HIGHER LEVEL

Mathematics
Analysis and Approaches
for the IB Diploma

IBRAHIM WAZIR
TIM GARRY

Published by Pearson Education Limited, 80 Strand, London, WC2R 0RL

www.pearsonglobalschools.com

Text © Pearson Education Limited 2019
Development edited by Jim Newall
Copy edited by Linnet Bruce
Proofread by Eric Pradel and Martin Payne
Indexed by Georgie Bowden
Designed by © Pearson Education Limited 2019
Typeset by © Tech-Set Ltd, Gateshead, UK
Original illustrations © Pearson Education Limited 2019
Illustrated by © Tech-Set Ltd, Gateshead, UK
Cover design by © Pearson Education Limited 2019
Cover images: Front: © **Getty Images:** Busà Photography
Inside front cover: **Shutterstock.com:** Dmitry Lobanov

The rights of Ibrahim Wazir and Tim Garry to be identified as the authors of this work have been asserted by them in accordance with the Copyright, Designs and Patents Act 1988.

First published 2019

24
IMP 12

British Library Cataloguing in Publication Data
A catalogue record for this book is available from the British Library

ISBN 978 0 435 19342 3

Printed in Slovakia by Neografia

Acknowledgements
The authors and publisher would like to thank the following individuals and organisations for their kind permission to reproduce copyright material.

Photographs

(Key: b-bottom; c-centre; l-left; r-right; t-top)

Getty Images: JPL/Moment/Getty Images 1, baxsyl/Moment/Getty Images 57, d3sign/Moment/Getty Images 107, Paul Biris/Moment/Getty Images 161, Image Source/Ditto/Getty Images 203, Westend61/Getty Images 235, Alberto Manuel Urosa Toledano/Moment/Getty Images 301, Jamil Caram/EyeEm/Getty Images 345, Chris eberhardt/Getty Images 379, J Broughton Photography/Moment/Getty Images 449, Howard Pugh (Marais)/Moment/Getty Images 519, aaaaimages/Moment/Getty Images 573, Ryan Trussler/Getty Images 627, Johner Images/Getty Images 697, Gabriel Perez/Moment/Getty Images 765, Witthaya Prasongsin/Moment/Getty Images 835. **Piet Mondrian:** Figure 4 Composition with Red, Blue and Yellow (1926) Piet Mondrian.

All other images © Pearson Education

We are grateful to the following for permission to reproduce copyright material:

Text

pages 912–913, Edge Foundation Inc.: What Kind of Thing Is a Number? A Talk with Reuben Hersh, Wed, Oct 24, 2018. Used with permission of Edge Foundation Inc.

Text extracts relating to the IB syllabus and assessment have been reproduced from IBO documents. Our thanks go to the International Baccalaureate for permission to reproduce its copyright.

This work has been developed independently from and is not endorsed by the International Baccalaureate (IB). International Baccalaureate® is a registered trademark of the International Baccalaureate Organization.

This work is produced by Pearson Education and is not endorsed by any trademark owner referenced in this publication.

Dedications

I dedicate this work to the memory of my parents and my brother, Saeed, who passed away during the early stages of work on this edition.

My special thanks go to my wife, Lody, for standing beside me throughout writing this book. She has been my inspiration and motivation for continuing to improve my knowledge and move my career forward. She is my rock, and I dedicate this book to her.

My thanks go to all the students and teachers who used the earlier editions and sent us their comments.

Ibrahim Wazir

In loving memory of my parents.

I wish to express my deepest thanks and love to my wife, Val, for her unflappable good nature and support – and for smiling and laughing with me each day. I am infinitely thankful for our wonderful and kind-hearted children – Bethany, Neil and Rhona. My love for you all is immeasurable.

Tim Garry

Contents

Introduction

IB Mathematics: Analysis and Approaches Higher Level **syllabus topics**

1. Number and Algebra
2. Functions
3. Geometry and Trigonometry
4. Statistics and Probability
5. Calculus

This textbook comprehensively covers all of the material in the syllabus for the two-year *Mathematics: Analysis and Approaches Higher Level* course of the International Baccalaureate (IB) Diploma Programme (DP). First teaching of this course starts in the autumn of 2019 with first exams occurring in May 2021. We, the authors, have strived to thoroughly explain and demonstrate the mathematical concepts and methods listed in the course syllabus.

Content

As you will see when you look at the table of contents, the five syllabus topics (see margin note) are fully covered, though some are split over different chapters in order to group the information as logically as possible. This textbook has been designed so that the chapters proceed in a manner that supports effective learning of the course content. Thus – although not essential – it is recommended that you read and study the chapters in numerical order. It is particularly important that you thoroughly review and understand all of the content in the first chapter, *Algebra and function basics*, before studying any of the other chapters.

Other than the final two chapters (**Theory of knowledge** and **Internal assessment**), each chapter has a set of **exercises** at the end of every section. Also, at the end of each chapter there is a set of **practice questions,** which are designed to expose you to questions that are more 'exam-like'. Many of the end-of-chapter practice questions are taken from past IB exam papers. Near the end of the book, you will find answers to all of the exercises and practice questions. There are also numerous **worked examples** throughout the book, showing you how to apply the concepts and skills you are studying.

The Internal assessment chapter provides thorough information and advice on the required **mathematical exploration component**. Your teacher will advise you on the timeline for completing your exploration and will provide critical support during the process of choosing your topic and writing the draft and final versions of your exploration.

The final chapter in the book will support your involvement in the **Theory of knowledge** course. It is a thought-provoking chapter that will stimulate you to think more deeply and critically about the nature of knowledge in mathematics and the relationship between mathematics and other areas of knowledge.

eBook

Included with this textbook is an eBook that contains a digital copy of the textbook and additional high-quality enrichment materials to promote your understanding of a wide range of concepts and skills encountered throughout the course. These materials include:

- Interactive *GeoGebra* **applets** demonstrating key concepts
- **Worked solutions** for all exercises and practice questions
- Graphical display calculator (**GDC**) support

To access the eBook, please follow the instructions located on the inside cover.

Information boxes

As you read this textbook, you will encounter numerous boxes of different colours containing a wide range of helpful information.

Learning objectives

You will find learning objectives at the start of each chapter. They set out the content and aspects of learning covered in the chapter.

Learning objectives

By the end of this chapter, you should be familiar with...

- different forms of equations of lines and their gradients and intercepts
- parallel and perpendicular lines
- different methods to solve a system of linear equations (maximum of three equations in three unknowns)

Key facts

Key facts are drawn from the main text and highlighted for quick reference to help you identify clear learning points.

 A function is **one-to-one** if each element y in the range is the image of exactly one element x in the domain.

Hints

Specific hints can be found alongside explanations, questions, exercises, and worked examples, providing insight into how to analyse/answer a question. They also identify common errors and pitfalls.

 If you use a graph to answer a question on an IB mathematics exam, you must include a clear and well-labelled sketch in your working.

Notes

Notes include general information or advice.

 Quadratic equations will be covered in detail in Chapter 2.

Examples

Worked examples show you how to tackle questions and apply the concepts and skills you are studying.

Example 1.5

Find x such that the distance between points $(1, 2)$ and $(x, -10)$ is 13 units.

Solution

$d = 13 = \sqrt{(x-1)^2 + (-10-2)^2} \Rightarrow 13^2 = (x-1)^2 + (-12)^2$

$\Rightarrow 169 = x^2 - 2x + 1 + 144 \Rightarrow x^2 - 2x - 24 = 0$

$\Rightarrow (x-6)(x+4) = 0 \Rightarrow x - 6 = 0 \text{ or } x + 4 = 0$

$\Rightarrow x = 6 \text{ or } x = -4$

How to use this book

This book is designed to be read by you – the student. It is very important that you read this book carefully. We have strived to write a readable book – and we hope that your teacher will routinely give you reading assignments from this textbook, thus giving you valuable time for productive explanations and discussions in the classroom. Developing your ability to read and understand mathematical explanations will prove to be valuable to your long-term intellectual development, while also helping you to comprehend mathematical ideas and acquire vital skills to be successful in the *Analysis and Approaches* HL course. Your goal should be understanding, not just remembering. You should always read a chapter section thoroughly before attempting any of the exercises at the end of the section.

Our aim is to support genuine inquiry into mathematical concepts while maintaining a coherent and engaging approach. We have included material to help you gain insight into appropriate and wise use of your GDC and an appreciation of the importance of proof as an essential skill in mathematics. We endeavoured to write clear and thorough explanations supported by suitable worked examples, with the overall goal of presenting sound mathematics with sufficient rigour and detail at a level appropriate for a student of HL mathematics.

For over 10 years, we have been writing successful textbooks for IB mathematics courses. During that time, we have received many useful comments from both teachers and students. If you have suggestions for improving this textbook, please feel free to write to us at globalschools@pearson.com. We wish you all the best in your mathematical endeavours.

Ibrahim Wazir and Tim Garry

Algebra and function basics

1

1 Algebra and function basics

Learning objectives

By the end of this chapter, you should be familiar with...
- different forms of equations of lines and their gradients and intercepts
- parallel and perpendicular lines
- different methods to solve a system of linear equations (maximum of three equations in three unknowns)
- the concept of a function and its domain, range and graph
- mathematical notation for functions
- composite functions
- characteristics of an inverse function and finding the inverse function $f^{-1}(x)$
- self-inverse functions
- transformations of graphs and composite transformations of graphs
- the graphs of the functions $y = |f(x)|$, $y = f(|x|)$ and $y = \dfrac{1}{f(x)}$

1.1 Equations and formulae

Equations, identities and formulae

You will encounter a wide variety of equations in this course. Essentially, an equation is a statement equating two algebraic expressions that may be true or false depending upon the value(s) substituted for the variable(s). Values of the variables that make the equation true are called **solutions** or **roots** of the equation. All of the solutions to an equation comprise the **solution set** of the equation. An equation that is true for all possible values of the variable is called an **identity**.

Many equations are often referred to as a **formula** (plural: formulae) and typically contain more than one variable and, often, other symbols that represent specific constants or **parameters** (constants that may change in value but do not alter the properties of the expression). Formulae with which you are familiar include: $A = \pi r^2$, $d = rt$, $d = \sqrt{(x_1 - x_2)^2 + (y_1 - y_2)^2}$ and $V = \dfrac{4}{3}\pi r^3$.

Whereas most equations that we encounter will have numerical solutions, we can solve a formula for one variable in terms of other variables – often referred to as changing the subject of a formula.

Example 1.1

(a) Solve for b in the formula $a^2 + b^2 = c^2$

(b) Solve for l in the formula $T = 2\pi\sqrt{\dfrac{l}{g}}$

(c) Solve for R in the formula $M = \dfrac{nR}{R + r}$

Solution

(a) $a^2 + b^2 = c^2 \Rightarrow b^2 = c^2 - a^2 \Rightarrow b = \pm\sqrt{c^2 - a^2}$

   If b is a length then $b = \sqrt{c^2 - a^2}$

(b) $T = 2\pi\sqrt{\dfrac{l}{g}} \Rightarrow \sqrt{\dfrac{l}{g}} = \dfrac{T}{2\pi} \Rightarrow \dfrac{l}{g} = \dfrac{T^2}{4\pi^2} \Rightarrow l = \dfrac{T^2 g}{4\pi^2}$

(c) $I = \dfrac{nR}{R + r} \Rightarrow I(R + r) = nR \Rightarrow IR + Ir = nR$

   $\Rightarrow IR - nR = -Ir \Rightarrow R(I - n) = -Ir$

   $\Rightarrow R = \dfrac{Ir}{n - I}$

Note that factorisation was required in solving for R in part (c).

Equations and graphs

Two important characteristics of any equation are the number of variables (unknowns) and the type of algebraic expressions it contains (e.g. polynomials, rational expressions, trigonometric, exponential). Nearly all of the equations in this course will have either one or two variables. In this chapter we will only discuss equations with algebraic expressions that are polynomials. Solutions for equations with a single variable consist of individual numbers that can be graphed as points on a number line. The **graph** of an equation is a visual representation of the equation's solution set. For example, the solution set of the one-variable equation containing quadratic and linear polynomials $x^2 = 2x + 8$ is $x \in \{-2, 4\}$. The graph of this one-variable equation (Figure 1.1) is depicted on a one-dimensional coordinate system, i.e. the real number line.

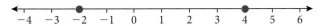

Figure 1.1 Graph of the solution set for the equation $x^2 = 2x + 8$

The solution set of a two-variable equation will be an **ordered pair** of numbers. An ordered pair corresponds to a location indicated by a point on a two-dimensional coordinate system, i.e. a **coordinate plane**. For example, the solution set of the two-variable **quadratic equation** $y = x^2$ will be an infinite set of ordered pairs (x, y) that satisfy the equation. Four ordered pairs in the solution set are shown in red in Figure 1.2. The graph of all the ordered pairs in the solution set forms a curve as shown in blue.

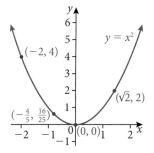

Figure 1.2 Graph of the solution set of the equation $y = x^2$

Quadratic equations will be covered in detail in Chapter 2.

Equations of lines

A one-variable **linear equation** in x can always be written in the form $ax + b = 0$, with $a \neq 0$, and it will have exactly one solution, namely $x = -\dfrac{b}{a}$. An example

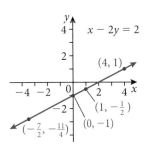

Figure 1.3 The graph of $x - 2y = 2$

of a two variable **linear equation in x and y** is $x - 2y = 2$. The graph of this equation's solution set (an infinite set of ordered pairs) is a **line** (Figure 1.3).

The **slope** or **gradient** m, of a non-vertical line is defined by the formula:

$$m = \frac{y_2 - y_1}{x_2 - x_1} = \frac{\text{vertical change}}{\text{horizontal change}}$$

Because division by zero is undefined, the slope of a vertical line is undefined.

Using the two points $\left(1, -\frac{1}{2}\right)$ and $(4, 1)$ we compute the slope of the line with

equation $x - 2y = 2$ to be $m = \dfrac{1 - \left(-\frac{1}{2}\right)}{4 - 1} = \dfrac{\frac{3}{2}}{3} = \dfrac{1}{2}$

If we solve for y we can rewrite the equation in the form $y = \frac{1}{2}x - 1$

Note that the coefficient of x is the slope of the line and the constant term is the y-coordinate of the point at which the line intersects the y-axis, that is, the y-intercept. There are several forms for writing linear equations.

general form	$ax + by + c = 0$	every line has an equation in this form if both a and $b \neq 0$
slope-intercept form	$y = mx + c$	m is the slope; $(0, c)$ is the y-intercept
point-slope form	$y - y_1 = m(x - x_1)$	m is the slope; (x_1, y_1) is a known point on the line
horizontal line	$y = c$	slope is zero; $(0, c)$ is the y-intercept
vertical line	$x = c$	slope is undefined; unless the line is the y-axis, no y-intercept

Table 1.1 Forms for equations of lines

Most problems involving linear equations and their graphs fall into two categories: (1) given an equation, determine its graph; and (2) given a graph, or some information about it, find its equation.

For lines, the first type of problem is often best solved by using the slope-intercept form. For the second type of problem, the point-slope form is usually most useful.

Example 1.2

Without using a GDC, sketch the line that is the graph of each linear equation written in general form.

(a) $5x + 3y - 6 = 0$ (b) $y - 4 = 0$ (c) $x + 3 = 0$

Solution

(a) Solve for y to write the equation in slope-intercept form.

$$5x + 3y - 6 = 0 \Rightarrow 3y = -5x + 6 \Rightarrow y = -\frac{5}{3}x + 2$$

The line has a y-intercept of $(0, 2)$ and a slope of $-\frac{5}{3}$

(b) The equation $y - 4 = 0$ is equivalent to $y = 4$, the graph of which is a horizontal line with a y-intercept of $(0, 4)$

(c) The equation $x + 3 = 0$ is equivalent to $x = -3$, the graph of which is a vertical line with no y-intercept; but, it has an x-intercept of $(-3, 0)$

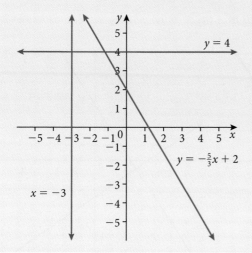

Example 1.3

(a) Find the equation of the line that passes through the point $(3, 31)$ and has a slope of 12. Write the equation in slope-intercept form.

(b) Find the linear equation in C and F knowing that $C = 10$ when $F = 50$, and $C = 100$ when $F = 212$. Solve for F in terms of C.

Solution

(a) Substitute $x_1 = 3$, $y_1 = 31$ and $m = 12$ into the point-slope form:

$$y - y_1 = m(x - x_1) \Rightarrow y - 31 = 12(x - 3) \Rightarrow y = 12x - 36 + 31$$
$$\Rightarrow y = 12x - 5$$

(b) The two points, ordered pairs (C, F), that are known to be on the line are $(10, 50)$ and $(100, 212)$. The variable C corresponds to the x variable and F corresponds to y in the definitions and forms stated above.

The slope of the line is $m = \dfrac{F_2 - F_1}{C_2 - C_1} = \dfrac{212 - 50}{100 - 10} = \dfrac{162}{90} = \dfrac{9}{5}$

Choose one of the points on the line, say $(10, 50)$, and substitute it and the slope into the point-slope form:

$$F - F_1 = m(C - C_1) \Rightarrow F - 50 = \frac{9}{5}(C - 10) \Rightarrow F = \frac{9}{5}C - 18 + 50$$

$$\Rightarrow F = \frac{9}{5}C + 32$$

The slope of a line is a convenient tool for determining whether two lines are parallel or perpendicular. The two lines shown in Figure 1.4 suggest that two distinct non-vertical lines are **parallel** if and only if their slopes are equal, $m_1 = m_2$.

The two lines shown in Figure 1.5 suggest that two non-vertical lines are perpendicular if and only if their slopes are negative reciprocals – that is, $m_1 = -\dfrac{1}{m_2}$, which is equivalent to $m_1 \cdot m_2 = -1$.

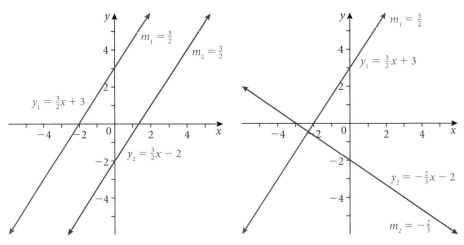

Figure 1.4 Parallel lines

Figure 1.5 Perpendicular lines

Distances and midpoints

Recall that absolute value is used to define the **distance** (always non-negative) between two points on the real number line. The distance between the points A and B on the real number line is $|B - A|$, which is equivalent to $|A - B|$.

The points A and B are the endpoints of a line segment that is denoted with the notation $[AB]$ and the length of the line segment is denoted AB. In Figure 1.6, the distance between A and B is $AB = |4 - (-2)| = |-2 - 4| = 6$.

Figure 1.6 The length of the line segment $[AB]$ is denoted by AB

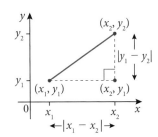

Figure 1.7 Distance between two points on a coordinate plane

The distance d between the two points (x_1, y_1) and (x_2, y_2) in the coordinate plane is
$$d = \sqrt{(x_1 - x_2)^2 + (y_1 - y_2)^2}$$

We can find the distance between two general points (x_1, y_1) and (x_2, y_2) on a coordinate plane using the definition for distance on a number line and Pythagoras' theorem. For the points (x_1, y_1) and (x_2, y_2), the horizontal distance between them is $|x_1 - x_2|$ and the vertical distance is $|y_1 - y_2|$. As illustrated in Figure 1.7, these distances are the lengths of two legs of a right-angled triangle whose hypotenuse is the distance between the points. If d represents the distance between (x_1, y_1) and (x_2, y_2), then by Pythagoras' theorem $d^2 = |x_1 - x_2|^2 + |y_1 - y_2|^2$. Because the square of any number is positive, the absolute value is not necessary to give us the **distance formula** for two-dimensional coordinates.

The coordinates of the **midpoint** of a line segment are the average values of the corresponding coordinates of the two endpoints.

 The midpoint of the line segment joining the points (x_1, y_1) and (x_2, y_2) in the coordinate plane is $\left(\dfrac{x_1 + x_2}{2}, \dfrac{y_1 + y_2}{2} \right)$

Example 1.4

(a) Show that the points $P(1, 2)$, $Q(3, 1)$ and $R(4, 8)$ are the vertices of a right-angled triangle.

(b) Find the midpoint of the hypotenuse of the triangle PQR.

Solution

(a) The three points are plotted and the line segments joining them are drawn in Figure 1.8. We can find the exact lengths of the three sides of the triangle by applying the distance formula.

$$PQ = \sqrt{(1 - 3)^2 + (2 - 1)^2} = \sqrt{4 + 1} = \sqrt{5}$$

$$QR = \sqrt{(3 - 4)^2 + (1 - 8)^2} = \sqrt{1 + 49} = \sqrt{50}$$

$$PR = \sqrt{(1 - 4)^2 + (2 - 8)^2} = \sqrt{9 + 36} = \sqrt{45}$$

$$(PQ)^2 + (PR)^2 = (QR)^2 \text{ because } (\sqrt{5})^2 + (\sqrt{45})^2 = 5 + 45 = 50 = (\sqrt{50})^2$$

The lengths of the three sides of the triangle satisfy Pythagoras' theorem, confirming that the triangle is a right-angled triangle.

(b) QR is the hypotenuse. Let the midpoint of QR be point M. Using the midpoint formula, $M = \left(\dfrac{3 + 4}{2}, \dfrac{1 + 8}{2} \right) = \left(\dfrac{7}{2}, \dfrac{9}{2} \right)$. This point is plotted in Figure 1.8

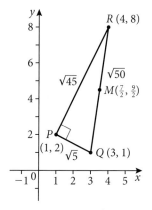

Figure 1.8 Diagram for Example 1.4

Example 1.5

Find x such that the distance between points $(1, 2)$ and $(x, -10)$ is 13 units.

Solution

$$d = 13 = \sqrt{(x - 1)^2 + (-10 - 2)^2} \Rightarrow 13^2 = (x - 1)^2 + (-12)^2$$

$$\Rightarrow 169 = x^2 - 2x + 1 + 144 \Rightarrow x^2 - 2x - 24 = 0$$

$$\Rightarrow (x - 6)(x + 4) = 0 \Rightarrow x - 6 = 0 \text{ or } x + 4 = 0$$

$$\Rightarrow x = 6 \text{ or } x = -4 \text{ (see Figure 1.9)}$$

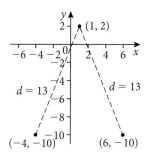

Figure 1.9 Graph for Example 1.5 showing the two points that are 13 units from $(1, 2)$

Systems of linear equations

Many problems involve sets of equations with several variables, rather than just a single equation with one or two variables. Such a set of equations is often called a set, or system, of **simultaneous equations** because we find the values for the variables that solve all of the equations simultaneously. In this section, we only consider sets of simultaneous equations containing linear equations; that is, **systems of linear equations**. We will take a brief look at four solution methods:

- graphical method (with technology)
- substitution method
- elimination method
- technology (without graphing)

We will first consider systems of two linear equations in two unknowns and then extend our methods to systems of three linear equations in three unknowns.

Graphical method

The graph of each equation in a **system of two linear equations in two unknowns** is a line. The graphical interpretation of such a system of equations corresponds to determining the point or points that lie on both lines. Two lines in a coordinate plane relate to one another in only one of three ways: (1) intersect at exactly one point, (2) intersect at all points on each line (i.e. the lines are identical, or coincident), or (3) the two lines do not intersect (i.e. the lines are parallel). These three possibilities are illustrated in Figure 1.10.

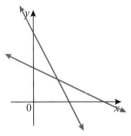 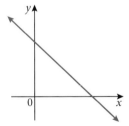 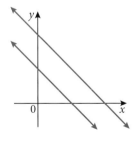

Intersect at exactly one point; exactly one solution

Identical – coincident lines; infinite solutions

Never intersect – parallel lines; no solution

Figure 1.10 Possible relationship between two lines in a coordinate plane

Although a graphical approach to solving a system of linear equations provides a helpful visual picture of the number and location of solutions, it can be tedious and inaccurate if done by hand. The graphical method is far more efficient and accurate when performed on a GDC.

Although the systems of two equations in two unknowns considered in this chapter contain only linear equations, it is important to mention that the graphical and substitution methods are effective for solving systems of two equations in two unknowns where not all of the equations are linear, e.g. one linear and one quadratic equation.

A system of equations that has no solution is **inconsistent**. A system of equations that has at least one solution is **consistent**.

Example 1.6

Use the graphical features of a GDC to solve each system of linear equations.

(a) $2x + 3y = 6$
$\quad 2x - y = -10$

(b) $7x - 5y = 20$
$\quad 3x + y = 2$

Solution

(a) Rewrite each equation in slope-intercept form, i.e. $y = mx + c$. This is a necessity if we use our GDC and is also very useful for graphing by hand.

$$2x + 3y = 6 \Rightarrow 3y = -2x + 6 \Rightarrow y = -\frac{2}{3}x + 2$$

and

$$2x - y = -10 \Rightarrow y = 2x + 10$$

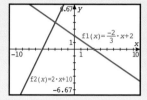  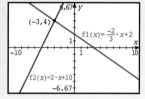

The intersection point and solution to the system of equations is $x = -3$ and $y = 4$, or $(-3, 4)$.

(b) $7x - 5y = 20 \Rightarrow 5y = 7x - 20 \Rightarrow y = \frac{7}{5}x - 4$

and

$$3x + y = 2 \Rightarrow y = -3x + 2$$

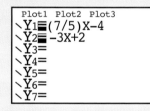 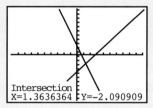

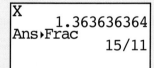 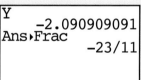

The solution to the system of equations is $x = \frac{15}{11}$ and $y = -\frac{23}{11}$, or $\left(\frac{15}{11}, -\frac{23}{11}\right)$.

Elimination method

To solve a system using the **elimination method**, we try to combine the two linear equations using sums or differences in order to eliminate one of the variables. Before combining the equations, we often need to multiply one or both of the equations by a suitable constant to produce coefficients for one of the variables that are equal (then subtract the equations), or that differ only in sign (then add the equations).

Example 1.7

Use the elimination method to solve each system of linear equations.

(a) $5x + 3y = 9$
$2x - 4y = 14$

(b) $x - 2y = 7$
$2x - 4y = 5$

Solution

(a) We can obtain coefficients for y that differ only in sign by multiplying the first equation by 4 and the second equation by 3. Then add the equations to eliminate the variable y.

$$5x + 3y = 9 \rightarrow 20x + 12y = 36$$
$$2x - 4y = 14 \rightarrow \underline{6x - 12y = 42}$$
$$26x = 78$$
$$x = \frac{78}{26}$$
$$x = 3$$

By substituting the value of 3 for x in either of the original equations we can solve for y.

$$5x + 3y = 9 \Rightarrow 5(3) + 3y = 9 \Rightarrow 3y = -6 \Rightarrow y = -2$$

The solution is $(3, -2)$

(b) To obtain coefficients for x that are equal, multiply the first equation by 2 and then subtract the equations to eliminate the variable x.

$$x - 2y = 7 \rightarrow 2x - 4y = 14$$
$$2x - 4y = 5 \rightarrow \underline{2x - 4y = 5}$$
$$0 = 9$$

Because it is not possible for 0 to equal 9, there is no solution. The lines on the graphs of the two equations are parallel. To confirm this, we can rewrite each of the equations in the form $y = mx + c$.

$$x - 2y = 7 \Rightarrow 2y = x - 7 \Rightarrow y = \frac{1}{2}x - \frac{7}{2} \text{ and}$$

$$2x - 4y = 5 \Rightarrow 4y = 2x - 5 \Rightarrow y = \frac{1}{2}x - \frac{5}{2}$$

Both equations have a slope of $\frac{1}{2}$, but different y-intercepts.

Therefore, the lines are parallel. This confirms that this system of linear equations has no solution.

Substitution method

The algebraic method that can be applied effectively to the widest variety of simultaneous equations, including non-linear equations, is the substitution method. Using this method, we choose one of the equations and solve for one of the variables in terms of the other variable. We then substitute this expression into the other equation to produce an equation with only one variable, which we can solve directly.

Example 1.8

Use the substitution method to solve each system of linear equations.

(a) $3x - y = -9$
    $6x + 2y = 2$

(b) $-2x + 6y = 4$
    $3x - 9y = -6$

Solution

(a) Solve for y in the top equation:

$$3x - y = -9 \Rightarrow y = 3x + 9$$

Substitute $3x + 9$ for y in the bottom equation:

$$6x + 2(3x + 9) = 2 \Rightarrow 6x + 6x + 18 = 2 \Rightarrow 12x = -16$$

$$\Rightarrow x = -\frac{16}{12} = -\frac{4}{3}$$

Now substitute $-\frac{4}{3}$ for x in either equation to solve for y.

$$3\left(-\frac{4}{3}\right) - y = -9 \Rightarrow y = -4 + 9 \Rightarrow y = 5$$

The solution is $x = -\frac{4}{3}, y = 5$; or $\left(-\frac{4}{3}, 5\right)$

(b) Solve for x in the top equation: $-2x + 6y = 4 \Rightarrow 2x = 6y - 4$

$$\Rightarrow x = 3y - 2$$

Substitute $3y - 2$ for x in the bottom equation:

$$3(3y - 2) - 9y = -6 \Rightarrow 9y - 6 - 9y = -6 \Rightarrow 0 = 0$$

This resulting equation $0 = 0$ is true for any values of x and y. The two equations are equivalent, and their graphs will produce identical lines, that is coincident lines. Therefore, the solution set consists of all points (x, y) lying on the line $-2x + 6y = 4$ $\left(\text{or } y = \frac{1}{3}x + \frac{2}{3}\right)$

Technology

As shown in Example 1.6 (a), we can use our GDC to graph the two lines whose equations constitute a system of two linear equations and then apply an 'intersection' command to find the ordered pair that solves the system. Alternatively, your GDC should have a simultaneous equation solver (or systems of equations solver) that can be used to solve a system of linear equations without graphing.

If you use a graph to answer a question on an IB mathematics exam, you must include a clear and well-labelled sketch in your working. Thus, on an exam there is often more effort involved in solving a system of linear equations by graphing with your GDC compared to solving the system using a simultaneous equation solver (or systems of linear equations solver) on your GDC.

Example 1.9

When flying in calm air (with no headwind or tailwind) with its propellers rotating at a particular rate, a small aeroplane has a speed of s kilometres per hour. With the propeller rotating at the same rate, the aeroplane flies 473 km in 2 hours as it flies against a headwind, and 887 km in 3 hours flying with the same wind as a tailwind. Find s and find the speed of the wind.

Solution

Let w represent the speed of the wind. Applying distance = rate × time gives the following two equations:

$$\begin{cases} 473 = 2(s - w) \\ 887 = 3(s + w) \end{cases} \Rightarrow \begin{cases} 473 = 2s - 2w \\ 887 = 3s + 3w \end{cases}$$

This system of equations can be solved with a simultaneous equation solver on a GDC.

$$\text{linSolve}\left(\begin{cases} 473=2\cdot s - 2\cdot w \\ 887=3\cdot s + 3\cdot w \end{cases}, \{s,w\}\right)$$
$$\left\{\frac{3139}{12}, \frac{355}{12}\right\}$$
$$\left\{\frac{3139}{12}, \frac{355}{12}\right\} \blacktriangleright \text{Decimal}$$
$$\{266.083, 29.5833\}$$

Therefore, $s \approx 266 \text{ km h}^{-1}$ and the wind speed is 29.6 km h^{-1}, accurate to 3 significant figures.

The graph of a linear equation in x, y and z, $ax + by + cz = d$, is a plane in a three-dimensional coordinate system. 'Linear' refers to the fact that the equation is a first-degree equation.

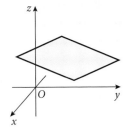

As with solving systems of two linear equations in two unknowns, **systems of three linear equations in three unknowns** also have three possible outcomes: a unique solution, an infinite number of solutions or no solution (inconsistent system). However, the graph of a linear equation in three unknowns is not a line, but a plane in three-dimensional space (covered in detail in Chapter 9). Hence, using a graphical approach is impractical for solving systems of three linear equations in three unknowns. Applying an elimination method is feasible, but the algebra required can be very tedious. So, in general, when solving a system of three linear equations in three unknowns it is best to use your GDC's simultaneous equation solver if a GDC is allowed, and an elimination method when a GDC is not allowed. The elimination method demonstrated in Example 1.10 (a) using **elementary row operations** is essentially the same as the elimination method shown earlier for systems of two linear equations, where equations are combined with each other, or each side of an equation is multiplied by a constant in order to isolate one of the variables. However, when using row operations to solve systems with more than two equations, we normally do not write the variables if all of the equations have the variables in the same order and the constant is on the other side of the equals sign.

Example 1.10

Solve the following system of equations in two ways:

$$\begin{cases} 2x + y - z = 2 \\ x + 3y + 2z = 1 \\ 2x + 4y + 6z = 6 \end{cases}$$

(a) using row operations (an elimination method), and

(b) using a simultaneous equation solver on a GDC.

Solution

(a) There are three types of elementary row operations: (1) multiply any row by a non-zero real number; (2) interchange any two rows; (3) add or subtract a multiple of one row to another row. The objective is to apply a sequence of row operations so that both the x and y terms are eliminated from one equation and the x term is eliminated from another of the three equations.

1. Subtract row 1 from row 3:

$$\left[\begin{array}{ccc|c} 2 & 1 & -1 & 2 \\ 1 & 3 & 2 & 1 \\ 0 & 3 & 7 & 4 \end{array}\right]$$

2. Multiply row 2 by 2 and subtract it from row 1:

$$\left[\begin{array}{ccc|c} 0 & -5 & -5 & 0 \\ 1 & 3 & 2 & 1 \\ 0 & 3 & 7 & 4 \end{array}\right]$$

3. Multiply row 3 by 5:

$$\left[\begin{array}{ccc|c} 0 & -5 & -5 & 0 \\ 1 & 3 & 2 & 1 \\ 0 & 15 & 35 & 20 \end{array}\right]$$

4. Multiply row 1 by 3 and add to row 3:

$$\left[\begin{array}{ccc|c} 0 & -5 & -5 & 0 \\ 1 & 3 & 2 & 1 \\ 0 & 0 & 20 & 20 \end{array}\right]$$

At this stage, we have eliminated both of the x and y terms from one equation and eliminated the x term from another equation. We can simplify the resulting system by one last step where we multiply row 1 by $-\dfrac{1}{5}$ and multiply row 3 by $\dfrac{1}{20}$, producing $\left[\begin{array}{ccc|c} 0 & 1 & 1 & 0 \\ 1 & 3 & 2 & 1 \\ 0 & 0 & 1 & 1 \end{array}\right]$ which is equivalent to the original system of equations. Writing each equation (row) with the variables we can solve for z directly, and for x and y by a process of substitution.

$$\begin{cases} 0x + y + z = 0 \\ x + 3y + 2z = 1 \\ 0x + 0y + z = 1 \end{cases}$$ Clearly, $z = 1$. Substituting this into the first

equation gives $y + 1 = 0$. Thus, $y = -1$. And, substituting into the second equation gives $x + 3(-1) + 2(1) = 1$. So, $x = 2$. Therefore, the system has the unique solution of $x = 2$, $y = -1$, $z = 1$.

(b) Using the systems of linear equations solver on a GDC.

$$\text{linSolve}\left(\begin{cases} 2 \cdot x + y - z = 2 \\ z + 3 \cdot y + 2 \cdot z = 2 \\ 2 \cdot x + 4 \cdot y + 6 \cdot z = 6 \end{cases}, \{x, y, z\}\right)$$

$$\{2, -1, 1\}$$

The sequence of row operations needed to produce one equation with both the x and y terms eliminated and another one of the three equations with the x term eliminated is not unique. Two people can carry out a different sequence of row operations while obtaining the correct solution.

13

Don't be confused when row operations on a system of three equations produces something like

$$\begin{bmatrix} 2 & -1 & 5 & | & 1 \\ 1 & 4 & -2 & | & -3 \\ 0 & 0 & 3 & | & 0 \end{bmatrix}$$

The bottom equation is $3z = 0$. So, $z = 0$. Thus, the system has the unique solution $\left(\frac{1}{8}, -\frac{3}{4}, 0\right)$

When using row operations to solve a system of three linear equations that has **no solution**, the equation (row) in which x and y have been eliminated will be an equation that is false for all values of z. For example, if after a sequence of row operations, we obtain $\begin{bmatrix} 2 & -1 & 5 & | & 1 \\ 1 & 4 & -2 & | & -3 \\ 0 & 0 & 0 & | & 2 \end{bmatrix}$ then the bottom equation is $0x + 0y + 0z = 2$ which is always false. Hence, the system has no solution.

If row operations on a system produces $\begin{bmatrix} 2 & -1 & 5 & | & 1 \\ 1 & 4 & -2 & | & -3 \\ 0 & 0 & 0 & | & 0 \end{bmatrix}$, then there are **infinite solutions** because the bottom equation is always true.

Exercise 1.1

1. Solve for the indicated variable in each formula.
   (a) $m(h - x) = n$, solve for x
   (b) $v = \sqrt{ab - t}$, solve for a
   (c) $A = \frac{h}{2}(b_1 + b_2)$, solve for b_1
   (d) $A = \frac{1}{2}r^2\theta$, solve for r
   (e) $\frac{f}{g} = \frac{h}{k}$, solve for k
   (f) $at = x - bt$, solve for t
   (g) $V = \frac{1}{3}\pi r^3 h$, solve for r
   (h) $F = \frac{g}{m_1 k + m_2 k}$, solve for k

2. Find the equation of the line that passes through the two given points. Write the line in slope-intercept form ($y = mx + c$), if possible.
   (a) $(-9, 1)$ and $(3, -7)$
   (b) $(3, -4)$ and $(10, -4)$
   (c) $(-12, -9)$ and $(4, 11)$
   (d) $\left(\frac{7}{3}, -\frac{1}{2}\right)$ and $\left(\frac{7}{3}, \frac{5}{2}\right)$
   (e) Find the equation of the line that passes through the point $(7, -17)$ and is parallel to the line with equation $4x + y - 3 = 0$. Write the line in slope-intercept form ($y = mx + c$).
   (f) Find the equation of the line that passes through the point $\left(-5, \frac{11}{2}\right)$ and is perpendicular to the line with equation $2x - 5y - 35 = 0$. Write the line in slope-intercept form ($y = mx + c$).

3. Find the exact distance between each pair of points and then find the midpoint of the line segment joining the two points.
   (a) $(-4, 10)$ and $(4, -5)$
   (b) $(-1, 2)$ and $(5, 4)$
   (c) $\left(\frac{1}{2}, 1\right)$ and $\left(-\frac{5}{2}, \frac{4}{3}\right)$
   (d) $(12, 2)$ and $(-10, 9)$

4. Find the value(s) of k so that the distance between the points is 5 units.
   (a) $(5, -1)$ and $(k, 2)$
   (b) $(-2, -7)$ and $(1, k)$

5. Show that the given points form the vertices of the indicated polygon.
   (a) Right-angled triangle: $(4, 0)$, $(2, 1)$ and $(-1, -5)$
   (b) Isosceles triangle: $(1, -3)$, $(3, 2)$ and $(-2, 4)$
   (c) Parallelogram: $(0, 1)$, $(3, 7)$, $(4, 4)$ and $(1, -2)$

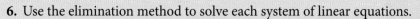

6. Use the elimination method to solve each system of linear equations.

   (a) $x + 3y = 8$
       $x - 2y = 3$

   (b) $x - 6y = 1$
       $3x + 2y = 13$

   (c) $6x + 3y = 6$
       $5x + 4y = -1$

   (d) $8x - 12y = 4$
       $-2x + 3y = 2$

   (e) $5x + 7y = 9$
       $-11x - 5y = 1$

7. Use the substitution method to solve each system of linear equations.

   (a) $2x + y = 1$
       $3x + 2y = 3$

   (b) $3x - 2y = 7$
       $5x - y = -7$

   (c) $2x + 8y = -6$
       $-5x - 20y = 15$

   (d) $\dfrac{x}{5} + \dfrac{y}{2} = 8$
       $x + y = 20$

   (e) $2x - y = -2$
       $4x + y = 5$

   (f) $0.4x + 0.3y = 1$
       $0.25x + 0.1y = -0.25$

8. Use your GDC to solve each system of two linear equations.

   (a) $3x + 2y = 9$
       $7x + 11y = 2$

   (b) $3.62x - 5.88y = -10.11$
       $0.08x - 0.02y = 0.92$

   (c) $2x - 3y = 4$
       $5x + 2y = 1$

9. Use row operations to solve each system of three linear equations.
   If the system has infinite solutions, simply write 'infinite solutions'.

   (a) $\begin{cases} 4x - y + z = -5 \\ 2x + 2y + 3z = 10 \\ 5x - 2y + 6z = 1 \end{cases}$

   (b) $\begin{cases} 4x - 2y + 3z = -2 \\ 2x + 2y + 5z = 16 \\ 8x - 5y - 2z = 4 \end{cases}$

   (c) $\begin{cases} 5x - 3y + 2z = 2 \\ 2x + 2y - 3z = 3 \\ x - 7y + 8z = -4 \end{cases}$

   (d) $\begin{cases} 3x - 2y + z = -29 \\ -4x + y - 3z = 37 \\ x - 5y + z = -24 \end{cases}$

10. Use your GDC to solve each system of three linear equations. If the
    system has infinite solutions, simply write 'infinite solutions'.

    (a) $\begin{cases} x - 3y - 2z = 8 \\ -2x + 7y + 3z = -19 \\ x - y - 3z = 3 \end{cases}$

    (b) $\begin{cases} 2x + 3y + 5z = 4 \\ 3x + 5y + 9z = 7 \\ 5x + 9y + 17z = 1 \end{cases}$

    (c) $\begin{cases} -x + 4y - 2z = 12 \\ 2x - 9y + 5z = -25 \\ -x + 5y - 4z = 10 \end{cases}$

    (d) $\begin{cases} 2x + 3y + 5z = 4 \\ 3x + 5y + 9z = 7 \\ 5x + 9y + 17z = 13 \end{cases}$

11. Find the value(s) of k such that the following system of equations
    has no solution.

    $x + y + (k - 1) = 2$
    $kx - z = -3$
    $6x + 2y - 3z = 1$

1.2 Definition of a function

A mapping illustrates how some values in the domain of a function are paired with values in the range of the function. Here is a mapping for the function $y = |x|$

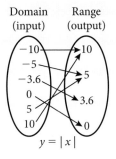

Domain (input) Range (output)

$y = |x|$

Many mathematical relationships concern how the value of one variable determines the value of a second variable. In general, suppose that the values of a particular **independent variable**, for example x, determine the values of a **dependent variable** y in such a way that for a specific value of x, a single value of y is determined. Then we say that y is a **function** of x and we write $y = f(x)$ (read y equals f of x) or $y = g(x)$, and so on, where the letter f or g, etc. represents the name of the function. For example:

- Period T is a function of length L: $T = 2\pi\sqrt{\dfrac{L}{g}}$

- Area A is a function of radius r: $A = \pi r^2$

- °F (degrees Fahrenheit) is a function of °C: $F = \dfrac{9}{5}C + 32$

- Distance d from the origin is a function of x: $d = |x|$

Other useful ways of representing a function include a graph of the equation on a **Cartesian coordinate system** (also called a rectangular coordinate system), a **table**, a **set of ordered pairs**, or a **mapping**.

The coordinate system for the graph of an equation has the independent variable on the horizontal axis and the dependent variable on the vertical axis.

The largest possible set of values for the independent variable (the **input** set) is called the **domain**, and the set of resulting values for the dependent variable (the **output** set) is called the **range**. In the context of a mapping, each value in the domain is mapped to its **image** in the range. All of the various ways of representing a mathematical function illustrate that its defining characteristic is that it is a rule by which each number in the domain determines a unique number in the range.

A **function** is a correspondence (mapping) between two sets X and Y in which each element of set X corresponds to (maps to) exactly one element of set Y. The domain is set X (independent variable) and the range is set Y (dependent variable).

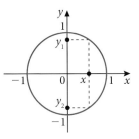

Figure 1.11 Graph of $x^2 + y^2 = 1$

Not all equations represent a function. The solution set for the equation $x^2 + y^2 = 1$ is the set of ordered pairs (x, y) on the circle of radius equal to 1 and centre at the origin (see Figure 1.11). If we solve the equation for y, we get $y = \pm\sqrt{1 - x^2}$. It is clear that any value of x between -1 and 1 will produce two different values of y (opposites). Since at least one value in the domain (x) determines more than one value in the range (y), the equation does not represent a function. A correspondence between two sets that does not satisfy the definition of a function is called a **relation**.

Alternative definition of a function

A function is a relation in which no two different ordered pairs have the same first coordinate. A vertical line intersects the graph of a function at no more than one point (vertical line test).

For many physical phenomena, we observe that one quantity depends on another. The word function is used to describe this dependence of one quantity on another – that is, how the value of an independent variable determines the value of a dependent variable. A common mathematical task is to find how to express one variable as a function of another variable.

Example 1.11

(a) Express the volume V of a cube as a function of the length e of each edge.

(b) Express the volume V of a cube as a function of its surface area S.

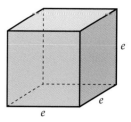

Figure 1.12 Cube for Example 1.11

Solution

(a) V as a function of e is $V = e^3$

(b) The surface area of the cube consists of six squares each with an area of e^2. Hence, the surface area is $6e^2$; that is, $S = 6e^2$. We need to write V in terms of S. We can do this by first expressing e in terms of S, and then substituting this expression for e in the equation $V = e^3$.

$$S = 6e^2 \Rightarrow e^2 = \frac{S}{6} \Rightarrow e = \sqrt{\frac{S}{6}}. \text{ Substituting, } V = \left(\sqrt{\frac{S}{6}}\right)^3 = \frac{\left(S^{\frac{1}{2}}\right)^3}{\left(6^{\frac{1}{2}}\right)^3}$$

$$= \frac{S^{\frac{3}{2}}}{6^{\frac{3}{2}}} = \frac{S \cdot S^{\frac{1}{2}}}{6 \cdot 6^{\frac{1}{2}}} = \frac{S}{6}\sqrt{\frac{S}{6}}$$

V as a function of S is $V = \dfrac{S}{6}\sqrt{\dfrac{S}{6}}$

Example 1.12

An offshore wind turbine is located at point W, 4 km offshore from the nearest point P on a straight coastline. A maintenance station is at point M, 3 km down the coast from P. An engineer is returning by a small boat from the wind turbine. He sails to point D that is located between P and M at an unknown distance x km from point P. From there, he walks to the maintenance station. The boat sails at 3 km hr^{-1} and the engineer can walk at 6 km hr^{-1}. Express the total time T (hours) for the trip from the wind turbine to the maintenance station as a function of x (km).

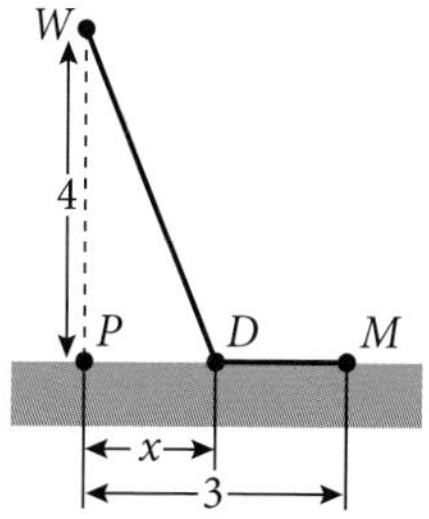

Figure 1.13 Diagram for Example 1.12

Solution

To get an equation for T in terms of x, use the fact that time $= \dfrac{\text{distance}}{\text{rate}}$

We then have

$$T = \frac{\text{distance } WD}{3} + \frac{\text{distance } DM}{6}$$

The distance WD can be expressed in terms of x using Pythagoras' theorem.

$$WD^2 = x^2 + 4^2 \Rightarrow WD = \sqrt{x^2 + 16}$$

To express T in terms of only the single variable x, note that $DM = 3 - x$
Then the total time T can be written in terms of x by the equation

$$T = \frac{\sqrt{x^2 + 16}}{3} + \frac{3 - x}{6} \text{ or } T = \frac{1}{3}\sqrt{x^2 + 16} + \frac{1}{2} - \frac{x}{6}$$

17

Domain and range of a function

The domain of a function may be stated explicitly, or it may be implied by the expression that defines the function. For most of this course, we can assume that functions are real-valued functions of a real variable. The domain and range will contain only real numbers or some subset of the real numbers. The domain of a function is the set of all real numbers for which the expression is defined as a real number, if not explicitly stated otherwise. For example, if a certain value of x is substituted into the algebraic expression defining a function and it causes division by zero or the square root of a negative number (both undefined in the real numbers) to occur, that value of x cannot be in the domain.

The domain of a function may also be implied by the physical context or limitations that exist in a problem. For example, in both functions derived in Example 1.11 the domain is the set of positive real numbers (denoted by $\mathbb{R}^+$) because neither a length (edge of a cube) nor a surface area (face of a cube) can have a value that is negative or zero. In Example 1.12 the domain for the function is $0 < x < 3$ because of the constraints given in the problem. Usually the range of a function is not given explicitly and is determined by analysing the output of the function for all values of the input (domain). The range of a function is often more difficult to find than the domain, and analysing the graph of a function is very helpful in determining it. A combination of algebraic and graphical analysis is very useful in determining the domain and range of a function.

In Chapter 8, we learn about functions where the variables can have values that are imaginary numbers.

Example 1.13

Find the domain of each function.

(a) $\{(-6, -3), (-1, 0), (2, 3), (3, 0), (5, 4)\}$

(b) Volume of a sphere: $V = \dfrac{4}{3}\pi r^3$

(c) $y = \dfrac{5}{2x - 6}$

(d) $y = \sqrt{3 - x}$

Solution

(a) The function consists of a set of ordered pairs. The domain of the function consists of all first coordinates of the ordered pairs. Therefore, the domain is the set $x \in \{-6, -1, 2, 3, 5\}$.

(b) The physical context tells us that a sphere cannot have a radius that is negative or zero. Therefore, the domain is the set of all real numbers r such that $r > 0$.

(c) Since division by zero is not defined for real numbers then $2x - 6 \neq 0$. Therefore, the domain is the set of all real numbers x such that $x \in \mathbb{R}$, $x \neq 3$.

(d) Since the square root of a negative number is not real, then $3 - x > 0$. Therefore, the domain is all real numbers x such that $x < 3$.

Example 1.14

Find the domain and range for the function $y = x^2$

Solution

Using algebraic analysis: Squaring any real number produces another real number. Therefore, the domain of $y = x^2$ is the set of all real numbers ($\mathbb{R}$). Since the square of any positive or negative number will be positive and the square of zero is zero, then the range is the set of all real numbers greater than or equal to zero.

Using graphical analysis: For the domain, focus on the x-axis and scan the graph from $-\infty$ to $+\infty$. There are no gaps or blank regions in the graph and the parabola will continue to get wider as x goes to either $-\infty$ or $+\infty$. Therefore, the domain is all real numbers. For the range, focus on the y-axis and scan from $-\infty$ to $+\infty$. The parabola will continue to increase as y goes to $+\infty$, but the graph does not go below the x-axis. The parabola has no points with negative y coordinates. Therefore, the range is the set of real numbers greater than or equal to zero.

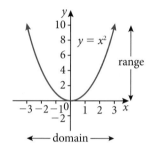

Figure 1.14 Graphical solution to Example 1.14

Description in words	Interval notation
Domain is any real number	Domain is $\{x : x \in \mathbb{R}\}$, or Domain is $x \in\ ]-\infty, \infty[$
Range is any number greater than or equal to zero	Range is $\{y : y \geqslant 0\}$, or Range is $y \in\ ]0, \infty[$

Table 1.2 Different ways of expressing the domain and range of $y = x^2$

The inequality $2 \leqslant x < 5$ can also be written as $[2, 5\ [$. The number 2 is included, but 5 is not. When determining the domain and range of a function, use both algebraic and graphical analysis. Do not rely too much on using just one approach. For graphical analysis of a function, producing a graph on your GDC that shows all the important features is essential.

Function notation

It is common practice to name a function using a single letter, with f, g and h commonly used. Given that the domain variable is x and the range variable is y, the symbol $f(x)$ denotes the unique value of y that is generated by the value of x.

Another notation – sometimes referred to as mapping notation – is based on the idea that the function f is the rule that maps x to $f(x)$ and is written $f : x \mapsto f(x)$. For each value of x in the domain, the corresponding unique value of y in the range is called the function value at x, or the image of x under f. The image of x may be written as $f(x)$ or as y. For example, for the function $f(x) = x^2$: '$f(3) = 9$', or 'if $x = 3$, then $y = 9$.'

Notation	Description in words
$f(x) = x^2$	The function f, in terms of x, is x^2; or, simply f of x equals x^2
$f : x \mapsto x^2$	The function f maps x to x^2
$f(3) = 9$	The value of the function f when $x = 3$ is 9; or, simply f of 3 equals 9
$f : 3 \mapsto 9$	The image of 3 under the function f is 9

Table 1.3 Function notation

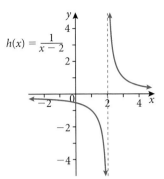

$h(x) = \dfrac{1}{x - 2}$

Figure 1.15 Diagram for Example 1.15

Example 1.15

Find the domain and range of the function $h{:}x \mapsto \dfrac{1}{x - 2}$

Solution

Using algebraic analysis: The function produces a real number for all x, except for $x = 2$ when division by zero occurs. Hence, $x = 2$ is the only real number not in the domain. Since the numerator of $\dfrac{1}{x - 2}$ can never be zero, the value of y cannot be zero. Hence, $y = 0$ is the only real number not in the range.

Using graphical analysis: A horizontal scan shows a gap at $x = 2$ dividing the graph of the equation into two branches that both continue indefinitely with no other gaps as $x \to \pm\infty$. Both branches are **asymptotic** (approach but do not intersect) to the vertical line $x = 2$. This line is a **vertical asymptote** and is drawn as a dashed line (it is not part of the graph of the equation). A vertical scan reveals a gap at $y = 0$ (x-axis) with both branches of the graph continuing indefinitely with no other gaps as $y \to \pm\infty$. Both branches are also asymptotic to the x-axis. The x-axis is a **horizontal asymptote**.

Both approaches confirm that the domain and range for $h{:}x \mapsto \dfrac{1}{x - 2}$ are:

domain: $\{x{:}x \in \mathbb{R},\, x \neq 2\}$ or $x \in\,]-\infty,\, 2[\cup\,]2,\, \infty[$

range: $\{y{:}y \in \mathbb{R},\, y \neq 0\}$ or $y \in\,]-\infty,\, 0[\cup\,]0,\, \infty[$

Example 1.16

Consider the function $f(x) = \sqrt{x + 4}$
(a) Find:
    (i) $f(7)$ (ii) $f(32)$ (iii) $f(-4)$
(b) Find the values of x for which f is undefined.
(c) State the domain and range of f.

Solution

(a) (i) $f(7) = \sqrt{7 + 4} = \sqrt{11} \approx 3.32$ (3 s.f.)
    (ii) $f(32) = \sqrt{32 + 4} = \sqrt{36} = 6$
    (iii) $f(-4) = \sqrt{-4 + 4} = \sqrt{0} = 0$

(b) $f(x)$ will be undefined (square root of a negative) when $x + 4 < 0$.
    Therefore, $f(x)$ is undefined when $x < -4$.

(c) It follows from the result in (b) that the domain of f is $\{x{:}x \geqslant -4\}$.
    The symbol $\sqrt{}$ stands for the **principal square root** that, by definition, can only give a result that is positive or zero. Therefore, the range of f is $\{y{:}y \geqslant 0\}$. The domain and range are confirmed by analysing the graph of the function.

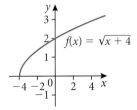

$f(x) = \sqrt{x + 4}$

Figure 1.16 Graph for the solution to Example 1.16 (c)

Example 1.17

Find the domain and range of the function $f(x) = \dfrac{1}{\sqrt{9 - x^2}}$

As Example 1.17 illustrates, it is dangerous to completely trust graphs produced on a GDC without also doing some algebraic thinking. It is important to check that the graph shown is comprehensive (shows all important features), and that the graph agrees with algebraic analysis of the function, for example. where the function should be zero, positive, negative, undefined, or increasing/decreasing without bound.

Solution

The graph of $y = \dfrac{1}{\sqrt{9 - x^2}}$ shown here, agrees with algebraic analysis indicating that the expression $\dfrac{1}{\sqrt{9 - x^2}}$ will be positive for all x, and is defined only for $-3 < x < 3$. Further analysis and tracing the graph reveals that $f(x)$ has

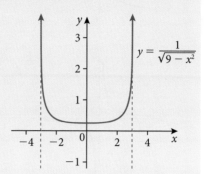

a minimum at $\left(0, \dfrac{1}{3}\right)$. The graph on the GDC is misleading in that it appears to show that the function has a maximum value of approximately $y \approx 2.8037849$. Can this be correct? A lack of algebraic thinking and over-reliance on a GDC could easily lead to a mistake. The graph abruptly stops its curve upwards because of low screen resolution. Function values should get quite large for values of x a little less than 3, because the value of $\sqrt{9 - x^2}$ will be small, making the fraction $\dfrac{1}{\sqrt{9 - x^2}}$ large.

Using a GDC to make a table for $f(x)$ or evaluating the function for values of x very close to -3 or 3 confirms that as x approaches -3 or 3, y increases without bound, i.e. y goes to $+\infty$. Hence, $f(x)$ has vertical asymptotes of $x = -3$ and $x = 3$. This combination of graphical and algebraic analysis leads to the conclusion that the domain of $f(x)$ is $\{x : -3 < x < 3\}$, and the range of $f(x)$ is $\left\{y : y \geqslant \dfrac{1}{3}\right\}$

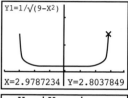

Figure 1.17 GDC screens for solution to Example 1.17

Exercise 1.2

1. **(i)** Match each equation to one of the graphs.

   (ii) State whether or not the equation represents any of the functions shown. Justify your answer. Assume that x is the independent variable and y is the dependent variable.

   (a) $y = 2x$

   (b) $y = -3$

   (c) $x - y = 2$

   (d) $x^2 + y^2 = 4$

   (e) $y = 2 - x$

   (f) $y = x^2 + 2$

   (g) $y^3 = x$

   (h) $y = \dfrac{2}{x}$

   (i) $x^2 + y = 2$

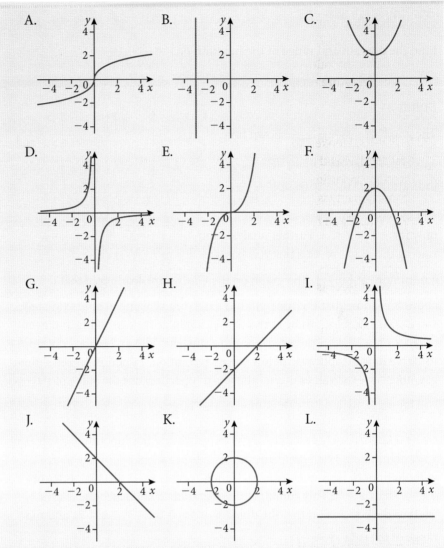

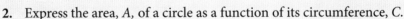

2. Express the area, A, of a circle as a function of its circumference, C.

3. Express the area, A, of an equilateral triangle as a function of the length, ℓ, of each of its sides.

4. A rectangular swimming pool with dimensions 12 metres by 18 metres is surrounded by a pavement of uniform width x metres. Find the area of the pavement, A, as a function of x.

5. In a right-angled isosceles triangle, the two equal sides have length x units and the hypotenuse has length h units. Write h as a function of x.

6. The pressure P (measured in kilopascals, kPa) for a particular sample of gas is directly proportional to the temperature T (measured in degrees kelvin, K) and inversely proportional to the volume V (measured in litres, L). With k representing the constant of proportionality, this relationship can be written in the form of the equation $P = k\dfrac{T}{V}$

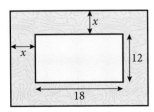

Figure 1.18 Diagram for question 4

(a) Find the constant of proportionality, k, if 150 L of gas exerts a pressure of 23.5 kPa at a temperature of 375 K.

(b) Using the value of k from part (a) and assuming that the temperature is held constant at 375 K, write the volume V as a function of pressure P for this sample of gas.

7. In physics, Hooke's law states that the force F (measured in newtons, N) needed to extend a spring by x units beyond its natural length is directly proportional to the extension x. Assume that the constant of proportionality is k (known as the spring constant for a particular spring).

(a) Write F as a function of x.

(b) A spring has a natural length of 12 cm and a force of 25 N stretches the spring to a length of 16 cm. Work out the spring constant k.

(c) What force is needed to stretch the spring to a length of 18 cm?

8. Find the domain of each of the following functions.

(a) $\{(-6.2, -7), (-1.5, -2), (0.7, 0), (3.2, 3), (3.8, 3)\}$

(b) Surface area of a sphere: $S = 4\pi r^2$

(c) $f(x) = \dfrac{2}{5}x - 7$ (d) $h{:}x \mapsto x^2 - 4$ (e) $g(t) = \sqrt{3 - t}$

(f) $h(t) = \sqrt[3]{t}$ (g) $f{:}x \mapsto \dfrac{6}{x^2 - 9}$ (h) $f(x) = \sqrt{\dfrac{1}{x^2} - 1}$

9. Do all linear equations represent a function? Explain.

10. Consider the function $h(x) = \sqrt{x - 4}$

(a) Find: (i) $h(21)$ (ii) $h(53)$ (iii) $h(4)$

(b) Find the values of x for which h is undefined.

(c) State the domain and range of h.

11. For each function below:

(i) find the domain and range of the function

(ii) sketch a comprehensive graph of the function, clearly indicating any intercepts or asymptotes.

(a) $f{:}x \mapsto \dfrac{1}{x - 5}$ (b) $g(x) = \dfrac{1}{\sqrt{x^2 - 9}}$ (c) $h(x) = \dfrac{2x - 1}{x + 2}$

(d) $p{:}x \mapsto \sqrt{5 - 2x^2}$ (e) $f(x) = \dfrac{1}{x} - 4$

Composition of functions

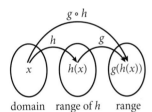

Consider the function in Example 1.16, $f(x) = \sqrt{x+4}$. When we evaluate $f(x)$ for a certain value of x in the domain, for example, $x = 5$, it is necessary to perform computations in two separate steps in a certain order.

$$f(5) = \sqrt{5+4} \Rightarrow f(5) = \sqrt{9} \quad \text{Step 1: compute the sum of } 5 + 4$$

$$\Rightarrow f(5) = 3 \quad \text{Step 2: compute the square root of 9}$$

Given that the function has two separate evaluation steps, $f(x)$ can be seen as a combination of two simpler functions that are performed in a specified order. According to how $f(x)$ is evaluated, the simpler function to be performed first is the rule of adding 4 and the second is the rule of taking the square root. If $h(x) = x + 4$ and $g(x) = \sqrt{x}$, then we can create (compose) the function $f(x)$ from a combination of $h(x)$ and $g(x)$ as follows:

$$f(x) = g(h(x))$$

$$= g(x + 4) \quad \text{Step 1: substitute } x+4 \text{ for } h(x) \text{ making } x+4 \text{ the argument of } g(x)$$

$$= \sqrt{x + 4} \quad \text{Step 2: apply the function } g(x) \text{ on the argument } x+4$$

We obtain the rule $\sqrt{x+4}$ by first applying the rule $x+4$ and then applying the rule $\sqrt{x}$. A function that is obtained from simpler functions by applying one after another in this way is called a **composite function**. $f(x) = \sqrt{x+4}$ is the **composition** of $h(x) = x+4$ followed by $g(x) = \sqrt{x}$. In other words, f is obtained by substituting h into g, and can be denoted in function notation by $g(h(x))$ – read 'g of h of x'.

Start with a number x in the domain of h and find its image $h(x)$. If this number $h(x)$ is in the domain of g, we then compute the value of $g(h(x))$. The resulting composite function is denoted as $(g \circ h(x))$. See Figure 1.19.

Example 1.18

If $f(x) = 3x$ and $g(x) = 2x - 6$, find:

(a) (i) $(f \circ g)(5)$ (ii) Express $(f \circ g)(x)$ as a single function rule (expression).

(b) (i) $(g \circ f)(5)$ (ii) Express $(g \circ f)(x)$ as a single function rule (expression).

(c) (i) $(g \circ g)(5)$ (ii) Express $(g \circ g)(x)$ as a single function rule (expression).

Solution

(a) (i) $(f \circ g)(5) = f(g(5)) = f(2 \cdot 5 - 6) = f(4) = 3 \cdot 4 = 12$

(ii) $(f \circ g)(x) = f(g(x)) = f(2x - 6) = 3(2x - 6) = 6x - 18$

Therefore, $(f \circ g)(x) = 6x - 18$

Check with result from (i): $(f \circ g)(5) = 6 \cdot 5 - 18 = 30 - 18 = 12$

(b) (i) $(g \circ f)(5) = g(f(5)) = g(3 \cdot 5) = g(15) = 2 \cdot 15 - 6 = 24$

(ii) $(g \circ f)(x) = g(f(x)) = g(3x) = 2(3x) - 6 = 6x - 6$

Therefore, $(g \circ f)(x) = 6x - 6$

Check with result from (i): $(g \circ f)(5) = 6 \cdot 5 - 6 = 30 - 6 = 24$

(c) (i) $(g \circ g)(5) = g(g(5)) = g(2 \cdot 5 - 6) = g(4) = 2 \cdot 4 - 6 = 2$

(ii) $(g \circ g)(x) = g(g(x)) = g(2x - 6) = 2(2x - 6) - 6 = 4x - 18$

Therefore, $(g \circ g)(x) = 4x - 18$

Check with result from (i): $(g \circ g)(5) = 4 \cdot 5 - 18 = 20 - 18 = 2$

> The notations $(g \circ h)(x)$ and $g(h(x))$ are both commonly used to denote a composite function where h is applied first then followed by applying g. Since you are reading this from left to right, it is easy to apply the functions in the incorrect order. It may be helpful to read $g \circ h$ as 'g following h' to highlight the order in which the functions are applied. Also, in either notation, $(g \circ h)(x)$ or $g(h(x))$, the function applied first is closest to the variable x.

It is important to notice that in parts (a)(ii) and (b)(ii) in Example 1.18, $f \circ g$ is not equal to $g \circ f$. At the start of this section, it was shown how the two functions $h(x) = x + 4$ and $g(x) = \sqrt{x}$ could be combined into the composite function $(g \circ h)(x)$ to create the single function $f(x) = \sqrt{x + 4}$. However, the composite function $(h \circ g)(x)$ (the functions applied in reverse order) creates a different function: $(h \circ g)(x) = h(g(x)) = h(\sqrt{x}) = \sqrt{x} + 4$. Since, $\sqrt{x} + 4 \neq \sqrt{x + 4}$ then $f \circ g$ is not equal to $g \circ f$. Is it always true that $f \circ g \neq g \circ f$? The next example will answer that question.

Example 1.19

Given $f : x \mapsto 3x - 6$ and $g : x \mapsto \frac{1}{3}x + 2$, find:

(a) $(f \circ g)(x)$ 　　　　　　　(b) $(g \circ f)(x)$

Solution

(a) $(f \circ g)(x) = f(g(x)) = f\left(\frac{1}{3}x + 2\right) = 3\left(\frac{1}{3}x + 2\right) - 6 = x + 6 - 6 = x$

(b) $(g \circ f)(x) = g(f(x)) = g(3x - 6) = \frac{1}{3}(3x - 6) + 2 = x - 2 + 2 = x$

Example 1.19 shows that it is possible for $f \circ g$ to be equal to $g \circ f$. You will learn in the next section that this occurs in some cases where there is a special relationship between the pair of functions. However, in general $f \circ g \neq g \circ f$.

Decomposing a composite function

In examples 1.18 and 1.19, we created a single function by forming the composite of two functions. As with the function $f(x) = \sqrt{x + 4}$, it is also important for us to be able to identify two functions that make up a composite function, in other words, to decompose a function into two simpler functions. When we are doing this it is very useful to think of the function that is applied first as the inside function, and the function that is applied second as the outside function. In the function $f(x) = \sqrt{x + 4}$, the inside function is $h(x) = x + 4$ and the outside function is $g(x) = \sqrt{x}$.

Example 1.20

Each of these functions is a composite function of the form $(f \circ g)(x)$. For each, find the two component functions f and g.

(a) $h: x \mapsto \dfrac{1}{x + 3}$ (b) $k: x \mapsto 2^{4x+1}$ (c) $p(x) = \sqrt[3]{x^2 - 4}$

Solution

(a) When we evaluate the function $h(x)$ for a certain x in the domain, we first evaluate the expression $x + 3$, and then evaluate the expression $\dfrac{1}{x}$. Hence, the inside function (applied first) is $y = x + 3$, and the outside function (applied second) is $y = \dfrac{1}{x}$. So the two component functions are $g(x) = x + 3$ and $f(x) = \dfrac{1}{x}$

(b) Evaluating $k(x)$ requires us to first evaluate the expression $4x + 1$, and then evaluate the expression 2^x. Hence, the inside function is $y = 4x + 1$, and the outside function is $y = 2^x$. The two composite functions are $g(x) = 4x + 1$ and $f(x) = 2^x$.

(c) Evaluating $p(x)$ requires us to perform three separate evaluation steps: squaring a number, subtracting four, and then taking the cube root. Hence, it is possible to decompose $p(x)$ into three component functions: $h(x) = x^2$, $g(x) = x - 4$ and $f(x) = \sqrt[3]{x}$. However, for our purposes it is best to decompose the composite function into only two component functions: $g(x) = x^2 - 4$, and $f(x) = \sqrt[3]{x}$.

Finding the domain of a composite function

It is important to note that in order for a value of x to be in the domain of the composite function $g \circ h$, two conditions must be met: (1) x must be in the domain of h, and (2) $h(x)$ must be in the domain of g. Likewise, it is also worth noting that $g(h(x))$ is in the range of $g \circ h$ only if x is in the domain of $g \circ h$. The next example illustrates these points – and also that, in general, the domains of $g \circ h$ and $h \circ g$ are not the same.

Example 1.21

Let $g(x) = x^2 - 4$ and $h(x) = \sqrt{x}$. Find:

(a) $(g \circ h)(x)$ and its domain and range

(b) $(h \circ g)(x)$ and its domain and range.

Solution

First, establish the domain and range for both g and h. For $g(x) = x^2 - 4$, the domain is $x \in \mathbb{R}$ and the range is $y \geqslant -4$. For $h(x) = \sqrt{x}$, the domain is $x \geqslant 0$ and the range is $y \geqslant 0$.

(a) $(g \circ h)(x) = g(h(x))$

$\quad\quad\quad = g(\sqrt{x})$ To be in the domain of $g \circ h$, $\sqrt{x}$ must be defined for $x \Rightarrow x \geqslant 0$

$\quad\quad\quad = (\sqrt{x^2}) - 4$ Therefore, the domain of $g \circ h$ is $x \geqslant 0$

$\quad\quad\quad = x - 4$ Since $x \geqslant 0$, then the range for $y = x - 4$ is $y \geqslant -4$.

Therefore, $(g \circ h)(x) = x - 4$, and its domain is $x \geqslant 0$, and its range is $y \geqslant -4$

(b) $(h \circ g)(x) = h(g(x))$ $g(x) = x^2 - 4$ must be in the domain of $h \Rightarrow x^2 - 4 \geqslant 0 \Rightarrow x^2 \geqslant 4$

$\quad\quad\quad = h(x^2 - 4)$ Therefore, the domain of $h \circ g$ is $x \leqslant -2$ or $x \geqslant 2$ and

$\quad\quad\quad = \sqrt{x^2 - 4}$ with $x \leqslant -2$ or $x \geqslant 2$, the range for $y = \sqrt{x^2 - 4}$ is $y \geqslant 0$

Therefore, $(h \circ g)(x) = \sqrt{x^2 - 4}$, and its domain is $x \leqslant -2$ or $x \geqslant 2$, and its range is $y \geqslant 0$

Exercise 1.3

1. Let $f(x) = 2x$ and $g(x) = \dfrac{1}{x - 3}, x \neq 0$

   Find the value of:

   (a) $(f \circ g)(5)$ (b) $(g \circ f)(5)$

   Find the function rule (expression) for:

   (c) $(f \circ g)(x)$ (d) $(g \circ f)(x)$

2. Let $f: x \mapsto 2x - 3$ and $g: x \mapsto 2 - x^2$

   Evaluate:

   (a) $(f \circ g)(0)$ (b) $(g \circ f)(0)$ (c) $(f \circ f)(4)$

   (d) $(g \circ g)(-3)$ (e) $(f \circ g)(-1)$ (f) $(g \circ f)(-3)$

   Find the expression for:

   (g) $(f \circ g)(x)$ (h) $(g \circ f)(x)$ (i) $(f \circ f)(x)$ (j) $(g \circ g)(x)$

3. For each pair of functions, find $(f \circ g)(x)$ and $(g \circ f)(x)$ and state the domain for each.

   (a) $f(x) = 4x - 1, g(x) = 2 + 3x$ **(b)** $f(x) = x^2 + 1, g(x) = -2x$

   (c) $f(x) = \sqrt{x + 1}, g(x) = 1 + x^2$ **(d)** $f(x) = \dfrac{2}{x + 4}, g(x) = x - 1$

   (e) $f(x) = 3x + 5, g(x) = \dfrac{x - 5}{3}$ **(f)** $f(x) = 2 - x^3, g(x) = \sqrt[3]{1 - x^2}$

   (g) $f(x) = \dfrac{2x}{4 - x}, g(x) = \dfrac{1}{x^2}$

   (h) $f(x) = \dfrac{2}{x + 3} - 3, g(x) = \dfrac{2}{x + 3} - 3$

   (i) $f(x) = \dfrac{x}{x - 1}, g(x) = x^2 - 1$

4. Let $g(x) = \sqrt{x - 1}$ and $h(x) = 10 - x^2$. Find:

   (a) $(g \circ h)(x)$ and its domain and range

   (b) $(h \circ g)(x)$ and its domain and range.

5. Let $f(x) = \dfrac{1}{x}$ and $g(x) = 10 - x^2$. Find:

   (a) $(f \circ g)(x)$ and its domain and range

   (b) $(g \circ f)(x)$ and its domain and range.

6. Determine functions g and h so that $f(x) = g(h(x))$

   (a) $f(x) = (x + 3)^2$ **(b)** $f(x) = \sqrt{x - 5}$ **(c)** $f(x) = 7 - \sqrt{x}$

   (d) $f(x) = \dfrac{1}{x + 3}$ **(e)** $f(x) = 10^{x+1}$ **(f)** $f(x) = \sqrt[3]{x - 9}$

   (g) $f(x) = |x^2 - 9|$ **(h)** $f(x) = \dfrac{1}{\sqrt{x - 5}}$

7. Find the domain for:

   (i) the function f **(ii)** the function g **(iii)** the composite function $f \circ g$

   (a) $f(x) = \sqrt{x}, g(x) = x^2 + 1$ **(b)** $f(x) = \dfrac{1}{x}, g(x) = x + 3$

   (c) $f(x) = \dfrac{3}{x^2 - 1}, g(x) = x + 1$ **(d)** $f(x) = 2x + 3, g(x) = \dfrac{x}{2}$

1.4 Inverse functions

Pairs of inverse functions

If we choose a number and cube it (raise it to the power of 3), and then take the cube root of the result, the answer is the original number. The same result would occur if we applied the two rules in the reverse order. That is, first take the cube root of a number and then cube the result; again, the answer is the original number.

Write each of these rules as a function with function notation. Write the cubing function as $f(x) = x^3$, and the cube root function as $g(x) = \sqrt[3]{x}$. Now, using what we know about composite functions and operations with radicals and powers, we can write what was described above in symbolic form.

Cube a number and then take the cube root of the result:

$$g(f(x)) = \sqrt[3]{x^3} = (x^3)^{\frac{1}{3}} = x^1 = x$$

For example, $g(f(-2)) = \sqrt[3]{(-2)^3} = \sqrt[3]{-8} = -2$

Take the cube root of a number and then cube the result:

$$f(g(x)) = \left(\sqrt[3]{x}\right)^3 = \left(x^{\frac{1}{3}}\right)^3 = x^1 = x$$

For example, $f(g(27)) = \left(\sqrt[3]{27}\right)^3 = (3)^3 = 27$

Because function g has this reverse (inverse) effect on function f, we call function g the **inverse** of function f. Function f has the same inverse effect on function g [$g(27) = 3$ and then $f(3) = 27$], making f the inverse function of g. The functions f and g are inverses of each other. The cubing and cube root functions are an example of a pair of **inverse functions**. The mapping diagram for functions f and g (Figure 1.20) illustrates the relationship for a pair of inverse functions where the domain of one is the range for the other.

You should already be familiar with pairs of **inverse operations**. Addition and subtraction are inverse operations. For example, the rule of adding six $(x + 6)$, and the rule of subtracting six $(x - 6)$, undo each other. Accordingly, the functions $f(x) = x + 6$ and $g(x) = x - 6$ are a pair of inverse functions. Multiplication and division are also inverse operations.

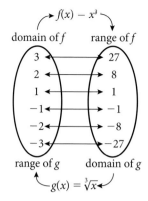

Figure 1.20 A mapping diagram for the cubing and cube root functions

The composite of two inverse functions is the function that always produces the same number that was first substituted into the function. This function is called the **identity function** because it assigns each number in its domain to itself and is denoted by $I(x) = x$.

If f and g are two functions such that $(f \circ g)(x) = x$ for every x in the domain of g and $(g \circ f)(x) = x$ for every x in the domain of f, then the function g is the **inverse** of the function f. The notation to indicate the function that is the inverse of function f is f^{-1}. Therefore, $(f \circ f^{-1})(x) = x$ and $(f^{-1} \circ f)(x) = x$

The domain of f must be equal to the range of f^{-1}, and the range of f must be equal to the domain of f^{-1}.

Remember that the notation $(f \circ g)(x)$ is equivalent to $f(g(x))$. It follows from the definition that if g is the inverse of f, then it must also be true that f is the inverse of g.

Do not mistake the -1 in the notation f^{-1} for a power. It is not a power. If a superscript of -1 is applied to the name of a function, as in f^{-1} or $\sin^{-1}$, then it denotes the function that is the inverse of the named function (e.g. f or sin). If a superscript of -1 is applied to an expression, as in 7^{-1} or $(2x + 5)^{-1}$, then it is a power and denotes the reciprocal of the expression.

In general, the functions $f(x)$ and $g(x)$ are a pair of inverse functions if the following two statements are true:

1 $g(f(x)) = x$ for all x in the domain of f

2 $f(g(x)) = x$ for all x in the domain of g

For a pair of inverse functions, f and g, the composite functions $f(g(x))$ and $g(f(x))$ are equal. Remember that this is not generally true for an arbitrary pair of functions.

Example 1.22

Given $h(x) = \dfrac{x - 3}{2}$ and $p(x) = 2x + 3$, show that h and p are inverse functions.

Solution

Since the domain and range of both $h(x)$ and $p(x)$ is the set of all real numbers, then:

For any real number x, $p(h(x)) = p\left(\dfrac{x-3}{2}\right) = 2\left(\dfrac{x-3}{2}\right) + 3$

$$= x - 3 + 3 = x$$

For any real number x, $h(p(x)) = h(2x + 3) = \dfrac{(2x+3)-3}{2} = \dfrac{2x}{2} = x$

Since $p(h(x)) = h(p(x)) = x$ then h and p are a pair of inverse functions.

A function is **one-to-one** if each element y in the range is the image of exactly one element x in the domain.

It is clear that both $f(x) = x^3$ and $g(x) = \sqrt[3]{x}$ satisfy the definition of a function because for both f and g every number in its domain determines exactly one number in its range. Since they are a pair of inverse functions then the reverse is also true for both; that is, every number in its range is determined by exactly one number in its domain. Such a function is called a **one-to-one function**. The phrase one-to-one is appropriate because each value in the domain corresponds to exactly **one** value in the range, and each value in the range corresponds to exactly **one** value in the domain.

The existence of an inverse function

Determining whether a function is one-to-one is very useful because the inverse of a one-to-one function will also be a function. Analysing the graph of a function is the most effective way to determine if a function is one-to-one. Let's look at the graph of the one-to-one function $f(x) = x^3$ shown in Figure 1.21. It is clear that as the values of x increase over the domain (from $-\infty$ to ∞), the function values are always increasing. A function that is always increasing, or always decreasing, throughout its domain is one-to-one and has an inverse function.

A function that is not one-to-one (always increasing or always decreasing) can be made so by restricting its domain.

The function $f(x) = x^2$ (Figure 1.22) is not one-to-one for all real numbers. However, the function $g(x) = x^2$ with domain $x \geqslant 0$ (Figure 1.23) is always increasing (one-to-one), and the function $h(x) = x^2$ with domain $x \leqslant 0$ (Figure 1.24) is always decreasing (one-to-one).

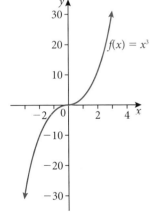

Figure 1.21 Graph of $f(x) = x^3$ which is increasing as x goes from $-\infty$ to ∞

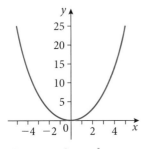

Figure 1.22 $f(x) = x^2$

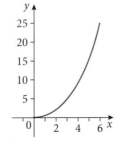

Figure 1.23 $g(x) = x^2, x \geqslant 0$

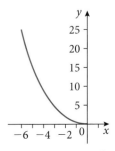

Figure 1.24 $h(x) = x^2, x \leqslant 0$

A function for which at least one element y in the range is the image of more than one element x in the domain is called a **many-to-one function**. Examples of many-to-one functions that we have already encountered are $y = x^2$, $x \in \mathbb{R}$ and $y = |x|$, $x \in \mathbb{R}$. As Figure 1.25 illustrates for $y = |x|$, a horizontal line exists that intersects a many-to-one function at more than one point. Thus, the inverse of a many-to-one function will not be a function.

If a function f is always increasing or always decreasing in its domain (i.e. it is monotonic), then f has an inverse f^{-1}.

No horizontal line can pass through the graph of a one-to-one function at more than one point.

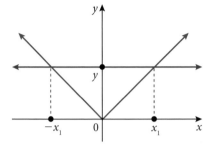

Figure 1.25 Graph of $y = |x|$; an example of a many-to-one function

Finding the inverse of a function

Example 1.23

The function f is defined for $x \in \mathbb{R}$ by $f(x) = 4x - 8$

(a) Determine if f has an inverse f^{-1}. If not, restrict the domain of f in order to find an inverse function f^{-1}

(b) Verify the result by showing that $(f \circ f^{-1})(x) = x$ and $(f^{-1} \circ f)(x) = x$

(c) Graph f and its inverse function f^{-1} on the same set of axes.

Solution

(a) Recognise that f is an increasing function for $(-\infty, \infty)$ because the graph of $f(x) = 4x - 8$ is a straight line with a constant slope of 4. Therefore, f is a one-to-one function and it has an inverse f^{-1}

(b) To find the equation for f^{-1}, start by switching the domain (x) and range (y) since the domain of f becomes the range of f^{-1} and the range of f becomes the domain of f^{-1}, as stated in the definition. Also, recall that $y = f(x)$.

$f(x) = 4x - 8$

$y = 4x - 8$ write $y = f(x)$

$x = 4y - 8$ interchange x and y (switch the domain and range)

$4y = x + 8$ solve for y (dependent variable) in terms of x (independent variable)

$$y = \frac{1}{4}x + 2$$

$$f^{-1}(x) = \frac{1}{4}x + 2 \quad \text{resulting equation is } y = f^{-1}(x)$$

Verify that f and f^{-1} are inverses by showing that $f(f^{-1}(x)) = x$ and $f^{-1}(f(x)) = x$

$$f\left(\frac{1}{4}x + 2\right) = 4\left(\frac{1}{4}x + 2\right) - 8 = x + 8 - 8 = x$$

$$f^{-1}(4x - 8) = \frac{1}{4}(4x - 8) + 2 = x - 2 + 2 - x$$

This confirms that $y = 4x - 8$ and $y = \frac{1}{4}x + 2$ are inverses of each other. Here is a graph of this pair of inverse functions.

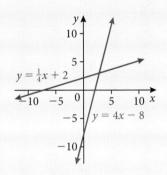

The method of interchanging domain (x) and range (y) to find the inverse function used in Example 1.23 also gives us a way for obtaining the graph of f^{-1} from the graph of f. Given the reversing effect that a pair of inverse functions have on each other, if $f(a) = b$ then $f^{-1}(b) = a$. Hence, if the ordered pair (a, b) is a point on the graph of $y = f(x)$, then the reversed ordered pair (b, a) must be on the graph of $y = f^{-1}(x)$. Figure 1.26 shows that the point (b, a) can be found by reflecting the point (a, b) about the line $y = x$. Therefore, the following statement can be made about the graphs of a pair of inverse functions.

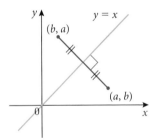

Figure 1.26 The point (b, a) is a reflection about the line $y = x$ of the point (a, b)

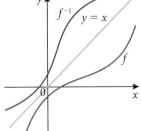

Figure 1.27 Graphs of f and f^{-1} are symmetric about the line $y = x$

The graph of f^{-1} is a reflection of the graph of f about the line $y = x$.

Example 1.24

The function f is defined for $x \in \mathbb{R}$ by $f{:}x \mapsto \dfrac{x^2 + 3}{x^2 + 1}$

(a) Determine if f has an inverse f^{-1}. If not, restrict the domain of f in order to find an inverse function f^{-1}.

(b) Graph f and its inverse f^{-1} on the same set of axes.

Solution

A graph of f produced on a GDC reveals that it is not monotonic over its domain $(-\infty, \infty)$. It is increasing for $(-\infty, 0]$, and decreasing for $[0, \infty)$. Therefore, f does not have an inverse f^{-1} for $x \in \mathbb{R}$. It is customary to restrict the domain to the 'largest' set possible. Hence, we can choose to restrict the domain to either $x \in (-\infty, 0]$

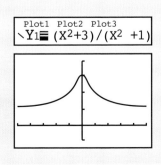

(making f an increasing function), or $x \in [0, \infty)$ (making f a decreasing function). Change the domain from $x \in \mathbb{R}$ to $x \in [0, \infty)$.

We use a method similar to that in Example 1.23 to find the equation for f^{-1}. First solve for x in terms of y and then interchange the domain (x) and range (y).

$$f:x \mapsto \frac{x^2 + 3}{x^2 + 1} \Rightarrow y = \frac{x^2 + 3}{x^2 + 1} \Rightarrow x^2 y + y = x^2 + 3 \Rightarrow x^2 y - x^2 = 3 - y$$

$$\Rightarrow x^2(y - 1) = 3 - y \Rightarrow x^2 = \frac{3 - y}{y - 1} \Rightarrow x = \pm\sqrt{\frac{3 - y}{y - 1}}$$

$$\Rightarrow y = \pm\sqrt{\frac{3 - x}{x - 1}}$$

Since we chose to restrict the domain of f to $x \in [0, \infty)$, the range of f^{-1} will be $y \in [0, \infty)$. Therefore, from the working above the resulting inverse function is $f^{-1}(x) = \sqrt{\frac{3 - x}{x - 1}}$. The graphs of f and f^{-1} show symmetry about the line $y = x$.

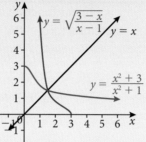

To find the inverse of a function f:

1. Determine if the function is one-to-one; if not, restrict the domain so that it is.
2. Replace $f(x)$ with y.
3. Solve for x in terms of y.
4. Interchange x and y.
5. Replace y with $f^{-1}(x)$.
6. The domain of f^{-1} is equal to the range of f; and the range of f^{-1} is equal to the domain of f.

Example 1.25

Consider the function $f:x \mapsto \sqrt{x + 3}$, $x \geqslant -3$

(a) Determine the inverse function f^{-1}

(b) Find the domain of f^{-1}

Solution

(a) Following the steps for finding the inverse of a function gives:

$y = \sqrt{x + 3}$ replace $f(x)$ with y

$y^2 = x + 3$ solve for x in terms of y; squaring both sides

$x = y^2 - 3$ solve for x

$y = x^2 - 3$ interchange x and y

Therefore, $f^{-1}: x \mapsto x^2 - 3$ replace y with $f^{-1}(x)$

(b) The domain explicitly defined for f is $x \geqslant -3$ and since the $\sqrt{\ }$ symbol stands for the principal square root (positive), then the range of f is all positive real numbers, i.e. $y \geqslant 0$. The domain of f^{-1} is equal to the range of f, therefore the domain of f^{-1} is $x \geqslant 0$.

Example 1.26

Consider the functions $f(x) = 2(x + 4)$ and $g(x) = \frac{1 - x}{3}$

(a) Find g^{-1} and state its domain and range.

(b) Solve the equation $(f \circ g^{-1})(x) = 2$

Solution

(a) $y = \dfrac{1 - x}{3}$ replace $f(x)$ with y

$x = \dfrac{1 - y}{3}$ interchange x and y

$3x = 1 - y$ solve for y

$y = -3x + 1$ solved for y

Therefore, $g^{-1}(x) = -3x + 1$ replace y with $g^{-1}(x)$

g is a linear function and its domain is $x \in \mathbb{R}$ and its range is $y \in \mathbb{R}$; therefore, for g^{-1} the domain is $x \in \mathbb{R}$ and the range is $y \in \mathbb{R}$.

(b) $(f \circ g^{-1})(x) = f(g^{-1}(x)) = f(-3x + 1) = 2$

$2[(-3x + 1) + 4] = 2$

$-6x + 2 + 8 = 2$

$-6x = -8$

$x = \dfrac{4}{3}$

Exercise 1.4

In questions 1–4, assume that f is a one-to-one function.

1. (a) If $f(2) = -5$, then what is $f^{-1}(-5)$?

   (b) If $f^{-1}(6) = 10$, then what is $f(10)$?

2. (a) If $f(-1) = 13$, then what is $f^{-1}(13)$?

   (b) If $f^{-1}(b) = a$, then what is $f(a)$?

3. If $g(x) = 3x - 7$, then what is $g^{-1}(5)$?

4. If $h(x) = x^2 - 8x$, with $x \geqslant 4$, then what is $h^{-1}(-12)$?

5. For each pair of functions, show algebraically and graphically that f and g are inverse functions by:

   (i) verifying that $(f \circ g)(x) = x$ and $(g \circ f)(x) = x$

   (ii) sketching the graphs of f and g on the same set of axes with equal scales on the x-axis and y-axis.

   Use your GDC to assist in making your sketches on paper.

   (a) $f{:}x \mapsto x + 6; \ g{:}x \mapsto x - 6$

   (b) $f{:}x \mapsto 4x; \ g{:}x \mapsto \dfrac{x}{4}$

   (c) $f{:}x \mapsto 3x + 9; \ g{:}x \mapsto \dfrac{1}{3}x - 3$

   (d) $f{:}x \mapsto \dfrac{1}{x}; \ g{:}x \mapsto \dfrac{1}{x}$

(e) $f\!:\!x \mapsto x^2 - 2, x \geqslant 0$; $g\!:\!x \mapsto \sqrt{x + 2}, x \geqslant -2$

(f) $f\!:\!x \mapsto 5 - 7x$; $g\!:\!x \mapsto \dfrac{5 - x}{7}$

(g) $f\!:\!x \mapsto \dfrac{1}{1 + x}$; $g\!:\!x \mapsto \dfrac{1 - x}{x}$

(h) $f\!:\!x \mapsto (6 - x)^{\frac{1}{2}}$; $g\!:\!x \mapsto 6 - x^2, x \geqslant 0$

(i) $f\!:\!x \mapsto x^2 - 2x + 3, x \geqslant 1$; $g\!:\!x \mapsto 1 + \sqrt{x - 2}, x \geqslant 2$

(j) $f\!:\!x \mapsto \sqrt[3]{\dfrac{x + 6}{2}}$; $g\!:\!x \mapsto 2x^3 - 6$

6. Find the inverse function f^{-1} and state its domain.

(a) $f(x) = 2x - 3$

(b) $f(x) = \dfrac{x + 7}{4}$

(c) $f(x) = \sqrt{x}$

(d) $f(x) = \dfrac{1}{x + 2}$

(e) $f(x) = 4 - x^2, x \geqslant 0$

(f) $f(x) = \sqrt{x - 5}$

(g) $f(x) = ax + b, a \neq 0$

(h) $f(x) = x^2 + 2x, x \geqslant -1$

(i) $f(x) = \dfrac{x^2 - 1}{x^2 + 1}, x \leqslant 0$

(j) $f(x) = x^3 + 1$

7. Determine if f has an inverse f^{-1}. If not, restrict the domain of f in order to find an inverse function. Graph f and its inverse f^{-1} on the same set of axes.

(a) $f(x) = \dfrac{2x + 3}{x - 1}$

(b) $f(x) = (x - 2)^2$

(c) $f(x) = \dfrac{1}{x^2}$

(d) $f(x) = 2 - x^4$

8. Use your GDC to graph the function $f(x) = \dfrac{2x}{1 + x^2}, x \in \mathbb{R}$. Find three intervals for which f is a one-to-one function (monotonic) and hence, will have an inverse f^{-1} on the interval. The union of all three intervals is all real numbers.

9. Use the functions $g(x) = x + 3$ and $h(x) = 2x - 4$ to find the indicated value or the indicated function.

(a) $(g^{-1} \circ h^{-1})(5)$

(b) $(h^{-1} \circ g^{-1})(9)$

(c) $(g^{-1} \circ g^{-1})(2)$

(d) $(h^{-1} \circ h^{-1})(2)$

(e) $g^{-1} \circ h^{-1}$

(f) $h^{-1} \circ g^{-1}$

(g) $(g \circ h)^{-1}$

(h) $(h \circ g)^{-1}$

10. The reciprocal function, $f(x) = \dfrac{1}{x}$, is its own inverse (self-inverse). Show that any function in the form $f(x) = \dfrac{a}{x + b} - b, a \neq 0$ is its own inverse.

1.5 Transformations of functions

Even when you use your GDC to sketch the graph of a function, it is helpful to know what to expect in terms of the location and shape of the graph – and even more so, if you're not allowed to use your GDC for a particular question. In this section, you will look at how certain changes to the equation of a function can affect, or **transform**, the location and shape of its graph. You will investigate three different types of **transformations** of functions that include how the graph of a function can be **translated**, **reflected** and **stretched** (or **shrunk**). Studying graphical transformations will help you to sketch and visualise many different functions efficiently. You will also take a closer look at two specific functions: the absolute value function, $y = |x|$, and the reciprocal function, $y = \dfrac{1}{x}$

> When analysing the graph of a function it is often convenient to express a function in the form $y = f(x)$. As we have done throughout this chapter, we can refer to a function such as $f(x) = x^2$ by the equation $y = x^2$.

Graphs of common functions

It is important to be familiar with the location and shape of a certain set of common functions. For example, from our previous knowledge about linear equations, we can determine the location of the linear function $f(x) = ax + b$. We know that the graph of this function is a line whose slope is a and whose y-intercept is $(0, b)$.

The eight graphs in Figure 1.28 represent some of the most commonly used functions in algebra. You should be familiar with the characteristics of the graphs of these common functions. This will help you predict and analyse the graphs of more complicated functions that are derived from applying one or more transformations to these simple functions.

> There are other important basic functions with which you should be familiar, for example, logarithmic and exponential functions, but you will learn about these in later chapters.

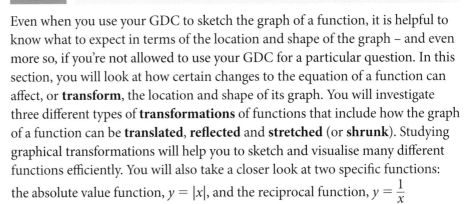

(a) Constant function (b) Identity function (c) Absolute value function

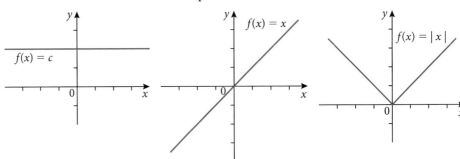

(d) Squaring function (e) Square root function (f) Cubing function

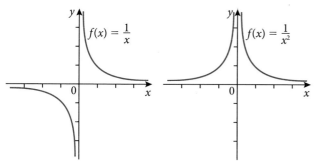

(g)   Reciprocal function   (h)   Inverse square function

Figure 1.28 Graphs of common functions

The word *inverse* can have different meanings in mathematics depending on the context. In section 1.4, 'inverse' is used to describe operations or functions that undo each other. However, 'inverse' is sometimes used to denote the **multiplicative inverse** (or **reciprocal**) of a number or function. This is how it is used in the names for the functions **g** and **h** shown above. The function in **g** is sometimes called the **reciprocal function**.

We will see that many functions have graphs that are a transformation (translation, reflection or stretch), or a combination of transformations, of one of these common functions.

Vertical and horizontal translations

Use your GDC to graph each of these functions: $f(x) = x^2$, $g(x) = x^2 + 3$ and $h(x) = x^2 - 2$. How do the graphs of g and h compare with the graph of f? The graphs of g and h appear to have the same shape – it's only the location, or position, that has changed compared to f. Although the curves (parabolas) appear to be getting closer together, their vertical separation at every value of x is constant.

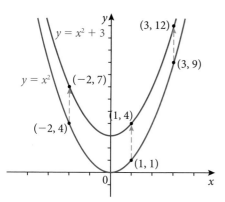

Figure 1.29 Translating $f(x) = x^2$ up 3 units.

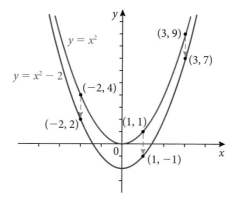

Figure 1.30 Translating $f(x) = x^2$ down 2 units.

Given $k > 0$:
- The graph of $y = f(x) + k$ is obtained by translating the graph of $y = f(x)$ up by k units.
- The graph of $y = f(x) - k$ is obtained by translating the graph of $y = f(x)$ down by k units.

As Figures 1.29 and 1.30 show, we can obtain the graph of $g(x) = x^2 + 3$ by translating (shifting) the graph of $f(x) = x^2$ **up** three units, and we can obtain the graph of $h(x) = x^2 - 2$ by translating the graph of $f(x) = x^2$ **down** two units.

Change function g to $g(x) = (x + 3)^2$ and change function h to $h(x) = (x - 2)^2$. Graph these two functions along with the original function $f(x) = x^2$ on your GDC. This time you can observe that the functions g and h can be obtained by a horizontal translation of f.

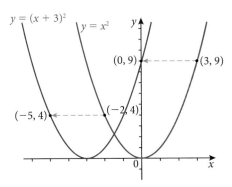

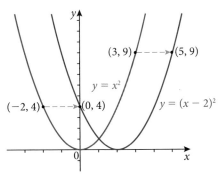

Figure 1.31 Translating $y = x^2$ left 3 units to produce $y = (x + 3)^2$

Figure 1.32 Translating $y = x^2$ right 3 units the graph of to produce the graph of $y = (x - 2)^2$

Given $k > 0$:
- The graph of $y = f(x - h)$ is obtained by translating the graph of $y = f(x)$ h units to the right.
- The graph of $y = f(x + h)$ is obtained by translating the graph of $y = f(x)$ h units to the left.

As Figures 1.31 and 1.32 show, we can obtain the graph of $g(x) = (x + 3)^2$ by translating the graph of $f(x) = x^2$ three units to the **left**, and we can obtain the graph of $h(x) = (x - 2)^2$ by translating the graph of $f(x) = x^2$ two units to the **right**.

Example 1.27

The diagrams show how the graph of $y = \sqrt{x}$ is transformed to the graph of $y = f(x)$ in three steps. For each diagram, (a) and (b), give the equation of the curve.

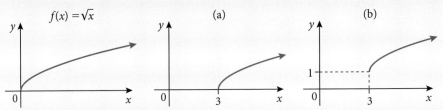

In Example 1.27, if the transformations were performed in reverse order, that is, the vertical translation followed by the horizontal translation, we would get the same final graph (in part (b)) with the same equation. The order in which we apply both a vertical and horizontal translation on a function does not make any difference. The translations are commutative. However, as we will see further in the chapter, it can make a difference how other sequences of transformations are applied. In general, transformations are not commutative.

Solution

To obtain the graph in (a), the graph of $y = \sqrt{x}$ is translated three units to the right. To produce the equation of the translated graph, -3 is added inside the argument of the function $y = \sqrt{x}$.

Therefore, the equation of the curve graphed in (a) is $y = \sqrt{x - 3}$

To obtain the graph in (b), the graph of $y = \sqrt{x - 3}$ is translated up one unit. To produce the equation of the translated graph, $+1$ is added outside the function. Therefore, the equation of the curve graphed in (b) is

$$y = \sqrt{x - 3} + 1 \text{ or } y = 1 + \sqrt{x - 3}$$

Example 1.28

Write the equation of the absolute value function shown by Figure 1.33

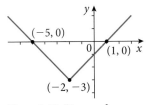

Figure 1.33 Diagram for Example 1.28

Solution

The graph shown is exactly the same shape as the graph of the equation $y = |x|$ but in a different position. Given that the vertex is $(-2, -3)$, it is clear that this graph can be obtained by translating $y = |x|$ two units left and then three units down. When we move $y = |x|$ two units left we get the graph of $y = |x + 2|$. Moving the graph of $y = |x + 2|$ down three units down gives us the graph of $y = |x + 2| - 3$. Therefore, the equation of the graph shown is $y = |x + 2| - 3$

We would get the same result if we applied the translations in the reverse order.

Reflections

Use your GDC to graph the two functions $f(x) = x^2$ and $g(x) = -x^2$. The graph of $g(x) = -x^2$ is a reflection in the x-axis of $f(x) = x^2$. This certainly makes sense because g is formed by multiplying f by -1, causing the y-coordinate of each point on the graph of $y = -x^2$ to be the negative of the y-coordinate of the point on the graph of $y = x^2$ that has the same x-coordinate.

Figures 1.34 and 1.35 show that the graph of $y = -f(x)$ is obtained by reflecting the graph of $y = f(x)$ in the x-axis.

The expression $-x^2$ is potentially ambiguous. It is accepted to be equivalent to $-(x)^2$. It is not equivalent to $(-x)^2$. For example, if you enter the expression -3^2 into your GDC it gives a result of -9, *not* $+9$. The expression -3^2 is consistently interpreted as 3^2 being multiplied by -1. The same as $-x^2$ is interpreted as x^2 being multiplied by -1.

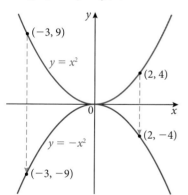

Figure 1.34 Reflecting $y = x^2$ in the x-axis

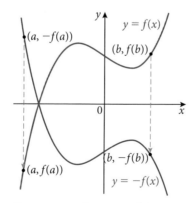

Figure 1.35 Reflecting $y = f(x)$ in the x-axis

Graph $f(x) = \sqrt{x - 2}$ and $g(x) = \sqrt{-x - 2}$. With $f(x) = x^2$ and $g(x) = -x^2$, g was formed by multiplying the entire function f by -1. However, for $f(x) = \sqrt{x - 2}$ and $g(x) = \sqrt{-x - 2}$, g is formed by multiplying the variable x by -1. In this case, the graph of $g(x) = \sqrt{-x - 2}$ is a reflection in the y-axis of $f(x) = \sqrt{x - 2}$. This makes sense if you recognise that the y-coordinate on the graph of $y = \sqrt{-x}$ will be the same as the y-coordinate on the graph of $y = \sqrt{x}$ if the value substituted for x in $y = \sqrt{-x}$ is the opposite of the x value in $y = \sqrt{x}$. For example, if $x = 9$ then $y = \sqrt{9} = 3$; and, if $x = -9$ then $y = \sqrt{-(-9)} = \sqrt{9} = 3$. Opposite values of x in the two functions produce the same y-coordinate for each.

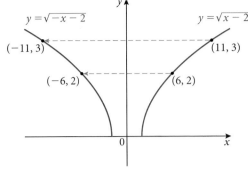

Figure 1.36 Reflecting $y = \sqrt{x-2}$ in the y-axis

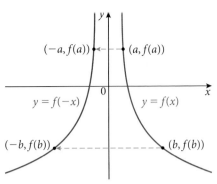

Figure 1.37 Reflecting $y = f(x)$ in the y-axis

The graph of $y = -f(x)$ is obtained by reflecting the graph of $y = f(x)$ in the x-axis.
The graph of $y = f(-x)$ is obtained by reflecting the graph of $y = f(x)$ in the y-axis.

Figures 1.36 and 1.37 illustrate how the graph of $y = f(-x)$ is obtained by reflecting the graph of $y = f(x)$ in the y-axis.

Example 1.29

For $g(x) = 2x^3 - 6x^2 + 3$, find:

(a) the function $h(x)$ that is the reflection of $g(x)$ in the x-axis

(b) the function $p(x)$ that is the reflection of $g(x)$ in the y-axis.

Solution

(a) Knowing that $y = -f(x)$ is the reflection of $y = f(x)$ in the x-axis, then $h(x) = -g(x) = -(2x^3 - 6x^2 + 3) \Rightarrow h(x) = -2x^3 + 6x^2 - 3$ will be the reflection of $g(x)$ in the x-axis. We can verify the result on the GDC – graphing the original equation $y = 2x^3 - 6x^2 + 3$ in bold style.

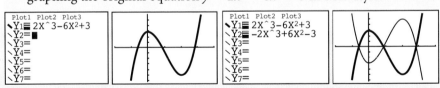

(b) Knowing that $y = f(-x)$ is the reflection of $y = f(x)$ in the y-axis, we need to substitute $-x$ in for x in $y = g(x)$.
Thus, $p(x) = g(-x) = 2(-x)^3 - 6(-x)^2 + 3 \Rightarrow p(x) = -2x^3 - 6x + 3$ will be the reflection of $g(x)$ in the y-axis. Again, we can verify the result on the GDC – graphing the original equation $y = 2x^3 - 6x^2 + 3$ in bold style.

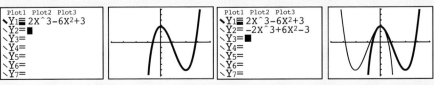

Non-rigid transformations: stretching and shrinking

Horizontal and vertical translations, and reflections in the x- and y-axes are called **rigid transformations** because the shape of the graph does not change – only its position is changed. **Non-rigid transformations** cause the shape of the original graph to change. The non-rigid transformations that you will study cause the shape of a graph to stretch or shrink in either the vertical or horizontal direction.

Vertical stretch or shrink

Graph the functions: $f(x) = x^2$, $g(x) = 3x^2$ and $h(x) = \frac{1}{3}x^2$. How do the graphs of g and h compare to the graph of f? Refer to figures 1.38 and 1.40. Clearly, the shape of the graphs of g and h is not the same as the graph of f. Multiplying the function f by a positive number greater than one, or less than one, has distorted the shape of the graph. For a certain value of x, the y-coordinate of $y = 3x^2$ is three times the y-coordinate of $y = x^2$. Therefore, the graph of $y = 3x^2$ can be obtained by **vertically stretching** the graph of $y = x^2$ by a factor of 3 (**scale factor** 3).

Likewise, the graph of $y = \frac{1}{3}x^2$ can be obtained by **vertically shrinking** the graph of $y = x^2$ by scale factor $\frac{1}{3}$

Figures 1.38 and 1.39 below show how multiplying a function by a positive number, a, greater than 1 causes a transformation in which the function stretches vertically by scale factor a. A point (x, y) on the graph of $y = f(x)$ is transformed to the point (x, ay) on the graph of $y = af(x)$.

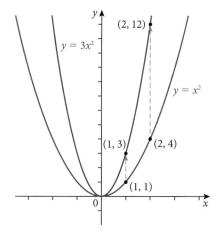

Figure 1.38 Vertical stretch of $y = x^2$ by scale factor 3

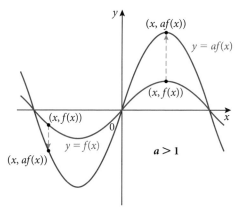

Figure 1.39 Vertical stretch of $y = f(x)$ by scale factor a when $a > 1$

Figures 1.40 and 1.41 below show how multiplying a function by a positive number, a, greater than 0 and less than 1 causes the function to shrink vertically by scale factor a. A point (x, y) on the graph of $y = f(x)$ is transformed to the point (x, ay) on the graph of $y = af(x)$.

If $a > 1$, then the graph of $y = af(x)$ is obtained by vertically stretching the graph of $y = f(x)$.

If $0 < a < 1$, then the graph of $y = af(x)$ is obtained by vertically shrinking the graph of $y = f(x)$.

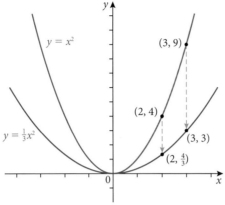

Figure 1.40 Vertical shrink of $y = x^2$ by scale factor $\dfrac{1}{3}$

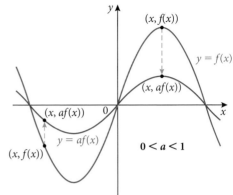

Figure 1.41 Vertical shrink of $y = f(x)$ by scale factor a, when $0 < a < 1$

Horizontal stretch or shrink

We will now investigate how the graph of $y = f(ax)$ is obtained from the graph of $y = f(x)$. Given $f(x) = x^2 - 4x$, find another function, $g(x)$, such that $g(x) = f(2x)$. We substitute $2x$ for x in the function f, giving $g(x) = (2x)^2 - 4(2x)$. For the purposes of our investigation, leave $g(x)$ in this form. On your GDC, graph these two functions, $f(x) = x^2 - 4x$ and $g(x) = (2x)^2 - 4(2x)$, using the indicated viewing window and graphing f in bold style (Figure 1.42).

Comparing the graphs of the two equations, we can see that $y = g(x)$ is not a translation or a reflection of $y = f(x)$. It is similar to the shrinking effect that occurs for $y = af(x)$ when $0 < a < 1$, except, instead of a vertical shrinking, the graph of $y = g(x) = f(2x)$ is obtained by horizontally shrinking the graph of $y = f(x)$. Given that it is a shrinking, the scale factor must be less than 1.

Consider the point $(4, 0)$ on the graph of $y = f(x)$. The point on the graph of $y = g(x) = f(2x)$ with the same y-coordinate and on the same side of the parabola is $(2, 0)$. The x-coordinate of the point on $y = f(2x)$ is the x-coordinate of the point on $y = f(x)$ multiplied by $\dfrac{1}{2}$. Use your GDC to confirm this for other pairs of corresponding points on $y = x^2 - 4x$ and $y = (2x)^2 - 4(2x)$ that have the same y-coordinate. The graph of $y = f(2x)$ can be obtained by horizontally shrinking the graph of $y = f(x)$ with scale factor $\dfrac{1}{2}$. This makes sense because if $f(2x_2) = (2x_2)^2 - 4(2x_2)$ and $f(x_1) = x_1^2 - 4x_1$ are to produce the same y-value, then $2x_2 = x_1$, and thus $x_2 = \dfrac{1}{2}x_1$. Figures 1.43 and 1.44 show how multiplying the x variable of a function by a positive number, a, greater than 1, causes the function to shrink horizontally by scale factor $\dfrac{1}{a}$. A point (x, y) on the graph of $y = f(x)$ is transformed to the point $\left(\dfrac{1}{a}x, y\right)$ on the graph of $y = f(ax)$.

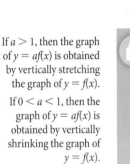

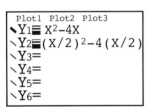

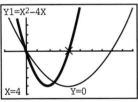

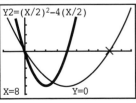

Figure 1.42 Graphs of $y = x^2 - 4x$ (in bold) and $y = \left(\dfrac{x}{2}\right)^2 - 4\left(\dfrac{x}{2}\right)$

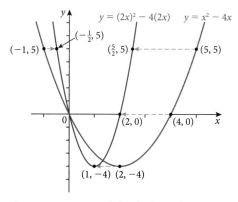

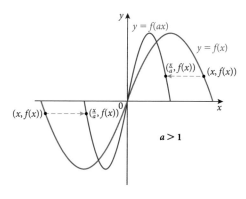

Figure 1.43 Horizontal shrink of $y = x^2 - 4x$ by scale factor $\frac{1}{2}$

Figure 1.44 Horizontal shrink of $y = f(x)$ by scale factor $\frac{1}{a}$, $a > 1$

If $0 < a < 1$, then the graph of the function $y = f(ax)$ is obtained by a horizontal stretching – rather than a shrinking – of the graph of $y = f(x)$ because the scale factor $\frac{1}{a}$ will be a value between 0 and 1 if $0 < a < 1$. Now, letting $a = \frac{1}{2}$ and, again using the function $f(x) = x^2 - 4x$, find $g(x)$, such that $g(x) = f\left(\frac{1}{2}x\right)$. Substitute $\frac{x}{2}$ for x in f, giving $g(x) = \left(\frac{x}{2}\right)^2 - 4\left(\frac{x}{2}\right)$. On your GDC, graph the functions f and g using the indicated viewing window with f in bold.

The graph of $y = \left(\frac{x}{2}\right)^2 - 4\left(\frac{x}{2}\right)$ is a horizontal stretching of the graph of

$y = x^2 - 4x$ by scale factor $\frac{1}{a} = \frac{1}{\frac{1}{2}} = 2$. For example, the point $(4, 0)$ on $y = f(x)$

has been moved horizontally to the point $(8, 0)$ on $y = g(x) = f\left(\frac{x}{2}\right)$.

Figures 1.45 and 1.46 below show how multiplying the x variable of a function by a positive number, a, greater than 0 and less than 1, causes the function to stretch horizontally by scale factor $\frac{1}{a}$. A point (x, y) on the graph of $y = f(x)$ is

transformed to the point $\left(\frac{1}{a}x, y\right)$ on the graph of $y = f(ax)$.

If $a > 1$, then the graph of $y = f(ax)$ is obtained by horizontally shrinking the graph of $y = f(x)$.

If $0 < a < 1$, then the graph of $y = f(ax)$ is obtained by horizontally stretching the graph of $y = f(x)$.

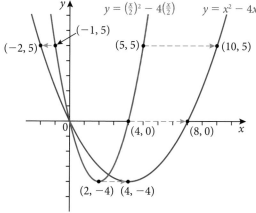

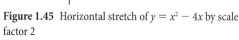

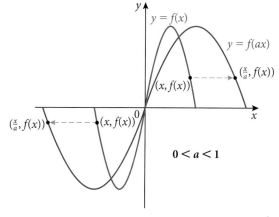

Figure 1.45 Horizontal stretch of $y = x^2 - 4x$ by scale factor 2

Figure 1.46 Horizontal stretch of $y = f(x)$ by scale factor $\frac{1}{a}$, $0 < a < 1$

Example 1.30

The graph of $y = f(x)$ is shown. Sketch the graph of each transformation.

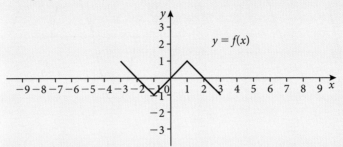

(a) $y = 3f(x)$

(b) $y = \frac{1}{3}f(x)$

(c) $y = f(3x)$

(d) $y = f\left(\frac{1}{3}x\right)$

Solution

(a) The graph of $y = 3f(x)$ is obtained by vertically stretching the graph of $y = f(x)$ with scale factor 3.

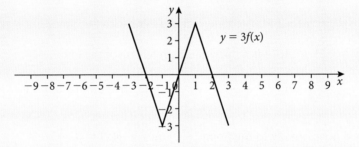

(b) The graph of $y = \frac{1}{3}f(x)$ is obtained by vertically shrinking the graph of $y = f(x)$ with scale factor $\frac{1}{3}$.

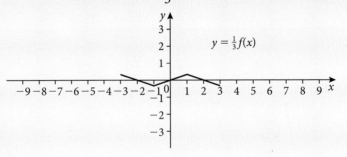

(c) The graph of $y = f(3x)$ is obtained by horizontally shrinking the graph of $y = f(x)$ with scale factor $\frac{1}{3}$.

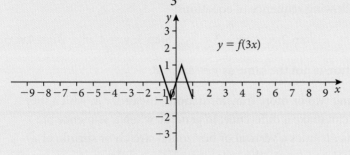

(d) The graph of $y = f\left(\frac{1}{3}\right)x$ is obtained by horizontally stretching the graph of $y = f(x)$ with scale factor 3.

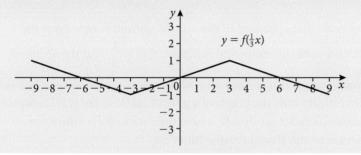

Example 1.31

Describe the sequence of transformations performed on the graph of $y = x^2$ to obtain the graph of $y = 4x^2 - 3$

Solution

Step 1: Start with the graph of $y = x^2$

Step 2: Vertically stretch $y = x^2$ by scale factor 4

Step 3: Vertically translate $y = 4x^2$ three units down

Step 1:

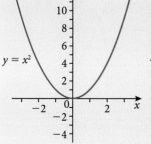

Step 2:

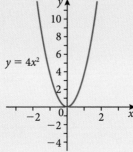

Step 3:

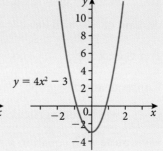

45

Note that in Example 1.31, a vertical stretch followed by a vertical translation does not produce the same graph if the two transformations are performed in reverse order. A vertical translation followed by a vertical stretch would generate the following sequence of equations:

Step 1: $y = x^2$ Step 2: $y = x^2 - 3$ Step 3: $y = 4(x^2 - 3) = 4x^2 - 12$

This final equation is not the same as $y = 4x^2 - 3$

When combining two or more transformations, the order in which they are performed can make a difference. In general, when a sequence of transformations includes a vertical or horizontal stretch or shrink, or a reflection through the x-axis, the order may make a difference.

Reciprocal and absolute value graphs

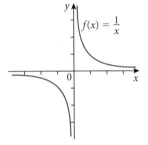

Figure 1.47 Reciprocal function

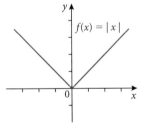

Figure 1.48 Absolute value function

Two of the functions that appeared in the set of common functions at the start of this section were the reciprocal function, $f(x) = \dfrac{1}{x}$, and the absolute value function.

We will now investigate how the graph of a given function, say $g(x)$, compares to that of a composite function $f(g(x))$, where the function f is either the reciprocal function or the absolute value function.

Example 1.32

Given that $f(x) = \dfrac{1}{x}$, $g(x) = -2x + 4$ and $h(x) = x^2 + 2x - 3$, sketch the graphs of the composite functions $f(g(x))$ and $f(h(x))$. Discuss the characteristics of the graphs of $f \circ g$ and $g \circ f$.

Solution

$$f(g(x)) = \frac{1}{g(x)} \Rightarrow y = \frac{1}{-2x + 4}$$

Clearly the reciprocal of g will be undefined wherever $g(x) = 0$, which means the domain of $\dfrac{1}{g(x)}$ will be $\{x : x \in \mathbb{R}, x \neq 2\}$.

Consequently, the graph of $\dfrac{1}{g(x)}$ will have a **vertical asymptote** with equation $x = 2$

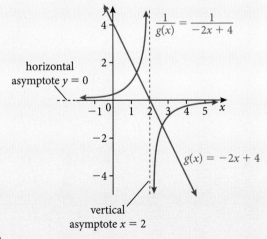

The graph of g illustrates that as x approaches the value of 2 ($x \to 2$) from the left side, the value of $g(x)$ is always positive but is converging to zero.

Therefore, as $x \to 2$ from the left (or, $x \to 2^-$), the values of $\dfrac{1}{g(x)}$ become increasingly large in the positive direction. We can express this behaviour symbolically by writing, as $x \to 2^-$, $\dfrac{1}{g(x)} \to +\infty$.

Similarly, as $x \to 2^+$, $\dfrac{1}{g(x)} \to -\infty$. Also, the x-axis ($y = 0$) is a **horizontal asymptote** for the graph of $\dfrac{1}{g(x)}$ because as the value of $g(x)$ becomes very large (either positive or negative), the value of $\dfrac{1}{g(x)}$ converges to zero; or, in symbols, as $x \to \pm\infty$, $\dfrac{1}{g(x)} \to 0$.

$$f(h(x)) = \frac{1}{h(x)} = \frac{1}{x^2 + 2x - 3}$$

$$= \frac{1}{(x + 3)(x - 1)}$$

Domain for $\dfrac{1}{h(x)}$ is

$\{x : x \in \mathbb{R}, x \neq -3, x \neq 1\}$

Since $h(x) = 0$ for $x = -3$ and $x = 1$ we anticipate that the graph of its reciprocal, $\dfrac{1}{h(x)}$, will have vertical asymptotes of $x = -3$ and $x = 1$.

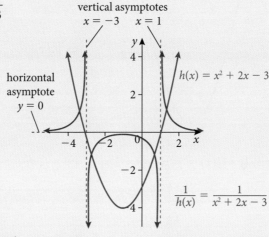

This is confirmed by the fact that as $x \to -3^-$, $\dfrac{1}{h(x)} \to +\infty$; as $x \to -3^+$, $\dfrac{1}{h(x)} \to -\infty$; and as $x^- \to 1$, $\dfrac{1}{h(x)} \to -\infty$; as $x \to 1^+$, $\dfrac{1}{h(x)} \to +\infty$.

The graph of $\dfrac{1}{h(x)}$ will also have a horizontal asymptote of $y = 0$ (x-axis) because as $x \to \pm\infty$, $\dfrac{1}{h(x)} \to 0$

In general, the line $x = c$ is a vertical asymptote of the graph of f if $f(x) \to \infty$ or $f(x) \to -\infty$ as x approaches c from either the left or from the right. The line $y = c$ is a horizontal asymptote of the graph of f if $f(x)$ approaches c as $x \to \infty$ or $x \to -\infty$

Example 1.33

Given $f(x) = |x|$ and using the same functions g and h from Example 1.32,

(a) graph the composite functions $f \circ g$ and $f \circ h$

(b) graph the composite functions $g \circ f$ and $h \circ f$

Solution

(a) $(f \circ g)(x) = f(-2x + 4) = |-2x + 4|$

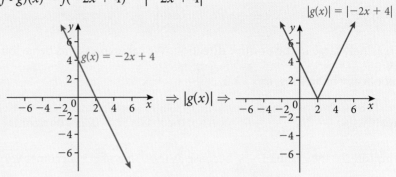

$(f \circ h)(x) = f(x^2 + 2x - 3) = |x^2 + 2x - 3|$

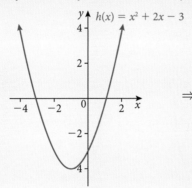 $\Rightarrow |h(x)| \Rightarrow$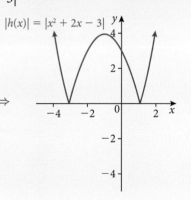

(b) $(g \circ f)(x) = g(|x|) = -2|x| + 4$

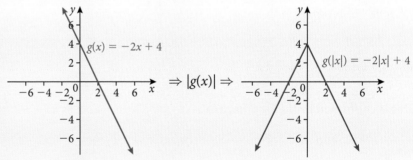

$(h \circ f)(x) = h(|x|) = |x|^2 + 2|x| - 3$

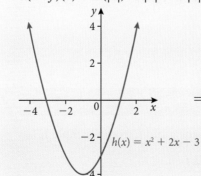 $\Rightarrow |h(x)| \Rightarrow$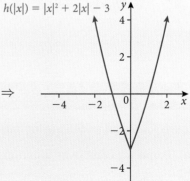

From part (a) of Example 1.33, you can see the change that occurs from the graph of a function to the graph of the **absolute value of the function**. Any portion of the graph of $g(x)$ or $h(x)$ that was below the x-axis gets reflected above the x-axis.

In part (b), you can see a change from the graph of a function to the graph of the **function of the absolute value**. Any portion of the graph of $g(x)$ or $h(x)$ that was left of the y-axis is eliminated, and any portion that was to the right of the y-axis is reflected to the left of the y-axis. Since the portion that was right of the y-axis remains, the resulting graph is always symmetric about the y-axis.

 Assume that a, h and k are positive real numbers.

Transformed function	Transformation performed on $y = f(x)$
$y = f(x) + k$	vertical translation k units up
$y = f(x) - k$	vertical translation k units down
$y = f(x - h)$	horizontal translation h units right
$y = f(x + h)$	horizontal translation h units left
$y = -f(x)$	reflection in the x-axis
$y = f(-x)$	reflection in the y-axis
$y = af(x)$	vertical stretch ($a > 1$) or shrink ($0 < a < 1$)
$y = f(ax)$	horizontal stretch ($0 < a < 1$) or shrink ($a > 1$)
$y = |f(x)|$	portion of the graph of $y = f(x)$ below x-axis is reflected above the x-axis
$y = f(|x|)$	symmetric about the y-axis; portion right of the y-axis is reflected in the y-axis

Table 1.4 Summary of transformations on the graphs of functions

Exercise 1.5

1. Sketch the graph of f, without a GDC or plotting points, by using your knowledge of some of the basic functions shown at the start of the Section 1.5.

   (a) $f{:}x \mapsto x^2 - 6$

   (b) $f{:}x \mapsto (x - 6)^2$

   (c) $f{:}x \mapsto |x| + 4$

   (d) $f{:}x \mapsto |x + 4|$

   (e) $f{:}x \mapsto 5 + \sqrt{x - 2}$

   (f) $f{:}x \mapsto \dfrac{1}{x - 3}$

   (g) $f{:}x \mapsto \dfrac{1}{(x + 5)^2} + 2$

   (h) $f{:}x \mapsto -x^3 - 4$

   (i) $f{:}x \mapsto -|x - 1| + 6$

   (j) $f{:}x \mapsto \sqrt{-x + 3}$

   (k) $f{:}x \mapsto 3\sqrt{x}$

   (l) $f{:}x \mapsto \dfrac{1}{2}x^2$

   (m) $f{:}x \mapsto \left(\dfrac{1}{2}x\right)^2$

   (n) $f{:}x \mapsto (-x)^3$

2. Write the equation for each graph.

(a)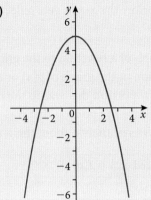

(b)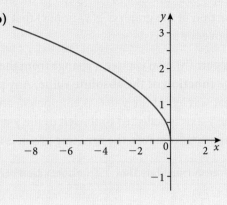

(c)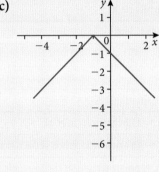

(d)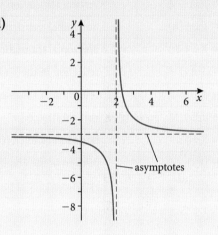

asymptotes

3. The graph of f is given. Sketch the graph of each transformed function.

   (a) $y = f(x) - 3$

   (b) $y = f(x - 3)$

   (c) $y = 2f(x)$

   (d) $y = f(2x)$

   (e) $y = -f(x)$

   (f) $y = f(-x)$

   (g) $y = 2f(x) + 4$

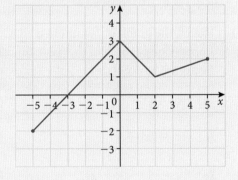

4. Specify a sequence of transformations to perform on the graph of $y = x^2$ to obtain the graph of the given function.

   (a) $g: x \mapsto (x - 3)^2 + 5$

   (b) $h: x \mapsto -x^2 + 2$

   (c) $p: x \mapsto \frac{1}{2}(x + 4)^2$

   (d) $f: x \mapsto [3(x - 1)]^2 - 6$

5. Without using your GDC, for each function $f(x)$, sketch the graph of:

   (i) $\dfrac{1}{f(x)}$ (ii) $|f(x)|$ (iii) $f(|x|)$

   Clearly label any intercepts or asymptotes.

   (a) $f(x) = \dfrac{1}{2}x - 4$

   (b) $f(x) = (x - 4)(x + 2)$

   (c) $f(x) = x^3$

Chapter 1 practice questions

1. The functions f and g are defined as $f{:}x \mapsto \sqrt{x - 3}$ and $g{:}x \mapsto x^2 + 2x$. The function $(f \circ g)(x)$ is defined for all $x \in \mathbb{R}$, except for the interval $]a, b[$.

   (a) Calculate the values of a and b

   (b) Find the range of $f \circ g$

2. Two functions g and h are defined as $g(x) = 2x - 7$ and $h(x) = 3(2 - x)$

   Find: **(a)** $g^{-1}(3)$

             (b) $(h \circ g)(6)$

3. Consider the functions $f(x) = 5x - 2$ and $g(x) = \dfrac{4 - x}{3}$

   (a) Find g^{-1}

   (b) Solve the equation $(f \circ g^{-1})(x) = 8$

4. The functions g and h are defined by $g{:}x \mapsto x - 3$ and $h{:}x \mapsto 2x$

   (a) Find an expression for $(g \circ h)(x)$

   (b) Show that $g^{-1}(14) + h^{-1}(14) = 24$

5. Use row operations to show that the following system of three linear equations has an infinite number of solutions.
   $$\begin{cases} 2x - y + 3z = 2 \\ 3x + y + 2z = -2 \\ -x + 2y - 3z = -4 \end{cases}$$

6. Use your GDC to solve the following system of three linear equations.
   $$\begin{cases} 3x - 2y + 5z = -3 \\ 2x + 6y - 4z = 20 \\ 4x - 3y + 5z = -5 \end{cases}$$

7. The diagram shows the graph of $y = f(x)$. It has maximum and minimum points at $(0, 0)$ and $(1, -1)$, respectively.

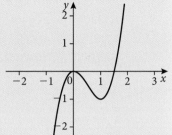

   (a) Copy the diagram, and then add in the graph of $y = f(x + 1) - \dfrac{1}{2}$

   (b) Find the coordinates of the minimum and maximum points of $y = f(x + 1) - \dfrac{1}{2}$

8. The diagram shows parts of the graphs of $y = x^2$ and $y = -\dfrac{1}{2}(x + 5)^2 + 3$

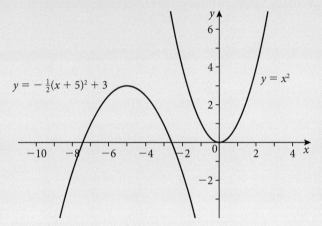

   The graph of $y = x^2$ may be transformed into the graph of

   $y = -\dfrac{1}{2}(x + 5)^2 + 3$ by these transformations:

   A reflection in the line $y = 0$, followed by a vertical stretch with scale factor k, followed by a horizontal translation of p units, followed by a vertical translation of q units.

   Write down the value of:

   (a) k (b) p (c) q

9. The function f is defined by $f(x) = \dfrac{4}{\sqrt{16 - x^2}}$, for $-4 < x < 4$

   (a) Without using a GDC, sketch the graph of f.

   (b) Write down the equation of each vertical asymptote.

   (c) Write down the range of the function f.

10. Let $g:x \mapsto \dfrac{1}{x}, x \neq 0$

  (a) Without using a GDC, sketch the graph of g.

  The graph of g is transformed to the graph of h by a translation of 4 units to the left and 2 units down.

  (b) Find an expression for the function h.

  (c) (i) Find the x- and y-intercepts of h.

  (ii) Write down the equations of the asymptotes of h.

  (iii) Sketch the graph of h.

11. Consider $f(x) = \sqrt{x + 3}$

  (a) Find:

  (i) $f(8)$ **(ii)** $f(46)$ **(iii)** $f(-3)$

  (b) Find the values of x for which f is undefined.

  (c) Let $g:x \mapsto x^2 - 5$. Find $(g \circ f)(x)$.

12. Let $g(x) = \dfrac{x - 8}{2}$ and $h(x) = x^2 - 1$

  (a) Find $g^{-1}(-2)$

  (b) Find an expression for $(g^{-1} \circ h)(x)$

  (c) Solve $(g^{-1} \circ h)(x) = 22$

13. Given the functions $f:x \mapsto 3x - 1$ and $g:x \mapsto \dfrac{4}{x}$, find the following:

  (a) f^{-1} **(b)** $f \circ g$ **(c)** $(f \circ g)^{-1}$ **(d)** $g \circ g$

14. (a) The diagram shows part of the graph of the function
  $$h(x) = \frac{a}{x - b}$$
  The curve passes through the point $A(-4, -8)$.
  The vertical line MN is an asymptote.

  Find the value of:
  (i) a **(ii)** b

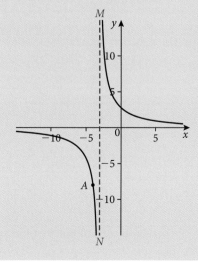

(b) The graph of $h(x)$ is transformed
as shown in the diagram.
The point A is transformed to
$A'(-4, 8)$.

Give a full geometric description
of the transformation.

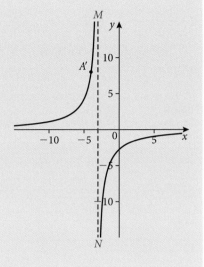

15. The graph of $y = f(x)$ is shown in the diagram.

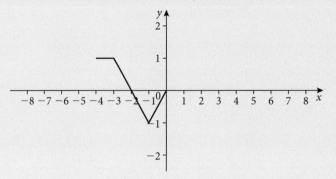

(a) Make two copies of the coordinate system as shown in the diagram
but without the graph of $y = f(x)$. On the first diagram sketch a
graph of $y = 2f(x)$, and on the second diagram sketch a graph of
$y = f(x - 4)$

(b) The point $A(-3, 1)$ is on the graph of $y = f(x)$. The point A' is
the corresponding point on the graph of $y = -f(x) - 1$. Find the
coordinates of A'.

16. The diagram shows
the graph of $y_1 = f(x)$.
The x-axis is a tangent
to $f(x)$ at $x = m$ and
$f(x)$ crosses the x-axis
at $x = n$.

On the same diagram,
sketch the graph of
$y_2 = f(x - k)$, where
$0 < k < n - m$ and
indicate the coordinates of the points of intersection of y_2 with the x-axis.

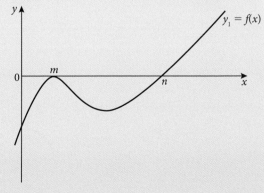

17. Given functions $f: x \mapsto x + 1$ and $g: x \mapsto x^3$, find the function $(f \circ g)^{-1}$

18. If $f(x) = \dfrac{x}{x + 1}$, for $x \neq -1$ and $g(x) = (f \circ f)(x)$, find:

    (a) $g(x)$ (b) $(g \circ g)(2)$

19. Let $f: x \mapsto \sqrt{\dfrac{1}{x^2} - 2}$. Find:

    (a) the set of real values of x for which f is real and finite

    (b) the range of f.

20. The function $f: x \mapsto \dfrac{2x + 1}{x - 1}$, $x \in \mathbb{R}$, $x \neq 1$. Find the inverse function, f^{-1}, clearly stating its domain.

21. The one-to-one function f is defined on the domain $x > 0$ by
    $$f(x) = \frac{2x - 1}{x + 2}$$

    (a) State the range, A, of f.

    (b) Obtain an expression for $f^{-1}(x)$, for $x \in A$.

22. The function f is defined by $f: x \mapsto x^3$

    Find an expression for $g(x)$ in terms of x in each of the following cases:

    (a) $(f \circ g)(x) = x + 1$ (b) $(g \circ f)(x) = x + 1$

23. (a) Find the largest set S of values of x such that the function
    $$f(x) = \frac{1}{\sqrt{3 - x^2}} \text{ takes real values.}$$

    (b) Find the range of the function f defined on the domain S.

24. Let f and g be two functions. Given that $(f \circ g)(x) = \dfrac{x + 1}{2}$ and $g(x) = 2x - 1$, find $f(x - 3)$.

25. The diagram shows part of the graph of $y = f(x)$ that passes through the points A, B, C and D.

Sketch, indicating clearly the images of A, B, C and D, the graphs of

(a) $y = f(x - 4)$ **(b)** $y = f(-3x)$

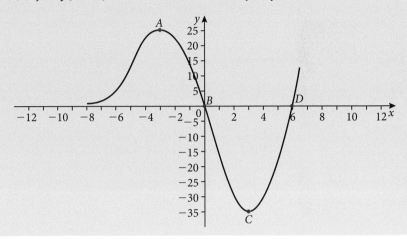

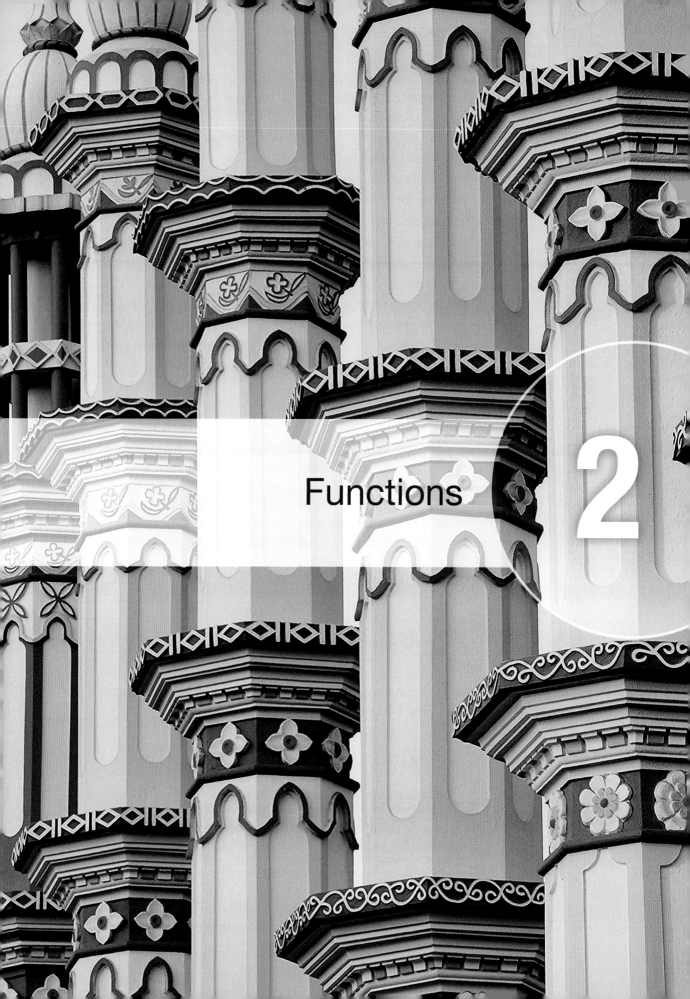

Functions

2

Learning objectives

By the end of this chapter, you should be familiar with...

- polynomial functions and their graphs
- zeros and factors of polynomial functions; roots of polynomial equations
- applying the factor and remainder theorems
- the sum and product of the roots of a polynomial equation
- quadratic functions, and different forms in which to express them
- finding characteristics of a parabola: axis of symmetry, x-intercepts and vertex
- solving quadratic equations and inequalities
- applying the quadratic formula and the discriminant of a quadratic equation
- rational functions and their graphs, and identifying all asymptotes
- solving a variety of equations and inequalities, both graphically and analytically
- partial fractions.

This chapter will focus on polynomial functions (which include quadratic functions) and rational functions. There are other function types that you need to be familiar with for this course. Chapter 3 will cover exponential functions and logarithmic functions, and Chapter 6 will focus on trigonometric functions. Along with polynomial and rational functions, this chapter will address solving polynomial equations, solving other types of equations and solving inequalities. The chapter closes with methods of manipulating rational expressions – techniques that will be very useful later in the course.

For this course, we restrict our study to polynomial functions with coefficients that are real numbers.

2.1 Polynomial functions

The most common type of algebraic function is a polynomial function. Here are some examples of polynomial functions:

$$f(x) = x^3, \ h(t) = -2t^2 + 16t - 24,$$
$$g(y) = y^5 + y^4 - 11y^3 + 7y^2 + 10y - 8$$

Definition of a polynomial function in the variable x

A polynomial function P is a function that can be expressed as

$$P(x) = a_n x^n + a_{n-1} x^{n-1} + \cdots + a_1 x + a_0, \ a_n \neq 0$$

where the non-negative integer n is the degree of the polynomial function.
The numbers $a_0, a_1, a_2, \ldots, a_n$ are real numbers and are the coefficients of the polynomial.
a_n is the leading coefficient, $a_n x^n$ is the leading term and a_0 is the constant term.

It is common practice to use subscript notation for coefficients of general polynomial functions, but for polynomial functions of low degree, the following simpler forms are often used.

Degree	Function form	Function name	Graph
Zero	$P(x) = a$	Constant function	Horizontal line
First	$P(x) = ax + b, a \neq 0$	Linear function	Line with slope a
Second	$P(x) = ax^2 + bx + c, a \neq 0$	Quadratic function	Parabola (∪-shape, 1 turn)
Third	$P(x) = ax^3 + bx^2 + cx + d, a \neq 0$	Cubic function	S-shape (2, 1 or no turns)

Table 2.1 Features of polynomial functions of low degree

To identify an individual term in a polynomial function, we use the function name correlated with the power of x contained in the term. For example, the polynomial function $f(x) = x^3 - 9x + 4$ has a cubic term of x^3, no quadratic term, a linear term of $-9x$, and a constant term of 4.

For each polynomial function $P(x)$ there is a corresponding **polynomial equation** $P(x) = 0$. When we solve polynomial equations, we often refer to the solutions as **roots**.

Approaches to finding zeros of various polynomial functions will be considered in the next three sections of this chapter.

If P is a function and c is a number such that $P(c) = 0$, then c is a **zero** of the function P (or of the polynomial P) and $x = c$ is a **root** of the equation $P(x) = 0$.

The use of the word root to denote the solution of a polynomial equation should not be confused with the use of the word in the context of square root, cube root, fifth root, etc.

Graphs of polynomial functions

The graph of a first-degree polynomial function (linear function), such as $P(x) = x$, is a line (Figure 2.1(a)). The graph of every second-degree polynomial function (quadratic function) is a parabola (Figure 2.1(b)).

The simplest type of polynomial function is one whose rule is given by a power of x. In Figure 2.1, the graphs of $P(x) = x^n$ for $n = 1, 2, 3, 4, 5$, and 6 are shown. As the figure suggests, the graph of $P(x) = x^n$ has the same general U-shape as $y = x^2$ when n is even, and the same general S-shape as $y = x^3$ when n is odd. However, as the degree n increases, the graphs of polynomial functions become flatter near the origin and steeper away from the origin.

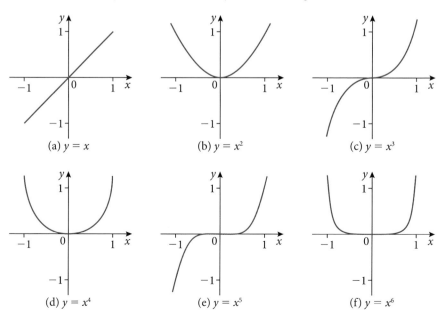

(a) $y = x$ (b) $y = x^2$ (c) $y = x^3$

(d) $y = x^4$ (e) $y = x^5$ (f) $y = x^6$

Figure 2.1 Graphs of $P(x) = x^n$ for increasing n

Another interesting observation is that depending on the degree of the polynomial function, its graph displays a certain type of symmetry.

The graph of $P(x) = x^n$ has rotational symmetry with respect to the origin when n is odd. These functions are called **odd functions**.

The graph of $P(x) = x^n$ has line symmetry with respect to the y-axis when n is even. These functions are called **even functions**.

We will meet formal definitions for odd and even functions in Chapter 6 when we investigate the graphs of the sine and cosine functions.

The graphs of polynomial functions that are not in the form $P(x) = x^n$ are more difficult to sketch. However, the graphs of all polynomial functions share these properties:

The third property of polynomial functions is referred to as the **end behaviour** because it describes how the curve behaves at the left and right ends (i.e. as $x \to -\infty$ and as $x \to +\infty$). The end behaviour of a polynomial function is determined by its degree and by the sign of its leading coefficient.

1 The graph is a smooth curve (i.e. it has no sharp, pointed turns − only smooth, rounded turns).

2 The graph is continuous (i.e. it has no breaks, gaps or holes).

3 The graph rises ($y \to \infty$) or falls ($y \to -\infty$) without bound as $x \to +\infty$ or $x \to -\infty$

4 The graph extends on forever both to the left ($-\infty$) and to the right ($+\infty$); domain is $\mathbb{R}$

5 The graph of a polynomial function of degree n has at most $(n-1)$ turning points.

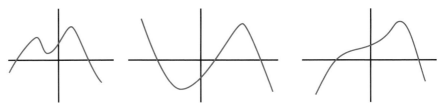

Figure 2.2 The graph of a polynomial function is a smooth, unbroken, continuous curve

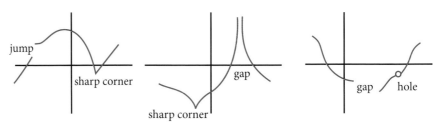

Figure 2.3 When a graph has jumps, gaps, holes or sharp corners, then it is not the graph of a polynomial function. None of these functions are polynomial functions

2.2 Quadratic functions

A linear function is a polynomial function of degree 1 that can be written in the general form $f(x) = ax + b$, where $a \neq 0$. Linear equations were briefly reviewed in Section 1.1. Any linear function will have a single solution (root) of $x = -\dfrac{b}{a}$. This is a formula that gives the zero of any linear polynomial.

In this section, we will focus on quadratic functions that are functions consisting of a second-degree polynomial that can be written in the form $f(x) = ax^2 + bx + c$, such that $a \neq 0$. You are probably familiar with the formula that gives the zeros of any quadratic polynomial, that is, the **quadratic formula**. We will also investigate other methods of finding zeros of quadratics and consider important characteristics of the graphs of quadratic functions.

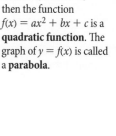

If a, b and c are real numbers, and $a \neq 0$, then the function $f(x) = ax^2 + bx + c$ is a **quadratic function**. The graph of $y = f(x)$ is called a **parabola**.

Each parabola is symmetric about a vertical line called its **axis of symmetry**. The axis of symmetry passes through a point on the parabola called the **vertex** of the parabola, as shown in Figure 2.4. If the leading coefficient, a, of the quadratic function is positive, then the parabola opens upwards (concave up) – and the y-coordinate of the vertex will be a **minimum value** for the function. If the leading coefficient is negative, then the parabola opens downwards (concave down) – and the y-coordinate of the vertex will be a **maximum value** for the function.

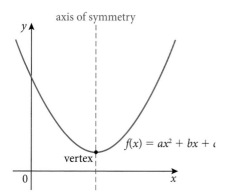

If $a > 0$ then the parabola opens upwards.

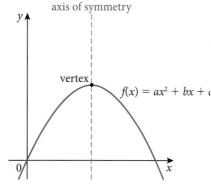

If $a < 0$ then the parabola opens downwards.

Figure 2.4 'Concave up' and 'concave down' parabolas

The graph of $f(x) = a(x - h)^2 + k$

From Section 1.5, we know that the graph of the equation $y = (x + 3)^2 + 2$ can be obtained by translating $y = x^2$ three units to the left and two units up. As we are familiar with the shape and position of the graph of $y = x^2$ and we know the two translations that transform $y = x^2$ to $y = (x + 3)^2 + 2$, we can sketch the graph of $y = (x + 3)^2 + 2$ (see Figure 2.5). We can also determine the axis of symmetry and the vertex of the graph. Figure 2.6 shows that the graph of $y = (x + 3)^2 + 2$ has an axis of symmetry of $x = -3$ and a vertex at $(-3, 2)$. The equation $y = (x + 3)^2 + 2$ can also be written as $y = x^2 + 6x + 11$.

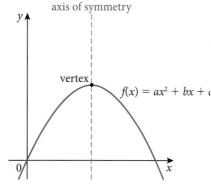

Figure 2.5 Translating $y = x^2$ to produce $y = (x + 3)^2 + 2$

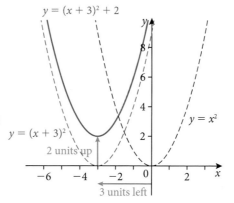

Figure 2.6 The axis of symmetry and the vertex for the graph of $y = (x + 3)^2 + 2$

As we can identify the vertex of the parabola easily when the equation is written as $y = (x + 3)^2 + 2$, we often refer to this as the vertex form of the quadratic equation, and $y = x^2 + 6x + 11$ as the general form.

Completing the square

When we are visualising and sketching quadratic functions, it is helpful to have them written in vertex form. We can rewrite a quadratic function written in general form into vertex form by completing the square.

For any real number p, the quadratic expression $x^2 + px + \left(\dfrac{p}{2}\right)^2$ is the square

of $\left(x + \dfrac{p}{2}\right)$. Convince yourself of this by expanding $\left(x + \dfrac{p}{2}\right)^2$. The technique

of completing the square is to add a constant to a quadratic expression to make it the square of a binomial expression. If the coefficient of the quadratic term (x^2) is

$+1$, the coefficient of the linear term is p, and the constant term is $\left(\dfrac{p}{2}\right)^2$, then

$x^2 + px + \left(\dfrac{p}{2}\right)^2 = \left(x + \dfrac{p}{2}\right)^2$ and the square is completed. Remember that

the coefficient of the quadratic term (leading coefficient) must be equal to $+1$ before completing the square.

If a quadratic function is written in vertex form, that is $f(x) = a(x - h)^2 + k$, with $a \neq 0$, then the graph of f has an axis of symmetry of $x = h$ and a vertex at (h, k).

$f(x) = a(x - h)^2 + k$ is sometimes referred to as the **standard form** of a quadratic function.

Example 2.1

Find the equation of the axis of symmetry and the coordinates of the vertex of the graph of $f(x) = x^2 - 8x + 18$ by rewriting the function in the form $f(x) = a(x - h)^2 + k$.

Solution

To complete the square and get the quadratic expression $x^2 - 8x + 18$ in the

form $x^2 + px + \left(\dfrac{p}{2}\right)^2$, the constant term needs to be $\left(\dfrac{-8}{2}\right)^2 = 16$. We need

to add 16, but also subtract 16, so that we are adding zero overall and, hence, not changing the original expression.

$f(x) = (x^2 - 8x + 16) - 16 + 18$ adding zero $(-16 + 16)$ to the right side

$f(x) = x^2 - 8x + 16 + 2$ $x^2 - 8x + 16$ fits the pattern

$$x^2 + px + \left(\dfrac{p}{2}\right)^2 \text{ with } p = -8$$

$f(x) = (x - 4)^2 + 2$ $x^2 - 8x + 16 = (x - 4)^2$

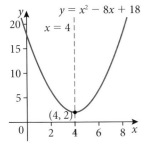

Figure 2.7 Graph for the solution to Example 2.1

Example 2.2

For the function $g : x \mapsto -2x^2 - 12x + 7$,

(a) find the axis of symmetry and the coordinates of the vertex of the graph of g

(b) indicate the transformations that can be applied to $y = x^2$ to obtain the graph of g

(c) find the minimum or the maximum value.

Solution

(a) $g : x \mapsto -2\left(x^2 + 6x - \dfrac{7}{2}\right)$

factorise so that coefficient of quadratic term is $+1$

$g : x \mapsto -2\left[(x^2 + 6x + 9) - 9 - \dfrac{7}{2}\right]$

$p = 6 \Rightarrow \left(\dfrac{p}{2}\right)^2 = 9$; hence, add $+9 - 9$ (zero)

$g : x \mapsto -2\left[(x + 3)^2 - \dfrac{18}{2} - \dfrac{7}{2}\right]$

$x^2 + 6x + 9 = (x + 3)^2$

$g : x \mapsto -2\left[(x + 3)^2 - \dfrac{25}{2}\right]$

$g : x \mapsto -2(x + 3)^2 + 25$

multiply through by -2 to remove outer brackets

$g : x \mapsto -2(x - (-3))^2 + 25$

express in vertex form $g : x \mapsto a(x - h)^2 + k$

The axis of symmetry of the graph of g is the vertical line $x = -3$ and the vertex is at $(-3, 25)$.

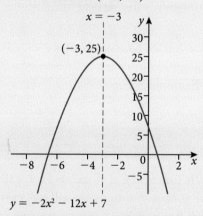

$y = -2x^2 - 12x + 7$

(b) Since $g : x \mapsto -2x^2 - 12x + 7 = -2(x + 3)^2 + 25$, we can obtain the graph of g by applying the following transformations (in the order given) on the graph of $y = x^2$: horizontal translation of 3 units left; reflection in the x-axis (parabola opening down); vertical stretch of factor 2; and a vertical translation of 25 units up.

(c) The parabola opens down because the leading coefficient is negative. Therefore, g has a maximum and no minimum value. The maximum value is 25 (y-coordinate of vertex) at $x = -3$.

Example 2.3 derives a general expression for the axis of symmetry and vertex of a quadratic function in the general form $f(x) = ax^2 + bx + c$ by completing the square.

Example 2.3

Find the axis of symmetry and the vertex for the general quadratic function $f(x) = ax^2 + bx + c$

Solution

$$f(x) = a\left(x^2 + \frac{b}{a}x + \frac{c}{a}\right) \qquad \text{factorise so that coefficient of } x^2 \text{ term is } +1$$

$$f(x) = a\left[x^2 + \frac{b}{a}x + \left(\frac{b}{2a}\right)^2 - \left(\frac{b}{2a}\right)^2 + \frac{c}{a}\right] \qquad p = \frac{b}{a} \Rightarrow \left(\frac{p}{2}\right)^2 = \left(\frac{b}{2a}\right)^2$$

$$f(x) = a\left[\left(x + \frac{b}{2a}\right)^2 - \frac{b^2}{4a^2} + \frac{c}{a}\right] \qquad x^2 + \frac{b}{a}x + \left(\frac{b}{2a}\right)^2 = \left(x + \frac{b}{2a}\right)^2$$

$$f(x) = a\left(x + \frac{b}{2a}\right)^2 - \frac{b^2}{4a} + c \qquad \text{multiply through by } a$$

$$f(x) = a\left(x - \left(-\frac{b}{2a}\right)\right)^2 + c - \frac{b^2}{4a} \qquad \text{express in vertex form}$$
$$f(x) = a(x - h)^2 + k$$

Check the results for Example 2.2 using the formulae for the axis of symmetry and vertex. For the function $g:x \mapsto -2x^2 - 12x + 7$:

$$x = -\frac{b}{2a} = -\frac{-12}{2(-2)} = -3 \Rightarrow \text{axis of symmetry is the vertical line } x = -3$$

$$c - \frac{b^2}{4a} = 7 - \frac{(-12)^2}{4(-2)} = \frac{56}{8} + \frac{144}{8} = 25 \Rightarrow \text{vertex has coordinates } (-3, 25)$$

These results agree with the results from Example 2.2.

This result leads to the following generalisation.
For the graph of the quadratic function $f(x) = ax^2 + bx + c$, the axis of symmetry is the vertical line with the equation $x = -\frac{b}{2a}$ and the vertex has coordinates $\left(-\frac{b}{2a}, c - \frac{b^2}{4a}\right)$

Zeros of a quadratic function

A specific value of x is a **zero** of a quadratic function $f(x) = ax^2 + bx + c$ if it is a solution (or **root**) to the equation $ax^2 + bx + c = 0$. As we will see, every quadratic function has two zeros, although it is possible for the same zero to occur twice (double zero, or double root). The x-coordinate of any point(s) where f crosses the x-axis (y-coordinate is zero) is a **real zero** of the function. A quadratic function can have no, one or two real zeros as Figure 2.8 illustrates. Finding all real zeros of a quadratic function requires you to solve quadratic equations of the form $ax^2 + bx + c = 0$. Although $a \neq 0$, it is possible for b or c to be equal to zero.

Figure 2.8 Different quadratic functions with different numbers of real zeros

one real zero

two real zeros

no real zeros

Now let us consider the general quadratic equation $ax^2 + bx + c = 0$ whose roots are $x = \alpha$ and $x = \beta$. We can write the quadratic equation with roots α and β as:

$$ax^2 + bx + c = (x - \alpha)(x - \beta) = 0$$

$$x^2 - \alpha x - \beta x + \alpha\beta = 0$$

$$x^2 - (\alpha + \beta)x + \alpha\beta = 0$$

In the next section, the factor theorem formally states the relationship between linear factors of the form $(x - \alpha)$ and the zeros for any polynomial.

Since the equation $ax^2 + bx + c = 0$ can also be written as $x^2 + \dfrac{b}{a}x + \dfrac{c}{a} = 0$, then:

$$x^2 - (\alpha + \beta)x + \alpha\beta = x^2 + \dfrac{b}{a}x + \dfrac{c}{a}$$

Equating coefficients of both sides, gives the results:

$$\alpha + \beta = -\dfrac{b}{a} \text{ and } \alpha\beta = \dfrac{c}{a}$$

For any quadratic equation in the form $ax^2 + bx + c = 0$, the **sum of roots** of the equation is $-\dfrac{b}{a}$ and the **product of the roots is** $\dfrac{c}{a}$.

Example 2.4

Given that α and β are the roots of the given equation, find the sum, $\alpha + \beta$, and product, $\alpha\beta$, of the roots.

(a) $x^2 - 5x + 3 = 0$ (b) $3x^2 + 4x - 7 = 0$

Solution

(a) For the equation $x^2 - 5x + 3 = 0$, $a = 1$, $b = -5$ and $c = 3$

Therefore, $\alpha + \beta = -\dfrac{b}{a} = -\dfrac{-5}{1} = 5$ and $\alpha\beta = \dfrac{c}{a} = \dfrac{3}{1} = 3$

(b) For the equation $3x^2 + 4x - 7 = 0$, $a = 3$, $b = 4$ and $c = -7$

Therefore, $\alpha + \beta = -\dfrac{b}{a} = -\dfrac{4}{3}$ and $\alpha\beta = \dfrac{c}{a} = -\dfrac{7}{3}$

If the sum and product of the roots of a quadratic equation are known then the equation can be written in the form: $x^2 - (\text{sum of roots})x + (\text{product of roots}) = 0$

Example 2.5

Given that α and β are the roots of the equation $2x^2 + 6x - 5 = 0$, find a quadratic equation whose roots are:

(a) $2\alpha, 2\beta$ (b) $\dfrac{1}{\alpha + 1}, \dfrac{1}{\beta + 1}$

Solution

For the equation $2x^2 + 6x - 5 = 0$, $a = 2$, $b = 6$ and $c = -5$

So, $\alpha + \beta = -\dfrac{b}{a} = -\dfrac{6}{2} = -3$ and $\alpha\beta = \dfrac{c}{a} = \dfrac{-5}{2}$

(a) Sum of the new roots $= 2\alpha + 2\beta = 2(\alpha + \beta) = 2(-3) = -6$

So, for the new equation $-\dfrac{b}{a} = -6$

Product of the new roots $= 2\alpha \cdot 2\beta = 4\alpha\beta = 4\left(-\dfrac{5}{2}\right) = -10$

So, for the new equation $\dfrac{c}{a} = -10$

The new equation can be written as $ax^2 + bx + c = 0$ or

$$x^2 + \frac{b}{a}x + \frac{c}{a} = 0$$

Therefore, the quadratic equation with roots 2α, 2β is

$$x^2 - (-6)x - 10 = 0 \Rightarrow x^2 + 6x - 10 = 0$$

(b)   Sum of the new roots $= \dfrac{1}{\alpha + 1} + \dfrac{1}{\beta + 1} = \dfrac{\beta + 1 + \alpha + 1}{(\alpha + 1)(\beta + 1)}$

$$= \frac{\alpha + \beta + 2}{\alpha\beta + \alpha + \beta + 1} = \frac{-3 + 2}{-\dfrac{5}{2} - 3 + 1} = \frac{-1}{-\dfrac{9}{2}} = \frac{2}{9}$$

So, for the new equation $-\dfrac{b}{a} = \dfrac{2}{9}$

Product of the new roots $= \dfrac{1}{\alpha + 1}\left(\dfrac{1}{\beta + 1}\right) = \dfrac{1}{\alpha\beta + \alpha + \beta + 1}$

$$= \frac{1}{-\dfrac{5}{2} - 3 + 1} = \frac{1}{-\dfrac{9}{2}} = -\frac{2}{9}$$

So, for the new equation $\dfrac{c}{a} = -\dfrac{2}{9}$

The new equation can be written as $x^2 + \dfrac{b}{a}x + \dfrac{c}{a} = 0$

Therefore, the quadratic equation with roots

$\dfrac{1}{\alpha + 1}, \dfrac{1}{\beta + 1}$ is $x^2 - \dfrac{2}{9}x - \dfrac{2}{9} = 0$ or $9x^2 - 2x - 2 = 0$

Example 2.6

Given that the roots of the equation $x^2 - 4x + 2 = 0$ are α and β, find the values of the following expressions.

(a) $\alpha^2 + \beta^2$ (b) $\dfrac{1}{\alpha^2} + \dfrac{1}{\beta^2}$

Solution

With $x^2 - 4x + 2 = 0$, $\alpha + \beta = -\dfrac{b}{a} = -\dfrac{-4}{1} = 4$ and $\alpha\beta = \dfrac{c}{a} = \dfrac{2}{1} = 2$

Both of the expressions $\alpha^2 + \beta^2$ and $\dfrac{1}{\alpha^2} + \dfrac{1}{\beta^2}$ need to be expressed in terms of $\alpha + \beta$ and $\alpha\beta$

(a) $\alpha^2 + \beta^2 = \alpha^2 + 2\alpha\beta + \beta^2 - 2\alpha\beta = (\alpha + \beta)^2 - 2\alpha\beta$

     Substituting the values for $\alpha + \beta$ and $\alpha\beta$, gives:

     $\alpha^2 + \beta^2 = 4^2 - 2 \cdot 2 = 16 - 4 = 12$

(b) $\dfrac{1}{\alpha^2} + \dfrac{1}{\beta^2} = \dfrac{\beta^2}{\alpha^2\beta^2} + \dfrac{\alpha^2}{\alpha^2\beta^2} = \dfrac{\alpha^2 + \beta^2}{(\alpha\beta)^2}$

     From part (a) we know that $\alpha^2 + \beta^2 = (\alpha + \beta)^2 - 2\alpha\beta$.

     Substituting this into the numerator gives:

$$\frac{1}{\alpha^2} + \frac{1}{\beta^2} = \frac{(\alpha + \beta)^2 - 2\alpha\beta}{(\alpha\beta)^2}$$

Then substituting the values for $\alpha + \beta$ and $\alpha\beta$ from above, gives

$$\frac{4^2 - 2 \cdot 2}{2^2} = \frac{12}{4} = 3$$

Therefore, $\dfrac{1}{\alpha^2} + \dfrac{1}{\beta^2} = 3$

The quadratic formula and the discriminant

The expression that is beneath the radical sign in the quadratic formula, $b^2 - 4ac$, determines whether the zeros of the quadratic function are real or not real (imaginary). Because it acts to 'discriminate' between the type of zeros, $b^2 - 4ac$ is called the **discriminant**. It is often labelled with the Greek letter Δ (upper case delta). The value of the discriminant can also indicate if the zeros are equal and if they are rational.

For the quadratic function $f(x) = ax^2 + bx + c, (a \neq 0)$ where a, b and c are real numbers:

If $\Delta = b^2 - 4ac > 0$, then f has two distinct real zeros, and the graph of f intersects the x-axis twice.

If $\Delta = b^2 - 4ac = 0$, then f has one real zero (double root), and the graph of f intersects the x-axis once (i.e. it is tangent to the x-axis).

If $\Delta = b^2 - 4ac < 0$, then f has two imaginary zeros, and the graph of f does not intersect the x-axis.

In the special case when a, b and c are integers and the discriminant is the square of an integer (a perfect square), the polynomial $ax^2 + bx + c$ has two distinct **rational zeros**.

If the zeros of a quadratic polynomial are rational – either two distinct zeros or two equal zeros (double zero/root) – then the polynomial can be factorised. That is, if $ax^2 + bx + c$ has rational zeros then $ax^2 + bx + c = (mx + n)(px + q)$ where m, n, p and q are rational numbers.

When the discriminant is zero, the solution of a quadratic function is $x = \dfrac{-b \pm \sqrt{b^2 - 4ac}}{2a}$ $= \dfrac{-b \pm \sqrt{0}}{2a} = -\dfrac{b}{2a}$. As mentioned, this solution of $-\dfrac{b}{2a}$ is called a double zero (or root), which can also be described as a **zero of multiplicity of 2**. If a and b are integers, then the zero $-\dfrac{b}{2a}$ will be rational. When we solve polynomial functions of higher degree later in this chapter, we will encounter zeros of higher multiplicity.

Remember that the **roots** of a polynomial equation are those values of x for which $P(x) = 0$. These values of x are called the **zeros** of the polynomial P.

Example 2.7

Use the discriminant to determine how many real roots each equation has. Confirm the result by graphing the corresponding quadratic function for each equation on your GDC.

(a) $2x^2 + 5x - 3 = 0$ (b) $4x^2 - 12x + 9 = 0$ (c) $2x^2 - 5x + 6 = 0$

Solution

(a) The discriminant is $\Delta = 5^2 - 4(2)(-3) = 49 > 0$. Therefore, the equation has two distinct real roots. This result is confirmed by the graph of the quadratic function $y = 2x^2 + 5x - 3$ that clearly intersects the x-axis twice.

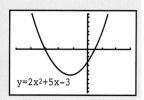
y=2x²+5x-3

Also, since $\Delta = 49$ is a perfect square, the two roots are rational and the quadratic polynomial $2x^2 + 5x - 3$ can be factorised:
$2x^2 + 5x - 3 = (2x - 1)(x + 3) = 0$

The two rational roots are $x = \dfrac{1}{2}$ and $x = -3$

(b) The discriminant is $\Delta = (-12)^2 - 4(4)(9) = 0$. Therefore, the equation has one rational root (a double root). The graph on the GDC of $y = 4x^2 - 12x + 9$ appears to intersect the x-axis at only one point. We can be more confident with this conclusion by investigating further – for example, tracing or looking at a table of values on the GDC.

Also, since the root is rational ($\Delta = 0$), the polynomial $4x^2 - 12x + 9$ can be factorised.

$$4x^2 - 12x + 9 = (2x - 3)(2x - 3) = \left[2\left(x - \frac{3}{2}\right)2\left(x - \frac{3}{2}\right)\right]$$

$$= 4\left(x - \frac{3}{2}\right)^2 = 0$$

There are two equal linear factors, which means there are two equal rational zeros − both equal to $\dfrac{3}{2}$ in this case.

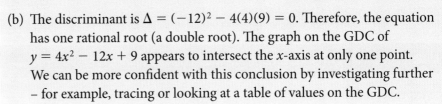

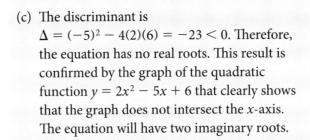

Figure 2.9 GDC screens for the solution to Example 2.7 (b)

(c) The discriminant is $\Delta = (-5)^2 - 4(2)(6) = -23 < 0$. Therefore, the equation has no real roots. This result is confirmed by the graph of the quadratic function $y = 2x^2 - 5x + 6$ that clearly shows that the graph does not intersect the x-axis. The equation will have two imaginary roots.

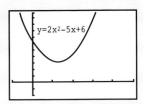

Example 2.8

For $4x^2 + 4kx + 9 = 0$, determine the value(s) of k so that the equation has:

(a) one real zero, (b) two distinct real zeros, and (c) no real zeros.

Solution

(a) For one real zero, $\Delta = (4k)^2 - 4(4)(9) = 0 \Rightarrow 16k^2 - 144 = 0$
$\Rightarrow 16k^2 = 144 \Rightarrow k^2 = 9 \Rightarrow k = \pm 3$

(b) For two distinct real zeros, $\Delta = (4k)^2 - 4(4)(9) > 0 \Rightarrow 16k^2 > 144$
$\Rightarrow k^2 > 9 \Rightarrow k < -3$ or $k > 3$

(c) For no real zeros, $\Delta = (4k)^2 - 4(4)(9) < 0 \Rightarrow 16k^2 < 144 \Rightarrow k^2 < 9$
$\Rightarrow k > -3$ and $k < 3 \Rightarrow -3 < k < 3$

The graph of $f(x) = a(x - p)(x - q)$

If a quadratic function is written in the form $f(x) = a(x - p)(x - q)$, we can easily identify the x-intercepts of the graph of f.

Consider that $f(p) = a(p - p)(p - q) = a(0)(p - q) = 0$ and that $f(q) = a(q - p)(q - q) = a(q - p)(0) = 0$. Therefore, the quadratic function $f(x) = a(x - p)(x - q)$ will intersect the x-axis at the points $(p, 0)$ and $(q, 0)$. We need to factorise in order to rewrite a quadratic function in the form $f(x) = ax^2 + bx + c$ to the form $f(x) = a(x - p)(x - q)$. Hence, $f(x) = a(x - p)(x - q)$ can be referred to as the factorised form of a quadratic function. As a parabola is symmetric, the x-intercepts $(p, 0)$ and $(q, 0)$ will be equidistant from the axis of symmetry (see Figure 2.9). As a result, the equation of the axis of symmetry and the x-coordinate of the vertex of the parabola can be found from finding the average of p and q.

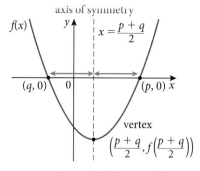

Figure 2.10 Features of the graph of a quadratic function

Example 2.9

Find the equation of each quadratic function from the graph in the form $f(x) = a(x - p)(x - q)$ and also in the form $f(x) = ax^2 + bx + c$

(a)

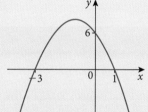

(b)

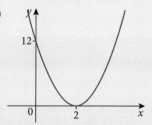

If a quadratic function is written in the form $f(x) = a(x - p)(x - q)$, with $a \neq 0$, then the graph of f has x-intercepts at $(p, 0)$ and $(q, 0)$, an axis of symmetry with equation $x = \dfrac{p + q}{2}$, and a vertex

at $\left(\dfrac{p + q}{2}, f\left(\dfrac{p + q}{2}\right)\right)$

Solution

(a) Since the x-intercepts are -3 and 1, $y = a(x + 3)(x - 1)$

The y-intercept is 6, so when $x = 0$, $y = 6$. Hence, $6 = a(0 + 3)(0 - 1) = -3a \Rightarrow a = -2$ ($a < 0$ agrees with the fact that the parabola is opening downwards). The function is $f(x) = -2(x + 3)(x - 1)$
Expanding to remove brackets: $f(x) = -2x^2 - 4x + 6$

(b) The function has one x-intercept at 2 (double root), so $p = q = 2$ and $y = a(x - 2)(x - 2) = a(x - 2)^2$. The y-intercept is 12, so when $x = 0$, $y = 12$. Hence, $12 = a(0 - 2)^2 = 4a \Rightarrow a = 3$ ($a > 0$ agrees with the parabola opening upwards). The function is $f(x) = 3(x - 2)^2$.
Expanding the brackets: $f(x) = 3x^2 - 12x + 12$

Example 2.10

The graph of a quadratic function intersects the x-axis at the points $(-6, 0)$ and $(-2, 0)$ and also passes through the point $(2, 16)$.

(a) Write the function in the form $f(x) = a(x - p)(x - q)$

(b) Find the coordinates of the vertex of the parabola.

(c) Write the function in the form $f(x) = a(x - h)^2 + k$

Solution

(a) The x-intercepts of $(-6, 0)$ and $(-2, 0)$ gives $f(x) = a(x + 6)(x + 2)$.
Since f passes through $(2, 16)$, then $f(2) = 16 \Rightarrow f(2) = a(2 + 6)(2 + 2)$

$$= 32a = 16 \Rightarrow a = \frac{1}{2}$$

Therefore, $f(x) = \frac{1}{2}(x + 6)(x + 2)$

(b) The x-coordinate of the vertex is the average of the x-coordinates of the intercepts. $x = \dfrac{-6 - 2}{2} = -4$

The y-coordinate of the vertex is $y = f(-4) = \frac{1}{2}(-4 + 6)(-4 + 2) = -2$

Hence, the coordinates of the vertex are $(-4, -2)$

(c) In vertex form, the quadratic function is $f(x) = \frac{1}{2}(x + 4)^2 - 2$

Exercise 2.2

1. For each of the quadratic functions f, find

   (i) the equation for the axis of symmetry and the coordinates of the vertex by algebraic methods

   (ii) the transformation(s) that can be applied to $y = x^2$ to obtain the graph of $y = f(x)$

   (iii) the minimum or maximum value of f.

   Check your results by using your GDC.

   (a) $f:x \mapsto x^2 - 10x + 32$ **(b)** $f:x \mapsto x^2 + 6x + 8$

   (c) $f:x \mapsto -2x^2 - 4x + 10$ **(d)** $f:x \mapsto 4x^2 - 4x + 9$

   (e) $f:x \mapsto \frac{1}{2}x^2 + 7x + 26$

2. Solve each quadratic equation by using factorisation.

   (a) $x^2 + 2x - 8 = 0$ **(b)** $x^2 = 3x + 10$

   (c) $6x^2 - 9x = 0$ **(d)** $3x^2 + 11x - 4 = 0$

   (e) $3x^2 + 18 = 15x$ **(f)** $9x - 2 = 4x^2$

3. Use the method of completing the square to solve each quadratic equation.

   (a) $x^2 + 4x - 3 = 0$ **(b)** $x^2 - 4x - 5 = 0$

   (c) $x^2 - 2x + 3 = 0$ **(d)** $2x^2 + 16x + 6 = 0$

   (e) $x^2 + 2x - 8 = 0$ **(f)** $-2x^2 + 4x + 9 = 0$

4. Let $f(x) = x^2 - 4x - 1$

   (a) Use the quadratic formula to find the zeros of the function.

   (b) Use the zeros to find the equation for the axis of symmetry of the parabola.

   (c) Find the minimum or maximum value of f.

5. Determine the number of real solutions to each equation.

   (a) $x^2 + 3x + 2 = 0$ (b) $2x^2 - 3x + 2 = 0$

   (c) $x^2 - 1 = 0$ (d) $2x^2 - \dfrac{9}{4}x + 1 = 0$

6. Find the value(s) of p for which the equation $2x^2 + px + 1 = 0$ has one real solution.

7. Find the value(s) of m for which the equation $x^2 + 4x + k = 0$ has two distinct real solutions.

8. The equation $x^2 - 4kx + 4 = 0$ has two distinct real solutions. Find the set of all possible values of k.

9. Find all possible values of m so that the graph of the function $g:x \mapsto mx^2 + 6x + m$ does not touch the x-axis.

10. Find the range of values of k such that $3x^2 - 12x + k > 0$ for all real values of x.

For question 10, consider what must be true about the zeros of the quadratic equation $y = 3x^2 - 12x + k$

11. Prove that the expression $x - 2 - x^2$ is negative for all real values of x.

12. Find a quadratic function in the form $y = ax^2 + bx + c$ that satisfies the given conditions.

    (a) The function has zeros of $x = -1$ and $x = 4$ and its graph intersects the y-axis at $(0, 8)$.

    (b) The function has zeros of $x = \dfrac{1}{2}$ and $x = 3$ and its graph passes through the point $(-1, 4)$.

13. Find the range of values for k in order for the equation $2x^2 + (3 - k)x + k + 3 = 0$ to have two imaginary solutions.

14. For what values of m does the function $f(x) = 5x^2 - kx + 2$ have two distinct real zeros?

15. The graph of a quadratic function passes through the points $\left(-\dfrac{17}{4}, \dfrac{55}{8}\right)$, $(-4, 4)$ and $\left(\dfrac{7}{4}, \dfrac{55}{8}\right)$. Express the function in the form $f(x) = ax^2 + bx + c$, where $a, b, c \in \mathbb{Z}$.

16. The maximum value of the function $f(x) = ax^2 + bx + c$ is 10. Given that $f(3) = f(-1) = 2$, find $f(2)$.

17. Find the values of x for which $4x + 1 < x^2 + 4$

18. Show that there is no real value t for which the equation
$2x^2 + (2 - t)x + t^2 + 3 = 0$ has real roots.

19. Show that the two roots of $ax^2 + bx + a = 0$ are reciprocals of each other.

20. Find the sum and product of the roots for each quadratic equation.

(a) $2x^2 + 6x - 5 = 0$ (b) $x^2 = 1 - 3x$

(c) $4x^2 - 6 = 0$ (d) $x^2 + ax - 2a = 0$

(e) $m(m - 2) = 4(m + 1)$ (f) $3x - \dfrac{2}{x} = 1$

21. The roots of the equation $2x^2 - 3x + 6 = 0$ are α and β. Find a quadratic equation with integer coefficients whose roots are $\dfrac{\alpha}{\beta}$ and $\dfrac{\beta}{\alpha}$.

22. If α and β are the roots of the equation $3x^2 + 5x + 4 = 0$, find the value of each expression.

(a) $\alpha^2 + \beta^2$ (b) $\dfrac{\alpha}{\beta} + \dfrac{\beta}{\alpha}$ (c) $\alpha^3 + \beta^3$

> For question 22 (c), factorise $\alpha^3 + \beta^3$ into a product of a binomial and a trinomial.

23. Consider the quadratic equation $x^2 + 8x + k = 0$ where k is a constant. Find both roots of the equation given that one root of the equation is three times the other. Find the value of k.

24. The roots of the equation $x^2 + x + 4 = 0$ are α and β.

(a) Without solving the equation, find the value of the expression $\dfrac{1}{\alpha} + \dfrac{1}{\beta}$.

(b) Find a quadratic equation whose roots are $\dfrac{1}{\alpha}$ and $\dfrac{1}{\beta}$.

25. If α and β are roots of the quadratic equation $5x^2 - 3x - 1 = 0$, find a quadratic equation with integer coefficients that has the roots:

(a) $\dfrac{1}{\alpha^2}$ and $\dfrac{1}{\beta^2}$ (b) $\dfrac{\alpha^2}{\beta}$ and $\dfrac{\beta^2}{\alpha}$

2.3 Zeros, factors and remainders

Finding the zeros of polynomial functions is a feature of many problems in algebra, calculus and other areas of mathematics. In our analysis of quadratic functions in Section 2.2, we saw the connection between the graphical and algebraic approaches to finding zeros. Information obtained from the graph of a function can be used to help find its zeros and, conversely, information about the zeros of a polynomial function can be used to help sketch its graph. Results and observations from the last section lead us to make some conclusions about real zeros of all polynomial functions.

If $P(x)$ is a polynomial function and c is a real number, then these statements are equivalent:
- $x = c$ is a zero of the function P
- $x = c$ is a solution (or root) of the polynomial equation $P(x) = 0$
- $(x - c)$ is a linear factor of the polynomial $P(x)$
- $(c, 0)$ is an x-intercept of the graph of the function P

Polynomial division

As with integers, finding the factors of polynomials is closely related to dividing polynomials. An integer n is **divisible** by another integer m if m is a factor of n. If n is not divisible by m, we can use the process of **long division** to find the quotient of the numbers and the remainder. For example, let's use division to divide 485 by 34.

Polynomial division is not a required algebraic method for this course; however, it can be a useful tool.

$$
\begin{array}{r}
14 \\
34\overline{)485} \\
34 \\
\hline
145 \\
136 \\
\hline
9
\end{array}
\qquad
\begin{array}{rl}
14 & \text{quotient} \\
\text{check:} \quad \times \ 34 & \text{divisor} \\
\hline
56 & \\
42 & \\
\hline
476 & \\
+ \quad 9 & \text{remainder} \\
\hline
485 & \text{dividend}
\end{array}
$$

The number 485 is the **dividend**, 34 is the **divisor**, 14 is the **quotient** and 9 is the **remainder**. The division process (or algorithm) stops when a remainder is less than the divisor. The procedure shown above for checking the division result may be expressed as

$$485 = 34 \times 14 + 9$$

or in words as

$$\text{dividend} = \text{divisor} \times \text{quotient} + \text{remainder}$$

The process of division for polynomials is similar to that for integers. If a polynomial $D(x)$ is a factor of polynomial $P(x)$, then $P(x)$ is divisible by $D(x)$. However, if $D(x)$ is not a factor of $P(x)$ then we can use a **long division algorithm for polynomials** to find a quotient polynomial $Q(x)$ and a remainder polynomial $R(x)$ such that $P(x) = D(x) \cdot Q(x) + R(x)$. In the same way that the remainder must be less than the divisor when dividing integers,

73

the remainder must be a polynomial of a lower degree than the divisor when dividing polynomials. Consequently, when the divisor is a linear polynomial (degree of 1), the remainder must be of degree 0, that is, a constant.

Example 2.11

Find the quotient $Q(x)$ and remainder $R(x)$ when $P(x) = 2x^3 - 5x^2 + 6x - 3$ is divided by $D(x) = x - 2$

Solution

$$
\begin{array}{r}
2x^2 - x + 4 \\
x - 2 \overline{)\,2x^3 - 5x^2 + 6x - 3} \\
\underline{2x^3 - 4x^2} \qquad \leftarrow 2x^2(x-2) \\
-x^2 + 6x \qquad \leftarrow \text{subtract} \\
\underline{-x^2 + 2x} \qquad \leftarrow -x(x-2) \\
4x - 3 \leftarrow \text{subtract} \\
\underline{4x - 8} \leftarrow 4(x-2) \\
5 \leftarrow \text{subtract}
\end{array}
$$

Thus, the quotient $Q(x)$ is $2x^2 - x + 4$ and the remainder is 5. Therefore, we can write:

$$2x^3 - 5x^2 + 6x - 3 = (x-2)(2x^2 - x + 4) + 5$$

This equation provides a means to check the result by expanding and simplifying the right side and verifying it is equal to the left side.

$$
\begin{aligned}
2x^3 - 5x^2 + 6x - 3 &= (x-2)(2x^2 - x + 4) + 5 \\
&= (2x^3 - x^2 + 4x - 4x^2 + 2x - 8) + 5 \\
&= 2x^3 - 5x^2 + 6x - 3
\end{aligned}
$$

Taking the identity $P(x) = D(x) \cdot Q(x) + R(x)$ and dividing both sides by $D(x)$ produces the equivalent identity $\dfrac{P(x)}{D(x)} = Q(x) + \dfrac{R(x)}{D(x)}$

Hence, the result for Example 2.11 could also be written as

$$\frac{2x^3 - 5x^2 + 6x - 3}{x - 2} = 2x^2 - x + 4 + \frac{5}{x - 2}$$

> A common error when performing long division with polynomials is to add rather than subtract during each cycle of the process.

Note that writing the result in this manner is the same as rewriting $17 = 5 \times 3 + 2$ as $\dfrac{17}{5} = 3 + \dfrac{2}{5}$, which we commonly write as the mixed number $3\dfrac{2}{5}$

If $P(x)$ and $D(x)$ are polynomials such that $D(x) \neq 0$, and the degree of $D(x)$ is less than or equal to the degree of $P(x)$, then there exist unique polynomials $Q(x)$ and $R(x)$ such that

$$P(x) = D(x) \cdot Q(x) + R(x)$$

$$\begin{array}{cccc}
\text{dividend} & \text{divisor} & \text{quotient} & \text{remainder}
\end{array}$$

and where $R(x)$ is either zero or of degree less than the degree of $D(x)$.

Remainder and factor theorems

As illustrated by Example 2.11, we commonly divide polynomials of higher degree by linear polynomials. Let's look at what happens to the division algorithm when the divisor $D(x)$ is a linear polynomial of the form $(x - c)$. Since the degree of the remainder $R(x)$ must be less than the degree of the divisor (degree of one in this case), the remainder will be a constant, simply written as R. Then the division algorithm for a linear divisor is the identity:

$$P(x) = (x - c) \cdot Q(x) + R$$

If we evaluate the polynomial function P at the number $x = c$, we obtain

$$P(c) = (c - c) \cdot Q(c) + R = 0 \cdot Q(c) + R = R$$

Thus, the remainder R is equal to $P(c)$, the value of the polynomial P at $x = c$. Because this is true for any polynomial P and any linear divisor $(x - c)$, we have the remainder theorem.

In the **remainder theorem**, if a polynomial function $P(x)$ is divided by $(x - c)$, then the remainder is the value $P(c)$.

Example 2.12

Find the remainder when $g(x) = 2x^3 + 5x^2 - 8x + 3$ is divided by $x + 4$

Solution

The linear polynomial $x + 4$ is equivalent to $x - (-4)$. Applying the remainder theorem, the required remainder is equal to the value of $g(-4)$.

$$g(-4) = 2(-4)^3 + 5\,(-4)^2 - 8(-4) + 3 = 2(-64) + 5(16) + 32 + 3$$
$$= -128 + 80 + 35 = -13$$

Therefore, when the polynomial function $g(x)$ is divided by $x + 4$ the remainder is -13.

A consequence of the remainder theorem is the factor theorem, which also follows intuitively from our discussion in the previous section about the zeros and factors of quadratic functions. It formalises the relationship between zeros and linear factors of all polynomial functions with real coefficients.

In the **factor theorem**, a polynomial function $P(x)$ has a factor $(x - c)$ if and only if $P(c) = 0$.

It is important to understand that the factor theorem is a **biconditional** statement of the form "A if and only if B". Such a statement is true in either direction; that is, "if A then B", and also "if B then A" — usually abbreviated A → B and B → A, respectively.

Sum and product of the roots of any polynomial equation

In the previous section, we found a way to express the sum and product of the roots of a quadratic equation, $ax^2 + bx + c = 0$, in terms of a, b and c. It is natural to wonder whether a similar method could be found for polynomial equations of degree greater than two.

Using the same approach as in the previous section for quadratic equations, let's consider the general cubic equation $ax^3 + bx^2 + cx + d = 0$ whose roots are $x - \alpha$, $x - \beta$ and $x = \gamma$. It follows that this general cubic equation can be written in the form $x^3 + \dfrac{b}{a}x^2 + \dfrac{c}{a}x + \dfrac{d}{a} = 0$. Applying the factor theorem, it can also be written in the form $(x - \alpha)(x - \beta)(x - \gamma) = 0$. Expanding the brackets gives

$$(x - \alpha)(x - \beta)(x - \gamma) = x^3 - \alpha x^2 - \beta x^2 - \gamma x^2 + \alpha\beta x + \beta\gamma x + \alpha\gamma x - \alpha\beta\gamma$$
$$= 0$$
$$x^3 - (\alpha + \beta + \gamma)x^2 + (\alpha\beta + \beta\gamma + \alpha\gamma)x - \alpha\beta\gamma = 0$$

Equating coefficients for $x^3 + \dfrac{b}{a}x^2 + \dfrac{c}{a}x + \dfrac{d}{a} = 0$ and

$x^3 - (\alpha + \beta + \gamma)x^2 + (\alpha\beta + \beta\gamma + \alpha\gamma)x - \alpha\beta\gamma = 0$ gives us the following results for the sum and product of the roots for any cubic equation.

$$\alpha + \beta + \gamma = -\frac{b}{a} \text{ and } \alpha\beta\gamma = -\frac{d}{a}$$

This result for the sum and product of the roots of any cubic equation looks very similar to that for any quadratic equation. The only difference is that the product of the roots, $\alpha\beta\gamma$, is the negative of the quotient $\dfrac{\text{constant term}}{\text{leading coefficient}}$

For the general quartic equation $ax^4 + bx^3 + cx^2 + dx + e = 0$ with roots α, β, γ and δ, the factored form of the equation expands as follows:

$$(x - \alpha)(x - \beta)(x - \gamma)(x - \delta)$$
$$= x^4 - (\alpha + \beta + \gamma + \delta)x^2 + (\alpha\beta + \alpha\gamma + \alpha\delta + \beta\gamma + \beta\delta + \gamma\delta)x -$$
$$(\alpha\beta\gamma + \alpha\beta\delta + \alpha\gamma\delta + \beta\gamma\delta) + \alpha\beta\gamma\delta = 0$$

Since this is equivalent to $x^4 + \dfrac{b}{a}x^3 + \dfrac{c}{a}x^2 + \dfrac{d}{a}x + \dfrac{e}{a} = 0$, then the sum and product of the roots for any quartic equation are $\alpha + \beta + \gamma + \delta = -\dfrac{b}{a}$ and $\alpha\beta\gamma\delta = \dfrac{e}{a}$.

These results for the sum and product of roots for polynomial equations of degree 2 (quadratic), degree 3 (cubic) and degree 4 (quartic) lead to the following result for any polynomial function of degree n that we state without a formal proof.

For the polynomial equation of degree n given by $P(x) = a_n x^n + a_{n-1}x^{n-1} + \cdots + a_1 x + a_0 = 0$, $a_n \neq 0$, the sum of the roots is $-\dfrac{a_{n-1}}{a_n}$ and the product of the roots is $\dfrac{(-1)^n a_0}{a_n}$

Example 2.13

Two roots of the equation $x^3 - 3x^2 + kx + 75 = 0$ are opposite in sign. Find the values of all the roots and the constant k.

Solution

Let the three unknown roots be represented by α, $-\alpha$ and β.

Then $\alpha - \alpha + \beta = 3 \Rightarrow \beta = 3$

and $\alpha(-\alpha)\beta = -75 \Rightarrow \alpha(-\alpha)(3) = -75$

$$\Rightarrow -3\alpha^2 = -75$$
$$\Rightarrow \alpha^2 = 25$$
$$\Rightarrow \alpha = \pm 5$$

Therefore, the three roots are 5, -5 and 3.

To find the value of k, write the cubic in factored form and expand.

$$(x - 3)(x + 5)(x - 5) = 0 \Rightarrow (x - 3)(x^2 - 25) = 0$$
$$\Rightarrow x^3 - 3x^2 - 25x + 75 = 0$$

Therefore, $k = -25$

Example 2.14

Consider the quartic equation $2x^4 + 3x^3 - 16x^2 + 15x - 4 = 0$. Given that the equation has a double root of $x = 1$, find the value of the other two roots.

Solution

There are other strategies, for example, using factors and polynomial division, but it is more efficient to apply what we know about the sum and product of the roots (zeros) of a polynomial equation.

Let x_3 and x_4 represent the two unknown roots.

From the fact that the sum of the roots is $-\dfrac{a_{n-1}}{a_n}$, then $1 + 1 + x_3 + x_4 = -\dfrac{3}{2}$

Thus, $x_3 + x_4 = -\dfrac{7}{2}$

Also, since the product of the roots is $\dfrac{(-1)^n a_0}{a_n}$, then $1 \cdot 1 \cdot x_3 x_4 = -\dfrac{4}{2}$

Thus, $x_3 x_4 = -2$

To find x_3 and x_4, we need to solve the system of equations $\begin{cases} x_3 + x_4 = -\dfrac{7}{2} \\ x_3 x_4 = -2 \end{cases}$

Solving for x_3 in the first equation gives $x_3 = -x_4 - \dfrac{7}{2}$

Substituting into the other equation gives $\left(-x_4 - \dfrac{7}{2}\right) x_4 = -2$

$$(x_4)^2 + \frac{7}{2} x_4 - 2 = 0$$
$$2(x_4)^2 + 7 x_4 - 4 = 0$$
$$(2x_4 - 1)(x_4 + 4) = 0$$
$$x_4 = \frac{1}{2} \text{ or } x_4 = -4$$

If $x_4 = \dfrac{1}{2}$, then $x_3 = -\dfrac{1}{2} - \dfrac{7}{2} = -4$. And, if $x_4 = -4$, then $x_3 = \dfrac{1}{2}$

Therefore, the other two zeros are $\dfrac{1}{2}$ and -4.

Exercise 2.3

1. In each question, two polynomials P and D are given.
   Use long division to divide $P(x)$ by $D(x)$, and express $P(x)$ in the form
   $P(x) = D(x) \cdot Q(x) + R(x)$.
   (a) $P(x) = 3x^2 + 5x - 5$, $D(x) = x + 3$
   (b) $P(x) = 3x^4 - 8x^3 + 9x + 5$, $D(x) = x - 2$
   (c) $P(x) = x^3 - 5x^2 + 3x - 7$, $D(x) = x - 4$
   (d) $P(x) = 9x^3 + 12x^2 - 5x + 1$, $D(x) = 3x - 1$

2. Given that $x - 1$ is a factor of the function
   $f(x) = 2x^3 - 17x^2 + 22x - 7$, factorise f completely.

3. Given that $2x + 1$ is a factor of the function
   $f(x) = 6x^3 - 5x^2 - 12x - 4$, factorise f completely.

4. Given that $x + \dfrac{2}{3}$ is a factor of the function
   $f(x) = 3x^4 + 2x^3 - 36x^2 + 24x + 32$, factorise f completely.

5. Find the quotient and the remainder for each rational expression.
   (a) $\dfrac{x^2 - 5x + 4}{x - 3}$
   (b) $\dfrac{x^3 + 2x^2 + 2x + 1}{x + 2}$
   (c) $\dfrac{9x^2 - x + 5}{3x^2 - 7x}$
   (c) $\dfrac{x^5 + 3x^3 - 6}{x - 1}$

6. Use the remainder theorem to find the remainder when $P(x)$ is divided
   by $(x - c)$ for the given value of c.
   (a) $P(x) = 2x^3 - 3x^2 + 4x - 7$, $c = 2$
   (b) $P(x) = x^5 - 2x^4 + 3x^2 - 20x + 3$, $c = -1$
   (c) $P(x) = 5x^4 + 30x^3 - 40x^2 + 36x + 14$, $c = -7$
   (d) $P(x) = x^3 - x + 1$, $c = \dfrac{1}{4}$

7. Given that $x = 2$ is a double root of the polynomial $x^4 - 5x^3 + 7x^2 - 4$,
   find all remaining zeros of the polynomial.

8. Find the values of k such that 3 is a zero of $f(x) = x^3 + x^2 + kx + 1$.

9. Find the values of a and b such that 1 and 4 are zeros of
   $f(x) = 2x^4 - 5x^3 - 14x^2 + ax + b$.

10. Find a polynomial with real coefficients satisfying the given conditions.
    (a) degree of 3; and zeros of -2, 1 and 4
    (b) degree of 4; and zeros of -1, 3 (multiplicity of 2) and -2
    (c) degree of 3; and 2 is the only zero (multiplicity of 3)

11. The polynomial $6x^3 + 7x^2 + ax + b$ has a remainder of 72 when divided by $(x - 2)$ and is exactly divisible (i.e. remainder is zero) by $(x + 1)$.

   (a) Find the values of a and b.

   (b) Show that $(2x - 1)$ is also a factor of the polynomial and, hence, find the third factor.

12. The polynomial $p(x) = (ax + b)^3$ leaves a remainder of -1 when divided by $(x + 1)$, and a remainder of 27 when divided by $(x - 2)$. Find the values of the real numbers a and b.

13. $(x - 2)$ and $(x + 2)$ are factors of $x^3 + ax^2 + bx + c$, and the remainder is 10 when divided by $(x - 3)$. Find the values of a, b and c.

14. When divided by $(x + 2)$, the expression $5x^3 - 3x^2 + ax + 7$ leaves a remainder of R. When the expression $4x^3 + ax^2 + 7x - 4$ is divided by $(x + 2)$ there is a remainder of $2R$. Find the value of the constant a.

15. Given that the roots of the equation $x^3 - 19x^2 + bx - 216 = 0$ are consecutive terms in a geometric sequence, find the value of b and solve the equation.

16. (a) Prove that when a polynomial $P(x)$ is divided by $(ax - b)$ the remainder is $P\left(\dfrac{b}{a}\right)$.

   (b) Hence find the remainder when $9x^3 - x + 5$ is divided by $(3x + 2)$.

17. Find the sum and product of the roots of the following equations.

   (a) $x^4 - \dfrac{2}{3}x^3 + 3x^2 - 2x + 5 = 0$

   (b) $(x - 2)^3 = x^4 - 1$

   (c) $\dfrac{3}{x^2 + 2} = \dfrac{2x^2 - x}{2x^5 + 1}$

18. If α, β and γ are the three roots of the cubic equation $ax^3 + bx^2 + cx + d = 0$, show that $\alpha\beta + \alpha\gamma + \beta\gamma = \dfrac{c}{a}$

19. One of the zeros of the equation $x^3 - 63x + 162 = 0$ is twice that of one of the other zeros. Find all three zeros.

20. Find the three zeros of the equation $x^3 - 6x^2 - 24x + 64 = 0$ given that they are consecutive terms in a geometric sequence.

For question 20, let the zeros be represented by $\dfrac{\alpha}{r}$, α, αr where r is the common ratio.

21. Find the value of k such that the zeros of the equation $x^3 - 6x^2 + kx + 10 = 0$ are in an arithmetic progression; that is, they can be represented by α, $\alpha + d$ and $\alpha + 2d$ for some constant d.

For question 21, use the result from question 20.

22. Find the value of k if the roots of the equation $x^3 + x^2 + 2x + k = 0$ are in geometric progression.

Another important category of functions is rational functions, which are functions in the form $R(x) = \dfrac{f(x)}{g(x)}$ where f and g are polynomials and the domain of the function R is the set of all real numbers except the real zeros of polynomial g in the denominator. Some examples of rational functions are:

$$p(x) = \frac{1}{x - 5} \qquad q(x) = \frac{x + 2}{(x + 3)(x - 1)} \qquad r(x) = \frac{x}{x^2 + 1}$$

The domain of p excludes $x = 5$, and the domain of q excludes $x = -3$ and $x = 1$. The domain of r is all real numbers because the polynomial $x^2 + 1$ has no real zeros.

Example 2.15

Find the domain and range of $h(x) = \dfrac{1}{x - 2}$. Sketch the graph of h.

Solution

Because the denominator is zero when $x = 2$, the domain of h is all real numbers except $x = 2$, i.e. $x \in \mathbb{R}$, $x \neq 2$. Determining the range of the function is a little less straightforward. It is clear that the function could never take on a value of zero because that will only occur if the numerator is zero. And since the denominator can have any value except zero it seems that the function values of h could be any real number except zero. To confirm this and to determine the behaviour of the function (and shape of the graph), some values of the domain and range (pairs of coordinates) are displayed in the tables.

A fraction is zero only if its numerator is zero.

x approaches 2 from the left

x	$h(x)$
-98	-0.01
-8	-0.1
0	-0.5
1	-1
1.5	-2
1.9	-10
1.99	-100
1.999	-1000

x approaches 2 from the right

x	$h(x)$
102	0.01
12	0.1
4	0.5
3	1
2.5	2
2.1	10
2.01	100
2.001	1000

The values in the tables provide clear evidence that the range of h is all real numbers except zero, i.e. $h(x) \in \mathbb{R}$, $h(x) \neq 0$. The values in the tables also show that as $x \to -\infty$, $h(x) \to 0$ from below (sometimes written $h(x) \to 0^-$) and as $x \to +\infty$, $h(x) \to 0$ from above ($h(x) \to 0^+$). It follows that the line with equation $y = 0$ (the x-axis) is a horizontal asymptote for the graph of h. As $x \to 2$ from the left (sometimes written $x \to 2^-$), $h(x)$ appears to decrease without bound, whereas as $x \to 2$ from the right ($x \to 2^+$), $h(x)$ appears to increase without bound. This indicates that the graph of h will have a vertical asymptote at $x = 2$. This behaviour is confirmed by the graph of h.

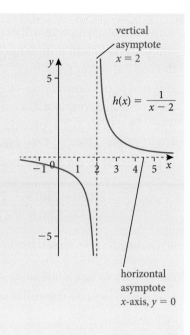

The line $y = c$ is a **horizontal asymptote** of the graph of the function f if at least one of the following statements are true:

- as $x \to +\infty$, then $f(x) \to c^+$
- as $x \to -\infty$, then $f(x) \to c^+$
- as $x \to +\infty$, then $f(x) \to c^-$
- as $x \to -\infty$, then $f(x) \to c^-$

The line $x = d$ is a **vertical asymptote** of the graph of the function f if at least one of the following statements are true:

- as $x \to d^+$, then $f(x) \to +\infty$
- as $x \to d^+$, then $f(x) \to -\infty$
- as $x \to d^-$, then $f(x) \to +\infty$
- as $x \to d^-$, then $f(x) \to -\infty$

Example 2.16

Consider the function $f(x) = \dfrac{3x^2 - 12}{x^2 + 3x - 4}$. Sketch the graph of f and identify any asymptotes and any x- or y-intercepts. Use your sketch to confirm the domain and range of the function.

Solution

First, completely factorise both the numerator and denominator.

$$f(x) = \frac{3x^2 - 12}{x^2 + 3x - 4} = \frac{3(x + 2)(x - 2)}{(x - 1)(x + 4)}$$

The x-intercepts will occur where the numerator is zero. Hence, the x-intercepts are $(-2, 0)$ and $(2, 0)$. A y-intercept will occur when $x = 0$.

$$f(0) = \frac{3(2)(-2)}{(-1)(4)} = 3, \text{ so the } y\text{-intercept is } (0, 3)$$

Any vertical asymptote will occur where the denominator is zero, that is, where the function is undefined. From the factored form of f we see that the vertical asymptotes are $x = 1$ and $x = -4$. We need to determine if the graph of f goes down ($f(x) \to -\infty$) or goes up ($f(x) \to \infty$) on either side of each vertical asymptote. It's easiest to do this by simply analysing what the sign of h will be as x approaches 1 and -4 from both the left and right. For example, as $x \to 1^-$ we can use a test value close to and to the left of 1 (e.g. $x = 0.9$) to check whether $f(x)$ is positive or negative to the left of 1.

$$f(x) = \frac{3(0.9 + 2)(0.9 - 2)}{(0.9 - 1)(0.9 + 4)} \Rightarrow \frac{(+)(-)}{(-)(+)} \Rightarrow f(x) > 0 \Rightarrow \text{as } x \to 1^-,$$

then $f(x) \to +\infty$ (rises)

As $x \to 1^+$ we use a test value close to and to the right of 1 (e.g. $x = 1.1$) to check whether $f(x)$ is positive or negative to the right of 1.

$$f(x) = \frac{3(1.1 + 2)(1.1 - 2)}{(1.1 - 1)(1.1 + 4)} \Rightarrow \frac{(+)(-)}{(+)(+)} \Rightarrow f(x) < 0 \Rightarrow \text{as } x \to 1^+,$$

then $f(x) \to -\infty$ (falls)

Conducting similar analysis for the vertical asymptote of $x = -4$ produces:

$$f(x) = \frac{3(-4.1 + 2)(-4.1 - 2)}{(-4.1 - 1)(-4.1 + 4)} \Rightarrow \frac{(-)(-)}{(-)(-)} \Rightarrow f(x) > 0 \Rightarrow x \to -4^-,$$

then $f(x) \to +\infty$ (rises)

$$f(x) = \frac{3(-3.9 + 2)(-3.9 - 2)}{(-3.9 - 1)(-3.9 + 4)} \Rightarrow \frac{(-)(-)}{(-)(+)} \Rightarrow f(x) < 0 \Rightarrow x \to -4^+,$$

then $f(x) \to -\infty$ (falls)

A horizontal asymptote (if it exists) is the value that $f(x)$ approaches as $x \to \pm\infty$. To find this value, we divide both the numerator and denominator by the highest power of x that appears in the denominator (x^2 for function f).

$$f(x) = \frac{\dfrac{3x^2}{x^2} - \dfrac{12}{x^2}}{\dfrac{x^2}{x^2} + \dfrac{3x}{x^2} - \dfrac{4}{x^2}} = \frac{3 - 0}{1 + 0 - 0} = 3$$

Hence, the horizontal asymptote is $y = 3$.

As we know the behaviour (rising or falling) of the function on either side of each vertical asymptote and that the graph will approach the horizontal asymptote as $x \to \pm\infty$, an accurate sketch can be made, as shown.

The further the number n is from 0, the closer the number $\frac{1}{n}$ is to 0. Conversely, the closer the number n is to 0, the further the number $\frac{1}{n}$ is from 0. These facts can be expressed mathematically using the concept of a limit expressed in limit notation as: $\lim\limits_{n \to \infty} \dfrac{1}{n} = 0$ and $\lim\limits_{n \to 0} \dfrac{1}{n} = \infty$.

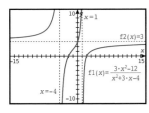

Figure 2.11 GDC screen for the solution to Example 2.16

Because the zeros of the polynomial in the denominator are $x = 1$ and $x = -4$, the domain of f is all real numbers except 1 and -4. From our analysis and from the sketch of the graph, it is clear that between $x = -4$ and $x = 1$ the function takes on all values from $-\infty$ to $+\infty$, therefore the range of f is all real numbers.

We are in the habit of cancelling factors in algebraic expressions, such as

$$\frac{x^2 - 1}{x - 1} = \frac{(x + 1)(x - 1)}{x - 1} = x + 1$$

However, the function $f(x) = \dfrac{x^2 - 1}{x - 1}$ and the function $g(x) = x + 1$ are not the same function. The difference occurs when $x = 1$. $f(1) = \dfrac{1^2 - 1}{1 - 1} = \dfrac{0}{0}$, which is undefined, and $g(1) = 1 + 1 = 2$. So, 1 is not in the domain of f but is in the domain of g. As we might expect, the graphs of the two functions appear identical, but upon closer inspection it is clear that there is a 'hole' (point discontinuity) in the graph of f at the point $(1, 2)$. Thus, f is a **discontinuous** function, but the polynomial function g is continuous. f and g are different functions. (see Figure 2.12)

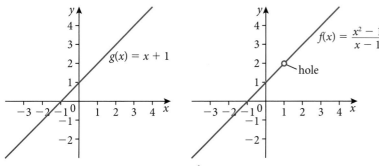

Try graphing $y = \dfrac{x^2 - 1}{x - 1}$ on your GDC and zooming in closely to the region around the point $(1, 2)$. Can you see the hole?

Figure 2.12 Point discontinuity for $f(x) = \dfrac{x^2 - 1}{x - 1}$, but not for $g(x) = x + 1$

In working with rational functions, we often assume that every linear factor that appears in both the numerator and in the denominator has been cancelled. Therefore, for a rational function in the form $\dfrac{f(x)}{g(x)}$, we can usually assume that the polynomial functions f and g have no common factors.

Example 2.17

Find any asymptotes for the function $p(x) = \dfrac{x^2 - 9}{x - 4}$

Solution

The denominator is zero when $x = 4$, thus the line with equation $x = 4$ is a vertical asymptote.

Although the numerator $x^2 - 9$ is not divisible by $x - 4$, it does have a larger degree. Some insight into the behaviour of function p may be gained by dividing $x - 4$ into $x^2 - 9$. Since the degree of the numerator is one greater than the degree of the denominator, the quotient will be a linear polynomial.

Recalling from the previous section that $\dfrac{P(x)}{D(x)} = Q(x) + \dfrac{R(x)}{D(x)}$, where Q and R are the quotient and remainder, we can rewrite $p(x)$ as a linear polynomial plus a fraction.

Since the denominator is in the form $(x - c)$ we can carry out the division efficiently by means of long division.

$$
\begin{array}{r}
x + 4 \\
x - 4 \overline{) x^2 + 0x - 9} \\
\underline{x^2 - 4x} \\
4x - 9 \\
\underline{4x - 16} \\
7
\end{array}
\qquad
\text{Hence, } p(x) = \frac{x^2 - 9}{x - 4} = x + 4 + \frac{7}{x - 4}
$$

As $x \to \pm\infty$, the fraction $\dfrac{7}{x - 4} \to 0$. This tells us about the end behaviour of function p, namely that the graph of p will get closer and closer to the line $y = x + 4$ as the values of x get further away from the origin.

Symbolically, this can be expressed as follows: as $x \to \pm\infty$, $p(x) \to x + 4$. We can graph both the rational function $p(x)$ and the line $y = x + 4$ on a GDC to visually confirm our analysis.

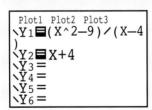
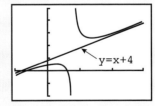

If a line is an asymptote of a graph but is neither horizontal nor vertical, it is called an **oblique asymptote** (sometimes called a slant asymptote). The graph of any rational function of the form $\dfrac{f(x)}{g(x)}$ where the degree of function f is one more than the degree of function g will have an oblique asymptote.

Using Example 2.17 as a model, we can set out a general procedure for analysing a rational function, leading to producing a sketch of its graph and determining its domain and range.

Analysing a rational function

$R(x) = \dfrac{f(x)}{g(x)}$, given functions f and g have no common factors

1. Factorise: Completely factorise both the numerator and denominator.

2. Intercepts: A zero of f will be a zero of R and hence, an x-intercept of the graph of R. The y-intercept is found by evaluating $R(0)$.

3. Vertical asymptotes: A zero of g will give the location of a vertical asymptote (if any). Then perform a sign analysis to see if $R(x) \to +\infty$ or $R(x) \to -\infty$ on either side of each vertical asymptote.

4. Horizontal asymptotes: Find the horizontal asymptote (if any) by dividing both f and g by the highest power of x that appears in g, and then letting $x \to \pm\infty$.

5. Oblique asymptotes: If the degree of g is one more than the degree of f, then the graph of R will have an oblique asymptote. Divide g into f to find the quotient $Q(x)$ and the remainder. The oblique asymptote will be the line with equation $y = Q(x)$.

6. Sketch of graph: Start by drawing dashed lines where the asymptotes are located. Use the information about the intercepts, whether $R(x)$ falls or rises on either side of a vertical asymptote, and additional points as needed to make an accurate sketch.

7. Domain and range: The domain of R will be all real numbers except the zeros of g. You need to study the graph carefully in order to determine the range. Often, but not always (as in Example 2.15), the value of the function at the horizontal asymptote will not be included in the range.

End behaviour of a rational function

Let R be the rational function given by

$$R(x) = \frac{f(x)}{g(x)} = \frac{a_n x^n + a_{n-1} x^{n-1} + \cdots + a_1 x + a_0}{b_m x^m + b_{m-1} x^{m-1} + \cdots + b_1 x + b_0} \quad \text{where functions}$$

f and g have no common factors. Then the following holds true:

If $n < m$, then the x-axis (line $y = 0$) is a horizontal asymptote for the graph of R.

If $n = m$, then the line $y = \dfrac{a_n}{b_m}$ is a horizontal asymptote for the graph of R.

If $n > m$, then the graph of R has no horizontal asymptote. However, if the degree of f is one more than the degree of g, then the graph of R will have an oblique asymptote.

Exercise 2.4

1. Sketch the graph of each rational function without the aid of your GDC. On your sketch, clearly indicate any x- or y- intercepts and any asymptotes (vertical, horizontal or oblique). Use your GDC to verify your sketch.

(a) $f(x) = \dfrac{1}{x + 2}$

(b) $g(x) = \dfrac{3}{x - 2}$

(c) $h(x) = \dfrac{1 - 4x}{1 - x}$

(d) $R(x) = \dfrac{x}{x^2 - 9}$

(e) $p(x) = \dfrac{2}{x^2 + 2x - 3}$

(f) $M(x) = \dfrac{x^2 + 1}{x}$

(g) $f(x) = \dfrac{x}{x^2 + 4x + 4}$

(h) $h(x) = \dfrac{x^2 + 2x}{x - 1}$

(i) $g(x) = \dfrac{2x + 8}{x^2 - x - 12}$

(j) $C(x) = \dfrac{x - 2}{x^2 - 4x}$

2. Use your GDC to sketch a graph of each function, and state the domain and range of the function.

(a) $f(x) = \dfrac{2x^2 + 5}{x^2 - 4}$

(b) $g(x) = \dfrac{x + 4}{x^2 + 3x - 4}$

(c) $h(x) = \dfrac{6}{x^2 + 6}$

(d) $r(x) = \dfrac{x^2 - 2x + 1}{x - 1}$

3. Use your GDC to sketch a graph of each function. Clearly label any x- or y- intercepts and any asymptotes.

(a) $f(x) = \dfrac{2x - 5}{2x^2 + 9x - 18}$

(b) $g(x) = \dfrac{x^2 + x + 1}{x - 1}$

(c) $h(x) = \dfrac{3x^2}{x^2 + x + 2}$

(d) $g(x) = \dfrac{1}{x^3 - x^2 - 4x + 4}$

4. If a, b and c are all positive, sketch the curve $y = \dfrac{x - a}{(x - b)(x - c)}$ for each of the following conditions:

(a) $a < b < c$ (b) $b < a < c$ (c) $b < c < a$

5. A drug is given to a patient and the concentration of the drug in the bloodstream is carefully monitored. At time $t \geqslant 0$ (in minutes after patient received the drug), the concentration, C, in milligrams per litre (mg/L) is given by the following function.

$$C(t) = \dfrac{25t}{t^2 + 4}$$

(a) Sketch a graph of the drug concentration (mg/L) versus time (min).

(b) When does the highest concentration of the drug occur, and what is it?

(c) What eventually happens to the concentration of the drug in the patient's bloodstream?

(d) How long does it take for the concentration to drop below 0.5 mg/L?

2.5 Solving equations and inequalities

We have studied some approaches to analysing and solving polynomial equations in this chapter. Some problems lead to equations with expressions that are not polynomials, for example, expressions with radicals, fractions, or absolute value. Problems in mathematics often involve inequalities rather than equations. We need to be familiar with effective methods for solving inequalities involving polynomials – and again, radicals, fractions, or absolute value.

Equations involving a radical

Example 2.18

Solve for x in the equation $\sqrt{3x + 6} = 2x + 1$

Solution

Squaring both sides gives $3x + 6 = (2x + 1)^2$

$$3x + 6 = 4x^2 + 4x + 1$$

$$4x^2 + x - 5 = 0$$

Factorising $(4x + 5)(x - 1) = 0$

$$x = -\frac{5}{4} \text{ or } x = 1$$

Check both solutions in the original equation

When $x = -\frac{5}{4}$, $\sqrt{3\left(-\frac{5}{4}\right) + 6} = 2\left(-\frac{5}{4}\right) + 1 \Rightarrow \sqrt{\frac{9}{4}} = -\frac{3}{2} \Rightarrow \frac{3}{2} \neq -\frac{3}{2}$

$$\Rightarrow x = -\frac{5}{4} \text{ is not a solution.}$$

When $x = 1$, $\sqrt{3(1) + 6} = 2(1) + 1 \Rightarrow \sqrt{9} = 3 \Rightarrow 3 = 3 \Rightarrow x = 1$ is the only solution.

If two quantities are equal, for example $a = b$, then it is certainly true that $a^2 = b^2$, and $a^3 = b^3$, and so on. However, the converse is not necessarily true. A simple example can illustrate this.

Consider the trivial equation $x = 3$. There is only one value of x that makes the equation true – and that is 3. Now if we take this equation and square both sides we transform it to the equation $x^2 = 9$. This transformed equation has two solutions, 3 and -3, so it is not equivalent to the original equation. By squaring both sides we gained an extra solution, often called an extraneous solution, that satisfies the transformed equation but not the original equation as occurred in Example 2.18. Whenever you raise both sides of an equation by a power it is imperative that you check all solutions in the original equation.

Every solution of the equation $a = b$ is also a solution of the equation $a^n = b^n$, but it is not necessarily true that every solution of $a^n = b^n$ is a solution of $a = b$.

Example 2.19

Solve for x in the equation $\sqrt{2x - 3} - \sqrt{x + 7} = 2$

Solution

Squaring both sides of the original equation will produce a messy expression on the left side, so it is better to rearrange the terms so that one side of the equation contains only a single radical term.

$$\sqrt{2x - 3} = 2 + \sqrt{x + 7}$$

$$(\sqrt{2x - 3})^2 = (2 + \sqrt{x + 7})^2$$

$$2x - 3 = 4 + 4\sqrt{x + 7} + x + 7$$

$$x - 14 = 4\sqrt{x + 7}$$

$$(x - 14)^2 = (4\sqrt{x + 7})^2 \qquad \text{squaring both sides again to}$$
$$\text{eliminate the radical}$$

$$x^2 - 28x + 196 = 16(x + 7)$$

$$x^2 - 44x + 84 = 0$$

$$(x - 2)(x - 42) = 0$$

$$x = 2 \ \text{ or } \ x = 42$$

checking both solutions in the original equation

when $x = 2$, $\sqrt{2(2) - 3} \overset{?}{=} 2 + \sqrt{2 + 7} \Rightarrow \sqrt{1} \overset{?}{=} 2 + \sqrt{9} \Rightarrow 1 \neq 5$

Thus, $x = 2$ is not a solution.

when $x = 42$, $\sqrt{2(42) - 3} \overset{?}{=} 2 + \sqrt{42 + 7} \Rightarrow \sqrt{81} \overset{?}{=} 2 + \sqrt{49} \Rightarrow 9 = 2 + 7$

Thus, $x = 42$ is a solution.

We can verify the single solution of $x = 42$ using a GDC by graphing the equation $y = \sqrt{2x - 3} - \sqrt{x + 7} - 2$ and looking for x-intercepts (zeros). Since we are restricted to real number solutions, the smallest possible value for x that can be substituted into the equation is $\dfrac{3}{2}$. This helps determine a suitable viewing window for the graph on the GDC.

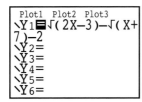 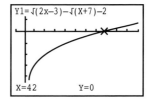

Figure 2.13 Verifying the single solution of $x = 42$ on a GDC

This verifies that $x = 42$ is the only solution to the equation $\sqrt{2x - 3} - \sqrt{x + 7} = 2$

Equations involving fractions

It is also possible for extraneous solutions to appear when solving equations with fractions.

Example 2.20

Find all real solutions of the equation $\dfrac{2x}{4 - x^2} + \dfrac{1}{x + 2} = 3$ and verify solution(s) with a GDC.

Solution

Multiply both sides of the equation by the least common denominator of the fractions, which is $4 - x^2$.

$$\frac{4 - x^2}{1} \cdot \frac{2x}{4 - x^2} + \frac{(2 - x)(2 + x)}{1} \cdot \frac{1}{x + 2} = 3(4 - x^2)$$

$$\text{factorising } 4 - x^2 \text{ gives } (2 - x)(2 + x)$$

$$2x + 2 - x = 12 - 3x^2$$

$$3x^2 + x - 10 = 0$$

$$(3x - 5)(x + 2) = 0$$

$$x = \frac{5}{3} \text{ or } x = -2$$

Clearly $x = -2$ cannot be a solution because that would cause division by zero in the original equation.

The GDC screen shows that the equation $y = \dfrac{2x}{4 - x^2} + \dfrac{1}{x + 2} - 3$ has an

x-intercept at $\left(\dfrac{5}{3}, 0\right)$, confirming the solution $x = \dfrac{5}{3}$.

> (!) Not only is it possible to gain an extraneous solution when solving certain equations, it is also possible to lose a correct solution by incorrectly dividing both sides of an equation by a common factor. For example, solve for x in the equation $4(x + 2)^2 = 3x(x + 2)$. Dividing both sides by $(x + 2)$ gives $4(x + 2) = 3x \Rightarrow 4x + 8 = 3x \Rightarrow x = -8$. However, there are two solutions, $x = -8$ and $x = -2$. The solution of $x = -2$ was lost because a factor of $x + 2$ was eliminated from both sides of the original equation. This is a common error to be avoided.

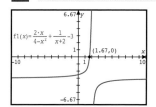

Figure 2.14 GDC screen for the solution to Example 2.20

Equations in quadratic form

In Section 2.2 we covered methods of solving quadratic equations. As the three previous examples illustrate, quadratic equations commonly appear in a range of mathematical problems. The methods of solving quadratics can sometimes be applied to other equations. An equation in the form $at^2 + bt + c = 0$, where t is an algebraic expression, is an equation in **quadratic form**. We can solve such equations by substituting for the algebraic expression and then applying an appropriate method for solving a quadratic equation.

Example 2.21

Find all real solutions of the equation $2m^4 - 5m^2 + 2 = 0$

Solution

The equation can be written as $2(m^2)^2 - 5(m^2) + 2 = 0$ showing it is quadratic in terms of m^2.

$$2m^4 - 5m^2 + 2 = 0 \qquad \text{substitute } t \text{ for } m^2$$

$$2t^2 - 5t + 2 = 0$$

$$(2t - 1)(t - 2) = 0$$

$$t = \frac{1}{2} \text{ or } t = 2$$

$$m^2 = \frac{1}{2} \text{ or } m^2 = 2 \qquad \text{substituting } m^2 \text{ for } t$$

$$m = \pm\sqrt{\frac{1}{2}} = \pm\frac{\sqrt{2}}{2} \text{ or } m = \pm\sqrt{2}$$

Check the solutions (which are in two pairs of opposites) by substituting them directly into the original equation. A value for m will be raised to the 4th and 2nd powers, thus we only need to check one value from each pair of opposites.

$$\text{when } m = \frac{\sqrt{2}}{2}, 2\left(\frac{\sqrt{2}}{2}\right)^4 - 5\left(\frac{\sqrt{2}}{2}\right)^2 + 2 = 0 \Rightarrow 2\left(\frac{1}{4}\right) - 5\left(\frac{1}{2}\right) + 2 = 0$$

$$\Rightarrow \frac{1}{2} - \frac{5}{2} + 2 = 0$$

$$\Rightarrow 0 = 0$$

$$\text{when } m = \sqrt{2}, 2(\sqrt{2})^4 - 5(\sqrt{2})^2 + 2 = 0 \Rightarrow 2(4) - 5(2) + 2 = 0$$

$$\Rightarrow 8 - 10 + 2 = 0$$

$$\Rightarrow 0 = 0$$

Therefore, the solutions to the equation are $m = \dfrac{\sqrt{2}}{2}, -\dfrac{\sqrt{2}}{2}, \sqrt{2}$ and $-\sqrt{2}$

A 4th degree polynomial equation in quadratic form

Example 2.22

Find all solutions, expressed exactly, to the equation $w^{\frac{1}{2}} = 4w^{\frac{1}{4}} - 2$

Solution

$$w^{\frac{1}{2}} - 4w^{\frac{1}{4}} + 2 = 0 \qquad \text{set the equation to zero}$$

$$\left(w^{\frac{1}{4}}\right)^2 - 4\left(w^{\frac{1}{4}}\right) + 2 = 0 \qquad \text{attempt to write in quadratic form } at^2 + bt + c = 0$$

$t^2 - 4t + 2 = 0$ make appropriate substitution; in this case, let $w^{\frac{1}{4}} = t$

$t = \dfrac{-(-4) \pm \sqrt{(-4)^2 - 4(1)(2)}}{2}$ trinomial does not factorise; apply quadratic formula

$t = \dfrac{4 \pm \sqrt{8}}{2} = \dfrac{4 \pm 2\sqrt{2}}{2}$

$t = 2 \pm \sqrt{2}$

$w^{\frac{1}{4}} = 2 \pm \sqrt{2}$ substituting $w^{\frac{1}{4}}$ back in for t

$w = (2 + \sqrt{2})^4$ or $w = (2 - \sqrt{2})^4$

$w = \left((2 + \sqrt{2})^2\right)^2$ or $w = \left((2 - \sqrt{2})^2\right)^2$

$w = (6 + 4\sqrt{2})^2$ or $w = (6 - 4\sqrt{2})^2$

$w = 68 + 48\sqrt{2} \approx 135.882$ or $w = 68 - 48\sqrt{2} \approx 0.117749$ (approx. values found with GDC)

```
68+48√2
              135.882251
68−48√2
          0.1177490061
```

It will be difficult to check these two solutions by substituting them directly into the original equation as we did in Example 2.21. It will be more efficient to use our GDC.

```
                                      0.117749
nSolve( w^(1/2) = 4·w^(1/4) −2, w )
                                      135.882
nSolve( w^(1/2) = 4·w^(1/4) −2, w )|w>1
```

Most GDC models have an equation solver. The main limitation of this GDC feature is that it will usually return only approximate solutions. However, even if exact solutions are required, approximate solutions from a GDC are still very helpful as a check of the exact solutions obtained algebraically.

Equations involving absolute value (modulus)

Equations involving absolute value occur in a range of different topics in mathematics. To solve an equation containing one or more absolute value expressions, we apply the definition that states that the absolute value of a real number a, denoted by $|a|$, is given by

$$|a| = \begin{cases} a & \text{if } a \geqslant 0 \\ -a & \text{if } a < 0 \end{cases}$$

$|a|$ also has the geometric interpretation of being the distance between the coordinate a and the origin on the real number line.

We will encounter equations in later chapters — for example, equations with logarithms and trigonometric functions — that will be in quadratic form.

Example 2.23

Use an algebraic approach to solve the equation $|2x + 7| = 13$. Check any solution(s) on a GDC.

Solution

The expression inside the absolute value symbols must be either 13 or -13, so $2x + 7$ equals 13 or -13. Hence, the given equation is satisfied if either

$$2x + 7 = 13 \quad \text{or} \quad 2x + 7 = -13$$
$$2x = 6 \qquad\qquad 2x = -20$$
$$x = 3 \qquad\qquad x = -10$$

The solutions are $x = 3$ and $x = -10$.

To check the solutions on a GDC, graph the equation $y = |2x + 7| - 13$ and confirm that $x = 3$ and $x = -10$ are the x-intercepts of the graph.

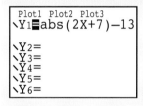 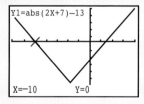 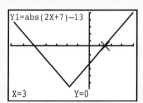

The x-intercepts of the graph of $y = |2x + 7| - 13$ agree with the solutions to the equation.

Example 2.24

Find algebraically the solution(s) to the equation $|2x - 3| = |7 - 3x|$. Check the solution(s) graphically.

Solution

There are four possibilities:

$$2x - 3 = 7 - 3x \qquad\qquad 2x - 3 = -(7 - 3x)$$
$$-(2x - 3) = 7 - 3x \qquad -(2x - 3) = -(7 - 3x)$$

The first and last equations are equivalent, and the second and third equations are also equivalent. So, it is only necessary to solve the first two equations.

$$2x - 3 = 7 - 3x \quad \text{or} \quad 2x - 3 = -(7 - 3x)$$
$$5x = 10 \qquad\qquad 2x - 3 = -7 + 3x$$
$$x = 2 \qquad\qquad 4 = x \Rightarrow x = 4$$

To check, we can graph the equations $y_1 = |2x - 3|$ and $y_2 = |7 - 3x|$ and confirm that the x-coordinates of their points of intersection agree with the solutions to the given equation.

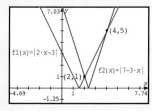

Solving inequalities

Working with inequalities is very important for many of the topics in this course. Four important properties for inequalities are given below.

For three real numbers a, b and c:
- If $a > b$ and $b > c$, then $a > c$.
- If $a > b$ and $c > 0$, then $ac > bc$.
- If $a > b$ and $c < 0$, then $ac < bc$.
- If $a > b$, then $a + c > b + c$.

Quadratic inequalities

In the topics covered in this course, you will need to be as proficient with solving inequalities as with solving equations. We have already solved some simple linear inequalities, and here we will consider strategies for other inequalities – particularly involving quadratic and absolute value expressions.

Example 2.25

Find the values of x that solve the inequality $x^2 > x$

Solution

It is possible to determine the solution set to this inequality by a method of trial and error, or simply using a mental process. That may be successful, but generally speaking it is a good idea to attempt to find the solution set by some algebraic method and then check, usually by means of a GDC. For this example, it is tempting to consider dividing both sides by x, but that cannot be done because it is not known whether x is positive or negative. Recall that when multiplying or dividing both sides of an inequality by a negative number it is necessary to reverse the inequality sign (3rd property of inequalities listed above). Instead, a better approach is to place all terms on one side of the inequality (with zero on the other side) and then try to factorise.

$$x^2 > x$$
$$x^2 - x > 0$$
$$x(x - 1) > 0$$

Now analyse the signs of the two different factors in a 'sign chart'.

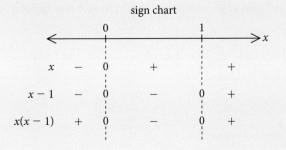

The solution set, $x < 0$ or $x > 1$, for Example 2.25 comprises two intervals that do not intersect (disjoint). It is incorrect to write the solution as $0 > x > 1$, or as $1 < x < 0$. Both of these formats imply that the solution set consists of the values of x between 0 and 1, but that is not the case. Only write the 'combined' inequality $a < x < b$ if $x > a$ and $x < b$ where the two intervals are intersecting between a and b.

The sign chart indicates that the product of the two factors, $x(x - 1)$, will be positive when x is less than 0 or greater than 1.

Therefore, the solution set is $x < 0$ or $x > 1$.

Inequalities with quadratic polynomials arise in many different contexts. Problems in which we need to analyse the value of the discriminant of a quadratic equation will usually require us to solve a quadratic inequality, as the next example illustrates.

Example 2.26

Given $f(x) = 3kx^2 - (k + 3)x + k - 2 = 0$, find the range of values of k for which f has no real zeros.

Solution

The quadratic function f will have no real zeros when its discriminant is negative. Since f is written in the form $ax^2 + bx + c = 0$, then, in terms of the parameter k, $a = 3k$, $b = -(k + 3)$, and $c = k - 2$. Substituting these values into the discriminant, we have the inequality:

$(-(k + 3))^2 - 4(3k)(k - 2) < 0$

$k^2 + 6k + 9 - 12k^2 + 24k < 0$

$-11k^2 + 30k + 9 < 0$ easier to factorise if leading coefficient is positive

$11k^2 - 30k - 9 > 0$ multiply both sides by -1; reverse inequality sign

$$k = \frac{-(-30) \pm \sqrt{(-30)^2 - 4(11)(-9)}}{2(11)} = \frac{30 \pm \sqrt{1296}}{22} = \frac{30 \pm 36}{22}$$

$$k = \frac{30 + 36}{22} = \frac{66}{22} = 3 \text{ or } k = \frac{30 - 36}{22} = -\frac{6}{22} = -\frac{3}{11}$$

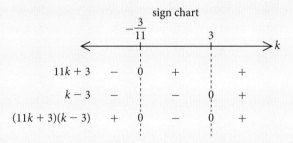

sign chart

The two rational zeros indicate $11k^2 - 30k - 9$ could have been factorised into $(11k + 3)(k - 3)$

The results of the sign chart indicate that the solution set to the inequality $(11k + 3)(k - 3) > 0$ is $k < -\frac{11}{3}$ or $k > 3$.

Therefore, any value of k such that $k < -\frac{11}{3}$ or $k > 3$ will cause the function f to have no real zeros.

Absolute value (modulus) inequalities

As mentioned before, absolute value can be used to indicate distance on the number line. For example, the equation $|x| = 3$ means that some number x is a distance of 3 units from the origin. The two solutions to this equation are $x = 3$ and $x = -3$. Consequently, the inequality $|x| < 3$ means that x lies at most 3 units from the origin, as shown in Figure 2.15.

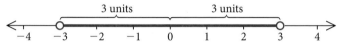

Figure 2.15 Graph of solution set for $|x| < 3$

This means that x lies between -3 and 3, that is, $-3 < x < 3$. Similarly, the inequality $|x| > 3$ means that x lies 3 or more units from the origin. This occurs if x is to the left of -3 (that is, $x < -3$) or if x lies to the right of 3 (that is, $x > 3$).

Properties of absolute value inequalities

For any real numbers x and c such that $c > 0$:

$|x| < c$ if and only if $-c < x < c$

$|x| > c$ if and only if $x < -c$ or $x > c$

Example 2.27

Solve for x in the inequality $|3x - 7| \geqslant 8$

Solution

Applying the second property for absolute value inequalities, we have

$$3x - 7 \leqslant -8 \text{ or } 3x - 7 \geqslant 8$$

$$3x \leqslant -1 \text{ or } 3x \geqslant 15$$

$$x \leqslant -\frac{1}{3} \text{ or } x \geqslant 5$$

Therefore, the solution set is the union of two half-open intervals $x \leqslant -\frac{1}{3}$ or $x \geqslant 5$, which can also be written in interval notation as $\left]-\infty, -\frac{1}{3}\right] \cup [5, \infty[$.

Example 2.28

Find the values of x which satisfy the inequality $\left|\dfrac{x}{x + 4}\right| < 2$

Solution

Applying the first property for absolute value inequalities gives

$$-2 < \frac{x}{x + 4} < 2$$

We cannot multiply both sides by $x + 4$ unless we take into account the two different cases:

when $x + 4$ is positive (inequality is not reversed), and

when $x + 4$ is negative (inequality sign is reversed).

Instead, let's solve the two inequalities in the 'combined' inequality separately by rearranging so that zero is on one side, and then analyse where the expression on the other side is zero, positive and negative. This is similar to the approach used in Example 2.27.

$$\frac{x}{x+4} > -2 \quad \text{and} \quad \frac{x}{x+4} < 2 \qquad \text{The word 'and' indicates intersection}$$

$$\frac{x}{x+4} + 2 > 0 \quad \text{and} \quad \frac{x}{x+4} - 2 < 0$$

$$\frac{x}{x+4} + \frac{2x+8}{x+4} > 0 \quad \text{and} \quad \frac{x}{x+4} - \frac{2x+8}{x+4} < 0$$

$$\frac{3x+8}{x+4} > 0 \quad \text{and} \quad \frac{-x-8}{x+4} < 0$$

$3x + 8$	$-$		$-$	0	$+$
$x + 4$	$-$	0	$+$		$+$
$\dfrac{3x+8}{x+4}$	$+$	X	$-$	0	$+$

$$x < -4 \cup x > -\frac{8}{3}$$

$-x - 8$	$+$	0	$-$		$-$
$x + 4$	$-$		$-$	0	$+$
$\dfrac{-x-8}{x+4}$	$-$	0	$+$	X	$-$

$$x < -8 \cup x > -4$$

$$\cap$$

The solution set for the original 'combined' inequality, $-2 < \dfrac{x}{x+4} < 2$, will be the intersection of the solution sets of the two separate inequalities graphed above on the number line. Thus, the solution set is $x < -8$ or $x > -\dfrac{8}{3}$.

A graphical check using a GDC can be effectively performed by graphing the equation $y = \left| \dfrac{x}{x+4} \right| - 2$ and observing where the graph is below the x-axis. The values of x for which this is true will correspond to the solution set for the inequality $\left| \dfrac{x}{x+4} \right| < 2$

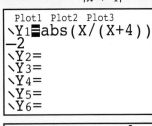

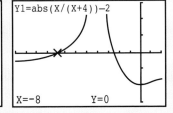

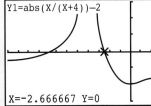

Algebraic and graphical methods

Example 2.29

Solve the inequality $|x - 4| > 2|x - 7|$

Solution

Method 1 – Algebraic

If $a > 0$, $b > 0$ and $a = b$, then $a^2 = b^2$. Since the expressions on both sides must be positive, we can square both sides and remove the absolute value signs.

$(x - 4)^2 > (2(x - 7))^2$

$x^2 - 8x + 16 > 4(x^2 - 14x + 49)$

$x^2 - 8x + 16 > 4x^2 - 56x + 196$

$0 > 3x^2 - 48x + 180$

$0 > x^2 - 16x + 60$

$(x - 10)(x - 6) < 0$

Therefore, the solution set is the open interval $6 < x < 10$.

Method 2 – Graphical

We can graph the equations $y_1 = |x - 4|$ and $y_2 = 2|x - 7|$ and use a GDC to determine for what values of x the graph of y_1 is above the graph of y_2.

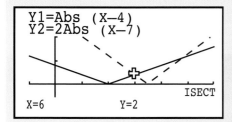
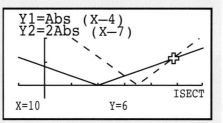

The equation $y_2 = 2|x - 7|$ has been graphed in a dashed style. By using the 'intersect' command on the GDC we find that the graph of y_1 is above the graph of y_2 for $6 < x < 10$. Therefore, the solution set is the open interval $6 < x < 10$.

Example 2.30

Find the values of x such that $\dfrac{x}{x + 8} \leqslant \dfrac{1}{x - 1}$. Solve algebraically.

Solution

As applied in previous examples, an effective algebraic approach is to rearrange the inequality so that both fractions are on the same side with zero on the other side. Then, combine the two fractions into one fraction and analyse where the fraction is zero, positive and negative.

$$\frac{x}{x+8} - \frac{1}{x-1} \leqslant 0$$

$$\frac{x(x-1) - (x+8)}{(x+8)(x-1)} \leqslant 0$$

$$\frac{x^2 - 2x - 8}{(x+8)(x-1)} \leqslant 0$$

$$\frac{(x+2)(x-4)}{(x+8)(x-1)} \leqslant 0$$

	-8		-2		1		4	
$x+2$	$-$		$-$	0 $+$		$+$		$+$
$x-4$	$-$		$-$		$-$		0 $+$	
$x+8$	$-$	0 $+$		$+$		$+$		$+$
$x-1$	$-$		$-$		$-$	0 $+$		$+$
$\dfrac{(x+2)(x-4)}{(x+8)(x-1)}$	$+$		$-$		$+$		$-$	$+$

Therefore, $\dfrac{x}{x+8} \leqslant \dfrac{1}{x-1}$ when $-8 < x \leqslant -2$ or $1 < x \leqslant 4$, which can also be expressed in interval notation as $]-8, -2] \cup \,] 1, 4]$

Exercise 2.5

1. Solve for x in each equation. If possible, find all real solutions and express them exactly. If this is not possible, solve using your GDC and approximate any solutions to 3 significant figures. Be sure to check answers and to recognise any extraneous solutions.

(a) $\sqrt{x+6} + 2x = 9$

(b) $\sqrt{x+7} + 5 = x$

(c) $\sqrt{7x+14} - 2 = x$

(d) $\sqrt{2x+3} - \sqrt{x-2} = 2$

(e) $\dfrac{5}{x+4} - \dfrac{4}{x} = \dfrac{21}{5x+20}$

(f) $\dfrac{x+1}{2x+3} = \dfrac{5x-1}{7x+3}$

(g) $\dfrac{1}{x} - \dfrac{1}{x+1} = \dfrac{1}{x+4}$

(h) $\dfrac{2x}{1-x^2} + \dfrac{1}{x+1} = 2$

(i) $x^4 - 2x^2 - 15 = 0$

(j) $2x^{\frac{2}{3}} - x^{\frac{1}{3}} - 15 = 0$

(k) $x^6 - 35x^3 + 216 = 0$

(l) $5x^{-2} - x^{-1} - 2 = 0$

(m) $|3x + 4| = 8$

(n) $|x + 6| = |3x - 24|$

(o) $|5x + 1| = 2x$

(p) $|x - 1| + |x| = 3$

(q) $\left|\dfrac{x + 1}{x - 1}\right| = 3$

(r) $\sqrt{x} - \dfrac{6}{\sqrt{x}} = 1$

(s) $\sqrt{4 - x} - \sqrt{6 + x} = \sqrt{14 + 2x}$

(t) $\dfrac{6}{x^2 + 1} = \dfrac{1}{x^2} + \dfrac{10}{x^2 + 4}$

(u) $x - \sqrt{x + 10} = 0$

(v) $6x - 37\sqrt{x} + 56 = 0$

2. Find the values of x that satisfy each inequality.

(a) $3x^2 - 4 < 4x$

(b) $\dfrac{2x - 1}{x + 2} \geqslant 1$

(c) $2x^2 + 8x \leqslant 120$

(d) $|1 - 4x| > 7$

(e) $|x - 3| > |x - 14|$

(f) $\left|\dfrac{x^2 - 4}{x}\right| \leqslant 3$

(g) $\dfrac{x}{x - 2} > \dfrac{1}{x + 1}$

(h) $\dfrac{4x - 1}{x^2 - 2x - 3} < 3$

3. Given $f(x) = 3kx^2 - (k + 3)x + k - 2 = 0$, find the range of values of k for which f has no real zeros.

4. Find the values of p for which the equation $px^2 - 3x + 1 = 0$ has

(a) one real solution   **(b)** two real solutions   **(c)** no real solutions.

5. Given $f(x) = x^2 + x(k - 1) + k^2$, find the range of values of k so that $f(x) > 0$ for all real values of x.

6. Show that both of the following inequalities are true for all real numbers m and n such that $m > 0$, $n > 0$.

(a) $m + \dfrac{1}{n} \geqslant 2$

(b) $(m + n)\left(\dfrac{1}{m} + \dfrac{1}{n}\right) \geqslant 4$

7. Find all of the solutions exactly to the equation $(x^2 + x)^2 = 5x^2 + 5x - 6$

8. If a, b and c are positive and unequal, show that
$(a + b + c)^2 < 3(a^2 + b^2 + c^2)$

9. Find the values of x that solve each inequality.

(a) $\left|\dfrac{2x - 3}{x}\right| < 1$

(b) $\dfrac{3}{x - 1} - \dfrac{2}{x + 1} < 1$

10. Provide a geometric or algebraic argument to show that
$|a + b| \leqslant |a| + |b|$ for all $a, b \in \mathbb{R}$.

2.6 Partial fractions

In arithmetic, when we add fractions we find the least common denominator, then multiply both the numerator and denominator of each term by what is needed to complete the common denominator. For example:

$$\frac{2}{3} + \frac{5}{7} = \frac{2}{3} \cdot \frac{7}{7} + \frac{5}{7} \cdot \frac{3}{3} = \frac{14 + 15}{21} = \frac{29}{21}, \text{ also}$$

$$\frac{2}{3} + \frac{5}{9} + \frac{1}{27} = \frac{2}{3} \cdot \frac{9}{9} + \frac{5}{9} \cdot \frac{3}{3} + \frac{1}{27} = \frac{18 + 15 + 1}{27} = \frac{34}{27}$$

Reversing the process is called expressing each compound fraction as **partial fractions**. Given, for example, the fraction $\frac{29}{21} = \frac{29}{3 \times 7}$, we express it as a sum of two fractions, one with denominator 3 and the other with denominator 7. Hence the name partial fractions.

The process of finding the partial fractions is a straightforward process. We write:

$$\frac{29}{3 \times 7} = \frac{a}{3} + \frac{b}{7} \text{ and then we solve for two integers } a \text{ and } b.$$

$$\frac{29}{3 \times 7} = \frac{a}{3} + \frac{b}{7} = \frac{7a + 3b}{21} \Rightarrow 7a + 3b = 29$$

Now by trial and error we can find that $a = 2$ and $b = 5$. Other answers are also possible $(-1, 12), (8, -9) \ldots$

Notice the situation in the second example. The lowest common multiple contains different powers of the same number. Consequently, when finding the partial fractions decomposition, you need to consider that all powers less than or equal to the highest one may be present. That is, when we set up the process of decomposing $\frac{34}{27}$ we set it up as:

$$\frac{34}{27} = \frac{a}{27} + \frac{b}{9} + \frac{c}{3}$$

And then we attempt to find the values of a, b, and c.

In algebra, we carry out that process on to addition of rational expressions. Once again we multiply the numerator and denominator of each term by what was missing from the denominator of that term.

> **Partial fractions decomposition**
> The method of partial fractions decomposition is extremely helpful in evaluating certain integrals as you will see in the chapter on integral calculus. With partial fractions decomposition, we are going to reverse the process and decompose a rational expression into two or more simpler proper rational expressions. This is a very useful skill in which a single fraction with a factorisable denominator is split into the sum of two or more fractions (partial fractions) whose denominators are the factors of the original denominator.
>
> For example: $\dfrac{12x - 1}{2x^2 - 5x - 3} = \dfrac{2}{2x + 1} + \dfrac{5}{x - 3}$

Example 2.31

Find the partial fraction decomposition of $\dfrac{x+1}{x^2+5x+6}$

Solution

$\dfrac{x+1}{x^2+5x+6} = \dfrac{x+1}{(x+2)(x+3)}$, and hence we will attempt to find two numbers

a and b such that $\dfrac{x+1}{x^2+5x+6} \equiv \dfrac{a}{x+2} + \dfrac{b}{x+3}$

Notice that we wrote this as an identity rather than an equality because it has to be true for all values of x.

$$\dfrac{x+1}{x^2+5x+6} \equiv \dfrac{a}{x+2} + \dfrac{b}{x+3} = \dfrac{a(x+3) + b(x+2)}{(x+2)(x+3)}$$

Since the denominators of these identical fractions are the same, their numerators must also be the same. That is,

$$x + 1 \equiv a(x+3) + b(x+2)$$

We have two methods of solution here.

Method 1

$$x + 1 \equiv a(x+3) + b(x+2) \Leftrightarrow x + 1 \equiv (a+b)x + (3a+2b)$$

For two polynomials to be identical, the coefficients of the same powers must be the same; that is, the coefficient of x on the left must be the same as the coefficient of x on the right, and similarly with the constant terms. Hence

$1 = a + b$ and $1 = 3a + 2b$. Now, solving the system with two equations will yield, $a = -1$ and $b = 2$ and hence

$$\dfrac{x+1}{x^2+5x+6} \equiv \dfrac{-1}{x+2} + \dfrac{2}{x+3}$$

Method 2

$$x + 1 \equiv a(x+3) + b(x+2)$$

Again, since this is an identity, the two sides must be the same for any choice of x. Hence, we can substitute any two numbers for x to get the value of each of a and b; specifically replacing x with -3 yields:

$$x + 1 \equiv a(x+3) + b(x+2) \Rightarrow -2 = -b \Rightarrow b = 2$$

Notice how the choice of -3 eliminated the term with a and allowed us to find b directly. Replacing x with -2 yields:

$$x + 1 \equiv a(x+3) + b(x+2) \Rightarrow -1 = a$$

This is of course the same result as above. Also notice here how the choice of -2 eliminated the term with b and allowed us to find a directly.

Note: This method is helpful in cases where there are no repeated factors, and is faster whenever applicable. We will discuss this in more detail later.

Example 2.32

Decompose the rational expression $\dfrac{5x^2 + 16x + 17}{2x^3 + 9x^2 + 7x - 6}$ into partial fractions.

Solution

$$\frac{5x^2 + 16x + 17}{2x^3 + 9x^2 + 7x - 6} \equiv \frac{5x^2 + 16x + 17}{(2x - 1)(x + 2)(x + 3)} \equiv \frac{a}{2x - 1} + \frac{b}{x + 2} + \frac{c}{x + 3}$$

Method 1

$5x^2 + 16x + 17$
$$\equiv a(x + 2)(x + 3) + b(2x - 1)(x + 3) + c(2x - 1)(x + 2)$$
$$\equiv (a + 2b + 2c)x^2 + (5a + 5b + 3c)x + 6a - 3b - 2c$$

This will lead to the following system:

$$\begin{cases} a + 2b + 2c = 5 \\ 5a + 5b + 3c = 16 \\ 6a - 3b - 2c = 17 \end{cases}$$

Using any method of our choice for solving systems of equations, we should have $a = 3$, $b = -1$, $c = 2$ and hence:

$$\frac{5x^2 + 16x + 17}{2x^3 + 9x^2 + 7x - 6} \equiv \frac{3}{2x - 1} - \frac{1}{x + 2} + \frac{2}{x + 3}$$

Method 2

$$5x^2 + 16x + 17 \equiv a(x + 2)(x + 3) + b(2x - 1)(x + 3) + c(2x - 1)(x + 2)$$

$$x = -2 \Rightarrow 5 = -5b \Rightarrow b = -1$$

$$x = -3 \Rightarrow 14 = 7c \Rightarrow c = 2$$

$$x = \frac{1}{2} \Rightarrow \frac{105}{4} = \frac{35}{4}a \Rightarrow a = 3$$

Partial fraction decomposition works only for proper rational expressions; that is, the degree of the numerator must be less than the degree of the denominator. If not, we must perform long division first, then perform the partial fraction decomposition on the rational part (the remainder over the divisor). After we have done the partial fraction decomposition, we just add back in the quotient part from the long division.

Linear factors: The partial fractions that we are decomposing the rational expression into must be proper. Hence, in each partial fraction, when the denominator is linear, the only thing that can be in the numerator is a constant. So, for every linear factor in the denominator, we will need a constant over that in the numerator. See Example 2.31 and Example 2.32.

Repeated linear factors: If the denominator of the rational expression contains repeated linear factors, then, following our discussion in the introduction, the process is as follows:

We need to include a factor in the expansion for each power possible. For example, if we have $(x - 1)^3$, we will need to include $(x - 1)$, $(x - 1)^2$, and

$(x - 1)^5$. Each of those $(x - 1)$ factors would have a constant term in the numerator because $(x - 1)$ is linear, no matter what power it is raised to.

For example:

$$\frac{13x^3 - 62x^2 + 101x - 58}{(x - 1)^3(2x - 5)} \equiv \frac{a}{(x - 1)^3} + \frac{b}{(x - 1)^2} + \frac{c}{x - 1} + \frac{d}{2x - 5}$$

Irreducible quadratic factors: If the rational expression we are decomposing contains irreducible quadratic factors in the denominator, then the numerator could have a linear term and/or a constant term. So, for every irreducible quadratic factor in the denominator, we will need a linear term and a constant term in the numerator.

For example: $\dfrac{-8x^3 + 15x^2 - 26x + 33}{(x - 1)^2(2x^2 + 5)} \equiv \dfrac{a}{(x - 1)^2} + \dfrac{b}{x - 1} + \dfrac{cx + d}{2x^2 + 5}$

Note: It may turn out that any of the numbers a, b, c, or d is zero.

In examinations, the denominator of any fraction to be decomposed into partial fractions will not have a degree greater than 3, and the degree of the numerator will be less than the degree of the denominator.

Example 2.33

Write $\dfrac{3x - 1}{x^2 + 4x + 4}$ as the sum of partial fractions.

Solution

The first step is to factorise the denominator.

$x^2 + 4x + 4 = (x + 2)^2$ the denominator has a repeated linear factor

$$\frac{3x - 1}{x^2 + 4x + 4} = \frac{3x - 1}{(x + 2)^2}$$

Because there are two (i.e. repeated) linear factors of $x + 2$ in the denominator of the original rational expression, then it must have a partial fraction with a denominator of $(x + 2)^2$, and it may also have a partial fraction with a denominator of $x + 2$.

So, we are looking for constants A and B such that

$$\frac{3x - 1}{(x + 2)^2} \equiv \frac{A}{x + 2} + \frac{B}{(x + 2)^2}$$

Multiplying both sides of the identity by $(x + 2)^2$ gives
$3x - 1 \equiv A(x + 2) + B$

Essentially the task is to find the unique values of A and B such that this equation is an identity, i.e. it is true for all values of x (all values of x for which the original fraction is defined, in this case $x \neq -2$). However, as you may recall, the cover-up method allows us to choose helpful values of x including such numbers. For example, in this case, if $x = -2$ then A is eliminated and the value of B can be found directly.

Let $x = -2$: $3x - 1 = A(x + 2) + B \Rightarrow 3(-2) - 1 = A \cdot 0 + B$
$$\Rightarrow B = -7$$

Let $x = 0$: $3x - 1 = A(x + 2) + B \Rightarrow 3 \cdot 0 - 1 = 2A - 7$
$$\Rightarrow 2A = 6$$
$$\Rightarrow A = 3$$

Therefore, $\dfrac{3x - 1}{x^2 + 4x + 4} = \dfrac{3}{x + 2} - \dfrac{7}{(x + 2)^2}$

Exercise 2.6

1. Decompose each rational expression into partial fractions.

   (a) $\dfrac{5x + 1}{x^2 + x - 2}$

   (b) $\dfrac{x + 4}{x^2 - 2x}$

   (c) $\dfrac{x + 2}{x^2 + 4x + 3}$

   (d) $\dfrac{5x^2 + 20x + 6}{x^3 + 2x^2 + x}$

   (e) $\dfrac{2x^2 + x - 12}{x^3 + 5x^2 + 6x}$

   (f) $\dfrac{4x^2 + 2x - 1}{x^3 + x^2}$

   (g) $\dfrac{3}{x^2 + x - 2}$

   (h) $\dfrac{5 - x}{2x^2 + x - 1}$

   (i) $\dfrac{3x + 4}{(x + 2)^2}$

   (j) $\dfrac{12}{x^4 - x^3 - 2x^2}$

   (k) $\dfrac{2}{x^3 + x}$

   (l) $\dfrac{x + 2}{x^3 + 3x}$

   (m) $\dfrac{3x + 2}{x^3 + 6x}$

   (n) $\dfrac{2x + 3}{x^3 + 8x}$

   (o) $\dfrac{x + 5}{x^3 - 5x^2 + 4x}$

Chapter 2 practice questions

1. Solve for x in the equation $x^2 - (a + 3b)x + 3ab = 0$

2. Find the values of x that solve this inequality:
   $$\frac{3x - 2}{5} + 3 \geqslant \frac{4x - 1}{3}$$

3. Find the value of c such that the vertex of the parabola $y = 3x^2 - 8x + c$ is $\left(\dfrac{4}{3}, -\dfrac{1}{3}\right)$.

4. The quadratic function $f(x) = ax^2 + bx + c$ has the following characteristics:
   - passes through the point $(2, 4)$
   - has a maximum value of 6 when $x = 4$
   - has a zero of $x = 4 + 2\sqrt{3}$

   Find the values of a, b and c.

5. If the roots of the equation $x^3 + 5x^2 + px + q = 0$ are ω, 2ω and $\omega + 3$, find the values of ω, p and q.

6. Find all values of m such that the equation $mx^2 - 2(m + 2)x + m + 2 = 0$
   (a) has two real roots
   (b) has two real roots (one positive and one negative).

7. $x - 1$ and $x + 1$ are factors of the polynomial $x^3 + ax^2 + bx + c$, and the polynomial has a remainder of 12 when divided by $x - 2$. Find the values of a, b and c.

8. Solve the inequality $|x| < 5|x - 6|$

9. Find the range of values for k in order for the equation $2x^2 + (3 - k)x + k + 3 = 0$ to have two imaginary solutions.

10. Consider the rational function $f(x) = \dfrac{2x^2 + 8x + 7}{x^2 + 4x + 5}$

    (a) Write $f(x)$ in the form $a - \dfrac{b}{(x + c)^2 + d}$

    (b) State the values of (i) $\lim\limits_{x \to +\infty} f(x)$, and (ii) $\lim\limits_{x \to -\infty} f(x)$.

    (c) State the coordinates of the minimum point on the graph of $f(x)$.

11. Find the values of k so that the equation $(k - 2)x^2 + 4x - 2k + 1 = 0$ has two distinct real roots.

12. When the function $f(x) = 6x^4 + 11x^3 - 22x^2 + ax + 6$ is divided by $(x + 1)$, the remainder is -20. Find the value of a.

13. The polynomial $p(x) - (ax + b)^3$ leaves a remainder of -1 when divided by $(x + 1)$, and a remainder of 27 when divided by $(x - 2)$. Find the values of the real numbers a and b.

14. The polynomial $f(x) = x^3 + 3x^2 + ax + b$ leaves the same remainder when divided by $(x - 2)$ as when divided by $(x + 1)$. Find the value of a.

15. When the polynomial $x^4 + ax + 3$ is divided by $(x - 1)$, the remainder is 8. Find the value of a.

16. The polynomial $x^3 + ax^2 - 3x + b$ is divisible by $(x - 2)$ and has a remainder 6 when divided by $(x + 1)$. Find the value of a and of b.

17. The polynomial $x^2 - 4x + 3$ is a factor of $x^3 + (a - 4)x^2 + (3 - 4a)x + 3$. Calculate the value of the constant a.

18. Consider $f(x) = x^3 - 2x^2 - 5x + k$. Find the value of k if $(x + 2)$ is a factor of $f(x)$.

19. The equation $kx^2 - 3x + (k + 2) = 0$ has two distinct real roots. Find the set of possible values of k.

20. Consider the equation $(1 + 2k)x^2 - 10x + k - 2 = 0$, $k \in \mathbb{R}$. Find the set of values of k for which the equation has real roots.

21. Find the range of values of m such that for all x the inequality $m(x + 1) \leqslant x^2$ is true.

22. Find the values of x for which $|5 - 3x| \leqslant |x + 1|$

23. Solve the inequality $x^2 - 4 + \dfrac{3}{x} < 0$

24. Solve the inequality $|x - 2| \geqslant |2x + 1|$

25. Let $f(x) = \dfrac{x + 4}{x + 1}, x \neq -1$ and $g(x) = \dfrac{x - 2}{x - 4}, x \neq 4$.

    Find the set of values of x such that $f(x) \leqslant g(x)$.

26. Solve the inequality $\left| \dfrac{x + 9}{x - 9} \right| \leqslant 2$

27. Find all values of x that satisfy the inequality $\dfrac{2x}{|x - 1|} < 1$

28. Express the fraction $\dfrac{2x - 5}{x^2 + x - 2}$ as the sum of two fractions.

29. Given that $\dfrac{2x - 48}{x^2 - 9} = \dfrac{A}{x + 3} + \dfrac{B}{x - 3}$, find the values of A and B.

30. Write $\dfrac{a - b}{(x - a)(x - b)}$ as the sum of partial fractions.

Sequences and series

3

Learning objectives

By the end of this chapter, you should be familiar with…

- working with arithmetic and geometric sequences; finding the sum of finite arithmetic and geometric sequences, and finding the sum of infinite geometric series
- using sigma notation
- using the binomial theorem for the expansion of $(a + b)^n$, $n \in \mathbb{N}$
- working with counting principles, including permutations and combinations.

The heights of consecutive bounces of a ball, compound interest, population growth, and Fibonacci numbers are only a few of the applications of sequences and series that we have seen in previous courses. In this chapter we will review these concepts, consolidate understanding, and take them one step further.

3.1 Sequences

Look at this pattern:

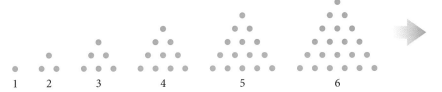

Figure 3.1 Sequence of dots in triangular arrays

The first term represents 1 dot, the second represents 3 dots, etc…
This pattern can be represented as a list of numbers written in a definite order:

$a_1 = 1$, $a_2 = 3$, $a_3 = 6$, …

The number a_1 is the first term, a_2 the second term, a_3 the third term, and so on. The nth term is a_n.

While the idea of a sequence of numbers, $a_1, a_2, a_3, \ldots$ is straightforward, it is useful to think of a sequence as a function. The sequence in Figure 3.1 can also be described in function notation as:

$f(1) = 1, f(2) = 3, f(3) = 6$, and so on, where the domain is $\mathbb{Z}^+$

Here are some more examples of sequences:

1 6, 12, 18, 24, 30

2 $3, 9, 27, \ldots, 3^k, \ldots$

3 $\left\{ \dfrac{1}{i^2}; i = 1, 2, 3, \ldots, 10 \right\}$

4 $\{b_1, b_2, \ldots, b_n, \ldots\}$, sometimes used with an abbreviation $\{b_n\}$

The first and third sequences are **finite** and the second and fourth are **infinite**. In the second and fourth sequences, we were able to define a rule that yields the nth number in the sequence (called the nth term) as a function of n, the term's

> If a_n is the nth term of a sequence, then a_{n-1} is the term before it and a_{n+1} is the term after.

number. In this sense you can think of a sequence, as a **function** that assigns a **unique** number (a_n) to each positive integer n.

Example 3.1

Find the first 5 terms and the 50th term of the sequence $\{b_n\}$ such that
$$b_n = 2 - \frac{1}{n^2}$$

Solution

Since we are given an explicit expression for the nth term as a function of its term number n, we only need to find the value of that function for the required terms:

$$b_1 = 2 - \frac{1}{1^2} = 1 \qquad b_2 = 2 - \frac{1}{2^2} = 1\frac{3}{4} \qquad b_3 = 2 - \frac{1}{3^2} = 1\frac{8}{9}$$

$$b_4 = 2 - \frac{1}{4^2} = 1\frac{15}{16} \qquad b_5 = 2 - \frac{1}{5^2} = 1\frac{24}{25} \qquad b_{50} = 2 - \frac{1}{50^2} = 1\frac{2499}{2500}$$

So, informally, a **sequence** is an **ordered set** of **real numbers**. That is, there is a first number, a second, and so on. The notation used for these sets is shown in Example 3.1. The way the function was defined in Example 3.1 is called the **explicit** definition of a sequence. There are other ways to define sequences, one of which is the **recursive** definition (also called the **inductive** definition). The following example will show you how this is used.

Example 3.2

Find the first 5 terms and the 20th term of the sequence $\{b_n\}$ such that $b_1 = 5$ and $b_n = 2(b_{n-1} + 3)$

Solution

The defining formula for this sequence is recursive. It allows us to find the nth term b_n if we know the preceding term b_{n-1}. Thus, we can find the second term from the first, the third from the second, and so on. Since we know the first term $b_1 = 5$, we can calculate the rest:

$$b_2 = 2(b_1 + 3) = 2(5 + 3) = 16$$
$$b_3 = 2(b_2 + 3) = 2(16 + 3) = 38$$
$$b_4 = 2(b_3 + 3) = 2(38 + 3) = 82$$
$$b_5 = 2(b_4 + 3) = 2(82 + 3) = 170$$

So, the first 5 terms of this sequence are 5, 16, 38, 82, and 170. However, to find the 20th term, we must first find all 19 preceding terms. This is one of the drawbacks of this type of definition, unless we can change the definition into explicit form. This can easily be done using a GDC (Figure 3.2).

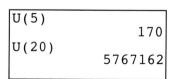

```
Plot1   Plot2   Plot3
 nMin=1
∴U(n)◼2(u(n−1)+3
)
 U(nMin)◼5▮
```

```
U(5)
                 170
U(20)
             5767162
```

Figure 3.2 GDC screens for Example 3.2

109

Example 3.3

A Fibonacci sequence is defined recursively as:

$$F_n = \begin{cases} 1 & n = 1 \\ 1 & n = 2 \\ F_{n-1} + F_{n-2} & n > 2 \end{cases}$$

(a) Find the first 10 terms of the sequence.

(b) Find the sum of the first 10 terms of the sequence.

(c) By observing that $F_1 = F_3 - F_2, F_2 = F_4 - F_3$, and so on, derive a formula for the sum of the first n Fibonacci numbers.

Solution

(a) 1, 1, 2, 3, 5, 8, 13, 21, 34, 55

(b) $S_1 = 1, S_2 = 2, S_3 = 4, S_4 = 7, S_5 = 12, S_6 = 20, S_7 = 33, S_8 = 54,$
$S_9 = 88, S_{10} = 143$

(c) Since $F_3 = F_2 + F_1$, then:

$$F_1 = F_3 - F_2$$
$$F_2 = F_4 - F_3$$
$$F_3 = F_5 - F_4$$
$$F_4 = F_6 - F_5$$
$$\vdots \qquad \vdots \qquad \vdots$$
$$F_n = F_{n+2} - F_{n+1}$$

$$S_n = F_{n+2} - F_2$$

Notice that $S_5 = 12 = F_7 - F_2 = 13 - 1$ and $S_8 = 54 = F_{10} - F_2 = 55 - 1$

Note: parts (a) and (b) can be made easy by using a spreadsheet:

	A	B	C	D
1	F(n)	S(n)		
2	1	1		
3	1	2		
4	2	4		
5	3	7		
6	5	12		
7	8	20		
8	13	33		
9	21	54		
10	34	88		
11	55	143		Let this cell be A2 + A3. Then copy it down
12	89	232		
13	144	376		
14	233	609		
15	377	986		Let this cell be B10 + A11. Then copy it down
16	610	1596		
17	987	2583		

Notice that not all sequences have formulae, either recursive or explicit. Some sequences are given only by listing their terms. There are two types that we will look at here: arithmetic and geometric sequences. We will look at them in the next two sections.

Exercise 3.1

1. Find the first 5 terms of each infinite sequence.

   (a) $s(n) = 2n - 3$

   (b) $g(k) = 2^k - 3$

   (c) $f(n) = 3 \times 2^{-n}$

   (d) $a_n = (-1)^n(2^n) + 3$

   (e) $\begin{cases} a_1 = 5 \\ a_n = a_{n-1} + 3; \quad \text{for } n > 1 \end{cases}$

   (f) $\begin{cases} b_1 = 3 \\ b_n = b_{n-1} + 2n; \quad \text{for } n \geqslant 2 \end{cases}$

2. Find the first 5 terms and the 50th term of each sequence.

   (a) $a_n = 2n - 3$

   (b) $b_n = 2 \times 3^{n-1}$

   (c) $u_n = (-1)^{n-1}\left(\dfrac{2n}{n^2 + 2}\right)$

   (d) $a_n = n^{n-1}$

   (e) $a_n = 2a_{n-1} + 5$ and $a_1 = 3$

   (f) $u_{n+1} = \dfrac{3}{2u_n + 1}$ and $u_1 = 0$

   (g) $b_n = 3b_{n-1}$ and $b_1 = 2$

   (h) $a_n = a_{n-1} + 2$ and $a_1 = -1$

3. Suggest a recursive definition for each sequence.

   (a) $\dfrac{1}{3}, \dfrac{1}{12}, \dfrac{1}{48}, \dfrac{1}{192}, \ \ldots$

   (b) $\dfrac{1}{2}a, \dfrac{2}{3}a^3, \dfrac{8}{9}a^5, \dfrac{32}{27}a^7, \ \ldots$

   (c) $a - 5k, 2a - 4k, 3a - 3k, 4a - 2k, 5a - k, \ \cdots$

4. Write down a possible formula that gives the nth term of each sequence.

   (a) $4, 7, 12, 19, \ldots$

   (b) $2, 5, 8, 11, \ldots$

   (c) $1, \dfrac{3}{4}, \dfrac{5}{9}, \dfrac{7}{16}, \dfrac{9}{25}, \ \ldots$

   (d) $\dfrac{1}{4}, \dfrac{3}{5}, \dfrac{5}{6}, 1, \dfrac{9}{8}, \ \ldots$

5. A sequence is defined as $a_n = \dfrac{F_{n+1}}{F_n}, n > 1$, where F_n is a member of the Fibonacci sequence.

   (a) Write down the first 10 terms of a_n

   (b) Show that $a_n = 1 + \dfrac{1}{a_{n-1}}$

6. A sequence is defined as:

   $$F_n = \dfrac{1}{\sqrt{5}}\left(\dfrac{(1 + \sqrt{5})^n - (1 - \sqrt{5})^n}{2^n}\right)$$

   (a) Find the first 10 terms of this sequence and compare them to Fibonacci numbers.

   (b) Show that $3 \pm \sqrt{5} = \dfrac{(1 \pm \sqrt{5})^2}{2}$

   (c) Use the result in **(b)** to verify that F_n satisfies the recursive definition of the Fibonacci sequence.

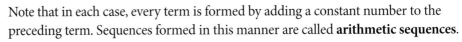

3.2 Arithmetic sequences

Here are some sequences and a recursive formula for each of them.

$7, 14, 21, 28, 35, 42, \ldots$ $\qquad a_1 = 7$ and $a_n = a_{n-1} + 7$, for $n > 1$

$2, 11, 20, 29, 38, 47, \ldots$ $\qquad a_1 = 2$ and $a_n = a_{n-1} + 9$, for $n > 1$

$39, 30, 21, 12, 3, -6, \ldots$ $\qquad a_1 = 39$ and $a_n = a_{n-1} - 9$, for $n > 1$

Note that in each case, every term is formed by adding a constant number to the preceding term. Sequences formed in this manner are called **arithmetic sequences**.

So, for the sequences above, 7 is the common difference for the first, 9 is the common difference for the second, and -9 is the common difference for the third.

This description gives us the recursive definition of the arithmetic sequence. It is possible, however, to find the explicit definition of the sequence.

Applying the recursive definition repeatedly will enable us to see the expression we are seeking:

$$a_2 = a_1 + d$$
$$a_3 = a_2 + d = a_1 + d + d = a_1 + 2d$$
$$a_4 = a_3 + d = a_1 + 2d + d = a_1 + 3d$$

So, we get to the nth term by adding d to a_1, $(n - 1)$ times.

This result is useful in finding any term of a sequence without knowing the previous terms.

> The arithmetic sequence can be looked at as a linear function as explained in the introduction to this chapter. In other words, for every increase of one unit in n, the value of the sequence will increase by d units. As the first term is a_1, the point $(1, a_1)$ belongs to this function. The constant increase d can be considered to be the gradient (slope) of this linear model, hence the nth term, the dependent variable in this case, can be found by using the point-slope form of the equation of a line:
>
> $$y - y_1 = m(x - x_1)$$
> $$a_n - a_1 = d(n - 1) \Leftrightarrow a_n = a_1 + (n - 1)d$$
>
> This agrees with our definition of an arithmetic sequence.

Definition of an arithmetic sequence

A sequence $a_1, a_2, a_3, \ldots$ is an arithmetic sequence if there is a constant d for which $a_n = a_{n-1} + d$ for all integers $n > 1$, where d is called the common difference of the sequence, and $d = a_n - a_{n-1}$ for all integers $n > 1$.

The general (nth) term of an arithmetic sequence, a_n with first term a_1 and common difference d may be expressed explicitly as $a_n = a_1 + (n - 1)d$.

Example 3.4

Find the nth term and the 50th term of the sequence $2, 11, 20, 29, 38, 47, \ldots$

Solution

This is an arithmetic sequence with first term 2 and common difference 9. Therefore:

$$a_n = a_1 + (n - 1)d = 2 + (n - 1) \times 9 = 9n - 7$$

$$\Rightarrow a_{50} = 9 \times 50 - 7 = 443$$

Example 3.5

(a) Find the recursive and the explicit forms of the definition of the sequence:

$$13, 8, 3, -2, \ldots$$

(b) Calculate the value of the 25th term.

Solution

(a) This is clearly an arithmetic sequence, with common difference -5.

Recursive definition: $\quad \begin{aligned} a_1 &= 13 \\ a_n &= a_{n-1} - 5 \end{aligned}$

Explicit definition: $\quad a_n = 13 - 5(n - 1) = 18 - 5n$

(b) $a_{25} = 18 - 5 \times 25 = -107$

Example 3.6

Find a definition for the arithmetic sequence whose first term is 5 and fifth term is 11.

Solution

Since the fifth term is given, using the explicit form, we have:

$$a_5 = a_1 + (5 - 1)d \Rightarrow 11 = 5 + 4d \Rightarrow d = \frac{3}{2}$$

This leads to the general term:

$$a_n = 5 + \frac{3}{2}(n - 1), \text{ or equivalently}$$

$$\begin{cases} a_1 = 5 \\ a_n = a_{n-1} + \dfrac{3}{2}, n > 1 \end{cases}$$

Example 3.7

Insert four arithmetic means between 3 and 7.

Solution

Since there are four means between 3 and 7, the problem can be reduced to a situation similar to Example 3.6, by considering the first term to be 3 and the sixth term to be 7. The rest is left as an exercise for you.

In a finite arithmetic sequence $a_1, a_2, a_3, \ldots, a_k$, the terms $a_2, a_3, \ldots, a_{k-1}$ are called **arithmetic means** between a_1 and a_k.

Exercise 3.2

1. Insert four arithmetic means between 3 and 7.

2. State whether or not each sequence is an arithmetic sequence. If it is, find the common difference and the 50th term. If it is not, say why not.

   (a) $a_n = 2n - 3$

   (b) $b_n = n + 2$

   (c) $c_n = c_{n-1} + 2$, and $c_1 = -1$

   (d) $u_n = 3u_{n-1} + 2$

   (e) $2, 5, 7, 12, 19, \ldots$

   (f) $2, -5, -12, -19, \ldots$

3. For each arithmetic sequence find:

   (i) the 8th term

   (ii) an explicit formula for the nth term

   (iii) a recursive formula for the nth term.

   (a) $-2, 2, 6, 10, \ldots$

   (b) $29, 25, 21, 17, \ldots$

   (c) $-6, 3, 12, 21, \ldots$

   (d) $10.07, 9.95, 9.83, 9.71, \ldots$

   (e) $100, 97, 94, 91, \ldots$

   (f) $2, \dfrac{3}{4}, -\dfrac{1}{2}, -\dfrac{7}{4}, \ldots$

4. Find five arithmetic means between 13 and -23.

5. Find three arithmetic means between 299 and 300.

6. In an arithmetic sequence, $a_5 = 16$ and $a_{14} = 42$.
   Find an explicit formula for the nth term of this sequence.

7. In an arithmetic sequence, $a_3 = -40$ and $a_9 = -18$.
   Find an explicit formula for the nth term of this sequence.

8. The first three terms and the last term are given for each sequence. Find the number of terms.

   (a) $3, 9, 15, \ldots, 525$

   (b) $9, 3, -3, \ldots, -201$

   (c) $3\dfrac{1}{8}, 4\dfrac{1}{4}, 5\dfrac{3}{8}, \ldots, 14\dfrac{3}{8}$

   (d) $\dfrac{1}{3}, \dfrac{1}{2}, \dfrac{2}{3}, \ldots, 2\dfrac{5}{6}$

   (e) $1 - k, 1 + k, 1 + 3k, \ldots, 1 + 19k$

9. Find five arithmetic means between 15 and -21.

10. Find three arithmetic means between 99 and 100.

11. In an arithmetic sequence, $a_3 = 11$ and $a_{12} = 47$.
    Find an explicit formula for the nth term of this sequence.

12. In an arithmetic sequence, $a_7 = -48$ and $a_{13} = -10$.
Find an explicit formula for the nth term of this sequence.

13. The 30th term of an arithmetic sequence is 147 and the common difference is 4. Find a formula for the nth term.

14. The first term of an arithmetic sequence is -7 and the common difference is 3. Is 9803 a term of this sequence? If so, which term?

15. The first term of an arithmetic sequence is 9689 and the 100th term is 8996. Show that the 110th term is 8926. Is 1 a term of this sequence? If so, which term?

16. The first term of an arithmetic sequence is 2 and the 30th term is 147. Is 995 a term of this sequence? If so, which term?

3.3 Geometric sequences

Examine the following sequences and the most likely recursive formula for each of them.

7, 14, 28, 56, 112, 224, ... $a_1 = 7$ and $a_n = a_{n-1} \times 2$, for $n > 1$

2, 18, 162, 1458, 13 122, ... $a_1 = 2$ and $a_n = a_{n-1} \times 9$, for $n > 1$

48, -24, 12, -6, 3, -1.5, ... $a_1 = 48$ and $a_n = a_{n-1} \times (-0.5)$, for $n > 1$

Note that in each case, every term is formed by multiplying a constant number with the preceding term. Sequences formed in this manner are called **geometric sequences.**

Thus, for the preceding sequences, 2 is the common ratio for the first, 9 is the common ratio for the second and -0.5 is the common ratio for the third.

This description gives us the recursive definition of the geometric sequence. It is possible, however, to find the explicit definition of the sequence.

Applying the recursive definition repeatedly will enable us to see the expression we are seeking:

$a_2 = a_1 \times r$

$a_3 = a_2 \times r = a_1 \times r \times r = a_1 \times r^2$

$a_4 = a_3 \times r = a_1 \times r^2 \times r = a_1 \times r^3$

We can see that we get to the nth term by multiplying r by a_1, $(n - 1)$ times.

This result is useful in finding any term of a sequence without knowing the previous terms.

Definition of a geometric sequence

A sequence a_1, a_2, a_3, is a **geometric sequence** if there is a constant r for which

$a_n = a_{n-1} \times r$

for all integers $n > 1$, where r is the **common ratio** of the sequence, and $r = a_n \div a_{n-1}$ for all integers $n > 1$.

nth term of a geometric sequence

The general (nth) term of a geometric sequence, a_n with common ratio r and first term a_1 may be expressed explicitly as

$a_n = a_1 \times r^{(n-1)}$

Example 3.8

(a) Find the geometric sequence with $a_1 = 2$ and $r = 3$

(b) Describe the sequence $3, -12, 48, -192, 768, \ldots$

(c) Describe the sequence $1, \dfrac{1}{2}, \dfrac{1}{4}, \dfrac{1}{8}, \ldots$

(d) Graph the sequence $a_n = \dfrac{1}{4} \cdot 3^{n-1}$

Solution

(a) The geometric sequence is $2, 6, 18, 54, \ldots, 2 \times 3^{n-1}$. Notice that the ratio of any two consecutive terms is 3.

(b) This is a geometric sequence with $a_1 = 3$ and $r = -4$. The nth term is $a_n = 3 \times (-4)^{n-1}$. Notice that when the common ratio is negative, the terms of the sequence alternate in sign.

(c) The nth term of this sequence is $a_n = 1 \times \left(\dfrac{1}{2}\right)^{n-1}$. Notice that the ratio of any two consecutive terms is $\dfrac{1}{2}$. Also, notice that the terms decrease in value.

(d) The terms of the sequence lie on the graph of the exponential function $y = \dfrac{1}{4} \cdot 3^{x-1}$

Example 3.9

At 8:00 a.m., 1000 mg of medicine is given to a patient. At the end of each hour, the amount of medicine in the patient's bloodstream is 60% of the amount present at the beginning of the hour.

(a) What portion of the medicine remains in the patient's bloodstream at 12 noon if no additional medication had been given?

(b) If a second dose of 1000 mg is given at 10:00 a.m., what is the total amount of the medication in the patient's bloodstream at 12 noon?

Solution

(a) Use the geometric model, as there is a constant multiple at the end of each hour. Hence, the amount at the end of any hour after giving the medicine is:

$$a_n = a_1 \times r^{n-1}, \text{ where } n \text{ is the number of hours.}$$

So, at 12 noon, $n = 5$ and $a_5 = 1000 \times 0.6^{(5-1)} = 129.6$ mg

(b) For the second dose, the amount of medicine at noon corresponds to $n = 3$:

$$a_3 = 1000 \times 0.6^{(3-1)} = 360$$

So, the amount of medicine is $129.6 + 360 = 489.6\,\text{mg}$

Compound interest

Compound interest is an example of a geometric sequence.

Interest compounded annually

When we borrow money we pay interest, and when we invest money we receive interest. Suppose an amount of €1000 is put into a savings account that has an annual interest rate of 6%. How much money will we have in the bank at the end of 4 years?

It is important to note that the 6% interest is given annually and is added to the savings account, so that in the following year it will also earn interest, and so on.

Time in years	Amount in the account (€)
0	1000
1	$1000 + 1000 \times 0.06 = 1000(1 + 0.06)$
2	$1000(1 + 0.06) + (1000(1 + 0.06)) \times 0.06 = 1000(1 + 0.06)(1 + 0.06) = 1000(1 + 0.06)^2$
3	$1000(1 + 0.06)^2 + (1000(1 + 0.06)^2) \times 0.06 = 1000(1 + 0.06)^2(1 + 0.06) = 1000(1 + 0.06)^3$
4	$1000(1 + 0.06)^3 + (1000(1 + 0.06)^3) \times 0.06 = 1000(1 + 0.06)^3(1 + 0.06) = \mathbf{1000(1 + 0.06)^4}$

Table 3.1 Compound interest

This appears to be a geometric sequence with five terms. You will notice that the number of terms is five, as both the beginning and the end of the first year are counted. (Initial value, when time $= 0$, is the first term.)

In general, if a **principal** of P euros is invested in an account that yields an annual interest rate r (expressed as a decimal), and this interest is added at the end of every year to the principal, then we can use the geometric sequence formula to calculate the **future value A,** which is accumulated after t years.

If we repeat the steps above, with

$A_0 = P =$ initial amount

$r =$ annual interest rate

$t =$ number of years

it becomes easier to develop the formula:

Time in years	Amount in the account
0	$A_0 = P$
1	$A_1 = P + Pr = P(1 + r)$
2	$A_2 = A_1(1 + r) = P(1 + r)^2$
$\vdots$	$\vdots$
t	$A_t = P(1 + r)^t$

Table 3.2 Compound interest formula

Notice that since we are counting from 0 to t, we have $t + 1$ terms, so we are using the geometric sequence formula:

$$a_n = a_1 \times r^{n-1} \Rightarrow A_t = A_1 \times (1 + r)^{t+1-1}$$

Interest compounded n times per year

Suppose that the principal P is invested as before but the interest is paid n times per year. Then $\dfrac{r}{n}$ is the interest paid every compounding period. Since every year we have n periods, for t years we have nt periods. The amount A in the account after t years is:

$$A = P\left(1 + \frac{r}{n}\right)^{nt}$$

Example 3.10

€1000 is invested in an account paying compound interest at a rate of 6% per annum. Calculate the amount of money in the account after 10 years if the compounding is:

(a) annual   (b) quarterly   (c) monthly.

Solution

(a) The amount after 10 years is:

$$A = 1000(1 + 0.06)^{10} = €1790.85$$

(b) The amount after 10 years quarterly compounding is:

$$A = 1000\left(1 + \frac{0.06}{4}\right)^{40} = €1814.02$$

(c) The amount after 10 years monthly compounding is:

$$A = 1000\left(1 + \frac{0.06}{12}\right)^{120} = €1819.40$$

Example 3.11

You invest €1000 at 6% per annum, compounded quarterly. How long will it take for this investment to increase to €2000?

Solution

Let $P = 1000$, $r = 0.06$, $n = 4$, and $A = 2000$ in the compound interest formula

$$A = P\left(1 + \frac{r}{n}\right)^{nt}$$

and then solve for t.

$$2000 = 1000\left(1 + \frac{0.06}{4}\right)^{4t} \Rightarrow 2 = 1.015^{4t}$$

Using a GDC, we can graph the functions $y = 2$ and $y = 1.015^{4t}$ and then find the intersection between their graphs.

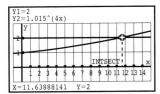

Figure 3.3 GDC screen for the solution to Example 3.11

It will take the €1000 investment 11.64 years to double to €2000. This translates into approximately 47 quarters.

We can check our work to see that this is accurate by using the compound interest formula:

$$A = 1000\left(1 + \frac{0.06}{4}\right)^{47} = €2013.28$$

In the next chapter you will learn how to solve Example 3.11 algebraically, using logarithms.

Example 3.12

You want to invest €1000. What annual interest rate is needed to make this investment grow to €2000 in 10 years if interest is compounded quarterly?

Solution

Let $P = 1000$, $n = 4$, $t = 10$ and $A = 2000$ in the compound interest formula

$$A = P\left(1 + \frac{r}{n}\right)^{nt}$$

and solve for r:

$$2000 = 1000\left(1 + \frac{r}{4}\right)^{40} \Rightarrow 2 = \left(1 + \frac{r}{4}\right)^{40}$$

$$\Rightarrow 1 + \frac{r}{4} = \sqrt[40]{2} \Rightarrow r = 4\left(\sqrt[40]{2} - 1\right) = 0.0699$$

So, at an annual rate of 7% compounded quarterly, the €1000 investment will grow to at least €2000 in 10 years.

We can check to see if our work is accurate by using the compound interest formula:

$$A = 1000\left(1 + \frac{0.07}{4}\right)^{40} = €2001.60$$

Population growth

The same formulae can be applied when dealing with population growth.

Example 3.13

The population of Baden in Austria grows at an annual rate of 0.35%. The population of Baden in 1981 was 23 140. What is the estimate of the population of Baden for 2025?

Solution

This situation can be modelled by a geometric sequence whose first term is 23 140 and common ratio 1.0035. Since we count the population of 1981 among the terms, the number of terms is 45.

In Chapter 4, more realistic population growth models will be explored and more efficient methods will be developed, including the ability to calculate interest that is continuously compounded.

2025 is equivalent to the 45th term in this sequence. The estimated population for Baden is therefore:

Population (2025) = a_{45} = 23 140(1.0035)^{44} = 26 985

Exercise 3.3

1. For each sequence:
   (i) determine whether the sequence is arithmetic, geometric, or neither.
   (ii) find the common difference for the arithmetic ones and the common ratio for the geometric ones.
   (iii) find the 10th term for each arithmetic or geometric sequence.

   (a) $3, 3^{a+1}, 3^{2a+1}, 3^{3a+1}, \ldots$
   (b) $a_n = 3n - 3$

   (c) $b_n = 2^{n+2}$
   (d) $c_n = 2c_{n-1} - 2$, and $c_1 = -1$

   (e) $u_n = 3u_{n-1}$, and $u_1 = 4$
   (f) 2, 5, 12.5, 31.25, 78.125, ...

   (g) 2, −5, 12.5, −31.25, 78.125, ...
   (h) 2, 2.75, 3.5, 4.25, 5, ...

   (i) $18, -12, 8, -\dfrac{16}{3}, \dfrac{32}{9}, \ldots$
   (j) 52, 55, 58, 61, ...

   (k) −1, 3, −9, 27, −81, ...
   (l) 0.1, 0.2, 0.4, 0.8, 1.6, 3.2, ...

   (m) 3, 6, 12, 18, 21, 27, ...
   (n) 6, 14, 20, 28, 34, ...

   (o) 2.4, 3.7, 5, 6.3, 7.6, ...

2. For each arithmetic or geometric sequence, find:
   (i) the 8th term
   (ii) an explicit formula for the nth term
   (iii) a recursive formula for the nth term.

   (a) −3, 2, 7, 12, ...
   (b) 19, 15, 11, 7, ...

   (c) −8, 3, 14, 25, ...
   (d) 10.05, 9.95, 9.85, 9.75, ...

   (e) 100, 99, 98, 97, ...
   (f) $2, \dfrac{1}{2}, -1, -\dfrac{5}{2}, \ldots$

   (g) 3, 6, 12, 24, ...
   (h) 4, 12, 36, 108, ...

   (i) 5, −5, 5, −5, ...
   (j) 3, −6, 12, −24, ...

   (k) 972, −324, 108, −36, ...
   (l) $-2, 3, -\dfrac{9}{2}, \dfrac{27}{4}, \ldots$

   (m) $35, 25, \dfrac{125}{7}, \dfrac{625}{49}, \ldots$
   (n) $-6, -3, -\dfrac{3}{2}, -\dfrac{3}{4}, \ldots$

   (o) 9.5, 19, 38, 76, ...
   (p) 100, 95, 90.25, ...

   (q) $2, \dfrac{3}{4}, \dfrac{9}{32}, \dfrac{27}{256}, \ldots$

3. Find four geometric means between 3 and 96.

4. Find three geometric means between 7 and 4375.

5. Find a geometric mean between 16 and 81.

6. Find four geometric means between 7 and 1701.

7. Find a geometric mean between 9 and 64.

8. The first term of a geometric sequence is 24 and the fourth term is 3. Find the fifth term and an expression for the nth term.

9. The first term of a geometric sequence is 24 and the third term is 6. Find the fourth term and an expression for the nth term.

10. The common ratio in a geometric sequence is $\frac{2}{7}$ and the fourth term is $\frac{14}{3}$. Find the third term.

11. Which term of the geometric sequence 6, 18, 54, … is 118 098?

12. The fourth term and the seventh term of a geometric sequence are 18 and $\frac{729}{8}$. Is $\frac{59\,049}{128}$ a term of this sequence? If so, which term is it?

13. The third term and the sixth term of a geometric sequence are 18 and $\frac{243}{4}$. Is $\frac{19\,683}{64}$ a term of this sequence? If so, which term is it?

14. Vitoria put €1500 into a savings account that pays 4% annual interest compounded semiannually. How much will her account hold 10 years later if she does not make any additional investments in this account?

15. At the birth of her daughter Jane, Charlotte deposited £500 into a savings account. The annual interest rate was 4% compounded quarterly. How much money will Jane have on her 16th birthday?

16. How much money should you invest now if you wish to have an amount of €4000 in your account after 6 years if interest is compounded quarterly at an annual rate of 5%?

17. In 2017, the population of a town in Switzerland was estimated to be 7554. How large would the town's population be in 2022 if it grows at a rate of 0.5% annually?

18. The common ratio in a geometric sequence is $\frac{3}{7}$ and the fourth term is $\frac{14}{3}$. Find the second term.

In a finite geometric sequence $a_1, a_2, a_3, …, a_k$, the terms $a_2, a_3, … a_k - 1$ are called **geometric means** between a_1 and a_k.

In questions 5 and 7, this is also called the **mean proportional**.

19. Which term of the geometric sequence 7, 21, 63, … is 137 781?

20. At her son Erik's birth, Astrid deposited £1000 into a savings account. The annual interest rate was 6% compounded quarterly. How much money will Erik have on his 18th birthday?

3.4 Series

In common usage, the word series is the same thing as sequence. But in mathematics, a series is the sum of terms in a sequence. For a sequence of values a_n, the corresponding series is the sequence of S_n with:

$$S_n = a_1 + a_2 + \ldots + a_{n-1} + a_n$$

If the terms are in an arithmetic sequence, then the sum is an **arithmetic series**.

Sigma notation

Most of the series we consider in mathematics are **infinite** series. This is to emphasise that the series contain an infinite number of terms. Any sum in the series S_k will be called a partial sum and is given by:

$$S_k = a_1 + a_2 + \ldots + a_{k-1} + a_k$$

For convenience, this partial sum is written using sigma notation:

$$S_k = \sum_{i=1}^{k} a_i = a_1 + a_2 + \ldots + a_{k-1} + a_k$$

Sigma notation is a concise and convenient way to represent long sums. The symbol $\sum$ is the Greek capital letter *Sigma* that refers to the initial letter of the word 'sum'. So this expression means the sum of all the terms a_i where i takes the values from 1 to k. We can also write $\sum_{i=m}^{n} a_i$ to mean the sum of the terms a_i where i takes the values from m to n. In such a sum, m is called the lower limit and n the upper limit.

This indicates ending with $i = n$
This indicates addition → $\sum_{i=m}^{n} a_i$
This indicates starting with $i = m$

For example, suppose we measure the heights of six children. We will denote their heights by x_1, x_2, x_3, x_4, x_5 and x_6.

The sum of their heights $x_1 + x_2 + x_3 + x_4 + x_5 + x_6$ is written more neatly, by using sigma notation, as $\sum_{i=1}^{6} x_i$.

The symbol $\sum$ means 'add up'. Underneath $\sum$ we see $i = 1$ and on top of it 6. This means that i is replaced by whole numbers starting at the bottom number, 1, until the top number, 6, is reached.

Thus $\sum_{i=3}^{6} x_i = x_3 + x_4 + x_5 + x_6$ and $\sum_{i=3}^{5} x_i = x_3 + x_4 + x_5$

So, the notation $\sum_{i=1}^{n} x_i$ tells us:

- to add the scores x_i
- where to start: x_1
- where to stop: x_n (where n is some integer)

Now take the heights of the children to be $x_1 = 112\,\text{cm}$, $x_2 = 96\,\text{cm}$, $x_3 = 120\,\text{cm}$, $x_4 = 132\,\text{cm}$, $x_5 = 106\,\text{cm}$, and $x_6 = 120\,\text{cm}$.

Then the total height (in cm) is

$$\sum_{k=1}^{6} x_k = x_1 + x_2 + x_3 + x_4 + x_5 + x_6$$
$$= 112 + 96 + 120 + 132 + 106 + 120 = 686\,\text{cm}$$

Notice that we have used k instead of i in the formula above. The i is what we call a dummy variable – any letter can be used.

$$\sum_{k=1}^{n} x_k = \sum_{i=1}^{n} x_i$$

Example 3.14

Write each series in full:

(a) $\sum_{i=1}^{5} i^4$ (b) $\sum_{r=3}^{7} 3^r$ (c) $\sum_{j=1}^{n} x_j p(x_j)$

Solution

(a) $\sum_{i=1}^{5} i^4 = 1^4 + 2^4 + 3^4 + 4^4 + 5^4$

(b) $\sum_{r=3}^{7} 3^r = 3^3 + 3^4 + 3^5 + 3^6 + 3^7$

(c) $\sum_{j=1}^{n} x_j p(x_j) = x_1 p(x_1) + x_2 p(x_2) + \ldots + x_n p(x_n)$

Example 3.15

Evaluate $\sum_{n=0}^{5} 2^n$

Solution

$\sum_{n=0}^{5} 2^n = 2^0 + 2^1 + 2^2 + 2^3 + 2^4 + 2^5 = 63$

Example 3.16

Write the sum $\frac{1}{2} - \frac{2}{3} + \frac{3}{4} - \frac{4}{5} + \ldots + \frac{99}{100}$ using sigma notation.

Solution

The numerator and denominator in each term are consecutive integers, so they take on the absolute value of $\frac{k}{k+1}$ or any equivalent form. The signs of the terms alternate and there are 99 terms. To take care of the sign, we use some power of (-1) that will start with a positive value. If we use $(-1)^k$, then the first term will be negative, hence we can use $(-1)^{k+1}$ instead. We can therefore write the sum as

$$(-1)^{1+1}\left(\frac{1}{2}\right) + (-1)^{2+1}\left(\frac{2}{3}\right) + (-1)^{3+1}\left(\frac{3}{4}\right) + \ldots + (-1)^{99+1}\left(\frac{99}{100}\right)$$

$$= \sum_{k=1}^{99} (-1)^{k+1}\left(\frac{k}{k+1}\right)$$

Properties of sigma notation

There are a number of useful results we can obtain when we use sigma notation.

1. For example, suppose we have a sum of constant terms:

$$\sum_{i=1}^{5} 2$$

What does this mean? If we write this out in full, we get:

$$\sum_{i=1}^{5} 2 = 2 + 2 + 2 + 2 + 2 = 5 \times 2 = 10$$

In general, if we sum a constant n times then we can write:

$$\sum_{i=1}^{n} k = k + k + \ldots + k = n \times k = nk$$

2. Suppose we have the sum of a constant multiplied by i. For example:

$$\sum_{i=1}^{5} 5i = 5 \times 1 + 5 \times 2 + 5 \times 3 + 5 \times 4 + 5 \times 5$$

$$= 5 \times (1 + 2 + 3 + 4 + 5) = 75$$

However, this can also be interpreted as:

$$\sum_{i=1}^{5} 5i = 5 \times 1 + 5 \times 2 + 5 \times 3 + 5 \times 4 + 5 \times 5$$

$$= 5 \times (1 + 2 + 3 + 4 + 5) = 5\sum_{i=1}^{5} i$$

which implies that:

$$\sum_{i=1}^{5} 5i = 5\sum_{i=1}^{5} i$$

In general, we can say:

$$\sum_{i=1}^{n} ki = k \times 1 + k \times 2 + \ldots + k \times n$$

$$= k \times (1 + 2 + \ldots + n)$$

$$= k\sum_{i=1}^{n} i$$

3. Suppose that we need to consider the summation of two different functions, such as:

$$\sum_{k=1}^{n}(k^2 + k^3) = (1^2 + 1^3) + (2^2 + 2^3) + \ldots + (n^2 + n^3)$$

$$= (1^2 + 2^2 + \ldots + n^2) + (1^3 + 2^3 + \ldots + n^3)$$

$$= \sum_{k=1}^{n} k^2 + \sum_{k=1}^{n} k^3$$

In general

$$\sum_{k=1}^{n}(f(k) + g(k)) = \sum_{k=1}^{n} f(k) + \sum_{k=1}^{n} g(k)$$

4. At times it is convenient to change powers, for example:

(i) $\displaystyle\sum_{i=1}^{k} a_i = a_1 + a_2 + \ldots + a_{k-1} + a_k$ is the same as $\displaystyle\sum_{i=0}^{k-1} a_{i+1}$

$$= a_1 + a_2 + \ldots + a_{k-1} + a_k$$

(ii) $\displaystyle\sum_{i=1}^{n} f(i) = \sum_{i=1}^{k} f(i) + \sum_{i=k+1}^{n} f(i)$ or equivalently $\displaystyle\sum_{i=k+1}^{n} f(i) = \sum_{i=1}^{n} f(i) - \sum_{i=1}^{k} f(i)$

Arithmetic series

In arithmetic series, we are concerned with adding the terms of arithmetic sequences. It is very helpful to be able to find an easy expression for the partial sums of such a series.

Let's start with an example:

Find the partial sum for the first 50 terms of the series

$3 + 8 + 13 + 18 + \ldots$

Write the 50 terms in ascending order and then in descending order underneath. Add the terms together as shown.

$$S_{50} = \quad 3 + \quad 8 + \quad 13 + \ldots + 248$$
$$S_{50} = 248 + 243 + 238 + \ldots + \quad 3$$
$$\overline{2\,S_{50} = 251 + 251 + 251 + \ldots + 251}$$

There are 50 terms in this sum, and hence

$$2\,S_{50} = 50 \times 251 \Rightarrow S_{50} = 6275$$

This reasoning can be extended to any arithmetic series in order to develop a formula for the nth partial sum S_n.

Let $\{a_n\}$ be an arithmetic sequence with first term a_1 and common difference d. We can construct the series in two ways: Forwards, by adding d to a_1 repeatedly, and backwards by subtracting d from a_n repeatedly. We get the following two expressions for the sum:

$$S_n = a_1 + a_2 + a_3 + \ldots + a_n = a_1 + (a_1 + d) + (a_1 + 2d) + \ldots + (a_1 + (n-1)d)$$

$$S_n = a_n + a_{n-1} + a_{n-2} + \ldots + a_1 = a_n + (a_n - d) + (a_n - 2d) + \ldots + (a_n - (n-1)d)$$

By adding, term by term vertically, we get:

$$S_n = a_1 + \qquad (a_1 + d) + (a_1 + 2d) + \ldots + (a_1 + (n-1)d)$$
$$S_n = a_n + \qquad (a_n - d) + (a_n - 2d) + \ldots + (a_n - (n-1)d)$$
$$2S_n = (a_1 + a_n) + (a_1 + a_n) + (a_1 + a_n) + \ldots + (a_1 + a_n)$$

Since there are n terms, we can reduce the expression above to:

$$2S_n = n(a_1 + a_n)$$

which can be reduced to:

$$S_n = \frac{n}{2}(a_1 + a_n)$$

which in turn can be changed to give an interesting perspective of the sum:

$$S_n = n\left(\frac{a_1 + a_n}{2}\right)$$

which is n times the average of the first and last terms!

If we substitute $a_1 + (n-1)d$ for a_n then we get an alternative formula for the sum:

$$S_n = \frac{n}{2}(a_1 + a_1 + (n-1)d) = \frac{n}{2}(2a_1 + (n-1)d)$$

> The partial sum S_n of an arithmetic series is given by one of the following:
>
> $S_n = \frac{n}{2}(a_1 + a_n)$, or $S_n = n\left(\frac{a_1 + a_n}{2}\right)$, or $S_n = \frac{n}{2}(2a_1 + (n-1)d)$

Example 3.17

Find the partial sum for the first 50 terms of the series $3 + 8 + 13 + 18 + \ldots$

Solution

Using the second formula for the sum we get:

$$S_{50} = \frac{50}{2}(2 \times 3 + (50 - 1)5) = 25 \times 251 = 6275$$

Using the first formula requires that we know the nth term.
So, $a_{50} = 3 + 49 \times 5 = 248$ which now can be used:

$$S_{50} = 25(3 + 248) = 6275$$

Example 3.18

You are given a sequence of figures as shown in the diagram.

P1   P2   P3

(a) P1 has six line segments. How many line segments are in P20?

(b) Is there a figure with 4401 segments? If so, which one? If not, why not?

(c) Find the total number of line segments in the first 880 figures.

Solution

(a) In each new figure, 5 line segments are added. This is an arithmetic sequence with first term 6, and common difference 5.

Using the nth term form:
$$P20 = 6 + 5(20 - 1) = 101$$

(b) The term whose value is 4401 satisfies the nth term form:
$$4401 = 6 + 5(n - 1), \text{ thus } n = \frac{4401 - 6}{5} + 1 = 880$$

Therefore, 4401 is the 880th figure.

(c) We use one of the formulae for the arithmetic series:
$$S_{880} = \frac{880}{2}(6 + 4401) = 1\,939\,080$$

$$\text{or } S_{880} = \frac{880}{2}(2 \times 6 + 5(880 - 1)) = 1\,939\,080$$

Geometric series

As is the case with arithmetic series, in several cases it is desirable to find a general expression for the nth partial sum of a geometric series.

Let us start with an example:

Find the partial sum for the first 20 terms of the series $3 + 6 + 12 + 24 + \ldots$

We express S_{20} in two different ways and subtract them:

$$S_{20} = 3 + 6 + 12 + \ldots + 1572864$$
$$2S_{20} = \quad\quad 6 + 12 + \ldots + 1572864 + 3\,145\,728$$
$$\overline{-S_{20} = 3 \quad\quad\quad\quad\quad\quad\quad\quad\quad - 3\,145\,728}$$
$$\Rightarrow S_{20} = 3\,145\,725$$

This reasoning can be extended to any geometric series in order to develop a formula for the nth partial sum S_n.

Let $\{a_n\}$ be a geometric sequence with first term a_1 and common ratio $r \neq 1$. We can construct the series in two ways as before, and, using the definition of the geometric sequence, $a_n = a_{n-1} \times r$, then:

$$S_n = a_1 + a_2 + a_3 + \ldots + a_{n-1} + a_n$$
$$rS_n = ra_1 + ra_2 + ra_3 + \ldots + ra_{n-1} + ra_n$$
$$\downarrow \quad \downarrow \quad\quad\quad\quad\quad \downarrow$$
$$= a_2 + a_3 + \ldots + a_{n-1} + a_n + ra_n$$

Now, we subtract the second expression from the first to get:

$$S_n - rS_n = a_1 - ra_n \Rightarrow S_n(1 - r) = a_1 - ra_n \Rightarrow S_n = \frac{a_1 - ra_n}{1 - r}, r \neq 1$$

This expression, however, requires that r, a_1, and a_n be known to find the sum. But, using the nth term expression developed earlier, we can simplify this sum formula to:

$$S_n = \frac{a_1 - ra_n}{1 - r} = \frac{a_1 - ra_1 r^{n-1}}{1 - r} = \frac{a_1(1 - r^n)}{1 - r}, r \neq 1$$

Partial sum of a geometric series
The partial sum, S_n, of n terms of a geometric sequence with common ratio r ($r \neq 1$) and first term a_1 is:

$$S_n = \frac{a_1(1 - r^n)}{1 - r} \quad \left[\text{equivalent to } S_n = \frac{a_1(r^n - 1)}{r - 1}\right]$$

Example 3.19

Find the partial sum for the first 20 terms of the series $3 + 6 + 12 + 24 + \ldots$

Solution

$$S_{20} = 3 \times \frac{(1 - 2^{20})}{1 - 2} = \frac{3(1 - 1\,048\,576)}{-1} = 3\,145\,725$$

Infinite geometric series

Consider the series $\sum_{k=1}^{n} 2\left(\frac{1}{2}\right)^{k-1} = 2 + 1 + \frac{1}{2} + \frac{1}{4} + \frac{1}{8} + \ldots$

Consider also finding the partial sums for 10, 20, and 100 terms. We are looking for are the partial sums of a geometric series:

$$\sum_{k=1}^{10} 2\left(\frac{1}{2}\right)^{k-1} = 2 \times \frac{1 - \left(\frac{1}{2}\right)^{10}}{1 - \frac{1}{2}} \approx 3.996$$

$$\sum_{k=1}^{20} 2\left(\frac{1}{2}\right)^{k-1} = 2 \times \frac{1 - \left(\frac{1}{2}\right)^{20}}{1 - \frac{1}{2}} \approx 3.999\,996$$

$$\sum_{k=1}^{100} 2\left(\frac{1}{2}\right)^{k-1} = 2 \times \frac{1 - \left(\frac{1}{2}\right)^{100}}{1 - \frac{1}{2}} \approx 4$$

As the number of terms increases, the partial sum appears to be approaching the number 4. This is no coincidence. In the language of limits,

$$\lim_{n \to \infty} \sum_{k=1}^{n} 2\left(\frac{1}{2}\right)^{k-1} = \lim_{n \to \infty} 2 \times \frac{1 - \left(\frac{1}{2}\right)^k}{1 - \frac{1}{2}} = 2 \times \frac{1 - 0}{\frac{1}{2}} = 4, \text{ since } \lim_{n \to \infty} \left(\frac{1}{2}\right)^n = 0$$

This type of problem allows us to extend the usual concept of a sum of a **finite** number of terms to make sense of sums in which an **infinite** number of terms are involved. Such series are called **infinite series**.

One thing to be made clear about infinite series is that they are not true sums! The associative property of addition of real numbers allows us to extend the definition of the sum of two numbers, such as $a + b$, to three or four, or n numbers, but not to an infinite number of numbers. For example, you can add any specific number of 5s together and get a real number, but if you add an infinite number of 5s together, you cannot get a real number. The remarkable thing about infinite series though, is that in some cases, such as the example above, the sequence of partial sums (which are true sums) approaches a finite limit L. The limit in our example is 4.

We write this as $\lim\limits_{n\to\infty} \sum\limits_{k=1}^{n} a_k = \lim\limits_{n\to\infty}(a_1 + a_2 + \dots + a_n) = L$

We say that the series **converges** to L, and it is convenient to define L as the **sum of the infinite series**. We use the notation:

$$\sum_{k=1}^{\infty} a_k = \lim_{n\to\infty} \sum_{k=1}^{n} a_k = L$$

We can therefore write the limit in the previous example:

$$\sum_{k=1}^{\infty} 2\left(\frac{1}{2}\right)^{k-1} = \lim_{n\to\infty} \sum_{k=1}^{n} 2\left(\frac{1}{2}\right)^{k-1} = 4$$

If the series does not have a limit, then it **diverges**.

We are now ready to develop a general rule for **infinite geometric series**. As we know, the sum of a geometric sequence is given by:

$$S_n = \frac{a_1 - r\,a_n}{1 - r} = \frac{a_1 - r\,a_1\,r^{n-1}}{1 - r} = \frac{a_1(1 - r^n)}{1 - r}, r \neq 1$$

If $|r| < 1$, then $\lim\limits_{n\to\infty} r^n = 0$ and hence:

$$\lim_{n\to\infty} S_n = S = \lim_{n\to\infty} \frac{a_1(1 - r^n)}{1 - r} = \frac{a_1}{1 - r}$$

We will call this **the sum of an infinite convergent geometric sequence**. In all other cases the series diverges. The proof is left as an exercise.

In the case of $\sum\limits_{k=1}^{\infty} 2\left(\frac{1}{2}\right)^{k-1} = \dfrac{2}{1 - \frac{1}{2}} = 4$ as already shown.

Sum of an infinite convergent geometric sequence
The sum, S_∞, of an infinite convergent geometric sequence with first term a_1 such that the common ratio, r, satisfies the condition $|r| < 1$ is given by:

$$S_\infty = \frac{a_1}{1 - r}$$

Example 3.20

A rational number is a number that can be expressed as a quotient of two integers. Show that $0.\overline{6} = 0.666\ldots$ is a rational number.

Solution

$0.\overline{6} = 0.666\ldots = 0.6 + 0.06 + 0.006 + 0.0006 + \ldots$

$$= \frac{6}{10} + \frac{6}{10} \cdot \frac{1}{10} + \frac{6}{10} \cdot \left(\frac{1}{10}\right)^2 + \frac{6}{10} \cdot \left(\frac{1}{10}\right)^3 + \ldots$$

This is an infinite geometric series with $a_1 = \frac{6}{10}$ and $r = \frac{1}{10}$, therefore:

$$0.\overline{6} = \frac{\frac{6}{10}}{1 - \frac{1}{10}} = \frac{6}{10} \cdot \frac{10}{9} = \frac{2}{3}$$

Example 3.21

A ball has elasticity such that, on each bounce, it bounces to 80% of its previous height. Find the total vertical distance travelled down and up by this ball when it is dropped from a height of 3 m and is allowed to keep bouncing until it comes to rest. Ignore friction and air resistance.

Solution

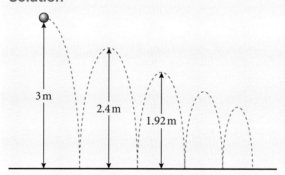

After the ball is dropped, the initial 3 m, it bounces up and down a distance of 2.4 m. On each bounce after the first bounce, the ball travels 0.8 times the previous height twice – once upwards and once downwards. So, the total vertical distance is given by

$$h = 3 + 2[2.4 + (2.4 \times 0.8) + (2.4 \times 0.8^2) + \ldots] = 3 + 2 \times l$$

The terms inside the square brackets form an infinite geometric series with $a_1 = 2.4$ and $r = 0.8$. The value of that quantity is:

$$l = \frac{2.4}{1 - 0.8} = 12$$

Hence the total distance required is $h = 3 + 2(12) = 27$ m

Applications of series to compound interest calculations

Annuities

An **annuity** is a sequence of equal periodic payments. If you are saving money by depositing the same amount at the end of each compounding period, the annuity is called an **ordinary annuity**. Using geometric series, you can calculate the **future value (FV)** of this annuity, which is the amount of money you have after making the last payment.

You invest €1000 at the end of each year for 10 years at a fixed annual interest rate of 6%, as shown in Table 3.3.

Year	Amount invested (€)	Future value (€)
10	1000	1000
9	1000	$1000(1 + 0.06)$
8	1000	$1000(1 + 0.06)^2$
$\vdots$		
1	1000	$1000(1 + 0.06)^9$

Table 3.3 Calculating the future value

The future value FV of this investment is the sum of the entries in the last column:

$$FV = 1000 + 1000(1 + 0.06) + 1000(1 + 0.06)^2 + \ldots + 1000(1 + 0.06)^9$$

This sum is a partial sum of a geometric series with $n = 10$ and $r = 1 + 0.06$
Hence:

$$FV = \frac{1000(1 - (1 + 0.06)^{10})}{1 - (1 + 0.06)} = \frac{1000(1 - (1 + 0.06)^{10})}{-0.06} = €13\,180.79$$

```
Compound Interest          Compound Interest
 I%  =6                      FV =13180.79494
 PV  =0
 PMT =-1000
 FV  =0
 P/V =1
 C/Y =1
```

Figure 3.4 This result can also be produced with a GDC

We can generalise the previous formula in the same manner. Let the periodic payment be R, and the periodic interest rate be i – that is, $i = \frac{r}{n}$.
Let the number of periodic payments be m.

Period	Amount invested	Future value
m	R	R
$m - 1$	R	$R(1 + i)$
$m - 2$	R	$R(1 + i)^2$
$\vdots$		
1	R	$R(1 + i)^{m-1}$

Table 3.4 Formula for calculating the future value

The future value *FV* is the sum of the entries in the last column:

$$FV = R + R(1 + i) + R(1 + i)^2 + \ldots + R(1 + i)^{m-1}$$

This is a partial sum of a geometric series with *m* terms and $r = 1 + i$. Hence:

$$FV = \frac{R(1 - (1 + i)^m)}{1 - (1 + i)} = \frac{R(1 - (1 + i)^m)}{-i} = R\left(\frac{(1 + i)^m - 1}{i}\right)$$

If the payment is made at the beginning of the period rather than at the end, then the annuity is called an **annuity due** and the future value after *m* periods will be slightly different.

Period	Amount invested	Future value
m	R	$R(1 + i)$
$m - 1$	R	$R(1 + i)^2$
$m - 2$	R	$R(1 + i)^3$
1	R	$R(1 + i)^m$

The future value of this investment is the sum of the entries in the last column:

$$FV = R(1 + i) + R(1 + i)^2 + \ldots + R(1 + i)^{m-1} + R(1 + i)^m$$

This is a partial sum of a geometric series with *m* terms and $r = 1 + i$.

$$FV = \frac{R(1 + i)(1 - (1 + i)^m)}{1 - (1 + i)} = \frac{R(1 + i - (1 + i)^{m+1})}{-i} = R\left(\frac{(1 + i)^{m+1} - 1}{i} - 1\right)$$

If the previous investment is made at the beginning of the year rather than at the end, then in 10 years we have:

$$FV = R\left(\frac{(1 + i)^{m+1} - 1}{i}\right) - 1 = 1000\left(\frac{(1 + 0.06)^{10+1} - 1}{0.06} - 1\right) = 13\,971.64$$

Exercise 3.4

1. Find the sum of the arithmetic sequence $11 + 17 + \ldots + 365$

2. Find the sum of this sequence:
$$2 - 3 + \frac{9}{2} - \frac{27}{4} + \ldots - \frac{177\,147}{1024}$$

3. Evaluate $\sum_{k=0}^{13}(2 - 0.3k)$

4. Evaluate $2 - \frac{4}{5} + \frac{8}{25} - \frac{16}{125} + \ldots$

5. Evaluate $\frac{1}{3} + \frac{\sqrt{3}}{12} + \frac{1}{16} + \frac{\sqrt{3}}{64} + \frac{3}{256} + \ldots$

6. Express each repeating decimal as a fraction:

   (a) $0.\overline{52}$ (b) $0.4\overline{53}$ (c) $3.01\overline{37}$

7. At the beginning of every month, Maggie invests $150 in an account that pays a 6% annual rate of interest. How much money will there be in the account after six years?

8. Find the sum of each series.

   (a) $9 + 13 + 17 + \ldots + 85$ (b) $8 + 14 + 20 + \ldots + 278$

   (c) $155 + 158 + 161 + \ldots + 527$

9. The kth term of an arithmetic sequence is $2 + 3k$. Find, in terms of n, the sum of the first n terms of this sequence.

10. For the arithmetic sequence that begins $17 + 20 + 23 + \ldots$, for what value of n will the partial sum S_n of the sequence exceed 678?

11. For the arithmetic sequence that begins $-18 - 11 - 4 \ldots$, for what value of n will the partial sum S_n of the sequence exceed 2335?

12. An arithmetic sequence has a as first term and $2d$ as common difference, i.e., $a, a + 2d, a + 4d, \ldots$. The sum of the first 50 terms is T. Another sequence, with first term $a + d$ and common difference $2d$, is combined with the first one to produce a new arithmetic sequence. Let the sum of the first 100 terms of the new combined sequence be S. If $2T + 200 = S$, find d.

13. Consider the arithmetic sequence $3, 7, 11, \ldots, 999$.

    (a) Find the number of terms and the sum of this sequence.

    (b) Create a new sequence by removing every third term, i.e., $11, 23, \ldots$. Find the sum of the terms of the remaining sequence.

14. The sum of the first 10 terms of an arithmetic sequence is 235 and the sum of the second 10 terms is 735. Find the first term and the common difference.

15. Use your GDC or a spreadsheet to evaluate each sum.

    (a) $\displaystyle\sum_{k=1}^{20}(k^2 + 1)$ (b) $\displaystyle\sum_{i=3}^{17}\left(\frac{1}{i^2 + 3}\right)$ (c) $\displaystyle\sum_{n=1}^{100}(-1)^n\frac{3}{n}$

16. Find the sum of the arithmetic series $13 + 19 + \ldots + 367$

17. Find the sum of the arithmetic series:
    $$2 - \frac{4}{3} + \frac{8}{9} - \frac{16}{27} + \ldots - \frac{4096}{177\,147}$$

18. Evaluate $\displaystyle\sum_{k=0}^{11}(3 + 0.2k)$

19. Evaluate $2 - \dfrac{4}{3} + \dfrac{8}{9} - \dfrac{16}{27} + \ldots$

20. Evaluate $\dfrac{1}{2} + \dfrac{\sqrt{2}}{2\sqrt{3}} + \dfrac{1}{3} + \dfrac{\sqrt{2}}{3\sqrt{3}} + \dfrac{2}{9} + \ldots$

21. Find the first four partial sums and the nth partial sum of each sequence.

    (a) $u_n = \dfrac{3}{5^n}$ (b) $v_n = \dfrac{1}{n^2 + 3n + 2}$ (c) $u_n = \sqrt{n + 1} - \sqrt{n}$

For question 21 part (b), show that
$$v_n = \frac{1}{n+1} - \frac{1}{n+2}$$

133

22. A ball is dropped from a height of 16 m. Every time it hits the ground it bounces to 81% of its previous height.

  (a) Find the maximum height it reaches after the 10th bounce.

  (b) Find the total distance travelled by the ball until it comes to rest. (Assume no friction and no loss of elasticity.)

23. The sides of a square are 16 cm in length. A new square is formed by joining the midpoints of the adjacent sides and then two of the resulting triangles are coloured, as shown.

  (a) If the process is repeated six more times, determine the total area of the shaded region.

  (b) If the process were to be repeated indefinitely, find the total area of the shaded region.

24. The largest rectangle in the diagram below measures 4 cm by 2 cm. Another rectangle is constructed inside it, measuring 2 cm by 1 cm. The process is repeated. The region surrounding every other inner rectangle is shaded, as shown.

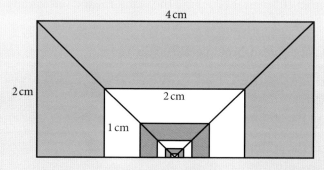

  (a) Find the total area for the three regions shaded already.

  (b) If the process were to be repeated indefinitely, find the total area of the shaded regions.

25. Find each sum.

  (a) $7 + 12 + 17 + 22 + \ldots + 337 + 342$

  (b) $9486 + 9479 + 9472 + 9465 + \ldots + 8919 + 8912$

  (c) $2 + 6 + 18 + 54 + \ldots + 3\,188\,646 + 9\,565\,938$

  (d) $120 + 24 + \dfrac{24}{5} + \dfrac{24}{25} + \ldots + \dfrac{24}{78\,125}$

3.5 The binomial theorem

A binomial is a polynomial with two terms. For example, $x + y$ is a binomial. In principle, it is easy to raise $x + y$ to any power, but raising it to high powers would be tedious. In this chapter, we will find a formula that gives the expansion of $(x + y)^n$ for any positive integer n. (The proof of the binomial theorem is given in Chapter 5 as optional material.)

Let's look at some particular cases of the expansion of $(x + y)^n$

$(x+y)^0 = 1$

$(x+y)^1 = x + y$

$(x+y)^2 = x^2 + 2xy + y^2$

$(x+y)^3 = x^3 + 3x^2y + 3xy^2 + y^3$

$(x+y)^4 = x^4 + 4x^3y + 6x^2y^2 + 4xy^3 + y^4$

$(x+y)^5 = x^5 + 5x^4y + 10x^3y^2 + 10x^2y^3 + 5xy^4 + y^5$

$(x+y)^6 = x^6 + 6x^5y + 15x^4y^2 + 20x^3y^3 + 15x^2y^4 + 6xy^5 + y^6$

There are several things that we notice after looking at the expansion:

- There are $n + 1$ terms in the expansion of $(x + y)^n$

- The degree of each term is n.

- The powers on x begin with n and decrease to 0.

- The powers on y begin with 0 and increase to n.

- The coefficients are symmetric.

For instance, notice how the powers of x and y behave in the expansion of $(x + y)^5$

The powers of x decrease:

$(x + y)^5 = x^{\boxed{5}} + 5x^{\boxed{4}}y + 10x^{\boxed{3}}y^2 + 10x^{\boxed{2}}y^3 + 5x^{\boxed{1}}y^4 + y^5$

The powers of y increase:

$(x + y)^5 = x^5 + 5x^4y^{\boxed{1}} + 10x^3y^{\boxed{2}} + 10x^2y^{\boxed{3}} + 5xy^{\boxed{4}} + y^{\boxed{5}}$

With these observations, we can now proceed to expand any binomial raised to power n: $(x + y)^n$. For example, leaving a blank for the missing coefficients, the expansion for $(x + y)^7$ can be written as:

$(x + y)^7 = \Box x^7 + \Box x^6y + \Box x^5y^2 + \Box x^4y^3 + \Box x^3y^4 + \Box x^2y^5 + \Box xy^6 + \Box y^7$

To finish the expansion, we need to determine these coefficients. In order to see the pattern, look at the coefficients of the expansion at the start of the section.

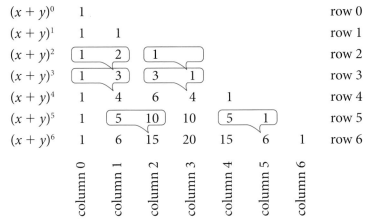

$$(x + y)^0 \quad 1 \qquad\qquad\qquad\qquad\qquad\qquad\qquad\qquad \text{row } 0$$
$$(x + y)^1 \quad 1 \quad 1 \qquad\qquad\qquad\qquad\qquad\qquad \text{row } 1$$
$$(x + y)^2 \quad \boxed{1 \quad 2} \quad \boxed{1} \qquad\qquad\qquad \text{row } 2$$
$$(x + y)^3 \quad \boxed{1 \quad 3} \quad \boxed{3 \quad 1} \qquad\qquad \text{row } 3$$
$$(x + y)^4 \quad 1 \quad 4 \quad 6 \quad 4 \quad 1 \qquad\qquad \text{row } 4$$
$$(x + y)^5 \quad 1 \quad \boxed{5 \quad 10} \quad 10 \quad \boxed{5 \quad 1} \qquad \text{row } 5$$
$$(x + y)^6 \quad 1 \quad 6 \quad 15 \quad 20 \quad 15 \quad 6 \quad 1 \qquad \text{row } 6$$

column 0, column 1, column 2, column 3, column 4, column 5, column 6

Pascal's triangle
Every entry in a row is the sum of the term directly above it and the entry to the left of it. When there is no entry, the value is considered zero.

A triangle like the one above is known as Pascal's triangle. The first and second terms in row 3 give you the second term in row 4, the third and fourth terms in row 3 give you the fourth term of row 4, the second and third terms in row 5 give the third term in row 6, and the fifth and sixth terms in row 5 give you the sixth term in row 6. So now we can state a key property of Pascal's triangle.

Take the last entry in row 5, for example; there is no entry directly above it, so its value is $0 + 1 = 1$.

From this property it is easy to find all the terms in any row of Pascal's triangle from the row above it. So, for the expansion of $(x + y)^7$, the terms are found from row 6 as follows:

$$0 \longrightarrow 1 \longrightarrow 6 \longrightarrow 15 \longrightarrow 20 \longrightarrow 15 \longrightarrow 6 \longrightarrow 1 \longrightarrow 0$$
$$\downarrow \qquad \downarrow \qquad \downarrow \qquad \downarrow \qquad \downarrow \qquad \downarrow \qquad \downarrow \qquad \downarrow$$
$$1 \qquad 7 \qquad 21 \qquad 35 \qquad 35 \qquad 21 \qquad 7 \qquad 1$$

So, $(x + y)^7 = x^7 + \boxed{7}x^6 y + \boxed{21}x^5 y^2 + \boxed{35}x^4 y^3 + \boxed{35}x^3 y^4 + \boxed{21}x^2 y^5 + \boxed{7}xy^6 + y^7$

Several sources use a slightly different arrangement for Pascal's triangle. The common usage considers the triangle as isosceles and uses the principle that every two entries add up to give the entry diagonally below them, as shown in the diagram.

$$1$$
$$1 \qquad 1$$
$$1 \qquad 2 \qquad 1$$
$$1 \qquad 3 \qquad 3 \qquad 1$$
$$1 \qquad 4 \qquad 6 \qquad 4 \qquad 1$$
$$1 \qquad 5 \qquad 10 \qquad 10 \qquad 5 \qquad 1$$

Example 3.22

Use Pascal's triangle to expand $(2k - 3)^5$

Solution

We can find the expansion by replacing x by $2k$ and y by -3 in the binomial expansion of $(x + y)^5$.

Using the fifth row of Pascal's triangle for the coefficients will give:

$$1(2k)^5 + 5(2k)^4(-3) + 10(2k)^3(-3)^2 + 10(2k)^2(-3)^3 + 5(2k)(-3)^4 + 1(-3)^5$$
$$= 32k^5 - 240k^4 + 720k^3 - 1080k^2 + 810k - 243$$

Pascal's triangle is a useful tool for finding the coefficients of the binomial expansion for relatively small values of n. It is not very efficient for large values of n. Imagine we want to evaluate $(x + y)^{20}$. Using Pascal's triangle, we would need the terms in the 19th row, and the 18th row and so on. This makes the process tedious and not practical.

Luckily, there is a formula we can use to find the coefficients of any Pascal's triangle row. This formula is the binomial formula, whose proof will appear in Chapter 5. Every entry in Pascal's triangle is denoted by $\binom{n}{r}$ or $_nC_r$ – this is also known as the binomial coefficient. In $_nC_r$, n is the row number and r is the column number. To understand the binomial coefficient, we need to understand what factorial notation means.

Factorial notation
The product of the first n positive integers is denoted by $n!$ and is called n **factorial**:
$$n! = n \times (n - 1) \times (n - 2) \times \cdots \times 3 \times 2 \times 1$$
We also define $0! = 1$.

This definition of the factorial makes many formulae involving the multiplication of consecutive positive integers shorter and easier to write. That includes the binomial coefficient.

The binomial coefficient
With n and r as non-negative integers such that $n \geq r$, the **binomial coefficient** $_nC_r$ $\left[\text{or } \binom{n}{r}\right]$ is defined by
$$_nC_r = \binom{n}{r} = \frac{n!}{r!(n - r)!}$$

$_nC_r$ is also written as ^nC_r

When simplified, $_nC_r$ can be written as $_nC_r = \dfrac{n(n - 1)(n - 2)\cdots(n - r + 1)}{r!}$.
For example $_nC_3 = \dfrac{n(n - 1)(n - 2)}{3!}$.

Example 3.23

Find the value of:

(a) $_7C_3$ (b) $_7C_4$ (c) $\binom{7}{0}$ (d) $\binom{7}{7}$

Solution

(a) $_7C_3 = \dfrac{7!}{3!(7 - 3)!} = \dfrac{7!}{3!4!} = \dfrac{1 \cdot 2 \cdot 3 \cdot 4 \cdot 5 \cdot 6 \cdot 7}{(1 \cdot 2 \cdot 3)(1 \cdot 2 \cdot 3 \cdot 4)} = \dfrac{5 \cdot 6 \cdot 7}{1 \cdot 2 \cdot 3} = 35,$

or using the other form of the expression for the binomial coefficient
$$_7C_3 = \frac{7 \cdot 6 \cdot 5}{3!} = \frac{210}{6} = 35$$

(b) $_7C_4 = \dfrac{7!}{4!(7 - 4)!} = \dfrac{7!}{4!3!} = \dfrac{1 \cdot 2 \cdot 3 \cdot 4 \cdot 5 \cdot 6 \cdot 7}{(1 \cdot 2 \cdot 3 \cdot 4)(1 \cdot 2 \cdot 3)} = \dfrac{5 \cdot 6 \cdot 7}{1 \cdot 2 \cdot 3} = 35$

(c) $\binom{7}{0} = \dfrac{7!}{0!(7 - 0)!} = \dfrac{7!}{0!7!} = \dfrac{1}{1} = 1$

(d) $\binom{7}{7} = \dfrac{7!}{7!(7 - 7)!} = \dfrac{7!}{7!0!} = \dfrac{1}{1} = 1$

Although the binomial coefficient $\binom{n}{r}$ appears as a fraction, all its results where n and r are non-negative integers are positive integers. Also notice the **symmetry** of the coefficient in Example 3.23. This is a property that you are asked to prove in the exercises:

$$_nC_r = {}_nC_{n-r}$$

Example 3.24

Calculate each binomial coefficient:

$$\binom{6}{0}, \quad \binom{6}{1}, \quad \binom{6}{2}, \quad \binom{6}{3}, \quad \binom{6}{4}, \quad \binom{6}{5}, \quad \binom{6}{6}$$

Solution

$$\binom{6}{0} = 1 \quad \binom{6}{1} = 6 \quad \binom{6}{2} = 15 \quad \binom{6}{3} = 20 \quad \binom{6}{4} = 15 \quad \binom{6}{5} = 6 \quad \binom{6}{6} = 1$$

The values in Example 3.24 are the entries in the 6th row of Pascal's triangle. We can write Pascal's triangle in the following manner:

$$\binom{0}{0}$$

$$\binom{1}{0} \quad \binom{1}{1}$$

$$\binom{2}{0} \quad \binom{2}{1} \quad \binom{2}{2}$$

$$\binom{3}{0} \quad \binom{3}{1} \quad \binom{3}{2} \quad \binom{3}{3}$$

$$\vdots \quad \vdots \quad \vdots \quad \vdots$$

$$\binom{n}{0} \quad \binom{n}{1} \quad \cdots \quad \cdots \quad \cdots \quad \cdots \quad \binom{n}{n}$$

Example 3.25

Calculate $_nC_{r-1} + {}_nC_r$ (This is called Pascal's rule.)

Solution

$$
\begin{aligned}
_nC_{r-1} + {}_nC_r &= \frac{n!}{(r-1)!(n-r+1)!} + \frac{n!}{r!(n-r)!} \\
&= \frac{n! \cdot r}{r \cdot (r-1)!(n-r+1)!} + \frac{n! \cdot (n-r+1)}{r!(n-r)! \cdot (n-r+1)} \\
&= \frac{n! \cdot r}{r!(n-r+1)!} + \frac{n! \cdot (n-r+1)}{r!(n-r+1)!} \\
&= \frac{n!(r+n-r+1)}{r!(n-r+1)!} \\
&= \frac{(n+1)!}{r!(n+1-r)!} \\
&= {}_{n+1}C_r
\end{aligned}
$$

If we read the result in Example 3.5 carefully, it says that the sum of the terms in the nth row, $(r-1)$th and rth columns, is equal to the entry in the $(n+1)$th row and rth column.

$$
\begin{array}{ccc}
 & (r-1)\text{th column} & r\text{th column} \\
n\text{th row} & {}_nC_{r-1} & + \quad {}_nC_r \\
 & & \| \\
(n+1)\text{th row} & & {}_{n+1}C_r
\end{array}
$$

That is, the two entries on the left are adjacent entries in the nth row of Pascal's triangle and the entry on the right is the entry in the $(n+1)$th row directly below the rightmost entry. This is precisely the principle behind Pascal's triangle!

Using the binomial theorem

We are now prepared to state the binomial theorem:

$$(x+y)^n = \binom{n}{0}x^n + \binom{n}{1}x^{n-1}y + \binom{n}{2}x^{n-2}y^2 + \binom{n}{3}x^{n-3}y^3 + \cdots + \binom{n}{n-1}xy^{n-1} + \binom{n}{n}y^n$$

or

$$(x+y)^n = {}_nC_0 x^n + {}_nC_1 x^{n-1}y + {}_nC_2 x^{n-2}y^2 + {}_nC_3 x^{n-3}y^3 + \cdots + {}_nC_n y^n$$

In a compact form, we can use sigma notation to express the theorem as follows:

$$(x+y)^n = \sum_{i=0}^{n} \binom{n}{i}x^{n-i}y^i = \sum_{i=0}^{n} {}_nC_i x^{n-i}y^i$$

Example 3.26

Use the binomial theorem to expand $(x+y)^7$

Solution

$$(x+y)^7 = \binom{7}{0}x^7 + \binom{7}{1}x^{7-1}y + \binom{7}{2}x^{7-2}y^2 + \binom{7}{3}x^{7-3}y^3 + \binom{7}{4}x^{7-4}y^4$$

$$+ \binom{7}{5}x^{7-5}y^5 + \binom{7}{6}x^{7-6}y^6 + \binom{7}{7}y^7$$

$$= x^7 + 7x^6y + 21x^5y^2 + 35x^4y^3 + 35x^3y^4 + 21x^2y^5 + 7xy^6 + y^7$$

Example 3.27

Find the expansion for $(2k-3)^5$

Solution

$$(2k-3)^5 = \binom{5}{0}(2k)^5 + \binom{5}{1}(2k)^4(-3) + \binom{5}{2}(2k)^3(-3)^2 + \binom{5}{3}(2k)^2(-3)^3$$

$$+ \binom{5}{4}(2k)(-3)^4 + \binom{5}{5}(-3)^5$$

$$= 32k^5 - 240k^4 + 720k^3 - 1080k^2 + 810k - 243$$

The proof of the binomial theorem is optional and will require mathematical induction. We will develop the proof in Chapter 5.

Why is the binomial theorem related to the number of combinations of n elements taken r at a time?

Consider evaluating $(x+y)^n$. In doing so, we must multiply $(x+y)$ by itself n times. As we know, one term has to be x^n. How do we get this term? x^n is the result of multiplying x in each of the n factors $(x+y)$ and that can happen in only one way. However, consider the term containing x^r. Having a power of r over the x means that the x in each of r factors must be multiplied, and the rest will be the $(n-r)$ y-terms. This can happen in ${}_nC_r$ ways. Hence the coefficient of the term $x^r y^{n-r}$ is ${}_nC_r$

Example 3.28

Find the term containing a^3 in the expansion $(2a - 3b)^9$

Solution

To find the term, we do not need to expand the whole expression.

Since $(x + y)^n = \sum_{i=0}^{n} {}_nC_i\, x^{n-i}y^i$, the term containing a^3 is the term where $n - i = 3$, i.e. when $i = 6$. So the required term is

$${}_9C_6(2a)^{9-6}(-3b)^6 = 84 \cdot 8a^3 \cdot 729b^6 = 489\,888a^3b^6$$

Example 3.29

Find the term independent of x in:

(a) $\left(2x^2 - \dfrac{3}{x}\right)^9$ (b) $\left(4x^3 - \dfrac{2}{x^2}\right)^5$

Solution

(a) 'Independent of x' means the term with no x variable, i.e. the constant term in the expansion of $\left(2x^2 - \dfrac{3}{x}\right)^9 = \sum_{k=0}^{9} {}_9C_k(2x^2)^k\left(-\dfrac{3}{x}\right)^{9-k}$

The constant term contains x^0. Thus it is the term where $(x^2)^k\left(\dfrac{1}{x}\right)^{9-k} = x^0$,

i.e. $2k = 9 - k \Rightarrow k = 3$, so the term is ${}_9C_3(2x^2)^3\left(-\dfrac{3}{x}\right)^6 = 84 \cdot 8x^6 \cdot \dfrac{729}{x^6}$

$$= 489\,888$$

(b) Similarly, $(x^3)^k\left(\dfrac{1}{x^2}\right)^{5-k} = x^0$, i.e., $3k = 10 - 2k \Rightarrow k = 2$, so the term is

$${}_5C_2(4x^3)^2\left(-\dfrac{2}{x^2}\right)^3 = -1280$$

```
(9C3)×(2³)×(-3)⁶
              489888
(5C2)×4²×(-2)³
              -1280
```

Example 3.30

Find the coefficient of b^6 in the expansion of $\left(2b^2 - \dfrac{1}{b}\right)^{12}$

Solution

The general term is $\dbinom{12}{i}(2b^2)^{12-i}\left(-\dfrac{1}{b}\right)^i = \dbinom{12}{i}(2)^{12-i}(b^2)^{12-i}\left(-\dfrac{1}{b}\right)^i$

$$= \dbinom{12}{i}(2)^{12-i}b^{24-2i}b^{-i}(-1)^i$$

$$= \dbinom{12}{i}(2)^{12-i}b^{24-3i}(-1)^i$$

$24 - 3i = 6 \Rightarrow i = 6$. So the coefficient in question is $\dbinom{12}{6}(2)^6(-1)^6 = 59\,136$

1. Use Pascal's triangle to expand each binomial.

(a) $(x + 2y)^5$ (b) $(a - b)^4$ (c) $(x - 3)^6$

(d) $(2 - x^3)^4$ (e) $(x - 3b)^7$ (f) $\left(2n + \dfrac{1}{n^2}\right)^6$

(g) $\left(\dfrac{3}{x} - 2\sqrt{x}\right)^4$

2. Evaluate each expression.

(a) $\dbinom{8}{3}$ (b) $\dbinom{18}{5} - \dbinom{18}{13}$ (c) $\dbinom{7}{4}\dbinom{7}{3}$

(d) $\dbinom{5}{0} + \dbinom{5}{1} + \dbinom{5}{2} + \dbinom{5}{3} + \dbinom{5}{4} + \dbinom{5}{5}$

(e) $\dbinom{6}{0} - \dbinom{6}{1} + \dbinom{6}{2} - \dbinom{6}{3} + \dbinom{6}{4} - \dbinom{6}{5} + \dbinom{6}{6}$

3. Use the binomial theorem to expand each expression.

(a) $(x - 2y)^7$ (b) $(2a - b)^6$ (c) $(x - 4)^5$

(d) $(2 + x^3)^6$ (e) $(3x - b)^7$ (f) $\left(2n - \dfrac{1}{n^2}\right)^6$

(g) $\left(\dfrac{2}{x} - 3\sqrt{x}\right)^4$ (h) $(1 + \sqrt{5})^4 + (1 + \sqrt{5})^4$

(i) $(\sqrt{3} + 1)^8 - (\sqrt{3} + 1)^8$

4. Consider the expression $\left(x - \dfrac{2}{x}\right)^{45}$

(a) Find the first three terms of this expansion.

(b) Find the constant term if it exists or justify why it doesn't exist.

(c) Find the last three terms of the expansion.

(d) Find the term containing x^3 if it exists or justify why it doesn't exist.

5. Prove that $_nC_k = {}_nC_{n-k}$ for all $n, k \in \mathbb{N}$ and $n \geqslant k$

6. Prove that for any positive integer n,

$$\dbinom{n}{1} + \dbinom{n}{2} + \dots + \dbinom{n}{n-1} + \dbinom{n}{n} = 2^n - 1.$$

 For question 6:
$2^n = (1 + 1)^n$

7. Consider all $n, k \in \mathbb{N}$ and $n \geqslant k$

(a) Verify that $k! = k(k - 1)!$

(b) Verify that $(n - k + 1)! = (n - k + 1)(n - k)!$

(c) Justify the steps given in the proof of $\dbinom{n}{r-1} + \dbinom{n}{r} = \dbinom{n+1}{r}$ in Example 3.25.

8. Find the value of the expression:

$$\dbinom{6}{0}\left(\dfrac{1}{3}\right)^6 + \dbinom{6}{1}\left(\dfrac{1}{3}\right)^5\left(\dfrac{2}{3}\right) + \dbinom{6}{2}\left(\dfrac{1}{3}\right)^4\left(\dfrac{2}{3}\right)^2 + \dots + \dbinom{6}{6}\left(\dfrac{2}{3}\right)^6$$

9. Find the value of the expression:

$$\binom{8}{0}\left(\frac{2}{5}\right)^8 + \binom{8}{1}\left(\frac{2}{5}\right)^7\left(\frac{3}{5}\right) + \binom{8}{2}\left(\frac{2}{5}\right)^6\left(\frac{3}{5}\right)^2 + \cdots + \binom{8}{8}\left(\frac{3}{5}\right)^8$$

10. Find the value of the expression:

$$\binom{n}{0}\left(\frac{1}{7}\right)^n + \binom{n}{1}\left(\frac{1}{7}\right)^{n-1}\left(\frac{6}{7}\right) + \binom{n}{2}\left(\frac{1}{7}\right)^{n-2}\left(\frac{6}{7}\right)^2 + \cdots + \binom{n}{n}\left(\frac{6}{7}\right)^n$$

11. Find the term independent of x in the expansion of $\left(x^2 - \dfrac{1}{x}\right)^6$

12. Find the term independent of x in the expansion of $\left(3x - \dfrac{2}{x}\right)^8$

13. Find the term independent of x in the expansion of $\left(2x - \dfrac{3}{x^3}\right)^8$

14. Find the first three terms in the expansion of $(1 + x)^{10}$ and use them to find an approximation to:

    (a) 1.01^{10} (b) 0.99^{10}

15. Show that $\begin{pmatrix} n \\ r-1 \end{pmatrix} + 2\begin{pmatrix} n \\ r \end{pmatrix} + \begin{pmatrix} n \\ r+1 \end{pmatrix} = \begin{pmatrix} n+2 \\ r+1 \end{pmatrix}$ and interpret your result on the entries in Pascal's triangle.

16. Express each repeating decimal as a fraction:

    (a) $0.\overline{7}$ (b) $0.3\overline{45}$ (c) $3.21\overline{29}$

17. Find the coefficient of x^6 in the expansion of $(2x - 3)^9$

18. Find the coefficient of x^3b^4 in the expansion of $(ax + b)^7$

19. Find the constant term in the expansion of $\left(\dfrac{2}{z^2} - z\right)^{15}$

20. Expand $(3n - 2m)^5$

21. Find the coefficient of r^{10} in the expansion of $(4 + 3r^2)^9$

22. In the expansion of $(2 - kx)^5$, the coefficient of x^3 is -1080. Find the constant k.

3.6 Counting principles

Simple counting problems

This section will introduce you to some of the basic principles of counting. In Chapter 5 you will apply some of this in justifying the binomial theorem and in Chapter 11 you will use these principles to tackle many probability problems. We will start with two examples.

Example 3.31

Each integer from 1 to 9 is written on a paper chip. The paper chips are placed in a box. One chip is chosen, the number is recorded, and the chip is put back in the box. This is then repeated for a second chip. The numbers on the chips are added. In how many ways can you get a sum of 8?

Solution

To solve this problem, count the different number of ways that a total of 8 can be obtained:

1st chip	1	2	3	4	5	6	7
2nd chip	7	6	5	4	3	2	1

From this list, it is clear that there are seven different ways of getting a sum of 8.

Example 3.32

Suppose now that the first chip is chosen and the number is recorded, but the chip is not put back in the box, then the second chip is drawn. In how many ways can you get a sum of 8?

Solution

To solve this problem, also count the different number of ways that a total of 8 can be obtained:

1st chip	1	2	3	5	6	7
2nd chip	7	6	5	3	2	1

From this list, it is clear that there six different ways of getting a sum of 8.

The difference between this and Example 3.31 is described by saying that the first random selection is done **with replacement**, while the second is done **without replacement**, which ruled out the drawing of two 4s.

Fundamental principle of counting

Examples 3.31 and 3.32 show you simple counting principles in which you can list every possible way that an event can happen. In many other cases, listing the ways an event can happen may not be feasible. In such cases we need to rely on counting principles. The most important of which is the fundamental principle of counting, also known as the multiplication principle. Consider the following examples.

Example 3.33

For making sandwiches, there are three kinds of bread and four kinds of cheese. Sandwiches can be made with or without pickles. How many different kinds of sandwich can be made?

Solution

There are three kinds of bread, each of which could be combined with one of four kinds of cheese. That would make $3 \times 4 = 12$ sandwiches. But, also, sandwiches can be made with or without pickles, thus 12 with pickles and 12 without. That is, there are $3 \times 4 \times 2 = 24$ possible sandwiches.

Example 3.34

How many 3-digit even numbers are there?

Solution

The first digit cannot be zero, since the number has to be a 3-digit number, so there are 9 possibilities for the hundreds digit. There is no condition on what the tens digit should be, so we have 10 possibilities, and to be an even number, the number must end with 0, 2, 4, 6, or 8. Therefore , there are $9 \times 10 \times 5 = 450$ 3-digit even numbers.

Examples 3.33 and 3.34 are examples of the fundamental principle of counting.

Fundamental Principle of Counting

The fundamental principle of counting states that if there are m ways an event can occur followed by n ways a second event can occur, then there are a total of $m \times n$ ways that the two can occur.

This principle can be extended to more than two events or processes:

If there are k events that can happen in $n_1, n_2, \ldots, n_k$ ways, then the whole sequence can happen in $n_1 \times n_2 \times \ldots \times n_k$ ways.

Example 3.35

A large school issues special coded ID cards that consist of two letters followed by three numerals. For example, AB 737 is such a code. How many different ID cards can be issued if the letters or numbers can be used more than once?

Solution

Since the letters can be used more than once, each letter position can be filled in 26 different ways. That is, the letters can be filled in $26 \times 26 = 676$ ways. Each number position can be filled in 10 different ways; hence the numerals can be filled in $10 \times 10 \times 10 = 1000$ different ways. So, the code can be formed in $676 \times 1000 = 676\,000$ different ways.

Permutations

One major application of the fundamental principle of counting is in determining the number of ways that n objects can be arranged. For example, we have five books we want to put on a shelf: Mathematics (M), Physics (P), English (E), Biology (B), and History (H). In how many ways can we do this?

Consider the positions that we want to place the books in as shown:

$$\underset{1\ \ 2\ \ 3\ \ 4\ \ 5}{\sqcup\ \sqcup\ \sqcup\ \sqcup\ \sqcup}$$

If we decide to put the Mathematics book in position 1, then there are 4 different ways of putting a book in position 2:

$$\underset{1\ \ 2\ \ 3\ \ 4\ \ 5}{\underline{M\ P\ \sqcup\ \sqcup\ \sqcup}} \quad \underset{1\ \ 2\ \ 3\ \ 4\ \ 5}{\underline{M\ E\ \sqcup\ \sqcup\ \sqcup}} \quad \underset{1\ \ 2\ \ 3\ \ 4\ \ 5}{\underline{M\ B\ \sqcup\ \sqcup\ \sqcup}} \quad \underset{1\ \ 2\ \ 3\ \ 4\ \ 5}{\underline{M\ H\ \sqcup\ \sqcup\ \sqcup}}$$

Since we can put any of the 5 books in the first position there will be $5 \times 4 = 20$ ways of shelving the first two books. Once we place the books in positions 1 and 2, the third book can be chosen from any one of three books left:

$$\underset{1\ \ 2\ \ 3\ \ 4\ \ 5}{\underline{M\ P\ E\ \sqcup\ \sqcup}} \quad \underset{1\ \ 2\ \ 3\ \ 4\ \ 5}{\underline{M\ P\ B\ \sqcup\ \sqcup}} \quad \underset{1\ \ 2\ \ 3\ \ 4\ \ 5}{\underline{M\ P\ H\ \sqcup\ \sqcup}}$$

Once we have selected three books, there are two books for the fourth position and only one way of placing the fifth book. So, the number of ways of arranging all 5 books is

$$5 \times 4 \times 3 \times 2 \times 1 = 120 = 5!$$

Number of permutations of n objects

The books example can be applied to n objects rather than only 5. The number of ways of filling in the first position can be done in n ways.

$$\underset{1\quad\ \ 2\quad\ \ 3\quad\ \ 4\qquad\ \ n}{\underline{\underset{\downarrow}{n}\ \ \underset{\downarrow}{n-1}\ \underset{\downarrow}{n-2}\ \underset{\downarrow}{n-3}\ \cdots\ \underset{\downarrow}{1}}}$$

Once the first position is filled, the second position can be filled by any of the $n - 1$ objects left, and hence, using the fundamental principle of counting, there will be $n \cdot (n - 1)$ different ways for filling the first two positions. Repeating the same procedure until the nth position is filled is therefore:

$$n \cdot (n - 1) \cdot (n - 2) \cdots 2 \cdot 1 = n!$$

Frequently, we are engaged in arranging a subset of the whole collection of objects rather than the entire collection. For example, suppose we want to shelve three of the books rather than all five of them. The discussion will be similar to the previous situation. However, we have to limit our search to the first three positions only – that is, the number of ways we can shelve three out of the five books is:

$$5 \times 4 \times 3 = 60$$

To change this product into factorial notation, we do the following:

$$5 \times 4 \times 3 = 5 \times 4 \times 3 \times \frac{2!}{2!} = \frac{5 \times 4 \times 3 \times 2 \times 1}{2!} = \frac{5!}{2!} = \frac{5!}{(5 - 3)!}$$

145

The number of permutations of n objects taken r at a time is:
$$^nP_r = {}_nP_r = P_r^n = P(n, r)$$
$$= \frac{n!}{(n-r)!}, n \geqslant r$$

This leads us to the result on the left.

To verify the formula, we can proceed in the same manner as with the permutation of n objects.

$$\underbrace{n}_{1} \underbrace{n-1}_{2} \underbrace{n-2}_{3} \underbrace{n-3}_{4} \cdots \underbrace{n-(r-1)}_{r}$$

When we arrive at the rth position, we would have used $r - 1$ objects already, and hence we are left with $n - (r - 1) = n - r + 1$ objects to fill this position. So, the number of ways of arranging n objects taken r at a time is:

$$_nP_r = n \cdot (n - 1) \cdot (n - 2) \cdots (n - r + 1)$$

Here again, to make the expression more manageable, we can write it in factorial notation:

$$_nP_r = n \cdot (n - 1) \cdot (n - 2) \cdots (n - r + 1)$$

$$= n \cdot (n - 1) \cdot (n - 2) \cdots (n - r + 1) \frac{(n - r)!}{(n - r)!}$$

$$= \frac{n \cdot (n - 1) \cdot (n - 2) \cdots (n - r + 1) \cdot (n - r)!}{(n - r)!} = \frac{n!}{(n - r)!}$$

Example 3.36

Fifteen drivers are taking part in a car race. In how many different ways can the top six positions be filled?

Solution

Since the drivers are all different, this is a permutation of 15 'objects' taken 6 at a time.

$$_{15}P_6 = \frac{15!}{(15 - 6)!} = 3\,603\,600$$

We can also calculate this easily using a GDC.

```
15P6
          3603600
15!÷9!
          3603600
```

Combinations

A **combination** is a selection of some or all of a number of different objects. It is an unordered collection of unique sizes. In a permutation, the order of occurrence of the objects or the arrangement is important, but in a combination, the order of occurrence of the objects is not important. In that sense, a combination of r objects out of n objects is a subset of the set of n objects.

For example, there are 24 permutations of three letters from the letters ABCD. However, there are only 4 combinations. Here is why:

ABC	ABD	ACD	BCD
ACB	ADB	ADC	BDC
BAC	BAD	CAD	CBD
BCA	BDA	CDA	CDB
CAB	DAB	DAC	DBC
CBA	DBA	DCA	DCB

For one combination, ABC for example, there are $3! = 6$ permutations. This is true for all combinations. So, the number of permutations is 6 times the number of combinations. That is:

$$_4P_3 = 3!\,_4C_3$$

where $_4C_3$ is the number of combinations of the 4 letters taken 3 at a time.

According to the previous result, we can write:

$$_4C_3 = \frac{_4P_3}{3!} = \frac{\dfrac{4!}{(4-3)!}}{3!} = \frac{4!}{3!(4-3)!}$$

The ISO notation for $_nC_r$ is $\binom{n}{r}$. In this book, we will use both notations interchangeably.

The last result can also be generalised to n elements combined r at a time.

Every subset of r objects (combination), gives rise to $r!$ permutations. So, if you have $_nC_r$ combinations, these will result in $r!\,_nC_r$ permutations. Therefore:

$$_nP_r = r!\,_nC_r \Leftrightarrow\ _nC_r = \frac{_nP_r}{r!} = \frac{\dfrac{n!}{(n-r)!}}{r!} = \frac{n!}{(n-r)!r!}$$

Example 3.37

A lottery has 45 numbers. If you buy a ticket, then you choose 6 of these numbers. In how many ways can you select 6 numbers from 45?

Solution

Since 6 numbers will be chosen and order is not an issue here, this is a combination case. The number of possible choices is:

$$\binom{45}{6} = 8\,145\,060$$

This can also be calculated using a GDC:

```
45 nCr 6
         8145060
```

Example 3.38

In some card games, a standard deck of 52 playing cards is used, and a 'hand' is made up of 5 cards.

(a) How many possible hands are there?

(b) How many possible hands are there with 3 diamonds and 2 hearts?

Solution

(a) A player can re-order the cards after receiving them, so the order is not important, and thus this is a combination of 52 cards taken 5 at a time:

$$\binom{52}{5} = 2\,598\,960$$

(b) Since there are 13 diamonds and we want 3 of them, there are $\binom{13}{3} = 286$ ways to get 3 diamonds. Since there are 13 hearts and we want 2 of them, there are $\binom{13}{2} = 78$ ways to get 2 hearts. Since we want them both to occur at the same time, we use the fundamental principle of counting and multiply 286 by 78 to get 22 308 possible hands.

```
52 nCr 5
            2598960
13 nCr 3
                286
13 nCr 2
                 78
```

Figure 3.5 GDC screen for solution to Example 3.38 (b)

Example 3.39

A code is made up of 6 different digits. How many possible codes are there?

Solution

Since there are 10 digits and we are choosing 6 of them, and since the order we use these digits makes a difference in the code, this is permutation case. The number of possible codes is:

$$_{10}P_6 = 151\,200$$

Exercise 3.6

1. Evaluate each expression.

   (a) $_5P_5$ (b) $5!$ (c) $_{20}P_1$ (d) $_8P_3$

2. Evaluate each expression.

   (a) $_5C_5$ (b) $_5C_0$ (c) $_{10}C_3$ (d) $_{10}C_7$

3. Evaluate each expression.

   (a) $\binom{7}{3} + \binom{7}{4}$ (b) $\binom{8}{4}$ (c) $\binom{10}{6} + \binom{10}{7}$ (d) $\binom{11}{7}$

4. Evaluate each expression.

(a) $\binom{8}{5} - \binom{8}{3}$ (b) $11 \cdot 10!$ (c) $\binom{10}{3} - \binom{10}{7}$ (d) $\binom{10}{1}$

5. State whether each equation below is true.

(a) $\dfrac{10!}{5!} = 2!$ (b) $(5!)^2 = 25!$ (c) $\binom{101}{8} = \binom{101}{93}$

6. You are buying a computer and have the following choices: three types of hard drive, two types of main processor, and four types of graphics card. How many different systems can you choose from?

7. You are going to a restaurant with a set menu. There are three starters, four main meals, two drinks, and three desserts. How many different choices are available for you to choose your meal from?

8. A school needs a teacher for each of three subjects: PE, Mathematics, and English. They have eight applicants for the PE position, three applicants for Mathematics and 13 applicants for English. How many different combinations of choices does the school have?

9. You are given a multiple-choice test where each question has four possible answers. The test is made up of 12 questions and you did not prepare for the test and are thus guessing at random. In how many ways could you answer all the questions on the test?

10. The test in question 9 is divided into two parts, the first six are true or false questions and the last six are multiple choice questions. In how many different ways can you answer all the questions on that test?

11. Passwords on a network are made up of two parts. One part consists of three letters, not necessarily different, the second part consists of five digits, also not necessarily different. How many passwords are possible on this network?

12. In question 11, how many 5-digit numbers can be made if the units digit cannot be 0?

13. Four couples are to be seated in a theatre row. Find how many different ways they can be seated if:

(a) no restrictions are made   (b) each couple sits together.

14. Five girls and three boys walk single file through a doorway. Find how many different orders they can walk through the doorway if:

(a) there are no constraints   (b) the girls must go first.

15. Write all the permutations of the letters in JANE.

16. Write all the permutations of the letters in MAGIE taken three at a time.

17. A computer code is made up of three letters followed by four digits.

    (a) How many possible codes are there?

    (b) 97 of the three-letter combinations cannot be used because they are offensive. How many codes are still possible?

18. A local club has 17 members: 10 females and 7 males. They have to elect three officers: president, deputy, and treasurer. Find how many ways to elect the officers if:

    (a) there are no restrictions

    (b) the president is to be a male

    (c) the deputy must be a male, the president can be any gender, but the treasurer must be a female

    (d) the president and deputy are of the same gender.

19. The research and development department for a computer manufacturer has 26 employees: eight mathematicians, 12 computer scientists, and six engineers. They need to select three employees to be leaders of the group. Find how many ways they can do this if:

    (a) the three officers are of the same specialisation

    (b) at least one of them must be an engineer

    (c) two of them must be mathematicians.

20. A combination lock has three numbers, each in the range 1 to 50.

    (a) How many different combinations are possible?

    (b) How many combinations do not have duplicates?

    (c) How many have the first and second numbers matching?

    (d) How many have exactly two of the numbers matching?

21. In how many ways can five couples be seated around a circle so that each couple sits together?

22. (a) How many subsets of $\{1, 2, 3, \ldots, 9\}$ have two elements?

    (b) How many subsets of $\{1, 2, 3, \ldots, 9\}$ have an odd number of elements?

23. Nine men and 12 women make up a community archery club. Four members are needed for an upcoming competition.

    (a) How many 4-member teams can they form?

    (b) How many of these 4-member teams have the same number of women and men?

(c) How many of these 4-member teams have more women than men?

Tim is the best male archer, and Gwen is the best female archer in the club. Only one of the two can be chosen to be on the team.

(d) Repeat parts (a)–(c) with these changes.

24. A shipment of 100 hard disks contains four defective disks. A sample of six disks is chosen for inspection.

   (a) How many different possible samples are there?

   (b) How many samples could contain all four defective disks?
   What percentage of the total number of possible samples is that?

   (c) How many samples could contain at least one defective disk?
   What percentage of the total number of possible samples is that?

25. There are three political parties represented in a parliament: ten conservatives, eight liberals, and four independents. A committee of six members needs to be set up.

   (a) How many different committees are possible?

   (b) How many committees with equal representation are possible?

26. How many ways are there for nine boys and six girls to stand in a line so that no two girls stand next to each other?

Chapter 3 practice questions

1. In an arithmetic sequence, the first term is 4, the 4th term is 19 and the nth term is 99. Find the common difference and the number of terms, n.

2. How much money should you invest now if you wish to have $3000 in your account after 6 years, if interest is compounded quarterly at an annual rate of 6%?

3. Two students, Nick and Maxine, decide to start preparing for their IB exams 15 weeks ahead of the exams. Nick starts by studying for 12 hours in the first week and plans to increase the amount by 2 hours per week. Maxine starts with 12 hours in the first week and decides to increase her time by 10% every week.

   (a) How many hours will each student study in week 5?

   (b) How many hours in total will each student study for the 15 weeks?

   (c) In which week will Maxine exceed 40 hours per week?

   (d) In which week will Maxine catch up with Nick in the number of hours spent studying per week?

4. Two diet schemes are available for people to lose weight. Plan A promises the patient an initial weight loss of 1000 grams the first month with a steady loss of an additional 80 grams every month after the first, for a maximum duration of 12 months.

   Plan B starts with a weight loss of 1000 grams the first month and an increase in weight loss by 6% more every subsequent month.

   (a) Write down the number of grams lost under Plan B in the second and third months.

   (b) Find the weight lost in the 12th month for each plan.

   (c) Find the total weight loss during a 12 month period under

       (i) Plan A   (ii) Plan B.

5. You start a savings plan to buy a car, where you invest €500 at the beginning of each year for 10 years. Your bank offers a fixed rate of 6% per year, compounded annually.

   Calculate, giving your answers to the nearest euro(€):

   (a) how much the first €500 is worth at the end of 10 years

   (b) the total value of your investment at the end of the 10 years.

6. The first three terms of an arithmetic sequence are 6, 9.5, and 13.

   (a) What is the 40th term of the sequence?

   (b) What is the sum of the first 103 terms of the sequence?

7. $\{a_n\}$ is defined as follows:
   $$a_n = \sqrt[3]{8 - a_{n-1}^3}$$

   (a) Given that $a_1 = 1$, evaluate a_2, a_3, and a_4. Describe $\{a_n\}$.

   (b) Given that $a_1 = 2$, evaluate a_2, a_3, and a_4. Describe $\{a_n\}$.

8. A marathon runner plans her training program for a 20 km race. On the first day she plans to run 2 km, then she wants to increase her distance by 500 m on each subsequent training day.

   (a) On which day of her training does she first run a distance of 20 km?

   (b) By the time she manages to run the 20 km distance, what is the total distance she would have run for the whole training program?

9. In a certain country, smartphones were first introduced in the year 2010. During the first year, 1600 people bought a smartphone. In 2011, the number of new participants was 2400, and in 2012 the new participants numbered 3600.

   (a) You notice that the trend follows a geometric sequence. Find the common ratio.

(b) Assuming that the trend continues:

　　(i) how many participants will join in 2022?

　　(ii) in what year would the number of new participants first exceed 50 000?

Between 2010 and 2012, the total number of participants reached 7600.

(c) What is the total number of participants between 2010 and 2022?

During this period, the total adult population remains approximately 800 000.

(d) Use this information to suggest a reason why this trend in growth would not continue.

10. In an arithmetic sequence, the first term is 25, the fourth term is 13, and the nth term is $-11\,995$. Find the common difference d and the number of terms n.

11. The midpoints M, N, P, and Q of the sides of a square of side 1 cm are joined to form a new square.

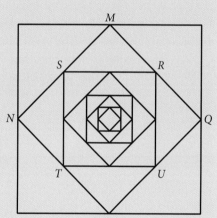

(a) Show that the side length of the square $MNPQ$ is $\dfrac{\sqrt{2}}{2}$

(b) Find the area of square $MNPQ$.

A new third square $RSTU$ is constructed in the same manner.

(c) **(i)** Find the area of $RSTU$.

　　(ii) Show that the areas of the squares are in a geometric sequence and find the common ratio.

The procedure continues indefinitely.

(d) **(i)** Find the area of the 10th square.

　　(ii) Find the sum of the areas of all the squares.

12. Aristede is a dedicated swimmer. He goes swimming once every week. He starts the first week of the year by swimming 200 metres. Each week after that he swims 20 metres more than the previous week. He does this for the whole year (52 weeks).

(a) How far does he swim in the final week?

(b) How far does he swim altogether?

13. The diagram shows three iterations of constructing squares in the following manner: A square of side 3 units is drawn, then it is divided into nine smaller squares and the middle square is shaded (below, left). Each of the unshaded squares is in turn divided into nine squares and the process is repeated. The area of the first shaded square is 1 unit.

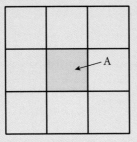 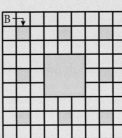 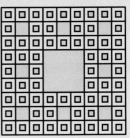

(a) Find the area of each of the squares A and B.

(b) Find the area of any small square in the third diagram.

(c) Find the area of the shaded regions in the second and third iterations.

(d) If the process was continued indefinitely, find the area left unshaded.

14. The table shows four series of numbers. One is an arithmetic series, one is a converging geometric series, one is a diverging geometric series, and the fourth is neither geometric nor arithmetic.

	Series	Type of series
(i)	$2 + 22 + 222 + 2222 + \ldots$	
(ii)	$2 + \dfrac{4}{3} + \dfrac{8}{9} + \dfrac{16}{27} + \ldots$	
(iii)	$0.8 + 0.78 + 0.76 + 0.74 + \ldots$	
(iv)	$2 + \dfrac{8}{3} + \dfrac{32}{9} + \dfrac{128}{27} + \ldots$	

(a) Copy and complete the table by stating the type of each series.

(b) Find the sum of the infinite geometric series.

15. Two IT companies offer apparently similar salary schemes for their new appointees. Kell offers a starting salary of €18,000 per year and an annual increase of €400 each year after the first. YBO offers a starting salary of €17,000 per year and an annual increase of 7% each year after the first.

(a) (i) Write down the salary paid in the 2nd and 3rd years for each company.

(ii) Calculate the total amount than an employee working for 10 years will accumulate over 10 years in each company.

(iii) Calculate the salary paid in the tenth year in each company.

(b) Tim works at Kell and Merijayne works at YBO.

    (i) When would Merijayne start earning more than Tim?

    (ii) What is the minimum number of years that Merijayne requires so that her total earnings exceed Tim's total earnings?

16. A theatre has 24 rows of seats. There are 16 seats in the first row, and each successive row increases by 2 seats.

    (a) Calculate the number of seats in the 24th row.

    (b) Calculate the number of seats in the whole theatre.

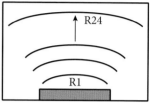

Figure 3.6 Diagram for question 16

17. The amount of €7000 is invested at 5.25% annual compound interest.

    (a) Write down an expression for the value of this investment after t full years.

    (b) Calculate the minimum number of years required for this amount to become €10,000.

    (c) For the same number of years as in part **(b)**, would an investment of the same amount be better if it were invested at a 5% rate compounded quarterly?

18. With S_n denoting the sum of the first n terms of an arithmetic sequence, we are given that $S_1 = 9$ and $S_2 = 20$.

    (a) Find the second term.

    (b) Calculate the common difference of the sequence.

    (c) Find the fourth term.

19. Consider an arithmetic sequence whose second term is 7. The sum of the first four terms of this sequence is 12. Find the first term and the common difference of the sequence.

20. Given that
$$(1 + x)^5(1 + ax)^6 \equiv 1 + bx + 10x^2 + \ldots + a^6x^{11},$$
find the values of $a, b \in \mathbb{Z}$.

21. In an arithmetic sequence of positive terms, a_n represents the nth term.

Given that $\dfrac{a_5}{a_{12}} = \dfrac{6}{13}$ and $a_1 \times a_3 = 32$, find $\displaystyle\sum_{i=1}^{100} a_i$

22. In an arithmetic sequence, $a_1 = 5$ and $a_2 = 13$.

   (a) Write down, in terms of n, an expression for the nth term, a_n.

   (b) Find n such that $a_n < 400$.

23. Find the coefficient of x^7 in the expansion of $(2 + 3x)^{10}$, giving your answer as a whole number.

24. The sum of the first n terms of an arithmetic sequence is $S_n = 3n^2 - 2n$. Find the nth term, u_n.

25. Mr Blue, Mr Black, Mr Green, Mrs White, Mrs Yellow and Mrs Red sit around a circular table for a meeting. Mr Black and Mrs White must not sit together.

   Calculate the number of different ways these six people can sit at the table without Mr Black and Mrs White sitting together.

26. Consider the arithmetic sequence $85, 78, 71, \ldots$
   Find the sum of its positive terms.

27. When we expand $\left(x + \dfrac{1}{kx^2}\right)^7$, the coefficient of x is $\dfrac{7}{3}$.
   Find all possible values of k.

28. The sum to infinity of a geometric sequence is $\dfrac{27}{2}$, and the sum of its first three terms is 13. Find the first term.

29. In how many ways can six different books be divided between two students so that each student receives at least one book?

30. Find the sum to infinity of the geometric series $-12 + 8 - \dfrac{16}{3} \ldots$

31. A geometric sequence is defined by $u_n = 3(4)^{n+1}$, $n \in \mathbb{Z}^+$, where u_n is the nth term.

   (a) Find the common ratio r.

   (b) Hence, find S_n, the sum of the first n terms of this sequence.

32. Consider the infinite geometric series:

$$1 + \left(\frac{3x}{5}\right) + \left(\frac{3x}{5}\right)^2 + \left(\frac{3x}{5}\right)^3 + \ldots$$

   (a) For what values of x does the series converge?

   (b) Find the sum of the series if $x = 1.5$

33. How many four-digit numbers are there that contain at least one digit 3?

34. Consider the arithmetic series $S_n = 2 + 5 + 8 + \ldots$

    (a) Find an expression for the partial sum S_n, in terms of n.

    (b) For what value of n is $S_n = 1365$?

35. Find the coefficient of x^3 when the binomial $\left(1 - \dfrac{1}{2}x\right)^8$ is expanded.

36. Find $\displaystyle\sum_{r=1}^{50} \ln(2^r)$, giving the answer in the form $a \ln 2$, where $a \in \mathbb{Q}$

37. Consider the sequence $\{a_n\}$ defined recursively by:
$$a_{n+1} = 3a_n - 2a_{n-1}, n \in \mathbb{Z}^+, \text{ with } a_0 = 1, a_1 = 2$$

    (a) Find $a_2, a_3,$ and a_4.

    (b) (i) Find the explicit form for a_n in terms of n.

    (ii) Verify that your answer to part (i) satisfies the given recursive definition.

38. The sum to infinity of a geometric sequence with all positive terms is 27, and the sum of the first two terms is 15. Find the value of:

    (a) the common ratio

    (b) the first term

39. The first four terms of an arithmetic sequence are $2, a - b, 2a + b + 7$, and $a - 3b$, where a and b are constants. Find a and b.

40. A school's Mathematics Club has eight members. A team of four will be chosen for an upcoming event. However, the two oldest members cannot both be chosen. Find the number of ways the team may be chosen.

41. Three consecutive terms of an arithmetic sequence are: $a, 1,$ and b. The terms $1, a,$ and b are consecutive terms of a geometric sequence. If $a \neq b$, find the value of a and of b.

42. The diagram opposite shows a sector AOB of a circle of radius 1 and centre O, where $A\widehat{O}B = \theta$.

    The lines $(AB_1), (A_1B_2),$ and (A_2B_3) are perpendicular to OB. A_1B_1 and A_2B_2 are arcs of circles with centre O.

    Calculate the sum to infinity of the arc lengths:

    $$AB + A_1B_1 + A_2B_2 + A_3B_3 + \ldots$$

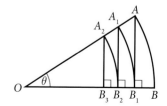

Figure 3.7 Diagram for question 42

43. The sum of the first n terms of a sequence is given by $S_n = 2n^2 - n$

(a) Find the first three terms of the sequence.

(b) Find an expression for the nth term of the sequence, giving your answer in terms of n.

44. (a) Expand $(2 + x)^5$, giving your answer in ascending powers of x.

(b) Hence, find the exact value of 2.01^5

45. You invest \$5000 at an annual compound interest rate of 6.3%.

(a) Write an expression for the value of this investment after t full years.

(b) Find the value of this investment at the end of five years.

(c) After how many full years will the value of the investment exceed \$10,000?

46. The sum of the first n terms of an arithmetic sequence $\{u_n\}$ is given by the formula $S_n = 4n^2 - 2n$. Three terms of this sequence, u_2, u_m and u_{32}, are consecutive terms in a geometric sequence. Find m.

47. The sum of the first 16 terms of an arithmetic sequence $\{u_n\}$ is 12. Find the first term and the common difference if the ninth term is zero.

48. (a) Write down the first four terms of the expansion of $(1 - x)^n$, with $n > 2$, in ascending powers of x.

(b) The absolute values of the coefficients of 2nd, 3rd, and 4th terms of the expansion in (a) are consecutive terms in an arithmetic sequence.

(i) Show that $n^3 - 9n^2 + 14n = 0$

(ii) Find the value of n.

49. From a group of five seniors and six juniors, four students are chosen.

(a) Determine how many possible groups can be chosen.

(b) Determine how many groups can be formed consisting of two seniors and two juniors.

(c) Determine how many groups can be formed consisting of at least one junior.

50. **(a)** Write down the full expansion of $(3 + x)^4$ in ascending powers of x.

(b) Find the exact value of 3.1^4

51. Find the number of ways in which seven different books can be given to three students, if the youngest student is to receive three books and the others receive two books each.

52. **(a)** Write down how many integers between 10 and 300 are divisible by 7.

(b) Express the sum of these integers in sigma notation.

(c) Find the sum above.

(d) Given an arithmetic sequence with first term 1000 and common difference -7, find the smallest n so that the sum of the first n terms of this sequence is negative.

53. Jim uses a pack of nine cards with integers $\{1, 2, 3, 4, 5, 6, 7, 8, 9\}$ written on them. Each card displays one of these integers. Jim is going to select four cards at random from this pack.

(a) Find the number of selections Jim could make if the largest integer drawn among the four cards is either a 5, a 6, or a 7.

(b) Find the number of selections Jim could make if at least two of the four integers drawn are even.

54. Three wives and their husbands are to sit on a bench for a photograph.

(a) Find the number of ways this can be done if the three wives want to sit together.

(b) Find the number of ways this can be done if the three wives will all sit apart.

55. Let $\{u_n\}$, $n \in \mathbb{Z}^+$, be an arithmetic sequence with first term a and common difference d, where $d \neq 0$. Let another sequence $\{v_n\}$, $n \in \mathbb{Z}^+$, be defined by $v_n = 2^{u_n}$.

(a) **(i)** Show that $\dfrac{v_{n+1}}{v_n}$ is a constant.

(ii) Write down the first term of the sequence $\{v_n\}$.

(iii) Write down a formula for v_n in terms of a, d, and n.

Let S_n be the sum of the first n terms of the sequence $\{v_n\}$.

(b)   **(i)** Find S_n in terms of a, d, and n.

   (ii) Find the values of d for which $\sum_{1}^{\infty} v_i$ exists.

You are now told that $\sum_{1}^{\infty} v_i$ does exist and is denoted by S_∞.

(iii) Write down S_∞ in terms of a and d.

(iv) Given that $S_\infty = 2^{a+1}$, find the value of d.

Exponential and logarithmic functions

4

4 Exponential and logarithmic functions

Learning objectives

By the end of this chapter, you should be familiar with...

- exponential functions and their graphs
- concepts of exponential growth and decay, and applications
- the nature and significance of the number e
- logarithmic functions and their graphs
- properties of logarithms
- solving equations involving exponential expressions
- solving equations involving logarithmic expressions.

This chapter examines exponential and logarithmic functions. A variety of functions have already been considered in this text: polynomial functions (linear, quadratic and cubic functions), functions with radicals (the square root function), rational functions (inverse and inverse square functions), and absolute value functions.

Exponential functions help us model a wide variety of physical phenomena. The natural exponential function (or simply, the exponential function), $f(x) = e^x$, is one of the most important functions in calculus. Exponential functions and their applications – especially to situations involving growth and decay – will be covered at length.

Logarithms, which were originally invented as a computational tool, lead to logarithmic functions. These functions are closely related to exponential functions and play an equally important part in calculus and a range of applications. We will learn that certain exponential and logarithmic functions are inverses of each other.

4.1 Exponential functions

Characteristics of exponential functions

The two equations $y = x^2$ and $y = 2^x$ are similar in that they both contain a base and a power. In $y = x^2$, the base is the variable x and the power is the constant 2. In $y = 2^x$, the base is the constant 2 and the power is the variable x.

The quadratic function $y = x^2$ is in the form '**variable base**$^{\text{constant power}}$', where the base is a variable and the power is an integer greater than or equal to zero (non-negative integer). Any function in this form is called a power function (another type of polynomial function).

The function $y = 2^x$ is in the form '**constant base**$^{\text{variable power}}$', where the base is a positive real number (not equal to one) and the power is a variable. Any function in this form is called an exponential function.

To illustrate a fundamental difference between exponential functions and power functions, consider the function values for $y = x^2$ and $y = 2^x$ when x is an integer from 0 to 10. Table 4.1 clearly shows how the values for the exponential function eventually increase at a significantly faster rate than the power function.

> To demonstrate just how quickly $y = 2^x$ increases, consider what would happen if you were able to repeatedly fold a piece of paper in half 50 times. A typical piece of paper is about five thousandths of a centimetre thick. Each time you fold the paper, the thickness of the paper doubles, so after 50 folds the thickness of the folded paper is the height of a stack of 2^{50} pieces of paper. The thickness of the paper after being folded 50 times would be $2^{50} \times 0.005$ cm which is more than 56 million kilometres (nearly 35 million miles)! Compare that with the height of a stack of 50^2 pieces of paper, which would be a meagre 12.5 centimetres – only 0.000125 kilometres.

x	$y = x^2$	$y = 2^x$
0	0	1
1	1	2
2	4	4
3	9	8
4	16	16
5	25	32
6	36	64
7	49	128
8	64	256
9	81	512
10	100	1024

Table 4.1 Contrast between power function and exponential function

Another important point to make is that power (and polynomial) functions can easily be defined, and computed, for any real number. For any power function $y = x^n$, where n is any positive integer, y is found by simply taking x and repeatedly multiplying it n times. Hence, x can be any real number. For example, for the power function $y = x^3$, if $x = \pi$, then $y = \pi^3 \approx 31.00619811\ldots$ Since a power function like $y = x^3$ is defined for all real numbers, we can graph it as a continuous curve so that every real number is the x-coordinate of some point on the curve. What about the exponential function $y = 2^x$? Can we compute a value for y for any real number x? Before we try, let's first consider x being any rational number and recall the laws of powers (see margin note).

> For $b > 0$ and $m, n \in \mathbb{Q}$ (rational numbers):
> $$b^m \cdot b^n = b^{m+n}$$
> $$\frac{b^m}{b^n} = b^{m-n}$$
> $$(b^m)^n = b^{mn}$$
> $$b^0 = 1$$
> $$b^{-m} = \frac{1}{b^m}$$

We also covered the definition of a rational exponent.

From these established facts, we are able to compute b^x ($b > 0$) when x is any rational number. For example, $b^{4.7} = b^{\frac{47}{10}}$ represents the 10th root of b raised to the 47th power $\left(\sqrt[10]{b^{47}}\right)$.

> For $b > 0$ and $m, n \in \mathbb{Z}$ (integers):
> $$b^{\frac{m}{n}} = \sqrt[n]{b^m} = \left(\sqrt[n]{b}\right)^m$$

Graphs of exponential functions

Using this definition of irrational powers, we can now construct a complete graph of any exponential function $f(x) = b^x$ such that b is a number greater than zero ($b \neq 1$) and x is any real number.

Example 4.1

Graph each exponential function on the same set of axes by plotting points.
$$f(x) = 3^x \text{ and } g(x) = \left(\frac{1}{3}\right)^x$$

Solution

Calculate values for each function for integral values of x from -3 to 3. Knowing that exponential functions are defined for all real numbers (not just integers), we can sketch a smooth curve, shown in the graph, filling in between the ordered pairs, shown in the table.

x	$f(x) = 3^x$	$g(x) = \left(\dfrac{1}{3}\right)^x$
-3	$\dfrac{1}{27}$	27
-2	$\dfrac{1}{9}$	9
-1	$\dfrac{1}{3}$	3
0	1	1
1	3	$\dfrac{1}{3}$
2	9	$\dfrac{1}{9}$
3	27	$\dfrac{1}{27}$

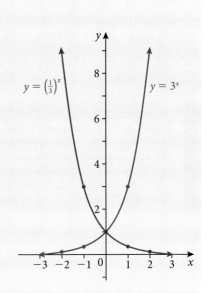

Remember that in Section 1.5 we established that the graph of $y = f(-x)$ is obtained by reflecting the graph of $y = f(x)$ in the y-axis. It is clear that the graph of function g is a reflection of function f in the y-axis. We can use some laws of powers to show that $g(x) = f(-x)$.

$$g(x) = \left(\frac{1}{3}\right)^x = \frac{1^x}{3^x} = \frac{1}{3^x} = 3^{-x} = f(-x)$$

It is useful to point out that both of the graphs, $y = 3^x$ and $y = \left(\dfrac{1}{3}\right)^x$, pass through the point $(0, 1)$ and have a horizontal asymptote of $y = 0$ (x-axis). The same is true for the graph of all exponential functions in the form $y = b^x$ given that $b \neq 1$. If $b = 1$, then $y = 1^x = 1$, and the graph is a horizontal line rather than a constantly increasing or decreasing curve.

If $b > 0$ and $b \neq 1$, an **exponential function** with base b is the function defined by $f(x) = b^x$
The **domain** of f is the set of real numbers ($x \in \mathbb{R}$) and the **range** of f is the set of positive real numbers ($y > 0$). The graph of f passes through $(0, 1)$, has the x-axis as a **horizontal asymptote**, and, depending on the value of the base of the exponential function b, will either be a continually increasing **exponential growth curve** or a continually decreasing **exponential decay curve**.

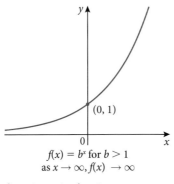

$f(x) = b^x$ for $b > 1$
as $x \to \infty$, $f(x) \to \infty$

f is an increasing function
exponential growth curve

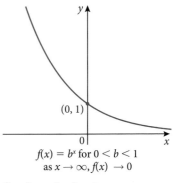

$f(x) = b^x$ for $0 < b < 1$
as $x \to \infty$, $f(x) \to 0$

f is a decreasing function
exponential decay curve

The graphs of all exponential functions will display a characteristic growth or decay curve. As we shall see, many natural phenomena exhibit exponential growth or decay. Also, the graphs of exponential functions behave asymptotically for either very large positive values of x (decay curve) or very large negative values of x (growth curve). This means that there will exist a horizontal line that the graph will approach but not intersect as either $x \to \infty$ or as $x \to -\infty$.

Transformations of exponential functions

Recalling from Section 1.5 how the graphs of functions are translated and reflected, we can efficiently sketch the graph of many exponential functions.

Example 4.2

Using the graph of $f(x) = 2^x$, sketch the graph of each function below. State the domain and range for each function and the equation of its horizontal asymptote.

(a) $g(x) = 2^x + 3$

(b) $h(x) = 2^{-x}$

(c) $p(x) = -2^x$

(d) $r(x) = 2^{x-4}$

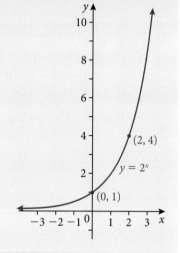

Solution

(a) The graph of $g(x) = 2^x + 3$ can be obtained by translating the graph of $f(x) = 2^x$ vertically three units up.

The horizontal asymptote for g is $y = 3$.

For function g, the domain is $x \in \mathbb{R}$ and the range is $y > 3$.

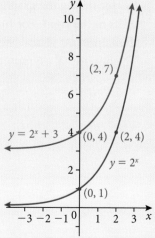

(b) The graph of $h(x) = 2^{-x}$ can be obtained by reflecting the graph of $f(x) = 2^x$ in the y-axis.

The horizontal asymptote is $y = 0$ (x-axis).

For function h, the domain is $x \in \mathbb{R}$ and the range is $y > 0$.

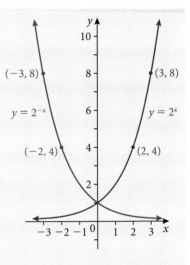

(c) The graph of $p(x) = -2^x$ can be obtained by reflecting the graph of $f(x) = 2^x$ in the y-axis. For function p, the domain is $x \in \mathbb{R}$ and the range is $y < 0$.

The horizontal asymptote is $y = 0$ (x-axis).

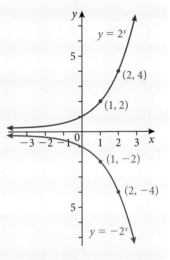

(d) The graph of $r(x) = 2^{x-4}$ can be obtained by translating the graph of $f(x) = 2^x$ four units to the right. For function r, the domain is $x \in \mathbb{R}$ and the range is $y > 0$. The horizontal asymptote is $y = 0$ (x-axis).

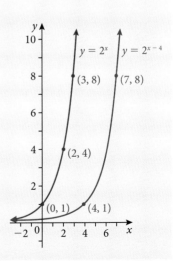

Note that for function p in part (c) of Example 4.2, the horizontal asymptote is an upper bound (no function value is equal to or greater than $y = 0$). Whereas, in parts (a), (b), and (d), the horizontal asymptote for each function is a lower bound (no function value is equal to or less than the y value of the asymptote).

4.2 Exponential growth and decay

Mathematical models of growth and decay

Exponential functions are well suited as a mathematical model for a wide variety of steadily increasing or decreasing phenomena of many kinds, including population growth (or decline), investment of money with compound interest, and radioactive decay. Recall from the previous chapter that the formula for finding terms in a geometric sequence (repeated multiplication by common ratio r) is an exponential function. Many instances of growth or decay occur exponentially (repeated multiplication by a growth or decay factor).

Exponential models are equations of the form $A(t) = A_0 b^t$, where $A_0 \neq 0$, $b > 0$, and $b \neq 1$. $A(t)$ is the amount after time t. $A(0) = A_0 b^0 = A_0 (1) = A_0$, so A_0 is called the **initial amount** (often the value at time $t = 0$). If $b > 1$, then $A(t)$ is an **exponential growth model**, and if $0 < b < 1$, then $A(t)$ is an **exponential decay model**. The value of b, the base of the exponential function, is often called the **growth or decay factor**.

Example 4.3

A sample count of bacteria in a culture indicates that the number of bacteria is doubling every hour. The estimated count at 15:00 was 12 000 bacteria.

(a) Find the estimated count three hours earlier (at 12:00).

(b) Write an exponential growth function for the number of bacteria at t hours after 12:00.

Solution

(a) Consider the time at 12:00 to be the starting, or initial, time and label it $t = 0$ hours. Then the time at 15:00 is $t = 3$. The amount of bacteria at any time t (in hours) will double after an hour so the growth factor, b, is 2. Therefore, $A(t) = A_0 (2)^t$. Knowing that $A(3) = 12\,000$, compute A_0. $12\,000 = A_0 (2)^3 \Rightarrow 12\,000 = 8A_0 \Rightarrow A_0 = 1500$. Therefore, the estimated count at 12:00 was 1500 bacteria.

(b) The growth function for the number of bacteria t hours after 12:00 is $A(t) = 1500(2)^t$.

Radioactive material decays at exponential rates. The **half-life** is the time it takes for a given amount of material to decrease to half of its original amount. An exponential function that models the activity of a radioactive source with a known value for the half-life, h, will be of the form $A(t) = A_0\left(\frac{1}{2}\right)^k$ where the growth factor is $\frac{1}{2}$ and k represents the number of half-lives that have occurred (the number of times that A_0 is multiplied by $\frac{1}{2}$). If t represents the amount of time, then the number of half-lives will be $\frac{t}{h}$.

For example, if the half-life of a certain material is 25 days and the amount of time that has passed since measuring the amount A_0 is 75 days, then the number of half-lives is $k = \frac{t}{h} = \frac{75}{25} = 3$, and activity is equal to $A_0\left(\frac{1}{2}\right)^3 = \frac{A_0}{8}$.

If a certain initial amount, A_0, of material decreases with a half-life of h, the amount of material that remains at time t is given by the exponential decay model $A(t) = A_0\left(\frac{1}{2}\right)^{\frac{t}{h}}$.
The time units for h and t must be the same.

Example 4.4

The half-life of radioactive carbon-14 is approximately 5730 years. What percentage of carbon-14 remains after 15 000 years?

Solution

The exponential decay model for the carbon-14 is $A(t) = A_0\left(\frac{1}{2}\right)^{\frac{t}{5730}}$.

After 15 000 years: $A(15\,000) = A_0\left(\frac{1}{2}\right)^{\frac{15000}{5730}} \approx 0.163\,A_0$. Thus, 16.3% of the carbon-14 remains.

Compound interest

Recall from Chapter 3 that exponential functions occur in calculating compound interest. If an initial amount of money P, called the **principal**, is invested at an interest rate r per time period, then after one time period the amount of interest is $P \times r$ and the total amount of money is $A = P + Pr = P(1 + r)$. If the interest is added to the principal, then the new principal is $P(1 + r)$, and the total amount after another time period is $A = P(1 + r)(1 + r) = P(1 + r)^2$. In the same way, after a third time period the amount is $A = P(1 + r)^3$. In general, after k periods the total amount is $A = P(1 + r)^k$, an exponential function with growth factor $1 + r$.

For example, if the amount of money in a bank account is earning interest at a rate of 6.5% per time period, then the growth factor is $1 + 0.065 = 1.065$. Is it possible for r to be negative? Yes, if an amount (not just money) is decreasing. For example, if the population of a town is decreasing by 12% per time period, then the decay factor is $1 - 0.12 = 0.88$.

For compound interest, if the annual interest rate is r and interest is compounded (number of times added in) n times per year, then for each time period the interest rate is $\frac{r}{n}$, and there are $n \times t$ time periods in t years.

Compound interest formula
The exponential function for calculating the amount of money after t years, $A(t)$, where P is the initial amount or principal, the annual interest rate is r, and the number of times interest is compounded per year is n, is given by $A(t) = P\left(1 + \frac{r}{n}\right)^{nt}$

Example 4.5

An initial amount of 1000 euros is deposited into an account earning 5.25% interest per year. Find the amount in the account after 8 years if interest is compounded annually, semiannually, quarterly, monthly, and daily.

Solution

We use the exponential function associated with compound interest with values of $P = 1000$, $r = 0.0525$ and $t = 8$ to compute the results below.

Compounding	n	Amount (€) after 8 years
Annually	1	$1000\left(1 + \dfrac{0.0525}{1}\right)^8 = 1505.83$
Semiannually	2	$1000\left(1 + \dfrac{0.0525}{2}\right)^{2(8)} = 1513.74$
Quarterly	4	$1000\left(1 + \dfrac{0.0525}{4}\right)^{4(8)} = 1517.81$
Monthly	12	$1000\left(1 + \dfrac{0.0525}{12}\right)^{12(8)} = 1520.57$
Daily	365	$1000\left(1 + \dfrac{0.0525}{365}\right)^{365(8)} = 1521.92$

Example 4.6

A new car is purchased for \$22,000. If the value of the car decreases (depreciates) at a rate of approximately 15% per year, what will be the approximate value of the car to the nearest whole dollar in 4.5 years?

Solution

The decay factor for the exponential function is $1 - r = 1 - 0.15 = 0.85$. In other words, after each year the car's value is 85% of what it was one year before. We use the exponential decay model $A(t) = A_0 b^t$ with values $A_0 = 22\,000$, $b = 0.85$ and $t = 4.5$.

$$A(4.5) = 22\,000\,(0.85)^{4.5} \approx 10\,588$$

In 4.5 years, the value of the car will be approximately \$10,588.

Exercise 4.1 & 4.2

1. (a) Write the equation for an exponential equation with base $b > 0$.

   (b) Given $b \neq 1$, state the domain and range of this function.

   (c) Sketch the general shape of the graph of this exponential function for each of two cases:

   (i) $b > 1$ (ii) $0 < b < 1$

2. Sketch a graph of each function below and state:

   (i)   the coordinates of any x-intercept(s) and y-intercept

   (ii) the equation of any asymptote(s)

   (iii) the domain and range.

   (a) $f(x) = 3^{x+4}$ (b) $g(x) = -2^x + 8$ (c) $h(x) = 4^{-x} - 1$

   (d) $p(x) = \dfrac{1}{2^x - 1}$ (e) $q(x) = 3(3^{-x}) - 3$ (f) $k(x) = 2^{-|x-2|} + 1$

3. A general exponential function is written in the form $f(x) = a\,(b)^{x-c} + d$. State the domain, range, y-intercept and the equation of the horizontal asymptote in terms of the parameters a, b, c and d.

4. Using your GDC and a graph viewing window with Xmin $= -2$, Xmax $= 2$, Ymin $= 0$ and Ymax $= 4$, sketch a graph for each exponential equation below on the same set of axes.

   (a) $y = 2^x$ (b) $y = 4^x$ (c) $y = 8^x$

   (d) $y = 2^{-x}$ (e) $y = 4^{-x}$ (f) $y = 8^{-x}$

5. Write equations that are equivalent to the equations in question 4 **(d)**, **(e)** and **(f)** but have a power of positive x rather than negative x.

6. If $1 < a < b$, then which is steeper: the graph of $y = a^x$ or $y = b^x$? Why?

7. The population of a city triples every 25 years. At time $t = 0$, the population is $100\,000$. Write a function for the population $P(t)$ as a function of t. Find the population after:

   (a) 50 years   (b) 70 years   (c) 100 years

8. An experiment involves a colony of bacteria in a solution. The number of bacteria doubles approximately every 3 minutes and the number of bacteria at the start of the experiment is 10^4. Write a function for the number of bacteria $N(t)$ as a function of t (in minutes). Find approximately how many bacteria there are after:

   (a) 3 minutes   (b) 9 minutes

   (c) 27 minutes   (d) one hour

9. A bank offers an investment account that will double your money in 10 years.

   (a) Express $A(t)$, the amount of money in the account after t years, in the form $A(t) = A_0\,(r)^t$.

   (b) If interest was added into the account just once at the end of each year (simple interest), then find the annual interest rate for the account (to 3 significant figures).

10. $10,000 is invested at an annual interest rate of 11%, compounded quarterly. Find the value of the investment after:

    (a) 5 years   (b) 10 years   (c) 15 years

11. A sum of $5000 is deposited into an investment account that earns interest at a rate of 9% per year, compounded monthly.

    (a) Write the function $A(t)$ that computes the value of the investment after t years.

    (b) Use your GDC to sketch a graph of $A(t)$ with values of t on the horizontal axis ranging from $t = 0$ years to $t = 25$ years.

    (c) Use the graph on your GDC to determine the minimum number of years (to the nearest whole year) for this investment to have a value greater than $20,000.

12. $10,000 is invested at an annual interest rate of 11% for a period of 5 years. Find the value of the investment for each compounding period.

    (a) annually   (b) monthly

    (c) daily   (d) hourly

13. Imagine a bank account that has the fantastic annual interest rate of 100%. Assume we deposit $1 into this account. Exactly one year later, how much will be in the account for each compounding period?

    (a) annually   (b) monthly

    (c) daily   (d) hourly

    (e) every minute

14. Each year for the past eight years, the population of deer in a certain national park increased at a steady rate of 3.2% per year. The present population is approximately 248 000.

    (a) What was the approximate number of deer one year ago?

    (b) What was the approximate number of deer eight years ago?

15. Radioactive carbon-14 has a half-life of 5730 years. The remains of an animal are found 20 000 years after it died. Approximately what percentage (to 3 significant figures) of the original amount of carbon-14 (when the animal was alive) would we expect to find?

16. The amount of a drug in the bloodstream of a patient decays exponentially with a half-life of 36 hours after entering the patient's bloodstream. An amount of the drug, A_0, is injected into the bloodstream at 12:00 on Monday. How much of the drug will be in the patient's bloodstream five days later (12:00 on Saturday)?

17. An open can is filled with 1000 ml of fluid that evaporates at a rate of 30% per week.

(a) Write a function, $A(w)$, that gives the amount of fluid after w weeks.

(b) Use your GDC to find how many weeks (to the nearest whole number) it will take for the volume of fluid to be less than 1 ml.

18. Why are exponential functions of the form $f(x) = b^x$ defined so that $b > 1$?

19. You are offered a highly paid job that lasts for just 30 days. How much would you get paid for the payment plan, I or II, that offers the larger salary?

I One dollar on the first day of the month, two dollars on the second day, three dollars on the third day, and so on (getting paid one dollar more each day), for 30 days.

II One cent ($0.01) on the first day, two cents ($0.02) on the second day, four cents on the third day, eight cents on the fourth day and so on (each day getting paid double from the previous day), for 30 days.

20. Each exponential function graphed here can be written in the form $f(x) = k(a)^x$. Find the value of a and k for each.

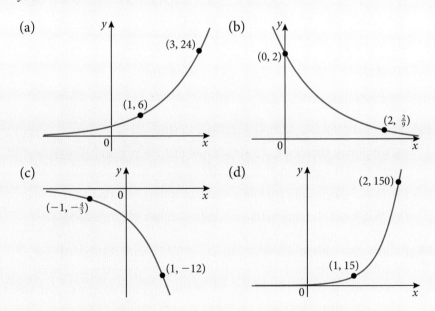

(a)

(b)

(c)

(d)

4.3 The number e

In an exponential function, $f(x) = b^x$, b is any positive constant and x is any real number. Graphs of $y = b^x$ for a few values where $b \geq 1$ are shown in Figure 4.1. As noted in Section 4.1, all the graphs pass through the point $(0, 1)$.

The question arises: what is the *best* number to choose for the base b? There is a good argument for $b = 10$ since we most commonly use a base 10 number system. Your GDC will have the expression 10^x as a built-in command. The base $b = 2$ is also plausible because a binary number system (base 2) is used in many processes, especially in computer systems. However, the most important base is an irrational number that is denoted by the letter e. As we will see, the value of e to 5 decimal places is 2.71828. The importance of e will be clearer when we get to calculus topics. The number π (another very useful irrational number) has a natural geometric significance as the ratio of circumference to diameter for any circle. The number e also occurs in a 'natural' manner. We will illustrate this in two different ways: first, by considering the **rate of change** of an exponential function, and second, by re-visiting compound interest and considering **continuous change** rather than **incremental change**.

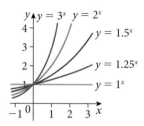

Figure 4.1 Graphs of $y = b^x$ for some values when $b \geq 1$

The constant e is of enormous mathematical significance and it appears 'naturally' in many mathematical processes. Jakob Bernoulli first observed e when studying sequences of numbers in connection to compound interest problems.

Rate of change (slope) of an exponential function

Since exponential functions (and associated logarithmic functions) are very important in calculus, the criteria we will use to determine the best value for b will be based on considering the slope of the curve $y = b^x$. In calculus, we are interested in the rate of change (the slope of the graph of functions). Our goal is to find a value for b such that the slope of the graph of $y = b^x$ at a particular value of x is equal to the function value y. We could investigate this by trial and error (and with a GDC this might prove fruitful) but it would not guarantee us an exact value and it could prove inefficient. Let's narrow our investigation to studying the slope of the curves at the point $(0, 1)$, which is convenient because it is shared by all the curves.

To obtain a good estimate for the value of e we will use the diagram in Figure 4.2, where the scales on the x- and y-axes are equal and $P(0, 1)$ is the y-intercept of the graph of $y = e^x$. Q is a point close to point P with coordinates (h, e^h). PR and RQ are parallel to the x- and y-axes, respectively, and they intersect at point $R(h, 1)$. The slope of the curve is always changing. It is not constant as with a straight line. As we will justify more thoroughly in our study of differential calculus in Chapter 12, the slope of a curve at a point will be equal to the slope of the line tangent to the curve at that point. PS is the tangent line to the curve at P, intersecting RQ at S. So, we are looking for the value e such that the slope of the tangent line PS is equal to 1. It follows that $\dfrac{RS}{PR} = 1$ and because $PR = h$ then $RS = h$. Since we have chosen Q close to P, we can

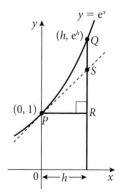

Figure 4.2 Graph of $y = e^x$; the slope of the tangent line PS is equal to 1

Table 4.2

h	$e = (1 + h)^{\frac{1}{h}}$
0.1	2.593742…
0.01	2.704814…
0.001	2.716924…
0.0001	2.718146…
0.00001	2.718268…
0.000001	2.718280…
0.0000001	2.718282…

Table 4.2 Values for $e \approx (1 + h)^{\frac{1}{h}}$ as h approaches zero (accuracy to 7 significant figures). To an accuracy of six significant figures, it appears that the value of e is 2.71828.

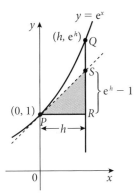

Figure 4.3 At $x = 0$, the rate of change of $y = e^x$ is equal to 1

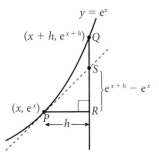

Figure 4.4 The rate of change of $y = e^x$ at a general value of x

assume that h is very small. Therefore, $RS \approx RQ$ and $\frac{RQ}{RS} \approx 1$. The value of $\frac{RQ}{RS}$ will get closer and closer to the value of 1 as h gets smaller (as Q gets closer to P). Since the y-coordinate of R is 1, then $RQ = e^h - 1$. Substituting h for RS and $e^h - 1$ for RQ into $\frac{RQ}{RS} \approx 1$, gives $\frac{e^h - 1}{h} \approx 1$. We wish to obtain an estimate for e so we multiply through by h to get $e^h - 1 \approx h$ leading to $e^h \approx h + 1$. To isolate e we raise both sides to the $\frac{1}{h}$ power, finally producing $e \approx (1 + h)^{\frac{1}{h}}$.

If h is made small enough, the expression above should give a good estimation of the value of e. Using the approximation $e \approx (1 + h)^{\frac{1}{h}}$, Table 4.2 shows values for e as h approaches zero.

e is defined as
$$e = \lim_{h \to 0} (1 + h)^{\frac{1}{h}}$$
The definition is read 'e equals the limit of $(1 + h)^{\frac{1}{h}}$ as h goes to zero.'

Geometrically speaking, as point Q gets closer to point P ($h \to 0$), and also closer to point S, we wanted the slope of the tangent line at $(0, 1)$, $\frac{RS}{PR}$, to be equal to 1. This is the same as saying that we wanted $\frac{e^h - 1}{h} \to 1$ as $h \to 0$ (see coloured triangle in Figure 4.3). The value of e approximated to increasing accuracy in Table 4.2 is the number that makes this happen. A non-geometrical way of describing this feature of the graph is to say that the **rate of change** (slope) of the function $y = e^x$ at $x = 0$ is equal to 1.

The rate of change of $y = e^x$ at a general value of x (Figure 4.4) can be obtained in the same way by fixing point P on the curve with coordinates (x, e^x) and a nearby point Q with coordinates $(x + h, e^{x+h})$. Then the rate of change of the function at point P is $\frac{e^{x+h} - e^x}{h}$ as $h \to 0$. We cannot evaluate the limit of $\frac{e^{x+h} - e^x}{h}$ as $h \to 0$ directly by substituting 0 for h. By applying some algebra and knowing that $\frac{e^h - 1}{h} \to 1$ as $h \to 0$, we can evaluate the required limit.

As $h \to 0$, $\frac{e^{x+h} - e^x}{h} = \frac{e^x e^h - e^x}{h} = \frac{e^x(e^h - 1)}{h} = e^x \left[\frac{e^h - 1}{h}\right] \to e^x \cdot 1 = e^x$

Therefore, for any value of x, the rate of change of the function $y = e^x$ is e^x. In other words, the rate of change of the function at any value in the domain (x) is equal to the corresponding value of the range (y). This is the amazing feature of $y = e^x$ that makes e the most useful and 'natural' choice for the base of an exponential function, and the irrational number $e \approx 2.71828\ldots$ is the only base for which this is true.

Continuously compounded interest

In Section 4.2 and in Chapter 3, we computed amounts of money resulting from an initial amount (principal) with interest being compounded (added in) at discrete intervals (e.g. yearly, monthly, daily). In the formula that we used, $A(t) = P\left(1 + \dfrac{r}{n}\right)^{nt}$, n is the number of times that interest is compounded per year. Instead of adding interest only at discrete intervals, let's investigate what happens if we try to add interest continuously; that is, let the value of n increase without bound ($n \to \infty$).

As described in question 13 in Exercise 4.1 & 4.2, consider investing just $1 at a very generous annual interest rate of 100%. How much will be in the account at the end of just one year? It depends on how often the interest is compounded; in other words, it depends on the value of n in the compound interest formula $A(t) = P\left(1 + \dfrac{r}{n}\right)^{nt}$. See Table 4.3 for results for different values of n.

As the number of compounding periods during the year, n, increases, the amount at the end of the year appears to approach a limiting value. We realise that as $n \to \infty$, the quantity of $\left(1 + \dfrac{1}{n}\right)^{n}$ approaches the number e. To twelve decimal places, e is 2.718281828459.

Compounding	n	After 1 year
Annually	1	2
Semiannually	2	2.25
Quarterly	4	2.44140625…
Monthly	12	2.61303529022…
Daily	365	2.71456748202…
Hourly	8760	2.71812669063…
Every minute	525 600	2.7182792154…
Every second	31 536 000	2.71828247254…

Table 4.3 Values for $\left(1 + \dfrac{1}{n}\right)^{n}$ as $n \to \infty$

e is also defined as:

$$e = \lim_{n \to \infty} \left(1 + \frac{1}{n}\right)^{n}$$

The definition is read 'e equals the limit of $\left(1 + \dfrac{1}{n}\right)^{n}$ as n goes to infinity.'

Note that the two definitions for the number e are equivalent. Take our first limit definition for e: $e = \lim\limits_{h \to 0} (1 + h)^{\frac{1}{h}}$. Letting $\dfrac{1}{h} = n$, it follows that $h = \dfrac{1}{n}$ and as

$h \to 0$ then $n \to \infty$. Substituting $\dfrac{1}{n}$ for h, n for $\dfrac{1}{h}$, and evaluating the limit as

$n \to \infty$ transforms $\lim\limits_{h \to 0} (1 + h)^{\frac{1}{h}}$ to $\lim\limits_{n \to \infty} \left(1 + \dfrac{1}{n}\right)^{n}$, which is our second limit definition for e.

As the number of compoundings, n, increases without bound, we approach continuous compounding where interest is being added continuously. In the formula for calculating amounts resulting from compound interest, letting $m = \dfrac{n}{r}$, produces:

$$A(t) = P\left(1 + \frac{r}{n}\right)^{nt} = P\left(1 + \frac{1}{m}\right)^{mrt} = P\left[\left(1 + \frac{1}{m}\right)^{m}\right]^{rt}$$

Continuous compound interest formula
An exponential function for calculating the amount of money after t years, $A(t)$, for interest compounded continuously, where P is the initial amount or principal and r is the annual interest rate, is given by $A(t) = Pe^{rt}$

Now if $n \to \infty$ and the interest rate r is constant, then $\frac{n}{r} = m \to \infty$. From the limit definition of e, we know that if $m \to \infty$, then $\left(1 + \frac{1}{m}\right)^{m} \to$ e. Therefore, for continuous compounding, it follows that $A(t) = P\left[\left(1 + \frac{1}{m}\right)^{m}\right]^{rt} = P[\text{e}]^{rt}$.

This result is part of the reason that e is the best choice for the base of an exponential function modelling change that occurs continually (e.g. radioactive decay) rather than in discrete intervals.

Example 4.7

The starting balance of an investment account is 1000 euros. The account earns interest at an annual rate of 5.25%. Assuming there are no further withdrawals or deposits (other than interest), find the total amount in the account after 10 years if the interest is added to the account:

(a) annually   (b) quarterly   (c) continuously

Solution

(a) $A(t) = P(1 + r)^t = 1000(1 + 0.0525)^{10} = 1669.10$ euros

(b) $A(t) = P\left(1 + \frac{r}{n}\right)^{nt} = 1000\left(1 + \frac{0.0525}{4}\right)^{4 \cdot 10} = 1684.70$ euros

(c) $A(t) = Pe^{rt} = 1000e^{0.0525\,(10)} = 1690.46$ euros

The **natural exponential function** is the function defined as $f(x) = e^x$
As with other exponential functions, the domain of the natural exponential function is the set of all real numbers $(x \in \mathbb{R})$, and its range is the set of positive numbers $(y > 0)$. The natural exponential function is often referred to as *the* exponential function.

The natural exponential function and continuous change

For many applications involving continuous change, the most suitable choice for a mathematical model is an exponential function with a base having the value of e.

The formula developed for continuously compounded interest does not apply only to applications involving adding interest to financial accounts. It can be used to model growth or decay of a quantity that is changing exponentially (repeated multiplication by a constant ratio, or growth/decay factor) and the change is continuous, or approaching continuous. Another version of a formula for continuous change that we will learn more about in calculus is stated below.

Continuous exponential growth/decay
If an initial quantity C (when $t = 0$) grows or decays continuously at a rate r over a certain time period, then the amount $A(t)$ after t time periods is given by the function $A(t) = Ce^{rt}$. If $r > 0$, then the quantity is increasing (growing). If $r < 0$, then the quantity is decreasing (decaying).

Example 4.8

A particular commercial aeroplane costs $150 million new. The aeroplane will lose value at a continuous rate. This is modelled by the continuous decay function $C(t) = 150e^{-0.053t}$ where $C(t)$ is the value of the aeroplane (in millions) after t years.

(a) How much (to the nearest million dollars) would the aeroplane be worth precisely five years after being purchased?

(b) If the aeroplane is purchased in 2020, what would be the first year that it is worth less than half of its original cost?

(c) Find the value of b (to 4 significant figures) for a discrete decay model, $D(t) = 150b^t$, so that $D(t)$ is a suitable model to describe the same decay as $C(t)$.

Solution

(a) $C(5) = 150e^{-0.053(5)} \approx 115$
The value is approximately $115 million after five years.

(b) Using a GDC, we graph the decay equation $y = 150e^{-0.053x}$ and the horizontal line $y = 75$ and determine the intersection point.

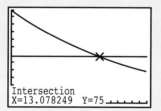

The x-coordinate of the intersection point is approximately 13.08. At the start of 2023, the aeroplane's value is not yet half of its original value. Therefore, the first year that the aeroplane is worth less than half of its original cost is 2024.

(c) One way to find the value of b so that $D(t) = 150b^t$ serves as a reasonable substitute for $C(t) = 150e^{-0.053t}$ is to compute some function values for $C(t)$ and use them to compute the relative change from one year to the next.

$$C(1) = 150e^{-0.053(1)} \approx 142.2570$$
$$C(2) = 150e^{-0.053(2)} \approx 134.9137$$
$$C(3) = 150e^{-0.053(3)} \approx 127.9495$$

Relative change from year 1 to year 2: $\dfrac{134.9137 - 142.2570}{142.2570} \approx -0.05162$

Compute relative change from year 2 to year 3 to make sure it agrees with the result above.

Relative change from year 2 to year 3: $\dfrac{127.9495 - 134.9137}{134.9137} \approx -0.05162$

The annual rate of decay, b, is the fraction of what remains after each year. Thus, $b = 1 - 0.05162 = 0.94838$; and to 4 significant figures the annual rate of decay is $b \approx 0.9484$. Therefore, the discrete decay model is $D(t) = 150(0.9484)^t$

To check that the two decay models give similar results for each year, we can use a GDC to display a table of values for both models side by side for easy comparison.

```
Plot1 Plot2 Plot3
\Y1■150e^(-.053X
)
\Y2■150(.9484)^X
\Y3=
\Y4=
\Y5=
```

X	Y1	Y2
0	150	150
1	142.26	142.26
2	134.91	134.92
3	127.95	127.96
4	121.34	121.34
5	115.08	115.09
6	109.14	109.15

X=0

Exercise 4.3

1. Sketch a graph of each function below and state:
   (i) the coordinates of any x-intercept(s) and y-intercept
   (ii) the equation of any asymptote(s)
   (iii) the domain and range.

   (a) $f(x) = e^{2x-1}$ (b) $g(x) = e^{(-x+1)}$ (c) $h(x) = -2e^x$

   (d) $p(x) = e^{x^2} - e$ (e) $h(x) = \dfrac{1}{1 - e^x}$ (f) $h(x) = e^{|x+2|} - 1$

2. (a) State a definition of the number e as a limit.

   (b) Evaluate $\left(1 - \dfrac{1}{n}\right)^n$ for $n = 100$, $n = 10\,000$ and $n = 1\,000\,000$.

   (c) To 5 significant figures, what appears to be the value of
   $\lim\limits_{n \to \infty}\left(1 - \dfrac{1}{n}\right)^n$?

   (d) How does this number relate to the number e?

3. Use your GDC to graph the curve $y = \left(1 + \dfrac{1}{x}\right)^x$ and the horizontal line $y = 2.72$. Use a graph window so that x ranges from 0 to 20 and y ranges from 0 to 3. Describe the behaviour of the graph of $y = \left(1 + \dfrac{1}{x}\right)^x$.
   Will it ever intersect the graph of $y = 2.72$? Explain.

4. Two different banks, Bank A and Bank B, offer accounts with exactly the same annual interest rate of 6.85%. However, the account from Bank A has the interest compounded monthly, whereas the account from Bank B compounds the interest continuously. To decide which bank to open an account with, you calculate the amount of interest you would earn after three years from an initial deposit of 500 euros in each bank's account. It is assumed that you make no further deposits and no withdrawals during the three years.

   (a) How much interest would you earn from each of the accounts?

   (b) Which bank's account earns more, and how much more?

5. Dina wishes to deposit $1000 into an investment account and then withdraw the total in the account in five years. She has the choice of two different accounts.

   Blue Star account: Interest is earned at an annual interest rate of 6.13% compounded weekly (52 weeks in a year).

   Red Star account: Interest is earned at an annual interest rate of 5.95% compounded continuously.

   (a) Which investment account will result in the greatest total at the end of five years?

   (b) What is the total after five years for this account?

   (c) How much more is it than the total for the other account?

6. Strontium-90 is a radioactive isotope of strontium. Strontium-90 decays according to the function $A(t) = Ce^{-0.0239t}$, where t is time in years and C is the initial amount of strontium-90 when $t = 0$. Find the percentage remaining of a sample of strontium-90 after:

   (a) 1 year   (b) 10 years   (c) 100 years   (d) 250 years

7. A radioactive substance decays in such a way that the mass (in kilograms) remaining after t days is given by the function $A(t) = 5e^{-0.0347t}$.

   (a) Find the mass at time $t = 0$.

   (b) What percentage of the initial mass remains after 10 days?

   (c) On your GDC and then on paper, draw a graph of the function $A(t)$ for $0 \leqslant t \leqslant 50$.

   (d) Use one of your graphs to approximate, to the nearest whole day, the half-life of the radioactive substance.

8. Which of the given interest rates and compounding periods would provide the best investment?

   Option A   8.5% per year, compounded semiannually

   Option B   8.25% per year, compounded quarterly

   Option C   8% per year, compounded continuously

9. In certain conditions, the bacteria that cause cholera, *Vibrios cholerae*, can grow rapidly in number. In a laboratory experiment, a culture of *Vibrios cholerae* is started with 20 bacteria. The bacteria's growth is modelled with the continuous growth model $A(t) = 20e^{0.068t}$ where $A(t)$ is the number of bacteria after t minutes.

   (a) Determine the value of r for the discrete growth model $B(t) = 20(r)^t$, so that $B(t)$ is equivalent to $A(t)$.

   (b) For both of these models, by what percentage does the number of bacteria grow each minute?

10. By comparing the graph of each of the following equations to the graph of $y = e^x$, determine if the slope of the tangent line at the point $(0, 1)$ for the graph of each equation is less than or equal to 1.

   (a) $y = 2^x$ (b) $y = \left(\dfrac{5}{2}\right)^x$ (c) $y = \left(\dfrac{11}{4}\right)^x$ (d) $y = 3^x$

11. £1000 is invested at 4.5% annual interest, compounded continuously.

   (a) How much money is in the account after 10 years? After 20 years?

   (b) Use your GDC to determine how many years (to the nearest tenth of a year) it takes for the initial investment to double to £2000.

   (c) If £5000 is invested at the same rate of interest and also compounded continuously, how many years (to the nearest tenth) would it take to double?

   (d) Are the answers to (b) and (c) the same or different? Why?

4.4   Logarithmic functions

In Example 4.7 (c), we used the equation $A(t) = 1000e^{0.075t}$ to compute the amount of money in an account after t years. Now suppose we wish to determine how much time, t, it takes for the initial investment of 1000 euros to double. To find this, we need to solve the following equation for t: $2000 = 1000e^{0.075t} \Rightarrow 2 = e^{0.075t}$. The unknown t is in the exponent. At this point in the book, we do not have an algebraic method to solve such an equation, but developing the concept of a logarithm will provide us with the means to do so.

The inverse of an exponential function

For $b > 1$, an exponential function with base b is increasing for all x, and for $0 < b < 1$ an exponential function is decreasing for all x. It follows from this that all exponential functions must be one-to-one. Recall from Section 1.4 that a one-to-one function passes both a vertical line test and a horizontal line test. We demonstrated that an inverse function would exist for any one-to-one function. Therefore, an exponential function with base b such that $b > 0$ and

$b \neq 1$ will have an inverse function, which is given in the Key fact box. Also recall from Section 1.4 that the domain of a function $f(x)$ is the range of its inverse function $f^{-1}(x)$, and, similarly, the range of $f(x)$ is the domain of $f^{-1}(x)$. The domain and range are switched around for a function and its inverse.

For $b > 0$ and $b \neq 1$, the **logarithmic function** $y = \log_b x$ (read 'logarithm with base b of x') is the inverse of the exponential function with base b.

$y = \log_b x$ if and only if $x = b^y$

The domain of the logarithmic function $y = \log_b x$ is the set of positive numbers $(x > 0)$ and its range is all real numbers $(y \in \mathbb{R})$.

Logarithmic expressions and equations

When evaluating logarithms, note that a logarithm is an exponent. This means that the value of $\log_b x$ is the power to which b must be raised to obtain x. For example, $\log_2 8 = 3$ because 2 must be raised to the power of 3 to obtain 8. That is, $\log_2 8 = 3$ if and only if $2^3 = 8$.

We can use the definition of a logarithmic function to translate a logarithmic equation into an exponential equation and vice versa. When doing this it is helpful to remember, as the definition stated, that in either form (logarithmic or exponential) the base is the same.

logarithmic equation

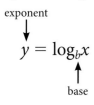

exponential equation

Example 4.9

Find the value of each logarithm.

(a) $\log_7 49$ (b) $\log_5\left(\dfrac{1}{5}\right)$ (c) $\log_6 \sqrt{6}$ (d) $\log_4 64$ (e) $\log_{10} 0.001$

Solution

We set each logarithmic expression equal to y and use the definition of a logarithmic function to obtain an equivalent equation in exponential form. We then solve for y by applying the logical fact that if $b > 0$, $b \neq 1$ and $b^y = b^k$, then $y = k$.

(a) Let $y = \log_7 49$, which is equivalent to the exponential equation $7^y = 49$. Since $49 = 7^2$, then $7^y = 7^2$. Therefore, $y = 2 \Rightarrow \log_7 49 = 2$

(b) Let $y = \log_5\left(\dfrac{1}{5}\right)$, which is equivalent to the exponential equation $5^y = \dfrac{1}{5}$. Since $\dfrac{1}{5} = 5^{-1}$, then $5^y = 5^{-1}$. Therefore, $y = -1 \Rightarrow \log_5\left(\dfrac{1}{5}\right) = -1$

(c) Let $y = \log_6 \sqrt{6}$, which is equivalent to the exponential equation $6^y = \sqrt{6}$. Since $\sqrt{6} = 6^{\frac{1}{2}}$, then $6^y = 6^{\frac{1}{2}}$. Therefore, $y = \dfrac{1}{2} \Rightarrow \log_6 \sqrt{6} = \dfrac{1}{2}$

(d) Let $y = \log_4 64$, which is equivalent to the exponential equation $4^y = 64$. Since $64 = 4^3$, then $4^y = 4^3$. Therefore, $y = 3 \Rightarrow \log_4 64 = 3$

(e) Let $y = \log_{10} 0.001$, which is equivalent to the exponential equation

$10^y = 0.001$. Since $0.001 = \dfrac{1}{1000} = \dfrac{1}{10^3} = 10^{-3}$, then $10^y = 10^{-3}$.

Therefore, $y = -3 \Rightarrow \log_{10} 0.001 = -3$

Example 4.10

Find the domain of the function $f(x) = \log_2(4 - x^2)$

Solution

From the definition of a logarithmic function, the domain of $y = \log_b x$ is $x > 0$, thus for $f(x)$ it follows that $4 - x^2 > 0 \Rightarrow (2 + x)(2 - x) > 0$ $\Rightarrow -2 < x < 2$. Hence, the domain is $-2 < x < 2$.

Properties of logarithms

As with all functions and their inverses, their graphs are reflections of each other in the line $y = x$. Figure 4.5 illustrates this relationship for exponential and logarithmic functions and also confirms the domain and range for the logarithmic function stated in the definition above.

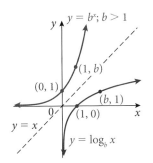

Figure 4.5 Reflection of $y = \log_b x$ in the line $y = x$

Notice that the points $(0, 1)$ and $(1, 0)$ are mirror images of each other in the line $y = x$. This corresponds to the fact that since $b^0 = 1$, then $\log_b 1 = 0$. Another pair of mirror image points $(1, b)$ and $(b, 1)$ highlight the fact that $\log_b b = 1$.

Notice also that since the x-axis is a horizontal asymptote of $y = b^x$, then the y-axis is a vertical asymptote of $y = \log_b x$.

In Section 1.4 we established that a function f and its inverse function f^{-1} satisfy the equations:

$$f^{-1}(f(x)) = x \qquad \text{for } x \text{ in the domain of } f$$
$$f(f^{-1}(x)) = x \qquad \text{for } x \text{ in the domain of } f^{-1}$$

When applied to $f(x) = b^x$ and $f^{-1}(x) = \log_b x$ these equations become:

$$\log_b(b^x) = x \qquad x \in \mathbb{R}$$
$$b^{\log_b x} = x \qquad x > 0$$

The logarithmic function with base 10 is called the **common logarithmic function**. On calculators and on your GDC, this function is denoted by **log**. The value of the expression $\log_{10} 1000$ is 3 because 10^3 is 3. Generally, for common logarithms (base 10) we omit writing the base of 10. Hence, if **log** is written with no base indicated, it is assumed to have a base of 10. For example, $\log 0.01 = -2$.

> **common logarithm:** $\qquad \log_{10} x = \log x$

As with exponential functions, the most widely used logarithmic function (and the other logarithmic function supplied on all calculators) is the logarithmic

Properties of logarithms I

For $b > 0$ and $b \neq 1$, the following statements are true:

1. $\log_b 1 = 0$ because $b^0 = 1$

2. $\log_b b = 1$ because $b^1 = b$

3. $\log_b(b^x) = x$ because $b^x = b^x$

4. $b^{\log_b x} = x$ because $\log_b x$ is the power to which b must be raised to get x

function with the base of e. This function is known as the **natural logarithmic function** and it is the inverse of the natural exponential function $y = e^x$. The natural logarithmic function is denoted by the symbol **ln,** and the expression $\ln x$ is read as 'the natural logarithm of x'.

natural logarithm: $\qquad \log_e x = \ln x$

Example 4.11

Evaluate each expression.

(a) $\log\left(\dfrac{1}{10}\right)$

(b) $\log(\sqrt{10})$

(c) $\log 1$

(d) $10^{\log 47}$

(e) $\log 50$

(f) $\ln e$

(g) $\ln\left(\dfrac{1}{e^3}\right)$

(h) $\ln 1$

(i) $e^{\ln 5}$

(j) $\ln 50$

Solution

(a) $\log\left(\dfrac{1}{10}\right) = \log(10^{-1}) = -1$

(b) $\log(\sqrt{10}) = \log\left(10^{\frac{1}{2}}\right) = \dfrac{1}{2}$

(c) $\log 1 = \log(10^0) = 0$

(d) $10^{\log 47} = 47$

(e) $\log 50 \approx 1.699$ [using GDC]

(f) $\ln e = 1$

(g) $\ln\left(\dfrac{1}{e^3}\right) = \ln(e^{-3}) = -3$

(h) $\ln 1 = \ln(e^0) = 0$

(i) $e^{\ln 5} = 5$

(j) $\ln 50 \approx 3.912$ [using GDC]

Example 4.12

Figure 4.6 shows the graph of the line $y = x$ and two curves. Curve A is the graph of the equation $y = \log x$. Curve B is the reflection of curve A in the line $y = x$.

(a) Write the equation for curve B.

(b) Write the coordinates of the y-intercept of curve B.

Solution

(a) Curve A is the graph of $y = \log x$, the common logarithm with base 10, which could also be written as $y = \log_{10} x$. Curve B is the inverse of $y = \log_{10} x$ since it is the reflection of it in the line $y = x$. Hence, the equation for curve B is the exponential equation $y = 10^x$.

(b) The y-intercept occurs when $x = 0$. For curve B, $y = 10^0 = 1$. Therefore, the y-intercept for curve B is $(0, 1)$.

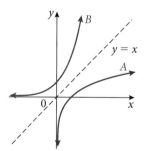

Figure 4.6 Graph for Example 4.12

The logarithmic function with base b is the inverse of the exponential function with base b. Therefore, it makes sense that the laws of powers should have corresponding properties involving logarithms. For example, the exponential property $b^0 = 1$ corresponds to the logarithmic property $\log_b 1 = 0$. We will state and prove three further important logarithmic properties that correspond to the following three laws of exponents.

- $b^m \cdot b^n = b^{m+n}$

- $\dfrac{b^m}{b^n} = b^{m-n}$

- $(b^m)^n = b^{mn}$

Properties of logarithms II

Given $M > 0$, $N > 0$ and k is any real number, then the following properties are true for logarithms with $b > 0$ and $b \neq 1$.

	Property	Description
1	$\log_b(MN) = \log_b M + \log_b N$	the log of a product is the sum of the logs of its factors
2	$\log_b\left(\dfrac{M}{N}\right) = \log_b M - \log_b N$	the log of a quotient is the log of the numerator minus the log of the denominator
3	$\log_b(M^k) = k\log_b M$	the log of a number raised to a power is the power times the log of the number

Any of these properties can be applied in either direction.

Proof

Property 1

Let $x = \log_b M$ and $y = \log_b N$

The corresponding exponential form of these two equations are

$b^x = M$ and $b^y = N$

Then, $\log_b(MN) = \log_b(b^x b^y) = \log_b(b^{x+y}) = x + y$

It's given that $x = \log_b M$ and $y = \log_b N$, hence $x + y = \log_b M + \log_b N$.

Therefore, $\log_b(MN) = \log_b M + \log_b N$

Property 2

Again, let $x = \log_b M$ and $y = \log_b N \Rightarrow b^x = M$ and $b^y = N$

Then, $\log_b\left(\dfrac{M}{N}\right) = \log_b\left(\dfrac{b^x}{b^y}\right) = \log_b(b^{x-y}) = x - y$

With $x = \log_b M$ and $y = \log_b N$, then $x - y = \log_b M - \log_b N$.

Therefore, $\log_b\left(\dfrac{M}{N}\right) = \log_b M - \log_b N$

Property 3

Let $x = \log_b M \Rightarrow b^x = M$

Now, take the logarithm of M^k and substitute b^x for M.

$\log_b(M^k) = \log_b[(b^x)^k] = \log_b(b^{kx}) = kx$

It's given that $x = \log_b M$, hence $kx = k \log_b M$

Therefore, $\log_b(M^k) = k \log_b M$

The notation $f(x)$ uses brackets not to indicate multiplication, but to indicate the argument of the function f. The symbol f is the name of a function, not a variable; it is not multiplying the variable x. Therefore, in general, $f(x + y)$ is not equal to $f(x) + f(y)$. Likewise, the symbol **log** is also the name of a function. Therefore, $\log_b(x + y)$ is not equal to $\log_b(x) + \log_b(y)$. Other mistakes to avoid include incorrectly simplifying quotients or powers of logarithms. Specifically,

$\dfrac{\log_b x}{\log_b y} \neq \log\left(\dfrac{x}{y}\right)$ and $(\log_b x)^k \neq k(\log_b x)$

Example 4.13

Use the properties of logarithms to write each logarithmic expression below as a sum, difference, and/or constant multiple of simple logarithms (logarithms without sums, products, quotients or powers).

(a) $\log_2(8x)$

(b) $\ln\left(\dfrac{3}{y}\right)$

(c) $\log(\sqrt{7})$

(d) $\log_b\left(\dfrac{x^3}{y^2}\right)$

(e) $\ln(5\,e^2)$

(f) $\log\left(\dfrac{m + n}{n}\right)$

Solution

(a) $\log_2(8x) = \log_2 8 + \log_2 x = 3 + \log_2 x$

(b) $\ln\left(\dfrac{3}{y}\right) = \ln 3 - \ln y$

(c) $\log(\sqrt{7}) = \left(\log 7^{\frac{1}{2}}\right) = \dfrac{1}{2}\log 7$

(d) $\log_b\left(\dfrac{x^3}{y^2}\right) = \log_b(x^3) - \log_b(y^2) = 3 \log_b x - 2 \log_b y$

(e) $\ln(5\,e^2) = \ln 5 + \ln(e^2) = \ln 5 + 2 \ln e = \ln 5 + 2(1) \approx 3.609$

(f) $\log\left(\dfrac{m + n}{m}\right) = \log(m + n) - \log m$ [remember $\log(m + n) \neq \log m + \log n$]

Example 4.14

Write each expression as the logarithm of a single quantity.

(a) $\log 6 + \log x$

(b) $\log_2 5 + 2 \log_2 3$

(c) $\ln y - \ln 4$

(d) $\log_b 12 - \dfrac{1}{2}\log_b 9$

(e) $\log_3 M + \log_3 N - 2 \log_3 P$

(f) $\log_2 80 - \log_2 5$

Solution

(a) $\log 6 + \log x = \log(6x)$

(b) $\log_2 5 + 2\log_2 3 = \log_2 5 + \log_2(3^2) = \log_2 5 + \log_2 9$
$$= \log_2(5 \cdot 9) = \log_2 45$$

(c) $\ln y - \ln 4 = \ln\left(\dfrac{y}{4}\right)$

(d) $\log_b 12 - \dfrac{1}{2}\log_b 9 = \log_b 12 - \log_b(9^{\frac{1}{2}}) = \log_b 12 - \log_b(\sqrt{9})$
$$= \log_b 12 - \log_b 3$$
$$= \log_b\left(\dfrac{12}{3}\right) = \log_b 4$$

(e) $\log_3 M + \log_3 N - 2\log_3 P = \log_3(MN) - \log_3(P^2) = \log_3\left(\dfrac{MN}{P^2}\right)$

(f) $\log_2 80 - \log_2 5 = \log_2\left(\dfrac{80}{5}\right) = \log_2 16 = 4$ [because $2^4 = 16$]

Change of base

The answer to Example 4.14 (f) was $\log_2 16$, which we can compute to be exactly 4 because we know that $2^4 = 16$. The answer to Example 4.13 (e) was $2 + \ln 5$, which we approximated to 3.609 using the natural logarithm function key (**ln**) on our GDC. But, what if we wanted to compute an approximate value for $\log_2 45$, the answer to Example 14 (b)? Some GDCs can evaluate only common logarithms (base 10) and natural logarithms (base e).

To evaluate logarithmic expressions and graph logarithmic functions to other bases we may need to apply a change of base formula (see margin note).

Let a, b and x be positive real numbers such that $a \neq 1$ and $b \neq 1$. Then $\log_b x$ can be expressed in terms of logarithms in any other base a, as follows:

$$\log_b x = \dfrac{\log_a x}{\log_a b}$$

Proof

$y = \log_b x \Rightarrow b^y = x$ convert from logarithmic form to exponential form

$\log_a x = \log_a(b^y)$ if $b^y = x$, then the log of each with same bases are equal

$\log_a x = y\log_a b$ applying the property $\log_b(M^k) = k\log_b M$

$y = \dfrac{\log_a x}{\log_a b}$ divide both sides by $\log_a b$

To apply the change of base formula, let $a = 10$ or $a = $ e. Then the logarithm of any base b can be expressed in terms of either common logarithms or natural logarithms. For example:

$$\log_2 x = \dfrac{\log x}{\log 2} \text{ or } \dfrac{\ln x}{\ln 2} \qquad \log_5 x = \dfrac{\log x}{\log 5} \text{ or } \dfrac{\ln x}{\ln 5}$$

$$\log_2 45 = \dfrac{\log 45}{\log 2} = \dfrac{\ln 45}{\ln 2} \approx 5.492 \quad \text{[using GDC]}$$

Example 4.15

Use the change of base formula and common or natural logarithms to evaluate each logarithmic expression. Start by making a rough mental estimate. Write your answer to 3 significant figures.

(a) $\log_3 30$ (b) $\log_9 6$

Solution

(a) The value of $\log_3 30$ is the power to which 3 is raised to obtain 30. We know that $3^3 = 27$ and $3^4 = 81$. Therefore, the value of $\log_3 30$ is between 3 and 4, and will be much closer to 3 than 4, perhaps around 3.1. Using the change of base formula and common logarithms, we obtain

$\log_3 30 = \dfrac{\log 30}{\log 3} \approx 3.096$. This agrees well with the mental estimate.

After computing the answer on a GDC, use the GDC to also check it by raising 3 to the answer and confirming that it gives a result of 30.

```
log(30)/log(3)
           3.095903274
3^Ans
                    30
■
```
```
ln(6)/ln(9)
           .8154648768
9^Ans
                     6
■
```

(b) The value of $\log_9 6$ is the power to which 9 is raised to obtain 6. We know that $9^{\frac{1}{2}} = \sqrt{9} = 3$ and $9^1 = 9$. Therefore, the value of $\log_9 6$ is between $\dfrac{1}{2}$ and 1, perhaps around 0.75. Using the change of base formula and natural logarithms, we obtain $\log_9 6 = \dfrac{\ln 6}{\ln 9} \approx 0.815$.

This agrees well with the mental estimate.

Exercise 4.4

1. Express each logarithmic equation as an exponential equation.

   (a) $\log_2 16 = 4$ (b) $\ln 1 = 0$ (c) $\log 100 = 2$

   (d) $\log 0.01 = -2$ (e) $\log_7 343 = 3$ (f) $\ln\left(\dfrac{1}{e}\right) = -1$

   (g) $\log 50 = y$ (h) $\ln x = 12$ (i) $\ln(x + 2) = 3$

2. Express each exponential equation as a logarithmic equation.

   (a) $2^{10} = 1024$ (b) $10^{-4} = 0.0001$ (c) $4^{-\frac{1}{2}} = \dfrac{1}{2}$

   (d) $3^4 = 81$ (e) $10^0 = 1$ (f) $e^x = 5$

   (g) $2^{-3} = 0.125$ (h) $e^4 = y$ (i) $10^{x+1} = y$

3. Find the exact value of each expression without using your GDC.

(a) $\log_2 64$

(b) $\log_4 64$

(c) $\log_2\left(\dfrac{1}{8}\right)$

(d) $\log_3(3^5)$

(e) $\log_{16} 8$

(f) $\log_{27} 3$

(g) $\log_{10}(0.001)$

(h) $\ln e^{13}$

(i) $\log_8 1$

(j) $10^{\log 6}$

(k) $\log_3\left(\dfrac{1}{27}\right)$

(l) $e^{\ln\sqrt{2}}$

(m) $\log 1000$

(n) $\ln(\sqrt{e})$

(o) $\ln\left(\dfrac{1}{e^2}\right)$

(p) $\log_3(81^{22})$

(q) $\log_4 2$

(r) $3^{\log_3 18}$

(s) $\log_5\left(\sqrt[3]{5}\right)$

(t) $10^{\log\pi}$

4. Use a GDC to evaluate each expression, correct to 3 significant figures.

(a) $\log\sqrt{3}$

(b) $\ln\sqrt{3}$

(c) $\log 25$

(d) $\log\left(\dfrac{1+\sqrt{5}}{2}\right)$

(e) $\ln(100^3)$

5. Find the domain of each function exactly.

(a) $y = \log(x-2)$

(b) $y = \ln(x^2)$

(c) $y = \log(x) - 2$

(d) $y = \log_7(8 - 5x)$

(e) $y = \sqrt{x+2} - \log_3(9 - 3x)$

(f) $y = \sqrt{\ln(1-x)}$

6. Find the domain and range of each function exactly.

(a) $y = \dfrac{1}{\ln x}$

(b) $y = |\ln(x-1)|$

(c) $y = \dfrac{x}{\log x}$

7. Find the equation of the function that is graphed in the form $f(x) = \log_b x$

(a)

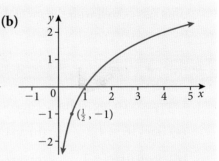

(b)

(c)

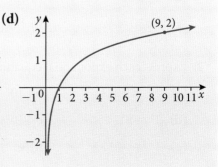

(d)

8. Use properties of logarithms to write each logarithmic expression as a sum, difference, and/or constant multiple of simple logarithms (logarithms without sums, products, quotients or powers).

(a) $\log_2(2m)$

(b) $\log\left(\dfrac{9}{x}\right)$

(c) $\ln(\sqrt[5]{x})$

(d) $\log_3(a\,b^3)$

(e) $\log[10x(1+r)^t]$

(f) $\ln\left(\dfrac{m^3}{n}\right)$

9. Write each expression in terms of $\log_b p$, $\log_b q$ and $\log_b r$.

(a) $\log_b(pqr)$

(b) $\log_b\left(\dfrac{p^2 q^3}{r}\right)$

(c) $\log_b(\sqrt[4]{pq})$

(d) $\log_b\sqrt{\dfrac{qr}{p}}$

(e) $\log_b\left(\dfrac{p\sqrt{q}}{r}\right)$

(f) $\log_b\left(\dfrac{(pq)^3}{\sqrt{r}}\right)$

10. Write each expression as the logarithm of a single quantity.

(a) $\log(x^2) + \log\left(\dfrac{1}{x}\right)$

(b) $\log_3 9 + 3\log_3 2$

(c) $4\ln y - \ln 4$

(d) $\log_b 12 - \dfrac{1}{2}\log_b 9$

(e) $\log x - \log y - \log z$

(f) $2\ln 6 - 1$

> ! For question 10 (f):
> $\ln(?) = 1$

11. Use the change of base formula and common or natural logarithms to evaluate each logarithmic expression. Give your answers to 3 significant figures.

(a) $\log_2 1000$

(b) $\log_{0.5} 40$

(c) $\log_6 40$

(d) $\log_5(0.75)$

12. Use the change of base formula to rewrite each function in terms of natural logarithms. Then graph the function on a GDC and use the graph to find, to 3 significant figures, the value of the function when $x = 20$.

(a) $f(x) = \log_2 x$

(b) $f(x) = \log_5 x$

13. Use the change of base formula to prove the following statement.

$$\log_b a = \dfrac{1}{\log_a b}$$

14. Show that $\log e = \dfrac{1}{\ln 10}$

189

15. The relationship between the number of decibels dB (one variable) and the intensity I of a sound in watts per square metre is given by the formula $dB = 10 \log\left(\dfrac{I}{10^{-16}}\right)$. Use properties of logarithms to write the formula in simpler form. Then find the number of decibels of a sound with an intensity of 10^{-4} watts per square metre.

16. (a) Given the exponential function $f(x) = 5(2)^x$, show that $\log f(x)$ varies linearly with x; that is, find the linear equation in terms of x that is equal to $\log f(x)$.

    (b) Prove that for any exponential function in the form $f(x) = ab^x$, the function $\log f(x)$ is linear and can be written in the form $\log f(x) = mx + c$. Find the constants m (slope) and c (y-intercept) in terms of $\log a$ and $\log b$.

4.5 Exponential and logarithmic equations

Solving exponential equations

At the start of Section 4.4 we wanted to find a way to determine how much time t (in years) it would take for an investment of 1000 euros to double if the investment earns interest at an annual rate of 7.5%. Since the interest is compounded continuously, we need to solve this equation:

$$2000 = 1000\,e^{0.075t} \Rightarrow 2 = e^{0.075t}$$

The equation has the variable t as part of the power. With the properties of logarithms established in Section 4.4, we now have a way to algebraically solve such equations. Along with these properties, we need to apply the logic that if two expressions are equal, then their logarithms with the same base must also be equal. That is, if $m = n$, then $\log_b m = \log_b n$.

Example 4.16

Solve for t in the equation $2 = e^{0.075t}$

Give your answer accurate to 3 significant figures.

Solution

$2 = e^{0.075t}$

$\ln 2 = \ln(e^{0.075t})$ take natural logarithm of both sides

$\ln 2 = 0.075t$ apply the property $\log_b(b^x) = x$

$t = \dfrac{\ln 2}{0.075} \approx 9.24$

Example 4.16 serves to illustrate a general strategy for solving exponential equations. To solve an exponential equation, first isolate the exponential expression and take the logarithm of both sides. Then apply a property of logarithms so that the variable is no longer in the power and can be isolated on one side of the equation. By taking the logarithm of both sides of an exponential equation, we are making use of the inverse relationship between exponential and logarithmic functions. Symbolically, this method can be represented as follows:

(i) If $b = 10$ or e: $y = b^x \Rightarrow \log_b y = \log_b b^x \Rightarrow \log_b y = x$

(ii) If $b \neq 10$ or e: $y = b^x \Rightarrow \log_a y = \log_a b^x \Rightarrow \log_a y = x \log_a b \Rightarrow x = \dfrac{\log_a y}{\log_a b}$

Example 4.17

Solve for x in the equation $3^{x-4} = 24$

Give the answer to 3 significant figures.

Solution

$3^{x-4} = 24$

$\log(3^{x-4}) = \log 24$	take common logarithm of both sides
$(x - 4)\log 3 = \log 24$	apply the property $\log_b(M^k) = k\log_b M$
$x - 4 = \dfrac{\log 24}{\log 3}$	divide both sides by $\log 3$ $\left[\text{note: } \dfrac{\log 24}{\log 3} \neq \log 8\right]$
$x = \dfrac{\log 24}{\log 3} + 4$	
$x \approx 6.89$	using GDC

 We could have used natural logarithms instead of common logarithms to solve the equation in Example 4.17. Using the same method but with natural logarithms we get

$x = \dfrac{\ln 24}{\ln 3} + 4 \approx 6.89$

Recall Example 3.11 in which we solved an exponential equation graphically because we did not yet have the tools to solve it algebraically. Let's now solve it using logarithms.

Example 4.18

You invest €1000 at 6% per annum compounded quarterly.

How long will it take this investment to increase to €2000?

Solution

Using the compound interest formula from Section 4.2, $A(t) = P\left(1 + \dfrac{r}{n}\right)^{nt}$, with $P = €1000$, $r = 0.06$ and $n = 4$, we need to solve for t when $A(t) = 2P$.

$2P = P\left(1 + \dfrac{0.06}{4}\right)^{4t}$ substitute $2P$ for $A(t)$

$2 = 1.015^{4t}$ divide both sides by P

$\ln 2 \doteq \ln(1.015^{4t})$ take natural logarithm of both sides

$\ln 2 = 4t \ln 1.015$ apply the property $\log_b(M^k) = k \log_b M$

$t = \dfrac{\ln 2}{4 \ln 1.015}$

$t \approx 11.6389$ evaluated on GDC

$\dfrac{\ln(2)}{4 \cdot \ln(1.015)}$	11.6389

The investment will double in 11.64 years, or about 11 years and 8 months.

Example 4.19

The bacteria that cause strep throat will grow at a rate of about 2.3% per minute. To the nearest whole minute, how long will it take for these bacteria to double in number?

Solution

Let t represent time in minutes and let A_0 represent the number of bacteria at $t = 0$.

Using the exponential growth model from Section 4.2, $A(t) = A_0 \, b^t$, the growth factor, b, is $1 + 0.023 = 1.023$ giving $A(t) = A_0 (1.023)^t$. The same equation would apply to money earning 2.3% annual interest with the money being added (compounded) once per year rather than once per minute.

So, our mathematical model assumes that the number of bacteria increase incrementally, with the number increasing by 2.3% at the end of each minute. To find the doubling time, find the value of t so that $A(t) = 2A_0$.

$2A_0 \doteq A_0 (1.023)^t$ substitute $2A_0$ for $A(t)$

$2 = 1.023^t$ divide both sides by A_0

$\ln 2 = \ln(1.023^t)$ take natural logarithm of both sides

$\ln 2 = t \ln 1.023$ apply the property $\log_b(M^k) = k \log_b M$

$t = \dfrac{\ln 2}{\ln 1.023} \approx 30.482$

The number of bacteria will double in about 30 minutes.

Alternative solution

What if we assumed continuous growth instead of incremental growth? We apply the continuous exponential growth model from Section 4.3 $C: A(t) = C e^{rt}$ with initial amount C, and $r = 0.023$

$2C = Ce^{0.023t}$ substitute $2C$ for $A(t)$

$2 = e^{0.023t}$ divide both sides by C

$\ln 2 = \ln(e^{0.023t})$ take natural logarithm of both sides

$\ln 2 = 0.023t$ apply the property $\log_b(b^x) = x$

$t = \dfrac{\ln 2}{0.023} \approx 30.137$

Continuous growth has a slightly shorter doubling time but, rounded to the nearest minute, it also gives an answer of 30 minutes.

Example 4.20

$1000 is deposited in an investment account that earns interest at an annual rate of 10%, compounded monthly. Calculate the minimum number of years needed for the amount in the account to exceed $4000.

Solution

We use the exponential function associated with compound interest,
$A(t) = P\left(1 + \dfrac{r}{n}\right)^{nt}$ with $P = 1000$, $r = 0.1$ and $n = 12$

$$4000 = 1000\left(1 + \frac{0.1}{12}\right)^{12t} \Rightarrow 4 = (1.008\overline{3})^{12t} \Rightarrow \log 4 = \log[(1.008\overline{3})^{12t}] \Rightarrow$$

$$\log 4 = 12t \log(1.008\overline{3}) \Rightarrow t = \frac{\log 4}{12 \log(1.008\overline{3})} \approx 13.92 \text{ years}$$

The minimum time needed for the account to exceed $4000 is 14 years.

Example 4.21

A 20-gram sample of radioactive iodine decays so that the mass remaining after t days is given by the equation $A(t) = 20\,e^{-0.087t}$, where $A(t)$ is measured in grams. After how many days (to the nearest whole day) are there only 5 grams remaining?

Solution

$$5 = 20\,e^{-0.087t} \Rightarrow \frac{5}{20} = e^{-0.087t} \Rightarrow \ln 0.25 = \ln(e^{-0.087t}) \Rightarrow \ln 0.25 = -0.087t$$

$$\Rightarrow t = \frac{\ln 0.25}{-0.087} \approx 15.93$$

After about 16 days there are only 5 grams remaining.

Example 4.22

Solve for x in the equation $3^{2x} - 18 = 3^{x+1}$. Express any answers in exact form.

Solution

The key to solving this equation is recognising that it can be written in quadratic form. In Section 2.5, we solved equations of the form $at^2 + bt + c = 0$, where t is an algebraic expression. This is not immediately obvious for this equation. We need to apply some laws of powers to show that the equation is quadratic for the expression 3^x.

$$3^{2x} - 18 = 3^{x+1}$$

$$(3^x)^2 - 3^1 \cdot 3^x - 18 = 0 \qquad \text{applying rules } b^{mn} = (b^m)^n \text{ and } b^{m+n} = b^m b^n$$

Substituting a single variable, for example y, for the expression 3^x clearly makes the equation quadratic in terms of 3^x. We solve first for y and then solve for x after substituting 3^x back for y.

$$y^2 - 3y - 18 = 0$$

$$(y + 3)(y - 6) = 0$$

$$y = -3 \quad \text{or} \quad y = 6$$

$$3^x = -3 \quad \text{or} \quad 3^x = 6$$

$3^x = -3$ has no solution $\qquad$ raising a positive number to a power cannot produce a negative number

$$3^x = 6$$

$$\ln(3^x) = \ln 6 \qquad \text{take logarithms of both sides}$$

$$x \ln 3 = \ln 6$$

Therefore, the one solution to the equation is exactly $x = \dfrac{\ln 6}{\ln 3}$.

There are a couple of common algebra errors to avoid in Example 4.22

- If $3^x = -3$, then it does not follow that $x = -1$. A power of -1 indicates reciprocal.

- If $x = \dfrac{\ln 6}{\ln 3}$, it does not follow that $x = \ln 2$. The rule $\log m - \log n = \log\left(\dfrac{m}{n}\right)$ does not apply to the expression $\dfrac{\ln 6}{\ln 3}$.

Solving logarithmic equations

A logarithmic equation is an equation where the variable appears within the argument of a logarithm. For example, $\log x = \dfrac{1}{2}$ or $\ln x = 4$. We can solve both of these logarithmic equations directly by applying the definition of a logarithmic function (Section 4.4):

$$y = \log_b x \text{ if and only if } x = b^y$$

The logarithmic equation $\log x = \dfrac{1}{2}$ is equivalent to the exponential equation $x = 10^{\frac{1}{2}} = \sqrt{10}$, which leads directly to the solution. Likewise, the equation $\ln x = 4$ is equivalent to $x = e^4 \approx 54.598$. Both of these equations could have been solved by means of another method that makes use of the following two facts:

if $a - b$ then $n^a = n^b$

$$b^{\log_b x} = x$$

To understand the second fact, remember that a logarithm is an exponent. The value of $\log_b x$ is the power to which b is raised to give x. And b is being raised to this value; hence the expression $b^{\log_b x}$ is equivalent to x. Therefore, another method for solving the logarithmic equation $\ln x = 4$ is to **exponentiate** both sides; that is, use the expressions on either side of the equals sign as powers for exponential expressions with equal bases. The base needs to be the base of the logarithm.

$$\ln x = 4 \Rightarrow e^{\ln x} = e^4 \Rightarrow x = e^4$$

In the context of solving equations, to **exponentiate** means to use two equal expressions as powers to construct another equality involving equal bases. For example: If $a = b$, then $m^a = m^b$.

Example 4.23

Solve for x in the equation $\log_3(2x - 5) = 2$

Solution

$\log_3(2x - 5) = 2$

Method 1	**Method 2**	
$2x - 5 = 3^2$	$y = \log_b x \Leftrightarrow x = b^y$	$3^{\log_3(2x-5)} = 3^2$ exponentiate with base $= 3$
$2x = 9 + 5$	$2x - 5 = 9$	
$x = 7$	$x = 7$	

Example 4.24

Solve for x in terms of k: $\log_2(5x) = 3 + k$

Solution

$\log_2(5x) = 3 + k \Rightarrow 2^{\log_2(5k)} = 2^{3+k}$ exponentiate both sides with base $= 2$

$5x = 2^3 \cdot 2^k$ law of powers $b^m \cdot b^n = b^{m+n}$ used 'in reverse'

$x = \dfrac{8}{5}(2^k)$

For some logarithmic equations it is necessary to first apply a property, or properties, of logarithms to simplify combinations of logarithmic expressions before solving.

Example 4.25

Solve for x in the equation $\log_2 x + \log_2(10 - x) = 4$

Solution

$\log_2 x + \log_2(10 - x) = 4$

$\log_2[x(10 - x)] = 4$ property of logarithms: $\log_b M + \log_b N = \log_b(MN)$

$$10x - x^2 = 2^4 \qquad \text{changing from logarithmic form to exponential form}$$
$$x^2 - 10x + 16 = 0$$
$$(x - 2)(x - 8) = 0$$
$$x = 2 \text{ or } x = 8$$

When solving logarithmic equations, you should be careful to always check if the original equation is a true statement when any solutions are substituted in for the variable. For Example 4.25, both of the solutions $x = 2$ and $x = 8$ produce true statements when substituted into the original equations. Sometimes 'extra' (extraneous) invalid solutions are produced (discussed in Section 2.5), as illustrated in the next example.

Example 4.26

Solve for x in the equation $\ln(x - 2) + \ln(2x - 3) = 2\ln x$

Solution

$$\ln(x - 2) + \ln(2x - 3) = 2\ln x$$
$$\ln[(x - 2)(2x - 3)] = \ln x^2 \qquad \text{properties of logarithms}$$
$$\ln(2x^2 - 7x + 6) = \ln x^2$$
$$e^{\ln(2x^2 - 7x + 6)} = e^{\ln x^2} \qquad \text{exponentiate both sides}$$
$$2x^2 - 7x + 6 = x^2$$
$$x^2 - 7x + 6 = 0$$
$$(x - 6)(x - 1) = 0 \qquad \text{factorise}$$
$$x = 6 \text{ or } x = 1$$

Substituting these two possible solutions indicates that $x = 1$ is not a valid solution. The reason is that if we try to substitute 1 for x in the original equation, we are not able to evaluate the expression $\ln(2x - 3)$ because we can only take the logarithm of a positive number. Therefore, $x = 6$ is the only solution. $x = 1$ is an extraneous solution that is not valid.

Solving, or checking the solutions to, a logarithmic equation on your GDC will help you avoid extraneous solutions. To solve Example 4.26 on your GDC, a useful approach is to first set the equation equal to zero. Then graph the expression (after setting it equal to y) and observe where the graph intersects the x-axis ($y = 0$).

Graphical solution for Example 4.26:

$$\ln(x - 2) + \ln(2x - 3) = 2\ln x \implies \ln(x - 2) + \ln(2x - 3) - 2\ln x = 0$$

Graph the equation $y = \ln(x - 2) + \ln(2x - 3) - 2\ln x$ on your GDC and find x-intercepts.

The graph intersects the x-axis only at $x = 6$ and not at $x = 1$. Hence, $x = 6$ is the only valid solution and $x = 1$ is an extraneous solution.

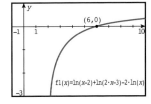

Figure 4.7 GDC graph of $y = \ln(x - 2) + \ln(2x - 3) - 2\ln x$

Exponential and logarithmic inequalities

In Section 2.5, we covered methods of solving a variety of inequalities. These methods can also be applied to solving inequalities involving exponential and logarithmic functions. It is important to consider the domain of any functions in the inequality, and to check any solutions in the original inequality in case any extraneous solutions occur.

Example 4.27

Find the solution set to the inequality: $2\log_3 x - 1 < 0$

Solution

Due to the domain of the logarithmic function, all solutions must be positive.

Method 1 (algebraic solution)

Rewrite the inequality as the equation $2\log_3 x - 1 = 0$ and find the exact solution.

$$2\log_3 x = 1 \implies \log_3 x = \frac{1}{2} \implies x = 3^{\frac{1}{2}} = \sqrt{3}$$

Substitute 'test' values, x_1 and x_2, into the original inequality such that $0 < x_1 < \sqrt{3}$ and $x_2 > \sqrt{3}$.

Let $x_1 = 1$: $\log_3 1 - 1 = 0 - 1 = -1 < 0$ (true)
Let $x_1 = 9$: $\log_3 9 - 1 = 2 - 1 = 1 \not< 0$ (false)

Therefore, the solution set is $0 < x < \sqrt{3}$

Method 2 (graphical solution)

Graph the equation $y = 2\log_3 x - 1$ on a GDC and use it to determine the portion of the graph that is less than zero (below the x-axis). But, how do we input the expression $\log_3 x$ on a GDC? We can use the change of base formula to write $\log_3 x = \dfrac{\log x}{\log 3}$

The y-axis is a vertical asymptote. The graph indicates that the solution set is $0 < x < 1.7320508$. Although the graphical method is efficient and effective it does not give an exact result.

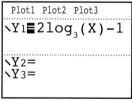

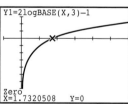

Figure 4.8 GDC graph of $y = 2\log_3 x - 1$

Example 4.28

Solve for x in the inequality $(e^x - 2)(e^x + 6) \leqslant 3e^x$

Solution

The fact that the left side is factorised is not helpful because the other side of the inequality is not zero. So we need to expand the left side and rearrange terms to get zero on the right side.

$(e^x - 2)(e^x + 6) \leqslant 3 e^x$

$(e^x)^2 + 4 e^x - 12 \leqslant 3 e^x$

$e^{2x} + e^x - 12 \leqslant 0$ now factorise this expression

$(e^x - 3)(e^x + 4) \leqslant 0$ find where each factor is zero and construct a sign chart

$e^x - 3 = 0 \implies e^x = 3 \implies x = \ln 3$

and

$e^x + 4 = 0 \implies e^x = -4 \implies$ no solution

Since $x = \ln 3$ is the only zero of the expression $(e^x - 3)(e^x + 4)$ we only need to test x values on either side of $x = \ln 3$. The factor $e^x + 4$ will always be positive.

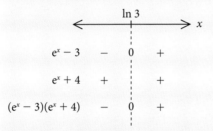

Therefore, the solution set is $x \leqslant \ln 3$

Exercise 4.5

1. Solve each equation for x. Give x accurate to 3 significant figures.

   (a) $10^x = 5$ **(b)** $4^x = 32$ **(c)** $8^{x-6} = 60$

   (d) $2^{x+3} = 100$ **(e)** $\left(\dfrac{1}{5}\right)^x = 22$ **(f)** $e^x = 15$

   (g) $10^x = e$ **(h)** $3^{2x-1} = 35$ **(i)** $2^{x+1} = 3^{x-1}$

   (j) $2 e^{10x} = 19$ **(k)** $6^{\frac{x}{2}} = 5^{1-x}$ **(l)** $\left(1 + \dfrac{0.05}{12}\right)^{12x} = 3$

2. Solve each equation for x. Give answers exactly.

   (a) $4^x - 2^{x+1} = 48$ **(b)** $2^{2x+1} - 2^{x+1} + 1 = 2^x$

   (c) $6^{2x+1} - 17(6^x) + 12 = 0$ **(d)** $3^{2x+1} + 3 = 10(3^x)$

3. \$5000 is invested in an account that pays 7.5% interest per year, compounded quarterly.

   (a) Find the amount in the account after 3 years.

   (b) How long will it take for the money in the account to double? Give the answer to the nearest quarter of a year.

For question 2 (a) write 4 as 2^2

4. Find how long it will take for an investment of €500 to triple in value when the interest is 8.5% per year, compounded continuously. Give the answer in number of years accurate to 3 significant figures.

5. A single bacterium begins a colony in a laboratory dish. If the colony doubles every hour, after how many hours does the colony first have more than one million bacteria?

6. Find the least number of years for an investment to double if interest is compounded annually with the following interest rates.

   (a) 3%   (b) 6%   (c) 9%

7. A new car purchased in 2015 decreases in value by 11% per year. When is the first year that the car is worth less than half of its original value?

8. Uranium-234 is a radioactive substance that has a half-life of 2.46×10^5 years.

   (a) Find the amount remaining from a 1 gram sample after 1000 years.

   (b) How long will it take a one gram sample to decompose until its mass is 700 milligrams (0.7 grams)? Give the answer in years, accurate to 3 significant figures.

9. The stray dog population in a town is growing exponentially with about 18% more stray dogs each year. In 2018, there are 16 stray dogs.

   (a) Find the projected population of stray dogs after 5 years.

   (b) When is the first year that the number of stray dogs is greater than 70?

10. A water tank initially contains 1000 litres of water. At time $t = 0$ minutes, a tap is opened and water flows out of the tank. The volume, V litres, that remains in the tank after t minutes is given by the exponential function $V(t) = 1000\,(0.925)^t$.

    (a) Find the value of V after ten minutes.

    (b) Find how long, to the nearest second, it takes for half of the initial amount of water to flow out of the tank.

    (c) The tank is considered 'empty' when only 5% of the water remains. From when the tap is first opened, how many whole minutes have passed before the tank can first be considered empty?

11. The mass m kilograms of a radioactive substance at time t days is given by $m = 5e^{-0.13t}$.

    (a) What is the initial mass?

    (b) How long does it take for the substance to decay to 0.5 kilograms? Give the answer in days, accurate to 3 significant figures.

12. Solve the following logarithmic equations for x.

Give exact answers and be sure to check for extraneous solutions.

(a) $\log_2(3x - 4) = 4$ (b) $\log(x - 4) = 2$

(c) $\ln x = -3$ (d) $\log_{16} x = \dfrac{1}{2}$

(e) $\log \sqrt{x + 2} = 1$ (f) $\ln(x^2) = 16$

(g) $\log_2(x^2 + 8) = \log_2 x + \log_2 6$ (h) $\log_3(x - 8) + \log_3 x = 2$

(i) $\log 7 - \log(4x + 5) + \log(2x - 3) = 0$

(j) $\log_3 x + \log_3(x - 2) = 1$ (k) $\log x^8 = (\log x)^4$

13. Solve each inequality.

(a) $5 \log_4 x + 2 > 0$

(b) $2 \log x^2 - 3 \log x < \log 8x - \log 4x$

(c) $(e^x - 2)(e^x - 3) < 2 e^x$

(d) $3 + \ln x > e^x$

Chapter 4 practice questions

1. Part of the graph of $y = 2 - \log_3(x + 1)$ is shown.

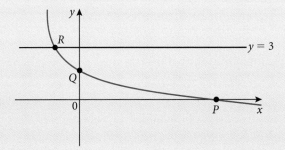

The graph intersects the x-axis at point P, the y-axis at point Q and the line $y = 3$ at point R.

Find:

(a) the x-coordinate of point P

(b) the y-coordinate of point Q

(c) the coordinates of point R.

2. The amount $A(t)$, in grams, of a certain radioactive substance remaining after t years, decays by the formula $A(t) = A_0 e^{-0.0045t}$, where A_0 is the initial amount.

(a) If 5 grams are left after 800 years, how many grams were present initially?

(b) What is the half-life of the substance?

3. Solve for x in the equation $\log_2(5x^2 - x - 2) = 2 + 2\log_2 x$

4. If $\log_2 4\sqrt{2} = x$, $\log_z y = 4$, and $y = 4x^2 - 2x - 6 + z$, find y.

5. Find the exact values of t for which $2e^{3t} - 7e^{2t} + 7e^t = 2$

6. Find the exact solution(s) to the equation $8e^2 - 2e \ln x = (\ln x)^2$

7. Find the exact value of x for each equation.

   (a) $\log_3 x - 4\log_x 3 + 3 = 0$ **(b)** $\log_2(x - 5) + \log_2(x + 2) = 3$

8. Express each as a single logarithm.

   (a) $2\log a + 3\log b - \log c$ **(b)** $3\ln x - \dfrac{1}{2}\ln y + 1$

9. A piece of wood is recovered from an ancient building during an archaeological excavation. The formula $A(t) = A_0 e^{-0.000124t}$ is used to determine the age of the wood, where A_0 is the activity of carbon-14 in any living tree, $A(t)$ is the activity of carbon in the wood being dated, and t is the age of the wood in years. For the ancient piece of wood, it is found that $A(t)$ is 79% of the activity of the carbon in a living tree. How old is the piece of wood, to the nearest 100 years?

10. The graph of the equation $y = \log_3(2x - 3) - 4$ intersects the x-axis at the point $(c, 0)$. Without using your GDC, find the exact value of c.

11. The graph of $y = b^x$, $b > 1$ is shown. On separate coordinate planes, sketch the graphs of:

   (a) $y = b^{-x}$ **(b)** $y = b^{1-x}$

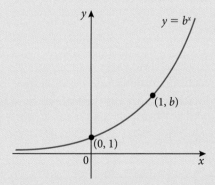

12. A radium isotope decays exponentially and its half-life is 1600 years.

   If A_0 represents the initial activity of radium in a sample and $A(t)$ represents the activity after t years, then $A(t) = A_0 e^{-kt}$.

   (a) Find the value of k accurate to 3 significant figures.

   (b) Find the percentage of the original amount of radium that will be remaining after 4000 years.

13. Solve the equation $e^{-x} - x + 1 = 0$

14. Find the set of values of x for which $|0.1x^2 - 2x + 3| < \log_{10} x$

15. Determine the values of x that satisfy the inequality $\dfrac{x e^x}{x^2 - 1} \geqslant 1$

16. (a) Solve the equation $2(4^x) + 4^{-x} = 3$

    (b) (i) Solve the equation $a^x = e^{2x+1}$, where $a > 0$, giving your answer for x in terms of a.

    (ii) For what value of a does the equation have no solution?

17. The solution of $2^{2x+3} = 2^{x+1} + 3$ can be expressed in the form $a + \log_2 b$ where $a, b \in \mathbb{Z}$. Find the value of a and b.

18. Solve $2(\ln x)^2 = 3\ln x - 1$ for x. Give your answers in exact form.

19. A sum of $100 is invested.

    (a) If the interest is compounded annually at a rate of 5% per year, find the total value V of the investment after 20 years.

    (b) If the interest is compounded monthly at a rate of $\dfrac{5}{12}$% per month, find the minimum number of months for the value of the investment to exceed V.

20. Solve the equation $9\log_5 x = 25\log_x 5$, expressing your answer in the form $5^{\frac{p}{q}}$, where $p, q \in \mathbb{Z}$.

21. Solve $|\ln(x + 3)| = 1$. Give your answers in exact form.

22. Solve the equation $\left| e^{2x} - \dfrac{1}{x + 2} \right| = 2$

23. An experiment is carried out in which the number n of bacteria in a liquid is given by the formula $n = 650\, e^{kt}$, where t is the time in minutes after the beginning of the experiment and k is a constant. The number of bacteria doubles every 20 minutes. Find the exact value of k.

24. The function f is defined for $x > 2$ by $f(x) = \ln x + \ln(x - 2) - \ln(x^2 - 4)$

    (a) Express $f(x)$ in the form $\ln\left(\dfrac{x}{x + a}\right)$

    (b) Find an expression for $f^{-1}(x)$

25. The function f is defined by $f : x \mapsto e^x - 1 - x$

    (a) Use your GDC to find the minimum value of f.

    (b) Prove that $e^x \geqslant 1 + x$ for all real values of x.

Proofs

5

Learning objectives

By the end of this chapter, you should be familiar with...

- the skills essential to read and practise abstract mathematical concepts: logic basics for proofs such as statements, negations of statements, and compound statements
- the concept of proof and several proof procedures: direct proofs, proofs with the contrapositive, contradiction or counter examples
- writing and understanding proof by mathematical induction.

In this chapter we will introduce you to the study of mathematical reasoning. We will give you an idea of what mathematical proofs are and will explain their importance. This will give you some reference to check to see if your proofs are correct.

One of the goals of this chapter is to introduce the language of mathematical argument. This means learning to read and evaluate mathematical statements critically and being able to write mathematical explanations in clear, logically correct language.

5.1 Basic laws and simple proofs

Logic basics

The most vital notion in mathematics is that of proof. In mathematics, it is the process of establishing the validity of a statement, especially by derivation from other statements in accordance with principles of reasoning.

Mathematicians use deductive reasoning to develop and extend a theory by drawing conclusions based on statements accepted as true. Proofs are essential in mathematical reasoning because they demonstrate that the conclusions are true. Generally speaking, a mathematical explanation for a conclusion has no value if the explanation cannot be supported by an acceptable proof. It is not enough to test the statement with a few examples.

A fundamental principle of **deductive** reasoning is the **law of the excluded middle:** every statement is either true or false, never both. The most basic true statements are **axioms**. They are assumptions that set forth the basic relations among the fundamental (undefined) objects of the theory.

Additional true statements are **definitions** that introduce more objects in terms of the fundamental ones.

Example 5.1

Prove that the sum of any two odd integers is an even integer.

Proof

We first need definitions of even and odd:

An integer is even if it can be written in the form $2k$, for some integer k.

An integer is odd if it can be written in the form $2k + 1$, for some integer k.

The proof follows straight from the definition.

Let m and n be any two odd integers. We want to show that $m + n$ is an even integer.

By definition, an integer is odd if it can be written in the form $2k + 1$ (or $2k - 1$) for some integer k. Thus, there exist integers r and s such that $m = 2r + 1$ and $n = 2s + 1$. Add $m + n$:

$$m + n = 2r + 1 + 2s + 1 = 2(r + s + 1)$$

Because $r + s + 1$ is an integer, this shows that $m + n$ has the form $2k$ and thus is an even integer.

There are several noteworthy comments:

- The word 'any' in the statement of the theorem means the proof must work regardless of which odd integers you choose. It is not good enough to simply select, for example, 3 and 7, then write $3 + 7 = 10$. This is an example, or test, of the theorem, not a mathematical proof.

- According to the definition, $m = 2r + 1$ and $n = 2s + 1$ together represent all possible pairs of odd integers since we did not restrict the values that r and s can take.

- The proof makes direct reference to the definition. If you know the definition of every word in the statement of a theorem, you will often discover a proof simply by writing down the definitions.

- The theorem itself did not mention any variables. The proof required a calculation for which these were essential. In this case, the variables r and s come from the definition of oddness. A great mistake is to think that the proof is nothing more than the calculation. This is the easy bit, and it means nothing without the supporting sentences.

The rules of logic are used to distinguish between valid and invalid mathematical arguments and to give precise meaning to mathematical statements. Because a major goal of this chapter is to help you learn how to understand and construct correct mathematical arguments, you will be introduced to logic.

 A **statement** (or **proposition**) is a sentence that asserts a fact that is either true or false, but not both.

Example 5.2

Which of these are statements?

(a) Vienna is the capital of Austria.

(b) 12 is a prime number.

(c) $2 + 3 = 5$

(d) Euler was a great writer.

(e) $2 + 3 = 6$

(f) $n + 1 \leqslant 15$

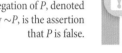

Solution

(a) Yes

(b) Yes

(c) Yes

(d) No (neither true nor false)

(e) Yes

(f) This is not a statement until n is assigned a value. This is an example of an open statement.

The area of logic that deals with statements (propositions) is called **propositional calculus** or **propositional logic**.

Let P be a **statement**. The negation of P, denoted by $\sim P$, is the assertion that P is false.

The statement $\sim P$ is read 'not P.' The truth value of the negation of P, $\sim P$, is the opposite of the truth value of P. $\sim P$ is also denoted as $\neg P$.

We use letters to denote statement variables (or propositional variables), that is, variables that represent statements, just as letters are used to denote numerical variables. If a statement is true, its truth value is 'true', denoted by T, and if it is a false statement, then its truth value is 'false', denoted by F.

For example, if P is '12 is a prime number' then $\sim P$ is '12 is not a prime number' or '12 is a composite number'. P is false, and thus $\sim P$ is true.

An important distinction must be made between a statement and the form of a statement. '12 is not a prime number' is a statement with a truth value of T. We use the form $\sim P$ to represent this statement. However, the form $\sim P$ itself has no truth value unless P is assigned to a specific statement, such as '12 is a prime number' in this case. The form such as $\sim P$ does not have a truth value. Instead, each form has a list of truth values that depend on the values assigned to its components. This list is displayed by presenting all possible combinations for the truth values of its components in a **truth table**. The value of $\sim P$ depends on the two possible values of P; its truth table is:

P	$\sim P$
T	F
F	T

Table 5.1 Truth table for statements P and not P.

This is the truth table for the negation of statement P. This table has a row for each of the two possible truth values of a statement P. Each row shows the truth value of $\sim P$ corresponding to the truth value of P in that row.

There is some similarity between logical statements and set theory. It gives you a visual tool to help increase your understanding of arguments. For example, when we talk about a statement P, it is like saying that an element belongs to a set.

'P: n is an even integer' is like saying '$n \in$ set of even integers', or even simpler, see the Venn diagram in Figure 5.1.

Here, E is the set of even integers and E' represents the complement of this set, where U is the universal set.

If P is true, then $n \in E$.

If P is false, then $\sim P$ is true and $n \in E'$

In other words, n is not an even integer.

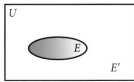

Figure 5.1 Venn diagram for statements P and not P

Example 5.3

Find the negation of each of the following statements and write them in simple English:

(a) *P*: Kevin is looking for a new car.

(b) *Q*: 142 is a perfect square.

(c) *R*: Tim walked his dogs 20 km on Sunday.

Solution

(a) ~*P*: It is not the case that Kevin is looking for a new car. This negation can be more simply expressed as 'Kevin is not looking for a new car.'

(b) ~*Q*: It is not the case that 142 is a perfect square. This negation can be more simply expressed as '142 is not a perfect square.'

(c) ~*R*: It is not the case that Tim walked his dogs 20 km on Sunday. This negation can be more simply expressed as 'Tim did not walk his dogs 20 km on Sunday.'

The negation of a statement can also be considered the result of the operation of the negation operator on a statement. The negation operator constructs a new statement from a single existing statement. The logical operators that are used to form new statements from two or more existing statements are also called connectives.

Let *P* and *Q* be statements.

The **conjunction** of *P* and *Q*, denoted by $P \wedge Q$, is the statement '*P* and *Q*'.

The conjunction $P \wedge Q$ is true when both *P* and *Q* are true, and is false otherwise.

The **disjunction** of *P* and *Q*, denoted by $P \vee Q$, is the statement '*P* or *Q*'.

$P \vee Q$ is true when at least one of *P* or *Q* is true.

P	*Q*	$P \wedge Q$
T	*T*	*T*
T	*F*	*F*
F	*T*	*F*
F	*F*	*F*

Table 5.2 Truth table for the conjunction of *P* and *Q*

P	*Q*	$P \vee Q$
T	*T*	*T*
T	*F*	*T*
F	*T*	*T*
F	*F*	*F*

Table 5.3 Truth table for the disjunction of *P* and *Q*

The conjunction and disjunction ideas are demonstrated in the diagram. For the conjunction to be true, the points have to be in the area occupied by both *P* and *Q*; for the disjunction, it suffices to be in any of them.

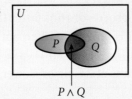

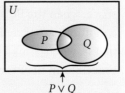

Important: In set theory, *P* and *Q* have some relationship, while in logical statements they do not have to be connected. The only thing that counts is their respective truth value.

Tables 5.2 and 5.3 display the truth tables of $P \wedge Q$ and $P \vee Q$. Each table has a row for each of the four possible combinations of truth values of P and Q. The four rows correspond to the pairs of truth values TT, TF, FT, and FF, where the first truth value in the pair is the truth value of P and the second truth value is the truth value of Q.

In logic, the word 'but' is sometimes used instead of 'and' in a conjunction. For example, the statement 'The car is running, but the back door is open' is another way of saying 'The car is running, and the back door is open.'

'While' and 'although' are also used.

Example 5.4

C is the statement '41 is composite' and M is '45 is a multiple of 5'. C is false, and M is true. Thus '41 is composite and 45 is a multiple of 5', written using connectives as $C \wedge M$ is a false statement, while '41 is composite or 45 is a multiple of 5', written using connectives as $C \vee M$ is true.

Also, the statement 'Either 41 is composite or 45 is not a multiple of 5' written using connectives as $C \vee {\sim}M$ is false.

Example 5.5

Find the conjunction and disjunction of the statements M and N, where M is the statement 'Tim's PC has more than 100 GB free hard disk space', and N is the statement 'The processor in Tim's PC runs faster than 2 GHz'.

Solution

The conjunction of these statements, $M \wedge N$, is the statement 'Tim's PC has more than 100 GB free hard disk space, and the processor in Tim's PC runs faster than 2 GHz.' This conjunction can be expressed more simply as 'Tim's PC has more than 100 GB free hard disk space, and its processor runs faster than 2 GHz.' For this conjunction to be true, both conditions given must be true. It is false when one or both of these conditions are false.

The disjunction of M and N, $M \vee N$, is the statement 'Tim's PC has at least 100 GB free hard disk space, or the processor in Tim's PC runs faster than 2 GHz.'

This statement is true when Tim's PC has at least 100 GB free hard disk space, when the PC's processor runs faster than 2 GHz, and when both conditions are true. It is only false when both of these conditions are false, that is, when Tim's PC has less than 100 GB free hard disk space and the processor in his PC runs at 2 GHz or slower.

Conditional statements

We will now look at several other important ways in which propositions can be combined. Specifically, we will see how statements can lead from one to another.

The statement $P \Rightarrow Q$ (also written as $P \rightarrow Q$) is called conditional because $P \Rightarrow Q$ asserts that Q is true on the condition that P holds. It is also called an **implication**.

The truth table for the conditional statement $P \Rightarrow Q$ is shown in Table 5.4.

P	Q	$P \Rightarrow Q$
T	T	T
T	F	F
F	T	T
F	F	T

Table 5.4 Truth table for $P \Rightarrow Q$

The conditional parallel is shown in the diagram, which demonstrates the idea of 'subset'. This is a very important tool that helps later in using contrapositives in proofs. $P \subseteq Q$ is similar to $P \Rightarrow Q$.

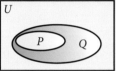

For $P \subseteq Q$ to be true, then when $x \in P$, x must belong to Q.

According to Table 5.4, there is only one way that $P \Rightarrow Q$ can be false: when P is true and Q is false. Thus, this truth table agrees with the way we appreciate promises: the only circumstances where a promise is broken is when the premise is true but the person making the promise fails to make the consequent action true.

Example 5.6

Consider the statement your teacher makes to you: 'If you get 80% or more on your final exam, then you will receive a course grade of 7.'

The only time you feel cheated is when you get more than 80% on the final but do not receive a grade 7, which is the second row in Table 5.4. Some individuals are confused with rows 3 and 4. How can a false statement lead to a true one? This situation is similar to you getting less than 80% but still receiving a grade 7. This may happen because your grade depends on many factors other than the final exam. Row 4 is equivalent to you getting less than 80% and not receiving a grade 7.

Definition of a conditional statement

Let P and Q be statements. The **conditional statement** $P \Rightarrow Q$ is the statement 'if P, then Q.'

The conditional statement $P \Rightarrow Q$ is false when P is true and Q is false, and true otherwise.

In the conditional statement $P \Rightarrow Q$, P is called the hypothesis (or antecedent or premise) and Q is called the conclusion (or consequence).

Conditional statements play such an essential role in mathematical reasoning; thus, a variety of terminology is used to express $P \Rightarrow Q$.

- If P then Q
- P implies Q
- P only if Q.
- Q when P
- Q is necessary for P
- P is sufficient for Q
- Q if P

As discussed earlier, $P \Rightarrow Q$, in set theory is similar to when $P \subseteq Q$. The first diagram below resembles the rows in the truth table corresponding to the cases where the truth value of $P \rightarrow Q$ is True. Points x, y, and z represent situations similar to rows 1, 3, and 4.

x is analogous to the first row: $x \in P$ and $x \in Q$ is true.

In the second diagram below, y is analogous to the third row: $y \notin P$ but still $y \in Q$.

z is similar to the fourth row: $z \notin P$ and $z \notin Q$.

u is comparable to the case where the truth value of the implication is False.

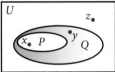

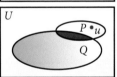

$u \in P$ but $u \notin Q$. That is $P \not\subseteq Q$. Here u is parallel to the second row of the table.

Example 5.7

Suppose you tell your younger brother 'If the concert is sold out, I will take you to the movies.' This promise is broken (the conditional sentence is false) only when the concert was sold out (the antecedent is true) and you did not take your brother to the movies (the consequent is false). This is row 2 of the truth table. In all other situations, the promise is true. If there were tickets left (row 3 and 4 of the table), we don't say the promise was broken, regardless of whether you decided to go to the movies. The promise is also kept in the situation where the concert is sold out and you went to the movies, which is row 1 of the table.

One curious consequence of the truth table for $P \Rightarrow Q$ is that a conditional sentence may be true even when there is no connection between the hypothesis and the conclusion. The reason for this is that the truth value of $P \Rightarrow Q$ depends only on the truth value of components P and Q, not on their interpretation. For this reason, all of the following are true:

- If $\cos \pi = 0$, then 12 is a prime number (row 4 of Table 5.4)
- $2\pi > 5 \Rightarrow 3 + 4 = 7$ (row 1)
- $2\pi < 5 \Rightarrow 3 + 4 = 7$ (row 3)

On the other hand, the following statement is false:

- $2\pi > 5 \Rightarrow 3 + 4 = 12$ (row 2)

Our truth table portrayal for $P \Rightarrow Q$ captures the same meaning for 'If ..., then...' type statements that you have always used in mathematics. For example, if we think of x as some fixed real number, we all know that

'If $x > 10$, then $x > 5$'

is a true statement, no matter which number x we have in mind.

Let's examine why we say this sentence is true for some specific values of x, where the hypothesis P is '$x > 10$' and the conclusion Q is '$x > 5$.'

When $x = 20$, both P and Q are true, as in row 1 of the truth table. If $x = 8$, this corresponds to row 3 of the table, and for $x = 2$ we have the situation in row 4. There is no case corresponding to row 2 because $P \Rightarrow Q$ is true. Note that when we say 'If P, then Q' is true, we don't claim that either P or Q is true. What we do say is that no matter what number we think of, *if* it's larger than 10, then it's also larger than 5.

A related concept is that of **equivalence** $P \Leftrightarrow Q$. P and Q are equivalent if $P \Rightarrow Q$ and $Q \Rightarrow P$. The truth table is given in the table blow.

P	Q	$P \Leftrightarrow Q$
T	T	T
T	F	F
F	T	F
F	F	T

Other expressions that translate into $P \Leftrightarrow Q$:
- P if and only if Q (sometimes written as P iff Q)
- P is necessary and sufficient to Q

Two important implications related to $P \Rightarrow Q$, its **converse**, and its **contrapositive**, will play a very important role in some types of proofs that will appear later in the chapter.

Example 5.8

Write both the converse and the contrapositive for each statement below. In each case, determine whether the given statement is true or false, state whether the converse is true or false, and state whether the contrapositive is true or false.

(a) If $5 < \sqrt{26}$, then 6 is a composite number.

(b) If $5 \geq \sqrt{26}$, then 6 is a composite number.

(c) If $5 < \sqrt{26}$, then 6 is a prime number.

(d) If $5 \geq \sqrt{26}$, then 6 is a prime number.

Solution

(a) Converse: If 6 is a composite number, then $5 < \sqrt{26}$
   Contrapositive: If 6 is a prime number, then $5 \geq \sqrt{26}$
   Statement is True. Row 1 of the truth table.
   Converse is True. Row 1 of the truth table.
   Contrapositive is True. Row 4 of the truth table.

(b) Converse: If 6 is a composite number, then $5 \geq \sqrt{26}$
   Contrapositive: If 6 is a prime number, then $5 < \sqrt{26}$
   Statement is True. Row 3 of the truth table.
   Converse is False. Row 2 of the truth table.
   Contrapositive is True. Row 3 of the truth table.

(c) Converse: If 6 is a prime number, then $5 < \sqrt{26}$
   Contrapositive: If 6 is a composite number, then $5 \geq \sqrt{26}$
   Statement is False. Row 2 of the truth table.
   Converse is True. Row 3 of the truth table.
   Contrapositive is False. Row 2 of the truth table.

(d) Converse: If 6 is a prime number, then $5 \geq \sqrt{26}$
   Contrapositive: If 6 is a composite number, then $5 < \sqrt{26}$
   Statement is True. Row 4 of the truth table.
   Converse is True. Row 4 of the truth table.
   Contrapositive is True. Row 1 of the truth table.

Example 5.8 suggests that the truth values of a conditional sentence and its contrapositive are the same, but there seems to be little connection between the truth values of a sentence and its converse. We describe the relationships in the following theorem.

For statements P and Q:

1 $P \Rightarrow Q$ is equivalent to its contrapositive $\sim Q \Rightarrow \sim P$
2 $P \Rightarrow Q$ is not equivalent to its converse $Q \Rightarrow P$

We can use truth tables to examine these statements.

P	Q	$P \Rightarrow Q$	$\sim P$	$\sim Q$	$Q \Rightarrow P$	$\sim Q \Rightarrow \sim P$
T	T	T	F	F	T	T
T	F	F	F	T	T	F
F	T	T	T	F	F	T
F	F	T	T	T	T	T

1 $P \Rightarrow Q$ is equivalent to its contrapositive $\sim Q \Rightarrow \sim P$ because column 3 in the truth table is identical to column 7.
2 $P \Rightarrow Q$ is not equivalent to its converse $Q \Rightarrow P$ because column 3 in the truth table differs from column 6 in rows 2 and 3.

The Venn diagram uses sets to imitate the contrapositive case.

When $x \notin Q$ (i.e. $\sim Q$) implies that x cannot be in P (i.e. $\sim P$), could only mean that $P \subset Q$, (i.e. $P \Rightarrow Q$)

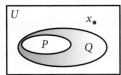

Valid arguments

Later in this chapter we will study proofs. Proofs in mathematics are **valid arguments** that establish the truth of mathematical statements. By an **argument**, we mean a **sequence of statements** that end with a conclusion. By **valid**, we mean that the **conclusion,** or final statement of the argument, must follow from the truth of the preceding statements, or premises, of the argument. That is, an argument is valid if and only if it is impossible for all the premises to be true and the conclusion to be false. To deduce new statements from statements we already have, we use rules of inference, which are templates for constructing valid arguments. Rules of inference are our basic tools for establishing the truth of statements.

Consider the following argument:

'If you buy a ticket, you can fly to Frankfurt tomorrow.'

'You buy a ticket.'

Therefore, 'You can fly to Frankfurt tomorrow.'

We would like to determine if this is a valid argument. That is, we would like to determine whether the conclusion 'You can fly to Frankfurt tomorrow' must be true when the premises 'If you buy a ticket' and 'If you buy a ticket, you can fly to Frankfurt tomorrow' are both true.

Before we discuss the validity of this particular argument, we will look at its form. Use P to represent 'You buy a ticket' and Q to represent 'You can fly to Frankfurt tomorrow.'

Then, the argument has the form

$$P \Rightarrow Q$$
$$\frac{P}{\therefore Q}$$

We can look at the truth table of the implication $P \Rightarrow Q$

When P is true, and $P \Rightarrow Q$ is true, there is only one case left for Q, which is true.

Another way of looking at it is to look for the truth table of $(P \wedge (P \Rightarrow Q)) \Rightarrow Q$

P	Q	$P \Rightarrow Q$	$P \wedge (P \Rightarrow Q)$	$(P \wedge (P \Rightarrow Q)) \Rightarrow Q$
T	T	T	T	T
T	F	F	F	T
F	T	T	F	T
F	F	T	F	T

Table 5.6 Truth table for $(P \wedge (P \Rightarrow Q)) \Rightarrow Q$

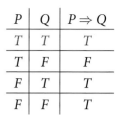

P	Q	$P \Rightarrow Q$
T	T	T
T	F	F
F	T	T
F	F	T

Table 5.5 Truth table for $P \Rightarrow Q$

Notice that this is a **tautology**. That is, it is always true. In particular, when P and $P \Rightarrow Q$ are both true, then Q must be true as is clear from the first row of the table.

What happens when we replace P and Q in this argument form by statements where either P or $P \Rightarrow Q$ is true? For example, suppose that P represents 'You have access to my grade book' and Q represents 'You can change your term grade' and that P is true, but $P \Rightarrow Q$ is false.

The argument we obtain by substituting these values of P and Q into the argument form is

'If you have access to my grade book, then you can change your term grade.'

'You have access to my grade book.'

∴ 'You can change your term grade.'

The argument we obtained is a valid argument, but because one of the premises, namely the first premise, is false, we cannot conclude that the conclusion is true. (Most likely, this conclusion is false.)

In our discussion, to analyse an argument, we replaced propositions by propositional variables. This changed an argument to an argument form. The validity of an argument follows from the validity of the form of the argument. We summarise the terminology used to discuss the validity of arguments with our definition of the key notions.

$\sim(P \Rightarrow Q)$ is logically equivalent to $P \wedge \sim Q$. That is, the negation of an implication is a conjunction and not an implication itself.

For example, if x is an integer, then the negation of the sentence 'If x is odd then x^2 is odd' is 'x is odd but x^2 is even.'

This can be seen from the truth table.

P	Q	$P \Rightarrow Q$	$\sim(P \Rightarrow Q)$	$\sim Q$	$P \wedge \sim Q$
T	T	T	F	F	F
T	F	F	T	T	T
F	T	T	F	F	F
F	F	T	F	T	F

An **argument** is a sequence of statements. All but the final statement in the argument are called premises and the final statement is called the conclusion. An argument is valid if the truth of all its premises implies that the conclusion is true.

An argument form in logic is a sequence of compound statements involving propositional variables. An argument form is valid no matter which particular propositions are substituted for the propositional variables in its premises; the conclusion is true if the premises are all true.

The tautology $(P \wedge (P \Rightarrow Q)) \Rightarrow Q$ is the basis of the rule of inference called **modus ponens**, or the **law of detachment**. (Modus ponens is Latin for 'mode that affirms'.) This tautology leads to the following valid argument form, which we have already seen in our initial discussion about arguments:

$$P \Rightarrow Q$$
$$\underline{P}$$
$$\therefore Q$$

Example 5.9

Suppose that the conditional statement 'If you score more than 80% on your final exam, you will receive a grade 7 for the whole term' and its hypothesis, 'You scored more than 80% on your final exam', are true. Then, by modus ponens, it follows that the conclusion of the conditional statement, 'You will receive a grade 7 for the whole term', is true.

A valid argument can still lead to an incorrect conclusion if one or more of its premises is false. For example:

Determine whether or not the argument given here is valid and determine whether its conclusion must be true.

'If $\sqrt{5} = 2.5$, then $5 = 5.0625$. We know that $\sqrt{5} = 2.5$. Consequently, $(\sqrt{5})^2 = 5 = 2.5^2 = 5.0625$.'

Solution Let P be the statement '$\sqrt{5} = 2.5$' and Q the statement '$5 = 5.0625$.'

The premises of the argument are $P \Rightarrow Q$ and P, and its conclusion is Q. This argument is valid because it is constructed by using modus ponens. However, one of its premises, $\sqrt{5} = 2.5$, is false. Consequently, we cannot conclude that the conclusion is true. Furthermore, note that the conclusion of this argument is false, because $5 \neq 5.0625$.

Table 5.7 gives you a list of possible valid arguments.

Valid arguments	
Argument	**Tautology**
$P \Rightarrow Q$ $\underline{P}$ $\therefore Q$	$(P \wedge (P \Rightarrow Q)) \Rightarrow Q$
$P \Rightarrow Q$ $\underline{\sim Q}$ $\therefore \sim P$	$(\sim Q \wedge (P \Rightarrow Q)) \Rightarrow \sim P$
$P \Rightarrow Q$ $\underline{Q \Rightarrow R}$ $\therefore P \Rightarrow R$	$((P \Rightarrow Q) \wedge (Q \Rightarrow R)) \Rightarrow (P \Rightarrow R)$
$P \vee Q$ $\underline{\sim P}$ $\therefore Q$	$(\sim P \wedge (P \vee Q)) \Rightarrow Q$
$P \vee Q$ $\underline{\sim P \vee R}$ $\therefore Q \vee R$	$((P \vee Q) \wedge (\sim P \vee R)) \Rightarrow (Q \vee R)$

Table 5.7 Some valid arguments

In the remaining sections of this chapter, we will introduce the required logic principle when it is needed.

1. Which of the following is a statement?

   (a) $4^{500} > 5^{400}$

   (b) It is difficult to find solutions to the equation $2x^3 - 3x^2 + 5x - 8 = 0$

   (c) $7238 + 8327$

   (d) $2874 = 67 \times 73 - 2017$

   (e) There is a prime number larger than 100^{100}

2. In each of the following, find value(s) of a for which the statement is true.

   (a) $3 < 4$ and $a = 7$ (b) $4 < 3$ and $a = 7$

   (c) $4 < 3$ or $a = 7$ (d) If $a = 7$ then $3 < 4$

   (e) If $a = 7$ then $3 > 4$ (f) If $3 > 4$ then $a = 7$

3. Identify the antecedent and the consequent for each of the following conditional sentences. Assume that a, b, and f represent some fixed sequence, integer, or function, respectively.

   (a) If quadrilaterals have three sides, then triangles have four sides.

   (b) If the moon is made of butter, then $\sqrt{3}$ is a rational number.

   (c) b divides 5 only if b divides 30.

   (d) The differentiability of f is sufficient for f to be continuous.

   (e) A sequence a is bounded whenever a is convergent.

   (f) A function f is bounded if f is integrable.

   (g) $3 + 2 = 5$ is necessary for $3 + 3 = 6$.

4. Write the converse and contrapositive of each statement in question 3.

5. What can you say about the truth value of Q when:

   (a) P is false and $P \Rightarrow Q$ is true?

   (b) P is true and $P \Rightarrow Q$ is true?

   (c) P is true and $P \Rightarrow Q$ is false?

6. Which of the following conditional sentences are true?

   (a) If a pentagon has five sides, then Canada is in Asia.

   (b) If a pentagon has six sides, then Canada is in Asia.

   (c) If $12 + 7 = 20$, then $7 + 7 = 14$.

   (d) If one interior angle of a right-angled triangle is 100°, then the other interior angle is 80°.

   (e) $2 + 1 = 3$ is sufficient to $3 > 7$.

5.2 Direct proofs

There are four typical methods for proving $P \Rightarrow Q$. In practice, extensive proofs will use quite a few of them.

- **Direct**: Assume P and logically deduce Q.
- **Contrapositive**: Assume $\sim Q$ and deduce $\sim P$. This is enough since the contrapositive $\sim Q \Rightarrow \sim P$ is logically equivalent to $P \Rightarrow Q$.
- **Contradiction**: Assume that P and $\sim Q$ are true and deduce a contradiction. Since $P \wedge \sim Q$ implies a contradiction, this shows that $P \wedge \sim Q$ must be false. Because $P \wedge \sim Q$ is equivalent to $\sim (P \Rightarrow Q)$, this is enough to conclude that $P \Rightarrow Q$ is true.
- **Induction**: This has a completely different spirit: we will consider it in Section 5.4.

We cannot define all terms nor prove all statements from previous ones. We begin with an initial set of statements, called **axioms** (or **postulates**), that are assumed to be true. We then derive theorems that are true in any situation where the axioms are true. The Pythagorean theorem, for example, is a theorem whose proof is based on the five axioms of Euclidean geometry. In a situation where the Euclidean axioms are not all true (which can happen), the Pythagorean theorem may not be true.

There must also be an initial set of **undefined terms** – concepts fundamental to the context of study. In geometry, the concept of a point is an undefined term. In this book, real numbers are not formally defined. While a precise definition of a real number could be given, doing so would take us far from our intended goals.

From the axioms and undefined terms, you can introduce new concepts (new definitions), and, finally, new theorems can be proved. The structure of a proof for a particular theorem depends greatly on the logical form of the theorem.

A direct proof of a conditional statement $P \Rightarrow Q$ is constructed when the first step is the assumption that P is true; subsequent steps are constructed using rules of inference, with the last step showing that Q must also be true. A direct proof shows that a conditional statement $P \Rightarrow Q$ is true by showing that if P is true, then Q must also be true, so that the combination P is true and Q is false never occurs. In a direct proof, we assume that P is true and use axioms, definitions, and previously proven theorems, together with rules of inference, to show that Q must also be true. You will find that direct proofs of many results are quite straightforward, with a fairly obvious sequence of steps leading from the hypothesis to the conclusion.

Recall the truth table for $P \Rightarrow Q$. Suppose that the hypothesis P is true and that $P \Rightarrow Q$ is true: that is, $P \Rightarrow Q$ is a theorem. We must be in the first row of the truth table, and so the conclusion Q is also true. In a direct proof we start by assuming the hypothesis (P) is true and make a logical argument ($P \Rightarrow Q$) which asserts that the conclusion (Q) is true. As such, it is often convenient to rewrite the statement of a theorem as an implication of the form $P \Rightarrow Q$. Example 5.10 is an example of a direct proof.

Example 5.10

Show that the product of two odd integers is an odd integer.

Proof

We can translate this into statements and connectives:

P: m and n are odd integers. This is the hypothesis.

Q: mn is an odd integer. This is what we need to show, the conclusion.

Let m and n be odd integers. We need to show that mn is an odd integer.

By definition, an integer is odd if it can be written in the form $2k + 1$ for some integer k. Hence there are integers r and s such that $m = 2r + 1$ and $n = 2s + 1$.

Now consider the product $mn = (2r + 1)(2s + 1)$

$$= 4rs + 2r + 2s + 1$$

$$= 2(2rs + r + s) + 1$$

Since $2rs + r + s$ is an integer, then mn is of the form $2k + 1$ and thus is an odd integer as requested.

Showing that $P \Rightarrow Q$ is true requires an argument. This is the proof.

An erroneous proof may lead to invalid results. In Example 5.10, a common mistake is to let
$m = 2r + 1$ and
$n = 2r + 1$, then
$mn = (2r + 1)(2r + 1)$
$= 2(2r^2 + 2r) + 1$
is odd. However, this takes care of only one situation, namely when $m = n$ and does not address the general case when $m \neq n$.

Example 5.11

Let $m, n \in \mathbb{Z}$. Prove that m and n are both odd if and only if (iff) the product mn is odd.

Proof

There are two statements here:

($\Rightarrow$) If m and n are both odd, then the product mn is odd.

($\Leftarrow$) If the product mn is odd, then both integers m and n are odd.

Most often when there are two directions we have to prove them separately.

($\Rightarrow$) This was proved in Example 5.10

($\Leftarrow$) If the product mn is odd, then there are two cases to consider:

i.   at least one of m or n is even

ii. m and n are both odd.

In the first case, say m is even. That is, $m = 2k$ for some integer k and thus $mn = 2kn$ is even, which cannot be true. Thus, we are only left with both numbers being odd. This kind of proof may be called proof by exhaustion. An alternative approach will be outlined in section 5.3.

Example 5.12

Prove the following statement: If $m, n \in \mathbb{Z}$, and m is divisible by n, then m is divisible by kn, where k is any integer.

Proof

This is a case where we can use a counter-example to prove that it is false.

Examples alone are not proofs of the truth of a statement, but counter-examples are enough to disprove a false statement.

For example, 18 is divisible by 3, and it is also divisible by 6, which is a multiple of 3. However, 18 is not divisible by 15, which is also a multiple of 3.

Example 5.13

Prove the following statement: If $m, n \in \mathbb{Z}$, and m is divisible by n, then m^2 is divisible by n^2.

Proof

If m is divisible by n, then $m = kn$ for some integer k. Then $m^2 = k^2 n^2$ and hence m^2 is divisible by n^2.

Example 5.14

Prove the following statement: If $n \in \mathbb{Z}$ and n is not a multiple of 5, then n^2 has only remainders 1 or 4 upon dividing by 5.

Proof

We use direct proof using proof by cases.

There are four cases: n remainders 1, 2, 3, or 4 upon dividing by 5. (n is not a multiple of 5.)

- If n has remainder 1, then $n = 5k + 1$ for some $k \in \mathbb{Z}$ and so $n^2 = 5(5k^2 + 2k) + 1$ has remainder 1.

- If n has remainder 2, then $n = 5k + 2$ for some $k \in \mathbb{Z}$ and so $n^2 = 5(5k^2 + 4k) + 4$ has remainder 4.

- If n has remainder 3, then $n = 5k + 3$ for some $k \in \mathbb{Z}$ and so $n^2 = 5(5k^2 + 6k + 5) + 4$ has remainder 4.

- If n has remainder 4, then $n = 5k + 4$ for some $k \in \mathbb{Z}$ and so $n^2 = 5(5k^2 + 8k + 3) + 1$ has remainder 1.

Therefore, n^2 has only remainders 1 or 4 upon dividing by 5.

5.3 Indirect proofs

Direct proofs lead from the premises of a theorem to the conclusion. They begin with the premises, continue with a sequence of deductions, and end with the conclusion. However, we will see that attempts at direct proofs often reach dead ends. We need other methods of proving theorems of the form for all x $(P(x) \Rightarrow Q(x))$. Proofs of theorems of this type are not direct proofs, that is, they do not start with the premises and end with the conclusion. They are called **indirect** proofs.

Proof by contradiction

One of the forms of indirect proof is the proof by **contradiction**. In a proof by contradiction, to prove that some statement P is true, we instead assume that P is false, then proceed to derive an impossible statement (a contradiction). This means that P cannot be false, and therefore P must be true.

Suppose we want to prove that a statement P is true. Furthermore, suppose that we can find a contradiction Q such that $\sim P \Rightarrow Q$ is true. Because Q is false, but $\sim P \Rightarrow Q$ is true, we can conclude that $\sim P$ is false, which means that P is true. This is row 4 of Table 5.4 as applied to $\sim P \Rightarrow Q$.

$\sim P$	Q	$\sim P \Rightarrow Q$
$\vdots$	$\vdots$	$\vdots$
F	F	T

We will show this by a few examples.

> De Morgan's laws are very important in negating conjunctions and disjunctions. We state them here without proof. You can use truth tables to show their validity.
>
> 1. $\sim(P \wedge Q) \Leftrightarrow \sim P \vee \sim Q$
> 2. $\sim(P \vee Q) \Leftrightarrow \sim P \wedge \sim Q$

Example 5.15

Consider the real numbers m and n. Prove that if m is rational and n is irrational, then $m + n$ is irrational.

Proof

We prove this by contradiction.

Assume that there is a rational number m and an irrational number n where the number $m + n$ is rational.

Since $m + n$ is rational, there exist two integers p and q where $q \neq 0$ such that $m + n = \dfrac{p}{q}$.

Similarly, since m is rational, for some integers r and s where $s \neq 0$, then $m = \dfrac{r}{s}$.

By rearranging the equations above and simplifying gives us the following:

$$m + n = \frac{p}{q} \Rightarrow n = \frac{p}{q} - m = \frac{p}{q} - \frac{r}{s} = \frac{ps - rq}{qs}$$

From this, we see that since $(ps - rq)$ and qs are integers with qs being nonzero as neither q nor s is zero, we can write n as the ratio of two integers, $(ps - rq)$ and qs. Thus, n must be rational. This contradicts our earlier assumption that n is irrational.

We have reached a contradiction, so the assumption must have been wrong. Therefore, if m is rational and n is irrational, then $m + n$ must be irrational.

When writing proofs by contradiction, we organise the proof as follows:

- Begin by saying that you're going to write a proof by contradiction. For example, you could write 'By contradiction, assume that ….' Then, clearly write out the negation of the statement you're trying to prove.

- Don't skip this step! It's important to write out the negation of what you want to prove. This tells the reader what assumptions you're making and it forces you to write out the contradiction. It therefore makes the proof easier to read and reduces the chances that you accidentally take the negation of the statement incorrectly.

- Starting with your assumption , continue to achieve something impossible, such as '1 = 0', or a number is 'both even and odd', or that a number is 'both rational and irrational'.

- State that you've reached a contradiction and, if it's not obvious, explain why it's a logical contradiction. This explains to the reader why your assumption couldn't possibly be right.

- Conclude the proof. For example, 'Therefore, our assumption must have been wrong, so […].' You can put whatever you'd like here as long as it explains why the contradiction you arrived at actually shows that the original assumption was incorrect.

Remember that $\sim(P \Rightarrow Q)$ is logically equivalent to $P \wedge \sim Q$

Example 5.16

Prove that $\sqrt{2}$ is irrational.

Proof

We will use contradiction for this proof.

Assume the statement to be false, that is, assume $\sqrt{2}$ is rational. Then, by definition of rational numbers, $\sqrt{2}$ can be written as a reduced fraction $\frac{m}{n}$ where the two integers m and n, with $n \neq 0$, have no common divisor except 1.

Now, $\sqrt{2} = \frac{m}{n} \Rightarrow (\sqrt{2})^2 = \frac{m^2}{n^2} \Rightarrow m^2 = 2n^2$

This indicates that m^2 is even. If m^2 is even, m must also be even. (This is assumed true here, but proved somewhere else.) Thus, $m = 2k$ for some integer k. This leads us to $m^2 = 4k^2$, and hence $2n^2 = m^2 = 4k^2 \Rightarrow n^2 = 2k^2$, thus n^2 is even, which in turn leads to n being even.

Thus, both m and n are even, and hence they have another common factor, 2, which contradicts the assumption that m and n have no common divisor other than 1.

Therefore, assuming $\sqrt{2}$ to be rational leads us to a contradiction and so $\sqrt{2}$ cannot be rational.

Example 5.17

Prove that for $n \in \mathbb{Z}$, if $3n + 2$ is odd, then n is odd.

Proof

We will use contradiction in this proof. Assume that $3n + 2$ is odd, and n not odd (n is even).

If n is even, then $n = 2k$ for some integer k.

Thus $3n + 2 = 3(2k) + 2 = 2(3k + 1)$. This means that $3n + 2$ is even. However, we know that $3n + 2$ is odd, and that is a contradiction. Therefore, the assumption that n is not odd is false and hence n must be odd.

Proof by contrapositive

Proof by contradiction is not the only way to perform an indirect proof; there is another technique, the proof by contrapositive.

A useful type of indirect proof is known as **proof by contraposition**. Proofs by contraposition make use of the fact that the conditional statement $P \Rightarrow Q$ is equivalent to its contrapositive, $\sim Q \Rightarrow \sim P$. This means that the conditional statement $P \Rightarrow Q$ can be proved by showing that its contrapositive, $\sim Q \Rightarrow \sim P$, is true. In a proof by contraposition of $P \Rightarrow Q$, we take $\sim Q$ as a premise, and using axioms, definitions, and previously proven theorems, together with rules of inference, we show that $\sim P$ must follow. The following examples show that proof by contraposition can succeed when we cannot easily find a direct proof.

In reading mathematical statements, we often see the following two quantifiers:

Universal quantifier, $\forall$ is read as 'for all', 'any', or 'for every' etc.

Existential quantifier, $\exists$ is read 'there exists' or 'there exists at least one'.

For example, for all real numbers n, $n^2 \geqslant n$ can be written as: $\forall n \in \mathbb{R}, n^2 \geqslant n$. Or, there is an integer smaller than π that can be written as: $\exists n \in \mathbb{Z}$ such that $n < \pi$.

To prove a statement of the form 'If A, then B,' do the following:

Form the contrapositive. In particular, negate A and B.

Prove directly that $\sim B$ implies $\sim A$.

Example 5.18

Prove that for any real number x, if x^2 is irrational, then x is irrational.

Proof

We will use contrapositive proof; we prove that if x is rational, then x^2 is rational.

Suppose that x is a rational number. That means that we can write $x = \dfrac{p}{q}$ where p and q are integers with $q \neq 0$. So, $x^2 = \dfrac{p^2}{q^2}$ where p^2 and q^2 are integers with $q^2 \neq 0$. Thus, we can write x^2 as the ratio of two integers. Consequently, x^2 is rational. Therefore, since the contrapositive of the statement 'if x^2 is irrational, then x is irrational' is true, the statement itself must be true.

Note that this proof can also be done using contradiction:

Assume x^2 is irrational and x rational. This means that we can write $x = \dfrac{p}{q}$ where p and q are integers with $q \neq 0$. Thus $x^2 = \dfrac{p^2}{q^2}$ where p^2 and q^2 are integers with $q^2 \neq 0$, which means that x^2 is rational. This is a contradiction and hence the assumption must be false and therefore if x^2 is irrational, x must be also irrational.

Remember that
$\sim(P \to Q)$ is $P \wedge \sim Q$

When writing a proof by contrapositive, organise the proof as follows:

- Start off by saying that you're going to prove the contrapositive of the statement in question. For example, you could say something like 'We will prove the contrapositive of this statement, namely, that ...' or 'By contraposition, we will instead prove that ...'

- Don't skip this step! It's important for several reasons. First, it tells the reader that in the proof you're not going to prove the original statement, but in its place, you're going to prove the contrapositive. Second, it forces you to write out the contrapositive of the statement that you're trying to prove, reducing the likelihood that you accidentally take the contrapositive incorrectly.

- With any proof method of your choice, prove the contrapositive of the statement. Often, you'll prove the contrapositive of the statement using a direct proof. This means that if you want to prove the statement 'If P is true, then Q is true,' you'll start off by assuming that Q is false and will prove that P is false.

Example 5.19

Prove the statement from example 5.11: m and n are both odd if and only if the product mn is odd.

Proof

There are two statements here:

($\Rightarrow$) Has been proven earlier.

($\Leftarrow$) If the product mn is odd, then both integers m and n are odd.

The negation of this is: the integers m and n are not both odd (i.e. at least one of m or n is even,) then product mn is even.

Without loss of generality, say m is even. That is $m = 2k$ for some integer k and thus $mn = 2kn$, which is even. Therefore, the contrapositive is true and hence the original statement 'If the product mn is odd, then both integers m and n are odd' must also be true.

Example 5.20

Prove that for any two integers m and n, if $m + n$ is odd, then exactly one of m or n is odd.

Proof

The statement of the problem is an implication of the form $P \to Q$, where:

P: The sum $m + n$ of integers m and n is odd.

Q: Exactly one of m or n is odd.

A direct proof would require that we assume P is true and logically deduce the truth of Q. The problem is that it may be hard to work with these statements, especially Q. The negation of Q is, however, easier:

$\sim Q$: m and n are both even or both odd.

$\sim P$: The sum $m + n$ of integers m and n is even.

Since $P \Rightarrow Q$ is logically equivalent to the simpler-seeming contrapositive $\sim Q \Rightarrow \sim P$, we choose to prove the latter. This is, after all, equivalent to proving the original implication.

There are two cases: m and n are both even, or both odd.

Case 1: Let $m = 2r$ and $n = 2s$ be even. Then $m + n = 2(r + n)$ is even.

Case 2: Let $m = 2r + 1$ and $n = 2s + 1$ be odd. Then $m + n = 2(r + n + 1)$ is even.

This proves the contrapositive and hence the original statement is true.

Example 5.21

Prove that for all integers a, b, and c, if $c \nmid a\, b$ then $c \nmid a$ and $c \nmid b$.

Proof

We need to find the contrapositive of the given statement.

First, negate '$c \nmid a$ and $c \nmid b$.' The negation is '$c \mid a$ or $c \mid b$.' (The 'and' becomes an 'or' because of De Morgan's law.) The initial hypothesis' negation is simply '$c \mid a\, b$.' Thus we are trying to prove '$c \mid a$ or $c \mid b \rightarrow c \mid ab$.'

Suppose that c divides a. Then $a = nc$ for some $n \in \mathbb{Z}$, and $ab = ncb = c(nb)$, so $c \mid ab$. Similarly, if $c \mid b$, then $b = mc$ for some $m \in \mathbb{Z}$, and $ab = amc = c(am)$, and so $c \mid ab$. Therefore, we have proven the result by contraposition.

In Example 5.21, a direct proof would be awkward (and obviously quite difficult), so contraposition is the best way to prove it.

Example 5.22

Prove that for $x \in \mathbb{Z}$, if $x^2 + 6x - 5$ is even, then x is odd.

Proof

Direct: If $x^2 + 6x - 5$ is even, then there exists an integer k such that $x^2 + 6x - 5 = 2k$.

Thus $x^2 + 6x = 2k + 5$ which makes $x^2 + 6x = x(x + 6)$ odd. Now, if x is even, $x(x + 6)$ is even, which cannot be true, thus x must be odd. Alternatively, we can say, if x is odd, so is $x + 6$ and their product is odd.

Contraposition: Suppose that x is even. Then we want to show that $x^2 + 6x - 5$ is odd. Since x is even we can write: $x = 2k$ for some $k \in \mathbb{Z}$, and so $x^2 + 6x - 5 = 4k^2 + 12k - 5 = 2(2k^2 + 6k - 2) - 1$

This is of the form $2n - 1$ and hence an odd number.

Example 5.23

Prove that for a positive real number a, if a is irrational then $\sqrt{a}$ is irrational.

Proof

Stated differently, we need to prove that $a \notin \mathbb{Q} \Rightarrow \sqrt{a} \notin \mathbb{Q}$.

We will use the contrapositive and attempt to prove $\sqrt{a} \in \mathbb{Q} \Rightarrow a \in \mathbb{Q}$.

If $\sqrt{a} \in \mathbb{Q}$, then there are two relatively prime integers m and n, with $n \neq 0$, such that $\sqrt{a} = \dfrac{m}{n}$ by the definition of rational numbers.

Thus $a = (\sqrt{a})^2 = \dfrac{m^2}{n^2}$, and since m and n, with $n \neq 0$, are integers, then m^2 and n^2, with $n^2 \neq 0$, are also integers.

So, a can be written as the quotient of two integers and hence it is a rational number. By proving the contrapositive, the statement itself is also true.

Exercise 5.2 & 5.3

1. Prove or disprove the given conjecture.

   (a) There is an even integer that can be expressed as the sum of three even integers.

   (b) Every even integer can be expressed as the sum of three even integers.

   (c) There is an odd integer that can be expressed as the sum of two odd integers.

   (d) Every odd integer can be expressed as the sum of three odd integers.

   (e) The sum of any three consecutive integers is divisible by 3.

   (f) The sum of any four consecutive integers is divisible by 4.

   (g) The product of any three consecutive integers is divisible by 6.

   (h) $a, b \in \mathbb{Z}$. If $a \mid b$ and $b \mid a$, then $a = b$.

2. The English mathematician Augustus De Morgan died in 1871. He once made the claim: 'I turned x years of age in the year x^2.'

   (a) Find the year he was born.

   (b) You have a friend who satisfies the same claim. How old would your friend have been in 2018? Justify that you are correct.

3. Prove that for an integer n, if $n \geqslant 2$, then $n! + 2$ is even.

4. Let $n \in \mathbb{Z}$.

   (a) Prove that $5n + 3$ is even if $7n - 2$ is odd.

   (b) Can you conclude anything about $7n - 2$ if $5n + 3$ is odd?

5. Let $m, n \in \mathbb{Z}$. Prove that for $m^2 + n^2$ to be even, it is necessary for m and n to be both even or both odd. (We also say, they have the same parity.)

6. Prove that if x and y are positive real numbers, then $\sqrt{x + y} \neq \sqrt{x} + \sqrt{y}$

7. Prove that if x and y are real numbers, then $xy = 0$ iff $x = 0$ or $y = 0$

8. (a) Write the statement 'There is an odd number that is a perfect square' using quantifiers.

   (b) Write its negation.

   (c) Is the original statement true? Justify.

9. **(a)** Write the statement 'Every positive integer is divisible by 13' using quantifiers, then write its negation.

   (b) Is the original statement true? Justify.

10. Given $a, b \in \mathbb{Z}$, prove that if $a^2(b^2 - 2b)$ is odd, then both a and b are odd.

11. Given $m, n \in \mathbb{Z}$, prove that if $25 \nmid mn$, then $5 \nmid m$ or $5 \nmid n$.

12. If $m, n \in \mathbb{Z}$ and $m + n$ is even, prove that $m^2 + n^2$ is even.

13. Prove that an integer n is even if and only if $n^2 + 2n + 9$ is odd.

14. If a is an odd integer, prove that $a^2 + 3a + 5$ is odd.

5.4 Mathematical induction

Domino effect

Apart from playing games of strategy, another familiar activity using dominoes is to place them on edge in lines, then topple the first domino, which falls on and topples the second, which topples the third, and so on, resulting in all of the dominoes falling. Arrangements of millions of dominoes have been made, which have taken many minutes to fall.

The Netherlands has hosted an annual domino toppling competition, called *Domino Day*, since 1986. The record, achieved in 2006, is 4 079 381 dominoes.

Similar phenomena of chains of small events each causing similar events which lead to an eventual grand result, by analogy, are called domino effects. The phenomenon also has some theoretical bearing to familiar applications like the amplifier, digital signal processing, and information processing.

Induction

In mathematics, we have a parallel in **mathematical induction**, which is a method for proving a statement that is maintained about every natural number. For example:

$$1 + 2 + 3 + \cdots + n = \frac{n(n + 1)}{2}$$

This claims that the sum of consecutive natural numbers from 1 to n is half the product of the last term, n, and the term after it, $n + 1$.

We want to prove that this will be true for $n = 1$, $n = 2$, $n = 3$, and so on. We can test the formula for any given number, say $n = 3$:

$$1 + 2 + 3 = \frac{1}{2} \cdot 3 \cdot 4 = 6, \text{ which is true.}$$

It is also true for $n = 4$:

$$1 + 2 + 3 + 4 = \frac{1}{2} \cdot 4 \cdot 5 = 10$$

But how do we prove this rule for every value of n?

We prove it using the principle of mathematical induction.

Statement 1: When a statement is true for the natural number $n = k$, then it is also true for its successor, $n = k + 1$;

and

Statement 2: If the statement is true for $n = 1$ then the statement is true for every natural number n.

When the statement is true for 1, then according to statement 1, it will also be true for 2. And that implies it will be true for 3, which implies it will be true for 4. And so on. It will thus be true for every natural number.

To prove a statement by induction, then, we must prove statements 1 and 2.

The hypothesis of statement 1: 'The statement is true for $n = k$' is called the **induction assumption**, or the induction hypothesis. It is what we assume when we prove a theorem by induction.

> The order of the steps varies from one source to another. We present you with both arrangements.

Example 5.24

Prove that the sum of the first n natural numbers is given by this formula:

$$1 + 2 + 3 + \cdots + n = \frac{n(n + 1)}{2}$$

Proof

We will call this statement $S(n)$, because it depends on n.

We will prove statements 1 and 2 for $S(n)$. First, we assume that the statement is true for $n = k$; that is, assume that $S(k)$ is true:

$$S(k): 1 + 2 + 3 + \cdots + k = \frac{k(k + 1)}{2} \tag{1}$$

This is the induction assumption. Assuming this, we must prove that $S(k + 1)$ is also true. That is, we must show:

$$S(k + 1): 1 + 2 + 3 + \cdots + k + (k + 1) = \frac{(k + 1)((k + 1) + 1)}{2} \tag{2}$$

To do that, add the next term $(k + 1)$ to both sides of the induction assumption, line (1), and then simplify:

$$\begin{aligned} S(k + 1): 1 + 2 + 3 + \cdots + k + (k + 1) &= \frac{k(k + 1)}{2} + (k + 1) \\ &= \frac{k(k + 1) + 2(k + 1)}{2} \\ &= \frac{(k + 1)(k + 2)}{2} \\ &= \frac{(k + 1)((k + 1) + 1)}{2} \end{aligned}$$

This is equation (2), which is the first thing we wanted to show.

Next, we must show that the statement is true for $n = 1$. We have:

$$S(1) : 1 = \frac{1(1 + 1)}{2}$$

The formula therefore is true for $n = 1$. We have now fulfilled both conditions of the principle of mathematical induction. $S(n)$ is therefore true for every natural number.

Example 5.25

In an investigation to find the sum of the first n positive odd integers, we can do the following: Investigate the sums of the first few odd integers and then try to come up with a conjecture. Then mathematical induction will provide us with a tool to prove the conjecture.

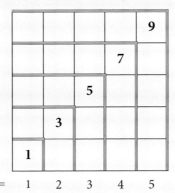

$$n = \quad 1 \quad 2 \quad 3 \quad 4 \quad 5$$

For $n = 1$, the sum is $1 = 1$

For $n = 2$, the sum is $1 + 3 = 4$

For $n = 3$, the sum is $1 + 3 + 5 = 9$

For $n = 4$, the sum is $1 + 3 + 5 + 7 = 16$

For $n = 5$, the sum is $1 + 3 + 5 + 7 + 9 = 25$

When we compare the number of integers being added with their sum, they are related; that is, the sum of n such integers is n^2.

n	1	2	3	4	5	6	...	n
sum	1	4	9	16	25	36	...	n^2

Solution

Let $S(n)$ denote the statement that the sum of the first n odd positive integers is n^2.

First, we must complete the basis step: show that $S(1)$ is true. Then we carry out the inductive step: show that $S(k + 1)$ is true whenever $S(k)$ is assumed true.

Basis step: $S(1)$, which means that the sum of the first odd integer is 1^2. This is obvious as the sum of 1 is 1.

Inductive step: Show that the implication $S(k) \Rightarrow S(k + 1)$ is true, regardless of the choice of k. Start with an assumption that $S(k)$ is true for any k:

$$1 + 3 + 5 + \ldots + (2k - 1) = k^2$$

Now, we must show that $S(k + 1)$ is true.

$$S(k + 1): 1 + 3 + 5 + \ldots + (2k + 1) = (k + 1)^2,$$

[the $(k + 1)$th odd integer is $2(k + 1) - 1 = 2k + 1$]

The left hand side can be written as

$$1 + 3 + 5 + \ldots + (2k - 1) + (2k + 1) = k^2 + (2k + 1) = k^2 + 2k + 1$$
$$= (k + 1)^2$$

Therefore $1 + 3 + 5 + \ldots + (2k + 1) = (k + 1)^2$, which is nothing but $S(k + 1)$. This shows that $S(k + 1)$ follows from $S(k)$.

Since $S(1)$ is true, and the implication $S(k) \Rightarrow S(k + 1)$ is true for all positive integers k, the mathematical induction principle shows that $S(n)$ is true for all positive integers n.

The nth odd positive integer is $2n - 1$. This is so because we are adding '2' a total of $n - 1$ times to 1. In other words, $1 + 2(n - 1) = 2n - 1$

In a proof by mathematical induction, we do not assume that $S(k)$ is true for all positive integers! We only show that if it is assumed that $S(k)$ is true, then $S(k + 1)$ is also true.

Example 5.26

Prove that $3^n < n!$ for all integers $n > 6$.

Proof

Let $S(n)$ be the statement that $3^n < n!$

Basis step: To prove this inequality the basis step must be $S(7)$.

Note that $S(6)$: $3^6 = 729 < 6! = 720$ is not true.

$S(7)$: $3^7 = 2187 < 7! = 5040$ is true.

Inductive step: Assume $S(k)$ is true. That is, assume that $3^k < k!$ is true. We must show that $S(k + 1)$ is also true. That is, we must show that $3^{k+1} < (k + 1)!$

On the assumption that $3^k < k!$, multiply both sides of this inequality by 3:

$3 \cdot 3^k < 3 \cdot k!$, and since $k > 6$, then $3 < k + 1$, hence

$3 \cdot 3^k < 3 \cdot k!$
$\qquad < (k + 1) \cdot k!$
$\qquad = (k + 1)!$
$\Rightarrow 3^{k+1} < (k + 1)!$

This shows that $S(k + 1)$ is true whenever $S(k)$ is true. This completes the inductive step of the proof.

Therefore, $3^n < n!$ for all integers $n > 6$.

When we use mathematical induction to prove a statement $S(n)$, we show that $S(1)$ is true. Then we know that $S(2)$ is true, since $S(1) \Rightarrow S(1 + 1)$. Further, we know that $S(3)$ is true, since $S(2) \Rightarrow S(2 + 1)$. Continuing along these lines, we see that $S(n)$ is true for every positive integer n.

Show that in an arithmetic sequence where $a_n = a_{n-1} + d$, the nth term can be given by the formula:

$$a_n = a_1 + (n - 1)d$$

Solution

Let $S(n)$ be the statement that $a_n = a_1 + (n - 1)d$

Basis step: To prove this formula the basis step must be $S(1)$.

$S(1)$: $a_1 = a_1 + (1 - 1)d = a_1$ is true.

Inductive step: Assume $S(k)$ is true. That is, assume that $a_k = a_1 + (k - 1)d$ is true. We must show that $S(k + 1)$ is also true. That is, we must show that $a_{k+1} = a_1 + (k + 1 - 1)d = a_1 + kd$

On the assumption that $a_k = a_1 + (k - 1)d$:

$a_{k+1} = a_k + d$ by definition of an arithmetic sequence, hence:

$$a_{k+1} = a_k + d = a_1 + (k - 1)d + d = a_1 + kd$$

This shows that $S(k + 1)$ is true whenever $S(k)$ is true. This completes the inductive step of the proof.

Therefore, $a_n = a_1 + (n - 1)d$ for all integers n.

Show that in an arithmetic series: $S_n = \dfrac{n}{2}(2a_1 + (n - 1)d)$

Solution

Notice here that we are using $P(n)$ rather than $S(n)$. The use of the name does not influence the method.

Let $P(n)$ be the statement that $S_n = \dfrac{n}{2}(2a_1 + (n - 1)d)$

Basis step: To prove this formula the basis step must be $P(1)$.

$$P(1): S_1 = \frac{1}{2}(2a_1 + (1 - 1)d) = a_1 \text{ is true. } (S_1 = a_1)$$

Inductive step: Assume $P(k)$ is true. i.e. assume that $S_k = \dfrac{k}{2}(2a_1 + (k - 1)d)$ is true. We must show that $P(k + 1)$ is also true. That is, we must show that

$$S_{k+1} = \frac{k + 1}{2}(2a_1 + (k + 1 - 1)d) = \frac{k + 1}{2}(2a_1 + kd)$$

on the assumption that $S_k = \dfrac{k}{2}(2a_1 + (k - 1)d)$

$S_{k+1} = S_k + a_{k+1}$ by definition of an arithmetic series, hence

$$S_{k+1} = \underbrace{S_k} + \overbrace{a_{k+1}} = \underbrace{\frac{k}{2}(2a_1 + (k-1)d)} + \overbrace{a_1 + kd}$$

By combining like terms and simplifying, this expression can be reduced to

$$S_{k+1} = \frac{k}{2} \cdot 2a_1 + \frac{k}{2}(k-1)d + a_1 + kd = (k+1)a_1 + \frac{k}{2}(k-1)d + kd$$

$$= \frac{(k+1)}{2} \cdot 2a_1 + \frac{k(k+1)}{2}d = \frac{k+1}{2}(2a_1 + kd)$$

This shows that $P(k + 1)$ is true whenever $P(k)$ is true. This completes the inductive step of the proof.

Therefore, $S_n = \frac{n}{2}(2a_1 + (n-1)d)$ for all integers n.

Example 5.29

Show that 3 divides $n^3 + 2n$ for all non-negative integers n.

Solution

Let $P(n)$ be the statement that '3 divides $n^3 + 2n$'

Basis step: To prove this formula the basis step must be $P(0)$.

$P(0)$: is true since $0^3 + 2(0) = 0$ is a multiple of 3.

If we are not convinced, we can try $P(1)$: $1^3 + 2(1) = 3$ is a multiple of 3.

Inductive step: Assume $P(k)$ is true. That is, assume that 3 divides $k^3 + 2k$. We must prove that $P(k + 1)$ is true, That is, 3 divides $(k + 1)^3 + 2(k + 1)$

Note that

$$(k + 1)^3 + 2(k + 1) = k^3 + 3k^2 + 3k + 1 + 2k + 2$$
$$= (k^3 + 2k) + 3k^2 + 3k + 1 + 2$$
$$= (k^3 + 2k) + 3(k^2 + k + 1)$$

Since both terms in this sum are multiples of 3 (the first by the induction hypothesis and the second because it is 3 times an integer), it follows that the sum is a multiple of 3. Hence $(k + 1)^3 + 2(k + 1)$ is a multiple of 3.

This shows that $P(k + 1)$ is true whenever $P(k)$ is true, which completes the inductive step of the proof.

Therefore, 3 divides $n^3 + 2n$ for all non-negative integers n.

Example 5.30

Using mathematical induction, show that for all non-negative integers n:

$$\binom{n}{0} + \binom{n}{1} + \binom{n}{2} + \ldots + \binom{n}{n-1} + \binom{n}{n} = 2^n$$

Solution

Let $P(n)$ be the statement that $\binom{n}{0} + \binom{n}{1} + \binom{n}{2} + \ldots + \binom{n}{n-1} + \binom{n}{n} = 2^n$

Basis step: To prove this formula the basis step must be $P(0)$.

$P(0)$: is true since $\binom{0}{0} = 2^0 = 1$ is true. Moreover, $P(1)$ is also true since

$\binom{1}{0} + \binom{1}{1} = 2^1$ is true.

Inductive step: Assume $P(k)$ is true. i.e. assume that:

$$\binom{k}{0} + \binom{k}{1} + \binom{k}{2} + \ldots + \binom{k}{k-1} + \binom{k}{k} = 2^k$$

$\binom{n}{r}$ is the same as $_nC_r$.

Recall from section 3.5 that $\binom{n}{r-1} + \binom{n}{r} = \binom{n+1}{r}$ which is the basis of Pascal's triangle. Using this fact, we can perform the following addition:

$$\binom{k}{0} + \binom{k}{1} + \binom{k}{2} + \ldots \ldots + \binom{k}{k-1} + \binom{k}{k} = 2^k$$

$$\underline{\binom{k}{0} + \binom{k}{1} + \binom{k}{2} + \ldots + \binom{k}{k-1} + \binom{k}{k} \qquad\qquad = 2^k}$$

$$\binom{k}{0} + \binom{k+1}{1} + \binom{k+1}{2} + \ldots \ldots + \binom{k+1}{k} + \binom{k}{k} = 2 \cdot 2^k$$

However, $\binom{k}{0} = \binom{k+1}{0} = \binom{k}{k} = \binom{k+1}{k+1} = 1$, so the last result can be written as

$$\binom{k+1}{0} + \binom{k+1}{1} + \binom{k+1}{2} + \ldots \ldots + \binom{k+1}{k} + \binom{k+1}{k+1}$$

$$= 2 \cdot 2^k = 2^{k+1}$$

This shows that $P(k+1)$ is true whenever $P(k)$ is true, which completes the inductive step of the proof.

Therefore, $\binom{n}{0} + \binom{n}{1} + \binom{n}{2} + \ldots + \binom{n}{n-1} + \binom{n}{n} = 2^n$ for all non-negative integers n.

Exercise 5.4

1. Find a formula for the sum of the first n even positive integers and prove it using mathematical induction.

2. Let $a_1, a_2, a_3, \ldots$ be a sequence defined by:

$$a_1 = 1, a_n = 3a_{n-1}; n \geqslant 1$$

Show that $a_n = 3^{n-1}$ for all positive integers n.

3. Let $a_1, a_2, a_3, \ldots$ be a sequence defined by:

$$a_1 = 1, a_n = a_{n-1} + 4; n \geqslant 2$$

Show that $a_n = 4n - 3$ for all positive integers $n > 1$.

4. Let $a_1, a_2, a_3, \ldots$ be a sequence defined by:

$$a_1 = 1, a_n = 2a_{n-1} + 1; n \geqslant 2$$

Show that $a_n = 2^n - 1$ for all positive integers $n > 1$.

5. Let $a_1, a_2, a_3, \ldots$ be a sequence defined by:

$$a_1 = \frac{1}{2}, a_n = a_{n-1} + \frac{1}{n(n+1)}; n \geqslant 2$$

Show that $a_n = \dfrac{n}{n+1}$ for all positive integers $n \geqslant 1$.

6. Find a formula for $\dfrac{1}{2} + \dfrac{1}{4} + \dfrac{1}{8} + \cdots + \dfrac{1}{2^n}$ and then use mathematical induction to prove your formula.

7. Show that $1 + 2 + 2^2 + \ldots + 2^n = 2^{n+1} - 1$, for all non-negative integers n.

8. Show, using mathematical induction, that in a geometric sequence $a_n = a_1 r^{n-1}$

9. Show, using mathematical induction, that in a geometric series

$$S_n = \frac{a - ar^n}{1 - r}$$

10. Prove that $2^n < n!$ for all positive integers larger than 3.

11. Prove that $2^n > n^2$ for all positive integers larger than 4.

12. Show that $1 \cdot 1! + 2 \cdot 2! + 3 \cdot 3! + \cdots + n \cdot n! = (n+1)! - 1$

13. Show that $\dfrac{1}{1 \cdot 2} + \dfrac{1}{2 \cdot 3} + \dfrac{1}{3 \cdot 4} + \cdots + \dfrac{1}{n \cdot (n+1)} = \dfrac{n}{n+1}$ for all positive integers n.

14. Show that $n^3 - n$ is divisible by 3 for all positive integers n.

15. Show that $n^5 - n$ is divisible by 5 for all positive integers n.

16. Show that $n^3 - n$ is divisible by 6 for all positive integers n.

17. Show that $n^2 + n$ is an even number for all integers n.

18. Show that $5^n - 1$ is divisible by 4 for all integers n.

19. Show that $\begin{pmatrix} a & 0 \\ 0 & b \end{pmatrix}^n = \begin{pmatrix} a^n & 0 \\ 0 & b^n \end{pmatrix}$ for every positive integer n and where a and b are real numbers.

20. Prove each of the following statements.

(a) $\displaystyle\sum_{i=1}^{n} (2i + 4) = n^2 + 5n$ for each positive integer n.

(b) $\displaystyle\sum_{i=1}^{n} (2 \cdot 3^{i-1}) = 3^n - 1$ for each positive integer n.

(c) $\displaystyle\sum_{i=1}^{n} \frac{1}{(2i - 1)(2i + 1)} = \frac{n}{2n + 1}$ for each positive integer n.

Chapter 5 practice questions

1. $\{a^n\}$ is defined as follows:

$$a_n = \sqrt[3]{8 - (a_{n-1})^3}$$

(a) Given that $a_1 = 1$, evaluate a_2, a_3, and a_4. Describe $\{a_n\}$.

(b) Given that $a_1 = 2$, evaluate a_2, a_3, and a_4. Describe $\{a_n\}$.

2. Use mathematical induction to prove that $5^n + 9^n + 2$ is divisible by 4, for $n \in \mathbb{Z}^+$.

3. The sum of the first n terms of an arithmetic sequence $\{u_n\}$ is given by the formula $S_n = 4n^2 - 2n$. Three terms of this sequence, u_2, u_m, and u_{32}, are consecutive terms in a geometric sequence. Find m.

4. Prove that $\forall n \in \mathbb{N}$, $1^3 + 2^3 + 3^3 + \cdots + n^3 = \dfrac{n^2(n + 1)^2}{4}$

5. Prove that $24 \mid (5^{2n} - 1)$ for all integers $n \geqslant 0$.

6. The Fibonacci sequence is defined as

$$F_1 = F_2 = 1 \text{ and } F_{n+1} = F_n + F_{n-1}$$

Prove that $F_1^2 + F_2^2 + F_3^2 + \cdots + F_n^2 = F_n F_{n+1}$

7. $n \in \mathbb{Z}^+$. Prove that $\dfrac{1}{2!} + \dfrac{2}{3!} + \dfrac{3}{4!} + \cdots + \dfrac{n}{(n + 1)!} = 1 - \dfrac{1}{(n + 1)!}$

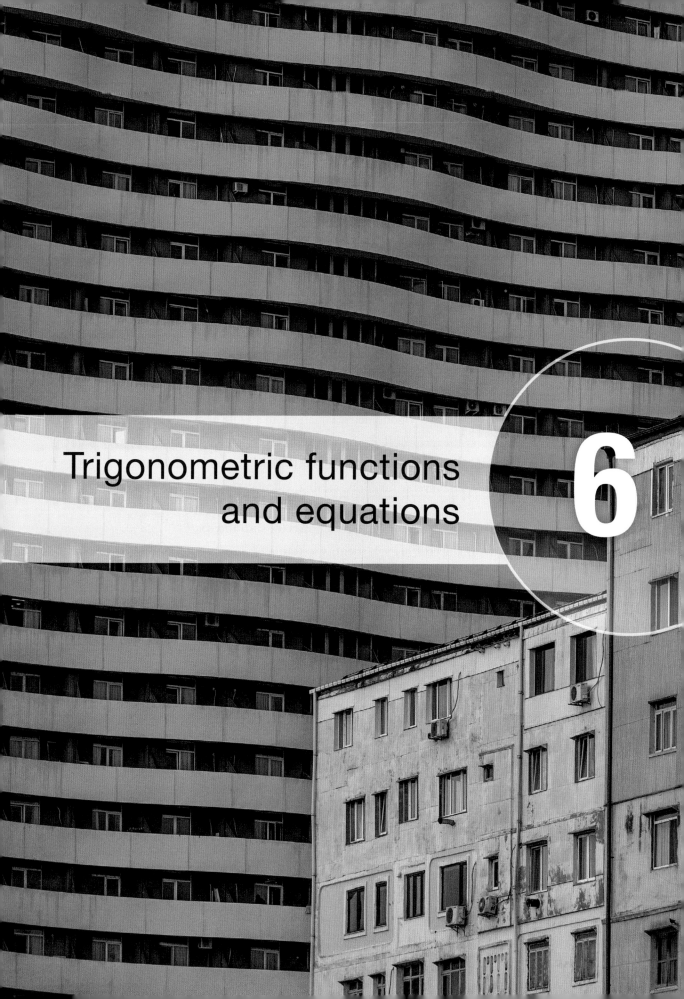

Trigonometric functions and equations

6

Learning objectives

By the end of this chapter, you should be familiar with...

- angles measured in radians
- computing the length of an arc and the area of a sector
- the unit circle and the definitions for $\sin\theta$, $\cos\theta$, and $\tan\theta$
- knowing exact values of $\sin\theta$, $\cos\theta$, and $\tan\theta$ for $\theta = 0, \dfrac{\pi}{6}, \dfrac{\pi}{4}, \dfrac{\pi}{3}, \dfrac{\pi}{2}$ and their multiples
- the Pythagorean identities and double angle identities for sine and cosine
- the relationships between $\sin\theta$, $\cos\theta$, and $\tan\theta$
- the graphs of $\sin\theta$, $\cos\theta$, and $\tan\theta$
- the amplitude and period for the graphs of $\sin\theta$ and $\cos\theta$
- composite functions of the form $a\sin(b(x + c)) + d$ and $a\cos(b(x + c)) + d$ and their graphs
- applying trigonometry to real-life problems
- solving trigonometric equations in a finite interval
- the reciprocal trigonometric ratios $\sec\theta$, $\csc\theta$, and $\cot\theta$
- the Pythagorean identities involving $\tan\theta$, $\sec\theta$, $\csc\theta$, and $\cot\theta$
- the inverse functions $\arcsin x$, $\arccos x$, $\arctan x$ and their domains, ranges and graphs
- the compound angle identities for $\sin\theta$ and $\cos\theta$
- the double angle identity for $\tan\theta$
- relationships between trigonometric functions and the symmetry of their graphs.

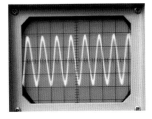

The oscilloscope shows the pressure of a sound wave versus time for a high-pitched sound. The graph is a repetitive pattern that can be expressed as the sum of different sine waves. A sine wave is any transformation of the graph of the trigonometric function $y = \sin x$ and takes the form $y = a\sin[b(x + c)] + d$.

Trigonometry developed from the use and study of triangles in surveying, navigation, architecture, and astronomy to find relationships between the side lengths of triangles and the measurement of angles. As a result, trigonometric functions were initially defined as functions of angles – that is, functions with angle measurements as their domains. With the development of calculus in the 17th century and the growth of knowledge in the sciences, the application of trigonometric functions grew to include a wide variety of periodic (repetitive) phenomena such as wave motion, vibrating strings, oscillating pendulums, alternating electrical current, and biological cycles. These applications of trigonometric functions require their domains to be sets of real numbers without reference to angles or triangles. Hence, trigonometry can be approached from two different perspectives – **functions of angles** or **functions of real numbers**. This chapter focuses on the latter – viewing trigonometric functions as defined in terms of a real number that is the **length of an arc** along the unit circle.

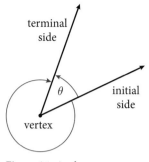

Figure 6.1 Angle components

6.1 Angles, circles, arcs and sectors

An **angle** in a plane is made by rotating a ray about its endpoint, called the vertex of the angle. The starting position of the ray is called the initial side and the position of the ray after rotation is called the terminal side of the angle (Figure 6.1). An angle with its vertex at the origin and its initial side on the positive x-axis is in **standard position** (Figure 6.2a). A **positive angle** is produced when a ray is rotated in an anticlockwise direction, and a **negative angle** when rotated in a clockwise direction.

Two angles in standard position that have the same terminal sides regardless of the direction or number of rotations are called **coterminal angles**. Greek letters are often used to represent angles, and the direction of rotation is indicated by an arc with an arrow at its endpoint. The x- and y-axes divide the coordinate plane into four quadrants (numbered with Roman numerals). Figure 6.2b shows a positive angle α and a negative angle β that are coterminal in quadrant III.

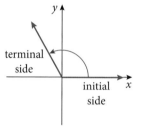

Figure 6.2a Standard position of an angle

Measuring angles: degree measure and radian measure

A unit of one degree (1°) is defined to be $\dfrac{1}{360}$ of one anticlockwise revolution about the vertex. There is another method of measuring angles that is more natural. Instead of dividing a full revolution into an arbitrary number of equal divisions (e.g. 360), consider an angle that has its vertex at the centre of a circle (a central angle) and subtends (or intercepts) a part of the circle, called an arc of the circle. Figure 6.3 shows three circles with radii of different lengths $(r_1 < r_2 < r_3)$ and the same central angle θ subtending (intercepting) the arc lengths s_1, s_2, and s_3. Regardless of the size of the circle (i.e. length of the radius), the ratio of arc length, s, to radius, r, for a given angle will be constant. For the angle θ in Figure 6.3, $\dfrac{s_1}{r_1} = \dfrac{s_2}{r_2} = \dfrac{s_3}{r_3}$. Because this ratio is an arc length divided by another length (radius), it is just an ordinary real number and has no units.

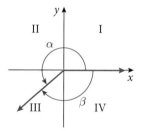

Figure 6.2b Coterminal angles

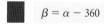
$$\beta = \alpha - 360$$

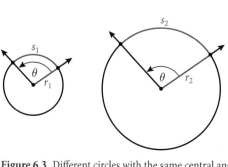

Figure 6.3 Different circles with the same central angle θ subtending different arcs, but the ratio of arc length to radius, $\dfrac{s}{r}$, remains constant.

237

Major and minor arcs

If a central angle is less than 180°, then the subtended arc is referred to as a minor arc. If a central angle is greater than 180°, then the subtended arc is referred to as a major arc.

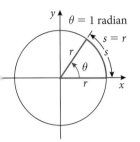

One radian is the measure of a central angle θ of a circle that subtends an arc s of the circle that is exactly the same length as the radius r of the circle. That is, when $\theta = 1$ radian, arc length = radius.

The ratio $\frac{s}{r}$ indicates how many radius lengths, r, fit into the length of arc s.

For example, if $\frac{s}{r} = 2$, then the length of s is equal to two radius lengths.

This accounts for the name **radian**.

The unit circle

When an angle is measured in radians it makes sense to draw it in standard position. It follows that the angle will be a central angle of a circle whose centre is at the origin. As Figure 6.3 illustrated, it makes no difference what size circle is used. The most practical circle to use is the circle with a radius of one unit, because the radian measure of an angle will simply be equal to the length of the subtended arc.

$$\text{Radian measure: } \theta = \frac{s}{r} \qquad \text{If } r = 1, \text{ then } \theta = \frac{s}{1} = s$$

The circle with a radius of one unit and centre at the origin $(0, 0)$ is called the unit circle (Figure 6.4). The equation for the unit circle is $x^2 + y^2 = 1$. Because the circumference of a circle with radius r is $2\pi r$, a central angle of one full anticlockwise revolution (360°) subtends an arc on the unit circle equal to 2π units. Hence, if an angle has a degree measure of 360°, its radian measure is exactly 2π. It follows that an angle of 180° has a radian measure of exactly π. This fact can be used to convert between degree measure and radian measure.

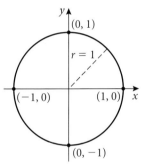

Figure 6.4 The unit circle

Conversion between degrees and radians

An angle with a radian measure of 1 has a degree measure of 57.3° (3 s.f.).

Knowing the four facts given in Example 6.1 can help you quickly convert mentally between degrees and radians for many common angles.

Example 6.1

Convert 30° and 45° to radian measure and sketch the corresponding arc on the unit circle. Use these results to convert 60° and 90° to radian measure.

Solution

Note that the 'degree' units cancel.

$$30° = 30°\left(\frac{\pi}{180°}\right) = \left(\frac{30°}{180°}\right)\pi = \frac{\pi}{6} \qquad 45° = 45°\left(\frac{\pi}{180°}\right) = \left(\frac{45°}{180°}\right)\pi = \frac{\pi}{4}$$

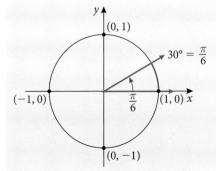

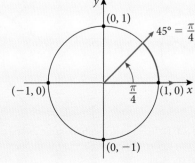

Since $60° = 2(30°)$ and $30° = \dfrac{\pi}{6}$, then $60° = 2\left(\dfrac{\pi}{6}\right) = \dfrac{\pi}{3}$

Similarly, $90° = 2(45°)$ and $45° = \dfrac{\pi}{4}$, so $90° = 2\left(\dfrac{\pi}{4}\right) = \dfrac{\pi}{2}$

Example 6.2

(a) Convert the following radian measures to degrees.
Express them exactly, if possible. Otherwise, express them accurate to 3 significant figures.

   (i) $\dfrac{4\pi}{3}$ (ii) $-\dfrac{3\pi}{2}$ (iii) 5 (iv) 1.38

(b) Convert the following degree measures to radians. Express them exactly.

   (i) $135°$ (ii) $-150°$ (iii) $175°$ (iv) $10°$

Be sure to set your GDC to degree mode or radian mode, as appropriate. As you progress further in mathematics (especially calculus) radian measure is far more useful.

Solution

(a) (i) $\dfrac{4\pi}{3} = 4\left(\dfrac{\pi}{3}\right) = 4(60°) = 240°$

   (ii) $-\dfrac{3\pi}{2} = -\dfrac{3}{2}(\pi) = -\dfrac{3}{2}(180°) = -270°$

   (iii) $5\left(\dfrac{180°}{\pi}\right) = 286°$ (3 s.f.)

   (iv) $1.38\left(\dfrac{180°}{\pi}\right) = 79.1°$ (3 s.f.)

(b) (i) $135° = 3(45°) = 3\left(\dfrac{\pi}{4}\right) = \dfrac{3\pi}{4}$

   (ii) $-150° = -5(30°) = -5\left(\dfrac{\pi}{6}\right) = -\dfrac{5\pi}{6}$

   (iii) $175°\left(\dfrac{\pi}{180°}\right) = 3.05$ (3 s.f.)

   (iv) $10°\left(\dfrac{\pi}{180°}\right) = 0.175$ (3 s.f.)

Because 2π is approximately 6.28 (3 s.f.), there are a little more than six radius lengths in one revolution, as shown in Figure 6.5.

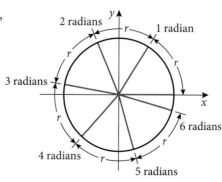

Figure 6.5 Arcs with lengths equal to the radius placed along the circumference of a circle.

Figure 6.6 shows all of the angles between 0° and 360° inclusive that are multiples of 30° or 45°, and their equivalent radian measure. You will benefit by being able to convert quickly between degree measure and radian measure for these common angles.

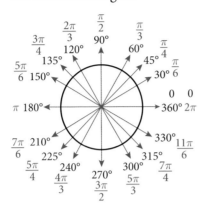

Figure 6.6 Degree and radian measure for common angles.

Arc length

For any angle θ, its radian measure is given by $\theta = \frac{s}{r}$. Simple rearrangement of this formula leads to another formula for computing arc length (see margin note).

For a circle of radius r, a central angle θ subtends an arc length s of the circle given by $s = r\theta$, where θ is in radian measure.

The units of the product $r\theta$ are equal to the units of r because in radian measure θ has no units.

Example 6.3

A circle has a radius of 10 cm. Find the length of the arc of the circle subtended by a central angle of 150°.

Solution

To use the formula $s = r\theta$, we must first convert 150° to radian measure.

$$150° = 150°\left(\frac{\pi}{180°}\right) = \frac{150\pi}{180} = \frac{5\pi}{6}$$

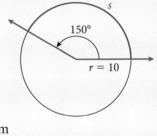

Substituting $r = 10$ cm into $s = r\theta$ gives:

$$s = 10\left(\frac{5\pi}{6}\right) = \frac{25\pi}{3} \approx 26.179\,94 \text{ cm}$$

The length of the arc is 26.2 cm (3 s.f.)

Example 6.4

The diagram shows a circle of centre O with radius $r = 6$ cm. Angle AOB subtends the minor arc AB such that the length of the arc is 10 cm. Find the measure of angle AOB in degrees, accurate to 3 significant figures.

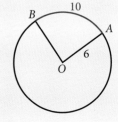

Solution

Rearrange the arc length formula, $s = r\theta$, giving $\theta = \frac{s}{r}$

Remember that the result for θ will be in radians.

Therefore, angle $AOB = \frac{10}{6}$ or $1.\dot{6}$ radians.

Now, we convert to degrees: $\frac{5}{3}\left(\frac{180°}{\pi}\right) \approx 95.492\,97°$

The degree measure of angle AOB is $95.5°$ (3 s.f.)

Geometry of a circle

inscribed circle of a regular polygon – the radius is perpendicular to the side of the polygon at the point of tangency

circumscribed circle of a regular polygon

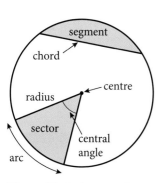
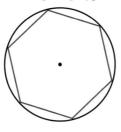
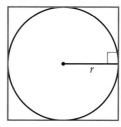

Figure 6.7 Circle terminology

Sector of a circle

Area of a sector
In a circle of radius r, the area of a sector with a central angle θ measured in radians is $A = \frac{1}{2}r^2\theta$

A **sector of a circle** is the region bounded by an arc of the circle and the two sides of a central angle (Figure 6.7). The ratio of the area of a sector to the area of the circle (πr^2) is equal to the ratio of the length of the subtended arc to the circumference of the circle ($2\pi r$). If s is the arc length and A is the area of the sector, we can write the following proportion:

$$\frac{A}{\pi r^2} = \frac{s}{2\pi r}$$

Solving for A gives:

$$A = \frac{\pi r^2 s}{2\pi r} = \frac{1}{2}rs$$

From the formula for arc length we have $s = r\theta$, with θ the radian measure of the central angle. Substituting $r\theta$ for s gives the area of a sector to be

$$A = \frac{1}{2}rs = \frac{1}{2}r(r\theta) = \frac{1}{2}r^2\theta$$

The formula for arc length, $s = r\theta$, and the formula for area of a sector, $A = \frac{1}{2}r^2\theta$, are true only when θ is in radians.

This result makes sense because, if the sector is the entire circle, $\theta = 2\pi$ and area $A = \frac{1}{2}r^2\theta = \frac{1}{2}r^2(2\pi) = \pi r^2$, which is the formula for the area of a circle.

Example 6.5

A circle of radius 9 cm has a sector whose central angle measures $\frac{2\pi}{3}$. Find the exact values of:

(a) the length of the arc subtended by the central angle

(b) the area of the sector.

Solution

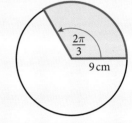

(a) $s = r\theta \Rightarrow s = 9\left(\frac{2\pi}{3}\right) = 6\pi$

The length of the arc is exactly 6π cm.

(b) $A = \frac{1}{2}r^2\theta \Rightarrow A = \frac{1}{2}(9)^2\left(\frac{2\pi}{3}\right) = 27\pi$

The area of the sector is exactly 27π cm^2.

Exercise 6.1

1. Convert each angle into radians.

   (a) 60° (b) 150° (c) −270° (d) 36°

   (e) 135° (f) 50° (g) −45° (h) 400°

2. Convert each angle into degrees. If possible, express exactly, otherwise express accurate to 3 significant figures.

   (a) $\frac{3\pi}{4}$ (b) $\frac{7\pi}{2}$ (c) 2 (d) $\frac{7\pi}{6}$

   (e) −2.5 (f) $\frac{5\pi}{3}$ (g) $\frac{\pi}{12}$ (h) 1.57

3. Find two angles (one positive and one negative) that are coterminal with the angle given in standard position below. If no units are given, assume the angle is in radian measure.

   (a) 30° (b) $\frac{3\pi}{2}$ (c) 175°

   (d) $-\frac{\pi}{6}$ (e) $\frac{5\pi}{3}$ (f) 3.25

4. Find the length of the arc s in each diagram.

(a)

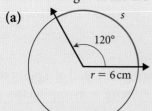

120°

$r = 6\,\text{cm}$

s

(b)

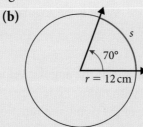

70°

$r = 12\,\text{cm}$

s

5. Find the angle θ in the diagram in both radians and degrees.

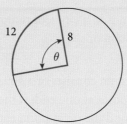

12

8

θ

6. Find the radius r of the circle in the diagram.

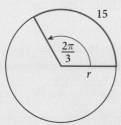

15

$\dfrac{2\pi}{3}$

r

7. Find the area of the sector in each diagram.

(a)

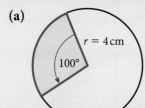

$r = 4\,\text{cm}$

100°

(b)

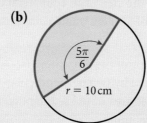

$\dfrac{5\pi}{6}$

$r = 10\,\text{cm}$

8. An arc of length 60 cm subtends a central angle α in a circle of radius 20 cm. Find the measure of α in both degrees and radians, accurate to 3 s.f.

9. Find the length of an arc that subtends a central angle of 2 radians in a circle of radius 16 cm.

10. The area of a sector of a circle with a central angle of 60° is 24 cm². Find the radius of the circle.

11. A bicycle with wheels 70 cm in diameter is travelling such that its wheels complete one and a half revolutions every second. That is, the angular velocity of each wheel is 1.5 revolutions per second.
    (a) What is the angular velocity of the bicycle's wheel in radians per second?
    (b) At what speed is the bicycle travelling along the ground? (This is the linear velocity of any point on the wheel that touches the ground.)

12. A bicycle with wheels 70 cm in diameter is travelling along a road at 25 km h^{-1}. What is the angular velocity of a bicycle wheel in radians per second?

13. Given that ω is the angular velocity in radians per second of a point on a circle with radius r cm, express the linear velocity, v, in centimetres per second, of the point as a function in terms of ω and r.

14. A chord of length 26 cm is in a circle of radius 20 cm. Find the length of the arc that the chord subtends.

15. A circular irrigation system consists of a 400 m pipe that is rotated around a central pivot point at one end. If the irrigation pipe makes one full revolution around the pivot point each day, how much area, in square metres, does it irrigate each hour?

16. (a) Find the radius of a circle circumscribed about a regular polygon of 64 sides, if one side measures 3 cm.

    (b) Calculate the difference between the circumference of the circle and the perimeter of the polygon.

17. What is the area of an equilateral triangle that has an inscribed circle with an area of 50π cm^2, and a circumscribed circle with an area of 200π cm^2?

18. The sector of the circle here is subtended by two perpendicular radii. The area of the segment is A square units.
Find an expression for the area of the circle in terms of A.

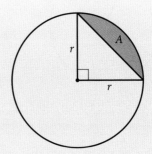

6.2 The unit circle and trigonometric functions

Several important functions can be described by mapping the coordinates of points on the real number line onto the points of the unit circle.

Suppose that the real number line is tangent to the unit circle at the point $(1, 0)$, and that zero on the number line matches with $(1, 0)$ on the circle, as shown in Figure 6.8. Because of the properties of circles, the real number line in this position will be perpendicular to the x-axis. The scales on the number line, the x-axis, and the y-axis need to be the same. Imagine that the real number line is flexible like a string and can wrap around the circle, with zero on the number line remaining fixed to the point $(1, 0)$ on the unit circle. When the top portion of the string moves along the circle, the wrapping is anticlockwise $(t > 0)$, and when the bottom portion of the string moves along the circle, the wrapping is clockwise $(t < 0)$. As the string wraps around the unit circle, each real number t on the string is mapped onto a point (x, y) on the circle. Hence, the real number line from 0 to t makes an arc of length t starting on the circle at $(1, 0)$ and ending at the point (x, y) on the circle. For example, since the circumference of the unit circle is 2π, the number $t = 2\pi$ will be wrapped anticlockwise around the circle to the point $(1, 0)$. Similarly, the number $t = \pi$ will be wrapped anticlockwise halfway around the circle to the point $(-1, 0)$ on the circle. And the number $t = -\dfrac{\pi}{2}$ will be wrapped clockwise one-quarter of the way around the circle to the point $(0, -1)$ on the circle. Note that each number t on the real number line is mapped (corresponds) to exactly one point on the unit circle, thereby satisfying the definition of a function – consequently this mapping is called a **wrapping function**.

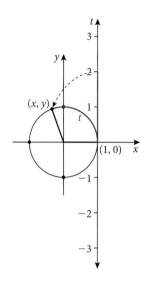

Figure 6.8 The wrapping function

Before we leave our mental picture of the string (representing the real number line) wrapping around the unit circle, consider any pair of points on the string that are exactly 2π units from each other. Let these two points represent the real numbers t_1 and $t_1 + 2\pi$. Because the circumference of the unit circle is 2π, these two numbers will be mapped to the same point on the unit circle. Furthermore, consider the infinite number of points whose distance from t_1 is any integer multiple of 2π, i.e. $t_1 + k \cdot 2\pi$, $k \in \mathbb{Z}$, and again all of these numbers will be mapped to the same point on the unit circle. Consequently, the wrapping function is not a one-to-one function. Output for the function (points on the unit circle) is unchanged by the addition of any integer multiple of 2π to any input value (a real number). Functions that behave in such a repetitive (or cyclic) manner are called periodic.

A function f such that $f(x) = f(x + p)$ is a **periodic function**. If p is the least positive constant for which $f(x) = f(x + p)$ is true, then p is called the **period** of the function.

We are surrounded by periodic functions, e.g. average daily temperature; sun rise and the day of the year; animal populations over many years; the height of tides and the position of the moon; and, an electrocardiogram.

Trigonometric functions

The x- and y-coordinates of the points on the unit circle can be used to define six **trigonometric functions**: the **sine**, **cosine**, **tangent**, **cosecant**, **secant** and **cotangent** functions. These are often abbreviated as **sin**, **cos**, **tan**, **csc** (or **cosec**), **sec** and **cot** respectively.

When the real number t is mapped to a point (x, y) on the unit circle, the value of the y-coordinate is assigned to the sine function; the x-coordinate is assigned to the cosine function; and the ratio of the two coordinates $\dfrac{y}{x}$ is assigned to

Let t be any real number and (x, y) a point on the unit circle to which t is mapped. Then the function definitions are:
$\sin t = y$
$\cos t = x$
$\tan t = \dfrac{y}{x}, x \neq 0$
$\csc t = \dfrac{1}{y}, y \neq 0$
$\sec t = \dfrac{1}{x}, x \neq 0$
$\cot t = \dfrac{x}{y}, y \neq 0$

When trigonometric functions are defined as circular functions based on the unit circle, radian measure is used. The values for the domain of the sine and cosine functions are real numbers that are arc lengths on the unit circle. The arc length on the unit circle subtends an angle in standard position whose radian measure is equivalent to the arc length (see Figure 6.9).

the tangent function. Sine, cosine, and tangent are often referred to as the **basic trigonometric functions**. Cosecant, secant, and cotangent are each a reciprocal of one of the basic trigonometric functions and are often referred to as the **reciprocal trigonometric functions**. All six are defined by means of the length of an arc on the unit circle, as shown in the Key fact box on the previous page.

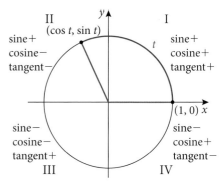

On the unit circle: $x = \cos t$, $y = \sin t$.

Figure 6.9 Signs of the trigonometric functions depend on the quadrant where the arc t terminates.

Evaluating the trigonometric functions for any value of t involves finding the coordinates of the point on the unit circle where the arc of length t will 'wrap to' (or terminate) starting at the point $(1, 0)$. It is useful to remember that an arc of length π is equal to one half of the circumference of the unit circle. All of the values for t in this example are positive, so the arc length will wrap along the unit circle in an anticlockwise direction.

Example 6.6

Evaluate the six trigonometric functions for each value of t.

(a) $t = 0$ (b) $t = \dfrac{\pi}{2}$ (c) $t = \pi$ (d) $t = \dfrac{3\pi}{2}$ (e) $t = 2\pi$

Solution

(a) An arc of length $t = 0$ has no length so it 'terminates' at the point $(1, 0)$. By definition:

$$\sin 0 = y = 0 \qquad\qquad \csc 0 = \frac{1}{y} = \frac{1}{0} \text{ is undefined}$$

$$\cos 0 = x = 1 \qquad\qquad \sec 0 = \frac{1}{x} = \frac{1}{1} = 1$$

$$\tan 0 = \frac{y}{x} = \frac{0}{1} = 0 \qquad\qquad \cot 0 = \frac{x}{y} = \frac{1}{0} \text{ is undefined}$$

(b) An arc of length $t = \dfrac{\pi}{2}$ is equivalent to one-quarter of the circumference of the unit circle (Figure 6.10) so it terminates at the point $(0, 1)$. By definition:

$$\sin \frac{\pi}{2} = y = 1 \qquad\qquad \csc \frac{\pi}{2} = \frac{1}{y} = 1$$

$$\cos \frac{\pi}{2} = x = 0 \qquad\qquad \sec \frac{\pi}{2} = \frac{1}{x} \text{ is undefined}$$

$$\tan \frac{\pi}{2} = \frac{y}{x} = \frac{1}{0} \text{ is undefined} \qquad\qquad \cot \frac{\pi}{2} = \frac{x}{y} = 0$$

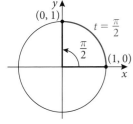

Figure 6.10 Arc length of $\dfrac{\pi}{2}$ or one-quarter of an anticlockwise revolution.

(c) An arc of length $t = \pi$ is equivalent to half of the circumference of the unit circle (Figure 6.11) so it terminates at the point $(-1, 0)$. By definition:

$$\sin \pi = y = 0 \qquad\qquad \csc \pi = \frac{1}{y} \text{ is undefined}$$

$$\cos \pi = x = -1 \qquad\qquad \sec \pi = \frac{1}{-1} = -1$$

$$\tan \pi = \frac{y}{x} = \frac{0}{-1} = 0 \qquad\qquad \cot \pi = \frac{x}{y} \text{ is undefined}$$

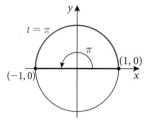

Figure 6.11 Arc length of π or half of an anticlockwise revolution.

(d) An arc of length $t = \frac{3\pi}{2}$ is equivalent to three-quarters of the circumference of the unit circle (Figure 6.12), so it terminates at the point $(0, -1)$. By definition:

$$\sin \frac{3\pi}{2} = y = -1 \qquad\qquad \csc \frac{3\pi}{2} = \frac{1}{y} = -1$$

$$\cos \frac{3\pi}{2} = x = 0 \qquad\qquad \sec \frac{3\pi}{2} = \frac{1}{x} \text{ is undefined}$$

$$\tan \frac{3\pi}{2} = \frac{y}{x} = \frac{-1}{0} \text{ is undefined} \qquad\qquad \cot \frac{3\pi}{2} = \frac{x}{y} = 0$$

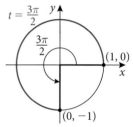

Figure 6.12 Arc length of $\frac{3\pi}{2}$ or three-quarters of an anticlockwise revolution.

(e) An arc of length $t = 2\pi$ terminates at the same point as an arc of length $t = 0$ (Figure 6.13), so the values of the trigonometric functions are the same as found in part (a):

$$\sin 0 = y = 0 \qquad\qquad \csc 0 = \frac{1}{y} \text{ is undefined}$$

$$\cos 0 = x = 1 \qquad\qquad \sec 0 = \frac{1}{x} = 1$$

$$\tan 0 = \frac{y}{x} = \frac{0}{1} = 0 \qquad\qquad \cot 0 = \frac{x}{y} \text{ is undefined}$$

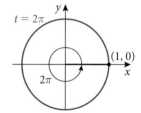

Figure 6.13 : Arc length of 2π or one full anticlockwise revolution.

From our previous discussion of periodic functions, we can conclude that all of the trigonometric functions are periodic. Given that the sine and cosine functions are generated directly from the wrapping function, the period of each of these functions is 2π. That is:

$$\sin t = \sin(t + k \cdot 2\pi), k \in \mathbb{Z} \text{ and } \cos t = \cos(t + k \cdot 2\pi), k \in \mathbb{Z}$$

Since the cosecant and secant functions are the respective reciprocals of sine and cosine, the period of cosecant and secant will also be 2π.

Initial evidence from Example 6.6 indicates that the period of the tangent function is π. That is,

$$\tan t = \tan(t + k \cdot \pi), k \in \mathbb{Z}$$

We will establish these results graphically in the next section. Also note that since the trigonometric functions are periodic (i.e. function values repeat) then they are not one-to-one functions. This is an important fact when establishing inverse trigonometric functions (see Section 6.6).

Domains of the six trigonometric functions

$f(t) = \sin t$ and $f(t) = \cos t$
domain: $\{t : t \in \mathbb{R}\}$
$f(t) = \tan t$ and $f(t) = \sec t$
domain:
$\{t : t \in \mathbb{R}, t \neq \frac{\pi}{2} + k\pi, k \in \mathbb{Z}\}$
$f(t) = \cot t$ and $f(t) = \csc t$
domain:
$\{t : t \in \mathbb{R}, t \neq k\pi, k \in \mathbb{Z}\}$

Evaluating trigonometric functions

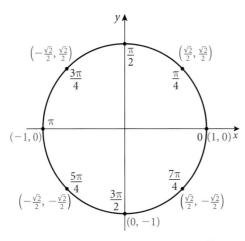

Figure 6.14 Arc lengths that are multiples of $\frac{\pi}{4}$ divide the unit circle into eight equally spaced points.

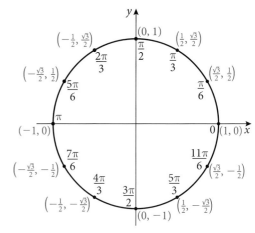

Figure 6.15 Arc lengths that are multiples of $\frac{\pi}{6}$ divide the unit circle into twelve equally spaced points.

The following four identities follow directly from the definitions for the trigonometric functions.

$$\tan t = \frac{\sin t}{\cos t} \quad \csc t = \frac{1}{\sin t}$$
$$\sec t = \frac{1}{\cos t} \quad \cot t = \frac{\cos t}{\sin t}$$

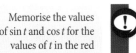

Memorise the values of $\sin t$ and $\cos t$ for the values of t in the red box in Table 6.1. These values can be used to derive the value of all six trigonometric functions for any multiple of $\frac{\pi}{6}, \frac{\pi}{4}, \frac{\pi}{3}$ or $\frac{\pi}{2}$.

You will find it very helpful to memorise the exact values of sine and cosine for numbers that are multiples of $\frac{\pi}{6}$ and $\frac{\pi}{4}$.

Use the unit circle diagrams shown in Figures 6.14 and 6.15 as a guide to help you do this and to visualise the location of the terminal points of different arc lengths. With the symmetry of the unit circle and a point's location in the coordinate plane telling us the sign of x and y (see Figure 6.9), we only need to remember the sine and cosine of common values of t in the first quadrant and on the positive x- and y-axes. These are organised in Table 6.1.

t	$\sin t$	$\cos t$	$\tan t$	$\csc t$	$\sec t$	$\cot t$
0	0	1	0	undefined	1	undefined
$\frac{\pi}{6}$	$\frac{1}{2}$	$\frac{\sqrt{3}}{2}$	$\frac{\sqrt{3}}{3}$	2	$\frac{2\sqrt{3}}{3}$	$\sqrt{3}$
$\frac{\pi}{4}$	$\frac{\sqrt{2}}{2}$	$\frac{\sqrt{2}}{2}$	1	$\sqrt{2}$	$\sqrt{2}$	1
$\frac{\pi}{3}$	$\frac{\sqrt{3}}{2}$	$\frac{1}{2}$	$\sqrt{3}$	$\frac{2\sqrt{3}}{3}$	2	$\frac{\sqrt{3}}{3}$
$\frac{\pi}{2}$	1	0	undefined	1	undefined	0

Table 6.1 The trigonometric functions evaluated for special values of t.

If t is not a multiple of $\frac{\pi}{6}, \frac{\pi}{4}, \frac{\pi}{3}$ or $\frac{\pi}{2}$, then the approximate values of the trigonometric functions for that number can be found using your GDC.

If s and t are coterminal arcs (i.e. they terminate at the same point), then the trigonometric functions of s are equal to those of t.

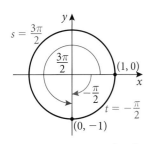

Figure 6.16 Coterminal angles $\frac{3\pi}{2}$ and $-\frac{\pi}{2}$

For example, the arcs $s = \dfrac{3\pi}{2}$ and $t = -\dfrac{\pi}{2}$ are coterminal (Figure 6.16).

Thus, $\sin\dfrac{3\pi}{2} = \sin\left(-\dfrac{\pi}{2}\right)$, $\tan\dfrac{3\pi}{2} = \tan\left(-\dfrac{\pi}{2}\right)$, and so on.

Exercise 6.2

1. (a) By knowing the ratios of sides in any triangle with angles measuring 30°, 60° and 90° (see figure), find the coordinates of the points on the unit circle where arcs of length $t = \dfrac{\pi}{6}$ and $t = \dfrac{\pi}{3}$ terminate in the first quadrant.

    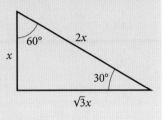

    (b) Using the result from (a) and applying symmetry about the unit circle, find the coordinates of the points on the unit circle corresponding to arcs whose lengths are

    $$\dfrac{2\pi}{3}, \dfrac{5\pi}{6}, \dfrac{7\pi}{6}, \dfrac{4\pi}{3}, \dfrac{5\pi}{3}, \dfrac{11\pi}{6}.$$

    Draw a large unit circle and label all of these points with their coordinates and the measure of the arc that terminates at each point.

2. Quadrant I of the unit circle shown below indicates angles in intervals of 10° and also indicates angles in radian measure of 0.5, 1 and 1.5. Use the figure and the definitions of the sine and cosine functions to approximate the function values to 1 decimal place in questions (a) to (h).

    Check your answers with your GDC, and make sure you are using the correct angle measure mode.

    (a) $\cos 50°$

    (b) $\sin 80°$

    (c) $\cos 1$

    (d) $\sin 0.5$

    (e) $\tan 70°$

    (f) $\cos 1.5$

    (g) $\sin 20°$

    (h) $\tan 1$

    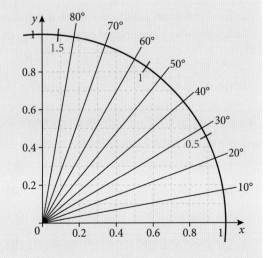

3. *t* is the length of an arc on the unit circle starting from $(1, 0)$. For each value of *t* below:

(i) State the quadrant in which the terminal point of the arc lies.

(ii) Find the coordinates of the terminal point (x, y) on the unit circle. Give exact values for *x* and *y* if possible, otherwise give the value accurate to 3 significant figures.

(a) $t = \dfrac{\pi}{6}$ **(b)** $t = \dfrac{5\pi}{3}$ **(c)** $t = \dfrac{7\pi}{4}$

(d) $t = \dfrac{3\pi}{2}$ **(e)** $t = 2$ **(f)** $t = -\dfrac{\pi}{4}$

(g) $t = -1$ **(h)** $t = -\dfrac{5\pi}{4}$ **(i)** $t = 3.52$

4. State the exact value (if possible) of the sine, cosine and tangent of the given real number.

(a) $\dfrac{\pi}{3}$ **(b)** $\dfrac{5\pi}{6}$ **(c)** $-\dfrac{3\pi}{4}$

(d) $\dfrac{\pi}{2}$ **(e)** $-\dfrac{4\pi}{3}$ **(f)** 3π

(g) $\dfrac{3\pi}{2}$ **(h)** $-\dfrac{7\pi}{6}$ **(i)** 1.25π

5. Use the periodic properties of the sine and cosine functions to find the exact value of $\sin x$ and $\cos x$ for each value of *x*:

(a) $x = \dfrac{13\pi}{6}$ **(b)** $x = \dfrac{10\pi}{3}$ **(c)** $x = \dfrac{15\pi}{4}$ **(d)** $x = \dfrac{17\pi}{6}$

6. Find the exact function values, if possible. Do not use your GDC.

(a) $\cos\dfrac{5\pi}{6}$ **(b)** $\sin 315°$ **(c)** $\tan\dfrac{3\pi}{2}$

(d) $\sec\dfrac{5\pi}{3}$ **(e)** $\csc 240°$

7. Find the exact function values, if possible. Otherwise, find the value accurate to 3 significant figures.

(a) $\sin 2.5$ **(b)** $\cot 120°$ **(c)** $\cos\dfrac{5\pi}{4}$

(d) $\sec 6$ **(e)** $\tan \pi$

8. Specify in which quadrant(s) an angle θ in standard position could be, given the stated conditions.

(a) $\sin \theta > 0$ **(b)** $\sin \theta > 0$ and $\cos \theta < 0$

(c) $\sin \theta < 0$ and $\tan \theta > 0$ **(d)** $\cos \theta < 0$ and $\tan \theta < 0$

(e) $\cos \theta > 0$ **(f)** $\sec \theta > 0$ and $\tan \theta > 0$

(g) $\cos \theta > 0$ and $\csc \theta < 0$ **(h)** $\cot \theta < 0$

6.3 Graphs of trigonometric functions

From the previous section we know that trigonometric functions are periodic; that is, their values repeat in a regular manner. The graphs of the trigonometric functions should provide a picture of this periodic behaviour. In this section we will graph the sine, cosine, and tangent functions, and transformations of the sine and cosine functions.

Graphs of the sine and cosine functions

Since the period of the sine function is 2π, we know that two values of t (domain) that differ by 2π will produce the same value for y (range). This means that any portion of the graph of $y = \sin t$ with a t-interval of length 2π (called one period or cycle of the graph) will repeat. Remember that the domain of the sine function is all real numbers, so one period of the graph of $y = \sin t$ will repeat indefinitely in the positive and negative directions. Therefore, in order to construct a complete graph of $y = \sin t$, we need to graph just one period of the function – for example, from $t = 0$ to $t = 2\pi$ – and then repeat the pattern in both directions.

We know from Section 6.2 that $\sin t$ is the y-coordinate of the terminal point on the unit circle corresponding to the real number t (Figure 6.18). In order to generate one period of the graph of $y = \sin t$, we need to record the y-coordinates of a point on the unit circle and the corresponding value of t as the point travels anticlockwise one revolution, starting from the point $(1, 0)$. These values are then plotted on a graph with t on the horizontal axis and y ($\sin t$) on the vertical axis. Figure 6.19 illustrates this process in a sequence of diagrams.

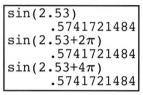

Figure 6.17 The period of $y = \sin x$ is 2π

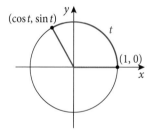

Figure 6.18 Coordinates of the terminal point of arc t give the values of $\cos t$ and $\sin t$.

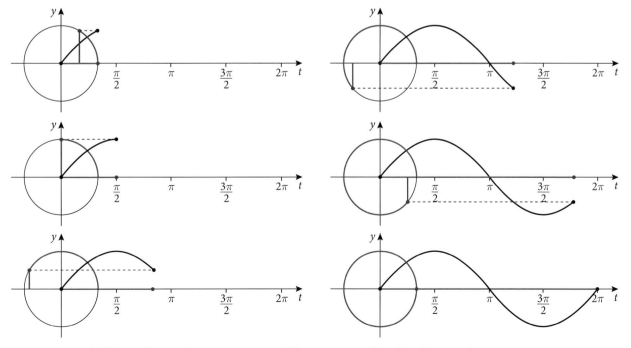

Figure 6.19 Graph of the sine function for $0 \leqslant t \leqslant 2\pi$ generated from a point travelling along the unit circle.

As the point (cos *t*, sin *t*) travels along the unit circle, the *x*-coordinate (cos *t*) goes through the same cycle of values as the *y*-coordinate (sin *t*) does. The only difference is that the *x*-coordinate begins at a different value in the cycle – when $t = 0$, $y = 0$, but $x = 1$. The result is that the graph of $y = \cos t$ is the exact same shape as $y = \sin t$ but it has been shifted $\frac{\pi}{2}$ units to the left. The graph of $y = \cos t$ for $0 \leqslant t \leqslant 2\pi$ is shown in Figure 6.20.

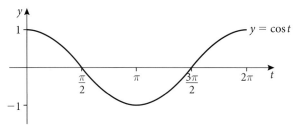

Figure 6.20 Graph of $y = \cos t$ for $0 \leqslant t \leqslant 2\pi$

The convention is to use the letter *x* to denote the variable in the domain of the function. Hence from here on, we will use the letter *x* rather than *t* and write the trigonometric functions as $y = \sin x$, $y = \cos x$, and $y = \tan x$.

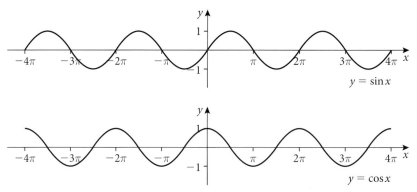

Figure 6.21 $y = \sin x$ and $y = \cos x$, $-4\pi \leqslant x \leqslant 4\pi$

A function is odd if, for each *x* in the domain of *f*, $f(-x) = -f(x)$. The graph of an odd function is symmetric with respect to the origin (rotational symmetry).

A function is even if, for each *x* in the domain of *f*, $f(-x) = f(x)$. The graph of an even function is symmetric with respect to the *x*-axis (line symmetry).

The sine function is odd because $\sin(-x) = -\sin(x)$, and the cosine function is even because $\cos(-x) = \cos(x)$.

Aside from their periodic behaviour, these graphs reveal further properties of the functions $y = \sin x$ and $y = \cos x$. Note that the sine function has a maximum value of $y = 1$ for all $x = \frac{\pi}{2} + k \cdot 2\pi$, $k \in \mathbb{Z}$, and has a minimum value of $y = -1$ for all $x = -\frac{\pi}{2} + k \cdot 2\pi$, $k \in \mathbb{Z}$. The cosine function has a maximum value of $y = 1$ for all $x = k \cdot 2\pi$, $k \in \mathbb{Z}$, and has a minimum value of $y = -1$ for all $x = \pi + k \cdot 2\pi$, $k \in \mathbb{Z}$. This also confirms that both functions have a domain of all real numbers and a range of $-1 \leqslant y \leqslant 1$.

Closer inspection of the graphs in Figure 6.21 shows that the graph of $y = \sin x$ has rotational symmetry about the origin – that is, it can be rotated 180° about the origin and it remains the same. This graph symmetry can be expressed with the identity: $\sin(-x) = -\sin x$. For example, $\sin\left(-\frac{\pi}{6}\right) = -\frac{1}{2}$ and $-\left[\sin\left(\frac{\pi}{6}\right)\right] = -\left[\frac{1}{2}\right] = -\frac{1}{2}$. A function that is symmetric about the origin is called an **odd function**.

The graph of $y = \cos x$ has line symmetry over the x-axis – that is, it remains the same when reflected over the x-axis. This graph symmetry can be expressed with the identity:

$\cos(-x) = \cos x$. For example, $\cos\left(-\dfrac{\pi}{6}\right) = \dfrac{\sqrt{3}}{2}$ and $\cos\dfrac{\pi}{6} = \dfrac{\sqrt{3}}{2}$.

A function that is symmetric about the y-axis is called an **even function**.

Graphs of transformations of the sine and cosine functions

In Section 1.5 we learned how to transform the graph of a function by horizontal and vertical translations, by reflections in the coordinate axes, and by stretching and shrinking – both horizontally and vertically.

Review of transformations of graphs of functions
Assume that a, b, c and d are real numbers.

To obtain the graph of:	From the graph of $y = f(x)$:
$y = f(x) + d$	Translate d units up for $d > 0$, d units down for $d < 0$
$y = f(x + c)$	Translate c units left for $c > 0$, c units right for $c < 0$
$y = -f(x)$	Reflect in the x-axis
$y = a f(x)$	Vertical stretch $(a > 1)$ or shrink $(0 < a < 1)$ of factor a
$y = f(-x)$	Reflect in the y-axis
$y = f(bx)$	Horizontal stretch $(0 < b < 1)$ or shrink $(b > 1)$ of factor $\dfrac{1}{b}$

In this section we will look at the composition of sine and cosine functions of the form $f(x) = a \sin[b(x + c)] + d$ and $f(x) = a \cos[b(x + c)] + d$

Example 6.7

Sketch the graph of each function on the interval $-\pi \leqslant x \leqslant 3\pi$.

(a) $f(x) = 2 \cos x$

(b) $g(x) = \cos x + 3$

(c) $h(x) = 2 \cos x + 3$

(d) $p(x) = \dfrac{1}{2}\sin x - 2$

Solution

(a) Since $a = 2$, the graph of $y = 2 \cos x$ is obtained by stretching the graph of $y = \cos x$ vertically by a factor of 2.

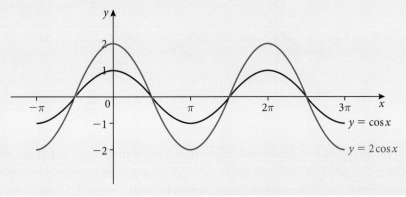

(b) Since $d = 3$, the graph of $y = \cos x + 3$ is obtained by translating 3 units up the graph of $y = \cos x$.

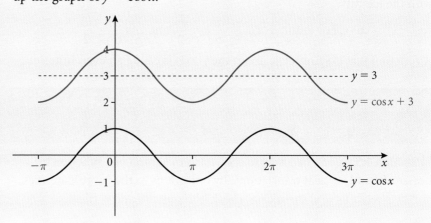

(c) We can obtain the graph of $y = 2\cos x + 3$ by combining both of the transformations to the graph of $y = \cos x$ performed in parts (a) and (b) – namely, a vertical stretch of factor 2 and a translation 3 units up.

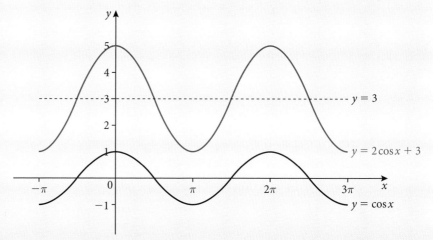

(d) The graph of $y = \dfrac{1}{2}\sin x - 2$ can be obtained by shrinking the graph of $y = \sin x$ vertically by a factor of $\dfrac{1}{2}$ and then translating it down 2 units.

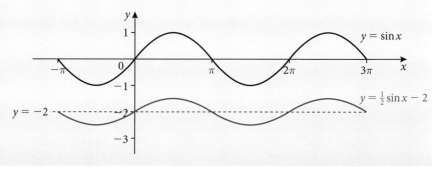

In (a), the graph of $y = 2\cos x$ has many of the same properties as the graph of $y = \cos x$: the same period, and the maximum and minimum values occur at the same x values. However, the graph ranges between -2 and 2 instead of -1

and 1. This difference is best described by referring to the graph's **amplitude**. The amplitude of $y = \cos x$ is 1 and the amplitude of $y = 2\cos x$ is 2. The amplitude (always positive) is not always equal to the maximum value. In (b), the amplitude of $y = \cos x + 3$ is 1; in (c), the amplitude of $y = 2\cos x + 3$ is 2; and in (d), the amplitude of $y = \frac{1}{2}\sin x - 2$ is $\frac{1}{2}$. For all three of these, the graphs oscillate about the horizontal mid-line $y = d$. How high and low the graph oscillates with respect to the mid-line is the graph's amplitude. With respect to the general form $y = af(x)$, changing the amplitude is equivalent to a vertical stretching or shrinking. Thus, we can give a more precise definition of amplitude in terms of the parameter a.

Amplitude of the graph of sine and cosine functions
The graphs of $f(x) = a\sin[b(x + c)] + d$ and $f(x) = a\cos[b(x + c)] + d$ have an amplitude equal to $|a|$.

Example 6.8

Waves are produced in a long tank of water. The depth of the water, d metres, at t seconds at a fixed location in the tank is modelled by the function $d(t) = M\cos\left(\frac{\pi}{2}t\right) + K$, where M and K are positive constants. The figure shows the graph of $d(t)$ for $0 \leqslant t \leqslant 12$, indicating that the point $(2, 5.1)$ is a minimum and the point $(8, 9.7)$ is a maximum.

(a) Find the value of K and the value of M.

(b) After $t = 0$, find the first time when the depth of the water is 9.7 metres.

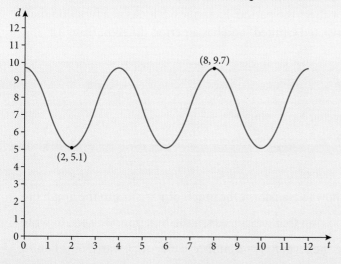

Solution

(a) The constant K is equivalent to the constant d in the general form $f(x) = a\cos[b(x + c)] + d$. To find the value of K and the equation of the horizontal mid-line, $y = K$, find the average of the function's maximum and minimum value: $K = \dfrac{9.7 + 5.1}{2} = 7.4$. The constant M is equivalent to the constant a whose absolute value is the amplitude. The amplitude is the difference between the function's maximum value and the mid-line: $|M| = 9.7 - 7.4 = 2.3$. So, $M = 2.3$ or $M = -2.3$

Try $M = 2.3$ by evaluating the function at one of the known values:

$$d(2) = 2.3\cos\left(\frac{\pi}{2}(2)\right) + 7.4 = 2.3\cos\pi + 7.4 = 2.3(-1) + 7.4 = 5.1$$

This agrees with the point $(2, 5.1)$ on the graph. Therefore, $M = 2.3$

(b) Maximum values of the function ($d = 9.7$) occur at values of t that differ by a value equal to the period. The graph shows that the difference in t-values from the minimum $(2, 5.1)$ to the maximum $(8, 9.7)$ is equivalent to one-and-a-half periods. Therefore, the period is 4 and the first time after $t = 0$ at which $d = 9.7$ is $t = 4$.

All four of the functions in Example 6.7 had the same period of 2π, but the function in Example 6.8 had a period of 4. Because $y = \sin x$ completes one period from $x = 0$ to $x = 2\pi$, it follows that $y = \sin bx$ completes one period from $bx = 0$ to $bx = 2\pi$. This implies that $y = \sin bx$ completes one period from $x = 0$ to $x = \frac{2\pi}{b}$. This agrees with the period for the function

$d(t) = 2.3\cos\left(\frac{\pi}{2}t\right) + 7.4$ in Example 6.8: period $= \dfrac{2\pi}{b} = \dfrac{2\pi}{\frac{\pi}{2}} = \dfrac{2\pi}{1} \cdot \dfrac{2}{\pi} = 4$

Note that the change in amplitude and vertical translation had no effect on the period. We should also expect that a horizontal translation of a sine or cosine curve should not affect the period. Example 6.9 looks at a function that is horizontally translated (shifted) and has a period different from 2π.

Example 6.9

Sketch the function $f(x) = \sin\left(2x + \dfrac{2\pi}{3}\right)$

Solution

To determine how to transform the graph of $y = \sin x$ to the graph of $y = \sin\left(2x + \dfrac{2\pi}{3}\right)$ so that we can sketch the function, we need to write the function in the form $f(x) = a\sin[b(x + c)] + d$. Clearly, $a = 1$ and $d = 0$, but we need to take out a common factor of 2 from $2x + \dfrac{2\pi}{3}$ to get $f(x) = \sin\left[2\left(x + \dfrac{\pi}{3}\right)\right]$. From the transformations studied in Chapter 1, we expect the graph of f is obtained by first translating the graph of $y = \sin x$ to the left $\dfrac{\pi}{3}$ units, then a horizontally shrinking by a factor of $\dfrac{1}{2}$. The following graphs show the two steps of transforming $y = \sin x$ to $y = \sin\left[2\left(x + \dfrac{\pi}{3}\right)\right]$.

(sidebar)

Period and horizontal translation (phase shift) of sine and cosine functions
Given that b is a positive real number,
$y = a\sin[b(x + c)] + d$
and
$y = a\cos[b(x + c)] + d$
have a **period of** $\dfrac{2\pi}{b}$ and a horizontal translation (**phase shift**) of $-c$.

A horizontal translation of a sine or cosine curve is often referred to as a **phase shift**. The equations
$y = \sin\left(x + \dfrac{\pi}{3}\right)$ and
$y = \sin\left[2\left(x + \dfrac{\pi}{3}\right)\right]$
both have a phase shift of $-\dfrac{\pi}{3}$.

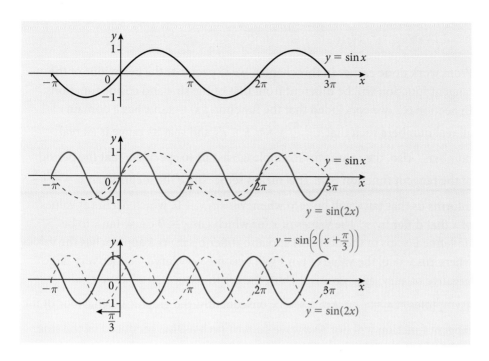

Example 6.10

Here is a graph of a function in the form $y = a \cos bx$.

(a) Write down the value of a.
(b) Calculate the value of b.

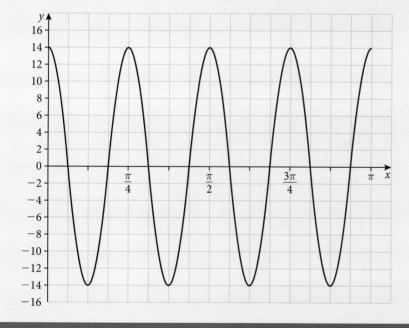

Solution

(a) The amplitude of the graph is 14. Therefore, $a = 14$.

(b) From inspecting the graph, we can see that the period is $\frac{\pi}{4}$.

$$\text{Period} = \frac{2\pi}{b} = \frac{\pi}{4} \Rightarrow b\pi = 8\pi \Rightarrow b = 8$$

Graph of the tangent function

From work done earlier in this chapter, we expect that the behaviour of the tangent function will be different from that of the sine and cosine functions. In Section 6.2, we concluded that the function $f(x) = \tan x$ has a domain of all real numbers such that $x \neq \dfrac{\pi}{2} + k\pi$, $k \in \mathbb{Z}$, and that its range is all real numbers. Also, the results for Example 6.6 led us to speculate that the period of the tangent function is π. This makes sense, since the identity $\tan x = \dfrac{\sin x}{\cos x}$ informs us that $\tan x$ will be zero whenever $\sin x = 0$, which occurs at values of x that differ by π. The values of x for which $\cos x = 0$ cause $\tan x$ to be undefined (gaps in the domain) and also differ by π. As x approaches the values where $\cos x = 0$, the value of $\tan x$ will become very large – either very large negative or very large positive. Thus, the graph of $y = \tan x$ has vertical asymptotes at $x = \dfrac{\pi}{2} + k\pi$, $k \in \mathbb{Z}$. Consequently, the graphical behaviour of the tangent function will not be a wave pattern such as that produced by the sine and cosine functions, but rather a series of separate curves that repeat every π units (Figure 6.22).

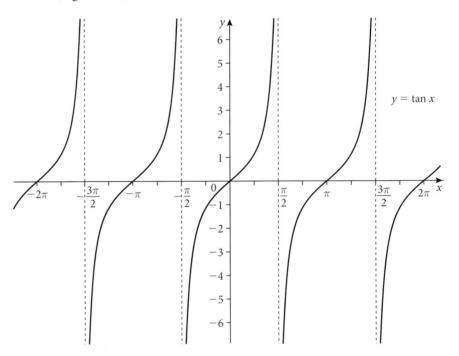

Figure 6.22 Graph of $y = \tan x$ for $-2\pi \leqslant x \leqslant 2\pi$

The graph gives clear confirmation that the period of the tangent function is π. Also, the graph of $y = \tan x$ has rotational symmetry about the origin – that is, it can be rotated 180° about the origin and it remains the same. Hence, like the sine function, tangent is an odd function and $\tan(-x) = -\tan x$.

By applying the symmetry of their graphs and a horizontal translation, we can establish the following relationships for the sine, cosine, and tangent functions.

$\sin(-\theta) = -\sin(\theta)$ because sine is an odd function (symmetric about the origin). The graph of $\sin(\pi - \theta)$ is a translation of the graph of $\sin(-\theta)$ π units to the left (half the period of sine), which makes it equal to the value of $\sin(\theta)$. Thus, $\sin(\pi - \theta) = \sin(\theta)$.

$\cos(-\theta) = \cos(\theta)$ because cosine is an even function (symmetric about the y-axis). The graph of $\cos(\pi - \theta)$ is a translation of the graph of $\cos(-\theta)$ π units to the left (half the period of cosine), which makes it equal to the value of $-\cos(\theta)$. Thus, $\cos(\pi - \theta) = -\cos(\theta)$.

$\tan(-\theta) = -\tan(\theta)$ because tangent is an odd function (symmetric about the origin). The graph of $\tan(\pi - \theta)$ is a translation of the graph of $\tan(-\theta)$ π units to the left, and because the period of tangent is π, $\tan(\pi - \theta) = -\tan(\theta)$.

Although the graph of $y = \tan x$ can undergo a vertical stretch or shrink, it does not have an amplitude since the tangent function has no maximum or minimum value. However, other transformations can affect the period of the tangent function.

Example 6.11

Sketch each function.

(a) $f(x) = \tan 2x$

(b) $g(x) = \tan\left[2\left(x - \dfrac{\pi}{4}\right)\right]$

Solution

(a) An equation in the form $y = f(bx)$ indicates a horizontal shrink of $f(x)$ by a factor of $\dfrac{1}{b}$. Hence, the period of $y = \tan 2x$ is $\dfrac{1}{2} \cdot \pi = \dfrac{\pi}{2}$

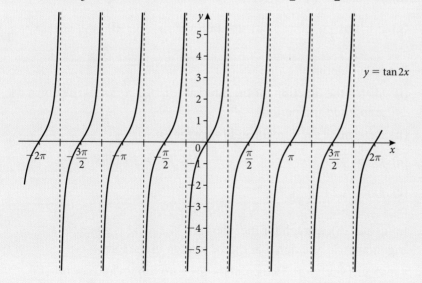

(b) The graph of $y = \tan\left[2\left(x - \dfrac{\pi}{4}\right)\right]$ is obtained by first translating the graph of $y = \tan x$ to the right $\dfrac{\pi}{4}$ units, and then a horizontal shrink by a factor of $\dfrac{1}{2}$. As for $f(x) = \tan 2x$ in part (a) the period of $g(x) = \tan\left[2\left(x - \dfrac{\pi}{4}\right)\right]$ is $\dfrac{\pi}{2}$.

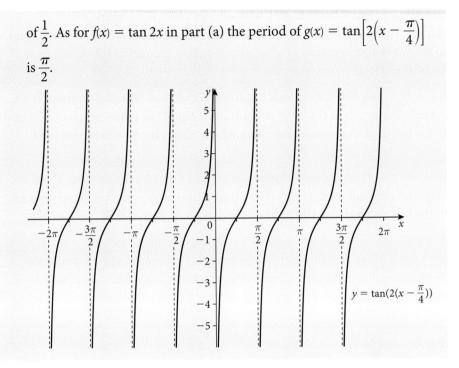

$$y = \tan\left(2\left(x - \tfrac{\pi}{4}\right)\right)$$

Exercise 6.3

1. Without using your GDC, sketch a graph of each equation on the interval $-\pi \leqslant x \leqslant 3\pi$.

   (a) $y = 2\sin x$ **(b)** $y = \cos x - 2$ **(c)** $y = \dfrac{1}{2}\cos x$

   (d) $y = \sin\left(x - \dfrac{\pi}{2}\right)$ **(e)** $y = \cos(2x)$ **(f)** $y = 1 + \tan x$

   (g) $y = \sin\left(\dfrac{x}{2}\right)$ **(h)** $y = \tan\left(x + \dfrac{\pi}{2}\right)$ **(i)** $y = \cos\left(2x - \dfrac{\pi}{4}\right)$

2. For each function:

   (i) Sketch the function for the interval $-\pi \leqslant x \leqslant 5\pi$. Write down its amplitude and period.

   (ii) Determine the domain and range for $f(x)$.

   (a) $f(x) = \dfrac{1}{2}\cos x - 3$ **(b)** $f(x) = 3\sin(3x) - \dfrac{1}{2}$

   (c) $f(x) = 1.2\sin\left(\dfrac{x}{2}\right) + 4.3$

3. In (a) and (b), a graph for the interval $0 \leqslant x \leqslant 12$ is given for a trigonometric function that can be written in the form

   $$y = A\sin\left(\dfrac{\pi}{4}x\right) + B.$$ Two points – one a minimum and the other a maximum – are indicated on the graph. Find the value of A and of B for each.

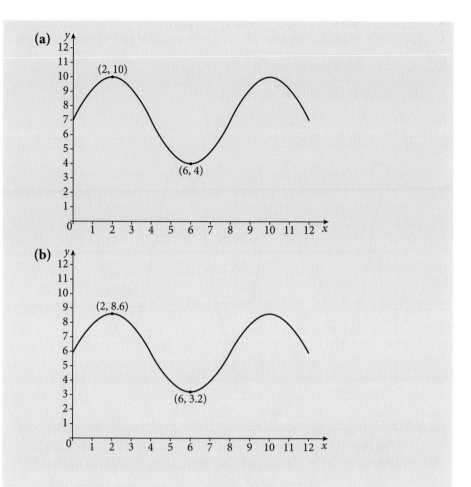

(a)

(2, 10)

(6, 4)

(b)

(2, 8.6)

(6, 3.2)

4. A graph for the interval $0 \leqslant x \leqslant 12$ is given for a trigonometric
   function that can be written in the form $y = A \cos\left(\dfrac{\pi}{4}x\right) + B$.

   Two points – one a minimum and the other a maximum – are indicated
   on the graph. Find the value of A and of B.

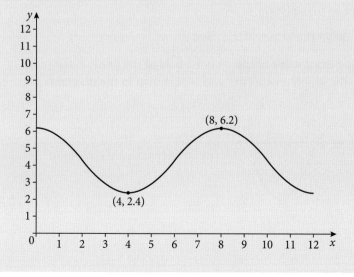

(8, 6.2)

(4, 2.4)

5. The graph of a function in the form $y = p \cos qx$ is given in the diagram.

   (a) Write down the value of p.

   (b) Calculate the value of q.

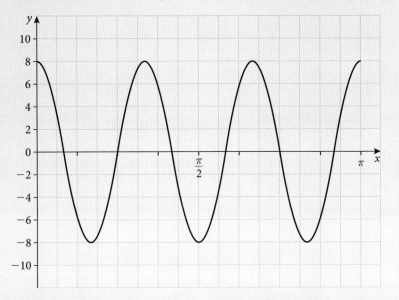

6. (a) With help from your GDC, sketch the graphs of the three reciprocal trigonometric functions $y = \csc x$, $y = \sec x$ and $y = \cot x$ for the interval $0 \leqslant x \leqslant 2\pi$. Include any vertical asymptotes as dashed lines.

   (b) State the domain and range for each of the three reciprocal trigonometric functions.

7. The diagram shows part of the graph of a function whose equation is in the form $y = a \sin(bx) + c$.

   (a) Write down the values of a, b, and c.

   (b) Find the exact value of the x-coordinate of the point P, the point where the graph crosses the x-axis as shown in the diagram.

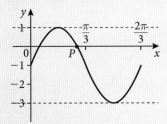

8. The graph represents $y = a \sin(x + b) + c$, where a, b, and c are constants. Find values for a, b, and c.

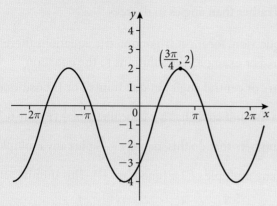

6.4 Trigonometric equations

The primary focus of this section is to give an overview of concepts and strategies for solving **trigonometric equations**. In general, we will look at finding solutions by means of applying algebraic techniques (analytic solution) and/or by analysing a graph (graphical solution). Some examples of trigonometric equations are:

$$\csc x = 2, \quad \sin^2 \theta + \cos^2 \theta = 1, \quad 2\cos(3x - \pi) = 1,$$

$$\sec^2 \alpha - 2\tan \alpha - 4 = 0, \quad \tan 2\theta = \frac{2\tan}{1 - \tan^2 \theta}$$

The equations $\sin^2 \theta + \cos^2 \theta = 1$ and $\tan 2\theta = \dfrac{2\tan}{1 - \tan^2 \theta}$ are examples of special equations called **identities** (Section 6.5). An identity is an equation that is true for all possible values of the variable. The other equations are true for only certain values or for none. Trigonometric identities will be covered in Section 6.5. We will be applying methods similar to that used to solve equations encountered earlier in this book.

The unit circle and exact solutions to trigonometric equations

When you are asked to solve a trigonometric equation, there are two important questions you need to consider:

- Is it possible, or required, to express any solution(s) exactly?
- For what interval of the variable are all solutions to be found?

263

With regard to the first question, exact solutions are only attainable, in most cases, if they are an integer multiple of $\frac{\pi}{6}$ or $\frac{\pi}{4}$. We are primarily interested in finding numerical solutions rather than angles in degrees.

With regard to the second question, for most trigonometric equations there are infinitely many solutions. For example, the solutions to the equation $\sin x = \frac{1}{2}$ are any number (arc or central angle) in quadrants I or II positioned so that the terminal point on the unit circle has a y-coordinate of $\frac{1}{2}$ (Figure 6.23). There are an infinite set of numbers that do this, namely $-\frac{\pi}{6}$ plus any multiple of 2π (quadrant I) or $\frac{5\pi}{6}$ plus any multiple of 2π (quadrant II). This infinite set is written as $x = \frac{\pi}{6} + k \cdot 2\pi$ or $x = \frac{5\pi}{6} + k \cdot 2\pi, k \in \mathbb{Z}$. However, for this course, the number of solutions to any trigonometric equation will be limited to a finite set by the fact that the solution set will always be restricted to a specified interval. For the equation $\sin x = \frac{1}{2}$, if the solution set is restricted to the interval $0 \leqslant x < 2\pi$, then the solutions are $x = \frac{\pi}{6}$ and $x = \frac{5\pi}{6}$. If the solution set is restricted to the interval $0 \leqslant x < \frac{\pi}{2}$, then there is just the one solution, $x = \frac{\pi}{6}$. Figure 6.23 shows the solutions to $\sin x = \frac{1}{2}$ for the interval $0 \leqslant x < 2\pi$, and Figure 6.24 illustrates how the graph of $y = \sin x$ can be used to locate the solutions for the equation $\sin x = \frac{1}{2}$ for different intervals of x.

When asked to solve a trigonometric equation, a solution interval will always be given, as in Example 6.12.

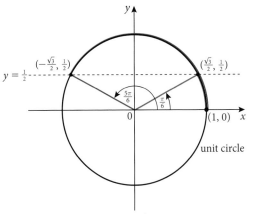

Figure 6.23 Solution to $\sin x = \frac{1}{2}, 0 \leqslant x < 2\pi$

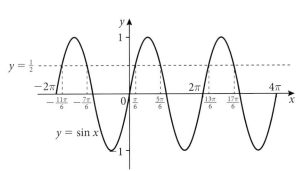

Figure 6.24 Points of intersection between $y = \sin x$ and $y = \frac{1}{2}$

Example 6.12

Find the exact solution(s) to the equation $\sin x \cos x = 2 \cos x$ for $-\pi < x < \pi$. Solve without a GDC (analytic solution).

Solution

There is a temptation to divide both sides by $\cos x$, but this can result in losing a solution to the equation. In fact, for this equation, both solutions would be lost. Instead, set the equation equal to zero and take out the common factor of $\cos x$.

$$\sin x \cos x - 2 \cos x = 0$$

$$\cos x (\sin x - 2) = 0$$

$$\cos x = 0 \text{ or } \sin x = 2$$

2 is outside the range of the sine function so there is no solution to $\sin x = 2$. Solutions to $\cos x = 0$ occur for arcs (angles) that terminate where the x-coordinate is 0. For the solution interval $-\pi < x < \pi$, this occurs where the unit circle intersects the y-axis as shown in the diagram. The analytic solution gives the exact solutions of $x = \dfrac{\pi}{2}$ and $x = -\dfrac{\pi}{2}$.

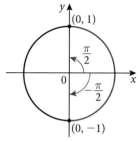

Figure 6.25 Diagram for solution to Example 6.12

Your GDC can be a very effective tool for searching for solutions graphically. However, it can be limited when exact solutions are required. The sequence of GDC images here shows a graphical solution for the equation in Example 6.12. Rather than finding the intersection of the separate graphs of $y = \sin x \cos x$ and $y = 2 \cos x$, the more efficient approach shown here is to find the zeros of the graph of $y = \sin x \cos x - 2 \cos x$.

If exact solutions are required, you need to first attempt an analytic solution, and then a graphical confirmation can be performed.

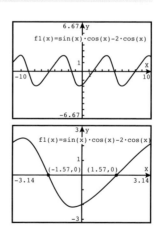

Figure 6.26 Graphical solution for the equation in Example 6.12

Example 6.13

Find the exact solution(s) to the equation $\tan \theta + 1 = 0$ for $0 \leqslant \theta < 360°$

Solution

Since the solution interval is expressed in degrees, it is necessary to give any solution as an angle in degree measure. Solutions to this equation are values of θ such that $\tan \theta = -1$.

Applying the identity $\tan \theta = \dfrac{\sin \theta}{\cos \theta}$, we have $\dfrac{\sin \theta}{\cos \theta} = -1$

The expression $\tan \theta + 1$ is not equivalent to $\tan (\theta + 1)$. It is a good habit to use brackets to make it absolutely clear what is, or is not, the argument of a function.

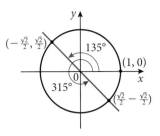

Figure 6.27 Diagram for solution to Example 6.13

We need to find any angles θ such that $\sin \theta$ and $\cos \theta$ have opposite signs. This occurs in quadrant II at $\theta = 135°$ and in quadrant IV at $\theta = 315°$ as shown in the diagram.

It is possible to arrive at exact answers that are not multiples of $\dfrac{\pi}{6}$ or $\dfrac{\pi}{4}$, as Example 6.14 illustrates.

Example 6.14

Find the exact solution(s) to the equation $\cos^2\!\left(x - \dfrac{\pi}{3}\right) = \dfrac{1}{2}$, for $0 \leqslant x < 2\pi$

Solution

The expression $\cos^2\!\left(x - \dfrac{\pi}{3}\right)$ can also be written as $\left[\cos\!\left(x - \dfrac{\pi}{3}\right)\right]^2$

The first step is to take the square root of both sides – remembering that every positive number has two square roots – which gives

$$\cos\!\left(x - \frac{\pi}{3}\right) = \pm\sqrt{\frac{1}{2}} = \pm\frac{1}{\sqrt{2}} = \pm\frac{\sqrt{2}}{2}$$

All of the odd integer multiples of $\dfrac{\pi}{4}\left(\ldots -\dfrac{3\pi}{4}, -\dfrac{\pi}{4}, 0, \dfrac{\pi}{4}, \dfrac{3\pi}{4}, \ldots\right)$ have a cosine equal to either $\dfrac{\sqrt{2}}{2}$ or $-\dfrac{\sqrt{2}}{2}$. That is, $x - \dfrac{\pi}{3} = \dfrac{\pi}{4} + k \cdot \dfrac{\pi}{2}$

Now, solve for x: $x = \dfrac{\pi}{4} + \dfrac{\pi}{3} + k \cdot \dfrac{\pi}{2} = \dfrac{7\pi}{12} + k \cdot \dfrac{6\pi}{12}$

The last step is to substitute in different integer values for k to generate all the possible values for x so that $0 \leqslant x < 2\pi$.

when $k = 0 : x = \dfrac{7\pi}{12}$

when $k = 1 : x = \dfrac{7\pi}{12} + \dfrac{6\pi}{12} = \dfrac{13\pi}{12}$

when $k = 2 : x = \dfrac{7\pi}{12} + \dfrac{12\pi}{12} = \dfrac{19\pi}{12}$

when $k = 3 : x = \dfrac{7\pi}{12} + \dfrac{18\pi}{12} = \dfrac{25\pi}{12}$

however $\dfrac{25\pi}{12} > 2\pi \ldots$ but,

when $k = -1 : x = \dfrac{7\pi}{12} - \dfrac{6\pi}{12} = \dfrac{\pi}{12}$. Therefore, there are four exact solutions

in the interval $0 \leqslant x \leqslant 2\pi$, and they are: $x = \dfrac{\pi}{12}, \dfrac{7\pi}{12}, \dfrac{13\pi}{12}$ and $\dfrac{19\pi}{12}$

As we did at the end of Example 6.12, check the solutions to trigonometric equations with your GDC. The sequence of GDC images verifies that $x = \dfrac{\pi}{12}$ is the first solution to the equation in Example 6.14.

When entering the equation $y = \cos^2\left(x - \dfrac{\pi}{3}\right)$ into your GDC (as shown in the first GDC image), you will have to enter it in the form $y = \left[\cos\left(x - \dfrac{\pi}{3}\right)\right]^2$.

Be aware that $\cos^2\left(x - \dfrac{\pi}{3}\right)$ is not equivalent to $\cos\left(x - \dfrac{\pi}{3}\right)^2$. The expression $\cos\left(x - \dfrac{\pi}{3}\right)^2$ indicates that the quantity $\left(x - \dfrac{\pi}{3}\right)$ is squared first and then the cosine of the resulting value is found. However, the expression $y = \cos^2\left(x - \dfrac{\pi}{3}\right)$ indicates that the cosine of $\left(x - \dfrac{\pi}{3}\right)$ is found first and then that value is squared.

Graphical solutions to trigonometric equations

If exact solutions are not required, then a graphical solution using your GDC is a very effective way to find approximate solutions to trigonometric equations. Unless instructed to do otherwise, you should give approximate solutions to an accuracy of 3 significant figures.

Example 6.15

Find all solutions to the equation $3\tan x = 2\cos x$ in the interval $0 \leqslant x < 2\pi$

Solution

Graph the equation $y = 3\tan x - 2\cos x$ and find all of its zeros (x-intercepts) in the interval $0 \leqslant x < 2\pi$. Because the domain of the tangent function is $\left\{x : x \in \mathbb{R},\ x \neq \dfrac{\pi}{2} + k\pi, k \in \mathbb{Z}\right\}$, we expect there to be gaps (and vertical asymptotes) in the graph at $x = \dfrac{\pi}{2}$ and at $x = \dfrac{3\pi}{2}$

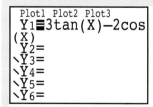 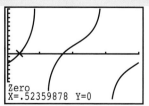 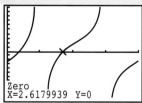

This sequence of GDC images indicates approximate solutions of $x \approx 0.524$ and $x \approx 2.62$, accurate to 3 significant figures.

A graphical approach is effective and appropriate when it is very difficult, or not possible, to find exact solutions.

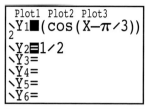

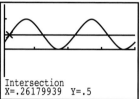

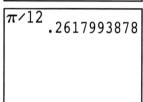

Figure 6.28 Solving the equation on a GDC

It is possible to solve the equation in Example 6.15 analytically. The exact solutions are $x = \dfrac{\pi}{6}$ and $x = \dfrac{5\pi}{6}$.

Example 6.16

The peak height h metres of ocean waves during a storm is given by the equation $h = 9 + 4 \sin\left(\dfrac{t}{2}\right)$ where t is the number of hours after midnight.
An alarm is triggered when the peak height goes above 12.5 metres. Find the value of t when the alarm first sounds.

Solution

Graph the equations $y = 9 + 4 \sin\left(\dfrac{x}{2}\right)$ and $y = 12.5$ and find the first point of intersection for $x > 0$.

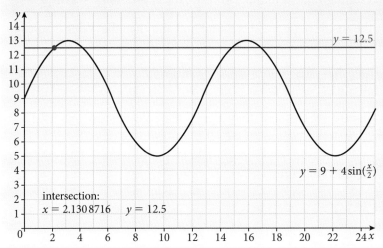

Using the Intersect command on a GDC indicates that the first point of intersection has an x-coordinate of approximately 2.13. Therefore, the alarm will first sound when $t \approx 2.13$.

Analytic solutions to trigonometric equations

An analytical approach requires you to devise a solution strategy using algebraic methods that you have applied to other types of equations – such as quadratic equations. Often, but not always, trigonometric equations that demand an analytic approach will result in exact solutions. Although our approach for equations in this section focuses on algebraic techniques, it is important to use graphical methods to support or confirm analytical solutions.

Example 6.17

Solve $2 \sin^2 x + \sin x = 0$, for $0 \leqslant x < 2\pi$

Solution

Factorising gives $\sin x(2 \sin x + 1) = 0$

$$\sin x = 0 \text{ or } \sin x = -\frac{1}{2}$$

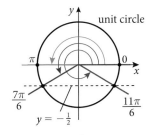

Figure 6.29 Solution to Example 6.17

Solutions to $\sin x = 0$ are where the angle is on the x-axis; and solutions to $\sin x = -\dfrac{1}{2}$ are angles in quadrants III and IV such that their intersection point with the unit circle has y-coordinate of $-\dfrac{1}{2}$

For $\sin x = 0$: $x = 0, \pi$ For $\sin x = -\dfrac{1}{2}$: $x = \dfrac{7\pi}{6}, \dfrac{11\pi}{6}$

Therefore, the solutions are $x = 0, \pi, \dfrac{7\pi}{6}, \dfrac{11\pi}{6}$

Example 6.18 illustrates how the application of a trigonometric identity can be helpful to rewrite an equation in a way that allows us to solve it algebraically. The next section will introduce many further trigonometric identities and examples of using them to assist in solving trigonometric equations.

Example 6.18

Solve $3\cos x + \cot x = 0$ for $0 \leqslant x \leqslant 2\pi$

Solution

Since the structure of this equation is such that an expression is set equal to zero, it would be ideal to be able to use the same algebraic technique as the previous example – that is, factorise it and solve for when each factor is zero. However, it is not possible to factorise the expression $3\cos x + \cot x$, and rewriting the equation as $3\cos x = -\cot x$ does not help either.

Are there any expressions in the equation for which we can substitute an equivalent expression that will make the equation accessible to an algebraic solution? We do not have any equivalent expressions for $\cos x$, but we do have an identity for $\cot x$. Since $\cot x$ is the reciprocal of $\tan x$ we know that $\cot x = \dfrac{\cos x}{\sin x}$. Let's substitute $\dfrac{\cos x}{\sin x}$ for $\cot x$.

$$3\cos x + \frac{\cos x}{\sin x} = 0$$

Noting that $\sin x \neq 0$, multiply both sides by $\sin x$: $\dfrac{3\sin x \cos x + \cos x}{\sin x} = 0$

A fraction equals zero when the numerator equals zero:
$3\sin x \cos x + \cos x = 0$

Factorise: $\cos x(3\sin x + 1) = 0$

$\cos x = 0$ or $\sin x = -\dfrac{1}{3}$

For $\cos x = 0$: $x = \dfrac{\pi}{2}, \dfrac{3\pi}{2}$

Although exact answers were not demanded in Example 6.17, given our knowledge of the unit circle and familiarity with the sine of common values (i.e. multiples of $\dfrac{\pi}{6}$ and $\dfrac{\pi}{4}$), we are able to give exact answers without any difficulty. It would have been acceptable to give approximate solutions using your GDC, but this would have required considerably more effort than providing exact solutions. Entering and graphing the equation $y = 2\sin^2 x + \sin x$ on your GDC would not be the most efficient or appropriate solution method, but if sufficient time is available it is an effective way to confirm your exact solutions. Remember that $\sin^2 x$ must be entered in a GDC as $(\sin x)^2$.

We know that solutions to $\cos x = 0$ are values of x giving the two exact solutions of $\dfrac{\pi}{2}$ and $\dfrac{3\pi}{2}$. Although we know solutions to $\sin x = -\dfrac{1}{3}$ are in quadrants III and IV, we do not know their exact values. So, we will need to use our GDC to find approximate solutions to

$\sin x = -\dfrac{1}{3}$ for $0 \leqslant x \leqslant 2\pi$

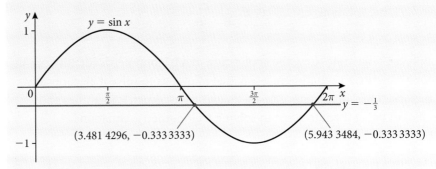

(3.481 4296, −0.333 3333) (5.943 3484, −0.333 3333)

Thus, for $\sin x = -\dfrac{1}{3}$: $x = 3.48$ or $x = 5.94$ (both to 3 s.f.)

Therefore, the full solution set for the equation is $x = \dfrac{\pi}{2}, \dfrac{3\pi}{2}; x \approx 3.48, 5.94$

Exercise 6.4

1. Find the exact solution(s) to each equation for $0 \leqslant x < 2\pi$. Verify your solution(s) with your GDC.

   (a) $\cos x = \dfrac{1}{2}$

   (b) $2 \sin x + 1 = 0$

   (c) $1 - \cot x = 0$

   (d) $\sqrt{3} = 2 \sin x$

   (e) $2 \sin^2 x = 1$

   (f) $4 \cos^2 x = 3$

   (g) $\tan^2 x - 1 = 0$

   (h) $4 \cos^2 x = 1$

   (i) $\tan x(\tan x + 1) = 0$

   (j) $\sin x \cos x = 0$

   (k) $5 - \sec x = 3$

   (l) $\csc^2 x = 2$

2. Use your GDC to find approximate solution(s) for $0 \leqslant x < 2\pi$. Express solutions to 3 significant figures.

   (a) $\sin x = 0.4$

   (b) $3 \cos x + 1 = 0$

   (c) $\tan x = 2$

   (d) $\sec 2x = 3.46$

   (e) $\cos(x - 1) = -0.38$

   (f) $3 \tan^2 x = 1$

   (g) $\csc(2x - 3) = \dfrac{3}{2}$

   (h) $3 \cot x = 10$

3. Given that k is any integer, list all of the possible values for x that are in the specified interval.

(a) $\dfrac{\pi}{2} + k \cdot \pi,\ -3\pi \leqslant x \leqslant 3\pi$

(b) $\dfrac{\pi}{6} + k \cdot 2\pi,\ -2\pi \leqslant x \leqslant 2\pi$

(c) $\dfrac{7\pi}{12} + k \cdot \pi,\ 0 \leqslant x < 2\pi$

(d) $\dfrac{\pi}{4} + k \cdot \dfrac{\pi}{4},\ 0 \leqslant x < 4\pi$

4. Find the exact solutions for the indicated interval. The interval will also indicate whether the solutions are given in degree or radian measure. Write a complete analytic solution.

(a) $\cos\left(x - \dfrac{\pi}{6}\right) = -\dfrac{1}{2},\ 0 \leqslant x < 2\pi$

(b) $\tan(\theta + \pi) = 1,\ -\pi \leqslant \theta \leqslant \pi$

(c) $\sin 2x = \dfrac{\sqrt{3}}{2},\ 0 \leqslant x < 360°$

(d) $\sin^2\left(\alpha + \dfrac{\pi}{2}\right) = \dfrac{3}{4},\ \dfrac{-\pi}{2} \leqslant \alpha \leqslant \dfrac{\pi}{2}$

(e) $2\cos^2\theta - 5\cos\theta - 3 = 0,\ 0 \leqslant \theta < 2\pi$

(f) $3\tan x = 2\cos x,\ 0 \leqslant x < 2\pi$

(g) $2\cos(3x + 24°) = \sqrt{2},\ 0 \leqslant x < 360°$

(h) $9\sec^2\theta = 12,\ 0 \leqslant \theta \leqslant \pi$

5. The number, N, of empty bird nests in a park is approximated by the function $N = 74 + 42\sin\left(\dfrac{\pi}{12}t\right)$, where t is the number of hours after midnight. Find the value of t when the number of empty nests first equals 90. Give the answer to 1 decimal place.

6. In Edinburgh, the number of hours of daylight, H, on day D is modelled by the function $H = 12 + 7.26\sin\left[\dfrac{2\pi}{365}(D - 80)\right]$ where D is the number of days after December 31 (e.g. January 1 is $D = 1$, January 2 is $D = 2$, and so on). Do not use your GDC for part (a).

(a) Which days of the year have 12 hours of daylight?

(b) Which days of the year have about 15 hours of daylight?

(c) How many days of the year have more than 17 hours of daylight?

7. Solve each equation for the stated solution interval. Find exact solutions when possible, otherwise give solutions to 3 significant figures. Verify solutions with your GDC.

   (a) $2\cos^2 x + \cos x = 0, 0 \leqslant x < 2\pi$

   (b) $2\sin^2 \theta - \sin \theta - 1 = 0, 0 \leqslant \theta < 2\pi$

   (c) $\tan^2 x - \tan x = 2, -90° \leqslant x \leqslant 90°$

   (d) $3\cos^2 x - 6\cos x = 2, -\pi < x \leqslant \pi$

   (e) $2\sin \beta = 3\cos \beta, 0 \leqslant \beta \leqslant 180°$

   (f) $\sin^2 x = \cos^2 x, 0 \leqslant x \leqslant \pi$

   (g) $\sec^2 x + 2\sec x + 4 = 0, 0 \leqslant x < 2\pi$

   (h) $\sin x \tan x = 3\sin x, 0 \leqslant x < 360°$

6.5 Trigonometric identities

You will recall that an identity is an equation that is true for all values of the variable for which the expressions in the equation are defined. Several trigonometric identities were introduced earlier in this chapter and a number of important new identities are presented and proved in this section.

Trigonometric identities are used in a variety of ways. For example, one of the reciprocal identities is applied whenever the cosecant, secant or cotangent function is evaluated on a calculator. The following uses of trigonometric identities will be illustrated in this section:

- evaluating trigonometric functions
- simplifying trigonometric expressions
- proving other trigonometric identities
- solving trigonometric equations.

First we will develop some further trigonometric identities that are organised into three groups: Pythagorean identities, compound angle identities, and double angle identities.

At the start of Section 6.4, it was stated that the equation $\sin^2 \theta + \cos^2 \theta = 1$ is an identity; that is, it's true for all possible values of θ. Let's prove that this is indeed the case.

Recall from Section 6.1 that the equation for the unit circle is $x^2 + y^2 = 1$. That is, the coordinates (x, y) of any point on the circle satisfy the equation $x^2 + y^2 = 1$. As we learned in Section 6.2, if θ is any real number that represents a central angle (in radian measure) of the unit circle that terminates at (x, y), then $x = \cos \theta$ and $y = \sin \theta$. Substituting directly into the equation for the circle gives $\sin^2 \theta + \cos^2 \theta = 1$. Therefore, the equation $\sin^2 \theta + \cos^2 \theta = 1$ is true for any real number θ.

Phrases such as 'prove the identity' and 'verify the identity' are often used. Both mean 'prove that the given equation is an identity'. We do this by performing a series of algebraic manipulations to show that the expression on one side of the equation can be transformed into the expression on the other side, or that both expressions can be transformed into some third expression. When verifying that an equation is an identity, you should not perform an operation to both sides of the equation; for example, multiplying both sides of the equation by a quantity. This can be done only if it is known that the two sides of the equation are equal, but this is exactly what we are trying to verify in the process of proving an identity.

Example 6.19

Prove that $1 + \tan^2 \theta = \sec^2 \theta$ is an identity.

Solution

There is more of an opportunity to perform algebraic manipulations on the left side than the right side. Thus, our task is to transform the expression $1 + \tan^2 \theta$ into the expression $\sec^2 \theta$.

$$1 + \tan^2 \theta = \sec^2 \theta$$ Using the identity $\tan \theta = \dfrac{\sin \theta}{\cos \theta}$, substitute $\dfrac{\sin^2 \theta}{\cos^2 \theta}$ for $\tan^2 \theta$

$$1 + \frac{\sin^2 \theta}{\cos^2 \theta} =$$ Find a common denominator.

$$\frac{\cos^2 \theta}{\cos^2 \theta} + \frac{\sin^2 \theta}{\cos^2 \theta} =$$

$$\frac{\cos^2 \theta + \sin^2 \theta}{\cos^2 \theta} =$$ Apply the Pythagorean identity $\sin^2 \theta + \cos^2 \theta = 1$

$$\frac{1}{\cos^2 \theta} =$$ Because $\dfrac{1}{\cos \theta} = \sec \theta$, then $\dfrac{1}{\cos^2 \theta} = \sec^2 \theta$.

$$\sec^2 \theta = \sec^2 \theta$$ QED

Another identity that can be proved in a similar way is
$1 + \cot^2 \theta = \csc^2 \theta$.

Pythagorean identities
$\sin^2 \theta + \cos^2 \theta = 1$
$1 + \tan^2 \theta = \sec^2 \theta$
$1 + \cot^2 \theta = \csc^2 \theta$

The Pythagorean identities are sometimes used in radical forms such as
$\sin \theta = \pm\sqrt{1 - \cos^2 \theta}$ or
$\tan \theta = \pm\sqrt{\sec^2 \theta - 1}$
where the sign ($+$ or $-$) depends on which quadrant θ is in.

Example 6.20

(a) Express $2\cos^2 x + \sin x$ in terms of $\sin x$ only.

(b) Solve the equation $2\cos^2 x + \sin x = -1$ for x in the interval $0 \leqslant x \leqslant 2\pi$, expressing your answer(s) exactly.

Solution

(a) $2\cos^2 x + \sin x = 2(1 - \sin^2 x) + \sin x$ Using Pythagorean identity:
$\qquad\qquad\qquad = 2 - 2\sin^2 x + \sin x$ $\cos^2 x = 1 - \sin^2 x$.

(b) $2\cos^2 x + \sin x = -1$

$2 - 2\sin^2 x + \sin x - 3 = 0$

$(2\sin^2 x - 3)(\sin x + 1) = 0$

$\sin x = \dfrac{3}{2}$ or $\sin x = -1$

For $\sin x = \dfrac{3}{2}$: no solution because $\dfrac{3}{2}$ is not in the range of sine function.

For $\sin x = -1$: $x = \dfrac{3\pi}{2}$.

Therefore, only one solution in $0 \leqslant x \leqslant 2\pi$: $x = \dfrac{3\pi}{2}$

Compound angle identities (sum and difference identities)

In this section we develop trigonometric identities known as the **compound angle identities** for sine, cosine, and tangent. These contain the expressions $\sin(\alpha + \beta)$, $\sin(\alpha - \beta)$, $\cos(\alpha + \beta)$, $\cos(\alpha - \beta)$, $\tan(\alpha + \beta)$, and $\tan(\alpha - \beta)$. We first find a formula for $\cos(\alpha + \beta)$.

You might wonder whether $\cos(\alpha + \beta) = \cos\alpha + \cos\beta$. Often it is easier to prove a mathematical statement false than to prove it true. One counterexample is sufficient to prove a statement false.

Let $\alpha = \dfrac{\pi}{3}$ and $\beta = \dfrac{\pi}{6}$. Does $\cos\left(\dfrac{\pi}{3} + \dfrac{\pi}{6}\right) = \cos\dfrac{\pi}{3} + \cos\dfrac{\pi}{6}$?

$\cos\left(\dfrac{\pi}{3} + \dfrac{\pi}{6}\right) = \cos\left(\dfrac{2\pi}{6} + \dfrac{\pi}{6}\right) = \cos\left(\dfrac{3\pi}{6}\right) = \cos\left(\dfrac{\pi}{2}\right) = 0$ and

$\cos\dfrac{\pi}{3} + \cos\dfrac{\pi}{6} = \dfrac{1}{2} + \dfrac{\sqrt{3}}{2} = \dfrac{1 + \sqrt{3}}{2}$

Thus, the answer is no; $\cos\left(\dfrac{\pi}{3} + \dfrac{\pi}{6}\right) \neq \cos\dfrac{\pi}{3} + \cos\dfrac{\pi}{6}$

Although $\cos(\alpha + \beta) = \cos\alpha + \cos\beta$ may be true for some values (e.g. it's true for $\alpha = \dfrac{\pi}{2}$ and $\beta = \dfrac{3\pi}{4}$), it's not true for all possible values of α and β, and therefore, it is not an identity.

Derivation of identity for the cosine of the sum of two numbers

> Greek letters such as α (alpha), β (beta), and θ (theta) are frequently used to name angles. In the development of the formula for $\cos(\alpha + \beta)$, α and β are arcs along the unit circle, but they could just as well represent the central angle (in radian measure) that cuts off (subtends) the arc.

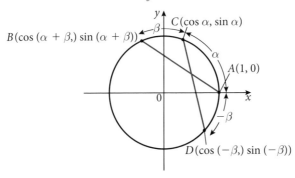

Figure 6.30 Two chords of the unit circle

To find a formula for $\cos(\alpha + \beta)$, we use Figure 6.30 showing the four points A, B, C, and D on the unit circle and the two chords AB and CD. The arc lengths α, β and $-\beta$ are marked. The coordinates of A, B, C, and D in terms of sines and cosines of the arcs are also indicated. The coordinates of point D are $(\cos(-\beta), \sin(-\beta))$, but we can apply the odd/even identities to write the coordinates of D more simply as $(\cos \beta, -\sin \beta)$. Note that the arc length from A to B is equal to the arc length from D to C because they both have a length equal to $\alpha + \beta$. Since equal arcs on a circle determine equal chords, it must follow that $AB = CD$. Using the respective coordinates for A, B, C, and D, we can express $AB = CD$ using the distance formula as

$$\sqrt{(\cos(\alpha + \beta) - 1)^2 + \sin^2(\alpha + \beta)} = \sqrt{(\cos \alpha - \cos \beta)^2 + (\sin \alpha + \sin \beta)^2}$$

Squaring both sides and expanding, gives:

$$\cos^2(\alpha + \beta) - 2\cos(\alpha + \beta) + 1 + \sin^2(\alpha + \beta)$$
$$= \cos^2 \alpha - 2\cos \alpha \cos \beta + \cos^2 \beta + \sin^2 \alpha + 2\sin \alpha \sin \beta + \sin^2 \beta$$
$$[\cos^2(\alpha + \beta) + \sin^2(\alpha + \beta)] - 2\cos(\alpha + \beta) + 1$$
$$= (\cos^2 \alpha + \sin^2 \alpha) + (\sin^2 \beta + \cos^2 \beta) - 2\cos \alpha \cos \beta + 2\sin \alpha \sin \beta$$

Applying the identity $\sin^2 \theta + \cos^2 \theta = 1$, we can replace three expressions with one:

$$1 - 2\cos(\alpha + \beta) + 1 = 1 + 1 - 2\cos \alpha \cos \beta + 2\sin \alpha \sin \beta$$

Subtracting 2 from each side and dividing both sides by -2, gives

$$\cos(\alpha + \beta) = \cos \alpha \cos \beta - \sin \alpha \sin \beta$$

This is the identity for the cosine of the sum of two numbers.

Now we can find exact values of a trigonometric function for numbers other than multiples of $\frac{\pi}{6}$ or $\frac{\pi}{4}$.

Example 6.21

Find the exact values for $\cos \dfrac{5\pi}{12}$

Solution

$$\frac{5\pi}{12} = \frac{\pi}{4} + \frac{\pi}{6}$$

Applying the identity $\cos(\alpha + \beta) = \cos \alpha \cos \beta - \sin \alpha \sin \beta$

with $\alpha = \dfrac{\pi}{4}$ and $\beta = \dfrac{\pi}{6}$, gives $\cos\left(\dfrac{\pi}{4} + \dfrac{\pi}{6}\right) = \cos\dfrac{\pi}{4}\cos\dfrac{\pi}{6} - \sin\dfrac{\pi}{4}\sin\dfrac{\pi}{6}$

$$= \left(\frac{\sqrt{2}}{2}\right)\left(\frac{\sqrt{3}}{2}\right) - \left(\frac{\sqrt{2}}{2}\right)\left(\frac{1}{2}\right)$$

$$= \frac{\sqrt{6}}{4} - \frac{\sqrt{2}}{4} = \frac{\sqrt{6} - \sqrt{2}}{4}$$

Therefore, $\cos\dfrac{5\pi}{12} = \dfrac{\sqrt{6} - \sqrt{2}}{4}$

Derivation of identity for the cosine of the difference of two numbers

We can use the identity for the cosine of the sum of two numbers and the fact that cosine is an even function and sine is an odd function to derive the formula for $\cos(\alpha - \beta)$.

Let's replace β with $-\beta$ in $\cos(\alpha + \beta) = \cos \alpha \cos \beta - \sin \alpha \sin \beta$.

$$\cos[\alpha + (-\beta)] = \cos \alpha \cos(-\beta) - \sin \alpha \sin(-\beta)$$

Substituting $-\sin \beta$ for $\sin(-\beta)$, and $\cos \beta$ for $\cos(-\beta)$, gives

$$\cos(\alpha - \beta) = \cos \alpha \cos \beta + \sin \alpha \sin \beta$$

This is the identity for the cosine of the difference of two numbers.

Example 6.22

Given that A and B are numbers representing arcs or angles that are in the first quadrant, and $\sin A = \dfrac{4}{5}$ and $\cos B = \dfrac{12}{13}$, find the exact values of

(a) $\cos(A + B)$ (b) $\cos(A - B)$

Solution

We are given the exact values for $\sin A$ and $\cos B$, but we also need exact values for $\sin B$ and $\cos A$ in order to use the sum and difference identities for cosine.

Since B is in the first quadrant then $\sin B > 0$ and rearranging one of the Pythagorean identities: $\sin B = \sqrt{1 - \cos^2 B} = \sqrt{1 - \left(\dfrac{12}{13}\right)^2} = \sqrt{\dfrac{25}{169}} = \dfrac{5}{13}$

Similarly, $\cos A = \sqrt{1 - \sin^2 A} = \sqrt{1 - \left(\dfrac{4}{5}\right)^2} = \sqrt{\dfrac{9}{25}} = \dfrac{3}{5}$

(a) Use the identity for the cosine of the sum of two numbers:

$$\cos(A + B) = \cos A \cos B - \sin A \sin B = \left(\dfrac{3}{5}\right)\left(\dfrac{12}{13}\right) - \left(\dfrac{4}{5}\right)\left(\dfrac{5}{13}\right) = \dfrac{16}{65}$$

(b) Use the identity for the cosine of the difference of two numbers:

$$\cos(A - B) = \cos A \cos B + \sin A \sin B = \left(\dfrac{3}{5}\right)\left(\dfrac{12}{13}\right) + \left(\dfrac{4}{5}\right)\left(\dfrac{5}{13}\right) = \dfrac{56}{65}$$

In Example 6.22, we obtained $\cos(A + B)$ and $\cos(A - B)$ without finding the actual values of A and B.

Derivation of identities for the sine of the sum and difference of two numbers

The identity $\cos(\alpha - \beta) = \cos \alpha \cos \beta + \sin \alpha \sin \beta$ can be used to derive an identity for $\sin(\alpha + \beta)$. Substituting $\frac{\pi}{2}$ for α and $(\alpha + \beta)$ for β, gives

$$\cos\left[\frac{\pi}{2} - (\alpha + \beta)\right] = \cos\left[\left(\frac{\pi}{2} - \alpha\right) - \beta\right] = \cos\left(\frac{\pi}{2} - \alpha\right)\cos \beta + \sin\left(\frac{\pi}{2} - \alpha\right)\sin \beta$$

Now using the co-function identities $\cos\left(\frac{\pi}{2} - x\right) = \sin x$ and $\sin\left(\frac{\pi}{2} - x\right) = \cos x$, we have

$$\sin(\alpha + \beta) = \sin \alpha \cos \beta + \cos \alpha \sin \beta$$

This is the identity for the sine of the sum of two numbers.

By replacing β with $-\beta$ in the identity $\sin(\alpha + \beta) = \sin \alpha \cos \beta + \cos \alpha \sin \beta$, we get

$$\sin(\alpha - \beta) = \sin \alpha \cos(-\beta) + \cos \alpha \sin(-\beta)$$

Applying the odd/even identities for $\cos(-\beta)$ and $\sin(-\beta)$, produces

$$\sin(\alpha - \beta) = \sin \alpha \cos \beta - \cos \alpha \sin \beta$$

This is the identity for the sine of the difference of two numbers.

Derivation of identities for the tangent of the sum and difference of two numbers

To produce an identity for $\tan(\alpha + \beta)$ in terms of $\tan \alpha$ and $\tan \beta$, we start with the fundamental identity that the tangent is the quotient of the sine and cosine:

$$\tan(\alpha + \beta) = \frac{\sin(\alpha + \beta)}{\cos(\alpha + \beta)}, \text{ given } \cos(\alpha + \beta) \neq 0$$

$$= \frac{\sin \alpha \cos \beta + \cos \alpha \sin \beta}{\cos \alpha \cos \beta - \sin \alpha \sin \beta}$$

So that the identity will involve $\tan \alpha$ and $\tan \beta$, we divide the numerator and denominator by $\cos \alpha \cos \beta$, with the assumption that $\cos \alpha \cos \beta \neq 0$.

$$= \frac{\dfrac{\sin \alpha \cos \beta}{\cos \alpha \cos \beta} + \dfrac{\cos \alpha \sin \beta}{\cos \alpha \cos \beta}}{\dfrac{\cos \alpha \cos \beta}{\cos \alpha \cos \beta} - \dfrac{\sin \alpha \sin \beta}{\cos \alpha \cos \beta}}$$

$$\tan(\alpha + \beta) = \frac{\tan \alpha + \tan \beta}{1 - \tan \alpha \tan \beta}$$

This is the identity for the tangent of the sum of two numbers.

If in this identity β is replaced with $-\beta$, we get

$$\tan[\alpha + (-\beta)] = \frac{\tan \alpha + \tan(-\beta)}{1 - \tan \alpha \tan(-\beta)}$$

Tangent is an odd function, so $\tan(-\beta) = -\tan\beta$. Making this substitution gives

$$\tan(\alpha - \beta) = \frac{\tan\alpha - \tan\beta}{1 + \tan\alpha\tan\beta}$$

This is the identity for the tangent of the difference of two numbers.

Compound angle identities

$\cos(\alpha + \beta) = \cos\alpha\cos\beta - \sin\alpha\sin\beta$ $\qquad$ $\cos(\alpha - \beta) = \cos\alpha\cos\beta + \sin\alpha\sin\beta$

$\sin(\alpha + \beta) = \sin\alpha\cos\beta + \cos\alpha\sin\beta$ $\qquad$ $\sin(\alpha - \beta) = \sin\alpha\cos\beta - \cos\alpha\sin\beta$

$\tan(\alpha + \beta) = \dfrac{\tan\alpha + \tan\beta}{1 - \tan\alpha\tan\beta}$ $\qquad$ $\tan(\alpha - \beta) = \dfrac{\tan\alpha - \tan\beta}{1 + \tan\alpha\tan\beta}$

The compound angle identities are also referred to as the 'sum and difference identities', or the 'addition and subtraction identities'.

Example 6.23

Given that $\tan(A + B) = \dfrac{1}{7}$ and $\tan A = 3$, find the value of $\tan B$.

Solution

$\tan(A + B) = \dfrac{\tan A + \tan B}{1 - \tan A\tan B}$ $\qquad$ Start with the identity for the tangent of the sum of two numbers

$\dfrac{1}{7} = \dfrac{3 + \tan B}{1 - 3\tan B}$ $\qquad$ Substitute the given values for $\tan A$ and $\tan(A + B)$ into the identity.

$21 + 7\tan B = 1 - 3\tan B$ $\qquad$ Cross-multiply and solve for $\tan B$

$10\tan B = -20$

$\tan B = -2$

Note that, similar to Example 6.22, we found the exact value of $\tan B$ without finding the actual value of B. In fact, we're not even certain which quadrant B is in, only that it must be in either quadrant II or IV since $\tan B < 0$.

Double angle identities

Is $\sin 2\theta = 2\sin\theta$ an identity? Clearly, it is not, as the counterexample $\theta = \dfrac{\pi}{6}$ shows.

$$\sin\left(2 \cdot \frac{\pi}{6}\right) = \sin\left(\frac{\pi}{3}\right) = \frac{\sqrt{3}}{2}, \text{ and } 2\sin\left(\frac{\pi}{6}\right) = 2\left(\frac{1}{2}\right) = 1$$

A direct consequence of the compound angle identities we have developed are formulae for $\sin 2\theta$, $\cos 2\theta$ and $\tan 2\theta$ – that is, **double angle identities**. For example, the formula for $\sin 2\theta$ can be derived by taking the identity for the sine of two numbers and by letting $\alpha = \beta = \theta$.

$$\sin 2\theta = \sin(\theta + \theta) = \sin\theta\cos\theta + \cos\theta\sin\theta = 2\cos\theta\sin\theta$$

Similarly, for $\cos 2\theta$ we have,

$$\cos 2\theta = \cos(\theta + \theta) = \cos\theta\cos\theta - \sin\theta\sin\theta = \cos^2\theta - \sin^2\theta$$

By applying the Pythagorean identity $\sin^2\theta + \cos^2\theta = 1$, we can write the double angle identity for $\cos 2\theta$ in two other useful ways.

$$\cos 2\theta = \cos^2\theta - \sin^2\theta = \cos^2\theta - (1 - \cos^2\theta) = 2\cos^2\theta - 1$$
$$\cos 2\theta = \cos^2\theta - \sin^2\theta = (1 - \sin^2\theta) - \sin^2\theta = 1 - 2\sin^2\theta$$

To derive the formula for expressing $\tan 2\theta$ in terms of $\tan\theta$, we take the same approach and start with the identity for the tangent of the sum of two numbers, and let $\alpha = \beta = \theta$.

$$\tan(\theta + \theta) = \frac{\tan\theta + \tan\theta}{1 - \tan\theta\tan\theta} = \frac{2\tan\theta}{1 - \tan^2\theta}$$

Now let's look at some further applications of the trigonometric identities we have established, especially for solving more sophisticated equations.

Double angle identities

$$\sin 2\theta = 2\sin\theta\cos\theta \qquad \cos 2\theta = \begin{cases} \cos^2\theta - \sin^2\theta \\ 2\cos^2\theta - 1 \\ 1 - 2\sin^2\theta \end{cases} \qquad \tan 2\theta = \frac{2\tan\theta}{1 - \tan^2\theta}$$

Example 6.24

Solve the equation $\cos 2x + \cos x = 0$ for $0 \leqslant x < 2\pi$ without a GDC.

Solution

The expression $\cos 2x + \cos x$ contains terms with only the cosine function but with different arguments, x and $2x$. To solve algebraically, we need both functions to have one argument, x (rather than $2x$). There are three different double angle identities for $\cos 2x$. It is best to have the equation in terms of one trigonometric function, so we choose to substitute $2\cos^2 x - 1$ for $\cos 2x$.

$$\cos 2x + \cos x = 0$$
$$\Rightarrow 2\cos^2 x - 1 + \cos x = 0$$
$$\Rightarrow 2\cos^2 x + \cos x - 1 = 0$$
$$(2\cos x - 1)(\cos x + 1) = 0$$
$$\Rightarrow \cos x = \frac{1}{2} \text{ or } \cos x = -1$$

for $\cos x = \frac{1}{2}$: $x = \frac{\pi}{3}, \frac{5\pi}{3}$

for $\cos x = -1$: $x = \pi$

Therefore, all of the solutions in the interval $0 \leqslant x < 2\pi$ are: $x = \frac{\pi}{3}, \pi, \frac{5\pi}{3}$

Example 6.25

Solve the equation $2 \sin 2x = 3 \cos x$ for $0 \leqslant x \leqslant \pi$

Solution

$2 \sin 2x = 3 \cos x$	
$2(2 \sin x \cos x) = 3 \cos x$	Using double angle identity for sine.
$4 \sin x \cos x = 3 \cos x$	Do not divide by $\cos x$; solution(s) may be eliminated.
$4 \sin x \cos x - 3 \cos x = 0$	Set equal to zero to prepare for solving by factorisation.
$\cos x(4 \sin x - 3) = 0$	Factorise.

$\cos x = 0$ or $\sin x = \dfrac{3}{4}$

For $\cos x = 0$: $x = \dfrac{\pi}{2}$

For $\sin x = \dfrac{3}{4}$: $x \approx 0.848$ or 2.29

Approximate solutions are found using the Intersect command on the GDC.
All solutions in interval $0 \leqslant x \leqslant \pi$ are: $x = \dfrac{\pi}{2}$; $x \approx 0.848$, 2.29

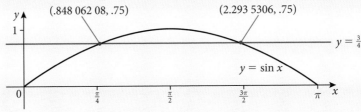

Example 6.26 illustrates how trigonometric identities can be applied to find exact values to trigonometric expressions.

A good approach to proving identities is to work on one side of the equation. Choosing the side that has an expression that is more complicated is often an efficient path to transform the expression to the one on the other side by means of algebraic manipulations and substitutions. .

Example 6.26

Given that $\cos x = \dfrac{1}{4}$ and that $0 < x < \dfrac{\pi}{2}$, find the exact values of

(a) $\sin x$ (b) $\sin 2x$

Solution

(a) Given $0 \leqslant x \leqslant \dfrac{\pi}{2}$ it follows that $\sin x > 0$, because the arc with length x will terminate in the first quadrant. The Pythagorean identity is useful when relating $\sin x$ and $\cos x$.

$$\sin^2 x = 1 - \cos^2 x \Rightarrow \sin x = \sqrt{1 - \cos^2 x}$$

$$\Rightarrow \sin x = \sqrt{1 - \left(\frac{1}{4}\right)^2} = \sqrt{\frac{15}{16}} = \frac{\sqrt{15}}{4}$$

(b) $\sin 2x = 2 \sin x \cos x = 2\left(\dfrac{\sqrt{15}}{4}\right)\left(\dfrac{1}{4}\right) = \dfrac{\sqrt{15}}{8}$

Example 6.27

Prove the identity $\dfrac{\cos A}{\cos A - \sin A} + \dfrac{\sin A}{\cos A + \sin A} = 1 + \tan 2A$

Solution

Although we could apply a double angle identity to $\tan 2A$ on the right side, it would not help simplify the expression. The left side appears better for simplification given that the common denominator of the two fractions is $\cos^2 A - \sin^2 A$, which is equivalent to $\cos 2A$.

$$\frac{\cos A}{\cos A - \sin A} \cdot \frac{\cos A + \sin A}{\cos A + \sin A} + \frac{\sin A}{\cos A + \sin A} \cdot \frac{\cos A - \sin A}{\cos A - \sin A} = \text{RHS}$$

$$\frac{\cos^2 A + \sin A \cos A}{\cos^2 A - \sin^2 A} + \frac{\sin A \cos A - \sin^2 A}{\cos^2 A - \sin^2 A} = \text{RHS}$$

$$\frac{\cos^2 A - \sin^2 A + 2 \sin A \cos A}{\cos^2 A - \sin^2 A} = \text{RHS}$$

$$\frac{\cos 2A + 2 \sin A \cos A}{\cos 2A} = \text{RHS} \qquad \text{Substitute } \cos 2A \text{ for } \cos^2 A - \sin^2 A$$

Observing that the right-hand side (RHS) has a term equal to 1 directs us to split the left side into two fractions, since one of the terms in the numerator is equal to the denominator.

$$\frac{\cos 2A}{\cos 2A} + \frac{2 \sin A \cos A}{\cos 2A} = \text{RHS}$$

$$1 + \frac{\sin 2A}{\cos 2A} = \text{RHS} \qquad \text{Substitute } \sin 2A \text{ for } 2 \sin A \cos A$$

$$1 + \tan 2A = 1 + \tan 2A \qquad \text{Applying tangent identity } \tan x = \frac{\sin x}{\cos x}$$

Reciprocal identities	$\csc\theta = \dfrac{1}{\sin\theta}$ $\sec\theta = \dfrac{1}{\cos\theta}$ $\cot\theta = \dfrac{1}{\tan\theta}$
Tangent and cotangent identities	$\tan\theta = \dfrac{\sin\theta}{\cos\theta}$ $\cot\theta = \dfrac{\cos\theta}{\sin\theta}$
Odd/even function identities	$\sin(-\theta) = -\sin\theta$ $\cos(-\theta) = \cos\theta$ $\tan(-\theta) = -\tan\theta$ $\sec(-\theta) = \cos\theta$ $\csc(-\theta) = -\csc\theta$ $\cot(-\theta) = -\tan\theta$
Symmetry/translation identities	$\sin\left(\dfrac{\pi}{2} - \theta\right) = \cos\theta$ $\sec\left(\dfrac{\pi}{2} - \theta\right) = \csc\theta$ $\tan\left(\dfrac{\pi}{2} - \theta\right) = \cot\theta$ $\cos\left(\dfrac{\pi}{2} - \theta\right) = \sin\theta$ $\csc\left(\dfrac{\pi}{2} - \theta\right) = \sec\theta$ $\cot\left(\dfrac{\pi}{2} - \theta\right) = \tan\theta$
Pythagorean identities	$\sin^2\theta + \cos^2\theta = 1$ $1 + \tan^2\theta = \sec^2\theta$ $1 + \cot^2\theta = \csc^2\theta$
Compound angle identities	$\sin(\alpha \pm \beta) = \sin\alpha\cos\beta \pm \cos\alpha\sin\beta$ $\cos(\alpha \pm \beta) = \cos\alpha\cos\beta \mp \sin\alpha\sin\beta$ $\tan(\alpha \pm \beta) = \dfrac{\tan\alpha \pm \tan\beta}{1 \mp \tan\alpha\tan\beta}$
Double angle identities	$\sin 2\theta = 2\cos\theta\sin\theta$ $\cos 2\theta = \begin{cases} \cos^2\theta - \sin^2\theta \\ 2\cos^2\theta - 1 \\ 1 - 2\sin^2\theta \end{cases}$ $\tan 2\theta = \dfrac{2\tan\theta}{1 - \tan^2\theta}$

Table 6.2 Summary of trigonometric identities

1. Use a compound angle identity to find the exact value of each expression.

   (a) $\cos\dfrac{7\pi}{12}$

   (b) $\sin 165°$

   (c) $\tan\dfrac{\pi}{12}$

   (d) $\sin\left(-\dfrac{5\pi}{12}\right)$

   (e) $\cos 255°$

   (f) $\cot 75°$

2. (a) Find the exact value of $\cos\dfrac{\pi}{12}$.

   (b) By writing $\cos\dfrac{\pi}{12}$ as $\cos\left(2 \cdot \dfrac{\pi}{24}\right)$ and using a double angle identity

   for cosine, find the exact value of $\cos\dfrac{\pi}{24}$.

3. Prove each co-function identity using the compound angle identities.

   (a) $\tan\left(\dfrac{\pi}{2} - \theta\right) = \cot \theta$

   (b) $\sin\left(\dfrac{\pi}{2} - \theta\right) = \cos \theta$

   (c) $\csc\left(\dfrac{\pi}{2} - \theta\right) = \sec \theta$

4. Given that $\sin x = \dfrac{3}{5}$ and that $0 < x < \dfrac{\pi}{2}$, find the exact values of:

   (a) $\cos x$

   (b) $\cos 2x$

   (c) $\sin 2x$

5. Given that $\cos x = -\dfrac{2}{3}$ and that $\dfrac{\pi}{2} < x < \pi$, find the exact values of:

   (a) $\sin x$

   (b) $\sin 2x$

   (c) $\cos 2x$

6. Find the exact values of $\sin 2\theta$, $\cos 2\theta$ and $\tan 2\theta$ subject to the given conditions.

   (a) $\sin \theta = \dfrac{2}{3}, \dfrac{\pi}{2} < \theta < \pi$

   (b) $\cos \theta = -\dfrac{4}{5}, \pi < \theta < \dfrac{3\pi}{2}$

   (c) $\tan \theta = 2, 0 < \theta < \dfrac{\pi}{2}$

   (d) $\sec \theta = -4, \csc \theta > 0$

7. Use a compound angle identity to write the given expression as a function of x alone.

   (a) $\cos(x - \pi)$

   (b) $\sin\left(x - \dfrac{\pi}{2}\right)$

   (c) $\tan(x + \pi)$

   (d) $\cos\left(x + \dfrac{\pi}{2}\right)$

8. Use identities to find an equivalent expression involving only sines and cosines, and then simplify it.

   (a) $\sec\theta + \sin\theta$

   (b) $\dfrac{\sec\theta \csc\theta}{\tan\theta \sin\theta}$

   (c) $\dfrac{\sec\theta + \csc\theta}{2}$

   (d) $\dfrac{1}{\cos^2\theta} + \dfrac{1}{\cot^2\theta}$

9. Simplify each expression.

   (a) $\cos\theta - \cos\theta\sin^2\theta$

   (b) $\dfrac{1 - \cos^2\theta}{\sin^2\theta}$

   (c) $\cos 2\theta + \sin^2\theta$

   (d) $\dfrac{\sin^2\theta}{\cos^2\theta} + \dfrac{1}{\cot^2\theta}$

   (e) $\sin(\alpha + \beta) + \sin(\alpha - \beta)$

   (f) $\dfrac{1 + \cos 2A}{2}$

   (g) $\cos(\alpha + \beta) + \cos(\alpha - \beta)$

   (h) $2\cos^2\theta - \cos 2\theta$

10. Prove each identity.

   (a) $\dfrac{\cos 2\theta}{\cos\theta + \sin\theta} = \cos\theta - \sin\theta$

   (b) $(1 - \cos\alpha)(1 + \sec\alpha) = \sin\alpha \tan\alpha$

   (c) $\dfrac{1 - \tan^2 x}{1 + \tan^2 x} = \cos 2x$

   (d) $\cos^4\theta - \sin^4\theta = \cos 2\theta$

   (e) $\cot\theta - \tan\theta = 2\cot 2\theta$

   (f) $\dfrac{\cos\beta - \sin\beta}{\cos\beta + \sin\beta} = \dfrac{\cos 2\beta}{1 + \sin 2\beta}$

   (g) $\dfrac{1}{\sec\theta(1 - \sin\theta)} = \sec\theta + \tan\theta$

   (h) $(\tan A - \sec A)^2 = \dfrac{1 - \sin A}{1 + \sin A}$

   (i) $\dfrac{\tan 2x \tan x}{\tan 2x - \tan x} = \sin 2x$

   (j) $\dfrac{\sin 2\theta - \cos 2\theta + 1}{\sin 2\theta + \cos 2\theta + 1} = \tan\theta$

   (k) $\dfrac{1 + \cos\alpha}{\sin\alpha} = 2\csc\alpha - \dfrac{\sin\alpha}{1 + \cos\alpha}$

   (l) $\dfrac{1 + \cos\beta}{\sin\beta} + \dfrac{\sin\beta}{1 + \cos\beta} = 2\csc\beta$

   (m) $\dfrac{\cot x - 1}{1 - \tan x} = \dfrac{\csc x}{\sec x}$

   (n) $\sin\left(\dfrac{\theta}{2}\right) = \pm\sqrt{\dfrac{1 - \cos\theta}{2}}$

For part (n), first prove that $\sin^2 x = \dfrac{1 - \cos 2x}{2}$, then make a suitable substitution for x. This identity is called the **half-angle identity** for sine.

11. Given the figure shown, find an expression in terms of x for the value of $\tan \theta$.

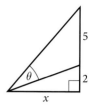

Figure 6.31 Diagram for question 11

12. Solve each equation for x in the given interval. Give answers exactly, if possible. Otherwise, give answers accurate to 3 significant figures.

(a) $2 \sin^2 x - \cos x = 1, 0 \leqslant x < 2\pi$

(b) $\sec^2 x = 8 \cos x, -\pi < x \leqslant \pi$

(c) $2 \cos x + \sin 2x = 0, -180° < x \leqslant 180°$

(d) $2 \sin x = \cos 2x, 0 \leqslant x < 2\pi$

(e) $\cos 2x = \sin^2 x, 0 \leqslant x < 2\pi$

(f) $2 \sin x \cos x + 1 = 0, 0 \leqslant x < 2\pi$

(g) $\cos^2 x - \sin^2 x = -\dfrac{1}{2}, 0 \leqslant x \leqslant \pi$

(h) $\sec^2 x - \tan x - 1 = 0, 0 \leqslant x < 2\pi$

(i) $\tan 2x + \tan x = 0, 0 \leqslant x < 2\pi$

(j) $2 \sin 2x \cos 3x + \cos 3x = 0, 0 \leqslant x \leqslant 180°$

13. Find an identity for $\sin 3x$ in terms of $\sin x$.

14. (a) By squaring $\sin^2 x + \cos^2 x$, prove that $\sin^4 x + \cos^4 x = \dfrac{1}{4}(\cos 4x + 3)$.

(b) Hence, or otherwise, solve the equation $\sin^4 x + \cos^4 x = \dfrac{1}{2}$ for $0 \leqslant x < 2\pi$.

6.6 Inverse trigonometric functions

In Section 1.4, we learned that if a function f is one-to-one then it has an inverse, f^{-1}. A defining characteristic of a one-to-one function is that it is always increasing or always decreasing in its domain. Also, recall that no horizontal line can pass through the graph of a one-to-one function at more than one point. It is evident that none of the trigonometric functions are one-to-one functions given their periodic nature. Therefore, the inverse of any of the trigonometric functions over their domain is not a function.

Defining the inverse sine function

Recall that the domain of $y = \sin x$ is all real numbers ($\mathbb{R}$) and its range is the set of all real numbers in the closed interval $-1 \leqslant y \leqslant 1$. The sine function is not one-to-one and hence its inverse is not a function, since more than one value of x corresponds to the same value of y. For example, $\sin\dfrac{\pi}{6} = \sin\dfrac{5\pi}{6} = \sin\dfrac{13\pi}{6} = \dfrac{1}{2}$. That is, for $y = \sin x$ there are an infinite number of ordered pairs with a y-coordinate of $\dfrac{1}{2}$ (see Figure 6.32).

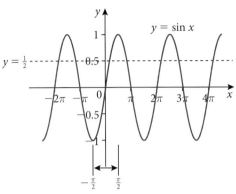

Figure 6.32 A horizontal line, $y = \dfrac{1}{2}$ shown here, intersects the graph of $y = \sin x$ more than once, indicating that the inverse of $y = \sin x$ is not a function. The portion of the graph (in red) from $-\dfrac{\pi}{2}$ to $\dfrac{\pi}{2}$ is used to define the inverse and only intersects a horizontal line once

A function that is not one-to-one can often be made so by restricting its domain. Consequently, even though there is no inverse function for the sine function for all $\mathbb{R}$, we can define the inverse sine function if we restrict its domain so that it is one-to-one (and passes the horizontal line test). We have an unlimited number of ways of restricting the domain but it seems sensible to select an interval of x including zero, and it is standard to restrict the domain to the largest set possible. Consider restricting the domain of $y = \sin x$ to the interval $-\dfrac{\pi}{2} \leqslant x \leqslant \dfrac{\pi}{2}$. In this interval, $y = \sin x$ is always increasing and takes on every value from -1 to 1 exactly once. Thus, the function $y = \sin x$ with domain $-\dfrac{\pi}{2} \leqslant x \leqslant \dfrac{\pi}{2}$ is one-to-one and its inverse is a function.

Thus, $\arcsin x$ (or $\sin^{-1} x$) is the number in the closed interval $\left[-\dfrac{\pi}{2}, \dfrac{\pi}{2}\right]$ whose sine is x. For example, $\arcsin\dfrac{1}{2} = \dfrac{\pi}{6}$ because the one number in the interval $\left[-\dfrac{\pi}{2}, \dfrac{\pi}{2}\right]$ whose sine is $\dfrac{1}{2}$ is $\dfrac{\pi}{6}$. Your GDC is programmed such that it will give the same result. If your GDC is in radian mode, it will give the approximate value of $\dfrac{\pi}{6}$ to several significant figures, and if it is in degree mode it will give the exact result of $30°$.

From the graphical symmetry of inverse functions, the graph of $y = \arcsin x$ is a reflection of $y = \sin x$ about the line $y = x$, as shown in Figures 6.33 and 6.34.

The equation $y = \arcsin x$ is interpreted as 'y is the arc whose sine is x', or 'y is the angle whose sine is x', or 'y is the real number whose sine is x'. Any GDC labels the inverse sine function as $\sin^{-1} x$. The symbols $y = \arcsin x$ and $y = \sin^{-1} x$ are both commonly used to indicate the inverse sine function, but a disadvantage of writing $y = \sin^{-1} x$ is that it can be confused with $y = (\sin x)^{-1}$ $= \dfrac{1}{\sin x} = \csc x$.

Inverse sine function
The inverse sine function, denoted by $y = \arcsin x$ or $y = \sin^{-1} x$, is the function with a domain of $-1 \leqslant x \leqslant 1$ and a range of $-\dfrac{\pi}{2} \leqslant y \leqslant \dfrac{\pi}{2}$ defined by $y = \arcsin x$ if and only if $x = \sin y$

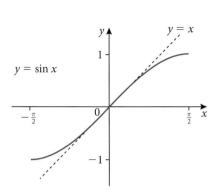

Figure 6.33 The graph of $y = \sin x$ with domain restricted to $-\dfrac{\pi}{2} \leqslant x \leqslant \dfrac{\pi}{2}$

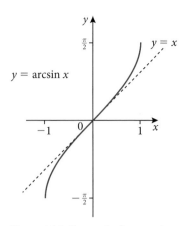

Figure 6.34 The graph of $y = \arcsin x$

Defining the inverse cosine and inverse tangent functions

The inverse cosine function and inverse tangent function can be defined by following a parallel procedure to that used for defining the inverse sine function. The graphs of $y = \cos x$ and $y = \tan x$ (Figures 6.35 and 6.36) show that neither function is one-to-one and consequently their inverses are not functions. Consider restricting the domain of the cosine function to the closed interval $0 \leqslant x \leqslant \pi$ (Figure 6.37) and restricting the domain of the tangent function to the open interval $-\dfrac{\pi}{2} < x < \dfrac{\pi}{2}$ (Figure 6.39). The interval for tangent cannot include the endpoints, $-\dfrac{\pi}{2}$ and $\dfrac{\pi}{2}$, because tangent is undefined for these values. For these domain restrictions, cosine and tangent will attain each of their function values exactly once. Hence, with these restrictions, both cosine and tangent will be one-to-one and their inverses will be functions.

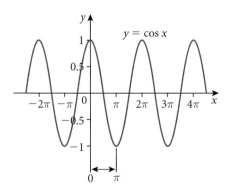

Figure 6.35 The graph of $y = \cos x$ with portion of the graph (in red) from 0 to π (inclusive) used to define its inverse

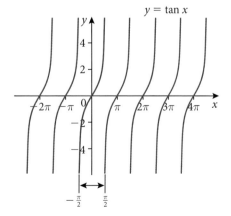

Figure 6.36 The graph of $y = \tan x$ with portion of the graph (in red) from $-\dfrac{\pi}{2}$ to $\dfrac{\pi}{2}$ (exclusive) used to define its inverse

Inverse cosine function
The inverse cosine function, denoted by $y = \arccos x$ or $y = \cos^{-1} x$, is the function with a domain of $-1 \leqslant x \leqslant 1$ and a range of $0 \leqslant y \leqslant \pi$ defined by $y = \arccos x$ if and only if $x = \cos y$

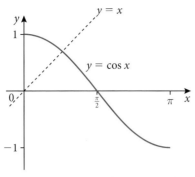

Figure 6.37 The graph of $y = \cos x$ with domain restricted to $0 \leqslant x \leqslant \pi$

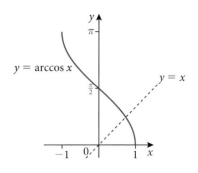

Figure 6.38 The graph of $y = \arccos x$

Inverse tangent function

The inverse tangent function, denoted by $y = \arctan x$ or $y = \tan^{-1} x$, is the function with a domain of $\mathbb{R}$ and a range of $-\dfrac{\pi}{2} < y < \dfrac{\pi}{2}$ defined by $y = \arctan x$ if and only if $x = \tan y$

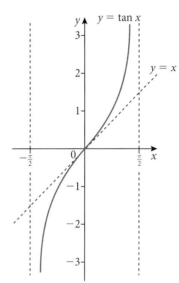

Figure 6.39 The graph of $y = \tan x$ with domain restricted to $-\dfrac{\pi}{2} < x < \dfrac{\pi}{2}$

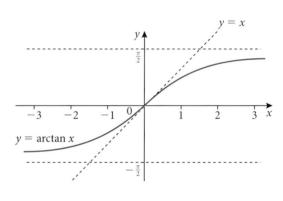

Figure 6.40 The graph of $y = \arctan x$

Example 6.28

Without using your GDC, find the exact value of each expression.

(a) $\arcsin\left(-\dfrac{\sqrt{3}}{2}\right)$ (b) $\arccos 1$ (c) $\arctan\sqrt{3}$ (d) $\arcsin\dfrac{3}{2}$

Solution

(a) The expression $\arcsin\left(-\dfrac{\sqrt{3}}{2}\right)$ can be interpreted as 'the number y such that $-\dfrac{\pi}{2} \leqslant y \leqslant \dfrac{\pi}{2}$ whose sine is $-\dfrac{\sqrt{3}}{2}$; or, 'the number in quadrant I or IV whose sine is $-\dfrac{\sqrt{3}}{2}$. We know sine

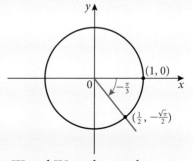

function values are negative in quadrants III and IV, so the number we are looking for is in quadrant IV.

The figure shows that the required number is $-\frac{\pi}{3}$. An angle of $-\frac{\pi}{3}$ in standard position intersects the unit circle at a point whose y-coordinate is $-\frac{\sqrt{3}}{2}$.

Therefore, $\arcsin\left(-\frac{\sqrt{3}}{2}\right) = -\frac{\pi}{3}$

(b) The range of the function $y = \arccos x$ is $0 \leqslant y \leqslant \pi$. Thus, we are looking for a number in quadrant I or II whose cosine is 1. The number must be 0, because an angle of measure 0 in standard position will intersect the unit circle at a point whose x-coordinate is 1. Therefore, $\arccos 1 = 0$.

(c) The range of the function $y = \arctan x$ is $-\frac{\pi}{2} < y < \frac{\pi}{2}$. So, we are looking for a number in quadrant I or IV such that $\frac{\text{sine}}{\text{cosine}}$ is equal to $\sqrt{3}$.

It must be in quadrant I because in quadrant IV tangent values are negative. Familiarity with the sine and cosine values for common angles helps us to recognise that the required ratio will be $\dfrac{\frac{\sqrt{3}}{2}}{\frac{1}{2}}$. The required number is $\frac{\pi}{3}$ because it is in the first quadrant with $\sin\frac{\pi}{3} = \frac{\sqrt{3}}{2}$ and $\cos\frac{\pi}{3} = \frac{1}{2}$. Therefore, $\arctan\sqrt{3} = \frac{\pi}{3}$

(d) The domain of the function $y = \arcsin x$ is $-1 \leqslant x \leqslant 1$, but $\frac{3}{2}$ is not in this interval. There is no number whose sine is $\frac{3}{2}$. Therefore, $\arcsin\frac{3}{2}$ is not defined.

Compositions of trigonometric and inverse trigonometric functions

Recall from Chapter 1 the following two properties for a pair of inverse functions.

$f(f^{-1}(x)) = x$ for all x in the domain of f^{-1}; and $f^{-1}(f(x)) = x$ for all x in the domain of f.

It follows that the following properties hold true for the inverse sine, cosine and tangent functions.

Inverse properties

If $-1 \leqslant \alpha \leqslant 1$, then $\sin(\arcsin \alpha) = \alpha$; and if $-\frac{\pi}{2} \leqslant \beta \leqslant \frac{\pi}{2}$, then $\arcsin(\sin \beta) = \beta$

If $-1 \leqslant \alpha \leqslant 1$, then $\cos(\arccos \alpha) = \alpha$; and if $0 \leqslant \beta \leqslant \pi$, then $\arccos(\cos \beta) = \beta$

If $\alpha \in \mathbb{R}$, then $\tan(\arctan \alpha) = \alpha$; and if $-\frac{\pi}{2} < \beta < \frac{\pi}{2}$, then $\arctan(\tan \beta) = \beta$

Note that the inverse property $\arcsin(\sin \beta) = \beta$ does not hold true when $\beta = \frac{3\pi}{4}$.

$\arcsin\left(\sin\frac{3\pi}{4}\right) =$

$\arcsin\left(\frac{\sqrt{2}}{2}\right) = \frac{\pi}{4}$ and

$\arcsin\left(\sin\frac{5\pi}{4}\right) =$

$\arcsin\left(-\frac{\sqrt{2}}{2}\right) = -\frac{\pi}{4}$.

The property $\arcsin(\sin \beta) = \beta$ is not valid for values of β outside the interval $-\frac{\pi}{2} \leqslant \beta \leqslant \frac{\pi}{2}$.

Similarly, the property $\arccos(\cos \beta) = \beta$ is not valid for values of β outside the interval $0 \leqslant \beta \leqslant \pi$; and $\arctan(\tan \beta) = \beta$ is not valid for values of β outside the interval $-\frac{\pi}{2} < \beta < \frac{\pi}{2}$.

Example 6.29

Find the exact values, if possible, for each expression.

(a) $\cos^{-1}\left(\cos\dfrac{4\pi}{3}\right)$ (b) $\tan(\arctan(-7))$ (c) $\sin(\arcsin\sqrt{3})$

Solution

(a) $\dfrac{4\pi}{3}$ is not in the range of the arccos function

$0 \leqslant \beta \leqslant \pi$.

However, using the symmetry of the unit

circle, we know that $\dfrac{4\pi}{3}$ has the same cosine

as $\dfrac{2\pi}{3}$ (see figure), which is in the interval

$0 \leqslant \beta \leqslant \pi$.

Thus, $\cos^{-1}\left(\cos\dfrac{4\pi}{3}\right) = \cos^{-1}\left(\cos\dfrac{2\pi}{3}\right) = \dfrac{2\pi}{3}$

(b) -7 is in the range of the tangent function (and in the domain of the arctangent function), so the inverse property applies. Therefore, $\tan(\arctan(-7)) = -7$

(c) $\sqrt{3}$ is not in the range of the sine function $-1 \leqslant \alpha \leqslant 1$, so $\arcsin\sqrt{3}$ is not defined. It follows that $\sin(\arcsin\sqrt{3})$ is not defined.

Example 6.30

Given that $C = \arctan 3 + \arcsin\left(\dfrac{5}{13}\right)$, find the exact value of $\cos C$.

Solution

Let $A = \arctan 3$ and $B = \arcsin\left(\dfrac{5}{13}\right)$. Thus, $C = A + B$ and a strategy for

finding $\cos C$ is to use the following compound angle identity:

$$\cos C = \cos(A + B) = \cos A \cos B - \sin A \sin B$$

We know that $\sin B = \dfrac{5}{13}$. We need to find exact values for $\cos A$, $\cos B$, and

$\sin A$. The range for $\arctan x$ is $-\dfrac{\pi}{2} < x < \dfrac{\pi}{2}$ and the range for $\arcsin x$ is

$-\dfrac{\pi}{2} \leqslant x \leqslant \dfrac{\pi}{2}$, and since $\tan A = 3 > 0$ and $\sin B = \dfrac{5}{13} > 0$, both A and B

are in quadrant I.

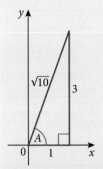

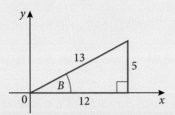

$$\sin A = \frac{3}{\sqrt{10}} = \frac{3\sqrt{10}}{10} \qquad\qquad \sin B = \frac{5}{13}$$

$$\cos A = \frac{1}{\sqrt{10}} = \frac{\sqrt{10}}{10} \qquad\qquad \cos B = \frac{12}{13}$$

Hence, $\cos C = \cos(A + B) = \cos A \cos B - \sin A \sin B$

$$= \left(\frac{\sqrt{10}}{10}\right)\left(\frac{12}{13}\right) - \left(\frac{3\sqrt{10}}{10}\right)\left(\frac{5}{13}\right)$$

$$= \frac{(12 - 15)\sqrt{10}}{130}$$

$$= \frac{-3\sqrt{10}}{130}$$

Therefore, $\cos C = \dfrac{-3\sqrt{10}}{130}$

Example 6.31

Find all solutions, accurate to 3 significant figures, to the equation $3 \sin 2\theta = 1$ in the interval $0 \leqslant \theta < 2\pi$.

Solution

A reasonable idea is to apply a double angle identity and substitute $2 \sin \theta \cos \theta$ for $\sin 2\theta$. Although a substitution like this proved to be an effective technique in Section 6.5, it is not always the best strategy. In this case, the transformed equation becomes $6 \sin \theta \cos \theta = 1$, which is difficult to solve. A better approach is:

$$3 \sin 2\theta = 1$$

$$\sin 2\theta = \frac{1}{3}$$

$$2\theta = \arcsin\left(\frac{1}{3}\right)$$

$$\theta = \frac{1}{2}\arcsin\left(\frac{1}{3}\right)$$

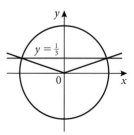

Figure 6.41 Diagram for solution to Example 6.31

There is one angle in quadrant I with a sine equal to $\frac{1}{3}$ and one angle in quadrant II with a sine equal to $\frac{1}{3}$ (Figure 6.41). None of the common angles has a sine equal to $\frac{1}{3}$, so we will need to use the inverse sine ($\sin^{-1}$) on a GDC to obtain an approximate answer. Since the range of the inverse sine function is $-\frac{\pi}{2} \leqslant y \leqslant \frac{\pi}{2}$, our GDC's result for $\sin^{-1}\left(\frac{1}{3}\right)$ will only give the angle (arc) in quadrant I. From the symmetry of the unit circle, we can obtain the angle in quadrant II by subtracting the angle in quadrant I from π.

```
sin⁻¹(1/3)
        .3398369095
.5*Ans
        .1699184547
Ans→A
        .1699184547
```

```
sin⁻¹(1/3)
        .3398369095
.5(π-Ans)
        1.400877872
Ans→B
        1.400877872
```

```
3sin(2A)
                1
3sin(2B)
                1
```

The GDC images below show the computation for both answers — and a check of the two answers.

Therefore, $\theta \approx 0.170$ or $\theta \approx 1.40$, accurate to 3 significant figures.

To an observer, the apparent size of an object depends on the distance from the observer to the object. The farther an object is from an observer, the smaller its apparent size. For example, although the sun's diameter is 400 times larger than that of the moon, the two objects appear to have the same diameter as viewed from the Earth. Thus, during a total solar eclipse, the moon blocks out the sun.

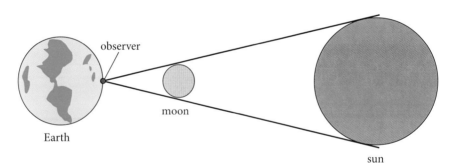
observer
moon
Earth
sun

Figure 6.42 On the surface of the Earth, the angle subtended by the moon and the sun is nearly the same. It is approximately 0.54 degrees for the moon and 0.52 degrees for the sun. The sun is 400 times wider than the moon and, coincidentally, 400 times further from the Earth than the moon.

Example 6.32

A painting that is 125 cm from top to bottom is hanging on the wall of a gallery such that its base is 250 cm from the floor. Pablo is standing x cm from the wall on which the painting is hung. Pablo's eyes are 170 cm from the floor and from where he stands the painting subtends an angle α degrees.

(a) Write a function for α in terms of x.

(b) Find α, accurate to 3 significant figures, for the following values of x:

(i) $x = 75$ cm (ii) $x = 125$ cm (iii) $x = 175$ cm

(c) Using a GDC, find to the nearest cm how far Pablo should stand from the wall so that the subtended angle α is a maximum.

Solution

(a) The figure shows α, the angle subtended by the painting, and β, the angle subtended by the part of the wall above eye level and below the painting. Let θ be the sum of these two angles. Hence, $\theta = \alpha + \beta$ and $\alpha = \theta - \beta$. From the compound angle identity for tangent, we have

$$\tan \alpha = \frac{\tan \theta - \tan \beta}{1 + \tan \theta \tan \beta}$$

From the right-angled triangles in the figure, we can determine that

$$\tan \beta = \frac{80}{x} \text{ and } \tan \theta = \frac{205}{x}$$

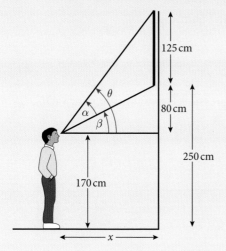

Substituting these into the expression for $\tan \alpha$, gives

$$\tan \alpha = \frac{\dfrac{205}{x} - \dfrac{80}{x}}{1 + \left(\dfrac{205}{x}\right)\left(\dfrac{80}{x}\right)}$$

$$\tan \alpha = \frac{\dfrac{125}{x}}{1 + \left(\dfrac{205}{x}\right)\left(\dfrac{80}{x}\right)} \cdot \frac{x^2}{x^2}$$

$$\tan \alpha = \frac{125x}{x^2 + 16400}$$

Therefore, $\alpha = \tan^{-1}\left(\dfrac{125x}{x^2 + 16400}\right)$

(b) (i) For $x = 75$ cm: $\alpha = \tan^{-1}\left(\dfrac{125 \cdot 75}{75^2 + 16400}\right) \approx \tan^{-1}(0.4256527)$
$\approx 23.06°$

   (ii) For $x = 125$ cm: $\alpha = \tan^{-1}\left(\dfrac{125 \cdot 125}{125^2 + 16400}\right) \approx \tan^{-1}(0.4879001)$
$\approx 26.01°$

   (iii) For $x = 175$ cm: $\alpha = \tan^{-1}\left(\dfrac{175 \cdot 175}{175^2 + 16400}\right) \approx \tan^{-1}(0.4651781)$
$\approx 24.95°$

(c) Graph the function found in (a). On a GDC, it will be entered as

$$y = \tan^{-1}\left(\frac{125x}{x^2 + 16400}\right).$$ Find the value of

x that gives the maximum value for y (subtended angle a) by either tracing or using a 'maximum' command on the calculator.

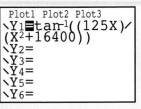

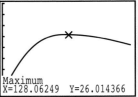

Therefore, if Pablo stands 128 cm away from the wall, the painting will subtend the widest possible angle at his eye. In other words, it will give him the 'best' view of the painting.

Exercise 6.6

1. Find the exact value of each expression without using your GDC.

   (a) $\arcsin 1$

   (b) $\arccos\left(\dfrac{1}{\sqrt{2}}\right)$

   (c) $\arctan(-\sqrt{3})$

   (d) $\arccos\left(-\dfrac{1}{2}\right)$

   (e) $\arctan 0$

   (f) $\arcsin\left(-\dfrac{\sqrt{3}}{2}\right)$

2. Find the exact value, if possible, without using your GDC, for each expression. Verify your result with your GDC.

   (a) $\sin^{-1}\left(\sin\dfrac{2\pi}{3}\right)$

   (b) $\cos^{-1}\left(\cos\dfrac{3}{2}\right)$

   (c) $\tan(\arctan 12)$

   (d) $\cos\left(\arccos\dfrac{2\pi}{3}\right)$

   (e) $\arctan\left(\tan\left(-\dfrac{3\pi}{4}\right)\right)$

   (f) $\sin(\arcsin \pi)$

   (g) $\sin\left(\arctan\dfrac{3}{4}\right)$

   (h) $\cos\left(\arcsin\left(\dfrac{7}{25}\right)\right)$

   (i) $\arcsin\left(\tan\dfrac{\pi}{3}\right)$

   (j) $\tan^{-1}\left(2\sin\dfrac{\pi}{3}\right)$

   (k) $\cos\left(\arctan\left(\dfrac{1}{2}\right)\right)$

   (l) $\cos(\sin^{-1}(0.6))$

   (m) $\sin\left(\arccos\left(\dfrac{3}{5}\right) + \arctan\left(\dfrac{5}{12}\right)\right)$

   (n) $\cos\left(\tan^{-1}3 + \sin^{-1}\left(\dfrac{1}{3}\right)\right)$

3. Rewrite each expression as an algebraic expression in terms of x.

   (a) $\cos(\arcsin x)$ **(b)** $\tan(\arccos x)$

   (c) $\cos(\tan^{-1} x)$ **(d)** $\sin(2\cos^{-1} x)$

   (e) $\tan\left(\dfrac{1}{2}\arccos x\right)$ **(f)** $\sin(\arcsin x + 2\arctan x)$

4. Show that $\arcsin\dfrac{4}{5} + \arcsin\dfrac{5}{13} = \arccos\dfrac{16}{65}$

5. Show that $\arctan\dfrac{1}{2} + \arctan\dfrac{1}{3} = \dfrac{\pi}{4}$

6. Find x if $\tan^{-1} x + \tan^{-1}(1 - x) = \tan^{-1}\left(\dfrac{4}{3}\right)$

7. Solve for x in the indicated interval.

   (a) $5\cos(2x) = 2,\ 0 \leqslant x \leqslant \pi$

   (b) $\tan\left(\dfrac{x}{2}\right) = 2,\ 0 < x \leqslant 2\pi$

   (c) $2\cos x - \sin x = 0,\ 0 < x \leqslant 2\pi$

   (d) $3\sec^2 x = 2\tan x + 4,\ 0 < x \leqslant 2\pi$

   (e) $2\tan^2 x - 3\tan x + 1 = 0,\ 0 \leqslant x \leqslant \pi$

   (f) $\tan x \csc x = 5,\ 0 < x \leqslant 2\pi$

   (g) $\tan 2x + 3\tan x = 0,\ 0 < x \leqslant 2\pi$

   (h) $2\cos^2 x - 3\sin 2x = 2,\ 0 \leqslant x \leqslant \pi$

8. An offshore lighthouse is located 2 km from a straight coastline. The lighthouse has a revolving light. Let θ be the angle that the beam of light from the lighthouse makes with the coastline, and P is the point on the coast that is the shortest distance from the lighthouse (see figure). Given that d is the distance in km from P to the point B where the beam of light is hitting the coast, express θ as a function of d. Sketch a complete graph of this function and indicate the portion of the graph that sufficiently represents the given situation.

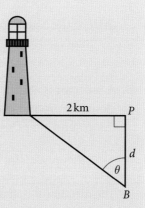

2 km

Chapter 6 practice questions

1. A toy on an elastic string is attached to the top of a doorway. It is pulled down and released, allowing it to bounce up and down. The length of the elastic string, L centimetres, is modelled by the function $L = 110 + 25 \cos(2\pi t)$ where t is time in seconds after release.

   (a) Find the length of the elastic string after 2 seconds.

   (b) Find the minimum length of the string.

   (c) Find the first time after release that the string is 85 cm.

   (d) What is the period of the motion?

2. Find the exact solution(s) to the equation $2 \sin^2 x - \cos x - 1 = 0$ for $0 \leqslant x \leqslant 2\pi$.

3. The diagram shows a circle of radius 6 cm. The perimeter of the shaded sector is 36 cm. Find the radian measure of the angle θ.

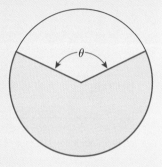

4. Consider the two functions $f(x) = \cos 4x$ and $g(x) = \cos\left(\dfrac{x}{2}\right)$

   (a) Write down:

      (i) the minimum value of the function f

      (ii) the period of g.

   (b) For the equation $f(x) = g(x)$, find the number of solutions in the interval $0 \leqslant x \leqslant \pi$.

5. A reflector is attached to the spoke of a bicycle wheel. As the wheel rolls along the ground, the distance, d centimetres, that the reflector is above the ground after t seconds is modelled by the function

$$d = p + q \sin\left(\frac{2\pi}{m} t\right), \text{ where } p, q \text{ and } m \text{ are constants}$$

   The distance d is at a maximum of 64 cm at $t = 0$ seconds and at $t = 0.5$ seconds, and is at a minimum of 6 cm at $t = 0.25$ seconds and at $t = 0.75$ seconds. Write down the value of:

   (a) p (b) q (c) m

6. Find all solutions to $1 + \sin 3x = \cos(0.25x)$ such that $x \in [0, \pi]$

7. Find all solutions to both trigonometric equations in the interval $x \in [0, 2\pi]$. Express the solutions exactly.

   (a) $2\cos^2 x + 5 \cos x + 2 = 0$

   (b) $\sin 2x - \cos x = 0$

8. The value of x is in the interval $\dfrac{\pi}{2} < x < \pi$ and $\cos^2 x = \dfrac{8}{9}$

   Without using your GDC, find the exact values for the following:

   (a) $\sin x$

   (b) $\cos 2x$

   (c) $\sin 2x$

9. The depth d metres of water in a harbour varies with the tides during each day. The first high (maximum) tide after midnight occurs at 5:00 am with a depth of 5.8 metres. The first low (minimum) tide occurs at 10:30 am with a depth of 2.6 metres.

   (a) Find a trigonometric function that models the depth, d, of the water t hours after midnight.

   (b) Find the depth of the water at 12 noon.

   (c) A large boat needs at least 3.5 metres of water to dock in the harbour. During what time interval after 12 noon can the boat dock safely?

10. Solve the equation $\tan^2 x + 2 \tan x - 3 = 0$ for $0 \leqslant x \leqslant \pi$. Give solutions exactly if possible, otherwise give to 3 significant figures.

11. The diagram shows a circle of centre O and radius 10 cm.

    The arc ABC subtends an angle of $\dfrac{3}{2}$ radians at the centre O.

    (a) Find the length of the arc ACB.

    (b) Find the area of the shaded region.

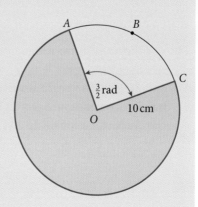

297

12. Consider the function $f(x) = \dfrac{5}{2}\cos\left(2x - \dfrac{\pi}{2}\right)$. For what values of k will the equation $f(x) = k$ have no solutions?

13. A portion of the graph of $y = k + a\sin x$ is shown below.

The graph passes through the points $(0, 1)$ and $\left(\dfrac{3\pi}{2}, 3\right)$.

Find the value of k and the value of a.

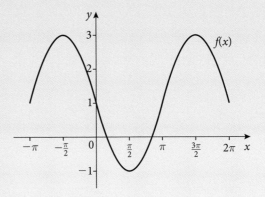

14. The angle α satisfies the equation $2\tan^2\alpha - 5\sec\alpha - 10 = 0$ where α is in the second quadrant. Find the exact value of $\sec\alpha$.

15. Triangles PTS and RTS have right angles at T, with angles α and β as shown in the diagram. Find the exact values of:

(a) $\sin(\alpha + \beta)$

(b) $\cos(\alpha + \beta)$

(c) $\tan(\alpha + \beta)$

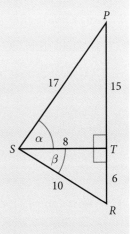

16. The diagram shows a right-angled triangle with legs of length 1 unit and 2 units. The angle at vertex P has a degree measure of $p°$. Find the exact values of $\sin 2p$ and $\sin 3p$.

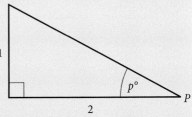

17. The obtuse angle B is such that $\tan B = -\dfrac{5}{12}$

Find the values of:

(a) $\sin B$

(b) $\cos B$

(c) $\sin 2B$

(d) $\cos 2B$

18. Given that $\tan 2\theta = \dfrac{3}{4}$, find the possible values of $\tan \theta$.

19. If $\sin(x - \alpha) = k \sin(x + \alpha)$, express $\tan x$ in terms of k and α.

20. Solve $\tan^2(2\theta) = 1$, in the interval $-\dfrac{\pi}{2} \leqslant \theta \leqslant \dfrac{\pi}{2}$

21. Let f be the function $f(x) = x \arccos x + \dfrac{1}{2}x$, for $-1 \leqslant x \leqslant 1$ and

g the function $g(x) = \cos 2x$, for $-1 \leqslant x \leqslant 1$

(a) Sketch the graph of f and of g.

(b) Write down the solution of the equation $f(x) = g(x)$.

(c) Write down the range of g.

22. Let ABC be a right-angled triangle, where $\widehat{C} = 90°$. The line segment AD bisects $B\widehat{A}C$, $BD = 3$, and $DC = 2$, as shown in the diagram. Find $D\widehat{A}C$.

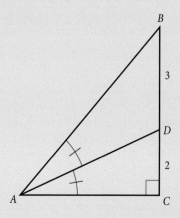

23. The diagram shows the cross-section of a water channel.

The equation that represents this boundary is

$y = 16\sec\left(\dfrac{\pi x}{36}\right) - 32$ where x and y are both measured in cm.

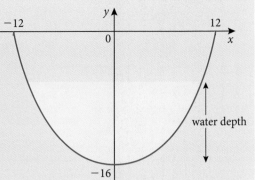

The top of the channel is level with the ground and has a width of 24 cm. The maximum depth of the channel is 16 cm.

Find the width of the water surface in the channel when the water depth is 10 cm. Give your answer in the form $a \arccos b$, where $a, b \in \mathbb{R}$.

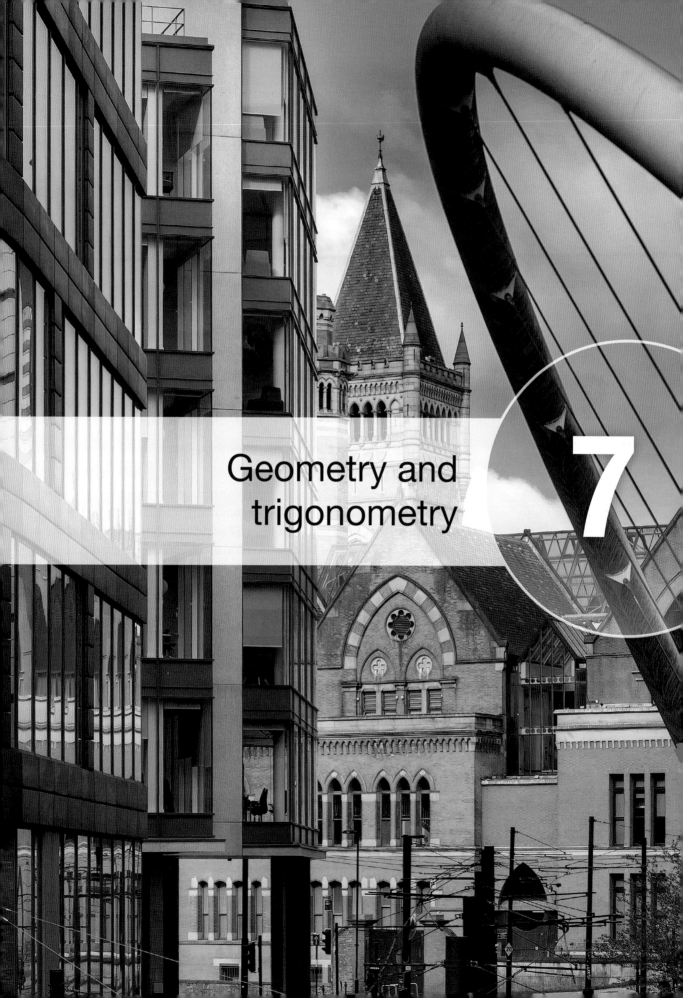

Geometry and trigonometry

7

7 Geometry and trigonometry

Learning objectives

By the end of this chapter, you should be familiar with...

- finding the distance between two points in 3-dimensional space

- finding the midpoint of a line segment in 3-dimensional space

- computing the volume and surface area of a solid such as a pyramid, cone, sphere, hemisphere or a solid made from a combination of these

- determining the size of an angle between two lines

- finding the sides and angles of a right-angled triangle using the sine, cosine and tangent ratios

- applying the sine rule and the cosine rule to find an unknown length or an angle

- computing the area of a triangle using the formula $\frac{1}{2}ab \sin C$

- solving problems involving 2-dimensional or 3-dimensional figures by means of right-angled and non-right-angled trigonometry

- solving problems involving compass bearings.

In this chapter, we cover some basic 3-dimensional geometry and trigonometry using right-angled triangles. This chapter may contain topics that you have studied before. Trigonometric functions will be defined in terms of the ratios of the sides of a right-angled triangle rather than in terms of an arc on the unit circle.

7.1 Measurements in three dimensions

Given the coordinates of two points $A(x_1, y_1)$ and $B(x_2, y_2)$ in a 2-dimensional plane, recall that the length AB of the segment $[AB]$ is the hypotenuse of a right-angled triangle where the length of one side of the right-angled triangle is the difference in the x-coordinates and the length of the other perpendicular side is the difference in the y-coordinates (Figure 7.1).

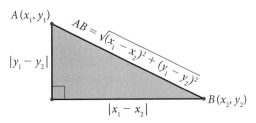

Figure 7.1 Distance between two points in a plane derived from the Pythagorean theorem

Thus, from the Pythagorean theorem, the distance, d, between the points with coordinates (x_1, y_1) and (x_2, y_2) is $d = \sqrt{(x_1 - x_2)^2 + (y_1 - y_2)^2}$.

The **midpoint** M of $[AB]$ is the point whose coordinates are the average of the x-coordinates and the average of the y-coordinates, respectively (Figure 7.2). Thus, the midpoint, M, of the line segment with endpoints (x_1, y_1) and (x_2, y_2) has coordinates $M\left(\dfrac{x_1 + x_2}{2}, \dfrac{y_1 + y_2}{2}\right)$.

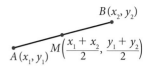

Figure 7.2 Midpoint of a line segment in a plane

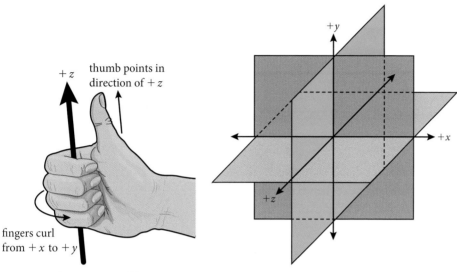

thumb points in direction of $+z$

fingers curl from $+x$ to $+y$

Figure 7.3 The orientation of the $+x$, $+y$ and $+z$ axes must follow the right-hand rule

Figure 7.4 3D coordinate system showing axes and the x-y plane, x-z plane and y-z plane

The rule for drawing the x, y and z axes of a 3D coordinate system is that the direction of the positive z axis points in the direction of the thumb on a right hand when the fingers curl in the direction from the positive x axis to the positive y axis, as illustrated in Figure 7.3. The three axes in Figure 7.4 also conform to the right-hand rule.

It is straightforward to extend the formulas for the distance between two points and the midpoint of a line segment from points in a 2-dimensional coordinate system to points in a 3-dimensional coordinate system (Figure 7.4). Consider a right rectangular prism (all six faces are rectangles: a cuboid) with a width of a units, a depth of b units and a height of c units (Figure 7.5). Consider a diagonal $[PQ]$ of the right rectangular prism. Applying the Pythagorean theorem twice – first to find the length of the diagonal of the bottom face and then to find the length, PQ, of the diagonal of the prism – gives $PQ = \sqrt{a^2 + b^2 + c^2}$

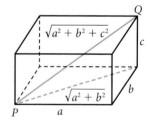

Figure 7.5 Pythagoras' theorem for three-dimensional space

A line segment in space with endpoints $P(x_1, y_1, z_1)$ and $Q(x_2, y_2, z_2)$ is the diagonal of a right rectangular prism as shown in Figure 7.6 and we can use Pythagoras' theorem, as we did for the prism in Figure 7.5, to find PQ.

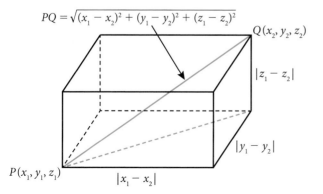

$$PQ = \sqrt{(x_1 - x_2)^2 + (y_1 - y_2)^2 + (z_1 - z_2)^2}$$

Figure 7.6 The distance between two points in space

The **distance**, d, between the points (x_1, y_1, z_1) and (x_2, y_2, z_2) is

$$d = \sqrt{(x_1 - x_2)^2 + (y_1 - y_2)^2 + (z_1 - z_2)^2}$$

The **midpoint** M of the line segment with endpoints (x_1, y_1, z_1) and (x_2, y_2, z_2) has coordinates

$$M\left(\frac{x_1 + x_2}{2}, \frac{y_1 + y_2}{2}, \frac{z_1 + z_2}{2}\right)$$

Example 7.1

Use the distance formula to show that the three points $F(1, -1, 3)$, $G(2, -4, 5)$, and $H(5, -13, 11)$ are collinear.

Solution

Find the exact distance between each of the three pairs of points. The three points are collinear (lie on the same line) if the sum of any two distances is equal to the third distance.

$$FG = \sqrt{(1 - 2)^2 + (-1 - (-4))^2 + (3 - 5)^2} = \sqrt{1 + 9 + 4} = \sqrt{14}$$

$$FH = \sqrt{(1 - 5)^2 + (-1 - (-13))^2 + (3 - 11)^2} = \sqrt{16 + 144 + 64} = 4\sqrt{14}$$

$$GH = \sqrt{(2 - 5)^2 + (-4 - (-13))^2 + (5 - 11)^2} = \sqrt{9 + 81 + 36} = 3\sqrt{14}$$

It is true that $FG + GH = FH$. Therefore, points F, G, and H are collinear.

Example 7.2

Given that three of the vertices of parallelogram $ABCD$ are $A(3, -1, 2)$, $B(1, 2, -4)$, and $C(-1, 1, 2)$, determine the coordinates of vertex D.

It is standard practice to label the vertices of a polygon in alphabetical order either clockwise or anti-clockwise.

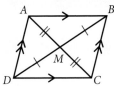

Solution

The diagonals of a parallelogram bisect each other. Thus, the diagonals AC and BD must have the same midpoint M.

$$M\left(\frac{3 - 1}{2}, \frac{-1 + 1}{2}, \frac{2 + 2}{2}\right) = M(1, 0, 2)$$

Let (x, y, z) be the coordinates of vertex D.

Then $\dfrac{x + 1}{2} = 1, \dfrac{y + 2}{2} = 0, \dfrac{z - 4}{2} = 2$.

Therefore, the coordinates of D are $(1, -2, 8)$.

3-dimensional solids: volumes and surface areas

Table 7.1 lists the solids and their respective formulae that you need to know for this course.

Solid	Volume	Surface area	Parameters
cuboid	$V = l \cdot w \cdot h$	$S = 2(l \cdot w + l \cdot h + w \cdot h)$	l = length; w = width; h = height
sphere	$V = \dfrac{4}{3}\pi r^3$	$S = 4\pi r^2$	r = radius
prism	$V = A_{base} \cdot h$	$S = 2 \cdot A_{base} + \sum\limits_{i=1}^{n} A_{lateral\,face}$	A_{base} = area of polygonal base $\sum\limits_{i=1}^{n} A_{lateral\,face}$ = sum of the n lateral faces, each of which is a rectangle
pyramid	$V = \dfrac{1}{3} A_{base} \cdot h$	$S = A_{base} + \sum\limits_{i=1}^{n} A_{lateral\,face}$	A_{base} = area of polygonal base $\sum\limits_{i=1}^{n} A_{lateral\,face}$ = sum of the n lateral faces, each of which is a triangle
cylinder	$V = \pi r^2 h$	$S = 2\pi r^2 + 2\pi rh$ or $= 2\pi r(r + h)$	r = radius; h = height
cone	$V = \dfrac{1}{3}\pi r^2 h$	$S = \pi r^2 + \pi rl$ where $l = \sqrt{r^2 + h^2}$	r = radius; h = height l = lateral height (or slant height)

Table 7.1 Volume and surface area formulae for different solids

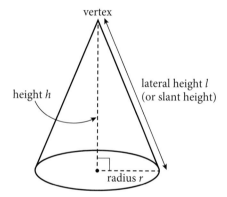

Figure 7.7 Features of a cone

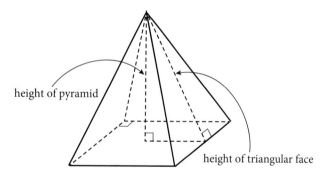

Figure 7.8 Features of a pyramid

Example 7.3

A triangular pyramid sits on top of a triangular right prism. The prism has a height of h. Find the height of the pyramid – in terms of h – so that the prism and the pyramid have the same volume.

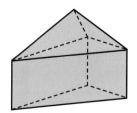

Figure 7.9 Diagram for Example 7.3

Solution

The volume of the prism is $V = A_{base} \cdot h$ and the volume of the pyramid is $V = \dfrac{1}{3} A_{base} \cdot h$. The two solids have the same base. Therefore, for the two solids to have the same volume, the height of the pyramid needs to be $3h$.

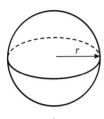

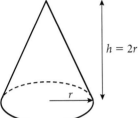

Figure 7.10 Diagrams for Example 7.4

Example 7.4

Compare a sphere and a cone. The radius of the sphere and the radius of cone are both r. The height of the cone is $2r$. Find the exact value of each ratio:

(a) The ratio of the volume of the sphere to the volume of the cone.

(b) The ratio of the surface area of the cone to the surface area of the sphere.

Solution

(a) The volume of the sphere is $V = \frac{4}{3}\pi r^3$

The volume of the cone is $V = \frac{1}{3}\pi r^2 h$. Since $h = 2r$, then

$$V = \frac{1}{3}\pi r^2(2r) = \frac{2}{3}\pi r^3$$

Thus, $\dfrac{V_{\text{sphere}}}{V_{\text{cone}}} = \dfrac{\frac{4}{3}\pi r^3}{\frac{2}{3}\pi r^3} = 2$

(b) The surface area of the cone is $S = \pi r^2 + \pi r l = \pi r^2 + \pi r\left(\sqrt{r^2 + h^2}\right)$

Substituting $h = 2r$, gives $S = \pi r^2 + \pi r\left(\sqrt{r^2 + (2r)^2}\right) = \pi r^2 + \pi r\left(\sqrt{5\, r^2}\right)$
$= \pi r^2(1 + \sqrt{5})$

The surface area of the sphere is $S = 4\pi r^2$

Thus, $\dfrac{S_{\text{cone}}}{S_{\text{sphere}}} = \dfrac{\pi r^2(1 + \sqrt{5})}{4\pi r^2} = \dfrac{1 + \sqrt{5}}{4}$

Exercise 7.1

For three line segments to form a triangle, their lengths must satisfy the **triangle inequality theorem** which states that the sum of the lengths of any two sides of a triangle must be greater than the length of the third side.

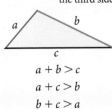

$a + b > c$
$a + c > b$
$b + c > a$

1. For each set of three points, determine whether they are the vertices of a scalene, isosceles, or equilateral triangle.
   (a) $(3, -1, 5), (-4, 0, 2), (2, 2, -1)$
   (b) $(-2, 4, -3), (4, -3, -2), (-3, -2, 4)$
   (c) $(4, 5, 0), (2, 6, 2), (2, 3, -1)$
   (d) $(a, b, c), (b, c, a), (c, a, b)$

2. Find the point on the y-axis that is a distance of $\sqrt{10}$ from the point $(1, 2, 3)$
   .

3. For each set of three points, determine whether or not they could be the vertices of a triangle.
   (a) $A(-1, 2, 3), B(1, 4, 5), C(5, 4, 0)$
   (b) $P(2, -3, 3), Q(1, 2, 4), R(3, -8, 2)$

4. Show that the points $(0, 7, 10), (-1, 6, 6)$, and $(-4, 9, 6)$ are the vertices of a right-angled isosceles triangle.

5. For each set of three points, determine whether they are collinear.
   (a) $(0, -1, -7), (2, 1, -9), (6, 5, -13)$
   (b) $(-2, 0, 4), (5, -1, 1), (4, -6, 3)$
   (c) $(1, 8, -4), (-3, 5, -1), (2, 7, 2)$
   (d) $(2, 3, 4), (-1, 2, -3), (-4, 1, -10)$

6. Find the distance of each of these points from:
       (i) the origin   **(ii)** the x-axis   **(iii)** the y-axis   **(iv)** the z-axis.
   (a) $(2, 6, -3)$
   (b) $(2, -\sqrt{3}, 3)$

7. $PQRS$ is a parallelogram. Given that three of the vertices are $P(6, -2, 4)$, $Q(2, 4, -8)$, and $R(-2, 2, 4)$, determine the coordinates of vertex S.

8. Consider the triangle with vertices $X(2, 2, 3)$, $Y(3, 7, 5)$, and $Z(1, 4, -2)$.
   (a) Show that triangle XYZ is isosceles.
   (b) Find the exact area of triangle XYZ.

9. A line segment connecting two antipodal (diametrically opposite) points on a sphere will pass through the centre of the sphere.
   Points $A(2, -7, -4)$ and $B(6, 1, 2)$ are a pair of antipodal points on a sphere. Find the exact surface area and exact volume of this sphere.

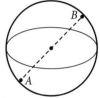

Figure 7.11 Diagram for question 9

10. A rectangular wooden box has dimensions: $62\,\text{cm} \times 44\,\text{cm} \times 20\,\text{cm}$. Find the length (to the nearest whole cm) of the longest piece of straight wire that can be placed completely inside the box.

11. A solid consists of a cone, cylinder and hemisphere joined together as shown in the diagram. Given the dimensions indicated in the diagram, find the volume and surface area of the solid, accurate to 3 significant figures.

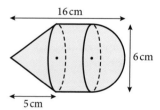

Figure 7.12 Diagram for question 11

12. The midpoints of the sides of a triangle are $(1, 5, -1)$, $(0, 4, -2)$, and $(2, 3, 4)$. Find the coordinates of the three vertices of the triangle.

13. A sphere with radius r is inscribed in a cylinder such that the sphere is tangent to the bases of the cylinder and touches the inside of the cylinder along a circle. Find:
   (a) the ratio of the sphere's volume to the cylinder's volume.
   (b) the ratio of the sphere's surface area to the cylinder's surface area.

14. A large building for storing grain consists of a cylinder with a cone on top (Figure 7.13). Given the dimensions indicated in the diagram, find the surface area and the volume of the building accurate to 3 significant figures. (The circular base is not included in the building's surface.)

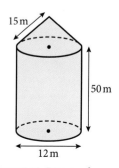

Figure 7.13 Diagram for question 14

307

15. A metal spike is made from a cube with a pyramid attached to one of the cube's faces, as shown in the diagram. Given the dimensions indicated in the diagram, find the exact volume and exact surface area of the spike.

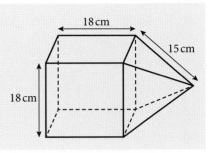

Right-angled triangles and trigonometric functions of acute angles

Right-angled triangles

In IB notation, the symbol $A\hat{B}C$ denotes the angle with its vertex at point B, with one side of the angle containing point A and the other side containing point C.

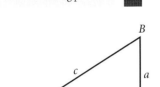

Figure 7.14 Conventional triangle notation

The conventional notation for triangles is to label the three vertices with capital letters, for example A, B, and C. The same capital letters can be used to represent the angles at these vertices, but we will often use a Greek letter, such as α (alpha), β (beta), or θ (theta) instead – as we did in Chapter 6. The corresponding lower-case letters, a, b, and c represent the lengths of the sides opposite the vertices. For example, b represents the length of the side opposite angle B – that is, the line segment AC (Figure 7.14).

Trigonometric functions of an acute angle

We can use properties of similar triangles and the definitions of the sine, cosine and tangent functions from Chapter 6 to define these functions in terms of the sides of a right-angled triangle.

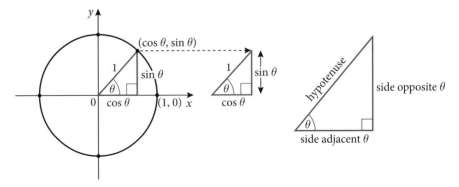

Figure 7.15 Trigonometric functions defined in terms of the sides of a right-angled triangle

$$\frac{\sin\theta}{1} = \frac{\text{opposite}}{\text{hypotenuse}}$$

$$\frac{\cos\theta}{1} = \frac{\text{adjacent}}{\text{hypotenuse}}$$

$$\frac{\tan\theta}{1} = \frac{\sin\theta}{\cos\theta} = \frac{\text{opposite}}{\text{adjacent}}$$

The right-angled triangles shown in Figure 7.15 are **similar triangles** because corresponding angles are the same size: each has a right angle and an acute angle of θ. It follows that the ratios of corresponding sides are equal, allowing us to write these three proportions involving the sine, cosine and tangent of the acute angle θ.

The definitions of the trigonometric functions in terms of the sides of a right-angled triangle follow directly from these three equations.

Let θ be an **acute angle** of a right-angled triangle. Then the sine, cosine and tangent functions of the angle θ are defined as:

$$\sin\theta = \frac{\text{side opposite angle }\theta}{\text{hypotenuse}} \qquad\qquad \csc\theta = \frac{\text{hypotenuse}}{\text{side opposite angle }\theta}$$

$$\cos\theta = \frac{\text{side adjacent angle }\theta}{\text{hypotenuse}} \qquad\qquad \sec\theta = \frac{\text{hypotenuse}}{\text{side adjacent angle }\theta}$$

$$\tan\theta = \frac{\text{side opposite angle }\theta}{\text{side adjacent angle }\theta} \qquad\qquad \cot\theta = \frac{\text{side adjacent angle }\theta}{\text{side opposite angle }\theta}$$

It follows that the trigonometric functions of an acute angle are positive.

It is important to understand that properties of similar triangles are the foundation of right-angled triangle trigonometry. Regardless of the size (i.e. lengths of the sides) of a right-angled triangle, so long as the angles don't change, the ratio of any two sides in the right-angled triangle will remain constant. All the right-angled triangles in Figure 7.16 have an acute angle with a measure of 30°. For each triangle, the ratio of the side opposite the 30° angle to the hypotenuse is exactly $\frac{1}{2}$.

In other words, the sine of 30° is always $\frac{1}{2}$. This agrees with results from the previous chapter that an angle of 30° is equivalent to $\frac{\pi}{6}$ in radian measure.

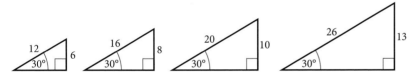

Figure 7.16 Corresponding ratios of a pair of sides for similar triangles are equal

Values of sine, cosine and tangent for common acute angles:

$$\sin 30° = \sin\frac{\pi}{6} = \frac{1}{2} \qquad \cos 30° = \cos\frac{\pi}{6} = \frac{\sqrt{3}}{2} \qquad \tan 30° = \tan\frac{\pi}{6} = \frac{\sqrt{3}}{3}$$

$$\sin 45° = \sin\frac{\pi}{4} = \frac{\sqrt{2}}{2} \qquad \cos 45° = \cos\frac{\pi}{4} = \frac{\sqrt{2}}{2} \qquad \tan 45° = \tan\frac{\pi}{4} = 1$$

$$\sin 60° = \sin\frac{\pi}{3} = \frac{\sqrt{3}}{2} \qquad \cos 60° = \cos\frac{\pi}{3} = \frac{1}{2} \qquad \tan 60° = \tan\frac{\pi}{3} = \sqrt{3}$$

Finding unknowns of right-angled triangles

We can use Pythagoras' theorem and trigonometric functions to find the size of any unknown side or angle. We will use trigonometric functions in two different ways – to find the length of a side, and to find the measure of an angle. Finding unknowns in right-angled triangles using the sine, cosine and tangent functions is essential to finding solutions to problems in fields such as astronomy, navigation, engineering, and architecture. In Section 7.4, we will see how trigonometry can also be used to find missing parts in triangles that are not right-angled triangles.

Observe that
$\sin 30° = \cos 60° = \frac{1}{2}$,
$\sin 60° = \cos 30° = \frac{\sqrt{3}}{2}$
and
$\sin 45° = \cos 45° = \frac{\sqrt{2}}{2}$.

Complementary angles (sum of 90°) have equal function values for sine and cosine. That is, for all angles x measured in degrees:
$\sin x = \cos(90° - x)$ or
$\sin(90° - x) = \cos x$.
As noted in Chapter 6, it is for this reason that sine and cosine are called co-functions.

Angles of depression and elevation

An imaginary line segment from an observation point O to a point P (representing the location of an object) is called the **line of sight** of P. If P is above O, the acute angle between the line of sight of P and a horizontal line passing through O is called the **angle of elevation** of P. If P is below O, the angle between the line of sight and the horizontal is called the **angle of depression** of P. This is illustrated in Figure 7.17.

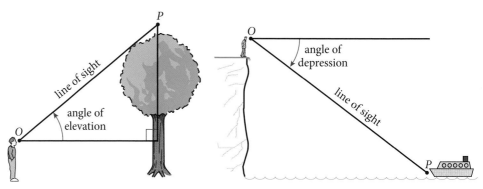

Figure 7.17 An angle of elevation or depression is always measured from the horizontal. Also, note that the angle of elevation from O to P is equal to the angle of depression from P to O

Example 7.5

Determine the lengths of the missing sides in triangle ABC given $c = 8.76$ cm, angle $A = 30°$, and the right angle is at C. Give exact answers when possible, otherwise give to an accuracy of 3 significant figures.

Solution

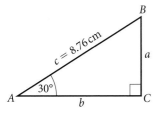

Figure 7.18 Solution to Example 7.5

Sketch triangle ABC indicating the known measurements.

From the definition of sine and cosine functions, we have:

$$\sin 30° = \frac{\text{opposite}}{\text{hypotenuse}} = \frac{a}{8.76} \quad a = 8.76 \sin 30°$$

$$a = 8.76\left(\frac{1}{2}\right) = 4.38$$

$$\cos 30° = \frac{\text{adjacent}}{\text{hypotenuse}} = \frac{b}{8.76}$$

$$b = 8.76 \cos 30°$$

$$b = 8.76\left(\frac{\sqrt{3}}{2}\right) \approx 7.586382537 \approx 7.59$$

Therefore $a = 4.38$ cm, $b \approx 7.59$ cm and, it's clear that angle $B = 60°$.

We can use Pythagoras' theorem to check our results for a and b.

$$a^2 + b^2 = c^2 \Rightarrow \sqrt{a^2 + b^2} = 8.76$$

Be aware that the result for a is exactly 4.38 cm (assuming measurements given for angle A and side c are exact), but the result for b can only be approximated. To reduce error when performing the check, we should use the most accurate value (i.e. most significant figures) for b possible. The most effective way to do this on your GDC is to use results that are stored to several significant figures.

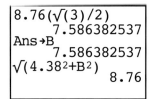

Figure 7.19 Using stored results on your GDC

Example 7.6

A man who is 183 cm tall casts a shadow 72 cm long on horizontal ground. What is the angle of elevation of the sun to the nearest tenth of a degree?

Solution

In Figure 7.20, the angle of elevation of the sun is labelled θ.

GDC computation in degree mode.

$$\tan\theta = \frac{183}{72}$$

$$\theta = \tan^{-1}\left(\frac{183}{72}\right)$$

$$\theta \approx 68.5°$$

The angle of elevation of the sun is approximately 68.5°

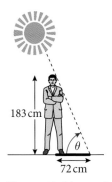

Figure 7.20 Diagram for Example 7.6

```
tan-1(183/72)
          68.52320902
```

Figure 7.21 GDC screen for the solution to Example 7.6

The notation for indicating the inverse of a function is a superscript of negative one. For example, the inverse of the tangent function is denoted as $\tan^{-1}$ on your GDC. The negative one is not an exponent, so it does not denote a reciprocal.

$$\tan^{-1} x \neq \frac{1}{\tan x}$$

Example 7.7

During a training exercise, an Air Force pilot is flying his jet at a constant altitude of 1200 metres. His task is to fire a missile at a target on the ground. At the moment he fires his missile, he is able to see the target at an angle of depression of 18.5°. If the missile travels in a straight line, what distance will the missile cover (to the nearest metre) from the jet to the target?

Solution

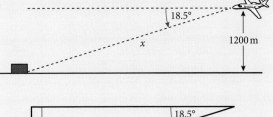

Draw a diagram to represent the information and let x be the distance that the missile travels from the jet to the target. A right-angled triangle can be 'extracted' from the diagram with one side 1200 metres, the angle

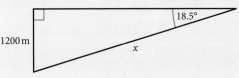

opposite that side is 18.5°, and the hypotenuse is x.

Applying the sine ratio, we can write the equation $\sin 18.5° = \dfrac{1200}{x}$

Then $x = \dfrac{1200}{\sin 18.5°} \approx 3781.85$

Hence, the missile travels approximately 3782 metres.

Example 7.8

A boat is sailing directly towards a cliff. The angle of elevation of a point on the top of the cliff and straight ahead of the boat increases from 10° to 15° as the ship sails a distance of 50 metres. Find the height of the cliff.

Solution

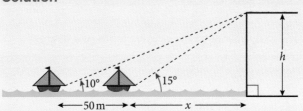

Draw a diagram that accurately represents the information, with the height of the cliff labelled h and the distance from the base of the cliff to the later position of the boat labelled x. There are two right-angled triangles that can be extracted from the diagram. From the smaller triangle, we have:

$$\tan 15° = \frac{h}{x} \Rightarrow h = x\tan 15°$$

From the larger triangle, we have:

$$\tan 10° = \frac{h}{x + 50} \Rightarrow h = (x + 50)\tan 10°$$

We can solve for x by setting the two expressions for h equal to each other. Then we can solve for h by substitution.

$$x\tan 15° = (x + 50)\tan 10°$$

$$x\tan 15° = x\tan 10° + 50\tan 10°$$

$$x(\tan 15° - \tan 10°) = 50\tan 10°$$

$$x = \frac{50\tan 10°}{\tan 15° - \tan 10°} \approx 96.225$$

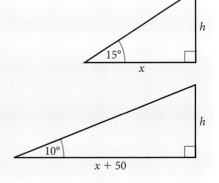

Substituting this value for x into $h = x\tan 15°$, gives:

$$h \approx 96.225\tan 15° \approx 25.783$$

Therefore, the height of the cliff is approximately 25.8 metres.

Example 7.9

Using a suitable right-angled triangle, find the exact minimum distance from the point $(8, 3)$ to the line with the equation $2x - y + 2 = 0$.

Solution

Graph the line with equation $2x - y + 2 = 0$. The minimum distance from the point $(8, 3)$ to the line is the length of the line segment drawn from the point perpendicular to the line. This minimum distance is labelled d in the diagram. d is also the height of the large yellow triangle formed by drawing vertical and horizontal line segments from $(8, 3)$ to the line.

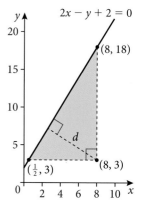

The area of the right-angled triangle is $A = \dfrac{1}{2}\left(\dfrac{15}{2}\right)(15) = \dfrac{225}{4}$

The area of the triangle can also be found by using the hypotenuse as the base and the distance d as the height. By Pythagoras' theorem, we have

$$\text{hypotenuse} = \sqrt{\left(\dfrac{15}{2}\right)^2 + 15^2} = \sqrt{\dfrac{1125}{4}} = \dfrac{\sqrt{225}\,\sqrt{5}}{\sqrt{4}} = \dfrac{15\,\sqrt{5}}{2}$$

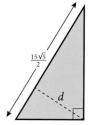

Thus, the area can also be expressed as $A = \dfrac{1}{2}\left(\dfrac{15\,\sqrt{5}}{2}\right)d$. We can solve for d by equating the two results for the area of the triangle.

Figure 7.22 Solution to Example 7.9

$$\dfrac{1}{2}\left(\dfrac{15\,\sqrt{5}}{2}\right)d = \dfrac{225}{4}$$

$$\dfrac{15\,\sqrt{5}}{4}\,d = \dfrac{225}{4}$$

$$d = \dfrac{225}{4} \cdot \dfrac{4}{15\,\sqrt{5}}$$

$$d = \dfrac{15}{\sqrt{5}} = \dfrac{15}{\sqrt{5}} \cdot \dfrac{\sqrt{5}}{\sqrt{5}} = \dfrac{15\,\sqrt{5}}{5} = 3\,\sqrt{5}$$

Therefore, the minimum distance from the point $(8, 3)$ to the line with equation $2x - y + 2 = 0$ is $3\,\sqrt{5}$ units.

Exercise 7.2

1. For each **(a)** to **(f)**

   (i) sketch a right-angled triangle corresponding to the acute angle θ

   (ii) find the exact value of the other five trigonometric functions associated with the angle

   (iii) use your GDC to find the degree measure of θ and the other acute angle (approximate to 3 significant figures).

   (a) $\sin\theta = \dfrac{3}{5}$ **(b)** $\cos\theta = \dfrac{5}{8}$ **(c)** $\tan\theta = 2$

   (d) $\cot\theta = \dfrac{1}{3}$ **(e)** $\sec\theta = \dfrac{11}{\sqrt{61}}$ **(f)** $\cos\theta = \dfrac{4\,\sqrt{65}}{65}$

2. Find the exact value of θ in degrees $(0 < \theta < 90°)$ and in radians $\left(0 < \theta < \dfrac{\pi}{2}\right)$ without using your GDC.

(a) $\cos\theta = \dfrac{1}{2}$

(b) $\sin\theta = \dfrac{\sqrt{2}}{2}$

(c) $\tan\theta = \sqrt{3}$

(d) $\csc\theta = \dfrac{2\sqrt{3}}{3}$

(e) $\cot\theta = 1$

(f) $\cos\theta = \dfrac{\sqrt{3}}{2}$

3. Find the values of x and y in each triangle. If possible, give an exact answer; otherwise, give the answer correct to 3 significant figures.

(a)

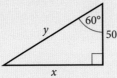

(b)

(c)

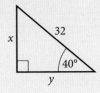

(d)

(e)

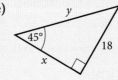

(f)

4. Find the size of the angles α and β in degrees. If possible, give an exact answer; otherwise, give your answer correct to 3 significant figures.

(a)

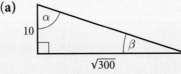

(b)

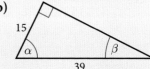

(c)

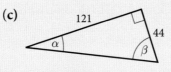

(d)

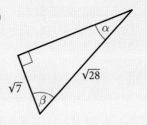

5. The tallest tree in the world is reputed to be a giant redwood named *Hyperion* located in Redwood National Park in California, USA. At a point 41.5 metres from the centre of its base and on the same elevation, the angle of elevation of the top of the tree is 70°. How tall is the tree? Give your answer to 3 significant figures.

6. The Eiffel Tower in Paris is 300 metres high (not including the antenna on top). What is the angle of elevation of the top of the tower from a point on the ground (assumed level) that is 125 metres from the centre of the tower's base?

7. A 1.62 m tall woman, standing 3 metres from a streetlight, casts a 2 m shadow. What is the height of the streetlight?

8. A pilot measures the angles of depression to two ships to be 40° and 52°. The pilot is flying at an elevation of 10 000 metres. Find the distance between the two ships.

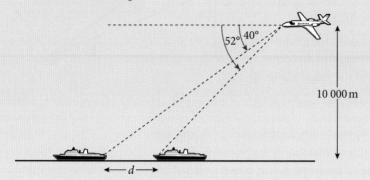

9. Find the size of all three angles in a triangle with sides of length 8 cm, 8 cm and 6 cm.

10. A boat is sighted from a 50-metre observation tower on the shoreline at an angle of depression of 4° moving directly towards the shore at a constant speed. Five minutes later the angle of depression of the boat is 12°. What is the speed of the boat in kilometres per hour?

11. Find the length of x indicated in Figure 7.23. Give your answer to 3 significant figures.

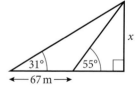

Figure 7.23 Diagram for question 11

12. A support wire for a tower is connected from an anchor point on level ground to the top of the tower. The straight wire makes a 65° angle with the ground at the anchor point. At a point 25 metres farther from the tower than the wire's anchor point and on the same side of the tower, the angle of elevation to the top of the tower is 35°. Find the wire length to the nearest tenth of a metre.

13. A 30-metre high building sits on top of a hill. The angles of elevation of the top and bottom of the building from the same spot at the base of the hill are measured to be 55° and 50° respectively. How high is the hill to the nearest metre?

14. The angle of elevation of the top of a vertical pole as seen from a point 10 metres away from the pole is double its angle of elevation as seen from a point 70 metres from the pole. Find the height (to the nearest tenth of a metre) of the pole above the level of the observer's eyes.

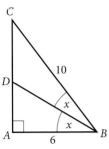

Figure 7.24 Diagram for question 15

15. Angle ABC of a right-angled triangle is bisected by segment BD. The lengths of sides AB and BC are given in Figure 7.24. Find the exact length of BD, expressing your answer in its simplest form.

16. In the diagram, $D\hat{E}C = C\hat{E}B = x°$ and $C\hat{D}E = B\hat{E}A = 90°$, $CD = 1$ unit, $DE = 3$ units. By writing $D\hat{E}A$ in terms of x, find the exact value of $\cos(D\hat{E}A)$.

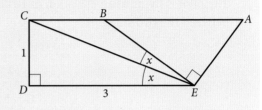

17. For any point with coordinates (p, q) and any line with equation $ax + by + c = 0$, find a formula in terms of a, b, c, p and q that gives the minimum (perpendicular) distance, d, from the point to the line.

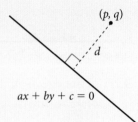

In question 18, first try expressing the formula using the tangent ratio.

18. Show that the length x in the diagram is given by the formula

$$x = \frac{d}{\cot\alpha - \cot\beta}$$

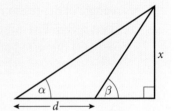

19. A spacecraft is travelling in a circular orbit 200 km above the surface of the Earth. Find the angle of depression (to the nearest degree) from the spacecraft to the horizon. Assume that the radius of the Earth is 6400 km. The 'horizontal' line through the spacecraft from which the angle of depression is measured will be parallel to a line tangent to the surface of the Earth directly below the spacecraft.

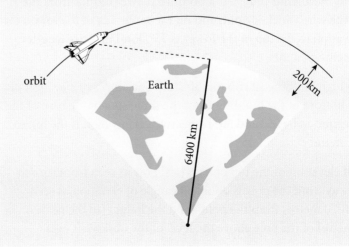

7.3 Trigonometric functions of any angle

In this section we will extend the trigonometric ratios to all angles, allowing us to solve problems involving any size angle.

Defining trigonometric functions for any angle in standard position

Consider the point $P(x, y)$ on the terminal side of an angle θ in standard position (Figure 7.25) such that r is the distance from the origin O to P. If θ is an acute angle, then we can construct a right-angled triangle POQ (Figure 7.26) by dropping a perpendicular from P to a point Q on the x-axis. It follows that:

$$\sin\theta = \frac{y}{r} \qquad \cos\theta = \frac{x}{r} \qquad \tan\theta = \frac{y}{x} \ (x \neq 0)$$

$$\csc\theta = \frac{r}{y} \ (y \neq 0) \qquad \sec\theta = \frac{r}{x} \ (x \neq 0) \qquad \cot\theta = \frac{x}{y} \ (y \neq 0)$$

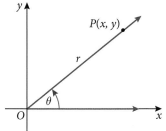

Figure 7.25 Angle θ in standard position

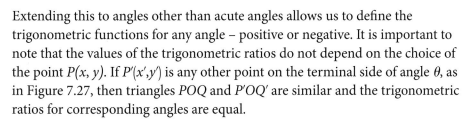

Figure 7.26 θ is an acute angle in $\triangle POQ$

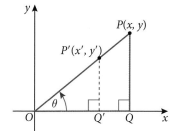

Figure 7.27 Similar right-angled triangles POQ and $P'OQ'$

Extending this to angles other than acute angles allows us to define the trigonometric functions for any angle – positive or negative. It is important to note that the values of the trigonometric ratios do not depend on the choice of the point $P(x, y)$. If $P'(x',y')$ is any other point on the terminal side of angle θ, as in Figure 7.27, then triangles POQ and $P'OQ'$ are similar and the trigonometric ratios for corresponding angles are equal.

Example 7.10

Find the sine, cosine and tangent of an angle α that contains the point $(-3, 4)$ on its terminal side when in standard position.

Solution

$$r = \sqrt{x^2 + y^2} = \sqrt{(-3)^2 + 4^2} = \sqrt{25} = 5$$

Then, $\sin\alpha = \dfrac{y}{r} = \dfrac{4}{5}$

$\cos\alpha = \dfrac{x}{r} = \dfrac{-3}{5} = -\dfrac{3}{5}$

$\tan\alpha = \dfrac{y}{x} = \dfrac{4}{-3} = -\dfrac{4}{3}$

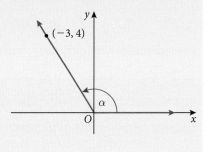

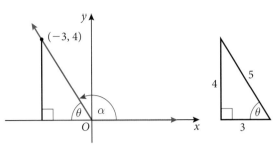

Figure 7.28 Reference triangle for computing trigonometric values for angle *a*

Note that for the angle α in Example 7.10, we can form a right-angled triangle by constructing a line segment from the point $(-3, 4)$ perpendicular to the x-axis, as shown in Figure 7.28. Clearly, $\theta = 180° - \alpha$. Furthermore, the values of the sine, cosine and tangent of the angle θ are the same as those for the angle α, except that the sign may be different.

Since all trigonometric functions are associated with either the x-coordinate (cosine), the y-coordinate (sine), or both x- and y-coordinates (tangent) of a point on the terminal side of the angle, then the sign of a trigonometric function will be positive or negative according to which quadrant it lies in, as shown in Figure 7.29.

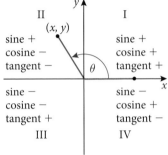

Figure 7.29 Signs of trigonometric function values in each quadrant

---

Example 7.11

Find the sine, cosine and tangent of the obtuse angle that measures 150°.

Solution

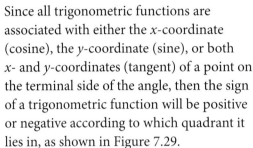

Figure 7.30 Solution to Example 7.11

The terminal side of the angle forms a 30° angle with the x-axis. The sine values for 150° and 30° will be exactly the same, and the cosine and tangent values will be the same but of opposite sign.

We know that $\sin 30° = \dfrac{1}{2}$, $\cos 30° = \dfrac{\sqrt{3}}{2}$

and $\tan 30° = \dfrac{\sqrt{3}}{3}$

Therefore, $\sin 150° = \dfrac{1}{2}$, $\cos 150° = -\dfrac{\sqrt{3}}{2}$ and $\tan 150° = -\dfrac{\sqrt{3}}{3}$

Example 7.11 illustrates three trigonometric identities for angles whose sum is 180° (i.e. a pair of supplementary angles). The following identities are true for any acute angle θ:

$$\sin(180° - \theta) = \sin\theta \qquad \cos(180° - \theta) = -\cos\theta \qquad \tan(180° - \theta) = -\tan\theta$$

$$\csc(180° - \theta) = \csc\theta \qquad \sec(180° - \theta) = -\sec\theta \qquad \cot(180° - \theta) = -\cot\theta$$

These identities are equivalent to ones in radian measure $(180° = \pi)$.

Example 7.12

Given $\sin\theta = \dfrac{5}{13}$, $90° < \theta < 180°$, find the exact values of $\cos\theta$ and $\tan\theta$.

Solution

θ is an angle in the 2nd quadrant. It follows from the definition $\sin\theta = \dfrac{y}{r}$ that with θ in standard position, there must a be a point on the terminal side of the angle that is 13 units from the origin ($r = 13$) and has a y-coordinate of 5, as shown in the diagram.

Thus, $x = \sqrt{13^2 - 5^2} = \sqrt{144} = 12$. Because θ is in the 2nd quadrant, the x-coordinate of the point is negative; thus, $x = -12$.

Therefore, $\cos\theta = \dfrac{-12}{13} = -\dfrac{12}{13}$, and $\tan\theta = \dfrac{5}{-12} = -\dfrac{5}{12}$.

Example 7.13

(a) Find the acute angle with the same sine ratio as (i) 135°, and (ii) 117°

(b) Find the acute angle with the same cosine ratio as (i) 300° and (ii) 342°

Solution

(a) (i) Angles in the 1st and 2nd quadrants have the same sine ratio. Hence, the identity $\sin(180° - \theta) = \sin\theta$. Since $180° - 135° = 45°$, then $\sin 135° = \sin 45°$

   (ii) Since $180° - 117° = 63°$, then $\sin 117° = \sin 63°$

(b) (i) Angles in the 1st and 4th quadrants have the same cosine ratio. Hence, the identity $\cos(360° - \theta) = \cos\theta$. Since $360° - 300° = 60°$, then $\cos 300° = \cos 60°$

   (ii) Since $360° - 342° = 18°$, then $\cos 342° = \cos 18°$

Areas of triangles

We are familiar with the standard formula for the area of a triangle, $\text{area} = \dfrac{1}{2} \times \text{base} \times \text{height}$ $\left(A = \dfrac{1}{2}bh\right)$, where the base, b, is a side of the triangle and the height, h, (or altitude) is a line segment perpendicular to the base (or the line containing it) and drawn to the vertex opposite the base, as shown in Figure 7.31.

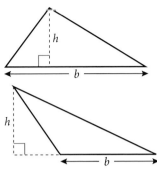

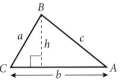

Figure 7.31 Area of a triangle $A = \frac{1}{2}bh$

Figure 7.32 An acute triangle

For a triangle with sides of lengths a and b and included angle C:

area of $\Delta = \frac{1}{2}ab\sin C$.

For any triangle labelled in the manner of the triangles in Figures 7.32 and 7.33, its area is given by any of the expressions.

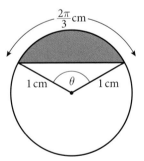

Figure 7.34 Diagram for Example 7.14

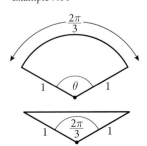

Figure 7.35 Solution for Example 7.14

If the lengths of two sides of a triangle and the measure of the angle between these sides (often called the included angle) are known, then the triangle is unique and has a fixed area. Hence, we should be able to calculate the area from just these measurements – two sides and the included angle. This calculation is quite straightforward when the triangle is a right-angled triangle and we know the lengths of the two legs on either side of the right angle.

Let's develop a general area formula that will apply to any triangle − right-angled, acute, or obtuse. For triangle ABC shown in Figure 7.32, suppose we know the lengths of the two sides a and b and the included angle C. If the height is h, then the area of ABC is $\frac{1}{2}bh$. From trigonometry, we know that $\sin C = \frac{h}{a}$, or $h = a\sin C$. Substituting $a\sin C$ for h gives area $= \frac{1}{2}bh = \frac{1}{2}b(a\sin C) = \frac{1}{2}ab\sin C$.

If the angle C is obtuse, then from Figure 7.33 we see that $\sin(180° − C) = \frac{h}{a}$.

So, the height is $h = a\sin(180° − C)$. However, $\sin(180° − C) = \sin C$. Thus, $h = a\sin C$ and, again, area $= \frac{1}{2}ab\sin C$.

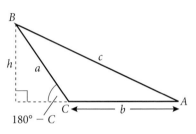

Figure 7.33 An obtuse triangle

area of $\Delta = \frac{1}{2}ab\sin C = \frac{1}{2}ac\sin B = \frac{1}{2}bc\sin A$

These three equivalent expressions will prove to be helpful for developing an important formula (the sine rule) for solving non-right-angled triangles in the next section.

Example 7.14

The circle shown has a radius of 1 cm and the central angle θ subtends an arc of length $\frac{2\pi}{3}$ cm. Find the area of the shaded region (a segment of the circle).

Solution

The formula for the area of a sector is $A = \frac{1}{2}r^2\theta$ where θ is the central angle in radians. Since the radius of the circle is 1, the length of the arc subtended by θ is the same as the size of θ. Thus, area of sector $= \frac{1}{2}(1)^2\left(\frac{2\pi}{3}\right) = \frac{\pi}{3}$ cm². The area of the triangle formed by the two radii and the chord is equal to $\frac{1}{2}(1)(1)\sin\left(\frac{2\pi}{3}\right) = \frac{1}{2}\left(\frac{\sqrt{3}}{2}\right) = \frac{\sqrt{3}}{4}$ cm²

We find the area of the shaded region (segment) by subtracting the area of the triangle from the area of the sector. Area $= \frac{\pi}{3} - \frac{\sqrt{3}}{4}$ or $\frac{4\pi - 3\sqrt{3}}{12}$ or approximately 0.614 cm² (3 s.f.)

Example 7.15

Show that it is possible to construct two different triangles with an area of 35 cm² that have sides measuring 8 cm and 13 cm. For each triangle, find the size of the (included) angle between the sides of 8 cm and 13 cm to the nearest tenth of a degree.

Solution

We can visualise the two different triangles with equal areas: one with an acute included angle (α) and the other with an obtuse included angle (β).

$$\text{area} = \frac{1}{2}(\text{side})(\text{side})(\text{sine of included angle}) = 35 \text{ cm}^2$$

$$\Rightarrow \frac{1}{2}(8)(13)(\sin\alpha) = 35$$

$$52\sin\alpha = 35$$

$$\sin\alpha = \frac{35}{52}$$

$$\alpha = \sin^{-1}\left(\frac{35}{52}\right)$$

$$= 42.3° \text{ (to the nearest tenth of a degree)}$$

Knowing that $\sin(180° - \alpha) = \sin\alpha$, the obtuse angle β is equal to $180° - 42.3° = 137.7°$

Therefore, there are two different triangles with sides 8 cm and 13 cm and an area of 35 cm², one with an included angle of 42.3° and the other with an included angle of 137.7°

Your GDC will give only the acute angle value for the $\sin^{-1}$ function.

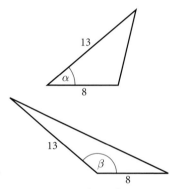
Figure 7.36 Solution for Example 7.15

Equations of lines and angles between two lines

Recall that the gradient m of a non-vertical line is defined as

$$m = \frac{y_2 - y_1}{x_2 - x_1} = \frac{\text{vertical change}}{\text{horizontal change}}$$

The equation of the line shown in Figure 7.37 has a gradient $m = \frac{1}{2}$ and a y-intercept of $(0, -1)$. So, the equation of the line is $y = \frac{1}{2}x - 1$. We can find the size of the acute angle θ between the line and the x-axis by using the tangent function (Figure 7.38).

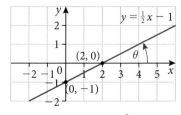

Figure 7.37 The line $y = \frac{1}{2}x - 1$

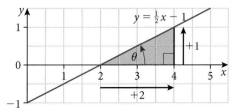

Figure 7.38 Angle between $y = \frac{1}{2}x - 1$ and x-axis

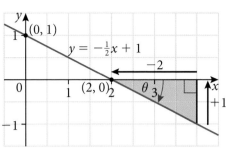

Figure 7.39 The line $y = -\frac{1}{2}x + 1$

$$\theta = \tan^{-1}(m) = \tan^{-1}\left(\frac{1}{2}\right) \approx 26.6°$$

Clearly, the gradient, m, of this line is equal to $\tan\theta$. If we know the angle between the line and the x-axis, and the y-intercept $(0, c)$, we can write the equation of the line in gradient-intercept form as

$$y = (\tan\theta)x + c$$

Before we can generalise for any non-horizontal line, let's look at a line with a negative gradient.

The gradient of the line is $-\frac{1}{2}$. In order for $\tan\theta$ to be equal to the gradient of the line, the angle θ must be the angle that the line makes with the x-axis in the positive direction, as shown in Figure 7.39. In this example, $\theta = \tan^{-1}(m) = \tan^{-1}\left(-\frac{1}{2}\right) \approx -26.6°$. Remember, a negative angle indicates a clockwise rotation from the initial side to the terminal side of the angle.

 If a line has a y-intercept of $(0, c)$ and makes an angle of θ with the positive direction of the x-axis, such that $-90° < \theta < 90°$, then the gradient of the line is $m = \tan\theta$ and the equation of the line is $y = (\tan\theta)x + c$.

The angle this line makes with any horizontal line will be θ.

We will now use triangle trigonometry to find the angle between any two intersecting lines – not just for a line intersecting the x-axis. Any pair of intersecting lines that are not perpendicular will have both an acute angle and an obtuse angle between them. When asked for an angle between two lines, the convention is to give the acute angle.

Example 7.16

Find the acute angle between the lines $y = 3x$ and $y = -x$

Solution

The angle between the line $y = 3x$ and the positive x-axis is α, and the angle between the line $y = -x$ and the positive x-axis is β.

$$\alpha = \tan^{-1}(3) \approx 71.565°$$

$$\beta = \tan^{-1}(-1) = -45°$$

The obtuse angle between the two lines is $\alpha - \beta \approx 71.565° - (-45°) \approx 116.565°$

Therefore, the acute angle θ between the two lines is $\theta \approx 180° - 116.565 \approx 63.4°$

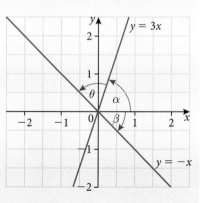

Example 7.17

Find the acute angle between the lines $y = 5x - 2$ and $y = \frac{1}{3}x - 1$

Solution

A horizontal line is drawn through the point of intersection.

The angle between $y = 5x - 2$ and this horizontal line is α, and the angle between $y = \frac{1}{3}x - 1$ and this horizontal line is β.

$\alpha = \tan^{-1}(5) \approx 78.690°$ and

$\beta = \tan^{-1}\left(\frac{1}{3}\right) \approx 18.435°$

The acute angle θ between the two lines is $\theta = \alpha - \beta \approx 78.690° - 18.435° \approx 60.3°$

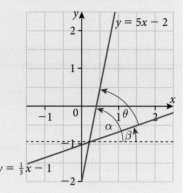

We can generalise the procedure for finding the angle between two lines as follows.

Given two non-vertical lines with equations of $y_1 = m_1x + c_1$ and $y_2 = m_2x + c_2$, the angle between the two lines is $\left|\tan^{-1}(m_1) - \tan^{-1}(m_2)\right|$. This angle may be acute or obtuse.

Example 7.18

(a) Find the exact equation of line L_1 that passes through the origin and makes an angle of $-60°$ (or $120°$) with the positive direction of the x-axis.

(b) The equation of line L_2 is $7x + y + 1 = 0$. Find the acute angle between L_1 and L_2.

Solution

(a) The equation of the line is given by
$y = (\tan\theta)x$

$\Rightarrow [\tan(-60°)]x = \left[\frac{\sin(-60°)}{\cos(-60°)}\right]x$

$= \left[\frac{-\dfrac{\sqrt{3}}{2}}{\dfrac{1}{2}}\right]x = (-\sqrt{3})x$

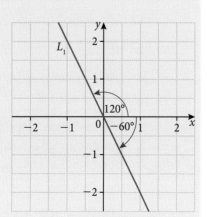

Therefore, the equation of L_1 is
$y = (-\sqrt{3})x$ or $y = -x\sqrt{3}$

Remember: $\tan(-60°) = \tan 120° = -\sqrt{3}$

(b) $L_2 : 7x + y + 1 = 0 \implies y = -7x - 1$

θ is the acute angle between L_1 and L_2.

$\theta = |\tan^{-1}(m_1) - \tan^{-1}(m_2)|$

$= |\tan^{-1}(-\sqrt{3}) - \tan^{-1}(-7)|$

$\implies \theta \approx -60° - (-81.870°) \approx -21.87°$

Therefore, the acute angle between the lines is 21.9° (3 s.f.)

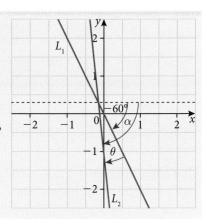

In many problems, it is necessary to calculate lengths and angles in three-dimensional structures. It is very important to analyse the three-dimensional diagram carefully and to extract any relevant triangles to find the missing angle(s) or length(s).

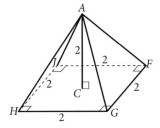

Figure 7.40 Diagram for Example 7.19

Example 7.19

The diagram shows a pyramid with a square base. It is a right pyramid so the line segment (the height) drawn from the top vertex A perpendicular to the base will intersect the square base at its centre C. Each side of the square base has a length of 2 cm and the height of the pyramid is also 2 cm. Find:

(a) the size of $A\widehat{G}F$

(b) the total surface area of the pyramid.

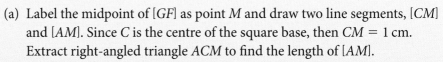

Solution

(a) Label the midpoint of $[GF]$ as point M and draw two line segments, $[CM]$ and $[AM]$. Since C is the centre of the square base, then $CM = 1$ cm. Extract right-angled triangle ACM to find the length of $[AM]$.

$AM = \sqrt{1^2 + 2^2} = \sqrt{5}$ $[AM]$ is perpendicular to $[GF]$

Extract right-angled triangle AMG and use the tangent ratio to find $A\widehat{G}M$.

$\tan(A\widehat{G}M) = \dfrac{\sqrt{5}}{1}$

$A\widehat{G}M = A\widehat{G}F = \tan^{-1}(\sqrt{5}) \approx 65.905°$

Therefore, $A\widehat{G}F \approx 65.9°$

(b) The total surface area comprises the square base plus four identical lateral faces that are all equilateral triangles. Triangle AGM is one-half the area of one of these triangular faces.

Area of triangle $AGM = \dfrac{1}{2}(1)(\sqrt{5}) = \dfrac{\sqrt{5}}{2} \implies$

Area of triangle $= AGF = 2\left(\dfrac{\sqrt{5}}{2}\right) = \sqrt{5}$

Surface area = area of square base + area of 4 lateral faces $= 2^2 + 4\sqrt{5}$
$= 4 + 4\sqrt{5} \approx 12.94 \text{ cm}^2$

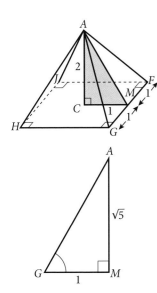

Figure 7.41 Solution to Example 7.19

1. In each diagram, find the exact value of the six trigonometric functions of the angle θ. Simplify your answers.

(a)

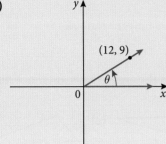

(b)

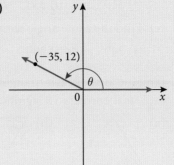

(c)

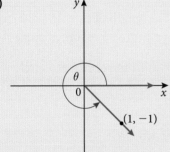

(d)

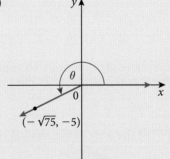

2. Without using your GDC, determine the exact values of all six trigonometric functions for each angle.

(a) $120°$ (b) $135°$ (c) $330°$ (d) $270°$ (e) $240°$

(f) $\dfrac{5\pi}{4}$ (g) $\dfrac{\pi}{6}$ (h) $\dfrac{7\pi}{6}$ (i) $-60°$ (j) $-\dfrac{3\pi}{2}$

(k) $\dfrac{5\pi}{3}$ (l) $-210°$ (m) $-\dfrac{\pi}{4}$ (n) π (o) 4.25π

3. Given that $\cos\theta = \dfrac{8}{17}$, $0° < \theta < 90°$, find the exact values of the other five trigonometric functions.

4. Given that $\tan\theta = -\dfrac{6}{5}$, $\sin\theta < 0$, find the exact values of $\sin\theta$ and $\cos\theta$.

5. Given that $\sin\theta = 0$, $\cos\theta < 0$, find the exact values of the other five trigonometric functions.

6. If $\sec\theta = 2$, $\dfrac{3\pi}{2} < \theta < 2\pi$, find the exact values of the other five trigonometric functions.

7. **(a)** Find the acute angle with the same sine ratio as
    (i) 150°, and **(ii)** 95°

    (b) Find the acute angle with the same cosine ratio as
    (i) 315°, and **(ii)** 353°

    (c) Find the acute angle with the same tangent ratio as
    (i) 240°, and **(ii)** 200°

8. Find the area of each triangle. Express the area exactly, or, if not possible, express it correct to 3 significant figures.

    (a) 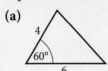   **(b)** 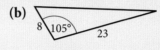   **(c)**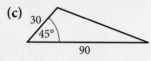

9. Triangle ABC has an area of 43 cm². The length of side AB is 12 cm and the length of side AC is 15 cm. Find the degree measure of angle A.

10. A chord AB subtends an angle of 120° at O, the centre of a circle with radius 15 centimetres. Find the area of **(a)** the sector AOB, and **(b)** the triangle AOB.

11. Find the area of the shaded region in each circle.

    (a) 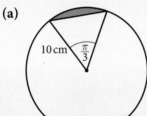   **(b)**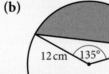

12. Two adjacent sides of a parallelogram have lengths a and b and the angle between these two sides is θ. Express the area of the parallelogram in terms of a, b and θ.

13. For the triangle shown, express y in terms of x.

    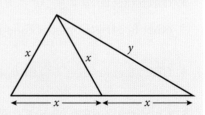

14. GJ bisects $F\widehat{G}H$ such that $F\widehat{G}J = H\widehat{G}J = \theta$. Express x in terms of h, f, and $\cos\theta$.

    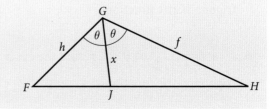

15. s is the length of each side of a regular polygon with n sides and r is the radius of the circumscribed circle. Show that $s = 2r \sin\left(\dfrac{180°}{n}\right)$

16. A triangle has two sides of lengths 6 cm and 8 cm and an included angle x.

(a) Express the area of the triangle as a function of x.

(b) State the domain and range of the function and sketch its graph for a suitable interval of x.

(c) Find the exact coordinates of the maximum point of the function. What type of triangle corresponds to this maximum? Explain why this triangle gives a maximum area.

17. A long metal rod is being carried down a hallway 3 metres wide. At the end of the hall there is a right-angled turn into a narrower hallway 2 metres wide. The angle that the rod makes with the outer wall is θ.

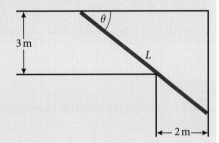

(a) Show that the length, L, of the rod is given by the function
$L(\theta) = 3 \csc \theta + 2 \sec \theta$

(b) On your GDC, graph the function L for the interval $0 < \theta < \dfrac{\pi}{2}$

(c) Using built-in features of your GDC, find the minimum value of the function L. Explain why this is the length of the longest rod that can be carried around the corner.

18. As viewed from the surface of the Earth at point A, the angle $D\widehat{A}E$ subtended by the full moon is $0.5182°$. Given that the distance from the Earth's surface to the moon's surface (AB) is approximately $383\,500$ km, find the radius, r, of the moon to 3 significant figures.

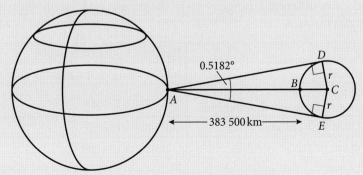

19. (a) Given that $\sin \theta = x$, find $\sec \theta$ in terms of x.

(b) Given that $\tan \beta = y$, find $\sin \beta$ in terms of y.

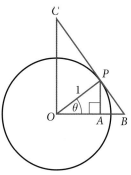

Figure 7.42 Diagram for question 20

20. Figure 7.42 shows the unit circle with angle θ in standard position. Segment BC is tangent to the circle at P and $B\widehat{O}C$ is a right angle. Each of the six trigonometric functions of θ is equal to the length of a line segment in the figure. For example, we know that $\sin\theta = AP$. For each of the five other trigonometric functions, find a line segment in the figure whose length equals the function value of θ.

21. Find the angle that the line through the given pair of points makes with the positive direction of the x-axis.

    (a) $(1, 4)$ and $(-1, 2)$ **(b)** $(-3, 1)$ and $(6, -5)$

    (c) $\left(2, \dfrac{1}{2}\right)$ and $(-4, -10)$

22 Find the acute angle between the two given lines.

    (a) $y = -2x$ and $y = x$ **(b)** $y = -3x + 5$ and $y = 2x$

7.4 The sine rule and the cosine rule

So far, we have used trigonometry to find an unknown angle or side of a right-angled triangle. We will now study methods for finding unknown lengths and angles in triangles that are not right-angled triangles. These general methods are effective for solving problems involving any kind of triangle – right-angled, acute or obtuse.

Possible triangles constructed from three given parts

We need to know at least three parts of a triangle to solve for other unknown parts. Different arrangements of the three known parts can be given. Before solving for unknown parts, it is helpful to know whether the three known parts determine a unique triangle, or more than one possible triangle. Table 7.2 summarises the five different arrangements of three parts and the number of possible triangles for each.

Known parts	Number of possible triangles
Three angles (AAA)	Infinite triangles (not possible to solve)
Three sides (SSS) (sum of any two must be greater than the third)	One unique triangle
Two sides and their included angle (SAS)	One unique triangle
Two angles and any side (ASA or AAS)	One unique triangle
Two sides and a non-included angle (SSA)	No triangle, one triangle or two triangles

Table 7.2 Possible triangles formed with three known parts

Arrangements ASA, AAS and SSA can be solved using the **sine rule**, whereas arrangements SSS and SAS can be solved using the **cosine rule**.

The sine rule

In Section 7.3, we showed that we can write three equivalent expressions for the area of any triangle for which we know two sides and the included angle.

Area of $\Delta = \frac{1}{2}\,ab \sin C = \frac{1}{2}\,ac \sin B = \frac{1}{2}\,bc \sin A$

Divide each expression by $\frac{1}{2}\,abc$:

$$\frac{\frac{1}{2}\,ab \sin C}{\frac{1}{2}\,abc} = \frac{\frac{1}{2}\,ac \sin B}{\frac{1}{2}\,abc} = \frac{\frac{1}{2}\,bc \sin A}{\frac{1}{2}\,abc}$$

We obtain the following three equivalent ratios – each containing the sine of an angle divided by the length of the side opposite the angle.

> If A, B and C are the angle measures of any triangle and a, b and c are, respectively, the lengths of the sides opposite these angles, then, according to the **sine rule**:
>
> $$\frac{\sin A}{a} = \frac{\sin B}{b} = \frac{\sin C}{c}$$
>
> Alternatively, the sine rule can also be written as $\dfrac{a}{\sin A} = \dfrac{b}{\sin B} = \dfrac{c}{\sin C}$

Finding unknowns given two angles and any side (ASA or AAS)

When we know two angles and any side of a triangle, we can use the sine rule to find any of the other angles or sides of the triangle.

Example 7.20

Find all the unknown angles and sides of triangle DEF shown in the diagram. Give all measurements correct to 1 decimal place.

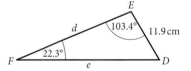

Figure 7.43 Diagram for Examle 7.20

Solution

The third angle of the triangle is

$$D = 180° - E - F = 180° - 103.4° - 22.3° = 54.3°$$

We can write an equation using the sine rule to solve for e

$$\frac{\sin 22.3°}{11.9} = \frac{\sin 103.4°}{e}$$

$$e = \frac{11.9 \sin 103.4°}{\sin 22.3°} \approx 30.507 \text{ cm}$$

We can write another equation using the sine rule to solve for d

$$\frac{\sin 22.3°}{11.9} = \frac{\sin 54.3°}{d}$$

$$d = \frac{11.9 \sin 54.3°}{\sin 22.3°} \approx 25.467 \text{ cm}$$

Therefore, the other parts of the triangle are:

$D = 54.3°$, $e \approx 30.5$ cm and $d \approx 25.5$ cm

 When using your GDC to find angles and lengths with the sine rule (or the cosine rule), remember to store intermediate answers on the GDC for greater accuracy. By not rounding until the final answer, you reduce the amount of rounding error.

Example 7.21

A tree on a sloping hill casts a shadow 45 metres down the slope of the hill. The gradient of the hill is $\dfrac{60}{12}$ and the angle of elevation of the sun is 35°. How tall is the tree, to the nearest tenth of a metre?

Solution

α is the angle that the hill makes with the horizontal. We can work out its size using the inverse tangent of $\dfrac{12}{60} = \dfrac{1}{5}$.

$\alpha = \tan^{-1}\left(\dfrac{1}{5}\right) \approx 11.3099°$

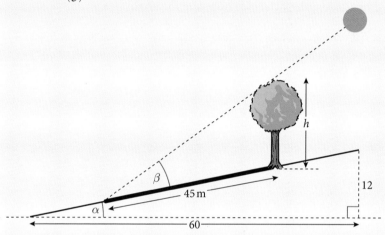

The height of the tree is h. The angle of elevation of the sun is the angle between the sun's rays and the horizontal. In the diagram, this angle of elevation is the sum of α and β. Thus, $\beta \approx 35° - 11.3099° \approx 23.6901°$

If a line segment is dropped from the base of the tree perpendicular to the length of 60 m, then we can sketch a right-angled triangle with $\alpha + \beta = 35°$ as one of its acute angles; the other acute angle – and the angle in the obtuse triangle opposite the side of 45 metres – must be 55°. We can apply the sine rule for the obtuse triangle to solve for h.

$$\frac{\sin 23.7°}{h} = \frac{\sin 55°}{45} \Rightarrow h = \frac{45 \sin 23.7°}{\sin 55°} \approx 22.0809$$

Therefore, the tree is approximately 22.1 metres tall.

Two sides and a non-included angle (SSA) – the ambiguous case

The SSA arrangement – two sides of a triangle and the size of an angle not between these two sides – can produce three different results: no triangle, one unique triangle, or two different triangles.

Example 7.22

Find all of the unknown angles and sides of triangle ABC where $a = 35$ cm, $b = 50$ cm and $A = 30°$. Give all measurements correct to 1 decimal place.

Solution

Here are the three parts we have in our attempt to construct triangle ABC:

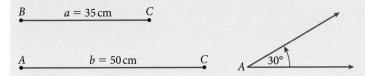

We attempt to construct the triangle, as shown. First draw angle A with its initial and terminal sides extended. Then measure off the known side $b = AC = 50$ on the terminal side. To construct side a (opposite angle A), we take point C as the centre and with radius $a = 35$ we draw an arc of a circle. The points on this arc are all possible positions for vertex B. Point B must be on the base line, so B can be located at any point of intersection of the circular arc and the base line. In this instance, with these particular measurements for the two sides and non-included angle, there are two points of intersection, which we label B_1 and B_2.

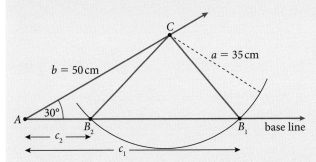

Therefore, we can construct two different triangles, triangle AB_1C and triangle AB_2C. Angle B_1 is acute and angle B_2 is obtuse. To complete the solution of this problem, we need to solve each of these triangles.

Solve triangle AB_1C:

We can solve for acute angle B_1 using the sine rule.

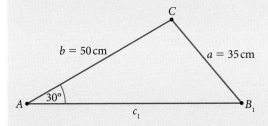

331

$$\frac{\sin 30°}{35} = \frac{\sin B_1}{50}$$

$$\sin B_1 = \frac{50 \sin 30°}{35} = \frac{50(0.5)}{35}$$

$$B_1 = \sin^{-1}\left(\frac{5}{7}\right) \approx 45.5847°$$

Then, $C \approx 180° - 30° - 45.5847° \approx 104.4153°$

With another application of the sine rule, we can solve for side c_1.

$$\frac{\sin 30°}{35} = \frac{\sin 104.4153°}{c_1}$$

$$c_1 \approx \frac{35 \sin 104.4153°}{\sin 30°} \approx \frac{35(0.96852)}{0.5} \approx 67.7964 \text{ cm}$$

Therefore, for triangle AB_1C: $B_1 \approx 45.6°$, $C \approx 104.4°$ and $c_1 \approx 67.8$ cm

Solve triangle AB_2C:

Solving for obtuse angle B_2 using the sine rule gives the same result as above, except we know that $90° < B_2 < 180°$. We also know that $\sin(180° - \theta) = \sin\theta$.

Thus, $B_2 = 180° - B_1 \approx 180° - 45.5847° \approx 134.4153°$

Then, $C \approx 180° - 30° - 134.4153° \approx 15.5847°$

Applying the sine rule again, we can solve for side c_2.

$$\frac{\sin 30°}{35} = \frac{\sin 15.5847°}{c_2}$$

$$c_2 \approx \frac{35 \sin 15.5847°}{\sin 30°} \approx \frac{35(0.26866)}{0.5}$$

$$\approx 18.8062 \text{ cm}$$

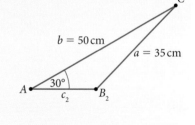

Therefore, for triangle AB_2C: $B_2 \approx 134.4°$, $C \approx 15.6°$ and $c_2 \approx 18.8$ cm

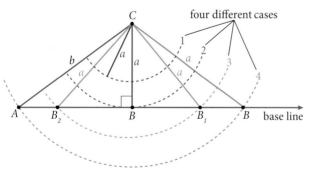

Figure 7.44 Four distinct cases for SSA when angle A is acute

We will now take a more general look and examine all the possible conditions and outcomes for the SSA arrangement. In general, we are given the lengths of two sides – call them a and b – and a non-included angle; for example, angle A that is opposite side a. From these measurements, we can determine the number of different triangles. Figure 7.44 shows the four different possibilities (or cases) when angle A is acute. The number of triangles depends on the length of side a.

In case 2, side a is perpendicular to the base line resulting in a single right-angled triangle, shown in Figure 7.45. In this case, $\sin A = \dfrac{a}{b}$ and $a = b \sin A$.

In case 1, a is shorter than it is in case 2, i.e. $b \sin A$. In case 3, which occurred in Example 7.22, a is longer than $b \sin A$, but less than b. And, in case 4, a is longer than b. These results are summarised in Table 7.3. Because the number of triangles may be none, one, or two, depending on the length of a (the side opposite the given angle), the SSA arrangement is called the **ambiguous case**.

Length of a	Number of triangles	Case number in Figure 7.44
$a < b \sin A$	No triangle	1
$a = b \sin A$	One right-angled triangle	2
$b \sin A < a < b$	Two triangles	3
$a > b$	One triangle	4

Table 7.3 Possible triangles formed with two known sides and an acute non-included angle

The situation is considerably simpler if angle A is obtuse rather than acute. Figure 7.46 shows that if $a > b$ then there is only one possible triangle, and if $a \leqslant b$ then no triangle is possible that contains angle A.

Example 7.23 uses the same SSA information given in Example 7.22 with the exception that side a is not fixed at 35 cm, but is allowed to vary.

Example 7.23

For triangle ABC, side $b = 50$ cm and angle $A = 30°$. Find the values for the length of side a that will produce: (i) no triangle, (ii) one triangle, (iii) two triangles.

Solution

Because this is an SSA arrangement (ambiguous case) and given A is an acute angle, then the number of different triangles that can be constructed depends on the length of a.

First calculate the value of $b \sin A$: $b \sin A = 50 \sin 30° = 50(0.5) = 25$ cm

Thus, if a is exactly 25 cm, triangle ABC is a right-angled triangle.
(i) When $a < 25$ cm, there is no triangle.
(ii) When $a = 25$ cm, or $a > 50$ cm, there is one unique triangle.
(iii) When 25 cm $< a < 50$ cm, there are two different possible triangles.

Example 7.24

The diagrams show two different triangles both satisfying the conditions:

$HK = 18$ cm, $JK = 15$ cm, $J\widehat{H}K = 53°$

(a) Calculate the size of $H\widehat{J}K$ in triangle 2.

(b) Calculate the area of triangle 1.

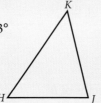
Triangle 1

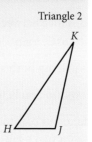
Triangle 2

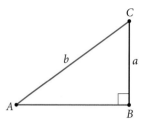

Figure 7.45 Case 2 for SSA: $a = b \sin A$, one right angle

Given the length of sides a and b and the non-included angle A is acute, the four cases and resulting triangles shown in Table 7.3 can occur.

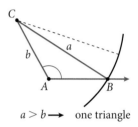

$a > b \longrightarrow$ one triangle

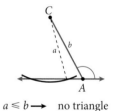

$a \leqslant b \longrightarrow$ no triangle

Figure 7.46 Angle A is obtuse

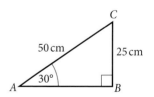

Figure 7.47 Diagram for solution to Example 7.23

Solution

From the sine rule, $\sin\left(\dfrac{H\hat{J}K}{18}\right) = \dfrac{\sin 53°}{15} \Rightarrow \sin(H\hat{J}K) = \dfrac{18\sin 53°}{15} \approx 0.95836$

$\Rightarrow \sin^{-1}(0.95836) \approx 73.408°$

However, $H\hat{J}K > 90° \Rightarrow H\hat{J}K \approx 180° - 73.408° \approx 106.592°$

Therefore in triangle 2, $H\hat{J}K \approx 107°$ (3 s.f.)

In triangle 1, $H\hat{J}K < 90° \Rightarrow H\hat{J}K \approx 73.408°$

$\Rightarrow H\hat{K}J \approx 180° - (73.408° + 53°) \approx 53.592°$

Area $= \dfrac{1}{2}(18)(15)\sin(53.592°) \approx 108.649\text{ cm}^2$

Therefore, the area of triangle 1 is approximately 109 cm² (3 s.f.)

The cosine rule

Two arrangements remain in our list (Table 7.2) of different ways to arrange three known parts of a triangle. If three sides of a triangle are known (SSS arrangement), or two sides of a triangle and the angle between them are known (SAS arrangement), then a unique triangle is determined. However, in both of these cases the sine rule cannot be used to work out the unknowns in the triangle.

For example, it is not possible to set up an equation using the sine rule to solve triangle PQR or triangle STU in Figure 7.48

Trying to solve $\triangle PQR$:

$\dfrac{\sin P}{4} = \dfrac{\sin R}{6} \Rightarrow$ two unknowns; cannot solve for angle P or angle R

Trying to solve $\triangle STU$:

$\dfrac{\sin 80°}{t} = \dfrac{\sin U}{13} \Rightarrow$ two unknowns; cannot solve for angle U or side t

We will need the **cosine rule** to solve triangles with SSS and SAS arrangements. To derive this law, we need to place a general triangle ABC in the coordinate plane so that one of the vertices is at the origin and one of the sides is on the positive x-axis. Figure 7.49 (top of next page) shows both an acute triangle ABC and an obtuse triangle ABC. In both cases, the coordinates of vertex A are $x = b\cos C$ and $y = b\sin C$. Because c is the distance from A to B, we can use the distance formula to write the following:

distance between $(b\cos C, b\sin C)$ and $(a, 0)$	$c = \sqrt{(b\cos C - a)^2 + (b\sin C - 0)^2}$
squaring both sides	$c^2 = (b\cos C - a)^2 + (b\sin C - 0)^2$
expand brackets	$c^2 = b^2\cos^2 C - 2ab\cos C + a^2 + b^2\sin^2 C$
take out common factor of b^2 from two terms	$c^2 = b^2(\cos^2 C + \sin^2 C) - 2ab\cos C + a^2$

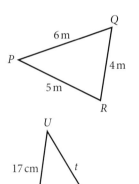

Figure 7.48 Two triangles that cannot be solved with the sine rule

apply trigonometric identity

$$c^2 = b^2 - 2ab\cos C + a^2$$

$$\cos^2\theta + \sin^2\theta = 1$$

rearrange terms

$$c^2 = a^2 + b^2 - 2ab\cos C$$

This equation gives one form of the cosine rule. Two other forms are obtained in a similar manner by having either vertex A or vertex B, rather than C, located at the origin.

It is helpful to understand the underlying pattern of the cosine rule when applying it to solve for parts of triangles. The pattern relies on choosing one particular angle of the triangle and then identifying the two sides that are adjacent to the angle and the one side that is opposite to it (Figure 7.50). The cosine rule can be used to solve for the chosen angle or the side opposite the chosen angle.

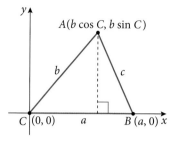

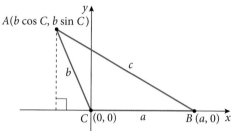

Figure 7.49 Deriving the cosine rule

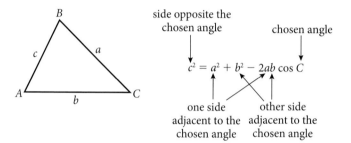

Figure 7.50 Applying the cosine rule

In any triangle ABC with corresponding sides a, b and c, the cosine rule states:

$$c^2 = a^2 + b^2 - 2ab\cos C$$
$$b^2 = a^2 + c^2 - 2ac\cos B$$
$$a^2 = b^2 + c^2 - 2bc\cos A$$

Finding unknowns given two sides and the included angle (SAS)

If we know two sides and the included angle, we can use the cosine rule to solve for the side opposite the given angle. Then it is best to solve for one of the two remaining angles using the sine rule.

Example 7.25

Find all of the unknown angles and sides of triangle STU. Give all measurements to 1 decimal place.

Solution

We first solve for side t opposite known angle $S\hat{T}U$ using the cosine rule.

$$t^2 = 13^2 + 17^2 - 2(13)(17)\cos 80°$$

$$t = \sqrt{13^2 + 17^2 - 2(13)(17)\cos 80°}$$

$$t \approx 19.5256$$

Now use the sine rule to solve for one of the other angles, say $T\hat{S}U$

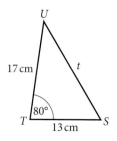

Figure 7.51 Diagram for Example 7.25

Example 8.1

Work out $\sqrt{-36} \cdot \sqrt{-49}$

Solution

First simplify each square root using rule 2:

$$\sqrt{-36} = \sqrt{36} \cdot \sqrt{-1} = 6 \cdot \sqrt{-1}$$

$$\sqrt{-49} = \sqrt{49} \cdot \sqrt{-1} = 7 \cdot \sqrt{-1}$$

And hence, using rule 1 with the other obvious rules

$$\sqrt{-36} \cdot \sqrt{-49} = 6 \cdot \sqrt{-1} \cdot 7 \cdot \sqrt{-1} = 42 \cdot \sqrt{-1} \cdot \sqrt{-1} = -42$$

To deal with quadratic formula expressions that consist of combinations of real numbers and square roots of negative numbers, we can apply the rules of binomials to numbers of the form

$$a + b\sqrt{-1}$$

where a and b are real numbers. For example, to add $(5 + 7\sqrt{-1})$ to $(2 - 3\sqrt{-1})$ we combine like terms as we do with polynomials:

$$(5 + 7\sqrt{-1}) + (2 - 3\sqrt{-1}) = 5 + 2 + 7\sqrt{-1} - 3\sqrt{-1}$$

$$= (5 + 2) + (7 - 3)\sqrt{-1} = 7 + 4\sqrt{-1}$$

Similarly, to multiply these numbers we use the binomial multiplication procedures

$$(5 + 7\sqrt{-1}) \cdot (2 - 3\sqrt{-1})$$

$$= 5 \cdot 2 + (7\sqrt{-1}) \cdot (-3\sqrt{-1}) + 5 \cdot (-3\sqrt{-1}) + (7\sqrt{-1}) \cdot 2$$

$$= 10 - 21 \cdot (\sqrt{-1})^2 - 15 \cdot \sqrt{-1} + 14 \cdot \sqrt{-1}$$

$$= 10 - 21 \cdot (-1) + (-15 + 14)\sqrt{-1} = 31 - \sqrt{-1}$$

Square roots of negative numbers were given the name imaginary numbers and Euler introduced the symbol i for $\sqrt{-1}$.

A **pure imaginary** number is a number of the form ki, where k is a real number and i, the **imaginary unit** is defined by $i^2 = -1$.

With this definition of i, a few interesting results are immediately apparent. For example,

$i^3 = i^2 \cdot i = -1 \cdot i = -i$

$i^4 = i^2 \cdot i^2 = (-1) \cdot (-1) = 1$

$i^5 = i^4 \cdot i = 1 \cdot i = i$

$i^6 = i^4 \cdot i^2 = i^2 = -1; i^7 = -i$, and finally $i^8 = 1$

This means we are able to evaluate any positive integer power of i using the following property

$i^{4n+k} = i^k, k = 0, 1, 2, 3$

So, for example

$i^{2122} = i^{2120+2} = i^2 = -1$, or $i^{2123} = i^{2120+3} = i^3 = -i$

In some cases, especially in engineering sciences, the number i is sometimes denoted as j.

Example 8.2

Simplify

(a) $\sqrt{-36} + \sqrt{-49}$

(b) $\sqrt{-36} \cdot \sqrt{-49}$

Solution

(a) $\sqrt{-36} + \sqrt{-49} = \sqrt{36}\sqrt{-1} + \sqrt{49}\sqrt{-1}$

$$= 6i + 7i = 13i$$

(b) $\sqrt{-36} \cdot \sqrt{-49} = 6i \cdot 7i = 42i^2$

$$= 42(-1) = -42$$

We do not define $i = \sqrt{-1}$ for a reason. It is the convention in mathematics that when we write $\sqrt{9}$ we mean the non-negative square root of 9, namely 3. We do not mean -3.

i does not belong to this category since we cannot say that i is the positive square root of -1, i.e., $i > 0$. If we do, then $-1 = i \times i > 0$, which is false, and if we say $i < 0$, then $-i > 0$, and $-1 = -i \times -i > 0$, which is also false. Actually $-i$ is also a square root of $\sqrt{-1}$ because $-i \times -i = i^2 = -1$.

With this in mind, we can use a convention which calls i the **principal** square root of -1 and write $i = \sqrt{-1}$.

Gauss introduced the idea of complex numbers by giving them the following definition.

A **complex number** is a number that can be written in the form $a + bi$, where a and b are real numbers and $i^2 = -1$. a is called the **real part** of the number and b is the **imaginary part**.

It is customary to denote complex numbers with the variable z. $z = 5 + 7i$ is the complex number with real part 5 and imaginary part 7, and $z = 2 - 3i$ has 2 as real part and -3 as imaginary.

Notation:

It is usual to write **Re(z)** for the real part of z and **Im(z)** for the imaginary part. So, $\text{Re}(2 + 3i) = 2$ and $\text{Im}(2 + 3i) = 3$

You can set up a GDC to do basic complex number operations.

```
Mode         :Comp
Frac Result  :d/c
Func Type    :Y=
Draw Type    :Connect
Derivative   :Off
Angle        :Deg
Complex Mode :a+bi
```

Algebraic structure of complex numbers

Gauss' definition of complex numbers triggers the following understanding of the set of complex numbers as an extension to our number sets in algebra.

The set of complex numbers, $\mathbb{C}$, is the set of ordered pairs of real numbers $\mathbb{C} = \{z = (x, y): x, y \in \mathbb{R}\}$, with the following additional structure.

Two complex numbers $z_1 = (x_1, y_1)$ and $z_2 = (x_2, y_2)$ are equal if their corresponding components are equal: $(x_1, y_1) = (x_2, y_2)$ if $x_1 = x_2$ and $y_1 = y_2$. That is, two complex numbers are equal if and only if their real parts are equal and their imaginary parts are equal.

This is equivalent to saying: $a + bi = c + di \Leftrightarrow a = c$ and $b = d$.

For example, if $2 - (y - 2)i = x + 3 + 5i$, then x must be -1 and y must be -3.

An interesting application of the way the equality works is in finding the square roots of complex numbers without a need for the trigonometric forms developed later in the chapter.

Find the square root(s) of $z = 5 + 12i$. Let the square root of z be $x + yi$, then

$(x + yi)^2 = 5 + 12i \Rightarrow x^2 - y^2 + 2xyi = 5 + 12i \Rightarrow x^2 - y^2 = 5$ and

$2xy = 12 \Rightarrow xy = 6 \Rightarrow y = \dfrac{6}{x}$, and when we substitute this value in $x^2 - y^2 = 5$, we have

$x^2 - \left(\dfrac{6}{x}\right)^2 = 5$. This simplifies to $x^4 - 5x^2 - 36 = 0$ which yields $x^2 = -4$ or $x^2 = 9, \Rightarrow x = \pm 3$,

This leads to $y = \pm 2$, that is, the two square roots of $5 + 12i$ are $3 + 2i$ or $-3 - 2i$.

```
(3+2i)²
            5+12i
(-3-2i)²
            5+12i
```

Figure 8.1 Finding the square roots of $5 + 12i$

Addition, subtraction and multiplication

Addition, subtraction and multiplication for complex numbers are defined as follows.

Addition and subtraction of complex numbers

$(x_1, y_1) + (x_2, y_2) = (x_1 + x_2, y_1 + y_2)$; or $(x_1, y_1) - (x_2, y_2) = (x_1 - x_2, y_1 - y_2)$.

This is equivalent to saying: $(a + bi) + (c + di) = (a + c) + (b + d)i$, or

$(a + bi) - (c + di) = (a - c) + (b - d)i$

Multiplication of complex numbers

$(x_1, y_1)(x_2, y_2) = (x_1 x_2 - y_1 y_2, x_1 y_2 + x_2 y_1)$

This is equivalent to using the binomial multiplication on $(a + bi)(c + di)$:

$(a + bi) \times (c + di) = ac + bdi^2 + adi + bci = ac - bd + (ad + bc)i$

Addition and multiplication of complex numbers inherit most of the properties of addition and multiplication of real numbers:

Commutativity: $z + w = w + z$ and $zw = wz$

Associativity: $z + (u + v) = (z + u) + v$ and $z(uv) = (zu)v$

Distributive property of multiplication: $z(u + v) = zu + zv$

Some complex numbers take up unique positions. For example, the complex number $(0, 0)$ has the properties of 0:

$(x, y) + (0, 0) = (x, y)$ and $(x, y)(0, 0) = (0, 0)$

It is therefore normal to identify it with 0. The symbol is exactly the same symbol used to identify the real 0. So, the real and complex zeros are the same number.

Another complex number of significance is $(1, 0)$. This number plays an important role in multiplication that stems from the following property:

$(x, y)(1, 0) = (x \cdot 1 - y \cdot 0, \; x \cdot 0 + y \cdot 1) = (x, y)$

For complex numbers, $(1, 0)$ behaves like the identity for multiplication for real numbers. Again, it is normal to write $(1, 0)$ as 1.

The third number of significance is $(0, 1)$. It has the notable characteristic of having a negative square. i.e.,

$(0, 1)(0, 1) = (0 \cdot 0 - 1 \cdot 1, \; 0 \cdot 1 + 1 \cdot 0) = (-1, 0)$

Using the definition above, $(0, 1) = 0 + 1\mathrm{i} = \mathrm{i}$. So, the last result should be no surprise to us since we know that

$$\mathrm{i} \times \mathrm{i} = -1 = (-1, 0)$$

Since (x, y) represents the complex number $x + y\mathrm{i}$, then every real number x can be written as $x + 0\mathrm{i} = (x, 0)$. The set of real numbers is therefore a subset of the set of complex numbers. They are the complex numbers whose imaginary part is 0. Similarly, pure imaginary numbers are of the form $0 + y\mathrm{i} = (0, y)$. They are the complex numbers whose real part is 0.

Example 8.3

Simplify each expression.

(a) $(4 - 5\mathrm{i}) + (7 + 8\mathrm{i})$ (b) $(4 - 5\mathrm{i}) - (7 + 8\mathrm{i})$ (c) $(4 - 5\mathrm{i})(7 + 8\mathrm{i})$

Solution

(a) $(4 - 5\mathrm{i}) + (7 + 8\mathrm{i}) = (4 + 7) + (-5 + 8)\mathrm{i} = 11 + 3\mathrm{i}$

(b) $(4 - 5\mathrm{i}) - (7 + 8\mathrm{i}) = (4 - 7) + (-5 - 8)\mathrm{i} = -3 - 13\mathrm{i}$

(c) $(4 - 5\mathrm{i})(7 + 8\mathrm{i}) = (4 \cdot 7 - (-5) \cdot 8) + (4 \cdot 8 + (-5) \cdot 7)\mathrm{i} = 68 - 3\mathrm{i}$

We can also use a GDC.

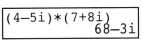

Figure 8.2 GDC screen for the solution to Example 8.3 (c)

Division

Multiplication can be used to perform division of complex numbers.

Example 8.4

Find the quotient $\dfrac{2 + 3\mathrm{i}}{1 + 2\mathrm{i}}$

Solution

Let $\dfrac{2 + 3\mathrm{i}}{1 + 2\mathrm{i}} = x + y\mathrm{i}$

Using multiplication and the equality of complex numbers

$$2 + 3\mathrm{i} = (1 + 2\mathrm{i})(x + y\mathrm{i}) \Leftrightarrow 2 + 3\mathrm{i} = x - 2y + (2x + y)\mathrm{i}$$

$$\Leftrightarrow \begin{cases} 2 = x - 2y \\ 3 = 2x + y \end{cases} \Rightarrow x = \frac{8}{5}, y = -\frac{1}{5}$$

We can also use a GDC.

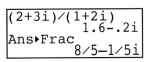

Figure 8.3 GDC screen for the solution to Example 8.4

Now, in general, $\dfrac{a + b\mathrm{i}}{c + d\mathrm{i}} = x + y\mathrm{i} \Leftrightarrow a + b\mathrm{i} = (x + y\mathrm{i})(c + d\mathrm{i})$

With the multiplication as described above

$$a + b\mathrm{i} = (cx - dy) + (dx + cy)\mathrm{i}$$

Again, by applying the equality of complex numbers property above, we get a system of two equations that can be solved

$$\begin{cases} cx - dy = a \\ dx + cy = b \end{cases} \Rightarrow x = \frac{ac + bd}{c^2 + d^2}; \, y = \frac{bc - ad}{c^2 + d^2}$$

The denominator $c^2 + d^2$ resulted from multiplying $c + d\mathrm{i}$ by $c - d\mathrm{i}$.

Conjugate

With every complex number $(a + b\mathrm{i})$ we associate another complex number $(a - b\mathrm{i})$ that is called its **conjugate**. The conjugate of number z is most often denoted with a bar over it, sometimes with an asterisk to the right of it, occasionally with an apostrophe and even less often with the plain word Conj, as in:

$$\bar{z} = z^* = z' = \mathrm{Conj}(z)$$

In this book, we will use z^* for the conjugate.

 In your own work, you can use any notation you feel comfortable with. You just need to understand that IB questions use the conjugate notation z^*.

The importance of the conjugate stems from the following property

$$(a + b\mathrm{i})(a - b\mathrm{i}) = a^2 - b^2\mathrm{i}^2 = a^2 + b^2$$

which is a non-negative real number. The product of a complex number and its conjugate is always a real number.

Example 8.5

Find the conjugate of z and verify that zz^* is a real number.

(a) $z = 2 + 3\mathrm{i}$ (b) $z = 5\mathrm{i}$ (c) $z = 11$

Solution

(a) $z^* = 2 - 3\mathrm{i}$

    $zz^* = (2 + 3\mathrm{i})(2 - 3\mathrm{i}) = 4 - 9\mathrm{i}^2 = 4 + 9 = 13$

(b) $z^* = -5\mathrm{i}$

    $zz^* = (5\mathrm{i})(-5\mathrm{i}) = -5\mathrm{i}^2 = (-5)(-1) = 5$

(c) $z^* = 11$

    $zz^* = 11 \times 11 = 121$

So, the method used in dividing two complex numbers can be achieved by multiplying the quotient by a fraction whose numerator and denominator are the conjugate $c - d\mathrm{i}$

$$\frac{a + b\mathrm{i}}{c + d\mathrm{i}} = \frac{a + b\mathrm{i}}{c + d\mathrm{i}} \cdot \frac{c - d\mathrm{i}}{c - d\mathrm{i}} = \frac{(a + b\mathrm{i})(c - d\mathrm{i})}{c^2 + d^2} = \frac{ac + bd}{c^2 + d^2} + \left(\frac{bc - ad}{c^2 + d^2}\right)\mathrm{i}$$

Example 8.6

Find each quotient and write your answer in Cartesian form.

(a) $\dfrac{4 - 5i}{7 + 8i}$

(b) $\dfrac{4 - 5i}{8i}$

(c) $\dfrac{4 - 5i}{7}$

Solution

(a) $\dfrac{4 - 5i}{7 + 8i} = \dfrac{4 - 5i}{7 + 8i} \cdot \dfrac{7 - 8i}{7 - 8i} = \dfrac{28 - 40 + (-32 - 35)i}{49 + 64} = -\dfrac{12}{113} - \dfrac{67}{113}i$

(b) $\dfrac{4 - 5i}{8i} = \dfrac{4 - 5i}{8i} \cdot \dfrac{-8i}{-8i} = \dfrac{-32i - 40}{64} = -\dfrac{5}{8} - \dfrac{1}{2}i$

(c) $\dfrac{4 - 5i}{7} = \dfrac{4}{7} - \dfrac{5}{7}i$

We can also use a GDC.

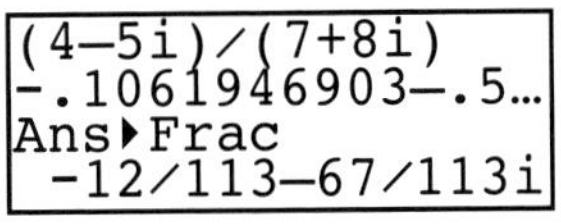

```
(4-5i)/(7+8i)
-.1061946903-.5…
Ans▸Frac
      -12/113-67/113i
```

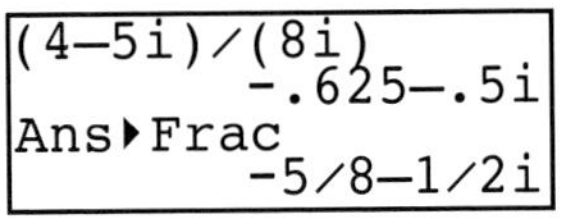

```
(4-5i)/(8i)
        -.625-.5i
Ans▸Frac
        -5/8-1/2i
```

Figure 8.4 GDC screens for the solution to Example 8.6

Example 8.7

Solve the system of equations and express your answer in Cartesian form.

$$(1 + i)z_1 - iz_2 = -3$$
$$2z_1 + (1 - i)z_2 = 3 - 3i$$

Solution

Multiply the first equation by 2, and the second equation by $(1 + i)$

$$\begin{cases} 2(1 + i)z_1 - 2iz_2 = -6 \\ 2(1 + i)z_1 + (1 + i)(1 - i)z_2 = (1 + i)(3 - 3i) \end{cases}$$

$$\Rightarrow \begin{cases} 2(1 + i)z_1 - 2iz_2 = -6 \\ 2(1 + i)z_1 + 2z_2 = 6 \end{cases}$$

By subtracting the last two equations, we get

$$(-2 - 2i)z_2 = -12$$

And hence, after multiplying the denominator by its conjugate

$$z_2 = \dfrac{-12}{-2 - 2i} = 3 - 3i$$

$$z_1 = \dfrac{-3 + i(3 - 3i)}{1 + i} = \dfrac{3}{2} + \dfrac{3}{2}i$$

Let z, z_1 and z_2 be complex numbers. They have these properties:

1 $(z^*)^* = z$

2 $z^* = z$ if and only if z is real.

3 $(z_1 + z_2)^* = z_1^* + z_2^*$ (the conjugate of the sum is the sum of conjugate(s)

4 $(-z)^* = -z^*$

5 $(z_1 \cdot z_2)^* = z_1^* \cdot z_2^*$ (the conjugate of the product is the product of conjugate(s)

6 $(z^{-1})^* = (z^*)^{-1}$, if $z \neq 0$.

Properties of conjugates

Conjugates have some important properties (see Key Fact box on the left).

Proof

(1) and (2) are obvious.

For example, $((a + bi)^*)^* = (a - bi)^* = a + bi$, and

$a - bi = a + bi \rightarrow 2bi = 0 \rightarrow b = 0$

(3) is proved by straightforward calculation

Let $z_1 = x_1 + iy_1$ and $z_2 = x_2 + iy_2$, then

$(z_1 + z_2)^* = ((x_1 + iy_1) + (x_2 + iy_2))^* = ((x_1 + x_2) + i(y_1 + y_2))^*$

$\qquad = (x_1 + x_2) - i(y_1 + y_2) = (x_1 - iy_1) + (x_2 - iy_2) = z_1^* + z_2^*$

(4) can now be proved using the above results

$(z + (-z))^* = 0^* = 0$

but, $(z + (-z))^* = z^* + (-z)^*$, so $z^* + (-z)^* = 0$, and $(-z)^* = -z^*$

Also (5) is proved by straightforward calculation,

$(z_1 \cdot z_2)^* = ((x_1 + iy_1) \cdot (x_2 + iy_2))^* = ((x_1 x_2 - y_1 y_2) + i(y_1 x_2 + x_1 y_2))^*$

$\qquad = (x_1 x_2 - y_1 y_2) - i(y_1 x_2 + x_1 y_2)$

$\qquad = (x_1 - iy_1) \cdot (x_2 - iy_2) = z_1^* \cdot z_2^*$

(6) Again, it is proved by straightforward calculation,

$(z^{-1})^* = \left(\dfrac{1}{x + iy}\right)^* = \left(\dfrac{1}{x + iy} \cdot \dfrac{x - iy}{x - iy}\right)^* = \left(\dfrac{x - iy}{x^2 + y^2}\right)^* = \dfrac{x + iy}{x^2 + y^2}$

$\qquad = \dfrac{x + iy}{x^2 + y^2} \cdot \dfrac{x - iy}{x - iy} = \dfrac{x^2 + y^2}{(x^2 + y^2)(x - iy)} = \dfrac{1}{x - iy} = (z^*)^{-1}$

The product also works for powers of complex numbers, i.e. $(z^2)^* = (z \times z)^* = z^* \times z^* = (z^*)^2$. This result can be generalised for any non-negative integer power n, i.e. $(z^n)^* = (z^*)^n$ and can be proved by mathematical induction.

Proof

The basis case, when $n = 0$, is obviously true $(z^0)^* = 1 = (z^*)^0$

Now assume $(z^k)^* = (z^*)^k$

$\qquad (z^{k+1})^* = (z^k z)^* = (z^k)^* z^* = (z^*)^k z^*$ (using the product rule)

Therefore, $(z^{k+1})^* = (z^*)^k z^* = (z^*)^{k+1}$

So, since the statement is true for $n = k$, it is also true for $n = k + 1$, and, by the principle of mathematical induction, it is true for all $n > 0$.

And finally, from (6),

$$(z(z^{-1}))^* = 1^* = 1$$

but, $(z(z^{-1}))^* = z^* (z^{-1})^*$, so $z^* (z^{-1})^* = 1$, and $(z^{-1})^* = \dfrac{1}{z^*} = (z^*)^{-1}$

Conjugate zeros of polynomials

In Chapter 2, you used the following result without proof.

If c is a root of a polynomial equation with real coefficients, then c^* is also a root.

We give the proof for $n = 3$, but the method is general.

$\qquad P(x) = ax^3 + bx^2 + dx + e$

Since c is a root of $P(x) = 0$, we have

$\qquad ac^3 + bc^2 + dc + e = 0$

$\qquad \Rightarrow (ac^3 + bc^2 + dc + e)^* = 0$, (since $0^* = 0$)

$\qquad \Rightarrow (ac^3)^* + (bc^2)^* + (dc)^* + e^* = 0$, (sum of conjugates theorem)

$\qquad \Rightarrow a(c^*)^3 + b(c^*)^2 + d(c^*) + e = 0$, (result of product conjugate)

$\qquad \Rightarrow (c^*)$ is a root of $P(x) = 0$

If c is a root of a polynomial equation with real coefficients, then c^* is also a root of the equation.

Complex numbers

Example 8.8

$1 + 2i$ is a zero of the polynomial $P(x) = x^3 - 5x^2 + 11x - 15$.
Find all other zeros.

Solution

Since the polynomial has real coefficients, then $1 - 2i$ is also a zero. Hence using the factor theorem, $P(x) = (x - (1 + 2i))(x - (1 - 2i))(x - c)$, where c is a real number to be found.

Now, $P(x) = (x^2 - 2x + 5)(x - c)$. c can either be found by division or by factorising by trial and error. In either case, $c = 3$

Exercise 8.1

1. Express each number in the form $a + bi$

   (a) $5 + \sqrt{-4}$ (b) $7 - \sqrt{-7}$ (c) -6

   (d) $-\sqrt{49}$ (e) $\sqrt{-81}$ (f) $-\sqrt{\dfrac{-25}{16}}$

2. Work out each calculation and express your answer in the form $a + bi$

   (a) $(-3 + 4i) + (2 - 5i)$ (b) $(-3 + 4i) - (2 - 5i)$

   (c) $(-3 + 4i)(2 - 5i)$ (d) $3i - (2 - 4i)$

   (e) $(2 - 7i)(3 + 4i)$ (f) $(1 + i)(2 - 3i)$

   (g) $\dfrac{3 + 2i}{2 + 5i}$ (h) $\dfrac{2 - i}{3 + 2i}$

   (i) $\left(\dfrac{2}{3} - \dfrac{1}{2}i\right) + \left(\dfrac{1}{3} + \dfrac{1}{2}i\right)$ (j) $\left(\dfrac{2}{3} - \dfrac{1}{2}i\right)\left(\dfrac{2}{3} + \dfrac{1}{2}i\right)$

   (k) $\left(\dfrac{2}{3} - \dfrac{1}{2}i\right) \div \left(\dfrac{1}{3} + \dfrac{1}{2}i\right)$ (l) $(2 + i)(3 - 2i)$

   (m) $\dfrac{1}{i}(3 - 7i)$ (n) $(2 + 5i) - (-2 - 5i)$

   (o) $\dfrac{13}{5 - 12i}$ (p) $\dfrac{12i}{3 + 4i}$

   (q) $3i\left(3 - \dfrac{2}{3}i\right)$ (r) $(3 + 5i)(6 - 10i)$

   (s) $\dfrac{39 - 52i}{24 + 10i}$ (t) $(7 - 4i)^{-1}$

   (u) $(5 - 12i)^{-1}$ (v) $\dfrac{3}{3 - 4i} + \dfrac{2}{6 + 8i}$

   (w) $\dfrac{(7 + 8i)(2 - 5i)}{5 - 12i}$ (x) $\dfrac{5 - \sqrt{-144}}{3 + \sqrt{-16}}$

3. Let $z = a + bi$. Find a and b if $(2 + 3i)z = 7 + i$

4. $(2 + yi)(x + i) = 1 + 3i$, where x and y are real numbers. Solve for x and y.

5. **(a)** Evaluate $(1 + \sqrt{3}\,i)^3$

   (b) Prove that $(1 + i\sqrt{3})^{6n} = 8^{2n}$, where $n \in \mathbb{Z}^+$

   (c) Hence find $(1 + i\sqrt{3})^{48}$

6. **(a)** Evaluate $(-\sqrt{2} + i\sqrt{2})^2$

   (b) Prove that $(-\sqrt{2} + i\sqrt{2})^{4k} = (-16)^k$, where $k \in \mathbb{Z}^+$

   (c) Hence find $(-\sqrt{2} + i\sqrt{2})^{46}$

7. If z is a complex number such that $|z + 4i| = 2|z + i|$, find the value of $|z|$.

8. Find the complex number z and write it in the form $a + bi$ when
   $$z = 3 + \frac{2i}{2 - i\sqrt{2}}$$

9. Find the values of the two real numbers x and y such that
   $(x + iy)(4 - 7i) = 3 + 2i$

10. Find the complex number z and write it in the form $a + bi$ when
    $i(z + 1) = 3z - 2$

11. Find the complex number z and write it in the form $a + bi$ when
    $$\frac{2 - i}{1 + 2i}\sqrt{z} = 2 - 3i$$

12. Find the values of the two real numbers x and y such that $(x + iy)^2 = 3 - 4i$

13. **(a)** Find the values of the two real numbers x and y such that
    $(x + iy)^2 = -8 + 6i$

    (b) Hence solve the equation $z^2 + (1 - i)z + 2 - 2i = 0$

14. If $z \in \mathbb{C}$, find all solutions to the equation $z^3 - 27i = 0$

15. Given that $z = \dfrac{1}{2} + 2i$ is a zero of the polynomial
    $f(x) = 4x^3 - 16x^2 + 29x - 51$, find the other zeros.

16. Find a polynomial function with integer coefficients and lowest possible
    degree that has $\dfrac{1}{2}, -1, 3 + i\sqrt{2}$ as zeros.

17. Find a polynomial function with integer coefficients and lowest possible
    degree that has $-2, -2, 1 + i\sqrt{3}$ as zeros.

18. Given that $z = 5 + 2i$ is a zero of the polynomial
    $f(x) = x^3 - 7x^2 - x + 87$, find the other zeros.

19. Given that $z = 1 - i\sqrt{3}$ is a zero of the polynomial
    $f(x) = 3x^3 - 4x^2 + 8x + 8$, find the other zeros.

20. Let $z \in \mathbb{C}$. If $\dfrac{z}{z^*} = a + bi$, show that $|a + bi| = 1$

21. Given that $z = (k + i)^4$ where k is a real number, find all values of k
    such that:

    (a) z is a real number   **(b)** z is purely imaginary.

22. Solve the system of equations

$$i z_1 + 2 z_2 = 3 - i$$
$$2 z_1 + (2 + i) z_2 = 7 + 2i$$

23. Solve the system of equations

$$i z_1 - (1 + i) z_2 = 3$$
$$(2 + i) z_1 + i z_2 = 4$$

8.2 The complex plane

Our definition of complex numbers as ordered pairs of real numbers enables us to look at them from a different perspective. Every ordered pair (x, y) determines a unique complex number $x + y$i, and vice versa. This correspondence is embodied in the geometric representation of complex numbers. Looking at complex numbers as points in the plane equipped with additional structure, changes the plane into what we call **complex plane**, or **Gauss plane**, or **Argand plane** (**diagram**). The complex plane has two axes, the horizontal axis is called the **real axis**, and the vertical axis is the **imaginary axis**. Every complex number $z = x + y$i is represented by a point (x, y) in the plane. The real part is measured along the real axis and the imaginary part along the imaginary axis.

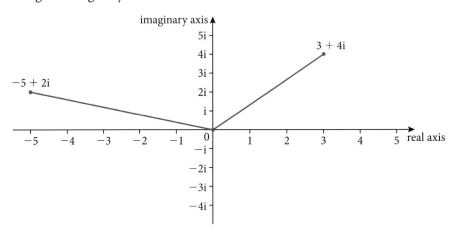

Figure 8.5 The complex numbers $3 + 4$i and $-5 + 2$i plotted in the complex plane

Let us consider the sum of two complex numbers

$$z_1 = x_1 + y_1 i \text{ and } z_2 = x_2 + y_2 i$$

As we have defined addition before

$$z_1 + z_2 = (x_1 + x_2) + (y_1 + y_2)i$$

This suggests that we consider complex numbers as vectors; that is, we regard the complex number $z = x + iy$ as a vector in Cartesian form whose terminal point is the complex number (x, y).

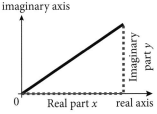

Figure 8.6 Real and imaginary parts of z

Since we are representing the complex numbers by vectors, that brings some analogies between the two sets. So, adding two complex numbers or subtracting them, or multiplying by a scalar, is similar in both sets.

If you are not familiar with vectors, it is advisable that you read section 9.1 first.

Example 8.9

Consider the complex numbers $z_1 = 3 + 4i$ and $z_2 = -5 + 2i$

Find

(a) $z_1 + z_2$
(b) $z_1 - z_2$

Solution

Draw a diagram.

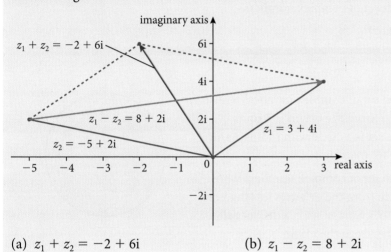

(a) $z_1 + z_2 = -2 + 6i$
(b) $z_1 - z_2 = 8 + 2i$

Note that the vector representing the sum, $-2 + 6i$, is the diagonal of the parallelogram with sides representing $3 + 4i$ and $-5 + 2i$, while the vector representing the difference is the second diagonal of the parallelogram.

The length of a vector also has a parallel in complex numbers. You recall that for a vector $v = \begin{pmatrix} x \\ y \end{pmatrix}$ the length of the vector is

$$|v| = \sqrt{x^2 + y^2}$$

For complex numbers, the modulus or absolute value (or magnitude) of the complex number $z = x + yi$ is

$$|z| = \sqrt{x^2 + y^2}$$

It follows immediately that since

$$z^* = x - yi \Rightarrow |z^*| = \sqrt{x^2 + (-y)^2} = \sqrt{x^2 + y^2}, \text{ then}$$
$$|z^*| = |z|$$

Also of interest is the following result:
$z \cdot z^* = (x + iy)(x - iy) = x^2 + y^2, |z^2| = x^2 + y^2, \text{ and } |z^*|^2 = x^2 + y^2$
$\Rightarrow z \cdot z^* = |z|^2 = |z^*|^2$

For example: $(3 + 4i)(3 - 4i) = 9 + 16 = 25 = \left(\sqrt{3^2 + 4^2}\right)^2$

Example 8.10

Calculate the modulus of each complex number.

(a) $z_1 = 5 - 6i$ (b) $z_2 = 12 + 5i$

Solution

(a) $|z_1| = |5 - 6i| = \sqrt{5^2 + 6^2} = \sqrt{61}$

(b) $|z_2| = |12 + 5i| = \sqrt{12^2 + 5^2} = \sqrt{169} = 13$

Example 8.11

Graph each set of complex numbers.

(a) $A = \{z : |z| = 3\}$ (b) $B = \{z : |z| \leq 3\}$

Solution

(a) A is the set of complex numbers whose distance from the origin is 3 units. So, the set is a circle with radius 3 and centre $(0, 0)$ as shown.

(b) B is the set of complex numbers whose distance from the origin is less than or equal to 3. So, the set is a disk of radius 3 and centre at the origin.

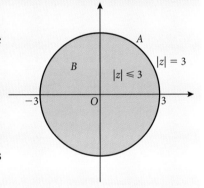

Another important property is the following result:
$$|z_1 z_2| = |z_1||z_2|$$

Proof

$$|z_1 z_2| = |(x_1 x_2 - y_1 y_2) + (x_1 y_2 + x_2 y_1)i| = \sqrt{(x_1 x_2 - y_1 y_2)^2 + (x_1 y_2 + x_2 y_1)^2}$$
$$= \sqrt{(x_1 x_2)^2 - 2x_1 x_2 y_1 y_2 + (y_1 y_2)^2 + (x_1 y_2)^2 + 2x_1 y_2 x_2 y_1 + (x_2 y_1)^2}$$
$$= \sqrt{(x_1 x_2)^2 + (y_1 y_2)^2 + (x_1 y_2)^2 + (x_2 y_1)^2}$$

But
$$|z_1||z_2| = \sqrt{x_1^2 + y_1^2}\sqrt{x_2^2 + y_2^2} = \sqrt{(x_1^2 + y_1^2)(x_2^2 + y_2^2)}$$
$$= \sqrt{(x_1 x_2)^2 + (y_1 y_2)^2 + (x_1 y_2)^2 + (x_2 y_1)^2}$$

And so, the result follows.

Example 8.12

Evaluate $|(3 + 4i)(5 + 12i)|$

Solution

$$|(3 + 4i)(5 + 12i)| = |3 + 4i||5 + 12i| = \sqrt{9 + 16}\sqrt{25 + 144} = 5 \times 13 = 65$$

or

$$|(3 + 4i)(5 + 12i)| = |-33 + 56i| = \sqrt{(-33)^2 + 56^2} = \sqrt{4225} = 65$$

Modulus-argument (trigonometric or polar) form of a complex number

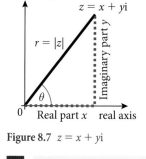

imaginary axis

Figure 8.7 $z = x + yi$

Every complex number $z = x + yi$ can be considered as an ordered pair (x, y). Hence using our knowledge of vectors, we can introduce a new form for representing complex numbers – the trigonometric form (also known as polar form).

The modulus-argument form uses the modulus of the complex number as its distance from the origin, $r \geq 0$, and θ the angle the 'vector' makes with the positive direction of the real axis.

Clearly $x = r\cos\theta$ and $y = r\sin\theta$; $r = \sqrt{x^2 + y^2}$; and $\tan\theta = \dfrac{y}{x}$.

Therefore $z = x + yi = r\cos\theta + (r\sin\theta)i = r(\cos\theta + i\sin\theta)$

The angle θ is called the **argument** of the complex number, arg(z). Arg(z) is not unique. However, all values differ by a multiple of 2π.

> The trigonometric form is called modulus-argument form by the IB. Also, this trigonometric form is abbreviated, for ease of writing, as follows:
> $z = x + yi = r(\cos\theta + i\sin\theta) = r\operatorname{cis}\theta$
> $\operatorname{cis}\theta$ stands for $\cos\theta + i\sin\theta$.

Example 8.13

Write each number in modulus-argument form.

(a) $z = 1 + i$ (b) $z = \sqrt{3} - i$ (c) $z = -5i$ (d) $z = 17$

Solution

(a) $r = \sqrt{1^2 + 1^2} = \sqrt{2}$; $\tan\theta = \dfrac{1}{1} = 1$,

hence, by observing the real and imaginary parts being positive, we can conclude that the argument must be $\theta = \dfrac{\pi}{4}$.

$\therefore z = \sqrt{2}\left(\cos\dfrac{\pi}{4} + i\sin\dfrac{\pi}{4}\right)$

$= \sqrt{2}\operatorname{cis}\dfrac{\pi}{4}$

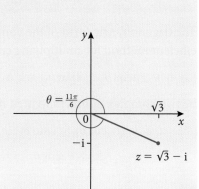

(b) $r = \sqrt{(\sqrt{3})^2 + (-1)^2} = \sqrt{4} = 2$;

$\tan\theta = \dfrac{-1}{\sqrt{3}}$. The real part is positive, the imaginary part is negative, and the point is therefore in the fourth quadrant, so $\theta = \dfrac{11\pi}{6}$.

$\therefore z = 2\left(\cos\dfrac{11\pi}{6} + i\sin\dfrac{11\pi}{6}\right)$

$= 2\operatorname{cis}\dfrac{11\pi}{6}$

We can also use $\theta = -\dfrac{\pi}{6}$

(c) $r = 5$ and $\theta = \dfrac{3\pi}{2}$ since it is on the

negative side of the imaginary axis.

$$\therefore z = 5\left(\cos\dfrac{3\pi}{2} + i\sin\dfrac{3\pi}{2}\right)$$

We can also use $\theta = -\dfrac{\pi}{2}$

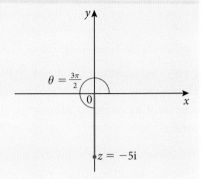

(d) $r = 17$ and $\theta = 0$

$$\therefore z = 17(\cos 0 + i\sin 0)$$

Example 8.14

Convert each complex number into its Cartesian form. Express your answer in exact form where possible.

(a) $z = 3\cos 150° + 3i\sin 150°$

(b) $z = 12\operatorname{cis}\dfrac{4\pi}{3}$

(c) $z = 6(\cos 50° + i\sin 50°)$

(d) $z = 15\left(\cos\dfrac{\pi}{2} + i\sin\dfrac{\pi}{2}\right)$

Solution

(a) $z = 3\left(\dfrac{-\sqrt{3}}{2}\right) + 3i\left(\dfrac{1}{2}\right) = \dfrac{-3\sqrt{3}}{2} + \dfrac{3i}{2}$

(b) $z = 12\cos\dfrac{4\pi}{3} + 12i\sin\dfrac{4\pi}{3} = 12 \cdot \dfrac{-1}{2} + 12i \cdot \dfrac{\sqrt{3}}{2} = -6 + 6i\sqrt{3}$

(c) $z = 6\cos 50° + 6i\sin 50° \approx 6 \cdot 0.643 + 6i \cdot 0.766 = 3.857 + 4.596i$

(d) $z = 15(0 + i) = 15i$

Multiplication

The analogy between complex numbers and vectors stops at multiplication. As you recall, multiplication of vectors is not well defined in the sense that there are two products – the scalar product, which is a scalar, not a vector, and the vector product (discussed in Chapter 9), which is a vector but is not in the plane! Complex number products are complex numbers!

The trigonometric form of the complex number offers a very interesting and efficient method for multiplying complex numbers.

Let $z_1 = r_1(\cos\theta_1 + i\sin\theta_1)$ and $z_2 = r_2(\cos\theta_2 + i\sin\theta_2)$

be two complex numbers written in polar form. Then

$$z_1 z_2 = (r_1(\cos\theta_1 + i\sin\theta_1))(r_2(\cos\theta_2 + i\sin\theta_2))$$

$$= r_1 r_2[(\cos\theta_1\cos\theta_2 - \sin\theta_1\sin\theta_2) + i(\sin\theta_1\cos\theta_2 + \sin\theta_2\cos\theta_1)]$$

Using the addition formulas for sine and cosine, we have the following formula:

$$z_1 z_2 = r_1 r_2[(\cos(\theta_1 + \theta_2)) + i(\sin(\theta_1 + \theta_2))]$$

This formula means: To multiply two complex numbers written in polar form, we multiply the moduli and add the arguments.

Example 8.15

Let $z_1 = 2 + 2i\sqrt{3}$ and $z_2 = -1 - i\sqrt{3}$

(a) Evaluate $z_1 z_2$ using their Cartesian forms.

(b) Evaluate $z_1 z_2$ using their polar forms and verify that the two results are the same.

Solution

(a) $z_1 z_2 = (2 + 2i\sqrt{3})(-1 - i\sqrt{3}) = (-2 + 6) + (-2\sqrt{3} - 2\sqrt{3})i = 4 - 4i\sqrt{3}$

(b) Converting both to polar form, we get

$$z_1 = 4\operatorname{cis}\frac{\pi}{3} \text{ and } z_2 = 2\operatorname{cis}\frac{4\pi}{3}, \text{ then}$$

$$z_1 z_2 = 4 \cdot 2\left(\operatorname{cis}\left(\frac{\pi}{3} + \frac{4\pi}{3}\right)\right) = 8\operatorname{cis}\left(\frac{5\pi}{3}\right) = 8\left(\cos\frac{5\pi}{3} + i\sin\frac{5\pi}{3}\right)$$

$$= 8\left(\frac{1}{2} + i\left(\frac{-\sqrt{3}}{2}\right)\right) = 4 - 4i\sqrt{3}$$

You may observe here that multiplying z_1 by z_2 resulted in a new number whose magnitude is twice that of z_1 and is rotated by an angle of $\frac{4\pi}{3}$. Alternatively you can see it as multiplying z_2 by z_1 which results in a complex number whose magnitude is 4 times that of z_2 and is rotated by an angle of $\frac{\pi}{3}$.

Example 8.16

Let $z_1 = -2 + 2i$ and $z_2 = 3\sqrt{3} - 3i$

Convert to polar form and multiply.

Solution

$$z_1 = 2\sqrt{2}\operatorname{cis}\frac{3\pi}{4} \text{ and } z_2 = \operatorname{cis}\frac{11\pi}{6}, \text{ then}$$

$$z_1 z_2 = 12\sqrt{2}\left(\operatorname{cis}\left(\frac{3\pi}{4} + \frac{11\pi}{6}\right)\right) = 12\sqrt{2}\operatorname{cis}\left(\frac{31\pi}{12}\right) = 12\sqrt{2}\operatorname{cis}\left(\frac{7\pi}{12} + 2\pi\right)$$

$$= 12\sqrt{2}\operatorname{cis}\left(\frac{7\pi}{12}\right) = 12\sqrt{2}\left(\cos\frac{7\pi}{12} + i\sin\frac{7\pi}{12}\right)$$

You can simplify this answer further to get an exact Cartesian form

$$z_1 z_2 = 12\sqrt{2}\left(\cos\frac{7\pi}{12} + i\sin\frac{7\pi}{12}\right) = 12\sqrt{2}\left(\cos\left(\frac{3\pi + 4\pi}{12}\right) + i\sin\frac{3\pi + 4\pi}{12}\right)$$

$$= 12\sqrt{2}\left(\cos\left(\frac{\pi}{4} + \frac{\pi}{3}\right) + i\sin\left(\frac{\pi}{4} + \frac{\pi}{3}\right)\right)$$

$$= 12\sqrt{2}\left(\left(\frac{\sqrt{2}}{2} \cdot \frac{1}{2} - \frac{\sqrt{2}}{2} \cdot \frac{\sqrt{3}}{2}\right) + i\left(\frac{\sqrt{2}}{2} \cdot \frac{1}{2} + \frac{\sqrt{2}}{2} \cdot \frac{\sqrt{3}}{2}\right)\right)$$

$$= 12\sqrt{2}\left(\frac{\sqrt{2} - \sqrt{6}}{4} + i\frac{\sqrt{2} + \sqrt{6}}{4}\right) = (6 - 6\sqrt{3}) + i(6 + 6\sqrt{3})$$

By comparing the Cartesian form of the product to the polar form, i.e., $12\sqrt{2}\left(\cos\frac{7\pi}{12} + i\sin\frac{7\pi}{12}\right)$ and

$12\sqrt{2}\left(\frac{\sqrt{2} - \sqrt{6}}{4} + i\frac{\sqrt{2} - \sqrt{6}}{4}\right)$, we can conclude that $\cos\frac{7\pi}{12} = \frac{\sqrt{2} - \sqrt{6}}{4}$ and $\sin\frac{7\pi}{12} = \frac{\sqrt{2} + \sqrt{6}}{4}$.

This observation gives us a way of using complex number multiplication in order to find exact values of some trigonometric functions.

If you set up your GDC to work with complex numbers in polar form, then it may look like this. You can also perform the polar form multiplication getting the Cartesian form directly.

Figure 8.8 Working with complex numbers in polar form on a GDC

imaginary axis

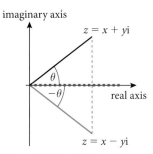

Figure 8.9 Graphically, a complex number and its conjugate are reflections of each other in the real axis

You may have noticed that the conjugate of a complex number $z = r(\cos\theta + i\sin\theta)$ is
$z^* = r(\cos\theta - i\sin\theta) = r(\cos(-\theta) + i\sin(-\theta))$.

Also,
$z \cdot z^* = r(\cos\theta + i\sin\theta) \cdot r(\cos\theta - i\sin\theta) = r^2(\cos^2\theta + \sin^2\theta) = r^2$

Graphically, a complex number and its conjugate are reflections of each other in the real axis.

Division of complex numbers

A similar approach gives us the rules for division of complex numbers.

Let $z_1 = r_1(\cos\theta_1 + i\sin\theta_1)$ and $z_2 = r_2(\cos\theta_2 + i\sin\theta_2)$

be two complex numbers written in trigonometric form. Then

$$\frac{z_1}{z_2} = \frac{r_1(\cos\theta_1 + i\sin\theta_1)}{r_2(\cos\theta_2 + i\sin\theta_2)} \cdot \frac{\cos\theta_2 - i\sin\theta_2}{\cos\theta_2 - i\sin\theta_2}$$

$$= \frac{r_1}{r_2}\left(\frac{(\cos\theta_1\cos\theta_2 + \sin\theta_1\sin\theta_2) + i(\sin\theta_1\cos\theta_2 - \sin\theta_2\cos\theta_1)}{\cos^2\theta_2 + \sin^2\theta_2}\right)$$

$$= \frac{r_1}{r_2}\left(\frac{(\cos\theta_1\cos\theta_2 + \sin\theta_1\sin\theta_2) + i(\sin\theta_1\cos\theta_2 - \sin\theta_2\cos\theta_1)}{1}\right)$$

Using the subtraction formulas for sine and cosine, we have
$$\frac{z_1}{z_2} = \frac{r_1}{r_2}[(\cos(\theta_1 - \theta_2)) + i(\sin(\theta_1 - \theta_2))]$$

This formula says: to divide two complex numbers written in trigonometric form, we divide the moduli and subtract the arguments.

In particular, if we take $z_1 = 1$ and $z_2 = z$, (i.e., $\theta_1 = 0$ and $\theta_2 = \theta$) and $z = r(\cos\theta + i\sin\theta)$, we obtain:
$$\frac{1}{z} = \frac{1}{r}(\cos(-\theta) + i\sin(-\theta)) = \frac{1}{r}(\cos(\theta) - i\sin(\theta))$$

Example 8.17

Let $z_1 = 1 + i$ and $z_2 = \sqrt{3} - i$

(a) Convert the complex numbers into polar form.

(b) Evaluate $\dfrac{1}{z_2}$

(c) Evaluate $\dfrac{z_1}{z_2}$

(d) Use the results from parts (a) to (c) to find the exact values of $\sin\dfrac{5\pi}{12}$ and $\cos\dfrac{5\pi}{12}$.

Solution

(a) $z_1 = \sqrt{2}\,\text{cis}\dfrac{\pi}{4}; z_2 = 2\text{cis}\dfrac{11\pi}{6} = 2\text{cis}\dfrac{-\pi}{6}$

(b) $\dfrac{1}{z_2} = \dfrac{1}{2}\text{cis}\left(-\dfrac{-\pi}{6}\right) = \dfrac{1}{2}\text{cis}\dfrac{\pi}{6}$

(c) We can find $\dfrac{z_1}{z_2}$ by multiplying z_1 by $\dfrac{1}{z_2}$, or by using division.

$$\frac{z_1}{z_2} = z_1 \cdot \frac{1}{z_2} = \left(\sqrt{2}\operatorname{cis}\frac{\pi}{4}\right) \cdot \left(\frac{1}{2}\operatorname{cis}\frac{\pi}{6}\right) = \frac{\sqrt{2}}{2}\operatorname{cis}\left(\frac{\pi}{4}+\frac{\pi}{6}\right) = \frac{\sqrt{2}}{2}\operatorname{cis}\left(\frac{5\pi}{12}\right), \text{ or}$$

$$\frac{z_1}{z_2} = \frac{\sqrt{2}\operatorname{cis}\dfrac{\pi}{4}}{2\operatorname{cis}\dfrac{-\pi}{6}} = \frac{\sqrt{2}}{2}\operatorname{cis}\left(\frac{\pi}{4}-\frac{-\pi}{6}\right) = \frac{\sqrt{2}}{2}\operatorname{cis}\left(\frac{5\pi}{12}\right)$$

(d) $\dfrac{z_1}{z_2} = \dfrac{1+i}{\sqrt{3}-i} \cdot \dfrac{\sqrt{3}+i}{\sqrt{3}+i} = \dfrac{\sqrt{3}-1+(\sqrt{3}+1)i}{4}$

$$\Rightarrow \frac{\sqrt{3}-1}{4} = \frac{\sqrt{2}}{2}\cos\frac{5\pi}{12} \Rightarrow \cos\frac{5\pi}{12} = \frac{\sqrt{3}-1}{4}\cdot\frac{2}{\sqrt{2}} = \frac{\sqrt{6}-\sqrt{2}}{4}$$

$$\Rightarrow \frac{\sqrt{3}+1}{4} = \frac{\sqrt{2}}{2}\sin\frac{5\pi}{12} \Rightarrow \sin\frac{5\pi}{12} = \frac{\sqrt{3}+1}{4}\cdot\frac{2}{\sqrt{2}} = \frac{\sqrt{6}+\sqrt{2}}{4}$$

Exercise 8.2

1. Write each complex number in polar form with argument θ, such that $0 < \theta < 2\pi$.

   (a) $2 + 2i$ (b) $\sqrt{3} + i$ (c) $2 - 2i$ (d) $\sqrt{6} - i\sqrt{2}$

   (e) $2 - 2i\sqrt{3}$ (f) $-3 + 3i$ (g) $4i$ (h) $-3\sqrt{3} - 3i$

   (i) $i + 1$ (j) -15 (k) $(4 + 3i)^{-1}$ (l) $i(3 + 3i)$

   (m) π (n) ei

2. Find $z_1 z_2$ and $\dfrac{z_1}{z_2}$ for each set of numbers.

   (a) $z_1 = \cos\dfrac{\pi}{2} + i\sin\dfrac{\pi}{2},\, z_2 = \cos\dfrac{\pi}{3} + i\sin\dfrac{\pi}{3}$

   (b) $z_1 = \cos\dfrac{5\pi}{6} + i\sin\dfrac{5\pi}{6},\, z_2 = \cos\dfrac{7\pi}{6} + i\sin\dfrac{7\pi}{6}$

   (c) $z_1 = \cos\dfrac{\pi}{6} + i\sin\dfrac{\pi}{6},\, z_2 = \cos\dfrac{2\pi}{3} + i\sin\dfrac{2\pi}{3}$

   (d) $z_1 = \cos\dfrac{13\pi}{12} + i\sin\dfrac{13\pi}{12},\, z_2 = \cos\dfrac{5\pi}{12} + i\sin\dfrac{5\pi}{12}$

   (e) $z_1 = 3\left(\cos\dfrac{3\pi}{4} + i\sin\dfrac{3\pi}{4}\right),\, z_2 = \dfrac{2}{3}\left(\cos\dfrac{4\pi}{3} + i\sin\dfrac{4\pi}{3}\right)$

   (f) $z_1 = 3\sqrt{2}\left(\cos\dfrac{5\pi}{4} + i\sin\dfrac{5\pi}{4}\right),\, z_2 = 2\left(\cos\dfrac{5\pi}{3} + i\sin\dfrac{5\pi}{3}\right)$

   (g) $z_1 = \cos135° + i\sin135°,\, z_2 = \cos90° + i\sin90°$

   (h) $z_1 = 3(\cos120° + i\sin120°),\, z_2 = 2(\cos240° + i\sin240°)$

   (i) $z_1 = \dfrac{5}{8}(\cos225° + i\sin225°),\, z_2 = \dfrac{\sqrt{3}}{2}(\cos330° + i\sin330°)$

   (j) $z_1 = 3\sqrt{2}(\cos315° + i\sin315°),\, z_2 = 2(\cos300° + i\sin300°)$

3. Write z_1 and z_2 in polar form, and then find the reciprocals $\dfrac{1}{z_1}, \dfrac{1}{z_2}$, the product $z_1 z_2$, and the quotient $\dfrac{z_1}{z_2}$

   (a) $z_1 = \sqrt{3} + i$ and $z_2 = 2 - 2i\sqrt{3}$

   (b) $z_1 = \sqrt{6} + i\sqrt{2}$ and $z_2 = 2\sqrt{3} - 6i$

   (c) $z_1 = 4\sqrt{3} + 4i$ and $z_2 = -3 - 3i$

   (d) $z_1 = \sqrt{3}\,i$ and $z_2 = -\sqrt{2} - i\sqrt{6}$

   (e) $z_1 = \sqrt{5} + i\sqrt{5}$ and $z_2 = 2i\sqrt{2}$

   (f) $z_1 = 1 + i\sqrt{3}$ and $z_2 = 2\sqrt{3}$

4. Consider the complex number z where $|z - i| = |z + 2i|$

   (a) Show that $\text{Im}(z) = -\dfrac{1}{2}$

   (b) Let z_1 and z_2 be the two possible values of z, such that $|z| = 1$

       (i) Sketch a diagram to show the points that represent z_1 and z_2 in the complex plane.

       (ii) Find $\arg(z_1)$ and $\arg(z_2)$

5. Use an Argand diagram to show that $|z + w| < |z| + |w|$

6. $z = \sqrt{3}\left(\cos\dfrac{2\pi}{3} + i\sin\dfrac{2\pi}{3}\right)$

   Express each complex number in Cartesian form.

   (a) $\dfrac{3}{\sqrt{3} + z}$ (b) $\dfrac{2z}{3 + z^2}$ (c) $\dfrac{3 - z^2}{3 + z^2}$

7. Find the modulus and argument of each complex number.

   $$z_1 = 2\sqrt{3} - 2i, \quad z_2 = 2 + 2i \quad \text{and} \quad z_3 = (2\sqrt{3} - 2i)(2 + 2i)$$

8. The numbers in question 7 represent the vertices of a triangle in the Argand diagram. Find the area of the triangle.

9. Identify, in the complex plane, the set of points that corresponds to each equation.

   (a) $|z| = 3$ (b) $z^* = -z$ (c) $z + z^* = 8$

   (d) $|z - 3| = 2$ (e) $|z - 1| + |z - 3| = 2$

10. Identify, in the complex plane, the set of points that corresponds to each inequality.

   (a) $|z| < 3$ (b) $|z - 3i| > 2$

8.3 Powers and roots of complex numbers

The formula established for the product of two complex numbers can be applied to derive a special formula for the nth power of a complex number.

Let $z = r(\cos\theta + i\sin\theta)$

So $z^2 = (r(\cos\theta + i\sin\theta))(r(\cos\theta + i\sin\theta))$

$\quad = r^2((\cos\theta\cos\theta - \sin\theta\sin\theta) + i(\sin\theta\cos\theta + \cos\theta\sin\theta))$

$\quad = r^2((\cos^2\theta - \sin^2\theta) + i(2\sin\theta\cos\theta)) = r^2(\cos 2\theta + i\sin 2\theta)$

Similarly,

$\quad z^3 = z \cdot z^2 = (r(\cos\theta + i\sin\theta))(r^2(\cos 2\theta + i\sin 2\theta))$

$\quad\quad = r^3(\cos(\theta + 2\theta) + i\sin(\theta + 2\theta)) = r^3(\cos 3\theta + i\sin 3\theta)$

In general we obtain the following theorem, named after the French mathematician Abraham de Moivre (1667−1754).

de Moivre's theorem

If $z = r(\cos\theta + i\sin\theta)$ and n is a positive integer, then
$z^n = (r(\cos\theta + i\sin\theta))^n = r^n(\cos n\theta + i\sin n\theta)$

In words: to find the nth power of any complex number written in trigonometric form, we take the nth power of the modulus and multiply the argument by n.

Proof of de Moivre's theorem

The proof of this theorem follows as an application of mathematical induction.
Let $P(n)$ be the statement $z^n = r^n(\cos n\theta + i\sin n\theta)$

Basis step: To prove this formula, the basis step must be $P(1)$.
$P(1)$: is true since $z^1 = r^1(\cos\theta + i\sin\theta)$, which is given!
[If you are not convinced, you can try $P(2)$: $z^2 = r^2(\cos 2\theta + i\sin 2\theta)$, which we showed above.]

Inductive step: Assume that $P(k)$ is true, i.e.,
$z^k = r^k(\cos k\theta + i\sin k\theta)$, we need to show that $P(k + 1)$ is also true.
i.e., we must show that $z^{k+1} = r^{k+1}(\cos(k + 1)\theta + i\sin(k + 1)\theta)$

Now
$z^{k+1} = z^k \cdot z = (r^k(\cos k\theta + i\sin k\theta))(r(\cos\theta + i\sin\theta))$ by assumption
$\quad = r^k r[(\cos k\theta\cos\theta - \sin k\theta\sin\theta) + i(\sin k\theta\cos\theta + \cos k\theta\sin\theta)]$
$\quad = r^{k+1}[\cos(k\theta + \theta) + i\sin(k\theta + \theta)]$ by addition formulas for sine and cosine
$\quad = r^{k+1}(\cos(k + 1)\theta + i\sin(k + 1)\theta)$

Therefore, by the principle of mathematical induction, since the theorem is true for $n = 1$, and whenever it is true for $n = k$, it was proved true for $n = k + 1$, then the theorem is true for all positive integers n.

As a matter of fact, de Moivre has stated 'his' formula only implicitly. Its standard form is due to Euler and was generalised by him to any real n.

De Moivre's theorem is valid for all real numbers n. However, the proof is beyond the scope of this course and this book and therefore we will accept the theorem as true.

Example 8.18

Find $(1 + i)^6$

Solution

First, we convert the number into polar form

$(1 + i) = \sqrt{2}\left(\cos\dfrac{\pi}{4} + i\sin\dfrac{\pi}{4}\right)$, now we can apply de Moivre's theorem

$(1 + i)^6 = \left[\sqrt{2}\left(\cos\dfrac{\pi}{4} + i\sin\dfrac{\pi}{4}\right)\right]^6 = (\sqrt{2})^6\left(\cos\left(6\cdot\dfrac{\pi}{4}\right) + i\sin\left(6\cdot\dfrac{\pi}{4}\right)\right)$

$\quad = 8\left(\cos\dfrac{3\pi}{2} + i\sin\dfrac{3\pi}{2}\right) = 8(-i) = -8i$

We can obtain the same answer using the binomial theorem.
$(1 + i)^6 = 1 + 6i + 15i^2 + 20i^3 + 15i^4 + 6i^5 + i^6$
$\quad = 1 + 6i - 15 - 20i + 15 + 6i - 1 = -8i$

When the powers get larger, applying the binomial theorem or Pascal's triangle will require more calculations.

Uses of de Moivre's theorem in problem solving

Several applications of this theorem prove very helpful in dealing with trigonometric identities and expressions.

- $z^{-1} = r^{-1}(\cos(-\theta) + i \sin(-\theta)) = \frac{1}{r}(\cos\theta - i \sin\theta)$
- $z^{-n} = (z^{-1})^n = (r^{-1}(\cos(-\theta) + i \sin(-\theta)))^n = r^{-n}(\cos(-n\theta) + i \sin(-n\theta))$

If we take the case when $r = 1$, then

- $z^n = \cos n\theta + i \sin n\theta$
- $z^{-n} = \cos(-n\theta) + i \sin(-n\theta) = \cos n\theta - i \sin n\theta$
- $z^n + z^{-n} = 2 \cos n\theta$
- $z^n - z^{-n} = 2i \sin n\theta$

These relationships are quite helpful in allowing us to write powers of $\cos\theta$ and $\sin\theta$ in terms of cosines and sines of multiples of θ.

Example 8.19

Find $\cos^3\theta$ in terms of first powers of the cosine function.

Solution

Start with $\left(z + \frac{1}{z}\right)^3 = (2\cos\theta)^3$

Expand the left-hand side:

$z^3 + 3z + \frac{3}{z} + \frac{1}{z^3} = 8\cos^3\theta$

$$\Rightarrow z^3 + \frac{1}{z^3} + 3\left(z + \frac{1}{z}\right) = 8\cos^3\theta$$

$$\Updownarrow \qquad\qquad \Updownarrow$$

$$\Rightarrow 2\cos3\theta + 3(2\cos\theta) = 8\cos^3\theta$$

$$\Rightarrow \cos^3\theta = \frac{1}{8}(2\cos3\theta + 3(2\cos\theta)) = \frac{1}{4}(\cos3\theta + 3\cos\theta)$$

Example 8.20

Find formulae for $\cos3\theta$ and $\sin3\theta$ in terms of $\sin\theta$ and $\cos\theta$.

Solution

For $n = 3$, using de Moivre's theorem and the binomial theorem, or Pascal's triangle, will give you formulae for sines and cosines of multiples of θ:

$$(\cos\theta + i \sin\theta)^3 = \cos3\theta + i \sin3\theta$$
$$= \cos^3\theta + 3i\cos^2\theta\sin\theta + 3i^2\cos\theta\sin^2\theta + i^3\sin^3\theta$$
$$= \cos^3\theta - 3\cos\theta\sin^2\theta + i(3\cos^2\theta\sin\theta - \sin^3\theta)$$

Comparing real and imaginary parts gives

$$\cos3\theta = \cos^3\theta - 3\cos\theta\sin^2\theta$$
$$\sin3\theta = 3\cos^2\theta\sin\theta - \sin^3\theta$$

Example 8.21

Simplify the expression $\dfrac{(\cos6\theta + i\sin6\theta)(\cos3\theta + i\sin3\theta)}{\cos4\theta + i\sin4\theta}$

Solution

Using de Moivre's theorem

$$\frac{(\cos6\theta + i\sin6\theta)(\cos3\theta + i\sin3\theta)}{\cos4\theta + i\sin4\theta} = \frac{(\cos\theta + i\sin\theta)^6(\cos\theta + i\sin\theta)^3}{(\cos\theta + i\sin\theta)^4}$$

$$= (\cos\theta + i\sin\theta)^5 = \cos5\theta + i\sin5\theta$$

*n*th roots of a complex number

De Moivre's theorem is an essential tool for finding *n*th roots of complex numbers.

An *n*th root of a given number z is a number w that satisfies the relation $w^n = z$.

For example, $w = 1 + i$ is a 6th root of $z = -8i$ because, as we have seen in Example 8.18, $(1 + i)^6 = -8i$; or $w = -\sqrt{3} + i$ is a 10th root of $512 + 512i\sqrt{3}$. This is also because $w^{10} = (-\sqrt{3} + i)^{10} = 512 + 512i\sqrt{3}$

To find the *n*th roots, apply the definition of an *n*th root.

Let $w = s(\cos\alpha + i\sin\alpha)$ be an *n*th root of $z = r(\cos\theta + i\sin\theta)$.
This means that $w^n = z$, i.e.

$$(s(\cos\alpha + i\sin\alpha))^n = r(\cos\theta + i\sin\theta)$$
$$\Rightarrow s^n(\cos n\alpha + i\sin n\alpha) = r(\cos\theta + i\sin\theta)$$

However, two complex numbers are equal if their moduli are equal; that is

$$s^n = r \Leftrightarrow s = \sqrt[n]{r} = r^{\frac{1}{n}}$$

Also

$$\cos n\alpha = \cos\theta \text{ and } \sin n\alpha = \sin\theta$$

Recall that both sine and cosine functions are periodic functions with period 2π, hence

$$\begin{cases} \cos n\alpha = \cos\theta \\ \sin n\alpha = \sin\theta \end{cases} \Rightarrow n\alpha = \theta + 2k\pi, k = 0\ 1, 2, ..., n-1$$

This leads to

$$\alpha = \frac{\theta + 2k\pi}{n} = \frac{\theta}{n} + \frac{2k\pi}{n}; k = 0, 1, 2, 3, ..., n-1$$

Let $z = r(\cos\theta + i\sin\theta)$ and let n be a positive integer.
z has n distinct *n*th roots:
$$z_k = \sqrt[n]{r}\left(\cos\left(\frac{\theta}{n} + \frac{2k\pi}{n}\right) + i\sin\left(\frac{\theta}{n} + \frac{2k\pi}{n}\right)\right)$$
Where $k = 1, 2, 3, ..., n-1$.

Notice that we stop the values of k at $n - 1$. This is because for values larger than or equal to n, principal arguments for these roots will be identical to those for $k = 0$ until $n - 1$.

Each of the n nth roots of z has the same modulus, $\sqrt[n]{r} = r^{\frac{1}{n}}$. Thus, all these roots lie on a circle in the complex plane whose radius is $\sqrt[n]{r} = r^{\frac{1}{n}}$. Also, since the arguments of consecutive roots differ by $\dfrac{2\pi}{n}$, the roots are equally spaced on this circle.

Example 8.22

Find the cubic roots of $z = -4\sqrt{2} + 4\mathrm{i}\sqrt{2}$

Solution

$r = 8$ and $\theta = \dfrac{3\pi}{4}$, so, the roots are

$$w = s\left(\cos\left(\frac{\theta}{n} + \frac{2k\pi}{n}\right) + \mathrm{i}\sin\left(\frac{\theta}{n} + \frac{2k\pi}{n}\right)\right)$$

$$= \sqrt[3]{8}\left(\cos\left(\frac{\frac{3\pi}{4}}{3} + \frac{2k\pi}{3}\right) + \mathrm{i}\sin\left(\frac{\frac{3\pi}{4}}{3} + \frac{2k\pi}{3}\right)\right)$$

$$= 2\left(\cos\left(\frac{\pi}{4} + \frac{2k\pi}{3}\right) + \mathrm{i}\sin\left(\frac{\pi}{4} + \frac{2k\pi}{3}\right)\right); k = 0, 1, 2$$

$$w_1 = 2\left(\cos\left(\frac{\pi}{4}\right) + \mathrm{i}\sin\left(\frac{\pi}{4}\right)\right)$$

$$w_2 = 2\left(\cos\left(\frac{\pi}{4} + \frac{2\pi}{3}\right) + \mathrm{i}\sin\left(\frac{\pi}{4} + \frac{2\pi}{3}\right)\right) = 2\left(\cos\left(\frac{11\pi}{12}\right) + \mathrm{i}\sin\left(\frac{11\pi}{12}\right)\right)$$

$$w_3 = 2\left(\cos\left(\frac{\pi}{4} + \frac{4\pi}{3}\right) + \mathrm{i}\sin\left(\frac{\pi}{4} + \frac{4\pi}{3}\right)\right) = 2\left(\cos\left(\frac{19\pi}{12}\right) + \mathrm{i}\sin\left(\frac{19\pi}{12}\right)\right)$$

Notice how the points are distributed equally around a circle with radius 2.

The difference between any two arguments is $\dfrac{2\pi}{3}$.

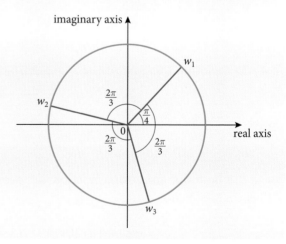

Notice that if we try to go beyond $k = 2$, we get back to w_1

$$w_4 = 2\left(\cos\left(\frac{\pi}{4} + \frac{6\pi}{3}\right) + \mathrm{i}\sin\left(\frac{\pi}{4} + \frac{6\pi}{3}\right)\right) = 2\left(\cos\left(\frac{\pi}{4} + 2\pi\right) + \mathrm{i}\sin\left(\frac{\pi}{4} + 2\pi\right)\right)$$

$$= 2\left(\cos\left(\frac{\pi}{4}\right) + \mathrm{i}\sin\left(\frac{\pi}{4}\right)\right) = w_1$$

Also, if we raise any of the roots to the third power, we will eventually get z. For example:

$$(w_2)^3 = \left[2\left(\cos\left(\frac{11\pi}{12}\right) + \mathrm{i}\sin\left(\frac{11\pi}{12}\right)\right)\right]^3 = 8\left(\cos\left(\frac{33\pi}{12}\right) + \mathrm{i}\sin\left(\frac{33\pi}{12}\right)\right)$$

$$= 8\left(\cos\left(\frac{11\pi}{4}\right) + \mathrm{i}\sin\left(\frac{11\pi}{4}\right)\right) = 8\left(\cos\left(\frac{3\pi}{4}\right) + \mathrm{i}\sin\left(\frac{3\pi}{4}\right)\right) = z$$

Example 8.23

Find the six sixth roots of $z = -64$ and graph them in the complex plane.

Solution

Here $r = 64$ and $\theta = \pi$. So, the roots are

$$w = s\left(\cos\left(\frac{\theta}{n} + \frac{2k\pi}{n}\right) + i\sin\left(\frac{\theta}{n} + \frac{2k\pi}{n}\right)\right)$$

$$= \sqrt[6]{64}\left(\cos\left(\frac{\pi}{6} + \frac{2k\pi}{6}\right) + i\sin\left(\frac{\pi}{6} + \frac{2k\pi}{6}\right)\right)$$

$$= 2\left(\cos\left(\frac{\pi}{6} + \frac{k\pi}{3}\right) + i\sin\left(\frac{\pi}{6} + \frac{k\pi}{3}\right)\right); k = 0, 1, 2, 3, 4, 5$$

$$w_1 = 2\left(\cos\left(\frac{\pi}{6}\right) + i\sin\left(\frac{\pi}{6}\right)\right)$$

$$w_2 = 2\left(\cos\left(\frac{\pi}{6} + \frac{\pi}{3}\right) + i\sin\left(\frac{\pi}{6} + \frac{\pi}{3}\right)\right) = 2\left(\cos\left(\frac{\pi}{2}\right) + i\sin\left(\frac{\pi}{2}\right)\right)$$

$$w_3 = 2\left(\cos\left(\frac{\pi}{6} + \frac{2\pi}{3}\right) + i\sin\left(\frac{\pi}{6} + \frac{2\pi}{3}\right)\right) = 2\left(\cos\left(\frac{5\pi}{6}\right) + i\sin\left(\frac{5\pi}{6}\right)\right)$$

$$w_4 = 2\left(\cos\left(\frac{\pi}{6} + \frac{3\pi}{3}\right) + i\sin\left(\frac{\pi}{6} + \frac{3\pi}{3}\right)\right) = 2\left(\cos\left(\frac{7\pi}{6}\right) + i\sin\left(\frac{7\pi}{6}\right)\right)$$

$$w_5 = 2\left(\cos\left(\frac{\pi}{6} + \frac{4\pi}{3}\right) + i\sin\left(\frac{\pi}{6} + \frac{4\pi}{3}\right)\right) = 2\left(\cos\left(\frac{3\pi}{2}\right) + i\sin\left(\frac{3\pi}{2}\right)\right)$$

$$w_6 = 2\left(\cos\left(\frac{\pi}{6} + \frac{5\pi}{3}\right) + i\sin\left(\frac{\pi}{6} + \frac{5\pi}{3}\right)\right) = 2\left(\cos\left(\frac{11\pi}{6}\right) + i\sin\left(\frac{11\pi}{6}\right)\right)$$

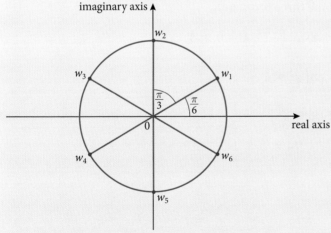

nth roots of unity

The rules we established can be applied to finding the nth roots of 1 (unity). Since 1 is a real number, in polar form it has a modulus of 1 and an argument of 0. We can write it as

$$1 = 1(\cos 0 + i\sin 0)$$

Now applying the rules above, 1 has n distinct nth roots given by

$$z_k = \sqrt[n]{r}\left(\cos\left(\frac{\theta}{n} + \frac{2k\pi}{n}\right) + i\sin\left(\frac{\theta}{n} + \frac{2k\pi}{n}\right)\right)$$

$$= \sqrt[n]{1}\left(\cos\left(\frac{0}{n} + \frac{2k\pi}{n}\right) + i\sin\left(\frac{0}{n} + \frac{2k\pi}{n}\right)\right)$$

$$= \cos\left(\frac{2k\pi}{n}\right) + i\sin\left(\frac{2k\pi}{n}\right); \; k = 0, 1, 2, ..., n - 1$$

Or in degrees

$$z_k = \cos\left(\frac{360k}{n}\right) + i\sin\left(\frac{360k}{n}\right); \; k = 0, 1, 2, ..., n - 1$$

Example 8.24

Find

(a) the square roots of unity

(b) the cube roots of unity

Solution

(a) Here $k = 2$, and therefore the two roots are

$$z_k = \cos\left(\frac{360k}{2}\right) + i\sin\left(\frac{360k}{2}\right); \; k = 0, 1$$

$$z_0 = \cos\left(\frac{0}{2}\right) + i\sin\left(\frac{0}{2}\right) = 1$$

$$z_1 = \cos\left(\frac{360°}{2}\right) + i\sin\left(\frac{360°}{2}\right) = \cos 180° + i\sin 180° = -1$$

(b) Here $k = 3$, and the three roots are

$$z_k = \cos\left(\frac{2k\pi}{3}\right) + i\sin\left(\frac{2k\pi}{3}\right); \; k = 0, 1, 3$$

$$z_0 = \cos\left(\frac{0}{3}\right) + i\sin\left(\frac{0}{3}\right) = 1$$

$$z_1 = \cos\left(\frac{2\pi}{3}\right) + i\sin\left(\frac{2\pi}{3}\right) = -\frac{1}{2} + i\frac{\sqrt{3}}{2}$$

$$z_2 = \cos\left(\frac{4\pi}{3}\right) + i\sin\left(\frac{4\pi}{3}\right) = -\frac{1}{2} - i\frac{\sqrt{3}}{2}$$

Euler's formula

The following results are proved in Chapter 16. Taylor's (Maclaurin's) series expansions for $\sin x$, $\cos x$ and e^x are

$$\sin x = x - \frac{x^3}{3!} + \frac{x^5}{5!} - \frac{x^7}{7!} + \cdots \qquad = \sum_0^\infty (-1)^n \frac{x^{2n+1}}{(2n + 1)!}$$

$$\cos x = 1 - \frac{x^2}{2!} + \frac{x^4}{4!} - \frac{x^6}{6!} + \cdots \qquad = \sum_0^\infty (-1)^n \frac{x^{2n}}{(2n)!}$$

$$e^x = 1 + x + \frac{x^2}{2!} + \frac{x^3}{3!} + \frac{x^4}{4!} + \cdots \qquad = \sum_0^\infty \frac{x^n}{n!}$$

Now add

$$\sin x + \cos x = 1 + x - \frac{x^2}{2!} - \frac{x^3}{3!} + \frac{x^4}{4!} + \frac{x^5}{5!} - \frac{x^6}{6!} - \frac{x^7}{7!} + \cdots$$

and compare the result to the expansion of e^x. The terms are similar, except for the signs of the terms. The signs in the sum alternate in a way where pairs of terms alternate. This property is typical of powers of i.

Look at $i, i^2, i^3, i^4, i^5, i^6, i^7, i^8 \ldots = i, \boxed{-1, -i}, \boxed{1, i}, \boxed{-1, -i}, 1, \cdots$

This suggests expanding e^{ix}

$$e^{ix} = 1 + ix + \frac{i^2 x^2}{2!} + \frac{i^3 x^3}{3!} + \frac{i^4 x^4}{4!} + \frac{i^5 x^5}{5!} + \frac{i^6 x^6}{6!} \cdots$$

$$= 1 + \frac{i^2 x^2}{2!} + \frac{i^4 x^4}{4!} + \frac{i^6 x^6}{6!} + \ldots + ix + \frac{i^3 x^3}{3!} + \frac{i^5 x^5}{5!} \cdots$$

$$= 1 - \frac{x^2}{2!} + \frac{x^4}{4!} - \frac{x^6}{6!} + \ldots + i\left(x + \frac{i^2 x^3}{3!} + \frac{i^4 x^5}{5!} + \ldots\right)$$

$$= 1 - \frac{x^2}{2!} + \frac{x^4}{4!} - \frac{x^6}{6!} + \ldots + i\left(x - \frac{x^3}{3!} + \frac{x^5}{5!} + \ldots\right)$$

$$= \cos x + i \sin x$$

Since, for any complex number
$z = x + iy = r(\cos\theta + i\sin\theta)$ and since $e^{i\theta} = \cos\theta + i\sin\theta$, then
$z = r(\cos\theta + i\sin\theta) = re^{i\theta}$.
This is known as Euler's formula.

Euler's form is also known as the exponential form of a complex number.

Example 8.25

Evaluate each complex number.

(a) $e^{i\pi}$ (b) $e^{i\frac{\pi}{2}}$

Solution

(a) $e^{i\pi} = \cos\pi + i\sin\pi = -1$ (b) $e^{i\frac{\pi}{2}} = \cos\frac{\pi}{2} + i\sin\frac{\pi}{2} = i$

$e^{i\pi} = \cos\pi + i\sin\pi = -1$ leads to Euler's remarkable formula $e^{i\pi} + 1 = 0$, which comprises the five most important mathematical constants: $e, \pi, i, 1$, and 0.

Example 8.26

Use Euler's formula to demonstrate de Moivre's theorem.

Solution

$$(r(\cos\theta + i\sin\theta))^n = (re^{i\theta})^n = r^n e^{in\theta}$$
$$= r^n(\cos n\theta + i\sin n\theta)$$

Example 8.27

Find the real and imaginary parts of each complex number.

(a) $z = 3e^{i\frac{\pi}{6}}$

(b) $z = 7e^{2i}$

Solution

(a) Since $|z| = 3$ and $\arg(z) = \dfrac{\pi}{6}$, $\operatorname{Re}(z) = 3\cos\dfrac{\pi}{6} = \dfrac{3\sqrt{3}}{2}$

and $\operatorname{Im}(z) = 3\sin\dfrac{\pi}{6} = \dfrac{3}{2}$

(b) Since $|z| = 7$ and $\arg(z) = 2$, $\operatorname{Re}(z) = 7\cos 2$ and $\operatorname{Im}(z) = 7\sin 2$

Example 8.28

Express $z = 5 + 5i$ in exponential (Euler) form.

Solution

$|z| = 5\sqrt{2}$ and $\tan\theta = \dfrac{5}{5} = 1 \Rightarrow \theta = \dfrac{\pi}{4}$, therefore $z = 5\sqrt{2}e^{i\frac{\pi}{4}}$

Example 8.29

Evaluate $(5 + 5i)^6$ and express your answer in Cartesian form.

Solution

Let $z = 5 + 5i$. From Example 8.28, $z = 5\sqrt{2}\,e^{i\frac{\pi}{4}}$, hence

$$z^6 = (5\sqrt{2}e^{i\frac{\pi}{4}})^6 = (5\sqrt{2})^6 e^{i\frac{\pi}{4}\times 6} = 125\,000e^{i\frac{3\pi}{2}} = -125\,000i$$

Alternatively,

$$(5 + 5i)^6 = \left(5\sqrt{2}\left(\cos\frac{\pi}{4} + i\sin\frac{\pi}{4}\right)\right)^6 = (5\sqrt{2})^6\left(\cos\frac{6\pi}{4} + i\sin\frac{6\pi}{4}\right)$$

$$= -125\,000i$$

Example 8.30

Simplify the expression $\dfrac{(\cos 6\theta + i\sin 6\theta)(\cos 3\theta + i\sin 3\theta)}{\cos 4\theta + i\sin 4\theta}$

Solution

$$\frac{(\cos 6\theta + i\sin 6\theta)(\cos 3\theta + i\sin 3\theta)}{\cos 4\theta + i\sin 4\theta} = \frac{e^{6i\theta} \cdot e^{3i\theta}}{e^{4i\theta}} = e^{5i\theta} = \cos 5\theta + i\sin 5\theta$$

Example 8.31

Use Euler's formula to find the cube roots of i.

Solution

$$i = e^{i\left(\frac{\pi}{2}+2k\pi\right)} \Rightarrow i^{\frac{1}{3}} = \left(e^{i\left(\frac{\pi}{2}+2k\pi\right)}\right)^{\frac{1}{3}} = e^{i\left(\frac{\pi}{6}+\frac{2k\pi}{3}\right)}; k = 0, 1, 2$$

Therefore,

$$z_0 = e^{i\left(\frac{\pi}{6}\right)} = \cos\frac{\pi}{6} + i\sin\frac{\pi}{6} = \frac{\sqrt{3}}{2} + \frac{i}{2}$$

$$z_1 = e^{i\left(\frac{\pi}{6}+\frac{2\pi}{3}\right)} = e^{i\left(\frac{5\pi}{6}\right)} = \cos\frac{5\pi}{6} + i\sin\frac{5\pi}{6} = -\frac{\sqrt{3}}{2} + \frac{i}{2}$$

$$z_2 = e^{i\left(\frac{\pi}{6}+\frac{4\pi}{3}\right)} = e^{i\left(\frac{3\pi}{2}\right)} = \cos\frac{3\pi}{2} + i\sin\frac{3\pi}{2} = -i$$

Euler's formula can also be helpful in finding nth roots of a complex number:

$$z = re^{i(\theta+2k\pi)}$$
$$\Rightarrow z_k = r^{\frac{1}{n}}e^{\left(\frac{\theta}{n} + \frac{2k\pi}{n}\right)},$$
$$k = 0, 1, \cdots, n-1$$

Exercise 8.3

1. Write each complex number in Cartesian form.
   (a) $z = 4e^{-i\frac{2\pi}{3}}$
   (b) $z = 3e^{2\pi i}$
   (c) $z = 3e^{0.5\pi i}$
   (d) $z = 4\text{cis}\left(\frac{7\pi}{12}\right)$ (exact value)
   (e) $z = 13e^{\frac{\pi i}{3}}$
   (f) $z = 3e^{1+\frac{\pi}{3}i}$

2. Write each complex number in exponential (Euler) form.
   (a) $2 + 2i$
   (b) $\sqrt{3} + i$
   (c) $\sqrt{6} - i\sqrt{2}$
   (d) $2 - 2i\sqrt{3}$
   (e) $-3 + 3i$
   (f) $4i$
   (g) $-3\sqrt{3} - 3i$
   (h) $i(3 + 3i)$
   (i) π
   (j) ei

3. Evaluate each complex number. Express in exact Cartesian form where possible.
   (a) $(1 + i)^{10}$
   (b) $(\sqrt{3} - i)^6$
   (c) $(3 + 3i\sqrt{3})^9$
   (d) $(2 - 2i)^{12}$
   (e) $(\sqrt{3} - i\sqrt{3})^8$
   (f) $(-3 + 3i)^7$
   (g) $(\sqrt{3} - i\sqrt{3})^{-8}$
   (h) $(-3\sqrt{3} - 3i)^{-7}$
   (i) $2(\sqrt{3} + i)^7$

4. Find each root and graph them in the complex plane.
   (a) The square roots of $4 + 4i\sqrt{3}$
   (b) The cube roots of $4 + 4i\sqrt{3}$
   (c) The fourth roots of -1
   (d) The sixth roots of i
   (e) The fifth roots of $-9 - 9i\sqrt{3}$

5. Solve each equation.
   (a) $z^5 - 32 = 0$
   (b) $z^8 + i = 0$
   (c) $z^3 + 4\sqrt{3} - 4i = 0$
   (d) $z^4 - 16 = 0$
   (e) $z^5 + 128 = 128i$
   (f) $z^6 - 64i = 0$

6. Use de Moivre's theorem to simplify each expression.

   (a) $(\cos(9\beta) + i\sin(9\beta))(\cos(5\beta) - i\sin(5\beta))$

   (b) $\dfrac{(\cos(6\beta) + i\sin(6\beta))(\cos(4\beta) + i\sin(4\beta))}{(\cos(3\beta) + i\sin(3\beta))}$

   (c) $(\cos(9\beta) + i\sin(9\beta))^{\frac{1}{3}}$

   (d) $\sqrt[n]{(\cos(2n\beta) + i\sin(2n\beta)}$

7. Use $e^{i\theta}$ to prove that $\cos(\alpha + \beta) = \cos\alpha\cos\beta - \sin\alpha\sin\beta$

8. Use de Moivre's theorem to show that:

   (a) $\cos 4\alpha = 8\cos^4\alpha - 8\cos^2\alpha + 1$

   (b) $\cos 5\alpha = 16\cos^5\alpha - 20\cos^3\alpha + 5\cos\alpha$

   (c) $\cos^4\alpha = \dfrac{1}{8}(\cos 4\alpha + 4\cos 2\alpha + 3)$

9. Let $z = \cos 2\alpha + i\sin 2\alpha$

   (a) Show that $z + \dfrac{1}{z} = 2\cos 2\alpha$ and that $2i\sin 2\alpha = z - \dfrac{1}{z}$

   (b) Find an expression for $\cos 2n\alpha$ and $\sin 2n\alpha$ in terms of z.

10. Let the cubic roots of 1 be 1, ω and ω^2. Simplify $(1 + 3\omega)(1 + 3\omega^2)$

11. (a) Show that the fourth roots of unity can be written as
    $$1, \beta, \beta^2, \text{ and } \beta^3$$

    (b) Simplify $(1 + \beta)(1 + \beta^2 + \beta^3)$

    (c) Show that $\beta + \beta^2 + \beta^3 = -1$

12. (a) Show that the fifth roots of unity can be written as
    $$1, \alpha, \alpha^2, \alpha^3 \text{ and } \alpha^4$$

    (b) Simplify $(1 + \alpha)(1 + \alpha^4)$

    (c) Show that $1 + \alpha + \alpha^2 + \alpha^3 + \alpha^4 = 0$

13. Show that $(1 + i\sqrt{3})^n + (1 - i\sqrt{3})^n$ is real and find its value for $n = 18$

14. Given that $z = (2a + 3i)^3$, and $a \in \mathbb{Z}^+$, find the values of a such that $\arg z = 135°$

Chapter 8 practice questions

1. Given that $z = x + yi$, find x and y such that $(1 - i)z = 1 - 3i$

2. ω is a complex solution of the equation $z^3 = 1$. If x and y are real numbers, evaluate:

   (a) $1 + \omega + \omega^2$

   (b) $(\omega x + \omega^2 y)(\omega y + \omega^2 x)$

3. **(a)** Expand and evaluate $(1 + i)^2$

   (b) Use mathematical induction to prove that $(1 + i)^{4n} = (-4)^n$, where $n \in \mathbb{Z}^+$

   (c) Hence or otherwise, find $(1 + i)^{32}$.

4. Consider the complex numbers $z_1 = \dfrac{\sqrt{6} - i\sqrt{2}}{2}$ and $z_2 = 1 - i$

   (a) Express z_1 and z_2 in polar form with modulus r and argument θ, where $r > 0$ and $-\dfrac{\pi}{2} \leqslant \theta \leqslant \dfrac{\pi}{2}$

   (b) Show that $\dfrac{z_1}{z_2} = \cos\dfrac{\pi}{12} + i\sin\dfrac{\pi}{12}$

   (c) Express $\dfrac{z_1}{z_2}$ in the form $a + bi$, where a and b are to be determined exactly in radical (surd) form. Hence, or otherwise, find the exact values of $\cos\dfrac{\pi}{12}$ and $\sin\dfrac{\pi}{12}$

5. Let $z_1 = a\left(\cos\dfrac{\pi}{3} + i\sin\dfrac{\pi}{3}\right)$ and $z_2 = b\left(\cos\dfrac{\pi}{4} + i\sin\dfrac{\pi}{4}\right)$

   Express $\left(\dfrac{z_1}{z_2}\right)^4$ in the form $z = x + yi$

6. Find the value of $|z|$ if z is a complex number such that $|z + 16| = 4\,|z + 1|$

7. Find the values of the real numbers a and b such that $(a + bi)(2 - i) = 5 - i$

8. Given that $z = (x + i)^2$, where $x \in \mathbb{R}^+$. If $\arg(z) = \dfrac{\pi}{3}$, find the exact value of x.

9. Write the complex number z in the form $z = a + bi$, with a and b real numbers if z satisfies

   $i(z + 2) = 1 - 2z$, where $i = \sqrt{-1}$

10. **(a)** Given that $z \in \mathbb{C}$, write $z^5 - 1$ as a product of a linear factor and a 4th degree factor.

    (b) Find the roots of $z^5 - 1 = 0$, giving your answers in modulus argument form $r\,\text{cis}\,\theta$ with

    $r > 0$ and $-\pi < \theta \leqslant \pi$

    (c) Factorise $z^4 + z^3 + z^2 + z + 1$ into two real quadratic factors.

11. **(a)** Write $8i$ in polar form.

    (b) Find the cube root of $8i$ that lies in the first quadrant and express your answer in:

    (i) polar form

    (ii) $a + bi$ form.

12. Consider the complex number $z = \dfrac{\left(\cos\dfrac{\pi}{3} + i\sin\dfrac{\pi}{3}\right)^3 \left(\cos\dfrac{\pi}{4} + i\sin\dfrac{\pi}{4}\right)^8}{\left(\cos\dfrac{\pi}{24} - i\sin\dfrac{\pi}{24}\right)^8}$

   (a) Find the modulus of z and argument of z.
       Express your answer in radians.

   (b) Show that z is a cube root of 1.

   (c) Simplify $(1 + 2z)(2 + z^2)$, and express your answer in the form
       $a + bi$, where a and b are exact real numbers.

13. Write the complex number z in $x + iy$ form, where $x, y \in \mathbb{Z}$ if z satisfies
    the equation

    $$\sqrt{z} = \frac{2}{1 - i} + 1 - 4i$$

14. (a) Consider the complex number $z = \cos\theta + i\sin\theta$.
        Prove, using mathematical induction, that for a positive integer n,
        $z^n = \cos n\theta + i\sin n\theta$.

    (b) Hence, show that $z^{-1} = \cos(-\theta) + i\sin(-\theta)$ and $z^n + z^{-n} = 2\cos n\theta$.

    (c) (i) Use the binomial theorem to expand $(z + z^{-1})^5$

        (ii) Hence, show that $\cos^5\theta = \dfrac{1}{16}(a\cos 5\theta + b\cos 3\theta + c\cos\theta)$,

        and find the values of a, b, and c.

15. Find the real numbers p and q that satisfy the equation

    $$2(p + iq) = q - ip - 2(1 - i)$$

16. $z^5 - 32 = 0$

    (a) Show that $z_1 = 2\left(\cos\dfrac{2\pi}{5} + i\sin\dfrac{2\pi}{5}\right)$ is one of the complex roots of
        this equation.

    (b) Find z_1^2, z_1^3, z_1^4, and z_1^5 giving your answer in modulus argument form.

    (c) Plot the points that represent z_1, z_1^2, z_1^3, z_1^4, and z_1^5 in the complex plane.

    (d) The point z_1^n is mapped to z_1^{n+1} by a composition of two linear
        transformations, where $n = 1, 2, 3, 4$. Give a full geometric
        description of the two transformations.

17. A complex number z is such that $|z| = |z - 3i|$

    (a) Show that the imaginary part of z is $\dfrac{3}{2}$

    (b) Let z_1 and z_2 be the two possible values of z, such that $|z| = 3$

        (i) Sketch a diagram to show the points that represent z_1 and z_2
             in the complex plane, where z_1 is in the first quadrant.

        (ii) Show that $\arg(z_1) = \dfrac{\pi}{6}$

        (iii) Find $\arg(z_2)$

    (c) Given that $\arg\left(\dfrac{z_1^k z_2}{2i}\right) = \pi$, find a value for k.

18. Find the value of integers a and b, such that $(a + i)(2 - bi) = 7 - i$

19. Consider the complex number $z = \cos\theta + i\sin\theta$

(a) Using de Moivre's theorem, show that $z^n + \dfrac{1}{z^n} = 2\cos n\theta$

(b) By expanding $\left(z + \dfrac{1}{z}\right)^4$ show that

$$\cos^4\theta = \frac{1}{8}(\cos 4\theta + 4\cos 2\theta + 3)$$

20. Consider the complex geometric series $e^{i\theta} + e^{2i\theta} + e^{3i\theta} + \ldots$

(a) Find an expression for z, the common ratio of this series.

(b) Show that $|z| < 1$

(c) Write down an expression for the sum to infinity of this series.

(d) (i) Express your answer to part (c) in terms of $\sin\theta$ and $\cos\theta$.

(ii) Hence show that

$$\cos\theta + \frac{1}{2}\cos 2\theta + \frac{1}{4}\cos 3\theta + \ldots = \frac{4\cos\theta - 2}{5 - 4\cos\theta}$$

21. a, b, and $c \in \mathbb{R}$.

The roots of equation $z^3 + az^2 + bz + c = 0$ are -2 and $(-3 + 2i)$. Find the value of a, b and c.

22. Find the complex number z with $|z| = 2\sqrt{5}$, and $\dfrac{25}{z} - \dfrac{15}{z^\star} = 1 - 8i$. Express your answer in $x + iy$ form.

23. Solve this simultaneous system of equations giving your answers in $x + iy$ form:

$$z_1 + 4z_2 = 29$$
$$iz_1 + z_2 = 3$$

24. (a) Solve the equation $x^2 - 4x + 8 = 0$. Denote its two roots by z_1 and z_2 and express them in exponential form with z_1 in the first quadrant.

(b) Find the value of $\dfrac{z_1^4}{z_2^2}$ and write it in the form $x + yi$

(c) Show that $z_1^4 = z_2^4$

(d) Find the value of $\dfrac{z_1}{z_2} + \dfrac{z_2}{z_1}$

(e) For what values of n is z_1 real?

25. (a) Show that $z = \cos\dfrac{2\pi}{7} + i\sin\dfrac{2\pi}{7}$ is a root of the equation $x^7 - 1 = 0$

(b) Show that $z^7 - 1 = (z - 1)(z^6 + z^5 + z^4 + z^3 + z^2 + z + 1)$ and deduce that $z^6 + z^5 + z^4 + z^3 + z^2 + z + 1 = 0$

(c) Show that $\cos\dfrac{2\pi}{7} + \cos\dfrac{4\pi}{7} + \cos\dfrac{6\pi}{7} = -\dfrac{1}{2}$

26. (a) Find the three roots of the equation $27z^3 + 8 = 0, z \in \mathbb{C}$.
    Give your answers in modulus-argument form.
    (b) The roots are represented by the vertices of a triangle in an Argand diagram. Find the area of this triangle and express it in the form $\dfrac{\sqrt{a}}{b}$

27. (a) Show that the complex number $2 + 3i$ is a root of the equation
    $$z^4 - 4z^3 + 17z^2 - 16z + 52 = 0$$
    (b) Hence, find all 4 roots of the equation.

28 (a) Let $z = a + bi$ and use this form to find all values of a and b such that $z^2 = i$
    (b) Hence, find the two solutions of $z^2 - (3 + i)z + (2 + i) = 0$

29. Find formulae for $\cos 4\theta$, $\sin 4\theta$, $\cos 5\theta$, and $\sin 6\theta$ in terms of $\cos \theta$ and $\sin \theta$, by using de Moivre's formula.

30. Compute the real and imaginary parts of $e^{\ln 2(1+i)}$

Vectors, lines, and planes

9

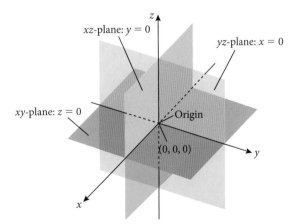

Figure 9.1 The coordinate planes divide space into 8 octants

Because we live in a three-dimensional world, it is essential that we study objects in three dimensions. In this section, we consider a three-dimensional coordinate system in which points are determined by ordered triplets.

We construct the coordinate system in the following manner: Choose three mutually perpendicular axes, as shown in Figure 9.1, to serve as our reference. The orientation of the system is right-handed in the sense that if you hold your right hand so that the fingers curl from the positive x-axis towards the positive y-axis, your thumb points along the z-axis (Figure 9.2).

Looking at it from a different perspective, if you are looking straight at the system, the yz-plane is the plane facing you, the xz-plane is perpendicular to it and extending out of the page towards you, and the xy-plane is the bottom part of that picture. The xy-, xz- and yz-planes are called the **coordinate planes**. Points in space are assigned coordinates in the same manner as in the plane. So, the point P is assigned the ordered triplet (x, y, z) to indicate that it is x, y, and z units from the yz-, xz-, and xy-planes, respectively (Figure 9.3).

Figure 9.2 The thumb points along the z-axis

In this chapter we will focus our study of vectors to 3-dimensional (3D) space. The good news is that many of the rules you will learn about vectors in 3D also apply to the vectors in 2D.

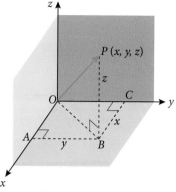

Figure 9.3 Point P is assigned the ordered triplet (x, y, z)

9.1 Vectors from a geometric viewpoint

Vectors can be represented geometrically by arrows in two- or three-dimensional space; the direction of the arrow specifies the direction of the vector, and the length of the arrow describes its magnitude. The first point in the arrow is called the **initial point** of the vector and the tip is called the **terminal point**. We shall denote vectors in lower case bold italic type (e.g. v) when using one letter to name the vector, and we will use $\overrightarrow{AB}$ to denote the vector from A to B. The handwritten notation will also be the latter.

If the initial point of a vector is at the origin, then the vector is said to be in standard position. It is also called the **position vector** of point P. The terminal point will have coordinates of the form (x, y, z). We call these coordinates the **components** of v, and we write: $v = \begin{pmatrix} x \\ y \\ z \end{pmatrix}$.

Vectors are also written as $v = (x, y, z)$

The length (magnitude) of a vector v is also known as its **modulus** or its **norm**, and it is written as $|v|$.

Looking at Figure 9.3 and using Pythagoras' theorem, we can show that the magnitude of a vector v, $|v| = \sqrt{x^2 + y^2 + z^2}$

Let $\overrightarrow{OP} = v$, then

$|v| = |\overrightarrow{OP}| = \sqrt{OB^2 + BP^2}$, since the triangle OBP is right-angled at B. Now, consider triangle OAB, which is right-angled at A:

$OB^2 = OA^2 + AB^2 = x^2 + y^2$, and therefore:

$$|v| = \sqrt{OB^2 + BP^2} = \sqrt{(x^2 + y^2) + z^2} = \sqrt{x^2 + y^2 + z^2}$$

Two vectors such as v and $\overrightarrow{AB}$ are equal (equivalent) if they have the same length (magnitude) and the same direction; we write $v = \overrightarrow{AB}$. Geometrically, two vectors are equal if they are translations of one another (Figure 9.4). In Figure 9.5 the four vectors are equal, even though they are in different positions.

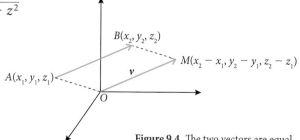

Figure 9.4 The two vectors are equal

381

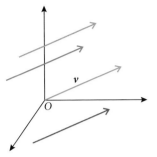

Figure 9.5 Free vectors

Because vectors are not affected by translation, the initial point of a vector v can be moved to any convenient position by making an appropriate translation.

Two vectors are said to be opposite if they have equal modulus but opposite direction (Figure 9.6).

If the initial and terminal points of a vector coincide, then the vector has length zero; we call this the **zero vector** and denote it by $\mathbf{0}$.

The zero vector does not have a specific direction, so we will agree that it can be assigned any convenient direction in a specific problem.

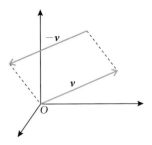

Figure 9.6 Opposite vectors

Addition and subtraction of vectors

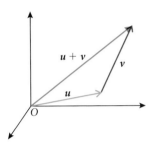

Figure 9.7 Triangular rule

According to the **triangular rule**, if u and v are vectors, then the sum $u + v$ is the vector from the initial point of u to the terminal point of v when the vectors are positioned so that the initial point of v is the terminal point of u, as shown in Figure 9.7.

Equivalently, $u + v$ is also the diagonal of the parallelogram whose sides are u and v, as shown in Figure 9.8.

The difference of the two vectors u and v can be dealt with in the same manner. So, the vector $w = u - v$ is a vector such that $u = v + w$.

In Figure 9.9, we see that the difference is along the diagonal joining the two terminal points of the vectors and in the direction from v to u.

When we discuss vectors, we will refer to real numbers as scalars.

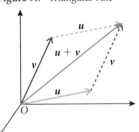

Figure 9.8 Vector sum as diagonal of a parallelogram

If k is a real positive number, ku is a vector of magnitude $k|u|$ and in the same direction as u. It follows that when k is negative, ku has magnitude $|k| \times |u|$ and is in the opposite direction to u. (Figure 9.10). This is also the condition for two vectors to be parallel.

 Two vectors are parallel if one vector is a scalar multiple of the other.

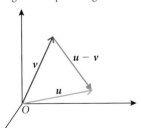

Figure 9.9 Difference of vectors

For example, the vector $\begin{pmatrix} -3 \\ 4 \\ -2 \end{pmatrix}$ is parallel to the vector $\begin{pmatrix} 4.5 \\ -6 \\ 3 \end{pmatrix}$ since

$$\begin{pmatrix} -3 \\ 4 \\ -2 \end{pmatrix} = -\frac{2}{3}\begin{pmatrix} 4.5 \\ -6 \\ 3 \end{pmatrix}$$

Components provide a simple way to perform several algebraic operations on vectors. First, by definition, we know that two vectors are equal if they have the same length and the same magnitude. So, if we choose to draw the two equal

vectors $u = \begin{pmatrix} u_1 \\ u_2 \\ u_3 \end{pmatrix}$ and $v = \begin{pmatrix} v_1 \\ v_2 \\ v_3 \end{pmatrix}$ from the origin, then their terminal points

must coincide, and hence $u_1 = v_1$, $u_2 = v_2$, and $u_3 = v_3$. So, we showed that equal vectors have the same components. The converse is also true; that is,

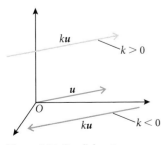

Figure 9.10 Parallel vectors

if $u_1 = v_1$, $u_2 = v_2$, and $u_3 = v_3$, then the two vectors are equal. The following results are also obvious from the simple geometry of similar figures:

If $u = \begin{pmatrix} u_1 \\ u_2 \\ u_3 \end{pmatrix}$ and $v = \begin{pmatrix} v_1 \\ v_2 \\ v_3 \end{pmatrix}$ and k is any real number, then $u + v = \begin{pmatrix} u_1 + v_1 \\ u_2 + v_2 \\ u_3 + v_3 \end{pmatrix}$

and $ku = \begin{pmatrix} ku_1 \\ ku_2 \\ ku_3 \end{pmatrix}$

If the initial point of the vector is not at the origin, then we can generalise the previous notation to any position.

If $\overrightarrow{AB}$ is a vector with initial point $A(x_1, y_1, z_1)$ and terminal point $B(x_2, y_2, z_2)$, then

$$\overrightarrow{AB} = \overrightarrow{OB} - \overrightarrow{OA} = \begin{pmatrix} x_2 - x_1 \\ y_2 - y_1 \\ z_2 - z_1 \end{pmatrix}$$

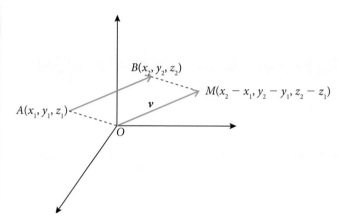

As illustrated in the diagram above, either by applying the distance formula or by using the equality of vectors v and $\overrightarrow{AB}$, $|\overrightarrow{AB}| = \sqrt{(x_2 - x_1)^2 + (y_2 - y_1)^2 + (z_2 - z_1)^2}$

Additionally, the following results can easily follow from properties of real numbers:

$u + v = v + u$

$(u + v) + w = u + (v + w)$

$k(u + v) = ku + kv.$

Example 9.1

Given the points $A(-2, 3, 5)$ and $B(1, 0, -4)$:

(a) find the components of vector $\overrightarrow{AB}$

(b) find the components of vector $\overrightarrow{BA}$

(c) find the components of vector $3\overrightarrow{AB}$

(d) find the components of vector $\overrightarrow{OA} + \overrightarrow{OB}$

(e) calculate $|\overrightarrow{AB}|$ and $|\overrightarrow{BA}|$

(f) calculate $|3\overrightarrow{AB}|$ and $|\overrightarrow{OA} + \overrightarrow{OB}|$

Solution

(a) $\overrightarrow{AB} = \overrightarrow{OB} - \overrightarrow{OA} = \begin{pmatrix} x_2 - x_1 \\ y_2 - y_1 \\ z_2 - z_1 \end{pmatrix} = \begin{pmatrix} 1 - (-2) \\ 0 - 3 \\ -4 - 5 \end{pmatrix} = \begin{pmatrix} 3 \\ -3 \\ -9 \end{pmatrix}$

(b) Since $\overrightarrow{BA}$ is the opposite of $\overrightarrow{AB}$, $\overrightarrow{BA} = \begin{pmatrix} -3 \\ 3 \\ 9 \end{pmatrix}$

(c) $3\overrightarrow{AB} = 3\begin{pmatrix} 3 \\ -3 \\ -9 \end{pmatrix} = \begin{pmatrix} 9 \\ -9 \\ -27 \end{pmatrix}$

(d) $\overrightarrow{OA} + \overrightarrow{OB} = \begin{pmatrix} -2 + 1 \\ 3 + 0 \\ 5 - 4 \end{pmatrix} = \begin{pmatrix} -1 \\ 3 \\ 1 \end{pmatrix}$

(e) $\left|\overrightarrow{AB}\right| = \sqrt{(x_2 - x_1)^2 + (y_2 - y_1)^2 + (z_2 - z_1)^2} = \sqrt{9 + 9 + 81} = 3\sqrt{11}$

$\left|\overrightarrow{BA}\right| = \sqrt{(x_2 - x_1)^2 + (y_2 - y_1)^2 + (z_2 - z_1)^2} = \sqrt{9 + 9 + 81} = 3\sqrt{11}$

(f) $\left|3\overrightarrow{AB}\right| = \sqrt{(x_2 - x_1)^2 + (y_2 - y_1)^2 + (z_2 - z_1)^2} = \sqrt{81 + 81 + 729} = 9\sqrt{11}$

$\left|\overrightarrow{OA} + \overrightarrow{OB}\right| = \left|\begin{pmatrix} -1 \\ 3 \\ 1 \end{pmatrix}\right| = \sqrt{1 + 9 + 1} = \sqrt{11}$

Note that $\left|3\overrightarrow{AB}\right| = 3\left|\overrightarrow{AB}\right|$

In general, $|\lambda v| = |\lambda|\,|v|$ – that is, the magnitude of a multiple of a vector is equal to the absolute multiple of the magnitude of the vector. For example, $|-3v| = 3|v|$.

Note also that $\left|\overrightarrow{OA} + \overrightarrow{OB}\right| = \sqrt{11} \neq \left|\overrightarrow{OA}\right| + \left|\overrightarrow{OB}\right|$

$$= \sqrt{4 + 9 + 25} + \sqrt{1 + 0 + 16} = \sqrt{38} + \sqrt{17}$$

Example 9.2

Find the coordinates of point M in terms of a constant so that the points M, $A(0, -1, 5)$, and $B(1, 2, 3)$ are collinear.

Solution

For the points to be collinear, it is enough to make $\overrightarrow{AM}$ parallel to $\overrightarrow{AB}$. If the two vectors are parallel, then one of them is a scalar multiple of the other.

Say $\overrightarrow{AM} = t\overrightarrow{AB}$

$$\overrightarrow{AM} = \begin{pmatrix} x \\ y + 1 \\ z - 5 \end{pmatrix} = t\begin{pmatrix} 1 \\ 3 \\ -2 \end{pmatrix} = \begin{pmatrix} t \\ 3t \\ -2t \end{pmatrix}$$

So, $x = t$, $y = 3t - 1$, and $z = 5 - 2t$

Unit vectors

A vector of length 1 is called a **unit vector**. So, in 2D space: the vectors $i = \begin{pmatrix} 1 \\ 0 \end{pmatrix}$ and $j = \begin{pmatrix} 0 \\ 1 \end{pmatrix}$ are unit vectors along the x- and y-axes, and in 3D space: the unit vectors along the axes are $i = \begin{pmatrix} 1 \\ 0 \\ 0 \end{pmatrix}, j = \begin{pmatrix} 0 \\ 1 \\ 0 \end{pmatrix}$ and $k = \begin{pmatrix} 0 \\ 0 \\ 1 \end{pmatrix}$.

The vectors i, j, and k are called the **base vectors** of the 3-space.

 The terms '2-space' and '3-space' are short forms for two-dimensional space and three-dimensional space respectively, also known as 2D and 3D. We will use 2D and 3D from here onwards.

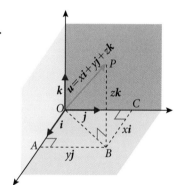

Figure 9.11 Base vectors in 3D

It follows that each vector in 3D space can be expressed uniquely in terms of i, j, and k as follows:

$$u = \begin{pmatrix} x \\ y \\ z \end{pmatrix} = \begin{pmatrix} x \\ 0 \\ 0 \end{pmatrix} + \begin{pmatrix} 0 \\ y \\ 0 \end{pmatrix} + \begin{pmatrix} 0 \\ 0 \\ z \end{pmatrix} = x\begin{pmatrix} 1 \\ 0 \\ 0 \end{pmatrix} + y\begin{pmatrix} 0 \\ 1 \\ 0 \end{pmatrix} + z\begin{pmatrix} 0 \\ 0 \\ 1 \end{pmatrix} = xi + yj + zk$$

So, in Example 9.1, $\overrightarrow{AB} = \begin{pmatrix} 3 \\ -3 \\ -9 \end{pmatrix} = 3i - 3j - 9k$

Unit vectors can be found in any direction, not only in the direction of the axes. For example, if we want to find the unit vector in the same direction as u, then we need to find a vector parallel to u, which has a magnitude of 1.

Since u has a magnitude of $|u|$, multiply this vector by its reciprocal, $\dfrac{1}{|u|}$, to give it a magnitude of 1. So, the unit vector v in the same direction as u is

$v = \dfrac{1}{|u|}u = \dfrac{u}{|u|}$. This is a unit vector since its length is 1.

Figure 9.12 u as a sum of multiples of the base vectors

Example 9.3

Find a unit vector in the direction of $v = i - 2j + 3k$

Solution

The length of the vector v is $\sqrt{1^2 + (-2)^2 + 3^2} = \sqrt{14}$, so the unit vector is

$\dfrac{1}{\sqrt{14}}(i - 2j + 3k) = \dfrac{i}{\sqrt{14}} - \dfrac{2j}{\sqrt{14}} + \dfrac{3k}{\sqrt{14}}$. To verify that this is a unit vector, we find its length:

$$\sqrt{\left(\dfrac{1}{\sqrt{14}}\right)^2 + \left(\dfrac{-2}{\sqrt{14}}\right)^2 + \left(\dfrac{3}{\sqrt{14}}\right)^2} = \sqrt{\dfrac{1}{14} + \dfrac{4}{14} + \dfrac{9}{14}} = 1$$

The unit vector plays another important role. It determines the direction of the given vector.

In 2D space, we can write the vector in a form that gives us its direction (in terms of the angle it makes with the horizontal axis, called the direction angle), and its magnitude.

In Figure 9.13, θ is the angle with the horizontal axis.

The unit vector v, in the same direction as u is:

$$v = 1\cos\theta\,i + 1\sin\theta\,j,$$

and from the results above,

$$v = \frac{1}{|u|}u \Rightarrow u = |u| \times v = |u|\cos\theta\,i + |u|\sin\theta\,j$$

$$= |u|(\cos\theta\,i + \sin\theta\,j)$$

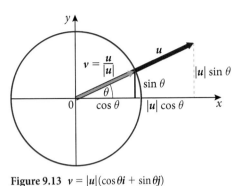

Figure 9.13 $v = |u|(\cos\theta\,i + \sin\theta\,j)$

Example 9.4

Find the vector with magnitude 2 that makes an angle of 60° with the positive x-axis.

Solution

$$v = |v|(\cos60°\,i + \sin60°\,j) = 2\left(\frac{1}{2}i + \frac{\sqrt{3}}{2}j\right) = i + \sqrt{3}\,j$$

Example 9.5

Find the magnitude and direction of the vector $v = 2\sqrt{3}\,i - 2j$

Solution

$$|v| = \sqrt{(2\sqrt{3})^2 + 4} = 4$$

$$\cos\theta = \frac{2\sqrt{3}}{4} = \frac{\sqrt{3}}{2},\ \sin\theta = \frac{-2}{4} = -\frac{1}{2} \Rightarrow \theta = -\frac{\pi}{6}$$

Example 9.6

(a) Find the unit vector that has the same direction as $v = i + 2j - 2k$

(b) Find a vector of length 6 that is parallel to $v = i - 2j + 3k$

Solution

(a) The vector v has magnitude $|v| = \sqrt{1 + 2^2 + (-2)^2} = 3$

so, the unit vector u in the same direction as v is

$$u = \frac{1}{3}v = \frac{1}{3}i + \frac{2}{3}j - \frac{2}{3}k$$

(b) Let w be the vector in question and u be the unit vector in the direction of v.

$$w = 6u = 6 \cdot \frac{1}{\sqrt{14}}(i - 2j + 3k) = \frac{6}{\sqrt{14}}i - \frac{12}{\sqrt{14}}j + \frac{18}{\sqrt{14}}k$$

Example 9.7

Example 9.7 introduces you to the vector equation of a line which you will learn more about in section 9.4.

r_1 and r_2 are the position vectors of two points A and B in space, and λ is a real number.

(a) Show that $r = (1 - \lambda)r_1 + \lambda r_2$ is the position vector of a point C on the straight line joining A and B.

(b) Consider the cases where $\lambda = 0, 1, \dfrac{1}{2}, -1, 2$, and $\dfrac{2}{3}$.

Solution

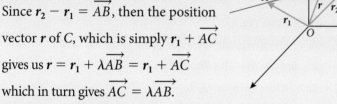

(a) Rewrite the equation

$$r = (1 - \lambda)r_1 + \lambda r_2 = r_1 + \lambda(r_2 - r_1).$$

Since $r_2 - r_1 = \overrightarrow{AB}$, then the position vector r of C, which is simply $r_1 + \overrightarrow{AC}$ gives us $r = r_1 + \lambda\overrightarrow{AB} = r_1 + \overrightarrow{AC}$ which in turn gives $\overrightarrow{AC} = \lambda\overrightarrow{AB}$.

This means that $\overrightarrow{AC}$ is parallel to $\overrightarrow{AB}$ and is a multiple of it.

(b) When $\lambda = 0$, then $r = r_1$ and C is at A.

When $\lambda = 1$, then $r = r_1 + \overrightarrow{AB}$ and C is at B.

When $\lambda = \dfrac{1}{2}$, then $r = r_1 + \dfrac{1}{2}\overrightarrow{AB}$ and C is the midpoint of AB.

When $\lambda = -1$, then $r = r_1 + -\overrightarrow{AB}$ and A is the midpoint of CB.

When $\lambda = 2$, then $r = r_1 + 2\overrightarrow{AB}$ and B is the midpoint of AC.

When $\lambda = \dfrac{2}{3}$, then $r = r_1 + \dfrac{2}{3}\overrightarrow{AB}$ and C is $\dfrac{2}{3}$ the way between A and B.

Exercise 9.1

1. Write each vector $\overrightarrow{AB}$ in component form.

   (a) $A\left(-\dfrac{3}{2}, -\dfrac{1}{2}, 1\right); B\left(1, -\dfrac{5}{2}, 1\right)$ (b) $A\left(-2, -\sqrt{3}, -\dfrac{1}{2}\right); B\left(1, \sqrt{3}, -\dfrac{1}{2}\right)$

   (c) $A(2, -3, 5); B(1, -1, 3)$ (d) $A(a, -a, 2a); B(-a, -2a, a)$

2. Given the coordinates of point P or Q and the components of $\overrightarrow{PQ}$, find the missing items.

   (a) $P\left(-\dfrac{3}{2}, -\dfrac{1}{2}, 1\right); \overrightarrow{PQ} = \begin{pmatrix} 1 \\ -\dfrac{5}{2} \\ 1 \end{pmatrix}$ (b) $Q\left(1, -\dfrac{5}{2}, 1\right); \overrightarrow{PQ} = \begin{pmatrix} -\dfrac{3}{2} \\ -\dfrac{1}{2} \\ 1 \end{pmatrix}$

   (c) $P(a, -2a, 2a); \overrightarrow{PQ} = \begin{pmatrix} -a \\ -2a \\ a \end{pmatrix}$

387

3. Find the coordinates of point M in terms of a constant so that the points M, A and B are collinear:

   (a) $A(0, 0, 5)$, $B(1, 1, 0)$ (b) $A(-1, 0, 1)$, $B(3, 5, -2)$

   (c) $A(2, 3, 4)$, $B(-2, -3, 5)$

4. Given the coordinates of the points A and B, find the coordinates of the point C such that B is the midpoint of line segment AC:

   (a) $A(3, -4, 0)$; $B(-1, 0, 1)$ (b) $A(-1, 3, 5)$; $B\left(-1, \frac{1}{2}, \frac{1}{3}\right)$

   (c) $A(1, 2, -1)$; $B(a, 2a, b)$

5. Given a triangle ABC and a point G such that
   $$\overrightarrow{GA} + \overrightarrow{GB} + \overrightarrow{GC} = 0$$
   find the coordinates of G for:

   (a) $A(-1, -1, -1)$; $B(-1, 2, -1)$; $C(1, 2, 3)$

   (b) $A(2, -3, 1)$; $B(1, -2, -5)$; $C(0, 0, 1)$

   (c) $A(a, 2a, 3a)$; $B(b, 2b, 3b)$; $C(c, 2c, 3c)$

6. Determine the fourth vertex D of the parallelogram $ABCD$ that has AB and BC as adjacent sides.

   (a) $A(\sqrt{3}, 2, -1)$; $B(1, 3, 0)$; $C(-\sqrt{3}, 2, -5)$

   (b) $A(\sqrt{2}\sqrt{3}, \sqrt{5})$; $B(3\sqrt{2}, -\sqrt{3}, \sqrt{5})$; $C(-2\sqrt{2}, \sqrt{3}, -3\sqrt{5})$

   (c) $A\left(-\frac{1}{2}, \frac{1}{3}, 0\right)$; $B\left(\frac{1}{2}, \frac{2}{3}, 5\right)$; $C\left(\frac{7}{2}, \frac{1}{3}, 1\right)$

7. Determine the values of m and n such that the vectors $v = \begin{pmatrix} m - 2 \\ m + n \\ -2m + n \end{pmatrix}$
   and $w = \begin{pmatrix} 2 \\ 4 \\ -6 \end{pmatrix}$ have the same direction.

8. Find a unit vector in the same direction as each vector.

   (a) $v = 2i + 2j - k$ (b) $v = 6i - 4j + 2k$ (c) $v = 2i - j - 2k$

9. Find a vector with the given magnitude and in the same direction as the given vector.

   (a) Magnitude 2, $v = 2i + 2j - k$

   (b) Magnitude 4, $v = 6i - 4j + 2k$

   (c) Magnitude 5, $v = 2i - j - 2k$

10. Let $u = i + 3j - 2k$ and $v = 2i + j$. Find:

    (a) $|u + v|$ (b) $|u| + |v|$ (c) $|-3u| + |3v|$

    (d) $\frac{1}{|u|}u$ (e) $\left|\frac{1}{|u|}u\right|$

11. Find the terminal points for each vector.

    (a) $w = 4i + 2j - 2k$, given the initial point $(-1, 2, -3)$

    (b) $v = 2i - 3j + k$, given the initial point $(-2, 1, 4)$

12. Find vectors that satisfy the stated conditions:

    (a) opposite direction of $u = \begin{pmatrix} -3 \\ 4 \end{pmatrix}$ and one-third the magnitude of u

    (b) magnitude 12 and same direction as $w = 4i + 2j - 2k$

    (c) of the form and $xi + yj - 2k$ and parallel to $w = i - 4j + 3k$.

13. Let u, v, and w be the vectors from each vertex of a triangle to the midpoint of the opposite side. Find $u + v + w$.

14. Find the scalar t (or show that there is none) so that the vector $v = ti - 2tj + 3tk$ is a unit vector.

15. Find the scalar t (or show that there is none) so that the vector $v = 2i - 2tj + 3tk$ is a unit vector.

16. Find the scalar t (or show that there is none) so that the vector $v = 0.5i - tj + 1.5tk$ is a unit vector.

17. The diagram shows a cube with a side length 8 units. O is the origin.

    (a) Find the position vectors of all the vertices.

    (b) L, M, and N are the midpoints of the respective edges. Find the position vectors of L, M, and N.

    (c) Show that $\overrightarrow{LM} + \overrightarrow{MN} + \overrightarrow{NL} = 0$

18. A triangular prism has side lengths $OA = 8$, $OB = 10$, and $OE = 12$. O is the origin.

    (a) Find the position vectors of C and D.

    (b) F and G are the midpoints of the edges AB and CD, respectively. Find their position vectors.

    (c) Find the vectors $\overrightarrow{AG}$ and $\overrightarrow{FD}$ and explain your results.

19. Find α such that $|\alpha i + (\alpha - 1)j + (\alpha + 1)k| = 2$

20. Let $a = \begin{pmatrix} 4 \\ -1 \\ 1 \end{pmatrix}$, $b = \begin{pmatrix} 1 \\ 1 \\ 1 \end{pmatrix}$, $c = \begin{pmatrix} -1 \\ 3 \\ 2 \end{pmatrix}$, $d = \begin{pmatrix} -3 \\ 0 \\ 1 \end{pmatrix}$

    Find the scalars α, β, and μ (or show that they cannot exist) such that $a = \alpha b + \beta c + \mu d$.

21. Repeat question 20 for $a = \begin{pmatrix} -1 \\ 1 \\ 5 \end{pmatrix}$, $b = \begin{pmatrix} 1 \\ 0 \\ 1 \end{pmatrix}$, $c = \begin{pmatrix} 3 \\ 2 \\ 0 \end{pmatrix}$, $d = \begin{pmatrix} 0 \\ 1 \\ 1 \end{pmatrix}$

22. Repeat question 20 for $a = \begin{pmatrix} 2 \\ 1 \\ -1 \end{pmatrix}$, $b = \begin{pmatrix} 1 \\ -1 \\ 0 \end{pmatrix}$, $c = \begin{pmatrix} 3 \\ 0 \\ 1 \end{pmatrix}$, $d = \begin{pmatrix} 4 \\ -1 \\ 1 \end{pmatrix}$

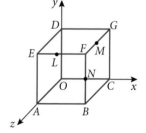

Figure 9.14 Diagram for question 17

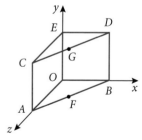

Figure 9.15 Diagram for question 18

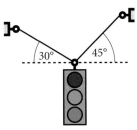

Figure 9.16 Diagram for question 24

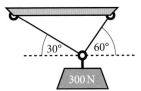

Figure 9.17 Diagram for question 25

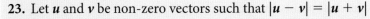

23. Let u and v be non-zero vectors such that $|u - v| = |u + v|$

    (a) What can you conclude about the parallelogram with u and v as adjacent sides?

    (b) Show that if $u = \begin{pmatrix} u_1 \\ u_2 \\ u_3 \end{pmatrix}$ and $v = \begin{pmatrix} v_1 \\ v_2 \\ v_3 \end{pmatrix}$, then $u_1 v_1 + u_2 v_2 + u_3 v_3 = 0$

24. A 125 N traffic light is hanging from two flexible cables. The magnitude of the force that each cable applies to the 'eye ring' holding the lights is called the cable tension. Find the cable tensions if the light is in equilibrium.

25. Find the tension in the cables used to hold a weight of 300 N as shown in the diagram.

9.2 Scalar (dot) product

Let $u = (u_1, u_2, u_3)$ and $v = (v_1, v_2, v_3)$. The dot product (scalar) is written as $u \times v$ and is defined as
$$u \cdot v = u_1 v_1 + u_2 v_2 + u_3 v_3$$
and:
$$u \cdot v = |u| \cdot |v| \cos\theta, \text{ where } \theta \text{ is the angle between the two vectors.}$$

Proof

Work out:
$$u^2 = u \times u = u_1 u_1 + u_2 u_2 + u_3 u_3 = u_1^2 + u_2^2 + u_3^2 = |u|^2$$

Then
$$|u - v|^2 = (u - v)(u - v) = u^2 + v^2 - 2u \cdot v = |u|^2 + |v|^2 - 2u \cdot v$$

Also, using the law of cosines,
$$|u - v|^2 = |u|^2 + |v|^2 - 2|u| \cdot |v| \cos\theta$$

By comparing these two results, we can conclude that $u \cdot v = |u| \cdot |v| \cos\theta$.

Conversely, using the law of cosines in Figure 9.18 gives
$$|u - v|^2 = |u|^2 + |v|^2 - 2|u| \cdot |v| \cos\theta$$

which in turn will give
$$\begin{aligned} 2|u| \cdot |v| \cos\theta &= |u|^2 + |v|^2 - |u - v|^2 \\ &= (u_1^2 + u_2^2 + u_3^2) + (v_1^2 + v_2^2 + v_3^2) \\ &\quad - ((u_1 - v_1)^2 + (u_2 - v_2)^2 + (u_3 - v_3)^2) \\ &= 2(u_1 v_1 + u_2 v_2 + u_3 v_3) \end{aligned}$$

Thus $|u| \cdot |v| \cos\theta = u_1 v_1 + u_2 v_2 + u_3 v_3$ and $u \cdot v = |u||v|\cos\theta$

The scalar product can also be used to find angles between vectors:
$$u \cdot v = |u||v|\cos\theta \Leftrightarrow \cos\theta = \frac{u \cdot v}{|u||v|}$$

Figure 9.18 Using the law of cosines

$\cos\theta = \dfrac{u \cdot v}{|u||v|} = \dfrac{u}{|u|} \cdot \dfrac{v}{|v|}$, where $\dfrac{u}{|u|}$ and $\dfrac{v}{|v|}$ are unit vectors in the direction of u and v respectively. That is, the cosine of the angle between two vectors is the dot product of the corresponding unit vectors.

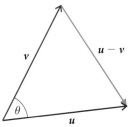

Example 9.8

Find the angle between the vectors $u = i - 2j + 2k$ and $v = -3i + 6j + 2k$

Solution

$$\cos\theta = \frac{u \cdot v}{|u||v|} = \frac{-3 - 12 + 4}{\sqrt{1 + 4 + 4}\sqrt{9 + 36 + 4}}$$

$$= \frac{-11}{21}$$

$$\Rightarrow \theta = \cos^{-1}\left(\frac{-11}{21}\right) \approx 2.12 \text{ radians}$$

If two vectors are perpendicular, then the dot product is zero.

This is because when the two vectors are perpendicular, the angle between them is $\pm90°$ and therefore

$$u \cdot v = |u||v|\cos90° = |u||v| \cdot 0 = 0$$

The base vectors of the coordinate system are obviously perpendicular:

$$i \cdot j = \begin{pmatrix} 1 \\ 0 \\ 0 \end{pmatrix} \cdot \begin{pmatrix} 0 \\ 1 \\ 0 \end{pmatrix} = 0, \text{ and similarly, } i \times k = 0, \text{ and } j \times k = 0$$

From the geometric definition of the dot product, we can see that for vectors of a given magnitude, the dot product measures the extent to which the vectors agree in direction. As the difference in direction increases from 0 to π, the dot product decreases:

If u and v have the same direction, then $\theta = 0$ and $u \cdot v = |u||v|\cos\theta = |u||v|$
This is the largest possible value for $u \cdot v$

If u and v have opposite directions, then $\theta = \pi$ and $u \cdot v = |u||v|\cos\pi = -|u||v|$
This is the least possible value for $u \cdot v$

If two vectors u and v are parallel, then
$u \cdot v = \pm|u||v|$

When the vectors are parallel, the angle between them is either 0° or 180°. And therefore

$$u \cdot v = |u||v|\cos0 = |u||v| \cdot 1 = |u||v|, \text{ or}$$

$$u \cdot v = |u||v|\cos180° = |u||v| \cdot (-1) = -|u||v|$$

Example 9.9

Determine which, if any, of these vectors are at right angles to each other (orthogonal).

$$u = 7i + 3j + 2k \quad v = -3i + 5j + 3k \quad w = i + k$$

Solution

$u \cdot v = 7(-3) + 3 \times 5 + 2 \times 3 = 0$; orthogonal vectors

$u \cdot w = 7 \times 1 + 3 \times 0 + 2 \times 1 = 9$; not orthogonal

$v \cdot w = -3 \times 1 + 5 \times 0 + 3 \times 1 = 0$; orthogonal vectors.

Example 9.10

The vertices of a triangle are $A(1, 2, 3)$, $B(-3, 2, 4)$ and $C(1, -4, 3)$. Show that the triangle is right-angled and find its area.

Solution

$$\overrightarrow{AB} = (-3 - 1)\boldsymbol{i} + (2 - 2)\boldsymbol{j} + (4 - 3)\boldsymbol{k} = -4\boldsymbol{i} + \boldsymbol{k}$$

$$\overrightarrow{AC} = (1 - 1)\boldsymbol{i} + (-4 - 2)\boldsymbol{j} + (3 - 3)\boldsymbol{k} = -6\boldsymbol{j}$$

$$\overrightarrow{BC} = (1 - (-3))\boldsymbol{i} + (-4 - 2)\boldsymbol{j} + (3 - 4)\boldsymbol{k} = 4\boldsymbol{i} - 6\boldsymbol{j} - \boldsymbol{k}$$

Since

$$\overrightarrow{AB} \cdot \overrightarrow{AC} = -4 \times 0 + 0 \times -6 + 1 \times 0 = 0,$$

the vectors are perpendicular.
So the triangle is right-angled at A.

The area of this triangle is half the product of the sides AB and AC.

$$\text{Area} = \frac{1}{2}|\overrightarrow{AB}||\overrightarrow{AC}|$$

$$= \frac{1}{2} \cdot \sqrt{(-4)^2 + 1} \cdot 6$$

$$= 3\sqrt{17}$$

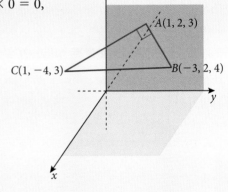

Proof for the key fact box

(a) $|\boldsymbol{u} \cdot \boldsymbol{v}| < |\boldsymbol{u}||\boldsymbol{v}|$

(b) $|\boldsymbol{u} + \boldsymbol{v}| < |\boldsymbol{u}| + |\boldsymbol{v}|$

(a) Since, $\boldsymbol{u} \cdot \boldsymbol{v} = |\boldsymbol{u}||\boldsymbol{v}|\cos\theta$, then $|\boldsymbol{u} \cdot \boldsymbol{v}| = ||\boldsymbol{u}||\boldsymbol{v}|\cos\theta| = |\boldsymbol{u}||\boldsymbol{v}||\cos\theta|$, and as $|\cos\theta| < 1$, then $|\boldsymbol{u} \cdot \boldsymbol{v}| < |\boldsymbol{u}||\boldsymbol{v}|$

(b) $|\boldsymbol{u} + \boldsymbol{v}|^2 = (\boldsymbol{u} + \boldsymbol{v}) \cdot (\boldsymbol{u} + \boldsymbol{v}) = \boldsymbol{u} \cdot \boldsymbol{u} + \boldsymbol{u} \cdot \boldsymbol{v} + \boldsymbol{v} \cdot \boldsymbol{u} + \boldsymbol{v} \cdot \boldsymbol{v}$

$$= |\boldsymbol{u}|^2 + 2(\boldsymbol{u} \cdot \boldsymbol{v}) + |\boldsymbol{v}|^2 \text{ since } \boldsymbol{u} \cdot \boldsymbol{v} = \boldsymbol{v} \cdot \boldsymbol{u}$$

but, $\boldsymbol{u} \cdot \boldsymbol{v} < |\boldsymbol{u} \cdot \boldsymbol{v}|$, since $\boldsymbol{u} \cdot \boldsymbol{v}$ may also be negative while $|\boldsymbol{u} \cdot \boldsymbol{v}|$ is not.

Also, $|\boldsymbol{u} \cdot \boldsymbol{v}| < |\boldsymbol{u}||\boldsymbol{v}|$, therefore

$$|\boldsymbol{u} + \boldsymbol{v}|^2 = |\boldsymbol{u}|^2 + 2(\boldsymbol{u} \cdot \boldsymbol{v}) + |\boldsymbol{v}|^2 < |\boldsymbol{u}|^2 + 2|\boldsymbol{u} \cdot \boldsymbol{v}| + |\boldsymbol{v}|^2$$

$$< |\boldsymbol{u}|^2 + 2|\boldsymbol{u}||\boldsymbol{v}| + |\boldsymbol{v}|^2 = (|\boldsymbol{u}| + |\boldsymbol{v}|)^2$$

Taking square roots, this implies that $|\boldsymbol{u} + \boldsymbol{v}| < |\boldsymbol{u}| + |\boldsymbol{v}|$

Direction angles, direction cosines

Figure 9.19 shows a non-zero vector $\boldsymbol{v}$. The angles α, β, and γ that the vector makes with the unit coordinate vectors are called the **direction angles** of $\boldsymbol{v}$, and $\cos\alpha$, $\cos\beta$, $\cos\gamma$ are called the **direction cosines**.

Let $\boldsymbol{v} = x\boldsymbol{i} + y\boldsymbol{j} + z\boldsymbol{k}$. Considering the right-angled triangles OAP, OCP and ODP, the hypotenuse in each of these triangles is OP – that is, $|\boldsymbol{v}|$. The side adjacent to an angle θ in a right triangle is related to it by

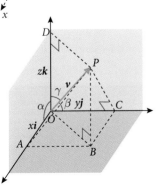

Figure 9.19 Direction angles of a vector

$$\cos\theta = \frac{adjacent}{hypotenuse} \Leftrightarrow adjacent = hypotenuse \cdot \cos\theta, \text{ so, in this case}$$

$x = |\boldsymbol{v}|\cos\alpha,\ y = |\boldsymbol{v}|\cos\beta,\ z = |\boldsymbol{v}|\cos\gamma$, and so

$$\boldsymbol{v} = (|\boldsymbol{v}|\cos\alpha)\,\boldsymbol{i} + (|\boldsymbol{v}|\cos\beta)\,\boldsymbol{j} + (|\boldsymbol{v}|\cos\gamma)\,\boldsymbol{k} = |\boldsymbol{v}|(\cos\alpha\,\boldsymbol{i} + \cos\beta\,\boldsymbol{j} + \cos\gamma\,\boldsymbol{k})$$

Taking the magnitude of both sides

$$|\boldsymbol{v}| = |\boldsymbol{v}|\sqrt{\cos^2\alpha + \cos^2\beta + \cos^2\gamma}$$

The discussion above leads to the following important result:

$$\cos^2\alpha + \cos^2\beta + \cos^2\gamma = 1$$

The sum of the squares of the direction cosines is always 1.

For a unit vector, the expression will be of the form

$\boldsymbol{u} = |\boldsymbol{u}|(\cos\alpha\,\boldsymbol{i} + \cos\beta\,\boldsymbol{j} + \cos\gamma\,\boldsymbol{k}) = \cos\alpha\,\boldsymbol{i} + \cos\beta\,\boldsymbol{j} + \cos\gamma\,\boldsymbol{k}$, (since $|\boldsymbol{u}| = 1$)

This means that for a unit vector its x-, y-, and z-coordinates are its direction cosines.

It is also important that you remember that $\cos\alpha = \dfrac{x}{|\boldsymbol{v}|}; \cos\beta = \dfrac{y}{|\boldsymbol{v}|}; \cos\gamma = \dfrac{z}{|\boldsymbol{v}|}$.

Example 9.11

Find the direction cosines of the vector $\boldsymbol{v} = 4\boldsymbol{i} - 2\boldsymbol{j} + 4\boldsymbol{k}$, and then approximate the direction angles to the nearest degree.

Solution

$$|\boldsymbol{v}| = \sqrt{4^2 + (-2)^2 + 4^2} = 6 \Rightarrow \boldsymbol{u} = \frac{\boldsymbol{v}}{|\boldsymbol{v}|} = \frac{2}{3}\boldsymbol{i} - \frac{1}{3}\boldsymbol{j} + \frac{2}{3}\boldsymbol{k}$$

$$\cos\alpha = \frac{2}{3},\ \cos\beta = -\frac{1}{3},\ \cos\gamma = \frac{2}{3}$$

$\alpha \approx 48°,\ \beta \approx 109°,\ \gamma \approx 48°$ to the nearest degree.

Example 9.12

Find the angle that a main diagonal of a cube with side a makes with the adjacent edges.

Solution

We can place the cube in a coordinate system such that three of its adjacent edges lie on the coordinate axes as shown. The diagonal, represented by the vector $\boldsymbol{v}$, has a terminal point (a, a, a). Hence

$$|\boldsymbol{v}| = \sqrt{a^2 + a^2 + a^2} = a\sqrt{3}$$

Take angle β, for example:

$$\beta = \cos^{-1}\left(\frac{a}{a\sqrt{3}}\right) = \cos^{-1}\left(\frac{1}{\sqrt{3}}\right) \approx 54.7°$$

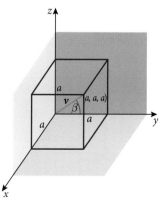

Figure 9.20 Solution to Example 9.12

393

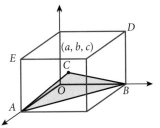

Figure 9.21 Diagram for Example 9.13

Example 9.13

The point C is at the centre of the rectangular box whose edges have measures a, b, and c. Find the size of angle ACB in terms of a, b, and c.

Solution

The point diagonally opposite to A is $D(0, b, c)$, so, $C\left(\dfrac{a}{2}, \dfrac{b}{2}, \dfrac{c}{2}\right)$.

Consequently

$$\overrightarrow{CA} = \left(a - \frac{a}{2}, 0 - \frac{b}{2}, 0 - \frac{c}{2}\right) = \left(\frac{a}{2}, -\frac{b}{2}, -\frac{c}{2}\right) \text{ and}$$

$$\overrightarrow{CB} = \left(0 - \frac{a}{2}, b - \frac{b}{2}, 0 - \frac{c}{2}\right) = \left(-\frac{a}{2}, \frac{b}{2}, -\frac{c}{2}\right)$$

$$\cos A\hat{C}B = \frac{\overrightarrow{CA} \cdot \overrightarrow{CB}}{|\overrightarrow{CA}||\overrightarrow{CB}|} = \frac{-\dfrac{a^2}{4} - \dfrac{b^2}{4} + \dfrac{c^2}{4}}{\sqrt{\dfrac{a^2}{4} + \dfrac{b^2}{4} + \dfrac{c^2}{4}}\sqrt{\dfrac{a^2}{4} + \dfrac{b^2}{4} + \dfrac{c^2}{4}}} = \frac{c^2 - a^2 - b^2}{a^2 + b^2 + c^2}$$

Exercise 9.2

1. Find the dot product and the angle between each pair of vectors.

   (a) $u = \begin{pmatrix} 3 \\ -2 \\ 4 \end{pmatrix}$, $v = 2i - j - 6k$ (b) $u = \begin{pmatrix} 2 \\ -6 \\ 0 \end{pmatrix}$, $v = \begin{pmatrix} -1 \\ 3 \\ 5 \end{pmatrix}$

   (c) $u = 3i - j$, $v = 5i + 2j$ (d) $u = i - 3j$, $v = 5j + 2k$

2. Find the dot product of each pair of vectors.

   (a) $|u| = 3$, $|v| = 4$, the angle between u and v is $\dfrac{\pi}{3}$

   (b) $|u| = 3$, $|v| = 4$, the angle between u and v is $\dfrac{2\pi}{3}$

3. State whether these vectors are orthogonal.
   If they are not orthogonal, is the angle acute?

   (a) $u = \begin{pmatrix} 2 \\ -6 \\ 4 \end{pmatrix}$, $v = \begin{pmatrix} -1 \\ 3 \\ 5 \end{pmatrix}$ (b) $u = 3i - 7j$, $v = 5i + 2j$

   (c) $u = i - 3j + 6k$, $v = 6j + 3k$

4. (a) Show that the vectors $v = -yi + xj$ and $w = yi - xj$ are both perpendicular to $u = xi + yj$.

   (b) Find two unit vectors that are perpendicular to $u = 2i - 3j$. Plot the three vectors in the same coordinate system.

5. For each vector v:
   (i) find the direction cosines
   (ii) show that they satisfy $\cos^2 \alpha + \cos^2 \beta + \cos^2 \gamma = 1$
   (iii) approximate the direction angles to the nearest degree.

   (a) $v = 2i - 3j + k$ (b) $v = i - 2j + k$

   (c) $v = 3i - 2j + k$ (d) $v = 3i - 4k$

6. Find a unit vector with direction angles $\dfrac{\pi}{3}, \dfrac{\pi}{4}, \dfrac{2\pi}{3}$.

7. Find a vector with magnitude 3 and direction angles $\dfrac{\pi}{4}, \dfrac{\pi}{4}, \dfrac{\pi}{2}$.

8. Determine m such that $\boldsymbol{u}$ and $\boldsymbol{v}$ are perpendicular.

(a) $\boldsymbol{u} = \begin{pmatrix} 3 \\ 5 \\ 0 \end{pmatrix}; \boldsymbol{v} = \begin{pmatrix} m-2 \\ m+3 \\ 0 \end{pmatrix}$
(b) $\boldsymbol{u} = \begin{pmatrix} 2m \\ m-1 \\ m+1 \end{pmatrix}; \boldsymbol{v} = \begin{pmatrix} m-1 \\ m \\ m-1 \end{pmatrix}$

9. Let $\boldsymbol{u} = \begin{pmatrix} -3 \\ 1 \\ 2 \end{pmatrix}, \boldsymbol{v} = \begin{pmatrix} 1 \\ 2 \\ 1 \end{pmatrix}$, and $\boldsymbol{w} = \boldsymbol{u} + m\boldsymbol{v}$

Determine the value of m so that the vectors $\boldsymbol{u}$ and $\boldsymbol{w}$ are orthogonal.

10. Let $\boldsymbol{u} = \begin{pmatrix} -2 \\ 5 \\ 4 \end{pmatrix}$, and $\boldsymbol{v} = \begin{pmatrix} 6 \\ -3 \\ 0 \end{pmatrix}$

Find, to the nearest degree, the measures of the angles between

(a) $\boldsymbol{u}$ and $\boldsymbol{v}$
(b) $\boldsymbol{u}$ and $(\boldsymbol{u} + \boldsymbol{v})$
(c) $\boldsymbol{v}$ and $(\boldsymbol{u} + \boldsymbol{v})$

11. Consider the three points $A(1, 2, -3)$, $B(3, 5, -2)$, and $C(m, 1, -10m)$. Determine m so that:

(a) A, B, and C are collinear
(b) $\overrightarrow{AB}$ and $\overrightarrow{AC}$ are perpendicular.

12. Consider the triangle with vertices $A(4, -2, -1)$, $B(3, -5, -1)$, and $C(3, 1, 2)$. Find the vector equations of each of its medians and then find the coordinates of its centroid (where the medians meet).

13. (a) Consider the tetrahedron $ABCD$ with vertices as shown in the diagram. Find, to the nearest degree, all the angles in the tetrahedron.

(b) Use the angles you found to calculate the total surface area of the tetrahedron.

(c) What angles does $\overrightarrow{DC}$ make with each of the coordinate axes?

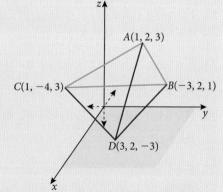

(d) Find $\left(\overrightarrow{DA} - \overrightarrow{DB}\right) \cdot \overrightarrow{AC}$

14. Find k such that the angle between the vectors $\begin{pmatrix} 3 \\ -k \\ -1 \end{pmatrix}$ and $\begin{pmatrix} 1 \\ -3 \\ k \end{pmatrix}$ is $\dfrac{\pi}{3}$.

15. Find k such that the angle between the vectors $\begin{pmatrix} k \\ 1 \\ 1 \end{pmatrix}$ and $\begin{pmatrix} 1 \\ k \\ 1 \end{pmatrix}$ is $\dfrac{\pi}{3}$.

16. Find x and y such that $\begin{pmatrix} 2 \\ x \\ y \end{pmatrix}$ is perpendicular to both $\begin{pmatrix} 3 \\ 1 \\ -1 \end{pmatrix}$ and $\begin{pmatrix} 4 \\ -1 \\ 2 \end{pmatrix}$.

17. Consider the vectors $\begin{pmatrix} 1-x \\ 2x-2 \\ 3+x \end{pmatrix}$ and $\begin{pmatrix} 2-x \\ 1+x \\ 1+x \end{pmatrix}$. Find the value(s) of x such that the two vectors are parallel.

18. In triangle ABC, $\overrightarrow{OA} = \begin{pmatrix} 2 \\ 3 \\ 1 \end{pmatrix}$, $\overrightarrow{OB} = \begin{pmatrix} 3 \\ 5 \\ 4 \end{pmatrix}$, and $\overrightarrow{BC} = \begin{pmatrix} -1 \\ 4 \\ 0 \end{pmatrix}$.

    (a) Find the size of $A\widehat{B}C$.

    (b) Find $\overrightarrow{AC}$ and use it to find measure of $B\widehat{A}C$.

19. Find the value(s) of b such that the vectors are orthogonal.

    (a) $\begin{pmatrix} b \\ 3 \\ 2 \end{pmatrix}$ and $\begin{pmatrix} 1 \\ b \\ 1 \end{pmatrix}$ (b) $\begin{pmatrix} 4 \\ -2 \\ 7 \end{pmatrix}$ and $\begin{pmatrix} b^2 \\ b \\ 0 \end{pmatrix}$

    (c) $\begin{pmatrix} b \\ 11 \\ -3 \end{pmatrix}$ and $\begin{pmatrix} 2b \\ -b \\ -5 \end{pmatrix}$ (d) $\begin{pmatrix} 2 \\ 5 \\ 2b \end{pmatrix}$ and $\begin{pmatrix} 6 \\ 4 \\ -b \end{pmatrix}$

20. If two vectors p and q are such that $|p| = |q|$, show that $p + q$ and $p - q$ are perpendicular. (This proves that the diagonals of a rhombus are perpendicular to each other.)

21. Shortly after take-off, an aeroplane is rising at a rate of $300 \, \text{m min}^{-1}$. It is heading north-west (on a bearing of $045°$) with an airspeed of $200 \, \text{km h}^{-1}$. Find the components of its velocity vector. The x-axis is in the east direction, the y-axis north and the z-axis is the elevation.

22. For what value of t is the vector $2ti + 4j - (10 + t)k$ perpendicular to the vector $i + tj + k$?

23. For what value of t is the vector $ti + 3j + 2k$ perpendicular to the vector $i + tj + k$?

24. For what value of t is the vector $4i - 2j + 7k$ perpendicular to the vector $t^2i + tj$?

25. Find the angle between the diagonal of a cube and a diagonal of one of the faces. Consider all possible cases.

26. Show that the vector $|a|b + |b|a$ bisects the angle between the two vectors a and b.

27. Let $u = i + mj + k$ and $v = 2i - j + nk$. Compute all values of m and n for which $u \perp v$ and $|u| = |v|$.

28. Show that $\dfrac{\pi}{4}, \dfrac{\pi}{6}, \dfrac{2\pi}{3}$ cannot be the direction angles of a vector.

29. A vector has direction angles $\alpha = \dfrac{\pi}{3}, \beta = \dfrac{\pi}{4}$
    Find the third direction angle, γ.

30. If all the direction angles of a vector are equal, what is the size of each angle?

31. The direction angles of a vector u are α, β, and γ. What are the direction angles of $-u$?

32. Find all possible values of a unit vector u that will be perpendicular to both $i + 2j + k$ and $3i - 4j + 2k$.

9.3 Vector (cross) product

In several applications of vectors there is a need to find a vector that is orthogonal to two given vectors. In this section we will discuss a type of vector multiplication that can be used for this purpose.

Let $u = \begin{pmatrix} u_1 \\ u_2 \\ u_3 \end{pmatrix}$ and $v = \begin{pmatrix} v_1 \\ v_2 \\ v_3 \end{pmatrix}$. The vector (cross) product is written as $u \times v$ and is defined as

$$u \times v = \begin{vmatrix} u_2 & u_3 \\ v_2 & v_3 \end{vmatrix} i - \begin{vmatrix} u_1 & u_3 \\ v_1 & v_3 \end{vmatrix} j + \begin{vmatrix} u_1 & u_2 \\ v_1 & v_2 \end{vmatrix} k$$

or using the properties of determinants, this definition is equivalent to

$$u \times v = \begin{vmatrix} i & j & k \\ u_1 & u_2 & u_3 \\ v_1 & v_2 & v_3 \end{vmatrix}$$

> In case you do not want to use matrix algebra, the definition of the cross product can be stated as:
> $$u \times v = (u_2 v_3 - v_2 u_3)i \\ - (u_1 v_3 - v_1 u_3)j \\ + (u_1 v_2 - v_1 u_2)k$$

Example 9.14

Given the vectors $u = 2i - 3j + k$ and $v = i + 3j - 2k$, find:

(a) $u \times v$ (b) $v \times u$ (c) $u \times u$

Solution

(a) $u \times v = \begin{vmatrix} i & j & k \\ 2 & -3 & 1 \\ 1 & 3 & -2 \end{vmatrix} = \begin{vmatrix} -3 & 1 \\ 3 & -2 \end{vmatrix} i - \begin{vmatrix} 2 & 1 \\ 1 & -2 \end{vmatrix} j + \begin{vmatrix} 2 & -3 \\ 1 & 3 \end{vmatrix} k$

$$= 3i + 5j + 9k$$

(b) $v \times u = \begin{vmatrix} i & j & k \\ 1 & 3 & -2 \\ 2 & -3 & 1 \end{vmatrix} = \begin{vmatrix} 3 & -2 \\ -3 & 1 \end{vmatrix} i - \begin{vmatrix} 1 & -2 \\ 2 & 1 \end{vmatrix} j + \begin{vmatrix} 1 & 3 \\ 2 & -3 \end{vmatrix} k$

$$= -3i - 5j - 9k$$

(c) $u \times u = \begin{vmatrix} i & j & k \\ 2 & -3 & 1 \\ 2 & -3 & 1 \end{vmatrix} = \begin{vmatrix} -3 & 1 \\ -3 & 1 \end{vmatrix} i - \begin{vmatrix} 2 & 1 \\ 2 & 1 \end{vmatrix} j + \begin{vmatrix} 2 & -3 \\ 2 & -3 \end{vmatrix} k = 0$

Note that $u \times v = -(v \times u)$

Determinants have many useful applications when we are dealing with vector products. Here are some of the properties, which we state without proof.

1 If two rows of a determinant are proportional, then the value of that determinant is zero.
So, for example, if $u = ai + bj + ck$, $v = mai + mbj + mck$, then

$$u \times v = \begin{vmatrix} i & j & k \\ a & b & c \\ ma & mb & mc \end{vmatrix} = 0$$

This result leads to an important property of vector products:
Two non-zero vectors are parallel if their cross product is zero.

2 If two rows of a determinant are interchanged, then the determinant's value is multiplied by (-1).
So, for instance, if $u = u_1 i + u_2 j + u_3 k$, $v = v_1 i + v_2 j + v_3 k$, then

$$u \times v = \begin{vmatrix} i & j & k \\ u_1 & u_2 & u_3 \\ v_1 & v_2 & v_3 \end{vmatrix} = -\begin{vmatrix} i & j & k \\ v_1 & v_2 & v_3 \\ u_1 & u_2 & u_3 \end{vmatrix} = -(v \times u)$$

Properties of the vector product

The following results are important.

1 $u \times (v \pm w) = (u \times v) \pm (u \times w)$

2 $u \times 0 = 0$

3 $u \times u = 0$

4 $i \times j = k; j \times k = i; k \times i = j$

5 $u \cdot (u \times v) = 0$ (i.e. $u \times v$ is orthogonal to u)

6 $v \cdot (u \times v) = 0$ (i.e. $u \times v$ is orthogonal to v)

Proof

To prove $i \times j = k$, we apply the definition.

$$i \times j = \begin{vmatrix} i & j & k \\ 1 & 0 & 0 \\ 0 & 1 & 0 \end{vmatrix} = k,$$

details are left as an exercise.

$$u \cdot (u \times v) = \begin{pmatrix} u_1 \\ u_2 \\ u_3 \end{pmatrix} \begin{pmatrix} \begin{vmatrix} u_2 & u_3 \\ v_2 & v_3 \end{vmatrix} \\ -\begin{vmatrix} u_1 & u_3 \\ v_1 & v_3 \end{vmatrix} \\ \begin{vmatrix} u_1 & u_2 \\ v_1 & v_2 \end{vmatrix} \end{pmatrix},$$

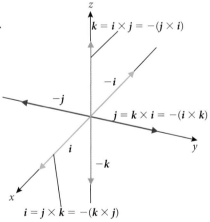

Figure 9.22 Vector product of two base vectors result in a third base vector

so that

$$u \cdot (u \times v) = u_1 \begin{vmatrix} u_2 & u_3 \\ v_2 & v_3 \end{vmatrix} - u_2 \begin{vmatrix} u_1 & u_3 \\ v_1 & v_3 \end{vmatrix} + u_3 \begin{vmatrix} u_1 & u_2 \\ v_1 & v_2 \end{vmatrix}$$

$$= u_1 u_2 v_3 - u_1 v_2 u_3 - u_2 u_1 v_3 + u_2 v_1 u_3 + u_3 u_1 v_2 - u_3 v_1 u_2$$

$$= 0$$

Properties 4 and 5 lead to an equivalent geometric definition of the cross product:

$(u \times v)$ is a vector perpendicular to both u and v and obeying the right-hand rule, and has the magnitude: $|u \times v| = |u||v|\sin\theta$

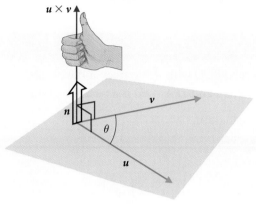
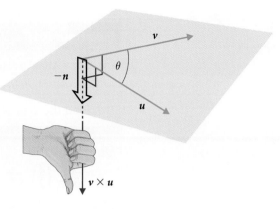

Remember that $u \times v = -(v \times u)$

Proof

The algebraic manipulation required for the proof is extensive, so we will keep the details away from this discussion:

$$|u \times v| = \sqrt{(u_2 v_3 - v_2 u_3)^2 + (u_1 v_3 - v_1 u_3)^2 + (u_1 v_2 - v_1 u_2)^2}$$

$$= \sqrt{(u_1^2 + u_2^2 + u_3^2)(v_1^2 + v_2^2 + v_3^2) - (u_1 v_1 + u_2 v_2 + u_3 v_3)^2}$$

$$= \sqrt{|u|^2 |u|^2 (u \cdot v)^2} = \sqrt{|u|^2 |v|^2 - (|u||v|\cos\theta)^2}$$

$$= |u||v|\sqrt{1 - \cos^2\theta} = |u||v|\sin\theta$$

The vector product gives you another method for finding the angle between two vectors.

$$|u \times v| = |u||v|\sin\theta \Leftrightarrow \sin\theta = \frac{|u \times v|}{|u||v|}$$

Example 9.15

Find a unit vector orthogonal to both vectors $u = 2i - 3j + k$ and $v = i + 3j - 2k$

Solution

$u \times v$ is orthogonal to both vectors.
$u \times v = 3i + 5j + 9k$

A unit vector in the same direction as $u \times v$ will also be orthogonal to both vectors. A unit vector is equal to the vector itself multiplied by the reciprocal of its magnitude (Section 9.1).

Find the magnitude of $u \times v$ first.
$|u \times v| = \sqrt{9 + 25 + 81} = \sqrt{115}$

The required unit vector is therefore $\dfrac{u \times v}{|u \times v|} = \dfrac{3i + 5j + 9k}{\sqrt{115}}$

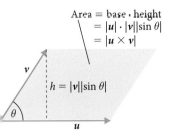

Area = base · height
$= |\boldsymbol{u}| \cdot |\boldsymbol{v}||\sin \theta|$
$= |\boldsymbol{u} \times \boldsymbol{v}|$

$h = |\boldsymbol{v}||\sin \theta|$

Figure 9.23 Magnitude of a vector product

The last result leads to the conclusion that the magnitude of the cross product is the area of the parallelogram that has $\boldsymbol{u}$ and $\boldsymbol{v}$ as adjacent sides.

Example 9.16

(a) Show that the quadrilateral $ABCD$ with its vertices at the points $A(3, 0, 2)$, $B(6, 2, 5)$, $C(1, 2, 2)$, and $D(4, 4, 5)$ is a parallelogram

(b) Find the area of the parallelogram.

Solution

(a) $\overrightarrow{AB} = \begin{pmatrix} 3 \\ 2 \\ 3 \end{pmatrix}, \overrightarrow{BD} = \begin{pmatrix} -2 \\ 2 \\ 0 \end{pmatrix}, \overrightarrow{CD} = \begin{pmatrix} 3 \\ 2 \\ 3 \end{pmatrix}, \overrightarrow{AC} = \begin{pmatrix} -2 \\ 2 \\ 0 \end{pmatrix}$

This implies that $\overrightarrow{AB} = \overrightarrow{CD}$ and $\overrightarrow{BD} = \overrightarrow{AC}$, which in turn means that the pairs of opposite sides of the quadrilateral are congruent and parallel. (We need only one pair).

Thus $ABDC$ is a parallelogram with AB and BD as adjacent sides.

(b) The area of the parallelogram is the magnitude of the cross product of $\overrightarrow{AB}$ and $\overrightarrow{BD}$

$$\overrightarrow{AB} \times \overrightarrow{BD} = \begin{vmatrix} \boldsymbol{i} & \boldsymbol{j} & \boldsymbol{k} \\ 3 & 2 & 3 \\ 02 & 2 & 0 \end{vmatrix} = \begin{pmatrix} -6 \\ -6 \\ 10 \end{pmatrix}$$

So the area of the parallelogram is

$$\left| \overrightarrow{AB} \times \overrightarrow{BD} \right| = \sqrt{36 + 36 + 100} = \sqrt{172} = 2\sqrt{43}$$

Example 9.17

Find the area of the triangle determined by the points $A(2, 2, 0)$, $B(-1, 0, 2)$, and $C(0, 4, 3)$.

Solution

The area of the triangle ABC is half the area of the parallelogram formed with AB and AC as its adjacent sides.

$$\overrightarrow{AB} = \begin{pmatrix} -3 \\ -2 \\ 2 \end{pmatrix} \text{ and } \overrightarrow{AC} = \begin{pmatrix} -2 \\ 2 \\ 3 \end{pmatrix}$$

So $\overrightarrow{AB} \times \overrightarrow{AC} = \begin{pmatrix} -10 \\ 5 \\ -10 \end{pmatrix}$, and hence

$$\text{area of triangle } ABC = \frac{1}{2} \left| \overrightarrow{AB} \times \overrightarrow{AC} \right| = \frac{1}{2}\sqrt{(-10)^2 + 5^2 + (-10)^2}$$

$$= \frac{1}{2}\sqrt{225} = \frac{15}{2}$$

The scalar triple product

Let $u = (u_1, u_2, u_3)$, $v = (v_1, v_2, v_3)$ and $w = (w_1, w_2, w_3)$. The scalar triple product is $u \cdot (v \times w)$
The component expression of this product is:

$$u \cdot (v \times w) = u \cdot \left(\begin{vmatrix} v_2 & v_3 \\ w_2 & w_3 \end{vmatrix} i - \begin{vmatrix} v_1 & v_3 \\ w_1 & w_3 \end{vmatrix} j + \begin{vmatrix} v_1 & v_2 \\ w_1 & w_2 \end{vmatrix} k \right)$$

$$= u_1 \begin{vmatrix} v_2 & v_3 \\ w_2 & w_3 \end{vmatrix} - u_2 \begin{vmatrix} v_1 & v_3 \\ w_1 & w_3 \end{vmatrix} + u_3 \begin{vmatrix} v_1 & v_2 \\ w_1 & w_2 \end{vmatrix}$$

$$= \begin{vmatrix} u_1 & u_2 & u_3 \\ v_1 & v_2 & v_3 \\ w_1 & w_2 & w_3 \end{vmatrix}$$

This product is very useful in its geometric interpretation as it is of great help in finding the equation of a plane later in the chapter.

Example 9.18

Calculate the scalar triple product of the vectors
$u = 2i - j - 5k$, $v = 2i + 5j - 5k$, $w = i + 4j + 3k$

Solution

$$u \cdot (v \times w) = \begin{vmatrix} 2 & -1 & -5 \\ 2 & 5 & -5 \\ 1 & 4 & 3 \end{vmatrix} = 66$$

Geometric interpretation

$|u \cdot (v \times w)|$ is the volume of the parallelepiped that has the three vectors as adjacent edges.

Proof

In Figure 9.24, $|v \times w|$ is the area of the parallelogram with sides v and w, which is also the base of the parallelepiped.

$$|u \cdot (v \times w)| = |u||v \times w|\cos\alpha$$

$$= |v \times w||u|\cos\alpha$$

As $|u|\cos(\alpha) = h$, the height of the parallelepiped, and $|v \times w|$ is the area of the base, the triple product's absolute value is the volume of the parallelepiped.

A direct consequence of the scalar triple product is that the volume of the parallelepiped is 0 if and only if the three vectors are coplanar. That is:

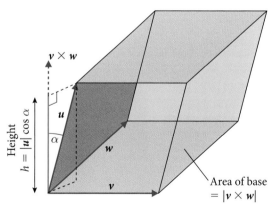

Figure 9.24 Parallelepiped formed by three vectors

If $u = (u_1, u_2, u_3)$, $v = (v_1, v_2, v_3)$ and $w = (w_1, w_2, w_3)$ are three vectors drawn from the same initial point, they lie in the same plane if:

$$u \cdot (v \times w) = \begin{vmatrix} u_1 & u_2 & u_3 \\ v_1 & v_2 & v_3 \\ w_1 & w_2 & w_3 \end{vmatrix} = 0$$

The parentheses in the scalar triple product are unnecessary, i.e.
$u \cdot (v \times w) = u \cdot v \times w$. This is so because $u \cdot v$ is a scalar.

401

Example 9.19

Consider the three vectors

$$u = 2i + j + mk, \; v = 3i + 2j + 3k, \; w = mi + 2j + k.$$

(a) Find the volume of the parallelepiped that has these vectors as sides.

(b) Show that these vectors can never be on the same plane.

Solution

(a) The volume of the parallelepiped is given by the absolute value of their scalar triple product:

$$\left| u \cdot (v \times w) \right| = \left| \begin{vmatrix} 2 & 1 & m \\ 3 & 2 & 3 \\ m & 2 & 1 \end{vmatrix} \right| = \left| -2m^2 + 9m - 11 \right|$$

(b) For the vectors to be coplanar, their scalar triple product must be zero. That is

$$-2m^2 + 9m - 11 = 0$$

However, since this is a quadratic equation, it can have real roots if $b^2 - 4ac > 0$. But $b^2 - 4ac = 81 - 88 = -7 < 0$, and thus the equation does not have any real roots and the three vectors can therefore never be coplanar.

Exercise 9.3

1. (a) Find the cross product using the definition: $i \times (i + j + k)$

   (b) Compare your answer to $(i \times i) + (i \times j) + (i \times k)$

2. Repeat question 1 for $j \times (i + j + k)$

3. Repeat question 1 for $k \times (i + j + k)$

4. Use the definition of vector products to verify

   $$u \times (v \pm w) = (u \times v) \pm (u \times w)$$

This is the distributive property of vector product over addition and subtraction.

5. Find $u \times v$ and check that it is orthogonal to both u and v.

   (a) $u = \begin{pmatrix} 2 \\ 3 \\ -2 \end{pmatrix}, v = \begin{pmatrix} -3 \\ 2 \\ 3 \end{pmatrix}$ (b) $u = 4i + 3j, v = -2j + 2k$

   (c) $u = \begin{pmatrix} 1 \\ 2 \\ -1 \end{pmatrix}, v = \begin{pmatrix} 4 \\ 1 \\ -3 \end{pmatrix}$ (d) $u = 5i + j + 2k, v = 3i + k$

6. Consider the vectors $u = 2i + j + mk$, $v = 3i + 2j + 3k$, and $w = mi + 2j + k$

   Find

   (a) $u \cdot (v \times w)$ (b) $w \cdot (u \times v)$ (c) $v \cdot (w \times u)$

7. Consider the vectors $u = \begin{pmatrix} 3 \\ 0 \\ 4 \end{pmatrix}$, $v = \begin{pmatrix} 1 \\ 2 \\ 8 \end{pmatrix}$ and $w = \begin{pmatrix} 2 \\ 5 \\ 6 \end{pmatrix}$
Find

   (a) $u \times (v \times w)$ (b) $(u \times v) \times w$ (c) $(u \times v) \times (v \times w)$

   (d) $(v \times w) \times (u \times v)$ (e) $(u \cdot w)v - (u \cdot v)w$ (f) $(w \cdot u)v - (w \cdot v)u$

8. Find a unit vector that is orthogonal to both

     $u = -6i + 4j + k$ and $v = 3i + j + 5k$

9. Find a unit vector that is normal (perpendicular) to the plane determined by the points $A(1, -1, 2)$, $B(2, 0, -1)$, and $C(0, 2, 1)$.

10. Find the area of the parallelogram that has u and v as adjacent sides

   (a) $u = 2i + 3k, v = i + 4j + 2k$ (b) $u = 3i + 4j + k, v = 3j - k$

11. Verify that the points $(2, -1, 1)$, $(5, 1, 4)$, $(0, 1, 1)$, and $(3, 3, 4)$ are the vertices of a parallelogram and find its area.

12. Show that the points $P(1, -1, 2)$, $Q(2, 0, 1)$, $R(3, 2, 0)$, and $S(5, 4, -2)$ are coplanar.

13. For what value(s) of m are these four points on the same plane?
     $A(m, 3, -2)$, $B(3, 4, m)$, $C(2, 0, -2)$, and $D(4, 8, 4)$

14. Find the area of the triangle with the given vertices.

   (a) $A(2, 6, -1)$, $B(1, 1, 1)$, $C(3, 5, 2)$

   (b) $A(3, 1, -2)$, $B(2, 5, 6)$, $C(6, 1, 8)$

15. Find $u \cdot (v \times w)$

   (a) $u = 3i - 2j + 2k, v = 5i + 2j - 2k, w = i + 2j + 6k$

   (b) $u = \begin{pmatrix} 2 \\ -1 \\ 3 \end{pmatrix}$, $v = \begin{pmatrix} 1 \\ 4 \\ 3 \end{pmatrix}$, $w = \begin{pmatrix} -3 \\ 2 \\ -2 \end{pmatrix}$

   (c) $u = \begin{pmatrix} 3 \\ 2 \\ 1 \end{pmatrix}$, $v = \begin{pmatrix} 1 \\ -3 \\ 1 \end{pmatrix}$, $w = \begin{pmatrix} 5 \\ 1 \\ 2 \end{pmatrix}$

16. Find the volume of the parallelepiped with u, v and w as adjacent edges.

   (a) $u = \begin{pmatrix} 3 \\ -5 \\ 3 \end{pmatrix}$, $v = \begin{pmatrix} 1 \\ 5 \\ -1 \end{pmatrix}$, $w = \begin{pmatrix} 3 \\ 2 \\ -3 \end{pmatrix}$

   (b) $u = 4i + 2j + 3k, v = 5i + 6j + 2k, w = 2i + 3j + 5k$

17. Determine whether or not the three given vectors are coplanar.

   (a) $u = \begin{pmatrix} 2 \\ -1 \\ 2 \end{pmatrix}$, $v = \begin{pmatrix} 4 \\ 1 \\ -1 \end{pmatrix}$, $w = \begin{pmatrix} 6 \\ -3 \\ 1 \end{pmatrix}$

   (b) $u = \begin{pmatrix} 4 \\ -2 \\ -1 \end{pmatrix}$, $v = \begin{pmatrix} 9 \\ -6 \\ -1 \end{pmatrix}$, $w = \begin{pmatrix} 6 \\ -6 \\ 1 \end{pmatrix}$

18. Find m such that the following vectors are coplanar; otherwise, show that it is not possible.

    (a) $u = \begin{pmatrix} 1 \\ m \\ 1 \end{pmatrix}, v = \begin{pmatrix} 3 \\ 0 \\ m \end{pmatrix}, w = \begin{pmatrix} 5 \\ -4 \\ 0 \end{pmatrix}$

    (b) $u = \begin{pmatrix} 2 \\ -3 \\ 2m \end{pmatrix}, v = \begin{pmatrix} m \\ -3 \\ 1 \end{pmatrix}, w = \begin{pmatrix} 1 \\ 3 \\ -2 \end{pmatrix}$

19. Consider the parallelepiped shown in the diagram.

    (a) Find the volume.

    (b) Find the area of the face determined by u and v.

    (c) Find the height of the parallelepiped from vertex D to the base.

    (d) Find the angle that w makes with the plane determined by u and v.

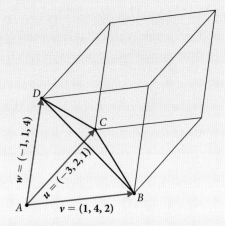

20. (a) The volume of a tetrahedron is $\frac{1}{3}$(base)(height).

    Use the results from question 19 to find the volume of tetrahedron $ABCD$. Compare this volume to the volume of the parallelepiped and make a general conjecture.

    (b) Use the results you have from part (a) to find the volume of the tetrahedron whose vertices are $A(0, 3, 1)$, $B(3, 2, -2)$, $C(2, 1, 2)$, and $D(4, -1, 4)$

21. What can you conclude about the angle between two non-zero vectors u and v if $u \cdot v = |u \times v|$?

22. Show that $|u \times v| = \sqrt{|u|^2|v|^2 - (u \cdot v)^2}$

23. Use the diagram at the right to show that the distance from a point P in space to a line L through two points A and B can be expressed as

    $$d = \frac{\left|\overrightarrow{AP} \times \overrightarrow{AB}\right|}{\left|\overrightarrow{AB}\right|}$$

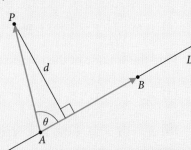

Use right triangle trigonometry to find d in terms of θ first.

24. Use the result from question 23 to find the distance from point A to the line through the points B and C:

    (a) $A(-2, 2, 3)$, $B(2, 2, 1)$, $C(-1, 4, -3)$

    (b) $A(5, 4)$, $B(3, 2)$, $C(1, 3)$

    (c) $A(2, 0, 1)$, $B(1, -2, 2)$, $C(3, 0, 2)$

25. Express $(\boldsymbol{u} + \boldsymbol{v}) \times (\boldsymbol{v} - \boldsymbol{u})$ in terms of $(\boldsymbol{u} \times \boldsymbol{v})$.

26. Express $(2\boldsymbol{u} + 3\boldsymbol{v}) \times (4\boldsymbol{v} - 5\boldsymbol{u})$ in terms of $(\boldsymbol{u} \times \boldsymbol{v})$.

27. Express $(m\boldsymbol{u} + n\boldsymbol{v}) \times (p\boldsymbol{v} + q\boldsymbol{u})$ in terms of $(\boldsymbol{u} \times \boldsymbol{v})$, where m, n, p, and q are scalars.

28. You are given a tetrahedron with its vertex at the origin and base ABC as shown in the diagram.

    (a) Find the area of the base ABC. Call it o.

    (b) Find the area of each face of the tetrahedron and call them a, b, and c.

    (c) Show that $o^2 = a^2 + b^2 + c^2$. This is sometimes called the 3D version of Pythagoras' theorem.

29. Find all vectors $\boldsymbol{v}$ such that $(-\boldsymbol{i} + 2\boldsymbol{j} + 3\boldsymbol{k}) \times \boldsymbol{v} = \boldsymbol{i} + 5\boldsymbol{j} - 3\boldsymbol{k}$. Otherwise, show that it is not possible.

30. Find all vectors $\boldsymbol{v}$ such that $(-\boldsymbol{i} + 2\boldsymbol{j} + 3\boldsymbol{k}) \times \boldsymbol{v} = \boldsymbol{i} + 5\boldsymbol{j}$. Otherwise, show that it is not possible.

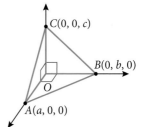

Figure 9.25 Diagram for question 28

9.4 Lines in space

In the same way that a plane can be defined by any three points, a straight line in space can be determined by any two points A and M that lie on it. Alternatively, the line can be determined by specifying a point on it and a direction given by a non-zero vector parallel to it. To investigate equations that describe lines in space, let's begin with a straight line L that passes through the point $A(x_0, y_0, z_0)$ and parallel to the vector $\boldsymbol{v} = a\boldsymbol{i} + b\boldsymbol{j} + z\boldsymbol{k}$, as shown in Figure 9.26. If L is the line that passes through A and is parallel to the non-zero vector $\boldsymbol{v}$, then L consists of all the points $M(x, y, z)$ for which the vector $\overrightarrow{AM}$ is parallel to $\boldsymbol{v}$.

In Section 9.1 we established that two vectors are parallel if one of them is a scalar multiple of the other. That is $\boldsymbol{v}$ is parallel to $\boldsymbol{u}$ if and only if $\boldsymbol{v} = t\boldsymbol{u}$ for some real number t.

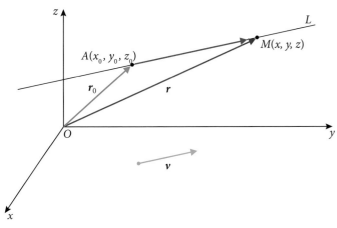

Figure 9.26 Any vector $\overrightarrow{AM}$ is parallel to $\boldsymbol{v}$

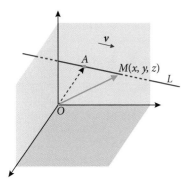

Figure 9.27 Vector AM is a multiple of v

The line that passes through the point $A(x_0, y_0, z_0)$ and parallel to the vector $v = \begin{pmatrix} a \\ b \\ c \end{pmatrix}$ has parametric equations:
$x = x_0 + at, y = y_0 + bt,$
$z = z_0 + ct$

This means that for the point M to be on L, $\overrightarrow{AM}$ must be a scalar multiple of v. That is

$$\overrightarrow{AM} = tv,$$

where t is a scalar. This equation can be written in coordinate form as

$$\begin{pmatrix} x - x_0 \\ y - y_0 \\ z - z_0 \end{pmatrix} = t \begin{pmatrix} a \\ b \\ c \end{pmatrix} = \begin{pmatrix} ta \\ tb \\ tc \end{pmatrix}$$

For two vectors to be equal, their components must be the same, so

$$x - x_0 = ta, \; y - y_0 = tb, \; z - z_0 = tc$$

This leads to the result:

$$x = x_0 + at, \; y = y_0 + bt, \; z = z_0 + ct$$

Example 9.20

(a) Find parametric equations of the line through $A(1, -2, 3)$ and parallel to
$$v = 5i + 4j - 6k$$

(b) Find parametric equations of the line through the points $A(1, -2, 3)$ and $B(2, 4, -2)$

Solution

(a) From the above theorem, $x = 1 + 5t, \; y = -2 + 4t, \; z = 3 - 6t$.

(b) We need to find a vector parallel to the given line. The vector $\overrightarrow{AB}$ provides a good choice. $\overrightarrow{AB} = \begin{pmatrix} 1 \\ 6 \\ -5 \end{pmatrix}$. So the equations are

$$x = 1 + t, \; y = -2 + 6t, \; z = 3 - 5t$$

Another set of equations could be

$$x = 2 + t, \; y = 4 + 6t, \; z = -2 - 5t$$

Other sets are possible by considering any vector parallel to $\overrightarrow{AB}$.

Vector equation of a line

An alternative route to interpreting the equation

$$\overrightarrow{AM} = tv$$

is to express it in terms of the position vectors r_0 of the fixed point A, and r, the position vector of M.

In Section 9.1 we discussed the difference of two vectors, which we can use here

$$\overrightarrow{AM} = \overrightarrow{OM} - \overrightarrow{OA}$$

$$\Rightarrow tv = r - r_0$$

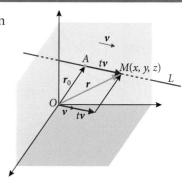

Figure 9.28 Position vector approach to equation of line

And hence we arrive at the vector equation of a line.

> The two approaches are very closely related. We can even say that the parametric equations are a detailed form of the vector equation.
>
> $r = r_0 + tv$
>
> $$\Leftrightarrow \begin{pmatrix} x \\ y \\ z \end{pmatrix} = \begin{pmatrix} x_0 \\ y_0 \\ z_0 \end{pmatrix} + t\begin{pmatrix} a \\ b \\ c \end{pmatrix} = \begin{pmatrix} x_0 \\ y_0 \\ z_0 \end{pmatrix} + \begin{pmatrix} ta \\ tb \\ tc \end{pmatrix} = \begin{pmatrix} x_0 + ta \\ y_0 + tb \\ z_0 + tc \end{pmatrix}$$
>
> $$\Leftrightarrow \begin{cases} x = x_0 + ta \\ y = y_0 + tb \\ z = z_0 + tc \end{cases}$$

The **vector equation** of the line $r = r_0 + tv$

r is the position vector of any point on the line, while r_0 is the position vector of a fixed point (A in this case) on the line and v is the vector parallel to the given line.

We can interpret vector equations in several ways. One of these has to do with displacement. That is, to reach point M from point O, we first arrive at A, and then go towards M along the line a multiple of v, tv.

Looking at Figure 9.29, for each value of t we describe a point on the line.

When $t > 0$, the points are in the same direction as v. When $t < 0$, the points are in the opposite direction.

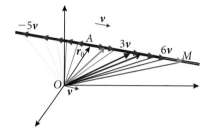

Figure 9.29 Points on a line as displacements

Example 9.21

Find a vector equation of the line that contains the point $(-1, 3, 0)$ and is parallel to $v = 3i - 2j + k$

Solution

From the discussion above, $r = (-i + 3j) + t(3i - 2j + k)$

When $t = 0$, the equation gives the point $(-1, 3)$. When $t = 1$, the equation gives $r = (-i + 3j) + (3i - 2j + k) = 2i + j + k$, a point shifted by $1v$ down the line.

Similarly, when $t = 3$

$r = (-i + 3j) + 3(3i - 2j + k) = 8i - 3j + 3k$, a point $3v$ down the line, etc.

Alternatively, the equation can be written as $r = (-1 + 3t)i + (3 - 2t)j + tk$.

This last form allows us to recognise the parametric equations of the line by simply reading the components of the vector on the right-hand side of the equation.

Example 9.22

Find a vector equation of the line passing through $A(2, 7)$ and $B(6, 2)$.

Solution

Let the vector $\overrightarrow{AB} = \begin{pmatrix} 6 - 2 \\ 2 - 7 \end{pmatrix} = \begin{pmatrix} 4 \\ -5 \end{pmatrix}$ be the vector giving the direction of the line, so $r = \begin{pmatrix} 2 \\ 7 \end{pmatrix} + t\begin{pmatrix} 4 \\ -5 \end{pmatrix}$, or equivalently

$r = 2i + 7j + t(4i - 5j)$

Example 9.23

Find parametric equations for the line through $A(-1, 1, 3)$ and parallel to the vector $v = 2i + 3j - k$.

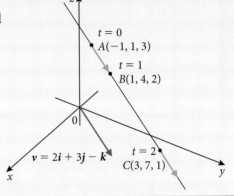

Solution

$x = -1 + 2t, y = 1 + 3t, z = 3 - t$

If we select a few points with their parametric values, we can see how the equation represents the line. For $t = 0$, as we expect, we are at point A; for $t = 1$, the point is B and for $t = 2$, the point is C. The arrows show the direction of increasing values of t.

Line segments

Sometimes, we would like to express a line segment in parameters. That is, to write the equation so that it describes the points making up the segment. For example, to parameterise the line segment between $A(3, 7, 1)$ and $B(1, 4, 2)$, we first find the direction vector

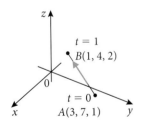

Figure 9.30 Line segments

$$\overrightarrow{AB} = \begin{pmatrix} -2 \\ -3 \\ 1 \end{pmatrix}$$

then we use point A as the fixed point on the line.

Thus, the parametric equations are

$$\begin{cases} x = 3 - 2t \\ y = 7 - 3t \\ z = 1 + t \end{cases}$$

Notice that when $t = 0$, the line starts at the point $A(3, 7, 1)$ and when $t = 1$, the line starts at $B(1, 4, 2)$. Therefore, to parameterise this segment we restrict the values of t to $0 < t < 1$. The new equation is then

$$x = 3 - 2t, y = 7 - 3t, z = 1 + t \quad 0 < t < 1$$

In general, to parameterise a line segment AB so that we represent the points included between the end points only, we can use the vector equation

$$r(t) = (1 - t)\overrightarrow{OA} + t\overrightarrow{OB}, \quad 0 < t < 1$$

In this parameterisation, when $t = 0$, $r = \overrightarrow{OA}$, and when $t = 1$, $r = \overrightarrow{OB}$. This way, r traces the segment AB from A to B for $0 < t < 1$.

The parameterisation with the vector equation can be expressed differently if we want to use parametric equations. If $A(x_1, y_1, z_1)$ and $B(x_2, y_2, z_2)$ are the endpoints of the segment, then

$$r(t) = (1 - t)\overrightarrow{OA} + t\overrightarrow{OB} = \overrightarrow{OA} - t\overrightarrow{OA} + t\overrightarrow{OB}$$
$$= \overrightarrow{OA} + t(\overrightarrow{OB} - \overrightarrow{OA}) = (x_1, y_1, z_1) + t(x_2 - x_1, y_2 - y_1, z_2 - z_1)$$
$$= (x_1 + t(x_2 - x_1), y_1 + t(y_2 - y_1), z_1 + t(z_2 - z_1))$$

So the parametric equations are

$$x = x_1 + t(x_2 - x_1), y = y_1 + t(y_2 - y_1), z = z_1 + t(z_2 - z_1), 0 < t < 1$$

Example 9.24

Parametrise the line segment through $A(2, -1, 5)$ and $B(4, 3, 2)$. Use the equation to find the midpoint of the segment.

Solution

$$r(t) = (1 - t)\overrightarrow{OA} + t\overrightarrow{OB}, \quad 0 < t < 1$$

$$= (1 - t)\begin{pmatrix} 2 \\ -1 \\ 5 \end{pmatrix} + t\begin{pmatrix} 4 \\ 3 \\ 2 \end{pmatrix} = \begin{pmatrix} 2 + 2t \\ -1 + 4t \\ 5 - 3t \end{pmatrix}$$

For the midpoint, $t = \dfrac{1}{2}$, and hence its coordinates are

$$r\left(\frac{1}{2}\right) = \left(2 + 2\left(\frac{1}{2}\right), -1 + 4\left(\frac{1}{2}\right), 5 - 3\left(\frac{1}{2}\right)\right) = \left(3, 1, \frac{7}{2}\right)$$

This method can be used to find points that divide the segment in any ratio; for example, $\dfrac{2}{3}$ of the way from A to B.

Equivalently, the parametric equations can be used

$$x = 2 + t(4 - 2), y = -1 + t(3 + 1), z = 5 + t(2 - 5)$$
$$x = 2 + 2t, y = -1 + 4t, z = 5 - 3t, 0 < t < 1$$

Symmetric (Cartesian) equations of lines

Another set of equations for a line is obtained by eliminating the parameter from the parametric equation.

If $a \neq 0$, $b \neq 0$, and $c \neq 0$, then the set of parametric equations can be re-arranged to yield the set of Cartesian (symmetric) equations:

$$\left.\begin{array}{l} x - x_0 = ta \Leftrightarrow \dfrac{x - x_0}{a} = t \\[2mm] y - y_0 = tb \Leftrightarrow \dfrac{y - y_0}{b} = t \\[2mm] z - z_0 = tc \Leftrightarrow \dfrac{z - z_0}{c} = t \end{array}\right\} \Leftrightarrow \dfrac{x - x_0}{a} = \dfrac{y - y_0}{b} = \dfrac{z - z_0}{c}$$

Notice that the coordinates (x_0, y_0, z_0) of the fixed point A on L appear in the numerators of the fractions, and that the components a, b, and c of a direction vector appear in the denominators of these fractions.

Example 9.25

Find the Cartesian equations of the line through $A(3, -7, 4)$ and $B(1, -4, -1)$.

Solution

In order to use the Cartesian equation, we find the vector v parallel to the line. Since A and B are two points that lie on the line, the vector $\overrightarrow{AB}$ will suffice. Thus, we let

$$v = \overrightarrow{AB} = (1 - 3)i + (-4 + 7)j + (-1 - 4)k = -2i + 3j - 5k$$

If we use A as the fixed point, then the Cartesian equations are

$$\frac{x - 3}{-2} = \frac{y + 7}{3} = \frac{z - 4}{-5}$$

Similarly, if we use B as the fixed point then

$$\frac{x - 1}{-2} = \frac{y + 4}{3} = \frac{z + 1}{-5}$$

Example 9.26

Let L be the line with Cartesian equations

$$\frac{x - 2}{3} = \frac{y + 1}{-2} = z - 4$$

Find a set of parametric equations for L.

Solution

Since the numbers in the denominators are the components of a vector parallel to L, then

$$u = 3i - 2j + k$$

The point $(2, -1, 4)$ lies on L.

Thus, a set of parametric equations of L is

$$x = 2 + 3t, y = -1 - 2t, z = 4 + t$$

A vector equation would be

$$r = \begin{pmatrix} 2 \\ -1 \\ 4 \end{pmatrix} + t\begin{pmatrix} 3 \\ -2 \\ 1 \end{pmatrix}$$

If any of the components a, b, or c is zero, then the Cartesian equations are written in a mixed form. For example, if $c = 0$, then we write

$$\frac{x - x_0}{a} = \frac{y - y_0}{b}, z = z_0$$

For example, the Cartesian set of equations for a line parallel to $2i - 3j$ through the point $(2, 1, -3)$ is

$$\frac{x - 2}{2} = \frac{y - 1}{-3}, z = -3$$

Intersecting, parallel, and skew straight lines

In the plane, lines can coincide, intersect, or be parallel. In space they can also be skew straight lines. Although these lines are not parallel, they do not intersect either. They lie in different planes.

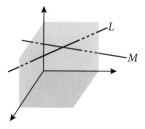

Figure 9.31 Two intersecting straight lines

How do we know whether or not two lines are parallel?

If the direction vectors are parallel, then the lines are. We can check to see if one of the vectors is a scalar multiple of the other. Alternatively, we can find the angle between them, and if it is 0° or 180°, then the lines are either parallel or coincident. The case for coincidence is always there, and we need to check it by examining a point on one of the lines to see whether it is also on the other line.

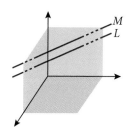

Figure 9.32 Two parallel lines

Example 9.27

Show that these two lines are parallel:

$L_1: x = 2 - 3t, y = t, z = -1 + 2t$

$L_2: x = 1 + 6s, y = 2 - 2s, z = 2 - 4s$

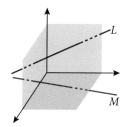

Figure 9.33 Two skew straight lines

Solution

Let l_1 be the vector parallel to L_1 and l_2 be the vector parallel to L_2.

$l_1 = -3i + j + 2k$ and $l_2 = 6i - 2j - 4k$

Now we can easily see that $l_2 = -2l_1$, and hence the vectors are parallel.

To check whether the lines coincide, examine the point $(2, 0, -1)$, which is on the first line, and see whether it lies on the second line, too.

If we choose $y = 0$, then $0 = 2 - 2s$, so $s = 1$; and when we substitute $s = 1$ into $x = 1 + 6s$ we find that x must be 7 in order for the point $(2, 0, -1)$ to be on L_2. Therefore, the lines cannot intersect, and their direction vectors are parallel, so they must be parallel.

Are the lines intersecting or skew?

If the direction vectors are not parallel, then the lines either intersect or are skew. The method starts by examining whether or not the lines intersect. If they do, then we can find the coordinates of the point of intersection; if they do not intersect, then we cannot find the coordinates of the point of intersection. Finding the coordinates of the point of intersection is a straightforward method that you already know: solving systems of equations.

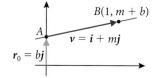

Figure 9.34 Vector equation of line in 2D

The vector equation of a line has an interesting application in proving a result that you already know – the condition for two lines to be perpendicular.

The slope-intercept form of the equation of the line is $y = mx + b$. We can consider the fixed point on the line to be its y-intercept, i.e., $r_0 = bj$, and another point $(1, m + b)$ on the line $\Rightarrow r = i + (m + b)j$. So, the direction vector of the line is

$v = r - r_0 = (i + (m + b)j) - bj = i + mj$.

Now, if you have two lines with slopes m_1 and m_2, their direction vectors can be written as $v_1 = i + m_1 j$ and $v_2 = i + m_2 j$.
For the two lines to be perpendicular, their direction vectors will also be perpendicular, and hence

$v_1 \cdot v_2 = (i + m_1 j) \cdot (i + m_2 j) = 1 + m_1 m_2 = 0 \Rightarrow m_1 m_2 = -1$

Example 9.28

The lines L_1 and L_2 have the equations:

$L_1: x = 1 + 4t, y = 5 - 4t, z = -1 + 5t$

$L_2: x = 2 + 8s, y = 4 - 3s, z = 5 + s$

Show that the lines are skew.

Solution

We first examine whether the lines are parallel. However, since the vector parallel to L_1 is $\boldsymbol{l}_1 = (4, -4, 5)$ and the vector parallel to L_2 is $\boldsymbol{l}_2 = (8, -3, 1)$ are not scalar multiples of each other, the vectors and consequently the lines are not parallel.

For the lines to intersect there should be some point $M(x_0, y_0, z_0)$ that satisfies the equations of both lines for some values of t and s. That is

$$x_0 = 1 + 4t = 2 + 8s; y_0 = 5 - 4t = 4 - 3s; z_0 = -1 + 5t = 5 + s$$

This leads to a set of three simultaneous equations in two unknowns: s and t.

By solving the first two equations:

$$\left.\begin{array}{r} 1 + 4t = 2 + 8s \\ 5 - 4t = 4 - 3s \end{array}\right\} \Rightarrow 6 = 6 + 5s \Rightarrow s = 0, t = \frac{1}{4}$$

For the system to be consistent, these values must satisfy the third equation, $-1 + \dfrac{5}{4} = 5 + 0$, which is false. Hence the system is inconsistent and the lines are skew.

Example 9.29

The lines L_1 and L_2 have the following parametric equations:

$L_1: x = 1 + 2t, y = 3 - 4t, z = -2 + 4t$

$L_2: x = 4 + 3s, y = 4 + s, z = -4 - 2s$

Show that the lines intersect.

Solution

We first examine whether the lines are parallel. However, since the vector parallel to L_1 is $\boldsymbol{l}_1 = \begin{pmatrix} 2 \\ -4 \\ 4 \end{pmatrix}$ and the vector parallel to L_2 is $\boldsymbol{l}_2 = \begin{pmatrix} 3 \\ 1 \\ -2 \end{pmatrix}$ are not scalar multiples of each other, the vectors and consequently the lines are not parallel.

For the lines to intersect there should be some point $M(x_0, y_0, z_0)$ that satisfies the equations of both lines for some values of t and s. That is

$$x_0 = 1 + 2t = 4 + 3s; y_0 = 3 - 4t = 4 + s; z_0 = -2 + 4t = -4 - 2s$$

This leads to a set of three simultaneous equations in two unknowns: s and t.

By solving the first two equations:

$$\left.\begin{array}{l} 1 + 2t = 4 + 3s \\ 3 - 4t = 4 + s \end{array}\right\} \Rightarrow 5 = 12 + 7s \Rightarrow s = -1, t = 0$$

For the system to be consistent, these values must satisfy the third equation,

$-2 + 4(0) = -4 - 2(-1) \Rightarrow -2 = -2$, which is a correct statement. Hence the two lines intersect.

The point of intersection can be found through substitution of the value of the parameter into the corresponding line equation:

$$L_1: (1, 3, -2) \text{ and also in } L_2: (4 - 3, 4 - 1, -4 - 2(-1)) = (1, 3, -2)$$

In vector form, finding the point of intersection, if it exists, follows a similar approach. For example, the vector equations of lines L_1 and L_2 are

$$L_1: r = \begin{pmatrix} 1 \\ 3 \\ -2 \end{pmatrix} + t\begin{pmatrix} 2 \\ -4 \\ 4 \end{pmatrix} \qquad L_2: r = \begin{pmatrix} 4 \\ 4 \\ -4 \end{pmatrix} + s\begin{pmatrix} 3 \\ 1 \\ -2 \end{pmatrix}$$

The condition of intersection is therefore

$$\begin{pmatrix} 1 \\ 3 \\ -2 \end{pmatrix} + t\begin{pmatrix} 2 \\ -4 \\ 4 \end{pmatrix} = \begin{pmatrix} 4 \\ 4 \\ -4 \end{pmatrix} + s\begin{pmatrix} 3 \\ 1 \\ -2 \end{pmatrix}$$

which leads to the same conclusion as in the case using parametric equations.

Application of lines to motion

The vector form of the equation of a line in space is more revealing when we think of the line as the path of an object, placed in an appropriate coordinate system, starting at position $A(x_0, y_0, z_0)$, and moving in the direction of v (Figure 9.35). Generally speaking, we find an object at an initial location A, represented by r_0.

The object moves on its path with a velocity vector $v = \begin{pmatrix} a \\ b \\ c \end{pmatrix}$.

The object's position at any point in time after the start can then be described by $r = r_0 + tv$.

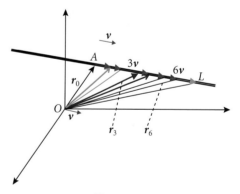

Figure 9.35 Vector equation as a path

Assuming the unit of time to be seconds, the equation tells us that for every second, the object moves a units in the x direction, b units in the y direction and c units in the z direction. So, for example, after 2 seconds we find the object at

$$r = r_0 + 2v.$$

The speed of the object is then $|v|$ in the v direction.

In general, we can write the vector equation in a slightly modified form

$$r(t) = r_0 + tv$$

$$= \quad\underbrace{r_0}_{\text{Initial position}} \quad + \quad \underbrace{t}_{\text{Time}} \quad \underbrace{|v|}_{\text{Speed}} \quad \underbrace{\frac{v}{|v|}}_{\text{Direction}}$$

In other words, the position of an object at time t is the initial position plus its *rate* $\times$ *time* (distance moved) in the direction $\boldsymbol{u} = \dfrac{\boldsymbol{v}}{|\boldsymbol{v}|}$ of its straight-line motion.

Let's clarify this in Example 9.30

Example 9.30

A model aeroplane is to fly directly from a platform at a reference point $(2, 1, 1)$ towards a point $(5, 5, 6)$ at a speed of $60\,\text{m}\,\text{min}^{-1}$.
What is the position of the aeroplane after 10 minutes, to the nearest metre?

Solution

The unit vector in the direction of the flight is $\boldsymbol{u} = \dfrac{3}{5\sqrt{2}}\boldsymbol{i} + \dfrac{4}{5\sqrt{2}}\boldsymbol{j} + \dfrac{5}{5\sqrt{2}}\boldsymbol{k}$

The position of the aeroplane at any time t is

$$\boldsymbol{r}(t) = \boldsymbol{r}_0 + t(speed)(\boldsymbol{u})$$

$$= (2\boldsymbol{i} + \boldsymbol{j} + \boldsymbol{k}) + (10)(60)\left(\dfrac{3}{5\sqrt{2}}\boldsymbol{i} + \dfrac{4}{5\sqrt{2}}\boldsymbol{j} + \dfrac{5}{5\sqrt{2}}\boldsymbol{k}\right)$$

$$= (2\boldsymbol{i} + \boldsymbol{j} + \boldsymbol{k}) + \left(\dfrac{360}{\sqrt{2}}\boldsymbol{i} + \dfrac{480}{\sqrt{2}}\boldsymbol{j} + \dfrac{600}{\sqrt{2}}\boldsymbol{k}\right)$$

So, the aeroplane is approximately at $(257, 340, 425)$, to the nearest metre.

Example 9.31

An object is moving in the plane of an appropriately fitted coordinate system such that its position is given by $\boldsymbol{r} = \begin{pmatrix} 3 \\ 1 \end{pmatrix} + t\begin{pmatrix} -2 \\ 3 \end{pmatrix}$, where t is the time in hours after start, and distances are measured in km.
(a) Find the initial position of the object.
(b) Show the position of the object on a graph at the start, after 1 hour, and after 3 hours.
(c) Find the velocity and speed of the object.

Solution

(a) Initial position is when $t = 0$. This is the point $(3, 1)$.

(b)

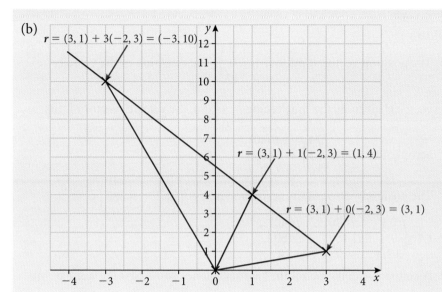

$r = (3, 1) + 3(-2, 3) = (-3, 10)$

$r = (3, 1) + 1(-2, 3) = (1, 4)$

$r = (3, 1) + 0(-2, 3) = (3, 1)$

(c) The velocity vector is $v = \begin{pmatrix} -2 \\ 3 \end{pmatrix}$, which means that every hour the object moves 2 units west and 3 units north.

The speed is $|v| = \sqrt{(-2)^2 + 3^2} = \sqrt{13}\ \mathrm{km\,h^{-1}}$

We can also express the velocity as $\sqrt{13}\ \mathrm{km\,h^{-1}}$ in the direction of $(-2, 3)$.

We can also express the direction in terms of the unit vector in the direction of v instead. That is, we can say that the speed is $\sqrt{13}\ \mathrm{km\,h^{-1}}$ in the direction of $\begin{pmatrix} \dfrac{-2}{\sqrt{13}} \\ \dfrac{3}{\sqrt{13}} \end{pmatrix}$

or equivalently at an angle of $\cos^{-1}\left(\dfrac{-2}{\sqrt{13}}\right) \approx 124°$ to the positive x-direction.

Example 9.32

At midday, an aeroplane A is passing in the vicinity of an airport at a height of 12 km and a speed of 800 km h^{-1}. The direction of the aeroplane is $\begin{pmatrix} 4 \\ 3 \\ 0 \end{pmatrix}$. Consider that $\begin{pmatrix} 1 \\ 0 \\ 0 \end{pmatrix}$ is a displacement of 1 km due east, $\begin{pmatrix} 0 \\ 1 \\ 0 \end{pmatrix}$ is a displacement of due north, and $\begin{pmatrix} 0 \\ 0 \\ 1 \end{pmatrix}$ is an altitude of 1 km.

(a) Using the airport as the origin, find the position vector r of the aeroplane t hours after midday.

(b) Find the position of the aeroplane 1 hour after midday.

(c) Another aeroplane B is heading towards the airport with velocity vector $\begin{pmatrix} -300 \\ -400 \\ 0 \end{pmatrix}$ from a location $\begin{pmatrix} 600 \\ 480 \\ 12 \end{pmatrix}$. Is there a danger of collision?

Solution

(a) The position vector at midday is $\begin{pmatrix} 0 \\ 0 \\ 12 \end{pmatrix}$. The direction of the velocity

vector is given by the unit vector $\dfrac{1}{5}\begin{pmatrix} 4 \\ 3 \\ 0 \end{pmatrix}$. So, the velocity vector of this

aeroplane is $800 \cdot \dfrac{1}{5}\begin{pmatrix} 4 \\ 3 \\ 0 \end{pmatrix} = \begin{pmatrix} 640 \\ 480 \\ 0 \end{pmatrix}$

The position vector of the aeroplane is: $r = \begin{pmatrix} 0 \\ 0 \\ 12 \end{pmatrix} + t\begin{pmatrix} 640 \\ 480 \\ 0 \end{pmatrix}$

(b) $r = \begin{pmatrix} 0 \\ 0 \\ 12 \end{pmatrix} + t\begin{pmatrix} 640 \\ 480 \\ 0 \end{pmatrix} = \begin{pmatrix} 640 \\ 480 \\ 12 \end{pmatrix}$

(c) A collision can happen if the two aeroplanes paths pass the same point
at the same time.

The position vector for the second aeroplane is

$r = \begin{pmatrix} 600 \\ 480 \\ 12 \end{pmatrix} + t\begin{pmatrix} -300 \\ -400 \\ 0 \end{pmatrix}$, if the two paths intersect, they may intersect

at instances corresponding to t_1 and t_2 and they should have the same
position, thus,

$\begin{pmatrix} 0 \\ 0 \\ 12 \end{pmatrix} + t_1\begin{pmatrix} 640 \\ 480 \\ 0 \end{pmatrix} = \begin{pmatrix} 600 \\ 480 \\ 12 \end{pmatrix} + t_2\begin{pmatrix} -300 \\ -400 \\ 0 \end{pmatrix}$

This gives rise to a set of three equations in two variables

$$\begin{cases} 640\,t_1 = 600 - 300\,t_2 \\ 480\,t_1 = 480 - 400\,t_2 \\ 12 = 12 \end{cases}$$

Solving the system of equations simultaneously will give $t_1 = \dfrac{6}{7}, t_2 = \dfrac{6}{35}$

This means that the aeroplanes' paths will cross at $(548.57, 411.43, 12)$.
However, there is no collision because aeroplane A will pass that point at
12:51 while aeroplane B will pass this point at 12.10.

Distance from a point to a line (optional)

2-dimensional space

There are several methods of
proving this theorem. We will
follow a vector approach.

The x- and y-intercepts of the line l are

$N\left(-\dfrac{c}{a}, 0\right)$ and $M\left(0, -\dfrac{c}{b}\right)$.

The equation of a line l is
written in the form
$ax + by + c = 0$.
The distance from a point
$P_0(x_0, y_0)$ to the line l is
given by

$d = \dfrac{|ax_0 + by_0 + c|}{\sqrt{a^2 + b^2}}$

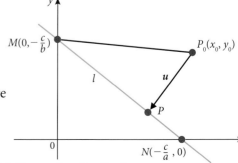

Figure 9.36 Distance from P_0 to line l

So, a vector parallel to l can be any vector in the direction of $\overrightarrow{NM} = \begin{pmatrix} \frac{c}{a} \\ -\frac{c}{b} \end{pmatrix}$

For convenience we will consider the vector $L = \begin{pmatrix} \frac{1}{a} \\ -\frac{1}{b} \end{pmatrix}$

Consider a vector in the direction of $\overrightarrow{P_0 P}$ perpendicular to l and vector $u = \begin{pmatrix} a \\ b \end{pmatrix}$
that is also perpendicular to l because $u \times L = 0$. So u is parallel to $\overrightarrow{P_0 P}$, then
in triangle MPP_0, the distance $\left|\overrightarrow{P_0 P}\right|$ is $\left|\overrightarrow{MP_0}\right| \cos(M\widehat{P_0}P)$.

$$\overrightarrow{MP_0} = \left(x_0, y_0 + \frac{c}{b}\right)$$

$$\left|\overrightarrow{P_0 P}\right| = \left\|\overrightarrow{MP_0}\right\| \cos(M\widehat{P_0}P) = \left\|\overrightarrow{MP_0}\right\| \cdot \frac{\overrightarrow{MP_0} \cdot \overrightarrow{P_0 P}}{\left|\overrightarrow{MP_0}\right| \cdot \left|\overrightarrow{P_0 P}\right|} = \frac{\left|\overrightarrow{MP_0} \cdot u\right|}{|u|}$$

$$= \left|\frac{\left(x_0, y_0 + \frac{c}{b}\right) \cdot (a, b)}{\sqrt{a^2 + b^2}}\right| = \frac{|a x_0 + b y_0 + c|}{\sqrt{a^2 + b^2}}$$

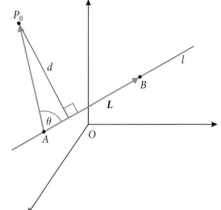

3-dimensional space

$$d = \left|\overrightarrow{AP_0}\right| \cdot \sin(\theta) = \left|\overrightarrow{AP_0}\right| \cdot \frac{\left|L \times \overrightarrow{AP_0}\right|}{|L| \cdot \left|\overrightarrow{AP_0}\right|} = \frac{\left|L \times \overrightarrow{AP_0}\right|}{|L|}$$

Where A is any point on line l and L is a vector parallel to l.

Figure 9.37 Distance from a point to a line in 3D

Example 9.33

Find the distance from the point $(1, 3)$ to the line with equation $2x - y = 7$.

Solution

The equation can be written as $2x - y - 7 = 0$ and hence the distance is
$$d = \frac{|2(1) - 3 - 7|}{\sqrt{2^2 + 1}} = \frac{8}{\sqrt{5}}$$

Example 9.34

Find the distance from the point $P(8, 1, -3)$ to the line containing $M(3, 0, 6)$ and $N(5, -2, 7)$.

Solution

$$\overrightarrow{MN} = (2, -2, 1) \text{ and } \overrightarrow{MP} = (5, 1, -9) \Rightarrow d = \frac{|(2, -2, 1) \times (5, 1, -9)|}{|(2, -2, 1)|}$$

$$= \frac{|(17, 23, 12)|}{\sqrt{2^2 + (-2)^2 + 1}} = \frac{\sqrt{17^2 + 23^2 + 12^2}}{3} = \frac{\sqrt{962}}{3}$$

Distance between two skew straight lines

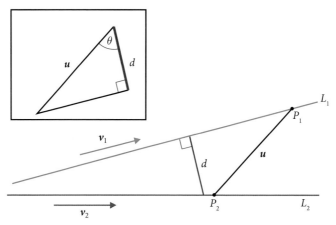

Figure 9.38 Two skew straight lines L_1 and L_2 are given with direction vectors v_1 and v_2

In Figure 9.38, two skew straight lines L_1 and L_2 are shown with direction vectors v_1 and v_2 respectively. We need to find the distance d, defined as the length of the common perpendicular, between them.

Consider any two fixed points P_1 on L_1 and P_2 on L_2. The distance d is the length of the orthogonal projection of vector $u = \overrightarrow{P_1 P_2}$ on a direction perpendicular to both v_1 and v_2. This direction can be determined by $v_1 \times v_2$. So,

$$d = \left\| \overrightarrow{P_1 P_2} \right| \cos \theta \right| = \left| \overrightarrow{P_1 P_2} \frac{\overrightarrow{P_1 P_2} \cdot (v_1 \times v_2)}{\left| \overrightarrow{P_1 P_2} \right| |v_1 \times v_2|} \right| = \left| \frac{\overrightarrow{P_1 P_2} \cdot (v_1 \times v_2)}{|v_1 \times v_2|} \right|$$

Example 9.35

Find the distance between the following skew lines:

$$L_1: r = \begin{pmatrix} 2 \\ 3 \\ 1 \end{pmatrix} + t \begin{pmatrix} 1 \\ 2 \\ -3 \end{pmatrix}; \ L_2: r = 4i + 2j + s(3i - j + k)$$

Solution

The two fixed points could be taken as $(2, 3, 1)$ and $(4, 2, 0)$ while the vectors are $v_1 = \begin{pmatrix} 1 \\ 2 \\ -3 \end{pmatrix}$ and $v_2 = \begin{pmatrix} 3 \\ -1 \\ 1 \end{pmatrix}$.

$$d = \left| \frac{(2, -1, -1) \cdot (-1, -10, -7)}{|(-1, -10, -7)|} \right| = \left| \frac{-2 + 10 + 7}{\sqrt{1 + 100 + 49}} \right| = \frac{15}{\sqrt{150}} = \frac{\sqrt{6}}{2}$$

The minimum distance could be found using other methods too. One of them would be to consider the line going from any point on L_1 to any point on L_2. This will give a parametric equation in s and t. Then, considering that this line will be perpendicular to both L_1 and L_2 ($u \cdot v_1 = 0$ and $u \cdot v_2 = 0$) enables us to set up a system of two equations that could be solved for s and t. Lastly, we get the distance between the points corresponding to the specific values we just established.

1. Find a vector equation, a set of parametric equations, and a set of Cartesian equations of the line containing the point A and parallel to the vector $\boldsymbol{u}$.

   (a) $A(-1, 0, 2)$, $\boldsymbol{u} = \begin{pmatrix} 1 \\ 5 \\ -4 \end{pmatrix}$

   (b) $A(3, -1, 2)$, $\boldsymbol{u} = \begin{pmatrix} 2 \\ 5 \\ -1 \end{pmatrix}$

   (c) $A(1, -2, 6)$, $\boldsymbol{u} = 3\boldsymbol{i} + 5\boldsymbol{j} - 11\boldsymbol{k}$

2. Find the equation of the line that passes through the points A and B.

   (a) $A(-1, 4, 2)$, $B(7, 5, 0)$

   (b) $A(4, 2, -3)$, $B(0, -2, 1)$

   (c) $A(1, 3, -3)$, $B(5, 1, 2)$

3. (a) Write the equation of the line through the points $(3, -2)$ and $(5, 1)$ in the form $\boldsymbol{r} = \boldsymbol{a} + t\boldsymbol{b}$.

   (b) Write the equation of the line through the points $(0, -2)$ and $(5, 0)$ in the form $\boldsymbol{r} = \boldsymbol{a} + t\boldsymbol{b}$.

4. The equation of a line in 2D space is given by $\boldsymbol{r} = \begin{pmatrix} 2 \\ 1 \end{pmatrix} + t\begin{pmatrix} 3 \\ -2 \end{pmatrix}$
   Write the equation in the form $ax + by = c$

5. Find the equation of a line through $(2, -3)$ that is parallel to the line with equation
   $$\boldsymbol{r} = 3\boldsymbol{i} - 7\boldsymbol{j} + \lambda(4\boldsymbol{i} - 3\boldsymbol{j})$$

6. Find the equation of a line through $(-2, 1, 4)$ and parallel to the vector
   $$3\boldsymbol{i} - 4\boldsymbol{j} + 7\boldsymbol{k}$$

7. Find the point of intersection of the two given lines, and if they do not intersect, explain why.

   (a) $L_1: \boldsymbol{r} = \begin{pmatrix} 2 \\ 2 \\ 3 \end{pmatrix} + t\begin{pmatrix} 1 \\ 3 \\ 1 \end{pmatrix}$; $L_2: \boldsymbol{r} = \begin{pmatrix} 2 \\ 3 \\ 4 \end{pmatrix} + s\begin{pmatrix} 1 \\ 4 \\ 2 \end{pmatrix}$

   (b) $L_1: \boldsymbol{r} = \begin{pmatrix} -1 \\ 3 \\ 1 \end{pmatrix} + t\begin{pmatrix} 4 \\ 1 \\ 0 \end{pmatrix}$; $L_2: \boldsymbol{r} = \begin{pmatrix} -13 \\ 1 \\ 2 \end{pmatrix} + s\begin{pmatrix} 12 \\ 6 \\ 3 \end{pmatrix}$

   (c) $L_1: \boldsymbol{r} = \boldsymbol{i} + 3\boldsymbol{j} + 5\boldsymbol{k} + t(7\boldsymbol{i} + \boldsymbol{j} - 3\boldsymbol{k})$;
       $L_2: \boldsymbol{r} = 4\boldsymbol{i} + 6\boldsymbol{j} + 7\boldsymbol{k} + s(-\boldsymbol{i} + 2\boldsymbol{k})$

   (d) $L_1: \begin{pmatrix} x \\ y \\ z \end{pmatrix} = \begin{pmatrix} 3 \\ 4 \\ 6 \end{pmatrix} + t\begin{pmatrix} -2 \\ 1 \\ -1 \end{pmatrix}$; $L_2: \begin{pmatrix} x \\ y \\ z \end{pmatrix} = \begin{pmatrix} 5 \\ -2 \\ 7 \end{pmatrix} + s\begin{pmatrix} -4 \\ 2 \\ -2 \end{pmatrix}$

8. Find the vector and parametric equations of each line:

   (a) through the points $(2, -1)$ and $(3, 2)$

   (b) through the point $(2, -1)$ and parallel to the vector $\begin{pmatrix} -3 \\ 7 \end{pmatrix}$

   (c) through the point $(2, -1)$ and perpendicular to the vector $\begin{pmatrix} -3 \\ 7 \end{pmatrix}$

   (d) with y-intercept $(0, 2)$ and in the direction of $2\boldsymbol{i} - 4\boldsymbol{j}$.

9. Consider the line with equation

$$\begin{pmatrix} x \\ y \\ z \end{pmatrix} = \begin{pmatrix} 3 \\ 4 \\ 6 \end{pmatrix} + t\begin{pmatrix} -2 \\ 1 \\ -1 \end{pmatrix}$$

(a) For what value of t does this line pass through the point $\left(0, \frac{11}{2}, \frac{9}{2}\right)$?

(b) Does the point $(-1, 4, 6)$ lie on this line?

(c) For what value of m does the point $\left(\frac{1-2m}{2}, 2m, 3\right)$ lie on the line?

10. Consider the following equations representing the paths of cars after starting time $t \geqslant 0$, where distances are measured in km and time in hours. For each car, determine the:

(i) starting position   (ii) velocity vector   (iii) speed.

(a) $r = \begin{pmatrix} 3 \\ -4 \end{pmatrix} + t\begin{pmatrix} 7 \\ 24 \end{pmatrix}$ (b) $\begin{pmatrix} x \\ y \end{pmatrix} = \begin{pmatrix} -3 \\ 1 \end{pmatrix} + t\begin{pmatrix} 5 \\ -12 \end{pmatrix}$

(c) $\begin{pmatrix} x \\ y \end{pmatrix} = \begin{pmatrix} 5 \\ -2 \end{pmatrix} + t\begin{pmatrix} 24 \\ -7 \end{pmatrix}$

11. Find the velocity vector, using the following data, for each racing car.

(a) direction $\begin{pmatrix} -3 \\ 4 \end{pmatrix}$ with a speed of $160 \, \text{km h}^{-1}$

(b) direction $\begin{pmatrix} 12 \\ -5 \end{pmatrix}$ with a speed of $170 \, \text{km h}^{-1}$

12. After leaving an intersection of roads located 3 km east and 2 km north of a city, a car is moving towards a traffic light 7 km east and 5 km north of the city at a constant speed of 30 km/h. Take the city as origin for an appropriate coordinate system.

(a) What is the velocity vector of the car?

(b) Write down the equation of the position of the car after t hours.

(c) When will the car reach the traffic light?

13. Consider the vectors $u = \begin{pmatrix} 1 \\ a \\ b \end{pmatrix}$, $v = i - 3j + 2k$ and $w = -2i + j - k$

(a) Find a and b so that u is perpendicular to both v and w.

(b) O is the origin, P a point whose position vector is v and Q has position vector w. Find the cosine of the angle between v and w.

(c) Hence find the sine of the angle and use it to find the area of the triangle OPQ.

14. The triangle ABC has vertices at the points $A(-1, 2, 3)$, $B(-1, 3, 5)$ and $C(0, -1, 1)$.

(a) Find the size of the angle θ between the vectors $\overrightarrow{AB}$ and $\overrightarrow{AC}$

(b) Hence, or otherwise, find the area of triangle ABC

Let L_1 be the line parallel to $\overrightarrow{AB}$ that passes through $D(2, -1, 0)$ and L_2 be the line parallel to $\overrightarrow{AC}$ that passes through $E(-1, 1, 1)$

(c) (i) Find the equations of the lines L_1 and L_2.

(ii) Hence show that L_1 and L_2 do not intersect.

15. Consider the points $A(1, 3, -17)$ and $B(6, -7, 8)$ that lie on line l.

 (a) Find an equation of line l, giving your answer in parametric form.

 (b) Point P is on l such that $\overrightarrow{OP}$ is perpendicular to l.
 Find the coordinates of P.

16. (a) Starting with the equation of a line in the form $mx + ny = p$, find a vector equation of the line.

 (b) Starting with a vector equation of a line where $r = r_0 + tv$, with

 $$r_0 = \begin{pmatrix} x_0 \\ y_0 \end{pmatrix}, v = \begin{pmatrix} a \\ b \end{pmatrix},$$

 (i) find an equation of the line in the form $mx + ny = p$.

 (ii) What is the relationship between the components of the direction vector $v = \begin{pmatrix} a \\ b \end{pmatrix}$ and the slope of the line?

17. Find a parametrisation for the line segment between points A and B.

 (a) $A(0, 0, 0)$, $B(1, 1, 3)$ 　　　　 (b) $A(-1, 0, 1)$, $B(1, 1, -2)$

 (c) $A(1, 0, -1)$, $B(0, 3, 0)$

18. Find a vector equation and a set of parametric equations of the line through the point $(0, 2, 3)$ and parallel to the line $r = (i - 2j) + 2tk$

19. Find a vector equation and a set of parametric equations of the line through the point $(1, 2, -1)$ and parallel to the line $r = t(2i - 3j + k)$

20. Find a vector equation and a set of parametric equations of the line through the origin and the point $A(x_0, y_0, z_0)$.

21. Find a vector equation and a set of parametric equations of the line through $(3, 2, -3)$ and perpendicular to:

 (a) the xz-plane 　　　　　　　 (b) the yz-plane.

22. Write a set of symmetric equations for the line through the origin and the point $A(x_0, y_0, z_0)$, $x_0, y_0, z_0 \neq 0$.

23. Determine whether the lines l_1 and l_2 are parallel, skew, or intersecting. If they intersect, find the coordinates of the point of intersection.

 (a) $l_1: x - 3 = 1 - y = \dfrac{z - 5}{2}$, $l_2: r = i + 4j + 2k + \lambda(j + k)$

 (b) $l_1: \begin{cases} x = -1 + s \\ y = 2 - 3s \\ z = 1 + 2s \end{cases}$; $l_2: \begin{cases} x = 2 - 2m \\ y = -1 + 6m \\ z = -4m \end{cases}$

 (c) $l_1: \dfrac{x - 3}{2} = \dfrac{1 + y}{4} = 2 - z$, $l_2: r = 3i + 2j - 2k + \lambda(2i + j + 2k)$

 (d) $l_1: x - 1 = \dfrac{y - 1}{3} = \dfrac{z + 4}{2}$, $l_2: 1 - x = -1 - y = \dfrac{z}{2}$

(e) $l_1: r = i + 2j + \lambda(-6i + 9j - 3k)$, $l_2: r = 2i + 3j + m(2i - 3j + k)$

(f) $\dfrac{x - 2}{5} = y - 1 = \dfrac{z - 2}{3}$ and $\dfrac{x + 4}{3} = \dfrac{7 - y}{3} = \dfrac{10 - z}{4}$

(g) $x = 1 + t, y = 2 - 2t, z = t + 5$ and
$x = 2 + 2m, y = 5 - 9m, z = 2 + 6m$

24. Find the point on the line $r = 2i + 3j + k + t(-3i + j + k)$ that is closest to the origin. (Use the parametric form and the distance formula and minimise the distance using derivatives.)

25. Find the point on the line $r = 4j + 5k + t(i - 3j + 2k)$ that is closest to the origin.

26. Find the point on the line $r = 5i + 2j + k + t(i - 3j + k)$ that is closest to the point $(-1, 4, 1)$.

9.4 Planes

To define or specify a plane is to identify it in such a way that makes it unique. One way is to set up an equation in a frame that will identify every point that belongs to the plane. There are many ways of specifying a plane.

A plane can be defined:

• by three non-collinear points
• by two intersecting straight lines
• to be perpendicular to a certain direction and at a specific distance from the origin, for example
• by being drawn through a given point and perpendicular to a given direction.

A direction, for our purposes, can be defined by a vector. In the case of a plane, the vector determining the direction is perpendicular to the plane and is said to be normal to the plane.

Equations of a plane

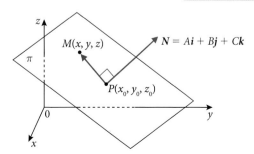

Figure 9.39 Normal to plane

The third and fourth ways of defining a plane are appropriate for deriving equations of a plane.

Cartesian (scalar) equation of a plane

Consider a plane π and a fixed point $P(x_0, y_0, z_0)$ on that plane. A vector $N = Ai + Bj + Ck$, called the **normal vector** to the plane, is perpendicular to the plane.

To find an equation for the plane, consider an arbitrary point $M(x, y, z)$ in space. Recalling that a line perpendicular to the plane is perpendicular to every line in the plane, we can conclude that for the point M to be on the plane, the vector $\mathbf{N}$ must be perpendicular to $\overrightarrow{PM}$.

Hence

$$\mathbf{N} \cdot \overrightarrow{PM} = 0 \text{ but } \overrightarrow{PM} = (x - x_0)\mathbf{i} + (y - y_0)\mathbf{j} + (z - z_0)\mathbf{k}, \text{ and}$$
$$\mathbf{N} \cdot \overrightarrow{PM} = 0 \Leftrightarrow (A\mathbf{i} + B\mathbf{j} + C\mathbf{k}) \cdot ((x - x_0)\mathbf{i} + (y - y_0)\mathbf{j} + (z - z_0)\mathbf{k}) = 0$$

Using the scalar product definition this can be simplified to

$$A(x - x_0) + B(y - y_0) + C(z - z_0) = 0$$

This is a Cartesian equation of a plane that passes through a point $P(x_0, y_0, z_0)$ and has a normal vector $\mathbf{N} = A\mathbf{i} + B\mathbf{j} + C\mathbf{k}$.

If $\mathbf{N}$ is normal to a given plane, then any vector parallel to $\mathbf{N}$ will be normal to the plane. Suppose we have chosen $3\mathbf{N}$ as our normal, then

$$3A(x - x_0) + 3B(y - y_0) + 3C(z - z_0) = 0$$
$$\Leftrightarrow A(x - x_0) + B(y - y_0) + C(z - z_0) = 0$$

Specifically, the unit vector $\mathbf{n}$ in the same direction as $\mathbf{N}$ is of particular importance, as we will see soon.

The equation can be simplified further:

$$A(x - x_0) + B(y - y_0) + C(z - z_0) = 0 \Leftrightarrow Ax + By + Cz = Ax_0 + By_0 + Cz_0,$$

and setting $Ax_0 + By_0 + Cz_0 = D$ will give us a more concise form of the equation

$$Ax + By + Cz = D$$

which is similar to the equation of a line in the plane, i.e., $Ax + By = C$

> In many sources, the equation of the plane is given in the form $Ax + By + Cz + D = 0$. This is the case when we set the quantity $Ax_0 + By_0 + Cz_0 = -D$. Each form has some advantage in using it. We will adhere to the previous form for reasons that will become clear.

Example 9.36

Write an equation for the plane that contains $(2, -3, 5)$ and has normal $\mathbf{N} = 2\mathbf{i} + \mathbf{j} - 3\mathbf{k}$.

Solution

A Cartesian equation for the plane is of the form

$$A(x - x_0) + B(y - y_0) + C(z - z_0) = 0$$
$$\Rightarrow 2(x - 2) + (y + 3) - 3(z - 5) = 0$$
$$\Rightarrow 2x + y - 3z = -14$$

Alternatively, since $\mathbf{N} = 2\mathbf{i} + \mathbf{j} - 3\mathbf{k}$ is the normal to the plane, then $2x + y - 3z = D$, but the line contains the point $(2, -3, 5)$, and thus

$$2(2) - 3 - 3(5) = D \Rightarrow -14 = D$$

therefore

$$2x + y - 3z = -14 \text{ as before.}$$

Example 9.37

Show that every equation of the form $Ax + By + Cz = D$ with $A^2 + B^2 + C^2 \neq 0$ represents a plane in space.

Solution

The equation $Ax + By + Cz = D$ is a linear equation in 3 variables, x, y, and z. This means that it has an infinite number of solutions, and hence we can be confident that there exist numbers x_0, y_0, z_0 such that $Ax_0 + By_0 + Cz_0 = D$.

Since the equation $Ax + By + Cz = D$ is also true, then

$$Ax + By + Cz = Ax_0 + By_0 + Cz_0 = D$$
$$\Leftrightarrow Ax + By + Cz - (Ax_0 + By_0 + Cz_0) = 0$$
$$\Leftrightarrow A(x - x_0) + B(y - y_0) + C(z - z_0) = 0$$

The last equation represents the equation of a plane through a fixed point $P(x_0, y_0, z_0)$ with a normal vector $N = Ai + Bj + Ck$. The condition $A^2 + B^2 + C^2 \neq 0$ guarantees that $N \neq 0$.

Vector equation of a plane

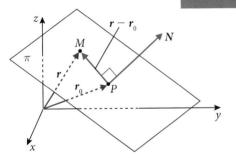

Figure 9.40 Normal to a plane is perpendicular to any vector in the plane

We can write the equation of the plane in vector notation. Use the same set up as before: the normal vector $N = Ai + Bj + Ck$, a fixed point $P(x_0, y_0, z_0)$ with a position vector $r_0 = x_0i + y_0j + z_0k$, and an arbitrary point $M(x, y, z)$ with a position vector $r = xi + yj + zk$.

The equation $A(x - x_0) + B(y - y_0) + C(z - z_0) = 0$ can be interpreted as the scalar product $N \cdot (r - r_0) = 0$ (Figure 9.40).

The normal N is perpendicular to $\overrightarrow{PM} = r - r_0$ and hence their dot product must be zero. Using the distributive property of the scalar product, we have

$$N \cdot (r - r_0) = 0 \Leftrightarrow N \cdot r - N \cdot r_0 = 0$$
$$\Leftrightarrow N \cdot r = N \cdot r_0$$

This is one form of the vector equation of a plane that passes through a point with position vector r_0 and has a normal N.

Notice here that
$$N \cdot r = Ax + By + Cz$$
and
$$N \cdot r_0 = Ax_0 + By_0 + Cz_0$$
$$= D,$$
which shows that $N \cdot r = N \cdot r_0$ is another way of stating $Ax + By + Cz = D$.

Unit vector equation of a plane

Figure 9.41 shows vector N as drawn from the origin O, along with the position vectors r and r_0.

The vector equation $N \cdot r = N \cdot r_0$ can be investigated further

$$N \cdot r = |N||r|\cos\theta_1 = |N|OR = |N|d$$

where $OR - d$ is the distance from the origin to the plane.

Also

$$N \cdot r_0 = |N||r_0|\cos\theta_2 = |N|d$$

In both cases, the result is of course the same. Both sides of the equation $N \cdot r = N \cdot r_0$ are equal to the same value: the magnitude of the normal multiplied by the distance from the origin.

Furthermore, if we divide either side by $|N|$, we get the distance from the origin to the plane.

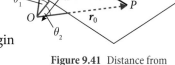

Figure 9.41 Distance from origin O to plane π

That is

$$N \cdot r = |N|d \Rightarrow d = \frac{N \cdot r}{|N|}$$

As well as

$$N \cdot r_0 = |N|d \Rightarrow d = \frac{N \cdot r_0}{|N|}$$

This last result gives us the basis for forming a new vector equation of the plane in terms of a unit vector perpendicular to it.

Let us call the unit vector normal to the plane n. So, using the results just established, we can write

$$d = \frac{N \cdot r}{|N|} = \frac{N \cdot r_0}{|N|}, \text{ which in turn can be simplified to}$$

$$\frac{N}{|N|} \cdot r = \frac{N}{|N|} \cdot r_0, \text{ and since } \frac{N}{|N|} = n, \text{ then obviously}$$

$$n \cdot r = n \cdot r_0 = d$$

This equation is very practical when we need to find the distance from the origin to the plane. The distance from the origin to a plane is the scalar product between the unit normal and the position vector of any point on the plane.

This will be shown in the following examples.

Example 9.38

Write a vector equation for the plane that contains $(2, -3, 5)$ and has normal $N = 2i + j - 3k$.

Solution

We apply the results of the previous discussion

$$N \cdot r = N \cdot r_0$$
$$\Rightarrow (2i + j - 3k) \cdot r = (2i + j - 3k) \cdot (2i - 3j + 5k)$$
$$\Rightarrow (2i + j - 3k) \cdot r = -14$$

Notice that this result can easily transfer into Cartesian form by expanding the scalar product on the left

$$(2i + j - 3k) \cdot r = -14$$
$$\Rightarrow (2i + j - 3k) \cdot (xi + yj + zk) = -14$$
$$\Rightarrow 2x + y - 3z = -14$$

Example 9.39

Show that the line l with equation $\mathbf{r} = 2\mathbf{i} - \mathbf{j} + 5\mathbf{k} + k(3\mathbf{i} + 2\mathbf{j} - 2\mathbf{k})$ is parallel to the plane P whose equation is $\mathbf{r} \times (2\mathbf{i} - 2\mathbf{j} + \mathbf{k}) = -3$ and find the distance between them.

Solution

Since $\mathbf{v} = 3\mathbf{i} + 2\mathbf{j} - 2\mathbf{k}$ is the direction vector of the line l, and since this vector is perpendicular to $\mathbf{N} = 2\mathbf{i} - 2\mathbf{j} + \mathbf{k}$, the normal to P, as $(3\mathbf{i} + 2\mathbf{j} - 2\mathbf{k}) \times (2\mathbf{i} - 2\mathbf{j} + \mathbf{k}) = 6 - 4 - 2 = 0$, then the line l must be parallel to plane P.

To find the distance between the line and the plane P, we may find the distance between a point on l, $(2, -1, 5)$ for example, and plane P. One way would be to consider a plane Q containing the given point and parallel to P. The equation of plane Q can be found using the last result:

$$\mathbf{N} \cdot \mathbf{r} = \mathbf{N} \cdot \mathbf{r}_0$$
$$\Rightarrow (2\mathbf{i} - 2\mathbf{j} + \mathbf{k}) \cdot \mathbf{r} = (2\mathbf{i} - 2\mathbf{j} + \mathbf{k}) \cdot (2\mathbf{i} - \mathbf{j} + 5\mathbf{k})$$
$$\Rightarrow (2\mathbf{i} - 2\mathbf{j} + \mathbf{k}) \cdot \mathbf{r} = 11$$

The distance from the origin to P is given by

$$d = \left| \frac{\mathbf{N} \cdot \mathbf{r}_0}{|\mathbf{N}|} \right| = \frac{3}{\sqrt{4 + 4 + 1}} = 1$$

while the distance from the origin to Q is

$$d = \left| \frac{\mathbf{N} \cdot \mathbf{r}_0}{|\mathbf{N}|} \right| = \frac{11}{\sqrt{4 + 4 + 1}} = \frac{11}{3}.$$ The distance between the two planes is

the sum of these two distances since the planes are on opposite sides of the origin (see note), and hence the required distance is $\frac{14}{3}$.

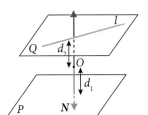

Figure 9.42 Solution to Example 9.39

The vector equation for P is $\mathbf{r} \times (2\mathbf{i} - 2\mathbf{j} + \mathbf{k}) = -3$, or $\mathbf{r} \times (-2\mathbf{i} + 2\mathbf{j} - \mathbf{k}) = 3$ and the equation for Q is $(2\mathbf{i} - 2\mathbf{j} + \mathbf{k}) \cdot \mathbf{r} = 11$. The normals to the two planes are opposite, and hence they are on opposite sides of the origin. If the two normals are in the same direction, then the distance between them will be the difference of the two distances from the origin.

Example 9.40

Show that the plane with vector equation $\mathbf{r} \times (2\mathbf{i} - 2\mathbf{j} + \mathbf{k}) = -3$ contains the line with equation

$$\mathbf{r} = \mathbf{i} + 3\mathbf{j} + \mathbf{k} + k(3\mathbf{i} + 2\mathbf{j} - 2\mathbf{k})$$

Solution

We have several ways available to us at this stage. One method is to check whether two points are common to the line and the plane. One point on the line is $(1, 3, 1)$.

Since $(1, 3, 1) \times (2, -2, 1) = 2 - 6 + 1 = -3$, then the point is on the plane. Another point on the line can be found by choosing any value for k, say $k = 1$. Thus another point has the position vector

$$\mathbf{r} = \mathbf{i} + 3\mathbf{j} + \mathbf{k} + 3\mathbf{i} + 2\mathbf{j} - 2\mathbf{k} = 4\mathbf{i} + 5\mathbf{j} - \mathbf{k}$$

Since $(4\boldsymbol{i} + 5\boldsymbol{j} - \boldsymbol{k}) \times (2\boldsymbol{i} - 2\boldsymbol{j} + \boldsymbol{k}) = 8 - 10 - 1 = -3$, this point will also lie on the plane and so the plane will contain the whole line.

Another method would be to check only one point and prove that the line is parallel to the plane, as was done in Example 9.39.

Example 9.41

Find the vector equation of the line through (1, 2, 3) that is perpendicular to the plane with vector equation $\boldsymbol{r} \times (2\boldsymbol{i} - 2\boldsymbol{j} + \boldsymbol{k}) = -3$ and find their point of intersection.

Solution

A vector parallel to the required line must be parallel to the normal vector to the plane. Hence a vector equation of the line is $\boldsymbol{r} = \begin{pmatrix} 1 \\ 2 \\ 3 \end{pmatrix} + t\begin{pmatrix} 2 \\ -2 \\ 1 \end{pmatrix}$.

To find the point of intersection, we consider any point on the line. Such a point would have the position vector $\begin{pmatrix} 1 + 2t \\ 2 - 2t \\ 3 + t \end{pmatrix}$. For this point to be on the plane, the following equation must be true

$$\begin{pmatrix} 1 + 2t \\ 2 - 2t \\ 3 + t \end{pmatrix} \cdot \begin{pmatrix} 2 \\ -2 \\ 1 \end{pmatrix} = -3; \text{ that is, } 2 + 4t - 4 + 4t + 3 + t = -3,$$

so $9t = -4$, and $t = -\dfrac{4}{9}$, giving the point of intersection of the line and the

plane as $\left(1 + 2 \times -\dfrac{4}{9}, 2 - 2 \times -\dfrac{4}{9}, 3 - \dfrac{4}{9}\right) = \left(\dfrac{1}{9}, \dfrac{26}{9}, \dfrac{23}{9}\right)$

Parametric form for the equation of a plane

Figure 9.43 shows that it is always possible to construct a parallelogram whose diagonal is $\boldsymbol{w}$ and whose sides are the non-parallel vectors $\boldsymbol{u}$ and $\boldsymbol{v}$ or their multiples.

For example, given the two non-parallel vectors $\boldsymbol{u} = \begin{pmatrix} 1 \\ 0 \\ -2 \end{pmatrix}$ and $\boldsymbol{v} = \begin{pmatrix} 3 \\ 1 \\ -9 \end{pmatrix}$,

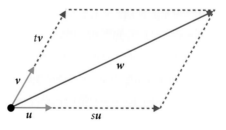

Figure 9.43 A vector in the plane is a linear combination of any two non-parallel vectors in the plane

Three coplanar vectors $\boldsymbol{u}$, $\boldsymbol{v}$, and $\boldsymbol{w}$ are given. If $\boldsymbol{u}$ and $\boldsymbol{v}$ are not parallel, then $\boldsymbol{w}$ can always be expressed as a linear combination of $\boldsymbol{u}$ and $\boldsymbol{v}$, i.e., it is always possible to find two scalars s and t such that $\boldsymbol{w} = s\boldsymbol{u} + t\boldsymbol{v}$.

we can always find the scalars s and t so that vector $\boldsymbol{w} = \begin{pmatrix} 2 \\ 1 \\ -7 \end{pmatrix}$ can be

expressed as a linear combination of $\boldsymbol{u}$ and $\boldsymbol{v}$.

Thus $\begin{pmatrix} 2 \\ 1 \\ -7 \end{pmatrix} = s\begin{pmatrix} 1 \\ 0 \\ -2 \end{pmatrix} + t\begin{pmatrix} 3 \\ 1 \\ -9 \end{pmatrix}$. To find s and t we solve the system of equations

$$\begin{cases} s + 3t = 2 \\ t = 1 \\ -2s - 9t = -7 \end{cases}$$

This system is consistent and yields the solution $s = -1$, and $t = 1$.

Thus $\boldsymbol{w} = -\boldsymbol{u} + \boldsymbol{v}$.

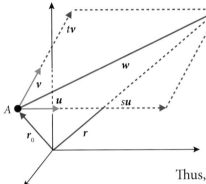

Now, consider the plane that is parallel to vectors $\boldsymbol{u}$ and $\boldsymbol{v}$ and contains the point A whose position vector is $\boldsymbol{r}_0$. As Figure 9.44 shows, the two vectors determine the direction of the plane and A 'fixes' it in space. So, if M is any point in this plane, then according to the previous theorem, $\overrightarrow{AM} = s\boldsymbol{u} + t\boldsymbol{v}$ where s and t are two scalars.

If $\boldsymbol{r}$ is the position vector of M then

$$\boldsymbol{r} = \boldsymbol{r}_0 + \overrightarrow{AM} = \boldsymbol{r}_0 + s\boldsymbol{u} + t\boldsymbol{v}$$

Figure 9.44 Parametric form of the equation of the plane

Thus, any equation of the form $\boldsymbol{r} = \boldsymbol{r}_0 + s\boldsymbol{u} + t\boldsymbol{v}$, where s and t are independent scalars, represents the equation of a plane parallel to $\boldsymbol{u}$ and $\boldsymbol{v}$ that contains the point with position vector $\boldsymbol{r}_0$.

Note that this equation is not unique. This is because we can start at any other fixed point on the plane other than A and we can choose any number of intersecting vectors in the plane other than $\boldsymbol{u}$ and $\boldsymbol{v}$. The parametric form of the equation of the plane is seldom needed or used.

Example 9.42

Find an equation of a plane with normal $\boldsymbol{q} = 2\boldsymbol{i} - 3\boldsymbol{j} + \boldsymbol{k}$ that contains the point $A(2, 1, 1)$. Use all forms you have learned.

Solution

The Cartesian equation

Consider the equation

$A(x - x_0) + B(y - y_0) + C(z - z_0) = 0 \Rightarrow 2(x - 2) - 3(y - 1) + (z - 1) = 0$,

and the equation would be: $2x - 3y + z - 2 = 0$

Or, start the equation

$Ax + By + Cz = D \Rightarrow 2x - 3y + z = D$

since the plane contains the point

$(2, 1, 1)$, then $2(2) - 3(1) + 1 = D$, and thus $D = 2$

Vector equations

We can also find the vector equation by applying

> It is easy to transform the vector equation into Cartesian by simply performing the dot product. The opposite is also true.

$$\boldsymbol{N} \cdot \boldsymbol{r} = \boldsymbol{N} \cdot \boldsymbol{r}_0 \Rightarrow \begin{pmatrix} 2 \\ -3 \\ 1 \end{pmatrix} \cdot \begin{pmatrix} x \\ y \\ z \end{pmatrix} = \begin{pmatrix} 2 \\ -3 \\ 1 \end{pmatrix} \cdot \begin{pmatrix} 2 \\ 1 \\ 1 \end{pmatrix} \Rightarrow \begin{pmatrix} 2 \\ -3 \\ 1 \end{pmatrix} \cdot \begin{pmatrix} x \\ y \\ z \end{pmatrix} = 2$$

Parametric equations

A parametric equation of this plane is not as straightforward and may not be the most efficient way of doing this problem. However, for the sake of giving an example, here is a way of doing it.

The parametric form requires that we have two vectors parallel to the plane. We may find the two vectors by considering that they have to be

perpendicular to $\begin{pmatrix} 2 \\ -3 \\ 1 \end{pmatrix}$. So, take a vector $\begin{pmatrix} 1 \\ 1 \\ z \end{pmatrix}$ and find z so that this vector is

perpendicular to $\begin{pmatrix} 2 \\ -3 \\ 1 \end{pmatrix}$.

$$\Rightarrow 2 - 3 + z = 0 \text{ and } z = 1$$

Do the same with $\begin{pmatrix} 1 \\ 0 \\ z \end{pmatrix}$, i.e., $2 + 0 + z = 0$ and $z = -2$. Therefore, two

vectors that are perpendicular to $\begin{pmatrix} 2 \\ -3 \\ 1 \end{pmatrix}$ are $\begin{pmatrix} 1 \\ 1 \\ 1 \end{pmatrix}$ and $\begin{pmatrix} 1 \\ 0 \\ -2 \end{pmatrix}$, and a parametric

equation of the plane is

$$\boldsymbol{r} = \boldsymbol{r}_0 + s\boldsymbol{u} + t\boldsymbol{v} \Rightarrow \boldsymbol{r} = \begin{pmatrix} 2 \\ 1 \\ 1 \end{pmatrix} + s\begin{pmatrix} 1 \\ 1 \\ 1 \end{pmatrix} + t\begin{pmatrix} 1 \\ 0 \\ -2 \end{pmatrix}$$

Note that the choice of the vectors is arbitrary and hence the parametric form is not unique.

Example 9.43

Find the equation of the plane that contains the points $A(1, 3, 0)$, $B(-2, 1, 2)$, and $C(1, -2, -1)$.

Solution

Consider any point $M(x, y, z)$ on this plane. For this point to belong to the plane, the following vectors must be coplanar: $\overrightarrow{AM}$, $\overrightarrow{AB}$, and $\overrightarrow{AC}$.

This means that the parallelepiped with these vectors as edges is flat; that is, it has zero volume. Since we know that the volume of the parallelepiped is the absolute value of the scalar triple product, we equate that value to zero and get the equation.

$$\overrightarrow{AM} = \begin{pmatrix} x - 1 \\ y - 3 \\ z \end{pmatrix}, \overrightarrow{AB} = \begin{pmatrix} -3 \\ -2 \\ 2 \end{pmatrix}, \overrightarrow{AC} = \begin{pmatrix} 0 \\ -5 \\ -1 \end{pmatrix}$$

$$\Rightarrow \overrightarrow{AM} \cdot (\overrightarrow{AB} \times \overrightarrow{AC}) = \begin{vmatrix} x - 1 & y - 3 & z \\ -3 & -2 & 2 \\ 0 & -5 & -1 \end{vmatrix} = 0$$

$$\Rightarrow 3(4x - y + 5z - 1) = 0$$

So, the Cartesian equation of the plane is $4x - y + 5z - 1 = 0$

We can also deduce the vector equation from this plane by writing it in scalar product form:

$$(4\mathbf{i} - \mathbf{j} + 5\mathbf{k}) \cdot (x\mathbf{i} + y\mathbf{j} + z\mathbf{k}) = 1$$

We can also find the vector form by thinking of the problem as a plane containing a fixed point and normal to a given vector.

We find the normal by computing the cross product of two vectors in the plane; in this case, we can take $\overrightarrow{AB}$ and $\overrightarrow{AC}$.

$$\overrightarrow{AB} \times \overrightarrow{AC} = \begin{vmatrix} \mathbf{i} & \mathbf{j} & \mathbf{k} \\ -3 & -2 & 2 \\ 0 & -5 & -1 \end{vmatrix} = 3(4\mathbf{i} - \mathbf{j} + 5\mathbf{k})$$

The equation of the plane is then

$$(4\mathbf{i} - \mathbf{j} + 5\mathbf{k}) \cdot (x\mathbf{i} + y\mathbf{j} + z\mathbf{k}) = (4\mathbf{i} - \mathbf{j} + 5\mathbf{k}) \cdot (\mathbf{i} + 3\mathbf{j})$$
$$(4\mathbf{i} - \mathbf{j} + 5\mathbf{k}) \cdot (x\mathbf{i} + y\mathbf{j} + z\mathbf{k}) = 1$$

That is, the same as above.

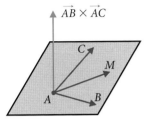

Figure 9.45 Solution to Example 9.43

Distance between a point and a plane

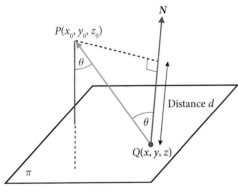

Figure 9.46 Distance from P to π

The distance between a point $P(x_0, y_0, z_0)$ and a plane with equation $Ax + By + Cz = D$ is given by

$$d = \frac{|Ax_0 + By_0 + Cz_0 - D|}{\sqrt{A^2 + B^2 + C^2}}$$

Let $Q(x, y, z)$ be any point on the plane and $\mathbf{N} = \begin{pmatrix} A \\ B \\ C \end{pmatrix}$ be a normal to the plane (Figure 9.46). The distance we are looking for is d. Then

$$d = \left\| \overrightarrow{QP} \right\| \cos\theta = \left| \left\| \overrightarrow{QP} \right\| \frac{\overrightarrow{QP} \cdot \mathbf{N}}{\left\| \overrightarrow{QP} \right\| |\mathbf{N}|} \right|$$

$$= \left| \frac{\overrightarrow{QP} \cdot \mathbf{N}}{|\mathbf{N}|} \right| = \left| \frac{(A, B, C) \cdot (x_0 - x, y_0 - y, z_0 - z)}{\sqrt{A^2 + B^2 + C^2}} \right|$$

$$= \frac{|A(x_0 - x) + B(y_0 - y) + C(z_0 - z)|}{\sqrt{A^2 + B^2 + C^2}}$$

$$= \frac{|Ax_0 + By_0 + Cz_0 - (Ax + By + Cz)|}{\sqrt{A^2 + B^2 + C^2}}$$

Since $Q(x, y, z)$ is on the plane, then $Ax + By + Cz = D$, so replacing this expression in the result above gives

$$d = \frac{|Ax_0 + By_0 + Cz_0 - D|}{\sqrt{A^2 + B^2 + C^2}}$$

This formula is similar to the distance between a point and a line in 2D space.

Example 9.44

Show that the line l with equation $r = 2i - j + 5k + k(3i + 2j - 2k)$ is parallel to the plane P whose equation is $r \cdot (2i - 2j + k) = -3$, and find the distance between them.

Solution

In Example 9.41 we showed that this line is parallel to the plane because it is perpendicular to the normal of the plane. To find the distance, we used a relatively complex approach. Now we can use the distance formula just established to find the required distance.

A point on the line is $(2, -1, 5)$ and the Cartesian equation of the plane is simply

$$2x - 2y + z = -3,$$

hence the distance is

$$d = \frac{|2(2) - 2(-1) + 1(5) - (-3)|}{\sqrt{4 + 4 + 1}} = \frac{14}{3}$$

Example 9.45

Find the distance between the two parallel planes:
$x + 2y - 2z = 3$ and $2x + 4y - 4z = 7$

Solution

It is enough to find the distance from one point on one of the planes to the other plane since all points are equidistant.

Take the point $(1, 1, z)$ on the first plane $1 + 2 - 2z = 3$, so $z = 0$.

Thus, the point is $(1, 1, 0)$ and the distance between the planes is

$$d = \frac{|2(1) + 4(1) - 4(0) - 7|}{\sqrt{4 + 16 + 16}} = \frac{1}{6}$$

Example 9.46

Find the distance between the two skew straight lines

$$L_1: r = \begin{pmatrix} 1 \\ 5 \\ -1 \end{pmatrix} + \lambda \begin{pmatrix} 4 \\ -4 \\ 5 \end{pmatrix} \text{ and } L_2: r = \begin{pmatrix} 2 \\ 4 \\ 5 \end{pmatrix} + \mu \begin{pmatrix} 8 \\ -3 \\ 1 \end{pmatrix}$$

Solution

We can reduce this problem to the type in the previous example by creating two planes that contain the given lines and are parallel to each other.

For the two planes to be parallel, they must be perpendicular to the same vector. Hence by finding the cross product of the direction vectors of the lines, we would have found a vector perpendicular to both.

$$l_1 \times l_2 = \begin{vmatrix} i & j & k \\ 4 & -4 & 5 \\ 8 & -3 & 1 \end{vmatrix} = 11i + 36j + 20k$$

Considering the point $(2, 4, 5)$ on L_2, the plane containing L_2 will be

$$\begin{pmatrix} 11 \\ 36 \\ 20 \end{pmatrix} \cdot \begin{pmatrix} x \\ y \\ z \end{pmatrix} = \begin{pmatrix} 11 \\ 36 \\ 20 \end{pmatrix} \cdot \begin{pmatrix} 2 \\ 4 \\ 5 \end{pmatrix}$$

or $11x + 36y + 20z = 266$ and the distance between $(1, 5, -1)$ on L_1 to this plane is

$$d = \frac{|11(1) + 36(5) + 20(-1) - 266|}{\sqrt{11^2 + 36^2 + 20^2}} = \frac{95}{\sqrt{1817}}$$

The angle between two planes

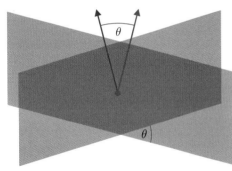

Figure 9.47 The angle between two planes is defined as the acute angle between them

The angle between two planes is defined as the acute angle between them (Figure 9.47).

Consider two planes P and Q with unit normals n_1 and n_2 respectively. Their vector equations are of the form

$$r \cdot n_1 = d_1; \, r \cdot n_2 = d_2$$

The angle between the planes is equal to the angle between the normal units n_1 and n_2 or the normals N_1 and N_2, and hence

$$\cos\theta = n_1 \cdot n_2 = \frac{N_1 \cdot N_2}{|N_1||N_2|}$$

For example, the angle between the planes with vector equations

$$r \times (i + j - 2k) = 3 \text{ and } r \times (2i - 2j + k) = 2$$

is given by

$$\cos\theta = \frac{(i + j - 2k) \cdot (2i - 2j + k)}{\sqrt{1 + 1 + 4} \cdot \sqrt{4 + 4 + 1}} = -\frac{2}{3\sqrt{6}}$$

This is the cosine of the obtuse angle between the two planes. The acute angle between them is $\cos^{-1}\left(\dfrac{2}{3\sqrt{6}}\right)$.

Example 9.47

Find the angle between the planes with equations:

$$2x - 3y + 5 = 0 \text{ and } 3x + y - z = 4$$

Solution

The two normals are $2\mathbf{i} - 3\mathbf{j}$ and $3\mathbf{i} + \mathbf{j} - \mathbf{k}$, and therefore the angle is given by

$$\cos\theta = \frac{(2\mathbf{i} - 3\mathbf{j}) \cdot (3\mathbf{i} + \mathbf{j} - \mathbf{k})}{\sqrt{13}\sqrt{11}} = \frac{3}{\sqrt{33}}$$

So, the angle between the planes is $\cos^{-1}\left(\dfrac{3}{\sqrt{33}}\right)$

The angle between a line and a plane

The angle between a line and a plane is defined as the angle θ formed by the line and its projection on the plane (Figure 9.48).

Consider the line l with equation $\mathbf{r} = \mathbf{r}_0 + \lambda\mathbf{a}$ and the plane with equation
$\mathbf{r} \cdot \mathbf{N} = \mathbf{D}$.

The acute angle ϕ between $\mathbf{N}$, the normal to the plane, and the line l can be found by using the law of cosines

$$\cos\phi = \left|\frac{\mathbf{a} \cdot \mathbf{N}}{||\mathbf{a}||\mathbf{N}||}\right|$$

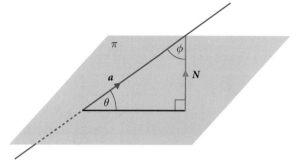

Figure 9.48 The angle between a line and a plane is defined as the angle θ formed by the line and its projection on the plane

If θ is the acute angle between the line and the plane then

$$\theta = \frac{\pi}{2} - \phi$$

Therefore, to find the angle between the line and the plane, we either:

- find the angle ϕ first and then find its complement, or

- since ϕ and θ are complements, then $\sin\theta = \cos\phi = \left|\dfrac{\mathbf{a} \cdot \mathbf{N}}{||\mathbf{a}||\mathbf{N}||}\right|$

For example, to find the angle between the line with equation
$\mathbf{r} = \mathbf{i} + 2\mathbf{j} - \mathbf{k} + \lambda(\mathbf{i} - \mathbf{j} + \mathbf{k})$ and the plane with equation $\mathbf{r} \times (2\mathbf{i} - \mathbf{j} + \mathbf{k}) = 3$
we find that

$$\sin\theta = \frac{(\mathbf{i} - \mathbf{j} + \mathbf{k}) \cdot (2\mathbf{i} - \mathbf{j} + \mathbf{k})}{\sqrt{3} \cdot \sqrt{6}} = \frac{4}{3\sqrt{2}}$$

$$\Rightarrow \theta = \sin^{-1}\frac{4}{3\sqrt{2}}$$

Example 9.48

Find the angle between the line with equations:
$\dfrac{x - 1}{2} = \dfrac{y - 2}{3} = \dfrac{z}{2}$, and the plane with equation $2x - y - z = 7$

Solution

The direction of the line is given by $a = (2, 3, 2)$ and the normal to the plane by $N = (2, -1, -1)$, and the angle is given by

$$\sin\theta = \left|\frac{(2, 3, 2) \cdot (2, -1, -1)}{\sqrt{17}.\sqrt{6}}\right| = \frac{1}{\sqrt{102}}$$

$$\Rightarrow \theta = \sin^{-1}\frac{1}{\sqrt{102}}$$

Line of intersection of two planes

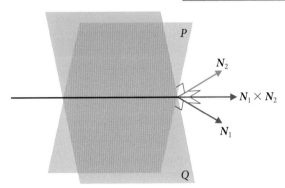

Figure 9.49 Line of intersection of two planes

Unless two planes are parallel they will intersect along a straight line. Consider two planes P and Q that have l as their line of intersection. Also let the planes have the vector equations

$$r \cdot N_1 = D_1 \text{ and } r \cdot N_2 = D_2$$

Since the line l lies in plane P then N_1, the normal to this plane, must be perpendicular to it. This is also true for N_2. Therefore, the direction of line l is perpendicular to both N_1 and N_2.

To find the line of intersection, we can use two methods:

1. the cross product of n_1 and n_2 as the direction of l and a specific point on the line,

2. the fact that all points on l must satisfy the equations of both planes; i.e., we solve a system of equations.

These methods are best demonstrated when n applied to a particular situation.

Example 9.49

Let the planes P and Q have the equations:

$$r \cdot (i + j - 3k) = 6 \text{ and } r \cdot (2i - j + k) = 4$$

Find the line of intersection using the two methods.

Solution

First method

To find a vector equation of the line of intersection, first we find the cross product of the two normals and then find a point on the line l.

$$(i + j - 3k) \times (2i - j + k) = 2i + 7j + 3k$$

To find a point on the line, we use the fact that the points on that line must satisfy both equations. So, consider the points on both planes that have the x-coordinate zero:

$$(0, y, z) \cdot (i + j - 3k) = 6 \text{ and } (0, y, z) \cdot (2i - j + k) = 4$$

$$\Rightarrow \begin{cases} y - 3z = 6 \\ -y + z = 4 \end{cases} \Rightarrow z = -5 \text{ and } y = -9$$

So, the vector equation of the line is: $r = \begin{pmatrix} 0 \\ -9 \\ -5 \end{pmatrix} + t\begin{pmatrix} 2 \\ 7 \\ 3 \end{pmatrix}$

Second method

The second method uses a system of equations to find the equation. The equations of the planes in Cartesian form are:

$$x + y - 3z = 6 \text{ and } 2x - y + z = 4$$

Since this system has to be solved simultaneously, and since there are two equations in three variables, we should consider one of these variables as a parameter and solve for the rest.

$$\begin{cases} x + y - 3z = 6 \\ 2x - y + z = 4 \end{cases} \Rightarrow 3x - 2z = 10 \Rightarrow z = -5 + \frac{3}{2}x; y = -9 + \frac{7}{2}x$$

Therefore, we either consider x to be the parameter or, for convenience purposes, we replace it by another parameter such as the following:

$$x = 2\lambda, y = 7\lambda - 9, z = 3\lambda - 5$$

This equation is equivalent to the one found using the first method.

If the equations of the planes are in parametric form it may not be necessary to convert them into Cartesian form. However, from Example 9.49, you may notice that it is sometimes more straightforward to follow the Cartesian method.

Example 9.50

Find the intersection between the two planes

$P: r = i + j + \lambda(2i - k) + \mu(i - j + k)$, and

$Q: r = 3i - k + s(i - j + 2k) + t(i + 2j - k)$

Solution

A point on P will have coordinates: $(1 + 2\lambda + \mu, 1 - \mu, -\lambda + \mu)$, while a point on Q will have coordinates $(3 + s + t, -s + 2t, -1 + 2s - t)$. For the collection of points on the intersection, coordinates must satisfy both equations, and hence

$$1 + 2\lambda + \mu = 3 + s + t$$

$$1 - \mu = -s + 2t$$

$$-\lambda + \mu = -1 + 2s - t$$

This system of three equations in four unknowns must have an infinite number of solutions if the planes intersect – the set of points that belong to the line of intersection. We can solve this system after re-arranging terms, which is best done by Gaussian elimination:

$$\begin{pmatrix} 2 & 1 & -1 & -1 & 2 \\ 0 & -1 & 1 & -2 & -1 \\ -1 & 1 & -2 & 1 & -1 \end{pmatrix} \Rightarrow \begin{pmatrix} 1 & 0 & 0 & -\frac{3}{2} & \frac{1}{2} \\ 0 & 1 & 0 & \frac{9}{2} & \frac{5}{2} \\ 0 & 0 & 1 & \frac{5}{2} & \frac{3}{2} \end{pmatrix}$$

The last result expresses λ, μ, and s in terms of t. To find the equation of the line we have to substitute these values in either of the two equations above. For example, it is easier to substitute for s in terms of t in the equation for Q. So, from the result above

$$s = \frac{3}{2} - \frac{5}{2}t$$

and the line will have the vector equation

$$r = 3i - k + \left(\frac{3}{2} - \frac{5}{2}t\right)(i - j + 2k) + t(i + 2j - k)$$

$$= \left(\frac{9}{2}i - \frac{3}{2}j + 2k\right) + t\left(-\frac{3}{2}i + \frac{9}{2}j - 6k\right)$$

If $\lambda = \frac{1}{2} + \frac{3}{2}t$ and $\mu = \frac{5}{2} - \frac{9}{2}t$ are substituted into the equation for P we will get the same result.

As we notice from Example 9.50, the process is long and complex even with many steps that are hidden to save space. Alternatively, the Cartesian solution may be more efficient.

P: point $(1, 1, 0)$ is on the plane, and $(2, 0, -1) \times (1, -1, 1) = (-1, -3, -2)$ is perpendicular to the plane, so the Cartesian equation is

$$(x - 1) + 3(y - 1) + 2z = 0, \text{ or } x + 3y + 2z = 4$$

Q: point $(3, 0, -1)$ is on the plane, and $(1, -1, 2) \times (1, 2, -1) = (-3, 3, 3)$ is perpendicular to the plane, so the Cartesian equation is

$$-3(x - 3) + 3y + 3(z + 1) = 0, \text{ or } x - y - z = 4$$

The intersection between the planes is the result of solving the following system

$$\begin{cases} x + 3y + 2z = 4 \\ x - y - z = 4 \end{cases} \Rightarrow x = 4 + m, y = -3m, z = 4m$$

This result compares to the previous one and appears to be more elegant.

Intersections of three planes

Three planes can intersect in three lines as shown in Figure 9.50.

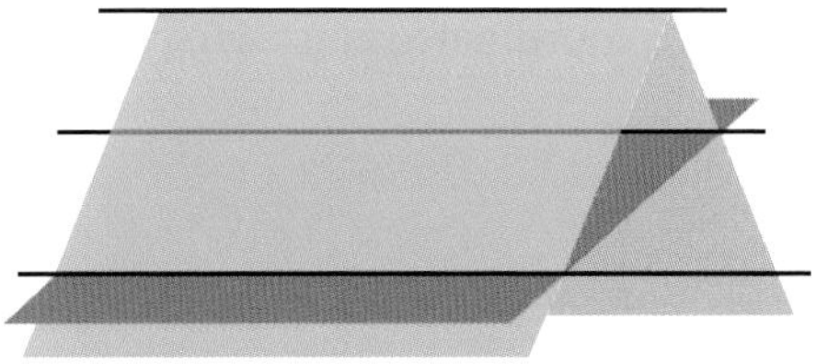

Figure 9.50 Three planes can intersect in three lines

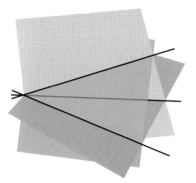

Figure 9.51 If the lines of intersection are not parallel, the three planes meet at one point

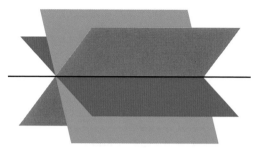

Figure 9.52 If the three planes pass through one straight line, there is an infinite number of solutions

The three lines of intersection are parallel. Hence, the system of equations they represent is inconsistent. If the lines of intersection are not parallel, then the three planes meet at one point as shown (Figure 9.51). This system is consistent with a unique solution. The three planes can also all pass through one straight line. In that case, the system is consistent with an infinite number of solutions (Figure 9.52).

Exercise 9.5

1. Which of the points $A(3, -2, -1)$, $B(2, 1, -1)$, and $C(1, 4, 0)$ lie in the plane $3x + 2y - 3z = 11$?

2. Which of the points $A(3, -2, -1)$, $B(2, 1, -2)$, and $C(1, 4, 0)$ lie in the plane $(i - 3j + k) \cdot (xi + yj + zk) = -6$?

3. Find an equation for the plane satisfying the given conditions. Give two forms for each equation, out of the three forms: Cartesian, vector, or parametric.

   (a) Contains the point $(3, -2, 4)$ and perpendicular to $2i - 4j + 3k$

   (b) Contains the point $(-3, 2, 1)$ and perpendicular to $2i + 3k$

   (c) Contains the point $(0, 3, 1)$ and perpendicular to $3k$

   (d) Contains the point $(3, -2, 4)$ and parallel to the plane $5x + y - 2z = 7$

   (e) Contains the point $(3, 0, 1)$ and parallel to the plane $y - 2z = 11$

   (f) Contains the point $(3, -2, 4)$ and the line $r = \begin{pmatrix} 1 \\ -2 \\ 5 \end{pmatrix} + t\begin{pmatrix} 2 \\ 1 \\ 2 \end{pmatrix}$

   (g) Contains the lines $r = \begin{pmatrix} 1 \\ -2 \\ 5 \end{pmatrix} + t\begin{pmatrix} 2 \\ 1 \\ 2 \end{pmatrix}$ and $r = \begin{pmatrix} 4 \\ 0 \\ 7 \end{pmatrix} + s\begin{pmatrix} 3 \\ 2 \\ 2 \end{pmatrix}$

   (h) Contains the point $(1, -3, 2)$ and the line given in parametric form: $x = 2t, y = 2 + t, z = -1 + 3t$

   (i) Contains the point $M(p, q, r)$ and perpendicular to the vector $\overrightarrow{OM}$

   (j) Contains the three points $(1, 2, 2)$, $(3, -1, 0)$, and $(7, 0, -2)$

   (k) Contains the three points $(2, -2, -2)$, $(3, -1, 3)$, and $(0, 1, 5)$

   (l) Contains the point $(1, -2, 3)$ and the line $x - 2 = y + 1 = \dfrac{z - 5}{3}$

   (m) Contains the two parallel lines
   $r = (1, -1, 5) + t(3i + 2j + 4k)$ and $r = (-3, 4, 0) + s(3i + 2j + 4k)$

   (n) Contains the point $(1, 1, 0)$ and parallel to the two lines
   $x = 2 + t, y = -1, z = t$ and $x = s, y = 2 - s, z = -1 + s$

437

4. Find the acute angle between the given lines or planes.

   (a) $3x + 4y - z = 1$ and $x - 2y = 3$

   (b) $4x - 7y + z = 3$ and $3x + 2y + 2z = 17$

   (c) $x = 4$ and $x + z = 4$

   (d) $x - 2y + 2z = 3$ and $x = 2 - 6t, y = 4 + 3t, z = 1 - 2t$

   (e) $(3i - k) \cdot \begin{pmatrix} x \\ y \\ z \end{pmatrix} = 4$ and $r = (2j + 3k) + m(-i + 2j - k)$

   (f) $x + y + z = 7$ and $z = 0$

5. Find the points of intersection of the given line and plane.

   (a) $r = 5i - 2k + \lambda(i - 3j + 4k)$ and
   $(i - 3j + 2k) \times (xi + yj + zk) = -35$

   (b) $r = \begin{pmatrix} 2 \\ 4 \\ 0 \end{pmatrix} + \mu \begin{pmatrix} 0 \\ -3 \\ 3 \end{pmatrix}$ and $4x - 2y + 3z - 30 = 0$

   (c) $x - 3 = \dfrac{y - 4}{5} = \dfrac{z - 6}{3}$ and $(2i - 4j + 6k) \cdot (xi + yj + zk) = 5$

   (d) $x = t, y = 4 - \dfrac{1}{3}t, z = 5 - \dfrac{5}{3}t$ and $3x - y + 2z = 6$

6. Find the line of intersection between the given planes.

   (a) $x = 10$ and $x + y + z = 3$

   (b) $2x - y + z = 5$ and $x + y - z = 4$

   (c) $\begin{pmatrix} 1 \\ -1 \\ -2 \end{pmatrix} \cdot \begin{pmatrix} x \\ y \\ z \end{pmatrix} = 1$ and $x - y - 2z = 5$

   (d) $r = \begin{pmatrix} 1 \\ 0 \\ 2 \end{pmatrix} + \lambda \begin{pmatrix} 1 \\ -2 \\ 0 \end{pmatrix} + \mu \begin{pmatrix} 3 \\ 2 \\ -8 \end{pmatrix}$ and $3x - y - z = 3$

7. Find a plane through $A(2, 1, -1)$ and perpendicular to the line of intersection of the planes $2x + y - z = 3$ and $x + 2y + z = 2$

8. Find a plane through the points $A(1, 2, 3)$ and $B(3, 2, 1)$ and perpendicular to the plane $r = (4i - j + 2k) \times (xi + yj + zk) = 7$

9. What point on the line through $(1, 2, 5)$ and $(3, 1, 1)$ is closest to the point $(2, -1, 5)$?

10. Find an equation of the plane that contains the line
    $r = (-i + 2j + 3k) + \lambda(i - 2j + k)$
    and is parallel to $\dfrac{x - 1}{3} = \dfrac{y + 2}{2} = \dfrac{z - 4}{4}$

11. Find an equation of the plane that contains the line
    $x = 1 + 2t, y = 1 + 2t, z = 2 - t$
    and is parallel to $x - 1 = \dfrac{y - 2}{2} = z - 7$

12. Show that the equation $\dfrac{x}{A} + \dfrac{y}{B} + \dfrac{z}{C} = 1$ is the equation of a plane.

13. Find the equation of a plane that contains the point $(4, -3, -1)$ and is perpendicular to the planes $2x - 3y + 4z = 5$ and $4x - 3z = 5$

14. Find the equation of a plane that contains the point $(2, 3, 0)$ and is perpendicular to the plane $2x - 3y + 4z = 5$ and parallel to the line $r(t) = (t - 3)i + (4 - 2t)j + (1 + t)k$

Chapter 9 practice questions

1. Consider vectors $u = -i + 2j$ and $v = 3i + 5j$, where i and j are unit vectors along the x-axis and y-axis respectively.

   (a) Find $u + 2v$ in terms of i and j.

   A vector w has the same direction as $u + 2v$ and has a magnitude of 26.

   (b) Find w in terms of i and j.

2. The circle shown on the right has centre O and radius 6.

   $$\overrightarrow{OA} = \begin{pmatrix} 6 \\ 0 \end{pmatrix}, \overrightarrow{OB} = \begin{pmatrix} -6 \\ 0 \end{pmatrix} \text{ and } \overrightarrow{OC} = \begin{pmatrix} 5 \\ \sqrt{11} \end{pmatrix}$$

   (a) Show that A, B, and C lie on the circle.

   (b) Find vector $\overrightarrow{AC}$.

   (c) Find the cosine of angle OAC.

   (d) Find the area of triangle ABC, giving your answer in the form $a\sqrt{11}$, where $a \in \mathbb{N}$.

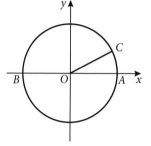

Figure 9.53 Diagram for question 2

3. Given vectors $u = 3i + 5j$ and $v = i - 2j$, find scalars a, b such that $a(u + v) = 8i + (b - 2)j$

4. Vector $\begin{pmatrix} 1 \\ 0 \end{pmatrix}$ represents a displacement due east, and vector $\begin{pmatrix} 0 \\ 1 \end{pmatrix}$ a displacement due north. Distances are in kilometres and time in hours.

   Two crews of workers are laying an underground cable in a north-south direction across a desert. At 06:00, each crew sets out from their base camp, which is situated at the origin $(0, 0)$. One crew is in vehicle T and the other in vehicle C.

   T has velocity vector $\begin{pmatrix} 18 \\ 24 \end{pmatrix}$, and C has velocity vector $\begin{pmatrix} 36 \\ -16 \end{pmatrix}$

   (a) Find the speed of each vehicle.

   (b) (i) Find the position vectors of each vehicle at 06:30

       (ii) Hence, or otherwise, find the distance between the vehicles at 06:30

   (c) At this time (06:30), C stops and its crew begins their day's work, laying cable in a northerly direction. T continues travelling in the same direction at the same speed until it is exactly north of C. The T crew then begins their day's work, laying cable in a southerly direction. At what time does the T crew begin laying cable?

(d) Each crew lays an average of 800 m of cable in an hour. If they work non-stop until their lunch break at 11:30, what is the distance between them at this time?

(e) How long would T take to return to base camp from its lunch-time position, assuming it travelled in a straight line and with the same average speed as on the morning journey? (Give your answer to the nearest minute.)

5. Vectors $\begin{pmatrix} 1 \\ 0 \end{pmatrix}$ and $\begin{pmatrix} 0 \\ 1 \end{pmatrix}$ represent displacements due ast and north respectively. The point $(0, 0)$ is the position of an airport. $r_1 = \begin{pmatrix} 16 \\ 12 \end{pmatrix} + t\begin{pmatrix} 12 \\ -5 \end{pmatrix}$ is the position vector of Aircraft 1. (t is the time in minutes since 12:00 and distance is in km.)

(a) Show that Aircraft 1:

    (i) is 20 km from the airport at 12:00

    (ii) has a speed of 13 km min^{-1}.

(b) Show that a Cartesian equation of the path of Aircraft 1 is:

    $5x + 12y = 224$.

Aircraft 2 has position vector $r_2 = \begin{pmatrix} 23 \\ -5 \end{pmatrix} + t\begin{pmatrix} 2.5 \\ 6 \end{pmatrix}$. ($t$ is the time in minutes since 12:00.)

(c) Find the angle between the paths of the two aircraft.

(d) (i) Find a Cartesian equation for the path of Aircraft 2.

    (ii) Hence find the coordinates of the point where the two paths cross.

(e) The two aircraft are flying at the same height. Will they collide?

6. A line passes through the point $(4, -1)$ and its direction is perpendicular to the vector $\begin{pmatrix} 2 \\ 3 \end{pmatrix}$. Find the equation of the line in the form $ax + by = p$, where a, b and p are integers to be determined.

7. In this question, the vector $\begin{pmatrix} 1 \\ 0 \end{pmatrix}$ represents a displacement due east and the vector $\begin{pmatrix} 0 \\ 1 \end{pmatrix}$ represents a displacement due north. Distances are in km.

The diagram shows the path of the oil-tanker *Aristides* relative to the port of Orto, which is situated at the point $(0, 0)$.

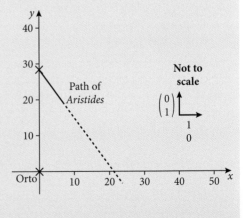

The position of the *Aristides* is given by the vector equation

$\begin{pmatrix} x \\ y \end{pmatrix} = \begin{pmatrix} 0 \\ 28 \end{pmatrix} + t\begin{pmatrix} 6 \\ -8 \end{pmatrix}$ at a time t hours after 12:00.

(a) Find the position of the *Aristides* at 13:00.

(b) Find:

    (i) the velocity vector   (ii) the speed of the *Aristides*.

Another ship, the cargo vessel *Boadicea*, is stationary, with position vector $\begin{pmatrix} 18 \\ 4 \end{pmatrix}$.

(d) Show that the two ships will collide, and find the time of collision.

To avoid collision, the *Boadicea* starts to move at 13:00 with velocity vector $\begin{pmatrix} 5 \\ 12 \end{pmatrix}$.

(e) Show that the position of the *Boadicea* for $t \geqslant 1$ is given by:

$$\begin{pmatrix} x \\ y \end{pmatrix} = \begin{pmatrix} 13 \\ -8 \end{pmatrix} + t\begin{pmatrix} 5 \\ 12 \end{pmatrix}$$

(f) Find how far apart the two ships are at 15:00.

8. Vectors $u = \begin{pmatrix} 2x \\ x + 3 \end{pmatrix}$ and $v = \begin{pmatrix} x + 1 \\ 5 \end{pmatrix}$ are perpendicular for two values of x.

(a) Write down the equation that these values of x must satisfy.

(b) Find all values of x.

9. Figure 9.54 shows the positions of towns O, A, B and X.
Town A is 240 km east and 70 km north of O.
Town B is 480 km east and 250 km north of O.
Town X is 339 km east and 238 km north of O.
An aeroplane flies at a constant speed of 300 km h^{-1} from O towards A.

(a) (i) Show that a unit vector in the direction of $\overrightarrow{OA}$ is $\begin{pmatrix} 0.96 \\ 0.28 \end{pmatrix}$

    (ii) Write down the velocity vector for the aeroplane in the form $\begin{pmatrix} v_1 \\ v_2 \end{pmatrix}$

    (iii) How long does it take the aeroplane to reach A?

At A, the aeroplane changes direction so it now flies towards B.
The angle between the original direction and the new direction is θ as shown in Figure 9.55. This diagram also shows the point Y, between A and B, where the aeroplane comes closest to X.

(b) Use the scalar product of two vectors to find the value of θ in degrees.

(c) (i) Write down the vector $\overrightarrow{AX}$.

    (ii) Show that the vector $n = \begin{pmatrix} -3 \\ 4 \end{pmatrix}$ is perpendicular to $\overrightarrow{AB}$.

    (iii) By finding the projection of $\overrightarrow{AX}$ in the direction of n, calculate the distance XY.

(d) How far is the aeroplane from A when it reaches Y?

10. Three of the coordinates of the parallelogram $STUV$ are $S(-2, -2)$, $T(7, 7)$ and $U(5, 15)$.

(a) Find the vector $\overrightarrow{ST}$ and hence the coordinates of V.

(b) Find a vector equation of the line UV in the form $r = p + \lambda d$ where $\lambda \in P$.

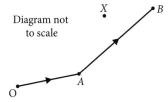

Diagram not to scale

Figure 9.54 Diagram for question 9

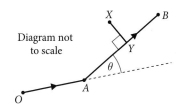

Diagram not to scale

Figure 9.55 Second diagram for question 9

(c) Show that the point E with position vector $\binom{1}{11}$ is on the line (UV), and find the value of λ for this point.

The point W has position vector $\binom{a}{17}$, $a \in \mathbb{R}$.

(d) (i) If $\left|\overrightarrow{EW}\right| = 2\sqrt{13}$, show that one value of a is -3 and find the other possible value of a.

    (ii) For $a = -3$, calculate the angle between $\overrightarrow{EW}$ and $\overrightarrow{ET}$.

11. Find the position vector of point P, the point of intersection of the two lines with vector equations

$$r_1 = \binom{5}{1} + \lambda\binom{3}{-2} \text{ and } r_2 = \binom{-2}{2} + t\binom{4}{1}$$

12. The diagram shows a parallelogram $OPQR$ in which

$$\overrightarrow{OP} = \binom{7}{3}, \overrightarrow{OQ} = \binom{10}{1}$$

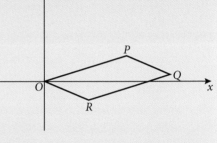

(a) Find the vector $\overrightarrow{OR}$.

(b) Use the scalar product of two vectors to show that
$$\cos O\widehat{P}Q = -\frac{15}{\sqrt{754}}$$

(c) (i) Explain why $\cos P\widehat{Q}R = -\cos O\widehat{P}Q$

    (ii) Hence show that $\sin P\widehat{Q}R = \dfrac{23}{\sqrt{754}}$

    (iii) Calculate the area of the parallelogram $OPQR$, giving your answer as an integer.

13. The diagram shows points A, B, and C, that are three vertices of a parallelogram $ABCD$. The point A has position vector $\binom{2}{2}$.

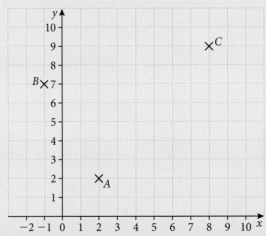

(a) Write down the position vectors of B and C.

(b) The position vector of point D is $\binom{d}{4}$. Find d.

(c) Find $\overrightarrow{BD}$

The line L passes through B and D.

(d) (i) Write down a vector equation of L in the form $\begin{pmatrix} x \\ y \end{pmatrix} = \begin{pmatrix} -1 \\ 7 \end{pmatrix} + t\begin{pmatrix} m \\ n \end{pmatrix}$

(ii) Find the value of t at point B.

(e) Let P be the point $(7, 5)$.
By finding the value of t at P, show that P lies on the line L.

(f) Show that $\overrightarrow{CP}$ is perpendicular to $\overrightarrow{BD}$.

14. Given are the points A, B, and C with position vectors $\begin{pmatrix} 4 \\ 2 \end{pmatrix}$, $\begin{pmatrix} 1 \\ -3 \end{pmatrix}$ and $\begin{pmatrix} -5 \\ -5 \end{pmatrix}$.

D is a point on the x-axis such that $ABCD$ forms a parallelogram.

(a) (i) Find $\overrightarrow{BC}$

(ii) Find the position vector of D.

(b) Find the angle between $\overrightarrow{BD}$ and $\overrightarrow{AC}$.

Line L_1 passes through A and is parallel to $i + 4j$. Line L_2 passes through B and is parallel to $2i + 7j$. A vector equation of L_1 is $r = (4i + 2j) + s(i + 4j)$.

(c) Write down a vector equation of L_2 in the form $r = b + tq$.

(d) The lines L_1 and L_2 intersect at point P. Find the position vector of P.

15. In this question, distance is in kilometres, time is in hours.

A balloon is moving at a constant height with a speed of 18 km h^{-1} in the direction of the vector $\begin{pmatrix} 3 \\ 4 \\ 0 \end{pmatrix}$. At time $t = 0$, the balloon is at point B with coordinates $(0, 0, 5)$.

(a) Show that the position vector b of the balloon at time t is given by

$$b = \begin{pmatrix} x \\ y \\ z \end{pmatrix} = \begin{pmatrix} 0 \\ 0 \\ 5 \end{pmatrix} + t\begin{pmatrix} 10.8 \\ 14.4 \\ 0 \end{pmatrix}$$

At time $t = 0$, a helicopter goes to deliver a message to the balloon. The position vector h of the helicopter at time t is given by

$$h = \begin{pmatrix} x \\ y \\ z \end{pmatrix} = \begin{pmatrix} 49 \\ 32 \\ 0 \end{pmatrix} + t\begin{pmatrix} -48 \\ -24 \\ 6 \end{pmatrix}$$

(b) (i) Write down the coordinates of the starting position of the helicopter.

(ii) Find the speed of the helicopter.

(c) The helicopter reaches the balloon at point R.

(i) Find the time the helicopter takes to reach the balloon.

(ii) Find the coordinates of R.

16. In this question the vector $\begin{pmatrix} 1 \\ 0 \end{pmatrix}$ represents a displacement due east and the vector $\begin{pmatrix} 0 \\ 1 \end{pmatrix}$ represents a displacement of 1 km north.

The diagram shows the positions of towns A, B and C in relation to an airport O, which is at the point $(0, 0)$. An aircraft flies over the three towns at a constant speed of $250 \, \text{km h}^{-1}$.

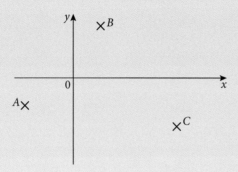

Town A is 600 km west and 200 km south of the airport.

Town B is 200 km east and 400 km north of the airport.

Town C is 1200 km east and 350 km south of the airport.

(a) (i) Find $\overrightarrow{AB}$.

   (ii) Show that the vector of length one unit in the direction of $\overrightarrow{AB}$ is $\begin{pmatrix} 0.8 \\ 0.6 \end{pmatrix}$

An aircraft flies over town A at 12:00, heading towards town B at $250 \, \text{km h}^{-1}$.

Let $\begin{pmatrix} p \\ q \end{pmatrix}$ be the velocity vector of the aircraft. Let t be the number of hours in flight after 12:00.

The position of the aircraft can be given by the vector equation

$$\begin{pmatrix} x \\ y \end{pmatrix} = \begin{pmatrix} -600 \\ -200 \end{pmatrix} + t \begin{pmatrix} p \\ q \end{pmatrix}.$$

(b) (i) Show that the velocity vector is $\begin{pmatrix} 200 \\ 150 \end{pmatrix}$

   (ii) Find the position of the aircraft at 13:00

   (iii) At what time is the aircraft flying over town B?

Over town B, the aircraft changes direction so it now flies towards town C. It takes five hours to travel the 1250 km between B and C. Over town A, the pilot noted that she had 17 000 litres of fuel left. The aircraft uses 1800 litres of fuel per hour when travelling at $250 \, \text{km h}^{-1}$. When the fuel gets below 1000 litres a warning light comes on.

(c) How far from town C will the aircraft be when the warning light comes on?

17. Given are the points $P(4, 1, -1)$, $Q(3, 3, 5)$, $R(1, 0, 2c)$, and $S(1, 1, 2)$.

   (a) Find the value of c so that the vectors $\overrightarrow{OR}$ and $\overrightarrow{PR}$ are orthogonal.

   For the remainder of the question, use the value of c found in part (a) for the coordinates of the point R.

   (b) Evaluate $\overrightarrow{PS} \times \overrightarrow{PR}$.

   (c) Find an equation of the line l passing through point Q and parallel to vector $\overrightarrow{PR}$.

   (d) Find an equation of the plane π which contains the line l and passes through the point S.

   (e) Find the shortest distance between the point P and the plane π.

18. Given are the points $A(1, 2, 1)$, $B(0, -1, 2)$, $C(1, 0, 2)$, and $D(2, -1, -6)$.

   (a) Find the vectors $\overrightarrow{AB}$ and $\overrightarrow{BC}$.

   (b) Calculate $\overrightarrow{AB} \times \overrightarrow{BC}$.

   (c) Hence, find the area of triangle ABC.

   (d) Find an equation of the plane P containing the points A, B, and C.

   (e) Find a set of parametric equations for the line through the point D and perpendicular to plane P.

   (f) Find the distance from the point D to plane P.

   (g) Find a unit vector perpendicular to plane P.

   (h) Find the coordinates of E, which is the reflection of D in the plane P.

19. (a) If $u = i + 2j + 3k$ and $v = 2i - j + 2k$, show that
   $$u \times v = 7i + 4j - 5k.$$

   (b) Let $w = \lambda u + \mu v$ where λ and μ are scalars. Show that w is perpendicular to the line of intersection of the planes
   $x + 2y + 3z = 5$ and $2x - y + 2z = 7$ for all values of λ and μ.

20. Given are the points $A(2, 1, -2)$, $B(2, -1, -1)$, and $C(1, 2, 2)$. O is the origin and $OAPBCQSR$ is a parallelepiped as shown in the diagram.

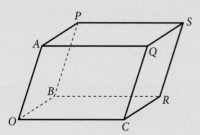

   (a) Find the coordinates of P, Q, R, and S.

   (b) Find a Cartesian equation for the plane $OAPB$.

   (c) Calculate the volume, V, of the parallelepiped given that
   $$V = \overrightarrow{OA} \times \overrightarrow{OB} \cdot \overrightarrow{OC}$$

21. Triangle ABC has vertices $A(-1, 2, 3)$, $B(-1, 3, 5)$ and $C(0, -1, 1)$.

    (a) Find the angle θ between vectors $\overrightarrow{AB}$ and $\overrightarrow{AC}$.

    (b) Hence, find the area of triangle ABC.

    Let l_1 be the line parallel to $\overrightarrow{AB}$ which passes through $D(2, -1, 0)$ and l_2 be the line parallel to $\overrightarrow{AC}$ which passes through $E(-1, 1, 1)$.

    (c) (i) Find the equations of the lines l_1 and l_2

        (ii) Hence, show that l_1 and l_2 do not intersect.

    (d) Find the shortest distance between l_1 and l_2.

22. (a) Solve the following system of linear equations

    $$x + 3y - 2z = -6$$
    $$2x + y + 3z = 7$$
    $$3x - y + z = 6$$

    (b) Find the vector $\mathbf{v} = (\mathbf{i} + 3\mathbf{j} - 2\mathbf{k}) \times (2\mathbf{i} + \mathbf{j} + 3\mathbf{k})$.

    (c) If $\mathbf{a} = \mathbf{i} + 3\mathbf{j} - 2\mathbf{k}$, $\mathbf{b} = 2\mathbf{i} + \mathbf{j} + 3\mathbf{k}$ and $\mathbf{u} = m\mathbf{a} + n\mathbf{b}$ where m, n are scalars, and $\mathbf{u} \neq 0$, show that $\mathbf{v}$ is perpendicular to $\mathbf{u}$ for all m and n.

    (d) The line l lies in the plane $3x - y + z = 6$, passes through the point $(1, -1, 2)$ and is perpendicular to $\mathbf{v}$. Find an equation of l.

23. Given are the points $A(1, 3, 1)$, $B(1, 2, 4)$, $C(2, 3, 6)$, and $D(5, -2, 1)$.

    (a) (i) Evaluate the vector product $\overrightarrow{AB} \times \overrightarrow{AC}$.

        (ii) Find the area of the triangle ABC.

    Let the plane containing the points A, B, C be π and the line passing through D perpendicular to π be L. P is the point of intersection of L and π.

    (b) (i) Find the Cartesian equation of π.

        (ii) Find the Cartesian equation of L.

    (c) Determine the coordinates of P.

    (d) Find the perpendicular distance from D to π.

24. Point $A(2, 5, -1)$ lies on line L, which is perpendicular to plane P with equation $x + y + z - 1 = 0$.

    (a) Find the Cartesian equation of the line L.

    (b) Find the point of intersection of the line L and plane P.

    (c) The point A is reflected in plane P. Find the coordinates of the image of A.

    (d) Find the distance from the point $B(2, 0, 6)$ to line L.

25. (a) Plane π_1 contains point $P(1, 2, 11)$. Vector $3\mathbf{i} - 4\mathbf{j} + \mathbf{k}$ is perpendicular to π_1. Find the Cartesian equation of π_1.

(b) Plane π_2 has equation $x + 3y - z = -4$.

(i) Show that point P also lies in the plane π_2.

(ii) Find a vector equation of the line of intersection of π_1 and π_2.

(c) Find the acute angle between π_1 and π_2.

26. A line l_1 has equation $\dfrac{x + 2}{3} = \dfrac{y}{1} = \dfrac{z - 9}{-2}$

(a) Let M be a point on l_1 with parameter μ. Express the coordinates of M in terms of μ.

(b) The line l_2 is parallel to l_1 and passes through $P(4, 0, -3)$.

(i) Write down an equation for l_2.

(ii) Express $\overrightarrow{PM}$ in terms of μ.

(c) The vector $\overrightarrow{PM}$ is perpendicular to l_1.

(i) Find the value of μ.

(ii) Find the distance between l_1 and l_2.

(d) Plane π_1 contains l_1 and l_2. Find an equation for π_1, giving your answer in the form $Ax + By + Cz = D$.

(e) Plane π_2 has equation $x - 5y - z = -11$. Verify that l_1 is the line of intersection of planes π_1 and π_2.

27. (a) Show that lines $\dfrac{x - 2}{1} = \dfrac{y - 2}{3} = \dfrac{z - 3}{1}$ and $\dfrac{x - 2}{1} = \dfrac{y - 3}{4} = \dfrac{z - 4}{2}$ intersect and find the coordinates of P, the point of intersection.

(b) Find the Cartesian equation of the plane π that contains the two lines.

(c) The point $Q(3, 4, 3)$ lies on π. Line L passes through the midpoint of $[PQ]$. Point S is on L such that $|\overrightarrow{PS}| = |\overrightarrow{QS}| = 3$, and the triangle PQS is normal to the plane π. Given that there are two possible positions for S, find their coordinates.

28. (a) Plane π_1 has equation $\mathbf{r} = \begin{pmatrix} 2 \\ 1 \\ 1 \end{pmatrix} + \lambda \begin{pmatrix} -2 \\ 1 \\ 8 \end{pmatrix} + \mu \begin{pmatrix} 1 \\ -3 \\ -9 \end{pmatrix}$

Plane π_2 has equation $\mathbf{r} = \begin{pmatrix} 2 \\ 0 \\ 1 \end{pmatrix} + s \begin{pmatrix} 1 \\ 2 \\ 1 \end{pmatrix} + t \begin{pmatrix} 1 \\ 1 \\ 1 \end{pmatrix}$

(i) For points that lie in π_1 and π_2, show that $\lambda = \mu$.

(ii) Hence, find a vector equation of the line of intersection of π_1 and π_2.

(b) Plane π_3 contains the line $\dfrac{2 - x}{3} = \dfrac{y}{-4} = z + 1$ and is perpendicular to $3\mathbf{i} - 2\mathbf{j} + \mathbf{k}$. Find the Cartesian equation of π_3.

(c) Find the intersection of π_1, π_2 and π_3.

29. Two planes have equations

$\pi_1: 4x + y + z = 8$ and $\pi_2: 4x + 3y - z = 0$

(a) Find the cosine of the angle between the two planes in the form $\sqrt{\dfrac{p}{q}}$ where $p, q \in \mathbb{Z}$.

Let L be the line of intersection of the two planes.

(b) (i) Show that L has direction $\begin{pmatrix} -1 \\ 2 \\ 2 \end{pmatrix}$

(ii) Show that the point $A(1, 0, 4)$ lies on both planes.

(iii) Write down a vector equation of L.

B is the point on π_1 with coordinates $(a, b, 1)$.

(c) Given the vector $\overrightarrow{AB}$ is perpendicular to L, find the value of a and the value of b.

(d) Show that $AB = 3\sqrt{2}$

The point P lies on L and $A\hat{B}P = 45°$.

(e) Find the coordinates of the two possible positions of P.

30. A line L has equation $\dfrac{x - 2}{p} = \dfrac{y - q}{2} = z - 1$, where $p, q \in \mathbb{R}$.

A plane π has equation $x + y + 3z = 9$.

(a) Show that L is not perpendicular to π.

(b) Given that L lies in the plane π, find the value of p and the value of q.

Consider the different case where the acute angle between L and π is θ where $\theta = \arcsin\left(\dfrac{1}{\sqrt{11}}\right)$

(c) (i) Show that $p = -2$.

(ii) If L intersects π at $z = -1$, find the value of q.

Statistics 10

Learning objectives

By the end of this chapter, you should be familiar with...

- concepts of population, sample, random sample, and frequency distribution of discrete and continuous data
- reliability of data sources and bias in sampling; recognising sampling techniques and their effectiveness
- interpreting outliers
- presenting data using frequency tables and diagrams, and box-and-whisker plots
- working with grouped data: mid-interval values, interval width, upper and lower interval boundaries, frequency histograms
- interpreting and calculating mean, median, mode; quartiles, percentiles
- interpreting and calculating range, interquartile range; variance, standard deviation
- interpreting and calculating cumulative frequency, cumulative frequency graphs, and using them to find median, quartiles, percentiles
- interpreting and calculating linear correlation of bivariate data
- interpreting and calculating linear regression models.

You will almost certainly encounter statistics, in one form or another, on a daily basis. Figure 10.1 shows an example.

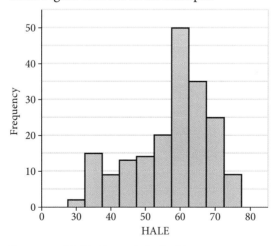

Figure 10.1 Worldwide population health

Data pertaining to worldwide population health in UN-member countries is widely available. Among the indicators reported is the **health-adjusted life expectancy** (HALE), which is based on life expectancy at birth, but includes an adjustment for time spent in poor health. It is most easily understood as the equivalent number of years in full health that a newborn can expect to live, based on current rates of ill-health and mortality. According to WHO (World Health Organization) rankings, years lost due to disability are substantially higher in poorer countries. Several factors contribute to this trend, including injury, blindness, paralysis, and the debilitating effects of tropical disease.

Of the countries ranked by WHO, Japan has the highest healthy life expectancy (75 years) and the lowest ranking country is Sierra Leone (29 years).

Reports similar to this one are commonplace in publications of several organisations, newspapers and magazines, and on the internet.

There are many questions we can ask about a report like this:

- How did the researchers collect the data?
- How can we be sure that these results are reliable?
- What conclusions should be drawn from this report?

The increased frequency with which statistical techniques are used in all fields, from business to agriculture to social and natural sciences, leads to the need for statistical literacy – familiarity with the goals and methods of these techniques – to be a part of any well-rounded educational programme.

Statistical methods for summary and analysis provide us with powerful tools for making sense out of the data we collect. In this chapter, we will start by introducing two basic components of most statistical problems – population and sample – and then discuss the methods of presenting and making sense of data.

In the language of statistics, one of the most basic concepts is sampling. In most statistical problems, we draw a specified number of measurements or data – a sample – from a much larger body of measurements, called the population. On the basis of our observations of the data in the well-chosen sample, we try to describe or predict the behaviour of the population.

A **population** is any entire collection of people, animals, plants, or things from which we may collect data. It is the entire group we are interested in, which we wish to describe or draw conclusions about. In order to make any generalisations about a population, a **sample** (that is meant to be representative of the population) is often studied. For each population there are many possible samples.

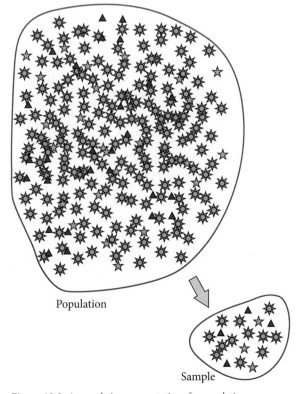

Population

Sample

Figure 10.2 A sample is representative of a population

For example, a study was conducted to examine the use of water for washing clothes in the households of an EU country.

1674 households were surveyed where the residents own and use a washing machine. The average amount of water used by each washing cycle was reported to be 58 minutes. The average time for each cycle was 42 minutes. It was also discovered that the amount of laundry done by a household every year is related to the household's income.

In this example, the population is household usage of water for washing, average wash cycle time, income, and so on. The sample is the set of measurements of 1674 households that took part in the study. Notice that the population and sample are the measurements and not the people. The households are 'experimental units' or subjects in this study.

In this chapter we will present some basic techniques in **descriptive statistics** – the branch of statistics concerned with describing sets of measurements, both samples and populations.

10.1 Graphical tools

A **variable** is a characteristic that changes or varies over time and/or for different objects under consideration.

Once we have collected a set of measurements, how can we display this set in a clear, understandable and readable form? First, we must be able to define what is meant by measurement or data, and to categorise the types of data we are likely to encounter. We begin by introducing some definitions of the new terms in the statistical language that we need to know.

For example, if you are measuring the height of adults in a certain area, the height is a variable that changes with time for an individual and from person to person. When a variable is actually measured, a set of measurements or **data** will result. The set of measurements is called a **data set**.

As the process of data collection begins, it becomes clear that often the amount of data collected is so large that it is difficult for the statistician to see the findings of the data. The statistician's objective is to summarise the data succinctly, bringing out the important characteristics of the numbers and values in such a way that a clear and accurate picture emerges.

When looking at statistical results, care must be taken with how data has been collected/measured. **Reliability** and **validity** are often used interchangeably when they are not related to statistics. In statistics, however, they refer to different properties of the statistical or experimental method.

Reliability is another term for consistency. For example, if one person takes the same personality test several times and always receives the same results, the test is reliable.

A test is valid if it measures what it is supposed to measure. If the results of the personality test claimed that a very shy person was in fact outgoing, the test would be invalid.

Reliability and validity are independent of each other. A measurement may be valid but not reliable, or reliable but not valid. Suppose your bathroom scale was reset to read 5 kg lighter. The weight it reads will be reliable (the same every time you step on it) but will not be valid, since it is not reading your actual weight.

Reliability or reproducibility in statistics refers to consistency in measurement: the capacity to produce the same result for two identical states, or, more practically, the closeness of the initial estimated value(s) to the subsequent estimated value(s).

Classification of variables

Numerical or categorical

When classifying data, there are two major classifications: numerical and categorical.

Numerical (quantitative) data: quantitative variables measure a numerical quantity or amount of each experimental unit. Quantitative data yields a numerical response.

For example: Yearly income of company presidents, the heights of students at school, the length of time it takes students to finish their lunch at school, and the total score you receive on exams are all numerical data.

There are two types of numerical data:

Discrete: responses that arise from counting. For example, the number of courses students take in a day.

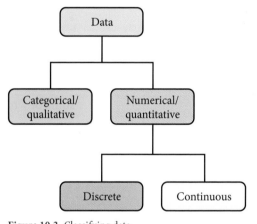

Figure 10.3 Classifying data

Continuous: responses that arise from measuring. For example, the time it takes a student to travel from home to school.

Categorical (qualitative data) data: qualitative variables measure a quality or characteristic of the experimental unit. Categorical data gives a qualitative response; that is, data is a kind or type, rather than a quantity.

For example: categorising students into first year IB or second year IB, into AI/SL, AI/HL, AA/HL, or AA/SL.

When data is first collected, there are some simple ways of beginning to organise it. These include an ordered array and the stem-and-leaf plot.

- Data in raw form (as collected):

  24, 26, 24, 21, 27, 27, 30, 41, 32, 38

- Data in an ordered array from smallest to largest (an ordered array is an arrangement of data in either ascending or descending order):

  21, 24, 24, 26, 27, 27, 30, 32, 38, 41

Suppose a consumer organisation was interested in studying weekly food and living expenses of college students. A survey of 80 students yielded the following expenses to the nearest euro:

38	50	55	60	46	51	58	64	50	49	48	65	58	61	65	53
39	51	56	61	48	53	59	65	54	54	54	59	65	66	47	49
40	51	56	62	47	55	60	63	60	59	59	50	46	45	54	47
41	52	57	64	50	53	58	67	67	66	65	58	54	52	55	52
44	52	57	64	51	55	61	68	67	54	55	48	57	57	66	66

Table 10.1 Weekly food and living expenses in euros for 80 college students

The first step in the analysis is a summary of the data, which should show the following information:

- What values of the variable have been measured?
- How often has each value occurred?

Such summaries can be done in many ways. The most useful are the frequency distribution and the histogram. There are other methods of presenting data, some of which we will discuss later.

Frequency distribution (table)

A **frequency distribution** is a table used to organise data; classes (groups) are the numerical intervals for a variable being studied, and each class in the table shows the frequency, or number of observations. Intervals are normally of equal size, must cover the range of the sample observations, and are non-overlapping (see Table 10.2).

There are some general rules for preparing frequency distributions that make it easier to summarise data and to communicate results.

Construction of a frequency distribution (table)

Rule 1: Intervals (classes) must be inclusive and non-overlapping; each observation must belong to one and only one class interval. Consider a frequency distribution for the living expenses of the 80 college students. If the frequency distribution contains the intervals '35–40' and '40–45', to which of these two classes would a person spending €40 belong?

The boundaries, or endpoints, of each class must be clearly defined. For our example, appropriate intervals would be '35 but less than 40' and '40 but less than 45'.

If classes are described with discrete limits such as '30–34', '35–39', '40–44', …, then the boundaries are midway between the neighbouring class limits/endpoints. That is, the classes above will be considered as '29.5 but less than 34.5', '34.5 but less than 39.5', '39.5 but less than 44.5' etc.… Here the boundaries are 29.5, 34.5, 39.5, 44.5. Each class width is 5.

Rule 2: Determine k, the number of classes. Practice and experience are the best guidelines for deciding the number of classes. In general, the number of classes could be between 5 and 10, but this is not an absolute rule. Practitioners use their judgement in these issues. If the number of classes is too few, some characteristics of the distribution will be hidden, and if too many, some characteristics will be lost with the detail.

Rule 3: Intervals should be the same width, w; the width is determined by the following:

$$\text{interval width} = \frac{\text{largest number} - \text{smallest number}}{\text{number of intervals}}$$

Both the number of intervals and the interval width should be rounded upwards, possibly to the next largest integer. The above formula can be used when there are no natural ways of grouping the data. If this formula is used, the interval width is generally rounded to a convenient whole number to provide for easy interpretation.

In the example of the weekly living expenses of students, a reasonable grouping with nice round numbers was that of '35 but less than 40' and '40 but less than 45', and so on.

Living expenses	Number of students	% of students
$35 \leqslant$ living expenses < 40	2	2.50
$40 \leqslant$ living expenses < 45	3	3.75
$45 \leqslant$ living expenses < 50	11	13.75
$50 \leqslant$ living expenses < 55	21	26.25
$55 \leqslant$ living expenses < 60	19	23.75
$60 \leqslant$ living expenses < 65	11	13.75
$65 \leqslant$ living expenses < 70	13	16.25
Total	**80**	**100.00**

Table 10.2 Frequency and percentage frequency distributions of weekly expenses

Grouping the data in a table like this one enables us to see some of its characteristics. For example, we can see that there are few students who spend as little as €35 to €45, while the majority of the students spend more than €45. Grouping the data will also cause some loss of detail, as we do not see from the table what the real values are in each class.

In Table 10.2, the impression we get is that the class midpoint, also known as the mid-interval value, will represent the data in that interval. For example, 37.5 will represent the data in the first class, while 62.5 will represent the data in the 60 to 65 class. 35 and 40 are known as the **interval boundaries**.

Graphically, we have a tool that helps us visualise the distribution: the **histogram**.

Histograms

A histogram is a graph that consists of vertical bars constructed on a horizontal line that is marked off with intervals for the variable being displayed.

The intervals correspond to those in a frequency distribution table. The height of each bar is proportional to the number of observations in that interval. The number of observations can also be displayed above the bars.

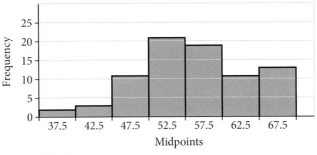

Figure 10.4 Histogram showing weekly living expenses

Here is an example of a histogram produced by a GDC.

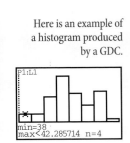

By looking at the histogram, it becomes clear that our observation above is true. From the histogram we also see that the distribution is not symmetric.

Cumulative and relative cumulative frequency distributions

A **cumulative frequency distribution** contains the total number of observations whose values are less than the upper limit for each interval. It is constructed by adding the frequencies of all frequency distribution intervals up to and including the present interval. A **relative cumulative frequency distribution** converts all cumulative frequencies to cumulative percentages.

A cumulative distribution and a relative (percentage) cumulative distribution for the weekly expenses data are shown in Table 10.3.

Living expenses, L	Number of students	Cumulative number of students	% of students	Cumulative % of students
$35 \leqslant L < 40$	2	2	2.50	2.50
$40 \leqslant L < 45$	3	5	3.75	6.25
$45 \leqslant L < 50$	11	16	13.75	20.00
$50 \leqslant L < 55$	21	37	26.25	46.25
$55 \leqslant L < 60$	19	56	23.75	70.00
$60 \leqslant L < 65$	11	67	13.75	83.75
$65 \leqslant L < 70$	13	80	16.25	100.00
	80		**100.00**	

Table 10.3 Cumulative frequency and cumulative relative frequency distributions of weekly expenses

Notice how each cumulative frequency is added to the frequency in the next interval to give you the next cumulative frequency. The same is true for the relative frequencies.

As we will see later, cumulative frequencies and their graphs help us analyse data that are given in group form.

Cumulative line graph/cumulative frequency graph

Sometimes called an **ogive**, this is a line that connects points that are the cumulative percentage of observations below the upper limit of each class in a cumulative frequency distribution.

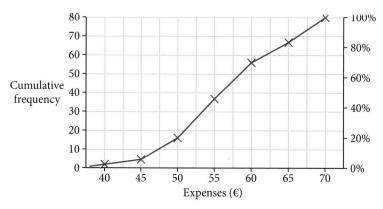

Figure 10.5 An ogive is a line that connects points that are the cumulative percentage of observations below the upper limit of each class

Notice how the height of each line at the upper boundary represents the cumulative frequency for that interval. For example, at 50 the height is 16 and at 60 it is 56.

Example 10.1

Here is some HALE data in raw form. Prepare a frequency table starting with 20 and a class interval of 5. Then draw a histogram of the data and a cumulative frequency graph.

29	36	40	44	48	52	54	56	59	60	61	61	62	63	64	66	68	71	72	73	63	64	66	68
31	36	41	44	49	52	54	57	59	60	61	62	62	64	64	66	68	71	72	75	63	64	66	68
33	36	41	44	49	52	55	57	59	60	61	62	62	64	65	66	69	71	72	35	38	43	47	71
34	37	41	45	49	53	55	58	59	60	61	62	63	64	65	66	69	71	73	36	40	44	48	71
34	37	42	45	50	53	55	58	59	60	61	62	63	64	65	67	70	71	73	50	54	56	59	72
35	37	42	45	50	53	55	58	59	60	61	62	63	64	65	67	70	71	73	51	54	56	59	72
35	37	43	46	50	54	55	58	59	60	61	62	63	64	65	67	70	71	73	60	60	61	62	73
35	38	43	46	50	54	55	58	59	60	61	62	63	64	65	67	70	72	73	60	61	61	62	73

Solution

We first sort the data and then make sure we count every number in one class only.

Life expectancy, l	Number of countries	Life expectancy, l	Number of countries
$25 \leqslant l < 30$	1	$55 \leqslant l < 60$	26
$30 \leqslant l < 35$	4	$60 \leqslant l < 65$	54
$35 \leqslant l < 40$	14	$65 \leqslant l < 70$	22
$40 \leqslant l < 45$	14	$70 \leqslant l < 75$	27
$45 \leqslant l < 50$	11	$75 \leqslant l < 80$	1
$50 \leqslant l < 55$	18		

This is the histogram created by a spreadsheet. Since all the classes have equal width, the height and the area give the same impression about the frequency of the class interval. For example, the $60 \leqslant l < 65$ class contains almost twice as much as the $55 \leqslant l < 60$ class, and the height of the histogram is also twice as high. So is the area. Similarly, the height of the $65 \leqslant l < 70$ class is double that of $45 \leqslant l < 50$.

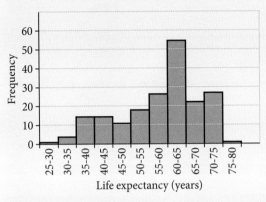

Life expectancy (years)

Life expectancy, l	Number of countries	Cumulative number of countries	Life expectancy, l	Number of countries	Cumulative number of countries
$25 \leqslant l < 30$	1	1	$55 \leqslant l < 60$	26	88
$30 \leqslant l < 35$	4	5	$60 \leqslant l < 65$	54	142
$35 \leqslant l < 40$	14	19	$65 \leqslant l < 70$	22	164
$40 \leqslant l < 45$	14	33	$70 \leqslant l < 75$	27	191
$45 \leqslant l < 50$	11	44	$75 \leqslant l < 80$	1	192
$50 \leqslant l < 55$	18	62			

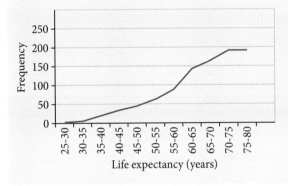

Life expectancy (years)

Sampling methods

Once a decision has been made to gather information by sampling, statisticians use a sampling technique that falls into one of two categories: statistical or non-statistical.

Both non-statistical and statistical sampling techniques are commonly used by decision makers. Regardless of which technique is used, the decision maker has the same objective – to obtain a sample that is a close representative of the population. There are some advantages to using a statistical sampling technique, but in many cases non-statistical sampling represents the only feasible way to sample.

Statistical sampling techniques use selection techniques based on chance selection. Advantages of using these techniques include their ability to represent the population they are drawn from. Disadvantages include cost of running them and time spent in the collection process.

Non-statistical sampling techniques use convenience, judgment, or other non-chance processes. Advantages include the ease of collection and minimal cost involved. Disadvantages include the great possibility that they don't represent the population they are drawn from, inclusion of bias, and non-reliability.

Statistical sampling techniques include:

- simple random sampling
- stratified random sampling
- systematic random sampling.

Simple random sampling: A method of selecting items from a population such that every possible sample of a specified size has an equal chance of being chosen. It is a method of sampling in which every sample element is selected only on the basis of chance, through a random process. The probability of selection in a true simple random sample is equal for each element. If a sample of 500 is selected from a population of 17 000 (a sampling frame of 17 000), then the probability of selection for each element is 500 ÷ 17 000, or about 0.03. Every element has an equal chance of being selected, just like the odds in a toss of a coin $\left(\frac{1}{2}\right)$ or a roll of a dice $\left(\frac{1}{6}\right)$.

Stratified random sampling is a statistical sampling method in which the population is divided into subgroups called strata, so that each population item belongs to only one stratum. The objective is to form strata such that the population values of interest within each stratum are as much alike as possible. Sample items are selected from each stratum using the simple random sampling method. For example, if you are interested in attitudes of voters in a certain city, you may divide your subjects by age group: under 25 years of age, 25 to 30, 30 to 40, and so on.

Systematic random sampling is a variant of simple random sampling. The first element is selected randomly from a list or from sequential files, and then every nth element is selected. This is a convenient method for drawing a random sample when the population elements are arranged sequentially. The total number of cases in the population is divided by the number of cases required for the sample. This division yields the sampling interval – the number of cases from one sampled case to another. If 50 cases are to be selected out of 1000, the **sampling interval** is 20 as every 20th case is selected.

Non-statistical sampling techniques include:

- quota sampling
- availability sampling.

Quota sampling is a sampling method that does not rely on probability, where elements are selected to ensure that the sample represents certain characteristics in proportion to their prevalence in the population. For example, suppose that you wish to sample adult residents of a town in a study of support for a tax increase to improve the town's schools. You know from historical data what the proportions of the town's residents are in terms of gender, race, age, and number of children. You think that each of these characteristics might influence support for new school taxes, so you want to be sure that the sample includes men, women, older people, younger people, and so on, in proportion to their numbers in the town population. Say that 48% of the town's adult residents are men and 52% are women, and that 60% are employed, 5% are unemployed, and 35% are out of the labour force. These percentages and the percentages corresponding to the other characteristics become the quotas for the sample. If you plan to include a total of 500 residents in your sample, 240 must be men (48% of 500), 260 must be women, 300 must be employed, and so on.

Availability sampling is where elements are selected based on convenience. Elements are selected for availability sampling because they're available or easy to find. Thus, this sampling method is also known as a haphazard, accidental, or convenience sample.

Exercise 10.1

1. Identify the experimental units, sensible population and sample on which each of the following variables is measured. Then indicate whether the variable is quantitative or qualitative.

   (a) Gender of a student

   (b) Number of errors on a final exam for 10th grade students

   (c) Height of a newborn child

   (d) Eye colour for children aged less than 14

   (e) Amount of time it takes to travel to work

   (f) Rating of a country's leader: excellent, good, fair, poor

   (g) Country of origin of students at international schools.

2. State what you expect the shapes of the distributions of the following variables to be: uniform, unimodal, bimodal, symmetric, etc. Explain why.

   (a) Number of goals scored by football players during last season

   (b) Weights of newborn babies in a major hospital during the course of 10 years

   (c) Number of countries visited by a student at an international school

   (d) Number of emails per week received by a high school student at your school.

3. Identify each variable as quantitative or qualitative:

   (a) Amount of time to finish your extended essay

   (b) Number of students in each section of IB Math SL

(c) Rating of your textbook as excellent, good, satisfactory, terrible

(d) Country of origin of each student in Math SL courses.

4. Identify each variable as discrete or continuous:

(a) Population of each country represented by SL students in your session of the exam

(b) Weight of IB Math SL exams printed every May since 1976

(c) Time it takes to mark an exam paper by an examiner

(d) Number of customers served at a bank counter

(e) Time it takes to finish a transaction at a bank counter

(f) Amount of sugar used in preparing your favourite cake.

5. Grade point averages (GPA) in several colleges are on a scale of 0–4. Here are the GPAs of 45 students at a certain college.

1.8	1.9	1.9	2.0	2.1	2.1	2.1	2.2	2.2	2.3	2.3	2.4	2.4	2.4	2.5
2.5	2.5	2.5	2.5	2.5	2.6	2.6	2.6	2.6	2.6	2.7	2.7	2.7	2.7	2.7
2.8	2.8	2.8	2.9	2.9	2.9	3.0	3.0	3.0	3.1	3.1	3.1	3.2	3.2	3.4

Draw a histogram, a relative frequency histogram, and a cumulative frequency graph. Describe the data in two to three sentences.

6. The following are the grades of an IB course with 40 students (two sections) on a 100-mark test. Use graphical methods you have learned so far to describe the grades.

61	62	93	94	91	92	86	87	55	56
63	64	86	87	82	83	76	77	57	58
94	95	89	90	67	68	62	63	72	73
87	88	68	69	65	66	75	76	84	85

7. The lengths of time (in months) between repeated speeding violations of 50 young drivers in a European country are given in the table below.

2.1	1.3	9.9	0.3	32.3	8.3	2.7	0.2	4.4	7.4
9.0	18.0	1.6	2.4	3.9	2.4	6.6	1.0	2.0	14.1
14.7	5.8	8.2	8.2	7.4	1.4	16.7	24.0	9.6	8.7
19.2	26.7	1.2	18.0	3.3	11.4	4.3	3.5	6.9	1.6
4.1	0.4	13.5	5.6	6.1	23.1	0.2	12.6	18.4	3.7

(a) Construct a histogram for the data.

(b) Would you describe the shape as symmetric?

(c) The law in this country requires that the driving licence be taken away if the driver repeats the violation within a period of 10 months. Use a cumulative frequency graph to estimate the fraction of drivers who may lose their licence.

8. To decide on the number of checkout counters needed to be open during rush hours in a supermarket, the management collected data from 60 customers for the time they spent waiting to be served. The times in minutes are given in the following table.

3.6	0.7	5.2	0.6	1.3	0.3	1.8	2.2	1.1	0.4
1.0	1.2	0.7	1.3	0.7	1.6	2.5	0.3	1.7	0.8
0.3	1.2	0.2	0.9	1.9	1.2	0.8	2.1	2.3	1.1
0.8	1.7	1.8	0.4	0.6	0.2	0.9	1.8	2.8	1.8
0.4	0.5	1.1	1.1	0.8	4.5	1.6	0.5	1.3	1.9
0.6	0.6	3.1	3.1	1.1	1.1	1.1	1.4	1.0	1.4

(a) Construct a relative frequency histogram for the times.

(b) Construct a cumulative frequency graph and estimate the number of customers who have to wait 2 minutes or more.

9. The histogram below shows the number of days in hospital spent by heart patients in a certain country's hospitals in the 2015–2017 period.

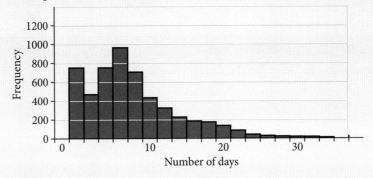

(a) Describe the data in a few sentences.

(b) Draw a cumulative frequency graph for the data.

(c) What percentage of the patients stayed less than 6 days?

10. János exercises on an almost daily basis. He records the length of time of his exercise on most of the days. Here is what he recorded for 2017.

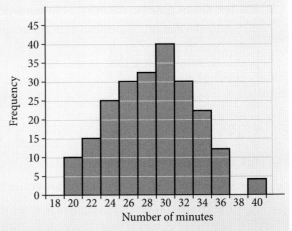

(a) What is the longest time he spent doing his exercises?

(b) What percentage of the time did he exercise more than 30 minutes?

(c) Draw a cumulative frequency graph for his exercise times.

11. Radar devices are installed at several locations on a main highway. Speeds, s, in $km\,h^{-1}$ of 400 cars travelling on that highway are measured and summarised in the following table.

Speed, s	$60 \leqslant s < 75$	$75 \leqslant s < 90$	$90 \leqslant s < 105$	$105 \leqslant s < 120$	$120 \leqslant s < 135$	Over 135
Frequency	20	70	110	150	40	10

(a) Construct a frequency table for the data.

(b) Draw a histogram to illustrate the data.

(c) Draw a cumulative frequency graph for the data.

(d) The speed limit in this country is $130\,km\,h^{-1}$. Use your graph in (c) to estimate the percentage of the drivers driving faster than this limit.

12. Electronic components used in the production of computers are manufactured in a factory and their lengths must be very accurate. Here are the lengths of a sample of 400 such components.

Length, l (mm)	$l < 5.00$	$5.00 \leqslant l < 5.05$	$5.05 \leqslant l < 5.10$	$5.10 \leqslant l < 5.15$	$5.15 \leqslant l < 5.20$	$l \geqslant 5.20$
Frequency	16	100	123	104	48	9

(a) Construct a cumulative relative frequency graph for the data.

(b) The components must have a length between 5.01 and 5.18 mm, and any component with a length above 5.18 has to be scrapped. Use your graph to estimate the percentage of components that must be scrapped from this production facility.

13. The time, t, in seconds, that 300 customers wait at a supermarket checkout are recorded in the table below.

Time, t	$t < 60$	$60 \leqslant t < 120$	$120 \leqslant t < 180$	$180 \leqslant t < 240$	$240 \leqslant t < 300$	$300 \leqslant t < 360$	$t \geqslant 360$
Frequency	12	15	42	105	66	45	15

(a) Draw a histogram of the data.

(b) Construct a cumulative frequency graph of the data.

(c) Use the cumulative frequency graph to estimate the waiting time that is exceeded by 25% of the customers.

10.2 Measures of central tendency

Summarising data can help us understand them, especially when the data set is large. This section presents several ways to summarise quantitative data by a typical value (a measure of location, such as the **mean**, **median**, or **mode**) and a measure of how well the typical value represents the list (a measure of spread, such as the **range**, **interquartile range**, or **standard deviation**). When looking at raw data, rather than looking at tables and graphs, it may be of interest to use summary measures to describe the data. The farthest we can reduce a set of data, and still retain any information at all, is to summarise the data with a single value. Measures of location do just that: they try to capture with a single number what is typical of the data. What single number is most representative of an entire list of numbers? We cannot say without defining 'representative' more precisely.

- The **median** of a set of n measurements is the value that falls in the middle position when the data are sorted in ascending order. In a histogram, the median is that middle value that divides the histogram into two equal areas.

- The **mode** of a set of data is the most common value among the data. It is rare that several data coincide exactly, unless the variable is discrete, or the measurements are reported with low precision.

- The **arithmetic mean** is commonly called the average. It is the sum of the data divided by the number of data.

When these measures are computed for a population they are called **parameters**. When they are computed from a sample they are called **statistics**.

A **statistic** is a descriptive measure computed from a sample of data. A **parameter** is a descriptive measure computed from an entire population of data.

Measures of central tendency provide information about a typical observation in the data or location in the data set.

The most common measure of central tendency is the arithmetic mean, usually referred to simply as the 'mean' or the 'average'.

Example 10.2

The following are the five closing prices of a stock market index for the first business week in November 2018. This is a sample of size $n = 5$ for the closing prices from the entire population: 2794.83, 2810.38, 2795.18, 2825.18, 2748.76.

What is the average closing price for the first business week in November 2018?

Solution

$$\text{average} = \frac{2794.83 + 2810.38 + 2795.18 + 2825.18 + 2748.76}{5} = 2794.87$$

This is called the sample mean. A second measure of central tendency is the median, which is the value in the middle position when the measurements are ordered from smallest to largest. The median of this data can be calculated only after if we sort them in ascending order:

2748.76   2794.83   **2795.18**   2810.38   2825.18

⇧

The **arithmetic mean** or **average** of a set of n measurements (data set) is equal to the sum of the measurements divided by n.

Notation:

The sample mean: $\bar{x} = \dfrac{\sum\limits_{i=1}^{n} x_i}{n} = \dfrac{x_1 + x_2 + x_3 + \cdots + x_n}{n}$, where n is the sample size. This is a statistic.

The population mean: $\mu = \dfrac{\sum\limits_{i=1}^{N} x_i}{N} = \dfrac{x_1 + x_2 + x_3 + \cdots + x_N}{N}$, where N is the population size. This is a parameter.

It is important to observe that you normally do not know the mean of the population μ and thus you usually estimate it with the sample mean $\bar{x}$.

In the previous example, we calculated the sample median by finding the third measurement to be in the middle position. However, in a different situation, where the number of measurements is even, the process is slightly different.

> The **median** of a set of n measurements is the value of x that falls in the middle position when the data are sorted in ascending order.

Let's say you took six tests last term and your marks, in ascending order, were: 52, 63, 74, 78, 80, 89.

There are two 'middle' observations here. To find the median, choose a value halfway between the two middle observations:

$$m = \frac{74 + 78}{2} = 76$$

Note that the position of the median can be given by $\dfrac{n + 1}{2}$. If this number ends with a decimal, you need to average the adjacent values.

In the stock exchange index case, with 5 observations, the position of the median is then at $\dfrac{5 + 1}{2} = 3$, which we found. In the grades example, the position of the median score is at $\dfrac{6 + 1}{2} = 3.5$, and hence we average the numbers at positions 3 and 4.

Although both the mean and median are good measures for the centre of a distribution, the median is less sensitive to extreme values or **outliers**. Look at these two sets of data:

Set 1:	52	63	74	78	80	89
Set 2:	12	63	74	78	80	89

Both have medians of 76, but the means are different.

Set 1 has a mean of:

$$\bar{x} = \frac{\sum x}{6} = \frac{436}{6} = 72.\dot{6}$$

But set 2 has a mean of:

$$\bar{x} = \frac{\sum x}{6} = \frac{396}{6} = 66$$

Clearly, the low outlier, 12, moves the mean towards it while leaving the median untouched. However, because the mean depends on every observation and uses all the information in the data, it is generally, wherever possible, the preferred measure of central tendency.

Example 10.3

The table lists the frequency distribution of 25 families in Lower Austria who were polled in a marketing survey to state the number of litres of milk consumed during a particular week.

Number of litres	Frequency	Relative frequency
0	2	0.08
1	5	0.20
2	9	0.36
3	5	0.20
4	3	0.12
5	1	0.04

Construct a frequency histogram and find the modal class, median and mean.

Solution

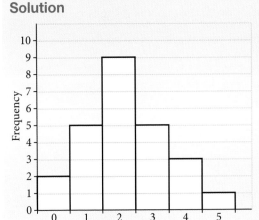

This histogram shows a relatively symmetric shape with a modal class at $x = 2$. The mean and median are not far from each other. The median is the 13th observation, which is 2, and the mean is calculated to be 2.2

> For lists, the mode is a most common (frequent) value. A list can have more than one mode. For histograms, a mode is a relative maximum.

> The symmetric shape of the histogram shows that the median, mean, and mode are all close together. This will be discussed further in the next section.

Shape of the distribution

An examination of the shape of a distribution will illustrate how the distribution is centred on the mean.

The shape of a distribution is said to be **symmetric** if the observations are balanced, or evenly distributed, about the mean. In a symmetric distribution, the mean, the median, and the mode are equal.

A distribution is **skewed** if the observations are not symmetrically distributed above and below the mean.

A **positively skewed** (or skewed to the right) distribution has a tail that extends to the right in the direction of positive values. A **negatively skewed** (or skewed to the left) distribution has a tail that extends to the left in the direction of negative values.

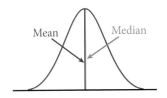

Figure 10.6 Symmetric distribution

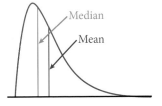

Figure 10.7 Positively skewed distribution

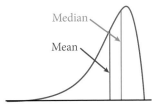

Figure 10.8 Negatively skewed distribution

Looking back at the WHO data, we see that the data are skewed to the left. Few countries have low life expectancies. The majority of countries have life expectancies between 50 and 65.

The average HALE is $\mu = \dfrac{\sum x}{n} = \dfrac{11028}{192} = 57.44$. Looking at the raw data,

it does not appear sensible to search for the mode, as there are four of them (59, 60, 61 and 62). However, after grouping the data into classes of length 5, we can see that the modal class is $60 \leqslant l < 65$.

The median: as there are 192 observations, which means that the median

is at $\dfrac{n+1}{2} = \dfrac{192+1}{2} = 96.5$, we take the average of the 96th and 97th

observations, which are both 60. So, the median is 60.

Knowing the median, we could say that a typical life expectancy is 60 years. How much does this really tell us? How well does this median describe the real situation? After all, not all countries have the same 60 years HALE. Whenever we find the centre of data, the next step is always to ask how well it actually summarises the data.

When we describe a distribution numerically, we always report a measure of its **spread** along with its centre.

Exercise 10.2

1. You are given eight measurements: 5, 4, 7, 8, 6, 6, 5, 7.

   (a) Find $\bar{x}$

   (b) Find the median.

   (c) Based on the previous results, are the data symmetric or skewed? Explain and support your conclusion with an appropriate graph.

2. You are given ten measurements: 5, 7, 8, 6, 12, 7, 8, 11, 4, 10.

   (a) Find $\bar{x}$

   (b) Find the median.

   (c) Find the mode.

3. The following table gives the number of cars owned by a sample of 50 typical households in a large city in Germany:

Number of cars	0	1	2	3
Number of households	12	24	8	6

   Find the average and the median number of cars. Which measure of central tendency is more appropriate here? Explain.

4. Ten businesses are listed below along with their 2017 revenue in millions of US dollars:

Company	Revenue ($ millions)	Company	Revenue ($ millions)
A	500 343	F	265 172
B	348 903	G	260 028
C	326 953	H	244 582
D	326 008	I	244 363
E	311 870	J	242 137

Calculate the mean and median of the revenues. Which measure is more appropriate in this case? Explain.

5. Even on a crucial examination, students tend to lose focus while writing their tests. In a psychology experiment, 20 students were given a 10-minute quiz and were observed for the number of seconds they spent 'on task'. Here are the results:

| 350 | 380 | 500 | 460 | 480 | 400 | 370 | 380 | 450 | 530 |
| 520 | 460 | 390 | 360 | 410 | 470 | 470 | 490 | 390 | 340 |

   (a) Find the mean and median of the time spent on task.

   (b) If you were writing a report to describe these times, which measure of central tendency would you use and why?

6. At 5:30 p.m. during the holiday season, a toy shop counted the number of items sold and the revenue collected for that day; the result was $n = 90$ toys with a total revenue of $\sum x = €4460$.

   (a) Find the average amount spent on each toy that day.

   Shortly before the shop closed at 6 p.m., two new purchases of €74 and €60 were made.

   (b) Calculate the new mean of the sales per toy that day.

7. A farmer has 144 bags of new potatoes weighing 2.15 kg each. He also has 56 potato bags from last year with an average weight of 1.80 kg. Find the mean weight of a bag of potatoes available for this farmer.

8. The following are the grades earned by 25 students on a 50-mark test in statistics.

   26, 27, 36, 38, 23, 26, 20, 35, 19, 24, 25, 27, 34,
   27, 26, 42, 46, 18, 22, 23, 24, 42, 46, 33, 40

   (a) Calculate the mean of the grades.

   (b) Draw a stem plot of the grades. Use the plot to estimate where the median is.

   (c) Draw a histogram of the grades.

   (d) Develop a cumulative frequency graph of the grades. Use your graph to estimate the mean.

9. The following are data concerning the injuries in road accidents in a certain country, classified by severity.

Year	Fatal	Serious	Slight
1970	758	7860	13 515
1975	699	6912	13 041
1980	644	7218	13 926
1985	550	6507	13 587
1990	491	5237	14 443
1995	361	4071	12 102
2000	297	3007	11 825
2005	264	2250	10 922

(a) Draw bar graphs for the total number of injuries and describe any patterns you observe.

(b) Draw pie charts for the different types of injuries for the years 1970, 1990, and 2005.

10. The data on the right report the car driver casualties in a certain district for 2017.

(a) Draw a histogram of the data.

(b) Estimate the mean of the data.

(c) Develop a cumulative frequency graph and use it to estimate the median of the data.

Age	Casualties
15–19	103
20–24	125
25–29	103
30–34	80
35–39	88
40–44	96
45–49	78
50–54	60
55–59	45
60–64	33
65–69	17
70–74	13
75–79	26

11. Use the data in question 9 of Exercise 10.1 to estimate the median and the mean of the number of days in hospital by heart patients.

12. Use the data in question 10 of Exercise 10.1 to estimate the median and the mean of the exercise time of János for 2017.

13. Use the data in question 11 of Exercise 10.1 to estimate the median and the mean speed of cars on the highway.

14. Use the data in question 12 of Exercise 10.1 to estimate the median and the mean length of components at this facility.

15. Use the data in question 13 of Exercise 10.1 to estimate the median and the mean of the waiting time for customers at this supermarket.

16. (a) Given that $\sum_{i=1}^{40} x_i = 1664$, find $\bar{x}$

    (b) Given that $\sum_{i=1}^{20} (x_i - 20) = 1664$, find $\bar{x}$

17. For a large class of 60 students, 12 points are added to each grade to boost scores on a relatively difficult test.

    (a) Knowing that $\sum (x + 12) = 4404$, find the mean score of this group of 60 students.

    (b) Another section of the class has 40 students and their average score is 67.4. Find the average of the whole group of 100 students.

10.3 Measures of variability

Measures of location summarise what is typical of elements of a list, but not every element is typical. Are all the elements close to each other? Are most of the elements close to each other? What is the biggest difference between elements? On average, how far are the elements from each other? The answers lie in the measures of spread or variability.

It is possible that two data sets have the same mean, but the individual observations in one set could vary more from the mean than do the observations in the second set. It takes more than the mean alone to describe data. Measures of variability (also called measures of dispersion or spread), which include the range, the variance, the standard deviation, and the coefficient of variation will help to summarise the data.

Range

The range of a data set is the difference between the largest and smallest observations. Consider the weekly expenses data given in Table 10.2 of Section 10.1. Also consider the same data when the largest value of 68 is replaced by 120. What is the range for these two sets of data?

	Expense data	Expense data with outlier
Minimum	38	38
Maximum	68	120
Range	30	82

Table 10.4 The range is the difference between the largest and smallest observations

Notice that the range is a single number, not an interval of values as you might think from its use in common speech. The maximum of the HALE data (page 457) is 75 and the minimum is 29, so the range is 46.

The range doesn't take into account how the data is distributed and is of course affected by extreme values (outliers) as illustrated in Table 10.4.

Variance and standard deviation

The most comprehensive measures of dispersion are those in terms of the average deviation from some location parameter.

Variance

The sample variance, s^2, is the sum of the squared differences between each observation and the sample mean, divided by the sample size minus 1.

$$s^2 = \frac{\sum_{i=1}^{n}(x_i - \bar{x})^2}{n - 1}$$

s^2 is called the unbiased estimate of the population variance σ^2 and is denoted as s_{n-1}^2. However, it is not required for this current IB syllabus. Discussion of the reason we define the sample variance in this manner is beyond the scope of this book. The use of $n - 1$ in the denominator has to deal with the use of the sample variance as an estimate of the population variance. Such an estimate has to be unbiased, and this sample variance is the most unbiased estimate of the population variance. However, the IB syllabus uses a different definition of the sample variance. The IB sample variance is listed as s_n^2 and is evaluated as follows:

$$s_n^2 = \frac{\sum_{i=1}^{n}(x_i - \bar{x})^2}{n}$$

It is obvious that $s_{n-1}^2 = \dfrac{\sum_{i=1}^{n}(x_i - \bar{x})^2}{n - 1} = \dfrac{n}{n - 1} \cdot \dfrac{\sum_{i=1}^{n}(x_i - \bar{x})^2}{n} = \dfrac{n}{n - 1} s_n^2.$

Or $s_n^2 = \dfrac{n - 1}{n} \cdot s_{n-1}^2$ in case you want to use s_x^2 from your GDC.

The population variance, σ^2, is the sum of the squared differences between each observation and the population mean divided by the population size, N.

$$\sigma^2 = \frac{\sum_{i=1}^{N}(x_i - \mu)^2}{N}$$

The variance is a measure of the variation about the mean squared. In order to make the units of the measure of variation the same as the data measurements, the square root is taken and the measure looked at is the standard deviation.

In your calculator, you should also be careful because the listed s_x in GDCs corresponds to s_{n-1}^2. So when you use your GDC, make sure you use what is called σ_x.

Standard deviation

The standard deviation measures the amount of spread around the mean.

The sample standard deviation, s_n, is the (positive) square root of the variance, and is defined as:

$$s_n = \sqrt{s_n^2} = \sqrt{\frac{\sum_{i=1}^{n}(x_i - \bar{x})^2}{n}}$$

$$s_{n-1} = \sqrt{s_{n-1}^2} = \sqrt{\frac{\sum_{i=1}^{n}(x_i - \bar{x})^2}{n-1}} = \sqrt{\frac{n}{n-1}} \cdot s_n, \text{ or } s_n = \sqrt{\frac{n-1}{n}} \cdot s_{n-1}$$

The population standard deviation is

$$\sigma = \sqrt{\sigma^2} = \sqrt{\frac{\sum_{i=1}^{N}(x_i - \mu)^2}{N}}$$

These are measures of variation about the mean.

- When $s = 0$, all the data takes on the same value and there is no variability about the mean.
- When s is large, there is a large amount of variability about the mean.

Consider the following example:

In business, investors invest their money in stocks whose prices fluctuate with market conditions. Stocks are considered risky if they have high fluctuations. Table 10.5 shows closing prices of two stocks traded on Vienna's stock market for the first seven business days in September 2017.

Even though the two stocks have similar central values, they do behave differently. It is obvious that stock B is more variable and it becomes more obvious when we calculate the standard deviations.

Stock A	Stock B
4	1
4.25	3
5	2.5
4.75	5
5.75	7
5.25	6.5
6	10
$\bar{x}_A = 5$	$\bar{x}_B = 5$
Median (A) = 5	Median (B) = 5

Table 10.5 Closing prices of two stocks traded on Vienna's stock market

The variances are:

$$s_A^2 = \frac{\sum_{i=1}^{7}(x_i - 5)^2}{7}$$

$$= \frac{(4-5)^2 + (4.25-5)^2 + (5-5)^2 + (4.75-5)^2 + (5.57-5)^2 + (5.25-5)^2 + (6-5)^2}{7} = 0.464$$

$$s_B^2 = \frac{\sum_{i=1}^{7}(x_i - 5)^2}{7}$$

$$= \frac{(1-5)^2 + (3-5)^2 + (2.5-5)^2 + (5-5)^2 + (7-5)^2 + (6.5-5)^2 + (10-5)^2}{7} = 8.21$$

This means that the standard deviations are: $s_A = 0.681$ and $s_B = 2.865$ Stock B is 4.2 times as variable as stock A.

You can use your GDC to do this.

When computing s_n^2 manually, you may find the following shortcuts useful:

$$s_n^2 = \frac{\sum_{i=1}^{n}(x_i - \bar{x})^2}{n} = \frac{\sum_{i=1}^{n}\left(x_i^2 - 2x_i\bar{x} + \bar{x}^2\right)}{n} = \frac{\sum_{i=1}^{n}x_i^2 - 2\sum_{i=1}^{n}x_i\bar{x} + \sum_{i=1}^{n}\bar{x}^2}{n}$$

$$= \frac{\sum_{i=1}^{n}x_i^2}{n} - \frac{2\bar{x}\sum_{i=1}^{n}x_i}{n} + \frac{\sum_{i=1}^{n}\bar{x}^2}{n} = \frac{\sum_{i=1}^{n}x_i^2}{n} - 2\bar{x}\sum_{i=1}^{n}\frac{x_i}{n} + \frac{n\bar{x}^2}{n} = \frac{\sum_{i=1}^{n}x_i^2}{n} - \bar{x}^2$$

Remember that once you have a good understanding of the standard deviation, you will rely on a GDC or software to do most of the calculation for you. Here is an example of a GDC output.

```
1-Var Stats        1-Var Stats
x̄=5                ↑n=7
Σx=35               minX=4
Σx²=178.25          Q1=4.25
Sx=.7359800722      Med=5
σx=.6813851439      Q3=5.75
↓n=7                maxX=6
                    ■
```

The s_x used by your GDC gives $s_{n-1} = \sqrt{\dfrac{\sum_{i=1}^{n}(x_i - \bar{x})^2}{n-1}}$ and $\sigma = \sqrt{\dfrac{\sum_{i=1}^{n}(x_i - \bar{x})^2}{n}}$ which is the s_n used in IB exams.

The screenshots also show that the GDC gives you $\sum x^2$ which can be used in case you want to find the variance by hand.

$$s_n^2 = \frac{\sum_{i=1}^{n}x_i^2}{n} - \bar{x}^2 = \frac{178.25}{7} - 5^2 = 0.464 \Rightarrow s_n = 0.681$$

The interquartile range and measures of non-central tendency

Percentiles and quartiles

Data must first be in ascending order.

Percentiles separate large ordered data sets into 100ths. The pth percentile is a number such that p percent of the observations are at or below that number.

Quartiles are descriptive measures that separate large ordered data sets into four quarters.

A test score in the 90th percentile indicates 90% of the tests scores were less than or equal to the score. An excellent performance! You scored in the upper 10% of all people taking the test.

The **first quartile**, Q_1, is another name for the 25th percentile. 25% of the observations are at or below the value of the first quartile. Q_1 is located in the $0.25(n + 1)$st position when the data is in ascending order. That is,

$$Q_1 = \frac{n + 1}{4}\text{th ordered observation}$$

The **third quartile**, Q_3, is another name for the 75th percentile. 75% of the observations are at or below the third quartile. Q_3 is located in the $0.75(n + 1)$st position when the data is in ascending order. That is,

$$Q_3 = \frac{3(n + 1)}{4}\text{th ordered observation}$$

The median, is the 50th percentile, also called the second quartile, Q_2.

A measure that helps to measure variability and is not affected by extreme values is the interquartile range. It avoids the problem of extreme values by just looking at the range of the middle 50% of the data.

Interquartile range

The interquartile range (IQR) measures the spread in the middle 50% of the data; it is the difference between the observations at the 25th and the 75th percentiles:

$$\text{IQR} = Q_3 - Q_1$$

If we consider the student expense data in Table 10.2, and once again look at that same data with the outlier 120 replacing the largest value 68, we have the following results:

A practical method to calculate the quartiles is to split the data into two halves at the median. (When n is odd, include the median in both halves.) The lower quartile is the median of the first half and the upper quartile is the median of the second half. For example, with the stocks data, {4, 4.25, 4.75, 5, 5.25, 5.75, 6}, $n = 7$, the median is the 4th observation, 5. The first quartile is then the median of {4, 4.25, 4.75, 5}, which is 4.5, and the third quartile is the median of {5, 5.25, 5.75, 6}, which is 5.5.

	Expense data	Expense data with outlier
Minimum	38	38
Q_1	50	50
Median	55	55
Q_3	61	61
Maximum	68	120
Range	30	82
IQR	11	11

Table 10.6 Expense data with outlier

The range doesn't take into account how the data is distributed and is of course affected by extreme values. We clearly saw that in Table 10.4. However, the IQR evidently does not have that problem.

Five-number summary

The five-number summary refers to the five descriptive measures: minimum, first quartile, median, third quartile, and the maximum.

Clearly, Minimum$< Q_1 <$Median $< Q_3 <$ Maximum

Box-and-whisker plot

Whenever we have a five-number summary, we can put the information together in one graphical display called a **box plot**, also called a **box-and-whisker** plot.

Let's make a box-and-whisker plot with the student expense data.

- Draw an axis spanning the range of the data; mark the numbers corresponding to the median, minimum, maximum, and the lower and upper quartiles.
- Draw a rectangle with the lower end at Q_1 and the upper end at Q_3.
- To help us consider outliers, mark the points corresponding to lower and upper fences. Mark them with a dashed line since they are not part of the box. The fences are constructed at the following positions:
  - Lower fence: $Q_1 - 1.5 \times IQR$ (in this case: $50 - 1.5\,(11) = 33.5$)
  - Upper fence: $Q_3 + 1.5 \times IQR$ (in this case: $61 + 1.5\,(11) = 77.5$)

Any point beyond the lower or upper fence is considered an **outlier**.

- Mark any outlier with an asterisk ($*$) on the graph.
- Extend horizontal lines called 'whiskers' from the ends of the box to the smallest and largest observations that are not outliers. In the first case these are 38 and 68, while in the second they are 38 and 67.

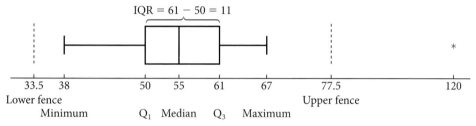

Figure 10.9 Outlier marked with an asterisk ($*$)

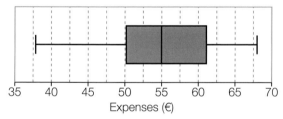

Figure 10.10 A box-and-whisker plot of the data created by a software package (Minitab)

As you see, the box contains the middle 50% of the data. The width of the box is the IQR. Now we know that the middle 50% of the students' expenses is €11. This seems a reasonable summary of the spread of the distribution, as you can see in the histogram below.

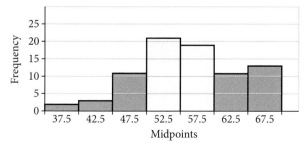

Figure 10.11 Histogram

Examples of GDC output for histograms and box-and-whisker plots:

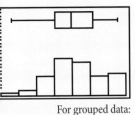

For grouped data:

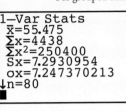

```
1-Var Stats
x̄=55.475
Σx=4438
Σx²=250400
Sx=7.2930954
σx=7.247370213
↓n=80
■
```

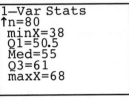

```
1-Var Stats
↑n=80
 minX=38
 Q1=50.5
 Med=55
 Q3=61
 maxX=68
```

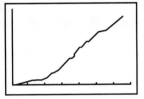

If you locate the IQR on the histogram, you can get another visual indication of the spread of the data.

This is a realistic ogive:

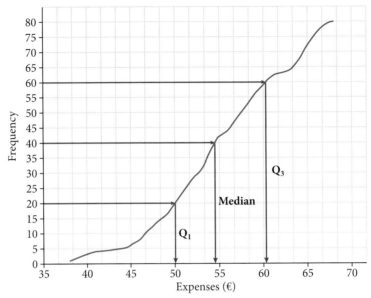

Figure 10.12 Ogive

Notice how we locate the first quartile. Since there are 80 observations, the first quartile is approximately at the $\frac{n+1}{4} = \frac{81}{4} \approx$ 20th position, which appears to be around 50.

The median is at the $\frac{n+1}{2} = \frac{81}{2} \approx$ 40th − 41st position, i.e., approximately at 55.

Similarly, the third quartile is at $\frac{3(n+1)}{4} = \frac{243}{4} \approx$ 61st, which happens here at approximately 61.

The calculation of the mean and variance for grouped data is essentially the same as for raw data. The difference lies in the use of frequencies to save typing (writing) all numbers. Here is a comparison:

Statistic	Raw data	Grouped data		Grouped data with intervals	
$\bar{x}$	$\bar{x} = \dfrac{\sum\limits_{\text{all } x} x}{n}$	$\bar{x} = \dfrac{\sum\limits_{\text{all } x} x_i \cdot f(x_i)}{n}$	$= \dfrac{\sum\limits_{\text{all } x} x_i \cdot f(x_i)}{\sum f(x_i)}$	$\bar{x} = \dfrac{\sum\limits_{\text{all } x} m_i \cdot f(m_i)}{n}$	$= \dfrac{\sum\limits_{\text{all } x} m_i \cdot f(m_i)}{\sum f(m_i)}$

Table 10.7 Data comparison

x_i	data point
$f(x_i)$	frequency of x_i
m_i	interval midpoint (mid-mark or mid-value)
$f(m_i)$	frequency of interval i
$\sum f(x_i), \sum f(m_i)$	total number of data points

For the grouped data reproduced here, this is how we estimate the mean and variance.

Living expenses, l	Midpoint m	Number of students $f(m)$	$m_i \times f(m_i)$	$(m_i - \bar{x})^2$	$(m_i - \bar{x})^2 \times f(m_i)$
$35 \leqslant l < 40$	37.5	2	75	344.5	688.9
$40 \leqslant l < 45$	42.5	3	127.5	183.9	551.6
$45 \leqslant l < 50$	47.5	11	522.5	73.3	806.0
$50 \leqslant l < 55$	52.5	21	1102.5	12.7	266.1
$55 \leqslant l < 60$	57.5	19	1092.5	2.1	39.4
$60 \leqslant l < 65$	62.5	11	687.5	41.5	456.2
$65 \leqslant l < 70$	67.5	13	877.5	130.9	1701.4
Totals		$\sum f(m_i)$ $= 80$	$\sum_{\text{all } x} m_i \cdot f(m_i)$ $= 4485$	$\sum_{\text{all } x} (m_i - \bar{x})^2 \cdot f(m_i)$ $= 4509.6$	
		Mean	$\dfrac{4485}{80} = 56.06$	**Variance**	$\dfrac{4509.6}{80} = 56.37$
				Standard deviation	7.51

Table 10.8 Estimating mean and variance

The numbers here are estimates of the mean and the variance, and eventually the standard deviation. As you will notice, they are not equal to the values we calculated earlier, but they are close. The reason for this is that by grouping we lose the detail in each interval. For example, the interval between 45 and 50 is represented by the mid interval 47.5. In essence, we are assuming that every number in the interval is equal to 47.5.

Example 10.4

Speed limits in some European cities are set to 50 km h^{-1}. Drivers in various cities react to such limits differently. The speeds of cars in Vienna, Brussels, and Stockholm are given in the table. Use box-and-whisker plots to compare the results.

Vienna	62	60	59	50	61	63	53	46	58	49	51	37	47	51	63	52	44	50	45	44
Brussels	64	61	63	57	49	49	46	58	45	60	51	36	65	45	47	46				
Stockholm	43	44	34	35	31	34	29	33	36	38	45	47	29	48	51	49	48			

Solution

Parallel box-and-whisker plots may be an appropriate tool to enable a comparison between the three data sets.

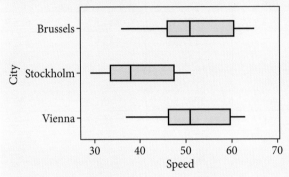

It appears that, on average, drivers in Brussels and Vienna tend to drive faster than the speed limit. The median in both cities is higher than 50, which means than more than 50% of the drivers in the two cities tend not to respect the speed limit. The variation in both cities is comparable with Brussels having a slightly wider range than Vienna. Almost all drivers in Stockholm appear to adhere to the 50 km h^{-1} limit. The median is around 40 km h^{-1} and the third quartile about 47, which means that more than 75% of the drivers in this city drive at a speed less than the 50 km h^{-1} limit.

Shape, centre, and spread

Statistics is about variation, so spread is an important fundamental concept. Measures of spread help us to precisely analyse what we do not know. If the values we are looking at are scattered very far from the centre, then the IQR and the standard deviation will be large. If these are large, then our central values will not represent the data very well. That is why we always report spread with any central value.

A practical way of seeing the significance of the standard deviation can be demonstrated with the following (optional) observations.

Empirical rule
If the data are close to being symmetric, as in Figure 10.13, then the following is true:

- The interval $\mu \pm \sigma$ contains approximately 68% of the measurements.
- The interval $\mu \pm 2\sigma$ contains approximately 95% of the measurements.
- The interval $\mu \pm 3\sigma$ contains approximately 99.7% of the measurements.

The empirical rule usually indicates if an observation is very far from the expected value or not. Take the following example:

I have recorded my car's fuel efficiency over the last 98 times that I have filled the tank with fuel. Here is the data expressing how many kilometres per litre the car travelled:

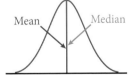

Figure 10.13 Symmetric distribution

Km/litre	Frequency	Km/litre	Frequency
6.0	1	10.0	14
7.0	1	10.5	7
7.5	4	11.0	9
8.0	8	11.5	5
8.5	14	12.0	1
9.0	21	12.5	2
9.5	11		

Table 10.9 Fuel efficiency data

The summary measures are:

Mean	9.454
σ	1.223
Median	9.25
Q_1	8.5
Q_3	10.125
IQR	1.625

Table 10.10 Summary measures

The histogram shows that the distribution is almost symmetric. The possible outlier has little effect on the mean and standard deviation. That is why the mean and median are almost the same. Looking at the box-and-whisker plot, we can see that there is one outlier. The confirmation is below:

$8.5 - 1.5 \times 1.625 = 6.06$ (3 s.f.); that is why 6 is considered an outlier.

$10.125 + 1.5 \times 1.625 = 12.6$ (3 s.f.); hence no outliers on this side.

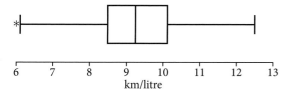

Figure 10.14 Fuel efficiency box-and-whisker plot

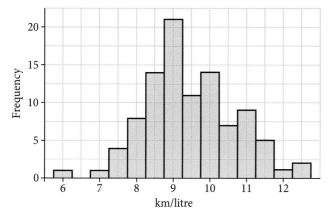

Figure 10.15 Fuel efficiency histogram

If we use the empirical rule, we can expect about 99.7% of the data to lie within three standard deviations of the mean; that is, $9.454 - 3 \times 1.223 = 5.8$ and $9.454 + 3 \times 1.223 = 13.1$. In fact, we see that all the data are within the specified interval, including the potential outlier.

What should you be able to tell about a quantitative variable?

You can report the shape of its distribution, and include a centre and a spread when describing a quantitative variable.

You can report on the central measure and which measure of spread as follows:

- If the shape is skewed, report the median and IQR. You may want to include the mean and standard deviation, but you should point out that the mean and median differ as this difference is a sign that the data are skewed. A histogram can help.

- If the shape is symmetric, report the mean and standard deviation. You may report the median and IQR as well.

- If there are clear outliers, report the data with and without the outliers. The differences may be revealing.

Example 10.5

The records of a large high school show the heights of their students (to the nearest cm) for the year 2016.

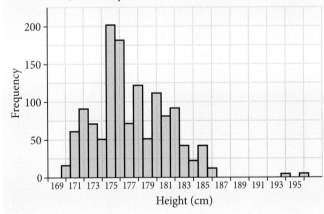

(a) Which statistics would best represent the data here? Why?

(b) Estimate the mean and standard deviation.

(c) Develop a cumulative frequency graph of the data.

(d) Use your result from part (c) to estimate the median, Q_1, Q_3, and IQR.

(e) Are there any outliers in the data? Why?

(f) Write a few sentences describing the distribution.

Solution

(a) The data appear to have outliers and are slightly skewed to the right. The most appropriate measure is the median since the mean is influenced by the extreme values.

(b) To calculate the mean and standard deviation, we will set up a table that will facilitate the calculation.

Height X_i	Number of students $f(x)$	$x_i \cdot f(x_i)$	$(x_i - \bar{x})^2$	$(x_i - \bar{x})^2 \times f(x_i)$
170	15	2550	51.84	777.6
171	60	10 260	38.44	2306.4
172	90	15 480	27.04	2433.6
⋮	⋮	⋮	⋮	⋮
194	2	388	282.24	564.5
196	3	588	353.44	1060.3
Totals	$\sum f(x_i)$ $= 1300$	$\sum\limits_{\text{all } x} x_i \cdot f(x_i)$ $= 230\,376$	$\sum\limits_{\text{all } x}(x_i - \bar{x})^2 \cdot f(x_i)$ $= 19\,927.6$	
	Mean	$\dfrac{230\,376}{1300} = 177.2$	**Variance**	$\dfrac{19\,927.4}{1300} = 15.33$
			Standard deviation	**3.92**

Note that using the alternative formula for the variance will also give the same result: (Due to rounding, answers will differ slightly.)

$$s_n^2 = \frac{\sum\limits_{i=1}^{n} x_i^2 \times f(x_i)}{n} - \bar{x}^2 = \frac{40\,845\,390}{1300} - 177.2123^2 = 15.3315$$

481

(c) To develop the cumulative frequency graph, we first need to develop the cumulative frequency table. This is done by accumulating the frequencies as shown below.

x	$f(x)$	Cum $f(x)$	x	$f(x)$	Cum $f(x)$
170	15	15	184	20	1245
171	60	75	185	40	1285
172	90	165	186	10	1295
173	70	235	187	0	1295
⋮	⋮	⋮	⋮	⋮	⋮
179	50	905	193	0	1295
180	110	1015	194	2	1297
181	80	1095	195	0	1297
182	90	1185	196	3	1300
183	40	1225			

The cumulative frequency table is constructed such that the cumulative frequency corresponding to any measurement is the number of observations that are less than or equal to its value. So, for example, the cumulative frequency corresponding to a height of 174 cm is 285, which consists of the 50 observations with height 174 and the 235 observations for heights less than 174.

The cumulative frequency graph plots the observations on the horizontal axis against their cumulative frequencies on the vertical axis as shown below.

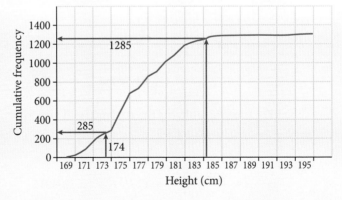

(d) The median is the observation between $\frac{1300}{2} = 650$th and 651st observations since the number is even. From the cumulative table, we see that the median is in the 176 interval. So the median is 176.

Q_1 is at $\frac{1301}{4} \approx 325$th observation. From the table, as 174 has a cumulative frequency of 285, and 175 has 485, then Q_1 has to be 175.

Also, Q_3 is at $\frac{3 \times 1301}{4} \approx 976$th observation. So, similarly, it is 180.

IQR = 180 − 175 = 5.

(e) To check for outliers, we calculate the lengths of the whiskers:

Lower fence: $175 - 1.5 \times 5 = 167.5$, which is lower than the minimum value, so there are no outliers on the left.

Upper fence: $180 + 1.5 \times 5 = 187.5$. So we have five outliers: two at 194 and three at 196.

(f) The distribution appears to be bimodal with two modes at 175 and 176. It is slightly skewed to the right with a few extreme values at 194 and 196. This is further confirmed with the fact that the mean of 177.2 is higher than the median of 176.

Here are the calculations using a GDC.

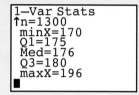
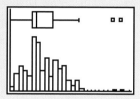

Exercise 10.3

1. The pulse rates of 15 patients chosen at random from visitors of a local clinic are given below.

   72, 80, 67, 68, 80, 68, 80, 56, 76, 68, 71, 76, 60, 79, 71

   (a) Calculate the mean and standard deviation of the pulse rate of the patients at the clinic.

   (b) Draw a box-and-whisker plot of the data and indicate the values of the different parts of the box.

   (c) Check if there are any outliers.

2. The numbers of passengers on 50 flights from Washington to London on a commercial airline are shown in this table:

165	173	158	171	177	156	178	210	160	164
141	127	119	146	147	155	187	162	185	125
163	179	187	174	166	174	139	138	153	142
153	163	185	149	154	154	180	117	168	182
130	182	209	126	159	150	143	198	189	218

   (a) Calculate the mean and standard deviation of the numbers of passengers on this airline between the two cities.

   (b) Set up a stem-and-leaf plot for the data and use it to find the median number of passengers.

   (c) Develop a cumulative frequency graph. Estimate the median, first and third quartiles. Draw a box-and-whisker plot.

(d) Find the IQR and use it to check whether there are any outliers.

(e) Use the empirical rule to check for outliers.

3. 100 students at a school took a practice IB exam using Paper 3. The paper was marked out of 60 marks. Here are the results:

Marks	0–9	10–19	20–29	30–39	40–49	50–60
No. of students	5	9	16	24	27	19

(a) Draw a cumulative frequency graph.

(b) Estimate the median and quartiles.

4. 130 first year IB students were given a placement test to decide whether they should go for SL or HL. The times, in minutes, for these students to finish the test are given in the table below:

Time, t	30–40	40–50	50–60	60–70	70–80	80–90	90–100	100–110	110–120
No. of students	8	12	24	29	19	16	12	8	2

(a) Draw a cumulative frequency graph.

(b) Estimate the median and the IQR.

(c) 20 students did not manage to finish the test after 120 minutes and had to hand it in uncompleted. Estimate the median finishing time for all 140 students.

5. The mean score of 26 students on a 40-mark paper is 22. The mean for another group of 84 other students is 32. Find the mean of the combined group of 110 students.

6. The scores on a 100-mark test of a sample of 80 students in a large school are given in the table.

Score	59–63	63–67	67–71	71–75	75–79	79–83	83–87
No. of students	6	10	18	24	10	8	4

(a) Find the mean and standard deviation of the scores of all students.

(b) A bonus of 13 points is to be added to these scores. What is the new value of the mean and standard deviation?

7. In a large theatre in London (1744 capacity), over a period of 10 years, there are 1000 performances of a particular production. The manager of the theatre group kept a record of the empty seats on the days it played, as shown in the table.

Number of empty seats	1–10	11–20	21–30	31–40	41–50	51–60	61–70	71–80	81–90	91–100
Days	15	50	100	170	260	220	90	45	30	20

(a) Copy and complete the following cumulative frequency table for the above information:

Number of empty seats	$x \leqslant 10$	$x \leqslant 20$	$x \leqslant 30$	$x \leqslant 40$	$x \leqslant 50$	$x \leqslant 60$	$x \leqslant 70$	$x \leqslant 80$	$x \leqslant 90$	$x \leqslant 100$
Days	15		165			815				1000

(b) Draw a cumulative frequency graph of this distribution. Use 1 unit on the vertical axis to represent 100 days and 1 unit on the horizontal axis to represent 10 seats.

(c) Use your graph to answer the following questions:
    (i) Find an estimate of the median number of empty seats.
    (ii) Find an estimate for the lower, upper quartile, and the IQR.
    (iii) The number of days when there were fewer than 35 empty seats were considered bumper days (lots of profit). How many days were considered bumper days?
    (iv) The highest 15% of the days with empty seats were categorised as loss days. What is the number of empty seats above which a day is claimed as a loss?

8. Aptitude tests sometimes involve jigsaw puzzles to test the ability of new applicants to perform precision assembly work in electronic instruments. One such company that produces computerised parts gave the following results for their job applicants:

Time to finish the puzzle, t (nearest second)	Number of applicants
$10 \leqslant t < 30$	16
$30 \leqslant t < 40$	24
$40 \leqslant t < 45$	22
$45 \leqslant t < 50$	26
$50 \leqslant t < 55$	38
$55 \leqslant t < 60$	36
$60 \leqslant t < 65$	32
$65 \leqslant t < 70$	18

(a) Draw a histogram of the data.

(b) Draw a cumulative frequency graph and estimate the median and IQR.

(c) Calculate the estimates of the mean and standard deviation of all such participants.

9. The heights (to the nearest cm) of baseball players at a school are given in the table:

Height	152	155	157	160	163	165	168	170	173	175	178	180	183	185	188	191	193
Frequency	2	6	9	7	5	20	18	7	12	5	11	8	9	4	2	4	1

(a) Find the five-number summary for this data.

(b) Display the data with a box-and-whisker plot and a histogram.

(c) Find the mean and standard deviation of the data.

(d) Describe the data with a few sentences.

(e) Draw a cumulative frequency graph and estimate the height of the player that is in the 90th percentile.

(f) 10 players' data were missing when the data was collected. The average height of the 10 players is 182. Find the average height of all the players, including the last 10.

10. Consider 10 data measures.

(a) If the mean of the first 9 measures is 12 and the 10th measure is 12, what is the mean of the 10 measures?

(b) If the mean of the first 9 measures is 11, and the 10th measure is 21, what is the mean of the 10 measures?

(c) If the mean of the first 9 measures is 11, and the mean of the 10 measures is 21, what is the value of the 10th measure?

11. Suppose that the mean of a set of data points is 30.

(a) It is discovered that a data point having a value of 25 was incorrectly entered as 15. What should be the revised value of the mean?

(b) Suppose an additional point of value 32 was added. Will this increase or decrease the value of the mean?

12. Half the values of a sample are equal to 20, one-sixth are equal to 40, and one-third are equal to 60. What is the sample mean?

13. The seven numbers 7, 10, 12, 17, 21, x, and y have a mean $\mu = 12$ and a variance $\sigma^2 = \dfrac{172}{7}$. Find x and y given that $x < y$.

14. A sample of 25 observations was taken out of a large population of measurements. If it is given that $\sum_{i=1}^{25} x_i = 278$ and $\sum_{i=1}^{n} x_i^2 = 3682$, estimate the mean and the variance of the population of measurements.

15. Use the data in question 9 of Exercise 10.1 to estimate the IQR and the standard deviation of the number of days in hospital by heart patients.

16. Use the data in question 10 of Exercise 10.1 to estimate the IQR and the standard deviation of the exercise time of János for 2017.

17. Use the data in question 11 of Exercise 10.1 to estimate the IQR and the standard deviation of the speed of cars on the highway.

18. Use the data in question 12 of Exercise 10.1 to estimate the IQR and the standard deviation of the length of components at this facility.

19. Use the data in question 13 of Exercise 10.1 to estimate the IQR and the standard deviation of the waiting time for customers at this supermarket.

10.4 Linear regression

Correlation and covariance

Scatter plot

The total time you devote preparing for an exam affects the score you obtain in that exam.

In general, the foot size of an adult is related to the height of that adult.

Smoking increases the chances of a heart attack.

Statements such as these concern the relationship between two variables. So far, we have considered how to describe the characteristics of one variable. In this section, we will look at relationships between two variables. This area of mathematics is known as **bivariate statistics**.

To study the relationship between two variables, we measure both variables on the same subjects. For example, if we are interested in the relationship between height and foot size, then for a group of individuals we record each person's height and foot size. This way we know which foot size goes with which height. Similarly, we can record the grades that students obtain along with the time they spent preparing for the exam. So, our data are sets of ordered pairs. These data allow us to study the link (association) between height and foot size or time spent studying and grade obtained. In fact, taller people tend to have larger foot sizes and the more you prepare for an exam the higher your grade is. We say that pairs of variables like these are **associated**.

Here are the scores of 10 students in an IB Chemistry SL class. The table shows the time they spent preparing for their test and the score they achieved.

Student	Tim	Joon	S-youn	Kevin	Steve	Niki	Henry	Anton	Cindy	Lucas
Hours	4	4.5	6	3.5	3	5	5.5	6.5	7	6.5
Score	65	80	83	61	55	79	85	89	92	95

Table 10.11 Student test score data

Here is a graph (scatter plot) of the data given in the table.

The horizontal axis shows the number of hours spent preparing and the vertical axis shows the scores achieved.

It appears that more hours spent preparing results in a higher score. We say that the scores on tests and the time preparing for them are

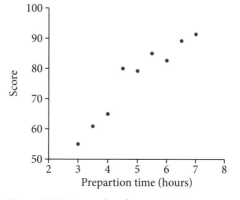

Figure 10.16 Scatterplot of preparation time vs score

Two variables measured on the same subjects are associated if specific values of one variable tend to occur in connection with particular values of the other variable.

487

associated. We call the time the **explanatory variable** (independent) and the grade the **response variable** (dependent). The students whose time and scores are recorded are the **subjects** of the **experiment/study**.

For instance, larger foot size tends to occur in taller individuals. Or, a higher rate of serious road accidents happens in connection with drivers who have a high level of alcohol concentration in their blood. We say that height and foot size are positively associated. Similarly, high alcohol levels are positively associated with involvement in serious road accidents. We can also claim that there is a negative association between time spent watching TV and scores achieved on weekly school tests for teenagers.

In our effort to study the nature of the relationship between two variables, we try to see how changes in one variable relate to changes in the other variable. For instance, we can look at how a person's height is related to their foot size. As discussed above, we call the first variable **explanatory** and the second the **response** variable. These are traditionally called **independent** and **dependent** variables, respectively.

The principles that guide our work on data are:

- Start with graphical display, and then explore numerical summaries.
- Look for overall patterns and deviations from those patterns.
- When the overall pattern is quite regular, use a mathematical model to describe it.

Graphical displays associated with one variable include histograms, box-and-whisker plots and others. In bivariate statistics the graphical tool we use is the **scatter plot**, or **scatter diagram**. In a scatter plot, each pair of observations is represented by a point on a grid. The horizontal component represents the explanatory variable and the vertical component represents the response variable.

Example 10.6

The data presented below is for 80 adults participating in a dieting program. The researchers believe that the metabolic rate (calories burned per 24 hours) is influenced by the lean body mass (k).

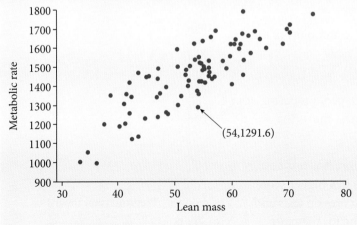

Does the scatter plot show that there is an association between the metabolic rate and lean mass?

You will observe that there is a positive association between these two variables; that is, the greater the lean mass, the higher the metabolic rate.

What to look for in a scatter plot

As a rule of thumb, when we examine a scatter plot, we look at the following characteristics:

- Overall pattern (form, direction and strength)
- Striking deviations from the pattern (outliers)

In Example 10.6, the form is roughly linear. That is, the points appear to cluster around a straight line. The direction, as mentioned earlier, appears to have a positive association. The strength is determined by how closely the points follow the form (this will be revisited later), even though some points stray away from the line. In Example 10.6, there do not appear to be any outliers.

 An outlier is an observation that falls outside the overall pattern of the relationship.

Example 10.7

The table below lists the fuel consumption in km/litre of 34 small cars during both city driving and highway driving. Draw a scatter plot of the data and comment on any patterns you observe.

City	7.3	8.5	8.5	7.3	7.7	5.1	4.7	4.3	7.3	3.8	3.8	6.4
Highway	10.2	11.9	11.9	10.7	10.7	8.5	6.8	6.8	9.8	6.4	5.5	9.4

City	5.1	9.4	6.8	5.5	8.5	8.5	6.4	11.1	5.1	9.0	8.1	8.1
Highway	7.3	11.9	9.8	8.1	11.1	12.4	9.8	13.7	8.1	12.4	11.5	11.9

City	6.8	7.7	6.8	7.7	10.7	9.8	8.5	7.7	6.0	25.6
Highway	9.8	11.1	9.8	9.8	13.7	13.2	12.4	11.1	9.4	28.2

Solution

Here is a scatter plot of the data.

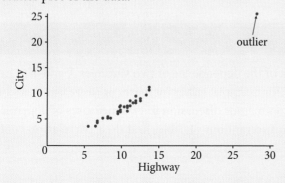

The data indicate that the fuel consumption in highway driving and city driving are positively associated. The relationship appears to be strong as the data are tightly clustered around a positively sloped line. However, we can see that there is one observation that is positioned quite far from the rest of the data. This observation is an outlier. Outliers in statistics are important. Sometimes they indicate a problem in the data being observed and sometimes they may have a special significance. In this case, the data corresponds to a 'hybrid' car, which uses battery power in addition to fuel, hence the high performance. This observation is not typical of the study and must be removed in order to get a clear indication of the nature of the relationship between the two variables. Here is an adjusted scatter plot without the data of the hybrid car.

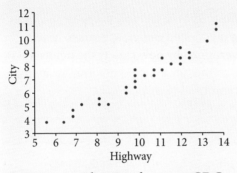

You can use either Excel or your GDC to produce scatter plots.

Here are samples of GDC output. Instructions differ depending on model.

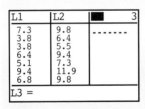

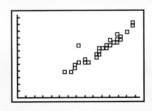

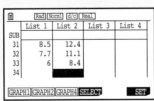

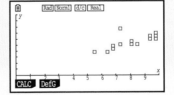

Covariance

Intuitively, we think of the dependence of two variables X and Y as implying that one variable, Y for example, either increases or decreases as X changes. In this book, we will confine our discussion to two measures of dependence: the **covariance** between two random variables and their **correlation coefficient**.

The scatter plots in Figures 10.17 to 10.19 show plots of variables X and Y, for samples of size 15.

In Figure 10.17, all points fall on a straight line. Obviously X and Y are dependent in this case. Suppose we know $E(X) = \mu_X$ and $E(Y) = \mu_Y$. Locate the point with coordinates (μ_X, μ_Y) and then locate any point (x_1, y_1), for example, and measure the deviations $(x_1 - \mu_X)$ and $(y_1 - \mu_Y)$. If the point is in the upper right corner, then both deviations are positive. Similarly, if the point is in the lower left corner, both deviations are negative.

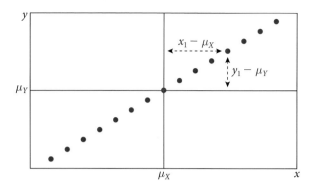

Figure 10.17 Scatter plot of Y against X

The product of the deviations $(x_1 - \mu_X)(y_1 - \mu_Y)$ is positive. This is a typical and extreme case of positive association. When the line representing the pattern in the data is positively sloped, the product of deviations of the mean is on average positive: that is $E((X - \mu_X)(Y - \mu_Y)) > 0$.

In Figure 10.18, the data follow a negatively sloped pattern. If the point is in the upper left corner, then the X-deviations are negative while the Y-deviations are positive. Similarly, if the point is in the lower right corner, the X-deviations are positive while the Y-deviations are negative. The product of the deviations $(x_1 - \mu_X)(y_1 - \mu_Y)$ is negative. These situations do not occur for the diagram in Figure 10.19 where little dependence (if any) exists between the variables.

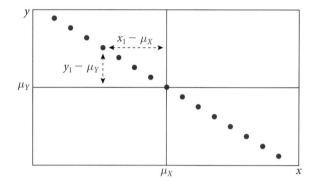

Figure 10.18 Scatter plot of Y against X

The deviations $(x_1 - \mu_X)$ and $(y_1 - \mu_Y)$ sometimes assume the same algebraic sign and sometimes opposite signs. Thus, the product $(x_1 - \mu_X)(y_1 - \mu_Y)$ will be positive sometimes and negative other times and the average may be close to zero.

The discussion above indicates that the average $E((X - \mu_X)(Y - \mu_Y))$ provides a measure of the linear dependence between X and Y. This quantity is called the covariance of X and Y.

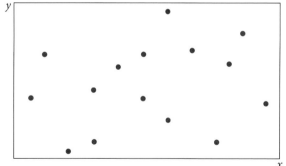

Figure 10.19 Scatter plot of Y against X

If X and Y are random variables with means μ_X and μ_Y, the **covariance** of X and Y is

$$\text{cov}(X,Y) = E[(X - \mu_X)(Y - \mu_Y)]$$

The larger the absolute value of the covariance of X and Y, the greater the linear dependence between X and Y. Positive values indicate that Y increases as X increases and negative values indicate that Y decreases as X increases. A zero value of the covariance indicates that the variables are linearly **uncorrelated** and that there is no linear association between X and Y.

Some facts worth knowing about covariance

1 A shortcut calculation formula can be helpful if you need to perform the calculations without using the built-in functions in your GDC or software:

$$
\begin{aligned}
\text{cov}(X, Y) &= \text{E}((X - \mu_X)(Y - \mu_Y)) \\
&= \text{E}(XY - X\mu_Y - \mu_X Y + \mu_X \mu_Y) \\
&= \text{E}(XY) - \text{E}(X\mu_Y) - \text{E}(\mu_X Y) + \text{E}(\mu_X \mu_Y) \\
&= \text{E}(XY) - \mu_Y \text{E}(X) - \mu_X \text{E}(Y) + \mu_X \mu_Y \\
&= \text{E}(XY) - \mu_Y \mu_X - \mu_X \mu_Y + \mu_X \mu_Y = \text{E}(XY) - \mu_X \mu_Y
\end{aligned}
$$

2 In fact, the above result leads to

$$
\text{cov}(X, X) = \text{E}(XX) - \mu_X \mu_X = \text{E}(X^2) - \mu_X^2 = \text{V}(X)
$$

V(X) is the variance of X.

3 If X and Y are **not independent**, then

$$
\text{V}(X + Y) = \text{V}(X) + 2\text{cov}(X, Y) + \text{V}(Y)
$$

4 If X and Y are **independent**, then

$$
\text{cov}(X, Y) = \text{E}(XY) - \mu_X \mu_Y = \text{E}(X)\text{E}(Y) - \mu_X \mu_Y = 0
$$

Consequently,

$$
\text{V}(X + Y) = \text{V}(X) + \text{V}(Y)
$$

Note that the converse of the theorem above is not true: if $\text{cov}(X,Y) = 0$, then X and Y are not necessarily independent.

Unfortunately, it is difficult to employ the covariance of X and Y as an absolute measure of association between variables because its value depends on the scales used.

In Example 10.7, the covariance of the data expressed as km/litre is 3.8.

However, if we change the scale from km/litre to mile/litre, then the covariance will be 1.49 even though the scatter plot does not indicate any change in the form nor the strength of association between the two variables.

This problem with covariance can be eliminated by 'standardising' its value by using the correlation coefficient, ρ, instead.

$$
\rho_{XY} = \frac{\text{cov}(X, Y)}{\sigma_X \sigma_Y}
$$

Since σ_X and σ_Y are both positive, the sign of the correlation coefficient is the same as that of the covariance.

Note that all models discussed concerning correlation and regression assume that data are samples that come from normal populations.

Correlation

A scatter plot is a useful tool that reveals the form, trend and strength of the association between two quantitative variables. In this book, we are only interested in linear relationships. As mentioned earlier, we say that a linear relationship is strong if the data are tightly packed around a line, and weak if they are widely dispersed around a line. Relying simply on the shape of the graph may be misleading though. Look at the scatter plots in Figures 10.20 and 10.21.

Figure 10.20 is a copy of the second graph in Example 10.7. The graph in Figure 10.21 gives the impression that the association is stronger than it is in Figure 10.20. This is due to the change in scale on the vertical axis. However, both scatter plots represent the same situation. We will need a more robust measure to support our first graphical impressions. This measure is the correlation coefficient.

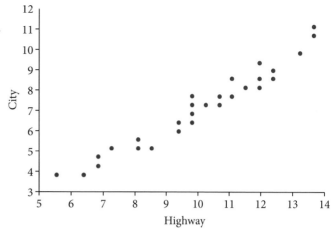

Figure 10.20 Copy of the second graph in Example 10.7

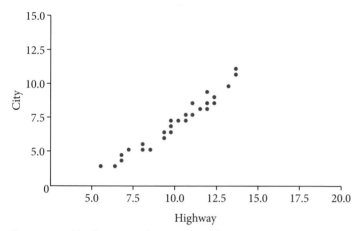

Figure 10.21 This figure gives the impression of a stronger association

Let us consider height and lean mass data collected from 130 twenty-year-olds. Here is the scatter plot.

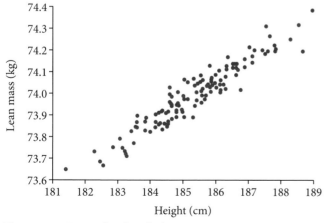

Figure 10.22 Scatter plot of weight data

Not surprisingly, the association between the two variables is strong. To measure the strength of this association, we use the correlation coefficient given by the following formula.

The correlation coefficient measures the strength and direction of the linear relationship between two quantitative variables.

For a set of data (x_i, y_i) of size n, the correlation coefficient is

$$R = \frac{1}{n-1}\sum\left(\frac{x_i - \bar{x}}{S_x}\right)\left(\frac{y_i - \bar{y}}{S_y}\right)$$

where $\bar{x}$ and $\bar{y}$ are the means of the variables and S_x and S_y are the standard deviations. Specific values of R are denoted by r.

R is also called the Pearson product-moment correlation coefficient. In fact, R is an unbiased estimate of the population coefficient, which is given by:

$$\rho = \frac{\text{cov}(X, Y)}{\sigma_x\sigma_y}$$

$$= \frac{1}{n}\sum\left(\frac{x_i - \mu_x}{\sigma_x}\right)\left(\frac{y_i - \mu_x}{\sigma_y}\right)$$

GDCs use r.

In exams, you will not be asked to calculate the coefficient by hand but to interpret the GDC result. There are several equivalent forms for the equation but it is not necessary, at this stage, to calculate any of them.

This formula is somewhat complex to put into practice; you will read the result from your calculator or computer output.

If we look at the formula, we see that the first component $\dfrac{x_i - \bar{x}}{S_x}$ is nothing but the standardised value for x_i. Similarly, the second component $\dfrac{y_i - \bar{y}}{S_y}$ is the standardised value for y_i. So, the correlation coefficient can be written as $R = \dfrac{\sum z_x z_y}{n-1}$. That is, the correlation coefficient is an average of the products of the standardised values of the two variables.

Whether we use the definition of r or ρ, it can be shown that they are equivalent. Hence, using your GDC will give you the correct value. If you are interested in seeing how to show their equivalence, here is one method.

Starting with ρ:

$$\rho = \frac{1}{n}\sum\left(\frac{x_i - \mu_y}{\sigma_y}\right)\left(\frac{y_i - \mu_y}{\sigma_y}\right) = \frac{1}{n}\sum\left(\frac{x_i - \mu_x}{\sqrt{\frac{\Sigma(x_i - \mu_x)^2}{n}}}\right)\left(\frac{y_i - \mu_y}{\sqrt{\frac{\Sigma(y_i - \mu_y)^2}{n}}}\right)$$

$$= \frac{1}{n}\sum\left(\frac{x_i - \mu_x}{\frac{1}{n}\sqrt{\Sigma(x_i - \mu_x)^2}}\right)\left(\frac{y_i - \mu_y}{\sqrt{\frac{\Sigma(y_i - \mu_y)^2}{n}}}\right) = \sum\frac{(x_i - \mu_x)(y_i - \mu_y)}{\sqrt{\Sigma(x_i - \mu_x)^2}\sqrt{\Sigma(y_i - \mu_y)^2}}$$

Starting with r:

$$R = \frac{1}{n-1}\sum\left(\frac{x_i - \bar{x}}{S_x}\right)\left(\frac{y_i - \bar{y}}{S_y}\right) = \frac{1}{n-1}\sum\left(\frac{x_i - x}{\sqrt{\frac{\Sigma(x_i - \bar{x})^2}{n-1}}}\right)\left(\frac{y_i - y}{\sqrt{\frac{\Sigma(y_i - \bar{y})^2}{n-1}}}\right)$$

$$= \frac{1}{n-1}\sum\left(\frac{x_i - \bar{x}}{\frac{1}{n-1}\sqrt{\Sigma(x_i - \bar{x})^2}}\right)\left(\frac{y_i - \bar{y}}{\sqrt{\Sigma(y_i - \bar{y})^2}}\right) = \sum\frac{(x_i - \bar{x})(y_i - \bar{y})}{\sqrt{\Sigma(x_i - \bar{x})^2}\sqrt{\Sigma(y_i - \bar{y})^2}}$$

Let's take the weight–height data and express it in pounds and inches instead, as shown in Figure 10.23.

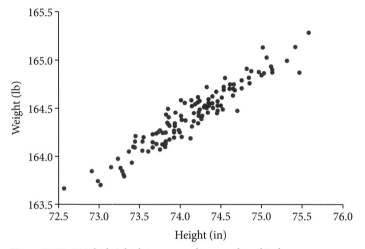

Figure 10.23 Weight-height data expressed in pounds and inches

Other than the scale on the axes being inches and pounds, the plot has the same form and direction and strength as the original one. Similarly, when we standardise the variables (Figure 10.24), we are subtracting a constant from each value and dividing by another constant.

Other than the centre of the data being at the origin, the form, direction, and strength appear to be the same.

This fact is verified by calculating the correlation coefficient for all three forms of the data. The result is always the same, 0.95 (software use).

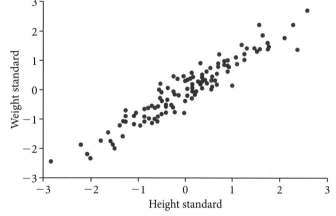

Figure 10.24 Standardisation of variables

```
LinReg
 y=ax+b
 a=1.108163254
 b=2.141125622
 r²=.8793206242
 r=.9377209735
```

```
           Rad Norm1 d/c Real
 LinearReg(ax+b)
  a   =1.10816325
  b   =2.14112562
  r   =0.93772097
  r²  =0.87932062
  MSe =0.60072376
 y=ax+b
                        COPY
```

Figure 10.25 For Example 10.7, the correlation can be read from your GDC

You may have observed in the technology output that r^2 is also reported. This measure is not required for your exam. However, it is an extremely useful and powerful tool. r^2 is known as the **coefficient of determination**. It reports the portion of variation in the response variable that can be explained by the variation in the explanatory variable. As such, r^2 can be expressed as a percentage. Using the data from Example 10.7, $r^2 = 0.879$, which can be interpreted as 'if all else is equal, then 88% of the variation in city consumption can be explained by variation in the highway consumption', i.e. on average, for cars with the same characteristics, if there is a 1 km/litre change in city consumption, we expect that 88% of this change can be explained by changes in the highway consumption. Using the data from Example 10.6, $r = 0.84$ and $r^2 = 0.7056$, which means that approximately 70.6% of the changes in the metabolic rate can be explained by changes in the lean mass.

Properties of the correlation coefficient

- The correlation coefficient is a measure of the strength of the linear association between two quantitative variables.
  - Do not apply correlation to non-quantitative data!
  - The coefficient does not prove a linear relationship. If there is a linear association, the coefficient will describe its strength.
- Outliers can distort the correlation coefficient. Special attention must be paid to such outliers.
- The correlation coefficient is always a number between -1 and $+1$. Values of R near 0 indicate weak or no relationship. Values close to $+1$ or -1 indicate strong association.
- R is independent of the units of measurement.
- R has no units and is not a percentage. For example, don't express a correlation of 0.85 as 85%.
- Correlation between two variables means that there is some association between them. It does not mean that one of them causes the other.

When there is no association, $\text{cov}(X, Y) = 0$.

Hence, $\rho = \dfrac{\text{cov}(X, Y)}{\sigma_x \sigma_y} = \dfrac{0}{\sigma_x \sigma_y} = 0$.

A proof for the values ± 1 is beyond the scope of this book.

So, correlation does not mean causation; that is, two variables can have a strong correlation without one of them being the cause of the changes in the other. For example, there may be a strong correlation between the amount of crude oil imported by country X and the rate of birth in country Y. That does not necessarily mean that the increase of oil imports causes an increase in birth rate. However, in some cases, there may be a causal relationship. For example, the increase in the average level of income in a certain country and the decrease of unemployment can have a strong negative correlation. This association is also causal. However, the task of proving that the relationship is causal comes with economics.

Example 10.8

The table below gives the data for a lab experiment involving the length (mm) of a metal alloy bar used in electronic equipment when it is exposed to heat.

Temp (°C)	40	45	50	55	60	65	70	75	80
Length	20	20.12	20.20	20.21	20.25	20.25	20.34	20.47	20.61

Draw a scatter plot. Comment on the strength of the relationship. Use both r and r^2.

Solution

Here is the scatter plot.

It appears that there is a relatively strong relationship.

This is confirmed by calculating the correlation coefficient. In this case, regardless of which formula we use (r or ρ), the correlation is approximately 0.95521. Using $r^2 = 91.2\%$ implies that 91.2% of the variation in the length can be explained by variation in the temperature.

Least squares regression

We have seen that correlation measures the strength and direction of a linear relationship between two quantitative variables. So, if we suspect from a scatter plot that the relationship is linear, then we need to summarise this linear behaviour; that is, we need to find an equation of a straight line that best fits the trend in the data. In this sub-section, we discuss how to find a **line of best fit** that describes the linear relationship between an explanatory and response variable when it exists.

Finding a line of best fit means finding a line that comes as close as possible to the points in the data set. Obviously, there is no straight line that contains all the points in the set.

Regression line

A **regression line** is a straight line that describes how a response variable changes with changes in an explanatory variable.

Let Y be the response variable and X be the explanatory variable. Since for the same value of the explanatory variable X we can expect several values of the response variable Y, our linear model enables us, on average, to predict the value of Y given a value of $X = x$, and hence we write the equation of the linear regression line in the form

$$E(Y) = \alpha + \beta x$$

This is to say, given a specific value of x, the expected value of Y is equal to $\alpha + \beta x$ where α is the value corresponding to $x = 0$, and β is the slope representing the rate with which the response variable changes with every change of one unit in the explanatory variable (gradient).

Note that the regression model can be stated 'formally' as $E(Y|X = x) = \alpha + \beta x$

In cases like this, our data are only samples from a population, and consequently we can only estimate the regression equation.

From the sample data we estimate the regression equation and we write our estimate as

$$y = bx + a$$

where b, the slope of the line, is an estimate of β and reflects how the response variable, Y, changes according to changes in the explanatory variable X. a is an estimate of α and is the value of the response variable corresponding to a zero value in X.

In the example of height–lean mass, the equation is

Lean mass (kg) $= 56.1 + 0.0966$ *Height (cm)*

That is, $b = 0.0966$ and $a = 56.1$

This means that *on average*, for every increase (decrease) of 1 cm in height, we predict an increase (decrease) of 0.0966 kg in weight. The interpretation of a is peculiar. As you know from algebra, a stands for the value of y (which is *Lean mass* in this case) corresponding to a zero value of x (which is height in this case). However, for this problem the interpretation is not ideal. It corresponds to a height of zero. The general rule is that if 0 is not included in the domain of the explanatory variable, then trying to interpret the intercept is pointless.

This issue has to do with what we call **extrapolation**. Extrapolation is the use of the regression line for predicting values far off the range of values of the explanatory variable x used to find the equation of that line. *Such predictions are often inaccurate.*

Why the least-squares regression line?

Let's take a simple example. The graph in Figure 10.26 represents a few points in a data set. The green line is the line of best fit. Take for example point (x_1, y_1). The point on the line $(x_1, \hat{y}_1)$ is the point whose y-coordinate $\hat{y}_1$ predicts the real y-coordinate, using the line of best fit. The distance $y_1 - \hat{y}_1$ is the error in this prediction. Similarly, $y_2 - \hat{y}_2$ and all other $y_i - \hat{y}_i$ represent the errors in their respective predictions. The line of best fit is the line that minimises the sum of all these errors. However, like the variance, some of these errors are positive and some are negative and thus they could cancel each other out. To avoid this, as we did with the variance, we try to minimise the squares of these errors. That is, the line of best fit is the line that minimises the sum $\Sigma(y_i - \hat{y}_i)^2$. Hence, it is called the **least-squares** line of regression $\hat{y} = bx + a$.

The process of finding the slope of such a line is beyond the scope of this book. Here are some of the many formulae for the slope and intercept:

$$b = \frac{\text{cov}(X, Y)}{\text{V}(X)} = \frac{\Sigma(x_i - \bar{x})(y_i - \bar{y})}{\Sigma(x_i - \bar{x})^2} = \frac{\Sigma x_i y_i - n\bar{x}\bar{y}}{\Sigma x_i^2 - n\bar{x}^2} = r\frac{s_y}{s_x}$$

Here, r is the correlation coefficient, $\bar{x}$, $\bar{y}$, s_x, and s_y are the means and standard deviations of the explanatory and response variables. The last form demonstrates the close relationship between the slope of the regression line and the correlation coefficient. One conclusion we can draw from this formula is that along a line of regression with slope b, a change of 1 standard deviation in the x direction will result in a change of r standard deviations in the y direction.

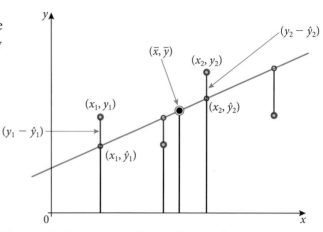

After estimating the slope, and using the fact that the line has to contain the point with coordinates $(\bar{x}, \bar{y})$, the intercept, a, can be found using $a = \bar{y} - b\bar{x}$.

Figure 10.26 Every regression line should contain the point $(\bar{x}, \bar{y})$

As you will notice from the equations, every regression line should contain the point $(\bar{x}, \bar{y})$ with the averages of the variables as coordinates.

Example 10.9

This scatter plot represents a random sample of IB students who went through four years of university, and is a comparison of their scores on the IB exams they took and their GPAs in their university studies (scale 1–4).

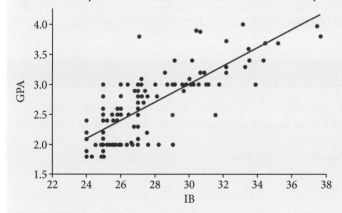

There appears to be a linear relationship. When we run a linear regression, the equation is:

GPA = $-1.51 + 0.151$ IB

This means that, on average, for every increase of 1 point in the total IB score, we expect an increase of 0.15 points in the GPA. If we want to predict the GPA of a student who scored 30 on an IB diploma, the model predicts, on average, a grade of:

GPA = $21.51 + 0.151(30) = 3.02$

The correlation coefficient of this relationship is $r = 0.758$, which is a relatively strong correlation. In addition, $r^2 = 57.5\%$. This means that changes in the IB score may help us explain 57.5% of the variation in the GPA.

Does that mean high IB scores cause high university grade averages? The answer is no. They only help predict the future university grade averages.

Features of the regression line

Exceptional cases of the regression line

If $r = 0$, the regression line is horizontal; its slope is zero.

If $r = 1$, all the points fall on a line with positive slope.

If $r = -1$, all the points fall on a line with negative slope.

- The **regression equation** can be used to predict the response variable according to values of the explanatory variable.
- The regression line must pass through the point $(\bar{x}, \bar{y})$.
- When the regression line is used for prediction and you substitute a specific value x_1 for the explanatory variable, the predicted value $\bar{y}_1$ of the response variable is an average value. For example, when we use the height–lean mass equation, Lean mass (kg) $= 56.1 + 0.0966$ Height (cm) to predict the weight corresponding to a height of 182 cm, the value we get (73.68 kg) is the average weight of twenty-year-old students of height 182 cm.

Estimating the value of r associated with a value of X that is larger than any of those observed, or smaller than any of those observed, is called **extrapolation**. Estimating the value of Y associated with a value of X that is within the range of the observed values of X but is not equal to any of the observed values of X is called **interpolation**.

Extrapolation is extremely suspect: without data in the range in which the estimate is wanted, there is no reason to believe that the relationship between X and Y is the same as it is in the region in which there are data.

Interpolation is sometimes reasonable when the scatter plot shows a strong relationship, especially if there are many data values near the value of X or Y at which the estimate is sought.

Example 10.10

Here are data for two variables. Draw the line of regression and indicate the distances, the sum of whose squares is minimised by the choice of the line of regression.

x	11	12	13	14	15	16	17
y	21	43	31	34	29	55	33

Solution

The scatter plot shows the data and line of regression. The red distances are those required.

The line has equation: $\hat{y} = 6.14 + 2.071x$

Look at the second table where we also introduced the value of each predicted y (Fit) and then calculated the distances (directed) whose squares were minimised.

x	y	Fit	Distance	Distance squared
11	21	28.92857	−7.92857	62.8622449
12	43	31	12	144
13	31	33.07143	−2.07143	4.290816327
14	34	35.14286	−1.14286	1.306122449
15	29	37.21429	−8.21429	67.4744898
16	55	39.28571	15.71429	246.9387755
17	33	41.35714	−8.35714	69.84183673

The minimum sum is 596.71. You can try to find any other line and you will notice that this is the minimum sum of the squares of distances.

Moreover, since $\bar{x} = 14$ and $\bar{y} = 35.14$, then:

$$35.14 = 6.14 + 2.071 \times 14$$

This indicates that the line contains the point $(\bar{x}, \bar{y})$.

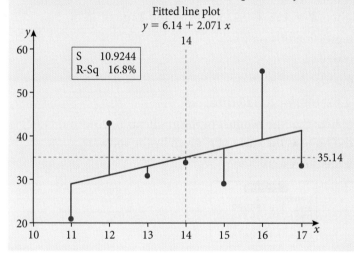

Fitted line plot
$y = 6.14 + 2.071\ x$

S   10.9244
R-Sq   16.8%

If you regress x on y instead, the equation of regression is $\hat{x} = dy + c$.

The resulting formulae for the slope and intercept are $d = r\dfrac{S_x}{S_y}$, and $c = \bar{x} - d\bar{y}$.

A remarkable relationship appears here between the gradients of the regression line and r.

For example, $b = r\dfrac{S_x}{S_y}$ and $d = r\dfrac{S_x}{S_y}$, and hence, $bd = r\dfrac{S_x}{S_y} \cdot r\dfrac{S_x}{S_y} = r^2$.

Note that in cases where the explanatory variable is 'not controlled', we can regress x on y instead, and the equation of regression is $\hat{x} = dy + c$.

The resulting formulae for the slope and intercept are

$$d = \frac{\text{cov}(X, Y)}{V(Y)} = \frac{\Sigma(x_i - \bar{x})(y_i - \bar{y})}{\Sigma(y_i - \bar{y})^2} = \frac{\Sigma x_i y_i - n\overline{xy}}{\Sigma y_i^2 - n\bar{y}^2} = r\frac{S_x}{S_y} \text{ and } c = \bar{x} - d\bar{y}$$

For example, $b = r\dfrac{S_y}{S_x}$ and $d = r\dfrac{S_x}{S_y}$,

Hence, $bd = r\dfrac{S_y}{S_x} \cdot r\dfrac{S_x}{S_y} = r^2$

Example 10.11

The following data represent the volume in cubic cm and weight in grams of a certain fruit studied by a biologist.

Volume (x)	223	236	242	226	223	221	233	222	222	218	232	223
Weight (y)	165	171	173	170	168	172	168	167	162	166	164	164

Obtain the least-squares regression line of y on x as well as the regression line of x on y. Use the model to predict the weight of a 230-cubic-cm fruit. Also, predict the volume of a 168 g fruit.

Solution

We will use software (you can use a GDC) for this calculation.

The least-squares regression of y on x is

$$Y = 115 + 0.233x$$

The predicted weight is $Y = 115 + 0.233(230) = 168.22$ g

The least squares regression of x on y is

$$X = 56.1 + 1.02y$$

The predicted volume is

$$X = 56.1 + 1.02(168) = 227.26 \text{ cubic cm}$$

You will also notice here that the product of the gradients (0.233) and (1.02) is 0.237, which is the same as the value of r^2 given by the software.

Using a GDC, here are the results:

```
LinReg
y=ax+b
a=.2327179047
b=114.7312151
r²=.2370923014
r=.4869212476
```

```
LinReg
y=ax+b
a=1.018796992
b=56.10150376
r²=.2370923014
r=.4869212476
```

Notice how the values of r and r^2 are the same.

Exercise 10.4

1. The following table lists the values of a response variable x against an explanatory variable y. Draw a scatter plot and comment on the strength of the relationship.

x	12	6	12	11	16	13	11	12	11	12	12	12	15	16	14	13	13	8	10	11
y	8	10	9	6	14	10	10	9	15	14	10	6	12	8	13	11	11	9	9	6

2. The data below represents the outcome of an experiment on a small car, relating fuel consumption to speed.

Speed km h^{-1}	60	65	70	75	80	85	90	95
Fuel consumption km/litre	16.9	16.8	15.9	15.9	14.4	14.3	13.2	14.3

Speed km h^{-1}	100	105	110	120	130	140	150
Fuel consumption km/litre	12.1	12.0	10.2	9.8	9.0	8.0	7.1

(a) Draw a scatter plot for the data.

(b) Describe the relationship and justify your choice of which variable is the explanatory variable and which is the response variable.

(c) Is the relationship strong? Explain your answer.

3. The tabular data is from World Bank statistics relating the Gross National Income per Capita (GNI/Cap) to Purchasing Power Parity (PPP) for some developed countries. The exchange rate adjusts so that an identical product in two different countries has the same price when expressed in the same currency. For example, a chocolate bar that sells for C$1.50 in a Canadian city should cost US$1.00 in a US city when the exchange rate between Canada and the US is 1.50 US$/C$. (Both chocolate bars cost US$1.00)

Country	GNI/Cap	PPP
NOR	85380	57130.0
CH	70350	49180.0
DK	58980	40140.0
SWE	49930	39600.0
NL	49720	42590.0
FIN	47170	37180.0
USA	47140	47020.0
AUT	46710	39410.0
BEL	45420	37840.0
D	43330	38170.0
F	42390	34440.0
JPN	42150	34790.0
SGP	40920	54700.0

(a) Draw a scatter plot of the data.

(b) Describe the relationship and justify your choice of which variable is the explanatory variable and which is the response variable.

(c) Is the relationship strong? Explain your answer.

4. In hotel management, it is necessary to estimate the electricity consumption in relation to the number of visitors. Here is the data for a large hotel.

Visitors	232	311	321	334	352	375	412	447	456	472	480	495	512
Consumption	237	278	270	303	298	328	387	390	376	402	431	430	432

(a) Draw a scatter plot of the data.

(b) Describe the relationship and justify your choice of which variable is the explanatory variable and which is the response variable.

(c) Is the relationship strong? Explain your answer.

5. Draw a regression model for questions 1 to 4 and interpret the slope of each.

6. To test the benefit of using an online tutoring course for exam preparation, 20 students were given a test before and after they took part in an experiment. The tests were similar and the scores before and after the experiment were recorded. The intention was to find how improved the scores were through participation in the experiment.

Student	1	2	3	4	5	6	7	8	9	10
Before	98	24	6	8	56	54	40	40	68	30
After	122	46	16	28	84	68	64	62	82	50

Student	11	12	13	14	15	16	17	18	19	20
Before	32	80	102	30	12	16	60	58	50	48
After	40	100	129	56	32	56	90	73	74	70

Analyse the data. For a student whose original score was 60, what do you expect, on average, the student's new score to be?

7. A large electronics company produces LCD monitors to be used in the computer industry. The monthly total cost of production over one year is given in the table below. (Number of units produced is in thousands and the cost is in 1000 euros.)

Number of units produced	16	31	57	76	13	25
Cost	1875	2586	3716	4712	1690	2191

Number of units produced	49	71	20	38	63	81
Cost	3319	4362	2005	2775	4116	4860

(a) Draw a scatter plot of the data.

(b) Write down the equation of the regression line representing the association between units of production and sales. Draw the regression line on your scatter plot.

(c) Interpret the slope of the regression line and comment on the strength of this association.

d) If the selling price of each unit during this year is 105 euros, what is the production level where the sales are equal to the cost?

8. The table below shows the scores for 12 students for IB Economics SL and IB Physics SL exams.

Economics	7	6	5	5	6	3	7	7	5	4	5	7
Physics	6	6	6	4	7	4	6	5	6	4	6	5

(a) Find the correlation coefficient and comment on your result.

(b) Find the regression equation that enables us to predict Economics scores from the Physics scores.

(c) What score in Economics would you expect for a candidate with a score of 4 in Physics?

9. Diamonds are usually priced according to weight. The carat is the usual measure of weight of a diamond. 1 carat is equivalent to 200 milligrams. Some experts use points as the measure instead. 1 point is equivalent to 2 milligrams. Therefore, every carat is equivalent to 100 points. So, a 0.5 carat diamond is worth 50 points.

Here is the data for 20 diamonds and their prices.

Points	73	103	106	21	31	100
Price (€)	5909	15260	13640	1287	2177	12837

Points	26	82	101	100	63	66
Price (€)	1911	6927	16143	10945	9117	6020

(a) Construct a scatter plot of the data. What type of trend do you observe?

(b) Write down the equation of a straight-line model relating the price to the number of points.

(c) Give a practical interpretation of the coefficients. If a practical interpretation is not possible, explain why.

(d) How well does the line fit the given data?

(e) Use the line you found to predict the price of a diamond with 63 points.

(f) Find the residual corresponding to your estimate in part (e).

10. 12 students in a graduating class study HL Economics and HL Physics. The marks they obtained on their mock exams in these subjects are given below:

Student	1	2	3	4	5	6	7	8	9	10	11	12
Economics	53	48	83	70	39	51	73	47	24	61	43	54
Physics	56	45	80	63	42	38	72	45	32	46	48	50

(a) Find the product moment correlation coefficient for the scores and write down its p-value.

(b) Interpret the p-value in the context of the question.

(c) Andrew obtained a grade of 64 in Economics. Use your model to predict his score in Physics.

(d) The same class sat mock exams in French SL and English HL and the correlation coefficient was 0.623. Using a 5% level of significance, determine whether the value indicates a positive association between the grades in French and English.

Chapter 10 practice questions

1. Given that μ is the mean of a data set $y_1, y_2, \ldots, y_{30}$. You know that $\sum_{i=1}^{30} y_i = 360$ and $\sum_{i=1}^{30} (y_i - \mu)^2 = 925$, find:

   (a) the value of μ

   (b) the standard deviation of the set.

2. Laura conducted a survey by asking students at her school about the time it takes each of them to get to school every morning. She scribbled the numbers on a piece of paper and, unfortunately, could not read the number of students who spend 40 minutes on their trip to school. The average number of minutes she had originally found was 34 minutes. Find out how many students spend 40 minutes on their trip.

Time in minutes	10	20	30	40	50
Number of students with this time	1	2	5	??	3

3. The following table gives 50 measurements of the time it took a certain reaction to be done in a laboratory experiment.

3.1	5.1	4.9	1.8	2.8	5.6	3.6	2.2	2.5	3.4
4.5	2.5	3.5	3.6	3.7	5.1	4.1	4.8	4.9	1.6
2.9	3.6	2.1	6.1	3.5	4.7	4	3.9	3.7	3.9
2.7	4.3	4	5.7	4.4	3.7	3.7	4.6	4.2	4
3.8	5.6	6.2	4.9	2.5	4.2	2.9	3.1	2.8	3.9

   (a) Construct a frequency table and histogram starting at 1.6 with an interval length of 0.5

   (b) What fraction of the measurements are less than 5.1?

   (c) Estimate, from your histogram, the median of this data set.

   (d) Estimate the mean and standard deviation using your frequency table.

   (e) Construct a cumulative frequency table.

   (f) From your cumulative frequency graph, estimate each of the numbers in the five-number summary.

4. In large cities around the world, governments offer parking facilities for public use. The histogram below gives a picture of the number of parking sites available with the capacity of each in a number of cities chosen at random.

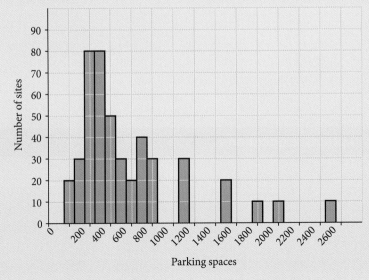

(a) Which statistics would best represent the data here? Why?

(b) Calculate the mean and standard deviation.

(c) Develop a cumulative frequency graph of the data.

(d) Use your result from part **(c)** to estimate the median, Q_1, Q_3, and IQR.

(e) Are there any outliers in the data? Why?

(f) Write a few sentences describing the distribution.

5. The box-and-whisker plots display the case prices of red wines produced in France, Italy, and Spain.

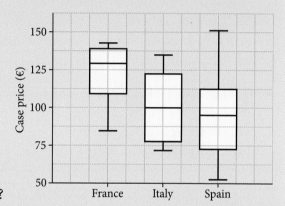

(a) Which country appears to produce the most expensive red wine? The least expensive?

(b) In which country are the red wines generally more expensive?

(c) Write a few sentences comparing the pricing of red wines in the three countries.

6.

112.72	53.55	54.12	54.33	58.79	59.26	60.39	62.45	52.22	52.52	52.58	52.85
54.06	51.34	51.93	52.09	52.14	52.24	52.24	52.53	53.50	51.82	51.93	52.00
52.78	52.82	50.28	50.49	51.28	51.28	51.52	51.62	52.40	52.43	49.83	50.46
50.95	51.07	51.11	49.45	49.45	49.73	49.76	49.93	50.19	50.32	50.63	48.64
49.79	50.19	50.62	50.96	49.09	49.16	49.29	49.74	49.74	49.75	49.84	49.76
52.90	52.91	53.40	52.18	52.57	52.72	50.56	50.87	50.90	49.32	49.70	

The table shows the times (in seconds) of the 71 male swimmers competing in the 100-metre race on the first day of the 2000 Summer Olympics in Sydney.

(a) Calculate the mean time and the standard deviation.

(b) Calculate the median and IQR.

(c) Explain the differences between these two sets of measures.

7. In a survey of universities in major cities in the world, the percentage of first-year students who graduate on time (some require 4 years and some 5 years) was reported. The summary statistics are given in the following table.

Number of universities surveyed	120
Mean percentage	69
Median percentage	70
Standard deviation	9.8
Minimum	42
Maximum	86
Range	44
Q_1	60.25
Q_3	75.75

(a) Is this distribution symmetric? Explain.

(b) Check for outliers.

(c) Create a box-and-whisker plot of the data.

(d) Describe the data in a short paragraph.

8. The International Heart Association studies, among other factors, the influence of the cholesterol level (in mg/dl) on the conditions of heart patients. In a study of 2000 subjects, the following cumulative relative frequency graph was recorded.

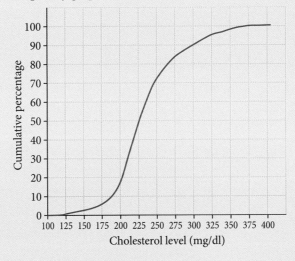

(a) Estimate the median cholesterol level of heart patients in the study.

(b) Estimate the first and third quartiles, and the 90th and the 10th percentiles.

(c) Estimate the IQR. Also estimate the number of patients in the middle 50% of this distribution.

(d) Create a box-and-whisker plot of the data.

(e) Give a short description of the distribution.

9. Many of the streets in Vienna, Austria, have a speed limit of $30 \, \text{km h}^{-1}$. On one Sunday evening the police registered the speed of cars passing an important intersection in order to give speeding tickets when drivers exceeded the limit. The following table is a random sample of 100 cars recorded that evening.

26	46	39	41	44	37	38	35	34	31
27	47	39	41	44	37	38	35	34	32
27	47	39	41	44	37	38	35	34	32
27	48	39	41	44	37	38	35	34	32
29	48	40	41	45	37	38	36	34	33
30	48	40	41	45	37	38	36	35	33
30	48	40	42	45	38	39	36	35	33
30	49	40	42	46	38	39	36	35	33
30	50	41	42	46	38	39	36	35	33
31	54	41	43	46	38	39	36	35	33

(a) Prepare a frequency table for the data.

(b) Draw a histogram of the data and describe its shape.

(c) Calculate, showing all work, the mean and standard deviation.

(d) Prepare a cumulative frequency table of the data.

(e) Find the median, Q_1, Q_3, and IQR.

(f) Are there any outliers in the data? Explain, using an appropriate diagram.

10. The following are the data collected from 50 industrial countries chosen at random in 2001. The data represent the per capita gasoline consumption in these countries. The Netherlands' consumption was at 1123 litres per capita while Italy stood at 2220 litres per capita.

2062	2076	1795	1732	2101	2211	1748	1239	1936	1658
1639	1924	2086	1970	2220	1919	1632	1894	1934	1903
1714	1689	1123	1671	1950	1705	1822	1539	1976	1999
2017	2055	1943	1553	1888	1749	2053	1963	2053	2117
1600	1795	2176	1445	1727	1751	1714	2024	1714	2133

(a) Calculate the mean, median, standard deviation, Q_1, Q_3, and IQR.

(b) Are there any outliers?

(c) Draw a box-and-whisker plot.

(d) What consumption levels are within 1 standard deviation from the mean?

(e) Germany, with a consumption level of 2758 litres per capita, was not included in the sample. What effect on the different statistics calculated would adding Germany have? Do not recalculate the statistics.

11. 90 students took part in an experiment where each reported the time, x, it took them to travel to school on a specific day, to the nearest minute. The teachers then reported back that the total travelling time for the course participants was $\sum x = 4460$ minutes.

(a) Find the mean number of minutes the students spent travelling to school that day.

Four students who were absent when the data was first collected reported that they spent 35, 39, 28, and 32 minutes, respectively.

(b) Calculate the new mean including these four students.

12. Two thousand students at a large university take the final Statistics examination, which is marked on a 100-scale, and the distribution of marks received is given in the table below:

Marks	1－10	11－20	21－30	31－40	41－50	51－60	61－70	71－80	81－90	91－100
Number of candidates	30	100	200	340	520	440	180	90	60	40

(a) Complete a copy of the table below so that it represents the cumulative frequency for each interval.

Marks	⩽10	⩽20	⩽30	⩽40	⩽50	⩽60	⩽70	⩽80	⩽90	⩽100
Number of candidates	30	130				1630				

(b) Draw a cumulative frequency graph of the distribution, using a scale of 1 cm for 100 students on the vertical axis and 1 cm for 10 marks on the horizontal axis.

(c) Use your graph to answer parts (i)–(iii).

   (i) Find an estimate for the median score.

   (ii) Candidates who scored less than 35 were required to retake the examination. How many candidates had to retake the exam?

   The highest-scoring 15% of candidates were awarded a distinction.

   (iii) Find the mark above which a distinction was awarded.

13. 100 physicians are taking part in a symposium. 72 of the physicians are male and 28 are female. The mean height of men is 179 cm and that for the women is 162 cm. Find the mean height of the 100 physicians.

14. Consider a population $x_1, x_2, ..., x_{25}$ such that $\sum_{i=1}^{25} x_i = 300$ and $\sum_{i=1}^{25} (x_i - \mu)^2 = 625$, where μ is the mean.

(a) Find the value of μ.

(b) Find the standard deviation of the population.

15. The table below lists the grades of students in a small class on a 50-points test.

Grade	10	20	30	40	50
Number of students with this grade	1	2	5	k	3

The mean grade is 34. Find the value of k.

16. Waiting times for 100 customers at a supermarket's cash counter were recorded and are shown in the table.

Waiting time, t (seconds)	Number of customers
$0 \leqslant t < 30$	5
$30 \leqslant t < 60$	15
$60 \leqslant t < 90$	33
$90 \leqslant t < 120$	21
$120 \leqslant t < 150$	11
$150 \leqslant t < 180$	7
$180 \leqslant t < 210$	5
$210 \leqslant t < 240$	3

(a) Estimate the mean waiting time for a customer.

(b) Set up a cumulative frequency table for these data.

(c) Use your table from part **(b)** to draw a cumulative frequency graph.

(d) Use your graph from part **(c)** to find estimates for the median and the first (lower) and third (upper) quartiles.

17. The diagram represents the lengths, in cm, of 80 plants grown in a laboratory.

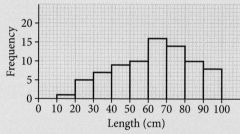

(a) How many plants have lengths in cm between

  (i) 50 and 60?

  (ii) 70 and 90?

(b) Calculate estimates for the mean and the standard deviation of the lengths of the plants.

(c) Explain what feature of the diagram suggests that the median is different from the mean.

(d) The following is an extract from the cumulative frequency table.

Length in cm	Cumulative frequency
<50	22
<60	32
<70	48
<80	62

Use the information in the table to estimate the median. Give your answer to 2 significant figures.

18. The table represents the weights, W, in grams, of 80 packets of roasted peanuts.

Weight (W)	Number of packets
$80 < W \leqslant 85$	5
$85 < W \leqslant 90$	10
$90 < W \leqslant 95$	15
$95 < W \leqslant 100$	26
$100 < W \leqslant 105$	13
$105 < W \leqslant 110$	7
$110 < W \leqslant 115$	4

(a) Use the midpoint of each interval to find an estimate for the standard deviation of the weights.

(b) Copy and complete the following cumulative frequency table for the above data.

Weight (W)	$W \leqslant 85$	$W \leqslant 90$	$W \leqslant 95$	$W \leqslant 100$	$W \leqslant 105$	$W \leqslant 110$	$W \leqslant 115$
Number of packets	5	15					80

(c) A cumulative frequency graph of the distribution is shown below, with a scale of 2 cm for 10 packets on the vertical axis and 2 cm for 5 grams on the horizontal axis.

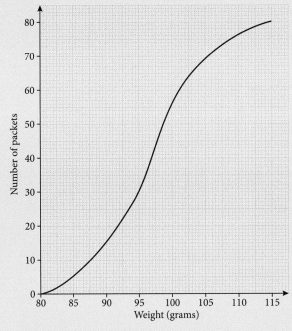

Use the graph to estimate:

(i) the median

(ii) the upper (third) quartile.

Give your answers to the nearest gram.

(d) Let W_1, W_2, ..., W_{80} be the individual weights of the packets, and let $\overline{W}$ be their mean. What is the value of the sum:

$$(W_1 - \overline{W}) + (W_2 - \overline{W}) + \ldots + (W_{79} - \overline{W}) + (W_{80} - \overline{W})?$$

(e) One of the 80 packets is selected at random. Given that its weight satisfies $5 < W \leqslant 110$, find the probability that its weight is greater than 100 grams.

19. A surveillance camera is mounted at a critical stretch of a highway and records the speed in km h^{-1} of cars passing a certain point. The table shows the record over a 5-minute interval.

Speed, s	Number of cars
$s \leqslant 60$	0
$60 < s \leqslant 70$	7
$70 < s \leqslant 80$	25
$80 < s \leqslant 90$	63
$90 < s \leqslant 100$	70
$100 < s \leqslant 110$	71
$110 < s \leqslant 120$	39
$120 < s \leqslant 130$	20
$130 < s \leqslant 140$	5
$s > 140$	0

(a) Estimate the mean speed of cars passing this point.

(b) The table below lists cumulative frequencies for the speeds above.

   (i) Write down the values of m and n.

   (ii) Construct a cumulative frequency graph to represent information in part (i).

Speed, s	Cumulative frequency
$s \leqslant 60$	0
$s \leqslant 70$	7
$s \leqslant 80$	32
$s \leqslant 90$	95
$s \leqslant 100$	m
$s \leqslant 110$	236
$s \leqslant 120$	n
$s \leqslant 130$	295
$s \leqslant 140$	300

(c) Use your graph in part (b) to determine:

   (i) the percentage of cars travelling at a speed in excess of 105 km h^{-1}

   (ii) the speed that is exceeded by 15% of the cars.

20. A taxi company has 200 taxi cabs. The cumulative frequency graph below shows the fares in dollars ($) taken by the cabs one morning.

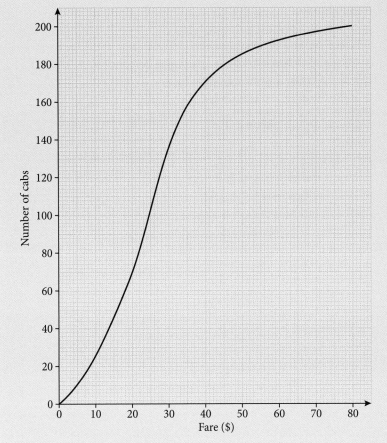

(a) Use the graph to estimate:

   (i) the median fare

   (ii) the number of cabs in which the fare taken is $35 or less.

The company charges 55 cents per kilometre for distance travelled. There are no other charges. Use the graph to answer the following.

(b) On that morning, 40% of the cabs travel less than *a* km. Find the value of *a*.

(c) What percentage of the cabs travelled more than 90 km that morning?

21. Three positive integers a, b, and c, where $a < b < c$, are such that their median is 11, their mean is 9, and their range is 10. Find the value of a.

22. A real estate agent keeps records of small houses sold in a suburb of Budapest, Hungary. The cumulative frequency table on the next page shows 100 houses that were sold in the second half of 2017. Prices are in thousands of euros.

Selling price, P (€1000)	$P \leqslant 100$	$P \leqslant 200$	$P \leqslant 300$	$P \leqslant 400$	$P \leqslant 500$
Total number of houses	12	58	87	94	100

(a) Draw a cumulative frequency graph for the information in the table.

(b) Use your graph to find the lower and upper quartiles as well as the interquartile range.

Below is the frequency distribution of the information above.

Selling price, P (€1000)	$0 < P \leqslant 100$	$100 < P \leqslant 200$	$200 < P \leqslant 300$	$300 < P \leqslant 400$	$400 < P \leqslant 500$
Number of houses	12	46	29	m	n

(c) Find the value of m and of n.

(d) Use mid-interval values to calculate an estimate for the mean selling price.

(e) Houses that sell for more than €350,000 are described as Luxury houses.

   (i) Use your graph to estimate the number of Luxury houses sold.

   (ii) Two Luxury houses are selected at random. Find the probability that both have a selling price of more than €400,000.

23. A student measured the diameters of 80 snail shells. His results are shown in the following cumulative frequency graph. The lower quartile (LQ) is 14 mm and is marked clearly on the graph.

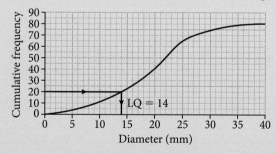

(a) On the graph, clearly mark the value of:

   (i) the median

   (ii) the upper quartile.

(b) Write down the interquartile range.

24. The cumulative frequency graph below shows the marks obtained in an examination by a group of 200 students.

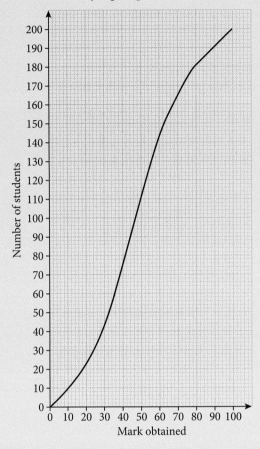

(a) Use the cumulative frequency graph to complete a copy of the frequency table below.

Mark (x)	$0 \leqslant x < 20$	$20 \leqslant x < 40$	$40 \leqslant x < 60$	$60 \leqslant x < 80$	$80 \leqslant x < 100$
Number of students	22				20

(b) Forty per cent of the students fail. Find the pass mark.

25. The cumulative frequency graph below shows the heights of 120 basketball players in centimetres.

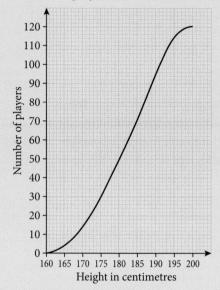

Use the graph to estimate:

(a) the median height

(b) the interquartile range.

26. Let a, b, c, and d be integers such that $a < b$, $b < c$ and $c = d$.

The mode of these four numbers is 11.

The range of these four numbers is 8.

The mean of these four numbers is 8.

Calculate the value of each of the integers a, b, c, and d.

27. A test marked out of 100 is taken by 800 students. The cumulative frequency graph for the marks is given below.

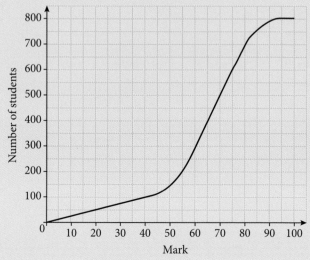

(a) Write down the number of students who scored 40 marks or less.

(b) The middle 50% of test results lie between marks a and b, where $a < b$. Find a and b.

28. The set of numbers $\{x, y, 10, 12, 16, 16, 18, 18\}$ has a mean of 13 and a variance σ^2 of 21. x and y are integers with $x < y$. Find x and y.

29. The following table gives the average yield of olives per tree, in kg, and the rainfall, in cm, for nine separate regions of Italy. You may assume that these data are a random sample from a bivariate normal distribution, with correlation coefficient ρ.

Rainfall (x)	11	10	15	13	7	18	22	20	28
Yield (y)	56	53	67	61	54	78	86	88	78

A scientist wishes to use these data to determine whether there is a positive correlation between rainfall and yield.

(a) Draw a scatter plot of the data and comment on its shape.

(b) Determine the product moment correlation coefficient for these data and comment about the strength of the relationship.

(c) Find the equation of the regression line of y on x and interpret its parameters.

(d) Hence, estimate the yield per tree in a tenth region where the rainfall was 19 cm.

(e) Determine the angle between the regression line of y on x and that of x on y. Give your answer to the nearest degree.

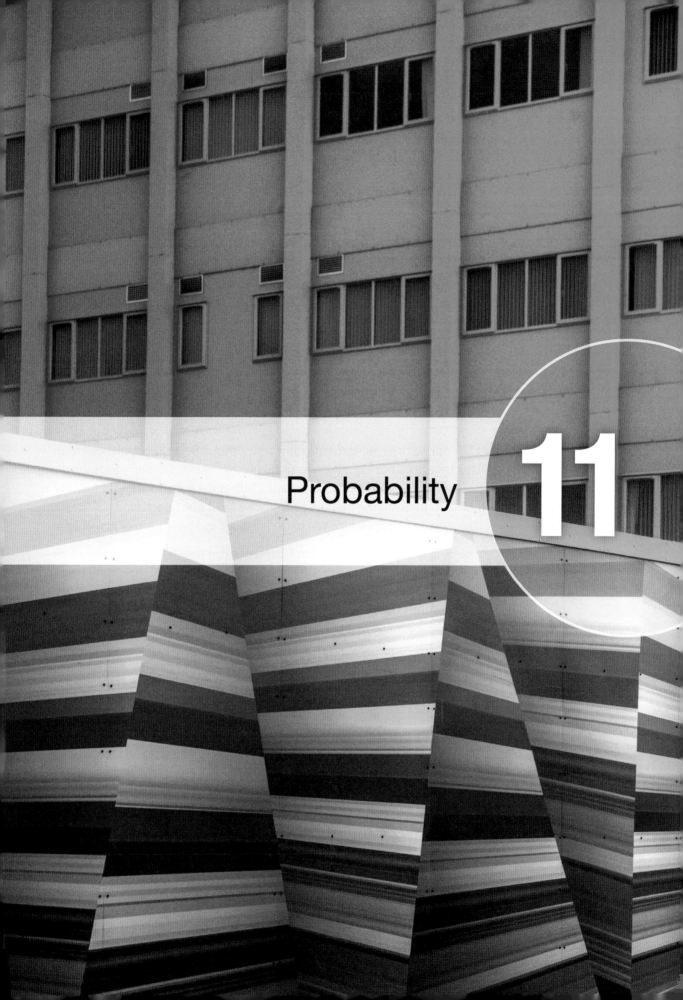

Probability

11

Learning objectives

By the end of this chapter, you should be familiar with...

- concepts of trial, outcome, equally likely outcomes, sample space (U), and event
- calculating the probability of an event A as $P(A) = n(A)/n(U)$
- the complementary events as A and A' (not A); $P(A) + P(A') = 1$
- combined events and using the formula
  $P(A \cup B) = P(A) + P(B) - P(A \cap B)$
- why $P(A \cap B) = 0$ for mutually exclusive events
- conditional probability and working with its definition as
  $P(A|B) = P(A \cap B)/P(B)$
- independent events; the definition $P(A|B) = P(A) = P(A|B')$
- using Venn diagrams, tree diagrams, and tables of outcomes to solve problems
- Using Bayes' theorem for a maximum of three events.

How can we use sample data to draw conclusions about the populations from which we drew our samples? The techniques we use in drawing conclusions are part of what we call inferential statistics. Inferential statistics uses **probability** as one of its tools. To use this tool properly, we must first understand how it works. This chapter will introduce you to the language and basic tools of probability.

The sample data we discussed in Chapter 10 can now be defined as **random variables**, whose values depend on the chance selection of the elements in the sample. Using probability as a tool, we will be able to create **probability distributions** that serve as models for random variables. We can then describe these using a mean and a standard deviation as we did in Chapter 10.

11.1 Randomness and probability

Probability is the study of randomness and uncertainty.

The reasoning in statistics rests on asking, 'How often would this method give a correct answer if I used it many, many times?' When we produce data by random sampling or by experiments, the laws of probability enable us to answer the question, 'What would happen if we did this many times?'

What does random mean? In ordinary speech, we use the word random to denote things that are unpredictable. Events that are **random** are not perfectly predictable, but they have long-term regularities that we can describe and quantify using probability. In contrast, **haphazard** events do not necessarily have long-term regularities. Take, for example, the flipping of an unbiased coin and observing the number of heads that appear. This is random behaviour.

There is a clear distinction between random and haphazard (chaotic) events. At first glance they might seem to be the same because neither of their outcomes can be anticipated with certainty.

When we flip a coin, there are only two outcomes, heads or tails. Figure 11.1 shows the results of the first 50 flips of an experiment where a coin was flipped 5000 times. Two trials are shown. The red graph shows the result of the experiment where the first flip was a head followed by a tail, making the proportion of heads to be 0.5 – the third flip was also a tail, so, the proportion of heads is 0.33, then 0.25. On the other hand, the other trial, shown in green, starts with a series of tails, then a head, which raises the proportion to 0.2, and so on.

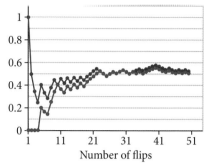

Figure 11.1 Probability of flipping a head

The proportion of heads is quite variable at first. However, in the long run, and as the number of flips increases, the proportion of heads stabilises around 0.5. We say that 0.5 is the **probability** of flipping a head.

It is important that you know that the proportion of heads in a small number of flips can be far from the probability. Probability describes only what happens in the long run. How a fair coin lands when it is flipped is an example of a random event. One cannot predict perfectly whether the coin will land on heads or tails. However, in repeated flips, the percentage of times the coin lands on heads will tend to settle down to a limit of 50%. The outcome of an individual flip is not perfectly predictable, but the long-term average behaviour is predictable. Thus, it is reasonable to consider the outcome of flipping a fair coin to be random.

Imagine the following scenario:

I drive to school every day. Close to the school, there is a traffic light. It appears that it is always red when I get there. I collected data over the course of one year (180 school days) and considered the green light to be a 'success'. Here is a partial table of the collected data.

Day	1	2	3	4	5	6	7	...
Light	red	green	red	green	red	red	red	...
% green	0	50	33.3	50	40	33.3	28.6	...

Table 11.1 Traffic light data

The first day it was red, so the proportion of success is 0% (0 out of 1), the second day it was green, so the cumulative frequency is now 50% (1 out of 2), the third day it was red again, so 33.3% (1 out of 3) and so on. As more data is collected, the new measurement becomes a smaller and a smaller fraction of the accumulated frequency, so, in the long run, the graph settles to the real chance of finding it green, which in this case is about 30%. The graph is shown in Figure 11.2

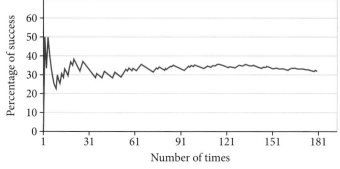

Figure 11.2 The probability of finding the traffic light green is about 30% over time

521

It you run a simulation for a longer period, you can see that it really stabilises around 30%. See Figure 11.3

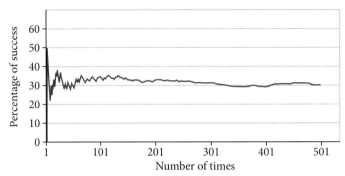

Figure 11.3 Further stabilisation at 30% over time

You should observe here that the randomness in the experiment is not in the traffic light itself, as it is controlled by a timer. In fact, if the system works well, it may turn green at the same time every day. The randomness of the event is the time I arrive at the traffic light.

If we ask for the probability of finding the traffic light green in the above example, our answer will be about 30%. We base our answer on knowing that, in the long run, the fraction of time that the traffic light was green is 30%. We could also say that the **long-run relative frequency** of green light settles down to about 30%.

Basic definitions

An **experiment** is the process by which an observation (or measurement) is obtained. A **random** (chance) **experiment** is an experiment where there is uncertainty concerning which of two or more possible outcomes will result.

Flipping a coin, rolling a dice and observing the number on the top surface, counting cars at a traffic light when it turns green, measuring daily rainfall in a certain area, and so on, are a few experiments in this sense of the word.

A description of a random phenomenon in the language of mathematics is called a **probability model**. For example, when we flip a coin, we cannot know the outcome in advance. What do we know? We are willing to say that the outcome will be either heads or tails. Because the coin appears to be balanced, we believe that each of these outcomes has probability 0.5. This description of flipping a coin has two parts:

- A list of possible outcomes
- A probability for each outcome.

The **sample space** U of a random experiment (or phenomenon) is the set of all possible outcomes.

The notation for sample space could also be S or any other letter.

For example, for one flip of a coin, the sample space is

$U = \{\text{heads, tails}\}$, or simply $U = \{\text{h, t}\}$

Example 11.1

Flip a coin twice (or two coins once) and record the results. What is the sample space?

Solution

$U = \{hh, ht, th, tt\}$

A **simple event** is the outcome we observe in a single repetition (trial) of an experiment.

An **event** is an **outcome** or a **set of outcomes** of a random experiment.

We can also look at the event as a subset of the sample space or as a collection of simple events.

Set theory provides a foundation for all of mathematics. The language of probability is much the same as the language of set theory. Logical statements can be interpreted as statements about sets.

Example 11.2

When rolling a standard six-sided dice, what are the sets of event A: 'observe an odd number', and event B: 'observe a number less than 5'?

Solution

Event A is the set $\{1, 3, 5\}$; event B is the set $\{1, 2, 3, 4\}$

Sometimes it helps to visualise an experiment using a **Venn diagram**. Figure 11.4 shows the outcomes of the dice-rolling experiment.

In general, in this book, we will use a rectangle to represent the sample space and closed curves to represent events.

To understand these definitions more clearly, let's look at the following example.

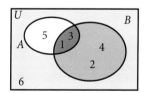

Figure 11.4 Venn diagram

Example 11.3

Flip a coin three times and record the results. Show the event 'observing two heads' on a Venn diagram.

Solution

The sample space is made up of 8 possible outcomes:

hhh, hht, tht, hth, htt, tth, thh, ttt.

Observing exactly two heads is an event with three elements: {hht, hth, thh}

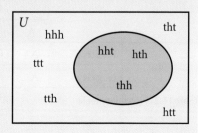

> **Some useful set theory results**
>
> Set operations have a number of properties, which are basic consequences of the definitions. Some examples are:
>
> - $A \cup B = B \cup A$
> - $(A')' = A$
> - $A \cap U = A$
> - $A \cup U = U$
> - $A \cap A' = \varnothing$
> - $A \cup A' = U$
>
> U is the sample space and $\varnothing$ is the empty set.
> Two valuable properties are known as De Morgan's laws, and state that
> $$(A \cup B)' = A' \cap B'$$
> $$(A \cap B)' = A' \cup B'$$
> And finally
> $$A \cap (B \cup C) = (A \cap B) \cup (A \cap C),$$
> $$A \cup (B \cap C) = (A \cup B) \cap (A \cup C)$$

Tree diagrams, tables and grids

In a two-stage experiment to check the blood types of patients, first we identify the type of the blood and then we classify the Rh factor as $+$ or $-$.

We can represent the events on a **tree diagram** (Figure 11.5)

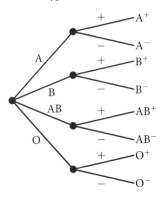

Figure 11.5 Blood types tree diagram

The sample space in this experiment is the set
$\{A^+, A^-, B^+, B^-, AB^+, AB^-, O^+, O^-\}$ as we can read from the last column.

The same simple events can also be arranged in a **probability table** or by using a 2-dimensional grid (Figure 11.6)

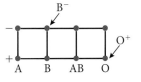

Figure 11.6 2D grid for the probability of blood types

		Blood type			
		A	**B**	**AB**	**O**
Rh factor	**Positive**	A^+	B^+	AB^+	O^+
	Negative	A^-	B^-	AB^-	O^-

Table 11.2 Probability table

Example 11.4

Two unbiased tetrahedral dice, with sides numbered
1 to 4, one blue and one yellow, are rolled.
List the elements of the following events when both
dice are thrown.

T = {3 is face down on at least one dice}

B = {the 3 on the blue dice is face down}

S = {sum of the face-down faces on both dice is a six}

Solution

We can use a sample space
diagram to help us.

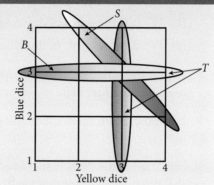

T = {(1, 3), (2, 3), (3, 3), (4, 3), (3, 2),
      (3, 1), (3, 4)}

B = {(1, 3), (2, 3), (3, 3)}

S = {(2, 4), (3, 3), (4, 2)}

Exercise 11.1

1. In a large school, a student is selected at random. Give a reasonable
   sample space for answers to each of the following questions:
   (a) Are you left-handed or right-handed?
   (b) What is your height in centimetres?
   (c) For how many minutes did you study last night?

2. You throw a coin and a standard six-sided dice and record the number
   and the face that appear in that order. For example, (5, h) represents a
   5 on the dice and a head on the coin. List the sample space.

3. You draw cards from a standard deck of 52 playing cards.
   (a) List the sample space if you draw one card at a time.
   (b) List the sample space if you draw two cards at a time.
   (c) How many outcomes do you have in each of the experiments above?

4. Tim carried out an experiment where he flipped 20 coins together and
   observed the number of heads showing. He repeated this experiment
   10 times and got the following results:
   11, 9, 10, 8, 13, 9, 6, 7, 10, 11
   (a) Use Tim's data to calculate the probability of obtaining a head.
   (b) Tim flipped the 20 coins for the 11th time. How many heads should
       he expect to observe?
   (c) If he flipped the coins 1000 times, how many heads should he expect
       to observe?

5. In the game 'Dungeons and Dragons', a four-sided dice with sides marked 1, 2, 3, and 4 is used. A character's intelligence is determined by rolling the dice twice and adding 1 to the sum of the two rolls.

   (a) What is the sample space for rolling the dice twice?

   (b) What is the sample space for a character's intelligence?

6. A box contains a yellow ball, a green ball, and a blue ball. You run an experiment where you draw a ball at random, look at its colour and then replace it and draw a second ball at random.

   (a) What is the sample space of this experiment?

   (b) What is the event of drawing yellow first?

   (c) What is the event of drawing the same colour twice?

7. Repeat the same exercise as in question 6, except this time without replacing the first ball.

8. Jane flips a coin three times and each time she notes whether it is heads or tails.

   (a) What is the sample space of this experiment?

   (b) What is the event that heads occur more often than tails?

9. Franz lives in Vienna. He and his family decided that their next vacation will be to either Italy or Hungary. If they go to Italy, they can fly, drive, or take the train. If they go to Hungary, they will drive or take a boat. Letting the outcome of the experiment be the location of their vacation and their mode of travel, list all the points in the sample space. Also list the sample space of the event 'fly to destination.'

10. A hospital codes patients according to whether they have or do not have health insurance, and according to their condition. The condition of the patient is rated as good (g), fair (f), serious (s), or critical (c). The hospital clerk marks a 0 for a non-insured patient and a 1 for an insured patient, and one of the above letters for the patient's condition. For example, (1, c) means an insured patient with critical condition.

   (a) List the sample space of this experiment.

   (b) What is the event: not insured, in serious or critical condition?

   (c) What is the event: patient in good or fair condition?

   (d) What is the event: patient has insurance?

11. A social study investigates people for different characteristics. One part of the study classifies people according to gender (G_1 = female, G_2 = male), drinking habits (K_1 = abstain, K_2 = drinks occasionally, K_3 = drinks frequently), and marital status (M_1 = married, M_2 = single, M_3 = divorced, M_4 = widowed).

   (a) List the elements of an appropriate sample space for observing a person in this study.

(b) Three events are defined as:

A = the person is a male, B = the person drinks, and C = the person is single

List the elements of each event A, B, and C.

(c) Interpret each event in the context of this situation:

(i) $A \cup B$ **(ii)** $A \cap C$ **(iii)** C' **(iv)** $A \cap B \cap C$ **(v)** $A' \cap B$

12. Cars leaving the highway can make a right turn (R), left turn (L), or go straight on (S). You are collecting data on traffic patterns at this intersection and you group your observations by taking four cars at a time every 5 minutes.

(a) List three outcomes in your sample space U. How many are there?

(b) List the outcomes in the event that all cars go in the same direction.

(c) List the outcomes that only two cars turn right.

(d) List the outcomes that only two cars go in the same direction.

13. You are collecting data on traffic at an intersection for vehicles leaving a highway. Your task is to collect information about the type of vehicle: truck (T), bus (B), or car (C). You are also recording whether the driver is wearing a seat belt (SY) or is not wearing seat belt (SN), as well as whether the headlights are on (O) or off (F).

(a) List the outcomes of your sample space, U.

(b) List the outcomes of the event SY (the driver is wearing a seat belt).

(c) List the outcomes of the event C (the vehicle is a car).

(d) List the outcomes of the event in $C \cap SY$, C', and $C \cup SY$.

14. Many electric systems use a built-in backup system so that the equipment using the system will work even if some parts fail. Such a system in given in the diagram.

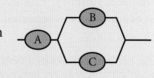

Two parts of this system are installed in parallel, so that the system will work if at least one of them works. If we code a working system by 1 and a failing system by 0, then one of the outcomes would be (1, 0, 1), which means parts A and C work while B failed.

(a) List the outcomes of your sample space, U.

(b) List the outcomes of the event X, that exactly 2 of the parts work.

(c) List the outcomes of the event Y, that at least 2 of the parts work.

(d) List the outcomes of the event Z, that the system functions.

(e) List the outcomes of the events:

(i) Z' **(ii)** $X \cup Z$ **(iii)** $X \cap Z$ **(iv)** $Y \cup Z$ **(v)** $Y \cap Z$

15. Your school library has five copies of George Polya's *How To Solve It*. Copies 1 and 2 are first edition, and copies 3, 4, and 5 are second edition. You are searching for a first edition book, and you will stop when you find a copy. For example, if you find copy 2 immediately, then the outcome is 2. Outcome 542 represents the outcome that a first edition was found on the third attempt.
   (a) List the outcomes of your sample space, U.
   (b) List the outcomes of the event A, that two books must be searched.
   (c) List the outcomes of the event B, that at least two books must be searched.
   (d) List the outcomes of the event C, that book 1 is found.

11.2 Probability assignments

There are a few theories of probability that assign meaning to statements like 'the probability that event A occurs is $p\%$.' In this book we will primarily examine only the **relative frequency theory**. In essence, we will follow the idea that probability is 'the long-run proportion of repetitions on which an event occurs.' This allows us to 'merge' two concepts into one:

Equally likely outcomes: In the theory of equally likely outcomes, probability has to do with symmetries and the indistinguishability of outcomes. If a given experiment or trial has n possible outcomes among which there is no preference, they are equally likely. Probabilities are between 0% and 100%. If an event consists of more than one possible outcome, the chance of the event is the number of ways it can occur, divided by the total number of things that could occur.

Frequency theory: Probability is the limit of the relative frequency with which an event occurs in repeated trials. Relative frequencies are always between 0% and 100%.

The probability of any event is the number of elements in an event A divided by the total number of elements in the sample space U.

$$P(A) = \frac{n(A)}{n(U)},$$

where $n(A)$ represents the number of outcomes in event A and $n(U)$ represents the total number of outcomes.

Probability rules

The basic rules of probability are:

Rule 1

Any probability is a number between 0 and 1; that is, the probability $P(A)$ of any event A satisfies $0 \leq P(A) \leq 1$. If the probability of an event is 0, the event never occurs. Likewise, if the probability is 1, it always occurs.

Remember that there is no such thing as negative probability or a probability greater than 1.

Rule 2

All possible outcomes together must have probability 1; that is, the probability of the sample space U is 1: $P(U) = 1$

Rule 3

If two events have no outcomes in common, the probability that one or the other occurs is the sum of their individual probabilities. Two events that have no outcomes in common, and hence can never occur together are called **disjoint** events or **mutually exclusive** events.

$$P(A \text{ or } B) = P(A) + P(B)$$

This is the **addition rule for mutually exclusive events.**

Rule 4

The event that contains the outcomes **not in A** is called the **complement** of A and is denoted by A'.

$$P(A') = 1 - P(A), \text{ or } P(A) = 1 - P(A').$$

No matter how large a chance you think an event has, there is no such thing as a probability larger than 1.

Example 11.5

When people create codes for their smartphones, the first digits follow distributions very similar to this one:

First digit	0	1	2	3	4	5	6	7	8	9
Probability	0.009	0.300	0.174	0.122	0.096	0.078	0.067	0.058	0.051	0.045

(a) Find the probabilities of the events:

$A = \{\text{first digit is 1}\}$

$B = \{\text{first digit is more than 5}\}$

$C = \{\text{first digit is an odd number}\}$

(b) Find the probability that the first digit is (i) 1 or greater than 5 (ii) not 1 (iii) an odd number or a number larger than 5.

Be careful with these rules: By the 'something has to happen' rule, the total of the probabilities of all possible outcomes must be 1. This is because they are disjoint, and their sum covers all the elements of the sample space. Suppose someone reports the following probabilities for students in your high school (4 years). The probability that a grade 1, 2, 3, or 4 student is chosen at random from the high school is 0.24, 0.24, 0.25, and 0.19 respectively with no other possibilities – you should know immediately that there is something wrong. These probabilities add up to 0.92. Similarly, if someone claims that these probabilities are 0.24, 0.28, 0.25, 0.26 respectively, there is also something wrong. These probabilities add up to 1.03, which is more than 1.

Solution

(a) From the table:

$P(A) = 0.300$

$P(B) = P(6) + P(7) + P(8) + P(9)$

$\quad = 0.067 + 0.058 + 0.051 + 0.045 = 0.221$

$P(C) = P(1) + P(3) + P(5) + P(7) + P(9)$

$\quad = 0.300 + 0.122 + 0.078 + 0.058 + 0.045$

$\quad = 0.603$

(b) (i) Since A and B are mutually exclusive, by the addition rule, the probability that the first digit is 1 or greater than 5 is

$\quad P(A \text{ or } B) = 0.300 + 0.221 = 0.521$

(ii) Using the complement rule, the probability that the first digit is not 1 is

$$P(A') = 1 - P(A) = 1 - 0.300 = 0.700$$

(iii) The probability that the first digit is an odd number or a number larger than 5:

$$P(B \text{ or } C) = P(1) + P(3) + P(5) + P(6) + P(7) + P(8) + P(9)$$
$$= 0.300 + 0.122 + 0.078 + 0.067 + 0.058 + 0.051 + 0.045$$
$$= 0.721$$

Notice here that $P(B \text{ or } C)$ is not the sum of $P(B)$ and $P(C)$ because B and C are not mutually exclusive.

Equally likely outcomes

In some cases, we are able to assume that individual outcomes are equally likely because of some balance in the experiment. Flipping a balanced coin renders heads or tails with an equally likely probability of 50%, and rolling a standard balanced dice gives the numbers from 1 to 6 as equally likely, with each having a probability of $\frac{1}{6}$.

Suppose that in Example 11.5 we consider all the digits to be equally likely to be selected, then our table would be:

First digit	0	1	2	3	4	5	6	7	8	9
Probability	0.1	0.1	0.1	0.1	0.1	0.1	0.1	0.1	0.1	0.1

Table 11.3 Probability if all digits are equally likely to be selected

$$P(A) = 0.1$$

$$P(B) = P(6) + P(7) + P(8) + P(9) = 4 \times 0.1 = 0.4$$

$$P(C) = P(1) + P(3) + P(5) + P(7) + P(9) = 5 \times 0.1 = 0.5$$

Also, by the complement rule, the probability that the first digit is not 1 is

$$P(A') = 1 - P(A) = 1 - 0.1 = 0.9$$

2-dimensional grids are also very helpful tools that are used to visualise 2-stage or sequential probability models. For example, consider rolling a normal unbiased six-sided dice twice. Figure 11.7 shows some events and how to use the grid in calculating their probabilities.

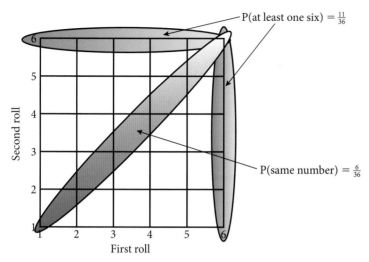

$P(\text{at least one six}) = \frac{11}{36}$

$P(\text{same number}) = \frac{6}{36}$

Figure 11.7 2D grid for calculating the probability of events when rolling a dice twice

If we are interested in the probability that at least one roll shows a 6, we count the points in the column corresponding to 6 on the first roll and the points in the row corresponding to 6 on the second roll, while keeping in mind that the point in the corner should not be counted twice.

If we are interested in the number showing on both rolls to be the same, then we count the points on the diagonal as shown.

If we are interested in the probability that the first roll shows a number larger than the second roll, then we pick the points below the diagonal.

Hence $P(\text{first number} > \text{second number}) = \dfrac{15}{36}$

Geometric probability

Some cases give rise to interpreting events as areas in the plane. Take, for example, shooting at a circular target at random. What is the probability of hitting the central part?

The probability of hitting the central part is given by

$$P = \frac{\pi \left(\dfrac{R}{4}\right)^2}{\pi R^2} = \frac{1}{16}$$

This calculation comes from the area of the small circle over the area of the whole target.

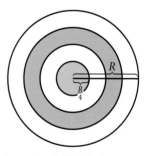

Figure 11.8 What is the probability of hitting the central part of the target?

Example 11.6

Lydia and Rania agreed to meet between 12:00 and 13:00. The first person to arrive will wait for at most 15 minutes. If the second person does not show up in this time, the first person will leave and they will arrange to meet at some other time. Assuming that their arrival times are random, what is the probability that they meet?

Solution

If Lydia arrives x minutes after 12:00 and Rania arrives y minutes after 12:00, then the condition for them to meet is $|x - y| \leq 15$, and $x \leq 60, y \leq 60$

Geometrically, the outcomes of their encounter region is given in the shaded region in the figure.

The area for each triangle is $\frac{1}{2}bh = \frac{1}{2}(45)^2$, so, the shaded area is $60^2 - 45^2$.

The probability they meet is therefore

$$\frac{60^2 - 45^2}{60^2} = \frac{7}{16}$$

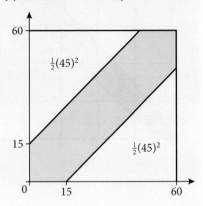

In an experiment where all outcomes are equally likely, the theoretical probability of an event A is given by

$$P(A) = \frac{n(A)}{n(U)}$$

where $n(A)$ is the number of outcomes that make up the event A, and $n(U)$ is the total number of outcomes in the sample space. The new ideas we want to discuss here involve the calculation of $n(A)$ and $n(U)$.

Probability calculation for equally likely outcomes using counting principles

Example 11.7

In a group of 18 students, 8 are females. Find the probability of choosing 5 students at random

(a) with all girls

(b) with three girls and two boys

(c) with at least one boy.

Solution

The total number of outcomes is the number of ways we can choose 5 out of the 18 students:

$$n(U) = {}_{18}C_5 = 8568$$

(a) This event will require that we select our group from among the 8 girls:

$$n(A) = {}_8C_5 = 56 \Rightarrow P(A) = \frac{56}{8568} = 0.0065$$

(b) This event will require that we select 3 out of the 8 girls, and at the same time we pick 2 out of the 10 boys. So, using the multiplication principle:

$$n(B) = {}_8C_3 \cdot {}_{10}C_2 = 56 \cdot 45 = 2520 \Rightarrow P(B) = \frac{2520}{8568} = 0.294$$

(c) This event can be approached in two ways:

To have at least 1 boy means that we can have 1, 2, 3, 4, or 5 boys. These are mutually exclusive, so the probability in question is the sum

Did you observe that ${}_8C_5 = {}_8C_3$? Why?

$$P(C) = \frac{{}_{10}C_1 \, {}_8C_4 + {}_{10}C_2 \, {}_8C_3 + {}_{10}C_3 \, {}_8C_2 + {}_{10}C_4 \, {}_8C_1 + {}_{10}C_5 \, {}_8C_0}{{}_{18}C_5}$$

$$= \frac{8512}{8568} = 0.9935, \text{ or}$$

Recognise that 'at least 1 boy' is the complement of no boys at all, i.e., 0 boys or all 5 girls.

$$P(C) = 1 - P(A) = 1 - 0.0065 = 0.9935$$

Example 11.8

A deck of playing cards has 52 cards. In a game, the player is given 5 cards.

(a) Find the probability of getting three cards of one denomination and two cards of another – (three 7s and two jacks for example).

This game can be played in two stages, First, the player is given 5 cards, and then the player can decide to exchange some of the cards. (The cards exchanged are discarded and not returned to the deck.)

A player was given the following hand: Q♠, Q♦, Q♥, 4♣, 9♠.
She decided to change the last two cards.

(b) Find the probability of

    (i)   three cards of one denomination and two cards of another

    (ii) having four queens.

Solution

(a) The sample space consists of all possible 5-card hands:

$$n(U) = \binom{52}{5} = {}_{52}C_5 = 2\,598\,960$$

Call the event of interest event A.

As there are 13 denominations in the deck, there are 13 choices for the first required denomination. Once a denomination is chosen, say 9, then there are ${}_4C_3$ ways of choosing three cards out of the four. Using the multiplication rule, there are $13 \cdot {}_4C_3$ ways of choosing three cards of the first denomination.

We are now left with 12 possible denominations for the second one, and each provides ${}_4C_2$ ways of getting two of the cards and, hence, using the multiplication rule, there are $12 \cdot {}_4C_2$ ways of choosing the cards for the second denomination. Again, using the multiplication rule, we will have $[13 \cdot {}_4C_3][12 \cdot {}_4C_2] = 3744$ ways of choosing the first and second denominations.

The requested probability is then

$$P(A) = \frac{3744}{2\,598\,960} \approx 0.00144$$

(b) (i) Since we have three queens, then we need only look for two cards of a different denomination. Now, there are only 47 cards left in the deck because we had 5 already. So the sample space has $n(U) = {}_{47}C_2 = 1081$ ways of getting the rest of the 5 cards. The other cards could be two 4s, two 9s or two of the rest of the 10 denominations.

We have ${}_3C_2 = 3$ ways of getting two 4s since 4♣ is already discarded. We also have ${}_3C_2 = 3$ ways to get two 9s. Or, for each of the other 10 denominations (no Q, no 4, and no 9), we have ${}_4C_2 = 6$ different ways of getting two of them; that is, we have $10 \cdot {}_4C_2 = 60$ different ways of getting two cards of the same denomination other than Q, 4, or 9.

So, the total number of ways of getting two cards of the same denomination is $3 + 3 + 60 = 66$ ways.

So, the required probability is $P(A) = \dfrac{n(A)}{n(U)} = \dfrac{66}{1081} \approx 0.0611$

(ii) To have four Qs we only have to look for one, and there is only one way of getting the missing Q♣. That leaves us with one card to be chosen from the 46 cards left.

Therefore $P(A) = \dfrac{46}{1081} \approx 0.0426$

Exercise 11.2

1. In a simple experiment, 20 chips with integers 1 to 20 inclusive were placed in a box and one chip was picked at random.
   (a) What is the probability that the number drawn is a multiple of 3?
   (b) What is the probability that the number drawn is not a multiple of 4?

2. The probability that an event A happens is 0.37
   (a) What is the probability that it does not happen?
   (b) What is the probability that it may or may not happen?

3. You are playing with an ordinary deck of 52 cards by drawing cards at random and looking at them.
   (a) Find the probability that the card you draw is
      (i)   the ace of hearts   (ii) the ace of hearts or any spade
      (iii) an ace or any heart   (iv) not a face card.
   (b) Now you draw the ten of diamonds and put it on the table and draw a second card. Find the probability that the second card
      (i)   is the ace of hearts   (ii) is not a face card.

(c) Now you draw the ten of diamonds and return it to the deck and draw a second card. Find the probability that the second card

(i) is the ace of hearts   (ii) is not a face card.

4. On Monday morning, my class wanted to know how many hours students spent studying on Sunday night. They stopped schoolmates at random as they arrived and asked each, 'How many hours did you study last night?' Here are the results of the sample they chose on Monday 15 January 2018.

Hours spent studying	0	1	2	3	4	5
Number of students	4	12	8	3	2	1

(a) Estimate the probability that a student selected at random spent less than three hours studying on Sunday night.

(b) Estimate the probability that a student studied for two or three hours.

(c) Estimate the probability that a student studied for less than six hours.

5. You throw a coin and a standard six-sided dice and record the number and the face that appear.

(a) Find the probability of getting a number larger than 3.

(b) Find the probability of getting a head and a 6.

6. A dice is constructed in such a way that a 1 has a chance of occurring twice as often as any other number.

(a) Find the probability that a 5 appears.

(b) Find the probability an odd number appears.

7. You are given two fair dice to roll in an experiment.

(a) Your first task is to report the numbers you observe.

(i) What is the sample space of your experiment?

(ii) What is the probability that the two numbers are the same?

(iii) What is the probability that the two numbers differ by 2?

(iv) What is the probability that the two numbers are not the same?

(b) Your second task is to report the sum of the numbers that appear.

(i) What is the probability that the sum is 1?

(ii) What is the probability that the sum is 9?

(iii) What is the probability that the sum is 8?

(iv) What is the probability that the sum is 13?

8. The blood types of people can be one of the four types: O, A, B, or AB. The distribution of people with these types differs from one group of people to another. Here are the distributions of blood types for randomly chosen people in the USA, China, and Russia.

Blood Type / Country	O	A	B	AB
USA	0.43	0.41	0.12	?
China	0.36	0.27	0.26	0.11
Russia	0.39	0.34	?	0.09

   (a) What is the probability of type AB in the USA?

   (b) Dirk lives in the USA and has type B blood. People with type B blood can receive blood only from people with type O or type B. What is the probability that a randomly chosen US citizen can donate blood to Dirk?

   (c) What is the probability of randomly choosing an American and a Chinese with type O blood?

   (d) What is the probability of randomly choosing an American, a Chinese and a Russian with type O blood?

   (e) What is the probability of randomly choosing an American, a Chinese and a Russian with the same blood type?

9. In each of the following situations, state whether or not the given assignment of probabilities to individual outcomes is legitimate. Give reasons for your answer.

   (a) A dice is loaded such that the probability of each face is according to the following assignment, where x is the number on the upper face and P(x) is its probability.

x	1	2	3	4	5	6
P(x)	0	$\frac{1}{6}$	$\frac{1}{3}$	$\frac{1}{3}$	$\frac{1}{6}$	0

   (b) Students at a school are categorised in terms of gender and whether or not they are diploma candidates.

   P(female, diploma candidate) = 0.57

   P(female, not a diploma candidate) = 0.23

   P(male, diploma candidate) = 0.43

   P(male, not a diploma candidate) = 0.18

   (c) Draw a card from a deck of 52 cards (x is the suit of the card and P(x) is its probability).

x	Hearts	Spades	Diamonds	Clubs
P(x)	$\frac{12}{52}$	$\frac{15}{52}$	$\frac{12}{52}$	$\frac{13}{52}$

10. In Switzerland, there are three official languages: German, French, and Italian. You choose a Swiss citizen at random and ask, 'What is your mother tongue?' Here is the distribution of responses:

Language	German	French	Italian	Other
Probability	0.58	0.24	0.12	?

(a) What is the probability that a Swiss person's mother tongue is not one of the official ones?

(b) What is the probability that a Swiss person's mother tongue is not German?

(c) What is the probability that you choose two Swiss people independently of each other and they both have German as their mother tongue?

(d) What is the probability that you choose two Swiss people independently of each other and they both have the same mother tongue?

11. The majority of email messages are now spam. Choose a spam email message at random. Here is the distribution of topics.

Topic	Adult	Financial	Health	Leisure	Products	Scams
Probability	0.165	0.142	0.075	0.081	0.209	0.145

(a) What is the probability of choosing a spam message and it does not concern any of these topics?

(b) Parents are usually concerned with spam messages that either have adult content or are scams. What is the probability that a randomly chosen spam email falls into one of the other categories?

12. Consider n to be a positive integer. Let $f(x) = \dfrac{{}_nC_{x+1}}{{}_nC_x}$, where x is also a positive integer. Determine the values of x (in terms of n) for which $f(x) < 1$.

13. Determine n for each of

(a) ${}_nC_2 = 190$

(b) ${}_nC_4 = {}_nC_8$

14. An experiment involves tossing a pair of dice, 1 white and 1 red, and recording the numbers that come up. Find the probability

(a) that the sum is greater than 8

(b) that a number greater than 4 appears on the white dice

(c) that a total of 5 or less appears.

15. Three books are picked from a shelf containing five novels, three science books and a thesaurus. Find the probability that

(a) the thesaurus is selected

(b) two novels and a science book are selected.

16. 5 cards are chosen at random from a deck of 52 cards. Find the probability that the 5 cards contain

(a) 3 kings   **(b)** 4 hearts and 1 diamond.

17. A class consists of 10 girls and 12 boys. A team of six members is to be chosen at random. Find the probability that the team contains

(a) one boy   **(b)** more boys than girls.

18. A committee of six people is to be chosen from a group of 15 people that contains two married couples.

(a) What is the probability that the committee will include both of the married couples?

(b) What is the probability that the committee will include the three youngest members in the group?

19. The computer department at your school has received a shipment of 25 printers, of which 10 are colour laser printers and the rest are black-and-white laser models. Six printers are selected at random to be checked for defects.

(a) Find the probability that exactly 3 of them are colour lasers.

(b) Find the probability that at least 3 are colour lasers.

20. The city of Graz just bought 30 buses from a certain company and put them into service in their public transport network. Shortly after the first week in service, 10 buses developed cracks on their instruments panel.

(a) How many ways are there to select a sample of six buses for a thorough inspection?

(b) What is the probability that half of the chosen buses have cracks?

(c) What is the probability that at least half of the chosen buses have cracks?

(d) What is the probability that at most half of the chosen buses have cracks?

21. A small factory has a 24-hour production facility. They employ 30 workers on the day shift, 22 workers on the evening shift, and 15 workers on the morning shift. A quality control consultant is to select 9 workers for in-depth interviews.

(a) What is the probability that all 9 come from the day shift?

(b) What is the probability that all 9 come from the same shift?

(c) What is the probability that at least two of the shifts are represented?

(d) What is the probability that at least one of the shifts is unrepresented?

22. **(a)** A box contains 8 chips numbered 1 to 8. Two are chosen at random and their numbers are added together. What is the probability that their sum is 7?

(b) A box contains 20 chips numbered 1 to 20. Two chips are chosen at random. What is the probability that the numbers differ by 3?

(c) A box contains 20 chips numbered 1 to 20. Two chips are chosen at random. What is the probability that the numbers on the two chips differ by more than 3?

23. Tim and Val want to meet for dinner at their favourite restaurant. They agreed on meeting between 20:00 and 21:00. The first person to arrive will order a salad and spend 30 minutes before ordering the main meal. If the second person does not arrive within the 30 minutes, the first person will pay the bill and leave. What is the probability they manage to eat dinner together at the restaurant?

24. Bus 48A and Tram 49 serve different routes in the city. They share one stop next to my house. They stop at this station every 20 minutes. Every stay is 3 minutes long. Assuming their arrivals are random, find the probability that both are at the stop at 12:00 on Monday.

25. During a dinner party, Magda plans on having her guests play charades. On slips of paper, Magda has written down the names of 8 films, 10 book titles, and 12 songs. The first guest chooses six slips of paper at random.

(a) What is the probability that two of each type get selected?

(b) What is the probability that all slips of paper are of the same type?

(c) What is the probability of only films and songs being selected?

26. Owen paints the faces of a wooden cube green. He cuts the cube into 1000 small cubes of equal size. He mixes the small cubes thoroughly. He draws one cube at random. Find the probability that the cube:

(a) has two faces coloured green

(b) has three coloured faces

(c) does not have a coloured face at all.

11.3 Operations with events

In Example 11.5, we considered the following events:

$B = \{$first digit is more than 5$\}$

$C = \{$first digit is an odd number$\}$

We also claimed that these two events are not mutually exclusive. This brings us to another concept for looking at combined events.

The intersection of two events B and C, denoted by the symbol $B \cap C$ or simply BC, is the event containing all outcomes common to B and C.

In Figure 11.9, $B \cap C = \{7, 9\}$ because these outcomes are in both B and C. Since the intersection has outcomes common to the two events B and C, they are not mutually exclusive.

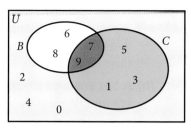

Figure 11.9 B and C are not mutually exclusive events

The probability of $B \cap C$ is $0.058 + 0.045 = 0.103$. The probability of B or C is not simply the sum of the two probabilities. How can we find the probability of B or C when they are not mutually exclusive? To answer this question, we need to define another operation.

Here $B \cup C = \{1, 3, 5, 6, 7, 8, 9\}$. In calculating the probability of $B \cup C$, we observe that the outcomes 7 and 9 are counted twice. To remedy the situation, if we decide to add the probabilities of B and C, we subtract one of the incidents of double counting. So,

$P(B \cup C) = 0.221 + 0.603 - 0.103 = 0.721$, which is the result we received with direct calculation. In general, we can state the following probability rule.

The union of two events B and C, $(B \cup C)$, is the event containing all the outcomes that belong to B or to C or to both.

For any two events A and B,
$P(A \cup B) = P(A) + P(B) - P(A \cap B)$.
As we see from Figure 11.10, $P(A \cap B)$ has been added twice, so the 'extra' one is subtracted to give the probability of $(A \cup B)$.

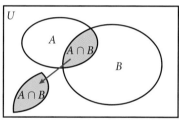

Figure 11.10 $P(A \cap B)$ has been added twice, so the 'extra' one is subtracted to give the probability of $(A \cup B)$

This general probability addition rule applies to the case of mutually exclusive events too. Consider any two mutually exclusive events A and B. The probability of A or B is given by

$P(A \cup B) = P(A) + P(B) - P(A \cap B) = P(A \cup B) = P(A) + P(B)$ since $P(A \cap B) = 0$.

Some useful results

1. $P(A \cap B) = P(B \cap A)$

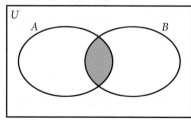

2. $P(A) = P(A \cap B) + P(A \cap B')$

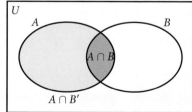

3. $P(B) = P(A \cap B) + P(A' \cap B)$

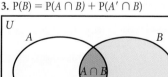

4. $P(A' \cap B') = 1 - P(A \cup B)$

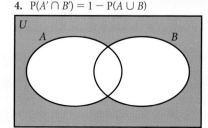

The simple multiplication rule

Consider the following situation: At a large school, 55% of the students are male. It is also known that the percentage of cyclists among males and females in this school is the same at 22%. What is the probability of selecting a student at random from this population and the student is a male cyclist?

Applying common sense only, we can think of the problem in the following manner. Since the proportion of cyclists is the same in both groups, cycling and gender are independent of each other in the sense that knowing that the student is a male does not influence the probability that he is a cyclist.

The chance that we pick a male student is 55%. From those 55% of the population, we know that 22% are cyclists, so, by simple arithmetic, the chance that we select a male cyclist is $0.22 \times 0.55 = 0.121\%$.

This is an example of the multiplication rule for independent events.

Two events A and B are independent if knowing that one of them occurs does not change the probability that the other occurs.
The multiplication rule for independent events is:
If two events A and B are **independent**, then
$P(A \cap B) = P(A) \times P(B)$

Example 11.9

There is a traffic light just before I get to school. The probability that I find the light green is 0.3.

Find the probability that I find it green on:

(a) two consecutive days (b) every school day one week.

Solution

(a) We can assume that my arrival and finding the light green is a random event and that if I find it green on one day, this does not influence how I find it the next day. In that case our calculation is very simple:

P(green the first and second day) = P(green first day) $\times$ P(green second day)

$$= 0.30 \times 0.30 = 0.09$$

(b) This rule can also be extended to more than two independent events.

P(green every day) = $0.3 \times 0.3 \times 0.3 \times 0.3 \times 0.3 = 0.00243$

Do not confuse independent with mutually exclusive. Mutually exclusive means that if one of the events occurs, then the other does not occur, while independent means that knowing one of the events occurs does not influence whether the other occurs.

Example 11.10

Computers bought from one company require repairs quite frequently. It is estimated that 17% of computers bought from one of these companies require one repair during the first month after purchase, 7% will need repairs twice during the first month, and 4% require three or more repairs.

(a) Find the probability that a computer chosen at random from this producer will need:

  (i) no repairs (ii) no more than one repair (iii) some repair.

(b) If you buy two such computers, finds the probability that

  (i) neither will require repair (ii) both will need repair.

Solution

(a) Since all of the events listed are disjoint, the addition rule can be used.

   (i) P(no repairs) = 1 – P(some repairs)

   $$= 1 - (0.17 + 0.07 + 0.04) = 1 - (0.28) = 0.72$$

   (ii) P(no more than one repair) = P(no repairs or one repair)

   $$= 0.72 + 0.17 = 0.89$$

   (iii) P(some repairs) = P(one or two or three or more repairs)

   $$= 0.17 + 0.07 + 0.04 = 0.28$$

(b) Since repairs on the two computers are independent from one another, the multiplication rule can be used. Use the probabilities of events from part (a) in the calculations.

   (i) P(neither will need repair) = (0.72)(0.72) = 0.5184

   (ii) P(both will need repair) = (0.28)(0.28) = 0.0784

Conditional probability

In probability, conditioning means incorporating new restrictions on the outcome of an experiment – updating probabilities to take into account new information. We will look at how conditional probability can be used to solve complicated problems.

Example 11.11

A public health department wants to study the exercise habits of high school students. They interview 768 students from grades 10 to 12 and ask them about their exercise habits. They categorised the students into the following three categories: frequent exercisers, occasional exercisers, and non-exercisers. The results are summarised in the table.

	Frequent	Occasional	Non-exerciser	Total
Male	127	73	214	414
Female	99	66	189	354
Total	226	139	403	768

If a student is selected at random from this study, find the probability that we select

(a) a female   (b) a male frequent exerciser   (c) a non-exerciser.

Solution

(a) $P(\text{Female}) = \dfrac{354}{768} = 0.461$

   So, 46.1% of our sample are females.

(b) Since we have 127 boys categorised as frequent exercisers, the chance of a male who exercises frequently will be

$$P(\text{male exercises frequently}) = \frac{127}{768} = 0.165$$

(c) $P(\text{non-exerciser}) = \frac{403}{768} = 0.525$

In Example 11.11, what if we know that the selected student is a girl? Does that influence the probability that the selected student is a non-exerciser? Yes it does.

Knowing that the selected student is a female changes our choices. The revised sample space is not made up of all students anymore. It is only the female students. The chance of finding a non-exerciser among the females is $\frac{189}{354} = 0.534$. That is, 53.4% of the females are non-exercisers as compared to the 52.5% of non-exercisers in the whole population.

This probability is called a conditional probability, and we write this as

$$P(\text{non-exerciser}|\text{Female}) = \frac{189}{354}$$

We read this as 'Probability of selecting a non-exerciser **given that** we have selected a female'.

The conditional probability of A given B, P(A|B), is the probability of the event A, updated on the basis of the knowledge that the event B occurred. Suppose that A is an event with probability $P(A) = p \neq 0$, and that $A \cap B = \emptyset$ (A and B are mutually exclusive). Then if we learn that B occurred, we know A did not occur, so we should revise the probability of A to be zero, $P(A|B) = 0$ (the conditional probability of A given B is zero).

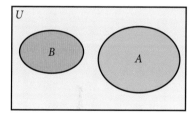

Figure 11.11 $P(A|B) = 0$

On the other hand, suppose that $A \cap B = B$ (B is a subset of A, so B implies A). Then if we learn that B occurred, we know A must have occurred as well, so we should revise the probability of A to be 100%, $P(A|B) = 1$ (the conditional probability of A given B is 100%).

Remember that the probability we assign to an event can change if we know that some other event has occurred. This idea is the key to understanding conditional probability.

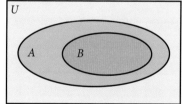

Figure 11.12 $P(A|B) = 1$

Imagine the following scenario:

You are playing cards and your opponent is about to give you a card. What is the probability that the card you receive is a queen?

There are 52 cards in the deck, and 4 of these cards are queens. So, assuming that the deck was thoroughly shuffled, the probability of receiving a queen is

$$P(\text{queen}) = \frac{4}{52} = \frac{1}{13}$$

This calculation assumes that you know nothing about any cards already dealt from the deck.

Suppose now that you are looking at the five cards you have in your hand, and one of them is a queen. You know nothing about the other 47 cards except that exactly three queens are among them. The probability of being given a queen as the next card, given what you know, is

$$P(\text{queen}|1 \text{ queen in hand}) = \frac{3}{47} \neq \frac{1}{13}$$

So, knowing that there is one queen among your five cards changes the probability of the next card being a queen.

Consider Example 11.11 again. We want to express the table frequencies as relative frequencies or probabilities. Our table will look like this:

	Frequent	**Occasional**	**Non-exerciser**
Male	0.165	0.095	0.279
Female	0.129	0.086	**0.246**

Table 11.3 Relative probabilities

To find the probability of selecting a student at random and finding that the student is a non-exerciser female, we look at the intersection of the female row with the non-exerciser column and find that this probability is 0.246.

Looking at this calculation from a different perspective, we can think about it in the following manner. We know that the percentage of females in our sample is 46.1 and, among those females, we found that 53.4% of those are non-exercisers. So, the percentage of female non-exercisers in the population is the 53.4% of those 46.1% females; that is, $0.534 \times 0.461 = 0.246$

In terms of events, this can be read as:

$$P(\text{non-exerciser}|\text{female}) \times P(\text{female}) = P(\text{female and non-exerciser}) \text{ or}$$
$$= P(\text{female} \cap \text{non-exerciser})$$

Given any events A and B, the probability that both events happen is given by

$P(A \cap B) = P(A|B) \times P(B)$

The previous discussion is an example of the **multiplication rule** of any two events A and B.

Example 11.12

Six green toys and four red toys (identical apart from colour) are placed in a container. A child is asked to select two toys at random. What is the probability that the child chooses two green toys?

Solution

To solve this problem, we can draw a tree diagram.

Each entry on the branches has a conditional probability. So, red on the second choice is, in fact, either Red|Red or Red|Green. We are interested in RR, so the probability is

$$P(RR) = P(R) \times P(R|R) = \frac{4}{10} \times \frac{3}{9} = 13.3\%$$

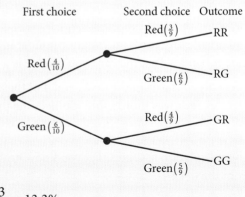

When $P(A \cap B) = P(A|B) \times P(B)$, and $P(B) \neq 0$, we can rearrange the multiplication rule to produce a definition of the conditional probability $P(A|B)$ in terms of the unconditional probabilities $P(A \cap B)$ and $P(B)$.

Why does this formula make sense?

First of all, note that it does agree with the intuitive answers we found above: if $A \cap B = \emptyset$, then $P(A \cap B) = 0$, so $P(A|B) = 0|P(B) = 0$; and if $A \cap B = B$, $P(A|B) = P(B)/P(B) = 100\%$.

Now, if we learn that B occurred, we can restrict attention to just those outcomes that are in B, and disregard the rest of U, so we have a new sample space that is just B (see Figure 11.13). For A to have occurred in addition to B, requires that $A \cap B$ occurred, so the conditional probability of A given B is $P(A \cap B)/P(B)$, just as we defined it above.

When $P(B) \neq 0$, the conditional probability of A given B is

$$P(A|B) = \frac{P(A \cap B)}{P(B)}$$

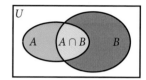

Figure 11.13 The conditional probability of A given B is $P(A \cap B)/P(B)$

Example 11.13

In an experiment to study the phenomenon of colour blindness, researchers collected information concerning 1000 people in a small town and categorised them according to colour blindness and gender. Here is a summary of the findings:

	Male (M)	Female (F)	Total
Colour blind (C)	40	2	42
Not colour blind (N)	470	488	958
Total	510	490	1000

What is the probability that a person is colour blind, given that the person is a woman?

Solution

To answer this question, we notice that we do not have to search the whole population for this event. We limit our search to the women. We have 490 women. As we only need to consider women, then when we search for colour-blindness, we only look for the women who are colour blind, i.e., the intersection. Here we only have two women. Therefore, the chance we get a colour blind person, given the person is a woman is

$$P(C|F) = \frac{P(C \cap F)}{P(F)} = \frac{n(C \cap F)}{n(F)} = \frac{2}{490} = 0.004$$

Notice here that we used the frequency rather than the probability. However, these are equivalent since dividing by $n(U)$ will transform the frequency into a probability.

$$\frac{n(C \cap F)}{n(F)} = \frac{\dfrac{n(C \cap F)}{n(U)}}{\dfrac{n(F)}{n(U)}} = \frac{P(C \cap F)}{P(F)} = P(C|F)$$

Example 11.14

One national airline is known for its punctuality. The probability that a regularly scheduled flight departs on time is $P(D) = 0.83$, the probability that it arrives on time is $P(A) = 0.92$, and the probability that it arrives and departs on time, $P(A \cap D) = 0.78$. Find the probability that a flight

(a) arrives on time, given that it departed on time

(b) departs on time, given that it arrived on time.

Solution

(a) The probability that a flight arrives on time given that it departed on time is
$$P(A|D) = \frac{P(A \cap D)}{P(D)} = \frac{0.78}{0.83} = 0.94$$

(b) The probability that a flight departs on time given that it arrived on time
$$P(D|A) = \frac{P(D \cap A)}{P(A)} = \frac{0.78}{0.92} = 0.85$$

Independence

Two events are **independent** if learning that one occurred does not affect the chance that the other occurred. That is, if $P(A|B) = P(A)$, and vice versa.

This means that if we apply our definition to the general multiplication rule, then
$$P(A \cap B) = P(A|B) \times P(B) = P(A) \times P(B)$$
which is the multiplication rule for independent events we studied earlier.

These results give us some helpful tools in checking the independence of events.

> Two events are independent if and only if either $P(A \cap B) = P(A) \times P(B)$ or $P(A|B) = P(A)$ Otherwise, the events are not independent.

Example 11.15

Take another look at Example 11.14. Are the events of arriving on time (A) and departing on time (D) independent?

Solution

We can answer this question in two different ways:

(a) $P(A) = 0.92$, and we found that $P(A|D) = 0.94$. Since the two values are not the same, then we can say that the two events are not independent.

(b) Alternatively, $P(A \cap D) = 0.78$ and $P(A) \times P(D) = 0.92 \times 0.83 = 0.76$
$$\neq P(A \cap D)$$

Example 11.16

If a doctor suspects that a patient is lactose intolerant, they might give the patient a breath test or a blood test, or both, in order to make a diagnosis.

A study found the following results in a particular country:

81% of the patients suspected to be lactose intolerant were given a breath test, 40% a blood test, and 25% both tests.

(a) Find the probability that a patient suspected to be lactose intolerant is given:

    (i) at least one test   (ii) exactly one test   (iii) no test.

(b) Are the events 'giving the patient a breath test' and 'giving the patient a blood test' independent?

Solution

A Venn diagram can help explain the solution.

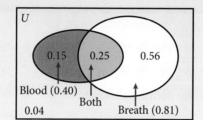

(a) (i) The probability that a driver receives a test means that he or she receives either a blood test, a breath test, or both tests. The probability can be calculated directly from the Venn diagram or by applying the addition rule. The diagram shows that if 81% receive the breath test and 25% are also given the blood test, then the remaining 56% do not receive a blood test. Similarly 15% of the blood test receivers do not get a breath test. So, the probability to receive a test is:

$0.56 + 0.25 + 0.15 = 0.96$

Also, if we apply the addition rule

P(Breath or Blood) = P(Breath) + P(Blood) − P(both)

                $= 0.81 + 0.40 − 0.25 = 0.96$

(ii) To receive exactly one test is to receive a blood test or a breath test, but not both. So, from the Venn diagram it is clear that this probability is $0.15 + 0.56 = 0.71$. To approach it differently, since we know that the union of the two events still contains the intersection, we can subtract the probability of the intersection from that of the union. Thus, $0.96 − 0.25 = 0.71$

(iii) To receive no test is equivalent to the complement of the union of the events. Hence, P(no test) = 1 − P(1 test) = 1 − 0.96 = 0.04

(b) To check for independence, we can use either of the two methods we tried before. Since all the necessary probabilities are given, we can use the product rule:

If they were independent, then

P(both tests) = P(Breath) × P(Blood) = 0.81 × 0.40 = 0.324, but

P(both tests) = 0.25. Therefore, the events of receiving a breath test and a blood test are not independent.

Example 11.17

Jane and Kate are long-time friends and frequently play tennis together. When Jane serves first, she wins 60% of the time; likewise, when Kate serves first, she wins 60% of the time. They alternate the person serving. They usually play for a prize, which is a chocolate bar. The first one who loses on her serve will have to buy the chocolate. Jane serves first.

(a) Find the probability that Jane pays on her second serve.

(b) Find the probability that Jane eventually pays for the chocolate.

(c) Find the probability that Kate pays for the chocolate.

Solution

A tree diagram can help in solving this problem. Let JW stand for Jane winning her serve, JL for Jane losing her serve and hence paying. KW and KL are defined similarly.

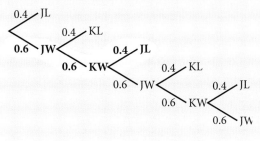

(a) For Jane to pay on her second serve, she should win her first serve. Kate must also win her first serve, and then Jane loses her second serve. See diagram above. The probability this happens is

$$P(JW) \cdot P(KW) \cdot P(JL) = 0.6 \times 0.6 \times 0.4 = 0.4 \cdot (0.6)^2 = 0.144$$

(b) For Jane to pay, she needs to be the first one to lose on her serve. This means she loses on the first serve or the second or the third, and so on. So, the probability that she pays is

$$P(J\ \text{pays}) = P(JL) + P(JW) \cdot P(KW) \cdot P(JL) + P(JW) \cdot P(KW) \cdot P(JW) \cdot P(KW) \cdot P(JL)$$

$$= 0.4 + 0.4 \cdot (0.6)^2 + 0.4 \cdot (0.6)^4 + \cdots$$

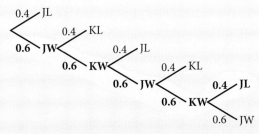

This appears to be the sum of an infinite geometric series with $(0.6)^2$ as common ratio, hence

$$P(J\ \text{pays}) = 0.4 + 0.4 \cdot (0.6)^2 + 0.4 \cdot (0.6)^4 + \cdots = \frac{0.4}{1 - (0.6)^2} = 0.625$$

(c) Obviously,

$$P(K \text{ pays}) = 1 - P(J \text{ pays}) = 1 - 0.625 = 0.375$$

This also gives us the opportunity to look at it differently. For Kate to pay, she needs to lose on her first serve, i.e., 0.6×0.4, or on her second, third, and so on,

$$P(K \text{ pays}) = 0.6 \times 0.4 + 0.6 \times (0.6)^2 \times 0.4 + 0.6 \times (0.6)^4 \times 0.4 + \cdots$$
$$= \frac{0.6 \times 0.4}{1 - (0.6)^2} = 0.375$$

Example 11.18

The figure here shows a target for a dart game. The radius of the board is 40 cm and it is divided into three regions as shown. You score 2 points if you hit the centre, 1 point for the middle region and 0 points for the outer region.

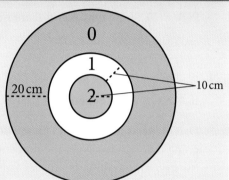

(a) What is the probability of scoring a 1 with one attempt?

(b) What is the probability of scoring a 2 with one attempt?

(c) How many attempts are necessary so that the probability of scoring at least one 2 is at least 50%?

Solution

(a) $P(1) = \dfrac{\pi(20^2 - 10^2)}{\pi(40^2)} = \dfrac{3}{16}$

(b) $P(2) = \dfrac{\pi(10^2)}{\pi(40^2)} = \dfrac{1}{16}$

(c) Let the number of attempts be n.

$$P(\text{at least one } 2) = 1 - P(\text{no } 2 \text{ in } n \text{ attempts}) = 1 - \left(\frac{15}{16}\right)^n$$

For this probability to be at least 50%, then

$$1 - \left(\frac{15}{16}\right)^n \geq 0.5 \Leftrightarrow \left(\frac{15}{16}\right)^n \leq 0.5$$

$$\Rightarrow n \ln\left(\frac{15}{16}\right) \leq \ln(0.5)$$

$$\Rightarrow n \geq \frac{\ln(0.5)}{\ln\left(\frac{15}{16}\right)} \quad \left\{ \text{since } \ln\left(\frac{15}{16}\right) < 0 \right\}$$

$$\Rightarrow n \geq 10.74$$

So, 11 attempts are required.

Exercise 11.3

1. Events A and B are given such that $P(A) = \frac{3}{4}$, $P(A \cup B) = \frac{4}{5}$, and $P(A \cap B) = \frac{3}{10}$. Find $P(B)$.

2. Events A and B are given such that $P(A) = \frac{7}{10}$, $P(A \cup B) = \frac{9}{10}$, and $P(A \cap B) = \frac{3}{10}$. Find
   (a) $P(B)$ (b) $P(B' \cap A)$ (c) $P(B \cap A')$
   (d) $P(B' \cap A')$ (e) $P(B|A')$

3. Events A and B are given such that $P(A) = \frac{1}{3}$, $P(A \cup B) = \frac{4}{9}$, and $P(B) = \frac{2}{9}$. Show that A and B are neither independent nor mutually exclusive.

4. Events A and B are given such that $P(A) = \frac{3}{7}$ and $P(A \cap B) = \frac{3}{10}$. If A and B are independent, find $P(A \cup B)$.

5. Driving tests in a certain city are not easy to pass the first time. After going through training, the percentage of new drivers passing the test the first time is 60%. If a driver fails the first test, the chance of passing it on a second try two weeks later is 75%. Otherwise, the driver has to retrain and take the test after 6 months. Find the probability that a randomly chosen new driver will pass the test without having to wait 6 months.

6. People with O^- blood type are universal donors. In other words, they can donate blood to individuals with any blood type. Only 8% of people have O^- blood.
   (a) One person randomly appears to give blood. What is the probability that the person does not have O^- blood?
   (b) Two people appear independently to give blood. Find the probability that
      (i) both have O^- blood
      (ii) at least one of them has O^- blood
      (iii) only one of them has O^- blood.
   (c) Eight people appear randomly to give blood. What is the probability that at least one of them has O^- blood?

7. PINs for smartphones usually consist of four digits that are not necessarily different.
   (a) How many possible PINs are there?
   (b) What is the probability that a PIN chosen at random does not start with a zero?

(c) What is the probability that a PIN contains at least one zero?

(d) Given a PIN with at least one zero, what is the probability that it starts with a zero?

8. An urn contains six red balls and two blue balls. We make two draws and each time we put the ball back after noting its colour.

   (a) What is the probability that at least one of the balls is red?

   (b) Given that at least one is red, what is the probability that the second one is red?

   (c) Given that at least one is red, what is the probability that the second one is blue?

9. Two dice are rolled and the numbers on the top face are observed.

   (a) List the elements of the sample space.

   (b) Let X represent the sum of the numbers observed. Copy and complete the following table:

x	2	3	4	5	6	7	8	9	10	11	12
$P(x)$		$\dfrac{1}{18}$									

   (c) **(i)** What is the probability that at least one dice shows a 6?

   (ii) What is the probability that the sum is at most 10?

   (iii) What is the probability that a dice shows 4 or the sum is 10?

   (iv) Given that the sum is 10, what is the probability that one of the dice is a 4?

10. A school has the following numbers categorised by class and gender:

	Grade 9	Grade 10	Grade 11	Grade 12	Total
Male	180	170	230	220	800
Female	200	130	190	180	700

   (a) What is the probability that a student chosen at random will be a female?

   (b) What is the probability that a student chosen at random is a male in Grade 12?

   (c) What is the probability that a female student chosen at random is a student in Grade 12?

   (d) What is the probability that a student chosen at random is female or in Grade 12?

   (e) What is the probability that a Grade 12 student chosen at random is a male?

   (f) Are gender and grade independent of each other? Explain.

11. Some young people do not like to wear glasses. A survey considered a large number of teenage students as to whether they needed glasses to correct their vision, and if they used the glasses when they needed to. Here are the results.

		Used glasses when needed	
		Yes	No
Need glasses for correct vision	Yes	0.41	0.15
	No	0.04	0.40

(a) Find the probability that a randomly chosen young person from this group
   (i) needs glasses
   (ii) needs glasses but does not use them.

(b) From those who need glasses, what is the probability that the student does not use them?

(c) Are the events of using and needing glasses independent?

12. Copy this table and fill in the missing entries.

P(A)	P(B)	Conditions for events A and B	P(A ∩ B)	P(A ∪ B)	P(A|B)
0.3	0.4	Mutually exclusive			
0.3	0.4	Independent			
0.1	0.5			0.6	
0.2	0.5		0.1		

13. In a large graduating class there are 100 students taking the IB exam. 40 students are doing Economics SL, 30 students are doing Physics SL, and 12 are doing both.

(a) A student is chosen at random. Find the probability that this student is doing Physics SL, given that they are doing Economics SL.

(b) Are doing Physics SL and Economics SL independent events?

14. A market chain in Germany accepts only Mastercard and Visa. It estimates that 21% of its customers use Mastercard, 57% use Visa, and 13% use both cards.

(a) What is the probability that a customer will have an acceptable credit card?

(b) What proportion of their customers has neither card?

(c) What proportion of their customers has exactly one acceptable card?

15. 132 of 300 patients at a hospital are signed up for a special exercise program that consists of a swimming class and an aerobics class. Each of these 132 patients takes at least one of the two classes. There are 78 patients in the swimming class and 84 in the aerobics class. Find the probability that a randomly chosen patient at this hospital is:

(a) not in the exercise program

(b) enrolled in both classes.

16. An ordinary unbiased 6-sided dice is rolled three times. Find the probability of rolling
    (a) three twos   (b) at least one two   (c) exactly one two.

17. An athlete is shooting arrows at a target. She has a record of hitting the centre 30% of the time. Find the probability that she hits the centre
    (a) with her second shot
    (b) exactly once with her first three shots
    (c) at least once with her first three shots.

18. Two unbiased dodecahedral (12 faces) dice, with faces numbered 1 to 12, are thrown. The scores are the numbers on the top side. Find the probability that
    (a) at least one 12 shows
    (b) the sum of the two dice is 12
    (c) there is a total score of at least 20
    (d) a total score of at least 20 is achieved, given that a 12 shows on one dice.

19. In question 18, two events are defined
    $A = \{$at least one of the numbers is a $10\}$
    $B = \{$the sum of the numbers is at most $15\}$
    Describe each event, list its elements and find the probability.
    (a) $A \cap B$ (b) $A \cup B$ (c) $(A \cap B')$ (d) $(A \cup B')$
    (e) $A' \cup B'$ (f) $A' \cap B'$ (g) $(A \cap B') \cup (A' \cap B)$

20. Consider any events A, B, and C. Prove each of the following:
    (a) $P(A \cap B) \geqslant P(A) + P(B) - 1$
    (b) $P(A \cup B \cup C) = P(A) + P(B) + P(C) - P(A \cap B) - P(A \cap C)$
    $$- P(B \cap C) + P(A \cap B \cap C)$$

21. Three fair 6-sided dice are rolled.
    (a) Find the probability that a triple is rolled.
    (b) Given that the roll is a sum of 8 or less, find the probability that a triple is rolled.
    (c) Find the probability that at least one six appears.
    (d) Given that the dice all have different numbers, find the probability that at least one six appears.

22. You are given four coins: one has two heads, one has two tails, and the other two are normal. You choose a coin at random and flip it. The result is tails. What is the probability that the opposite face is heads?

23. George and Kassanthra play a game in which they roll two unbiased six-sided dice. The first one who rolls a sum of 6 wins. Kassanthra rolls the dice first.

    (a) What is the probability that Kassanthra wins on her second roll?

    (b) What is the probability that George wins on his second roll?

    (c) What is the probability that Kassanthra wins?

24. A small repair shop for washing machines has the following demand for their services:.

    On 10% of the days, they have no requests, one request on 30% of the days, and two requests 50% of the time.

    (a) On Monday, what is the chance of more than two requests?

    (b) What is the chance of no requests for a whole week? (5-day week)

25. A small class has 5 boys and 6 girls. A group of four students are to be selected at random to interview a new school director.

    (a) Find the probability that the group contains at least one boy.

    (b) Find the probability that the majority of the group is girls.

    (c) Given that the group contains at least one boy, what is the chance that the boys are in the majority?

26. A construction company are bidding on three projects: B_1, B_2, and B_3. From previous experience they have the following probabilities of winning the bids: $P(B_1) = 0.22$, $P(B_2) = 0.25$, and $P(B_3) = 0.28$. Winning the bids are not independent from each other. The joint probabilities are given below.

	B_1	B_2	B_3
B_1		0.11	0.05
B_2	0.11		0.07
B_3	0.05	0.07	

    Also, $P(B_1 \cap B_2 \cap B_3) = 0.01$. Find each probability.

    (a) $P(B_1 \cup B_2)$ (b) $P(B'_1 \cap B'_2)$ (c) $P(B'_1 \cap B'_2) \cup B_3$

    (d) $P(B'_1 \cap B'_2 \cap B_3)$ (e) $P(B_2 \cap B_3 | B_1)$ (f) $P(B_2 \cup B_3 | B_1)$

27. Circuit boards used in electronic equipment go through more than one layer of inspection. The process of finding faults in the solder joints on these boards is highly subjective and prone to disagreements among inspectors. In a batch of 20 000 joints, Nick found 1448 faulty joints while David found 1502 faulty ones. All in all, among both inspectors 2390 joints were judged to be faulty. Find the probability that a randomly chosen joint is

    (a) judged to be faulty by neither of the two inspectors

    (b) judged to be defective by David but not Nick.

11.4 Bayes' theorem

Lie detectors (polygraphs), drug and alcohol tests, and disease screening tests are among the many applications where the results are often cautiously scrutinised. Tests with high precision rates are open for error. Bayes' theorem helps us understand and analyse the results of such tests.

Example 11.19

Suppose I have 9 indistinguishable boxes containing blue and red balls as shown below. They are of three types: Type A boxes contain one red and one blue ball; B boxes contain one blue and two red balls; C boxes contain one blue and three red balls. I mix up the boxes and choose one at random, and then I pick a ball from that box, also at random. I show you the ball. What is the probability that you can guess the box type it was drawn from if the ball is red?

A B C

Let R represent a red ball and A, B, or C representing box types. Find
(a) $P(A|R)$ (b) $P(B|R)$ (c) $P(C|R)$.

Solution

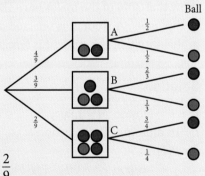

(a) Recalling information from conditional probability:

$$P(A|R) = \frac{P(A \cap R)}{P(R)}$$

Since we already know $P(A)$ and $P(R|A)$, then

$$P(A \cap R) = P(R|A) \cdot P(A) = \frac{4}{9} \cdot \frac{1}{2} = \frac{2}{9}$$

However, to find $P(R)$, we need to see what constitutes the event R. As we see in the figure, A, B, and C are mutually exclusive, and hence

$$R = (R \cap A) \cup (R \cap B) \cup (R \cap C) \Rightarrow$$
$$P(R) = P(R \cap A) + P(R \cap B) + P(R \cap C), \text{ and so}$$
$$P(R) = P(R \cap A) + P(R \cap B) + P(R \cap C)$$
$$= P(R|A) \cdot P(A) + P(R|B) \cdot P(B) + P(R|C) \cdot P(C)$$
$$= \frac{1}{2} \cdot \frac{4}{9} + \frac{2}{3} \cdot \frac{3}{9} + \frac{3}{4} \cdot \frac{2}{9} = \frac{11}{18}$$

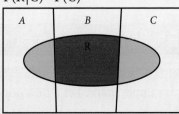

and now

$$P(A|R) = \frac{P(A \cap R)}{P(R)} = \frac{\frac{2}{9}}{\frac{11}{18}} = \frac{4}{11}$$

(b) Similarly

$$P(B|R) = \frac{P(B \cap R)}{P(R)} = \frac{\frac{2}{3} \cdot \frac{3}{9}}{\frac{11}{18}} = \frac{4}{11}$$

(c) $$P(C|R) = \frac{P(C \cap R)}{P(R)} = \frac{\frac{3}{4} \cdot \frac{2}{9}}{\frac{11}{18}} = \frac{3}{11}$$

So, A or B would be the more likely source of a red ball than C.

Conditional probability typically deals with the probability of an event when you have information about something that happened earlier. In conditional probability, we assume that the first event is known (selecting a box) and ask for the probability of the second event (colour of ball). Bayes' theorem deals with a reverse situation. It assumes that the second event is known (red ball) and asks for the probability of the first event.

Let us look at another example to see how we apply Bayes' theorem.

Example 11.20

60% of the students at a university are male and 40% are female. Records show that 30% of the males have IB diplomas while 75% of the females have IB diplomas. A student is selected and found to have a diploma. What is the probability that the student is female?

Solution

Use F = female, M = male, I = IB diploma. The question asks for $P(F|I)$.

$$P(F|I) = \frac{P(F \cap I)}{P(I)} = \frac{P(F) \cdot P(I|F)}{P(I)}$$

Since $I = (F \cap I) \cup (M \cap I)$, then

$$P(I) = 0.40 \times 0.75 + 0.60 \times 0.30$$
$$= 0.48, \text{ and so}$$

$$P(F|I) = \frac{P(F \cap I)}{P(I)} = \frac{0.40 \times 0.75}{0.48} = 0.625$$

Another interpretation of this number is the proportion of females among the IB holders.

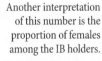

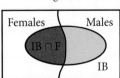

Bayes' theorem – simple case

Let U be a sample space with E_1 and E_2 mutually exclusive events that **partition** this sample space. Let F be a non-empty event in this sample space.

Then

$$P(E_1|F) = \frac{P(E_1 \cap F)}{P(F)} = \frac{P(E_1 \cap F)}{P(E_1 \cap F) + P(E_2 \cap F)}$$

$$= \frac{P(E_1) \cdot P(F|E_1)}{P(E_1) \cdot P(F|E_1) + P(E_2) \cdot P(F|E_2)}$$

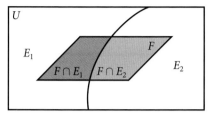

Figure 11.14 Let U be a sample space with E_1 and E_2 mutually exclusive events that **partition** this sample space. Let F be a non-empty event in this sample space

The probability of an event in terms of its mutually exclusive and exhaustive subsets is usually called the total probability:

$$P(F) = P(E_1) \cdot P(F|E_1) + P(E_2) \cdot P(F|E_2)$$

Sometimes, this rule is stated differently

$$P(E|F) = \frac{P(E \cap F)}{P(F)} = \frac{P(E \cap F)}{P(E \cap F) + P(E' \cap F)} = \frac{P(E) \cdot P(F|E)}{P(E) \cdot P(F|E) + P(E') \cdot P(F|E')}$$

Where E' is the complement of E.

Example 11.21

Lie detectors are not considered reliable in court. They are, however, administered to employees in sensitive positions. One such test gives a positive reading (the person is lying) when a person is lying 88% of the time, and a negative reading (the person is telling the truth) when a person is telling the truth 86% of the time. In a security-related question, the vast majority of subjects have no reason to lie, so 99% of the subjects will tell the truth. An employee produces a positive response on the polygraph. What is the probability that this employee is in fact telling the truth?

Solution

Let + denote the event of the test being positive (subject is lying), L the person is lying, and T the person is telling the truth.

Hence we have

$P(+|L) = 0.88$, $P(-|L) = 0.12$, $P(-|T)$
$\quad = 0.86$, $P(+|T) = 0.14$

$P(+) = P(+ \cap L) + P(+ \cap T)$
$\quad = P(+|L) \cdot P(L) + P(+|T) \cdot P(T)$
$\quad = 0.88 \times 0.01 + 0.14 \times 0.99$

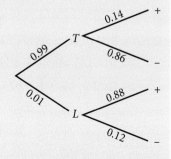

And finally

$$P(T|+) = \frac{P(|+[T] \cdot P(T)}{P(|+[L] \cdot P(L) + P(|+[T] \cdot P(T)} = \frac{0.14 \times 0.99}{0.88 \times 0.01 + 0.14 \times 0.99}$$
$$= 0.94$$

Thus, in screening this population of mostly innocent people, 94% of the positive polygraph reading will be misleading.

General rule

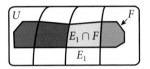

Figure 11.15 Diagram for the general rule

Let U be a sample space with $E_1, E_2, \ldots, E_n$ mutually exclusive events that partition this sample space. Let F be a non-empty event in this sample space. Then

$$P(E_i|F) = \frac{P(E_i \cap F)}{P(F)} = \frac{P(E_i \cap F)}{P(E_1 \cap F) + P(E_2 \cap F) + \ldots + P(E_n \cap F)}$$
$$= \frac{P(E_i) \cdot P(F|E_i)}{P(E_1) \cdot P(F|E_1) + P(E_2) \cdot P(F|E_2) + \ldots + P(E_n) \cdot P(F|E_n)}$$

Example 11.22

A paper factory produces high-quality paper using two machines. As with any process, the paper is checked for quality. 98% of the paper produced by machine A conforms to accepted standards. 97.5% of the paper produced by machine B conforms to standards. Machine A produces 60% of the paper in the factory. You pick some paper at random and it does not conform to the standards. What is the probability it was produced by machine A?

Solution

Let the non-conforming paper be called N. Since the paper is produced by these two machines, then the non-conforming paper has to be from the output of one of these machines; that is

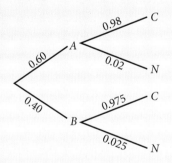

$$N = (N \cap A) \cup (N \cap B)$$

$$P(N) = P(N \cap A) + P(N \cap B)$$
$$= P(N|A) \cdot P(A) + P(N|B) \cdot P(B)$$
$$= 0.02 \times 0.60 + 0.025 \times 0.40 = 0.022$$

So, using Bayes' theorem:

$$P(A|N) = \frac{P(A \cap N)}{P(N)} = \frac{P(N|A) \cdot P(A)}{P(N)} = \frac{0.02 \times 0.60}{0.022} = 54.5\%$$

Example 11.23

In many countries the law requires that a driver's licence is withdrawn if the driver is found to have more than 0.05% blood alcohol concentration (BAC). The police use a test that will correctly identify a driver with more than 0.05% BAC 99% of the time, and will correctly identify a driver with 0.05% or less BAC 99% of the time. Assume that 0.5% of the drivers in a city drive under the influence of alcohol. Work out the probability that, given a positive test, a driver's BAC is, in fact, more than 0.05%.

Solution

Let D be the event of being over the limit and N indicate being under the limit. Let $+$ be the event of a positive test. We need to know $P(D|+)$.

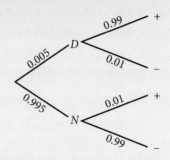

The question is to know the proportion of the drunk drivers who test positive out of all of those who test positive.

$P(D)$ is the probability that the driver is over the limit, regardless of any other information. This is 0.005, since 0.5% of the drivers drink and drive.

$P(N)$ is the probability that the driver is under the limit. This is $1 - P(D)$, or 0.995.

$P(+|D)$ is the probability that the test is positive, given that the driver is over the limit. This is 0.99, since the test is 99% accurate.

$P(+|N)$ is the probability that the test is positive, given that the driver is not over the limit. This is 0.01, since the test will produce a false positive for 1% of drivers who are under the limit.

$P(+)$ is the probability of a positive test event, regardless of other information. This is 0.0149 or 1.49%, which is found by adding the probability that a true positive result will appear (99% × 0.5% = 0.495%) plus the probability that a false positive will appear (1% × 99.5% = 0.995%).

Given this information, we can compute the $P(D|+)$ of a driver who tested positive being a driver with a BAC over the limit:

$$P(D|+) = \frac{P(D \cap +)}{P(+)} = \frac{P(+|D)P(D)}{P(+|D)P(D) + P(+|N)P(N)}$$

$$= \frac{0.99 \times 0.005}{0.99 \times 0.005 + 0.01 \times 0.995} = 0.3322$$

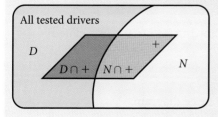

Even with the high precision of the test, the probability that a driver who tested positive in fact did have a BAC over the limit is only about 33%, so it is more likely that the driver is not over the limit.

Example 11.24

A computer manufacturer receives hard disks from three different suppliers. Marco supplies 40% of the disks, Berto supplies 25%, while Lukas supplies the rest. Disks from Marco have a defective rate of 4%, those from Berto 3%, while Lukas' disks have a 5% defective rate.

(a) A disk is checked at random. What is the probability that it is defective?

(b) A disk is checked and found to be defective. What is the probability that it was supplied by Lukas?

Solution

Let M represent Marco, B represent Berto, and L represent Lukas. Also let D represent defective items and G the non-defective ones. A tree diagram may be helpful here.

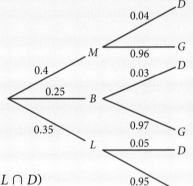

(a) A defective disk could be supplied by any of the three suppliers, hence:

$$P(D) = P(M \cap D) + P(R \cap D) + P(L \cap D)$$

$$= P(M) \cdot P(D|M) + P(B) \cdot P(D|B) + P(L) \cdot P(D|L)$$

$$= 0.4 \times 0.04 + 0.25 \times 0.03 + 0.35 \times 0.05$$

$$= 0.041$$

(b) This is a Bayes' theorem application. We are reversing the order of events:

$$P(L|D) = \frac{P(L \cap D)}{P(D)}$$

$$= \frac{P(L) \cdot P(D|L)}{P(M) \cdot P(D|M) + P(B) \cdot P(D|B) + P(L) \cdot P(D|L)}$$

$$= \frac{0.35 \times 0.05}{0.041} = 0.427$$

Exercise 11.4

1. In a sample space, U, there are the following events and associated probabilities:

$$P(E) = \frac{2}{3}, \; P(A|E) = \frac{3}{50}, \text{ and } P(A|E') = \frac{1}{25}$$

(a) Represent the information using a tree diagram.

(b) Find P(A)

(c) Find P($E|A$)

2. The rate of prostate cancer among men in 2002 in a certain country was approximately 26 cases per 100 000 people. Diagnosing this type of cancer saves the lives of about 70% of those treated.

   At a hospital, the probability of diagnosing a person with this cancer correctly is 78% and the probability of wrongly diagnosing a person without this cancer as having the disease is 6%.

   (a) What is the probability of diagnosing a person as having this cancer?

   (b) What percentage of those with a positive diagnosis, do, in fact, have cancer?

3. Police in a small town plan to enforce speed limits by installing radar detectors at the two main town entrances: east end and west end. Traffic statistics show that 40% of the cars entering the town use the east entrance and the rest use the west one. The east entrance detectors are operated 40% of the time and the west entrance are operated 60% of the time. Assuming that the proportion of speeding drivers is the same at both entrances, find the probability that:

   (a) a speeding driver is spotted passing one of the traps

   (b) a speeding driver who was spotted has come from the west end.

4. Coloured balls are placed in three boxes as shown in this table.

	Box		
	1	2	3
Green balls	4	8	6
Red balls	6	2	8
Blue balls	10	6	6

   A box is selected at random from which a ball is randomly drawn.

   (a) What is the probability that the ball is green?

   (b) Given that the ball is green, what is the probability that it was drawn from box 2?

5. Two coins are in your pocket. When flipped, one of the coins is biased with 0.6 probability of landing heads while the other coin is unbiased. You select one of the coins at random and flip it.

   (a) What is the probability it lands on heads?

   (b) Given that it lands on tails, what is the probability that it was the unbiased coin?

6. When answering a question on a multiple-choice test, a student is given five choices, one of which is correct. The test is designed so that the choices are very close and the probability of getting the correct answer, when you know the material, is 0.6. In a class where 70% of the students are well prepared, a randomly chosen student answers the question correctly. What is the probability that the student really knew the material?

7. Nigel is a student at Wigley College. To avoid coming late to his morning classes, he usually sets his alarm clock. 85% of the time he manages to remember to set his alarm. When the alarm goes off, he manages to get to his morning classes 90% of the time. If the alarm is not set, he still manages to get up and get to class 60% of the time.

   (a) What percentage of the days does he manage to get to his morning classes?

   (b) Given that Nigel made it to class one day, what is the chance that he did so without having set his alarm?

8. Cheung plays tennis. A player has two serves. If the first serve is successful, the game continues. If the first serve is not successful, the player is given another chance. If the second serve fails, then the player loses the point.

   Cheung is successful with his first serve 60% of the time and 95% of the time with his second serve. When his first serve is successful he goes on to win the point 75% of the time, and when it takes him two serves, he wins the point 50% of the time.

   (a) What is the probability that Cheung wins the point?

   (b) Cheung wins a point. What is the probability that he succeeded with the first serve?

9. In Bratislava, Slovakia, conventional wisdom says that days in February are snowy or fine. 80% of the time a fine day follows a fine day. 40% of the time a snowy day is followed by a fine day.

   The forecast for 1 February to be a fine day is 0.75

   (a) Find the probability that 2 February is fine.

   (b) Given that 2 February turns out to be snowy, what is the probability that 1 February was a fine day?

10. It is known that 33% of people over the age of 50 around the world have some kind of arthritis. A test has been developed to detect arthritis in individuals. This test was given to a large group of individuals with confirmed cases and a positive test result was achieved in 87% of the cases. That same test gave a positive test to 4% of individuals that do not have it.

   This test is given to an individual at random and it tests positive What is the probability that the individual has this disease?

11. A high school has a large graduating class. The table shows how the students are categorised according to their college plans and gender.

	Local university	University abroad
Male	51%	4%
Female	16%	29%

5% of these students are in the IB Chemistry HL class. 72% of that class will attend local university and the rest are going abroad.

(a) What percentage of the graduating class are Chemistry HL students planning on studying locally?

(b) What percentage of the students that are not Chemistry HL are going to study at a local university?

(c) Among the students studying locally, what percentage are Chemistry HL students?

(d) Among the students studying abroad, what percentage are Chemistry HL students?

12. When Olympic athletes are tested for illegal drug use (doping), the results of a single test are used to ban the athlete from competition. In an experiment on 1000 athletes, 100 were using the drug testosterone. During the medical experiment, the available test would positively identify 50% of the users. It would also falsely identify 9% of the non-users as users.

If an athlete tests positive, what is the probability that the athlete is really doping?

13. An engineering company employs three engineers who are responsible for cost estimates of new projects. Antonio makes 30% of the estimates, Richard makes 20%, and Sarah 50%. Like all estimates, there are usually some errors. The record of percentages of serious errors that cost the company thousands of euros shows Antonio's at 3%, Richard's at 2%, and Sarah's at 1%. Which of the three engineers is probably responsible for most of the serious errors?

14. At a small airport, when an aircraft comes within 10 km of the runway, radar detects it and generates an alarm signal 99% of the time. If an aircraft is not present, the radar generates a (false) alarm, with probability 0.10. We assume that an aircraft is present with probability 0.05.

(a) What is the probability that the radar gives an alarm signal?

(b) Given that there is no alarm signal, what is the probability that an aircraft is there?

15. You enter a chess tournament where your probability of winning a game is 0.5 against half the players (novices), 0.4 against a quarter of the players (experienced), and 0.3 against the remaining quarter of the players (masters). You play a game against a randomly chosen opponent.

(a) What is the probability of winning?

(b) Given that you won, what is the probability that the game was against a master?

16. Driving tests in a certain city are relatively easy to pass the first time you take them. After going through training, the percentage of new drivers passing the test the first time is 80%. If a driver fails the first test, there is a chance of passing it on a second, harder test, two weeks later; 50% of the second-chance drivers pass the test. If the second test is unsuccessful, a third attempt is given a week later and 30% of the participants pass it. Otherwise, the driver has to retrain and take the test after 1 year.

(a) Find the probability that a randomly chosen new driver will pass the test in three attempts or less.

(b) Find the probability that a randomly chosen new driver that passed the test, did so on the second attempt.

17. A school has 250 employees categorised by role and gender in the table.

	Teaching	Administrative	Support
Male	84	14	52
Female	56	26	18

An employee is randomly selected. Let A be the event that the employee is an administrative staff member, T teaching staff, S support, M male, and F female.

(a) Write down the probabilities:

    (i) $P(F)$ (ii) $P(F \cap T)$ (iii) $P(F \cup A')$ (iv) $P(F'|A)$

(b) Which events are independent of F, and which are mutually exclusive to F? Justify your choices.

(c) Given that 90% of teachers own cars, 80% of the administrative staff own cars, and 30% of the support staff own cars, find

    (i) the probability that a staff member chosen at random owns a car

    (ii) the probability that the staff member is teaching staff, knowing that the randomly chosen staff member owns a car.

18. Car insurance companies categorise drivers as high risk, medium risk, and low risk.

20% of the drivers insured by First Insurance are high risk, and 50% are medium risk. The company's actuaries estimate the chance that each class of drivers will have at least one accident in the coming 12 months as follows: high risk, 6%; medium, 3%; and low, 1%.

(a) Find the probability that a randomly chosen driver is a high-risk driver who will have an accident in a 12-month period.

(b) Find the probability that a randomly chosen driver insured by this company will have an accident in the next 12 months.

(c) A customer has a claim for an accident. What is the probability that the customer is a high-risk driver?

Chapter 11 practice questions

1. Two independent events A and B are given such that
   $P(A) = k$, $P(B) = k + 0.3$ and $P(A \cap B) = 0.18$
   (a) Find k (b) Find $P(A \cup B)$ (c) Find $P(A'|B')$

2. Many airport authorities test prospective employees for drug use. This procedure has plenty of opponents who claim that this procedure creates difficulties for some people and that it prevents others from getting these jobs, even if they are not drug users. The claim depends on the fact that these tests are not 100% accurate. To test this claim, assume that a test is 98% accurate in that it identifies a person as a user or non-user 98% of the time. Each job applicant takes this test twice. The tests are done at separate times and are designed to be independent of each other. Find the probability that:
   (a) a non-user fails both tests
   (b) a drug user is detected (i.e., they fail at least one test)
   (c) a drug user passes both tests.

3. Communications satellites are difficult to repair when something goes wrong. One satellite uses solar energy and it has two systems that provide electricity. The main system has a probability of failure of 0.002. It has a backup system that works independently of the main one and has a failure rate of 0.01. What is the probability that the systems do not fail at the same time?

4. In a group of 200 students taking the IB examination, 120 take Spanish, 60 take French, and 10 take both.
   (a) If a student is selected at random, find the probability that the student:
   (i)   takes either French or Spanish
   (ii) takes either French or Spanish but not both
   (iii) does not take any French or Spanish.
   (b) Given that a student takes Spanish, what is the probability that the student takes French?

5. In a factory producing computer disk drives, there are three machines that work independently to produce one of the components. In any production process, machines are not 100% fault free. The production after one batch from each machine is listed in the table.

	Defective	Non-defective
Machine I	6	120
Machine II	4	80
Machine III	10	150

(a) A component is chosen at random from the batches. Find the probability that the chosen component is:
   (i) from machine I
   (ii) a defective component from machine II
   (iii) non-defective or from machine I
   (iv) from machine I given that it is defective.

(b) Is the quality of the component dependent on the machine used?

6. At a school, the students are organising a lottery to raise money for their community. The lottery tickets consist of small coloured envelopes containing a small note. The note either says: 'You have won a prize!' or 'Sorry, try another ticket.' The envelopes have several colours. They have 70 red envelopes that contain two prizes, and the rest (130 tickets) contain four other prizes.

(a) You want to help this class and you buy a ticket hoping that it does not have a prize. Additionally, you don't like the red colour. You pick your ticket at random by closing your eyes. What is the probability that your ticket does not have a prize?

(b) You are surprised – you picked a red envelope. What is the probability that you did not win a prize?

7. Two events A and B have the conditions:

$P(A|B) = 0.30$, $P(B|A) = 0.60$, $P(A \cap B) = 0.18$

(a) Find $P(B)$

(b) Are A and B independent? Explain.

(c) Find $P(B \cap A')$

8. In several ski resorts in Austria and Switzerland, the local sports authorities use senior high school students as ski instructors to help deal with the surge in demand during vacations. However, to become an instructor, you have to pass a test and must be a senior at your school.

Here are the results of a survey of 120 students in a Swiss school who are training to become instructors. In this group, there are 70 boys and 50 girls. 74 students took the test; 32 boys and 16 girls passed the test, the rest, including 12 girls, failed the test. 10 of the students, including 6 girls, were too young to take the ski test.

(a) Copy and complete the table.

	Boys	Girls
Passed the ski test	32	16
Failed the ski test		12
Training, but did not take the test yet		
Too young to take the test		

(b) Find the probability that:

    (i) a student chosen at random has taken the test

    (ii) a girl chosen at random has taken the test

    (iii) a randomly chosen boy and randomly chosen girl have both passed the ski test.

9. Two events A and B are such that $P(A) = \dfrac{9}{16}$, $P(B) = \dfrac{3}{8}$, $P(A|B) = \dfrac{1}{4}$. Find the probability that:

    (a) both events will happen

    (b) only one of the events will happen

    (c) neither event will happen.

10. Martina plays tennis. When she serves, she has a 60% chance of succeeding with her first serve and continuing the game. She has a 95% chance on the second serve. Of course, if both serves are not successful, she loses the point.

    (a) Find the probability that she misses both serves.

    If Martina succeeds with the first serve, her chances of gaining the point against Steffy is 75%; if she is only successful with the second serve, her chances against Steffy for that point go down to 50%.

    (b) Find the probability that Martina wins a point against Steffy.

11. For events X and Y, $P(X) = 0.6$, $P(Y) = 0.8$ and $P(X \cup Y) = 1$. Find:

    (a) $P(X \cap Y)$ **(b)** $P(X' \cup Y')$.

12. In a survey, 100 managers were asked, 'Do you prefer to watch the news or play sport?' Of the 46 men in the survey, 33 said they would choose sport, while 29 women made this choice.

	Men	Women	Total
Watch the news			
Play sport	33	29	
Total	46		100

Find the probability that:

(a) a manager selected at random prefers to watch the news

(b) a manager prefers to watch the news, given that the manager is a man.

13. The Venn diagram shows a sample space U and events X and Y.

$n(U) = 36$, $n(X) = 11$, $n(Y) = 6$
and $n(X \cup Y)' = 21$

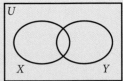

(a) Copy the diagram and shade the region $(X \cup Y)'$

(b) Find:

    (i) $n(X \cap Y)$ (ii) $P(X \cap Y)$

(c) Are events X and Y mutually exclusive? Explain why or why not.

14. In a survey of 200 people, 90 of whom were female, it was found that 60 people were unemployed, including 20 males.

(a) Copy and complete the table, using this information.

	Males	Females	Totals
Unemployed			
Employed			
Totals			200

(b) If a person is selected at random from this group of 200, find the probability that this person is:

    (i) an unemployed female

    (ii) a male, given that the person is employed.

15. A box contains 10 white chips, 10 red chips, and 6 green chips. Two chips are drawn at random from the box without replacement. What is the probability that they are of different colours?

16. The events B and C are dependent, where C is the event 'a student takes Chemistry', and B is the event 'a student takes Biology'. It is known that $P(C) = 0.4$, $P(B|C) = 0.6$, $P(B|C') = 0.5$

(a) With the help of a tree diagram, calculate the probability that a student takes Biology.

(b) Given that a student takes Biology, what is the probability that the student takes Chemistry?

17. Two fair dice are rolled and the number showing on each is noted. Find the probability that:

(a) the sum of the numbers is less than or equal to 7

(b) at least one dice shows a 3

(c) at least one dice shows a 3, given that the sum is less than 8.

18. For events A and B, the probabilities are $P(A) = \dfrac{3}{11}$, $P(B) = \dfrac{4}{11}$. Calculate the value of $P(A \cap B)$ if:

(a) $P(A \cup B) = \dfrac{6}{11}$

(b) events A and B are independent.

19. Consider events A and B such that $P(A) \neq 0$, $P(A) \neq 1$, $P(B) \neq 0$, and $P(B) \neq 1$.

    In each of the situations below, state whether A and B are mutually exclusive, independent, or neither.

    (a) $P(A|B) = P(A)$

    (b) $P(A \cap B) = 0$

    (c) $P(A \cap B) = P(A)$

20. Sophia is a student at an IB school.

    The probability that she will be woken by her alarm clock is $\dfrac{7}{8}$.

    If she is woken by her alarm clock, the probability that she will be late for school is $\dfrac{1}{4}$.

    If she is not woken by her alarm clock the probability that she will be late for school is $\dfrac{3}{5}$.

    Let W be the event 'Sophia is woken by her alarm clock'.

    Let L be the event 'Sophia is late for school'.

    (a) Draw a tree diagram for Sophia being late or on time.

    (b) Calculate the probability that Sophia will be late for school.

    (c) Given that Sophia is late for school, what is the probability that she was woken by her alarm clock?

21. A packet of seeds contains 40% radish seeds and 60% beans seeds. The probability of germination for a radish seed is 0.9, and for a bean seed it is 0.8. A seed is chosen at random from the packet.

    (a) Copy and complete the probability tree diagram.

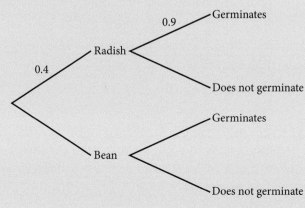

(b) **(i)** Calculate the probability that the chosen seed is a radish seed and germinates.

(ii) Calculate the probability that the chosen seed germinates.

(iii) Given that the seed germinates, calculate the probability that it is a radish seed.

22. The table shows the subjects studied by 210 students at a college.

	Year 1	Year 2	Totals
History	50	35	85
Science	15	30	45
Art	45	35	80
Totals	110	100	210

(a) A student from the college is selected at random.

Let A be the event 'the student studies Art'.

Let B be the event 'the student is in Year 2'.

(i) Find P(A).

(ii) Find the probability that the student is a Year 2 Art student.

(iii) Are the events A and B independent? Justify your answer.

(b) Given that a History student is selected at random, calculate the probability that the student is in Year 1.

(c) Two students are selected at random from the college. Calculate the probability that one student is in Year 1, and the other in Year 2.

23. A bag contains nine paper slips: two blank slips, three slips with odd numbers and four with even numbers. A slip is chosen at random from the bag and is not replaced. A second slip is chosen. Find the probability of choosing one even and one odd number in any order.

24. Two independent events X and Y are given with $P(X \cap Y) = 0.3$ and $P(X \cap Y') = 0.3$. Find $P(X \cup Y)$.

25. Sophia walks to school every day. If it is not raining, the probability that she is late is 0.2. If it is raining, the probability that she is late is $\frac{2}{3}$. The probability that it rains on any particular day is 0.25

Last Friday Sophia was late.

Find the probability that it was raining on that day.

26. The probability that Kier leaves his umbrella in any place he visits is $\frac{1}{3}$.

After visiting two friends in succession, he finds he has left his umbrella at one of his friends' places. What is the probability that he left his umbrella at the second friend's place?

27. Two girls, Catherine and Lucy, play a game in which they take turns throwing an unbiased six-sided dice. The first one to roll a 5 wins the game. Catherine is the first to roll.

    (a) Find the probability that:

        (i) Lucy wins on her first roll

        (ii) Catherine wins on her second roll

        (iii) Catherine wins on her nth roll.

    (b) Let p be the probability that Catherine wins the game.
Show that $p = \dfrac{1}{6} + \dfrac{25}{36}p$.

    (c) Find the probability that Lucy wins the game.

    (d) Suppose that they play the game six times. Find the probability that Catherine wins more games than Lucy.

28. A bag contains 22 red balls and 3 white balls. Three balls are selected at random, one after the other, without replacement.

    (a) The first two balls are white. What is the probability that the third ball is red?

    (b) What is the probability that exactly two of the three balls are red?

29. Roberto travels to school in a neighbouring town by bus every weekday from Monday to Friday. The probability that he catches the 8 a.m. bus on Friday is 0.66. The probability that he catches the 8 a.m. bus on any other weekday is 0.75. A weekday is chosen at random.

    (a) Find the probability that he catches the bus on that day.

    (b) Given that he catches the 8 a.m. bus on that day, find the probability that the chosen day is Friday.

30. Eric and Harriet play a game by throwing a dice in turn. If the dice shows a 3, 4, 5, or 6, the player who threw the dice wins the game. If the dice shows a 1 or 2, the other player has the next throw. Eric plays first, and the game continues until there is a winner.

    (a) Write down the probability that Eric wins on his first throw.

    (b) Calculate the probability that Harriet wins on her first throw.

    (c) Calculate the probability that Eric wins the game.

31. Bag A contains 2 red and 3 green balls.

    (a) Two balls are chosen at random from the bag without replacement. Find the probability that 2 red balls are chosen.

    Bag B contains 4 red and n green balls.

    (b) Two balls are chosen without replacement from this bag. If the probability that two red balls are chosen is $\dfrac{2}{15}$, show that $n = 6$.

A standard dice with six faces is rolled. If a 1 or 6 is obtained, two balls are chosen from bag *A*, otherwise two balls are chosen from bag *B*.

(c) Calculate the probability that two red balls are chosen.

(d) Given that two red balls are chosen, find the probability that a 1 or a 6 was obtained on the dice.

32. Six balls numbered 1, 2, 2, 3, 3, 3 are placed in a bag. Balls are taken one at a time from the bag at random and the number noted. Throughout the question a ball is always replaced before the next ball is taken.

(a) A single ball is taken from the bag. Let *X* denote the value shown on the ball. Find $E(X)$.

(b) Three balls are taken from the bag. Find the probability that:

    (i) the total of the three numbers is 5

    (ii) the median of the three numbers is 1.

(c) Ten balls are taken from the bag. Find the probability that less than four of the balls are numbered 2.

(d) Find the least number of balls that must be taken from the bag for the probability of taking out at least one ball numbered 2 to be greater than 0.95

Another bag also contains balls numbered 1, 2 or 3. Eight balls are to be taken from this bag at random. It is calculated that the expected number of balls numbered 1 is 4.8, and the variance of the number of balls numbered 2 is 1.5

(e) Find the least possible number of balls numbered 3 in this bag.

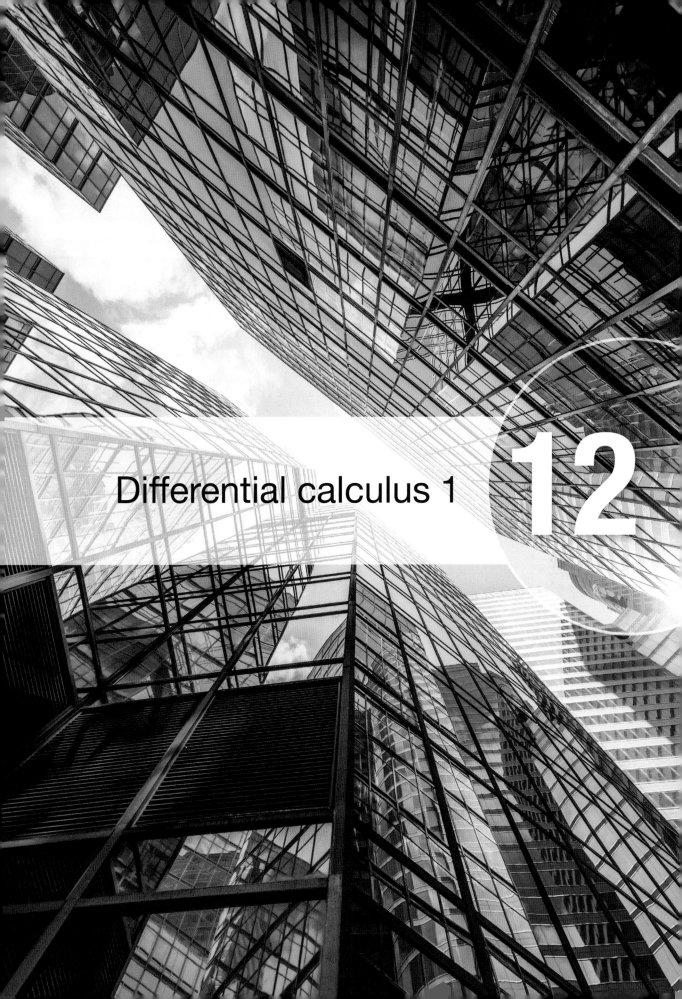

Differential calculus 1

Figure 12.1 shows a distance–time graph for a 50-kilometre bicycle ride that included going up and then down a steep hill. There are four time intervals labelled A, B, C, and D. The cyclist's speed is the lowest in interval B. It is the highest in interval C. The cyclist's speed is about the same in intervals A and D. The shape of the distance–time graph gives information about the cyclist's speed during a certain interval and at a particular moment (instant) during the ride.

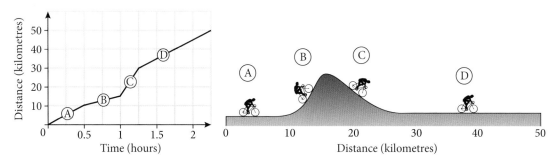

Figure 12.1 A distance–time graph for a cyclist

Calculus is the branch of mathematics that was developed to analyse and model change – such as velocity and acceleration. We can also apply it to study change in the context of slope, area, volume, and a wide range of other concepts that allow us to model real-life phenomena more precisely. Although mathematical techniques that we have previously studied dealt with many of these concepts, the ability to model change was restricted. Consider the curve in Figure 12.2 that illustrates the motion of an object by indicating the distance (y metres) travelled after a certain amount of time (t seconds). Without calculus, we can only compute the **average velocity** between two different times. With calculus, we can find the velocity of an object at a particular instant, known as its **instantaneous velocity** (Figure 12.3). The starting point for our study of calculus is the idea of a limit.

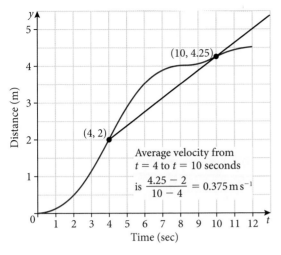

Figure 12.2 Average velocity from a distance–time graph

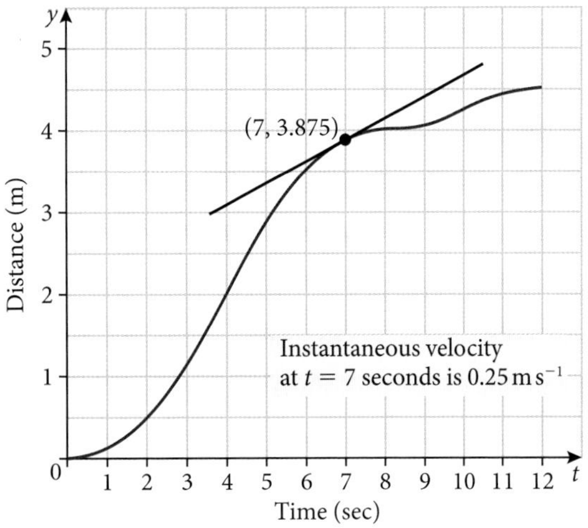

Figure 12.3 Instantaneous velocity from a distance–time graph

Limits of functions

A **limit** is one of the ideas that distinguish calculus from algebra, geometry, and trigonometry. The notion of a limit is a fundamental concept of calculus. Limits are not new to us. We often use the idea of a limit in many non-mathematical situations. We have already used mathematical limits in this book – finding the sum of an infinite geometric series and computing the irrational number e.

In Chapter 3 we established that if the sequence of partial sums for an infinite series **converges** to a finite number L, we say that the infinite series has a sum of L. We used limits to confirm algebraically that the infinite series

$2 + 1 + \dfrac{1}{2} + \dfrac{1}{4} + \dfrac{1}{8} + \cdots$ has a sum of 4. As part of the algebra for this,

we reasoned that as the value of n increases in the positive direction without

bound ($n \to +\infty$) the expression $\left(\dfrac{1}{2}\right)^n$ converges to zero – in other words,

the **limit** of $\left(\dfrac{1}{2}\right)^n$ as n goes to positive infinity is zero. This result is expressed

using limit notation as $\lim\limits_{n \to \infty} \left(\dfrac{1}{2}\right)^n = 0$. It is beyond the requirements of this

course to establish a precise formal definition of a limit, but a closer look at justifying a couple of limits can lead us to a useful informal definition.

In calculus we are interested in limits of functions of real numbers. Although many of the limits of functions that we will encounter can only be approached and not actually reached, this is not always the case.

For example, if asked to evaluate the limit of the function $f(x) = \dfrac{x}{2} - 1$ as x

approaches 6, then we evaluate the function for $x = 6$. Since $f(6) = 2$, then

$\lim\limits_{x \to 6}\left(\dfrac{x}{2} - 1\right) = 2$. However, it is more common that we are unable to evaluate

the limit of $f(x)$ as x approaches some number c because $f(c)$ does not exist.

Example 12.1

Find each limit using your GDC to analyse the behaviour of the function.

(a) $\lim\limits_{x \to 0} \dfrac{\sin x}{x}$

(b) $\lim\limits_{x \to 0} \dfrac{\cos x - 1}{x}$

Solution

(a) We are not able to evaluate this limit by direct substitution because when $x = 0$, $\dfrac{\sin x}{x} = \dfrac{0}{0}$ and is therefore undefined. We use our GDC (in radian mode) to analyse the behaviour of the function $y = \dfrac{\sin x}{x}$ as x approaches zero from the right side and the left side.

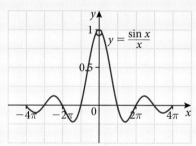

```
Plot1  Plot2  Plot3
\Y1≣ sin(X)/X
```

```
Y1(-0.05)
        .9995833854
Y1(0.05)
        .9995833854
1-Ans
     4.166145864E-4
```

```
Y1(-0.002)
        .9999993333
Y1(0.002)
        .9999993333
1-Ans
     6.66666655E-7
```

Although there is no point on the graph of $y = \dfrac{\sin x}{x}$ corresponding to $x = 0$, it is clear from the graph that as x approaches zero (from either direction) the value of $\dfrac{\sin x}{x}$ converges to 1.

We can get the value of $\dfrac{\sin x}{x}$ arbitrarily close to 1 depending on our choice of x. If we want $\dfrac{\sin x}{x}$ to be within 0.001 of 1, we choose $x = \pm 0.05$ giving $\dfrac{\sin 0.05}{0.05} \approx 0.999583$ and $1 - 0.999583 = 0.000417 < 0.001$; and if we want $\dfrac{\sin x}{x}$ to be within 0.000001 of 1, then we choose $x = \pm 0.002$ giving $\dfrac{\sin 0.002}{0.002} \approx 0.9999993333$ and $1 - 0.9999993333 = 0.0000006667 < 0.000001$; and so on. Therefore, $\lim\limits_{x \to 0} \dfrac{\sin x}{x} = 1$.

(b) As with $y = \dfrac{\sin x}{x}$, substituting $x = 0$ into the function $y = \dfrac{\cos x - 1}{x}$ produces $\dfrac{0}{0}$. The graph of $y = \dfrac{\cos x - 1}{x}$ shows that the function approaches 0 as x tends to 0.

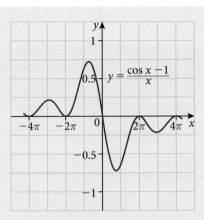

A table produced on a GDC also shows that the function approaches zero from both directions.

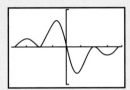

Therefore, $\lim\limits_{x \to 0} \dfrac{\cos x - 1}{x} = 0$

Functions do not necessarily converge to a finite value at every point. It is possible that a limit does not exist.

Example 12.2

Find $\lim\limits_{x \to 0} \dfrac{1}{x^2}$, if it exists.

Solution

As x approaches zero, the value of $\dfrac{1}{x^2}$ becomes increasingly large in the positive direction. The graph of the function indicates that we can make the values of $y = \dfrac{1}{x^2}$ arbitrarily large by choosing x close enough to zero. Therefore, the values of $y = \dfrac{1}{x^2}$ do not approach a finite number, so $\lim\limits_{x \to 0} \dfrac{1}{x^2}$ does not exist.

$y = \dfrac{1}{x^2}$

If $f(x)$ becomes arbitrarily close to a unique finite number L as x approaches c from either side, then the limit of $f(x)$ as x approaches c is L. The notation for indicating this is $\lim\limits_{x \to c} f(x) = L$.

When a function $f(x)$ becomes arbitrarily close to a finite number L, we say that $f(x)$ **converges** to L.

It is important to be able to apply some basic algebraic manipulation in order to evaluate the limits of some functions algebraically rather than by conjecturing from a graph or table. The following five properties of limits are also useful.

Properties of limits

Let a and b be real numbers, and let f and g be functions with the limits $\lim_{x \to a} f(x) = L$ and $\lim_{x \to a} g(x) = K$.

Constant: $\lim_{x \to a} b = b$

Scalar multiple: $\lim_{x \to a} [b \cdot f(x)] = b \cdot L$

Sum or difference: $\lim_{x \to a} [f(x) \pm g(x)] = L \pm K$

Product: $\lim_{x \to a} [f(x) \cdot g(x)] = L \cdot K$

Quotient: $\lim_{x \to a} \left[\dfrac{f(x)}{g(x)} \right] = \dfrac{L}{K}$, where $K \neq 0$

Often when trying to determine the limit of a quotient by direct substitution, we may get a meaningless fraction such as $\dfrac{0}{0}$ or $\dfrac{\infty}{\infty}$. Such an expression is called an **indeterminate form** because we cannot use it to determine the desired limit. When we are confronted with an indeterminate form we may be able to perform some algebraic manipulation to the quotient to get it into a form that allows the limit to be evaluated by direct substitution and/or applying known limits. Another technique, l'Hôpital's rule, can also be applied to limits of expressions in indeterminate form. l'Hôpital's rule will be covered in Chapter 13.

Example 12.3

Evaluate each limit algebraically.

(a) $\lim_{x \to \infty} \dfrac{5x - 3}{x}$

(b) $\lim_{p \to 0} (3x^2 - 4px + p^2)$

(c) $\lim_{h \to 0} \dfrac{[(x + h)^2 - 6] - (x^2 - 6)}{h}$

(d) $\lim_{x \to \infty} \dfrac{3x^2 - 5 - 1}{2x^2 + 1}$

Solution

(a) Split the fraction into two terms: $\lim_{x \to \infty} \dfrac{5x - 3}{x} = \lim_{x \to \infty} \left(\dfrac{5x}{x} - \dfrac{3}{x} \right)$

Apply the property: $\lim_{x \to a} [f(x) - g(x)] = L - K$

So, $\lim_{x \to \infty} \left(\dfrac{5x}{x} - \dfrac{3}{x} \right) = \lim_{x \to \infty} 5 - \lim_{x \to \infty} \dfrac{3}{x} = 5 - 0 = 5$

Therefore, $\lim_{x \to \infty} \dfrac{5x - 3}{x} = 5$

(b) Apply the property: $\lim_{x \to a} [f(x) \pm g(x)] = L \pm K$

$\lim_{p \to 0} (3x^2 - 4px + p^2) = \lim_{p \to 0} 3x^2 - \lim_{p \to 0} 4px + \lim_{p \to 0} p^2 = 3x^2 - 0 + 0 = 3x^2$

Therefore, $\lim_{p \to 0} (3x^2 - 4px + p^2) = 3x^2$

(c) $\displaystyle\lim_{h\to0}\frac{[(x+h)^2-6]-(x^2-6)}{h} = \lim_{h\to0}\frac{x^2+2xh+h^2-6-x^2+6}{h}$

$$= \lim_{h\to0}\frac{2xh+h^2}{h}$$

$$= \lim_{h\to0}\frac{h(2x+h)}{h}$$

$$= \lim_{h\to0}2x + \lim_{h\to0}h$$

$$= 2x + 0 = 2x$$

Therefore, $\displaystyle\lim_{h\to0}\frac{[(x+h)^2-6]-(x^2-6)}{h}=2x$

(d) Divide the numerator and denominator by the largest power of x.

$$\lim_{x\to\infty}\frac{3x^2+5x-1}{2x^2+1} = \lim_{x\to\infty}\frac{\dfrac{3x^2}{x^2}+\dfrac{5x}{x^2}-\dfrac{1}{x^2}}{\dfrac{2x^2}{x^2}+\dfrac{1}{x^2}}$$

$$= \lim_{x\to\infty}\frac{3+\dfrac{5}{x}-\dfrac{1}{x^2}}{2+\dfrac{1}{x^2}}$$

$$= \frac{\displaystyle\lim_{x\to\infty}3 + \lim_{x\to\infty}\frac{5}{x} - \lim_{x\to\infty}\frac{1}{x^2}}{\displaystyle\lim_{x\to\infty}2 + \lim_{x\to\infty}\frac{1}{x^2}}$$ By applying the product and quotient properties of limits

$$= \frac{3+0-0}{2+0}$$

Therefore, $\displaystyle\lim_{x\to\infty}\frac{3x^2+5x-1}{2x^2+1}=\frac{3}{2}$

The limits in parts (b) and (c) show that in some cases the limit of a function is itself a function.

We can relate the limit in Example 12.3 part (d) to the end behaviour of rational functions covered in Section 2.4. Since $\displaystyle\lim_{x\to\infty}\frac{3x^2+5x-1}{2x^2+1}=\frac{3}{2}$, the rational function $y=\dfrac{3x^2+5x-1}{2x^2+1}$ will have a horizontal asymptote at $y=\dfrac{3}{2}$. In other words as $x\to+\infty$ or $x\to-\infty$ the function approaches the value of $y=\dfrac{3}{2}$.

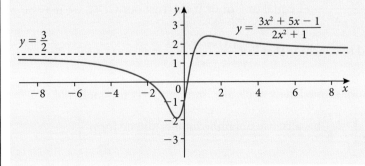

So far we have evaluated limits by guessing and checking with the help of a GDC. This process led us to conclude that $\lim\limits_{x \to 0} \dfrac{\sin x}{x} = 1$ and $\lim\limits_{x \to 0} \dfrac{\cos x - 1}{x} = 0$. It was reasonable to take this approach since it is not possible to perform algebraic manipulations on these expressions as we did with the expressions in Example 12.3. However, if possible, we should always try to use analytical methods to evaluate a limit as illustrated in Example 12.4.

Example 12.4

(a) Estimate the value of $\lim\limits_{x \to 0} \dfrac{\sqrt{x^2 + 4} - 2}{x^2}$ by evaluating the function

$$f(x) = \frac{\sqrt{x^2 + 4} - 2}{x^2} \text{ for } x = \pm 0.1, \pm 0.01, \pm 0.001, \pm 0.0001, \pm 0.00001,$$
$$\pm 0.000001, \pm 0.0000001, \pm 0.00000001$$

(b) Using algebra and the properties of limits, evaluate $\lim\limits_{x \to 0} \dfrac{\sqrt{x^2 + 4} - 2}{x^2}$

Solution

(a) A GDC that displays results to an accuracy of ten significant figures gives the following results.

x	$f(x)$
± 0.1	0.2498439450
± 0.01	0.2499984375
± 0.001	0.2499999843
± 0.0001	0.2499999763
± 0.00001	0.2500000207
± 0.000001	0.2500222251
± 0.0000001	0
± 0.00000001	0

The GDC results in the table seem unusual. Initially, as x approaches zero from either direction the function values appear to be approaching $\dfrac{1}{4}$, but then as the function is evaluated for values even closer to zero, the function values continue to decrease to zero.

Is $\lim\limits_{x \to 0} \dfrac{\sqrt{x^2 + 4} - 2}{x^2}$ equal to $\dfrac{1}{4}$ or 0? If we trust our GDC, we may be tempted to conclude that $\lim\limits_{x \to 0} \dfrac{\sqrt{x^2 + 4} - 2}{x^2} = 0$.

(b) We cannot immediately apply the limit property for quotients,

$\lim\limits_{x \to a} \left[\dfrac{f(x)}{g(x)} \right] = \dfrac{L}{K}$ because we obtainhe indeterminate form $\dfrac{0}{0}$.

We need to use the algebraic technique of multiplying numerator and denominator by the conjugate of the expression in the numerator. This will rationalise the numerator and may lead to an equivalent expression for which we can apply the quotient property for limits.

$$\lim_{x \to 0} \frac{\sqrt{x^2 + 4} - 2}{x^2} = \lim_{x \to 0} \frac{\sqrt{x^2 + 4} - 2}{x^2} \cdot \frac{\sqrt{x^2 + 4} + 2}{\sqrt{x^2 + 4} + 2}$$

$$= \lim_{x \to 0} \frac{\left(\sqrt{x^2 + 4}\right)^2 - 2^2}{x^2\left(\sqrt{x^2 + 4} + 2\right)}$$

$$= \lim_{x \to 0} \frac{x^2 + 4 - 4}{x^2\left(\sqrt{x^2 + 4} + 2\right)}$$

$$= \lim_{x \to 0} \frac{\cancel{x^2}}{\cancel{x^2}\left(\sqrt{x^2 + 4} + 2\right)}$$

$$= \lim_{x \to 0} \frac{1}{\sqrt{x^2 + 4} + 2}$$

$$= \frac{\lim_{x \to 0} 1}{\lim_{x \to 0}\left(\sqrt{x^2 + 4} + 2\right)} = \frac{1}{\sqrt{4} + 2} = \frac{1}{4}$$

Therefore, $\lim_{x \to 0} \dfrac{\sqrt{x^2 + 4} - 2}{x^2} = \dfrac{1}{4}$

A GDC will sometimes give a false value because of memory limitations. As $\sqrt{x^2 + 4}$ is very close to 2 when x is very small, a GDC will eventually consider $\sqrt{x^2 + 4}$ to be equal to $2.00\,000\,000 \dots$ (to as many digits as the GDC is capable of computing) when x is sufficiently small. Your GDC is a very powerful tool, but like any tool it does have its limitations.

Exercise 12.1

1. Evaluate each limit algebraically and then confirm your result using a table or graph on your GDC.

(a) $\displaystyle\lim_{n \to \infty} \frac{1 + 4n}{n}$

(b) $\displaystyle\lim_{h \to 0}(3x^2 + 2hx + h^2)$

(c) $\displaystyle\lim_{d \to 0} \frac{(x + d)^2 - x^2}{d}$

(d) $\displaystyle\lim_{x \to 3} \frac{x^2 - 9}{x - 3}$

2. Investigate the limit of each expression below (if it exists) as $x \to \infty$, by evaluating the expression for the following values of x: 10, 50, 100, 1000, 10000 and 1000000. Hence, make a conjecture for the value of each limit.

(a) $\displaystyle\lim_{x \to \infty} \frac{3x + 2}{x^2 - 3}$

(b) $\displaystyle\lim_{x \to \infty} \frac{5x - 6}{2x + 5}$

(c) $\displaystyle\lim_{x \to \infty} \frac{3x^2 + 2}{x - 3}$

For part 3(c), multiply numerator and denominator by the conjugate of the numerator.

For part 3(e), rewrite $\tan x$ as $\frac{\sin x}{\cos x}$ and apply the property $\lim_{x\to a}[f(x) \cdot g(x)] = L \cdot K$

For part 3(f), rewrite $\frac{\sin 3\theta}{\theta}$ as $3\left(\frac{\sin 3\theta}{3\theta}\right)$ and apply $\lim_{x\to 0}\frac{\sin x}{x} = 1$

3. Find each limit, if it exists.

(a) $\lim_{x\to 4}\dfrac{x-4}{x^2-16}$

(b) $\lim_{x\to 1}\dfrac{x^2+x-2}{x^2-1}$

(c) $\lim_{x\to 0}\dfrac{\sqrt{2+x}-\sqrt{2}}{x}$

(d) $\lim_{x\to\infty}\dfrac{x^3-1}{4x^3-3x+1}$

(e) $\lim_{x\to 0}\dfrac{\tan x}{x}$

(f) $\lim_{\theta\to 0}\dfrac{\sin 3\theta}{\theta}$

4. Use the graphing or table capabilities of your GDC to investigate the values of the expression $\left(1+\dfrac{1}{c}\right)^c$ as c increases without bound (i.e. $c\to\infty$). Explain the significance of the result.

5. It is known that the line $y=3$ is a horizontal asymptote for the function $f(x)$. State the value of each limit.

(a) $\lim_{x\to\infty} f(x)$

(b) $\lim_{x\to-\infty} f(x)$

6. It is known that the line $x=a$ is a vertical asymptote for the function $g(x)$ and $g(x)>0$. What conclusion can be made about $\lim_{x\to a} g(x)$?

7. State the equations of all horizontal and vertical asymptotes for each function. Confirm your results by graphing the function on your GDC.

(a) $f(x)=\dfrac{3x-1}{1+x}$

(b) $g(x)=\dfrac{1}{(x-2)^2}$

(c) $h(x)=\dfrac{1}{x-a}+b$

(d) $k(x)=\dfrac{2x^2-3}{x^2-9}$

(e) $d(x)=\dfrac{5-3x}{x^2-5x}$

(f) $p(x)=\dfrac{x^2-4}{x-4}$

8. For each limit, (i) use your GDC to estimate the limit, (ii) use analytical methods to evaluate the limit.

(a) $\lim_{x\to 2}\dfrac{\sqrt{x^2+5}-3}{x^2-2x}$

(b) $\lim_{x\to+\infty}\dfrac{4x-1}{\sqrt{x^2+2}}$

9. Show that $\lim_{h\to 0}\dfrac{\sqrt{x+h}-\sqrt{x}}{h}=\dfrac{1}{2\sqrt{x}}$

10. Show that $\lim_{h\to 0}\dfrac{\dfrac{1}{x+h}-\dfrac{1}{x}}{h}=-\dfrac{1}{x^2}$

12.2 The derivative of a function: definition and basic rules

Tangent lines and the slope (gradient) of a curve

Any linear function can be written in the form $y=mx+c$. This is the slope-intercept form for a linear equation where m is the slope (or gradient) of the graph and c is the y-coordinate of the point at which the graph intersects the y-axis (i.e. the y-intercept). The value of the slope m, defined as

$$m = \frac{y_2 - y_1}{x_2 - x_1} = \frac{\text{vertical change}}{\text{horizontal change}},$$ will be the same for any pair of points,

(x_1, y_1) and (x_2, y_2), on the line. An essential characteristic of the graph of a linear function is that it has a constant slope. This is not true for the graphs of non-linear functions.

Consider a boy walking up the side of a pitched roof (Figure 12.4).

At any point along the line segment PQ the boy is experiencing a slope of $\frac{3}{4}$.

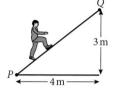

Figure 12.4 Slope of a straight line

Now consider the boy walking up the curve shown in Figure 12.5 that passes through the three points A, B, and C. As the boy walks along the curve from A to C, he will experience a steadily increasing slope. The slope is continually changing from one point to the next along the curve. Therefore, it is incorrect to say that a non-linear function, whose

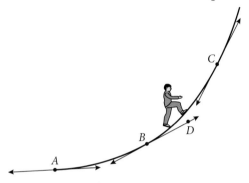

Figure 12.5 Slope of a curve

graph is a curve, has a slope – it has infinitely many slopes. We need a means to determine the slope of a non-linear function at a specific point on its graph.

Imagine if the slope of the curve in Figure 12.5 stopped increasing (remained constant) after point B. From that point on, the boy would move along a line with a slope equal to the slope of the curve at point B. This line – containing point D in the diagram – touches the curve just once, at point B. Line BD is **tangent** to the curve at point B. Therefore, finding the slope of the line that is tangent to a curve at a certain point will give us the slope of the curve at that point.

Finding the slope of a curve at a point – or better – finding a rule (function) that gives us the slope at any point on the curve is very useful information in many applications. The slope of a line, or of a curve at a point, is a measure of how fast variable y is changing as variable x changes. The slope represents the rate of change of y with respect to x. To find the slope of a tangent line we first need to clarify what it means to say that a line is tangent to a curve at a point. Then we can establish a method to find the tangent line at a point.

The three graphs in Figure 12.6 show different configurations of tangent lines. A tangent line may cross or intersect the graph at one or more points.

The slope (gradient) of a curve at a point is the slope of the line that is tangent to the curve at that point.

The word 'curve' can often mean the same as 'function', even if the function is linear.

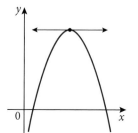
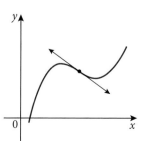
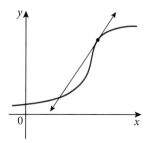

Figure 12.6 Different configurations of lines tangent to a curve

583

For many functions, the graph has a
tangent at every point. Informally, a
function is said to be smooth if it has this
property. Any linear function is certainly
smooth, since the tangent at each point
coincides with the original graph. However,
some graphs are not smooth at every point.
Consider the point (0, 0) on the graph of the
function $y = |x|$ (Figure 12.7). Zooming

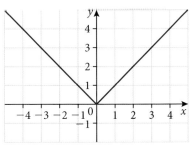

Figure 12.7 $y = |x|$

in on (0, 0) will always produce a V-shape rather than smoothing out to appear
more and more linear. Therefore, there is no tangent to the graph at this point.

One way to find the tangent line of a graph at a particular point is to make
a visual estimate. Figure 12.8 shows the distance–time graph for an object's
motion. The slope at any point (t, y) on the curve will give us the rate of
change of the distance y with respect to time t, in other words the object's
instantaneous velocity at time t. In the figure, an estimate of the line tangent to
the curve at (5, 3) has been drawn. Reading from the graph, the slope appears to
be $\frac{4}{6} = \frac{2}{3}$. In other words, the object has a velocity of approximately 0.667 m s^{-1}
at the instant when $t = 5$ seconds.

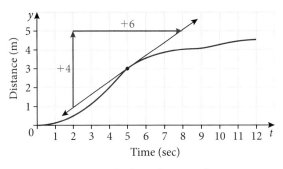

Figure 12.8 Estimating the slope of a tangent line

A more precise method of finding tangent lines makes use of a **secant line** and
a **limit process**. Suppose that f is any smooth function, so the tangent to its
graph exists at all points. A secant line (or chord) is drawn through the point
for which we are trying to find a tangent to f and a second point on the graph
of f, as shown in Figure 12.9 (a). If P is the point of tangency with coordinates
$(x, f(x))$, then choose a point Q to be a horizontal distance of h units away.
Hence, the coordinates of point Q are $(x + h, f(x + h))$ and the slope of the

secant line (PQ) is $m_{sec} = \dfrac{f(x + h) - f(x)}{(x + h) - x} = \dfrac{f(x + h) - f(x)}{h}$

The right side of this equation is often referred to as a difference quotient. The
numerator is the change in y, and the denominator h is the change in x. The
limit process of achieving better and better approximations for the slope of
the tangent at P consists of finding the slope of the secant PQ as Q moves ever
closer to P, as shown in the graphs in Figure 12.9 (b) and Figure 12.9 (c).
In doing so, the value of h will approach zero.

12. Show that there are two points at which the function $h(x) = 2x^2 - x^4$ has a maximum value, and one point at which h has a minimum value. Find the coordinates of these three points, indicating whether each is a maximum or minimum.

13. The normal to the curve $y = x^{\frac{1}{2}} + x^{\frac{1}{3}}$ at the point $(1, 2)$ meets the axes at $(a, 0)$ and $(0, b)$. Find a and b.

14. The displacement, s metres, of a car, t seconds after leaving a fixed point A, is given by $s(t) = 10t - \dfrac{1}{2}t^2$

    (a) Calculate the velocity when $t = 0$.

    (b) Calculate the value of t when the velocity is zero.

    (c) Calculate the displacement of the car from A when the velocity is zero.

15. A ball is thrown vertically upwards from ground level such that its height h at t seconds is given by $h = 14t - 4.9t^2$

    (a) Write expressions for the ball's velocity and acceleration.

    (b) Find the maximum height the ball reaches and the time it takes to reach the maximum.

    (c) At the moment the ball reaches its maximum height, what is the ball's velocity and acceleration?

16. Find the exact coordinates of the inflection point on the curve $y = x^3 + 12x^2 - x - 12$.

17. You are given the function $f(x) = 2\cos x - 3$. At the point on the curve where $x = \dfrac{\pi}{3}$, find:

    (a) the equation of the line tangent to f

    (b) the equation of the line normal to f.

    Express both equations exactly.

18. A manufacturer produces closed cylindrical cans of radius r cm and height h cm. Each can has a total surface area of 54π cm^2.

    (a) Solve for h in terms of r, and hence find an expression for the volume, V cm^3, of each can in terms of r.

    (b) Find the value of r for which the cans have their maximum possible volume.

19. The curve $y = ax^2 + bx + c$ has a maximum point at $(2, 18)$ and passes through the point $(0, 10)$. Find a, b, and c.

20. For the function $f(x) = \frac{1}{2}x^2 - 5x + 3$, find:

(a) the equation of the tangent line at $x = -2$

(b) the equation of the normal line at $x = -2$.

21. Consider the function $f(x) = x^4 - x^3$.

(a) Find the coordinates of any maximum or minimum points. Identify each as relative or absolute.

(b) State the domain and range of f.

(c) Find the coordinates of any inflection point(s).

(d) Sketch the function, clearly indicating any maximum, minimum, or inflection points.

22. Evaluate each limit.

(a) $\lim\limits_{x \to \infty} \dfrac{2 - 3x + 5x^2}{8 - 3x^2}$

(b) $\lim\limits_{x \to 0} \dfrac{\sqrt{x + 4} - 2}{x}$

(c) $\lim\limits_{x \to 1} \dfrac{x^3 - 1}{x - 1}$

(d) $\lim\limits_{h \to 0} \dfrac{\sqrt{(x + h) + 2} - \sqrt{x + 2}}{h}$

23. Find the derivative $f'(x)$ for each function.

(a) $f(x) = \dfrac{x^2 - 4x}{\sqrt{x}}$

(b) $f(x) = x^3 - 3\sin x$

(c) $f(x) = \dfrac{1}{x} + \dfrac{x}{2}$

(d) $f(x) = \dfrac{7}{3x^{13}}$

24. A point (p, q) is on the graph of $y = x^3 + x^2 - 9x - 9$, and the line tangent to the graph at (p, q) passes through the point $(4, -1)$. Find p and q.

25. For what values of c, such that $c \geq 0$, is the line $y = -\dfrac{1}{12}x + c$ normal to the graph of $y = x^3 + \dfrac{1}{3}$?

26. Find the points on the curve $y = \dfrac{1}{3}x^3 - x$ where the tangent line is parallel to the line $y = 3x$.

27. At what point does the line that is normal to the graph of $y = x - x^2$ at the point $(1, 0)$ intersect the graph of the curve a second time?

28. If $f(x) = \sqrt{x + 2}$, find $f'(x)$ by first principles.

29. An object moves along a line according to the position function $s(t) = t^3 - 9t^2 + 24t$. Find the positions of the object when:

    (a) its velocity is zero

    (b) its acceleration is zero.

30. A particle moves along a straight line in the time interval $0 \leqslant t \leqslant 2\pi$ such that its displacement from the origin O is s metres, given by the function $s = t + \sin t$.

    (a) Find the value(s) of t in the interval $0 \leqslant t \leqslant 2\pi$ when the particle changes direction.

    (b) Show that the particle always remains on the same side of the origin O.

    (c) Find the value(s) of t in the interval $0 \leqslant t \leqslant 2\pi$ when the acceleration of the particle is zero.

    (d) Sketch a graph of the displacement of the particle from O for $0 \leqslant t \leqslant 2\pi$, and state the maximum value of s in this interval.

31. The curve whose equation is $y = ax^3 + bx^2 + cx + d$ has a point of inflection at $(-1, 4)$, a turning point when $x = 2$, and passes through the point $(3, -7)$. Find the values of a, b, c, and d, and the y-coordinate of the turning point.

32. Find the stationary values of the function $f(x) = 1 - \dfrac{9}{x^2} + \dfrac{18}{x^4}$ and determine their nature.

33. (a) Find the equation of the tangent to the curve $y = \dfrac{1}{x}$ at the point $(1, 1)$

    (b) Find the equation of the tangent to the curve $y = \cos x$ at the point $\left(\dfrac{\pi}{2}, 0\right)$

    (c) Deduce that $\dfrac{1}{x} > \cos x$ for $0 \leqslant x \leqslant \dfrac{\pi}{2}$

34. Show that there is just one tangent to the curve $y = x^3 - x + 2$ that passes through the origin. Find its equation and the coordinates of the point of tangency.

35. The displacement, s metres, of a moving body B from a fixed point O at time t seconds is given by $s = 50t - 10t^2 + 1000$

    (a) Find the velocity of B in $m\,s^{-1}$

    (b) Find its maximum displacement from O.

36. The diagram shows a sketch of the graph of $y = f'(x)$ for $a \leqslant x \leqslant b$

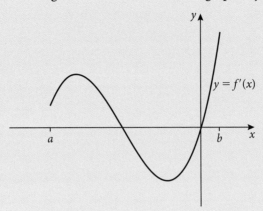

Sketch the graph of $y = f(x)$ for $a \leqslant x \leqslant b$, given that $f(0) = 0$ and $f(x) \geqslant 0$ for all x. On your graph you should clearly indicate any minimum or maximum points, or points of inflection.

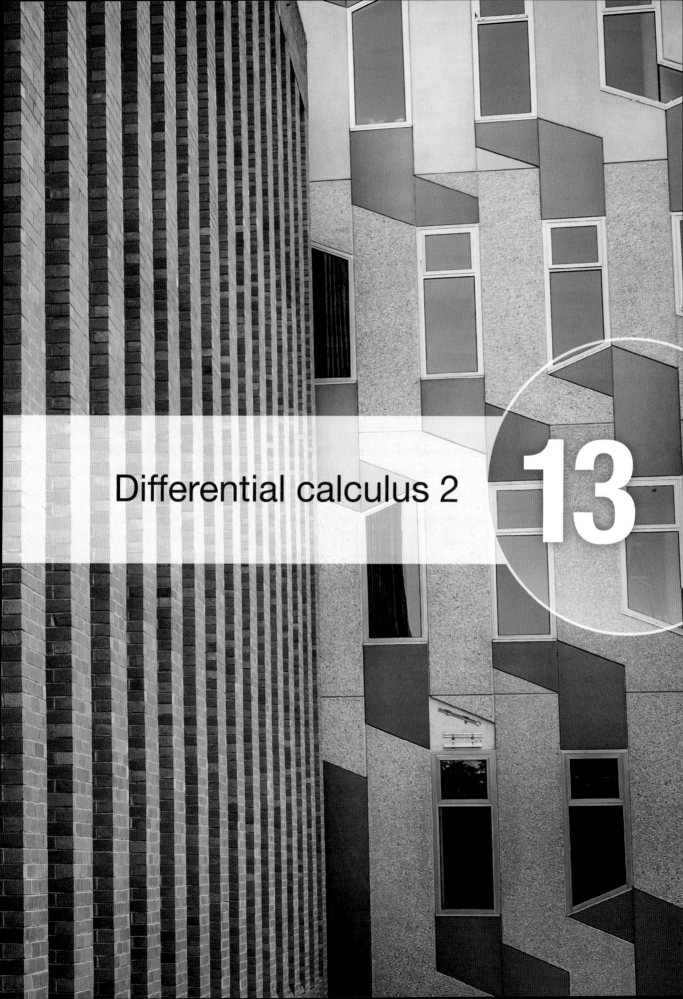

Differential calculus 2

13

Learning objectives

By the end of this chapter, you should be familiar with...

- finding the derivative of a composite function
- finding the derivative of a function in the form of a product or quotient
- finding the derivative of trigonometric and inverse trigonometric functions
- finding the derivative of exponential and logarithmic functions
- finding the rate of change of a variable that is defined implicitly (implicit differentiation)
- finding the rate of change of a variable from the relationship between two or more rates of change (related rates)
- solving problems requiring a solution that is an optimum – that is, a maximum or minimum (optimisation)
- using l'Hôpital's rule to evaluate a limit that is in indeterminate form.

Chapter 12 established some fundamental concepts and techniques of differential calculus, involving differentiation of functions. This chapter expands our set of differentiation rules and techniques and extends the applications introduced in Chapter 12, particularly using methods of finding extrema in the context of finding an optimum solution and solving problems involving more than one rate of change.

13.1 Derivatives of composite functions, products and quotients

The chain rule

We know how to differentiate functions such as $f(x) = x^3 + 2x - 3$ and $g(x) = \sqrt{x}$, but how do we differentiate the composite function $g(f(x)) = \sqrt{x^3 + 2x - 3}$?
We use the **chain rule** to compute the derivative of the composite of two functions. Because most functions that we encounter in applications are composites of other functions, it can be argued that the chain rule is the most important and most widely used rule of differentiation.

Table 13.1 shows some examples of functions that we can differentiate with the rules that we learned in Chapter 12, and further examples of functions that are best differentiated with the chain rule.

Differentiate without the chain rule	Differentiate with the chain rule
$y = \cos x$	$y = \cos 3x$
$y = 3x^2 + 5x$	$y = \sqrt{3x^2 + 5x}$
$y = \dfrac{1}{3x^2}$	$y = \dfrac{1}{3x^2 + x}$
$y = \sin x$	$y = \sin^2 x$

Table 13.1 Examples of functions that can be differentiated with the chain rule

The chain rule says that, given two functions, the derivative of their composite is the product of their derivatives – remembering that a derivative is a rate of change of one quantity (variable) with respect to another quantity (variable). For example, the function $y = 8x + 6 = 2(4x + 3)$ is the composite of the functions $y = 2u$ and $u = 4x + 3$. Note that the function y is in terms of u, and the function u is in terms of x. How are the derivatives of these three functions related?

$\dfrac{dy}{dx} = 8$, $\dfrac{dy}{du} = 2$ and $\dfrac{du}{dx} = 4$. Since $8 = 2 \cdot 4$, the derivatives are related such that $\dfrac{dy}{dx} = \dfrac{dy}{du} \cdot \dfrac{du}{dx}$. In other words, rates of change multiply.

Again, if we think of derivatives as rates of change, we can illustrate the relationship $\dfrac{dy}{dx} = \dfrac{dy}{du} \cdot \dfrac{du}{dx}$ with a practical example.

Consider the pair of levers in Figure 13.1 with lever endpoints U and U' connected by a segment that can shrink and stretch but always remains horizontal. Hence, points U and U' are always the same distance u from the ground.

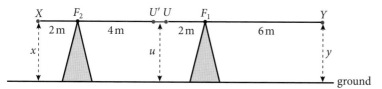

Figure 13.1 Two levers with horizontal connection between U' and U

As point Y moves down (Figure 13.2), points U and U' move up, and point X moves down, all at different rates. Let dy, du and dx represent the change in distance from the ground for the points Y, U and X, respectively. Because $YF_1 = 6$ and $UF_1 = 2$, then if point Y moves such that $dy = 3$, then $du = 1$. Since $U'F_2 = 4$ and $XF_2 = 2$, then if point U' moves so that $du = 2$, then $dx = 1$.

Hence, $\dfrac{dy}{du} = 3$ and $\dfrac{du}{dx} = 2$

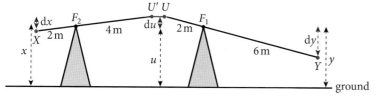

Figure 13.2 dx, du and dy represent the change in distance from the ground for X, U and Y

Combining these two results, we can see that for every 6 units that Y's distance changes, X's distance will change 1 unit. That is, $\dfrac{dy}{dx} = 6$. Therefore, we can write $\dfrac{dy}{dx} = \dfrac{dy}{du} \cdot \dfrac{du}{dx} = 3 \cdot 2 = 6$. In other words, the rate of change of y with respect to x is the product of the rate of change of y with respect to u and the rate of change of u with respect to x.

Example 13.1

The polynomial function $y = 16x^4 - 8x^2 + 1 = (4x^2 - 1)^2$ is the composite of $y = u^2$ and $u = 4x^2 - 1$. Use the chain rule to find $\dfrac{dy}{dx}$, the derivative of y with respect to x.

Solution

$$y = u^2 \quad \Rightarrow \quad \frac{dy}{du} = 2u$$

$$u = 4x^2 - 1 \quad \Rightarrow \quad \frac{du}{dx} = 8x$$

Applying the chain rule: $\dfrac{dy}{dx} = \dfrac{dy}{du} \cdot \dfrac{du}{dx} = 2u \cdot 8x$

$$= 2(4x^2 - 1) \cdot 8x$$

$$= 64x^3 - 16x$$

In this particular case, we could have differentiated the function in expanded form by differentiating term by term rather than differentiating the factorised form by the chain rule.

$\dfrac{dy}{dx} = \dfrac{d}{dx}(16x^4 - 8x^2 + 1) = 64x^3 - 16x$. It is not always easier to differentiate powers of polynomials by expanding and then differentiating term by term. For example, it is far easier to find the derivative of $y = (3x + 5)^8$ by using the chain rule.

Composite functions are often written using nested function notation. For example, the notation $f(g(x))$ denotes a function composed of functions f and g such that g is the inside function and f is the outside function. For the composite function $y = (4x^2 - 1)^2$ in Example 13.1, the inside function is $g(x) = 4x^2 - 1$ and the outside function is $f(u) = u^2$. Looking again at the solution for Example 13.1, we see that we can choose to express and work out the chain rule in function notation rather than Leibniz notation.

For $y = f(g(x)) = (4x^2 - 1)^2$ $y = f(u) = u^2, \ u = g(x) = 4x^2 - 1$

Leibniz notation	**Function notation**
$\dfrac{dy}{dx} = \dfrac{dy}{du} \cdot \dfrac{du}{dx} = 2u \cdot 8x$	$\dfrac{d}{dx}[f(g(x))] = f'(u) \cdot g'(x) = 2u \cdot 8x$
$= 2(4x^2 - 1) \cdot 8x$	$= f'(g(x)) \cdot g'(x) = 2(4x^2 - 1) \cdot 8x$
$= 64x^3 - 16x$	$= 64x^3 - 16x$

This leads us to formally state the chain rule in two different notations.

The chain rule needs to be applied carefully. Consider the function notation form for the chain rule $\dfrac{d}{dx}[f(g(x))] = f'(g(x)) \cdot g'(x)$. Although it is the product of two derivatives, it is important to point out that the first derivative involves the function f differentiated at $g(x)$ and the second is function g differentiated at x.

The chain rule
If $y = f(u)$ is a function in terms of u and $u = g(x)$ is a function in terms of x, then the function $y = f(g(x))$ is differentiated as follows.
In Leibniz form:
$$\frac{dy}{dx} = \frac{dy}{du} \cdot \frac{du}{dx}$$
In function notation form:
$$\frac{dy}{dx} = \frac{d}{dx}[f(g(x))] = f'(g(x)) \cdot g'(x)$$

The chain rule is easy to remember in Leibniz form, $\dfrac{dy}{dx} = \dfrac{dy}{du} \cdot \dfrac{du}{dx}$, but you should remember that they are not fractions. The expressions $\dfrac{dy}{dx}, \dfrac{dy}{du}$ and $\dfrac{du}{dx}$ are derivatives or, more precisely, limits, and although du and dx essentially represent very small changes in the variables u and x, we cannot guarantee that they are non-zero.

The function notation form of the chain rule offers a very useful way of saying the rule in words and, thus, a very useful structure for applying it.

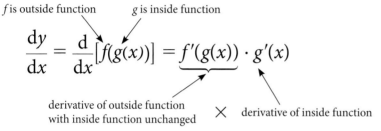

The chain rule in words:

$$\left(\begin{array}{c} \text{derivative of} \\ \text{composite} \end{array} \right) = \left(\begin{array}{c} \text{derivative of outside function} \\ \text{with inside function unchanged} \end{array} \right) \times \left(\begin{array}{c} \text{derivative of} \\ \text{inside function} \end{array} \right)$$

Although this is taking some liberties with mathematical language, the mathematical interpretation of the phrase 'with inside function unchanged' is that the derivative of the outside function f is evaluated at $g(x)$, the inside function.

Example 13.2

Differentiate each function by applying the chain rule.

(a) $y = \cos 3x$

(b) $y = \sqrt{3x^2 + 5x}$

(c) $y = \dfrac{1}{3x^2 + x}$

(d) $y = \sin^2 x$

(e) $y = \sin x^2$

(f) $y = \sqrt[3]{(7 - 5x)^2}$

Solution

(a) Start by 'decomposing' the composite function into the outside function and the inside function. $y = f(g(x)) = \cos 3x$, so the outside function is $f(u) = \cos u$ and the inside function is $g(x) = 3x$

In Leibniz form: $\dfrac{dy}{dx} = \dfrac{dy}{du} \cdot \dfrac{du}{dx} = (-\sin u) \cdot 3 = -3\sin(3x)$

Or, alternatively, in function notation form:

$\dfrac{dy}{dx} = f'(g(x)) \cdot g'(x) = \underbrace{[-\sin(3x)]} \cdot 3 = -3\sin(3x)$

derivative of outside function
with inside function unchanged $\qquad \times \qquad$ derivative of inside function

The chain rule is our most important rule of differentiation. It is an indispensable tool in differential calculus. Forgetting to apply the chain rule when it needs to be applied, or applying it improperly, is a common source of errors in calculus computations. It is important to understand it, practise it and master it.

The chain rule acquired its name because we use it to take derivatives of composites of functions by 'chaining' together their derivatives. A function could be the composite of more than two functions. When a function is the composite of three functions then we take the product of three derivatives 'chained' together. For example, if $y = f(u)$, $u = g(v)$ and $v = h(x)$, then the derivative of the function $y = f(g(h(x)))$ is $\dfrac{dy}{dx} = \dfrac{dy}{du} \cdot \dfrac{du}{dv} \cdot \dfrac{dv}{dx}$

(b) $y = f(g(x)) = \sqrt{3x^2 + 5x}$

So the outside function is $f(u) = \sqrt{u} = u^{\frac{1}{2}} \Rightarrow f'(u) = \dfrac{1}{2}u^{-\frac{1}{2}}$

and the inside function is $g(x) = 3x^2 + 5x$

$$\frac{dy}{dx} = f'(g(x)) \cdot g'(x) = \frac{1}{2}(3x^2 + 5x)^{-\frac{1}{2}} \cdot (6x + 5)$$

$$= \frac{6x + 5}{2(3x^2 + 5x)^{\frac{1}{2}}} \quad \text{or} \quad \frac{6x + 5}{2\sqrt{3x^2 + 5x}}$$

(c) $y = f(g(x)) = \dfrac{1}{3x^2 + x}$

So the outside function is $f(u) = \dfrac{1}{u} = u^{-1} \Rightarrow f'(u) = -u^{-2}$

and the inside function is $g(x) = 3x^2 + x$

$$\frac{dy}{dx} = f'(g(x)) \cdot g'(x) = -(3x^2 + x)^{-2} \cdot (6x + 1)$$

$$= -\frac{6x + 1}{(3x^2 + x)^2}$$

(d) The expression $\sin^2 x$ is an abbreviated way of writing $(\sin x)^2$.

$y = f(g(x)) = \sin^2 x = (\sin x)^2$

So the outside function is $f(u) = u^2 \Rightarrow f'(u) = 2u$

and the inside function is $g(x) = \sin x$

$$\frac{dy}{dx} = f'(g(x)) \cdot g'(x) = 2\sin x \cdot \cos x$$

$$= 2\sin x \cos x$$

(e) The expression $\sin x^2$ is equivalent to $\sin(x^2)$, and is not $(\sin x)^2$.

$y = f(g(x)) = \sin(x^2)$

So the outside function is $f(u) = \sin u \Rightarrow f'(u) = \cos u$

and the inside function is $g(x) = x^2$

$$\frac{dy}{dx} = f'(g(x)) \cdot g'(x) = \cos(x^2) \cdot 2x$$

$$= 2x\cos(x^2)$$

(f) First change from radical (surd) form to rational exponent form.

$y = f(g(x)) = (7 - 5x)^{\frac{2}{3}}$

So the outside function is $f(u) = u^{\frac{2}{3}} \Rightarrow f'(u) = \dfrac{2}{3}u^{-\frac{1}{3}}$

and the inside function is $g(x) = 7 - 5x$

$$\frac{dy}{dx} = f'(g(x)) \cdot g'(x) = \frac{2}{3}(7 - 5x)^{-\frac{1}{3}} \cdot (-5)$$

$$= -\frac{10}{3(\sqrt[3]{7 - 5x})}$$

Aim to write a function in a way that eliminates any confusion regarding the argument of the function. For example, write $\sin(x^2)$ rather than $\sin x^2$; $1 + \ln x$ rather than $\ln x + 1$; $5 + \sqrt{x}$ rather than $\sqrt{x} + 5$; $\ln(4 - x^2)$ rather than $\ln 4 - x^2$.

Example 13.3

Find the derivative of the function $y = (2x + 3)^3$

(a) by expanding the binomial and differentiating term by term

(b) by the chain rule.

Solution

(a) $y = (2x + 3)^3 = (2x + 3)(2x + 3)^2$

$$= (2x + 3)(4x^2 + 12x + 9)$$

$$= 8x^3 + 24x^2 + 18x + 12x^2 + 36x + 27$$

$$= 8x^3 + 36x^2 + 54x + 27$$

$$\frac{dy}{dx} = 24x^2 + 72x + 54$$

(b) $y = f(g(x)) = (2x + 3)^3 \Rightarrow y = f(u) = u^3; u = g(x) = 2x + 3$

$$\Rightarrow f'(u) = 3u^2; g'(x) = 2$$

$$\frac{dy}{dx} = \frac{dy}{du} \cdot \frac{du}{dx} = 3u^2 \cdot 2 = 6u^2$$

$$= 6(2x + 3)^2$$

$$= 6(4x^2 + 12x + 9)$$

$$= 24x^2 + 72x + 54$$

The product rule

With the differentiation rules that we have learned so far, we can differentiate some functions that are products. For example, we can differentiate the function $f(x) = (x^2 + 3x)(2x - 1)$ by expanding and then differentiating the polynomial term by term. In doing so, we are applying the sum and difference, constant multiple, and power rules from Section 12.2.

$$f(x) = (x^2 + 3x)(2x - 1) = 2x^3 + 5x^2 - 3x$$

$$f'(x) = 2\frac{d}{dx}(x^3) + 5\frac{d}{dx}(x^2) - 3\frac{d}{dx}(x)$$

$$f'(x) = 6x^2 + 10x - 3$$

The sum and difference rule states that the derivative of a sum/difference of two functions is the sum/difference of their derivatives. Is the derivative of the product of two functions the product of their derivatives? Let's try an example.

$$f(x) = (x^2 + 3x)(2x - 1)$$

$$f'(x) = \frac{d}{dx}(x^2 + 3x) \cdot \frac{d}{dx}(2x - 1)$$

$$f'(x) = (2x + 3) \cdot 2$$

$$f'(x) = 4x + 6$$

633

This is not the same result that we obtained earlier.

So the derivative of a product of two functions is not the product of their derivatives. However, there are many products, such as $y = (4x - 3)^3(x - 1)^4$ and $f(x) = x^2\sin x$, for which it is either difficult or impossible to write the function as a polynomial.

The product rule

If y is a function in terms of x that can be expressed as the product of two functions u and v that are also in terms of x, then the product $y = uv$ can be differentiated using the **product rule.**

$$\frac{dy}{dx} = \frac{d}{dx}(uv) = u\frac{dv}{dx} + v\frac{du}{dx}$$

or, equivalently, if $y = f(x) \cdot g(x)$, then

$$\frac{dy}{dx} = \frac{d}{dx}[f(x) \cdot g(x)] = f(x) \cdot g'(x) + g(x) \cdot f'(x)$$

An intuitive justification of the product rule can be provided by considering the product rule written in the form $\dfrac{dy}{dx} = \dfrac{d}{dx}(uv) = u\dfrac{dv}{dx} + v\dfrac{du}{dx}$ and analysing the relationship between the functions u, v, and y when there is a small change in the variable x. Recall that the definition of the derivative (Section 12.2),

is essentially the limit of $\dfrac{\text{change in } y}{\text{change in } x}$ as the 'change in x' goes to zero.

Let δx (read 'delta x') and δy represent small changes in x and y, respectively.

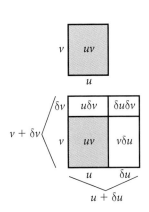

Figure 13.3 A graphic representation of the product rule

As $\delta x \to 0$, then $\dfrac{\delta y}{\delta x} \to \dfrac{dy}{dx}$, the derivative of y with respect to x. Any small change in x, δx, will cause small changes, δu and δv, in the values of functions u and v, respectively. Since $y = uv$, these changes will also cause a small change, δy, in the value of function y.

Now, consider the rectangles in Figure 13.3. The area of the first smaller rectangle is $y = uv$.

The values of u and v then increase by δu and δv, respectively.

The area of the larger rectangle is $y + \delta y = uv + u\delta v + v\delta u + \delta u\delta v$.

The product uv changes by the amount $\delta y = u\delta v + v\delta u + \delta u\delta v$.

Dividing through by δx gives $\dfrac{\delta y}{\delta x} = u\dfrac{\delta v}{\delta x} + v\dfrac{\delta u}{\delta x} + \delta u\dfrac{\delta v}{\delta x}$

Let $\delta x \to 0$ and $\delta u \to 0$, then:

$$\frac{\delta y}{\delta x} = u\frac{\delta v}{\delta x} + v\frac{\delta u}{\delta x} + \delta u\frac{\delta v}{\delta x} \implies \frac{dy}{dx} = u\frac{dv}{dx} + v\frac{du}{dx} + 0 \cdot \frac{dv}{dx}$$

Giving $\dfrac{dy}{dx} = u\dfrac{dv}{dx} + v\dfrac{du}{dx}$, the product rule.

Example 13.4

Find the derivative of the function $y = (x^2 + 3x)(2x - 1)$

(a) by expanding the binomial and differentiating term by term

(b) by the product rule.

Solution

(a) Expanding gives $y = (x^2 + 3x)(2x - 1) = 2x^3 + 5x^2 - 3x$

Therefore, $\dfrac{dy}{dx} = 6x^2 + 10x - 3$

(b) Let $u(x) = x^2 + 3x$ and $v(x) = 2x - 1$

Then $y = u(x) \cdot v(x)$ or simply $y = uv$

Using the product rule (in Leibniz form),

$$\frac{dy}{dx} = \frac{d}{dx}(uv) = u\frac{dv}{dx} + v\frac{du}{dx} = (x^2 + 3x) \cdot 2 + (2x - 1) \cdot (2x + 3)$$

$$= (2x^2 + 6x) + (4x^2 + 4x - 3)$$

$$= 6x^2 + 10x - 3$$

The result obtained from using the product rule agrees with the derivative obtained from differentiating the expanded polynomial.

Example 13.5

Given $y = x^2 \sin x$, find $\dfrac{dy}{dx}$

Solution

Let $y = f(x) \cdot g(x) = x^2 \sin x \Rightarrow f(x) = x^2$ and $g(x) = \sin x$

By the product rule (function notation form),

$$\frac{dy}{dx} = \frac{d}{dx}[f(x) \cdot g(x)] = f(x) \cdot g'(x) + g(x) \cdot f'(x)$$

$$= x^2 \cdot \cos x + (\sin x) \cdot 2x$$

$$\frac{dy}{dx} = x^2 \cos x + 2x \sin x$$

As with the chain rule, it is very helpful to remember the structure of the product rule in words.

first factor \qquad second factor

$$\frac{dy}{dx} = \frac{d}{dx}[f(x) \cdot g(x)] = f(x) \cdot g'(x) + g(x) \cdot f'(x)$$

Product of two functions, i.e. factors $=$ first factor $\times$ derivative of second factor $+$ second factor $\times$ derivative of first factor

Example 13.6

Find an equation of the tangent to the curve $y = \sin x \cos(2x)$ at the point where $x = \dfrac{\pi}{6}$.

Solution

To find the slope of the tangent we need to find the derivative of $y = \sin x \cos(2x)$. To do this, we will have to use more than one of the differentiation rules. First, we need the product rule because the function consists of the two factors $\sin x$ and $\cos(2x)$. The second factor is a composite of cosine and $2x$ so we also need the chain rule. In essence, the application of the chain rule will be nested within the product rule.

Applying the product rule to the entire function:

$$\frac{dy}{dx} = \sin x \frac{d}{dx}(\cos(2x)) + \cos(2x)\frac{d}{dx}\sin x$$

Using the chain rule for $\dfrac{d}{dx}(\cos(2x))$:

$$\frac{dy}{dx} = \sin x(-2\sin(2x)) + \cos(2x)\cos x$$

$$= -2\sin x \sin(2x) + \cos x \cos(2x)$$

At $x = \dfrac{\pi}{6}, \dfrac{dy}{dx} = -2\sin\left(\dfrac{\pi}{6}\right)\sin\left(2\cdot\dfrac{\pi}{6}\right) + \cos\left(\dfrac{\pi}{6}\right)\cos\left(2\cdot\dfrac{\pi}{6}\right)$

$$= -2\sin\left(\frac{\pi}{6}\right)\sin\left(\frac{\pi}{3}\right) + \cos\left(\frac{\pi}{6}\right)\cos\left(\frac{\pi}{3}\right)$$

$$= -2\left(\frac{1}{2}\right)\left(\frac{\sqrt{3}}{2}\right) + \left(\frac{\sqrt{3}}{2}\right)\left(\frac{1}{2}\right) = -\frac{\sqrt{3}}{4}$$

Hence, the slope of the tangent is $-\dfrac{\sqrt{3}}{4}$

Find the y-coordinate of the tangent point:

At $x = \dfrac{\pi}{6}, y = \sin\left(\dfrac{\pi}{6}\right)\cos\left(2\cdot\dfrac{\pi}{6}\right) = \sin\left(\dfrac{\pi}{6}\right)\cos\left(\dfrac{\pi}{3}\right) = \left(\dfrac{1}{2}\right)\left(\dfrac{1}{2}\right) = \dfrac{1}{4}$

So the tangent point is $\left(\dfrac{\pi}{6}, \dfrac{1}{4}\right)$

Using point-slope form for a linear equation, gives

$$y - \frac{1}{4} = -\frac{\sqrt{3}}{4}\left(x - \frac{\pi}{6}\right) \Rightarrow y = -\frac{\sqrt{3}}{4}x + \frac{\pi\sqrt{3}}{24} + \frac{1}{4} \text{ or } y = -\frac{\sqrt{3}}{4}x + \frac{6 + \pi\sqrt{3}}{24}$$

Therefore, an equation for the tangent to $y = \sin x \cos(2x)$ at $x = \dfrac{\pi}{6}$ is

$$y = -\frac{\sqrt{3}}{4}x + \frac{6 + \pi\sqrt{3}}{24}$$

Using a GDC can give a quick visual check for this result. $\left[\dfrac{\pi}{6} \approx 0.52359878\right]$

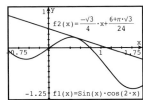

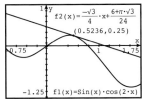

Figure 13.4 GDC screens for the solution to Example 13.6

The quotient rule

Just as the derivative of the product of two functions is not the product of their derivatives, the derivative of a quotient of two functions is not the quotient of their derivatives.

The quotient rule

If y is a function in terms of x that can be expressed as the quotient of two functions u and v that are also in terms of x, then the quotient $y = \frac{u}{v}$ can be differentiated using the **quotient rule**:

$$\frac{dy}{dx} = \frac{d}{dx}\left(\frac{u}{v}\right) = \frac{v\frac{du}{dx} - u\frac{dv}{dx}}{v^2}$$

or, equivalently, if $y = \frac{f(x)}{g(x)}$, then $\frac{dy}{dx} = \frac{d}{dx}\left[\frac{f(x)}{g(x)}\right] = \frac{g(x) \cdot f'(x) - f(x) \cdot g'(x)}{[g(x)]^2}$

Since subtraction is not commutative, make sure you set up the numerator of the quotient rule correctly.

As with the chain rule and the product rule, it is helpful to recognise the structure of the quotient rule by remembering it in words.

$$\binom{\text{derivative of}}{\text{quotient}} = \frac{(\text{denominator}) \times \binom{\text{derivative of}}{\text{numerator}} - (\text{numerator}) \times \binom{\text{derivative of}}{\text{denominator}}}{(\text{denominator})^2}$$

Note that we could have proved the quotient rule by writing the quotient $\frac{f(x)}{g(x)}$ as the product $f(x)[g(x)]^{-1}$ and applying the product rule and chain rule. As some of the examples here show, the derivative of a quotient can also be found by means of the product rule and/or the chain rule.

Example 13.7

For each function, find its derivative (i) by the quotient rule, and (ii) by another method.

(a) $g(x) = \dfrac{5x - 1}{3x^2}$ (b) $h(x) = \dfrac{1}{2x - 3}$ (c) $f(x) = \dfrac{3x - 2}{2x - 5}$

Solution

(a) (i) $g(x) = y = \dfrac{u}{v} = \dfrac{5x - 1}{3x^2}$

$$g'(x) = \frac{dy}{dx} = \frac{v\dfrac{du}{dx} - u\dfrac{dv}{dx}}{v^2} = \frac{3x^2 \cdot 5 - (5x - 1) \cdot 6x}{(3x^2)^2}$$

$$= \frac{15x^2 - 30x^2 + 6x}{9x^4} = \frac{3x(-5x + 2)}{9x^4}$$

$$= \frac{-5x + 2}{3x^3}$$

637

Notice that the results for parts (i) and (ii) are equivalent.

(ii) Using algebra, split the numerator:

$$g(x) = \frac{5x - 1}{3x^2} = \frac{5x}{3x^2} - \frac{1}{3x^2} = \frac{5}{3x} - \frac{1}{3x^2} = \frac{5}{3}x^{-1} - \frac{1}{3}x^{-2}$$

Now, differentiate term by term using the power rule.

$$g'(x) = \frac{5}{3}\frac{d}{dx}(x^{-1}) - \frac{1}{3}\frac{d}{dx}(x^{-2})$$

$$= \frac{5}{3}(-x^{-2}) - \frac{1}{3}(-2x^{-3})$$

$$= -\frac{5}{3x^2} + \frac{2}{3x^3}$$

(b) (i) $y = \dfrac{f(x)}{g(x)} = \dfrac{1}{2x - 3} \Rightarrow f(x) = 1$ and $g(x) = 2x - 3$

Using the quotient rule (function notation form),

$$\frac{dy}{dx} = \frac{d}{dx}\left[\frac{f(x)}{g(x)}\right] = \frac{g(x) \cdot f'(x) - f(x) \cdot g'(x)}{[g(x)]^2}$$

$$= \frac{(2x - 3) \cdot 0 - 1 \cdot (2)}{(2x - 3)^2}$$

$$= \frac{-2}{(2x - 3)^2}$$

(ii) $y = f(g(x)) = \dfrac{1}{2x - 3} = (2x - 3)^{-1}$

The outside function is $f(u) = u^{-1} \Rightarrow f'(u) = -u^{-2}$

and the inside function is $g(x) = 2x - 3$

Using the chain rule (function notation form):

$$\frac{dy}{dx} = f'(g(x)) \cdot g'(x) = -(2x - 3)^{-2} \cdot 2$$

$$= -\frac{2}{(2x - 3)^2}$$

(c) (i) $f(x) = y = \dfrac{u}{v} = \dfrac{3x - 2}{2x - 5}$

$$f'(x) = \frac{dy}{dx} = \frac{v\dfrac{du}{dx} - u\dfrac{dv}{dx}}{v^2} = \frac{(2x - 5) \cdot 3 - (3x - 2) \cdot 2}{(2x - 5)^2}$$

$$= \frac{6x - 15 - 6x + 4}{(2x - 5)^2}$$

$$= \frac{-11}{(2x - 5)^2}$$

(ii) Re-write $f(x)$ as a product and apply the product rule (with chain rule embedded)

$$f(x) = y = \frac{3x - 2}{2x - 5} = (3x - 2)(2x - 5)^{-1}$$

for $y = uv$, $u = 3x - 2$ and $v = (2x - 5)^{-1}$

$v = (2x - 5)^{-1}$ is a composite function, so we need to use the chain rule to find $\dfrac{dv}{dx}$

$$f'(x) = \frac{d}{dx}(uv) = u\frac{dv}{dx} + v\frac{du}{dx}$$

$$= (3x - 2) \cdot \frac{d}{dx}[(2x - 5)^{-1}] + (2x - 5)^{-1} \cdot 3$$

Applying the chain rule for $\dfrac{d}{dx}[(2x - 5)^{-1}]$

$$= (3x - 2)[-(2x - 5)^{-2} \cdot 2] + 3(2x - 5)^{-1}$$

$$= (-6x + 4)(2x - 5)^{-2} + 3(2x - 5)^{-1}$$

Taking out the highest common factor of $(2x - 5)^{-2}$

$$= (2x - 5)^{-2}[(-6x + 4) + 3(2x - 5)]$$

$$= (2x - 5)^{-2}[-6x + 4 + 6x - 15]$$

$$f'(x) = \frac{-11}{(2x - 5)^2}$$

As Example 13.7 demonstrates, it can be worth manipulating the quotient before differentiating as this may allow other more efficient differentiation techniques to be used.

Higher derivatives

If $y = f(x)$ is a function of x then, in general, the derivative will be some other function of x. As we have learned, the derivative indicates the rate of change of $f(x)$ with respect to x. The second derivative is an effective tool in verifying maxima, minima, and inflection points. In general, $\dfrac{d^2y}{dx^2}$ will also be a function of x and may be differentiated to give the third derivative of y with respect to x, denoted by $\dfrac{d^3y}{dx^3}$.

The nth derivative of y with respect to x is denoted by $\dfrac{d^n y}{dx^n}$. If the notation $f(x)$ is used, the first, second and third derivatives are written as $f'(x), f''(x),$ and $f'''(x)$, respectively. The fourth, fifth, and higher derivatives are denoted using a superscript number written in brackets. For example, $f^{(4)}(x)$ represents the fourth derivative of the function f with respect to x.

The process of computing the nth derivative of a function can be very tedious and can be achieved only by computing the successive derivatives in turn.

It is worthwhile to attempt to simplify the function $\dfrac{dy}{dx}$ before differentiating to find $\dfrac{d^2y}{dx^2}$, and in turn try to simplify this result before computing $\dfrac{d^3y}{dx^3}$, and so on.

The function $p(x) = \dfrac{3x^2}{5x - 1}$ initially looks similar to the function g in Example 13.7, part (a), (because they are reciprocals). However, it is not possible to split the denominator and express as two fractions.

Recognise that $\dfrac{3x^2}{5x - 1}$ is not equivalent to $\dfrac{3x^2}{5x} - \dfrac{3x^2}{1}$. Hence, in order to differentiate $p(x) = \dfrac{3x^2}{5x - 1}$ we would apply either the quotient rule, or the product rule with the function rewritten as $p(x) = 3x^2(5x - 1)^{-1}$, and use the chain rule to differentiate the factor $(5x - 1)^{-1}$.

639

Example 13.8

Given $y = \dfrac{1}{x}$, find a formula for the nth derivative $\dfrac{d^n y}{dx^n}$

Solution

Let's take successive derivatives of the function to see if we can spot a pattern, and then formulate a conjecture for the formula.

$$y = \frac{1}{x} = x^{-1}$$

$$\frac{dy}{dx} = (-1)x^{-2} = \frac{-1}{x^2}$$

$$\frac{d^2 y}{dx^2} = (-2)(-1)x^{-3} = \frac{2}{x^3}$$

$$\frac{d^3 y}{dx^3} = (-3)(-2)(-1)x^{-4} = \frac{-6}{x^4}$$

$$\frac{d^4 y}{dx^4} = (-4)(-3)(-2)(-1)x^{-4} = \frac{24}{x^5}$$

$$\frac{d^5 y}{dx^5} = (-5)(-4)(-3)(-2)(-1)x^{-4} = \frac{-120}{x^6}$$

Note that the sign of the result alternates; negative when n is odd, and positive when n is even. Thus, we need to incorporate the expression $(-1)^n$ into our formula since the successive values of $(-1)^n$ are $-1, 1, -1, 1, \ldots$.

Another factor needs to be $n!$ (n factorial) because $n! = n(n-1) \cdots 2 \cdot 1$. The last piece of the formula is that the power of x in the denominator is one more than the value of n.

Therefore, $\dfrac{d^{(n)} y}{dx^{(n)}} = \dfrac{(-1)^n n!}{x^{n+1}}$

Exercise 13.1

1. Find the derivative of each function.

   (a) $y = (3x - 8)^4$ (b) $y = \sqrt{1-x}$ (c) $y = \sin x \cos x$

   (d) $y = 2\sin\left(\dfrac{x}{2}\right)$ (e) $y = (x^2 + 4)^{-2}$ (f) $y = \dfrac{x+1}{x-1}$

   (g) $y = \dfrac{1}{\sqrt{x+2}}$ (h) $y = \cos^2 x$ (i) $y = x\sqrt{1-x}$

   (j) $y = \dfrac{1}{3x^2 - 5x + 7}$ (k) $y = \sqrt[3]{2x+5}$ (l) $y = (2x-1)^3(x^4 + 1)$

   (m) $y = \dfrac{\sin x}{x}$ (n) $y = \dfrac{x^2}{x+2}$ (o) $y = \sqrt[3]{x^2}\cos x$

2. Find the equation of the line that is tangent to the given curve at the specified value of x. Express the equation exactly in the form $y = mx + c$.

   (a) $y = (2x^2 - 1)^3$ at $x = -1$ **(b)** $y = \sqrt{3x^2 - 2}$ at $x = 3$

   (c) $y = \sin 2x$ at $x = \pi$ **(d)** $y = \dfrac{x^3 + 1}{2x}$ at $x = 1$

3. An object moves along a line so that its position, s, relative to a starting point at any time $t \geqslant 0$ is given by $s(t) = \cos(t^2 - 1)$.

   (a) Find the velocity of the object as a function of t.

   (b) What is the object's velocity at $t = 0$?

   (c) In the interval $0 < t < 2.5$, find any times (values of t) for which the object is stationary.

   (d) Describe the object's motion during the interval $0 < t < 2.5$

4. Find the equation of (i) the tangent, and (ii) the normal to the curve at the given point.

   (a) $y = \dfrac{2}{x^2 - 8}$ at $(3, 2)$

   (b) $y = \sqrt{1 + 4x}$ at $\left(2, \dfrac{2}{3}\right)$

   (c) $y = \dfrac{x}{x + 1}$ at $\left(1, \dfrac{1}{2}\right)$

5. Consider the trigonometric curve $y = \sin\left(2x - \dfrac{\pi}{2}\right)$

   (a) Find $\dfrac{dy}{dx}$ and $\dfrac{d^2y}{dx^2}$

   (b) Find the exact coordinates of any inflection points for the curve in the interval $0 < x < \pi$.

6. A curve has equation $y = x(x - 4)^2$

   (a) For this curve, find:

        (i) the x-intercepts

        (ii) the coordinates of the maximum point

        (iii) the coordinates of the point of inflection.

   (b) Use your answers to part **(a)** to sketch a graph of the curve for $0 \leqslant x \leqslant 4$, clearly indicating the features found in part **(a)**.

7. Consider the function $f(x) = \dfrac{x^2 - 3x + 4}{(x + 1)^2}$

   (a) Show that $f'(x) = \dfrac{5x - 11}{(x + 1)^3}$

   (b) Show that $f''(x) = \dfrac{-10x + 38}{(x + 1)^4}$

   (c) Does the graph of f have an inflection point at $x = 3.8$? Explain.

8. Find the first and second derivatives of the function $f(x) = \dfrac{x - a}{x + a}$

9. Given $y = \dfrac{1}{1-x}$, find a formula for the nth derivative $\dfrac{d^n y}{dx^n}$

10. The graph of the function $g(x) = \dfrac{8}{4 + x^2}$ is called the witch of Agnesi.

   (a) Find the exact coordinates of any extreme values or inflection points.

   (b) Determine all values of x for which
      (i) $g(x) < 0$ (ii) $g(x) = 0$ (iii) $g(x) > 0$

   (c) Find (i) $\lim\limits_{x \to -\infty} g(x)$, and (ii) $\lim\limits_{x \to +\infty} g(x)$

   (d) Sketch the graph of g.

11. Use the product rule to prove the constant multiple rule for differentiation. That is, show that $\dfrac{d}{dx}(c \cdot f(x)) = c \cdot \dfrac{d}{dx}(f(x))$ for any constant c.

12. If $y = x^4 - 6x^2$, show that y, $\dfrac{dy}{dx}$, and $\dfrac{d^2 y}{dx^2}$ are all negative on the interval $0 < x < 1$, but that $\dfrac{d^3 y}{dx^3}$ is positive on the same interval.

13.2 Derivatives of trigonometric and exponential functions

Derivatives of trigonometric functions

The derivative is a rule that gives us the slope of the tangent to the graph of a function at a particular point. Thus, we can use a function's derivative to deduce the behaviour of its graph. Conversely, we can gain insight about the derivative of a function from the shape of its graph.

In Chapter 12, we used the limit definition of the derivative (first principles) to determine that the derivative of $\sin x$ is $\cos x$, and that the derivative of $\cos x$ is $-\sin x$. We could have made a very confident conjecture for the derivative of $\sin x$ by analysing its graph.

Start with the graph of $f(x) = \sin x$ (Figure 13.5). The graph of $y = \sin x$ is periodic, with period 2π, so the same will be true of its derivative, which gives the slope at each point on the graph. Therefore, it's only necessary for us to consider the portion of the graph in the interval $0 \leqslant x \leqslant 2\pi$.

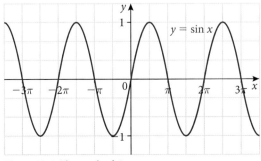

Figure 13.5 The graph of $\sin x$

Figure 13.6 shows two sets of axes having equal scales on the *x*- and *y*-axes and corresponding *x*-coordinates aligned vertically. On the top set of axes, $y = \sin x$ is graphed with tangents drawn at nine selected points. The points are chosen such that the slopes of the tangents at those points, in order, appear to be equal to 1, $\frac{1}{2}$, 0, $-\frac{1}{2}$, -1, $-\frac{1}{2}$, 0, $\frac{1}{2}$, 1. The values of these slopes are then plotted in the bottom graph, with the *y*-coordinate of each point indicating the slope of the curve for that particular *x* value. Hence, the points in the bottom set of axes should be on the graph of the derivative of $y = \sin x$.

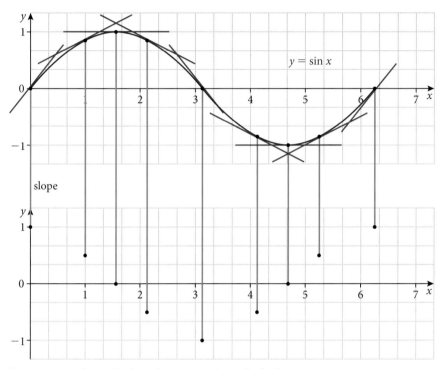

Figure 13.6 Analysing the slope of tangents to the graph of $y = \sin x$

Clearly the points representing the slope of the tangents to $y = \sin x$ plotted in Figure 13.6 are tracing out the graph of $y = \cos x$.

Although we will use this informal approach to conjecture the derivatives for $y = e^x$ and $y = \ln x$, it does not always work so smoothly. For example, let's analyse the graph of $y = \tan x$ in an attempt to guess its derivative.

We can use the GDC command that evaluates the derivative of a function at a specified point to graph the value of the derivative at all points on a graph. We used this technique to confirm the result in Example 12.9 part (b). The GDC screen images show the graph of $y = \tan x$ and then the GDC graphing its derivative (in bold) on the same set of axes. Although, as pointed out in Section 12.3, in general it is incorrect to graph a function and its derivative on the same pair of axes (units on the vertical axis will not be the same), it is helpful in seeing the connection between the graph of a function and that of its derivative.

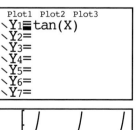

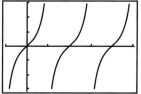

Figure 13.7 Using a GDC to graph $y = \tan x$

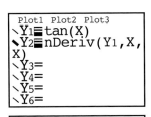

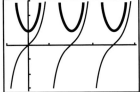

Figure 13.8 Using a GDC to graph the derivative (in bold)

The graph of the derivative of $\tan x$ is always above the x-axis, meaning that the derivative is always positive. This clearly agrees with the fact that the tangent function, except for where it is undefined, is always increasing (moving upwards) as the values of x increase. However, the shape of the graph does not bring to mind an easy conjecture for a rule for the derivative of $\tan x$.

Rather than use the limit definition for finding the derivative of $\tan x$, let's write $\tan x$ as $\dfrac{\sin x}{\cos x}$ and use the quotient rule.

$$\frac{d}{dx}(\tan x) = \frac{d}{dx}\left(\frac{\sin x}{\cos x}\right) = \frac{\cos x \dfrac{d}{dx}(\sin x) - \sin x \dfrac{d}{dx}(\cos x)}{\cos^2 x}$$

$$= \frac{\cos x \cos x - \sin x(-\sin x)}{\cos^2 x}$$

$$= \frac{\cos^2 x + \sin^2 x}{\cos^2 x}$$

$$= \frac{1}{\cos^2 x}$$

$$= \sec^2 x$$

Therefore, $\dfrac{d}{dx}(\tan x) = \sec^2 x$

Similarly, it can be shown that $\dfrac{d}{dx}(\cot x) = -\csc^2 x$

To find the derivative of $\sec x$ we can use the chain rule.

$$\frac{d}{dx}(\sec x) = \frac{d}{dx}\left(\frac{1}{\cos x}\right) = \frac{d}{dx}[(\cos x)^{-1}]$$

The table lists the derivatives of the six trigonometric functions.

$f(x)$	$f'(x)$
$\sin x$	$\cos x$
$\cos x$	$-\sin x$
$\tan x$	$\sec^2 x$
$\cot x$	$-\csc^2 x$
$\sec x$	$\sec x \tan x$
$\csc x$	$-\csc x \cot x$

Applying the chain rule: $\dfrac{d}{dx}(\sec x) = -(\cos x)^{-2}(-\sin x)$

$$= \frac{\sin x}{\cos^2 x}$$

$$= \frac{1}{\cos x} \cdot \frac{\sin x}{\cos x}$$

$$= \sec x \tan x$$

Therefore, $\dfrac{d}{dx}(\sec x) = \sec x \tan x$

Similarly, it can be shown that $\dfrac{d}{dx}(\csc x) = -\csc x \cot x$

Example 13.9

Find the derivative of each function.

(a) $y = \cos(\sqrt{x})$

(b) $y = \dfrac{x^3}{\sin x}$

(c) $y = x^2 \tan(3x)$

(d) $y = \sec^2(3x)$

Solution

(a) Applying the chain rule: $\dfrac{dy}{dx} = \dfrac{d}{dx}[\cos(\sqrt{x})] = -\sin(\sqrt{x}) \cdot \dfrac{d}{dx}(\sqrt{x})$

$$= -\sin(\sqrt{x}) \cdot \left(\dfrac{d}{dx}x^{\frac{1}{2}}\right)$$

Applying the power rule: $\qquad\qquad = -\sin(\sqrt{x}) \cdot \left(\dfrac{1}{2}x^{-\frac{1}{2}}\right)$

Therefore, $\dfrac{dy}{dx} = -\dfrac{\sin(\sqrt{x})}{2\sqrt{x}}$

(b) **Method 1** (quotient rule)

Applying the quotient rule:

$$\dfrac{dy}{dx} = \dfrac{d}{dx}\left(\dfrac{x^3}{\sin x}\right) = \dfrac{\sin x \cdot \dfrac{d}{dx}(x^3) - x^3 \cdot \dfrac{d}{dx}(\sin x)}{\sin^2 x}$$

Therefore, $\dfrac{dy}{dx} = \dfrac{3x^2\sin x - x^3\cos x}{\sin^2 x}$

Method 2 (product rule and chain rule)

Rewriting as a product: $\dfrac{dy}{dx} = \dfrac{d}{dx}\left(\dfrac{x^3}{\sin x}\right) = \dfrac{d}{dx}[x^3 \cdot (\sin x)^{-1}]$

Applying the product rule: $= x^3 \cdot \dfrac{d}{dx}[(\sin x)^{-1}] + (\sin x)^{-1} \cdot \dfrac{d}{dx}(x^3)$

$$= x^3[-(\sin x)^{-2}\cos x] + (\sin x)^{-1}(3x^2)$$

Taking out the common factor of $(\sin x)^{-2}$:

$$= (\sin x)^{-2}[-x^3\cos x + 3x^2\sin x]$$

Therefore, $\dfrac{dy}{dx} = \dfrac{3x^2\sin x - x^3\cos x}{\sin^2 x}$

(c) $\dfrac{dy}{dx} = \dfrac{d}{dx}[x^2\tan(3x)]$

Applying the product rule: $= x^2 \cdot \dfrac{d}{dx}(\tan(3x)) + \tan(3x) \cdot \dfrac{d}{dx}(x^2)$

Applying the chain rule for $\dfrac{d}{dx}(\tan(3x))$: $= x^2(3\sec^2(3x)) + (\tan(3x))(2x)$

$$= 3x^2\sec^2(3x) + 2x\tan(3x)$$

(d) $\dfrac{dy}{dx} = \dfrac{d}{dx}[\sec^2(3x)] = \dfrac{d}{dx}[(\sec(3x))^2]$

Applying the chain rule: $= 2\sec(3x) \cdot \dfrac{d}{dx}(\sec(3x))$

Applying the chain rule again: $= 2\sec(3x) \cdot \left(\sec(3x)\tan(3x) \cdot \dfrac{d}{dx}(3x)\right)$

$$= 2\sec(3x) \cdot (\sec(3x)\tan(3x) \cdot 3)$$

$$= 6\sec^2(3x)\tan(3x)\left[\text{equivalent to } \dfrac{6\sin(3x)}{\cos^3(3x)}\right]$$

Example 13.10

The motion of a particle moving along a straight line in the interval

$0 < t < 12$ (t in seconds) is given by the function $s(t) = \sin\left(\dfrac{t}{2}\right) - \cos\left(\dfrac{t}{2}\right) + 1$,

where s is the particle's displacement in centimetres from the origin O. The particle's displacement is negative when it is to the left of O, and positive when it is to the right of O.

(a) Find the exact time and displacement when the particle is (i) furthest to the right and (ii) furthest to the left during the interval $0 < t < 12$.

(b) Find the particle's exact maximum velocity to the right and at what exact time it occurs.

Solution

Displacement can only be a maximum or minimum when velocity is zero, i.e. $v(t) = 0$. Similarly, velocity can only be a maximum or minimum when acceleration is zero, i.e. $a(t) = 0$. So find the first and second derivatives of $s(t)$ giving us the velocity function, $v(t)$, and acceleration function, $a(t)$, respectively.

(a) $v(t) = s'(t) = \dfrac{\mathrm{d}}{\mathrm{d}x}\left[\sin\left(\dfrac{t}{2}\right) - \cos\left(\dfrac{t}{2}\right) + 1\right] = \dfrac{1}{2}\cos\left(\dfrac{t}{2}\right) + \dfrac{1}{2}\sin\left(\dfrac{t}{2}\right)$

Solve: $\dfrac{1}{2}\cos\left(\dfrac{t}{2}\right) + \dfrac{1}{2}\sin\left(\dfrac{t}{2}\right) = 0$

$\sin\left(\dfrac{t}{2}\right) = -\cos\left(\dfrac{t}{2}\right)$

$\dfrac{\sin\left(\dfrac{t}{2}\right)}{\cos\left(\dfrac{t}{2}\right)} = \tan\left(\dfrac{t}{2}\right) = -1$, given that $\cos\left(\dfrac{t}{2}\right) \neq 0$

$\tan\left(\dfrac{t}{2}\right) = -1$ when $\dfrac{t}{2} = \dfrac{3\pi}{4} + k \cdot \pi, k \in \mathbb{Z}$

Thus, $t = \dfrac{3\pi}{2} + k \cdot 2\pi, k \in \mathbb{Z}$

For $0 < t < 12$, $t = \dfrac{3\pi}{2}$ or $t = \dfrac{7\pi}{2}$

(i) Checking the sign (direction) of the particle's velocity just before and after these two times will show if they are maximum or minimum values.

Test values are $t = \pi$ and 2π for $t = \dfrac{3\pi}{2}$

$v(\pi) = \dfrac{1}{2}\cos\left(\dfrac{\pi}{2}\right) + \dfrac{1}{2}\sin\left(\dfrac{\pi}{2}\right) = 0 + \dfrac{1}{2} \cdot 1 = \dfrac{1}{2} > 0$

So the displacement is increasing before $t = \dfrac{3\pi}{2}$

$v(2\pi) = \dfrac{1}{2}\cos\left(\dfrac{2\pi}{2}\right) + \dfrac{1}{2}\sin\left(\dfrac{2\pi}{2}\right) = \dfrac{1}{2}(-1) < 0$

So the displacement is decreasing after $t = \dfrac{3\pi}{2}$

Hence, $s\left(\dfrac{3\pi}{2}\right) = \sin\left(\dfrac{3\pi}{4}\right) - \cos\left(\dfrac{3\pi}{4}\right) + 1 = \dfrac{\sqrt{2}}{2} - \left(-\dfrac{\sqrt{2}}{2}\right) + 1$

$= 1 + \sqrt{2}$ is a maximum

Therefore, the particle is furthest to the right (maximum displacement) at $t = \dfrac{3\pi}{2}$ seconds when its displacement is $1 + \sqrt{2}$ cm.

(ii) Test values are $t = 3\pi$ and 4π for $t = \dfrac{7\pi}{2}$

$$v(3\pi) = \frac{1}{2}\cos\left(\frac{3\pi}{2}\right) + \frac{1}{2}\sin\left(\frac{3\pi}{2}\right) = 0 + \frac{1}{2}(-1) < 0$$

So the displacement is decreasing before $t = \dfrac{7\pi}{2}$

$$v(4\pi) = \frac{1}{2}\cos(2\pi) + \frac{1}{2}\sin(2\pi) = \frac{1}{2}(1) + 0 > 0$$

So the displacement is increasing after $t = \dfrac{7\pi}{2}$

Hence, $s\left(\dfrac{7\pi}{2}\right) = \sin\left(\dfrac{7\pi}{4}\right) - \cos\left(\dfrac{7\pi}{4}\right) + 1 = -\dfrac{\sqrt{2}}{2} - \left(\dfrac{\sqrt{2}}{2}\right) + 1$

$$= 1 - \sqrt{2} \text{ is a minimum}$$

So the particle is furthest to the left (minimum displacement) at

$t = \dfrac{7\pi}{2}$ seconds when its displacement is $1 - \sqrt{2}$ cm.

(b) $a(t) = v'(t) = \dfrac{d}{dx}\left[\dfrac{1}{2}\cos\left(\dfrac{t}{2}\right) + \dfrac{1}{2}\sin\left(\dfrac{t}{2}\right)\right] = -\dfrac{1}{4}\sin\left(\dfrac{t}{2}\right) + \dfrac{1}{4}\cos\left(\dfrac{t}{2}\right)$

Solve: $\dfrac{1}{4}\cos\left(\dfrac{t}{2}\right) - \dfrac{1}{4}\sin\left(\dfrac{t}{2}\right) = 0$

$$\sin\left(\frac{t}{2}\right) = \cos\left(\frac{t}{2}\right)$$

$$\frac{\sin\left(\dfrac{t}{2}\right)}{\cos\left(\dfrac{t}{2}\right)} = \tan\left(\frac{t}{2}\right) = 1, \text{ given that } \cos\left(\frac{t}{2}\right) \neq 0$$

$\tan\left(\dfrac{t}{2}\right) = 1$ when $\dfrac{t}{2} = \dfrac{\pi}{4} + k \cdot \pi, k \in \mathbb{Z}$

Thus, $t = \dfrac{\pi}{2} + k \cdot 2\pi, k \in \mathbb{Z}$.

For $0 < t < 12, t = \dfrac{\pi}{2}$ or $t = \dfrac{5\pi}{2}$

To find the maximum velocity (moving right, speed > 0), let's evaluate the velocity at all critical points; that is, at endpoints for the time interval, $t = 0$ and $t = 12$, and where the acceleration is zero, $t = \dfrac{\pi}{2}$ and $t = \dfrac{5\pi}{2}$.

$$v(0) = \frac{1}{2}\cos(0) + \frac{1}{2}\sin(0) = \frac{1}{2}v\left(\frac{\pi}{2}\right) = \frac{1}{2}\cos\left(\frac{\pi}{4}\right) + \frac{1}{2}\sin\left(\frac{\pi}{4}\right)$$

$$= \frac{\sqrt{2}}{4} + \frac{\sqrt{2}}{4} = \frac{\sqrt{2}}{2} \approx 0.707$$

$$v\left(\frac{5\pi}{2}\right) = \frac{1}{2}\cos\left(\frac{5\pi}{4}\right) + \frac{1}{2}\sin\left(\frac{5\pi}{4}\right) = -\frac{\sqrt{2}}{4} - \frac{\sqrt{2}}{4} = -\frac{\sqrt{2}}{2} \approx -0.707$$

$$v(12) = \frac{1}{2}\cos(6) + \frac{1}{2}\sin(6) \approx -0.424$$

Therefore, the particle has a maximum velocity of $\dfrac{\sqrt{2}}{2}$ cm s^{-1} when $t = \dfrac{\pi}{2}$ seconds.

A graph of the displacement function $s(t) = \sin\left(\frac{t}{2}\right) - \cos\left(\frac{t}{2}\right) + 1$ gives a good visual confirmation of our results.

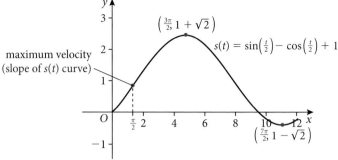

Figure 13.9 $s(t) = \sin\left(\frac{t}{2}\right) - \cos\left(\frac{t}{2}\right) + 1$

> You may be tempted to find the derivative of e^x by applying the rule for differentiating powers
> $$\frac{d}{dx}(x^n) = nx^{n-1}$$ but this only applies if a variable is raised to a constant power. An exponential function, such as $y = e^x$, is a constant raised to a variable power, so the power rule does not apply.

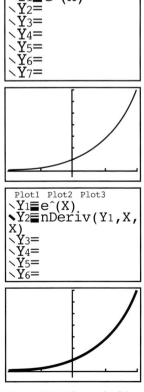

Figure 13.12 The graph of the derivative of e^x appears to be identical to e^x itself

Derivatives of exponential functions

An exponential function with base b is defined as $f(x) = b^x$, $b > 0$ and $b \neq 1$. The graph of f passes through $(0, 1)$, has the x-axis as a horizontal asymptote, and, depending on the value of the base b of the exponential function, will either be a continually increasing exponential growth curve (Figure 13.10) or a continually decreasing exponential decay curve (Figure 13.11).

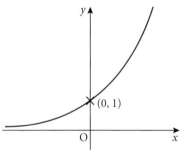

Figure 13.10 As $x \to \infty$, $f(x) \to \infty$, f is an increasing function, exponential growth curve

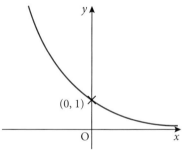

Figure 13.11 As $x \to \infty$, $f(x) \to 0$, f is a decreasing function, exponential decay curve

In Chapter 4, we learned that the exponential function is a particularly important function for modelling exponential growth and decay. The number e was defined in Section 4.3 as the limit of $\left(1 + \frac{1}{x}\right)^x$ as $x \to \infty$. Although the method was not successful in coming up with a conjecture for the derivative of the tangent function, let's try to guess the derivative of e^x by using a GDC to graph its derivative (Figure 13.12).

The graph of the derivative of e^x appears to be identical to e^x itself! This is a very interesting result, but one that we will see fits in exactly with the nature of exponential growth and decay.

The derivative of the exponential function is the exponential function. More precisely, the slope of the graph of $f(x) = e^x$ at any point (x, e^x) is equal to the y-coordinate of the point.

If $f(x) = e^x$, then $f'(x) = e^x$. Or, in Leibniz notation $\frac{d}{dx}(e^x) = e^x$

Example 13.11

Differentiate each function.

(a) $y = e^{2x+\ln x}$ (b) $y = \sqrt{x^2 + e^{4x}}$ (c) $y = \dfrac{e^x - e^{-x}}{e^x + e^{-x}}$

Solution

(a) Because $e^{2x+\ln x} = e^{2x}e^{\ln x}$ and $e^{\ln x} = x$, then $e^{2x+\ln x} = xe^{2x}$

$$\frac{dy}{dx} = \frac{d}{dx}(e^{2x+\ln x}) = \frac{d}{dx}(xe^{2x})$$

Applying the product rule: $= x \cdot \dfrac{d}{dx}(e^{2x}) + e^{2x} \cdot \dfrac{d}{dx}(x)$

Therefore, $\dfrac{dy}{dx} = 2xe^{2x} + e^{2x}$

(b) $\dfrac{dy}{dx} = \dfrac{d}{dx}(\sqrt{x^2 + e^{4x}}) = \dfrac{d}{dx}\left[(x^2 + e^{4x})^{\frac{1}{2}}\right]$

Applying the power rule and chain rule: $= \dfrac{1}{2}(x^2 + e^{4x})^{-\frac{1}{2}} \cdot \dfrac{d}{dx}(x^2 + e^{4x})$

$$= \frac{2x + 4e^{4x}}{2\sqrt{x^2 + e^{4x}}}$$

Therefore, $\dfrac{dy}{dx} = \dfrac{x + 2e^{4x}}{\sqrt{x^2 + e^{4x}}}$

(c) $\dfrac{dy}{dx} = \dfrac{d}{dx}\left(\dfrac{e^x - e^{-x}}{e^x + e^{-x}}\right)$

Using the quotient rule:

$$= \frac{(e^x + e^{-x}) \cdot \dfrac{d}{dx}(e^x - e^{-x}) - (e^x - e^{-x}) \cdot \dfrac{d}{dx}(e^x + e^{-x})}{(e^x + e^{-x})^2}$$

$$= \frac{(e^x + e^{-x})(e^x + e^{-x}) - (e^x - e^{-x})(e^x - e^{-x})}{(e^x + e^{-x})^2}$$

$$= \frac{(e^{2x} + 2e^x e^{-x} + e^{-2x}) - (e^{2x} - 2e^x e^{-x} + e^{-2x})}{(e^x + e^{-x})^2}$$

$$= \frac{4e^x e^{-x}}{(e^x + e^{-x})^2}$$

Therefore, $\dfrac{dy}{dx} = \dfrac{4}{(e^x + e^{-x})^2}$

What about exponential functions with bases other than e? We will now differentiate the general exponential function $f(x) = b^x$, $b > 1$, $b \neq 0$, using the limit definition of a derivative.

Definition of derivative: $\dfrac{d}{dx}(b^x) = \lim_{h \to 0} \dfrac{b^{x+h} - b^x}{h}$

Reverse of $a^m \cdot a^n = a^{m+n}$: $\qquad = \lim_{h \to 0} \dfrac{b^x \cdot b^h - b^x}{h}$

Factorising: $\qquad = \lim_{h \to 0} \dfrac{b^x(b^h - 1)}{h}$

$$= b^x \cdot \lim_{h \to 0} \frac{b^h - 1}{h}$$

So b^x is not affected by the value of h.

$\lim\limits_{h\to 0}\dfrac{b^h-1}{h}$ is equivalent to the slope of the graph of $f(x)=b^x$ at $x=0$; that is, $f'(0)$. Therefore, the derivative of the general exponential function $f(x)=b^x$ is $b^x\cdot f'(0)$. Although the value of $f'(0)$ is a constant, it depends on the value of the base b.

Applying the chain rule gives us the means to determine the value of $f'(0)$ in terms of b for the function $f(x)=b^x$. We can then state the rule for the derivative of the general exponential function $f(x)=b^x$.

We can use the laws of logarithms to write b^x in terms of e^x. Recall from Section 4.5 that $b^{\log_b x}=x$, and if $b=e$ then $e^{\ln x}=x$. Hence, $b^x=e^{x\ln b}$ because $e^{x\ln b}=e^{\ln(b^x)}=b^x$. We can now find the derivative of b^x by applying the chain rule to its equivalent expression $e^{x\ln b}$.

$$y=f(g(x))=e^{x\ln b}$$

The outside function is $f(u)=e^u \qquad f'(u)=e^u$

The inside function is $g(x)=x\ln b \qquad g'(x)=\ln b \ (\ln b\text{ is a constant})$

$$\frac{dy}{dx}=f'(g(x))\cdot g'(x)=e^{x\ln b}\cdot \ln b$$

$$=b^x\ln b$$

Therefore, $\dfrac{d}{dx}(b^x)=b^x\ln b$.

This result agrees with the fact that $\dfrac{d}{dx}(e^x)=e^x$. Using this general rule, $\dfrac{d}{dx}(b^x)=b^x\ln b$, then $\dfrac{d}{dx}(e^x)=e^x\ln e$. Since $\ln e=1$ then $\dfrac{d}{dx}(e^x)=e^x$.

Earlier we established that the derivative of the general exponential function $f(x)=b^x$ is $b^x\cdot f'(0)$, where $f'(0)$ is the slope of the graph at $x=0$. From our result above, we can see that for a specific base b the slope of the curve $y=b^x$ when $x=0$ is $\ln b$ because $b^0\ln b=\ln b$.

The first GDC screen image shows the value of $f'(0)$ for $b=2$, 3, and $\dfrac{1}{2}$. Evaluating $\ln 2$, $\ln 3$, and $\ln\left(\dfrac{1}{2}\right)$ confirms that $f'(0)$ is equal to $\ln b$.

For $b>0$ and $b\neq 1$, if $f(x)=b^x$, then $f'(x)=b^x\ln b$.

Or, in Leibniz notation:
$\dfrac{d}{dx}(b^x)=b^x\ln b$.

Be careful to distinguish between the power rule, $\dfrac{d}{dx}(x^n)=nx^{n-1}$, where the base is a variable and the exponent is a constant, and the rule for differentiating exponential functions, $\dfrac{d}{dx}(b^x)=b^x\ln b$, where the base is a constant and the exponent is a variable.

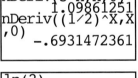

```
nDeriv(2^X,X,0)
         .6931472361
nDeriv(3^X,X,0)
         1.09861251
nDeriv((1/2)^X,X
,0)
         -.6931472361
```

```
ln(2)
         .6931471806
ln(3)
         1.098612289
ln(1/2)
         -.6931471806
```

Figure 13.13 $f'(0)$ is equal to $\ln b$

Example 13.12

Find the equation of the tangent to the curve $y=2^x$ at the point where $x=3$. Express the equation of the line exactly in the form $y=mx+c$.

Solution

We first find the derivative of $y=2^x$ and then evaluate it at $x=3$ to find the slope of the tangent.

$$y'=\frac{d}{dx}(2^x)=2^x(\ln 2)$$

$$y'(3)=2^3(\ln 2)=8\ln 2=\ln 2^8=\ln 256$$

$$m=\ln 256$$

Find the y-coordinate of the tangent point, $y(3) = 2^3 = 8$

The point is $(3, 8)$

Substituting into the point-slope form for a linear equation, gives

$$y - y_1 = m(x - x_1)$$
$$y - 8 = \ln 256(x - 3)$$

Therefore, the equation of the tangent line is $y = (\ln 256)x + 8 - 3\ln 256$

The GDC image confirms the result.

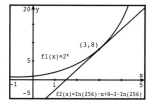

Figure 13.14 GDC screen for the solution to Example 13.12

Example 13.13

Find the coordinates of the point P lying on the graph of $y = 5^x$ such that the tangent to the curve at P passes through the origin.

Solution

Let $P = (x_0, y_0)$ be a point on the graph of $y = 5^x$. Since $\dfrac{dy}{dx} = 5^x(\ln 5)$,

the slope of the tangent line to the curve at P is given by $\dfrac{dy}{dx} = 5^{x_0}(\ln 5)$.

Substituting into the point-slope form for a linear equation:

$$y - y_0 = 5^{x_0}(\ln 5)(x - x_0)$$

If the line passes through the origin, then $(0, 0)$ must satisfy the equation.

$$0 - y_0 = 5^{x_0}(\ln 5)(0 - x_0)$$

$$-y_0 = 5^{x_0}(\ln 5)(-x_0)$$

But $y_0 = 5^{x_0}$, so $-5^{x_0} = 5^{x_0}(\ln 5)(-x_0)$

$$x_0 = \frac{5^{x_0}}{5^{x_0}\ln 5} = \frac{1}{\ln 5}$$

Then $y_0 = 5^{\frac{1}{\ln 5}}$

$$(y_0)^{\ln 5} = \left(5^{\frac{1}{\ln 5}}\right)^{\ln 5}$$

$$(y_0)^{\ln 5} = 5$$

$y_0 = e$ because $e^{\ln x} = x$

Therefore, the point P on the graph of $y = 5^x$ has coordinates $\left(\dfrac{1}{\ln 5}, e\right)$.

As a check, find the equation of the tangent to $y = 5^x$ at this point.

Since $\dfrac{dy}{dx} = 5^{x_0}(\ln 5)$ the slope is $5^{\frac{1}{\ln 5}}(\ln 5)$, but we showed above that $5^{\frac{1}{\ln 5}} = e$.

So, the slope is equivalent to $e \ln 5$. Substituting in the point-slope form:

$$y - e = e\ln 5\left(x - \frac{1}{\ln 5}\right)$$

$$y = e(\ln 5)x$$

651

1. Find the derivative of each function.

   (a) $y = x^2 e^x$

   (b) $y = 8^x$

   (c) $y = \tan e^x$

   (d) $y = \dfrac{x}{1 + \cos x}$

   (e) $y = \dfrac{e^x}{x}$

   (f) $y = \dfrac{1}{3}\sec^3 2x - \sec 2x$

   (g) $y = 4^{-x}$

   (h) $y = \cos x \tan x$

   (i) $y = \dfrac{x}{e^x - 1}$

   (j) $y = 4\cos(\sin 3x)$

   (k) $y = 2^{x+1}$

   (l) $y = \dfrac{1}{\csc x - \sec x}$

2. Find the equation of the tangent to the given curve at the specified value of x. Express the equation exactly in the form $y = mx + c$.

   (a) $y = \sin x$ at $x = \dfrac{\pi}{3}$

   (b) $y = x + e^x$ at $x = 0$

   (c) $y = 4\tan 2x$ at $x = \dfrac{\pi}{8}$

3. Consider the function $g(x) = x + 2\cos x$ on the interval $0 \leqslant x \leqslant 2\pi$.

   (a) find the exact x-coordinates of any stationary points

   (b) determine whether each stationary point is a maximum, minimum, or neither and give a brief explanation.

4. Find the coordinates of any stationary points on the curve $y = x - e^x$. Classify any such points as a maximum, minimum, or neither and explain.

5. Find the coordinates of any stationary points for each function on the interval $0 \leqslant x \leqslant 2\pi$. Indicate whether a stationary point is a maximum, minimum, or neither.

   (a) $f(x) = 4\sin x - \cos 2x$

   (b) $g(x) = \tan x(\tan x + 2)$

6. Find the equation of the normal to the curve $y = 3 + \sin x$ at the point where $x = \dfrac{\pi}{2}$.

7. Consider the function $f(x) = e^x - x^3$

   (a) Find $f'(x)$ and $f''(x)$

   (b) Find the x-coordinates (accurate to 3 significant figures) for any points where $f'(x) = 0$.

   (c) Indicate the intervals for which $f(x)$ is increasing, and indicate the intervals for which $f(x)$ is decreasing.

   (d) For the values of x found in part (b), state whether that point on the graph of f is a maximum, minimum, or neither.

   (e) Find the x-coordinate of any inflection point(s) for the graph of f.

   (f) Indicate the intervals for which $f(x)$ is concave up, and indicate the intervals for which $f(x)$ is concave down.

8. Show that the curves $y = e^{-x}$ and $y = e^{-x}\cos x$ are tangent at each point common to both curves. Sketch the two curves over the interval $-\dfrac{\pi}{2} \leqslant x \leqslant \dfrac{3\pi}{2}$

9. A particle moves in a straight line such that its displacement, s, is given by $s(t) = 4\cos t - \cos 2t$. If the particle comes to rest after T seconds, where $T > 0$, find:

   (a) the particle's acceleration at time T

   (b) the maximum speed of the particle for $0 < t < T$.

10. Find an equation for a line that is tangent to the graph of $y = e^x$ and passes through the origin.

11. Consider the exponential function $f(x) = 2^x$

   (a) Find $f'(x)$

   (b) Find the equation of the tangent to the graph of f at the point $(0, 1)$.

   (c) Explain why the graph of f has no stationary points.

12. Consider the function $h(x) = \dfrac{x^2 - 3}{e^x}$

   (a) Find the exact coordinates of any stationary points.

   (b) Determine whether each stationary point is a maximum, minimum, or neither.

   (c) State the value that the function approaches as
      (i) $x \to \infty$ (ii) $x \to -\infty$

   (d) Write down the equation of any asymptotes for the graph of $h(x)$.

   (e) Make an accurate sketch of the curve, indicating any extrema and points where the graph intersects the x- and y-axes.

13. Given $y = \sin x$, $\dfrac{dy}{dx} = \sin(x + a)$, $\dfrac{d^2y}{dx^2} = \sin(x + b)$, and $\dfrac{d^3y}{dx^3} = \sin(x + c)$, find:

   (a) the values of a, b, and c (b) a formula for $\dfrac{d^{(n)}y}{dx^{(n)}}$

14. (a) Find the first three derivatives of $y = xe^x$.

   (b) Suggest a formula for $\dfrac{d^{(n)}}{dx^{(n)}}(xe^x)$ that is true for all positive integers n.

   (c) Prove that your formula is true by using mathematical induction.

Implicit differentiation

An equation such as $3x - 2y - 8 = 0$ is said to define y as a function of x because it satisfies the definition of a function in that each value of x (domain) determines (corresponds to) a unique value of y (range). We can manipulate the equation in order to solve for y in terms of x, giving $y = \frac{3}{2}x - 4$. In this form, in which y is alone on one side of the equation, the equation is said to define y **explicitly** as a function of x. In the original form of the equation, $3x - 2y - 8 = 0$, the function is said to define y **implicitly** as a function of x.

If we wish to find the derivative of y with respect to x, $\frac{dy}{dx}$, when y is defined implicitly as function of x, we can often solve for y and then differentiate using one of the rules that we have established. For example, if we are asked to find $\frac{dy}{dx}$ for the equation $xy = 1$, we can write y explicitly as a function of x and then differentiate.

$$xy = 1$$

$$y = \frac{1}{x} = x^{-1}$$

$$\frac{dy}{dx} = \frac{d}{dx}(x^{-1}) = -x^{-2} = -\frac{1}{x^2}$$

Most of the functions that we have encountered so far can be described by expressing one variable explicitly in terms of another variable; for example, $y = \cos(2x)$ or $y = \sqrt{1 - x^2}$. But how do we find the derivative of y for an equation where we are not able to solve for y explicitly?

For example, we cannot solve the equation

$$x^3 + y^3 - 9xy = 0$$

for y in terms of x (Figure 13.15). However, there may exist one or more functions f such that if $y = f(x)$ then the equation

$$x^3 + [f(x)]^3 - 9x[f(x)] = 0$$

holds for all values of x in the domain of f. Hence, the function f is defined implicitly by the given equation.

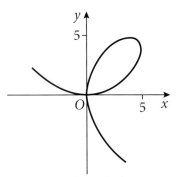

Figure 13.15 The graph of $x^3 + y^3 - 9xy = 0$ (called a *folium*, Latin for 'leaf')

With the assumption that the equation $x^3 + y^3 - 9xy = 0$ defines y as at least one differentiable function of x (Figure 13.16), we can find the derivative of y with respect to x using **implicit differentiation**.

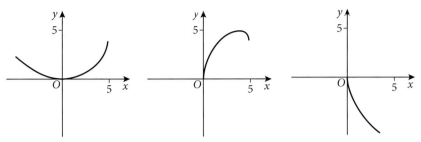

Figure 13.16 Although the equation $x^3 + y^3 - 9xy = 0$ is not a function, we can see that the graph of the equation can be separated into the graphs of three separate functions (they each pass the vertical line test for a function) – this demonstrates that the equation implicitly defines y as three functions of x.

Initially we differentiate term by term, with respect to x:

$$\frac{d}{dx}(x^3) + \frac{d}{dx}(y^3) - \frac{d}{dx}(9xy) = \frac{d}{dx}(0)$$

We can differentiate the first and last terms easily, and we can apply the constant rule to the third term:

$$3x^2 + \frac{d}{dx}(y^3) - 9\frac{d}{dx}(xy) = 0$$

Differentiating the second and third terms is a little more complicated, requiring the use of the chain rule and also the product rule for the third term. If y is defined implicitly as a function of x, then y^3 is also a (composite) function of x. Thus, applying the appropriate rules:

$$3x^2 + 3y^2 \cdot \frac{d}{dx}(y) - 9\left(x \cdot \frac{d}{dx}(y) + y \cdot \frac{d}{dx}(x)\right) = 0$$

$$3x^2 + 3y^2 \cdot \frac{dy}{dx} - 9\left(x \cdot \frac{dy}{dx} + y\right) = 0$$

$$3x^2 + 3y^2\frac{dy}{dx} - 9x\frac{dy}{dx} - 9y = 0$$

Now we solve the equation for $\frac{dy}{dx}$

$$\frac{dy}{dx}(3y^2 - 9x) = -3x^2 + 9y$$

$$\frac{dy}{dx} = \frac{-3x^2 + 9y}{3y^2 - 9x}$$

Therefore, $\frac{dy}{dx} = \frac{-x^2 + 3y}{y^2 - 3x}$

The process of implicit differentiation has given us a formula for $\frac{dy}{dx}$ that is the slope of the curve at any point (except where there is a vertical tangent and slope is undefined) and it is in terms of both x and y. This is not unexpected since we can see from the graph of the equation (Figure 13.16) that it is possible for two or three different points on the curve to have the same x-coordinate.

Thus, the slope of the curve (given by $\frac{dy}{dx}$) will depend on the values of both x and y, and not x alone as with functions where y is explicitly defined in terms of x.

In this section, it is assumed that y is defined as an implicit differentiable function of x (or more than one differentiable function as in the example above) so that the technique of implicit differentiation can be applied.

Process of implicit differentiation

1. Differentiate both sides of the equation, term by term, with respect to x. The chain rule must be applied for any terms containing y.
2. Collect all terms containing $\dfrac{dy}{dx}$ on one side of the equation and all other terms on the other side.
3. Take out $\dfrac{dy}{dx}$ as a factor.
4. Solve for $\dfrac{dy}{dx}$ by dividing both sides by the factor multiplying $\dfrac{dy}{dx}$.
5. Simplify the result, if possible.

Example 13.14

Consider the equation for the unit circle, $x^2 + y^2 = 1$, which is a relation, not a function.

(a) Solve for y and write all equations that express y as a function of x.
Find $\dfrac{dy}{dx}$ for each of these functions.

(b) Find $\dfrac{dy}{dx}$ by implicit differentiation.

(c) Find the equation of the tangent to the unit circle at the point $\left(-\dfrac{1}{2}, \dfrac{\sqrt{3}}{2}\right)$.

Solution

(a) Solving for y produces two equations, each defining y as a function of x.

$$x^2 + y^2 = 1 \implies y^2 = 1 - x^2 \implies y = \sqrt{1 - x^2} \text{ and } y = -\sqrt{1 - x^2}$$

Differentiating each of these with respect to x gives,

$$\frac{dy}{dx} = \frac{d}{dx}(\sqrt{1 - x^2}) = \frac{d}{dx}\left[(1 - x^2)^{\frac{1}{2}}\right] = \frac{1}{2}(1 - x^2)^{-\frac{1}{2}}(-2x)$$

$$\implies \frac{dy}{dx} = \frac{-x}{\sqrt{1 - x^2}}$$

$$\frac{dy}{dx} = \frac{d}{dx}(-\sqrt{1 - x^2}) = \frac{d}{dx}-\left[(1 - x^2)^{\frac{1}{2}}\right] = -\frac{1}{2}(1 - x^2)^{-\frac{1}{2}}(-2x)$$

$$\implies \frac{dy}{dx} = \frac{x}{\sqrt{1 - x^2}}$$

For the function $y = \sqrt{1 - x^2}$ we have $\dfrac{dy}{dx} = \dfrac{-x}{\sqrt{1 - x^2}}$

Since $y = \sqrt{1 - x^2}$, then $\dfrac{dy}{dx} = -\dfrac{x}{y}$

For the function $y = -\sqrt{1 - x^2}$ we have $\dfrac{dy}{dx} = \dfrac{x}{\sqrt{1 - x^2}}$

Since $y = -\sqrt{1 - x^2} \implies -y = \sqrt{1 - x^2}$, then $\dfrac{dy}{dx} = -\dfrac{x}{y}$

(b) Differentiate both sides term by term:

$$\frac{d}{dx}(x^2) + \frac{d}{dx}(y^2) = \frac{d}{dx}(1)$$

Apply the chain rule to differentiate y^2:

$$2x + 2y\frac{dy}{dx} = 0$$

$$2y\frac{dy}{dx} = -2x$$

solving for $\frac{dy}{dx}$

$$\frac{dy}{dx} = \frac{-2x}{2y}$$

Therefore, $\frac{dy}{dx} = -\frac{x}{y}$

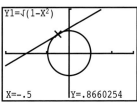

(c) At the point $\left(-\frac{1}{2}, \frac{\sqrt{3}}{2}\right)$ the slope of the tangent line is $\frac{dy}{dx} = -\left(\dfrac{-\frac{1}{2}}{\frac{\sqrt{3}}{2}}\right)$

$$= \frac{1}{\sqrt{3}} = \frac{\sqrt{3}}{3}$$

Substituting into the point-slope form:

$$y - \frac{\sqrt{3}}{2} = \frac{\sqrt{3}}{3}\left(x + \frac{1}{2}\right)$$

$$y = \frac{\sqrt{3}}{3}x + \frac{\sqrt{3}}{6} + \frac{\sqrt{3}}{2}$$

$$y = \frac{\sqrt{3}}{3}x + \frac{2\sqrt{3}}{3}$$

We can get a visual check by graphing the unit circle and the tangent on a GDC. In order to graph the complete unit circle on a GDC, we need to graph both functions found in part (a).

Figure 13.17 GDC screens for the solution to Example 13.14

Example 13.14 illustrates that even when it is possible to solve an equation explicitly for y in terms of x, it may be more efficient to find $\frac{dy}{dx}$ by implicit differentiation.

Example 13.15

(a) Find the points on the graph of $x^2 + 4xy + 13y^2 = 9$ at which the tangent is horizontal.

(b) Determine whether each point is a maximum, minimum, or neither.

Solution

(a) We need to find $\frac{dy}{dx}$, which we do by implicit differentiation.

Differentiate both sides term by term:

$$\frac{d}{dx}(x^2) + 4\frac{d}{dx}(xy) + 13\frac{d}{dx}(y^2) = \frac{d}{dx}(9)$$

Apply chain and product rules:

$$2x + 4\left(x\frac{d}{dx}(y) + y\frac{d}{dx}(x)\right) + 13\left(2y\frac{d}{dx}(y)\right) = 0$$

Collect terms containing $\dfrac{dy}{dx}$ on one side: $4x\dfrac{dy}{dx} + 26y\dfrac{dy}{dx} = -2x - 4y$

Take out $\dfrac{dy}{dx}$ as a factor: $\dfrac{dy}{dx}(4x + 26y) = -2x - 4y$

Solve for $\dfrac{dy}{dx}$: $\dfrac{dy}{dx} = \dfrac{-2x - 4y}{4x + 26y} = \dfrac{-x - 2y}{2x + 13y}$

To find horizontal tangents, solve $\dfrac{dy}{dx} = 0$

$$\dfrac{-x - 2y}{2x + 13y} = 0$$

$$-x - 2y = 0$$

$$y = -\dfrac{x}{2}$$

Of course, there are an infinite number of ordered pairs (x, y) that satisfy the equation $y = -\dfrac{x}{2}$. But the only ordered pairs that we want are ones that are on the curve $x^2 + 4xy + 13y^2 = 9$. So we substitute $-\dfrac{x}{2}$ for y

and solve to find x coordinates of points on the curve where $\dfrac{dy}{dx} = 0$.

$$x^2 + 4xy + 13y^2 = 9$$

$$x^2 + 4x\left(-\dfrac{x}{2}\right) + 13\left(-\dfrac{x}{2}\right)^2 = 9$$

$$x^2 - 2x^2 + \dfrac{13}{4}x^2 = 9$$

Multiply both sides by 4: $4x^2 - 8x^2 + 13x^2 = 36$

$$9x^2 = 36$$

$$x^2 = 4 \Rightarrow x = 2 \text{ or } x = -2$$

y coordinates: for $x = 2$, $y = -\dfrac{2}{2} = -1$;

for $x = -2$, $y = -\left(\dfrac{-2}{2}\right) = 1$

Therefore, the tangents to the curve at $(2, -1)$ and $(-2, 1)$ are horizontal.

(b) It is very difficult to determine the nature of the points $(2, -1)$ and $(-2, 1)$ by testing the sign of the derivative on either side of each point.

Since $\dfrac{dy}{dx}$ is in terms of both x and y we need an explicit equation for y in terms of x to find the y coordinate – but no explicit equation for y exists. It is also impossible to graph the curve $x^2 + 4xy + 13y^2 = 9$ on a GDC to see its shape. Find the second derivative, $\dfrac{d^2y}{dx^2}$, and apply the second derivative test (Section 13.3).

$$\dfrac{d^2y}{dx^2} = \dfrac{d}{dx}\left(\dfrac{-x - 2y}{2x + 13y}\right)$$

Apply quotient rule:

$$= \frac{(2x + 13y)\left[\frac{d}{dx}(-x - 2y)\right] - (-x - 2y)\left[\frac{d}{dx}(2x + 13y)\right]}{(2x + 13y)^2}$$

$$= \frac{(2x + 13y)\left(-1 - 2\frac{dy}{dx}\right) + (x + 2y)\left(2 + 13\frac{dy}{dx}\right)}{(2x + 13y)^2}$$

Substitute for $\frac{dy}{dx}$:

$$= \frac{(2x + 13y)\left(-1 - 2\left(\frac{-x - 2y}{2x + 13y}\right)\right) + (x + 2y)\left(2 + 13\left(\frac{-x - 2y}{2x + 13y}\right)\right)}{(2x + 13y)^2}$$

$$= \frac{2x + 13y}{2x + 13y} \cdot \frac{(2x + 13y)\left(-1 + \frac{2x + 4y}{2x + 13y}\right) + (x + 2y)\left(2 - \frac{13x + 26y}{2x + 13y}\right)}{(2x + 13y)^2}$$

$$= \frac{(2x + 13y)(-2x - 13y + 2x + 4y) + (x + 2y)(4x + 26y - 13x - 26y)}{(2x + 13y)^3}$$

$$= \frac{(2x + 13y)(-9y) + (x + 2y)(-9x)}{(2x + 13y)^3}$$

$$\therefore \frac{d^2y}{dx^2} = -\frac{9x^2 + 36xy + 117y^2}{(2x + 13y)^3} = \frac{-9(x^2 + 4xy + 13y^2)}{(2x + 13y)^3}$$

Now applying the second derivative test for both points where $\frac{dy}{dx} = 0$, we have

for $(2, -1)$, $\frac{d^2y}{dx^2} = \frac{-9(2^2 + 4(2)(-1) + 13(-1)^2)}{(2(2)^2 + 13(-1))^3} = \frac{81}{125} > 0$

$(2, -1)$ is a minimum

for $(-2, 1)$, $\frac{d^2y}{dx^2} = \frac{-9((-2)^2 + 4(-2)(1) + 13(1)^2)}{(2(-2)^2 + 13(1))^3} = -\frac{3}{343} < 0$

$(-2, 1)$ is a maximum

Even though it is not possible to graph the curve $x^2 + 4xy + 13y^2 = 9$ on a GDC, it is possible to find graphing software that can graph this equation. Figure 13.18 confirms our results for parts (a) and (b) of Example 13.15.

We still need to determine how to differentiate other important non-algebraic functions, namely logarithmic functions and inverse trigonometric functions.

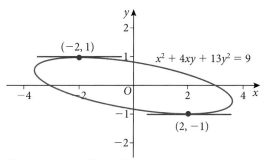

Figure 13.18 Graphing software confirms our results

659

Derivatives of logarithmic functions

We have explored how we can often form a strong conjecture for the derivative of a function by analysing the shape of the function's graph with the aid of some features of a GDC. Let's take this informal approach for finding the derivative for the natural logarithm function, $y = \ln x$, and then check our conjecture by deriving $\frac{d}{dx}(\ln x)$ by means of implicit differentiation.

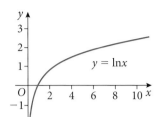

Figure 13.19 $y = \ln x$

The graph of $y = \ln x$ (Figure 13.19) is a particularly straightforward one. Its x-intercept is $(1, 0)$, and since its domain is all positive real numbers, it has no y-intercept. The y-axis is an asymptote, and the graph rises steadily, though less steeply, as $x \to \infty$. There is neither an upper nor a lower bound, so its range is all real numbers.

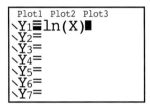

Use a GDC to view a graph of $y = \ln x$, a graph of its derivative, and to construct a table of ordered pairs, with x and the value of the derivative at x (as computed by the GDC).

In the table, each value in the Y_2 column is the slope of the curve (derivative) at the particular x value for $y = \ln x$. From the graph of the derivative, and especially from the table, we conjecture that the derivative of $\ln x$ is $\frac{1}{x}$. This agrees with the fact that for $x > 0$, the slope of the graph of $y = \ln x$ is always positive and as x increases the slope decreases.

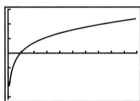

The inverse of $y = \ln x$ is $y = e^x$. Knowing this and that $\frac{d}{dx}(e^x) = e^x$, we can use implicit differentiation to confirm our conjecture.

$y = \ln x$

Use the inverse function relationship: $e^y = x$

Differentiate implicitly: $\frac{d}{dx}(e^y) = \frac{d}{dx}(x)$

$$e^y \frac{dy}{dx} = 1$$

$$\frac{dy}{dx} = \frac{1}{e^y}$$

Substitute x for e^y: $\frac{dy}{dx} = \frac{1}{x}$

Therefore, $\frac{d}{dx}(\ln x) = \frac{1}{x}$

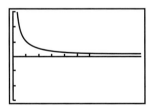

Figure 13.20 Using a GDC to construct a table of ordered pairs

What about the derivative of a logarithmic function with a base b rather than e? That is, logarithmic functions other than the natural logarithmic function.

To find the derivative of $\log_b x$ with any base $(b > 0, b \neq 1)$, we can use the change of base formula (Section 5.4) for logarithms to express $\log_b x$ in terms of the natural logarithm, $\ln x$, and then differentiate.

If $f(x) = \ln x$, then
$$f'(x) = \frac{1}{x}$$
Or, in Leibniz notation
$$\frac{d}{dx}(\ln x) = \frac{1}{x}$$

Apply change of base formula: $\log_b x = \frac{\ln x}{\ln b}$

Differentiate both sides: $\dfrac{d}{dx}(\log_b x) = \dfrac{d}{dx}\left(\dfrac{\ln x}{\ln b}\right) = \dfrac{d}{dx}\left(\dfrac{1}{\ln b} \cdot \ln x\right)$

$\dfrac{1}{\ln b}$ is a constant: $\qquad\qquad\qquad\qquad = \dfrac{1}{\ln b} \cdot \dfrac{d}{dx}(\ln x)$

$\qquad\qquad\qquad\qquad\qquad\qquad\qquad = \dfrac{1}{\ln b} \cdot \dfrac{1}{x}$

Therefore, $\dfrac{d}{dx}(\log_b x) = \dfrac{1}{x \ln b}$

Example 13.16

(a) Given $g(x) = \dfrac{1 + x}{1 - x}$, find $g'(x)$.

(b) Hence, find $f'(x)$ for $f(x) = \ln\left(\dfrac{1 + x}{1 - x}\right)$.

(c) Show that $f(x)$:

   (i)   is an odd function

   (ii) has no stationary points

   (iii) has one point of inflection, and give its coordinates.

Solution

(a) Apply the quotient rule: $g'(x) = \dfrac{(1 - x)\dfrac{d}{dx}(1 + x) - (1 + x)\dfrac{d}{dx}(1 - x)}{(1 - x)^2}$

$\qquad\qquad\qquad\qquad\qquad\quad = \dfrac{1 - x + 1 + x}{(1 - x)^2}$

$\qquad\therefore\ g'(x) = \dfrac{2}{(1 - x)^2}$

(b) Apply $\dfrac{d}{dx}(\ln x) = \dfrac{1}{x}$ and the chain rule: $f'(x) = \dfrac{d}{dx}\left[\ln\left(\dfrac{1 + x}{1 - x}\right)\right]$

$\qquad\qquad\qquad\qquad\qquad\qquad\qquad\qquad = \dfrac{1}{\dfrac{1 + x}{1 - x}} \cdot \dfrac{d}{dx}\left(\dfrac{1 + x}{1 - x}\right)$

Substitute result from part (a): $\qquad = \left(\dfrac{1 - x}{1 + x}\right)\left(\dfrac{2}{(1 - x)^2}\right)$

$\qquad\qquad\qquad\qquad\qquad\qquad\qquad\qquad = \dfrac{1}{1 + x} \cdot \dfrac{2}{1 - x}$

$\qquad\therefore\ f'(x) = \dfrac{2}{1 - x^2}$

(c) (i) In Section 6.3 we stated that a function f is odd if, for each x in the domain of f, $f(-x) = -f(x)$ and its graph is symmetric about the origin. This symmetry leads to the fact that the graph of the derivative of an odd function is symmetric about the y-axis – it is an even function. A function f is even if $f(-x) = f(x)$. Thus, it will suffice to show that $f'(x)$ is even in order to show that $f(x)$ is odd.

$\qquad\qquad f'(-x) = \dfrac{2}{1 - (-x)^2} = \dfrac{2}{1 - x^2} = f(x)$

Therefore, $f'(x)$ is even and it follows that $f(x)$ is odd.

(ii) A stationary point for a function can only occur where its derivative is zero.

$f'(x) = \dfrac{2}{1 - x^2} \neq 0$ because a rational expression can only equal zero when its numerator is zero. Therefore, $f(x)$ has no stationary points.

(iii) To find any inflection points, we start by finding where the second derivative is zero.

Apply the power and chain rules instead of the quotient rule:

$$f''(x) = \frac{\mathrm{d}}{\mathrm{d}x}\left(\frac{2}{1 - x^2}\right) = 2\frac{\mathrm{d}}{\mathrm{d}x}[(1 - x^2)^{-1}]$$

$$= 2[-(1 - x^2)^{-2}(-2x)]$$

$$f''(x) = \frac{4x}{(1 - x^2)^2} = 0 \text{ when } x = 0$$

To confirm that an inflection point does occur at $x = 0$ we need to show that the concavity of the graph of f changes at $x = 0$ ($f''(x)$ changes sign). Because $f(x)$ is defined only for $-1 < x < 1$, we choose $x = -\dfrac{1}{2}$ and $x = \dfrac{1}{2}$ as test points.

$$f''\left(-\frac{1}{2}\right) = \frac{4\left(-\frac{1}{2}\right)}{\left(1 - \left(-\frac{1}{2}\right)^2\right)^2} = -\frac{32}{9} < 0$$

$$f''\left(\frac{1}{2}\right) = \frac{4\left(\frac{1}{2}\right)}{\left(1 - \left(\frac{1}{2}\right)^2\right)^2} = \frac{32}{9} > 0$$

Since $f''(x)$ changes sign (and $f(x)$ changes concavity) at $x = 0$, f has an inflection point there. $f(0) = \ln\left(\dfrac{1 + 0}{1 - 0}\right) = \ln(1) = 0$. Therefore, the inflection point is at $(0, 0)$.

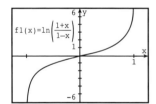

Figure 13.21 Solution to Example 13.16

Example 13.17

Find the equation of the tangent to the graph of $y = \log_{10}(x^3)$ at the point when $x = 4$. Express the equation exactly with any logarithms being expressed as natural logarithms.

Solution

Apply $\dfrac{\mathrm{d}}{\mathrm{d}x}(\log_b x) = \dfrac{1}{x \ln b}$ and chain rule:

$$\frac{\mathrm{d}y}{\mathrm{d}x} = \frac{\mathrm{d}}{\mathrm{d}x}[\log_{10}(x^3)] = \frac{1}{x^3 \ln 10} \cdot \frac{\mathrm{d}}{\mathrm{d}x}(x^3)$$

$$= \frac{1}{x^3 \ln 10} \cdot 3x^2$$

$$\frac{\mathrm{d}y}{\mathrm{d}x} = \frac{3}{x \ln 10}$$

Alternatively, we could use laws of logarithms to write $y = \log_{10}(x^3) = 3\log_{10}x$
and then $\dfrac{dy}{dx} = 3\dfrac{d}{dx}(\log_{10}x) = \dfrac{3}{x\ln 10}$, avoiding the use of the chain rule.

Using the change of base formula:

when $x = 4$, $\dfrac{dy}{dx} = \dfrac{3}{4\ln 10}$ and $y = \log_{10}(4^3) = \log_{10}64 = \dfrac{\ln 64}{\ln 10}$

Thus the tangent intersects the curve at the point $\left(4, \dfrac{\ln 64}{\ln 10}\right)$ and has a

slope of $\dfrac{3}{4\ln 10}$. Substituting into the point-slope form for a linear equation:

$$y - \dfrac{\ln 64}{\ln 10} = \dfrac{3}{4\ln 10}(x - 4)$$

$$y = \dfrac{3x}{4\ln 10} - \dfrac{3}{\ln 10} + \dfrac{\ln 64}{\ln 10}$$

$$y = \dfrac{3x}{4\ln 10} + \dfrac{-3 + \ln 64}{\ln 10}$$

Graphing the curve $y = \log_{10}(x^3)$ and the computed tangent line on a GDC gives a visual confirmation that the equation of the tangent line is correct.

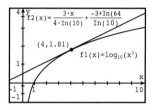

Figure 13.22 Solution to Example 13.17

Derivatives of inverse trigonometric functions

We have established that the derivative of the non-algebraic (transcendental) function $f(x) = \ln x$ is the algebraic function $f'(x) = \dfrac{1}{x}$. The same is true for the inverse trigonometric functions – they are transcendental but their derivatives are algebraic. The inverse trigonometric functions were discussed in Section 6.6. We will now use implicit differentiation to find the derivatives of the inverse functions for sine ($\arcsin x$), cosine ($\arccos x$), and tangent ($\arctan x$) (Figure 13.23).

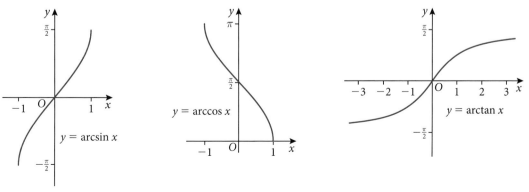

Figure 13.23 Inverse functions for sine, cosine and tangent

663

Recall from Chapter 6 that the notations $y = \arcsin x$ and $y = \sin^{-1} x$ are synonymous, but we will generally use $y = \arcsin x$.

Given the smooth shape of their graphs, we will assume that the functions $y = \arcsin x$, $y = \arccos x$, and $y = \arctan x$ are differentiable except where a vertical tangent exists. Since $y = \arcsin x$ and $y = \arccos x$ have vertical tangents at $x = -1$ and $x = 1$, they are differentiable throughout the interval $-1 < x < 1$. $y = \arctan x$ is differentiable for all real numbers.

Recall the definition of the arcsine function,

$$y = \arcsin x$$

So $\sin y = x$ for $-\dfrac{\pi}{2} \leqslant y \leqslant \dfrac{\pi}{2}$

Differentiating $\sin y = x$ implicitly with respect to x:

Differentiate both sides: $\dfrac{d}{dx}(\sin y) = \dfrac{d}{dx}(x)$

Use implicit differentiation: $(\cos y)\dfrac{dy}{dx} = 1$

Divide by $\cos y$: $\dfrac{dy}{dx} = \dfrac{1}{\cos y}$

That is, $\dfrac{d}{dx}(\arcsin x) = \dfrac{1}{\cos y}$

Dividing by $\cos y$ in the last step is allowed because $\cos y \neq 0$ for the interval in which $y = \arcsin x$ is differentiable; that is, $-\dfrac{\pi}{2} < y < \dfrac{\pi}{2}$ (quadrants I and IV). In fact, $\cos y > 0$ for $-\dfrac{\pi}{2} < y < \dfrac{\pi}{2}$. From the identity $\sin^2 x + \cos^2 x = 1$ we have $\cos x = \pm\sqrt{1 - \sin^2 x}$. Since $\cos y > 0$ we can replace $\cos y$ with $\sqrt{1 - \sin^2 y}$ and because $\sin y = x$ we get $\cos y = \sqrt{1 - x^2}$.

Therefore, $\dfrac{d}{dx}(\arcsin x) = \dfrac{1}{\sqrt{1 - x^2}}$

We can apply a similar process to find the derivative of the $\arccos x$ function, obtaining the result

$$\dfrac{d}{dx}(\arccos x) = -\dfrac{1}{\sqrt{1 - x^2}}$$

Although the domain for the inverse sine and inverse cosine functions is the fairly narrow closed interval $-1 \leqslant x \leqslant 1$, and they are differentiable on the open interval $-1 < x < 1$, the inverse tangent function is defined and differentiable for all real numbers. To find $\dfrac{d}{dx}(\arctan x)$, we follow a similar procedure to that for $\dfrac{d}{dx}(\arcsin x)$.

The definition of the inverse tangent (arctan) function is

$$y = \arctan x \qquad \qquad \tan y = x \text{ for } -\dfrac{\pi}{2} \leqslant y \leqslant \dfrac{\pi}{2}$$

Differentiating $\tan y = x$ implicitly with respect to x gives

Differentiate both sides: $\dfrac{d}{dx}(\tan y) = \dfrac{d}{dx}(x)$

Use implicit differentiation: $(\sec^2 y)\dfrac{dy}{dx} = 1$

Divide by $\sec^2 y$:

$$\frac{dy}{dx} = \frac{1}{\sec^2 y}$$

Apply identity $1 + \tan^2 y = \sec^2 y$:

$$\frac{dy}{dx} = \frac{1}{1 + \tan^2 y}$$

Therefore, $\tan y = x$

$$\frac{d}{dx}(\arctan x) = \frac{1}{1 + x^2}$$

We can find the derivatives for the inverse secant, inverse cosecant, and inverse cotangent functions by implicit differentiation. They are included in the list on the right but are not necessary for this course.

Derivatives of the inverse trigonometric functions

$$\frac{d}{dx}(\arcsin x) = \frac{1}{\sqrt{1 - x^2}}$$

$$\frac{d}{dx}(\text{arccsc}\, x) = -\frac{1}{x\sqrt{x^2 - 1}}$$

$$\frac{d}{dx}(\arccos x) = -\frac{1}{\sqrt{1 - x^2}}$$

$$\frac{d}{dx}(\text{arcsec}\, x) = \frac{1}{x\sqrt{x^2 - 1}}$$

$$\frac{d}{dx}(\arctan x) = \frac{1}{1 + x^2}$$

$$\frac{d}{dx}(\text{arccot}\, x) = -\frac{1}{1 + x^2}$$

Example 13.18

Find the derivative of each equation.

(a) $y = \cos^{-1}(e^{2x})$

(b) $y = x \arcsin 2x + \frac{1}{2}\sqrt{1 - 4x^2}$

(c) $\ln(x + y) = \arctan\left(\frac{x}{y}\right)$

Solution

(a) Use the chain rule and $\dfrac{d}{dx}(\arccos x) = -\dfrac{1}{\sqrt{1 - x^2}}$

$$\frac{dy}{dx} = \frac{d}{dx}[\cos^{-1}(e^{2x})] = -\frac{1}{\sqrt{1 - (e^{2x})^2}} \cdot \frac{d}{dx}(e^{2x})$$

Use the chain rule again:

$$= -\frac{1}{\sqrt{1 - e^{4x}}} \cdot e^{2x} \cdot 2$$

$$\frac{dy}{dx} = -\frac{2e^{2x}}{\sqrt{1 - e^{4x}}}$$

(b) $\dfrac{dy}{dx} = \dfrac{d}{dx}\left(x \arcsin 2x + \dfrac{1}{2}(1 - 4x^2)^{\frac{1}{2}}\right)$

$$= x\frac{d}{dx}(\arcsin 2x) + \arcsin 2x \frac{d}{dx}(x) + \frac{1}{2} \cdot \frac{1}{2}(1 - 4x^2)^{-\frac{1}{2}}\frac{d}{dx}(1 - 4x^2)$$

$$= x\left(\frac{1}{\sqrt{1 - (2x)^2}}\frac{d}{dx}(2x)\right) + \arcsin 2x + \frac{-8x}{4\sqrt{1 - 4x^2}}$$

$$= \frac{2x}{\sqrt{1 - (2x)^2}} + \arcsin 2x + \frac{-2x}{\sqrt{1 - 4x^2}}$$

$$\frac{dy}{dx} = \arcsin 2x$$

(c) Differentiate both sides implicitly: $\dfrac{d}{dx}[\ln(x + y)] = \dfrac{d}{dx}\left[\arctan\left(\dfrac{x}{y}\right)\right]$

Use the chain rule, $\dfrac{d}{dx}(\arctan x) = \dfrac{1}{1 + x^2}$, and the quotient rule:

$$\frac{1}{x + y}\left(1 + \frac{dy}{dx}\right) = \frac{1}{1 + \dfrac{x^2}{y^2}}\left(\frac{y - x\dfrac{dy}{dx}}{y^2}\right)$$

$$\frac{1 + \dfrac{dy}{dx}}{x + y} = \frac{y - x\dfrac{dy}{dx}}{x^2 + y^2}$$

$$x^2 + y^2 + \frac{dy}{dx}x^2 + \frac{dy}{dx}y^2 = xy + y^2 - \frac{dy}{dx}x^2 - \frac{dy}{dx}xy$$

$$\frac{dy}{dx}(2x^2 + xy + y^2) = xy - x^2$$

$$\frac{dy}{dx} = \frac{xy - x^2}{2x^2 + xy + y^2}$$

Example 13.19

A painting that is 175 cm from top to bottom is hanging on the wall of a gallery such that its base is 225 cm above the eye level of an observer.
How far from the wall should the observer stand so that the angle subtended at his eye by the painting is a maximum?

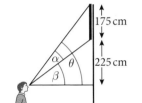

Figure 13.24 Diagram for Example 13.19

Solution

Change all lengths from centimetres to metres.

$$\tan\theta = \frac{4}{x} \text{ and } \tan\beta = \frac{\dfrac{9}{4}}{x}$$

Because $0 < \theta < \dfrac{\pi}{2}$ and $0 < \beta < \dfrac{\pi}{2}$, we have $\theta = \arctan\dfrac{4}{x}$ and $\beta = \arctan\left(\dfrac{\dfrac{9}{4}}{x}\right)$

Substituting these values of θ and β into the equation $\alpha = \theta - \beta$ gives

$$\alpha = \arctan\frac{4}{x} - \arctan\left(\frac{\dfrac{9}{4}}{x}\right)$$

Differentiating with respect to x gives

$$\frac{d\alpha}{dx} = \frac{d}{dx}\left[\arctan(4x^{-1}) - \arctan\left(\frac{9}{4}x^{-1}\right)\right]$$

$$= \frac{1}{1 + (4x^{-1})^2}(-4x^{-2}) - \frac{1}{1 + \left(\dfrac{9}{4}x^{-1}\right)^2}\left(-\frac{9}{4}x^{-2}\right)$$

$$= \frac{-4}{x^2 + 16} + \frac{\dfrac{9}{4}}{x^2 + \dfrac{81}{16}}$$

$$= \frac{-4}{x^2 + 16} + \frac{36}{16x^2 + 81}$$

Setting $\dfrac{d\alpha}{dx} = 0$:

$$36(x^2 + 16) - 4(16x^2 + 81) = 0$$

$$-28x^2 + 252 = 0$$

$x^2 = \dfrac{252}{28} = 9 \Rightarrow x = \pm 3$, however $x \neq -3$

We use the first derivative test to determine if the angle α is a maximum when $x = 3$.

When $x = 2$, $\dfrac{d\alpha}{dx} = \dfrac{7}{145} > 0$

When $x = 4$, $\dfrac{d\alpha}{dx} = -\dfrac{49}{2696} < 0$

Hence, the angle α has an absolute maximum value at $x = 3$. Therefore, the observer should stand 3 metres away from the wall to get the best view of the painting.

Exercise 13.3

1. Find the derivative of y with respect to x, $\dfrac{dy}{dx}$, by implicit differentiation.

   (a) $x^2 + y^2 = 16$
   (b) $x^2 y + xy^2 = 6$

   (c) $x = \tan y$
   (d) $x^2 - 3xy^2 + y^3 x - y^2 = 2$

   (e) $\dfrac{x}{y} - \dfrac{y}{x} = 1$
   (f) $xy\sqrt{x + y} = 1$

   (g) $x + \sin y = xy$
   (h) $x^2 y^3 = x^4 - y^4$

   (i) $xy + e^y = 0$
   (j) $(x + 2)^2 + (y + 3)^2 = 25$

   (k) $x = \arctan(x) - y$
   (l) $y + \sqrt{xy} = 3x^3$

2. Find the lines that are (i) tangent and (ii) normal to the curve at the given point.

   (a) $x^3 - xy - 3y^2 = 0$, at $(2, -2)$ (b) $16x^4 + y^4 = 32$, at $(1, 2)$

   (c) $2xy + \pi \sin y = 2\pi$, at $\left(1, \dfrac{\pi}{2}\right)$ (d) $\sqrt[3]{xy} = 14x + y$, at $(2, -32)$

3. For the circle $x^2 + y^2 = r^2$, show that the tangent at any point (x_1, y_1) on the circle is perpendicular to the line that passes through (x_1, y_1) and the centre of the circle.

4. Consider the equation $x^2 + xy + y^2 = 7$

   (a) Find the two points where the curve intersects the x-axis. Show that the tangents to the curve at these two points are parallel.

   (b) Find any points where the tangent to the curve is parallel to the x-axis.

   (c) Find any points where the tangent to the curve is parallel to the y-axis.

5. The line that is normal to the curve $x^2 + 2xy - 3y^2 = 0$ at $(1, 1)$ intersects the curve at another point. What is this point?

6. Find $\dfrac{dy}{dx}$ and $\dfrac{d^2y}{dx^2}$ for each equation.

   (a) $4x^2 + 9y^2 = 36$ (b) $xy = 2x - 3y$

7. Consider the equation $xy^3 = 1$. Find $\dfrac{dy}{dx}$ and $\dfrac{d^2y}{dx^2}$ by two different methods:

   (a) Solve for y in terms of x and differentiate explicitly.

   (b) Differentiate implicitly.

8. The graph of the equation $x^2 + y^2 = (2x^2 + 2y^2 - x)^2$ is a type of curve called a cardioid. A cardioid is a heart-shaped curve generated by a fixed point on a circle as it rolls around another circle having the same radius. Find the equation of the tangent to this particular cardioid at the point $\left(0, \dfrac{1}{2}\right)$

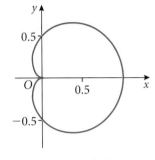

9. Find the derivative of y with respect to x, $\dfrac{dy}{dx}$.

   (a) $y = \ln(x^3 + 1)$ (b) $y = \ln(\sin x)$

   (c) $y = \log_5\sqrt{x^2 - 1}$ (d) $y = \ln\sqrt{\dfrac{1 + x}{1 - x}}$

   (e) $y = \sqrt{\log_{10}x}$ (f) $y = \ln\left(\dfrac{a - x}{a + x}\right)$

   (g) $y = \ln(e^{\cos x})$ (h) $y = \dfrac{1}{\log_3 x}$

   (i) $y = x\ln(x) - x$ (j) $y = \ln(ax) - (\ln b)\log_b x$

10. Find the equation of the line tangent to the graph of $y = \log_2 x$ at the point $x = 8$.

    (a) Express the equation exactly.

    (b) Can you find a way to graph $y = \log_2 x$ on your GDC in order to check your answer?

11. Given $y = \sqrt{\dfrac{x^2 - 1}{x^2 + 1}}$ we could find $\dfrac{dy}{dx}$ by applying the chain rule and the quotient rule. However, it is much easier to first take the natural logarithm of both sides, use the properties of logarithms to simplify as much as possible, and then differentiate implicitly to find $\dfrac{dy}{dx}$.

    This technique is called logarithmic differentiation. Use this technique to show that $\dfrac{dy}{dx} = \dfrac{2x}{(x^2 - 1)^{\frac{1}{2}}(x^2 + 1)^{\frac{3}{2}}}$

12. Find the x-coordinate, between 0 and 1, of the point of inflection on the graph of the function $f(x) = x^2\ln(x^2)$. Express your answer exactly.

13. (a) Given $g(x) = \dfrac{\ln x}{x}$, find expressions for $g'(x)$ and $g''(x)$.

(b) Show that g has an absolute maximum at $x = e$, and state the maximum value of g.

14. Find the derivative of y with respect to x, $\dfrac{dy}{dx}$.

(a) $y = \arctan(x + 1)$

(b) $y = \sin^{-1}\left(\dfrac{x}{\sqrt{1 + x^2}}\right)$

(c) $y = \arccos\left(\dfrac{3}{x^2}\right)$

(d) $\ln y = x \arctan x$

15. Given that $f(x) = \arcsin x + \arccos x$, find $f'(x)$. What can you conclude about the function f?

16. Show that if a is a constant:

(a) $\dfrac{d}{dx}\left[\arctan\left(\dfrac{x}{a}\right)\right] = \dfrac{a}{a^2 + x^2}$

(b) $\dfrac{d}{dx}\left[\arcsin\left(\dfrac{x}{a}\right)\right] = \dfrac{1}{\sqrt{a^2 - x^2}}$

17. Find the equation of the line tangent to the curve $y = 4x\arctan 2x$ at the point on the curve where $x = \dfrac{1}{2}$. Express the equation exactly in the form $y = mx + c$, where m and c are constants.

18. Consider the function $f(x) = \arcsin(\cos x)$ with domain of $0 \leqslant x < \pi$.

(a) Prove that f is a linear function.

(b) Express the function exactly in the form $f(x) = ax + b$, where a and b are constants.

19. A 3-metre tall statue is on top of a column such that the bottom of the statue is 2 metres above the eye level of a person viewing the statue. How far from the base of the column should the person stand to get the best view of the statue – that is, so that the angle subtended at the observer's eye by the statue is a maximum?

20. A particle moves along the x-axis so that its displacement, s (in metres), from the origin at any time $t \geqslant 0$ (in seconds) is given by $s(t) = \arctan\sqrt{t}$.

(a) Find the exact velocity of the particle **(i)** at $t = 1$ second, and **(ii)** at $t = 4$ seconds.

(b) Find the exact acceleration of the particle **(i)** at $t = 1$ second, and **(ii)** at $t = 4$ seconds.

(c) Describe the motion of the particle.

(d) What is the limiting displacement of the particle as t approaches infinity?

13.4 Related rates

Another important use of the chain rule is to find the rates of change of two or more variables that are changing with respect to time. Calculus provides us with the tools and techniques to solve problems where quantities (variables) are changing rather than static.

When a stone is thrown into a pond of water, a circular pattern of ripples is formed. In this situation, we can observe an ever-widening circle moving across the water. As the circular ripple moves across the water, the radius r of the circle, its circumference C, and its area A all increase as a function of time t. Not only are these quantities (variables) functions of time, but their values at any particular time t are related to one another by familiar formulae such as $C = 2\pi r$ and $A = \pi r^2$. Thus, their rates of change are also related to one another.

Example 13.20

A stone is thrown into a pond causing ripples in the form of concentric circles that move away from the point of impact at a rate of $20\ \text{cm s}^{-1}$.

(a) When a circular ripple has a radius of 50 cm, find:
  (i) the rate of change of the circle's circumference
  (ii) the rate of change of the circle's area.

(b) Repeat part (a) when the radius is 100 cm.

Solution

In calculus, a derivative represents a rate of change of one variable with respect to another variable. If the circles are moving outwards at a rate of $20\ \text{cm s}^{-1}$, then the rate of change of the radius is $20\ \text{cm s}^{-1}$, and in the notation of calculus we write $\dfrac{dr}{dt} = 20$

(a) (i) Knowing that the relationship between the radius, r, and the circumference, C, is $C = 2\pi r$, and that the rate of change of the radius with respect to time is $\dfrac{dr}{dt} = 20$, we can use the chain rule to find the rate of change of the circumference with respect to time, i.e. $\dfrac{dC}{dt}$.

$$\frac{dC}{dt} = \frac{dC}{dr} \cdot \frac{dr}{dt}$$

We need to find $\dfrac{dC}{dr}$, the rate of change (derivative) of circumference with respect to the radius. This rate can be derived from the relationship between the variables.

Given $C = 2\pi r$, differentiate both sides with respect to r:

$$\frac{d}{dr}(C) = \frac{d}{dr}(2\pi r)$$

Use implicit differentiation on the left hand side: $\dfrac{dC}{dr} = 2\pi$

Since the circumference C is a linear function of the radius r $(C = 2\pi r)$, the derivative $\dfrac{dC}{dr}$ is a constant.

We now substitute in for $\dfrac{dC}{dr}$ and $\dfrac{dr}{dt}$ to find the rate of change of the circumference with respect to time, $\dfrac{dC}{dt}$.

$$\frac{dC}{dt} = \frac{dC}{dr} \cdot \frac{dr}{dt} \Rightarrow \frac{dC}{dt} = 2\pi \cdot 20 = 40\pi\,\mathrm{cm\,s^{-1}}$$

The rate of change of a circular ripple's circumference is constant (40π). Therefore, the rate of change of the circumference is $40\pi\,\mathrm{cm\,s^{-1}}$ when the radius is 50 cm.

(ii) Similarly, to find the rate of change of the area with respect to time, $\dfrac{dA}{dt}$, we can use the chain rule to write

$$\frac{dA}{dt} = \frac{dA}{dr} \cdot \frac{dr}{dt}$$

Find $\dfrac{dA}{dr}$ from the formula $A = \pi r^2$, which relates variables A and r.

Differentiate both sides with respect to r: $\dfrac{d}{dr}(A) = \dfrac{d}{dr}(\pi r^2)$

Use implicit differentiation on the left hand side: $\dfrac{dA}{dr} = \pi(2r) = 2\pi r$

Since the area A is a nonlinear function of the radius r $(A = \pi r^2)$, the derivative $\dfrac{dA}{dr}$ is not a constant but has different values depending on the value of r.

We substitute in for $\dfrac{dA}{dr}$ and $\dfrac{dr}{dt}$ to find the rate of change of the area with respect to time, $\dfrac{dA}{dt}$.

$$\frac{dA}{dt} = \frac{dA}{dr} \cdot \frac{dr}{dt}$$

$$\frac{dA}{dt} = 2\pi r \cdot 20 = 40\pi r$$

Thus, the rate of change of A with respect to time, $\dfrac{dA}{dt}$, is a linear function in terms of the radius r.

When the radius is 50 cm,

$$\frac{dA}{dt} = 40\pi \cdot 50 = 2000\pi\,\mathrm{cm^2\,s^{-1}} \approx 6280\,\mathrm{cm^2\,s^{-1}}\ (\approx 0.628\,\mathrm{m^2\,s^{-1}})$$

(b) (i) The rate of change of the circumference is $40\pi\,\mathrm{cm\,s^{-1}}$

(ii) When the radius is 100 cm,

$$\frac{dA}{dt} = 40\pi \cdot 100 = 4000\pi\,\mathrm{cm^2\,s^{-1}} \approx 12600\,\mathrm{cm^2\,s^{-1}}\ (\approx 1.26\,\mathrm{m^2\,s^{-1}})$$

There is a slightly different method to determine $\dfrac{dC}{dt}$.
We can find the rate by differentiating implicitly with respect to time, t, both sides of the equation $C = 2\pi r$, that gives the relationship between the two changing quantities (variables).

Given $C = 2\pi r$, differentiate both sides with respect to t:

$$\frac{d}{dt}(C) = \frac{d}{dt}(2\pi r)$$

Use implicit differentiation:

$$\frac{dC}{dt} = 2\pi \frac{dr}{dt}$$

$$\frac{dC}{dt} = 2\pi \cdot 20$$
$$= 40\pi\,\mathrm{cm\,s^{-1}}$$

Note that when $r = 100$ cm the area is changing at twice the rate it was when $r = 50$ cm. It is important to include the appropriate units when an answer is a rate of change. For example, $\mathrm{cm\,s^{-1}}$, $\mathrm{m^2\,s^{-1}}$, litres $\mathrm{s^{-1}}$, etc.

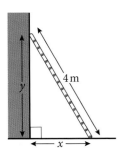

Figure 13.26 Diagram for Example 13.21

Example 13.21

A 4-metre ladder stands upright against a vertical wall. The foot of the ladder is pulled away from the wall at a constant rate of 0.75 m s^{-1}. Find how fast the top of the ladder is coming down the wall at the instant it is:

(a) 3 metres above the ground

(b) 1 metre above the ground. Give answers to 3 significant figures.

Solution

Let x and y represent the distances of the foot and top of the ladder, respectively, from the bottom of the wall. From Pythagoras' theorem, we have

$$x^2 + y^2 = 16$$

Given that the ladder is being pulled away at a rate of 0.75 m s^{-1}, then

$$\frac{dx}{dt} = 0.75 = \frac{3}{4}$$

So we know the rate $\dfrac{dx}{dt}$, and we need to find $\dfrac{dy}{dt}$ when $y = 3$ and when $y = 1$.

Rather than starting with the chain rule and writing an equation relating the different rates, let's use the chain rule by differentiating implicitly with respect to time the equation relating the relevant variables x and y.

$$\frac{d}{dt}(x^2 + y^2) = \frac{d}{dt}(16)$$

$$2x\frac{dx}{dt} + 2y\frac{dy}{dt} = 0$$

$$\frac{dy}{dt} = \left(-\frac{x}{y}\right)\frac{dx}{dt}$$

(a) We know $\dfrac{dx}{dt} = \dfrac{3}{4}$, so to find $\dfrac{dy}{dt}$ when $y = 3$ m, we find the corresponding value for x.

$$x^2 + y^2 = 16 \Rightarrow x = \sqrt{16 - y^2} \text{ for } y = 3: x = \sqrt{16 - 3^2} = \sqrt{7}$$

Hence, when $y = 3$: $\dfrac{dy}{dt} = -\dfrac{\sqrt{7}}{3} \cdot \dfrac{3}{4} = -\dfrac{\sqrt{7}}{4} \approx -0.661 \text{ m s}^{-1}$

(b) For $y = 1$: $x = \sqrt{16 - 1^2} = \sqrt{15}$

Hence, when $y = 1$: $\dfrac{dy}{dt} = -\dfrac{\sqrt{15}}{1} \cdot \dfrac{3}{4} = -\dfrac{3\sqrt{15}}{4} \approx -2.90 \text{ m s}^{-1}$

It makes sense that $\dfrac{dy}{dt}$ is negative because the distance y decreases as the ladder slides down.

Example 13.22

In Example 13.21, how fast is the angle between the ladder and the ground changing when $y = 2$ m?

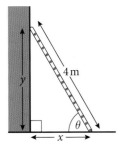

Figure 13.27 Diagram for Example 13.22

Solution

We know $\dfrac{dx}{dt} = \dfrac{3}{4}$ and we want to find $\dfrac{d\theta}{dt}$. We need a relationship, true at any instant, between the variables θ and x. Several trigonometric ratios could be used, but perhaps the most straightforward is:

$$x = 4\cos\theta$$

Now we differentiate implicitly with respect to t and solve for $\dfrac{d\theta}{dt}$.

$$\frac{d}{dt}(x) = \frac{d}{dt}(4\cos\theta)$$

$$\frac{dx}{dt} = -4\sin\theta\frac{d\theta}{dt}$$

$$\frac{d\theta}{dt} = -\frac{1}{4\sin\theta}\frac{dx}{dt}$$

When $y = 2$ we find that $\sin\theta = \dfrac{1}{2}$. Substituting for $\sin\theta$ and $\dfrac{d\theta}{dt}$:

$$\frac{d\theta}{dt} = -\frac{1}{4\left(\dfrac{1}{2}\right)}\cdot\frac{3}{4} = -\frac{3}{8}$$

Therefore, the angle is decreasing at a rate of $\dfrac{3}{8}$ radians s^{-1} (or approximately 21.5° s^{-1}).

Example 13.23

Consider a conical tank as shown in the diagram. Its radius at the top is 4 metres and its height is 8 metres. The tank is being filled with water at a rate of 2 m³ min⁻¹. How fast is the water level rising when it is 5 metres deep?

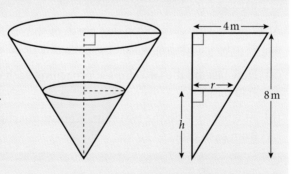

Solution

We know the rate of change of the volume with respect to time $\dfrac{dV}{dt} = 2$ m³ min⁻¹ and we need to find the rate of change of the height of the water level with respect to time, call it $\dfrac{dh}{dt}$.

Not including t, there are three variables involved in this problem: V, r, and h. The formula for the volume of a cone will give us an equation that relates all of these variables.

$$V = \frac{1}{3}\pi r^2 h$$

673

We differentiate this equation to get the rate $\frac{dr}{dt}$ in our result. We need to either find $\frac{dr}{dt}$ (which is possible) or eliminate r from the equation by solving for it in terms of one of the other variables and substitute. By using similar triangles we can write a proportion involving r and h.

$$\frac{r}{h} = \frac{4}{8} \Rightarrow r = \frac{h}{2}$$

Hence, $V = \frac{1}{3}\pi\left(\frac{h}{2}\right)^2 h$

$$V = \frac{\pi}{12}h^3$$

Differentiate implicitly with respect to t and solve for $\frac{dr}{dt}$:

$$\frac{dV}{dt} = \frac{\pi}{12} \cdot 3h^2\frac{dh}{dt} \Rightarrow \frac{dV}{dt} = \frac{\pi}{4}h^2\frac{dh}{dt} \Rightarrow \frac{dh}{dt} = \frac{4}{\pi h^2}\frac{dV}{dt}$$

Substitute $h = 5$ and $\frac{dV}{dt} = 2$:

$$\frac{dh}{dt} = \frac{4}{\pi(5)^2} \cdot 2 = \frac{8}{25\pi} \approx 0.102 \text{ m min}^{-1} \text{ (or } 10.2 \text{ cm min}^{-1})$$

Therefore, the water level is rising at a rate of 0.102 m min^{-1} when the water level is at 5 m.

> Be careful not to substitute in known quantities too early in the process of solving a related rates problem. Substitute the known values of any variables and any rates of change after differentiation. For example, in Example 13.23, h remained a variable (it is a quantity that is changing over time) until the last stage of the solution when we substituted $h = 5$. If we substituted earlier into $V = \frac{\pi}{12}h^3$, we would have obtained $\frac{dV}{dt} = 0$, which is obviously wrong.

Example 13.24 involves two rates of change.

Example 13.24

At 12 noon, ship A is 65 km due north of ship B. Ship A sails south at a rate of 14 km h^{-1}, and ship B sails west at a rate of 16 km h^{-1}.

(a) How fast are the two ships approaching each other $1\frac{1}{2}$ hours later at 1:30 p.m.?

(b) At what time do the two ships stop approaching and begin moving away from each other?

Solution

Let a and b be the distances that ships A and B are from the intersection of the ships' paths. Let c be the distance between the two ships. Since a is decreasing and b is increasing, we know that

$$\frac{da}{dt} = -14 \text{ km h}^{-1} \text{ and}$$

$$\frac{db}{dt} = 16 \text{ km h}^{-1}.$$

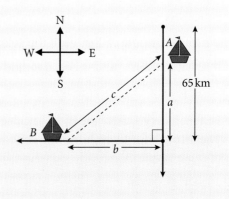

(a) The three variables are related by the equation

$$c^2 = a^2 + b^2$$

Differentiating implicitly with respect to t:

$$2c\frac{dc}{dt} = 2a\frac{da}{dt} + 2b\frac{db}{dt}$$

The rate at which the ships are approaching is $\frac{dc}{dt}$. Solve for $\frac{dc}{dt}$:

$$\frac{dc}{dt} = \frac{a\frac{da}{dt} + b\frac{db}{dt}}{c}$$

Substitute $\frac{da}{dt} = -14$ and $\frac{db}{dt} = 16$

$$\frac{dc}{dt} = \frac{-14a + 16b}{c}$$

The distances a and b are both functions of time; thus, they can be written in terms of t as

$$a = 65 - 14t \text{ and } b = 16t$$

Evaluating these expressions when $t = 1\frac{1}{2}$, gives $a = 44$, $b = 24$ and $c = \sqrt{44^2 + 24^2} \approx 50.12$. Substituting these values into the expression for $\frac{dc}{dt}$ gives

$$\frac{dc}{dt} \approx \frac{-14(44) + 16(24)}{50.12} \approx -4.629$$

Therefore, at 1:30 p.m. the distance between the two ships is decreasing at a rate of approximately -4.63 km h^{-1}.

(b) The time at which the two ships will stop approaching each other and begin to move away is when the value of $\frac{dc}{dt}$ changes from negative to positive. So we need to find when $\frac{dc}{dt} = 0$.

$$\frac{dc}{dt} = \frac{-14a + 16b}{c} = 0$$

$$-14a + 16b = 0$$

Substitute in $a = 65 - 14t$ and $b = 16t$:

$$-14(65 - 14t) + 16(16t) = 0$$

$$452t - 910 = 0$$

$$t = \frac{910}{452} \approx 2.013$$

Therefore, just moments after 3:30 p.m., the two ships will stop approaching and start moving away from each other.

In question 1, since volume is decreasing, $\dfrac{dV}{dt}$ is negative.

In question 2, $V = \dfrac{4}{3}\pi r^3$

Exercise 13.4

1. A water tank is in the shape of an inverted cone. Water is being drained from the tank at a constant rate of 2 m³ min⁻¹. The height of the tank is 8 m, and the diameter of the top of the tank is 6 m. When the depth of the water is 5 m, find, in units of cm min⁻¹:

   (a) the rate of change of the water level

   (b) the rate of change of the radius of the surface of the water.

2. A spherical balloon is being inflated at a constant rate of 240 cm³ s⁻¹.

   (a) At what rate is the radius increasing when the radius is 8 cm?

   (b) At what rate is the radius increasing after 5 seconds?

3. Oil is dripping from a car engine onto a garage floor making a growing circular stain. The radius, r, of the stain is increasing at a constant rate of 1 cm h⁻¹. When the radius is 4 cm, find:

   (a) the rate of change of the circumference of the stain

   (b) the rate of change of the area of the stain.

4. A hot air balloon is rising straight up from a level field at a constant rate of 50 m min⁻¹. An observer is standing 150 m from the point on the ground where the balloon was launched. Let θ be the angle between the ground and the point at which the observer is standing (angle of elevation of the balloon). What is the rate of change of θ (in radians min⁻¹) when the height of the balloon is 250 m?

5. Jenny is flying a kite at a constant height above level ground of 72 m. The wind carries the kite horizontally away at a rate of 6 m s⁻¹. How fast must Jenny let out the string at the moment when the kite is 120 m away from her?

6. A 5-foot tall girl is walking towards a 20-foot lamp post at a constant rate of 6 ft s⁻¹. The light from the lamp post causes the girl to cast a shadow. How fast is the tip of her shadow moving?

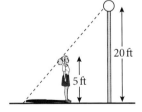

Figure 13.28 Diagram for question 6

7. Two cars start from a point A at the same time. One travels west at 60 km h⁻¹ and the other travels north at 35 km h⁻¹. How fast is the distance between them increasing 3 hours later?

8. A point moves along the curve $y = \sqrt{x^2 + 1}$ in such a way that $\dfrac{dx}{dt} = 4$. Find $\dfrac{dy}{dt}$ when $x = 3$.

9. A horizontal trough is 4 m long, 1.5 m wide, and 1 m deep. Its cross-section is an isosceles triangle. Water is flowing into the trough at a constant rate of 0.03 m³ s⁻¹. Find the rate at which the water level is rising 25 seconds after the water started flowing into the trough.

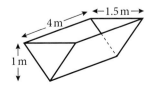

Figure 13.29 Diagram for question 9

10. The radius of a sphere is increasing at the constant rate of 3 mm s^{-1}. How fast is the volume changing when the surface area is 10 mm^2?

In question 10, surface area of a sphere $= 4\pi r^2$

11. Two roads, named A and B, intersect each other at an angle of 60°. Two cars, one on road A travelling at 40 km h^{-1} and the other on road B travelling at 50 km h^{-1}, are approaching the intersection. At a certain moment, the two cars are both 2 km from the intersection. How fast is the distance between them changing?

12. The diagonal of a cube is increasing at a rate of 8 cm s^{-1}. How fast is the length of a side of the cube increasing?

13. A point P is moving along a circle with equation $x^2 + y^2 = 100$ at a constant rate of 3 units s^{-1}. How fast is the projection of P on the x-axis moving when P is 5 units above the x-axis?

14. A jet is flying at a constant speed at an altitude of 10 000 m on a path that will take it directly over an observer on the ground. At a given instant, the observer determines that the angle of elevation of the jet is $\dfrac{\pi}{3}$ radians and is increasing at a constant rate of $\dfrac{1}{60}$ radians s^{-1}. Find the speed of the jet.

15. A television cameraman is filming a car race from a platform that is 40 metres from the racing track, following a car that is moving at 288 km h^{-1}. How fast, in degrees per second, will the camera be turning:
    (a) when the car is directly in front of the camera
    (b) a half second later? Give your answers to the nearest whole degree.

16. An aeroplane is flying due east at 640 km h^{-1} and climbing vertically at a rate of 180 m min^{-1}; an airport tower is tracking the aeroplane. Determine how fast the distance between the aeroplane and the tower is changing when the aeroplane is 5 km above the ground over a point exactly 6 km due west of the tower. Express the answer in km h^{-1}.

13.5 Optimisation

Many problems in science and mathematics involve finding the maximum or minimum value (**optimum** value) of a function over a specified or implied domain. The development of calculus in the 17th century was motivated to a large extent by maxima and minima (**optimisation**) problems. One such problem led Pierre de Fermat (1601–1665) to develop his principle of least time: a ray of light will follow the path that takes the least (or minimum) time. The solution to Fermat's principle lead to Snell's law, or the law of refraction. The solution is found by applying techniques of differential calculus, which can also be used to solve other optimisation problems involving ideas such as least cost, maximum profit, minimum surface area and greatest volume.

We have learned the theory of how to use the derivative of a function to locate points where the function has a maximum or minimum (i.e. extreme) value. It is important to remember that if the derivative of a function is zero at a certain point, it does not necessarily follow that the function has an extreme value (relative or absolute) at that point – it only ensures that the function has a horizontal tangent (stationary point) at that point. An extreme value may occur where the derivative is zero or at the endpoints of the function's domain.

Figure 13.30 shows the graph of $f(x) = x^4 - 8x^3 + 18x^2 - 16x - 2$. The derivative of $f(x)$ is $f'(x) = 4x^3 - 24x^2 + 36x - 16 = 4(x - 4)(x - 1)^2$. The function has horizontal tangents at both $x = 1$ and $x = 4$, since the derivative is zero at these points. However, an extreme value (absolute minimum) occurs only at $x = 4$. It is important to confirm – graphically or algebraically – the precise nature of a point on a function where the derivative is zero. Some different algebraic methods for confirming that a value is a maximum or minimum will be illustrated in the examples that follow.

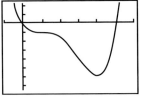

Figure 13.30 Graph of $f(x)$

It is also useful to know that one can often find extreme values (extrema) without calculus (e.g. using a minimum command on a graphics calculator). Calculator or computer technology can be very helpful in modelling, solving, or confirming solutions to optimisation problems. However, it is important to learn how to apply algebraic methods of differentiation to optimisation problems because it may be the only efficient way to obtain an accurate solution.

Let's start with a relatively straightforward example. We can use the steps in the solution to develop a general strategy that can be applied to more sophisticated problems.

Example 13.25

Find the maximum area of a rectangle inscribed in an isosceles right-angled triangle whose hypotenuse is 20 cm long.

Solution

Step 1: Draw an accurate diagram. Let the base of the rectangle be x cm and the height y cm.

The area of the rectangle is $A = xy \, \text{cm}^2$.

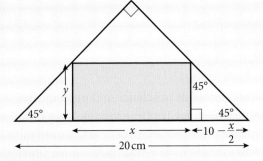

Step 2: Express the area as a function in terms of only one variable.

It can be deduced from the diagram that $y = 10 - \dfrac{x}{2}$.

Therefore, $A(x) = x\left(10 - \dfrac{x}{2}\right) = 10x - \dfrac{x^2}{2}$

x must be positive; from the diagram it is clear that x must be less than 20.

Step 3: Find the derivative of the area function and find for what value(s) of x it is zero.

$$A'(x) = 10 - x$$

$A'(x) = 0$ when $x = 10$

Step 4: Analyse $A(x)$ at $x = 10$ and also at the endpoints of the domain, $x = 0$ and $x = 20$

The second derivative test (Chapter 12) provides information about the concavity of a function. The second derivative is $A''(x) = -1$ and since $A''(x)$ is always negative, $A(x)$ is always concave down, indicating that $A(x)$ has a maximum at $x = 10$.

$A(0) = 0$ and $A(20) = 0$. Therefore $A(x)$ has an absolute maximum at $x = 10$.

Example 13.26

Two vertical posts, with heights of 7 m and 13 m, are secured by a rope going from the top of one post to a point on the ground that is between the posts and then to the top of the other post (see diagram). The distance between the two posts is 25 m. Where should the point at which the rope touches the ground be located so that the least amount of rope is used?

Solution

Step 1: Draw an accurate diagram. Draw the posts as line segments PQ and TS and the point where the rope touches the ground is labeled R. The optimum location of point R can be given as a distance from the base of the shorter post, QR, or from the taller post, SR. It is decided to give the answer as

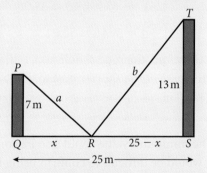

the distance from the shorter post, and this is labeled x. There are two other important unknown quantities – the lengths of the two portions of the rope, PR and TR. These are labeled a and b, respectively.

Step 2: The quantity to be minimised is the length L of the rope which is the sum of a and b. From Pythagoras' theorem, $a = \sqrt{x^2 + 49}$ and $b = \sqrt{(25 - x)^2 + 169}$. Therefore, the function for length (L) can be expressed in terms of the single variable x as

$$L(x) = \sqrt{x^2 + 49} + \sqrt{(25 - x)^2 + 169}$$
$$= \sqrt{x^2 + 49} + \sqrt{x^2 - 50x + 625 + 169}$$
$$= \sqrt{x^2 + 49} + \sqrt{x^2 - 50x + 794}$$

From the given information and diagram, the domain of $L(x)$ is $0 \leqslant x \leqslant 25$

General strategy for solving optimisation problems

Step 1: Draw a diagram that accurately illustrates the problem. Label all known parts of the diagram. Using variables, label the important unknown quantity (or quantities) (for example, x for base and y for height in Example 13.25).

Step 2: For the quantity that is to be optimised (area in Example 13.25), express this quantity as a function in terms of a single variable. From the diagram and/or information provided, determine the domain of this function.

Step 3: Find the derivative of the function from Step 2, and determine where the derivative is zero. The value (or values) of the variable when the derivative is zero, along with any domain endpoints, are the critical values to be tested (In Example 13.25: $x = 0$, $x = 10$ and $x = 20$).

Step 4: Using algebraic (e.g. second derivative test) or graphical (e.g. GDC) methods, analyse the nature (maximum, minimum, or neither) of the points at the critical values for the optimised function. Make sure you answer the precise question that was asked in the problem.

Step 3: To facilitate differentiation, express $L(x)$ using fractional exponents.

$$L(x) = (x^2 + 49)^{\frac{1}{2}} + (x^2 - 50x + 794)^{\frac{1}{2}}$$

Apply the chain rule for differentiation:

$$\frac{dL}{dx} = \frac{1}{2}(x^2 + 49)^{-\frac{1}{2}}(2x) + \frac{1}{2}(x^2 - 50x + 794)^{-\frac{1}{2}}(2x - 50)$$

$$= \frac{x}{\sqrt{x^2 + 49}} + \frac{x - 25}{\sqrt{x^2 - 50x + 794}}$$

By setting $\frac{dL}{dx} = 0$, we obtain

$$x\sqrt{x^2 - 50x + 794} = -(x - 25)\sqrt{x^2 + 49}$$

$$x^2(x^2 - 50x + 794) = (25 - x)^2(x^2 + 49)$$

$$x^4 - 50x^3 + 794x^2 = x^4 - 50x^3 + 674x^2 - 2450x + 30\,625$$

$$120x^2 + 2450x - 30\,625 = 0$$

$$5(4x - 35)(6x + 175) = 0$$

$$x = \frac{35}{4} \quad \text{or} \quad x = -\frac{175}{6}$$

Step 4: Since $x = -\frac{175}{6}$ is not in the domain for $L(x)$, then the critical values are $x = 0$, $x = \frac{35}{4}$ and $x = 25$. Simply evaluate $L(x)$ for these critical values.

$$L(0) = 49 + \sqrt{794} \approx 35.18$$

$$L(25) = \sqrt{674} + 13 \approx 38.96$$

$$L\left(\frac{35}{4}\right) = 5\sqrt{41} \approx 32.02$$

Therefore, the rope should touch the ground at a distance of 8.75 m from the base of the shorter post to give a minimum rope length of approximately 32.02 m.

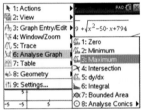

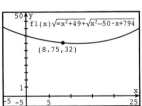

Figure 13.31 Performing steps 3 and 4 graphically

The minimum value could also be confirmed from the graph of $L(x)$, but it would be difficult to confirm using the second derivative test because of the tedious algebra required. From this example, we can see that applied optimisation problems can involve a high level of algebra. If you have access to suitable graphing technology, you could perform Steps 3 and 4 graphically rather than algebraically.

In both Examples 13.25 and 13.26, the extreme value occurred at a point where the derivative was zero. Although this often happens, an extreme value may also occur at the endpoint of the domain.

Example 13.27

A 4-metre length of wire is to be used to form a square and a circle. How much of the wire should be used to make the square and how much should be used to make the circle in order to enclose the greatest area?

Solution

Step 1: Let x = length of each edge of the square, and r = radius of the circle.

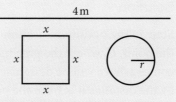

Step 2: The total area is given by $A = x^2 + \pi r^2$. The task is to write the area A as a function of a single variable. Therefore it is necessary to express r in terms of x or vice-versa and perform a substitution.

The perimeter of the square is $4x$ and the circumference of the circle is $2\pi r$. The total amount of wire is 4 m which gives:

$$4 = 4x + 2\pi r \implies 2\pi r = 4 - 4x \implies r = \frac{2(1 - x)}{\pi}$$

Substitute for r:

$$A(x) = x^2 + \pi\left[\frac{2(1 - x)}{\pi}\right]^2 = x^2 + \frac{4(1 - x)^2}{\pi} = \frac{1}{\pi}[(\pi + 4)x^2 - 8x + 4]$$

Because the square's perimeter is $4x$, then the domain for $A(x)$ is $0 \leqslant x \leqslant 1$

Step 3: Differentiate the function $A(x)$, set equal to zero, and solve.

$$\frac{d}{dx}\left(\frac{1}{\pi}[(\pi + 4)x^2 - 8x + 4]\right)$$

$$= \frac{1}{\pi}[2(\pi + 4)x - 8] = 0$$

$$2(\pi + 4)x - 8 = 0 \implies \implies x = \frac{4}{\pi + 4} \approx 0.5601$$

The critical values are $x = 0$, $x \approx 0.5601$ and $x = 1$

Step 4: Evaluate $A(x)$: $A(0) \approx 1.273$, $A(0.5601) \approx 0.5601$ and $A(1) = 1$

Therefore, the maximum area occurs when $x = 0$, which means all the wire is used for the circle.

Figure 13.32 Graph of the area function showing a maximum of 1.27 for the left endpoint of the domain at $x = 0$

Example 13.28

A pipeline needs to be constructed to link an offshore drilling rig to an onshore refinery depot. The oil rig is located at a distance (perpendicular to the coast) of 140 km from the coast. The depot is located inland at a perpendicular distance of 60 km from the coast. For modelling purposes, the coastline is assumed to follow a straight line. The rate at which crude oil is pumped through the pipeline varies according to several variables, including pipe dimensions, materials, temperature, and so on. On average, oil flows through the offshore section of the pipeline at a rate of 9 km h^{-1} and through the onshore section at a rate of 5 km h^{-1}. Assume that both sections of pipeline can travel straight from one point to another. At what point should the pipeline intersect with the coastline in order for the oil to take a minimum amount of time to flow from the rig to the depot?

13 Differential calculus 2

Solution

Step 1: The optimum location of the point, C, where the pipeline comes ashore will be designated by the distance it is from the point on the coast that is a minimum distance (perpendicular) from the rig, R (140 km).

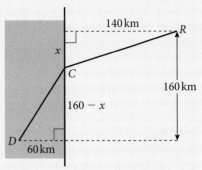

The distance from R to C is $\sqrt{x^2 + 140^2}$ and the distance from D (depot) to C is $\sqrt{(160 - x)^2 + 60^2}$.

Step 2: The quantity to be minimised is time, so it is necessary to express the total time it takes the oil to flow from R to D in terms of a single variable.

$$\text{time} = \frac{\text{distance}}{\text{rate}}$$

$$\text{time (offshore)} = \frac{\sqrt{x^2 + 19\,600}}{9}$$

$$\text{time (onshore)} = \frac{\sqrt{x^2 - 320x + 29\,200}}{5}$$

The function for time T in terms of x is:

$$T(x) = \frac{\sqrt{x^2 + 19\,600}}{9} + \frac{\sqrt{x^2 - 320x + 29\,200}}{5}$$

and the domain for $T(x)$ is $0 \leqslant x \leqslant 160$

Steps 3 and 4: The algebra for finding the derivative of $T(x)$ is similar to that of Step 3 in Example 13.26.

Use a GDC to find the value of x that produces a minimum for $T(x)$.

Therefore, the optimum point for the pipeline to intersect with the coast is approximately 134.9 km from the point on the coast nearest to the drilling rig.

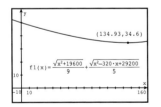

Figure 13.33 Solution to Example 13.28

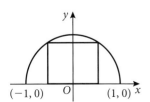

Figure 13.34 Diagram for question 1

Exercise 13.5

1. Find the dimensions of the rectangle with maximum area that is inscribed in a semicircle with radius 1 cm. Two vertices of the rectangle are on the semicircle and the other two vertices are on the x-axis, as shown in Figure 13.34.

2. A rectangular piece of aluminium is to be rolled to make a cylinder with open ends (a tube). Regardless of the dimensions of the rectangle, the perimeter of the rectangle must be 40 cm. Find the dimensions (length and width) of the rectangle that gives a maximum surface area for the cylinder.

3. Find the minimum distance from the graph of the function $y = \sqrt{x}$ to the point $\left(\frac{3}{2}, 0\right)$.

4. A rectangular box has height h cm, width x cm and length $2x$ cm. It is designed to have a volume equal to 1 litre (1000 cm³).

   (a) Show that $h = \dfrac{500}{x^2}$ cm.

   (b) Find an expression for the total surface area, s cm², of the box in terms of x.

   (c) What dimensions of the box will produce the minimum surface area?

5. The shape in Figure 13.35 consists of a rectangle $ABCD$ and two semicircles on either end. The rectangle has an area of 100 cm². If x represents the length of the rectangle AB, find the value of x that makes the perimeter of the entire figure a minimum.

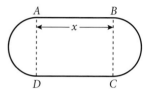

Figure 13.35 Diagram for question 5

6. Two vertical posts, with heights 12 metres and 8 metres, are 10 metres apart on horizontal ground (see Figure 13.36). A rope that stretches is attached to the top of both posts and is stretched down so that it touches the ground at point A between the two posts. The distance from the base of the taller post to point A is represented by x and the angle between the two sections of rope is θ. What value of x makes θ a maximum?

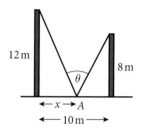

Figure 13.36 Diagram for question 6

7. A ladder is to be carried horizontally down an L-shaped hallway. The first section of the hallway is 2 metres wide, then there is a right-angled turn into a 3-metre wide section of the hallway. What is the longest ladder that can be carried around the corner?

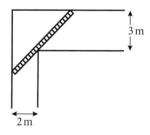

Figure 13.37 Diagram for question 7

8. Erica is walking from the wildlife observation tower (T) to the Big Desert Park office (O). The tower is 7 km due west and 10 km due south of the office. There is a road that goes to the office that Erica can get to if she walks 10 km due north from the tower. Erica can walk at a rate of 2 km h⁻¹ through the sandy terrain of the park, but she can walk a faster rate of 5 km h⁻¹ on the road. To what point A on the road should Erica walk in order to take the least time to walk from the tower to the office? Find the value of d such that point A is d km from the office.

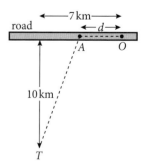

Figure 13.38 Diagram for question 8

9. Two vertices of a rectangle are on the x-axis, and the other two vertices are on the curve $y = \dfrac{8}{x^2 + 4}$. Find the maximum area of the rectangle.

10. A ship sailing due south at 16 km h⁻¹ is 10 kilometres north of a second ship going due west at 12 km h⁻¹. Find the minimum distance between the two ships.

11. Find the height, h, and the base radius, r, of the largest right circular cylinder that can be made by cutting it away from a sphere with a radius of R.

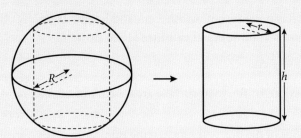

12. Nadia is standing at point A that is a km away in the countryside from a straight road XY. She wishes to reach the point Y where the distance from X to Y is b km. Her speed on the road is r km hr^{-1} and her speed travelling across the countryside is c km hr^{-1}, such that $r > c$. She wishes to reach Y as quickly as possible. Find the position of point P where she joins the road.

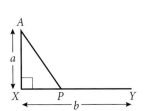

Figure 13.39 Diagram for question 12

13. A cone of height h and radius r is constructed from a circle with radius 10 cm by removing a sector AOC of arc length x cm and then connecting the edges OA and OC. What arc length x will produce the cone of maximum volume, and what is the volume?

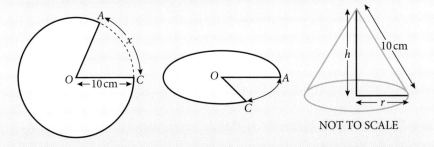

NOT TO SCALE

14. Point P is a units above the line AB, and point Q is b units below line AB. The velocity of light is u units sec^{-1} above AB and v units sec^{-1} below AB, and $u > v$. The angles α and β are the angles that a ray of light makes with a perpendicular to line AB above and below AB, respectively. Show that the following relationship must hold true:

$$\frac{\sin \alpha}{\sin \beta} = \frac{u}{v}$$

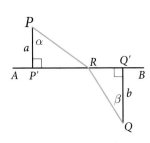

Figure 13.40 Diagram for question 14

13.6 l'Hôpital's rule

Rational functions were covered in Section 2.4, where we used an informal approach to evaluate limits of rational expressions to help determine different kinds of asymptotes related to the graph of a rational function. In Section 12.1, when differentiating from first principles, we encountered limits of rational expressions. Now that we have acquired some techniques for finding derivatives, we have access to a very useful and efficient technique for evaluating the limits of rational expressions (fractions) that are in what is referred to as **indeterminate form**. The two most common forms of a rational expression that are considered indeterminate are when the numerator and denominator are either both zero, or both increase without bound (i.e. go to infinity).

In Example 12.4, we performed some sophisticated algebraic operations to obtain the result: $\lim\limits_{x\to 0}\dfrac{\sqrt{x^2+4}-2}{x^2}=\dfrac{1}{4}$. Substituting zero for x in an attempt to evaluate the limit, the expression $\dfrac{\sqrt{x^2+4}-2}{x^2}$ produces the indeterminate form $\dfrac{0}{0}$. Not all limits of rational expressions in indeterminate form can be evaluated successfully with algebraic methods. Limits of rational expressions that are of indeterminate form are addressed by l'Hôpital's rule.

l'Hôpital's rule

Let f and g be functions whose derivatives can be found at any value in an open interval $\,]\,a,b\,[$, except possibly at some value c where $a < c < b$. Assume that $g'(x) \neq 0$, except possibly at c. Suppose that $\lim\limits_{x\to c} f(x) = 0$ and $\lim\limits_{x\to c} g(x) = 0$; or $\lim\limits_{x\to c} f(x) = \pm\infty$ and $\lim\limits_{x\to c} g(x) = \pm\infty$

(That is, the expression $\dfrac{f(x)}{g(x)}$ is in indeterminate form of $\dfrac{0}{0}$ or $\dfrac{\infty}{\infty}$.)

Then $\lim\limits_{x\to c}\dfrac{f(x)}{g(x)} = \lim\limits_{x\to c}\dfrac{f'(x)}{g'(x)}$ provided the limit on the right-hand side exists (or is infinite).

l'Hôpital's rule states simply that, given the right conditions, the limit of a quotient of functions is equal to the limit of the quotient of their derivatives. It is important to first verify the conditions regarding the limits of f and g before applying l'Hôpital's rule.

When you are applying l'Hôpital's rule, make sure that you differentiate the numerator and denominator separately. Do not use the quotient rule for differentiation.

The notation $x \to 2^+$ indicates that the value of x is approaching ever closer to 2 from the right (e.g. 3, 2.5, 2.25, 2.125, etc); and $x \to 2^-$ indicates that the value of x is approaching 2 from the left (e.g. 0.5, 1, 1.3, 1.5, etc).

Example 13.29

For each limit, use your GDC to conjecture a result, then find the limit using l'Hôpital's rule.

(a) $\lim\limits_{x\to 0}\left(\dfrac{x}{1-e^x}\right)$ (b) $\lim\limits_{x\to\frac{\pi}{2}}\left(\dfrac{\sec x}{1+\tan x}\right)$ (c) $\lim\limits_{x\to 0}(e^x+x)^{\frac{1}{x}}$

Solution

(a) To visualise $\lim\limits_{x\to 0}\left(\dfrac{x}{1-e^x}\right)$ we graph $f(x) = \dfrac{x}{1-e^x}$ using a GDC.

Although $x = 0$ is not in the domain of f, the graph appears to pass through the point $(0, -1)$ implying that $\lim\limits_{x\to 0}\left(\dfrac{x}{1-e^x}\right) = -1$.

Since $\lim\limits_{x\to 0} x = 0$ and $\lim\limits_{x\to 0}(1-e^x) = 0$, $\lim\limits_{x\to 0}\left(\dfrac{x}{1-e^x}\right)$ is in the indeterminate form $\dfrac{0}{0}$, and l'Hôpital's rule applies. Differentiating the numerator and denominator separately and evaluating the limit gives:

$$\lim_{x\to 0}\left(\frac{x}{1-e^x}\right) = \lim_{x\to 0}\left(\frac{1}{-e^x}\right) = \frac{1}{-1} = -1$$

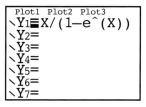

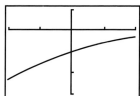

Figure 13.41 Solution to Example 13.29 (a)

685

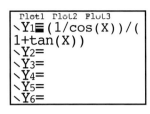

Figure 13.42 Solution to Example 13.29 (b)

$$\frac{\sec x \tan x}{\sec^2 x}$$
simplifies to $\sin x$

(b) Instead of viewing a graph of $f(x) = \dfrac{\sec x}{1 + \tan x}$ to conjecture a value for

$\displaystyle\lim_{x \to \frac{\pi}{2}}\left(\dfrac{\sec x}{1 + \tan x}\right)$, use a GDC to construct a table of function values near

$x = \dfrac{\pi}{2} \approx 1.5708$

The values in the table show that the function appears to be approaching 1 from either direction.

The values of $\sec x$ vanish to $+\infty$ when $x \to \dfrac{\pi}{2}$ from the left

$\left(\text{i.e. } x \to \dfrac{\pi}{2}^{-}\right)$ and vanish to $-\infty$ when $x \to \dfrac{\pi}{2}^{+}$.

Similarly, $\displaystyle\lim_{x \to \frac{\pi}{2}^{-}}(1 + \tan x) = +\infty$ and $\displaystyle\lim_{x \to \frac{\pi}{2}^{+}}(1 + \tan x) = -\infty$

So, when approaching $\dfrac{\pi}{2}$ from the left we have $\dfrac{+\infty}{+\infty}$ and $\dfrac{-\infty}{-\infty}$ when approaching from the right.

l'Hôpital's rule also applies to one-sided limits. Applying the rule to the right-hand limit gives

$$\lim_{x \to \frac{\pi}{2}^{+}}\left(\frac{\sec x}{1 + \tan x}\right) = \lim_{x \to \frac{\pi}{2}^{+}}\left(\frac{\sec x \tan x}{\sec^2 x}\right) = \lim_{x \to \frac{\pi}{2}^{+}} \sin x = 1$$

The left-hand limit is also 1, therefore the two-sided limit is equal to 1,

i.e. $\displaystyle\lim_{x \to \frac{\pi}{2}}\left(\dfrac{\sec x}{1 + \tan x}\right) = 1$

(c) To visualise $\displaystyle\lim_{x \to 0}(e^x + x)^{\frac{1}{x}}$ graph $f(x) = (e^x + x)^{\frac{1}{x}}$ using a GDC.

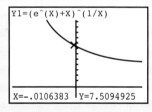
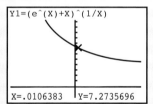

Tracing on the graph indicates that as $x \to 0$, the function approaches a value between 7.2735 and 7.5094. The exact value of the limit is not clear.

$\displaystyle\lim_{x \to 0}(e^x + x)^{\frac{1}{x}}$ is in the indeterminate form 1^{∞}. However, by taking the logarithm of both sides of $f(x) = (e^x + x)^{\frac{1}{x}}$ and then taking the limit, we can change the indeterminate form to $\dfrac{0}{0}$, for which we can apply l'Hôpital's rule.

$$\ln\left[f(x)\right] = \ln\left[(e^x + x)^{\frac{1}{x}}\right] = \frac{1}{x}\ln(e^x + x) = \frac{\ln(e^x + x)}{x}$$

Thus, $\ln\left[f(x)\right] = \dfrac{\ln(e^x + x)}{x}$, and taking the limit as $x \to 0$ of both sides produces

$$\lim_{x \to 0}\ln\left[f(x)\right] = \lim_{x \to 0}\frac{\ln(e^x + x)}{x}$$

The right-hand side is in the form $\frac{0}{0}$; apply l'Hôpital's rule:

$$= \lim_{x \to 0} \left(\frac{e^x + \dfrac{1}{e^x} + x}{1} \right) = \frac{e^0 + 1}{e^0 + 0} = 2$$

Hence, $\lim_{x \to 0} \ln[f(x)] = 2$

Since $f(x) = (e^x + x)^{\frac{1}{x}}$, then $\lim_{x \to 0}(e^x + x)^{\frac{1}{x}} = \lim_{x \to 0} f(x)$

Apply the rule $e^{\ln a} = a$: $= \lim_{x \to 0} e^{\ln f(x)}$

Use the result $\lim_{x \to 0} \ln[f(x)] = 2 = \lim_{x \to 0} e^2$

Therefore, $\lim_{x \to 0}(e^x + x)^{\frac{1}{x}} = e^2$

$e^2 \approx 7.38$ (to 3 s.f.), so the limit is within the range estimated from the graph on the GDC.

l'Hôpital's rule should not be applied if the limit is not in indeterminate form.

For example, consider the following limit: $\lim_{x \to 0} \left(\dfrac{\sin x}{x + 1} \right)$. The limit is not indeterminate because $\dfrac{\sin(0)}{0 + 1} = \dfrac{0}{1}$. Hence, the application of l'Hôpital's rule produces an incorrect result. This result is given by l'Hôpital's rule:

$\lim_{x \to 0} \left(\dfrac{\sin x}{x + 1} \right) = \lim_{x \to 0} \dfrac{\cos x}{1} = \dfrac{\cos(0)}{1} = \dfrac{1}{1} = 1$. The correct result can be obtained

simply from direct substitution: $\lim_{x \to 0} \left(\dfrac{\sin x}{x + 1} \right) = \dfrac{\sin(0)}{0 + 1} = \dfrac{0}{1} = 0$.

If, after applying l'Hôpital's rule, the quotient of the derivatives remains in indeterminate form, the rule can be applied more than once.

Example 13.30

Find $\lim_{x \to 1} \left(\dfrac{1 - x + \ln x}{x^3 - 3x + 2} \right)$

Solution

Substituting $x = 1$ into the rational expression gives $\dfrac{1 - 1 + \ln 1}{1^3 - 3 \cdot 1 + 2} = \dfrac{0}{0}$

So the limit is in the indeterminate form $\dfrac{0}{0}$ and we can apply l'Hôpital's rule.

$$\lim_{x \to 1} \left(\frac{1 - x + \ln x}{x^3 - 3x + 2} \right) = \lim_{x \to 1} \left(\frac{-1 + \dfrac{1}{x}}{3x^2 - 3} \right)$$

Substituting $x = 1$ again gives the indeterminate form $\dfrac{0}{0}$, so we apply

l'Hôpital's rule again, producing an expression that can be evaluated for $x = 1$.

$$\lim_{x \to 1} \left(\frac{1 - x + \ln x}{x^3 - 3x + 2} \right) = \lim_{x \to 1} \left(\frac{-1 + \dfrac{1}{x}}{3x^2 - 3} \right) = \lim_{x \to 1} \left(\frac{-\dfrac{1}{x^2}}{6x} \right) = -\frac{1}{6}$$

Exercise 13.6

1. Use l'Hopital's rule to find the value of each limit.

   (a) $\lim\limits_{x \to 0}\left(\dfrac{1 - \cos x}{x^2}\right)$ (b) $\lim\limits_{x \to 1}\left(\dfrac{x - 1}{\sqrt{x^2 + 3} - 2}\right)$ (c) $\lim\limits_{x \to 1}\left(\dfrac{1}{\ln x} - \dfrac{1}{x - 1}\right)$

2. Evaluate each limit.

   (a) $\lim\limits_{x \to 1}\left(\dfrac{x^2 - 1}{x^2 - 4x + 3}\right)$ (b) $\lim\limits_{x \to 0}\left(\dfrac{\sqrt[3]{1 - x} - 1}{x}\right)$

   (c) $\lim\limits_{x \to 0}\left(\dfrac{x - \sin x}{x^3}\right)$ (d) $\lim\limits_{x \to 0}\left(\dfrac{\frac{1}{x} - \cot x}{x}\right)$

   (e) $\left(\lim\limits_{x \to \infty}\dfrac{\ln(x + 1)}{\log_2 x}\right)$ (f) $\lim\limits_{x \to 0}\left(\dfrac{\ln(1 + x^2)}{\ln(1 - x^2)}\right)$

   (g) $\lim\limits_{x \to 0}\left(\dfrac{2 + x^2 - 2\cos x}{e^x + e^{-x} - 2\cos x}\right)$

3. Given $f(x) = (1 + x)^{\frac{1}{x}}$, find $\lim\limits_{x \to \infty} f(x)$

In question 3, start by taking the natural logarithm of both sides, converting the right side to the indeterminate form $\dfrac{0}{0}$. You can then use l'Hôpital's rule.

Chapter 13 practice questions

1. The diagram shows the graph of $y = f(x)$.

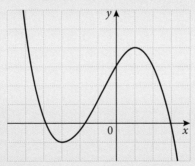

   Sketch the graph of $y = f'(x)$.

2. The diagram shows part of the graph of the function
   $f : x \mapsto -x^3 - 2x^2 + 8x$

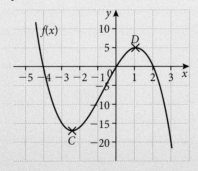

   The graph intersects the x-axis at $(-4, 0)$, $(0, 0)$ and $(2, 0)$.
   There is a minimum point at C and a maximum point at D.

(a) The function may also be written in the form
$f: x \mapsto -x(x - a)(x - b)$, where $a < b$.
Write down the value of:

  (i) a (ii) b

(b) Find:

  (i) $f'(x)$

  (ii) the exact values of x at which $f'(x) = 0$

  (iii) the value of the function at D.

(c) (i) Find the equation of the tangent to the graph of f at $(0, 0)$.

  (ii) This tangent cuts the graph of f at another point. Give the x-coordinate of this point.

3. In a controlled experiment, a tennis ball is dropped from the top observation deck (447 metres high) of the CN Tower in Toronto. The tennis ball's velocity is given by $v(t) = 66 - 66e^{-0.15t}$, where v is in metres per second and t is in seconds.

Find the value of v when:

(a) (i) $t = 0$ (ii) $t = 10$

(b) (i) Find an expression for the acceleration, a, as a function of t.

  (ii) What is the value of a when $t = 0$?

(c) (i) As t becomes large, what value does v approach?

  (ii) As t becomes large, what value does a approach?

  (iii) Explain the relationship between the answers to parts (i) and (ii).

4. (a) Given the function $f(x) = x^3 + 7x^2 + 8x - 3$, identify any points as a relative maximum or minimum, and find their exact coordinates.

(b) Find the exact coordinates of any inflection point(s).

5. Consider the function $g(x) = 2 + \dfrac{1}{e^{3x}}$

(a) (i) Find $g'(x)$

  (ii) Explain briefly how this shows that $g(x)$ is a decreasing function for all values of x (i.e. that $g(x)$ always decreases in value as x increases).

Let P be the point on the graph of g where $x = -\dfrac{1}{3}$

(b) Find an expression in terms of e for:

  (i) the y-coordinate of P

  (ii) the gradient of the tangent to the curve at P.

(c) Find the equation of the tangent to the curve at P, giving your answer in the form $y = mx + c$.

6. Consider the function f given by $f(x) = \dfrac{2x^2 - 13x + 20}{(x-1)^2}$, $x \neq 1$

   (a) Show that
   $$f'(x) = \frac{9x - 27}{(x-1)^3}, \quad x \neq 1$$
   The second derivative is given by $f''(x) = \dfrac{72 - 18x}{(x-1)^4}$, $x \neq 1$

   (b) Using values of $f'(x)$ and $f''(x)$, explain why a minimum must occur at $x = 3$.

   (c) There is a point of inflection on the graph of f. Write down the coordinates of this point.

7. Differentiate with respect to x:

   (a) $y = \dfrac{1}{(2x+3)^2}$ (b) $y = e^{\sin 5x}$ (c) $y = \tan^2(x^2)$

8. The curve with equation $y = Ax + B + \dfrac{C}{x}$, $x \in \mathbb{R}$, $x \neq 0$, has a minimum at $P(1, 4)$ and a maximum at $Q(-1, 0)$. Find the value of each of the constants A, B, and C.

9. Find $\dfrac{dy}{dx}$ and $\dfrac{d^2y}{dx^2}$ at the point $(1, 1)$ on the curve $x^3 + y^3 = 2$

10. Differentiate with respect to x:

    (a) $y = \dfrac{x}{e^x - 1}$ (b) $y = e^x \sin 2x$ (c) $y = (x^2 - 1)\ln(3x)$

11. The normal to the curve $y = x^2 - 4x$ at the point $(3, -3)$ intersects the x-axis at point P and the y-axis at point Q. Find the equation of the normal and the coordinates of P and Q.

12. Let $y = h(x)$ be a function of x for $0 \leqslant x \leqslant 6$. The graph of h has an inflection point at P, and a maximum point at M.

    Partial sketches of the curves of $h'(x)$ and $h''(x)$ are shown.

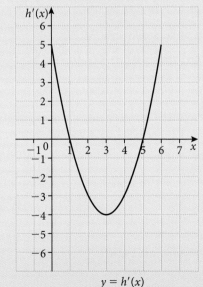

$y = h'(x)$

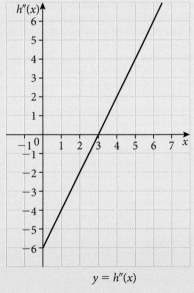

$y = h''(x)$

    (a) Write down the x-coordinate of P, and justify your answer.

(b) Write down the x-coordinate of M, and justify your answer.

(c) Given that $h(3) = 0$, sketch the graph of h. On the sketch, mark the points P and M.

13. Find the equation of the normal to the curve $x^2 + xy + y^2 - 3y = 10$ at the point $(2, 3)$.

14. A cylinder is to be made with an exact volume of $128\pi \, \text{cm}^3$. What should be the height h and the radius r of the cylinder's base so that cylinder's surface area is a minimum?

15. A rectangle has its base on the x-axis and its upper two vertices on the parabola $y = 12 - x^2$, as shown in the diagram. What is the largest area that the rectangle can have, and what are its dimensions (i.e. length and width)?

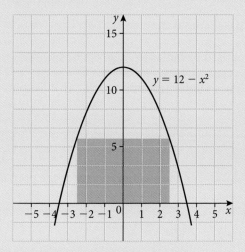

16. The diagram shows the graph of a function $y = f(x)$. At which point on the graph:

(a) are $f'(x)$ and $f''(x)$ both negative

(b) is $f'(x)$ negative and $f''(x)$ positive

(c) is $f'(x)$ positive and $f''(x)$ negative?

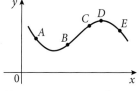

Figure 13.43 Diagram for question 16

17. Find the equation of the normal to the curve with equation $y = \dfrac{2x - 1}{x + 2}$ at the point $(-3, 7)$

18. Find the equation of **(a)** the tangent, and **(b)** the normal to the curve $y = \ln(4x - 3)$ at the point $(1, 0)$.

19. Let $f(x) = x^2 \ln x$

(a) Find the exact coordinates of any stationary points. Indicate whether each is a maximum or minimum (and absolute or relative).

(b) Find the exact coordinates of any inflection points.

20. Determine the constant a such that the function $f(x) = x^2 + \dfrac{a}{x}$

    (a) has a local minimum at $x = 2$

    (b) has a local minimum at $x = -3$.

    (c) Show that the function cannot have a local maximum for any value of a.

21. A line passes through the point $(3, 2)$ and intersects both the x-axis and the y-axis forming a triangular region in the first quadrant bounded by the x-axis, the y-axis, and the line. Find the equation of such a line that creates a triangle of minimum area.

22. Find the equation of both the tangent and normal to the curve $y = x \tan x$ at the point where $x = \dfrac{\pi}{4}$.

23. A very important function in statistics is the equation for the standard normal curve (mean $= 0$, standard deviation $= 1$) given by $f(x) = \dfrac{e^{-\frac{x^2}{2}}}{\sqrt{2\pi}}$

    (a) Find the coordinates of any stationary points and of any inflection points.

    (b) What happens when $x \to \infty$, and when $x \to -\infty$? Give the equation for any asymptotes.

    (c) Sketch a graph of $f(x)$ and indicate the location of any of the points found in part (a).

24. Let f be the function given by $f(x) = 2\ln(x^2 + 3) - x$

    (a) Find the x-coordinate of each maximum and minimum point of f. Justify your answer(s).

    (b) Find the x-coordinate of each inflection point of f. Justify your answer(s).

25. The rate at which cars on a road pass a certain point is known as the flow rate and is in units of cars per hour. The flow rate, F, of a certain road is given by $F(x) = \dfrac{2x}{18 + 0.015x^2}$, where x is the speed of the traffic in kilometres per hour. What speed will maximise the flow rate on the road?

26. Giving a reason, state whether or not the following argument is correct. Using l'Hopital's rule, $\displaystyle\lim_{x \to \pi^-}\left(\dfrac{\sin x}{1 - \cos x}\right) = \lim_{x \to \pi^-}\left(\dfrac{\cos x}{\sin x}\right) = -\infty$

27. Differentiate $y = \arccos(1 - 2x^2)$ with respect to x, and simplify your answer.

28. Find the gradient of the tangent to the curve $3x^2 + 4y^2 = 7$ at the point where $x = 1$ and $y > 0$

29. A normal to the graph of $y = \arctan(x - 1)$, for $x > 0$, has equation $y = -2x + c$, where $c \in \mathbb{R}$. Find the value of c.

30. The line $y = 16x - 9$ is a tangent to the curve $y = 2x^3 + ax^2 + bx - 9$ at the point $(1, 7)$. Find the values of a and b.

31. Let $y = \sin(kx) - kx\cos(kx)$, where k is a constant.

Show that $\dfrac{dy}{dx} = k^2 x \sin(kx)$

32. A curve has equation $xy^3 + 2x^2y = 3$. Find the equation of the tangent to this curve at the point $(1, 1)$.

33. The function f is defined by

$$f(x) = \frac{x^2 - x + 1}{x^2 + x + 1}$$

(a) (i) Find an expression for $f'(x)$, simplifying your answer.

(ii) The tangents to the curve of $f(x)$ at points A and B are parallel to the x-axis. Find the coordinates of A and of B.

(b) (i) Sketch the graph of $y = f'(x)$.

(ii) Find the x-coordinates of the three points of inflection on the graph of f.

(c) Find the range of:

(i) f

(ii) the composite function $f \circ f$.

34. Air is pumped into a spherical ball which expands at a rate of $8\,\text{cm}^3$ per second ($8\,\text{cm}^3\,\text{s}^{-1}$). Find the exact rate of increase of the radius of the ball when the radius is $2\,\text{cm}$.

35. The function f is defined by $f(x) = \dfrac{x^2}{2^x}$, for $x > 0$

(a) (i) Show that $f'(x) = \dfrac{2x - x^2 \ln 2}{2^x}$

(ii) Obtain an expression for $f''(x)$, simplifying your answer as far as possible.

(b) (i) Find the exact value of x satisfying the equation $f'(x) = 0$

(ii) Show that this value gives a maximum value for $f(x)$.

(c) Find the x-coordinates of the two points of inflection on the graph of f.

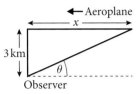

Figure 13.44 Diagram for question 36

36. An aeroplane is flying at a constant speed and at a constant altitude of 3 km in a straight line that will take it directly over an observer at ground level. At a given instant, the observer notes that the angle θ is $\frac{1}{3}\pi$ radians and is increasing at $\frac{1}{60}$ radians per second. Find the speed, in kilometres per hour, at which the aeroplane is moving towards the observer.

37. A curve has equation $f(x) = \dfrac{a}{b + e^{-cx}}$, $a \neq 0$, $b > 0$, $c > 0$

   (a) Show that $f''(x) = \dfrac{ac^2 e^{-cx}(e^{-cx} - b)}{(b + e^{-cx})^3}$

   (b) Find the coordinates of the point on the curve where $f''(x) = 0$

   (c) Show that this is a point of inflection.

38. The point $P(1, p)$, where $p > 0$, lies on the curve $2x^2 y + 3y^2 = 16$

   (a) Calculate the value of p.

   (b) Calculate the gradient of the tangent to the curve at P.

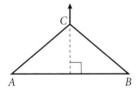

Figure 13.45 Diagram for question 39

39. The diagram shows an isosceles triangle ABC with $AB = 10$ cm and $AC = BC$. The vertex C is moving in a direction perpendicular to AB with speed 2 cm per second.

   Calculate the rate of increase of the angle $C\widehat{A}B$ at the moment the triangle is equilateral.

40. Find the equation of the normal to the curve $x^3 + y^3 - 9xy = 0$ at the point $(2, 4)$.

41. The function f is given by $f(x) = \dfrac{x^5 + 2}{x}$, $x \neq 0$. There is a point of inflection on the graph of f at the point P. Find the coordinates of P.

42. An experiment is carried out in which the number n of bacteria in a liquid is given by the formula $n = 650\,e^{kt}$, where t is the time in minutes after the beginning of the experiment and k is a constant. The number of bacteria doubles every 20 minutes. Find:

   (a) the exact value of k

   (b) the rate at which the number of bacteria is increasing when $t = 90$.

43. Let f be the function $f(x) = \ln(3x + 1)$ defined for $x > -\dfrac{1}{3}$.

(a) Find $f'(x)$.

(b) Find the equation of the normal to the curve $y = f(x)$ at the point where $x = 2$.

Give your answer in the form $y = ax + b$ where $a, b \in \mathbb{R}$.

44. Let $y = x \arcsin x$, $x \in\ ]-1, 1[$. Show that $\dfrac{d^2y}{dx^2} = \dfrac{2 - x^2}{(1 - x^2)^{\frac{3}{2}}}$.

45. Given that $e^{xy} - y^2 \ln x = e$ for $x \geqslant 1$, find $\dfrac{dy}{dx}$ at the point $(1, 1)$.

46. The function f is defined by $f(x) = \dfrac{2x}{x^2 + 6}$ for $x \geqslant b$, $b \in \mathbb{R}$.

(a) Show that $f'(x) = \dfrac{12 - 2x^2}{(x^2 + 6)^2}$

(b) Hence find the smallest exact value of b for which the inverse function f^{-1} exists. Justify your answer.

47. Consider the curve with equation $x^2 + xy + y^2 = 3$

(a) Find, in terms of k, the gradient of the curve at the point $(-1, k)$.

(b) Given that the tangent to the curve is parallel to the x-axis at this point, find the value of k.

48. André wants to get from point A located in the sea to point Y located on a straight stretch of beach. P is the point on the beach nearest to A such that $AP = 2$ km and $PY = 2$ km. André swims in a straight line to a point Q located on the beach and then runs to Y.

When André swims, he covers 1 km in $5\sqrt{5}$ minutes. When he runs, he covers 1 km in 5 minutes.

(a) If $PQ = x$ km, $0 \leqslant x \leqslant 2$, find an expression for the time T minutes taken by André to reach point Y.

(b) Show that $\dfrac{dT}{dx} = \dfrac{5\sqrt{5}x}{\sqrt{x^2 + 4}} - 5$

(c) (i) Solve $\dfrac{dT}{dx} = 0$

(ii) Use the value of x found in part (i) to determine the time, T minutes, taken for André to reach point Y.

(iii) Show that $\dfrac{d^2T}{dx^2} = \dfrac{20\sqrt{5}}{(x^2 + 4)^{\frac{3}{2}}}$ and hence show that the time found in part (ii) is a minimum.

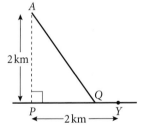

Figure 13.46 Diagram for question 48

695

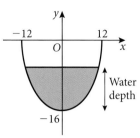

Figure 13.47 Diagram for question 49

49. The diagram shows the boundary of the cross-section of a water channel.

The equation that represents this boundary is $y = 16 \sec\left(\dfrac{\pi x}{36}\right) - 32$ where x and y are both measured in cm.

The top of the channel is level with the ground and has a width of 24 cm. The maximum depth of the channel is 16 cm.

Find the width of the water surface in the channel when the water depth is 10 cm. Give your answer in the form $a \arccos b$, where $a, b \in \mathbb{R}$.

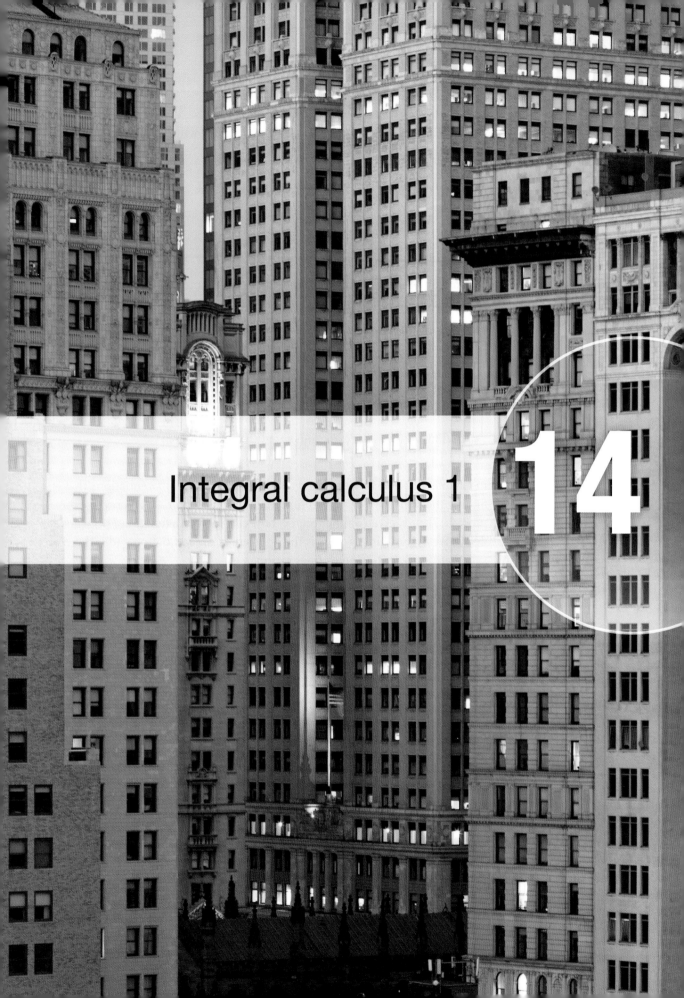

Integral calculus 1

14

In Chapters 12 and 13 we learned about the process of differentiation. That is, finding the derivative of a given function. In this chapter we will reverse the process. That is, given a function $f(x)$ how can we find a function $F(x)$ whose derivative is $f(x)$? This process is the opposite of differentiation and is therefore called **antidifferentiation** or **integration**.

14.1 Antiderivative

An **antiderivative** of the function $f(x)$ is a function $F(x)$ such that
$$\frac{d}{dx}F(x) = F'(x) = f(x)$$
wherever $f(x)$ is defined.

For instance, let $f(x) = x^2$. It is not difficult to discover an antiderivative of $f(x)$. Keep in mind that this is a power function. Since the power rule reduces the power of the function by 1, we examine the derivative of x^3: $\frac{d}{dx}(x^3) = 3x^2$

This derivative, however, is 3 times $f(x)$. To 'compensate' for the 'extra' 3 we have to multiply by $\frac{1}{3}$ so that the antiderivative is $\frac{1}{3}x^3$. Now $\frac{d}{dx}\left(\frac{1}{3}x^3\right) = x^2$, and therefore $\frac{1}{3}x^3$ is an antiderivative of x^2.

Table 14.1 shows some examples of functions, each paired with one of its antiderivatives. The diagrams show the relationship between the derivative and the integral as opposite operations.

Function $f(x)$	Antiderivative $F(x)$
1	x
x	$\dfrac{x^2}{2}$
$3x^2$	x^3
x^4	$\dfrac{x^5}{5}$
$\cos x$	$\sin x$
$\cos 2x$	$\dfrac{1}{2}\sin 2x$
e^x	e^x
$\sin x$	$-\cos x$

Table 14.1 Examples of functions paired to antiderivatives

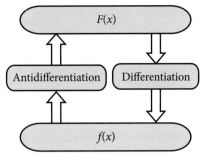

Figure 14.1 The relationship between a derivative and its integral

Example 14.1

Given the function $f(x) = 3x^2$. Find an antiderivative of $f(x)$.

Solution

$F_1(x) = x^3$ is one such antiderivative because $\dfrac{d}{dx}(F_1(x)) = 3x^2$

The following functions are also antiderivatives because the derivative of each one of them is also $3x^2$.

$$F_2(x) = x^3 + 27 \quad F_3(x) = x^3 - \pi \quad F_4(x) = x^3 + \sqrt{5}$$

Indeed, $F(x) = x^3 + c$ is an antiderivative of $f(x) = 3x^2$ for any constant c.

This is simply because

$$(F(x) + c)' = F'(x) + c' = F'(x) + 0 = f(x)$$

Thus we can say that any single function $f(x)$ has many antiderivatives, whereas a function has only one derivative.

If $F(x)$ is an antiderivative of $f(x)$, then so is $F(x) + c$ for any choice of the constant c. This statement is an indirect conclusion of one of the results of the mean value theorem. Two functions with the same derivative on an interval differ only by a constant on that interval.

Let $F(x)$ and $G(x)$ be any antiderivatives of $f(x)$; that is, $F'(x) = G'(x)$.

Take $H(x) = F(x) - G(x)$ and any two numbers x_1 and x_2 in the interval $[a, b]$ such that $x_1 < x_2$, then

$$H(x_2) - H(x_1) = (x_2 - x_1)H'(c) = (x_2 - x_1) \cdot (F'(c) - G'(c))$$
$$= (x_2 - x_1) \cdot 0 = 0 \Rightarrow H(x_1) = H(x_2)$$

which means $H(x)$ is a constant function. Hence $H(x) = F(x) - G(x) = $ constant. That is, any two antiderivatives of a function differ by a constant.

The **mean value theorem** states that a function $H(x)$, continuous over an interval $[a, b]$ and differentiable over $]a, b[$ satisfies:

$H(b) - H(a)$
$= (b - a)H'(c)$
for some $c \in \,]a, b[$

Note that if we differentiate an antiderivative of $f(x)$, we obtain $f(x)$. Thus

$$\frac{d}{dx}\left(\int f(x)\,dx\right) = f(x)$$

The expression $\int f(x)\,dx$ is called an **indefinite integral** of $f(x)$. The function $f(x)$ is called the **integrand**, and the constant c is called the **constant of integration**.

The integral symbol $\int$ is a medieval S, used by Leibniz as an abbreviation for the Latin word *summa* ('sum').

We think of the combination $\int [\ ]\,dx$ as a single symbol; we fill in the blank with the formula of the function whose antiderivative we seek. We may regard the differential dx as specifying the independent variable x both in the function $f(x)$ and in its antiderivatives. This is true for any independent variable, say t, with the notation adjusted appropriately. Thus

$$\frac{d}{dt}\left(\int f(t)\,dt\right) = f(t)$$

and

$$\int f(t)dt = F(t) + c$$

are equivalent statements.

Notation

The notation

$$\int f(x)dx = F(x) + c \qquad\qquad (1)$$

where c is an arbitrary constant, means that $F(x) + c$ is an antiderivative of $f(x)$.

Equivalently, $F(x)$ satisfies the condition that

$$\frac{d}{dx}F(x) = F'(x) = f(x) \qquad\qquad (2)$$

for all x in the domain of $f(x)$.

It is important to note that that **(1)** and **(2)** are just different notations to express the same fact. For example

$$\int x^2\,dx = \frac{1}{3}x^3 + c \text{ is equivalent to } \frac{d}{dx}\left(\frac{1}{3}x^3\right) = x^2$$

Derivative formula	Equivalent integration formula
$\dfrac{d}{dx}(x^3) = 3x^2$	$\int 3x^2\,dx = x^3 + c$
$\dfrac{d}{dx}(\sqrt{x}) = \dfrac{1}{2\sqrt{x}}$	$\int \dfrac{1}{2\sqrt{x}}\,dx = \sqrt{x} + c$
$\dfrac{d}{dt}(\tan t) = \sec^2 t$	$\int \sec^2 t\,dt = \tan t + c$
$\dfrac{d}{dv}\left(v^{\frac{3}{2}}\right) = \dfrac{3}{2}v^{\frac{1}{2}}$	$\int \dfrac{3}{2}v^{\frac{1}{2}}\,dv = v^{\frac{3}{2}} + c$

Table 14.2 Derivative formulae and their equivalent integration formulae

Basic integration formulae

Many basic integration formulae can be obtained directly from their companion differentiation formulae. Some of the most important are given in Table 14.3.

The integral sign and differential serve as delimiters, adjoining the integrand on the left and right, respectively. In particular we do not write $\int dx\,f(x)$ when we mean $\int f(x)dx$

	Derivation formula	Integration formula
1	$\dfrac{d}{dx}(x) = 1$	$\int dx = x + c$
2	$\dfrac{d}{dx}(x^{n+1}) = (n+1)x^n,\ n \neq -1$	$\int x^n\,dx = \dfrac{x^{n+1}}{n+1} + c,\ n \neq -1$
3	$\dfrac{d}{dx}(\sin x) = \cos x$	$\int \cos x\,dx = \sin x + c$
4	$\dfrac{d}{dv}(\cos v) = -\sin v$	$\int \sin v\,dv = -\cos v + c$
5	$\dfrac{d}{dt}(\tan t) = \sec^2 t$	$\int \sec^2 t\,dt = \tan t + c$
6	$\dfrac{d}{dv}(e^v) = e^v$	$\int e^v\,dv = e^v + c$
7	$\dfrac{d}{dx}(\ln|x|) = \dfrac{1}{x}$	$\int \dfrac{1}{x}\,dx = \ln|x| + c$
8	$\dfrac{d}{dx}\left(\dfrac{a^x}{\ln a}\right) = a^x$	$\int a^x\,dx = \dfrac{1}{\ln a}a^x + c$
9	$\dfrac{d}{dx}(\arcsin x) = \dfrac{1}{\sqrt{1 - x^2}}$	$\int \dfrac{dx}{\sqrt{1 - x^2}} = \arcsin x + c$

10	$\dfrac{d}{dx}(\arccos x) = \dfrac{-1}{\sqrt{1-x^2}}$	$\displaystyle\int \dfrac{-dx}{\sqrt{1-x^2}} = \arccos x + c = -\arcsin x + c$
11	$\dfrac{d}{dx}(\arctan x) = \dfrac{1}{1+x^2}$	$\displaystyle\int \dfrac{dx}{1+x^2} = \arctan x + c$
12	$\dfrac{d}{dt}\sec t = \tan t \sec t$	$\displaystyle\int \tan t \sec t\,dt = \sec t + c$
13	$\dfrac{d}{dx}\cot x = -\operatorname{cosec}^2 x$	$\displaystyle\int \operatorname{cosec}^2 t\,dt = -\cot t + c$
14	$\dfrac{d}{dt}\operatorname{cosec} t = -\cot t \operatorname{cosec} t$	$\displaystyle\int \cot t \operatorname{cosec} t\,dt = -\operatorname{cosec} t + c$

Table 14.3 Many basic integration formulae can be obtained directly from their companion differentiation formulae

Formula 7 is a special case of the 'power' rule shown in formula 2, but needs some modification.

If we are asked to integrate $\dfrac{1}{x}$, we may attempt to do it using the power rule:

$$\int \frac{1}{x}\,dx = \int x^{-1}\,dx = \frac{1}{(-1)+1}x^{(-1)+1} + c = \frac{1}{0}x^0 + c, \text{ which is undefined.}$$

However, the solution is found by observing what we learned in Chapter 13:

$$\frac{d}{dx}(\ln x) = \frac{1}{x}, x > 0, \text{ implies } \int \frac{1}{x}\,dx = \ln x + c, x > 0.$$

The function $\dfrac{1}{x}$ is differentiable for $x < 0$ too. So, we must be able to find its integral.

The solution lies in the chain rule.

If $x < 0$, then we can write $x = -u$ where $u > 0$. Then $dx = -du$, and

$$\int \frac{1}{x}\,dx = \int \frac{1}{-u}(-du) = \int \frac{1}{u}\,du = \ln u + c, u > 0$$

But $u = -x$, therefore when $x < 0$:

$$\int \frac{1}{x}\,dx = \ln u + c = \ln(-x) + c, \text{ and combining the two results, we have}$$

$$\int \frac{1}{x}\,dx = \ln|x| + c, x \neq 0$$

Suppose that $f(x)$ and $g(x)$ are differentiable functions and k is a constant. Then:

A constant factor can be moved through an integral sign; that is,

$$\int kf(x)\,dx = k\int f(x)\,dx$$

An antiderivative of a sum (difference) is the sum (difference) of the antiderivatives; i.e.,

$$\int (f(x) \pm g(x))\,dx = \int f(x)\,dx \pm \int g(x)\,dx$$

Example 14.2

Evaluate

(a) $\displaystyle\int 3\cos x\,dx$ (b) $\displaystyle\int(x^3 + x^2)\,dx$

Solution

(a) $\displaystyle\int 3\cos x\,dx = 3\int \cos x\,dx = 3\sin x + c$

(b) $\displaystyle\int(x^3 + x^2)\,dx = \int x^3\,dx + \int x^2\,dx = \frac{x^4}{4} + \frac{x^3}{3} + c$

Sometimes it is useful to rewrite the integrand in a different form before performing the integration.

Example 14.3

Evaluate

(a) $\int \dfrac{t^3 - 3t^5}{t^5} \, dt$ (b) $\int \dfrac{x + 5x^4}{x^2} \, dx$

Solution

(a) $\int \dfrac{t^3 - 3t^5}{t^5} \, dt = \int \dfrac{t^3}{t^5} \, dt - \int \dfrac{3t^5}{t^5} \, dt = \int t^{-2} \, dt - \int 3 \, dt = \dfrac{t^{-1}}{-1} - 3t + c$

$$= -\dfrac{1}{t} - 3t + c$$

(b) $\int \dfrac{x + 5x^4}{x^2} \, dx = \int \dfrac{x}{x^2} \, dx + \int \dfrac{5x^4}{x^2} \, dx = \int \dfrac{1}{x} \, dx + \int 5x^2 \, dx$

$$= \ln|x| + \dfrac{5x^3}{3} + c$$

Integration by simple substitution – change of variables

In this section we will study substitution, a technique that can often be used to transform complex integration problems into simpler ones.

The method of substitution depends on our understanding of the chain rule as well as the use of variables in integration. Two facts to recall:

When we find an antiderivative, we can use any other variable.

That is, $\int f(u) \, du = F(u) + c$, where u is a dummy variable in the sense that it can be replaced by any other variable.

Using the chain rule $\dfrac{d}{dx}(F(u(x))) = F'(u(x)) \cdot u'(x)$

Which can be written in integral form as $\int F'(u(x)) \cdot u'(x) \, dx = F(u(x)) + c$

Or equivalently, since $F(x)$ is an antiderivative of $f(x)$,

$\int f(u(x)) \cdot u'(x) \, dx = F(u(x)) + c$

For our purposes it will be useful and simpler to let $u(x) = u$ and to write $\dfrac{du}{dx} = u'(x)$ in its differential form as $du = u'(x) \, dx$ or simply $du = u' \, dx$.

We can now write the integral as

$$\int f(u(x)) \cdot u'(x) \, dx = \int f(u) \, du = F(u(x)) + c$$

Example 14.4 demonstrates how the method works.

Example 14.4

Evaluate

(a) $\int (x^3 + 2)^{10} \cdot 3x^2 \, dx$ (b) $\int \tan x \, dx$ (c) $\int \cos 5x \, dx$

(d) $\int \cos x^2 \cdot x \, dx$ (e) $\int e^{3x+1} \, dx$

Solution

(a) To integrate this function, it is simplest to make the substitution $u = x^3 + 2$, and so $du = 3x^2 \, dx$. Now we can write the integral as

$$\int (x^3 + 2)^{10} \cdot 3x^2 \, dx = \int u^{10} \, du = \frac{u^{11}}{11} + c = \frac{(x^3 + 2)^{11}}{11} + c$$

(b) This integrand has to be rewritten first and then we make the substitution:

$$\int \tan x \, dx = \int \frac{\sin x}{\cos x} \, dx = \int \frac{1}{\cos x} \cdot \sin x \, dx$$

We now let $u = \cos x \Rightarrow du = -\sin x \, dx$, and

$$\int \tan x \, dx = \int \frac{1}{\cos x} \cdot \sin x \, dx = \int \frac{1}{u} \cdot (-du) = -\int \frac{1}{u} \, du = -\ln|u| + c$$

This last result can be then expressed in two ways:

$$\int \tan x \, dx = -\ln|\cos x| + c, \text{ or}$$

$$\int \tan x \, dx = -\ln|\cos x| + c = \ln|(\cos x)^{-1}| + c = \ln\left|\frac{1}{(\cos x)}\right| + c$$

$$= \ln|\sec x| + c$$

(c) We let $u = 5x$, then $du = 5dx \Rightarrow dx = \frac{1}{5} du$, and so

$$\int \cos 5x \, dx = \int \cos u \cdot \frac{1}{5} \, du = \frac{1}{5} \int \cos u \, du = \frac{1}{5} \sin u + c$$

$$= \frac{1}{5} \sin 5x + c$$

Another method can be applied here:

The substitution $u = 5x$ requires $du = 5dx$. As there is no factor of 5 in the integrand, and since 5 is a constant, we can multiply and divide by 5 so that we group the 5 and dx to form the du required by the substitution:

$$\int \cos 5x \, dx = \frac{1}{5} \int \cos x \cdot 5dx = \frac{1}{5} \int \cos u \, du = \frac{1}{5} \sin u + c$$

$$= \frac{1}{5} \sin 5x + c$$

The main challenge in using the substitution rule is to think of an appropriate substitution. You should try to select u to be a part of the integrand whose differential is also included (except for the constant). In Example 14.4 (a), we selected u to be $(x^3 + 2)$ knowing that $du = 3x^2 \, dx$. Then we compensated for the absence of 3. Finding the right substitution is a subtle art, which you will acquire with practice. It is often the case that your first guess may not work.

(d) By letting $u = x^2$, $du = 2x\,dx$ and so

$$\int \cos x^2 \cdot x\,dx = \frac{1}{2}\int \cos x^2 \cdot 2x\,dx = \frac{1}{2}\int \cos u\,du$$

$$= \frac{1}{2}\sin u + c = \frac{1}{2}\sin x^2 + c$$

(e) $\displaystyle\int e^{3x+1}\,dx = \frac{1}{3}\int e^{3x+1}\,3dx = \frac{1}{3}\int e^u\,du = \frac{1}{3}e^u + c = \frac{1}{3}e^{3x+1} + c$

In integration, multiplying by a constant inside the integral and compensating for that with the reciprocal outside the integral depends on formula 2 from Table 14.3.

However, we cannot do this with a variable.

For example,

$$\int e^x\,dx = \frac{1}{2x}\int e^x 2x\,dx$$

is **not** valid because $2x$ is not a constant.

Example 14.5

Evaluate each integral.

(a) $\displaystyle\int e^{-3x}\,dx$ (b) $\displaystyle\int \sin^2 x\cos x\,dx$ (c) $\displaystyle\int 2\sin(3x-5)\,dx$

(d) $\displaystyle\int e^{mx+n}\,dx$ (e) $\displaystyle\int x\sqrt{x}\,dx$ and $F(1) = 2$

Solution

(a) Let $u = -3x$, then $du = -3dx$

$$\int e^{-3x}\,dx = -\frac{1}{3}\int e^{-3x}(-3\,dx) = -\frac{1}{3}\int e^u\,du = -\frac{1}{3}e^u + c$$

$$= -\frac{1}{3}e^{-3x} + c$$

(b) Let $u = \sin x \Rightarrow du = \cos x\,dx$, and hence

$$\int \sin^2 x\cos x\,dx = \int u^2\,du = \frac{1}{3}u^3 + c = \frac{1}{3}\sin^3 x + c$$

(c) Let $u = 3x - 5$, then $du = 3dx$

$$\int 2\sin(3x-5)\,dx = 2\cdot\frac{1}{3}\int \sin(3x-5)\,3\,dx = \frac{2}{3}\int \sin u\,du$$

$$= -\frac{2}{3}\cos u + c = -\frac{2}{3}\cos(3x-5) + c$$

(d) Let $u = mx + n$, then $du = m\,dx$

$$\int e^{mx+n}\,dx = \frac{1}{m}\int e^{mx+n}\,m\,dx = \frac{1}{m}\int e^u\,du = \frac{1}{m}e^u + c = \frac{1}{m}e^{mx+n} + c$$

(e) $F(x) = \displaystyle\int x\sqrt{x}\,dx = \int x^{\frac{3}{2}}\,dx = \frac{x^{\frac{5}{2}}}{\left(\frac{5}{2}\right)} + c = \frac{2}{5}x^{\frac{5}{2}} + c$, but $F(1) = 2$

$$F(1) = \frac{2}{5}1^{\frac{5}{2}} + c = \frac{2}{5} + c = 2 \Rightarrow c = \frac{8}{5}$$

Therefore $F(x) = \dfrac{2}{5}x^{\frac{5}{2}} + \dfrac{8}{5}$

Examples 14.4 and 14.5 make it clear that Table 14.2 is limited in scope because we cannot use the integrals directly to evaluate composite functions. We therefore need to revise some of the derivative formulae.

	Derivative formula	Integration formula
1	$\dfrac{d}{dx}(u(x)) = u'(x) \Rightarrow du = u'(x)dx$	$\displaystyle\int du = u + c$
2	$\dfrac{d}{dx}\left(\dfrac{u^{n+1}}{n+1}\right) = u^n u'(x),\, n \neq -1 \Rightarrow d\left(\dfrac{u^{n+1}}{n+1}\right) = u^n u'(x)\,dx$	$\displaystyle\int u^n\, du = \dfrac{u^{n+1}}{n+1} + c,\, n \neq -1$
3	$\dfrac{d}{dx}(\sin(u)) = \cos(u)u'(x) \Rightarrow d(\sin(u)) = \cos(u)u'(x)\,dx$	$\displaystyle\int \cos u\, du = \sin u + c$
4	$\dfrac{d}{dx}(\cos(u)) = -\sin(u)u'(x) \Rightarrow d(\cos(u)) = -\sin(u)u'(x)dx$	$\displaystyle\int \sin u\, du = -\cos u + c$
5	$\dfrac{d}{dt}(\tan u) = \sec^2 u\, u'(t) \Rightarrow d(\tan u) = \sec^2 u\, u'(t)\,dt$	$\displaystyle\int \sec^2 u\, du = \tan u + c$
6	$\dfrac{d}{dx}(e^u) = e^u u'(x)\,dx \Rightarrow d(e^u) = e^u u'(x)\,dx$	$\displaystyle\int e^u\, du = e^u + c$
7	$\dfrac{d}{dx}(\ln|u|) = \dfrac{1}{u}u'(x) \Rightarrow d(\ln|u|) = \dfrac{1}{u}u'(x)\,dx$	$\displaystyle\int \dfrac{1}{u}\, du = \ln|u| + c$
8	$\dfrac{d}{dx}(\arcsin u) = \dfrac{1}{\sqrt{1-u^2}}u'(x) \Rightarrow d(\arcsin u) = \dfrac{1}{\sqrt{1-u^2}}u'(x)\,dx$	$\displaystyle\int \dfrac{du}{\sqrt{1-u^2}} = \arcsin u + c$
9	$\dfrac{d}{dx}(\arctan u) = \dfrac{1}{1+u^2}u'(x) \Rightarrow d(\arctan u) = \dfrac{1}{1+u^2}u'(x)\,dx$	$\displaystyle\int \dfrac{du}{1+u^2} = \arctan u + c$

Table 14.4 More advanced derivative and integration formulae.

Example 14.6

Evaluate each integral.

(a) $\displaystyle\int \sqrt{6x + 11}\, dx$

(b) $\displaystyle\int (5x^3 + 2)^8 x^2\, dx$

(c) $\displaystyle\int \dfrac{x^3 - 2}{\sqrt[5]{x^4 - 8x + 13}}\, dx$

(d) $\displaystyle\int \sin^4(3x^2)\cos(3x^2)\, x\, dx$

Solution

(a) We let $u = 6x + 11$ and calculate du:

$$u = 6x + 11 \Rightarrow du = 6\, dx$$

Since du contains the factor 6, the integral is not in the form $\int f(u)\, du$. However, here we can use one of two approaches.

Introduce the factor 6, as we have done before; that is,

$$\int \sqrt{6x + 11}\, dx = \dfrac{1}{6}\int \sqrt{6x + 11}\, 6\, dx$$

$$= \dfrac{1}{6}\int \sqrt{u}\, du = \dfrac{1}{6}\int u^{\frac{1}{2}}\, du$$

$$= \frac{1}{6} \frac{u^{\frac{3}{2}}}{\frac{3}{2}} + c = \frac{2}{18} u^{\frac{3}{2}} + c$$

$$= \frac{1}{9}(6x + 11)^{\frac{3}{2}} + c$$

Or since $u = 6x + 11 \Rightarrow du = 6\,dx \Rightarrow dx = \dfrac{du}{6}$, then

$$\int \sqrt{6x + 11}\ dx = \int \sqrt{u}\ \frac{du}{6} = \frac{1}{6} \int u^{\frac{1}{2}}\,du,$$ then we follow the same steps as before.

(b) We let $u = 5x^3 + 2$, so $du = 15x^2\,dx$. This means that we need to introduce the factor 15 into the integrand

$$\int (5x^3 + 2)^8 x^2\,dx = \frac{1}{15} \int (5x^3 + 2)^8\,15x^2\,dx$$

$$= \frac{1}{15} \int u^8\,du = \frac{1}{15} \frac{u^9}{9} + c$$

$$= \frac{1}{135}(5x^3 + 2)^9 + c$$

(c) We let $u = x^4 - 8x + 13 \Rightarrow du = (4x^3 - 8)dx = 4(x^3 - 2)dx$

$$\int \frac{x^3 - 2}{\sqrt[5]{x^4 - 8x + 13}}\,dx = \frac{1}{4} \int \frac{4(x^3 - 2)dx}{\sqrt[5]{x^4 - 8x + 13}} = \frac{1}{4} \int \frac{du}{u^{\frac{1}{5}}}$$

$$= \frac{1}{4} \int u^{\frac{1}{5}}\,du = \frac{1}{4} \frac{u^{\frac{4}{5}}}{\frac{4}{5}} + c$$

$$= \frac{5}{16}(x^4 - 8x + 13)^{\frac{4}{5}} + c$$

(d) We let $u = \sin(3x^2) \Rightarrow du = \cos(3x^2)6x\,dx$ using the chain rule.

$$\int \sin^4(3x^2)\cos(3x^2)\,x\,dx = \frac{1}{6} \int \sin^4(3x^2)\cos(3x^2)\,6x\,dx$$

$$= \frac{1}{6} \int u^4\,du = \frac{1}{6} \frac{u^5}{5} + c$$

$$= \frac{1}{30} \sin^5(3x^2) + c$$

Exercise 14.1

1. Find the most general antiderivative of each function.

   (a) $f(x) = x + 2$

   (b) $f(t) = 3t^2 - 2t + 1$

   (c) $g(x) = \dfrac{1}{3} - \dfrac{2}{7}x^3$

   (d) $f(t) = (t - 1)(2t + 3)$

   (e) $g(u) = u^{\frac{2}{5}} - 4u^3$

   (f) $f(x) = 2\sqrt{x} - \dfrac{3}{2\sqrt{x}}$

   (g) $h(\theta) = 3\sin\theta + 4\cos\theta$

   (h) $f(t) = 3t^2 - 2\sin t$

   (i) $f(x) = \sqrt{x}(2x - 5)$

   (j) $g(\theta) = 3\cos\theta - 2\sec^2\theta$

(k) $h(t) = e^{3t-1}$

(l) $f(t) = \dfrac{2}{t}$

(m) $h(u) = \dfrac{t}{3t^2 + 5}$

(n) $h(\theta) = e^{\sin\theta}\cos\theta$

(o) $f(x) = (3 + 2x)^2$

2. Find f.

   (a) $f''(x) = 4x - 15x^2$

   (b) $f''(x) = 1 + 3x^2 - 4x^3, f'(0) = 2, f(1) = 2$

   (c) $f''(t) = 8t - \sin t$

   (d) $f'(x) = 12x^3 - 8x + 7, f(0) = 3$

   (e) $f'(\theta) = 2\cos\theta - \sin(2\theta)$

3. Evaluate each integral.

   (a) $\displaystyle\int x(3x^2 + 7)^5\,dx$

   (b) $\displaystyle\int \dfrac{x}{(3x^2 + 5)^4}\,dx$

   (c) $\displaystyle\int 2x^2\sqrt[4]{5x^3 + 2}\,dx$

   (d) $\displaystyle\int \dfrac{(3 + 2\sqrt{x})^5}{\sqrt{x}}\,dx$

   (e) $\displaystyle\int t^2\sqrt{2t^3 - 7}\,dt$

   (f) $\displaystyle\int \left(2 + \dfrac{3}{x}\right)^5\left(\dfrac{1}{x^2}\right)dx$

   (g) $\displaystyle\int \sin(7x - 3)dx$

   (h) $\displaystyle\int \dfrac{\sin(2\theta - 1)}{\cos(2\theta - 1) + 3}\,d\theta$

   (i) $\displaystyle\int \sec^2(5\theta - 2)d\theta$

   (j) $\displaystyle\int \cos(\pi x + 3)dx$

   (k) $\displaystyle\int \sec 2t \tan 2t\,dt$

   (l) $\displaystyle\int x\,e^{x^2+1}\,dx$

   (m) $\displaystyle\int \sqrt{t}\,e^{2t\sqrt{t}}\,dt$

   (n) $\displaystyle\int \dfrac{2}{\theta}(\ln\theta)^2\,d\theta$

4. Evaluate each integral.

   (a) $\displaystyle\int t\sqrt{3 - 5t^2}\,dt$

   (b) $\displaystyle\int \theta^2\sec^2\theta^3\,d\theta$

   (c) $\displaystyle\int \dfrac{\sin\sqrt{t}}{2\sqrt{t}}\,dt$

   (d) $\displaystyle\int \tan^5 2t\sec^2 2t\,dt$

   (e) $\displaystyle\int \dfrac{dx}{\sqrt{x}(\sqrt{x} + 2)}$

   (f) $\displaystyle\int \sec^5 2t\tan 2t\,dt$

   (g) $\displaystyle\int \dfrac{x + 3}{x^2 + 6x + 7}\,dx$

   (h) $\displaystyle\int \dfrac{k^3 x^3}{\sqrt{a^2 - a^4 x^4}}\,dx$

   (i) $\displaystyle\int 3x\sqrt{x - 1}\,dx$

   (j) $\displaystyle\int \csc^2\pi t\,dt$

   (k) $\displaystyle\int \sqrt{1 + \cos\theta}\sin\theta\,d\theta$

   (l) $\displaystyle\int t^2\sqrt{1 - t}\,dt$

   (m) $\displaystyle\int \dfrac{r^2 - 1}{\sqrt{2r - 1}}\,dr$

   (n) $\displaystyle\int \dfrac{e^{x^2} - e^{-x^2}}{e^{x^2} + e^{-x^2}}x\,dx$

Integration by parts

Although differentiation and integration are strongly linked, finding derivatives is very different from finding integrals. With the derivative rules available, we are able to find the derivative of just about any function we can think of. By contrast, we can compute antiderivatives for a rather small number of functions. Thus far, we have developed a set of basic integration formulas, most of which followed directly from the related differentiation formulas seen in Table 14.2.

In some cases, using substitution helps us reduce the difficulty of evaluating integrals by expressing them in familiar forms. However, there are many cases where simple substitution will not help. For example, we cannot evaluate

$$\int x \cos x \, dx$$

using the methods we have learned so far. However, we can evaluate this integral using integration by parts.

Recall the product rule for differentiation:

$$\frac{d}{dx}(u(x)v(x)) = u'(x)v(x) + u(x)v'(x),$$

which gives rise to the differential form

$$d(u(x)v(x)) = v(x)d(u(x)) + u(x)d(v(x))$$

and for convenience, we will write

$$d(uv) = v\,du + u\,dv$$

If we integrate both sides of this equation, we get

$$\int d(uv) = \int v\,du + \int u\,dv \Leftrightarrow uv = \int v\,du + \int u\,dv$$

Solving this equation for $\int u\,dv$, we get

$$\int u\,dv = uv - \int v\,du$$

This rule is **integration by parts**.

Example 14.7

Evaluate $\int x \cos x \, dx$

Solution

First, observe that we cannot evaluate this as it stands: it is not one of our basic integrals and no substitution can help either.

We need to make a clever choice of u and dv so that the integral on the right side is one that makes the evaluation easier.

We need to choose u (to differentiate) and dv (to integrate), thus we let

$u = x$, and $dv = \cos x\, dx$

Then $du = dx$, and $v = \sin x$. (We will introduce c at the end of the process.)

It can help to organise our work in table form:

$u = x \qquad du = dx$
$dv = \cos x\, dx \quad v = \sin x$

This gives us

$$\int \underset{u}{x}\ \underset{dv}{\underbrace{\cos x\, dx}} \ =\ \int u\, dv \ =\ uv \ -\ \int v\, du$$

$$=\ x\sin x \ -\ \int \sin x\, dx$$

$$=\ x\sin x \ +\ \cos x + c$$

To verify the result, simply differentiate the right-hand side.

$$\frac{d}{dx}(x\sin x + \cos x + c) = \sin x + x\cos x - \sin x + 0 = x\cos x$$

There are three other choices of u and dv that we can make in this problem.

(i) Let

$u = \cos x \quad du = -\sin x\, dx$
$dv = x\, dx \qquad v = \dfrac{x^2}{2}$

$\Rightarrow \int x\cos x\, dx = \dfrac{x^2}{2}\cos x + \int \dfrac{x^2}{2}\sin x\, dx$

This new integral is worse than the one we started with.

(ii) Let

$u = x\cos x \quad du = (\cos x - x\sin x)\, dx$
$dv = dx \qquad\quad v = x$

$\Rightarrow$

$\int x\cos x\, dx = x^2\cos x - \int x(\cos x - x\sin x)\, dx$

Again, this new integral is worse than the one we started with.

(iii) Let

$u = 1 \qquad du = 0$
$dv = x\cos x\, dx \quad v = ??$

This is obviously a bad choice since we still don't know how to integrate $dv = x\cos x\, dx$

The objective of integration by parts is to move from an integral $\int u\, dv$ that we can't see how to evaluate to an integral $\int v\, du$ that we can evaluate. So, keep in mind that integration by parts does not necessarily work all the time, and that we have to develop enough experience with such a process in order to make the correct choice for u and $v\, du$.

Example 14.8

Evaluate

(a) $\int xe^{-x}\, dx$

(b) $\int \ln x\, dx$

(c) $\int x^2 \ln x\, dx$

Solution

(a) $\left.\begin{array}{ll} u = x & du = dx \\ dv = e^{-x}\,dx & v = -e^{-x} \end{array}\right] \Rightarrow \int xe^{-x}\,dx = -xe^{-x} + \int e^{-x}\,dx$

$$= -xe^{-x} - e^{-x} + c$$

(b) $\left.\begin{array}{ll} u = \ln x & du = \dfrac{dx}{x} \\ dv = dx & v = x \end{array}\right] \Rightarrow \int \ln x\,dx = x\ln x - \int \cancel{x}\,\dfrac{dx}{\cancel{x}}$

$$= x\ln x - x + c$$

(c) Since x^2 is easier to integrate than $\ln x$, and the derivative of $\ln x$ is also simpler than $\ln x$ itself, we make the following substitutions.

$\left.\begin{array}{ll} u = \ln x & du = \dfrac{dx}{x} \\ dv = x^2\,dx & v = \dfrac{x^3}{3} \end{array}\right] \Rightarrow \int x^2 \ln x\,dx = \dfrac{x^3}{3}\ln x - \int \dfrac{x^{\cancel{3}2}}{3}\,\dfrac{dx}{\cancel{x}}$

$$= \dfrac{x^3}{3}\ln x - \int \dfrac{1}{3}x^2\,dx$$

$$= \dfrac{x^3}{3}\ln x - \dfrac{1}{9}x^3 + c$$

Sometimes we need to use integrations by parts more than once.

Example 14.9

Evaluate $\int x^2 \sin x\,dx$

Solution

Since $\sin x$ is equally easy to integrate or differentiate while x^2 is easier to differentiate, we make the following substitution

$\left.\begin{array}{ll} u = x^2 & du = 2x\,dx \\ dv = \sin x\,dx & v = -\cos x \end{array}\right] \Rightarrow \int x^2 \sin x\,dx = -x^2\cos x + 2\int x\cos x\,dx$

This first step simplified the original integral. However, the right-hand side still needs further integration. Here again, we use Integration by parts.

$\left.\begin{array}{ll} u = 2x & du = 2\,dx \\ dv = \cos x\,dx & v = \sin x \end{array}\right] \Rightarrow \int 2x\cos x\,dx = 2x\sin x - 2\int \sin x\,dx$

$$= 2x\sin x + 2\cos x + c$$

Combining the two results, we can now write

$$\int x^2 \sin x\,dx = -x^2\cos x + 2\int x\cos x\,dx$$

$$= -x^2\cos x + 2x\sin x + 2\cos x + c$$

When applying integration by parts more than once, we need to be careful not to change the nature of the substitution in successive applications. For instance, in Example 14.9, the first substitution was $u = x^2$ and $dv = \sin x\,dx$. In the second step, if we had switched the substitution to $u = \cos x$ and $dv = 2x\,dx$, we would have obtained

$\int x^2 \sin x\,dx = -x^2\cos x$ $+ x^2\cos x + \int x^2 \sin x\,dx$ $= \int x^2 \sin x\,dx$

This undoes the previous integration and returns to the original integral.

Example 14.10

Evaluate $\int x^2 e^x \, dx$

Solution

Since e^x is equally easy to integrate or differentiate, while x^2 is easier to differentiate, we make the following substitution

$$\left.\begin{array}{ll} u = x^2 & du = 2x \, dx \\ dv = e^x \, dx & v = e^x \end{array}\right] \Rightarrow \int x^2 e^x \, dx = x^2 e^x - \int 2x e^x \, dx$$

This first step simplified the original integral. However, the right-hand side still needs further integration. Here again, we use integration by parts.

$$\left.\begin{array}{ll} u = 2x & du = 2 \, dx \\ dv = e^x \, dx & v = e^x \end{array}\right] \Rightarrow \int 2x e^x \, dx = 2x e^x - 2 \int e^x \, dx$$
$$= 2x e^x - 2e^x + c$$

Hence

$$\int x^2 e^x \, dx = x^2 e^x - \int 2x e^x \, dx$$
$$= x^2 e^x - 2x e^x + 2e^x + c$$

Using integration by parts to find unknown integrals

Integrals like the one in Example 14.11 often occur in electricity problems. Their evaluation requires repeated applications of integration by parts followed by algebraic manipulation.

Example 14.11

Evaluate $\int \cos x \, e^x \, dx$

Solution

Let

$$\left.\begin{array}{ll} u = e^x & du = e^x \, dx \\ dv = \cos x \, dx & v = \sin x \end{array}\right] \Rightarrow \int \cos x e^x \, dx = e^x \sin x - \int \sin x \, e^x \, dx$$

The second integral is of the same nature, so use integration by parts again:

$$\left.\begin{array}{ll} u = e^x & du = e^x \, dx \\ dv = \sin x \, dx & v = -\cos x \end{array}\right] \Rightarrow \int \sin x e^x \, dx = -e^x \cos x + \int \cos x \, e^x \, dx$$

Hence:

$$\int \cos x\, e^x\, dx = e^x \sin x - \int \sin x e^x\, dx$$

$$= e^x \sin x - \left(-e^x \cos x + \int \cos x e^x\, dx\right)$$

$$= e^x \sin x + e^x \cos x - \int \cos x e^x\, dx$$

Now, the unknown integral appears on both sides of the equation, thus

$$\int \cos x\, e^x\, dx + \int \cos x\, e^x\, dx = e^x \sin x + e^x \cos x$$

$$\Rightarrow 2\int \cos x\, e^x\, dx = e^x \sin x + e^x \cos x$$

$$\Rightarrow \int \cos x\, e^x\, dx = \frac{e^x \sin x + e^x \cos x}{2} + c$$

Example 14.12

Evaluate $\int x \ln x\, dx$

Solution

$$\left.\begin{array}{ll} u = \ln x & du = \dfrac{dx}{x} \\[2mm] dv = x\, dx & v = \dfrac{x^2}{2} \end{array}\right] \Rightarrow \int x \ln x\, dx = \frac{x^2}{2}\ln x - \int \frac{x^{2\,1}}{2}\frac{dx}{x}$$

$$= \frac{x^2}{2}\ln x - \int \frac{x\, dx}{2} = \frac{x^2}{2}\ln x - \frac{x^2}{4} + c$$

Alternatively, we could have used a different substitution

$$\left.\begin{array}{ll} u = x \ln x & du = (\ln x + 1)dx \\[1mm] dv = dx & v = x \end{array}\right] \Rightarrow \int x \ln x\, dx = x^2 \ln x - \int x(\ln x + 1)dx$$

$$= x^2 \ln x - \int x \ln x\, dx - \int x\, dx$$

Adding $\int x \ln x\, dx$ to both sides and integrating $\int x\, dx$ we get

$$\int x \ln x\, dx + \int x \ln x\, dx = x^2 \ln x - \frac{x^2}{2} + c$$

$$\Rightarrow 2\int \mathrm{O}:: x \ln x\, dx = x^2 \ln x - \frac{x^2}{2} + c$$

$$\Rightarrow \int x \ln x\, dx = \frac{1}{2}\left(x^2 \ln x - \frac{x^2}{2} + c\right) = \frac{x^2 \ln x}{2} - \frac{x^2}{4} + c$$

The constant c is arbitrary, and hence it is unimportant whether we use $\frac{c}{2}$ or c in our final answer.

1. Evaluate each integral.

   (a) $\int x^2 e^{-x^3} dx$

   (b) $\int x^2 e^{-x} dx$

   (c) $\int x^2 \cos 3x \, dx$

   (d) $\int x^2 \sin ax \, dx$

   (e) $\int \cos x \ln(\sin x) \, dx$

   (f) $\int x \ln x^2 \, dx$

   (g) $\int x^2 \ln x \, dx$

   (h) $\int x^2(e^x - 1) \, dx$

   (i) $\int x \cos \pi x \, dx$

   (j) $\int e^{3t} \cos 2t \, dt$

   (k) $\int \arcsin x \, dx$

   (l) $\int x^3 e^x \, dx$

   (m) $\int e^{-2x} \sin 2x \, dx$

   (n) $\int \sin(\ln x) dx$

   (o) $\int \cos(\ln x) \, dx$

   (p) $\int \ln(x + x^2) \, dx$

   (q) $\int e^{kx} \sin x \, dx$

   (r) $\int x \sec^2 x \, dx$

   (s) $\int \sin x \sin 2x \, dx$

   (t) $\int x \arctan x \, dx$

   (u) $\int \dfrac{\ln x}{\sqrt{x}} \, dx$

2. In one scene of the movie *Stand and Deliver*, the teacher shows his students how to evaluate $\int x^2 \sin x \, dx$ by setting up a chart similar to this:

	$\sin x$	
x^2	$-\cos x$	$+$
$2x$	$-\sin x$	$-$
2	$\cos x$	$+$

   (a) Multiply across each row and add the result.

   (b) The integral is

   $$\int x^2 \sin x \, dx = -x^2 \cos x + 2x \sin x + 2 \cos x + c$$

   Explain why the method works for this problem.

3. Use the result of question 2 to evaluate each integral.

   (a) $\int x^4 \sin x \, dx$

   (b) $\int x^5 \cos x \, dx$

   (c) $\int x^4 e^x \, dx$

4. Show that the method used in question **2** will not work with $\int x^2 \ln x \, dx$

5. Show that
   $\int x^n e^x \, dx = x^n e^x - n \int x^{n-1} e^x \, dx$, then use this reduction formula to
   show that $\int x^4 e^x \, dx = ax^4 e^x + bx^3 e^x + cx^2 e^x + dx e^x + fe^x + g$, where
   $a, b, c, \ldots, g$ are to be determined.

6. Show that $\int x^n \ln x \, dx = \dfrac{x^{n+1}}{n+1} \ln x - \dfrac{x^{n+1}}{(n+1)^2} + c$

7. Show that $\int e^{mx} \cos nx \, dx = \dfrac{e^{mx}(m \cos nx + n \sin nx)}{m^2 + n^2} + c$

8. Show that $\int e^{mx} \sin nx \, dx = \dfrac{e^{mx}(m \sin nx - n \cos nx)}{m^2 + n^2} + c$

More methods of integration

In Section 14.2, we looked at a very powerful method for integration that has a wide range of applications. However, integration by parts does not work for all situations. In some cases, even when it does work, it may not be the most efficient method. In this section we will consider a few trigonometric integrals and some substitutions related to trigonometric functions or their inverses.

These trigonometric identities will prove very helpful:

1. $\cos^2\theta + \sin^2\theta = 1$

2. $\sin^2\theta = \dfrac{1 - \cos 2\theta}{2}$

3. $\cos^2\theta = \dfrac{1 + \cos 2\theta}{2}$

4. $\sec^2\theta = 1 + \tan^2\theta$

Example 14.13

Evaluate

(a) $\displaystyle\int \sin^2 x \, dx$ (b) $\displaystyle\int \cos^4\theta \, d\theta$

Solution

(a) We can use identity 2 from the list above:

$$\int \sin^2 x \, dx = \int \frac{1 - \cos 2x}{2} \, dx = \frac{1}{2}\int (1 - \cos 2x)dx$$

$$= \frac{1}{2}\left(x - \frac{1}{2}\sin 2x\right) + c$$

(b) Using identity 3:

$$\int \cos^4\theta \, d\theta = \int\left(\frac{1 + \cos 2\theta}{2}\right)^2 d\theta = \frac{1}{4}\int(1 + 2\cos 2\theta + \cos^2 2\theta)d\theta$$

$$= \frac{1}{4}\int\left(1 + 2\cos 2\theta + \frac{1 + \cos 4\theta}{2}\right)d\theta$$

$$= \frac{1}{8}\left(2\theta + 2\sin 2\theta + \theta + \frac{1}{4}\sin 4\theta\right) + c$$

$$= \frac{1}{32}(12\theta + 8\sin 2\theta + \sin 4\theta) + c$$

Here is a list how to find some integrals.

In exam papers, any non-standard cases will be accompanied by a recommended substitution.

Integral	How to find it
$\int \sin^m x \cos^n x \, dx$	If m is odd then break $\sin^m x$ into $\sin x$ and $\sin^{m-1} x$, use the substitution $u = \cos x$ and change the integral into the form $\int \cos^p x \sin x \, dx = \int u^p \, du$. Similarly if n is odd.
$\int \tan^m x \sec^n x \, dx$	If m and n are odd, break off a term for $\sec x \tan x$ and express the integrand in terms of $\sec x$ since $d(\sec x) = \sec x \tan x \, dx$
$\int \tan^n x \, dx$	Write the integrand as $\int \tan^{n-2} x \tan^2 x \, dx$, replace $\tan^2 x$ with $\sec^2 x - 1$ and then use $u = \tan x$
$\int \sec^n x \, dx$	If n is even, factor a $\sec^2 x$ out and write the rest in terms of $\tan^2 x + 1$. If n is odd, factor a $\sec^3 x$ out. Here, integration by parts may be useful.

Table 14.5 How to find some integrals

Example 14.14

Evaluate $\int \sec x \, dx$

Solution

This integral is evaluated using a clever multiplication by an atypical factor:

$$\int \sec x \, dx = \int \sec x \frac{\tan x + \sec x}{\tan x + \sec x} \, dx = \int \frac{\sec x \tan x + \sec^2 x}{\tan x + \sec x} \, dx$$

Now use the substitution $u = \sec x + \tan x \Rightarrow du = (\sec x \tan x + \sec^2 x) dx$, hence

$$\int \sec x \, dx = \int \frac{\sec x \tan x + \sec^2 x}{\tan x + \sec x} \, dx = \int \frac{du}{u}$$

$$= \ln|u| + c = \ln|\tan x + \sec x| + c$$

Example 14.15

Evaluate $\int \sec^3 x \, dx$

Solution

This can be evaluated using integration by parts and some of the results we used earlier.

$$u = \sec x \qquad du = \sec x \tan x \, dx$$
$$dv = \sec^2 x \, dx \qquad v = \tan x$$

Hence,

$$\int \sec^3 x \, dx = \sec x \tan x - \int \tan x \sec x \tan x \, dx$$

$$= \sec x \tan x - \int \sec x \tan^2 x \, dx$$

$$= \sec x \tan x - \int \sec x \sec^2 x - 1 \, dx$$

$$= \sec x \tan x - \int \sec^3 x \, dx + \int \sec x \, dx$$

Adding $\int \sec^3 x \, dx$ to both sides

$$2 \int \sec^3 x \, dx = \sec x \tan x + \int \sec x \, dx$$

$$= \sec x \tan x + \ln|\sec x + \tan x|$$

And finally

$$\int \sec^3 x \, dx = \frac{\sec x \tan x + \ln|\sec x + \tan x|}{2} + c$$

Example 14.16

Evaluate $\int \sin^3 x \cos^3 x \, dx$

Solution

This integral can be evaluated by separating either a cosine or a sine, then writing the rest of the expression in terms of sine or cosine.

We will separate a cosine here

$$\int \sin^3 x \cos^3 x \, dx = \int \sin^3 x \cos^2 x \cos x \, dx$$

$$= \int \sin^3 x (1 - \sin^2 x) \cos x \, dx$$

$$= \int (\sin^3 x - \sin^5 x) \cos x \, dx$$

Now we let

$u = \sin x \Rightarrow du = \cos x \, dx$, and hence

$$\int \sin^3 x \cos^3 x \, dx = \int (\sin^3 x - \sin^5 x) \cos x \, dx$$

$$= \int (u^3 - u^5) du = \frac{u^4}{4} - \frac{u^6}{6} + c$$

$$= \frac{\sin^4 x}{4} - \frac{\sin^6 x}{6} + c$$

Exercise 14.3

1. Evaluate each integral.

(a) $\int \sin^3 t \cos^2 t \, dt$

(b) $\int \sin^3 t \cos^3 t \, dt$

(c) $\int \sin^3 3\theta \cos 3\theta \, d\theta$

(d) $\int \frac{1}{t^2} \sin^5\left(\frac{1}{t}\right) \cos^2\left(\frac{1}{t}\right) dt$

(e) $\int \frac{\sin^3 x}{\cos^2 x} \, dx$

(f) $\int \tan^5 3x \sec^2 3x \, dx$

(g) $\int \theta \tan^3(\theta^2) \sec^4(\theta^2) \, d\theta$

(h) $\int \frac{1}{\sqrt{t}} \tan^3 \sqrt{t} \sec^3 \sqrt{t} \, dt$

(i) $\int \tan^4(5t) \, dt$

(j) $\int \frac{dt}{1 + \sin t}$

(k) $\int \frac{d\theta}{1 + \cos \theta}$

(l) $\int \frac{1 + \sin t}{\cos t} \, dt$

(m) $\int \frac{\sin x - 5 \cos x}{\sin x + \cos x} \, dx$

(n) $\int \frac{\sec \theta \tan \theta}{1 + \sec^2 \theta} \, d\theta$

(o) $\int \frac{\arctan t}{1 + t^2} \, dt$

(p) $\int \frac{1}{(1 + t^2)\arctan t} \, dt$

(q) $\int \frac{dx}{x\sqrt{1 - (\ln x)^2}}$

(r) $\int \sin^3 x \, dx$

(s) $\int \frac{\sin^3 x}{\sqrt{\cos x}} \, dx$

(t) $\int \frac{\sin^3 \sqrt{x}}{\sqrt{x}} \, dx$

(u) $\int \cos t \cos^3(\sin t) \, dt$

(v) $\int \frac{\cos \theta + \sin 2\theta}{\sin \theta} \, d\theta$

 For part (j), multiply the integrand by $\dfrac{1 - \sin t}{1 - \sin t}$

For part (m), find numbers a and b such that you can replace $\sin x - 5 \cos x$ with an expression involving terms $a(\sin x + \cos x)$ and $b(\cos x - \sin x)$

2. Evaluate each integral.

(a) $\int t \sec t \tan t \, dt$

(b) $\int \frac{\cos x}{2 - \sin x} \, dx$

(c) $\int e^{-2x} \tan(e^{-2x}) \, dx$

(d) $\int \frac{\sec(\sqrt{t})}{\sqrt{t}} \, dt$

(e) $\int \frac{dt}{1 + \cos 2t}$

(f) $\int \sqrt{1 - 9x^2} \, dx$

(g) $\int \frac{dx}{(x^2 + 4)^{3/2}}$

(h) $\int \sqrt{4 + t^2} \, dt$

(i) $\int \frac{3e^t \, dt}{4 + e^{2t}}$

(j) $\int \frac{1}{\sqrt{9 - 4x^2}} \, dx$

(k) $\int \frac{1}{\sqrt{4 + 9x^2}} \, dx$

(l) $\int \frac{\cos x}{\sqrt{1 + \sin^2 x}} \, dx$

(m) $\int \frac{x}{\sqrt{4 - x^2}} \, dx$

(n) $\int \frac{x}{x^2 + 16} \, dx$

(o) $\int \frac{\sqrt{4 - x^2}}{x^2} \, dx$

(p) $\int \frac{dx}{(9 - x^2)^{3/2}}$

 Parts (f) to –(v) will need trigonometric substitution.

717

(q) $\int x\sqrt{1 + x^2}\ dx$

(r) $\int e^{2x}\sqrt{1 + e^{2x}}\ dx$

(s) $\int e^x\sqrt{1 - e^{2x}}\ dx$

(t) $\int \dfrac{e^x\ dx}{\sqrt{e^{2x} + 9}}$

(u) $\int \dfrac{\ln x}{\sqrt{x}}\ dx$

(v) $\int \dfrac{x^3}{(x + 2)^2}\ dx$

3. The integral $\int \dfrac{x}{x^2 + 9}\ dx$ can be evaluated either by trigonometric substitution or by direct substitution. Do both, and reconcile the results.

4. The integral $\int \dfrac{x^2}{x^2 + 9}\ dx$ can be evaluated either by trigonometric substitution or by rewriting the numerator as $(x^2 + 9) - 9$. Do it both ways and reconcile the results.

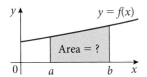

Figure 14.2 How do we find the area?

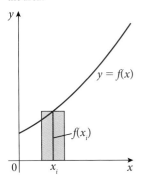

Figure 14.3 Dividing the base interval into subintervals

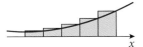

Figure 14.4 The total area of the rectangles can be viewed as an approximation

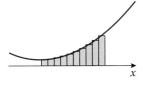

Figure 14.5 As n increases, the approximations get better

14.4 Area and the definite integral

The function $f(x)$ is continuous and non-negative on an interval $[a, b]$. How do we find the area between the graph of $f(x)$ and the interval $[a, b]$ on the x-axis? (Figure 14.2)

We divide the base interval $[a, b]$ into n equal subintervals, and over each subinterval construct a rectangle that extends from the x-axis to any point on the curve $y = f(x)$ that is above the subinterval; the particular point does not matter – it can be above the centre, above one endpoint, or above any other point in the subinterval. In Figure 14.3 it is at the centre.

For each n, the total area of the rectangles can be viewed as an approximation to the exact area in question. As n increases, these approximations will get better and better and will eventually approach the exact area as a limit. See Figures 14.3–14.5

A traditional approach would be to study how the choice of where to put the rectangular strip does not affect the approximation as the number of intervals increases. We can construct inscribed rectangles that, at the start, give us an underestimate of the area (Figure 14.6). On the other hand we can construct circumscribed rectangles that, at the start, overestimate the area (Figure 14.7).

As the number of intervals increases, the difference between the overestimates and the underestimates will approach 0.

Figures 14.8 and 14.9 show n inscribed and circumscribed rectangles and Figure 14.10 shows the difference between the overestimates and underestimates.

Figure 14.10 shows that as the number n increases, the difference between the estimates will approach 0. Because we set up our rectangles by choosing a point inside the interval, the areas of the rectangles will lie between the overestimates and underestimates, and hence, as the difference between the extremes approaches zero, the rectangles we construct will give the area of the region required.

If we consider the width of each interval to be $\triangle x$, then the area of any rectangle is given as

$$A_i = f(x_i^*)\triangle x$$

The total area of the rectangles so constructed is

$$A_n = \sum_{i=0}^{n} f(x_i^*)\triangle x$$

where x_i^* is an arbitrary point within any subinterval $[x_{i-1}, x_i]$, $x_0 = a$, and $x_n = b$.

In the case of a function $f(x)$ that has both positive and negative values on $[a, b]$, it is necessary to consider the signs of the areas in the following sense.

On each subinterval, we have a rectangle with width $\triangle x$ and height $f(x^*)$. If $f(x^*) > 0$, then this rectangle is above the x-axis; if $f(x^*) < 0$, then this rectangle is below the x-axis. We will consider the sum defined above as the sum of the signed areas of these rectangles. That means the total area on the interval is the sum of the areas above the x-axis minus the sum of the areas of the rectangles below the x-axis.

We are now ready to look at a loose definition of the definite integral:

If $f(x)$ is a continuous function defined for $a \leqslant x \leqslant b$, we divide the interval $[a, b]$ into n subintervals of equal width $\triangle x = \dfrac{(b-a)}{n}$. We let $x_0 = a$, and $x_n = b$ and we choose $x_1^*, x_2^*, \ldots, x_n^*$ in these subintervals, so that x_i^* lies in the ith subinterval $[x_{i-1}, x_i]$. Then the definite integral of $f(x)$ from a to b is

$$\int_a^b f(x)\,dx = \lim_{n\to\infty} \sum_{i=1}^{n} f(x_i^*)\triangle x$$

In the notation $\int_a^b f(x)\,dx$, a and b are called the limits of integration: a is the lower limit and b is the upper limit.

Because we have assumed that $f(x)$ is continuous, it can be proved that the limit definition above always exists and gives the same value no matter how we choose the points x_i^*. If we take these points at the centre, at two thirds the distance from the lower endpoint or at the upper endpoint, the value is the same. This why we will state the definition of the integral from now on as

$$\int_a^b f(x)\,dx = \lim_{n\to\infty} \sum_{i=1}^{n} f(x_i)\triangle x$$

For a more rigorous treatment of the definition of definite integrals using Riemann sums, refer to university calculus books. Such a treatment is beyond the scope of the IB syllabus and this book.

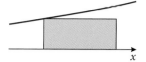

Figure 14.6 Underestimation of area

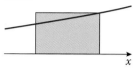

Figure 14.7 Overestimation of area

Figure 14.8 n inscribed rectangles

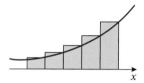

Figure 14.9 n circumscribed rectangles

Figure 14.10 difference between over- and under-estimates

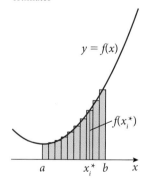

Figure 14.11 Area of each circumscribed rectangle

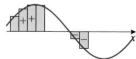

Figure 14.12 Areas above and below the x-axis

Calling the area under the function an integral is no coincidence. To make the point, let us take the following example:

Example 14.17

Find the area $A(x)$ between the graph of the function $f(x) = 3$ and the interval $[-1, x]$, and find the derivative $A'(x)$ of this area function.

Solution

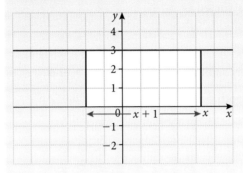

The area in question is

$$A(x) = 3(x - (-1)) = 3x + 3$$

$$A'(x) = 3 = f(x)$$

Example 14.18

Find the area $A(x)$ between the graph of the function $f(x) = 3x + 2$ and the interval $\left[-\dfrac{2}{3}, x\right]$, and find the derivative $A'(x)$ of this area function.

Solution

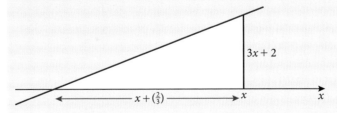

The area in question is

$$A(x) = \frac{1}{2}\left(x + \frac{2}{3}\right)(3x + 2) = \frac{1}{6}(3x + 2)^2$$

since this is the area of a triangle. Hence

$$A'(x) = \frac{1}{6} \times 2(3x + 2) \times 3 = 3x + 2 = f(x)$$

Example 14.19

Find the area $A(x)$ between the graph of the function $f(x) = x + 2$ and the interval $[-1, x]$, and find the derivative $A'(x)$ of this area function.

Solution

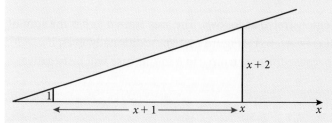

This is a trapezium, so the area is

$$A(x) = \frac{1}{2}(1 + (x + 2))(x + 1) = \frac{1}{2}(x^2 + 4x + 3), \text{ and}$$

$$A'(x) = \frac{1}{2} \times (2x + 4) = x + 2 = f(x)$$

Note that in every case, $A'(x) = f(x)$

That is, the derivative of the area function $A(x)$ is the function whose graph forms the upper boundary of the region. It can be shown that this relation is true not only for linear functions but for all continuous functions. Thus, to find the area function $A(x)$, we can look instead for a particular function whose derivative is $f(x)$. This is, of course, nothing but the antiderivative of $f(x)$.

So, intuitively, as we have seen above, we define the area function as

$$A(x) = \int_a^x f(t) \, dt, \text{ that is } A'(x) = f(x)$$

This is the trigger to the **fundamental theorem of calculus**.

We will now look at some of the properties of the definite integral.

Basic properties of the definite integral

$$\int_a^b f(x) \, dx = -\int_b^a f(x) \, dx$$

When we defined the definite integral $\int_a^b f(x) \, dx$, we implicitly assumed that $a < b$. When we reverse a and b, then $\triangle x$ changes from $\frac{(b - a)}{n}$ to $\frac{(a - b)}{n}$. Therefore the result above follows.

$$\int_a^b f(x) \, dx = 0$$

When $a = b$, then $\triangle x = 0$, and so, the result above follows.

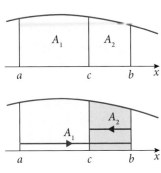

$$\int_a^b c \, dx = c(b - a)$$

$$\int_a^b [f(x) \pm g(x)] \, dx = \int_a^b f(x) \, dx \pm \int_a^b g(x) \, dx$$

$$\int_a^b c f(x) \, dx = c \int_a^b f(x) \, dx, \text{ where } c \text{ is any constant.}$$

$$\int_a^b f(x) \, dx = \int_a^c f(x) \, dx + \int_c^b f(x) \, dx$$

This property can be demonstrated as follows. The area from a to b is the sum of the two areas, that is $A(x) = A_1 + A_2$ (Figure 14.13). Additionally, even if $c > b$ the relationship holds because the area from c to b in this case will be negative.

Figure 14.13 $A(x) = A_1 + A_2$

Average value of a function

From statistics, the average value of a variable is $\bar{x} = \dfrac{\sum\limits_{i=1}^{n} \bar{x}_i}{n}$

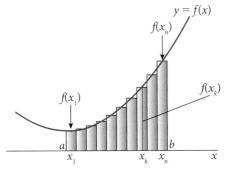

Figure 14.14 $y = f(x)$ partitioned into n subintervals

We can also think of the average value of a function in the same manner. Consider a continuous function $f(x)$ defined over a closed interval $[a, b]$. We partition this interval into n subintervals of equal length in a fashion similar to the previous discussion. Each interval has a length

$$\triangle x = \frac{b - a}{n}$$

The average value of $f(x)$ can be defined as

$$av(f) = \frac{f(x_1) + f(x_2) + \cdots + f(x_n)}{n}, \text{ or written in sigma notation}$$

$$av(f) = \frac{\sum\limits_{k=1}^{n} f(x_k)}{n} = \frac{1}{n} \sum\limits_{k=1}^{n} f(x_k)$$

However,

$$\triangle x = \frac{b - a}{n} \Rightarrow \frac{1}{n} = \frac{\triangle x}{b - a}, \text{ hence}$$

$$av(f) = \frac{1}{n} \sum\limits_{k=1}^{n} f(x_k) = \frac{\triangle x}{b - a} \sum\limits_{k=1}^{n} f(x_k)$$

This leads us to the following definition of the average value of a function $f(x)$ over an interval $[a, b]$:

The average (mean value) of an integrable function $f(x)$ over an interval $[a, b]$ is given by

$$av(f) = \frac{1}{b - a} \int_a^b f(x) \, dx$$

Max–min inequality

If f_{max} and f_{min} represent the maximum and minimum values of a non-negative, continuous, differentiable function $f(x)$ over an interval $[a, b]$, then the area under the curve lies between the area of the rectangle with base $[a, b]$ and f_{min} as height, and the rectangle with f_{max} as height.

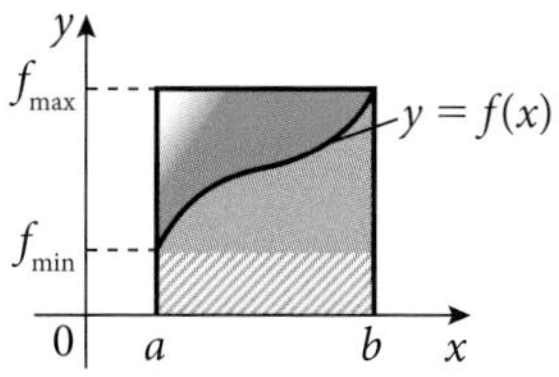

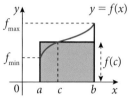

Figure 14.15 Max-min inequality

That is

$$(b - a) f_{min} \leqslant \int_a^b f(x)\, dx \leqslant (b - a) f_{max}$$

With the assumption that $b > a$, this in turn is equivalent to

$$f_{min} \leqslant \frac{1}{b - a} \int_a^b f(x)\, dx \leqslant f_{max}$$

Using the intermediate value theorem, we can ascertain that there is at least one point $c \in [a, b]$ where

$$f(c) = \frac{1}{b - a} \int_a^b f(x)\, dx$$

Figure 14.16 Average value

The value $f(c)$ in this theorem is, in fact, the average value of the function.

The first fundamental theorem of integral calculus

Our understanding of the definite integral as the area under the curve for $f(x)$ helps us establish the basis for the fundamental theorem of integral calculus.

In the definition of definite integral, we'll make the upper limit a variable, say x. Then we will call the area between a and x, $A(x)$; that is,

$$A(x) = \int_a^x f(t)\, dt$$

Consequently,

$$A(x + h) = \int_a^{x+h} f(t)\, dt$$

Now, if we want to find the derivative of $A(x)$, we evaluate

$$\lim_{h \to 0} \frac{A(x + h) - A(x)}{h}$$

Using the properties of definite integrals discussed earlier, we have

$$A(x + h) - A(x) = \int_a^{x+h} f(t)\, dt - \int_a^x f(t)\, dt$$

$$= \int_x^a f(t)\, dt + \int_a^{x+h} f(t)\, dt$$

$$= \int_x^{x+h} f(t)\, dt$$

723

Therefore

$$\lim_{h \to 0} \frac{A(x + h) - A(x)}{h} = \lim_{h \to 0} \frac{\int_x^{x+h} f(t)\,dt}{h} = \lim_{h \to 0} \frac{1}{h} \int_x^{x+h} f(t)\,dt$$

Looking at this result and what we established about the average value of $f(x)$ over the interval $[x, x + h]$, we can conclude that there is a point $c \in [x, x + h]$ such that

$$f(c) = \frac{1}{h} \int_x^{x+h} f(t)\,dt$$

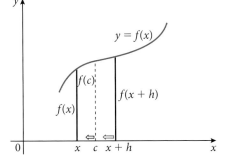

Figure 14.17 $f(c)$ approaching $f(x)$

What happens to c as h approaches 0? As h approaches 0, $x + h$ must approach x. This means, we are 'squeezing' c between x and a number approaching x. So, c must also approach x. That is

$$f(c) = f(x), \text{ and consequently}$$

$$\lim_{h \to 0} \frac{A(x + h) - A(x)}{h} = \lim_{h \to 0} \frac{1}{h} \int_x^{x+h} f(t)\,dt = f(c) = f(x)$$

This last equation is stating that

$$\frac{d}{dx}(A(x)) = A'(x) = \frac{d}{dx}\left(\int_a^x f(t)\,dt \right) = f(x)$$

It is important to remember that $\int_a^x f(t)\,dt$ is a function of x.

This very powerful statement is called the **first fundamental theorem of integral calculus**. In essence, it says that the processes of integration and differentiation are inverses of one another.

Example 14.20

Find each derivative.

(a) $\dfrac{d}{dx} \displaystyle\int_{-e}^x \sec^2 t\,dt$

(b) $\dfrac{d}{dx} \displaystyle\int_0^x \dfrac{dt}{1 + t^4}$

(c) $\dfrac{d}{dx} \displaystyle\int_x^\pi \dfrac{1}{1 + t^4}\,dt$

(d) $\dfrac{d}{dx} \displaystyle\int_0^{2x+x^3} \dfrac{1}{1 + t^4}\,dt$

(e) $\dfrac{d}{dx} \displaystyle\int_x^{2x+x^3} \dfrac{1}{1 + t^4}\,dt$

Solution

(a) This is a direct application of the fundamental theorem:

$$\frac{d}{dx} \int_{-e}^x \sec^2 t\,dt = \sec^2 x$$

(b) This is also straightforward:

$$\frac{d}{dx} \int_0^x \frac{dt}{1 + t^4} = \frac{1}{1 + x^4}$$

(c) We need to rewrite the expression before we perform the calculation.

$$\frac{d}{dx}\int_x^\pi \frac{1}{1+t^4}\,dt = \frac{d}{dx}\int_\pi^x -\frac{1}{1+t^4}\,dt = -\frac{d}{dx}\int_x^\pi \frac{1}{1+t^4}\,dt = \frac{-1}{1+x^4}$$

(d) This is a function of x, and the upper limit is a function of x, which makes $\int_0^{2x+x^3} \frac{1}{1+t^4}\,dt$ a composite of $\int_0^u \frac{1}{1+t^4}\,dt$ and $u = 2x + x^3$. So, we have to use the chain rule.

$$\frac{d}{dx}\int_0^{2x+x^3} \frac{1}{1+t^4}\,dt = \left(\frac{d}{du}\int_0^u \frac{1}{1+t^4}\,dt\right)\left(\frac{du}{dx}\right)$$

$$= \frac{1}{1+u^4}\cdot\frac{du}{dx}$$

$$= \frac{1}{1+(2x+x^3)^4}\cdot(2+3x^2)$$

$$= \frac{2+3x^2}{1+(2x+x^3)^4}$$

(e) Again, we need to rewrite the integral before evaluation

$$\frac{d}{dx}\int_x^{2x+x^3} \frac{1}{1+t^4}\,dt = \frac{d}{dx}\left(\int_x^k \frac{1}{1+t^4}\,dt + \int_k^{2x+x^3} \frac{1}{1+t^4}\,dt\right)$$

$$= \frac{2+3x^2}{1+(2x+x^3)^4} - \frac{1}{1+x^4}$$

The second fundamental theorem of integral calculus

Recall that $A(x) = \int_a^x f(t)\,dt$. If $F(x)$ is any antiderivative of $f(x)$, then applying what we learned earlier

$\quad F(x) = A(x) + c$ where c is an arbitrary constant.

Now

$\quad F(b) = A(b) + c = \int_a^b f(t)\,dt + c$, and

$\quad F(a) = A(a) + c = \int_a^a f(t)\,dt + c = 0 + c$, and hence

$\quad F(b) - F(a) = \int_a^b f(t)\,dt + c - c$

$$= \int_a^b f(t)\,dt$$

The second fundamental theorem of calculus states:
$$\int_a^b f(t)\,dt = F(b) - F(a)$$

725

The theorem is also known as the **evaluation theorem**. Also, since we know that $F'(x)$ is the rate of change in $F(x)$ with respect to x, and that $F(b) - F(a)$ is the change in y when x changes from a to b, we can reformulate the theorem in words to read:

The integral of a rate of change is the **total change**:

$$\int_a^b F'(x)\,dx = F(b) - F(a)$$

Here are a few instances where this applies:

- If $V'(t)$ is the rate at which a liquid flows into or out of a container at time t, then $\int_{t_1}^{t_2} V'(t)\,dt = V(t_2) - V(t_1)$ is the change in the amount of liquid in the container between time t_1 and t_2.

- If the rate of growth of a population is $n'(t)$, then $\int_{t_1}^{t_2} n'(t)\,dt = n(t_2) - n(t_1)$ is the increase (or decrease) in population during the period from t_1 to t_2.

This theorem has many other applications in calculus and several other fields. It is a very powerful tool that allows us to deal with problems of area, volume, and work. In this book, we will apply it to finding areas between functions and volumes of revolution as well in displacement problems.

Notation

We will use the following notation in evaluating definite integrals. If we know that $F(x)$ is an antiderivative of $f(x)$, then we will write

$$\int_a^b f(t)\,dt = F(x)\Big|_a^b$$
$$= F(b) - F(a)$$

Example 14.21

Evaluate each integral

(a) $\int_{-1}^{3} x^5\,dx$

(b) $\int_0^4 \sqrt{x}\,dx$

(c) $\int_\pi^{2\pi} \cos\theta\,d\theta$

(d) $\int_1^2 \frac{4 + u^2}{u^3}\,du$

Solution

(a) $\int_{-1}^{3} x^5\,dx = \frac{x^6}{6}\Big|_{-1}^{3} = \frac{3^6}{6} - \frac{1}{6} = \frac{364}{3}$

(b) $\int_0^4 \sqrt{x}\,dx = \frac{2}{3}x^{\frac{3}{2}}\Big|_0^4 = \frac{2}{3}4^{\frac{3}{2}} - 0 = \frac{16}{3}$

(c) $\int_\pi^{2\pi} \cos\theta\,d\theta = \sin\theta\Big|_\pi^{2\pi} = 0 - 0 = 0$

(d) $\int_1^2 \frac{4 + u^2}{u^3}\,du = \int_1^2 \left(\frac{4}{u^3} + \frac{1}{u}\right)du = 4\cdot\frac{u^{-2}}{-2} + \ln|u|\Big|_1^2$

$$= -2u^{-2} + \ln u\Big|_1^2$$

$$= (-2\cdot 2^{-2} + \ln 2) - (-2\cdot 1 + \ln 1)$$

$$= -\frac{1}{2} + \ln 2 + 2 = \frac{3}{2} + \ln 2$$

Using substitution with the definite integral

In Section 14.1, we discussed the use of substitution to evaluate integrals in cases that are not easily recognised. We established that

$$\int f(u(x)) \cdot u'(x)\, dx = \int f(u)du = F(u(x)) + c$$

When evaluating definite integrals by substitution, two methods are available.

- Evaluate the indefinite integral first, revert to the original variable, then use the fundamental theorem. For example, to evaluate

$$\int_0^{\frac{\pi}{3}} \tan^5 x \sec^2 x\, dx$$

we find the indefinite integral

$$\int \tan^5 x \sec^2 x\, dx = \int u^5\, du = \frac{1}{6}u^6 = \frac{1}{6}\tan^6 x$$

then we use the fundamental theorem

$$\int_0^{\frac{\pi}{3}} \tan^5 x \sec^2 x\, dx = \frac{1}{6}\tan^6 x\Big|_0^{\frac{\pi}{3}} = \frac{1}{6}(\sqrt{3})^6 = \frac{27}{6} = \frac{9}{2}$$

- Or we can use the following substitution rule for definite integrals

$$\int_a^b f(u(x))u'(x)\, dx = \int_{u(a)}^{u(b)} f(u)\, du$$

Proof:

If $F(x)$ is an antiderivative of $f(x)$, then by the fundamental theorem

$$\int_a^b f(u(x))u'(x)\, dx = F(u(x))\Big|_a^b = F(u(b)) - F(u(a))$$

Also

$$\int_{u(a)}^{u(b)} f(u)\, du = F(u)\Big|_{u(a)}^{u(b)} = F(u(b)) - F(u(a))$$

Therefore, to evaluate

$$\int_0^{\frac{\pi}{3}} \tan^5 x \sec^2 x\, dx$$

let $u = \tan x \Rightarrow u\left(\frac{\pi}{3}\right) = \sqrt{3}$, $u(0) = 0$, and so

$$\int_0^{\frac{\pi}{3}} \tan^5 x \sec^2 x\, dx = \int_0^{\sqrt{3}} u^5\, du = \frac{1}{6}u^6\Big|_0^{\sqrt{3}} = \frac{9}{2}$$

Example 14.22

Evaluate $\int_2^6 \sqrt{4x+1}\,dx$

Solution

Let $u = 4x + 1$, then $du = 4dx$. The limits of integration are $u(2) = 9$, and $u(6) = 25$. Therefore

$$\int_2^6 \sqrt{4x+1}\,dx = \frac{1}{4}\int_9^{25} \sqrt{u}\,du = \frac{1}{4}\left(\frac{2}{3}u^{3/2}\right)\Big|_9^{25}$$

$$= \frac{1}{6}(125 - 27) = \frac{49}{3}$$

Note that, using this method, we do not return to the original variable of integration. We simply evaluate the new integral between the appropriate values of u.

Notice that the substitution $u = 4x + 1$ stretched the interval $[2, 6]$ by a factor of 4, and shifted it by 1 unit to the right. But the areas are the same.

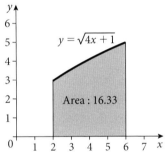

Figure 14.18 The area under the curve $y = \sqrt{4x+1}$ between $x = 2$ and $x = 6$

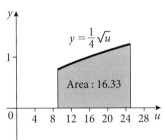

Figure 14.19 The area under the curve $y = \frac{1}{4}\sqrt{u}$ between $u = 9$ and $u = 25$

Exercise 14.4

1. Evaluate each integral.

(a) $\int_{-2}^1 (3x^2 - 4x^3)\,dx$

(b) $\int_2^7 8\,dx$

(c) $\int_1^5 \frac{2}{t^3}\,dt$

(d) $\int_2^2 (\cos t - \tan t)\,dt$

(e) $\int_1^7 \frac{2x^2 - 3x + 5}{\sqrt{x}}\,dx$

(f) $\int_0^\pi \cos\theta\,d\theta$

(g) $\int_0^\pi \sin\theta\,d\theta$

(h) $\int_3^1 (5x^4 + 3x^2)\,dx$

(i) $\int_1^3 \frac{u^5 + 2}{u^2}\,du$

(j) $\int_1^e \frac{2\,dx}{x}$

(k) $\int_1^3 \frac{2x}{x^2 + 2}\,dx$

(l) $\int_1^3 (2 - \sqrt{x})^2\,dx$

(m) $\int_0^{\frac{\pi}{4}} 3\sec^2\theta\,d\theta$

(n) $\int_0^1 (8x^7 + \sqrt{\pi})\,dx$

(o) (i) $\int_0^2 |3x|\,dx$ (ii) $\int_{-2}^0 |3x|\,dx$ (iii) $\int_{-2}^2 |3x|\,dx$

(p) $\int_0^{\frac{\pi}{2}} \sin 2x\,dx$

(q) $\int_1^9 \frac{1}{\sqrt{x}}\,dx$

(r) $\int_{-2}^2 (e^x - e^{-x})\,dx$

(s) $\int_{-1}^1 \frac{dx}{1 + x^2}$

(t) $\int_0^{\frac{1}{2}} \dfrac{dx}{\sqrt{1-x^2}}$

(u) $\int_{-1}^1 \dfrac{dx}{\sqrt{4-x^2}}$

(v) $\int_{-2}^0 \dfrac{dx}{4+x^2}$

2. Evaluate each integral.

(a) $\int_0^4 \dfrac{x^3\,dx}{\sqrt{x^2+1}}$

(b) $\int_1^{\sqrt{e}} \dfrac{\sin(\pi \ln x)}{x}\,dx$

(c) $\int_e^{e^2} \dfrac{dt}{t \ln t}$

(d) $\int_{-1}^2 3x\sqrt{9-x^2}\,dx$

(e) $\int_{-\frac{\pi}{3}}^{\frac{2\pi}{3}} \dfrac{\sin x}{\sqrt{3+\cos x}}\,dx$

(f) $\int_e^{e^2} \dfrac{\ln x}{x}\,dx$

(g) $\int_1^{\sqrt{3}} \dfrac{\sqrt{\arctan x}}{1+x^2}\,dx$

(h) $\int_1^{\sqrt{e}} \dfrac{dx}{x\sqrt{1-(\ln x)^2}}$

(i) $\int_{-\ln 2}^{\ln 2} \dfrac{e^{2x}}{e^{2x}+9}\,dx$

(j) $\int_{\ln 2}^{\ln(2/\sqrt{3})} \dfrac{e^{-2x}\,dx}{\sqrt{1-e^{-4x}}}$

(k) $\int_0^{\frac{\pi}{4}} \sqrt{\tan x}\,\sec^2 x\,dx$

(l) $\int_0^{\sqrt{\pi}} 7x\cos x^2\,dx$

(m) $\int_{\pi^2}^{4\pi^2} \dfrac{\sin\sqrt{x}}{\sqrt{x}}\,dx$

(n) $\int_0^1 \dfrac{\sqrt{3}\,x}{\sqrt{4-3x^4}}\,dx$

(o) $\int_0^{\frac{2}{\sqrt{3}}} \dfrac{dx}{9+4x^2}$

(p) $\int_1^{\sqrt{2}} \dfrac{x\,dx}{3+x^4}$

(q) $\int_0^{\frac{\pi}{6}} (1-\sin 3t)\cos 3t\,dt$

(r) $\int_0^{\frac{\pi}{4}} e^{\sin 2\theta}\cos 2\theta\,d\theta$

(s) $\int_0^{\frac{\pi}{8}} (3+e^{\tan 2t})\sec^2 2t\,dt$

(t) $\int_0^{\sqrt{\ln \pi}} 4t\,e^{t^2}\sin(e^{t^2})dt$

3. Find the average value of each function over the given interval.

(a) x^4, $[1,2]$

(b) $\cos x$, $\left[0, \dfrac{\pi}{2}\right]$

(c) $\sec^2 x$, $\left[\dfrac{\pi}{6}, \dfrac{\pi}{4}\right]$

(d) e^{-2x}, $[0,4]$

(e) $\dfrac{e^{3x}}{1+e^{6x}}$, $\left[\dfrac{-\ln 3}{6}, 0\right]$

4. Find the indicated derivative.

(a) $\dfrac{d}{dx}\int_2^x \dfrac{\sin t}{t}\,dt$

(b) $\dfrac{d}{dt}\int_t^3 \dfrac{\sin x}{x}\,dx$

(c) $\dfrac{d}{dx}\int_{x^2}^0 \dfrac{\sin t}{t}\,dt$

(d) $\dfrac{d}{dx}\int_0^{x^2} \dfrac{\sin u}{u}\,du$

(e) $\dfrac{d}{dt}\displaystyle\int_{-\pi}^{t}\dfrac{\cos y}{1+y^2}\,dy$

(f) $\dfrac{d}{dx}\displaystyle\int_{ax}^{bx}\dfrac{dt}{5+t^4}$

(g) $\dfrac{d}{d\theta}\displaystyle\int_{\sin\theta}^{\cos\theta}\dfrac{1}{1-x^2}\,dx$

(h) $\dfrac{d}{dx}\displaystyle\int_{5}^{x^{\frac{1}{4}}}e^{t^4+3t^2}\,dt$

5. Does the function $F(x) = \displaystyle\int_{0}^{2x-x^2}\cos\!\left(\dfrac{1}{1+t^2}\right)dt$ have an extreme value?

6. (a) Find $\displaystyle\int_{0}^{k}\dfrac{dx}{3x+2}$, giving your answer in terms of k.

(b) Given that $\displaystyle\int_{0}^{k}\dfrac{dx}{3x+2} = 1$, calculate the value of k.

7. Given that $p, q \in \mathbb{N}$, show that

$$\int_{0}^{1} x^p(1-x)^q\,dx = \int_{0}^{1} x^q(1-x)^p\,dx$$

Do not attempt to evaluate the integrals.

8. Given that $k \in \mathbb{N}$, evaluate each integral:

(a) $\displaystyle\int x(1-x)^k\,dx$

(b) $\displaystyle\int_{0}^{1} x(1-x)^k\,dx$

9. Let $F(x) = \displaystyle\int_{3}^{x}\sqrt{5t^2+2}\,dt$, find:

(a) $F(3)$ **(b)** $F'(3)$ **(c)** $F''(3)$

10. Show that the function

$$f(x) = \int_{x}^{3x}\dfrac{dt}{t}$$

is constant over the set of positive real numbers.

14.5   Integration by method of partial fractions

In this section, we will integrate rational functions with polynomial denominators. For example, if we find the indefinite integral $\displaystyle\int\dfrac{x+1}{x^2+5x+6}\,dx$, we first decompose the integrand into partial fractions (Section 2.6) and then the integration process is straightforward.

$$\dfrac{x+1}{x^2+5x+6} \equiv \dfrac{a}{x+2} + \dfrac{b}{x+3}$$

After solving for a and b we can perform the integration:

$$\int \frac{x+1}{x^2+5x+6} \, dx = \int \left(\frac{-1}{x+2} + \frac{1}{x+3} \right) dx = -\ln|x+2| + \ln|x+3| + c$$

$$= \ln\left| \frac{x+3}{x+2} \right| + c$$

Example 14.23

Find the indefinite integral $\int \dfrac{3x-1}{x^2+4x+4} \, dx$

Solution

Using partial fractions will make integration easier.

From Example 2.33 we know that

$$\frac{3x-1}{x^2+4x+4} \equiv \frac{3}{x+2} - \frac{7}{(x+2)^2}$$

Hence the integral can be rewritten as

$$\int \frac{3x-1}{x^2+4x+4} \, dx = \int \frac{3}{x+2} \, dx - \int \frac{7}{(x+2)^2} \, dx$$

These two integrals can be found by inspection, giving

$$\int \frac{3x-1}{x^2+4x+4} \, dx = 3\ln|x+2| + \frac{7}{x+2} + C$$

Example 14.24

Find the indefinite integral $\int \dfrac{2}{x^3+2x^2+2x} \, dx$

Solution

Factorising the denominator and separating fractions as we did in chapter 2, we have:

$$\frac{2}{x^3+2x^2+2x} \equiv \frac{1}{x} - \frac{x+2}{x^2+2x+2}$$

Hence, we can write the integral as

$$\int \frac{2}{x^3+2x^2+2x} \, dx = \int \frac{dx}{x} - \int \frac{x+2}{x^2+2x+2} \, dx$$

$$= \int \frac{dx}{x} - \int \frac{x+1+1}{x^2+2x+2} \, dx$$

$$= \int \frac{dx}{x} - \frac{1}{2}\int \frac{2x+2}{x^2+2x+2} \, dx - \int \frac{dx}{(x+1)^2+1}$$

$$= \ln|x| - \frac{1}{2}\ln(x^2+2x+2) - \arctan(x+1) + C$$

Example 14.25

Find the indefinite integral $\int \dfrac{5x^2 + 16x + 17}{2x^3 + 9x^2 + 7x - 6} dx$

Solution

From Example 2.32 we have

$$\int \frac{5x^2 + 16x + 17}{2x^3 + 9x^2 + 7x - 6} dx = \int \frac{3}{2x - 1} dx - \int \frac{1}{x + 2} dx + \int \frac{2}{x + 3} dx$$

$$= \frac{3}{2} \ln|2x - 1| - \ln|x + 2| + 2 \ln|x + 3| + c$$

For IB Mathematics HL exams, the original fraction to be rewritten as partial fractions will be such that the degree of the denominator will not be greater than 2; factors will be at most two distinct linear terms, and the degree of the numerator will be less than the degree of the denominator.

The last case is unlikely since it would require factorisation of a cubic, unless it is given in factorised form.

A few cases of partial fractions

Denominator is quadratic – it factorises into two distinct linear factors, and the numerator $p(x)$ is a constant or linear

$$\frac{p(x)}{(ax + b)(cx + d)} \equiv \frac{A}{ax + b} + \frac{B}{cx + d}$$

Denominator is quadratic – it factorises into two repeated linear factors, and the numerator $p(x)$ is a constant or linear

$$\frac{p(x)}{(ax + b)^2} \equiv \frac{A}{ax + b} + \frac{B}{(ax + b)^2}$$

Denominator is cubic – it factorises into three distinct linear factors, and the numerator $p(x)$ is a constant, linear, or quadratic

$$\frac{p(x)}{(ax + b)(cx + d)(ex + f)} = \frac{A}{ax + b} + \frac{B}{cx + d} + \frac{C}{ex + f}$$

Remember that a consequence of the fundamental theorem of algebra is that any polynomial with real coefficients can only have factors that are linear or quadratic.

Exercise 14.5

1. Evaluate each integral.

(a) $\int \dfrac{5x + 1}{x^2 + x - 2} dx$

(b) $\int \dfrac{x + 4}{x^2 - 2x} dx$

(c) $\int \dfrac{x + 2}{x^2 + 4x + 3} dx$

(d) $\int \dfrac{5x^2 + 20x + 6}{x^3 + 2x^2 + x} dx$

(e) $\int \dfrac{2x^2 + x - 12}{x^3 + 5x^2 + 6x} dx$

(f) $\int \dfrac{4x^2 + 2x - 1}{x^3 + x^2} dx$

(g) $\int \dfrac{3}{x^2 + x - 2} dx$

(h) $\int \dfrac{5 - x}{2x^2 + x - 1} dx$

(i) $\displaystyle\int \frac{3x + 4}{(x + 2)^2}\,dx$

(j) $\displaystyle\int \frac{12}{x^4 - x^3 - 2x^2}\,dx$

(k) $\displaystyle\int \frac{2}{x^3 + x}\,dx$

(l) $\displaystyle\int \frac{x + 2}{x^3 + 3x}\,dx$

(m) $\displaystyle\int \frac{3x + 2}{x^3 + 6x}\,dx$

(n) $\displaystyle\int \frac{2x + 3}{x^3 + 8x}\,dx$

14.6 Areas

We have seen how the area between a curve defined by $y = f(x)$ and the x-axis can be computed by the integral $\displaystyle\int_a^b f(x)\,dx$ on an interval $[a, b]$ where $f(x) \geqslant 0$. In this section, we shall use integration to find the area of more general regions between curves.

Areas between curves of functions of the form $y = f(x)$ and the x-axis

If the function $y = f(x)$ is always above the x-axis, finding the area is a straightforward computation of the integral $\displaystyle\int_a^b f(x)\,dx$.

Find the area between the curve $f(x) = x^3 - x + 1$ and the x-axis over the interval $[-1, 2]$

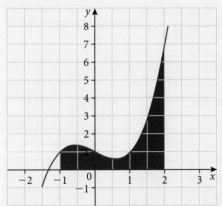

Solution

This area is:

$$\int_{-1}^{2}(x^3 - x + 1)\,dx = \left[\frac{x^4}{4} - \frac{x^2}{2} + x\right]_{-1}^{2} = (4 - 2 + 2) - \left(\frac{1}{4} - \frac{1}{2} - 1\right) = 2\frac{1}{4}$$

$$\int_{-1}^{2} (x^3 - x + 1)\,dx$$
$$\frac{21}{4}$$

Figure 14.20 Using a GDC to find the area

Using our GDC, the area in Example 14.26 is found by simply choosing the MATH menu, then the $\int dx$ menu item, then typing in the function with the integration limits.

In some cases, we will have to adjust how to work. This is the case when the graph intersects the x-axis. Since we are interested in the area bounded by the curve and the interval $[a, b]$ on the x-axis, we do not want the two areas to cancel each other. This is why we have to split the process into subintervals where we take the absolute values of the areas found and add them.

Example 14.27

Find the area under the curve $f(x) = x^3 - x - 1$ and the x-axis over the interval $[-1, 2]$

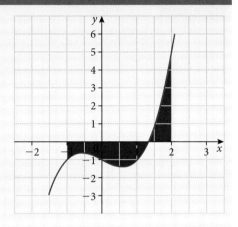

Solution

As we see from the diagram, a part of the graph is below the x-axis, and its area will be negative. If we try to integrate this function without paying attention to the intersection with the x-axis, here is what we get:

$$\int_{-1}^{2} (x^3 - x - 1)\, dx = \frac{x^4}{4} - \frac{x^2}{2} - x \Big|_{-1}^{2} = (4 - 2 - 2) - \left(\frac{1}{4} - \frac{1}{2} + 1\right)$$

$$= -\frac{3}{4}$$

This integration has to be split before we start. However, this is a function where we cannot find the intersection point. So, we either use a GDC to find the intersection or we just take the absolute values of the different parts of the region. This is done by integrating the absolute value of the function:

$$Area = \int_{a}^{b} |f(x)|\, dx$$

$\int_{-1}^{2} |x^3-x-1|\, dx$

$\qquad\qquad 3.614515769$

Figure 14.21 It is best to use a GDC

As we said earlier, this is not easy to find given the difficulty with the x-intercept. It is best if we use a GDC.

Hence, $Area = \int_{-1}^{2} |x^3 - x - 1|\, dx \approx 3.6145$

Example 14.28

Find the area enclosed by the graph of the function $f(x) = x^3 - 4x^2 + x + 6$ and the x-axis.

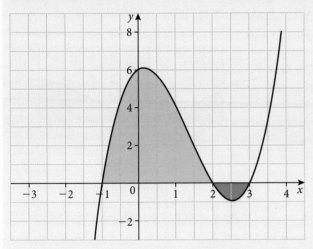

Solution

This function intersects the x-axis at three points where $x = -1$, 2, and 3. To find the area, we split it into two and then add the absolute values:

$$Area = \int_{-1}^{3} |f(x)|\, dx = \int_{-1}^{2} f(x)\, dx + \int_{2}^{3} (-f(x))\, dx$$

$$= \int_{-1}^{2} (x^3 - 4x^2 + x + 6)\, dx + \int_{2}^{3} (-x^3 + 4x^2 - x - 6)\, dx$$

$$= \frac{x^4}{4} - \frac{4x^3}{3} + \frac{x^2}{2} + 6x \Big|_{-1}^{2} + \left. -\frac{x^4}{4} + \frac{4x^3}{3} - \frac{x^2}{2} - 6x \right|_{2}^{3}$$

$$= \frac{45}{4} + \frac{7}{12} = \frac{71}{6}$$

Area between curves

In some practical problems, we may have to compute the area between two curves. Let $f(x)$ and $g(x)$ be functions such that $f(x) \geqslant g(x)$ on the interval $[a, b]$ (Figure 14.22). We do not insist that both functions are non-negative.

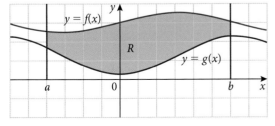

Figure 14.22 Area between two curves

735

To find the area of the region R between the curves from $x = a$ to $x = b$, we subtract the area between the lower curve $g(x)$ and the x-axis from the area between the upper curve $f(x)$ and the x-axis; that is

$$\text{Area of } R = \int_a^b f(x)\,dx - \int_a^b g(x)\,dx = \int_a^b [f(x) - g(x)]\,dx$$

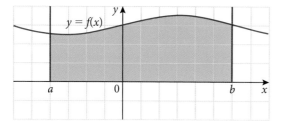 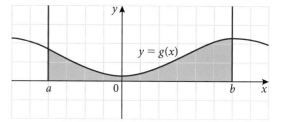

Figure 14.23 Areas under functions f and g

 If $f(x)$ and $g(x)$ are functions such that $f(x) \geqslant g(x)$ on the interval $[a, b]$, then the area between the two curves is given by $A = \sum_a^b [f(x) - g(x)]\,dx$

This fact applies to all functions, not only positive functions. These facts are used to define the area between curves.

Example 14.29

Find the area of the region between the curves $y = x^3$ and $y = x^2 - x$ on the interval $[0, 1]$.

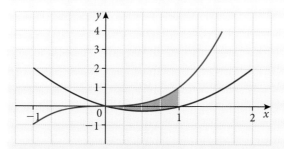

Solution

$y = x^3$ appears to be higher than $y = x^2 - x$ with one intersection at $x = 0$. Thus, the required area is

$$A = \int_0^1 [x^3 - (x^2 - x)]\,dx = \frac{x^4}{4} - \frac{x^3}{3} + \frac{x^2}{2}\Big|_0^1 = \frac{5}{12}$$

In some cases we must be very careful how we calculate the area. This is the case where the two functions intersect at more than one point.

Example 14.30

Find the area of the region bounded by the curves $y = x^3 + 2x^2$ and $y = x^2 + 2x$.

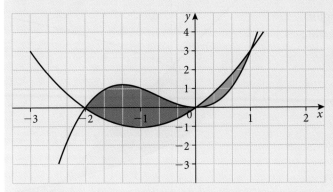

Solution

The two curves intersect when:

$$x^3 + 2x^2 = x^2 + 2x \Rightarrow x^3 + x^2 - 2x = 0 \Rightarrow x(x + 2)(x - 1) = 0$$

That is, when $x = -2, 0,$ or 1

The area is equal to:

$$A = \int_{-2}^{0} [x^3 + 2x^2 - (x^2 + 2x)]\, dx + \int_{0}^{1} [x^2 + 2x - (x^3 + 2x^2)]\, dx$$

$$= \int_{-2}^{0} [x^3 + x^2 - 2x]\, dx + \int_{0}^{1} [-x^2 + 2x - x^3]\, dx$$

$$= \left[\frac{x^4}{4} + \frac{x^3}{3} - x^2\right]_{-2}^{0} + \left[-\frac{x^4}{4} - \frac{x^3}{3} + x^2\right]_{0}^{1}$$

$$= 0 - \left[\frac{16}{4} - \frac{8}{3} - 4\right] + \left[-\frac{1}{4} - \frac{1}{3} + 1\right] - 0 = \frac{37}{12}$$

This discussion leads us to stating the general expression we should use in evaluating areas between curves.

If $f(x)$ and $g(x)$ are functions that are continuous on the interval $[a, b]$, then the area between the two curves is given by

$$A = \int_{a}^{b} |f(x) - g(x)|\, dx$$

We can do this on our GDC.

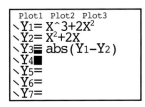

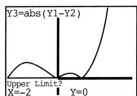

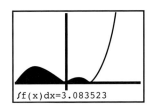

Figure 14.24 Using a GDC to find the area between two curves

Areas along the *y*-axis

To find the area enclosed by
$y = 1 - x$ and $y^2 = x + 1$,
it is best to treat the region
between them by regarding
x as a function of *y*
(Figure 14.25).

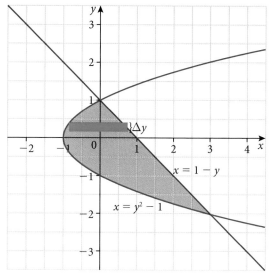

Figure 14.25 Area between two curves expressed by regarding *x* as a function of *y*

The area of the shaded
region can be calculated
using the integral:

$$A(y) = \int_{-2}^{1} |(1 - y) - (y^2 - 1)|\, dy$$

$$= \int_{-2}^{1} |2 - y - y^2|\, dy = \left| 2y - \frac{y^2}{2} - \frac{y^3}{3} \right|_{-2}^{1} = \frac{9}{2}$$

If we used *y* as a function of *x*, then the calculation would involve calculating
the area by dividing the interval into two: $[-1, 0]$ and $[0, 3]$.

In the first part, the area is enclosed between $y = \sqrt{x + 1}$ and $y = -\sqrt{x + 1}$,
and the area in the second part is enclosed by $y = 1 - x$ and $y = -\sqrt{x + 1}$:

$$A(x) = 2\int_{-1}^{0} \sqrt{x + 1}\, dx + \int_{0}^{3} \left((1 - x) - (-\sqrt{x + 1}) \right) dx$$

Exercise 14.6

1. Find the area of the region bounded by the given curves. Sketch the
   region and then compute the required area.

   (a) $y = x + 1, y = 7 - x^2$ **(b)** $y = \cos x, y = x - \frac{\pi}{2}, x = -\pi$

   (c) $y = 2x, y = x^2 - 2$ **(d)** $y = x^3, y = x^2 - 2, x = 1$

   (e) $y = x^6, y = x^2$ **(f)** $y = 5x - x^2, y = x^2$

   (g) $y = 2x - x^3, y = x - x^2$ **(h)** $y = \sin x, y = 2 - \sin x$ (one period)

   (i) $y = \frac{x}{2}, y = \sqrt{x}, x = 9$ **(j)** $y = \frac{x^4}{10}, y = 3x - x^3$

   (k) $y = \frac{1}{x}, y = \frac{1}{x^3}, x = 8$

   (l) $y = 2\sin x, y = \sqrt{3}\tan x, -\frac{\pi}{4} \leqslant x \leqslant \frac{\pi}{4}$

(m) $y = x - 1$, $y^2 = 2x + 6$ **(n)** $x = 2y^2$, $x = 4 + y^2$

(o) $4x + y^2 = 12$, $y = x$ **(p)** $x - y = 7$, $x = 2y^2 - y + 3$

(q) $x = y^2$, $x = 2y^2 - y - 2$

(r) $y = x^3 + 2x^2$, $y = x^3 - 2x$, $x = -3$, and $x = 2$

(s) $y = \sec^2 x$, $y = \sec x \tan x$, $x = -\dfrac{\pi}{3}$, and $x = \dfrac{\pi}{6}$

(t) $y = x^3 + 1$, $y = (x + 1)^2$

(u) $y = x^3 + x$, $y = 3x^2 - x$

(v) $y = 3 - \sqrt{x}$, $y = \dfrac{2\sqrt{x} + 1}{2\sqrt{x}}$

2. Find the area of the shaded region.

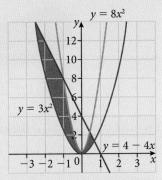

3. Find the area of the region enclosed by $y = e^x$, $x = 0$, and the tangent to $y = e^x$ at $x = 1$

4. Find the area of the inside of the 'loop' in the graph of the curve $y^2 = x^4(x + 3)$

5. Find the area enclosed by the curve $y^2 = 2x^2 - 4x^4$

6. Find the area of the region enclosed by $x = 3y^2$ and $x = 12y - y^2 - 5$

7. Find the area of the region enclosed by $y = (x - 2)^2$ and $y = x(x - 4)^2$

8. Find a value for $m > 0$ such that the area under the graph of $y = e^{2x}$ over the interval $[0, m]$ is 3 square units.

9. Find the area of the region bounded by $y = x^3 - 4x^2 + 3x$ and the x-axis.

14.7   Volumes with integrals

The underlying principle for finding the area of a plane region is to divide the region into thin strips, approximate the area of each strip by the area of a rectangle, and then add the approximations and take the limit of the sum to produce an integral for the area. The same strategy can be used to find the volume of a solid.

The idea is to divide the solid that stretches over an interval $[a, b]$ into thin slices, approximate the volume of each slice, add the approximations, and take the limit of the sum to produce an integral of the volume.

We start by taking cross-sections perpendicular to the x-axis, as shown in Figure 14.26. Each slice will be approximated by a solid whose volume will be equal to the product of its base times its height (Figure 14.27).

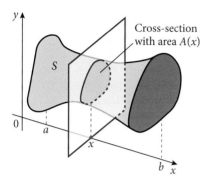

Figure 14.26 Taking a cross-section perpendicular to the x-axis

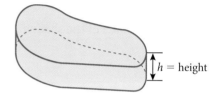

Figure 14.27 Volume = area of base × height

If we call the volume of the slice v_i and the area of its base $A(x)$, then

$$v_i = A(x_i) \cdot h = A(x_i) \cdot \Delta x_i$$

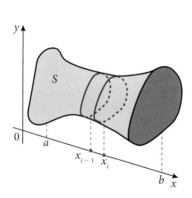

Figure 14.28 Subintervals of $[a, b]$

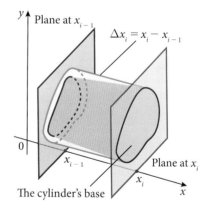

Figure 14.29 A cylindrical strip

Using this approximation, the volume of the whole solid can be found by

$$V \approx \sum_{i=1}^{n} A(x_i) \triangle x_i$$

Taking the limit as n increases and the widths of the subintervals approach zero yields the definite integral

$$V = \lim_{n \to \infty} \sum_{i=1}^{n} A(x_i) \triangle x_i = \int_{a}^{b} A(x) \, dx$$

If we place the solid along the y-axis and take the cross-sections perpendicular to that axis, we will arrive at a similar expression for the volume of the solid:

$$V = \lim_{n \to \infty} \sum_{i=1}^{n} A(y_i) \triangle y_i = \int_{a}^{b} A(y) \, dy$$

Example 14.31

Consider the solid formed when the graph of the parabola $y = \sqrt{2x}$ over $[0, 4]$ is rotated around the x-axis through an angle of 2π radians as shown in the diagrams.

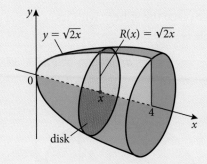

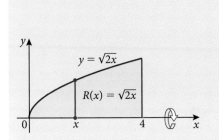

Solution

The cross-section is a circular disk whose radius is $y = \sqrt{2x}$. Therefore

$$A(x) = \pi R^2 = \pi(\sqrt{2x})^2 = 2\pi x$$

The volume is then

$$V = \int_0^4 A(x)\, dx = \int_0^4 2\pi x\, dx = 2\pi\left]\frac{x^2}{2}\right]_0^4 = 16\pi \text{ cubic units.}$$

If the region bounded by a closed interval $[a, b]$ on the x-axis and a function $f(x)$ is rotated about the x-axis, the volume of the resulting **solid of revolution** is given by:

$$V = \int_a^b \pi(f(x))^2\, dx$$

If the region bounded by a closed interval $[c, d]$ on the y-axis and a function $g(y)$ is rotated about the y-axis, the volume of the resulting solid of revolution is given by:

$$V = \int_c^d \pi(g(y))^2\, dy$$

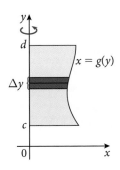 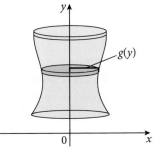

Example 14.32 is a special case of the general process for finding volumes of solids of revolution.

741

Example 14.32

Find the volume of a sphere with radius $R = a$.

Solution

If we place the sphere with its centre at the origin, then the equation of the circle is

$$x^2 + y^2 = a^2 \Rightarrow y = \pm \sqrt{a^2 - x^2}$$

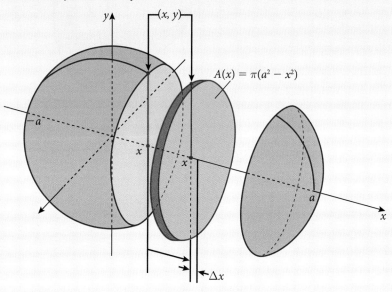

The cross-section of the sphere, perpendicular to the x-axis, is a circular disk with radius y; so the area is

$$A(x) = \pi R^2 = \pi y^2 = \pi \left(\sqrt{a^2 - x^2}\right)^2 = \pi(a^2 - x^2)$$

So, the volume of the sphere is

$$V = \int_{-a}^{a} \pi(a^2 - x^2)\,dx = \pi\left[a^2 x - \frac{x^3}{3}\right]_{-a}^{a}$$

$$= \pi\left(a^3 - \frac{a^3}{3}\right) - \pi\left(-a^3 + \frac{a^3}{3}\right) = \pi\left(2a^3 - 2\frac{a^3}{3}\right) = \frac{4\pi a^3}{3}$$

If we want to rotate the right-hand region of the circle around the y-axis, then the cross-section of the sphere, perpendicular to the y-axis, is a circular disk with radius x. Solving the equation for x instead:

$$x^2 + y^2 = a^2 \Rightarrow x = \pm \sqrt{a^2 - y^2}, \text{ and hence the area is}$$

$A(y) = \pi R^2 = \pi x^2 = \pi \left(\sqrt{a^2 - y^2}\right)^2 = \pi(a^2 - y^2)$, and the volume of the sphere is

$$V = \int_{-a}^{a} \pi(a^2 - y^2)\,dy = \pi\left[a^2 y - \frac{y^3}{3}\right]_{-a}^{a} = \pi\left(a^3 - \frac{a^3}{3}\right) - \pi\left(-a^3 + \frac{a^3}{3}\right)$$

$$= \pi\left(2a^3 - 2\frac{a^3}{3}\right) = \frac{4\pi a^3}{3}$$

The same result as given in Example 14.32

Example 14.33

Find the volume of the solid generated when the region enclosed by $y = \sqrt{3x}$, $x = 3$, and $y = 0$ is

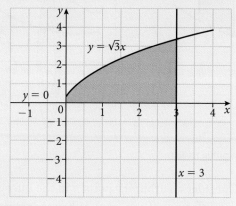

Solution

$$V = \int_0^3 \pi \left(f(x)\right)^2 dx$$

$$= \pi \int_0^3 (\sqrt{3x})^2 \, dx$$

$$= 3\pi \left[\frac{x^2}{2}\right]_0^3 = \frac{27\pi}{2}$$

Example 14.34

Find the volume of the solid generated when the region enclosed by $y = \sqrt{3x}$, $y = 3$, and $x = 0$ is revolved about the y-axis.

Solution

Here, we first find x as a function of y.

$y = \sqrt{3x} \Rightarrow x = \frac{y^2}{3}$, the interval on the y-axis is $[0, 3]$

So, the volume required is

$$V = \int_0^3 \pi \left(\frac{y^2}{3}\right)^2 dy = \frac{\pi}{9} \int_0^3 y^4 \, dy = \frac{\pi}{9}\left[\frac{y^5}{5}\right]_0^3 = \frac{27\pi}{5}$$

Washers

Consider the region R between two curves, $y = f(x)$ and $y = g(x)$, from $x = a$ to $x = b$ where $f(x) > g(x)$. Rotating R about the x-axis generates a solid of revolution S. How do we find the volume of S?

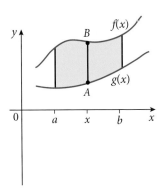 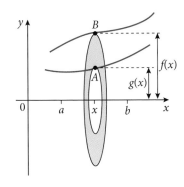 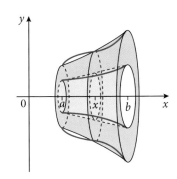

Figure 14.30 Generating washers

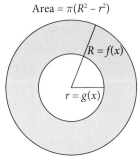

Area = $\pi(R^2 - r^2)$

Figure 14.31 Area of a typical washer

Consider an arbitrary point x in the interval $[a, b]$. The segment AB represents the difference $f(x) - g(x)$. When we rotate this slice, the cross-section perpendicular to the x-axis is going to look like a washer whose area is:

$$A = \pi(R^2 - r^2) = \pi\big((f(x))^2 - (g(x))^2\big)$$

So, the volume of S is

$$V = \int_a^b A(x)\,dx = \pi \int_a^b \big((f(x))^2 - (g(x))^2\big)\,dx$$

If we are rotating about the y-axis, a similar formula applies.

$$V = \pi \int_c^d \big((p(y))^2 - (q(y))^2\big)\,dy$$

To understand the washer more, we can think of it in the following manner. Let P be the solid generated by rotating the curve $y = f(x)$ and Q be the solid generated by rotating the curve $y = g(x)$. Then S can be found by removing the solid of revolution generated by $y = f(x)$ from the solid of revolution generated by $y = g(x)$

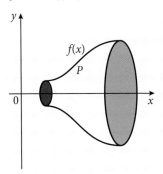 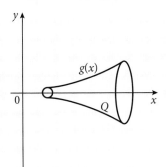 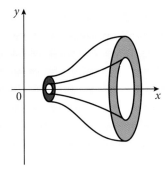

Volume of S = volume of P − volume of Q, which justifies the formula:

$$V = \pi \int_a^b (f(x))^2\,dx - \pi \int_a^b (g(x))^2\,dx = \pi \int_a^b \big((f(x))^2 - (g(x))^2\big)\,dx$$

Example 14.35

The region in the first quadrant between $f(x) = 6 - x^2$ and $h(x) = \dfrac{8}{x^2}$ is rotated about the x-axis. Find the volume of the generated solid.

Solution

The rotated region is shown in the diagram. $f(x)$ is larger than $h(x)$ in this interval. Moreover, the two curves intersect at:

$$\frac{8}{x^2} = 6 - x^2 \Rightarrow x = \sqrt{2}, x = 2$$

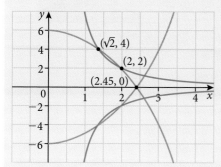

Hence the volume of the solid of revolution is:

$$V = \pi \int_{\sqrt{2}}^{2} \left((6 - x^2)^2 - \left(\frac{8}{x^2}\right)^2 \right) dx$$

$$= \pi \int_{\sqrt{2}}^{2} \left(x^4 - 12x^2 + 36 - \frac{64}{x^4} \right) dx$$

$$= \pi \left[\frac{x^5}{5} - 4x^3 + 36x + \frac{64}{3x^3} \right]_{\sqrt{2}}^{2}$$

$$= \frac{736 - 512\sqrt{2}}{15} \pi$$

An alternative method: volumes by cylindrical shells

Consider the region R under the curve $y = f(x)$. Rotate R about the y-axis. We divide R into vertical strips, each of width $\triangle x$ (Figure 14.32). When we rotate a strip around the y-axis, we generate a cylindrical shell of thickness $\triangle x$ and height $f(x)$ (Figure 14.33). To understand how we get the volume, we cut the shell vertically as shown and unfold it. The resulting rectangular parallelepiped has length $2\pi x$, height $f(x)$, and thickness $\triangle x$ (Figure 14.34).

So, the volume of this shell is

$$\triangle v_i = length \times height \times thickness = (2\pi x) \times f(x) \times \triangle x$$

The volume of the whole solid is the sum of the volumes of these shells as the number of shells increases, and consequently

$$V = \lim_{n \to \infty} \sum_{i=1}^{n} \triangle v_i = \lim_{\triangle x \to 0} \sum (2\pi x) \times f(x) \times \triangle x = 2\pi \int_a^b x f(x) dx$$

In many problems involving rotation about the y-axis, this would be more accessible than the disk-washer method.

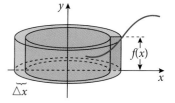

Figure 14.32 Divide R into vertical strips of width $\triangle x$

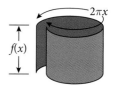

Figure 14.33 Cylindrical shell

Figure 14.34 Resulting rectangular parallelepiped

Example 14.36

Find the volume of the solid generated when we rotate the region under

$$f(x) = \frac{2}{1+x^2}, x = 0, \text{ and } x = 3 \text{ about}$$

the y-axis.

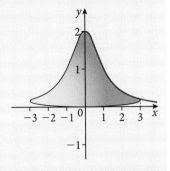

Solution

Using the shell method, we have

$$V = 2\pi \int_0^3 x \times \frac{2}{1+x^2} dx$$

$$= 2\pi \int_0^3 \frac{2x}{1+x^2} dx = 2\pi \int_0^{10} \frac{du}{u}$$

$$= 2\pi [\ln u]_1^{10} = 2\pi \ln 10$$

Exercise 14.7

1. Find the volume of the solid obtained by rotating the region bounded by the given curves about the x-axis. Sketch the region, the solid, and a typical disk.

   (a) $y = 3 - \dfrac{x}{3}, y = 0, x = 2, x = 3$

   (b) $y = 2 - x^2, y = 0$

   (c) $y = \sqrt{16 - x^2}, y = 0, x = 1, x = 3$

   (d) $y = \dfrac{3}{x}, y = 0, x = 1, x = 3$

   (e) $y = 3 - x, y = 0, x = 0$

   (f) $y = \sqrt{\sin x}, y = 0, 0 \leqslant x \leqslant \pi$

   (g) $y = \sqrt{\cos x}, y = 0, -\dfrac{\pi}{2} \leqslant x \leqslant \dfrac{\pi}{3}$

   (h) $y = 4 - x^2, y = 0$

   (i) $y = x^3 + 2x + 1, y = 0, x = 1$

   (j) $y = -4x - x^2, y = x^2$

   (k) $y = \sec x, x = \dfrac{\pi}{4}, x = \dfrac{\pi}{3}, = 0$

   (l) $y = 1 - x^2, y = x^3 + 1$

(m) $y = \sqrt{36 - x^2}, y = 4$

(n) $x = \sqrt{y}, y = 2x$

(o) $y = \sin x, y = \cos x, x = \dfrac{\pi}{4}, x = \dfrac{\pi}{2}$

(p) $y = 2x^2 + 4, y = x, x = 1, x = 3$

(q) $y = \sqrt{x^4 + 1}, y = 0, x = 1, x = 3$

(r) $y = 16 - x, y = 3x + 12, x = -1$

(s) $y = \dfrac{1}{x}, y = \dfrac{5}{2} - x$

2. Find the volume resulting from a rotation of the region shown in the diagram about:

   (a) the x-axis

   (b) the y-axis.

3. Find the volume of the solid obtained by rotating the region bounded by the given curves about the y-axis. Sketch the region, the solid, and a typical disk/shell.

   (a) $y = x^2, y = 0, x = 1, x = 3$

   (b) $y = x, y = \sqrt{9 - x^2}, x = 0$

   (c) $y = x^3 - 4x^2 + 4x, y = 0$

   (d) $y = \sqrt{3x}, x = 5, x = 11, y = 0$

   (e) $y = x^2, y = \dfrac{2}{1 + x^2}$

   (f) $y = \sqrt{x^2 + 2}, x = 3, y = 0, x = 0$

   (g) $y = \dfrac{7x}{\sqrt{x^3 + 7}}, x = 3, y = 0$

   (h) $y = \sin x, y = \cos x, x = \dfrac{\pi}{4}, x = \dfrac{\pi}{2}$

   (i) $y = 2x^2 + 4, y = x, x = 1, x = 3$

   (j) $y = \sin(x^2), y = 0, x = 0, x = \sqrt{\pi}$

   (k) $y = 5 - x^3, y = 5 - 4x$

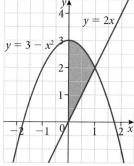

Figure 14.35 Diagram for question 2

747

14.8 Modelling linear motion

So far, our mathematical models considered the motion of an object only along a straight line. For example, projectile motion (e.g. a ball being thrown) is often modelled by a position function that simply gives the height (displacement) of the object. In this way, we are modelling the motion as if it was restricted to a vertical line.

In this section, we will again analyse the motion of an object as if its motion takes place along a straight line in space. This makes sense only if the mass (and thus, size) of the object is not taken into account. Hence, the object is modelled by a particle whose mass is considered to be zero. This study of motion, without reference either to the forces that cause it or to the mass of the object, is known as **kinematics**.

Displacement and total distance travelled

Recall from Chapter 13 that given time t, displacement s, velocity v, and acceleration a, we have:

$$v = \frac{ds}{dt} \text{ and } a = \frac{dv}{dt}, \text{ also } a = \frac{d}{dt}\left(\frac{ds}{dt}\right) = \frac{d^2s}{dt^2}$$

It is important to understand the difference between displacement and distance travelled. Consider a couple of simple examples of an object moving along the x-axis.

Assume that the object does not change direction during the interval $0 \leqslant t \leqslant 5$. If the position of the object at $t = 0$ is $x = 2$, and at $t = 5$ its position is $x = -3$, then its displacement, or change in position, is -5 because the object changed its position by 5 units in the negative x-direction. This can be calculated by: (final position) − (initial position) = $-3 - 2 = -5$

However, the distance travelled would be the absolute value of displacement, calculated by |final position − initial position| = $|-3 - 2| = 5$.

Assume that another object's initial and final positions are the same as in the first example; that is, at $t = 0$ its position is $x = 2$, and at $t = 5$ its position is $x = -3$. However, the object changed direction in that it first travelled to the left (negative velocity) from $x = 2$ to $x = -5$ during the interval $0 \leqslant t \leqslant 3$, and then travelled to the right (positive velocity) from $x = -5$ to $x = -3$. The object's displacement is -5, the same as in the first example because its net change in position is just the difference between its final and initial positions. However, it's clear that the object has travelled further than in the first example. But, we cannot calculate it the same way as we did in the first example. We will have to make a separate calculation for each interval where the direction changed. Hence, total distance travelled = $|-5 - 2| + |-3 - (-5)| = 7 + 2 = 9$.

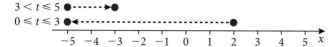

Figure 14.36 Travelled distances

♦ The **velocity** $v = \dfrac{ds}{dt}$ of a particle is a measure of how fast it is moving and of its direction of motion relative to a fixed point.

♦ The **speed** $|v|$ of a particle is a measure of how fast it is moving that does not indicate direction. Thus, speed is the magnitude of velocity and is always positive.

♦ The **acceleration** $a = \dfrac{dv}{dt}$ of a particle is a measure of how fast its velocity is changing.

Example 14.37

The displacement s of a particle on the x-axis, relative to the origin, is given by the position function $s(t) = -t^2 + 6t$ where s in centimetres and t is in seconds.

(a) Find a function for the particle's velocity $v(t)$ in terms of t. Graph the functions $s(t)$ and $v(t)$ on separate axes.

(b) Find the particle's position at the following times: $t = 0, 1, 3,$ and 6 seconds

(c) Find the particle's displacement for the following intervals: $0 \leqslant t \leqslant 1$, $1 \leqslant t \leqslant 3, 3 \leqslant t \leqslant 6,$ and $0 \leqslant t \leqslant 6$

(b) Find the particle's total distance travelled for the following intervals: $0 \leqslant t \leqslant 1, 1 \leqslant t \leqslant 3, 3 \leqslant t \leqslant 6,$ and $0 \leqslant t \leqslant 6$

Solution

(a) $v(t) = \dfrac{d}{dt}(-t^2 + 6t) = -2t + 6$

Position function: $s(t) = -t^2 + 6t$

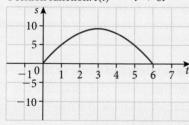

Velocity function: $v(t) = s'(t) = -2t + 6$

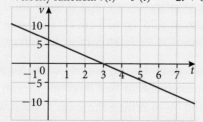

(b) The particle's position at:

$t = 0$ is $s(0) = -(0)^2 + 6(0) = 0$ cm

$t = 1$ is $s(1) = -(1)^2 + 6(1) = 5$ cm

$t = 3$ is $s(3) = -(3)^2 + 6(3) = 9$ cm

$t = 6$ is $s(6) = -(6)^2 + 6(6) = 0$ cm

(c) The particle's displacement for the interval:

$$0 \leqslant t \leqslant 1: \triangle \text{ position} = s(1) - s(0) = 5 - 0 = 5 \text{ cm}$$

$$1 \leqslant t \leqslant 3: \triangle \text{ position} = s(3) - s(1) = 9 - 5 = 4 \text{ cm}$$

$$3 \leqslant t \leqslant 6: \triangle \text{ position} = s(6) - s(3) = 0 - 9 = -9 \text{ cm}$$

$$0 \leqslant t \leqslant 6: \triangle \text{ position} = s(6) - s(0) = 0 - 0 = 0 \text{ cm}$$

This last result makes sense considering the particle moved to the right 9 cm then at $t = 3$ it turned around and moved to the left 9 cm, ending where it started – thus, no change in net position.

(d) The particle's total distance travelled for the interval:

$$0 \leqslant t \leqslant 1 \text{ is } |s(1) - s(0)| = |5 - 0| = 5$$

$$1 \leqslant t \leqslant 3 \text{ is } |s(3) - s(1)| = |9 - 5| = 4$$

$$3 \leqslant t \leqslant 6 \text{ is } |s(6) - s(3)| = |0 - 9| = |-9| = 9$$

The object's motion changed direction (velocity= 0) at $t = 3$

$$0 \leqslant t \leqslant 6 \text{ is } |s(3) - s(0)| + |s(6) - s(3)| = |9 - 0| + |0 - 9|$$
$$= 9 + 9 = 18$$

Since differentiation of the position function gives the velocity function $\left(\text{i.e. } v = \dfrac{ds}{dt}\right)$, we expect that the inverse of differentiation (integration) will lead us in the reverse direction – that is, from velocity to position. When velocity is constant, we can find the displacement with the formula:

$$\text{displacement} = \text{velocity} \times \text{change in time}$$

If we drove a car at a constant velocity of 50 km h^{-1} for 3 hours, then our displacement (same as distance travelled in this case) is 150 km. If a particle travelled to the left on the x-axis at a constant rate of -4 units s^{-1} for 5 seconds, then the particle's displacement is -20 units.

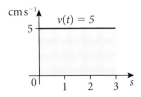

Figure 14.37 Velocity-time graph

The velocity–time graph (Figure 14.37) depicts an object's motion with a constant velocity of 5 cm s^{-1} for $0 \leqslant t \leqslant 3$. Clearly, the object's displacement is $5 \text{ cm s}^{-1} \times 3 \text{ s} = 15 \text{ cm}$ for this interval.

The area under the velocity curve for a certain interval is equal to the displacement for that interval. We can argue that just as the total area can be found by summing the areas of narrow rectangular strips, the displacement can be found by summing small displacements $(v \cdot \triangle t)$. Consider:

$$\text{displacement} = \text{velocity} \times \text{change in time} \Rightarrow s = v \cdot \triangle t \Rightarrow s = v \cdot dt$$

We already know that when $f(x) \geqslant 0$, the definite integral $\displaystyle\int_a^b f(x)\ dx$ gives the area between $y = f(x)$ and the x-axis from $x = a$ to $x = b$. And if $f(x) \leqslant 0$, then $\displaystyle\int_a^b f(x)\ dx$ gives a number that is the opposite of the area between $y = f(x)$ and the x-axis from a to b.

Given that $v(t)$ is the velocity function for a particle moving along a line, then:

$\int_a^b v(t)\, dt$ gives the displacement from $t = a$ to $t = b$

$\int_a^b |v(t)|\, dt$ gives the total distance travelled from $t = a$ to $t = b$

Let's apply integration to find the displacement and distance travelled for the two intervals $3 \leqslant t \leqslant 6$ and $0 \leqslant t \leqslant 6$ in Example 14.37

For $3 \leqslant t \leqslant 6$:

$$\text{Displacement} = \int_3^6 (-2t + 6)\, dt = -t^2 + 6t \Big|_3^6 = 0 - 9 = -9$$

$$\text{Distance travelled} = \int_3^6 |(-2t + 6)|\, dt = |-t^2 + 6t|\Big|_3^6 = |0 - 9| = 9$$

For $0 \leqslant t \leqslant 6$:

$$\text{Displacement} = \int_0^6 (-2t + 6)\, dt = -t^2 + 6t \Big|_0^6 = 0$$

$$\begin{aligned}
\text{Distance travelled} &= \int_0^3 |(-2t + 6)|\, dt + \int_3^6 |(-2t + 6)|\, dt \\
&\qquad \text{(particle changed direction at } t = 3) \\
&= |-t^2 + 6t|\Big|_0^3 + |-t^2 + 6t|\Big|_3^6 = 9 + 9 = 18
\end{aligned}$$

Example 14.38

The function $v(t) = \sin(\pi t)$ gives the velocity in m s^{-1} of a particle moving along the x-axis.

(a) Determine when the particle is moving to the right, to the left, and stopped. If it stops, determine if it changes direction at that time.

(b) Find the particle's displacement for the time interval $0 \leqslant t \leqslant 3$.

(c) Find the particle's total distance travelled for the time interval $0 \leqslant t \leqslant 3$.

Note when using a GDC or your computer, you do not need to separate the integrals as we did here.

Solution

(a) $v(t) = \sin(\pi t) = 0 \Rightarrow \sin(k \cdot \pi) = 0$ for $k \in \mathbb{Z} \Rightarrow \pi t = k\pi \Rightarrow t = k, k \in \mathbb{Z}$

for $0 \leqslant t \leqslant 3$, $t = 0, 1, 2, 3$. Therefore, the particle is stopped at $t = 0, 1, 2, 3$.

Since $t = 0$ and $t = 3$ are endpoints of the interval, the particle can change direction only at $t = 1$ or $t = 2$.

$v\left(\dfrac{1}{2}\right) = \sin\left(\pi \cdot \dfrac{1}{2}\right) = 1;\ v\left(\dfrac{3}{2}\right) = \sin\left(\pi \cdot \dfrac{3}{2}\right) = -1$

$\Rightarrow$ direction changes at $t = 1$

$v\left(\dfrac{3}{2}\right) = \sin\left(\pi \cdot \dfrac{3}{2}\right) = -1;\ v\left(\dfrac{5}{2}\right) = \sin\left(\pi \cdot \dfrac{5}{2}\right) = 1$

$\Rightarrow$ direction changes again at $t = 2$

(b) displacement $= \int_0^3 \sin(\pi t)\, dt = -\frac{1}{\pi}\cos(\pi t)\Big|_0^3$

$$= -\frac{1}{\pi}\cos(3\pi) - \left(-\frac{1}{\pi}\cos(0)\right) = \frac{2}{\pi} \approx 0.637 \text{ metres}$$

(c) total distance travelled $= \int_0^1 |\sin(\pi t)|\, dt + \int_1^2 |\sin(\pi t)|\, dt + \int_2^3 |\sin(\pi t)|\, dt$

$$= \left|\frac{2}{\pi}\right| + \left|-\frac{2}{\pi}\right| + \left|\frac{2}{\pi}\right| = \frac{6}{\pi} \approx 1.91 \text{ metres}$$

Note that in Example 14.39, the position function is not known precisely. The position function can be obtained by finding the antiderivative of the velocity function.

$$s(t) = \int v(t)\, dt = \int \sin(\pi t)\, dt = -\frac{1}{\pi}\cos(\pi t) + C$$

We can determine the constant of integration c only if we know the particle's initial position (or position at any other specific time). However, the particle's initial position will not affect displacement or distance travelled for any interval.

Position and velocity from acceleration

If we can obtain position from velocity by applying integration, then we can also obtain velocity from acceleration by integrating. Consider the next example.

Example 14.39

The motion of a falling parachutist is modelled as linear motion by considering that the parachutist is a particle moving along a line whose positive direction is vertically downwards. The parachute is opened at $t = 0$, at which time the parachutist's position is $s = 0$. According to the model, the acceleration function for the parachutist's motion for $t > 0$ is given by:

$$a(t) = -54e^{-1.5t}$$

(a) At the moment the parachute opens, the parachutist has a velocity of 42 m s^{-1}. Find the velocity function of the parachutist for $t > 0$. What does the model say about the parachutist's velocity as $t \to \infty$?

(b) Find the position function of the parachutist for $t > 0$.

Solution

(a) $v(t) = \int a(t)\, dt = \int (-54e^{-1.5t})\, dt$

$$= -54\left(\frac{1}{-1.5}\right)e^{-1.5t} + C$$

$$= 36e^{-1.5t} + C$$

Since $v = 42$ when $t = 0$, then $42 = 36e^0 + C \Rightarrow 42 = 36 + C \Rightarrow C = 6$

Therefore, after the parachute opens ($t > 0$) the velocity function is
$v(t) = 36e^{-1.5t} + 6$

Since $\lim\limits_{t \to \infty} e^{-1.5t} = \lim\limits_{t \to \infty} \dfrac{1}{e^{1.5t}} = 0$, then as $t \to 0$, $\lim\limits_{t \to \infty} v(t) = 6\,\text{m\,s}^{-1}$

(b) $s(t) = \displaystyle\int v(t)\,dt = \int (36e^{-1.5t} + 6)\,dt$

$$= 36\left(\frac{1}{-1.5}\right)e^{-1.5t} + 6t + C$$

$$= -24e^{-1.5t} + 6t + C$$

Since $s = 0$ when $t = 0$, then $0 = -24e^0 + 6(0) + C \Rightarrow 0 = -24 + C$
$\Rightarrow C = 24$

Therefore, after the parachute opens ($t > 0$), the position function is
$s(t) = -24e^{-1.5t} + 6t + 24$

The limit of the velocity as $t \to \infty$, for a falling object, is called the **terminal velocity** of the object. While the limit $t \to \infty$ is never attained as the parachutist eventually lands on the ground, the velocity gets close to the terminal velocity very quickly. For example, after just 8 seconds, the velocity is
$v(8) = 36e^{-1.5(8)} + 6$
$\approx 6.0002\,\text{m\,s}^{-1}$

Uniformly accelerated motion

Motion under the effect of gravity in the vicinity of Earth (or other planets) is an important case of rectilinear motion. This is called **uniformly accelerated motion**.

If a particle moves with constant acceleration along the s-axis, and if we know the initial speed and position of the particle, then it is possible to have specific formulas for the position and speed at any time t.

Assume acceleration is constant; that is, $a(t) = a$, $v(0) = v_0$ and $s(0) = s_0$.

$v(t) = \displaystyle\int a(t)dt = at + c$; however, we know that $v(0) = v_0$, so

$v(0) = v_0 = a \times 0 + c \Rightarrow c = v_0$, hence $v(t) = at + v_0$

$s(t) = \displaystyle\int v(t)dt = \int (at + v_0)dt = \frac{1}{2}at^2 + v_0 t + c$, and, as above, substituting $s(0) = s_0$ into the equation, we have

$s(t) = \dfrac{1}{2}at^2 + v_0 t + s_0$

When this is applied to the free-fall model (s-axis vertical), then

$v(t) = -gt + v_0$ and

$s(t) = -\dfrac{1}{2}gt^2 + v_0 t + s_0$, where $g = 9.8\,\text{m\,s}^{-2}$

Example 14.40

A ball is hit directly upwards from a point 2 m above the ground with initial velocity of 45 m s^{-1}. How high will the ball travel?

Solution

$$v(t) = -9.8t + 45$$

$$s(t) = -\frac{1}{2}(9.8)\,t^2 + 45t + 2 = -4.9t^2 + 45t + 2$$

The ball will rise until $v(t) = 0, \Rightarrow 0 = -9.8t + 45, \Rightarrow t \approx 4.6\,\text{s}$

At this time

$$s(4.6) = -4.9(4.6)^2 + 45(4.6) + 2 \approx 105.32\,\text{m}$$

Exercise 14.8

1. The velocity of a particle along a rectilinear path is given by each equation for $v(t)$ in m s^{-1}. Find both the net distance and the total distance it travels between the times $t = a$ and $t = b$.

   (a) $v(t) = t^2 - 11t + 24$, $a = 0$, $b = 10$

   (b) $v(t) = t - \dfrac{1}{t^2}$, $a = 0.1$, $b = 1$

   (c) $v(t) = \sin 2t$, $a = 0$, $b = \dfrac{\pi}{2}$

   (d) $v(t) = \sin t + \cos t$, $a = 0$, $b = \pi$

   (e) $v(t) = t^3 - 8\,t^2 + 15t$, $a = 0$, $b = 6$

   (f) $v(t) = \sin\left(\dfrac{\pi t}{2}\right) + \cos\left(\dfrac{\pi t}{2}\right)$, $a = 0$, $b = 1$

2. The acceleration of a particle along a rectilinear path is given by each equation for $a(t)$ in m s^{-2} and the initial velocity v_0 in m s^{-1} is also given. Find the velocity of the particle as a function of t, and both the net distance and the total distance travelled between times $t = a$ and $t = b$.

   (a) $a(t) = 3$, $v_0 = 0$, $a = 0$, $b = 2$

   (b) $a(t) = 2t - 4$, $v_0 = 3$, $a = 0$, $b = 3$

   (c) $a(t) = \sin t$, $v_0 = 0$, $a = 0$, $b = \dfrac{3\pi}{2}$

   (d) $a(t) = \dfrac{-1}{\sqrt{t + 1}}$, $v_0 = 2$, $a = 0$, $b = 4$

   (e) $a(t) = 6t - \dfrac{1}{(t + 1)^3}$, $v_0 = 2$, $a = 0$, $b = 2$

3. The velocity and initial position of an object moving along a coordinate line are given. Find the position of the object at time t.

   (a) $v = 9.8t + 5$, $s(0) = 10$

   (b) $v = 32t - 2$, $s(0.5) = 4$

   (c) $v = \sin \pi t$, $s(0) = 0$

   (d) $v = \dfrac{1}{t + 2}$, $t > -2$, $s(-1) = \dfrac{1}{2}$

4. The acceleration is given as well as the initial velocity and initial position of an object moving on a coordinate line. Find the position of the object at time t.

(a) $a = e^t$, $v(0) = 20$, $s(0) = 5$

(b) $a = 9.8$, $v(0) = -3$, $s(0) = 0$

(c) $a = -4\sin 2t$, $v(0) = 2$, $s(0) = -3$

(d) $a = \dfrac{9}{\pi^2}\cos\dfrac{3t}{\pi}$, $v(0) = 0$, $s(0) = -1$

5. An object moves with a speed of $v(t)$ m s^{-1} along the s-axis. Find the displacement and the distance travelled by the object during the given time interval.

(a) $v(t) = 2t - 4$; $0 \leqslant t \leqslant 6$

(b) $v(t) = |t - 3|$; $0 \leqslant t \leqslant 5$

(c) $v(t) = t^3 - 3t^2 + 2t$; $0 \leqslant t \leqslant 3$

(d) $v(t) = \sqrt{t} - 2$, $0 \leqslant t \leqslant 3$

6. An object moves with an acceleration $a(t)$ m s^{-2} along the s-axis. Find the displacement and the distance travelled by the object during the given time interval.

(a) $a(t) = t - 2$, $v_0 = 0$, $1 \leqslant t \leqslant 5$

(b) $a(t) = \dfrac{1}{\sqrt{5t + 1}}$, $v_0 = 2$, $0 \leqslant t \leqslant 3$

(c) $a(t) = -2$, $v_0 = 3$, $1 \leqslant t \leqslant 4$

7. The velocity of an object moving along the s-axis is $v = 9.8t - 3$

(a) Find the object's displacement between $t = 1$ and $t = 3$ given that $s(0) = 5$

(b) Find the object's displacement between $t = 1$ and $t = 3$ given that $s(0) = -2$

(c) Find the object's displacement between $t = 1$ and $t = 3$ given that $s(0) = s_0$

8. The displacement s metres of a moving object from a fixed point O at time t seconds is given by $s(t) = 50t - 10t^2 + 1000$.

(a) Find the velocity of the object in m s^{-1}.

(b) Find its maximum displacement from O.

9. A particle moves along a line so that its speed v at time t is given by

$$v(t) = \begin{cases} 5t & 0 \leqslant t < 1 \\ 6\sqrt{t} - \dfrac{1}{t} & t \geqslant 1 \end{cases}$$

where t is in seconds and v is in cm s^{-1}. Estimate the time(s) at which the particle is 4 cm from its starting position.

10. A projectile is fired vertically upwards with an initial velocity of 49 m s^{-1} from a platform 150 m high.

   (a) How long will it take the projectile to reach its maximum height?

   (b) What is the maximum height of the projectile?

   (c) How long will it take the projectile to pass its starting point on the way down?

   (d) What is the velocity of the projectile when it passes the starting point on the way down?

   (e) How long will it take the projectile to hit the ground?

   (f) What will its speed be at impact?

Chapter 10 practice questions

1. The graph in Figure 14.38 represents the function

$$f: x \mapsto p \cos x,\ p \in \mathbb{N}.$$

   Find:

   (a) the value of p

   (b) the area of the shaded region.

2. The diagram in Figure 14.39 shows part of the graph of $y = e^{\frac{x}{2}}$.

   (a) Find the coordinates of the point P, where the graph meets the y-axis.

   The shaded region between the graph and the x-axis, bounded by $x = 0$ and $x = \ln 2$, is rotated through 360° about the x-axis.

   (b) Write down an integral that represents the volume of the solid obtained.

   (c) Show that this volume is π.

3. The diagram in Figure 14.40 shows part of the graph of $y = \dfrac{1}{x}$.
   The area of the shaded region is 2 units.

   Find the exact value of a.

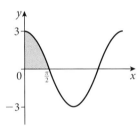

Figure 14.38 Graph for question 1

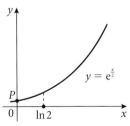

Figure 14.39 Diagram for question 2

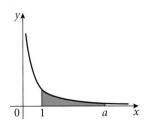

Figure 14.40 Diagram for question 3

4. (a) Find the equation of the tangent to the curve $y = \ln x$ at the point $(e, 1)$, and verify that the origin is on this line.

   (b) Show that $(x \ln x - x)' = \ln x$

   (c) The diagram shows the region enclosed by the curve $y = \ln x$, the tangent in part (a), and the line $y = 0$.

   Use the result of part (b) to show that the area of this region is $\frac{1}{2}e - 1$

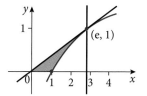

Figure 14.41 Diagram for question 4

5. The main runway at Concordville airport is 2 km long. An aeroplane, landing at Concordville touches down at point T and immediately starts to slow down. The point A is at the southern end of the runway. A marker is located at point P on the runway.

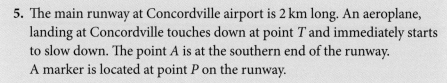

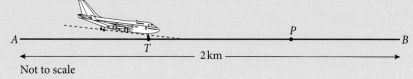

   As the aeroplane slows down, its distance, s, from A, is given by

   $$s = c + 100t - 4t^2$$

   where t is the time in seconds after touchdown, and c metres is the distance of T from A.

   (a) The aeroplane touches down 800 m from A, (i.e. $c = 800$).

   (i) Find the distance travelled by the aeroplane in the first 5 seconds after touchdown.

   (ii) Write down an expression for the velocity of the aeroplane at time t seconds after touchdown, and hence find the velocity after 5 seconds.

   The aeroplane passes the marker at P with a velocity of 36 m s^{-1}. Find:

   (iii) how many seconds after touchdown it passes the marker

   (iv) the distance from P to A.

   (b) Show that if the aeroplane touches down before reaching point P, it can stop before reaching the northern end, B, of the runway.

6. (a) Sketch the graph of $y = \pi \sin x - x$, $-3 \leqslant x \leqslant 3$, on millimetre square paper, using a scale of 2 cm per unit on each axis.

   Label and number both axes and indicate clearly the approximate positions of the x-intercepts and the local maximum and minimum points.

   (b) Find the solution of the equation $\pi \sin x - x = 0$, $x > 0$.

   (c) Find the indefinite integral $\int (\pi \sin x - x)dx$ and hence, or otherwise, calculate the area of the region enclosed by the graph, the x-axis, and the line $x = 1$.

757

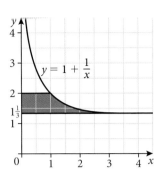

Figure 14.42 Diagram for question 7

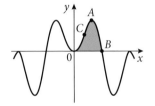

Figure 14.43 Diagram for question 9

7. Figure 14.42 shows the graph of the function $y = 1 + \dfrac{1}{x}, 0 < x \leqslant 4$.

Find the exact value of the area of the shaded region.

8. Note that radians are used throughout this question.

  (a) (i) Sketch the graph of $y = x^2 \cos x$, for $0 \leqslant x \leqslant 2$, making clear the approximate positions of the positive intercept, the maximum point, and the endpoints.

  (ii) Write down the approximate coordinates of the positive x-intercept, the maximum point and the endpoints.

  (b) Find the exact value of the positive x-intercept for $0 \leqslant x \leqslant 2$.

  Let R be the region in the first quadrant enclosed by the graph and the x-axis.

  (c) (i) Shade R on your sketch.

  (ii) Write down an integral which represents the area of R.

  (d) Evaluate the integral in part **(c) (ii)**, either by using a graphic display calculator or by using:

  $$\frac{\mathrm{d}}{\mathrm{d}x}(x^2 \sin x + 2x \cos x - 2 \sin x) = x^2 \cos x.$$

9. Note that radians are used throughout this question.
The function f is given by $f(x) = (\sin x)^2 \cos x$

Figure 14.43 shows part of the graph of $y = f(x)$.

The point A is a maximum point, the point B lies on the x-axis, and the point C is a point of inflection.

  (a) Give the period of f.

  (b) From consideration of the graph of $y = f(x)$, find the range of f, accurate to 1 significant figure.

  (c) (i) Find $f'(x)$

  (ii) Hence, show that at the point A, $\cos x = \sqrt{\dfrac{1}{3}}$

  (iii) Find the exact maximum value.

  (d) Find the exact value of the x-coordinate at the point B.

  (e) (i) Find $\int f(x)\mathrm{d}x$

  (ii) Find the area of the shaded region in the diagram.

  (f) Given that $f''(x) = 9(\cos x)^3 - 7 \cos x$, find the x-coordinate at the point C.

10. Note that radians are used throughout this question.

   (a) Draw the graph of $y = \pi + x \cos x$, $0 \leqslant x \leqslant 5$, on millimetre square graph paper, using a scale of 2 cm per unit. Make clear:

   (i) the integer values of x and y on each axis

   (ii) the approximate positions of the x-intercepts and the turning points.

   (b) Without the use of a calculator, show that π is a solution of the equation $\pi + x \cos x = 0$

   (c) Find another solution of the equation $\pi + x \cos x = 0$ for $0 \leqslant x \leqslant 5$, giving your answer to 6 significant figures.

   (d) Let R be the region enclosed by the graph and the axes for $0 \leqslant x \leqslant \pi$. Shade R on your diagram, and write down an integral which represents the area of R.

   (e) Evaluate the integral in part (d) to an accuracy of 6 significant figures. If considered necessary, you can make use of the result
   $$\frac{\mathrm{d}}{\mathrm{d}x}(x \sin x + \cos x) = x \cos x$$

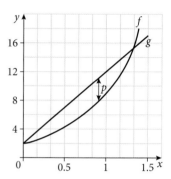

Figure 14.44 Diagram for question 11

11. Figure 14.44 shows the graphs of $f(x) = 1 + \mathrm{e}^{2x}$ and $g(x) = 10x + 2$, $0 \leqslant x \leqslant 1.5$

   (a) (i) Write down an expression for the vertical distance p between the graphs of f and g.

   (ii) Given that p has a maximum value for $0 \leqslant x \leqslant 1.5$, find the value of x at which this occurs.

   The graph of $y = f(x)$ only is shown Figure 14.45.
   When $x = a$, $y = 5$.

   (b) (i) Find $f^{-1}(x)$

   (ii) Hence, show that $a = \ln 2$

   (c) The region shaded in Figure 14.45 is rotated through 360° about the x-axis. Write down an expression for the volume obtained.

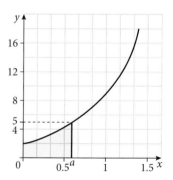

Figure 14.45 Second diagram for question 11

12. The area of the enclosed region shown in Figure 14.46 is defined by
   $$y \geqslant x^2 + 2, y \leqslant ax + 2, \text{ where } a > 0$$

   This region is rotated through 360° about the x-axis to form a solid of revolution. Find, in terms of a, the volume of this solid of revolution.

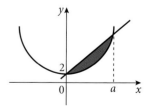

Figure 14.46 Diagram for question 12

759

13. Using the substitution $u = \frac{1}{2}x + 1$, or otherwise, find the integral

$$\int x\sqrt{\frac{1}{2}x + 1}\, dx$$

14. A particle moves along a straight line. When it is a distance s from a fixed point, where $s > 1$, the velocity v is given by $v = \frac{3s + 2}{2s - 1}$

Find the acceleration when $s = 2$.

15. The area between the graph of $y = e^x$ and the x-axis from $x = 0$ to $x = k$ $(k > 0)$ is rotated through $360°$ about the x-axis. In terms of k and e, find the volume of the solid generated.

16. Find the real number $k > 1$ for which $\displaystyle\int_1^k \left(1 + \frac{1}{x^2}\right) dx = \frac{3}{2}$

17. The acceleration, $a(t)$ m s^{-2}, of a fast train during the first 80 seconds of motion is given by

$$a(t) = -\frac{1}{20}t + 2$$

where t is the time in seconds. If the train starts from rest at $t = 0$, find the distance travelled by the train in the first minute.

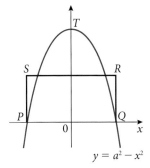

18. In Figure 14.47, PTQ is an arc of the parabola $y = a^2 - x^2$, where a is a positive constant and $PQRS$ is a rectangle. The area of rectangle $PQRS$ is equal to the area between the arc PTQ of the parabola and the x-axis.

$y = a^2 - x^2$

Figure 14.47 Diagram for question 18

Find, in terms of a, the dimensions of the rectangle.

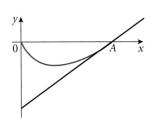

Figure 14.48 Diagram for question 19

19. Consider the function $f_k(x) = \begin{cases} x \ln x - kx & x > 0 \\ 0 & x = 0 \end{cases}$, where $k \in \mathbb{N}$

(a) Find the derivative of $f_k(x)$, $x > 0$.

(b) Find the interval over which $f(x)$ is increasing.

The graph of the function $f_k(x)$ is shown in Figure 14.48.

(c) (i) Show that the stationary point of $f_k(x)$ is at $x = e^{k-1}$.

   (ii) One x-intercept is at $(0, 0)$. Find the coordinates of the other x-intercept.

(d) Find the area enclosed by the curve and the x-axis.

(e) Find the equation of the tangent to the curve at A.

(f) Show that the area of the triangular region created by the tangent and the coordinate axes is twice the area enclosed by the curve and the x-axis.

(g) Show that the x-intercepts of $f_k(x)$ for consecutive values of k form a geometric sequence.

20. Consider the graphs of the functions $f(x) = a - |x - a|$ and $g(x) = |x - a|$, where $a > 0$. Find the value of a if the two graphs enclose an area of 12.5 square units.

21. The equation of motion of a particle with mass m subjected to a force kx can be written as $kx = mv\dfrac{\mathrm{d}v}{\mathrm{d}x}$, where x is the displacement and v is the velocity. When $x = 0$, $v = v_0$. Find v, in terms of v_0, k, and m, when $x = 2$.

22. (a) Sketch and label the graphs of $f(x) = e^{-x^2}$ and $g(x) = e^{-x^2} - 1$ for $0 \leqslant x \leqslant 1$, and shade the region A that is bounded by the graphs and the y-axis.

(b) Let the x-coordinate of the point of intersection of the curves $y = f(x)$ and $y = g(x)$ be p. Without finding the value of p, show that $\dfrac{p}{2} <$ area of region $A < p$.

(c) Find the value of p correct to 4 decimal places.

(d) Express the area of region A as a definite integral and calculate its value.

23. Let $f(x) = x \cos 3x$

(a) Use integration by parts to show that
$$\int f(x)\,\mathrm{d}x = \frac{1}{3}x \sin 3x + \frac{1}{9}\cos 3x + c$$

(b) Use your answer to part **(a)** to calculate the exact area enclosed by $f(x)$ and the x-axis in each of the following cases. Give your answers in terms of π.

(i) $\dfrac{\pi}{6} \leqslant x \leqslant \dfrac{3\pi}{6}$ **(ii)** $\dfrac{3\pi}{6} \leqslant x \leqslant \dfrac{5\pi}{6}$ **(iii)** $\dfrac{5\pi}{6} \leqslant x \leqslant \dfrac{7\pi}{6}$

(c) Given that the above areas are the first three terms of an arithmetic sequence, find an expression for the total area enclosed by $f(x)$ and the x-axis for $\dfrac{\pi}{6} \leqslant x \leqslant \dfrac{(2n + 1)\pi}{6}$, where $n \in \mathbb{Z}$. Give your answers in terms of n and π.

24. A particle is moving along a straight line so that t seconds after passing through a fixed point O on the line, its velocity $v(t)$ m s^{-1} is given by $v(t) = t \sin\left(\dfrac{\pi}{3}t\right)$.

(a) Find the values of t for which $v(t) = 0$, given that $0 \leqslant t \leqslant 6$.

(b) (i) Write down a mathematical expression for the total distance travelled by the particle in the first six seconds after passing through O.

(ii) Find this distance.

25. A particle is projected along a straight-line path. After t seconds, its velocity v in metres per second is given by $v = \dfrac{1}{2 + t^2}$

    (a) Find the distance travelled in the first second.

    (b) Find an expression for the acceleration at time t.

26. Figure 14.49 shows the shaded region R enclosed by the graph of $y = 2x\sqrt{1 + x^2}$, the x-axis, and the vertical line $x = k$.

    (a) Find $\dfrac{dy}{dx}$

    (b) Using the substitution $u = 1 + x^2$ or otherwise, show that
    $$\int 2x\sqrt{1 + x^2}\, dx = \frac{2}{3}(1 + x^2)^{\frac{3}{2}} + c$$

    (c) Given that the area of R equals 1, find the value of k.

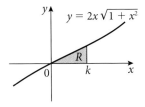

Figure 14.49 Diagram for question 26

27. A particle moves in a straight line with velocity in metres per second, at time t seconds, given by $v(t) = 6t^2 - 6t$, $t \geq 0$.

    Calculate the total distance travelled by the particle in the first two seconds of motion.

28. A particle moves in a straight line. Its velocity $v\,\text{m s}^{-1}$ after t seconds is given by $v = e^{-\sqrt{t}} \sin t$.

    Find the total distance travelled in the time interval $[0, 2\pi]$.

29. The temperature $T\,^\circ\text{C}$ of an object in a room, after t minutes, satisfies the differential equation $\dfrac{dT}{dt} = k(T - 22)$, where k is a constant.

    (a) Solve the differential equation showing that $T = Ae^{kt} + 22$, where A is a constant.

    (b) When $t = 0$, $T = 100$, and when $t = 15$, $T = 70$.

    (i) Use this information to find the values of A and k.

    (ii) Hence, find the value of t when $T = 40$.

30. Consider the function $f(x) = \dfrac{1}{x^2 + 5x + 4}$

    (a) Sketch the graph of the function, indicating the equations of the asymptotes, intercepts, and extreme values.

    (b) Find $\displaystyle\int_0^1 f(x)\, dx$ and express it in the form $\ln k$.

    (c) Sketch the graph of $f(|x|)$ and hence determine the area of the region between this graph, the x-axis, and the lines $x = -1$, and $x = 1$.

31. Use the substitution $u = x + 2$ to find $\displaystyle\int \frac{x^3\,dx}{(x+2)^2}$

32. (a) On the same axes, sketch the graphs of the functions, $f(x)$ and $g(x)$, where

$$f(x) = 4 - (1-x)^2, \text{ for } -2 \leqslant x \leqslant 4$$
$$g(x) = \ln(x+3) - 2, \text{ for } -3 \leqslant x \leqslant 5$$

(b) (i) Write down the equation of any vertical asymptotes.

(ii) State the x-intercept and y-intercept of $g(x)$.

(c) Find the values of x for which $f(x) = g(x)$.

(d) Let A be the region where $f(x) \geqslant g(x)$ and $x \geqslant 0$.

(i) On your graph, shade the region A.

(ii) Write down an integral that represents the area of A.

(iii) Evaluate this integral.

(e) In the region A, find the maximum vertical distance between $f(x)$ and $g(x)$.

33. Consider the equation $\dfrac{dy}{d\theta} = \dfrac{y}{e^{2\theta} + 1}$

(a) Use the substitution $x = e^\theta$ to show that $\displaystyle\int \frac{dy}{y} = \int \frac{dx}{x(x^2 + 1)}$

(b) Find $\displaystyle\int \frac{dx}{x(x^2 + 1)}$

(c) Hence, find y in terms of θ, if $y = \sqrt{2}$ when $\theta = 0$

34. Figure 14.50 shows part of the graph of $y = \dfrac{(\ln x)^2}{x}$, $x > 0$

(a) Find the extreme points of the curve.

(b) The region R is enclosed by the curve, the x-axis, and the line $x = e$. Find the area of the region R.

(c) Find the volume of the solid formed when the region R is rotated through 2π about the x-axis.

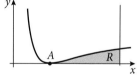

Figure 14.50 Diagram for question 34

35. (a) The functions f and g are defined by:

$$f(x) = \frac{e^x + e^{-x}}{2},\, x \in \mathbb{R} \qquad g(x) = \frac{e^x - e^{-x}}{2},\, x \in \mathbb{R}$$

(i) Show that $\dfrac{1}{4f(x) - 2g(x)} = \dfrac{e^x}{e^{2x} + 3}$

(ii) Use the substitution $u = e^x$ to find $\displaystyle\int_0^{\ln 3} \frac{1}{4f(x) - 2g(x)}\,dx$

Give your answer in the form $\dfrac{\pi\sqrt{a}}{b}$, where $a, b \in \mathbb{Z}^+$.

(b) Let $h(x) = nf(x) + g(x)$ where $n \in \mathbb{R}, n > 1$

    (i) By forming a quadratic equation in e^x, solve the equation $h(x) = k$, where $k \in \mathbb{R}^+$.

    (ii) Hence, or otherwise, show that the equation $h(x) = k$ has two real solutions, provided that $k > \sqrt{n^2 - 1}$ and $k \in \mathbb{R}^+$.

(c) Let $t(x) = \dfrac{g(x)}{f(x)}$

    (i) Show that $t'(x) = \dfrac{[f(x)]^2 - [g(x)]^2}{[f(x)]^2}$ for $x \in \mathbb{R}$.

    (ii) Hence show that $t'(x) > 0$ for $x \in \mathbb{R}$.

The normal distribution

The most important type of continuous random variable is the **normal** random variable. The probability density function of a normal random variable x is determined by two parameters: the mean or expected value μ, and the standard deviation σ of the variable.

The normal probability density function is a bell-shaped density curve that is symmetric about the mean μ. Its variability is measured by σ. The larger the value of σ, the more variability there is in the curve – that is, the higher the probability of finding values of the random variable further away from the mean. Figure 15.22 represents three different normal density functions with the same mean but different standard deviations. Note how the curves flatten as σ increases. This is because the area under the curve has to stay equal to 1.

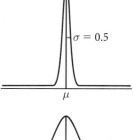
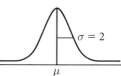
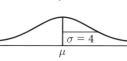

Figure 15.22 Three different normal density functions with the same mean but different standard deviations

The probability density function for a normally distributed random variable x is

$$f(x) = \frac{1}{\sigma\sqrt{2\pi}} e^{-\frac{(x-\pi)^2}{2\sigma^2}} = \frac{1}{\sigma\sqrt{2\pi}} e^{-\frac{1}{2}\left(\frac{x-\pi}{\sigma}\right)^2} \text{ for } -\infty < x < \infty$$

where μ and σ^2 are any number such that $-\infty < \mu < \infty$ and $0 < \sigma^2 < \infty$ and where e and π are the constants, e = 2.71828..... and π = 3.14159....

When a variable is normally distributed, we write:

$$X \sim N(\mu, \sigma^2)$$

Although we will not make direct use of the formula, it is interesting to note its properties because they help us understand how the normal distribution works.

The graph of a normal probability distribution is shown in Figure 15.23. The mean or expected value locates the centre of the distribution and the distribution is symmetric about this mean. Since the total area under the curve is 1, the symmetry of the curve implies that the area to the right of the mean and the area to the left are both equal to 0.5. Large values of σ tend to reduce the height of the curve and increase the spread, and small values of σ increase the height to compensate for the narrowness of the distribution.

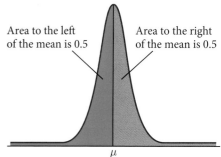

Area to the left of the mean is 0.5   Area to the right of the mean is 0.5

Figure 15.23 Normal probability distribution

So, the normal distribution is fully determined by its mean μ and its standard deviation, σ. Changing μ without changing σ moves the normal curve along the horizontal axis without changing its spread. The standard deviation σ controls the spread of the curve. You can also locate the standard deviation by eye on the curve. One σ to the right or left of the mean μ marks the point where the curvature of the curve changes. That is, as you move right from the mean, at the point where $x = \mu + \sigma$, the curve changes its curvature from downwards to upwards, and similarly as you move one σ to the left from the mean.

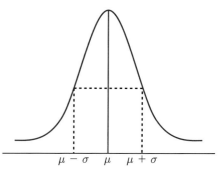

$\mu - \sigma$ μ $\mu + \sigma$

Figure 15.24 One σ to the right or left of the mean μ marks the point where the curvature of the curve changes.

803

Although there are many normal curves, they all have common properties.

Figure 15.25 illustrates this rule. Later in this section, you will learn how to find these areas from a table or from your GDC.

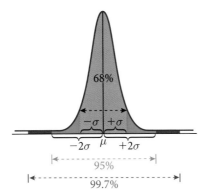

Figure 15.25 Empirical rule

Example 15.19

Heights of young German men between 18 and 19 years of age follow a distribution that is approximately normal, with mean 181 cm and a standard deviation of approximately 8 cm. Describe this population of young men.

Solution

According to the empirical rule, we find that approximately 68% of those young men have a height between 173 cm and 189 cm, 95% of them between 165 cm and 197 cm, and 99.7% between 157 cm and 205 cm. You can say that only 0.15% are taller than 205 cm or shorter than 157 cm.

As the empirical rule suggests, all normal distributions are the same if we measure in units of size σ about the mean μ as centre. Changing to these units is called **standardising**. To standardise a value, measure how far it is from the mean and express that distance in terms of σ.

The quantity $x - \mu$ tells us how far a value is from the mean; dividing by σ then tells us how many standard deviations that distance is equal to.

The standardising process is a transformation of the normal curve. For discussion purposes, assume the mean μ to be positive. The transformation $x - \mu$ shifts the graph back μ units. So, the new centre is shifted from μ back μ units. That is, the new centre is 0.

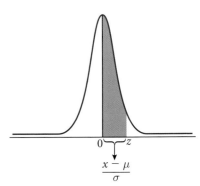

Figure 15.26 Transformation of the normal curve

Dividing by σ is going to scale the distances from the mean and express everything in terms of σ. So, a point that is one standard deviation from the mean is going to be 1 unit above the new mean; that is, it will be represented by $+1$. Now, if you look at the empirical rule discussed earlier, points that are within one standard deviation from the mean will be within a distance of 1 in the new distribution. Instead of being at $\mu + \sigma$ and $\mu - \sigma$, they will be at $0 + 1$ and $0 - 1$ respectively; that is, -1 and $+1$.

The new distribution created by this transformation is called the **standard normal distribution**. It has a mean of 0 and a standard deviation of 1. It is a very helpful distribution because it will enable us to read the areas under any normal distribution through the standardisation process.

Since linear transformations can transform all normal functions to standard, this becomes a very convenient and efficient way of finding the area under any normal distribution.

The proof that the mean and the variance of the standard normal variable are 0 and 1 respectively is straightforward.

Let $z = \dfrac{x - \mu}{\sigma}$ be the standard variable corresponding to a normal variable x.

$$E(z) = E\left(\frac{x - \mu}{\sigma}\right) = E\left(\frac{1}{\sigma}(x - \mu)\right) = \frac{1}{\sigma}E(x - \mu) = \frac{1}{\sigma}(\mu - \mu) = 0$$

$$V(z) = V\left(\frac{x - \mu}{\sigma}\right) = V\left(\frac{1}{\sigma}(x - \mu)\right) = \frac{1}{\sigma^2}V(x - \mu) = \frac{1}{\sigma^2}V(x) = \frac{1}{\sigma^2}\sigma^2 = 1$$

The probability density function for the standard normal distribution is

$$f(z) = \frac{1}{\sqrt{2\pi}}e^{-\frac{1}{2}z^2}$$

for $-\infty < z < \infty$

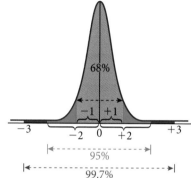

Figure 15.27 The standard normal distribution

Example 15.20

Using the data in Example 15.19, work out the z-score of a young German man with a height of

(a) 192 cm   (b) 175 cm.

Solution

(a) z-score:

$$z = \frac{x - \mu}{\sigma} = \frac{192 - 181}{8} = 1.375$$

or 1.375 standard deviations above the mean.

(b) $z = \dfrac{x - \mu}{\sigma} = \dfrac{175 - 181}{8} = -0.75$

or 0.75 standard deviations below the mean.

To find the probability that a normal variable x lies in the interval a to b, we need to find the area under the normal curve $N(\mu, \sigma^2)$ between the points a and b. However, there is an infinitely large number of normal curves – one for each mean and standard deviation. A separate table of areas for each of these curves is obviously not practical. Instead, we use one table for the standard normal distribution that gives us the required areas.

When we standardise a and b, we get two standard numbers z_1 and z_2 such that the area between z_1 and z_2 is the same as the area we need.

In this example, we are interested in the proportion of young German men whose height is between 175 cm and 192 cm.

To find the required area, we can use software or a GDC as shown in Figure 15.28.

With a GDC, we do not need to standardise our variables. However, there are cases where we need to understand standardisation in order to use it in solving some problems where the mean or the standard deviation, or both, are not given.

If we want to use the standard normal, our commands will be the same, but we do not need to include the mean and standard deviation. They are the default.

If we need the probability that a young man is taller than 175 cm, we can also read it either by looking at the distribution with the original data or by standardising.

```
normalcdf(175,19
2,181,8)
        .6888069418
```

```
normalcdf(-.75,1
.375)
        .6888069418
```

Figure 15.28 GDC output

Example 15.21

The age of graduate students in engineering programs throughout the USA is normally distributed with mean $\mu = 24.5$ and standard deviation $\sigma = 2.5$

A student is chosen at random.

(a) What is the probability that the student is younger than 26 years old?

(b) What proportion of students are older than 23.7 years?

(c) What percentage of students are between 22 and 28 years old?

(d) What percentage of the ages falls within

    (i)   1 standard deviation of the mean

    (ii) 2 standard deviations of the mean

    (iii) 3 standard deviations of the mean?

Solution

Let X = age of students, then $X \sim N(\mu = 24.5, \sigma^2 = 6.25)$

(a) We can either standardise and then read the table for the area left of 0.6 or use a GDC:

$$P\left(z < \frac{26 - 24.5}{2.5}\right) = P(z < 0.6) = 0.7257$$

Notice here that we put 0 as a lower limit. We can put a number as a lower limit far enough from the mean to make sure we are receiving the correct cumulative distribution.

```
Normal C.D
p    =0.72574688
z:Low=-9.8
z:Up =0.6
```

Figure 15.29 GDC screen for Example 15.21 (a)

(b) This can be done in a similar way:

$$P(x > 23.7) = P\left(z > \frac{23.7 - 24.5}{2.5}\right) = -0.32, \text{ so by symmetry we know}$$

$$P(z > -0.32) = P(z < 0.32) = 0.6255$$

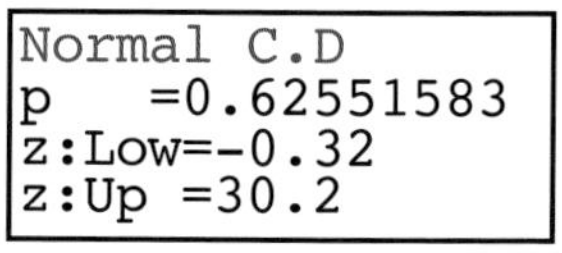

Figure 15.30 GDC screen for Example 15.21 (b)

(c) $P(22 < X < 28) = P\left(\frac{22 - 24.5}{2.5} < z < \frac{28 - 24.5}{2.5}\right) = P(-1 < z < 1.4)$

Find the area to the left of 1.4 and to the left of -1 and subtract them.

$$P(-1 < z < 1.4) = 0.9192 - 0.1587 = 0.7606 = 76.06\%$$

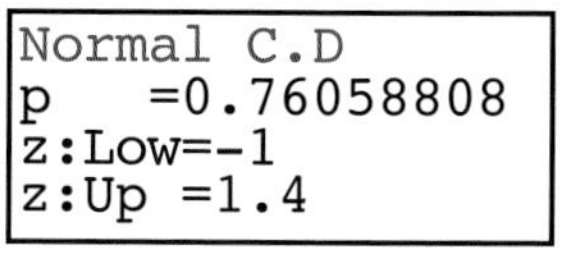

Figure 15.31 GDC screen for Example 15.21 (c)

(d) This is the empirical rule. Find what percentage of the approximately normal data will lie within 1, 2 and 3 standard deviations.

(i) $P(-1 \leqslant z \leqslant 1) = 0.6826$

(ii) $P(-2 \leqslant z \leqslant 2) = 0.9544$

(iii) $P(-3 \leqslant z \leqslant 3) = 0.9973$

These are the exact values corresponding to the empirical rule's 68%, 95% and 99.7%.

The inverse normal distribution

Another type of problem arises when we are given a cumulative probability and would like to find the value in our data that has this cumulative probability. For example, what age marks the 95th percentile? That is, what age is higher than or equal to 95% of the population? To answer this question, we need to reverse our steps. So far, we are given a value and then we look for the area corresponding to it. Now, we are given the area and we have to look for the number. That is why this is called the **inverse normal distribution**. Again, the approach is to find the standard inverse normal number and then to de-standardise it. That is, to find the value from the original data that corresponds to the z-value at hand.

For example, if we need to know what z-score the third quartile Q_3 is, we need to look up 0.75. The z-score corresponding to Q_3 is 0.6745 (Figure 15.32).

Suppose we want to find the z-score that leaves an area of 0.915 below it.

The z-score corresponding to 0.915 is 1.3722. That is

$$P(Z < 1.3722) = 0.915 \text{ (Figure 15.33)}$$

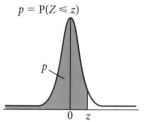

Figure 15.32 The z-score corresponding to Q_3 is 0.6745

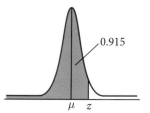

Figure 15.33
$P(Z < 1.3722) = 0.915$

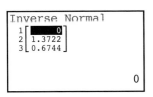

```
Inverse Normal
1[      0]
2[1.3722]
3[0.6744]

              0
```

```
Inverse Normal
  xInv=194.158829
```

Figure 15.34 Using invNorm

We could also use a GDC. The process is similar to the normal calculation, but choosing invNorm instead.

So, 95% of the young German men are shorter than 194.16 cm.

Example 15.22

The average time it takes fast trains to travel between London and Paris is 2 hours 15 minutes with a standard deviation of 4 minutes. Assume a normal distribution.

(a) What is the probability that a randomly chosen trip will take longer than 2 hours 20 minutes?

(b) What is the probability that a randomly chosen trip will take less than 2 hours 10 minutes?

(c) What is the IQR of the length of a trip?

Solution

We will do each problem using a GDC.

(a) $\mu = 2.25$ and $\sigma = 0.0667$

2 hours 20 minutes = 2.333, using our GDC. (We use z here for demonstration only. We don't need to standardise when using a GDC.)

$$P(x > 2.333) = P\left(z > \frac{2.333 - 2.25}{0.0667}\right) = P(z > 1.244) = 0.1067$$

(b) 2 hours 10 minutes = 2.167

$$P(x < 2.167) = P\left(z < \frac{2.167 - 2.25}{0.067}\right) = P(z < -1.244) = 0.106718$$

(c) To find the IQR, we need to find Q_1 and Q_3.

Q_1 is the number that leaves 25% of the data before it. Q_3 is the number that leaves 75% of the data before it. So, we need to find the inverse normal variable that has an area of 0.25 or 0.75 before it.

Using a GDC and the inverse normal, we find $Q_1 = 2.205$ and $Q_3 = 2.295$.

IQR = $2.295 - 2.205 = 0.090$ of an hour (5.4 minutes)

Example 15.23

The age at which babies develop the ability to walk can be described by a normal distribution model. It is known that 5% of the babies learn how to walk by the age of 10 months, while 25% need more than 13 months. Find the mean and standard deviation of the distribution.

Solution

There are several approaches to this problem. Here is one.

Look at the distance between 10 and 13 months in two different ways. First, 10 and 13 months are 3 months apart. When standardised, the respective z-scores are -1.645 and 0.674. The z-scores are 2.319 standard deviations apart. So, 3 months must be the same as 2.319 standard deviations.

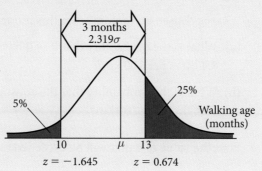

To find the z-score for the lowest 5% and the highest 25%, we can use a GDC.

Here is the calculation:

$2.319\,\sigma = 3$, $\sigma = 1.294$, and

$$z = \frac{x - \mu}{\sigma} \Rightarrow 0.674 = \frac{13 - \mu}{1.294} \Rightarrow \mu = 12.128, \text{ or alternatively}$$

$$-1.645 = \frac{10 - \mu}{\sigma} \Rightarrow \mu - 1.645\sigma = 10$$

$$0.674 = \frac{13 - \mu}{\sigma} \Rightarrow \mu + 0.674\sigma = 13$$

Solving the system of two equations in μ and σ will give the same result.

Exercise 15.4

1. The time it takes to change the batteries of your GDC is approximately normal with mean 50 hours and standard deviation of 7.5 hours.

   Find the probability that your newly equipped GDC will last

   (a) at least 50 hours   (b) between 50 and 75 hours

   (c) less than 42.5 hours   (d) between 42.5 and 57.5 hours

   (e) more than 65 hours   (f) 47.5 hours.

2. Find each of the following probabilities.

   (a) $p(|z| < 1.2)$

   (b) $p(|z| > 1.4)$

   (c) $p(X < 3.7)$, where $X \sim N(3, 3)$

   (d) $p(X > -3.7)$, where $X \sim N(3, 3)$

3. A car manufacturer introduces a new model that has a fuel consumption of 11.4 litres per 100 km in urban areas. Tests show that this model has a standard deviation of 1.26. The distribution is assumed to be normal.

   A car is chosen at random from this model.

   (a) What is the probability that it will have consumption less than 8.4 litres per 100 km?

   (b) What is the probability that the consumption is between 8.4 and 14.4 litres per 100 km?

4. Find the value of z that will be exceeded only 10% of the time.

5. Find the value of $z = z_0$ such that 95% of the values of z lie between $-z_0$ and $+z_0$.

6. The scores on a public school's examination are normally distributed with a mean of 550 and a standard deviation of 100.

   (a) What is the probability that a randomly chosen student from this population scores below 400?

   (b) What is the probability that a student will score between 450 and 650?

   (c) What score should you have in order to be in the 90th percentile?

   (d) Find the IQR of this distribution.

7. A company producing and packaging sugar for home consumption put labels on their sugar bags noting the weight to be 500 g. Their machines are known to fill the bags with weights that are normally distributed with a standard deviation of 5.7 g. A bag that contains less than 500 g is considered to be underweight.

   (a) The company decides to set their machines to fill the bags with a mean of 512 g. What fraction will be underweight?

   (b) The company wants the percentage of underweight bags to be a maximum of 4%. What should the mean be?

   (c) The company decides that they do not want to set the mean as high as 512 g, but instead at 510 g. What standard deviation gives them a maximum of 4% underweight bags?

8. In a large school, heights of students who are 13 years old are normally distributed with a mean of 151 cm and a standard deviation of 8 cm.

   Find the probability that a randomly chosen child is:

   (a) shorter than 166 cm

   (b) within 6 cm of the average.

9. The time it takes Kevin to get to school every day is normally distributed with a mean of 12 minutes and a standard deviation of 2 minutes. Estimate the number of days when Kevin takes:

 (a) longer than 17 minutes

 (b) less than 10 minutes

 (c) between 9 and 13 minutes.

There are 180 school days in Kevin's school year.

10. X has a normal distribution with mean 16. Given that the probability that X is less than 16.56 is 64%, find the standard deviation σ of this distribution.

11. X has a normal distribution with mean 91. Given that the probability that X is larger than 104 is 24.6%, find the standard deviation σ of this distribution.

12. X has a normal distribution with variance of 9. Given that the probability that X is more than 36.5 is 2.9%, find the mean μ of this distribution.

13. X has a normal distribution with standard deviation of 32. Given that the probability that X is more than 63 is 87.8%, find the mean μ of this distribution.

14. X has a normal distribution with variance of 25. Given that the probability that X is less than 27.5 is 0.312, find the mean μ of this distribution.

15. X has a normal distribution such that the probability that X is larger than 14.6 is 93.5% and $P(x > 29.6) = 2.2\%$. Find the mean μ and the standard deviation σ of this distribution.

16. $X \sim N(\mu, \sigma^2)$. $P(X > 19.6) = 0.16$ and $P(X < 17.6) = 0.012$. Find μ and σ.

17. $X \sim N(\mu, \sigma^2)$. $P(X > 162) = 0.122$ and $P(X < 56) = 0.0276$. Find μ and σ.

18. Wooden poles produced for electricity networks in rural areas have lengths that are normally distributed.

2% of the poles are rejected because they are considered too short, and 5% are rejected because they are too long.

 (a) Find the mean and standard deviation of these poles if the acceptable range is between 6.3 m and 7.5 m.

 (b) In a randomly selected sample of 20 poles, find the probability of finding 2 rejected poles.

19. Bottles of mineral water sold by a company are advertised to contain 1 litre of water. The company adjusts its filling process to fill the bottles with an average of 1012 ml to ensure that there is a minimum of 1 litre. The process follows a normal distribution with standard deviation of 5 ml.

    (a) Find the probability that a randomly chosen bottle contains more than 1010 ml.

    (b) Find the probability that a bottle contains less than the advertised volume.

    (c) In a shipment of 10 000 bottles, what is the expected number of under-filled bottles?

20. Cholesterol plays a major role in a person's heart health. High blood cholesterol is a major risk factor for coronary heart disease and stroke. The level of cholesterol in the blood is measured in milligrams per decilitre (mg/dL). According to the WHO, in general, less than 200 mg/dL is a desirable level, 200 to 239 is borderline high, and above 240 is a high risk level and the person with this level has more than twice the risk of heart disease as a person with less than 200 mg/dL.

    In a certain country, it is known that the average cholesterol level of the adult population is 184 mg/dL with a standard deviation of 22 mg/dL. It can be modelled by a normal distribution.

    (a) What percentage do you expect to be borderline high?

    (b) What percentage do you consider are high risk?

    (c) Estimate the interquartile range of the cholesterol levels in this country.

    (d) Above what value are the highest 2% of adults' cholesterol levels in this country?

21. A manufacturer of car tyres claims that its winter tyres can be described by a normal model with an average life of 52 000 km and a standard deviation of 4000 km.

    (a) If you buy a set of tyres from this manufacturer, is it reasonable for you to hope they last more than 64 000 km?

    (b) What percentage of these tyres do you expect to last less than 48 000 km?

    (c) What percentage of these tyres do you expect to last between 48 000 km and 56 000 km?

    (d) What is the IQR of the life of this type of tyre?

    (e) The company wants to guarantee a minimum life for these tires. They will refund customers whose tyres last less than a specific distance. What should their minimum life guarantee be so that they do not end up refunding more than 2% of their customers?

22. Chicken eggs are graded by size for the purpose of sales. In Europe, modern egg sizes are defined as follows: very large eggs have a mass of 73 g or more, large eggs are between 63 and 73 g, medium eggs are between 53 g and 63 g, and small eggs are less than 53 g.

   (a) Mature hens (older than 1 year) produce eggs with an average mass of 67 g. 98% of the eggs produced by mature hens are 53 g or above. What is the standard deviation if the egg production can be modelled by a normal distribution?

   (b) Young hens produce eggs with a mean of 51 g. Only 28% of their eggs exceed 53 g. What is the standard deviation?

   (c) A farmer finds that 7% of his farm's eggs are small, and 12% are very large. Estimate the mean and standard deviation of these eggs.

23. A machine produces bearings with diameters that are normally distributed with mean 3.0005 cm and standard deviation 0.0010 cm. Specifications require the bearing diameters to lie in the interval 3.000 ± 0.0020 cm. Those outside the interval are considered scrap and must be disposed of. What percentage of the production will be scrap?

24. A soft-drink machine can be regulated so that it discharges an average μ ml per bottle. The amount of fill is normally distributed with a standard deviation 9 ml.

   (a) Give the setting for μ so that 237 ml bottles will overflow only 1% of the time.

   (b) The standard deviation σ of the machine can be adjusted to the required levels when needed. What is the largest value of σ that will allow the actual amount dispensed to fall within 30 ml of the mean with probability at least 95%?

25. The speeds of cars on a main highway are approximately normal. Data collected at a certain point show that 95% of the cars travel at a speed less than $140 \, \text{km h}^{-1}$, and 10% travel at a speed less than $90 \, \text{km h}^{-1}$.

   (a) Find the average speed and the standard deviation for the cars travelling on that specific stretch of the highway.

   (b) Find the proportion of cars that travel at speeds exceeding $110 \, \text{km h}^{-1}$.

26. The random variable X is normally distributed and
$$P(X \leqslant 10) = 0.670; P(X \leqslant 12) = 0.937$$
   Find E(X).

27. A machine is set to produce bags of salt, whose weights are distributed normally, with a mean of 110 g and standard deviation of 1.142 g. If the weight of a bag of salt is less than 108 g, the bag is rejected. With these settings, 4% of the bags are rejected.

    The settings of the machine are altered, and it is found that 7% of the bags are rejected.

    (a) (i) If the mean has not changed, find the new standard deviation, correct to 3 decimal places.

    The machine is adjusted to operate with this new value of the standard deviation.

    (ii) Find the value, correct to 2 decimal places, at which the mean should be set so that only 4% of the bags are rejected.

    (b) With the new settings from part **(a)**, it is found that 80% of the bags of salt have a weight that lies between A g and B g, where A and B are symmetric about the mean. Find the values of A and B, giving your answers correct to 2 decimal places.

15.5 Expectation algebra

The expected value of a linear function of X

Example 15.24

You have a large box containing equal numbers of chips of two types.

You draw one chip and record the number. Find the expected value and variance of the number you record when the chips are numbered:

(a) 0 and 1

(b) 0 and 2

(c) 0 and 3.

Solution

(a) Since there is an equal chance of drawing 0 or 1, then

$$E(X) = \sum xp(x) = 0 \cdot \frac{1}{2} + 1 \cdot \frac{1}{2} = \frac{1}{2}$$

$$\text{Var}(X) = \sum x^2 p(x) - (E(X))^2 = 0^2 \cdot \frac{1}{2} + 1^2 \cdot \frac{1}{2} - \left(\frac{1}{2}\right)^2 = \frac{1}{4}$$

(b) Since there is an equal chance of drawing 0 or 2, then

$$E(Y) = \sum y p(y) = 0 \cdot \frac{1}{2} + 2 \cdot \frac{1}{2} = 1$$

$$\text{Var}(Y) = \sum y^2 p(y) - (E(Y))^2 = 0^2 \cdot \frac{1}{2} + 2^2 \cdot \frac{1}{2} - (1)^2 = 1$$

Note here that $Y = 2X$, and $E(Y) = 2E(X)$, while $\text{Var}(Y) = 4\text{Var}(X)$

(c) Since there is an equal chance of drawing 0 or 3, then

$$E(Y) = \sum y p(y) = 0 \cdot \frac{1}{2} + 3 \cdot \frac{1}{2} = 3 \cdot E(X)$$

$$\text{Var}(Y) = \sum y^2 p(y) - (E(Y))^2 = 0^2 \cdot \frac{1}{2} + 3^2 \cdot \frac{1}{2} - \left(\frac{3}{2}\right)^2 = \frac{9}{4}$$

$$= 9 \cdot \text{Var}(X)$$

The expected value of a linear function of X

Proof

Discrete case:

$E(aX + b) = aE(X) + b$

$$E(aX + b) = \sum (ax + b)p(x) = \sum (axp(x) + bp(x))$$

$$= \sum axp(x) + \sum bp(x) = a\sum xp(x) + b\sum p(x)$$

$$= aE(X) + b(1) = aE(X) + b$$

Continuous case:

$$E(aX + b) = \int (ax + b)p(x)dx = \int (axp(x) + bp(x))dx$$

$$= \int axp(x)dx + \int bp(x)dx = a\int xp(x)dx + b\int p(x)dx$$

$$= aE(X) + b(1) = aE(X) + b$$

We defined the variance of a random variable X as:

For the discrete case:

$$\sigma^2 = E((X - \mu)^2) = \sum (x - \mu)^2 \cdot p(x), \text{ we will call it Var}(X).$$

We also found a short-cut formula for the variance

$$\sigma^2 = \sum (x - \mu)^2 \cdot p(x) = \sum x^2 \cdot p(x) - \mu^2 = \sum x^2 \cdot p(x) - [E(X)]^2$$

For the continuous case:

$$\sigma^2 = E((X - \mu)^2) = \int (x - \mu)^2 \cdot p(x)dx, \text{ and the shortcut is}$$

$$\sigma^2 = \int (x - \mu)^2 \cdot p(x)dx = \int x^2 \cdot p(x)dx - \mu^2 = \int x^2 \cdot p(x)dx - [E(X)]^2$$

The variance of a linear function of X

$\text{Var}(aX + b) = a^2\text{Var}(X)$

Proof

For the discrete case:

Let $Y = aX + b$, which means that the random variable Y takes values $y = ax + b$ with the same probability as $p(x)$ since a and b are constants.

$$\text{Var}(aX + b) = \text{Var}(Y) - \sum(Y - E(Y))^2 p(y)$$
$$= \sum(aX + b - aE(X) - b)^2 p(x)$$
$$= \sum(a(X - E(X)))^2 p(x)$$
$$= a^2 \sum(X - E(X))^2 p(x) = a^2 \text{Var}(X)$$

The proof for the continuous case is similar to the discrete case.

Linear combinations of random variables

In this section, we present some results whose proofs go beyond the scope of the HL course and this book.

Example 15.25

You have a large box containing equal numbers of chips with the numbers 0 and 1 on them.

You draw one chip, record the number and return it to the box, then draw another chip and record the number.

(a) Find the expected value and variance of the sum of the numbers you record.

(b) Find the expected value and variance of the difference of the numbers you record.

Solution

(a) Since there is an equal chance of drawing 0 or 1, then the probability that the chip number is 0 or 1 is $\frac{1}{2}$.

The random variable $Z = X_1 + X_2$ in question is the sum of the two numbers. The values and their probabilities are summarised below.

z	sample points		$p(z)$
0	(0, 0)	$\frac{1}{2} \cdot \frac{1}{2}$	$\frac{1}{4}$
1	(1, 0), (0, 1)	$\frac{1}{2} \cdot \frac{1}{2} + \frac{1}{2} \cdot \frac{1}{2}$	$\frac{1}{2}$
2	(1, 1)	$\frac{1}{2} \cdot \frac{1}{2}$	$\frac{1}{4}$

$$E(Z) = \sum zp(z) = 0 \cdot \frac{1}{4} + 1 \cdot \frac{1}{2} + 2 \cdot \frac{1}{4} = 1$$

$$\text{Var}(Z) = \sum z^2 p(z) - (E(Z))^2 = 0^2 \cdot \frac{1}{4} + 1^2 \cdot \frac{1}{2} + 2^2 \cdot \frac{1}{4} - 1^2 = \frac{1}{2}$$

(b) Since there is an equal chance of drawing 0 or 1, then the probability that the chip number is 0 or 1 is $\frac{1}{2}$.

The random variable $Z = X_1 - X_2$ in question is the difference of the two numbers. The values and their probabilities are summarised below.

z	sample points		$p(z)$
0	$(0, 0), (1, 1)$	$\frac{1}{2} \cdot \frac{1}{2} + \frac{1}{2} \cdot \frac{1}{2}$	$\frac{1}{2}$
1	$(1, 0)$	$\frac{1}{2} \cdot \frac{1}{2}$	$\frac{1}{4}$
-1	$(0, 1)$	$\frac{1}{2} \cdot \frac{1}{2}$	$\frac{1}{4}$

$$E(Z) = \sum zp(z) = 0 \cdot \frac{1}{2} + 1 \cdot \frac{1}{4} - 1 \cdot \frac{1}{4} = 0$$

$$\text{Var}(Z) = \sum z^2 p(z) - (E(Z))^2 = 0^2 \cdot \frac{1}{2} + 1^2 \cdot \frac{1}{4} + (-1)^2 \cdot \frac{1}{4} - 0^2 = \frac{1}{2}$$

Let X and Y be any two random variables, then

1. $E(aX \pm bY) = aE(X) \pm bE(Y)$

and if the two variables are independent, then

2. $\text{Var}(aX \pm bY) = a^2 \text{Var}(X) + b^2 \text{Var}(Y)$

Note that we add the variances regardless of whether the variables are added or subtracted.

Looking back at Example 15.25, you will notice that:

$$E(X) = E(Y) = \frac{1}{2}, \text{ and}$$

$$E(X + Y) = 1 = \frac{1}{2} + \frac{1}{2} = E(X) + E(Y), \text{ and also}$$

$$E(X - Y) = 0 = \frac{1}{2} - \frac{1}{2} = E(X) - E(Y), \text{ and additionally}$$

$$\text{Var}(X) = \text{Var}(Y) = \frac{1}{4}, \text{ and}$$

$$\text{Var}(X \pm Y) = \frac{1}{2} = \frac{1}{4} + \frac{1}{4} = \text{Var}(X) + \text{Var}(Y)$$

Example 15.26 demonstrates the theorem applied to the case where $a = b = 1$.

Example 15.26

Consider the two random variables X and Y, where X is the number showing when we roll a tetrahedral die and Y is the number showing when we roll a cubical die. Work out:

(a) $E(X)$ and $V(X)$

(b) $E(Y)$ and $V(Y)$

(c) $E(X + Y)$

Solution

Here are their probability distributions:

x	1	2	3	4
$p(x)$	$\frac{1}{4}$	$\frac{1}{4}$	$\frac{1}{4}$	$\frac{1}{4}$

y	1	2	3	4	5	6
$p(y)$	$\frac{1}{6}$	$\frac{1}{6}$	$\frac{1}{6}$	$\frac{1}{6}$	$\frac{1}{6}$	$\frac{1}{6}$

$x + y$	2	3	4	5	6	7	8	9	10
	(1, 1)	(1, 2)	(1, 3)	(1, 4)	(1, 5)	(1, 6)	(2, 6)	(3, 6)	(4, 6)
		(2, 1)	(3, 1)	(4, 1)	(2, 4)	(2, 5)	(3, 5)	(4, 5)	
			(2, 2)	(2, 3)	(4, 2)	(3, 4)	(4, 4)		
				(3, 2)	(3, 3)	(4, 3)			
$p(x + y)$	$\frac{1}{4}\cdot\frac{1}{6}$	$2\left(\frac{1}{4}\cdot\frac{1}{6}\right)$	$3\left(\frac{1}{4}\cdot\frac{1}{6}\right)$	$4\left(\frac{1}{4}\cdot\frac{1}{6}\right)$	$4\left(\frac{1}{4}\cdot\frac{1}{6}\right)$	$4\left(\frac{1}{4}\cdot\frac{1}{6}\right)$	$3\left(\frac{1}{4}\cdot\frac{1}{6}\right)$	$2\left(\frac{1}{4}\cdot\frac{1}{6}\right)$	$\left(\frac{1}{4}\cdot\frac{1}{6}\right)$

This can be summarised as

$x + y$	2	3	4	5	6	7	8	9	10
$p(x + y)$	$\frac{1}{24}$	$\frac{2}{24}$	$\frac{3}{24}$	$\frac{4}{24}$	$\frac{4}{24}$	$\frac{4}{24}$	$\frac{3}{24}$	$\frac{2}{24}$	$\frac{1}{24}$

(a) $E(X) = \frac{5}{2}$, $V(X) = \frac{5}{4}$

(b) $E(Y) = \frac{7}{2}$, $V(Y) = \frac{329}{12}$

(c) $E(X + Y) = 6$

Interesting application I

If several observations of the same random variable are examined, then the results above have to be applied with great care.

1. If $X_1, \ldots, X_n$ are observations of the same random variable X, then

$$E(a_1 X_1 \pm a_2 X_2 \pm \ldots \pm a_n X_n) = a_1 E(X_1) \pm a_2 E(X_2) \pm \ldots \pm a_n E(X_n)$$

$$= a_1 E(X) \pm a_2 E(X) \pm \ldots \pm a_n E(X)$$

$$= (a_1 \pm a_2 \pm \ldots \pm a_n)E(X)$$

Special cases:

$$E(X_1 + X_2) = E(X) + E(X) = 2E(X)$$

also, $E(2X) = E(X + X) = E(X) + E(X) = 2E(X)$

which is also a special case of

$$E(aX + b) = aE(X) + b, \text{ when } a = 2 \text{ and } b = 0$$

This result can be generalised to

$$E(X_1 + X_2 + \ldots + X_n) = nE(X), \text{ and}$$

$$E(nX) = nE(X)$$

2. If $X_1, \ldots, X_n$ are independent observations of the same random variable X, then
$$\text{Var}(a_1 X_1 \pm a_2 X_2 \pm \ldots \pm a_n X_n) = a_1^2 \text{Var}(X) + a_2^2 \text{Var}(X) + \ldots + a_n^2 \text{Var}(X)$$
$$= (a_1^2 + a_2^2 + \ldots + a_n^2)\text{Var}(X)$$

Special cases:
$$\text{Var}(X_1 + X_2) = \text{Var}(X) + \text{Var}(X) = 2\text{Var}(X)$$

however $\text{Var}(2X) = \text{Var}(X + X) \neq \text{Var}(X) + \text{Var}(X) = 2\text{Var}(X)$

because X and X are not independent.
$$\text{Var}(2X) = 2^2 \text{Var}(X) = 4\text{Var}(X)$$

which is a special case of $\text{Var}(aX + b) = a^2\text{Var}(X)$ where $a = 2$ and $b = 0$.

Here too, the results can be generalised
$$\text{Var}(X_1 + X_2 + \ldots + X_n) = n\text{Var}(X)$$

while $\text{Var}(nX) = n^2 \text{Var}(X)$

Example 15.27

(a) Throw an unbiased cubic dice and define the random variable as the number on the upper side of the dice. Calculate the expected value and variance of this random variable.

(b) Throw two unbiased cubic dice and define the random variable as the sum of the numbers on the upper side of each dice. Calculate the expected value and variance of this random variable.

(c) Throw one dice and define the random variable as twice the number on the upper side of the dice. Calculate the expected value and variance of this random variable.

Solution

(a) Here is the probability distribution of the related random variables:

X	1	2	3	4	5	6	Σ
$P(X)$	$\frac{1}{6}$	$\frac{1}{6}$	$\frac{1}{6}$	$\frac{1}{6}$	$\frac{1}{6}$	$\frac{1}{6}$	1
$E(X)$	$\frac{1}{6}$	$\frac{2}{6}$	$\frac{3}{6}$	$\frac{4}{6}$	$\frac{5}{6}$	$\frac{6}{6}$	$\frac{7}{2}$
$E(X^2)$	$\frac{1}{6}$	$\frac{4}{6}$	$\frac{9}{6}$	$\frac{16}{6}$	$\frac{25}{6}$	$\frac{36}{6}$	$\frac{91}{6}$
$\text{Var}(X)$			$\frac{91}{6}$	$-$	$\left(\frac{7}{2}\right)^2$	$=$	$\frac{35}{12}$

(b)

$Y = X_1 + X_2$	2	3	4	5	6	7	8	9	10	11	12	Σ
$P(Y)$	$\frac{1}{36}$	$\frac{2}{36}$	$\frac{3}{36}$	$\frac{4}{36}$	$\frac{5}{36}$	$\frac{6}{36}$	$\frac{5}{36}$	$\frac{4}{36}$	$\frac{3}{36}$	$\frac{2}{36}$	$\frac{1}{36}$	1
$E(Y)$	$\frac{2}{36}$	$\frac{6}{36}$	$\frac{12}{36}$	$\frac{20}{36}$	$\frac{30}{36}$	$\frac{42}{36}$	$\frac{40}{36}$	$\frac{36}{36}$	$\frac{30}{36}$	$\frac{22}{36}$	$\frac{12}{36}$	$\frac{252}{36} = 7$
$E(Y^2)$	$\frac{4}{36}$	$\frac{18}{36}$	$\frac{48}{36}$	$\frac{100}{36}$	$\frac{180}{36}$	$\frac{294}{36}$	$\frac{320}{36}$	$\frac{324}{36}$	$\frac{300}{36}$	$\frac{242}{36}$	$\frac{144}{36}$	$\frac{1974}{36}$
$\text{Var}(Y)$								$\frac{1974}{36}$	$-$	$(7)^2$	$=$	$\frac{35}{6}$

819

(c)

$Y = 2X$	2	4	6	8	10	12	$\sum$
$P(Y)$	$\dfrac{1}{6}$	$\dfrac{1}{6}$	$\dfrac{1}{6}$	$\dfrac{1}{6}$	$\dfrac{1}{6}$	$\dfrac{1}{6}$	1
$E(Y)$	$\dfrac{2}{6}$	$\dfrac{4}{6}$	$\dfrac{6}{6}$	$\dfrac{8}{6}$	$\dfrac{10}{6}$	$\dfrac{12}{6}$	$\dfrac{42}{6} = 7$
$E(Y^2)$	$\dfrac{4}{6}$	$\dfrac{16}{6}$	$\dfrac{36}{6}$	$\dfrac{64}{6}$	$\dfrac{100}{6}$	$\dfrac{144}{6}$	$\dfrac{364}{6}$
$\mathrm{Var}(Y)$			$\dfrac{364}{6}$	$-$	$(7)^2$	$=$	$\dfrac{70}{6}$

Note that:

$$E(X_1 + X_2) = E(2X) = 7 = 2 \times \frac{7}{2} = 2E(X)$$

$$\mathrm{Var}(X_1 + X_2) = \frac{35}{6} = 2\left(\frac{35}{12}\right) = 2\mathrm{Var}(X)$$

$$\mathrm{Var}(2X) = \frac{70}{6} = \frac{35}{3} = 4\left(\frac{35}{12}\right) = 4\mathrm{Var}(X)$$

Example 15.28

In a multiple-choice quiz of 10 questions, each question provides 4 choices, one of which is correct. A student is guessing on all questions.

(a) Find the expected value and variance of the number of questions answered correctly by the student.

(b) Set up a table showing the probability distribution of the number of questions answered correctly by the student.

(c) Use your table to calculate the expected number and variance of the number of questions answered correctly by the student.

(d) The teacher will give a score of 3 marks for each question answered correctly and will not penalise wrong answers. Find the expected score and variance of the scores of the guessing student.

(e) Set up a table for the distribution of scores of the student and use it to calculate the expected value and variance of the scores.

Solution

(a) This is a binomial distribution with $n = 10$ and probability of success $p = 0.25$.

$$E(X) = np = 10\,(0.25) = 2.5$$

$$\mathrm{Var}(X) = npq = 10(0.25)(0.75) = 1.875$$

(b), (c)

X	0	1	2	3	4	5	6	7	8	9	10	Total
$p(x)$	0.06	0.19	0.28	0.25	0.15	0.06	0.02	0	0	0	0	1
$xp(x)$	0	0.19	0.56	0.75	0.58	0.29	0.1	0.02	0	0	0	2.5
$x^2p(x)$	0	0.19	1.13	2.25	2.34	1.46	0.58	0.15	0.02	0	0	8.125
							$\mathrm{Var}(X) = \sum x^2 p(x) - (\mathrm{E}(X))^2$					1.875

Note that the expected value and variance agree completely with the theoretical values found in part (a).

(d) Let $Y = 3X$ be the variable representing the score for each question, then

$$\mathrm{E}(Y) = 3\mathrm{E}(X) = 7.5$$

$$\mathrm{Var}(Y) = 9\mathrm{Var}(X) = 16.875$$

(e)

$Y = 3X$	0	3	6	9	12	15	18	21	24	27	30	Total
$p(y)$	0.06	0.19	0.28	0.25	0.15	0.06	0.02	0	0	0	0	1
$yp(y)$	0	0.56	1.69	2.25	1.75	0.88	0.29	0.06	0.01	0	0	7.5
$y^2p(y)$	0	1.69	10.1	20.3	21	13.1	5.26	1.36	0.22	0.02	0	73.125
							$\mathrm{Var}(Y) = \sum y^2 p(y) - (\mathrm{E}(Y))^2$					16.875

Note that the expected value and variance agree completely with the theoretical values found in part (d).

Interesting application II

Find $\mathrm{E}(\overline{X})$ and $\mathrm{Var}(\overline{X})$ in terms of n and μ.

Since $\overline{X} = \dfrac{\sum X_i}{n} = \dfrac{X_1 + X_2 + \ldots + X_n}{n}$,

and since $\mathrm{E}(X_1) = \mathrm{E}(X_2) = \cdots = \mathrm{E}(X_n) = \mu$, then

$$\mathrm{E}(\overline{X}) = \mathrm{E}\left(\frac{\sum X_i}{n}\right) = \mathrm{E}\left(\frac{X_1 + X_2 + \ldots + X_n}{n}\right)$$

$$= \mathrm{E}\left(\frac{1}{n}(X_1 + X_2 \ldots + X_n)\right) = \frac{1}{n}\mathrm{E}(X_1 + X_2 + \ldots + X_n)$$

$$= \frac{1}{n}(\mathrm{E}(X_1) + \mathrm{E}(X_2) + \ldots + \mathrm{E}(X_n)) = \frac{1}{n}(\mu + \mu + \ldots + \mu)$$

$$= \frac{1}{n} \cdot n\mu = \mu$$

Also, since $\mathrm{Var}(X_1) = \mathrm{Var}(X_2) = \cdots = \mathrm{Var}(X_n) = \sigma^2$, then

$$\mathrm{Var}(\overline{X}) = \mathrm{Var}\left(\frac{\sum X_i}{n}\right) = \mathrm{Var}\left(\frac{X_1 + X_2 + \ldots + X_n}{n}\right)$$

$$= \mathrm{Var}\left(\frac{1}{n}(X_1 + X_2 + \ldots + X_n)\right) = \frac{1}{n^2}\mathrm{Var}(X_1 + X_2 + \ldots + X_n)$$

$$= \frac{1}{n^2}(\text{Var}(X_1) + \text{Var}(X_2) + \ldots + \text{Var}(X_n)) = \frac{1}{n^2}(\sigma^2 + \sigma^2 + \ldots + \sigma^2)$$

$$= \frac{1}{n^2} \cdot n\sigma^2 = \frac{\sigma^2}{n}$$

These two results are of great importance when dealing with sampling distributions, confidence intervals and hypothesis testing.

Example 15.29

In a multiple-choice quiz of 10 questions, each question provides 4 choices, one of which is correct. A correct answer is worth 3 marks. A randomly selected group of 36 students who are not familiar with the topic are all guessing on all questions.

(a) Find the expected mean score for this group.

(b) Find the variance of the mean scores for this group.

(c) You are told that the distribution of scores is normal. What is the probability that a student in this group scores at least 9 marks?

(d) Under the same conditions as above, what is the probability that this group's mean is at least 9 marks?

Solution

(a) As we found before, $E(\overline{X}) = \mu = 7.5$

(b) Similarly $\text{Var}(\overline{X}) = \frac{\sigma^2}{n} = \frac{16.875}{36} = 0.46875$

(c) This is an individual observation's probability under a normal distribution with mean 7.5 and variance 16.875

$P(X \geq 9) = 0.3575$. This is the area under N(7.5, 16.875)

(d) This is an average value. The probability uses a normal distribution with mean 7.5 and a variance 0.46875

$P(\overline{X} \geq 9) = 0.000687$. This is the area under N(7.5, 0.46875)

Linear combinations of normally distributed random variables

A very significant property of normally distributed random variables is that a linear function of one of them, or a linear combination of several, is also normally distributed.

In particular:

- If X is normally distributed with a mean μ and a variance σ^2, i.e., $X \sim (\mu, \sigma^2)$, then $Y = aX + b$ is also normally distributed such that $Y \sim N(a\mu + b, a^2\sigma^2)$

- If X and Y are two normally distributed random variables, then $Z = aX \pm bY$ is also normally distributed with the following results:

$$X \sim (\mu_x, \sigma_x^2),\ Y \sim (\mu_y, \sigma_y^2) \Rightarrow Z \sim (\mu_x \pm \mu_y, \sigma_x^2 + \sigma_y^2)$$

Example 15.30

Test scores in an HL class are to be 'curved' as follows: every student will receive 5 marks, which are then added to twice the score on the test itself. Given that the test scores are normally distributed with an average of 35 and a standard deviation of 7 marks, find:

(a) the mean and standard deviation of the 'curved' score

(b) the probability that a student receives a score of at least 65 after curving.

Solution

(a) Let X be the raw score on the test, and hence $Y = 2X + 5$ will be the curved score.

$$E(Y) = 2 \times 35 + 5 = 75$$
$$\sigma = \sqrt{\mathrm{Var}(2X + 5)} = \sqrt{2^2 \mathrm{Var}(X)} = 2\sqrt{49} = 14$$

(b) $P(Y \geqslant 65) = 0.7625$

Example 15.31

Wooden barrels are traditionally used to store pickled cucumber in some European countries. To hold the wood together, steel rims are fixed around them to keep them tight. To keep the steel tight around the wood, the rings are slightly smaller in diameter than the barrel, heated so that they expand slightly before being fitted, then allowed to cool.

The diameter of one type of barrel is known to have a normal distribution with mean of 56 cm and a standard deviation of 0.20 cm. The rims are constructed so that they yield a diameter that is also normally distributed with a mean of 55.70 cm and a standard deviation of 0.30 cm when at room temperature. The rims are heated so that the diameter increases by 1.5%.

(a) What is the probability that a randomly chosen rim will fit around a randomly chosen barrel without heating?

(b) What is the probability that a randomly chosen rim will fit around a randomly chosen barrel with heating?

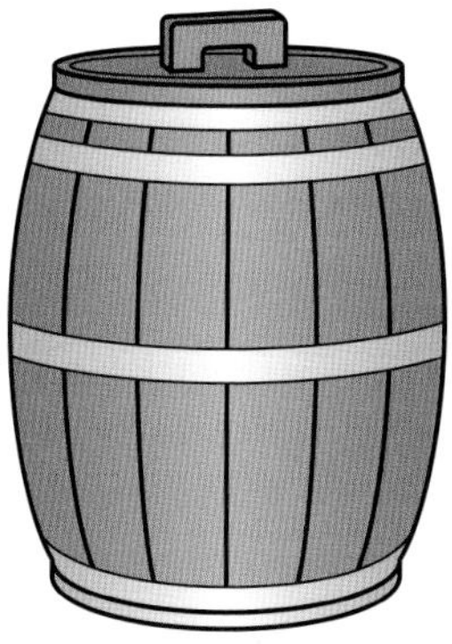

Figure 15.35 Wooden barrel

Solution

Let the barrel diameter be B and the rim diameter be R.
Therefore $B \sim N(56, 0.04)$ and $R \sim N(55.7, 0.09)$

(a) Before heating: For a rim to fit around a barrel, the rim's diameter must be larger than the barrel's diameter, i.e., $R - B > 0$.

Hence, to find the probability, we need to consider the distribution of the random variable $(R - B)$. Since R and B are randomly chosen, they are independent random variables and the new variable, call it $Y = R - B$ will also be normal.

$$E(Y) = E(R) - E(B) = 55.90 - 56 = -0.03, \text{ and}$$

$$\text{Var}(Y) = \text{Var}(R) + \text{Var}(B) = 0.09 + 0.04 = 0.13$$

Therefore $Y \sim N(-0.03, 0.13)$, and hence

$$P(R - B > 0) = 0.409$$

(b) After heating, the diameter of the rim becomes $1.015R$. For a heated rim to fit around a barrel, the rim's diameter must be larger than the barrel's diameter, i.e.,

$$1.015R - B > 0$$

Hence, if we want to find the probability, we need to consider the distribution of the random variable $(1.015R - B)$. Since R and B are randomly chosen, they are independent random variables and the new variable, call it $H = 1.015R - B$ will also be normal.

$$E(H) = 1.015E(R) - E(B) = 56.5355 - 56 = .5355, \text{ and}$$

$$\text{Var}(H) = 1.015^2\text{Var}(R) + \text{Var}(B) = 0.0927 + 0.04 = 0.133$$

Therefore $H \sim N(0.5355, 0.133)$, and hence

$$P(1.015R - B > 0) = 0.929$$

Summary of formulae

Formula	Note
$E(X) = \sum_{\text{all } x} xp(x)$	Discrete
$E(X) = \int_{\text{all } x} xp(x)\,dx$	Continuous
$E(aX + b) = aE(X) + b$, with $a, b \in \mathbb{R}$	
$\text{Var}(aX + b) = a^2\,\text{Var}(X)$	
$E(aX \pm bY) = aE(X) \pm bE(Y)$	
$E(XY) = E(X)E(Y)$	Independent
$\text{Var}(aX \pm bY) = a^2\text{Var}(X) + b^2\text{Var}(Y)$	Independent
$E(a_1X_1 \pm a_2X_2 \pm \ldots \pm a_nX_n) = a_1E(X_1) \pm a_2E(X_2) \pm \ldots \pm a_nE(X_n)$	
$\text{Var}(a_1X_1 \pm a_2X_2 \pm \ldots \pm a_nX_n) = a_1^2\,\text{Var}(X_1) + a_2^2\,\text{Var}(X_2) + \ldots + a_n^2\text{Var}(X_n)$	Independent
$E(X_1 + X_2 + \ldots + X_n) = nE(X)$, and $E(nX) = nE(X)$	
$\text{Var}(X_1 + X_2 + \ldots + X_n) = n\text{Var}(X)$	
$\text{Var}(nX) = n^2\text{Var}(X)$	

Table 15.9 Summary of formulae

Exercise 15.5

1. A discrete random variable X has the following probability distribution.

x	0	1	2	3	4
$p(X = x)$	0.1296	0.3456	0.3456	0.1536	0.0256

   (a) Find $p(X \geq 2)$, $p(1 \leq X \leq 3)$
   (b) Calculate $E(X)$ and $\text{Var}(X)$
   (c) Let $Y = 9 - 2X$. Calculate $E(Y)$ and $\text{Var}(Y)$.

2. A random variable X has the following probability distribution.

x	11	12	13	14	15
$p(X = x)$	0.25	0.2	0.35	k	0.07

   (a) Find the value of k and draw a histogram to represent the distribution.
   (b) Find $p(12 < X \leq 14)$, $p(X \geq 14)$
   (c) Find $E(X)$ and $\text{Var}(X)$.
   (d) If $Y = 2X$, find $E(Y)$ and $\text{Var}(Y)$ in two ways:
      (i) using what you learned in this chapter
      (ii) creating a table for all possible values of Z and then performing the calculations.
   (e) If $Z = X_1 + X_2$, where X_1 and X_2 are randomly chosen independent values of X, find $E(Z)$ and $\text{Var}(Z)$ in two ways:
      (i) using what you learned in this chapter
      (ii) creating a table for all possible values of Z and then performing the calculations.

3. Two unbiased dice, a cubical one and a tetrahedral one, are tossed together. The number that each dice lands on is the score.
   (a) Set up the probability distribution tables for the scores on each dice.
   (b) Calculate the mean and variance of each of the two variables.
   (c) Set up the probability distribution table for the sum of scores on both dice.
   (d) Calculate the mean and variance of the sum of scores in two ways:
      (i) by using the table you created in part (c)
      (ii) by using what you learned in this chapter.

4. We run an experiment where 36 cubical unbiased dice are thrown simultaneously and the average score is calculated. Suppose the experiment is repeated a large number of times. Calculate the expected value of the average score of the 36 dice and their standard deviation.

5. The probability distribution for a random variable M is given below.

m	1	2	3	4	5
$P(M = m)$	$50k + k^2 - 5$	$35k - 2k^2 - 3$	$6k^2 + 10k - 1$	$32k - 3$	$5k^2 + 12k - 1$

(a) Calculate

    (i) k (ii) $E(M)$ (iii) $Var(M)$

(b) If $N = 2M_1 + 3M_2$, where M_1 and M_2 are randomly chosen values of M, find:

    (i) $E(N)$ (ii) $Var(N)$

6. Two independent random variables X and Y have the properties:

    $E(X) = 3$, $Var(X) = 2$; $E(Y) = 7$, $Var(Y) = 1$

Calculate:

(a) $E(X + Y)$, $Var(X + Y)$ (b) $E(X - Y)$, $Var(X - Y)$

(c) $E(2X + 3Y)$, $Var(2X + 3Y)$ (d) $E(2X - 3Y)$, $Var(2X - 3Y)$

7. Two independent random variables X and Y have the properties:

    $E(X^2) = 9$, $Var(X) = 2$; $E(Y^2) = 16$, $Var(Y) = 3$

Calculate:

(a) $E(X + Y)$, $Var(X + Y)$ (b) $E(X - Y)$, $Var(X - Y)$

(c) $E(2X + 3Y)$, $Var(2X + 3Y)$ (d) $E(2X - 3Y)$, $Var(2X - 3Y)$

8. Two independent random variables X and Y have the properties:

    $E(X^2) = 12$, $Var(X) = 5$; $E(Y^2) = 6$, $Var(Y) = 2$

Calculate:

(a) $E(2X + Y)$, $Var(2X + Y)$ (b) $E(X - 3Y)$, $Var(X - 3Y)$

(c) $E(2X + 3Y)$, $Var(2X + 3Y)$ (d) $E(2X - 3Y)$, $Var(2X - 3Y)$

9. Aluminium pipes are produced for an industrial process by two machines. One machine produces 60% of the pipes, each with length 1.0 m, and the second machine produces 40% of the pipes, each with length 0.95 m. All pipes are collected in a central storage location.

(a) Find the expected length and variance of a pipe.

An instrument uses two of these pipes joined together in its production.

(b) Construct a table showing all possible lengths of the joined pipes and use the table to find the expected length and variance of the joined pipes. Use the theorems you learned in this chapter to consolidate your results.

(c) Another instrument uses three of these pipes. Repeat the calculations for part (b). To help you out with the table, here is a part of it:

l = length	2.85		3.05	3.15
p(l)	0.064	0.288		

10. Juice dispensers use juice concentrate mixed with an amount of water. A machine that dispenses apple juice uses, on average, 40 ml of juice concentrate and 260 ml of water for a 300 ml serving of apple juice. The volume of concentrate from this machine has a normal distribution with mean of 40 ml and a standard deviation of 5 ml, and the volume of water has a mean of 260 ml and a standard deviation of 8 ml.

(a) What is the probability that a serving from this dispenser will contain more than 305 ml?

(b) You can get a double serving from this machine. The machine will deal with the order as if it was two separate single servings. So, it produces two servings, one after the other. What is the probability that the amount you receive is less than 590 ml?

(c) A different dispenser deals with the double serving differently. It will simply double the amount of concentrate and the amount of water. What is the probability that the amount you receive is less than 590 ml?

11. A ballpoint pen has an internal chamber filled with ink, which is dispensed at the tip during use by the rolling action of a small metal sphere. Some pens have a small sphere with diameter 0.9 mm. The sphere must be held in place by a metal container. The metal spheres are produced by a machine and their diameters have a normal distribution with mean 0.9 mm and standard deviation of 0.05 mm. The containers are produced by different machines. The diameter of the opening of the container is normally distributed with a mean of 0.8 mm and standard deviation of 0.006 mm. The containers that are too large cannot hold the spheres and those that are too small do not allow enough ink to flow.

Figure 15.36 Ballpoint pen

The difference in diameters must not be smaller than 0.003 mm and not larger than 0.008 mm. One sphere and one container are usually chosen at random to assemble into a pen. What is the probability that they will match?

Chapter 15 practice questions

1. Residents of a small town have savings that are normally distributed with a mean of $3000 and a standard deviation of $500.

   (a) What percentage of townspeople have savings greater than $3200?

   (b) Two townspeople are chosen at random. What is the probability that both of them have savings between $2300 and $3300?

   (c) The percentage of townspeople with savings less than d dollars is 74.22%. Find the value of d.

2. A box contains 35 red discs and 5 black discs. A disc is selected at random and its colour noted. The disc is then replaced in the box.

   (a) In eight such selections, find the probability that a black disc is selected:

   (i) exactly once   (ii) at least once.

   (b) The process of selecting and replacing is carried out 400 times. What is the expected number of black discs that would be drawn?

3. The graph shows a normal curve for the random variable X, with mean μ and standard deviation σ.

   It is known that $P(X \geqslant 12) = 0.1$

   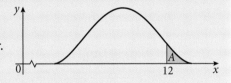

   (a) The shaded region A is the region under the curve where $X \geqslant 12$. Write down the area of the shaded region A.

   It is also known that $P(X \leqslant 8) = 0.1$

   (b) Find the value of μ, explaining your method in full.

   (c) Show that $\sigma = 1.56$, correct to 3 significant figures.

   (d) Find $P(X \leqslant 11)$.

4. The lifespan of a particular species of insect is normally distributed with a mean of 57 hours and a standard deviation of 4.4 hours.

   The probability that the lifespan of an insect of this species lies between 55 and 60 hours is represented by the shaded area in the diagram. This diagram represents the standard normal curve.

   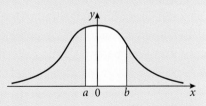

   (a) Write down the values of a and b.

   (b) Find the probability that the lifespan of an insect of this species is:

   (i) more than 55 hours   (ii) between 55 and 60 hours.

90% of the insects die after *t* hours.

(c) Represent this information on a standard normal curve diagram, similar to the one given in part (a), indicating clearly the area representing 90%.

(d) Find the value of *t*.

5. Intelligence quotient (IQ) in a certain population is normally distributed with a mean of 100 and a standard deviation of 15.

(a) What percentage of the population has an IQ between 90 and 125?

(b) If two people are chosen at random from the population, what is the probability that both have an IQ greater than 125?

6. Bags of cement are labelled as weighing 25 kg. The bags are filled by a machine and the actual weights are normally distributed with mean 25.7 kg and standard deviation 0.50 kg.

(a) What is the probability a bag selected at random will weigh less than 25.0 kg?

In order to reduce the number of underweight bags (bags weighing less than 25 kg) to 2.5% of the total, the mean is increased without changing the standard deviation.

(b) Show that the increased mean is 26.0 kg.

It is decided to purchase a more accurate machine for filling the bags. The requirements for this machine are that only 2.5% of bags be under 25 kg and that only 2.5% of bags be over 26 kg.

(c) Calculate the mean and standard deviation that satisfy these requirements.

The cost of the new machine is $5000. Cement sells for $0.80 per kg.

(d) Compared to the cost of operating with a 26 kg mean, how many bags must be filled in order to recover the cost of the new equipment?

7. The mass of packets of a breakfast cereal is normally distributed with a mean of 750 g and standard deviation of 25 g.

(a) Find the probability that a packet chosen at random has mass:
   (i)   less than 740 g
   (ii) at least 780 g
   (iii) between 740 g and 780 g.

(b) Two packets are chosen at random. What is the probability that both packets have a mass that is less than 740 g?

(c) The mass of 70% of the packets is more than *x* g. Find the value of *x*.

8. In a country called Tallopia, the height of adults is normally distributed with a mean of 187.5 cm and a standard deviation of 9.5 cm.

    (a) What percentage of adults in Tallopia are taller than than 197 cm?

    (b) A standard doorway in Tallopia is designed so that 99% of adults have a space of at least 17 cm over their heads when going through the doorway. Find the height of a standard doorway in Tallopia. Give your answer to the nearest cm.

9. It is claimed that the masses of a population of lions are normally distributed with a mean mass of 310 kg and a standard deviation of 30 kg.

    (a) Calculate the probability that a lion selected at random will have a mass of 350 kg or more.

    (b) The probability that the mass of a lion lies between a and b is 0.95, where a and b are symmetric about the mean. Find the values of a and b.

10. Reaction times of human beings are normally distributed with a mean of 0.76 seconds and a standard deviation of 0.06 seconds.

    The graph is that of the standard normal curve. The shaded area represents the probability that the reaction time of a person chosen at random is between 0.70 and 0.79 seconds.

    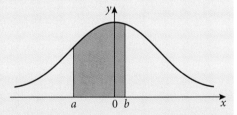

    (a) Write down the value of a and of b.

    (b) Calculate the probability that the reaction time of a person chosen at random is:

      (i) greater than 0.70 seconds

      (ii) between 0.70 and 0.79 seconds.

    Three per cent of the population have a reaction time less than c seconds.

    (c) (i) Represent this information on a diagram similar to the one above. Indicate clearly the area representing 3%.

      (ii) Find c.

11. The speeds of cars at a certain point on a straight road are normally distributed with mean μ and standard deviation σ. 15% of the cars travelled at speeds greater than 90 km h^{-1}, and 12% of them at speeds less than 40 km h^{-1}. Find μ and σ.

12. Bag A contains 2 red balls and 3 green balls. Two balls are chosen at random from the bag without replacement. Let X denote the number of red balls chosen. The table shows the probability distribution for X.

x	0	1	2
$P(X = x)$	$\dfrac{3}{10}$	$\dfrac{6}{10}$	$\dfrac{1}{10}$

   (a) Calculate $E(X)$, the mean number of red balls chosen.

   Bag B contains 4 red balls and 2 green balls. Two balls are chosen at random from bag B.

   (b) (i) Draw a tree diagram to represent this information, including the probability of each event.

   (ii) Hence find the probability distribution for Y, where Y is the number of red balls chosen.

   A standard dice with six faces is rolled. If a 1 or 6 is obtained, two balls are chosen from bag A, otherwise two balls are chosen from bag B.

   (c) Calculate the probability that two red balls are chosen.

   (d) Given that two red balls are obtained, find the conditional probability that a 1 or 6 was rolled on the dice.

13. Ball bearings are used in engines in large quantities. A car manufacturer buys these bearings from a factory. They agree on the following terms: The car company chooses a sample of 50 ball bearings from the shipment. If they find more than 2 defective bearings, the shipment is rejected. It is a fact that the factory produces 4% defective bearings.

   (a) What is the probability that the sample is free of defects?

   (b) What is the probability that the shipment is accepted?

   (c) What is the expected number of defective bearings in the sample of 50?

14. It is estimated that 2.3% of the cherry tomato fruits produced at a certain farm are considered to be small and cannot be sold for commercial purposes. The farmers have to separate such fruits and use them for domestic consumption instead.

   (a) 12 tomatoes are randomly selected from the produce.

   (i) Calculate the probability that 3 are not fit for selling.

   (ii) Calculate the probability that at least 4 are not fit for selling.

   (b) It is known that the sizes of such tomatoes are normally distributed with a mean of 3 cm and a standard deviation of 0.5 cm. Tomatoes that are categorised as large will have to be larger than 2.5 cm. What proportion of the produce is large?

15. A factory has a machine designed to produce 1 kg bags of sugar. It is found that the average weight of sugar in the bags is 1.02 kg. Assuming that the weights of the bags are normally distributed, find the standard deviation if 1.7% of the bags weigh below 1 kg.

Give your answer correct to the nearest 0.1 gram.

16. The continuous random variable X has probability density function $f(x)$ where

$$f_k(x) = \begin{cases} e - ke^{kx} & 0 \leqslant x \leqslant 1 \\ 0 & \text{otherwise} \end{cases}$$

(a) Show that $k = 1$.

(b) What is the probability that the random variable X has a value that lies between $\frac{1}{4}$ and $\frac{1}{2}$? Give your answer in terms of e.

(c) Find the mean and variance of the distribution. Give your answers exactly, in terms of e.

The random variable X above represents the lifetime, in years, of a certain type of battery.

(d) Find the probability that a battery lasts more than six months.

(e) A calculator is fitted with three of these batteries. Each battery fails independently of the other two. Find the probability that at the end of six months:

(i) none of the batteries has failed

(ii) exactly one of the batteries has failed.

17. The lifetime of a particular component of a solar cell is Y years, where Y is a continuous random variable with probability density function

$$f(y) = \begin{cases} 0 & y < 0 \\ 0.5e^{-y/2} & y \geqslant 0 \end{cases}$$

(a) Find the probability, correct to 4 significant figures, that a given component fails within six months.

Each solar cell has three components that work independently and the cell will continue to run if at least two of the components continue to work.

(b) Find the probability that a solar cell fails within six months.

18. Ian and Karl have been chosen to represent their countries in the Olympic discus throw. Assume that the distance thrown by each athlete is normally distributed. The mean distance thrown by Ian in the past year was 60.33 m with a standard deviation of 1.95 m.

(a) In the past year, 80% of Ian's throws have been longer than x metres. Find x, correct to 2 decimal places.

(b) In the past year, 80% of Karl's throws have been longer than 56.52 m. The mean distance of his throws was 59.39 m. Find the standard deviation of his throws, correct to 2 decimal places.

(c) This year, Karl's throws have a mean of 59.50 m and a standard deviation of 3.00 m. Ian's throws still have a mean of 60.33 m and standard deviation 1.95 m. In a competition, an athlete must have at least one throw of 65 m or more in the first round to qualify for the final round. Each athlete is allowed three throws in the first round.

    (i) Determine which of these two athletes is more likely to qualify for the final round on their first throw.

    (ii) Find the probability that both athletes qualify for the final round.

19. A company buys 44% of its stock of bolts from manufacturer A and the rest from manufacturer B. The diameters of the bolts produced by each manufacturer follow a normal distribution with a standard deviation of 0.16 mm.

The mean diameter of the bolts produced by manufacturer A is 1.56 mm. 24.2% of the bolts produced by manufacturer B have a diameter less than 1.52 mm.

(a) Find the mean diameter of the bolts produced by manufacturer B.

A bolt is chosen at random from the company's stock.

(b) Show that the probability that the diameter is less than 1.52 mm is 0.312, to 3 significant figures.

(c) The diameter of the bolt is found to be less than 1.52 mm. Find the probability that the bolt was produced by manufacturer B.

(d) Manufacturer B makes 8000 bolts in one day. It makes a profit of $1.50 on each bolt sold, on the condition that its diameter measures between 1.52 mm and 1.83 mm. Bolts whose diameters measure less than 1.52 mm must be discarded at a loss of $0.85 per bolt.

Bolts whose diameters measure over 1.83 mm are sold at a reduced profit of $0.50 per bolt.

Find the expected profit for manufacturer B, to the nearest $100.

20. Roger uses public transport to go to school each morning. The time he waits each morning for the transport is normally distributed with a mean of 15 minutes and a standard deviation of 3 minutes.

(a) On a specific morning, what is the probability that Roger waits more than 12 minutes?

(b) During a particular week (Monday to Friday), find the probability that:

    (i) Roger's total waiting time does not exceed 65 minutes

    (ii) he waits less than 12 minutes on at least three days of the week

    (iii) his average daily waiting time is more than 13 minutes.

21. The weights of male nurses in a hospital are known to be normally distributed with mean $\mu = 72$ kg and standard deviation $\sigma = 7.5$ kg. The hospital has a lift (elevator) with a maximum recommended load of 450 kg. Six male nurses enter the lift. Calculate the probability p that their combined weight exceeds the maximum recommended load.

22. The weights, X kg, of male birds of a certain species are normally distributed with mean 4.5 kg and standard deviation 0.2 kg. The weights, Y kg, of female birds of this species are normally distributed with mean 2.5 kg and standard deviation 0.15 kg.

    (a) (i) Find the mean and variance of $2Y - X$.

    (ii) Find the probability that the weight of a randomly chosen male bird is more than twice the weight of a randomly chosen female bird.

    (b) Two randomly chosen male birds and three randomly chosen female birds are placed together on a weighing machine for which the recommended maximum weight is 16 kg. Find the probability that this maximum weight is exceeded.

23. A shop sells apples and pears. The weights, in grams, of the apples may be assumed to have a $N(200, 15^2)$ distribution and the weights of the pears, in grams, may be assumed to have a $N(120, 10^2)$ distribution.

    (a) Find the probability that the weight of a randomly chosen apple is more than double the weight of a randomly chosen pear.

    (b) A shopper buys 3 apples and 4 pears. Find the probability that the total weight is greater than 1000 grams.

24. (a) The random variable Y is such that

    $$E(2Y + 3) = 6 \text{ and } Var(2 - 3Y) = 11$$

    Calculate:

    (i) $E(Y)$ (ii) $Var(Y)$ (iii) $E(Y^2)$.

    (b) Independent random variables R and S are such that $R \sim N(5, 1)$ and $S \sim N(8, 2)$.

    The random variable V is defined by $V = 3S - 4R$.
    Calculate $P(V > 5)$.

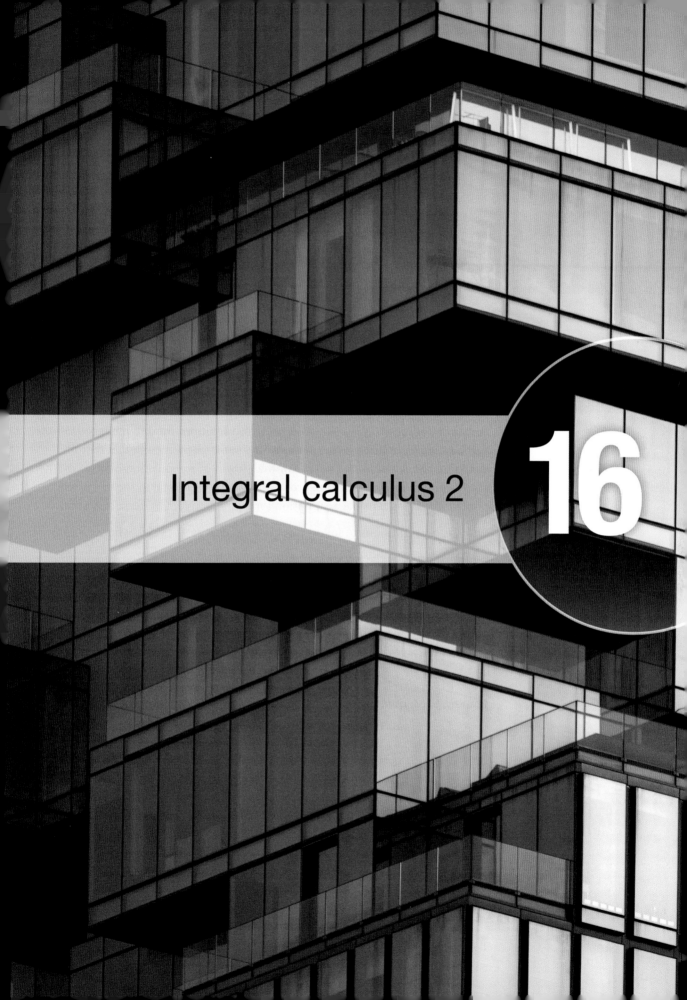

Integral calculus 2

16

Possibly one of the most significant applications of calculus is differential equations. Scientists often use calculus to find solutions for differential equations that have arisen in the process of modelling some phenomenon that they are studying. For example, one model for the growth of a population is based on the assumption that the population grows at a rate proportional to the size of the population. It is a sensible assumption for a population of bacteria or animals under ideal conditions to follow an equation such as $\dfrac{\mathrm{d}P}{\mathrm{d}t} = kP$, where t stands for time and P for the number of individuals, bacteria, or animals at any time t.

Although it is often impossible to find an explicit formula for the solution of a differential equation, we will see that graphical and numerical approaches provide the necessary information too.

In this chapter, you will study equations involving an unknown function and its derivative(s). These are called differential equations. Differential equations come in a wide variety of forms, and many different procedures – analytical, graphical and numerical – exist for finding their solutions.

16.1 Differential equations

In general, a **differential equation** is an equation that contains an unknown function and one or more of its derivatives. The **order** of a differential equation is the order of the highest derivative that occurs in the equation. A first order differential equation is an equation that involves an unknown function and its first derivative. For example, when we consider any of the differential equations

$$y' = x^4, \quad y' + 2ty = \sin t, \quad \frac{\mathrm{d}y}{\mathrm{d}x} = y + 2x, \quad \text{or} \quad \frac{\mathrm{d}y}{\mathrm{d}t} = ky$$

it is understood that y is an unknown function of x or t.

Here are some examples:

$$x\frac{dy}{dx} + y\frac{dy}{dx} - y = 0$$ first order differential equation $$F\left(x, y, \frac{dy}{dx}\right) = 0$$

$$\frac{d^2y}{dx^2} + 3\frac{dy}{dx} - 5y = 0$$ second order differential equation $$F\left(x, y, \frac{dy}{dx}, \frac{d^2y}{dx^2}\right) = 0$$

$$\frac{dy}{dt} + y\sin t - e^{\cos t} = 0$$ first order differential equation $$F\left(t, y, \frac{dy}{dt}\right) = 0$$

For this course, we only study **first order differential equations**. In a first order differential equation, the first derivative, $\frac{dy}{dx}$, of the unknown function can be isolated on one side of the equation. Hence, a simpler general form for first order differential equations is

$$\frac{dy}{dx} = F(x, y)$$

where $\frac{dy}{dx}$ is expressed as a function in terms of x and y. Note that the first order differential equations can all be rewritten in this form. For example,

$$x\frac{dy}{dx} + y\frac{dy}{dx} - y = 0 \Rightarrow \frac{dy}{dx} = \frac{y}{x + y}$$

Solution of a differential equation

A function f is called a **solution** of a differential equation if the equation is satisfied for all values of x in some interval when $y = f(x)$ and its derivatives are substituted into the equation. Thus, $y = f(x) = x - \frac{1}{x}$ is a solution of

$$x\frac{dy}{dx} + y = 2x$$ over the set of non-zero real numbers.

Since $\frac{dy}{dx} = 1 + \frac{1}{x^2} \Rightarrow x\left(1 + \frac{1}{x^2}\right) + x - \frac{1}{x} = x + \frac{1}{x} + x - \frac{1}{x} = 2x$, and thus

$$x\frac{dy}{dx} + y = 2x$$ for all values over this interval.

In algebra, we usually seek the unknown variable values that satisfy an equation such as $3x^2 - 2x - 5 = 0$. By contrast, in solving a differential equation, we are looking for the unknown functions $y = y(x)$ for which an identity such as $y'(x) = 3x^2$, $y(x)$ holds on some interval of real numbers. Usually, we want to find all solutions of the differential equation, if achievable.

Example 16.1

Verify that $y(x) = Ce^{x^3}$ is a solution to the differential equation $\frac{dy}{dx} = 3x^2y$

Solution

Since C is a constant in $y(x) = Ce^{x^3}$, then

$$\frac{dy}{dx} = C(3x^2 e^{x^3}) = 3x^2(Ce^{x^3}) = 3x^2y$$

837

Consequently, every function $y(x)$ of the form $y(x) = Ce^{x^3}$ satisfies – and thus is a solution of – the given differential equation $\dfrac{dy}{dx} = 3x^2 y$ for all real x.

In fact, $y(x) = Ce^{x^3}$ defines an infinite family of different solutions to this differential equation, one for each choice of the arbitrary constant C, as shown in the diagram.

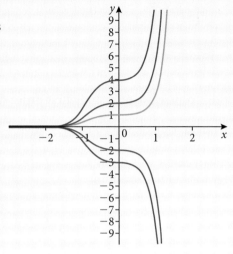

A differential equation may use symbols for the independent and dependent variables other than x and y. For simplicity, we will use x and y while we are developing theory and solution methods for differential equations. Also note that we are using F to represent a two-variable function that when set equal to $\dfrac{dy}{dx}$ is the differential equation, and f to represent the unknown function whose slope at the point (x, y) is $\dfrac{dy}{dx}$

Example 16.2

Verify that $y(x) = -\dfrac{1}{2x^4 + 3}$ is a solution to the differential equation

$$\frac{dy}{dx} = 8x^3 y^2$$

over the interval $]-\infty, \infty[$.

Solution

Note that the denominator in $y(x)$ is never zero and that $y(x)$ is differentiable everywhere. Furthermore, for all real numbers x

$$\frac{d}{dx}y(x) = \frac{d}{dx}\left(-\frac{1}{2x^4 + 3}\right) = \frac{8x^3}{(2x^4 + 3)^2} = 8x^3\left(-\frac{1}{2x^4 + 3}\right)^2 = 8x^3 y^2$$

Thus $y(x) = -\dfrac{1}{2x^4 + 3}$ is a solution to the given differential equation.

Example 16.3

The solution of a differential equation is the (initially unknown) function $y = f(x)$ whose derivative is $\dfrac{dy}{dx}$. Find a solution to the differential equation

$$\frac{dy}{dx} = \frac{1}{x + 1}, \quad x \neq -1$$

Solution

Every solution of this equation is an anti-derivative of $\dfrac{1}{x+1}$

$$y = \int \frac{1}{x+1}\,dx = \ln|x+1| + c, \quad x \neq -1$$

So, the solution of the differential equation

$\dfrac{dy}{dx} = \dfrac{1}{x+1}$ is the explicitly defined function

$y = \ln|x+1| + c$, where c is an arbitrary constant. This is called a general solution because it is not a single function, but an infinite 'family' of functions dependent on the constant c. The diagram shows some members of the family of solutions.

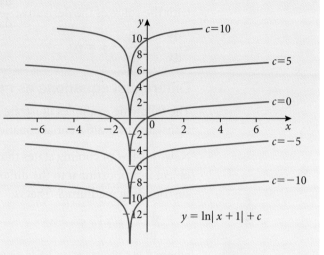

$$y = \ln|x+1| + c$$

When applying differential equations, we are usually not as interested in finding a family of solutions (the general solution) as we are in finding a solution that satisfies some additional requirement. In many physical problems we need to find the particular solution that satisfies a condition of the form $y(x_0) = y_0$. This is called an **initial condition**, and the problem of finding a solution of the differential equation that satisfies the initial condition is called an **initial-value problem**. For instance, in Example 16.3, if we are given the initial conditions that $y = 5$ when $x = 0$ then we can solve for c, giving $c = 5$ and the particular solution of $y = \ln|x+1| + 5$

Sometimes the solution of a differential equation will be expressed as an implicitly defined function. For example, the general solution to the equation

$$x\frac{dy}{dx} + y\frac{dy}{dx} - y = 0 \text{ is } \ln y = \frac{x}{y} + c$$

It is an equation relating x and y and implies a function exists that defines y as a function of x.

To verify that this is a solution, we differentiate: applying implicit differentiation and the product rule:

$$\frac{d}{dx}(\ln y) = \frac{d}{dx}\left(\frac{x}{y} + c\right) \Rightarrow \frac{1}{y}\frac{dy}{dx} = \frac{d}{dx}(xy^{-1}) + \frac{d}{dx}(c)$$

$$\Rightarrow \frac{1}{y}\frac{dy}{dx} = y^{-1} + x\left(-y^{-2}\frac{dy}{dx}\right) + 0 \Rightarrow \frac{1}{y}\frac{dy}{dx} = \frac{1}{y} - \frac{x}{y^2}\frac{dy}{dx}$$

$$\Rightarrow y^2\left(\frac{1}{y}\frac{dy}{dx}\right) = y^2\left(\frac{1}{y} - \frac{x}{y^2}\frac{dy}{dx}\right) \Rightarrow x\frac{dy}{dx} + y\frac{dy}{dx} - y = 0$$

In general, we wish to find the explicit solution of a differential equation written in the form $y = f(x)$ where f is a known function. However, it is sometimes not possible to solve for y. In such a case we must settle for an implicit solution written in the form $g(y) = f(x)$ where g and f are known functions and $g(y) \neq y$

839

Therefore, for any real number c the function $\ln y = \frac{x}{y} + c$ is a solution, in implicit form, to the differential equation $x\frac{dy}{dx} + y\frac{dy}{dx} - y = 0$. This means that the coordinates x and y of any point on the curve $\ln y = \frac{x}{y} + c$ combined with the value of the derivative $\frac{dy}{dx}$ at that point will solve the equation

$$x\frac{dy}{dx} + y\frac{dy}{dx} - y = 0$$

Differential equations as mathematical models

The following examples illustrate typical cases where scientific principles are translated into differential equations.

Newton's law of cooling states that the rate of change of the temperature T of an object is proportional to the difference between T and the temperature of the surrounding medium S. That is,

$$\frac{dT}{dt} = k(T - S)$$

where k is a constant and S is usually considered constant.

Population growth rate in cases where the birth and death rates are not variable is proportional to the size of the population. That is,

$$\frac{dP}{dt} = kP$$

where k is a constant.

We will learn how to solve such problems.

Separable differential equations

In this section, we look at separable differential equations, which are also known as **variables separable** differential equations.

The first order differential equation:

$$\frac{dy}{dx} = F(x, y)$$

is called variables separable when the function $F(x, y)$ can be factored into a product or quotient of two functions such as

$$\frac{dy}{dx} = p(x)q(y) \text{ or } \frac{dy}{dx} = \frac{g(x)}{h(y)}$$

In such cases, the variables x and y can be separated by writing

$$\frac{dy}{q(y)} = p(x)dx \text{ or } h(y)dy = g(x)dx$$

and then simply integrating both sides with respect to x:

$$\int \frac{dy}{q(y)} = \int p(x)dx + c \text{ or } \int h(y)dy = \int g(x)dx + c$$

You need to remember that $h(y)$ is a continuous function of y alone and $g(x)$ is a continuous function of x alone. This also applies to $q(y)$ and $p(x)$.

We also may say that the method of solution is separation of variables.

Here are some examples of differential equations that are separable:

Original differential equation	Rewritten with variables separated
$(x^2 + 4)y' = 3xy$	$\dfrac{dy}{y} = \dfrac{3x}{x^2 + 4}dx$
$\dfrac{3xe^y y'}{1 + e^{2y}} = 5$	$\dfrac{3e^y}{1 + e^{2y}}dy = \dfrac{5}{x}dx$
$\dfrac{dy}{dx} = xy + 4$	Not separable
$x^2\dfrac{dy}{dx} + y^2 = xy^2$	$\dfrac{1}{y^2}dy = \dfrac{x-1}{x^2}dx$

Table 16.1 Examples of differential equations that are separable

Example 16.4

Find the general solution of the differential equation
$$y' - 9x^2 y^2 = 5y^2$$

Solution

We first factor the equation to separate the variables
$$\frac{dy}{dx} = 5y^2 + 9x^2 y^2 \Rightarrow \frac{dy}{dx} = y^2(5 + 9x^2) \Rightarrow \frac{dy}{y^2} = (5 + 9x^2)dx$$

$$\Rightarrow -\frac{1}{y} = 5x + 3x^3 + c \Rightarrow y = \frac{-1}{5x + 3x^3 + c}$$

This is a general solution for the differential equation. In this case we are able to express the function in explicit form.

Example 16.5

Solve the differential equation
$$\frac{dy}{dx} = \frac{3x^2 y}{1 + 4y^2}$$

Solution

With very few steps, we can separate the variables
$$\frac{1 + 4y^2}{y}dy = 3x^2 dx$$

and now we can integrate both sides
$$\int \frac{1 + 4y^2}{y}dy = \int 3x^2 dx \Leftrightarrow \int \left(\frac{1}{y} + 4y\right)dy = \int 3x^2 dx$$

$$\ln|y| + 2y^3 = x^3 + c$$

It is not always obvious if a differential equation is separable. Some algebraic manipulation is needed to confirm that the differential equation can, in fact, be written in the form $\dfrac{dy}{dx} = p(x)q(y)$ or $\dfrac{dy}{dx} = \dfrac{g(x)}{h(y)}$. For example,

$$\frac{dy}{dx} = \frac{3}{xy} - \frac{x^2}{y}$$ is separable because it can be written as

$$\frac{dy}{dx} = \frac{1}{y}\left(\frac{3}{x} - x^2\right);$$

and $\dfrac{\tan x}{y}\dfrac{dy}{dx} = \dfrac{2}{\ln y}$ is also separable because it can be written as

$$\frac{dy}{dx} = \frac{2y}{\ln y}\cot x.$$

However, the equations $\dfrac{dy}{dx} = x^2 + y^2$ and $\dfrac{dy}{dx} = 1 + xy$ are not separable.

For every value of the arbitrary constant c, this defines an exact but implicit solution $y(x)$ as it cannot be written in an explicit form $y = f(x)$.

Some of the solution curves for a few values of c are shown in the diagram.

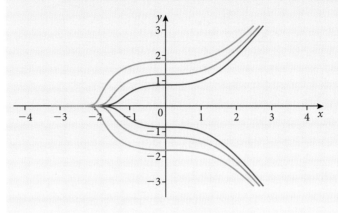

To solve equations by separation of variables:

- Write the differential equation in the standard form $\dfrac{dy}{dx} = f(x, y)$

- Can we separate the variables? That is, is $\dfrac{dy}{dx} = g(x)h(y)$ or $\dfrac{dy}{dx} = \dfrac{p(x)}{q(y)}$?

- If so, separate the variables to get $\dfrac{dy}{h(y)} = g(x)dx$ or $q(y)dy = p(x)dx$

- Integrate both parts, to get $\int \dfrac{dy}{h(y)} = \int g(x)dx + c$ or $\int q(y)dy = \int p(x)dx + c$

- Evaluate the integrals if you can and don't forget the arbitrary constant. Even though we have two integrals, one on the left and one on the right, we can combine both arbitrary constants into one.

- If possible, resolve the resulting equation with respect to y, to get the equation in explicit form $y = f(x)$

Example 16.6

Find the general solution of the differential equation

$$x^2 y \frac{dy}{dx} = x + 1, x > 0, y > 0$$

Solution

The equation is separable because we can rearrange the equation as:

$$\frac{dy}{dx} = \frac{1}{y}\left(\frac{x + 1}{x^2}\right)$$

Now separate the variables and integrate:

$$y\,dy = \frac{x + 1}{x^2}\,dx \Rightarrow \int y\,dy = \int \frac{x + 1}{x^2}\,dx$$

$$\Rightarrow \frac{1}{2}y^2 = \ln x - \frac{1}{x} + c$$

This is the general solution of the differential equation in implicit form. In this case, we can also find the explicit form:

$$y = \sqrt{2\ln x - \frac{2}{x} + c}$$

Example 16.7

Solve the initial value problem

$$\frac{dy}{dx} = \frac{y}{x+1}; \quad y(1) = 4$$

Solution

This is a variables separable type, so separate the variables and integrate:

$$\frac{dy}{dx} = \frac{y}{x+1} \Rightarrow \frac{dy}{y} = \frac{dx}{x+1} \Rightarrow \int \frac{dy}{y} = \int \frac{dx}{x+1}$$

$$\ln|y| = \ln|x+1| + c \Rightarrow |y| = e^{\ln|x+1|+c} = e^{\ln|x+1|}e^c = |x+1|e^c$$

Now, since c is an arbitrary constant, we can replace e^c with a constant C, and our solution becomes: $|y| = C|x+1|$

Using the initial condition: $4 = C|1+1| \Rightarrow C = 2$

and the particular solution $|y| = 2|x+1|$

that is $y = \pm 2(x+1)$

Example 16.8

Solve the initial value problem

$$\frac{dy}{dt} = (e^{y-t})\frac{1+t^2}{\cos y}; \quad y(0) = 0$$

Solution

This differential equation can be separated, but needs some work:

$$\frac{dy}{dt} = e^y e^{-t}\frac{1+t^2}{\cos y} \Rightarrow e^{-y}\cos y\, dy = e^{-t}(1+t^2)dt$$

Both sides need integration by parts

$$\int e^{-y}\cos y\, dy = \int e^{-t}(1+t^2)dt$$

$$\Rightarrow \frac{1}{2}e^{-y}(\cos y - \sin y) = e^{-t}(t^2 + 2t + 3) + c$$

with initial conditions applied

$$\frac{1}{2}e^{-y}(\cos y - \sin y) = e^{-t}(t^2 + 2t + 3) + c$$

$$\Rightarrow \frac{1}{2}e^{-0}(\cos 0 - \sin 0) = e^{-0}(0^2 + 2(0) + 3) + c$$

$$\frac{1}{2} = 3 + c \Rightarrow c = \frac{5}{2}$$

Therefore the particular solution is

$$\frac{1}{2}e^{-y}(\cos y - \sin y) = e^{-t}(t^2 + 2t + 3) + \frac{5}{2}$$

$$\Rightarrow e^{-y}(\cos y - \sin y) = 2e^{-t}(t^2 + 2t + 3) + 5$$

Notice here that our solution cannot be expressed explicitly.

Example 16.9

Solve the differential equation $x\,dx + e^{x+y}\cos y\,dy = 0$

Solution

The equation can be separated with a few steps:

$$x\,dx = e^y\cos y\,dy \Rightarrow e^y\cos y\,dy = -xe^{-x}\,dx$$

$$\Rightarrow \int e^y\cos y\,dy = -\int xe^{-x}\,dx$$

Using integration by parts and simplifying we get the implicit general solution:

$$\Rightarrow \frac{1}{2}e^y(\sin y + \cos y) = e^{-x}(x + 1) + c$$

Applied problems involving variables separable differential equations

Example 16.10

Find the general solution of the population growth model

$$\frac{dP}{dt} = kP$$

Solution

We can separate the variables easily: $\dfrac{dP}{P} = k\,dt$

Now integrate both sides to get $\int \dfrac{dP}{P} = \int k\,dt \Rightarrow \ln|P| = kt + c$

where c is an arbitrary constant. This last equation can be simplified to render an explicit expression for P:

$$\ln|P| = kt + c \Rightarrow |P| = e^{kt+c} = e^{kt}e^c = Ae^{kt}$$

where we replaced e^c with A. Thus $P = Ae^{kt}$

This is the general solution and all solutions to this problem will be in this form.

If the constant k is positive, the model describes population growth (Figure 16.1); if it is negative, it is decay (Figure 16.2).

If the population growth model in Example 16.10 had the additional initial value that at t_0 the population is P_0, then this particular population satisfies

$$P = A e^{kt}$$

and hence

$$P_0 = A e^{kt_0} \Rightarrow A = \frac{P_0}{e^{kt_0}} = P_0 e^{-kt_0}$$

and the solution to the initial value problem is

$$P = A e^{kt} = P_0 e^{-kt_0} e^{kt} = P_0 e^{k(t-t_0)}$$

There is a very important special case when $t_0 = 0$, the solution becomes

$$P = P_0 e^{k(t-t_0)} = P_0 e^{kt}$$

which is the usual growth model that starts at time $t = 0$ with initial population P_0.

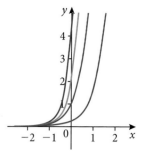

Figure 16.1 Population growth

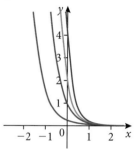

Figure 16.2 Population decay

Example 16.11

A cold object is placed in warmer medium that is kept at a constant temperature S. The rate of change of the temperature $T(t)$ with respect to time is proportional to the difference between the surrounding medium and the object and hence it satisfies

$$\frac{dT}{dt} = k(S - T) \text{ and } T(0) = T_0$$

where $k > 0$ and $T_0 < S$; that is, the initial temperature is less than the temperature of the surrounding medium. Find the solution to the initial value problem.

Solution

The variables can be separated, so:

$$\frac{dT}{dt} = k(S - T) \Leftrightarrow \frac{dT}{S - T} = k \, dt$$

Integrate and find the general solution first:

$$\int \frac{dT}{S - T} = \int k \, dt \Rightarrow -\ln|S - T| = kt + c_1 \Rightarrow \ln|S - T| = -kt - c_1$$

where c_1 is an arbitrary constant. Now since we know that the temperature T is less than the surrounding temperature, then $\ln|S - T| = \ln(S - T)$

The general solution then is

$$\ln(S - T) = -kt - c_1 \Rightarrow S - T = e^{-kt-c_1} \Rightarrow T = S - e^{-kt-c_1}$$

The initial condition implies

$$T = S - e^{-kt-c_1} \Rightarrow T_0 = S - e^{0-c_1} \Rightarrow e^{-c_1} = S - T_0 \Rightarrow c_1 = -\ln(S - T_0)$$

Therefore, substituting this value in the general solution

$$\ln(S - T) = -kt - c_1 = -kt + \ln(S - T_0)$$

$$\Rightarrow \ln(S - T) - \ln(S - T_0) = -kt \Rightarrow \ln\left(\frac{S - T}{S - T_0}\right) = -kt$$

$$\Rightarrow \frac{S - T}{S - T_0} = e^{-kt} \Rightarrow S - T = (S - T_0)e^{-kt} \Rightarrow T = S - (S - T_0)e^{-kt}$$

This is an example of what is called limited growth. This is so because the maximum value that T can achieve is S. For example, if a can of soda is taken out of a fridge and left in a room with constant temperature of 21°C, then the temperature of the soda will increase to eventually reach the room temperature.

In fact, since $k > 0$ and S is a constant, then

$$T = S - (S - T_0)e^{-kt} \Rightarrow \frac{dT}{dt} = (S - T_0)e^{-kt}$$

Also, since $T_0 < S$, then

$$\frac{dT}{dt} = (S - T_0)e^{-kt} > 0$$

The temperature will always increase.
As time passes,

$$\lim_{t\to\infty} e^{-kt} = 0 \Rightarrow \lim_{t\to\infty} T = \lim_{t\to\infty}\left(S - (S - T_0)e^{-kt}\right) = S$$

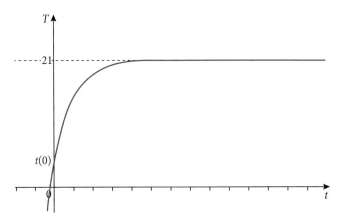

Figure 16.3 The temperature climbs up to 21 but does not exceed it

Example 16.12

The rate of decay of a substance y at any time t is directly proportional to the amount of y and also directly proportional to the amount of another substance x. The constant of proportionality is $-\frac{1}{2}$ and the value of x at any time t is given by $x = \dfrac{4}{(1+t)^2}$

(a) Given the initial conditions that $y = 10$ when $t = 0$, find y as an explicit function of t.

(b) Determine the amount of the substance remaining as t becomes very large.

Solution

(a) The rate of decay of substance y is proportional to the product xy, and with the constant of proportionality having a value of $-\frac{1}{2}$ and $x = \dfrac{4}{(1+t)^2}$, this gives

$$\frac{dy}{dt} = -\frac{1}{2}\left(\frac{4}{(1+t)^2}\right)y,$$

Separating variables gives:

$$\frac{dy}{y} = \frac{-2}{(1+t)^2}\,dt \Rightarrow \int \frac{dy}{y} = -2\int \frac{1}{(1+t)^2}\,dt$$

$$\Rightarrow \ln y = \frac{2}{1+t} + c$$

This is the implicit form. It can be expressed explicitly:

$$e^{\ln y} = e^{\frac{2}{1+t}+c} \Rightarrow y = e^{\frac{2}{1+t}}e^c = Ae^{\frac{2}{1+t}}$$

Now, $y = 10$ when $t = 0$, so:

$$10 = Ae^{\frac{2}{1+0}} \Rightarrow A = 10e^{-2};$$

Therefore the explicit form is $y = Ae^{\frac{2}{1+t}} = 10e^{-2}e^{\frac{2}{1+t}} = 10e^{\frac{-2t}{1+t}}$

(b) $\displaystyle\lim_{t\to\infty} \frac{-2t}{1+t} = -2 \Rightarrow y = 10e^{-2} = 1.36$ (3 s.f.)

In Example 16.13, we will find an explicit solution by separating the variables for a relatively straightforward first order differential equation, but one whose solution will prove useful in developing another solution method.

Example 16.13

Find the general solution to the differential equation $\dfrac{dy}{dx} = -2xy$

Solution

Separate the variables

$$\frac{dy}{y} = -2xdx$$

Note there is no solution when $y = 0$.

$$\int \frac{dy}{y} = -\int 2xdx \Rightarrow \ln|y| = -x^2 + c \Rightarrow e^{\ln|y|} = e^{-x^2+c}$$

$$\Rightarrow |y| = e^c e^{-x^2} = Ae^{-x^2}, A = e^c > 0$$

$$\Rightarrow y = \pm Ae^{-x^2}$$

We can include both of these solutions, and also the solution $y = 0$, by giving the general solution as

$$y = Ce^{-x^2}$$

with no restrictions on the constant C.

It is helpful to recognise that the explicit solution $y = Ce^{-x^2}$ in Example 16.13 defines a family of curves in the xy-plane. Some of these curves, with the corresponding value of C, are shown in Figure 16.4.

To determine a specific curve from this family we must impose an initial condition on the solution.

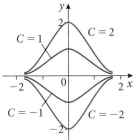

Figure 16.4 Family of curves

Logistic differential equations

Example 16.10 looked at the population growth model. In that model, we used the fact that the rate of change of a variable y (P in the example) is proportional to the value of y. We observed that the differential equation is of the form $\frac{dy}{dt} = ky$ and found that the general solution was $y = Ae^{kt}$.

Exponential growth is unlimited and does not provide an accurate model for the growth of a population over a long time period. To obtain a more realistic model we need to take account of the fact that as the population increases, quite a few factors will begin to affect the growth rate. For example, there will be increased competition for the limited resources that are available, increases in disease, and overcrowding of the limited available space, all of which would serve to slow the growth rate. In order to model this situation mathematically, we amend the differential equation leading to the modest exponential growth law we introduced by addition of a term that slows the growth down as the population increases. If we consider a closed environment (neglecting factors such as immigration and emigration), then the rate of change of population can be modelled by the differential equation

$$\frac{dy}{dt} = (b - d)y$$

where b and d denote the birth and death rates.

In a more general situation, the increased competition as the population grows will result in a corresponding increase in the death rate. One way to take this into account is to assume that the death rate is directly proportional to the present population, and that the birth rate remains constant. The resulting initial-value problem leading to the population growth can now be written as

$$\frac{dy}{dt} = (b_0 - d_0 y)y$$

where b_0 and d_0 are positive constants representing specific birth and death rates. It is useful to write the differential equation in the equivalent form

$$\frac{dy}{dt} = k\left(1 - \frac{y}{L}\right)y \quad (1)$$

where $k = b_0$, and $L = \dfrac{b_0}{d_0}$.

Equation (1) is called the **logistic equation**, and the corresponding population model is called the **logistic model**. The differential equation (1) is separable and can be solved without difficulty. The constant L is called **the carrying capacity** of the population. The graph of function y is called the **logistic curve** (Figure 16.5).

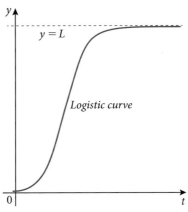

Figure 16.5 Logistic curve

From equation (1), that if $y < L$, then $\dfrac{dy}{dt} > 0$ and the population increases, whereas if $y > L$, then $\dfrac{dy}{dt} < 0$ and the population decreases. We can therefore interpret L as representing the maximum population that the environment can sustain.

Example 16.14

Solve the logistic differential equation $\dfrac{dy}{dt} = ky\left(1 - \dfrac{y}{L}\right)$

Solution

First separate the variables:

$$\frac{dy}{dt} = ky\left(1 - \frac{y}{L}\right) \Rightarrow \frac{dy}{y\left(1 - \frac{y}{L}\right)} = kdt$$

Then integrate both sides using partial fractions:

$$\int \frac{dy}{y\left(1 - \frac{y}{L}\right)} = \int kdt \Rightarrow \int \left(\frac{1}{y} + \frac{1}{L - y}\right)dy = \int kdt$$

$$\Rightarrow \ln|y| - \ln|L - y| = kt + c$$

$$\Rightarrow \ln\left|\frac{L - y}{y}\right| = -kt - c$$

$$\Rightarrow \frac{L - y}{y} = \pm e^{-c}e^{-kt} = be^{-kt}$$

$$\Rightarrow y = \frac{L}{1 + be^{-kt}}$$

From Example 16.14, we can conclude that all solutions of the logistic differential equation can be put in the general form

$$y = \frac{L}{1 + be^{-kt}}$$

849

Example 16.15

An official wildlife commission releases 20 deer into a protected forest. After 5 years, the deer population is 208. The commission believes that the environment cannot support more than 400 deer.

(a) Write a model for the deer population in terms of the number of years, t.

(b) Use the model to estimate the deer population after 10 years.

(c) Discuss the long-term trend of the deer population.

Solution

The growth of the deer population p is

$$\frac{dp}{dt} = kp\left(1 - \frac{p}{400}\right), \quad 20 \leqslant p \leqslant 400$$

where t is the number of years.

(a) We know that $L = 400$. Thus, the solution of the equation is of the form

$$p = \frac{400}{1 + be^{-kt}}$$

Since $p(0) = 20$, we can solve for b

$$20 = \frac{400}{1 + be^{-0}} \Rightarrow b = 19$$

Also, because $p = 208$ when $t = 5$, we can solve for k

$$208 = \frac{400}{1 + 19e^{-5k}} \Rightarrow k \approx 0.6049$$

Therefore the model for this deer population is $p = \dfrac{400}{1 + 19e^{-0.6049t}}$

(b) To estimate the deer population after 10 years, substitute 10 for t in the model:

$$p = \frac{400}{1 + 19e^{-0.6049 \times 10}} \approx 383$$

(c) As t increases without bound, the denominator in the model will approach 1 and so:

$$\lim_{t \to \infty} p = \lim_{t \to \infty} \frac{400}{1 + 19e^{-0.6049t}} = 400$$

Homogeneous differential equations

Some differential equations that are not separable in x and y can be made separable by a change of variables. This is true for differential equations of the form $y' = f(x, y)$, where f is a **homogeneous function**. The function given by $f(x, y)$ is **homogeneous of degree n** if $f(tx, ty) = t^n f(x, y)$, where n is an integer.

For example

- $f(x, y) = 2x^2y - 5x^3 + 7xy^2$ is homogeneous of degree 3 because

$$f(tx, ty) = 2(tx)^2(ty) - 5(tx)^3 + 7(tx)(ty)^2$$
$$= t^3(2x^2y - 5x^3 + 7xy^2) = t^3 f(x, y)$$

- $f(x, y) = 2x^2y - 5x^2y^2 + 7xy$ is not homogeneous because

$$f(tx, ty) = t^2(2tx^2y - 5t^2x^2y^2 + 7xy) \neq t^n(2x^2y - 5x^2y^2 + 7xy)$$

- $f(x, y) = \dfrac{4x}{7y}$ is a homogeneous function of degree 0 because

$$f(tx, ty) = \frac{4tx}{7ty} = t^0\frac{x}{y}$$

A differential equation is homogeneous if it is of the form
$M(x, y)\mathrm{d}x + N(x, y)\mathrm{d}y = 0$
where M and N are homogeneous functions of the same degree.

Equivalently, when a first-order differential equation can be written as $y' = f(x, y)$, then it is homogeneous iff the function f has the property $f(tx, ty) = f(x, y)$ for all $t \neq 0$. This condition is equivalent to saying that f can be expressed as a function of y/x, because if f is homogeneous, then

$$f(x, y) = f\left(x, x\left(\frac{y}{x}\right)\right) = f\left(1, \frac{y}{x}\right) = g\left(\frac{y}{x}\right), \quad (x \neq 0)$$

For example

- $(3x^2 - 2xy)\mathrm{d}x + 2y^2 = 0$ is homogeneous of degree 2.
- $(2x^2 + 3)\mathrm{d}x + 3y^2\,\mathrm{d}y = 0$ is not a homogeneous differential equation.

A homogeneous differential equation can be transformed into a variables separable differential equation by substituting $y = vx$ (or $x = uy$), where v is a differentiable function of x. Using this substitution, $\mathrm{d}y = v\mathrm{d}x + x\mathrm{d}v$ and, with a few steps of algebraic manipulation, it is possible to express the resulting equation as a separable equation in x and v, as seen in Example 16.16.

Example 16.16

Solve the differential equation $xyy' = 3x^2 + y^2$

Solution

This equation is homogeneous since:

$$y' = f(x, y) = \frac{3x^2 + y^2}{xy} \Rightarrow f(tx, ty) = \frac{3(tx)^2 + (ty)^2}{(tx)(ty)} = \frac{3x^2 + y^2}{xy} = y'$$

Several approaches are possible, one of which is:

$$y' = \frac{\mathrm{d}y}{\mathrm{d}x} = \frac{3x^2 + y^2}{xy} \Rightarrow \frac{v\mathrm{d}x + x\mathrm{d}v}{\mathrm{d}x} = \frac{3x^2 + v^2x^2}{x(vx)} = \frac{3 + v^2}{v}$$

$$\Rightarrow v^2\,\mathrm{d}x + xv\mathrm{d}v = 3\mathrm{d}x + v^2\,\mathrm{d}x \Rightarrow v\mathrm{d}v = \frac{3}{x}\mathrm{d}x$$

$$\Rightarrow \int v\mathrm{d}v = \int \frac{3}{x}\mathrm{d}x + c$$

$$\Rightarrow \frac{v^2}{2} = 3\ln|x| + c \Rightarrow v^2 = 6\ln|x| + 2c$$

By substituting the value of v back and simplifying, $y^2 = 6x^2(6\ln|x| + 2c)$

Example 16.17

Find the general solution of $(x^2 - y^2)dx + 3xydy = 0$

Solution

Since $(x^2 - y^2)$ and $3xy$ are both homogeneous of degree 2, we can use the method suggested above.

Substitute $y = vx$ and $dy = vdx + xdv$ into the given equation

$$(x^2 - y^2)dx + 3xydy = 0 \Rightarrow (x^2 - x^2v^2)dx + 3x^2v(vdx + xdv) = 0$$

$$\Rightarrow x^2(1 + 2v^2)dx + 3x^3vdv = 0$$

Divide by x^2, simplify and integrate

$$(1 + 2v^2)dx + 3xvdv = 0 \Rightarrow \int \frac{dx}{x} = -3\int \frac{vdv}{1 + 2v^2}$$

$$\ln|x| = -\frac{3}{4}\ln(1 + 2v^2) + C_1 \Rightarrow 4\ln|x| = -3\ln(1 + 2v^2) + 4C_1$$

Now let $4C_1 = \ln C$ and simplify

$$\ln x^4 = \ln(C(1 + 2v^2)^{-3}) \Rightarrow x^4 = C(1 + 2v^2)^{-3}$$

Substitute $v = \dfrac{y}{x}$ into this equation and simplify

$$x^4 = C\left(1 + 2\frac{y^2}{x^2}\right)^{-3} \Rightarrow x^4\left(1 + 2\frac{y^2}{x^2}\right)^3 = C \Rightarrow (x^2 + 2y^2)^3 = Cx^2$$

Exercise 16.1

1. **(a)** Show that every member of the family of functions $f(x) = \frac{1}{x}(\ln x + k)$

   is a solution of the differential equation $x^2\dfrac{dy}{dx} + xy - 1 = 0$

   (b) Write the equation in the form $\dfrac{dy}{dx} = F(x, y)$, and illustrate **(a)** by

   graphing two members of the family on a common screen.

   (c) Find a solution that satisfies the initial conditions $y(1) = 2$ and $y(2) = 1$.

2. Solve each differential equation.

   (a) $y'e^{y-x} = 1$ **(b)** $\dfrac{dy}{dx} = y^2x + x$

   (c) $e^{x-y}dy = xdx$ **(d)** $y' = xy^2 - x - y^2 + 1$

   (e) $(xy \ln x)y' = (y + 1)^2$ **(f)** $\dfrac{dy}{dx} = \dfrac{1 + 2y^2}{y \sin x}$

(g) $(1 + \tan y)y' = x^2 + 1$

(h) $\dfrac{dy}{dt} = \dfrac{te^t}{y\sqrt{y^2 + 1}}$

(i) $y \sec\theta \, dy = e^y \sin^2\theta \, d\theta$

(j) $\dfrac{dy}{dx} = e^x(1 + y^2)$

(k) $\dfrac{dy}{dx} = x\sqrt{\dfrac{1 - y^2}{1 - x^2}}$, $y(0) = 0$

(l) $y'(1 + e^x) = e^{x-y}$, $y(1) = 0$

(m) $(y + 1)dy = (x^2y - y)dx$, $y(3) = 1$

(n) $xy' - y = 2x^2y$, $y(1) = 1$

(o) $xy\,dx + e^{-x^2}(y^2 - 1)dy = 0$, $y(0) = 1$

3. Solve each initial value problem.

(a) $x^{-3}dy = 4y\,dx$, $y(0) = 3$

(b) $\dfrac{dy}{dx} = xy$, $y(0) = 1$

(c) $y' - xy^2 = 0$, $y(1) = 2$

(d) $y' - y^2 = 0$, $y(2) = 1$

(e) $\dfrac{dy}{dx} - e^y = 0$, $y(0) = 1$

(f) $\dfrac{dy}{dx} = y^{-2}x + y^{-2}$, $y(0) = 1$

(g) $x\,dy - y^2\,dx = -dy$, $y(0) = 1$

(h) $y^2\,dy - x\,dx = dx - dy$, $y(0) = 3$

(i) $yy' = xy^2 + x$, $y(0) = 0$

(j) $y' = \dfrac{xy - y}{y + 1}$, $y(2) = 1$

(k) $\dfrac{dy}{dx} = x\sqrt{\dfrac{1 - y^2}{1 - x^2}}$, $y(0) = 0$

(l) $y'(1 + e^x) = e^{x-y}$, $y(1) = 0$

(m) $(y + 1)dy = (x^2y - y)dx$, $y(3) = 1$

(n) $xy' - y = 2x^2y$, $y(1) = 1$

(o) $xy\,dx + e^{-x^2}(y^2 - 1)dy = 0$, $y(0) = 1$

4. Solve each differential equation.

(a) $2x\dfrac{dy}{dx} = x + y$

(b) $(x + y)dy = (x - y)dx$

(c) $(x^2 - y^2)y' = xy$

(d) $x\,dy - \left(2xe^{-\frac{y}{x}} + y\right)dx = 0$, $y(1) = 0$

(e) $\left(x\sec\dfrac{y}{x} + y\right)dx = x\,dy$, $y(1) = 0$

(f) $(x^2 + y^2)dx = (xy - x^2)dy$

(g) $y\,dx = (x + \sqrt{xy})dy$

5. The temperature T, in Celsius, of a kettle in a room satisfies the differential equation $\dfrac{dT}{dt} = m(T - 21)$, where t is in minutes and m is a constant.

(a) Solve the differential equation, showing that $T = Ce^{mt} + 21$, where C is an arbitrary constant.

(b) Given that $T(0) = 99$, and $T(15) = 69$, find:

    (i) the exact values of m and C

    (ii) the value of t when $T = 39$.

6. Studies indicate that there may be as few as 3200 tigers (*Panthera tigris*) left in the wild. The WWF arranged for the release of 25 tigers into a wildlife preserve. After two years there were 39 tigers in the preserve. The preserve has a capacity of 200 tigers.

   (a) Write a logistic equation that models the population of tigers in the preserve.

   (b) Find the population in the preserve after five years.

   (c) When will the population in the preserve reach 100?

7. The logistic equation $P(t) = \dfrac{2100}{1 + 29\,e^{-0.75t}}$ models the growth of a certain population.

   (a) Find the initial population.

   (b) Determine when the population will reach 50% of its carrying capacity.

   (c) Write a logistic differential equation that has the solution $P(t)$.

8. The number of bacteria in a certain culture grows at a rate that is proportional to the number of bacteria present.

   If the number increased from 500 to 2000 in two hours, determine

   (a) the number present after 12 hours

   (b) the doubling time.

16.2 First order linear differential equations – use of integrating factor

A first order differential equation has only the first derivative $\dfrac{dy}{dx}$ in the equation. A differential equation is linear when both $\dfrac{dy}{dx}$ and y appear only to the first power. The standard form for a first order linear differential equation is

$$\frac{dy}{dx} + P(x)y = Q(x)$$

It can also be written as $y' + P(x)y = Q(x)$.

We will now look at a method of solving first order linear differential equations of this form.

We will approach the solution using three cases of increasing complexity:

Case 1

$P(x)$ is a constant, $P(x) = p$ and $Q(x) = 0$

The equation has constant coefficients and is homogeneous.

$$\frac{dy}{dx} + py = 0 \Leftrightarrow \frac{dy}{y} = -p\,dx$$

This leads to the solution with direct integration:

$$\ln|y| = -px + c \Rightarrow y = Ce^{-px}$$

Case 2

$P(x)$ is a constant, $P(x) = p$ and $Q(x) \neq 0$

The equation has constant coefficients and is non-homogeneous.

$$\frac{dy}{dx} + py = Q(x)$$

where p is a constant. It is not possible to separate the variables. However, the left-hand side is a truncated part of the derivative of a product, and so we can find another function, u, that enables us to write the left-hand side as

$$\frac{dy}{dx}u + y\frac{du}{dx},$$

which is equivalent to

$$\frac{dy}{dx}u + y\frac{du}{dx} = \frac{d(y \cdot u)}{dx}$$

This can only happen if $u = e^{px}$ because

$$\frac{d}{dx}(y \cdot u) = \frac{d}{dx}(y \cdot e^{px}) = \frac{dy}{dx}e^{px} + y\frac{d}{dx}(e^{px}) = \frac{dy}{dx}e^{px} + y \cdot p \cdot e^{px}$$

Thus, if we multiply both sides of the differential equation by e^{px}, which is called an **integrating factor**, we get:

$$e^{px}\left(\frac{dy}{dx} + py\right) = e^{px}Q(x) \Rightarrow e^{px}\frac{dy}{dx} + pe^{px}y = e^{px}Q(x)$$

$$\Rightarrow \frac{d}{dx}(e^{px}y) = e^{px}Q(x),$$

and upon integration:

$$e^{px}y = \int e^{px}Q(x)dx + C,$$

where $\int e^{px}Q(x)dx$ can be integrated using familiar methods of integration. This finally leads to the general solution

$$y = \frac{1}{e^{px}}\int e^{px}Q(x)dx + C = e^{-px}\int e^{px}Q(x)dx + Ce^{-px}$$

The term e^{px} is called an integrating factor for the differential equation because it allows us, after multiplication, to solve the differential equation by integration.

Example 16.18

Find all solutions to

$$\frac{dy}{dx} + 5y = 3x$$

Solution

Multiply both sides of the equation by e^{5x}:

$$e^{5x}\frac{dy}{dx} + 5e^{5x}y = 3xe^{5x} \Leftrightarrow d(e^{5x}y) = 3xe^{5x}dx$$

$$\Leftrightarrow \int d(e^{5x}y) = \int 3xe^{5x}dx \Rightarrow e^{5x}y = \int 3xe^{5x}dx$$

The right-hand side can be integrated by parts to give

$$\frac{3x}{5}e^{5x} - \frac{3}{25}e^{5x} + C \Rightarrow e^{5x}y = \frac{3x}{5}e^{5x} - \frac{3}{25}e^{5x} + C$$

$$\Rightarrow y = \frac{3x}{5} - \frac{3}{25} + Ce^{-5x}$$

We can check this answer by differentiating it:

$$y = \frac{3x}{5} - \frac{3}{25} + Ce^{-5x} \Rightarrow \frac{dy}{dx} = \frac{3}{5} - 5Ce^{-5x}, \text{ thus}$$

$$\frac{dy}{dx} + 5y = \frac{3}{5} - 5Ce^{-5x} + 5\left(\frac{3x}{5} - \frac{3}{25} + Ce^{-5x}\right)$$

$$= \frac{3}{5} - 5Ce^{-5x} + 3x - \frac{3}{5} + 5Ce^{-5x} = 3x$$

Case 3

$P(x)$ is not a constant and $Q(x) \neq 0$.

The equation has variable coefficients and is non-homogeneous.

$$\frac{dy}{dx} + P(x)y = Q(x)$$

We first note the following facts:

From the fundamental theorem, Chapter 14: $\dfrac{d}{dx}\int P(x)dx = P(x)$ **(1)**

Using the chain rule: $\dfrac{d}{dx}e^{\int P(x)dx} = e^{\int P(x)dx} \cdot \dfrac{d}{dx}\int P(x)dx = P(x)e^{\int P(x)dx}$ **(2)**

Now consider the general differential equation

$$\frac{dy}{dx} + P(x)y = Q(x)$$

Multiply both sides by the integrating factor $e^{\int P(x)dx}$. Then we have:

$$e^{\int P(x)dx}\frac{dy}{dx} + P(x)e^{\int P(x)dx}y = e^{\int P(x)dx}f(x)$$

From **(2)**:

$$\frac{d}{dx}\left(ye^{\int P(x)dx}\right) = y\frac{d}{dx}\left(e^{\int P(x)dx}\right) + \frac{dy}{dx}e^{\int P(x)dx} = P(x)e^{\int P(x)dx}y + \frac{dy}{dx}e^{\int P(x)dx}$$

and by comparing the last two equations, we have

$$\frac{d}{dx}\left(ye^{\int P(x)dx}\right) = e^{\int P(x)dx}Q(x)$$

And so, by integration

$$ye^{\int P(x)dx} = \int e^{\int P(x)dx} Q(x)\,dx + C \Rightarrow y = \frac{1}{e^{\int P(x)dx}}\left(\int e^{\int P(x)dx} Q(x)\,dx + C\right)$$

$$\Rightarrow y = e^{-\int P(x)dx}\int e^{\int P(x)dx} Q(x)\,dx + Ce^{-\int P(x)dx}$$

Example 16.19

Find all solutions to $2x\dfrac{dy}{dx} + 8y = 5x^4$

Solution

This equation can be transformed into the standard form of a linear equation by dividing through by $2x$:

$$\frac{dy}{dx} + \frac{4}{x}y = \frac{5}{2}x^3$$

Here $P(x) = \dfrac{4}{x} \Rightarrow \int P(x)dx = \int \dfrac{4}{x}dx = 4\ln x = \ln x^4$, and $e^{\int P(x)dx} = e^{\ln x^4} = x^4$

So, our integrating factor is x^4.

Multiplying by x^4 we obtain:

$$x^4\,dy + 4x^3 y\,dx = \frac{5}{2}x^7\,dx \Leftrightarrow d(x^4 y) = \frac{5}{2}x^7\,dx$$

Finally, by integration we have

$$x^4 y = \frac{5x^8}{16} + C \Rightarrow y = \frac{5x^4}{16} + Cx^{-4}$$

Although the expression for the general solution given above looks quite complex, the basic steps for solving a first order linear differential equation by means of an integrating factor are relatively modest. Here are some practical steps to follow:

- If the differential equation is given as

$$f(x)\frac{dy}{dx} + g(x) = h(x)$$

  rewrite it in the form

$$\frac{dy}{dx} + P(x)y = Q(x), \text{ where } P(x) = \frac{g(x)}{f(x)}, \text{ and } Q(x) = \frac{h(x)}{f(x)}.$$

- Find the integrating factor $e^{\int P(x)dx}$ by first finding $\int P(x)dx$.
- Multiply both sides of the equation by the integrating factor.
- By integration, the left side is $ye^{\int P(x)dx}$. Evaluate the right side: $\int e^{\int P(x)dx} Q(x)$
- Write down the general solution

$$y = \frac{1}{e^{\int P(x)dx}}\left(\int e^{\int P(x)dx} Q(x)\,dx + C\right)$$

Example 16.20

Find the initial value solution of $y' + y\tan x = \cos^2 x$, $y(0) = 2$

Solution

$P(x) = \tan x$, $Q(x) = \cos^2 x$

Integrating factor:

$$e^{\int \tan x \, dx} = e^{\ln(\sec x)} = \sec x$$

Now multiply the equation by the integrating factor and simplify:

$$\sec x \left(\frac{dy}{dx} + y\tan x = \cos^2 x \right) \Rightarrow \sec x dy + y\sec x \tan x dx = \sec x \cos^2 x dx$$

$$\Rightarrow \sec x dy + y\sec x \tan x dx = \cos x dx \Rightarrow d(y\sec x) = \cos x dx$$

$$\Rightarrow y\sec x = \int \cos x dx \Rightarrow y\sec x = \sin x + c$$

$$\Rightarrow y = \frac{\sin x + c}{\sec x} = (\sin x + c)\cos x$$

$y(0) = 2 \Rightarrow (\sin 0 + c)\cos 0 = 2 \Rightarrow c = 2$

Therefore, the solution is $y = (\sin x + 2)\cos x$

Example 16.21

Find the solution to:
$$\frac{dy}{dt} = \frac{2t}{1 + t^2}y + \frac{2}{1 + t^2}, \, y(0) = 0.4$$

Solution

$$\frac{dy}{dt} - \frac{2t}{1 + t^2}y = \frac{2}{1 + t^2}$$

Integrating factor:

$$e^{\int -\frac{2t}{1+t^2}y} = e^{-\ln(1+t^2)} = e^{\ln\left(\frac{1}{1+t^2}\right)} = \frac{1}{1 + t^2}$$

Multiply by the integrating factor and simplify:

$$\left(\frac{dy}{dt} - \frac{2t}{1 + t^2}y = \frac{2}{1 + t^2} \right)\frac{1}{1 + t^2}$$

$$\Rightarrow \frac{1}{1 + t^2}dy - y\frac{2t}{1 + t^2}\frac{1}{1 + t^2}dt = \frac{2}{1 + t^2}\frac{1}{1 + t^2}dt$$

$$\Rightarrow d\left(y \cdot \frac{1}{1 + t^2}\right) = \frac{2}{(1 + t^2)^2}dt$$

Integrate to find the solution:

$$y \cdot \frac{1}{1 + t^2} = \int \frac{2}{(1 + t^2)^2} \, dt \Rightarrow y = (1 + t^2)\left(\tan^{-1} t + \frac{t}{1 + t^2} + c\right)$$

With $y(0) = 0.4$, $c = 0.4$ and the particular solution will be

$$y = (1 + t^2)\left(\tan^{-1} t + \frac{t}{1 + t^2} + 0.4\right)$$

Example 16.22

Solve the following initial value problem for $x > 0$

$$2x\frac{dy}{dx} - y = x + 1, \, y(2) = 4$$

Solution

Rewrite by dividing by $2x$:

$$\frac{dy}{dx} - \frac{1}{2x} \cdot y = \frac{x + 1}{2x} \Rightarrow dy - \frac{1}{2x} \cdot y \, dx = \frac{x + 1}{2x} dx$$

Integrating factor:

$$e^{\int -\frac{1}{2x} dx} = e^{-\frac{\ln x}{2}} = x^{-\frac{1}{2}}$$

Multiply by the integrating factor and simplify:

$$x^{-\frac{1}{2}} dy - x^{-\frac{1}{2}}\frac{1}{2x} \cdot y \, dx = x^{-\frac{1}{2}}\frac{x + 1}{2x} dx$$

$$x^{-\frac{1}{2}} dy - \frac{1}{2x^{\frac{3}{2}}} \cdot y \, dx = x^{-\frac{1}{2}}\left(\frac{1}{2} + \frac{1}{2x}\right) dx$$

$$x^{-\frac{1}{2}} y = \int\left(\frac{1}{2}x^{-\frac{1}{2}} + \frac{1}{2}x^{-\frac{3}{2}}\right) dx = x^{-\frac{1}{2}} - x^{-\frac{1}{2}} + c$$

$$\Rightarrow y = \frac{x^{\frac{1}{2}} - x^{-\frac{1}{2}} + c}{x^{-\frac{1}{2}}} = x - 1 + cx^{\frac{1}{2}}$$

$$y(2) = 4 \Rightarrow 4 = 2 - 1 + c \cdot 2^{\frac{1}{2}} \Rightarrow c = \frac{3}{\sqrt{2}}$$

Therefore, $y = x - 1 + \frac{3}{\sqrt{2}}\sqrt{x}$

Application to electric circuits

An electromotive force (usually supplied by a battery or generator) produces a potential difference of $E(t)$ volts (V) and a current of $I(t)$ amperes (A) at time t. The circuit in Figure16.6 also contains a resistor with a resistance of R ohms (Ω) and an inductor with an inductance of L henries (H).

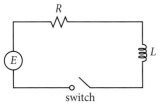

Figure 16.6 Electric circuit

Ohm's law gives the potential difference across the resistor as RI. The potential difference across the inductor is $L\left(\dfrac{\mathrm{d}I}{\mathrm{d}t}\right)$. One of Kirchhoff's laws says that the sum of the voltage drops is equal to the supplied voltage $E(t)$. Thus, we have

$$L\frac{\mathrm{d}I}{\mathrm{d}t} + RI = E(t)$$

which is a first order linear differential equation. The solution gives the current I at time t.

Example 16.23

Suppose that in the simple circuit of Figure 16.6 the resistance is $12\,\Omega$ and the inductance is $4\,\mathrm{H}$. The power supply provides a constant potential difference of $60\,\mathrm{V}$ and the switch is closed when $t = 0$ so the current starts with $I(0) = 0$. Find:

(a) $I(t)$

(b) the current after 1 second

(c) the limiting value of the current.

Solution

(a) The situation is modelled by the differential equation $L\dfrac{\mathrm{d}I}{\mathrm{d}t} + RI = E(t)$.

In this case $L = 4$, $R = 12$, and $E(t) = 60$.

$$L\frac{\mathrm{d}I}{\mathrm{d}t} + RI = E(t) \Rightarrow 4\frac{\mathrm{d}I}{\mathrm{d}t} + 12I = 60 \Rightarrow \frac{\mathrm{d}I}{\mathrm{d}t} + 3I = 15, \quad I(0) = 0$$

The integrating factor is $e^{\int 3\mathrm{d}t} = e^{3t}$

Now solve the linear equation:

$$e^{3t}\frac{\mathrm{d}I}{\mathrm{d}t} + 3e^{3t}I = 15e^{3t} \Rightarrow \frac{\mathrm{d}}{\mathrm{d}t}(e^{3t}I) = 15e^{3t} \Rightarrow I = \frac{1}{e^{3t}}(5e^{3t} + C)$$

With the initial condition of $I(0) = 0$, we have $0 = 5 + C$, so $C = -5$

Therefore, $I(t) = \dfrac{1}{e^{3t}}(5e^{3t} - 5) = 5(1 - e^{-3t})$

(b) After 1 second, the current is $I(1) = 5(1 - e^{-3}) \approx 4.75\,\mathrm{A}$

(c) The limiting value of the current is given by

$$\lim_{t \to \infty} I(t) = \lim_{t \to \infty} 5(1 - e^{-3t}) = 5\,\mathrm{A}$$

Example 16.24

Suppose that in the circuit in Figure 16.6, the resistance and inductance remain constant. A generator produces a variable potential difference of $E(t) = 120\sin 60t$ volts and the switch is closed when $t = 0$ so the current starts with $I(0) = 0$. Find $I(t)$.

Solution

The differential equation is now:

$$L\frac{dI}{dt} + RI = E(t) \Rightarrow 4\frac{dI}{dt} + 12I = 120\sin 60t$$

$$\Rightarrow \frac{dI}{dt} + 3I = 30\sin 60t, \quad I(0) = 0$$

Integrating factor is the same as before:

$$e^{3t}\frac{dI}{dt} + 3e^{3t}I = 30e^{3t}\sin 60t \Rightarrow \frac{d}{dt}(e^{3t}I) = 30\,e^{3t}\sin 60t$$

$$\Rightarrow I = \frac{1}{e^{3t}}\left(e^{3t}\left(\frac{10\sin 60t}{401} - \frac{200\cos 60t}{401}\right) + C\right)$$

$$= \frac{10}{401}(\sin 60t - 2\cos 60t) + Ce^{-3t}$$

With the initial condition of $I(0) = 0$, we have $0 = \frac{10}{401}(-2) + C \Rightarrow C = \frac{20}{401}$

Therefore,

$$I(t) = \frac{10}{401}(\sin 60t - 2\cos 60t) + \frac{20}{401}e^{-3t} = \frac{10}{401}(\sin 60t - 2\cos 60t + 2e^{-3t})$$

Application to mixture solutions

Example 16.25

A tank contains 50 litres of a solution composed of 90% water and 10% hydrochloric acid. A second solution containing 50% water and 50% hydrochloric acid is being poured into the tank at a rate of 4 litres per minute. At the same time, the tank is being drained at a rate of 5 litres per minute.

(a) How much acid will there be in the tank after t minutes?

(b) How much acid will there be in the tank after 10 minutes?

Solution

Let y be the number of litres of acid in the tank at any time t. Since acid is 10% of the contents, the number of litres in the tank is $y = 5$ when $t = 0$. Since the tank loses 1 litre of solution per minute, the number of litres of solution in the tank at any time t is $50 - t$, and as the tank loses 5 litres of solution per minute, it must lose $\frac{5}{50 - t}y$ litres of acid every minute.

Furthermore, as the tank is also gaining 2 litres of acid per minute, the rate of change of acid in the tank is

$$\frac{dy}{dt} = 2 - \frac{5}{50 - t}y \Rightarrow \frac{dy}{dt} + \frac{5}{50 - t}y = 2$$

(a) This is a linear differential equation with $P(t) = \dfrac{5}{50 - t}$

The integrating factor is $\dfrac{1}{(50 - t)^5}$

So, the general solution is $y = \dfrac{50 - t}{2} + C(50 - t)^5$

Because $y = 5$ when $t = 0$, we have $C = -\dfrac{20}{50^5}$

Thus, the amount of acid in the tank after t minutes is given by:

$$y = \frac{50 - t}{2} - 20\left(\frac{50 - t}{50}\right)^5$$

(b) When $t = 10$, the amount of acid is

$$y = \frac{50 - 10}{2} - 20\left(\frac{50 - 10}{50}\right)^5 \approx 13.45 \text{ litres}$$

Exercise 16.2

1. Determine whether or not each differential equation is linear.

   (a) $3\dfrac{dy}{dx} + xy = y^2$

   (b) $x^3\dfrac{dy}{dx} - 3y + 2x^2 = 0$

   (c) $5xy' = 3x - 2y$

   (d) $2y\dfrac{dy}{dx} = \sin 2x$

2. Find the general solution of $x\dfrac{dy}{dx} - 2y = x^2$

3. Find the particular solution of $(x^2 + 1)\dfrac{dy}{dx} + xy = (1 - 2x)\sqrt{x^2 + 1}$

   given that $y = 2$ when $x = 1$

4. Find the general solution to the equation $\dfrac{dy}{dx} = x - y$

5. Find the particular solution to $(1 + \sin x)\dfrac{dy}{dx} - y\cos x = (1 + \sin x)^4$
   given $y(0) = 1$

6. Solve each differential equation.

   (a) $e^x\dfrac{dy}{dx} + 2e^x y = 1$

   (b) $x\dfrac{dy}{dx} + 3y = \dfrac{\sin x}{x^3}$

   (c) $(x - 1)^3\dfrac{dy}{dx} + 4(x - 1)^2 = x + 1$

   (d) $\sin x\dfrac{dy}{dx} + y\cos x = \tan x$

   (e) $xy' - 2y = x^2$

   (f) $y' = x - y$

   (g) $xy' + y = \sqrt{x}$

   (h) $\sin x\dfrac{dy}{dx} + y\cos x = \sin(x^2)$

   (i) $(1 + t)\dfrac{du}{dt} + u = 1 + t, \quad t > 0$

(j) $x\dfrac{dy}{dx} + 2y = e^{x^2}$

(k) $\cos x \dfrac{dy}{dx} - \sin 2x = y \sin x, \quad -\dfrac{\pi}{2} < x < \dfrac{\pi}{2}$

(l) $x\dfrac{dy}{dx} + y = e^{-x} - xy, \quad x > 0$

7. Solve each initial value problem.

(a) $\dfrac{dy}{dx} + 2y = x, \quad y(0) = 1$

(b) $x\dfrac{dy}{dx} - 2y = x^3 \sec x \tan x, \quad y\!\left(\dfrac{\pi}{3}\right) = 2$

(c) $x^2 y' + 2xy = \ln x, \quad y(1) = 2$

(d) $t\dfrac{du}{dt} = t^2 + 3u, t > 0, \quad u(2) = 4$

(e) $xy' = y + x^2 \sin x, \quad y(\pi) = 0$

(f) $\dfrac{y'}{2y} - x = \dfrac{x e^{x^2}}{y}, \quad y(0) = 3$

8. The diagram shows a circuit containing a power source with potential difference E(t), a capacitor with a capacitance of C farads (F), and a resistor with a resistance of R ohms (Ω). The potential difference across the capacitor is Q/C, where Q is the charge (in coulombs). In this case, Kirchhoff's law gives

$$RI + \dfrac{Q}{C} = E(t)$$

Given that $I = \dfrac{dQ}{dt}$, we have $R\dfrac{dQ}{dt} + \dfrac{1}{C}Q = E(t)$

Suppose the resistance is 5 Ω, the capacitance is 0.05 F, a battery gives a constant potential difference of 60 V, and the initial charge is $Q(0) = 0$. Find the charge and the current at time t.

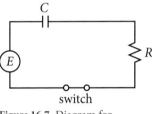

Figure 16.7 Diagram for question 8

16.3 Numerical solutions: Euler's method

Slope fields

Often the primary objective when solving a first order differential equation is to find an explicit solution. However, many differential equations used in mathematical models cannot be solved by means of an analytical method. For such equations, we must resort to graphical and/or numerical methods. These can be carried out by hand or by technology, and provide us with rough qualitative information about the graph of a solution to a differential equation.

A first order differential equation in the form $\dfrac{dy}{dx} = F(x, y)$ specifies the slope of the **solution curve** $y = f(x)$ at each point in the xy-plane where F is defined. We can use this fact to draw a short line segment whose slope is $F(x, y)$ at any point (x, y) in the plane. A plot of these line segments showing the slope (or direction) of the solution curve is a called a **slope field** (or direction field) for the first order differential equation. As a rule, the segments are drawn at representative points evenly spaced in both directions.

Figure 16.8 shows a slope field for the equation $\dfrac{dy}{dx} = x - y$.

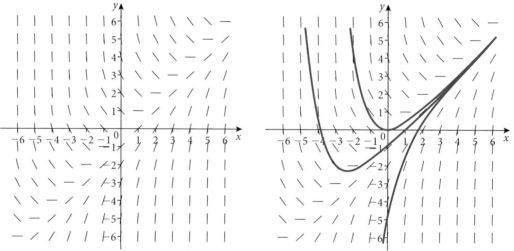

Figure 16.8 Slope field for the equation $\dfrac{dy}{dx} = x - y$

Figure 16.9 Solutions to a differential equation can be sketched

As you can imagine, it can be quite tedious to draw a slope field by hand. In practice, slope fields are easily generated by suitable graphing technology.

Solutions to a differential equation can be sketched by drawing curves that are at each point tangent to the line segment at that point. Thus, a family of solution curves can be produced. To use a slope field to sketch a particular solution, all we need to know is one point (an initial condition) that the solution curve passes through.

In this course, you will not be asked to draw slope fields. They are mentioned here as they help you understand the method used in the following procedure.

Euler's method

In cases where we are interested in finding a particular solution, the numerous line segments of a slope field could be distracting. Euler's method developed below enables us to approximate a single solution curve. The method is based almost entirely on the idea of a slope field.

Consider the initial-value problem:

$$\frac{dy}{dx} = x + y, \quad y(2) = 0$$

Figure 16.10 shows the slope field for the differential equation $\dfrac{dy}{dx} = x + y$.

An approximation to the particular solution can be sketched by drawing a smooth curve through the point (2, 0) that follows the slopes in the slope field.

Let $y(x)$ represent the solution curve. To approximate a value of y for a specific value of x, for example when $x = 3$, we could make an educated guess from the sketch of y made with the aid of the slope field. But if we want a more accurate approximation, we need to use a more refined method. The simplest numerical method is called **Euler's method**, after the prolific 18th-century mathematician who first devised this computational method to help him calculate the orbit of the moon.

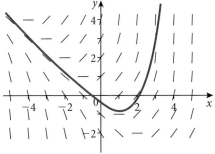

Figure 16.10 Slope field for the equation $\dfrac{dy}{dx} = x + y$ with an approximation to the solution curve sketched in

Euler's method uses the basic idea behind the construction of slope fields to find numerical approximations to solutions of differential equations. Let's illustrate the method with the initial-value problem that we have just been considering.

We know from the differential equation that the slope of the solution curve is 2 at the point (2, 0) because $\dfrac{dy}{dx} = x + y = 2 + 0 = 2$. Hence, the tangent to the solution curve at (2, 0) has the equation: $y - 0 = 2(x - 2) \Rightarrow y = 2x - 4$. We can use this tangent line as a rough approximation to the solution curve (see Figure 16.11). This approximation clearly becomes less accurate as we move away from the point of tangency (2, 0).

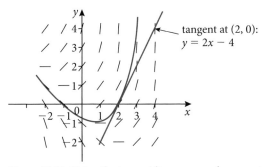

tangent at (2, 0):
$y = 2x - 4$

Figure 16.11 Using the tangent line as a rough approximation to the solution curve

Euler's method improves this approximation by moving a short horizontal distance (the **step size** h) along this tangent line and then changing direction according to the slope field. In this way we build an approximation to the curve by attaching little line segments together, each having the slope of the solution curve at its starting point.

In general, after being presented with an initial value problem $\dfrac{dy}{dx} = F(x, y)$,

$y(x_0) = y_0$, we choose a step size h. Starting at the point (x_0, y_0), for the interval $x_0 \leqslant x \leqslant x_0 + h$, we approximate the solution curve with the tangent line; that is, the line with slope $F(x_0, y_0)$. This takes us as far as the point (x_1, y_1), whose coordinates are calculated as follows:

$$x_1 = x_0 + h, \; y_2 = y_1 + hF(x_1, y_1)$$

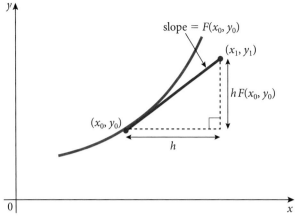

Figure 16.12 Approximating the solution curve for $\dfrac{dy}{dx} = F(x, y)$

Repeating this process, we get an approximation to the solution curve consisting of line segments joining the points (x_0, y_0), (x_1, y_1), (x_2, y_2), ...

Each computed value y_n is an estimate of the corresponding true solution y at $x = x_n$. The accuracy of the estimates depends on the choice of the step size h and the overall number of steps (iterations). Decreasing the step size while increasing the number of steps leads to increasingly more accurate estimates for solution values.

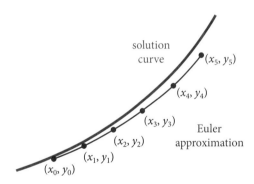

Figure 16.13 Euler approximation to a solution curve

Euler's numerical method

For the differential equation $\dfrac{dy}{dx} = F(x, y)$ with the initial condition $y(x_0) = y_0$, the recursive formulas for generating the coordinates of the unknown $(n + 1)$st point (x_{n+1}, y_{n+1}) from the known nth point (x_n, y_n) on the approximate solution curve (Euler approximation) are:

$x_{n+1} = x_n + h$, $y_{n+1} = y_n + hF(x_n, y_n)$, for $n = 1, 2, 3, ... N$,

where h, the step size, is a constant; and N is the total number of steps (iterations).

Let's now apply Euler's method to solve the equation at the start of this section.

Example 16.26

For the differential equation $\dfrac{dy}{dx} = x + y$, such that $y(2) = 0$, use Euler's method with a step value of 0.2 to find an approximate value of y when $x = 3$, giving your answer to 2 decimal places.

Solution

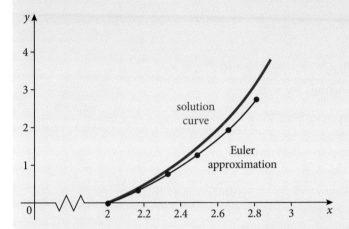

We use Euler's method to build an approximation to the true solution curve starting at $x = 2$ and finishing at $x = 3$ by piecing together five short segments. We are given that $h = 0.2$, $x_0 = 2$, $y_0 = 0$ and $F(x, y) = x + y$. Using the appropriate formulas for x_n and y_n and iterating five times:

$x_1 = x_0 + h = 2 + 0.2 = 2.2$ $\qquad$ $y_1 = y_0 + hF(x_0, y_0) = 0 + 0.2(2 + 0)$
$$= 0.4$$

$x_2 = x_1 + h = 2.2 + 0.2 = 2.4$ $\qquad$ $y_2 = y_1 + hF(x_1, y_1) = 0.4 + 0.2(2.2 + 0.4)$
$$= 0.92$$

$x_5 = x_4 + h = 2.8 + 0.2 = 3$ $\qquad$ $y_5 = y_4 + hF(x_4, y_4) = 2.4208 + 0.2(2.8 +$
$$2.4208) = 3.46496$$

This process leads to an approximate (2 decimal places) value of $y \approx 3.46$ when $x = 3$.

Because we will perform most of the calculations for each iteration on our GDC, it is often sufficient to simply display relevant results for each iteration in a table.

n	x_n	y_n	$hF(x_n, y_n)$	x_{n+1}	y_{n+1}
0	2	0	0.4	2.2	0.4
	↓	↓			↓
4	2.8	2.4208	1.04416	3.0	3.46496

The first order differential equation in Example 16.26 is linear, and hence can be solved using an integrating factor. Given $y(2) = 0$, the particular solution is $y = 3e^{x-2} - x - 1$.
To 3 significant figures, the true value of $y(3)$ is 4.15. Thus, our approximation of 3.46 has an error of approximately 16.6%. Using a program on our GDC or a spreadsheet, we could easily decrease the step size (increasing the number of steps) in order to improve the accuracy of the approximation. For example, if we used a step size of $h = 0.01$ (requiring 100 iterations) we would get an estimate of 4.11 (3 s.f.) reducing the error to less than 1%.

A numerical method like Euler's is especially useful when applied to a differential equation that cannot be solved by any known analytical methods.

Example 16.27

Given that $\dfrac{dy}{dx} = \dfrac{x+1}{xy+2}$ and $y = 1$ when $x = 0$, use Euler's method with step size $h = 0.25$ to approximate the value of y when $x = 1$. Give the approximation to 3 significant figures.

Solution

We have that $x_0 = 0$, $y_0 = 1$, $h = 0.25$ and $F(x, y) = \dfrac{x+1}{xy+2}$.

Thus the recursive formula for y_n is:

$$y_{n+1} = y_n + h\,F(x, y) = y_n + (0.25)\frac{x_n + 1}{x_n y_n + 2} \Rightarrow y_{n+1} = y_n + \frac{x_n + 1}{4x_n y_n + 8}$$

$n = 0$: $\quad x_1 = x_0 + h = 0 + 0.25 = 0.25$

$$y_1 = y_0 + \frac{x_0 + 1}{4x_0 y_0 + 8} = 1 + \frac{0 + 1}{4(0)(1) + 8} = \frac{9}{8} = 1.125$$

$n = 1$: $\quad x_2 = x_1 + h = 0.25 + 0.25 = 0.5$

$$y_2 = y_1 + \frac{x_1 + 1}{4x_1 y_1 + 8} = 1.125 + \frac{0.25 + 1}{4(0.25)(1.125) + 8} \approx 1.261986$$

$n = 2$: $\quad x_3 = x_2 + h = 0.5 + 0.25 = 0.75$

$$y_3 = y_2 + \frac{x_2 + 1}{4x_2 y_2 + 8} = 1.261986 + \frac{0.5 + 1}{4(0.5)(1.261986) + 8} \approx 1.404518$$

$n = 3$: $\quad x_4 = x_3 + h = 0.75 + 0.25 = 1$

$$y_4 = y_3 + \frac{x_3 + 1}{4x_3 y_3 + 8} = 1.404518 + \frac{0.75 + 1}{4(0.75)(1.404518) + 8} \approx 1.547801$$

Therefore, the approximate value of y when $x = 1$ is $y \approx 1.55$.

Exercise 16.3

In all the exercises for this section, a GDC, CAS, or computer is recommended.

1. Use Euler's method with step size 0.1 to estimate $y(0.5)$, where $y(x)$ is the solution of the initial-value problem $y' = y + xy$, $y(0) = 1$

2. (a) Use Euler's method to calculate $y(1)$, where $y(x)$ is the solution of the initial value problem

   $$\frac{dy}{dx} = 6x^2 - 3x^2 y, \quad y(0) = 3$$

   Use these step sizes:

   (i) $h = 0.1$ **(ii)** $h = 0.01$ **(iii)** $h = 0.001$

(b) Verify that $y = e^{-x^3} + 2$ is the exact solution of the differential equation.

(c) Find the errors in using Euler's method to compute $y(1)$ with the step sizes in part **(a)**.
What happens to the error when the step size is divided by 10?

3. Use Euler's method with step size 0.5 and 4 iterations to compute the approximate y-values of the solution of the initial-value problem

$$y' = 3x - 2y + 1, \quad y(1) = 2$$

4. Use Euler's method with step size 0.2 to estimate $y(1)$, where $y(x)$ is the solution of the initial-value problem $\dfrac{dy}{dx} = x + y^2, \quad y(0) = 0$

5. Use Euler's method with step size 0.1 to estimate $y(0.5)$, where $y(x)$ is the solution of the initial-value problem $\dfrac{dy}{dx} = x^2 + y^2, \quad y(0) = 1$

6. Use Euler's method to find approximate values of the solution of the given initial value problem at $t = 0.5, 1.5,$ and 3, with step size:

   (i) $h = 0.1$ **(ii)** $h = 0.05$ **(iii)** $h = 0.01$

   (a) $\dfrac{dy}{dt} = y(3 - ty), \quad y(0) = 0.5$

   (b) $y' = 5 - 3\sqrt{y}, \quad\quad y(0) = 2$

   (c) $\dfrac{dy}{dt} = \dfrac{4 - ty}{1 + y^2}, \quad\quad y(0) = -2$

7. Copy and complete the table using an exact solution of the differential equation and two approximations obtained using Euler's method to approximate the particular solution of the equation. Use $h = 0.2$ and $h = 0.1$ and give your answer to 4 decimal places.

$$\dfrac{dy}{dx} - y = \cos x, \quad y(0) = 0$$

x	0	0.2	0.4	0.6	0.8	1.0
$y(x)$ [exact]						
$y(x)$ [$h = 0.2$]						
$y(x)$ [$h = 0.1$]						

Have you ever considered how your calculator computes values for certain functions? For functions such as $f(x) = 3x^2 - 5x + 8$, $g(x) = \dfrac{x^4 + 2x}{-4x^3 + x^2 - 6}$, and $h(x) = \sqrt{7x - 3}$ the method of evaluation is fairly straightforward because these are algebraic functions. Algebraic functions can be expressed as a finite number of sums, differences, multiples, quotients, and radicals involving x^n. Polynomial functions, rational functions, and functions involving radicals are examples of algebraic functions.

But how does your calculator compute values for a function such as e^x? This is an example of a transcendental function. A transcendental function is non-algebraic – it cannot be expressed as a finite number of sums, differences, multiples, quotients, and radicals involving x^n. Other familiar transcendental functions include the trigonometric and logarithmic functions.

So, in order to compute values of transcendental functions such as e^x, the manufacturers of the calculator had to decide on a computational algorithm.

Local linear approximation

When solving certain problems, it can be useful, and sometimes necessary, to approximate nonlinear and non-algebraic functions by simpler ones. For example, the equation that estimates the period of a swinging pendulum may be greatly simplified if we use the fact that $\sin\theta \approx \theta$ when θ is close to 0 in radians. We saw in Chapters 12 and 13 that if a function is differentiable at a number x_0, then the tangent line to the graph of f through the point $(x_0, f(x_0))$ will closely approximate the function for values of x near x_0 (Figure 16.13).

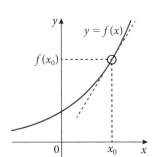

Figure 16.13 Tangent approximates the function values near x_0

To capture this intuitive awareness methodically, assume that the function f can be differentiated at x_0 and recall that the equation of the tangent to the graph of the function f through the point $(x_0, f(x_0))$ is $y - f(x_0) = f'(x_0)(x - x_0)$. Since this line closely approximates the graph of f for values of x near x_0, it follows that

$$f(x) \approx f(x_0) + f'(x_0)(x - x_0)$$

Example 16.28

(a) Show that $\sin\theta \approx \theta$ if θ is close to 0 (θ is in radians).

(b) Use the approximation from (a) to approximate $\sin 2°$, and compare your approximation to the result you get from your GDC.

Solution

(a) $f(\theta) = \sin\theta \Rightarrow f'(\theta) = \cos\theta$. Thus at $\theta_0 = 0$, the approximation above is
$$f(\theta) \approx f(0) + f'(0)(\theta - 0) = \sin 0 + \cos 0 \times \theta = \theta.$$
The diagram shows both the graph $f(\theta) = \sin\theta$ and its linear approximation $y = \theta$.

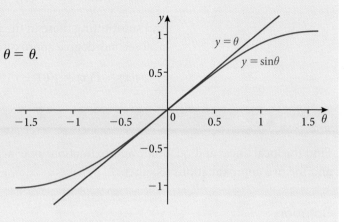

(b) In the approximation $\sin\theta \approx \theta$, θ is in radians, so, we must convert the angle into radians first:
$$2° = 2\left(\frac{\pi}{180}\right) \approx 0.0349066$$

Using GDC: $\sin 2° \approx 0.0348995$

Maclaurin and Taylor polynomials

We learned that the local linear approximation of a function f at a point x_0 is

$$f(x) \approx f(x_0) + f'(x_0)(x - x_0)$$

In this formula, the approximating function is $P(x) = f(x_0) + f'(x_0)(x - x_0)$, which is a first-degree polynomial with a value and first derivative value equal to those of f at x_0. That is:

$$P(x_0) = f(x_0) + f'(x_0)(x_0 - x_0) = f(x_0), \text{ and}$$
$$P'(x) = 0 + f'(x_0)(1 - 0) = f'(x_0) \Rightarrow P'(x_0) = f'(x_0)$$

However, as we move away from x_0 the accuracy of this approximation will decrease.

In order to increase this accuracy, we can consider higher degree polynomials whose value, and the values of its higher derivatives, agree with those of f at $x = x_0$.

To illustrate this idea, we will consider approximations of function f at $x = 0$.

Take, for example, a second-degree polynomial $P(x) = c_0 + c_1 x + c_2 x^2$

If the values of this polynomial with its first and second derivatives agree with those of f, then the polynomial has the same gradient and same **concavity** at $x = 0$, leading to a better approximation:

We want $P(0) = f(0)$, $P'(0) = f'(0)$, and $P''(0) = f''(0)$. Thus, we have the following set of equations:

$$P(x) = c_0 + c_1 x + c_2 x^2, \quad P(0) = c_0 = f(0)$$
$$P'(x) = c_1 + 2c_2 x, \quad\quad\ P'(0) = c_1 = f'(0)$$
$$P''(x) = 2c_2, \quad\quad\quad\quad P''(0) = 2c_2 = f''(0)$$

871

Hence it follows that

$$c_0 = f(0), \quad c_1 = f'(0), \quad c_2 = \frac{f''(0)}{2}$$

And substituting these in the polynomial equation will yield the formula for the local second-degree approximation at $x = 0$.

$$f(x) \approx P(x) = f(0) + f'(0)x + \frac{f''(0)}{2}x^2$$

Example 16.29

Find the local linear and quadratic approximations of e^x at $x = 0$, and graph e^x and the two approximations together.

Solution

Let $f(x) = e^x \Rightarrow f'(x) = f''(x) = e^x$

and hence $f(0) = f'(0) = f''(0) = e^0 = 1$

Thus, the approximations are:

$P(x) = f(0) + f'(0)x = 1 + x$,

and $P(x) = f(0) + f'(0)x + \dfrac{f''(0)}{2}x^2$

$\quad = 1 + x + \dfrac{x^2}{2}$

The diagram shows, as expected, that the quadratic approximation is more accurate than the linear one near $x = 0$.

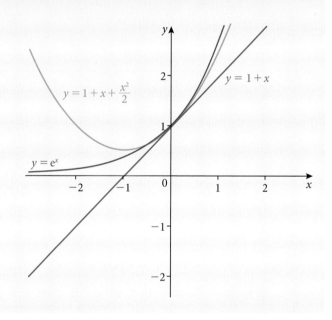

Given a function f that can be differentiated n times at $x = x_0$, we can find a polynomial P of degree n with the property that the value of P and its first n derivatives agree with those of f at x_0.

It is natural to ask if we can improve the accuracy of the second-degree approximation by using polynomials of higher degree.

This leads us to consider the following general problem.

Again, we begin with the case where $x_0 = 0$. Thus, the polynomial in question has the form:

$$P(x) = c_0 + c_1 x + c_2 x^2 + c_3 x^3 + \ldots + c_n x^n \tag{1}$$

Doing similar calculations as above:

$$P(x) = c_0 + c_1 x + c_2 x^2 + c_3 x^3 \ldots + c_n x^n \qquad P(0) = c_0 = f(0)$$
$$P'(x) = c_1 + 2c_2 x + 3c_3 x^2 \ldots + nc_n x^{n-1} \qquad P'(0) = c_1 = f'(0)$$
$$P''(x) = 2c_2 + 3 \cdot 2 \cdot c_3 x \ldots + n(n-1)c_n x^{n-2} \qquad P''(0) = 2c_2 = f''(0)$$
$$P'''(x) = 3 \cdot 2 \cdot c_3 \ldots + n(n-1)(n-2)c_n x^{n-3} \qquad P'''(0) = 3 \cdot 2 \cdot c_3 = f'''(0)$$
$$\vdots \qquad\qquad\qquad\qquad \vdots$$
$$P^{(n)}(x) = n(n-1)(n-2)\ldots(1)c_n \qquad P^{(n)}(0) = n(n-1)(n-2)\ldots(1)c_n = f^{(n)}(0)$$

Hence it follows

$$c_0 = f(0), \quad c_1 = f'(0), \quad c_2 = \frac{f''(0)}{2}, \quad c_3 = \frac{f'''(0)}{3 \cdot 2}, \quad \dots c_n = \frac{f^{(n)}(0)}{n(n-1)\dots}$$

That is

$$c_0 = f(0), \quad c_1 = f'(0), \quad c_2 = \frac{f''(0)}{2!}, \quad c_3 = \frac{f'''(0)}{3!}, \quad \dots c_n = \frac{f^{(n)}(0)}{n!}$$

And substituting back into **(1)**:

$$P(x) = f(0) + f'(0)x + \frac{f''(0)}{2!}x^2 + \frac{f'''(0)}{3!}x^3 + \dots + \frac{f^{(n)}(0)}{n!}x^n \qquad \textbf{(2)}$$

The polynomial in **(2)** is called the nth Maclaurin polynomial for f.

If f can be differentiated n times at 0, then we define the nth Maclaurin polynomial for f to be

$$P_n(x) = f(0) + f'(0)x + \frac{f''(0)}{2!}x^2 + \frac{f'''(0)}{3!}x^3 + \dots + \frac{f^{(n)}(0)}{n!}x^n \qquad \textbf{(3)}$$

This polynomial has the property that its value and the values of its first n derivatives agree with the values of f and its first n derivatives at $x = 0$.

Note that the polynomials $P_1(x)$ and $P_2(x)$ are respectively the local linear and quadratic approximations of f at $x = 0$.

Example 16.30

Find the Maclaurin polynomials P_0, P_1, P_2, P_3, and P_n for e^x and graph the first four together.

Solution

In order to apply **(2)**, we need to find the n derivatives of e^x.
This is straightforward since they are all equal to e^x:

$$f(0) = 1, \quad f'(0) = 1, \quad \frac{f''(0)}{2!} = \frac{1}{2!}, \quad \frac{f'''(0)}{3!} = \frac{1}{3!}, \quad \dots c_n = \frac{f^{(n)}(0)}{n!} = \frac{1}{n!}$$

Consequently, the polynomials are:

$$P_0(x) = f(0) = 1$$

$$P_1(x) = f(0) + f'(0)x = 1 + x$$

$$P_2(x) = f(0) + f'(0)x + \frac{f''(0)}{2!}x^2 = 1 + x + \frac{x^2}{2}$$

$$P_3(x) = f(0) + f'(0)x + \frac{f''(0)}{2!}x^2 + \frac{f'''(0)}{3!}x^3 = 1 + x + \frac{x^2}{2} + \frac{x^3}{6}$$

$$P_n(x) = f(0) + f'(0)x + \frac{f''(0)}{2!}x^2 + \frac{f'''(0)}{3!}x^3 + \dots + \frac{f^{(n)}(0)}{n!}x^n$$

$$= 1 + x + \frac{x^2}{2!} + \frac{x^3}{3!} + \dots + \frac{x^n}{n!}$$

Internal assessment

Internal assessment (IA) is an important component of the Analysis and Approaches HL course and contributes 20% to your final grade. It is a significant part of the overall assessment for the course and should be taken seriously. It should also be pointed out that your work in completing the IA component differs in important ways from the written exams (external assessment) for the course.

- Unlike written examinations, you do *not* perform IA work under strict time constraints.

- You have some freedom to decide which mathematical topic you wish to explore.

- Your IA work involves writing about mathematics, not just using mathematical procedures.

- Regular discussion with, and feedback from, your teacher will be essential.

- You should endeavour to explore a topic in which you have a genuine personal interest.

- You will be rewarded for evidence of creativity, curiosity, and independent thinking.

Mathematical exploration

To satisfy the IA component, you are required to complete a piece of written work on a mathematical topic that you choose in consultation with your teacher. This piece of written work is formally referred to as the mathematical exploration. It will be referred to simply as the 'exploration' throughout this chapter. Your primary objective is to *explore* a mathematical topic in which you are *genuinely interested* and that is at an *appropriate level* for the course. A fundamental aspect of your exploration must be the *use of mathematics* in a manner that clearly demonstrates your knowledge and understanding of the relevant mathematics. Your teacher may provide you with a list of ideas (or 'stimuli') to help you in the process of finding a suitable topic.

It is your responsibility to determine whether or not you are sufficiently interested in a particular topic – and it is your teacher's responsibility to determine if an exploration of the topic can be conducted at a mathematical level that is suitable for the course. Your teacher will help you determine if an exploration of a certain topic can potentially address the five assessment criteria satisfactorily. Your exploration should be approximately 12 to 20 pages long with double line spacing.

See the list of 200 ideas included in the eBook. You may find a suitable topic in the list, or the list may help you find or develop your own ideas for a mathematical topic to explore.

Internal assessment

Your exploration will be assessed by your teacher according to the following five criteria.

A Presentation

This criterion assesses the organisation and coherence of the exploration.
A well-organised exploration has an introduction, a rationale (which includes a brief explanation of why the topic was chosen), a description of the aim of the exploration, and a conclusion.

B Mathematical communication

This criterion assesses to what extent you are able to:

- use appropriate mathematical language (notation, symbols, terminology)
- clearly define key terms, variables, and parameters
- use multiple forms of mathematical representation, such as formulae, diagrams, tables, charts, graphs, and models
- apply a deductive approach in general, and present any proofs in a logical manner.

C Personal engagement

This criterion assesses the extent to which you engage with the exploration and present it in such a way that clearly shows *your own personal approach*. Personal engagement may be recognised in several different ways. These may include – but are not limited to – thinking independently and/or creatively, addressing personal interest, presenting mathematical ideas in your own words and diagrams, developing your own ideas and testing them, and creating your own examples to illustrate important results.

D Reflection

This criterion assesses how well you *review, analyse,* and *evaluate* the exploration. Although reflection may be seen in the conclusion to the exploration, you should also give evidence of reflective thought throughout the exploration. Reflection can be demonstrated by consideration of limitations and/or extensions, commenting on what you've learned, or comparing different mathematical methods and approaches.

E Use of mathematics

This criterion assesses to what extent and how well you use mathematics in your exploration. The mathematical working in your exploration needs to be *sufficiently sophisticated* and *rigorous*. The chosen topic should involve mathematics in the Analysis and Approaches HL syllabus or at a similar level. Sophistication and rigour can include understanding and use of challenging mathematical concepts, looking at a problem from different perspectives, mathematical arguments expressed clearly in a logical manner, or seeing underlying structures to link different areas of mathematics.

Your exploration will earn a score out of a total of 20 possible marks. The five criteria do not contribute equally to the overall score for your exploration. For example, criterion E (Use of mathematics) is 30% of the overall score, whereas crtieria C (Personal engagement) and D (Reflection) contribute 15% each.

It is very important that you familiarise yourself with the assessment criteria for the Analysis and Approaches HL exploration and refer to them while you are writing your exploration. The achievement levels for each criteria and associated descriptors are as follows:

A Presentation	
0	The exploration does not reach the standard described by the descriptors below.
1	The exploration has some coherence or some organisation.
2	The exploration has some coherence and shows some organisation.
3	The exploration is coherent and well organised.
4	The exploration is coherent, well organised, and concise.

B Mathematical communication	
0	The exploration does not reach the standard described by the descriptors below.
1	The exploration contains some relevant mathematical communication that is partially appropriate.
2	The exploration contains some relevant appropriate mathematical communication.
3	The mathematical communication is relevant, appropriate, and is mostly consistent.
4	The mathematical communication is relevant, appropriate, and consistent throughout.

C Personal engagement	
0	The exploration does not reach the standard described by the descriptors below.
1	There is evidence of some personal engagement.
2	There is evidence of significant personal engagement.
3	There is evidence of outstanding personal engagement.

D Reflection	
0	The exploration does not reach the standard described by the descriptors below.
1	There is evidence of limited reflection.
2	There is evidence of meaningful reflection.
3	There is substantial evidence of critical reflection.

E Use of mathematics	
0	The exploration does not reach the standard described by the descriptors below.
1	Some relevant mathematics is used. Limited understanding is demonstrated.
2	Some relevant mathematics is used. The mathematics explored is partially correct. Some knowledge and understanding are demonstrated.
3	Relevant mathematics commensurate with the level of the course is used. The mathematics explored is correct. Some knowledge and understanding are demonstrated.

4	Relevant mathematics commensurate with the level of the course is used. The mathematics explored is correct. Good knowledge and understanding are demonstrated.
5	Relevant mathematics commensurate with the level of the course is used. The mathematics explored is correct and demonstrates sophistication or rigour. Thorough knowledge and understanding are demonstrated.
6	Relevant mathematics commensurate with the level of the course is used. The mathematics explored is precise and demonstrates sophistication and rigour. Thorough knowledge and understanding are demonstrated.

Guidance

Conducting an in-depth individual exploration into the mathematics of a particular topic can be an interesting and very rewarding experience. It is important to take all stages of your work on the exploration seriously – not only because it is worth 20% of your final grade for the course but also because of the opportunity to pursue your own personal interests without the pressure of examination conditions. The exploration should *not* be approached as simply an extended homework assignment. The task of writing the exploration will require you to analyse, think, write, edit, and use mathematics in a readable and focused manner. Hopefully, it will also be enjoyable, thought-provoking, and satisfying, and it should give you the opportunity to gain a deeper appreciation for the beauty, power, and usefulness of mathematics.

Although it is required that your exploration is completely your own work, you should consult with your teacher on a regular basis. You are allowed to work collaboratively with fellow students, but this should be limited to the following: selecting a topic, finding resources, understanding relevant mathematical knowledge and skills, and receiving peer feedback on your writing. While you are encouraged to *talk* through your ideas with others, it is not appropriate for you to *work* with others on your exploration. Your teacher should provide support and advice during the planning and writing stages of your exploration. Both you and your teacher will need to verify the authenticity of your exploration.

Any text, diagrams, images, mathematical working, or ideas that are not your own must be cited where they appear in your exploration. Otherwise, all of the work connected with your exploration must be your own. Your exploration must provide the reader with the exact sources of quotations, ideas, and points of view with a complete and accurate bibliography. There are a number of acceptable bibliographic styles. Whichever style you choose, it must include all relevant source information and be applied consistently. Group work is not allowed. Also, if you are writing an extended essay for mathematics, you are not allowed to submit the same or similar piece of work for the exploration – and you should not write about the same mathematical topic for both.

In organising a successful exploration, consider the following suggestions:

1. Select a topic in which you are *genuinely interested*. Include a brief explanation in the early part of your exploration about why you chose your topic – including why you find it interesting.

Your teacher will provide oral and/or written advice on a draft of your exploration pertaining to how it can be improved. Your teacher will also write thorough and descriptive comments on the final version of your exploration to assist IB moderators in confirming the criteria scores they've awarded.

Warning: Failure to properly cite any text, diagrams, images, mathematical working, or ideas that are not your own may result in your exploration being reviewed for malpractice, which could have serious consequences.

If you are uncertain about the formatting and style of citations and a bibliography (not the same thing), then you should consult with teacher(s) at your school who have expertise in this area – such as an English teacher or librarian. A bibliography is required but it does not replace the need for appropriate citations (inline or footnotes) at the pertinent location in the exploration.

2. Consult with your teacher to confirm that the topic is at the *appropriate level of mathematics* – namely, that it is at the same or similar level of the mathematics in the HL syllabus.

3. Find as much *information* about the topic as possible. Although information found on websites can be very helpful, try to also find information in books, journals, textbooks, and other printed material.

4. Although there is no requirement that you present your exploration to your classmates, it should be written so that they can follow it without trouble. Your exploration needs to be *logically organised* and use appropriate mathematical terminology and notation.

5. The most important aspects of your exploration should be about *mathematical communication and using mathematics*. Although other aspects of your topic – for example, historical, personal, cultural – can be discussed, be careful to keep focus on the mathematical features.

6. Two of the assessment criteria – Personal engagement and Reflection – are about *what you think about the topic* you are exploring. Don't hesitate to pose your own relevant and insightful questions as part of your exploration – and then to address these questions using mathematics at a suitably sophisticated level along with sufficient written commentary.

7. Although your teacher will expect and require you to work independently, you are allowed to *consult with your teacher* – and your teacher is allowed to give you advice and feedback to a certain extent while you are working on your exploration. It is especially important to check with your teacher that any *mathematics in your exploration is correct*. Your teacher will not give mathematical answers or corrections but can indicate where any errors have been made or where improvement is needed.

Mathematical exploration - HL student checklist

- Is your exploration written entirely by yourself? Have you avoided simply replicating work and ideas from sources you found? ☐ Yes ☐ No

- Have you strived to apply your personal interest, develop your own ideas, and use critical thinking skills during your exploration? ☐ Yes ☐ No

- Did you refer to the five assessment criteria while writing your exploration? ☐ Yes ☐ No

- Does your exploration focus on good mathematical communication – and does it read like an article from a mathematical journal? ☐ Yes ☐ No

- Does your exploration have a clearly identified introduction and conclusion? ☐ Yes ☐ No

- Have you provided appropriate citation for any ideas, mathematical working, images, graphs, etc. that are not your own at the point they appear in your exploration? ☐ Yes ☐ No

- Not including the bibliography, is your exploration 12 to 20 pages? ☐ Yes ☐ No

- Are graphs, tables, and diagrams sufficiently described and labelled? ☐ Yes ☐ No

- To the best of your knowledge, have you used mathematics that is at the same level, or similar, to that studied in Analysis and Approaches HL? ☐ Yes ☐ No

- Have you attempted to discuss mathematical ideas, and use mathematics, with a sufficient level of sophistication and rigour? ☐ Yes ☐ No

- Are formulae, graphs, tables, and diagrams in the main body of text? (Preferably no full-page graphs, and no separate appendices.) ☐ Yes ☐ No

- Have you used technology – such as a GDC, spreadsheet, mathematics software, drawing and word-processing software – to enhance mathematical communication? ☐ Yes ☐ No

- Have you used appropriate mathematical language (notation, symbols, terminology) and defined key terms? ☐ Yes ☐ No

- Is the mathematics in your exploration performed precisely and accurately? ☐ Yes ☐ No

- Has calculator/computer notation and terminology been used? ($y = x^2$, not $y = x^2$; π, not <pi>; $\approx$, not 'approximately equal to'; $|x|$, not abs(x); etc) ☐ Yes ☐ No

- Have you included reflective and explanatory comments about the topic being explored throughout your exploration? ☐ Yes ☐ No

Finding, developing, and choosing a topic for your exploration

It is fair to say that the *most important stage* of completing your exploration is determining the mathematical topic you are going to investigate, write about, and apply. Your exploration is much more likely to be successful – and gratifying – if it focuses on a mathematical topic in which you have a genuine interest, is at a suitable level for the Analysis and Approaches HL course, and for which you are confident that you can discuss and use the relevant mathematics in a manner that demonstrates thorough knowledge and understanding. There is no single approach for determining an exploration topic that is guaranteed to be successful for all students. Your teacher will provide helpful advice and support. Your teacher may supply you with a short list of some broad stimuli to start the process of finding a much narrower topic. Many teachers have found that starting with a sufficiently narrow topic is often more successful than starting with a very broad topic that requires a significant effort to reduce to the extent that it can be explored in less than 20 pages (double spaced).

 Avoid choosing a topic that is too broad and/or too complicated.

In the eBook for this textbook you will find a list of 200 mathematical topics. Some of the topics in the list are broad but many are already quite narrow in scope. It is possible that some of these 200 topics could be the focus of an exploration, while others will require you to investigate further to develop a narrower focus to explore. Do not restrict yourself to the topics in the list. This list is only the tip of the iceberg with regard to potential topics for your exploration. Reading through this list may stimulate you to think of some other topic(s) that you may find interesting to explore. Many of the items in the list may be unfamiliar to you. A quick search on the internet should give you a better idea what each is about and help you determine if you're interested enough to investigate further – and to see if it might be a suitable topic for your exploration.

Theory of knowledge

At the start of his wonderful book *Nature's Numbers,* the mathematician Ian Stewart writes:

> 'We live in a universe of patterns. Every night the stars move in circles across the sky. The seasons cycle at yearly intervals. No two snowflakes are ever exactly the same, but they all have sixfold symmetry. Tigers and zebras are covered in patterns of stripes, leopards and hyenas are covered in patterns of spots. Intricate trains of waves march across the oceans; very similar trains of sand dunes march across the desert. Coloured arcs of light adorn the sky in the form of rainbows, and a bright circular halo sometimes surrounds the moon on winter nights. Spherical drops of water fall from clouds.'

We could add to Stewart's list. Wallpaper is patterned (there are surprisingly only 17 different distinct groups of possible patterns); buildings often exhibit mirror symmetry and their structure is carefully proportioned; the digital traces on memory sticks or hard drives are patterned in a way that makes them suitable for storing data; mechanical devices such as clocks and engines depend on symmetry and patterning for their smooth movement; the day is divided into equal parts that are represented using angles or digits; music possesses horizontal and vertical symmetries – and human behaviour is patterned.

It is no accident that the world is full of patterns. Symmetry in a building is not only easy on the eye but it ensures that the design is simple. Pattern is a labour-saving strategy. The same plan can be used for each window, or the plan for one side of a building can be used in reverse for the other side. These informational shortcuts can be found both in the man-made world and in nature. The same blueprint for generating twig patterns can be used for bigger branches, or one plan can be used for all the petals in a flower. It is a sort of design efficiency. The wealth of patterns in the world is a series of cost-effective solutions to problems – and that is why these patterns are worth studying.

Mathematics is one way in which human beings formally study patterns. While the natural sciences study patterns by going out into the world, collecting examples and analysing them, mathematics studies patterns in the abstract. Mathematics in its purest form is not fieldwork or experiment. Its raw materials are abstract structures specified by symbols, and mathematicians arrive at conclusions through their manipulation. In this sense, mathematics is a little 'other-wordly' – a characteristic that makes it interesting from a ToK perspective. It means that in some sense, mathematics is more like an art than a science. There is in this suggestion more than a hint of a deep reliance on creativity and imagination. A comparison with the arts and the sciences is instructive and reveals the truly special place that mathematics occupies in human knowledge.

In this chapter, we will investigate mathematics using the basic structure of the knowledge framework: Perspectives, methods and tools, and the link to the individual.

Under 'Perspectives', we will look at the orientation of mathematics within the academy. There are a number of key questions to be answered here:

- What is mathematics about?
- How should we think of mathematics: as a human construction or something in the world?
- Why is mathematics useful?

Under 'Methods and Tools' we will discuss exactly what mathematicians do – how they arrive at mathematical knowledge and what counts as facts and truth in mathematics. This is where we unpack the key conceptual building blocks of mathematical thought.

The final section deals with mathematics and the individual. What is the link between mathematics and supposedly subjective phenomena such as beauty? How reliable are our mathematical intuitions? Is mathematics a personal journey or is it something that we collaborate on?

On the way, we will have fun with infinite numbers, self-similar patterns and security codes. While it might be removed from the physical world, the world of mathematics is just as fascinating, if not more so. Enjoy!

What role does mathematics play in your life?

Perspectives

Mathematics and number

As a first definition, let's say that mathematics is the formal study of patterns. In this section we will see how far this basic idea takes us.

Imagine a simple pattern in the world – a set of similar objects, for example, a field of animals. Let's say that the animals are of the same kind – they are cows. To recognise that a group of different things all belong to the same kind is already remarkable. It means ignoring all the things that mark out individual animals and focusing only on what they have in common. Grouping a set of things together by common characteristics is a powerful technique in the sciences. If such a classification is effective, it might yield understanding, generalisations and predictions. We call groups that have these properties **natural kinds** – it is something that might be expected to happen in biology. But mathematics goes one step further. Suppose that we make a mark 'I' on a clay tablet for every cow in the field. We end up with a mark 'IIIIIII'. What we have done now is to abstract away everything about the animals in the field: the fact that they are animals, that they are cows, that they are eating grass. What is left is their number.

So, the simplest pattern that we can deal with abstractly is number. In a somewhat magical way, the inscriptions of the tablet **represent** the cows in the field. They are a convenient stand-in for the real world. If we want to find out what happens when we remove 'III' cows from the field. We can either move them physically or we simply separate the 'cow' symbols: 'IIIIII   III'. Manipulating the symbols is clearly easier to perform. Mathematics manipulates representations rather than the real world because it is easier.

We do not know if something like this story is accurate at the beginning of the long history of mathematics. But we do know that imprints on a Sumerian clay tablet led eventually to the astounding sophistication of the proof of Fermat's last theorem and to modern algebra, analysis, and geometry. Mathematics has been shaped by the job it is expected to perform and through countless quirks of culture. Improvised methods designed to deliver a temporary solution to an unforeseen problem become permanent. If they work well, they get passed on and take on a life of their own. Less good solutions eventually fall into disuse in a sort of Darwinian selection of competing ideas. We could call histories like this **cultural evolution**.

But has the counting of cows in a field really got anything to do with modern mathematics? Let's examine the example more closely. We add an 'I' on the tablet for each cow in the field, subject to two strict rules: no cow should be 'counted' more than once and all the cows in the field are counted. Although these rules are quite natural to us, they are mathematically sophisticated. Mathematically, we are establishing a mapping between the marks on the tablet and the cows in the field that is a **one-to-one correspondence**. This means a mapping links a mark to a unique cow (injective) and that all cows in the field are linked (surjective). While these early users of mathematics might not have understood it quite in these terms, they nonetheless needed to use these properties when counting. But there is something else at work here. The compound symbol 'IIIIIII' stands for the whole field of cows. It is a property of the whole set. It expresses the size of the set or its **cardinality**. The counting of cows in a field has a lot to do with the deep nature of mathematics itself.

Indeed, there are three more ideas illustrated by this simple example. The first is the power of numbers to create ordering: I II III IIII is such an ordering. This is called the **ordinal** property of number. Second, it illustrates the special place of sets and mappings in mathematics. We focused on the set of cows and the set of marks on the tablet. Third, we counted the first set by establishing a one-to-one correspondence with the second. This is a technique that works with any sets, including those that have infinitely many members. Mathematics is truly about sets and the mappings between them.

By representing the real world by marks bearing a special relation to their targets, human beings initiated perhaps the most extraordinary technical advance in their history: the invention of symbolic representation. Manipulating symbols is easier than manipulating objects in the world. Moreover, symbols allow this information to be communicated over distance and time. But the most powerful feature of symbols is that they can be used to represent states of affairs that are not physically present. Symbols can represent past worlds, possible worlds, and desired future worlds. Symbols allow us to tell stories, write histories, and make plans. Symbols that do not actually correspond with the world are called **counterfactuals**. They describe 'what if' situations. What if the Allies had lost World War II? What if we add sulfuric acid to copper? What if we wake up one morning to discover that we have been transformed into a giant insect? What if parallel lines could actually meet? What if there was a solution to the equation $x^2 = -1$? The power of symbolic representation is that it allows us to build abstract worlds – virtual realities where the 'what if' conditions are true.

There is a sense in which the world of mathematics is one such virtual universe, containing all manner of exciting and weird things. Mathematicians discuss 11-dimensional hypercubes, infinite sets of numbers, infinite numbers, surfaces that turn you from being right-handed to left-handed as you traverse them, spaces where the angles of a triangle add up to more than 180 degrees, spaces where parallel lines diverge, systems where the order of the operation matters (where $A * B$ is not the same as $B * A$), vectors in infinite-dimensional space, series that go on forever, and geometric figures that are self-similar called fractals (where you can take a small piece of the original figure then enlarge it and it looks identical – truly identical – to the original). And all this started with the making of a simple mark on a clay tablet.

Mathematics uses symbols to describe these amazing structures in the basic language of sets and the mappings between them. Because symbols are abstract and not limited to representing things in the world, mathematicians can use their imaginations to create a virtual reality following its own rule system unhindered by what the world is really like, a **counterfactual world**. In this world, mathematicians can explore the patterns they encounter.

Yet mathematics is remarkably useful in this world. From building bridges to controlling strategy in football, mathematics lies at the heart of the modern world. If mathematics really is so other-worldly, how come it has so much to say about this one?

This is an important question that motivates much of what follows.

If symbolic representation is the most significant technical advance in history, what would you put in second place?

Purpose: mathematics for its own sake

ToK uses the map metaphor; knowledge is taken to be like a map that is used for a particular purpose, such as solving a particular problem or answering a question. The map is a simplified picture of the world and its simplicity is its strength. It ensures that we get the job done with the least cognitive cost. If this is right, then it is natural to ask about the purpose of this particular map. What problems does it solve or what questions does it answer? There seem to be two categories: those questions that occur strictly within the virtual reality of mathematics itself (mathematics for its own sake) and those that occur in the world outside (mathematics as a tool). These categories broadly correspond to two subdivisions of mathematics that are often two different departments within a university: pure mathematics and applied mathematics.

Let's start with pure mathematics. A typical example of a problem in this category is how to solve a particular type of equation.

An example of a problem in pure mathematics might be how to solve the equation

$$(1) \quad x^3 - 2x^2 - x + 2 = 0$$

The task is to find a value for x that satisfies the equation. In books like this, there are many such equations and, in this context, they often have simple integer solutions. An initial strategy might be to try a value for x to see if it fits. If we try $x = 0$, then equation (1) gives us:

$$0^3 - 2 \cdot 0^2 - 0 + 2 = 0, \text{ i.e. } 2 = 0$$

which is clearly not true. So, we can say that $x = 0$ is not a solution to the equation.

But if we try $x = 1$, then equation (1) gives us:

$$1^3 - 2 \cdot 1^2 - 1 + 2 = 0$$

In other words, $1 - 2 - 1 + 2 = 0$ is true. So, $x = 1$ is a solution to the equation.

The trick now, as you know, is to factor out $(x - 1)$ from equation (1) to give:

$$(2) \quad (x - 1)(x^2 - x - 2) = 0$$

We can now try to find values of x that make the second bracket in (2) equal to 0. This can be done either by trying out hopeful values of x (2 seems to be a good bet, for example) or using the quadratic formula. We end up with $x = 2$ or $x = -1$

The equation therefore has three solutions: $x = 1$ or $x = -1$ or $x = 2$

The history of these problems illustrates the great attraction of pure mathematics. Certainly, these problems were of interest from the 7th century in what is now the Middle East – the home of algebra. The great 11th century Persian mathematician and poet Omar Khayyam wrote a treatise about similar so-called cubic equations and realised they could have more than one solution. By the 16th century, cubic equations were of public interest. In Italy, contests were held to showcase the ability of mathematicians to solve cubic equations, often with a great deal of money at stake. One such contest took place in 1635 between Antonio Fior and Niccolò Tartaglia. Fior was a student of Scipione del Ferro, who had found a method for solving equations of the type $x^3 + ax = b$, which is known as the 'unknowns and cubes problem' (where a and b are given numbers).

Del Ferro kept his method secret until just before his death when he passed the method on to his student. Fior began to boast that he knew how to solve cubics. Tartaglia also announced that he had been able to solve a number of cubic equations successfully. Fior immediately challenged Tartaglia to a contest. Each was to give the other a set of 30 problems and put up a sum of money. The person who had solved the most after 30 days would take all the money.

Tartaglia had produced a method to solve a different type of cubic $x^3 + ax^2 = b$. Fior was confident that his ability to solve cubic equations would defeat Tartaglia and submitted 30 problems of the 'unknowns and cubes' type, but Tartaglia submitted a variety of different problems. Although Tartaglia could not initially solve the 'unknowns and cubes' type of equation, he worked hard and discovered a method to solve this type of problem. He then managed to solve all of Fior's problems in less than two hours. In the meantime, Fior had made little progress with Tartaglia's problems and it was obvious who was the winner. Tartaglia did not take Fior's money though; the honour of winning was enough.

Tartaglia represents the essence of the pure mathematician: someone who is intrigued by puzzles and has a deep desire to solve them. It is the problem itself that is the motivation, not possible real-world applications.

What other knowledge is worth pursuing for its own sake?

A modern example is the solution of Fermat's conjecture by Andrew Wiles. The French mathematician Pierre de Fermat wrote the conjecture in 1627 as a short observation in his copy of *The Arithmetics of Diophantus*.

The conjecture is that the equation

$$A^n + B^n = C^n$$

where A, B, C are positive integers and $n > 2$ has no solution. Despite a large number of attempts to prove it, the conjecture remained unproved for 358 years until Wiles published his successful proof in 1995. The proof is way beyond the scope of this book, but there have been a number of interesting books and TV programmes made about it, including Simon Singh *Fermat's Last Theorem* (1997) and the BBC Horizon programme *Fermat's Last Theorem* (1996). As mathematician Roger Penrose remarked, '*QED: how to solve the greatest mathematical puzzle of your age. Lock self in room. Emerge 7 years later'*.

Purpose: mathematical models

Unlike pure mathematics, which is about the solution of exclusively mathematical puzzles, applied mathematics is about solving real-world problems. The mathematics it produces can be just as interesting from an insider's viewpoint as the problems of pure mathematics (and often the two are inseparable), but a piece of applied mathematics is judged by whether it can be usefully applied in the world.

Here is an example of applied mathematics at work. This is a problem that could have been posed in this book or, indeed (and this is the point), in a physics course.

A stone is dropped down a 30 m well. How long will it take the stone to reach the bottom of the well, neglecting the effect of air resistance?

The typical way to solve this type of problem is to use what we call a **mathematical model**. The essence of mathematical modelling is to produce a description of the problem where the main physical features become variables in an equation which is then solved and translated back into the real world.

To model the situation above:

We know that the acceleration due to gravity is 9.8 m s⁻², and we also know that the distance travelled *s* is given by the equation:

$$s = \frac{1}{2}at^2, \text{ where } a = \text{acceleration and } t = \text{time}$$

So we substitute the known values into the equation and get:

$$30 = \frac{1}{2}(9.8)t^2$$

Rearranging the equation gives us:

$$\frac{60}{9.8} = t^2, \text{ so } t = \sqrt{\frac{60}{9.8}} = 2.47 \text{ seconds (3 s.f.)}$$

There are a number of points to make about the process here that are typical of mathematical models.

(1) The model neglects factors that are known to operate in the real-world situation. There are two big assumptions made: that the stone will not experience air resistance, which will act as a significant drag force, and that the acceleration due to gravity is constant.

(2) The model appeals to a law of nature. In this case, the law of acceleration due to gravity.

(3) The model uses values for constants that are established empirically. In this case, the acceleration due to gravity at the Earth's surface.

We know that neither of the assumptions in (1) is true. The effect of air resistance can be highly significant. We know that if you have the misfortune to fall from an airplane above 100 m or so, the height does not matter – the speed of impact with the ground will be the same, around 150 km h⁻¹, because of the effect of air resistance (of course, it matters how you fall). The changing strength of gravitational force is a less important factor for normal wells.

But if we are dealing with a well that is 4000 km deep, then this factor would be significant. The point is that the model is actually fictional (it even breaks a major law of physics). It could never be true in the sense of exactly corresponding to reality. However, it is a sort of idealisation that we accept because the model provides an approximation to the behaviour of the stone (although not such a good one for deeper wells) and more importantly it gives us understanding of the system. If we were to make the modelling assumptions more realistic, the mathematics in the model would become too complicated to solve easily. Points (2) and (3) show us that the actual content of the model depends on something outside mathematics – namely some well-established results in physics. The mathematics is only a tool, albeit an important one. A model is a mathematical map – a simplified picture of reality that is useful.

Another beautiful example is the Lotka–Volterra model of prey–predator population dynamics in biology. This model was proposed by Alfred Lotka in 1925 and independently by Vito Volterra in 1926.

The model assumes a closed environment where there are only two species, prey and predator, and no other factors. The rate of growth of prey is assumed to be a constant proportion A of the population. The rate at which predators eat prey is B, which is assumed to be a constant proportion of the product of predators and prey. The death rate of predators, C, is assumed to be a constant proportion of the population, and there is a rate of generation of new predators, D, dependent on the product of prey and predators.

These modelling assumptions give rise to a pair of coupled differential equations:

$$(1) \quad \frac{dx}{dt} = Ax - Bxy$$

$$(2) \quad \frac{dy}{dt} = -Cy + Dxy$$

A modern computer package gives the following evolution of prey and predators over time:

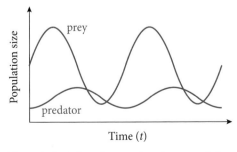

Figure 1 Evolution of prey and predator populations over time

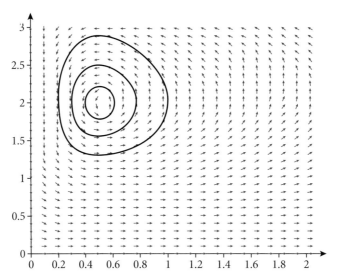

Figure 2 A phase space diagram. Number of prey (in units of 1000) on the x-axis, number of predators on the y-axis

It is interesting to look at a phase space diagram that represents each point (x, y) as a combination of numbers of prey and predators. Here the evolution of the system over time appears as a closed loop around the stationary point $\left(\dfrac{C}{D}, \dfrac{A}{B}\right)$, which is an 'attractor' of the dynamical system. (You could try to prove that this is a stationary point – it is not hard.) The position of an orbit around the attractor depends on the initial numbers of prey and predator. Notice that starting the model with too great a population of prey could end up with an extinction of predators (Figure 2) because the very high prey numbers leads to overpopulation of predators for whom there is not enough prey left to eat. The system itself is a nice example of circular causality.

As with the previous example, the modelling assumptions ensure that the mathematics of the model remains tractable, but the cost is that the model is not realistic. It is assumed that the prey do not die from natural causes or that the predators do not come into existence except through the provision of food. There is no competition between either prey or predators. Nonetheless, the model provides some important and powerful insights about the nature of population dynamics. As the model becomes more sophisticated and more factors are taken into consideration, not only does the mathematics become rapidly more difficult, but we lose sight of clear trends in the model (such as orbits around stationary points in phase space). We gain accuracy but lose understanding. This is a characteristic of both models and maps. A map that is as detailed as the territory it depicts is no use to anyone. It is precisely the simplification (literally what makes it false) that makes it useful. Virginia Woolf said about art, '*Art is not a copy of the world; one of the damn things is enough*', and the same could be said about models.

The distinction between pure and applied mathematics becomes blurred in the hands of someone like the great Carl Friedrich Gauss (1777–1855). He was perhaps happiest in the realm of number theory which he called the 'queen of mathematics', and the idea that queens stay in their rarified towers and do not dirty their hands in the ways of the world was perhaps not so far from his thinking. He found great satisfaction in working with patterns and sequences of numbers. It is the same Gauss who, as a young man, enabled astronomers to rediscover the minor planet Ceres after they had lost it in the glare of the sun, by calculating its orbit from the scant data that had been collected on its initial discovery in 1801 and then predicting where in the sky it would be found more than a year later. This feat immediately brought Gauss to the attention of the scientific community. His skills as a number theorist presented him with the opportunity of solving a very real scientific problem.

Who would have guessed that recent work in prime number theory would give rise to a system of encoding data that is used by banks all over the world? The system is called 'dual key cryptography'. The key to the code is a very large number that is the product of two primes. The bank holds one of the primes and the client's computer the other. The key can be made public because in order for it to work it has to be split up into its component prime factors. This task is virtually impossible for large numbers. For example, present computer programs would take longer than the 13.8 billion years since the big bang to find the two prime factors of the number:

25 195 908 475 657 893 494 027 183 240 048 398 571 429 282 126 204 032 027 777 137 836 043 662 020 707 595 556 264 018 525 880 784 406 918 290 641 249 515 082 189 298 559 149 176 184 502 808 489 120 072 844 992 687 392 807 287 776 735 971 418 347 270 261 896 375 014 971 824 691 165 077 613 379 859 095 700 097 330 459 748 808 428 401 797 429 100 642 458 691 817 195 118 746 121 515 172 654 632 282 216 869 987 549 182 422 433 637 259 085 141 865 462 043 576 798 423 387 184 774 447 920 739 934 236 584 823 824 281 198 163 815 010 674 810 451 660 377 306 056 201 619 676 256 133 844 143 603 833 904 414 952 634 432 190 114 657 544 454 178 424 020 924 616 515 723 350 778 707 749 817 125 772 467 962 926 386 356 373 289 912 154 831 438 167 899 885 040 445 364 023 527 381 951 378 636 564 391 212 010 397 122 822 120 720 357

But this number is indeed of the form of the product of two large primes. If you know one of them, it takes an ordinary computer a fraction of a second to do the division and find the other.

Just as pure research in the natural sciences produced results that could also be used for technological or engineering applications, so in mathematics, problems motivated purely from within the most abstract recesses of the subject (pure mathematics) give rise to very useful techniques for solving problems with strong applications in the world outside of mathematics. Mathematicians often practise their art as art for its own sake. They are motivated by the internal beauty and elegance of their subject. Nevertheless, it often happens that pure mathematics created for no other purpose than solving internal mathematical problems turns out to have some extraordinary and very practical applications.

Can you think of an example of a model that does not represent the world well but is nonetheless useful?

What other examples are there of pure research that end up having immense practical benefit?

Constructivist view of mathematics

Having thought a little about what the purpose of mathematics could be, let's move on to the question of whether it is best thought of as an invention or as something out there in the world.

Broadly speaking, the **constructivist** views mathematics as a human invention. The vision we had of mathematics as a vast virtual reality limited only by the imagination and the rules that are installed there is a constructivist view. However, we are then bound to ask why mathematics has so many useful applications in the real world. Why is mathematics important when it comes to building bridges, doing science and medicine, economics and even playing basketball? Chess is also a game invented by humans, but it does not have very much use in the outside world. Constructivism cannot account for the success of mathematics in the outside world.

On this view, mathematics is what might be called a **social fact**. A social fact is true by virtue of the role that it plays in our social lives. Social facts do have real causal power in the world. That a particular piece of paper is money is a social fact that does make things happen. That piece of paper acquires its status ultimately from a whole set of social agreements. In the end, social facts are produced by **language acts** – performances that change the social world. A language act would be a registry officer saying 'I pronounce you married'. The use of language in a **performative** manner creates social facts. Social facts are no less real or definite than those about the natural world. The statement 'John is married' is definitely either true or false. One is reminded of the story about the little boy who, when asked by his grandmother what day it will be tomorrow, replies, 'Let's wait and see'. Social facts do not require us to wait and see. They rely on social agreements, not on empirical evidence.

The mathematician Reuben Hersh argues for a type of constructivism that he calls **Humanism**. For Hersh, numbers and other mathematical objects are social facts. Hersh defends this view on the Edge website:

'[Mathematics] … is neither physical nor mental, it's social. It's part of culture, it's part of history, it's like law, like religion, like money, like all those very real things, which are real only as part of collective human consciousness. Being part of society and culture, it's both internal and external: internal to society and culture as a whole, external to the individual, who has to learn it from books and in school. That's what math is.'

Hersh called his theory of mathematics humanism because it's saying that mathematics is something human. 'There's no math without people. Many people think that ellipses and numbers and so on are there whether or not any people know about them; I think that's a confusion.'

Hersh points out that we do use numbers to describe physical reality and that this seems to contradict the idea that numbers are a social construction.

It is important to note here that we use numbers in two distinct ways: as nouns and as adjectives. When we say nine apples, nine is an adjective.

If it's an objective fact that there are nine apples on the table, that's just as objective as the fact that the apples are red, or that they're ripe or anything else about them; that's a fact. The problem occurs when we make a subconscious switch to 'nine' as an abstract noun in the sort of problems we deal with in Mathematics class. Hersh thinks that this is not really the same nine. They are connected, but the number nine is an abstract object as part of a number system. It is a result of our mathematics game – our deduction from axioms. It is a human creation.

Hersh sees a political and pedagogical dimension to his thinking about mathematics. He thinks that a humanistic vision of mathematics chimes in with more progressive politics. How can politics enter mathematics? As soon as we think of mathematics as a social construction then the exact arrangements by which this comes about – the institutions that build and maintain it – become important. These arrangements are political. Particularly interesting for us here is how a different view of mathematics can bring about changes in teaching and learning.

'Humanism sees mathematics as part of human culture and human history. It's hard to come to rigorous conclusions about this kind of thing, but I feel it's almost obvious that Platonism and Formalism are anti-educational, and interfere with understanding, and Humanism at least doesn't hurt and could be beneficial. Formalism is connected with rote, the traditional method, which is still common in many parts of the world. Here's an algorithm; practise it for a while; now here's another one. That's certainly what makes a lot of people hate mathematics (…) There are various kinds of Platonists. Some are good teachers, some are bad. But the Platonist idea, that, as my friend Phil Davis puts it, Pi is in the sky, helps to make mathematics intimidating and remote. It can be an excuse for a pupil's failure to learn, or for a teacher's saying "some people just don't get it". The humanistic philosophy brings mathematics down to earth, makes it accessible psychologically, and increases the likelihood that someone can learn it, because it's just one of the things that people do.'

There is a possibility that the arguments explored in this section might cast light on an aspect of mathematics learning that has seemed puzzling – why it is that mathematical ability is seen to be closely correlated with a certain type of intelligence. There is a widespread view that mathematics polarises society into two distinct groups: those who can do it and those who cannot. Those who cannot do it often feel the stigma of failure and that there is an exclusive club whose membership they have been denied. Those who can do it often find themselves labelled as 'nerds' or as people who are, in some sense, socially deficient. Is Hersh correct in attributing this to a formalistic or Platonic view? Is he right to suggest that if mathematics is just a meaningless set of formal exercises, then it will not be valued by society? If we deny that mathematics is out there to be discovered, it takes the stigma away from the particular individual who does not make the discovery. It is interesting to speculate on how consequences in the classroom flow from a humanist view of mathematics.

What are the strengths and weaknesses of mathematical humanism?

Platonic view of mathematics

One way to explain why mathematics applies so well to things like bridges and planets is simply to take mathematics as being out there in the world, independent of human beings. As with other things in the natural world, it is our task to discover it (literally to 'lift the cover'). This is called the **Platonic** view because the philosopher Plato (427–347 BC) took the view that mathematical objects belonged to the real world, underlying the world of appearances in which we lived. Mathematical objects such as perfect circles and numbers existed in this real world; circles on Earth were mere inferior shadows. Many mathematicians have at least some sympathy with this view. They talk about mathematical objects as though they had an existence independent of us and that we are accountable to mathematical truths in the same way as we are accountable to physical facts about the universe. They feel that there really is a mathematical world out there and that they are trying to discover truths about it, much like natural science discovers truth about the physical world.

This view is itself not entirely without problems. In ToK we might want to ask: 'If mathematics is out there in the world, where is it?' We do not see circles, triangles, $\sqrt{2}\,\pi$, i, e, and other mathematical objects obviously floating around in the world. We have to do a great deal of work to find them through inference and abstraction.

While this might be true, there is some evidence that mathematics is hidden not too far below the surface of our reality. Take prime numbers as an example. The Platonist might want to try to find them somewhere in nature. One place where she might start is in Tennessee. In the summer of 2016, the forests were alive with a cicada that exploits a property of prime numbers for its own survival. These cicadas have a curious life cycle. They stay in the ground for 13 years. Then they emerge and enjoy a relatively brief period courting and mating before laying eggs in the ground and dying. There is another species of cicada that has the same cycle and no fewer than 12 types that have a cycle of 17 years. There are, to add to the puzzle, none that have cycles of 12, 14, 15, 16 or 18 years. The clue is that 13 and 17 are prime numbers. There is a predator wasp that has evolved to have a similar life cycle. But if a predator had a life cycle of 6 years, the prey and the predator would only meet every $6 \times 17 = 112$ years. Whereas, if the cicada had a life cycle of 12 years, the prey and predator would meet every cicada cycle. Nature has discovered prime numbers through the cicada life cycles by evolutionary trial and error.

The relationship of nature to geometry was explored by the Scottish biologist D'Arcy Wentworth Thompson in his magnificent book of 1917, *On Growth and Form*. He explored the formation of shells and the wings of dragonflies, and examined the skeletons of dinosaurs through the eyes of a civil engineer constructing bridges and wondered about the formation of bee cells and the arrangement of sunflower seeds.

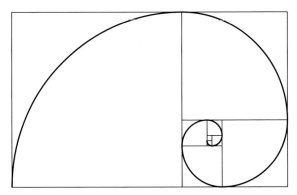

Figure 3 Spirals in nature

Many spirals in nature are formed, like the one in Figure 3, from the sequence:

1, 1, 2, 3, 5, 8, 13, 21, 34, …

This is called the Fibonacci sequence after the Italian mathematician Leonardo Pisano Bigolio (1170–1250), known as Fibonacci. The Fibonacci sequence is related to the golden number φ. The interested reader is referred to the many excellent sources on the internet.

Do you think that the mathematics teaching you have experienced reflects a Platonist or constructivist view of mathematics?

The methods and tools of mathematics

The language and concepts of mathematics

Knowledge in mathematics is like a map representing some aspect of the world. Like other areas of knowledge, it possesses a specialised vocabulary naming important concepts to build this map. Unlike some areas, this vocabulary is very precisely defined. This makes sense. If the world of mathematics is populated by some rather esoteric objects that are literally like nothing on Earth, then it is very important that these objects are precisely specified.

The other chapters in this book are all about establishing and using this very special vocabulary and becoming fluent in the methods that connect mathematical concepts into meaningful mathematical sentences. We will not spend too much time on these matters here, but there are a few aspects to highlight.

Notation

Since mathematical objects are abstract and we cannot point to them, we have to represent them with symbols. But the symbol and the idea are different things – there is a danger that we confuse them. Take representations of fractions. The symbols $\frac{1}{3}$, $\frac{2}{6}$, $\frac{3}{9}$, 0.3333… all represent the same number despite appearing to be quite different. (Perhaps the infinite number of ways of representing fractions is one of the reasons why some students have so much difficulty with them.) Some symbols such as $\frac{1}{0}$, $\sin^{-1}(1.2)$ or $\log(-2)$ have no meaning at all. More worrying is that an expression such as 'the smallest real number larger than 1' doesn't actually mean anything either. This is because there is no smallest real number larger than 1. (Think carefully about this.)

In a similar vein, the fact that there are different conventions for writing mathematics does not mean that the mathematics is different. Some conventions represent the number $\frac{3}{10}$ by the decimal 0.3, others by 0,3. Either way, the mathematics is the same and these do not really count as different mathematical cultures. Carl Friedrich Gauss, one of the greatest mathematicians of all time, said '*non notations, sed notions*' – not notations but notions.

Algebra

A staple method used in mathematics is the substitution of letters for numbers. In fact, mathematicians use letters for many sorts of mathematical objects, not just numbers. The reason is that they want to make generalised statements. By using a letter, they do not have to commit to making a statement about a specific number, but instead can make one about all numbers of a particular kind at once. This is a very powerful tool.

This is illustrated by a worked example. Imagine we want to prove that if we add an odd number to another odd number we get an even number. We hope to show that this is true for any choice of odd numbers. We could proceed by trying out different pairs of odd numbers and checking that the result is even:

$1 + 3 = 4$ even

$5 + 7 = 12$ even

$13 + 9 = 22$ even

$131 + 257 = 388$ even

You can see that this method will not serve as a proof because we would have to check every possible pair of odd numbers and, since this set is infinite, we would never finish. What we need is to define a general odd number without committing to a particular one. For example, we can define 'odd' by being 'one more than an even number'.

If k is an even number, then we can write $k = 2j$ for some whole number j.

If m is an odd number, then we can write $m = 2j + 1$ for some whole number j.

All we have to do now is to add two of these general odd numbers together.

So, we want to take two odd numbers, let's say $m = 2j + 1$ and $n = 2i + 1$ where j and i are whole numbers. There is a subtlety here because we use different letters j and i for the whole numbers in the expressions above because we want to allow m and n the possibility of being different odd numbers.

If we used the same letter, say j, in the expressions for m and n then we would be making our odd numbers equal and we would only have proved that if we add together two equal odd numbers, then the result is even.

Now we have to use some symbolic rules.

$$m + n = (2j + 1) + (2i + 1)$$

We can remove the brackets and rearrange to give:

$$m + n = 2j + 2i + 2$$

Finally, we can use the fact that 2 is a common factor of all terms in the expression to place it outside a bracket.

$$m + n = 2(j + i + 1)$$

But j, i and 1 are all whole numbers so $j + i + 1$ is also a whole number. Technically, this comes from the fact that the whole numbers are **closed under the operation of addition** because they form an important structure called a **group**. Let's call this whole number p.

So, we have that $m + n = 2p$. But this is precisely the definition of an even number that we started with. An even number is 2 times a whole number. This proves that any two odd numbers added together gives an even number.

The big chain of reasoning above is called a proof. It is immensely powerful because it covers an infinite number of situations. There is an infinite number of possible pairs of odd numbers to which the result applies. This is the power and beauty of using letters for numbers — a practice that was developed in Baghdad and Damascus about 1000 years ago. In one sense, mathematicians have a god-like ability when it comes to dealing with infinite sets.

Proof

Proof is the central concept in mathematics because it guarantees mathematical truth. When something is proved, we can say that it is true.

This type of truth is independent of place and time. In contrast to the science of the day, the mathematical truths of Pythagoras are just as true today as they were then – indeed his famous theorem is still taught today as can be seen in this book. But the science of the time has long been rejected. There were four chemical elements in the 4th century BC, and Aristotle thought that the heart was the organ for thinking. Actually, we do not have to go far back in time to find textbooks in the natural sciences that contain statements that we would dispute today. The truths of the natural sciences are always subject to revision, but mathematical truths are eternal.

But there is something even more striking about mathematical truths — that is, mathematical statements that have been proved. A statement such as 'odd + odd = even' has such power that we can say that it is certain. This is not just a matter of confidence – we are not talking about psychological certainty here.

It is certain because it cannot be otherwise. The negation of a mathematical truth (like 'odd + odd = odd') is to utter a self-contradiction or absurdity. Let's reflect on the power of this statement. This means that there is no possible world in which 'odd + odd = odd' (given the standard meanings of these terms). A story that makes this statement is describing a world that is self-contradictory – that is, an absurd and unintelligible world. Such a story is just not credible. But this means that mathematics is really radically different from other areas of knowledge, including the natural sciences. It is not a contradiction to say that the moon is composed of green cheese. There could be universes where this is true, but it just happens not to be in ours. Mathematics deals in what we call **necessary truths**, while the sciences deal mainly in **contingent truth**.

This is something that students of ToK should think about carefully. What is it about mathematical truth that makes it immune to revision and provides the basis for certainty and makes the negation of a mathematical truth a contradiction?

Recall that the constructivist sees mathematics as a big abstract game played by human beings according to invented rules. The hero of *The Glass Bead Game*, a novel by the German writer Hermann Hesse, must learn music, mathematics, and cultural history to play the game. On this view, mathematics is just like the glass bead game. There are parallels we can draw between a game like chess and mathematical proof. First, chess is played on a special board with pieces that can move in a particular way. The pieces must be set up on the board in a particular fashion before the game can begin. The same is true of mathematical proof. It starts with a collection of statements in mathematical language called **axioms**. They themselves cannot be proved. They are simply taken as self-evidently true and form the starting point for mathematical reasoning.

Once the game is set up, we can start playing. A move in chess means transforming the position of the pieces on the board by applying one of the game's rules that govern movement. Typically in chess, a move involves the movement of only one piece. (Can you think of an exception?) If the state of the pieces before the move was legitimate and the move was made according to the rules of the game, then the state of the pieces after the move is also legitimate. The same is true of a mathematical proof. One applies the rules (these are rules of algebra typically) to a line in the proof to get the next line. The whole proof is a chain of such moves.

Finally, the chess game ends. Either one of the players has achieved checkmate, or a stalemate (a draw) has been agreed. Similarly, a mathematical proof has an end. This is a point where the proof arrives at the required result at the end of the chain of reasoning. This result is called a **theorem**.

Once a proof of a mathematical statement is produced, we have a logical duty to believe the result, however unlikely. This is illustrated with a famous example.

Many people do not believe that $1 = 0.99999999\ldots$
(The three dots indicate that the 9's continue indefinitely).

The proof is straightforward.

Let	$x = 0.9999999\ldots$
Then	$10x = 9.9999999\ldots$
Subtract both equations	$10x - x = 9.99999999\ldots - 0.9999999\ldots$
This implies	$9x = 9$

Giving $x = 1$ as required.

$0.999999\ldots$ really does look very different to 1 but if the proof works then we are forced to believe that they are the same.

Are you happy with every stage of this proof?

Sets

A set is a collection of elements that can themselves be sets. They can be combined in various ways to produce new sets. The concepts of a set and membership of a set are **primitive**. This means that they cannot be explained in terms of more simple ideas. These seem to be rather modest beginnings on which to build the complexities of modern mathematics. Nevertheless, in the 20th century there were a number of projects that were designed to do just that: reduce the whole of mathematics to set theory. The most important work here was by Quine, von Neumann and Zermelo, and Bertrand Russell and Alfred North Whitehead in the three volumes of their *Principia Mathematica* of 1910–1913. Starting out with the notion of the empty set and the idea that no set can be a member of itself, we can construct the whole number system.

Mappings between sets

Once we have established sets in our mathematical universe, we want to do something useful with them. One of the most important ideas in the whole of mathematics is that of a mapping. A mapping is a rule that associates every member of a set with a member of a second set. This is what we were doing when we started this chapter by counting cows. We set up a one-to-one correspondence between a set of numbers and a set of cows.

Infinite sets

Consider the function $f(x) = 2x$ defined over the natural numbers.

Clearly it sets up a one-to-one correspondence between the set of natural numbers and the set of even numbers (check this yourself). So, this means that there are as many even numbers as there are natural numbers.

This is rather strange because we would think intuitively that there were more natural numbers than even numbers – they are after all the result of taking away an infinite number of odd numbers from the original set. But we are saying that the set that is left over has as many members as the original set. This strangeness is characteristic of infinite sets (indeed it can be used to define what we mean by infinite). Infinite sets can be put in a one-to-one correspondence with a proper subset of themselves.

But the story doesn't stop here. Using sets and mappings we can show that there are many different types of infinity. The set of natural numbers contains the smallest type of infinity, usually denoted by $\aleph_0$, which we call 'aleph nought'. In the 19th century, the German mathematician Georg Cantor showed by an ingenious argument that the number of numbers between 0 and 1 is a bigger type of infinity than aleph nought.

It turns out that there is an infinity of different types of infinity — a whole hierarchy of infinities, in fact — and this probably does not surprise you anymore, there are more infinities than finite cardinal numbers.

The methods and concepts of mathematics, therefore, are quite unlike anything to be found in the sciences, although they do seem to bear a strong resemblance to the arts in terms of the setting of the rules of the game and the use of the imagination. This is something we will explore in the next section.

What is it about the difference between the methods of the natural sciences and mathematics that accounts for the radical difference in types of knowledge produced?

Mathematics and the knower

English poet John Keats said, *"Beauty is truth, truth beauty – that is all / Ye know on earth, and all ye need to know."*

In this section we will see how mathematics impinges on our personal thinking about the world. One of the more surprising aspects of mathematics is the two-way link to the arts and beauty.

Beauty by the numbers

There is a long-held view that we find certain things beautiful because of their special proportions or some other intrinsic mathematical feature. This is the thinking that has inspired architects since the times of ancient Egypt and generations of painters, sculptors, musicians, and writers. Mathematics seems to endow beauty with a certain eternal objectivity. Things are beautiful because of the mathematical relationships between their parts. Moreover, this is a very public beauty because it can be dissected and discussed.

Let's take the example of the builders of the Parthenon. They were deeply interested in symmetry and proportion. In particular, they were interested in how to divide a line so that the proportion of the shorter part to the longer part is the same as that of the longer part to the whole. You can check that you get the quadratic equation $x^2 + x - 1 = 0$. One solution to this equation is the golden ratio $x = \dfrac{-1 + \sqrt{5}}{2} = 0.61803398875\ldots = \varphi$.

This proportion features significantly in the design of the Parthenon and many other buildings of the period. Since it is also related to the Fibonacci sequence, you will find φ turning up anywhere where there are spirals. It is used quite self-consciously in painting (Piet Mondrian, for example) and in music (particularly the music of Debussy). There are those who go as far as saying that it is present in the proportions of the perfect human figure and that we have a predisposition towards this ratio.

See if you can spot the connection between the golden ratio and the Fibonacci sequence. Hint: write down a difference equation for generating the sequence.

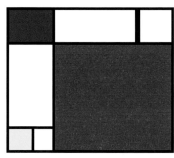

Figure 4 *Composition with Red, Blue and Yellow* (1926) Piet Mondrian. The proportions of some of the rectangles in this painting is φ

Beauty in numbers

Keats also put it the other way around: the beautiful is the true. Could we allow ourselves to be guided to truth in mathematics because of the beauty of the equations? This is a position taken by surprisingly many mathematicians. They look for beauty and elegance as an indicator of truth. Many mathematical physicists were guided in the 20th century by considerations of beauty and elegance.

Einstein suggested that the most incomprehensible thing about the universe was that it was comprehensible. From a ToK point of view, the most incomprehensible thing about the universe is that it is comprehensible in the language of mathematics. Galileo wrote, *'Philosophy is written in this grand book, the universe… It is written in the language of mathematics, and its characters are triangles, circles and other geometric figures…'*

Perhaps what is more puzzling is not just that we can describe the universe in mathematical terms, but that the mathematics we need to do this is mostly simple, elegant, and even beautiful.

To illustrate this, let's look at some of the famous equations of physics. Most people will be familiar with Einstein's field equations and Maxwell's equations.

$$R_{\mu\nu} - Rg_{\mu\nu} = \frac{8\pi G}{c^4} T_{\mu\nu}$$

EINSTEIN'S FIELD EQUATION for general relativity

Figure 5 Einstein's field equation

1. $\nabla \cdot \mathbf{D} = \rho_V$

2. $\nabla \cdot \mathbf{B} = 0$

3. $\nabla \times \mathbf{E} = -\dfrac{\partial \mathbf{B}}{\partial t}$

4. $\nabla \times \mathbf{H} = \dfrac{\partial \mathbf{D}}{\partial t} + \mathbf{J}$

Figure 6 Maxwell's equations

It is perplexing that the whole crazy complex universe can be described by such simple, elegant, and even beautiful equations. It seems that our mathematics fits the universe rather well. It is difficult to believe that mathematics is just a mind game that we humans have invented.

But the argument from simplicity and beauty goes further. Symmetry in the underlying algebra led mathematical physicists to propose the existence of new fundamental particles, which were subsequently discovered. In some cases, beauty and elegance of the mathematical description have even been used as evidence of truth. The physicist Paul Dirac said, *'It seems that if one is working from the point of view of getting beauty in one's equations, and if one has really a sound insight, one is on a sure line of progress'*.

Dirac's own equation for the electron must rate as one of the most profoundly beautiful of all. Its beauty lies in the extraordinary neatness of the underlying mathematics – it all seems to fit so perfectly together:

$$(i\partial\!\!\!/ - m)\psi = 0$$

Figure 7 Dirac's equation of the electron

The physicist and mathematician Palle Jorgensen wrote:

> *'[Dirac] … liked to use his equation for the electron as an example stressing that he was led to it by paying attention to the beauty of the math, more than to the physics experiments.'*

It was because of the structure of the mathematics in particular that there were two symmetrical parts to the equation — one representing a negatively charged particle (the electron) and the other a similar particle but with a positive charge — that scientists were led to the discovery of the positron. It seems fair to say that the mathematics did really come first here.

We will leave the last word on this subject to Dirac himself, writing in Scientific American in 1963:

> 'I think there is a moral to this story, namely that it is more important to have beauty in one's equations than to have them fit experiment.'

By any standards this is an extraordinary statement for a mathematical physicist to make.

Mathematics and personal intuitions

Sometimes our intuition can let us down badly when it comes to making judgments of probability. Here is an example to illustrate how we might have to correct our intuition by careful mathematical reasoning.

Consider the following case. There is a rare genetic disease among the population. Very few people have the disease. As a precaution, a test has been developed to detect whether particular individuals have the disease. Although the test is quite good, it is not perfect — it is only 99% accurate. Person X takes the test and it shows positive. The question for your mathematical intuition is: 'What is the probability that X actually has the disease?' (You should recognise this as being a problem of conditional probability.)

Think about this for a moment before we continue.

Many of the students (and teachers) we have worked with in the past give the same answer: the probability that X actually has the disease given a positive test result is about 99%. Did you say the same? If you did, then your mathematical intuition let you down – very badly.

Let's put some numbers into the problem to illustrate this. For the sake of simplicity, assume that the country in which the test takes place has a population of 10 million. We are told that the disease is very rare. Assume that only 100 people in the whole country have the disease. We are told that the test is 99% accurate so, of the 100 cases of the disease the test would show positive in 99 cases and negative in one. So far so good.

Now consider the 9 999 900 people who don't have the disease. In 99% of these cases the test does its job and records a negative result. In 1% of the cases however it gets it wrong and produces a positive result. 1% of 9 999 900 is 99 999. So, of the whole population tested there would be a total of $99\,999 + 99 = 100\,098$ positive results. But of these only 99 have the disease. Therefore, the probability of having the disease given a positive result is $\frac{99}{100\,098} = 0.0989\%$ or about 1 in 1000. This is quite a big difference from the 990 in 1000 that we expected intuitively. That is out by a whopping 99 000%.

What went wrong with intuition here?

Mathematics and personal qualities

There are undoubtedly special qualities well-suited to doing mathematics. There are a host of great mathematicians from Archimedes, Euclid, Hypatia, through to Andrew Wiles, Grigori Perelman, and Maryam Mirzakhani, who contributed significantly to the area. Maryam was the first woman to receive the Fields medal (the equivalent of the Nobel prize in mathematics). Although mathematics is collaborative in the sense that mathematicians build on the work of others and take on the challenges that the area itself has recognised as being important, it is nevertheless largely a solitary pursuit. It requires great depth of thought, imaginative leaps, careful and sometimes laborious computations, innovative ways of solving very hard problems, and, most of all, great persistence. Mathematicians need to develop their intuition and their nose for a profitable strategy. They are guided by emotion and by hunches — they are a far cry from the stereotype of the coldly logical thinker who is closer to computer than human.

Conclusion

We have seen that mathematics is really one of the crowning achievements of human civilisation. Its ancient art has been responsible for some of the most extraordinary intellectual journeys taken by humankind, and its methods have allowed the building of great cities, and the production of great art, and it has been the language of great science.

From a ToK perspective, mathematics, with its absolute and unchanging notion of necessary truth, makes a good contrast to the natural sciences with their reliance on observation of the external world, experimental method, and provisional nature of its results.

Two countering arguments should be set against this view of mathematics. The idea that the axioms of mathematics (the rules of the game) are arbitrary both deprives mathematics of its status as something independent of human beings, and makes it vulnerable to the charge that its results cannot ever be entirely relevant to the world outside mathematics.

Platonists would certainly argue that mathematics is out there in the universe, with or without human beings. They would argue that it is built into the structure of the cosmos – a fact that explains why the laws of the natural sciences lend themselves so readily to mathematical expression.

Both views produce challenging questions in ToK. The constructivist is a victim of the success of mathematics in fields such as the natural sciences. She has to account for why mathematics is so supremely good at describing the outside world to which, according to this view, it should ultimately be blind. The Platonist, on the other hand, finds it hard to identify mathematical structures embedded in the world or has a hard time explaining why they are there.

We have seen how mathematics is closely integrated into artistic thinking; perhaps because both are abstract areas of knowledge indirectly linked to the world and not held to account through experiment and observation, but instead, open to thought experiment and leaps of imagination. Mathematics can challenge our intuitions and can push our cognitive resources as individual knowers. Infinity is not something that the human mind can fathom in its entirety. Instead, mathematics gives us the tools to deal with it in precisely this unfathomed state. We can be challenged by results that seem counter to our intuition, but ultimately, the nature of mathematical proof is that it forces us to accept them nonetheless. In turn, individuals can, through their insight and personal perspectives, make ground-breaking contributions that change the direction of mathematics forever. The history of mathematics is a history of great thinkers building on the work of previous generations to do ever more powerful things using ever more sophisticated tools.

The Greek thinkers of the 4th century BC thought that mathematics lay at the core of human knowledge. They thought that mathematics was one of the few areas in which humans could apprehend the eternal forms only accessible to pure unembodied intellect. They thought that in mathematics they could glimpse the very framework on which the world and its myriad processes rested. Maybe they were right.

Answers

Chapter 1

Exercise 1.1

1. (a) $x = h - \dfrac{n}{m}$

(b) $a = \dfrac{v^2 + t}{b}$

(c) $b_1 = \dfrac{2A}{h} - b_2$

(d) $r = \sqrt{\dfrac{2A}{\theta}}$

(e) $k = \dfrac{gh}{f}$

(f) $t = \dfrac{x}{a + b}$

(g) $r = \sqrt[3]{\dfrac{3V}{\pi h}}$

(h) $k = \dfrac{g}{F(m_1 + m_2)}$

2. (a) $y = -\dfrac{2}{3}x - 5$

(b) $y = -4$

(c) $y = \dfrac{5}{4}x + 6$

(d) $x = \dfrac{7}{3}$

(e) $y = -4x + 11$

(f) $y = -\dfrac{5}{2}x - 7$

3. (a) (i) 17 **(ii)** $\left(0, \dfrac{5}{2}\right)$

(b) (i) $\sqrt{40}$ **(ii)** $(2, 3)$

(c) (i) $\dfrac{\sqrt{82}}{3}$ **(ii)** $\left(-1, \dfrac{7}{6}\right)$

(d) (i) $\sqrt{533}$ **(ii)** $\left(1, \dfrac{11}{2}\right)$

4. (a) $k = 1$ or 9 **(b)** $k = -11$ or -3

5. (a) $(\sqrt{5})^2 + (\sqrt{45})^2 = (\sqrt{50})^2$

(b) sides are: $\sqrt{29}, \sqrt{29}, \sqrt{58}$

(c) sides are: $\sqrt{45}, \sqrt{10}, \sqrt{45}, \sqrt{10}$

6. (a) $(5, 1)$ **(b)** $\left(4, \dfrac{1}{2}\right)$ **(c)** $(3, -4)$

(d) no solution **(e)** $(-1, 2)$

7. (a) $(-1, 3)$ **(b)** $(-3, -8)$

(c) lines are coincident; solution set is all points on the line
$y = -\dfrac{1}{4}x - \dfrac{3}{4}$

(d) $\left(\dfrac{20}{3}, \dfrac{40}{3}\right)$ **(e)** $\left(\dfrac{1}{2}, 3\right)$ **(f)** $(-5, 10)$

8. (a) $(5, -3)$ **(b)** $(14.1, 10.4)$ **(c)** $\left(\dfrac{11}{19}, -\dfrac{18}{19}\right)$

9. (a) $(-1, 3, 2)$ **(b)** $(5, 8, -2)$

(c) $\left(\dfrac{21}{32}, \dfrac{3}{32}, -\dfrac{1}{2}\right)$ **(d)** $(-7, 3, -2)$

10. (a) $(4, -2, 1)$ **(b)** no solution

(c) $(-2, 4, 3)$ **(d)** infinite solutions

11. $k = \dfrac{4}{3}$

Exercise 1.2

1. (a) G **(b)** L **(c)** H **(d)** K **(e)** J

(f) C **(g)** A **(h)** I **(i)** F

2. $A = \dfrac{C^2}{4\pi}$ **3.** $A = \dfrac{l^2\sqrt{3}}{8}$

4. $A = 4x^2 + 60x$ **5.** $h = x\sqrt{2}$

6. (a) 9.4 **(b)** $V = \dfrac{3525}{P}$

7. (a) $F = kx$ **(b)** 6.25 **(c)** $37.5\,\text{N}$

8. (a) $\{-6.2, -1.5, 0.7, 3.2, 3.8\}$ **(b)** $r > 0$

(c) $\mathbb{R}$ **(d)** $\mathbb{R}$ **(e)** $t \leqslant 3$

(f) $\mathbb{R}$ **(g)** $x \neq \pm 3$

(h) $-1 \leqslant x \leqslant 1$ and $x \neq 0$

9. no, $x = c$ is a vertical line

10. (a) (i) $\sqrt{17}$ **(ii)** 7 **(iii)** 0

(b) $x < 4$

(c) domain: $x \geqslant 4$, range: $h(x) \geqslant 0$

11. (a) (i) domain $\{x : x \in \mathbb{R}, x \neq 5\}$, range $\{y : y \in \mathbb{R}, y \neq 0\}$

(ii) y-intercept $\left(0, -\dfrac{1}{5}\right)$, vertical asymptote $x = 5$,
horizontal asymptote $y = 0$

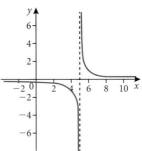

(b) (i) domain $\{x : x < -3, x > 3\}$, range $\{y : y > 0\}$
(ii) vertical asymptotes $x = -3$ and $x = 3$

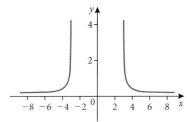

(c) (i) domain $\{x : x \in \mathbb{R}, x \neq -2\}$, range $\{y : y \in \mathbb{R}, y \neq 2\}$
(ii) y-intercept $\left(0, -\dfrac{1}{2}\right)$, vertical asymptote $x = -2$,
horizontal asymptote $y = 2$

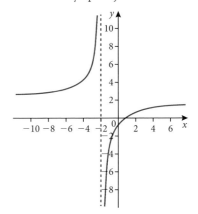

(d) (i) domain $\left\{x : -\dfrac{\sqrt{10}}{2} \leqslant x \leqslant \dfrac{\sqrt{10}}{2}\right\}$,
range $\left\{y : 0 < y \leqslant \sqrt{5}\right\}$

(ii) y-intercept $(0, \sqrt{5})$, x-intercepts $\left(-\dfrac{\sqrt{10}}{2}, 0\right)$
and $\left(\dfrac{\sqrt{10}}{2}, 0\right)$

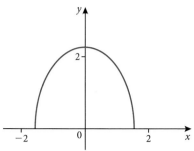

(e) (i) domain $\{x : x \in \mathbb{R}, x \neq 0\}$, range $\{y : y \in \mathbb{R}, y \neq -4\}$
  (ii) vertical asymptote $x = 0$, horizontal asymptote
    $y = -4$

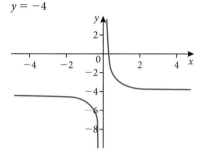

Exercise 1.3

1. (a) $(f \circ g)(5) = 1$ (b) $(g \circ f)(5) = \dfrac{1}{7}$

   (c) $(f \circ g)(x) = \dfrac{2}{x - 3}$ (d) $(g \circ f)(x) = \dfrac{1}{2x - 3}$

2. (a) 1 (b) -7 (c) 7
   (d) -47 (e) -1 (f) -79
   (g) $1 - 2x^2$ (h) $-4x^2 + 12x - 7$ (i) $4x - 9$
   (j) $-x^4 + 4x^2 - 2$

3. (a) $(f \circ g)(x) = 12x + 7$, domain: $x \in \mathbb{R}$;
      $(g \circ f)(x) = 12x - 1$, domain: $x \in \mathbb{R}$
   (b) $(f \circ g)(x) = 4x^2 + 1$, domain: $x \in \mathbb{R}$;
      $(g \circ f)(x) = -2x^2 - 2$, domain: $x \in \mathbb{R}$
   (c) $(f \circ g)(x) = \sqrt{x^2 + 2}$, domain: $x \in \mathbb{R}$;
      $(g \circ f)(x) = x + 2$, domain: $x \geqslant -1$
   (d) $(f \circ g)(x) = \dfrac{2}{x + 3}$, domain: $x \in \mathbb{R}, x \neq -3$;
      $(g \circ f)(x) = -\dfrac{x + 2}{x + 4}$, domain: $x \in \mathbb{R}, x \neq -4$
   (e) $(f \circ g)(x) = x$, domain: $x \in \mathbb{R}$;
      $(g \circ f)(x) = x$, domain: $x \in \mathbb{R}$
   (f) $(f \circ g)(x) = 1 + x^2$, domain: $x \in \mathbb{R}$;
      $(g \circ f)(x) = \sqrt[3]{-x^6 + 4x^3 - 3}$, domain: $x \in \mathbb{R}$
   (g) $(f \circ g)(x) = \dfrac{2}{4x^2 - 1}$, domain: $x \neq 0, x \neq \pm\dfrac{1}{2}$;
      $(g \circ f)(x) = \dfrac{(4 - x)^2}{4x^2}$, domain: $x \neq 0, x \neq 4$
   (h) $(f \circ g)(x) = x$, domain: $x \neq -3$;
      $(g \circ f)(x) = x$, domain: $x \neq -3$
   (i) $(f \circ g)(x) = \dfrac{x^2 - 1}{x^2 - 2}$, domain: $x \neq \pm\sqrt{2}$;
      $(g \circ f)(x) = \dfrac{2x - 1}{(x - 1)^2}$, domain: $x \neq 1$

4. (a) $(g \circ h)(x) = \sqrt{9 - x^2}$, domain: $-3 \leqslant x \leqslant 3$,
      range: $y > 0$
   (b) $(h \circ g)(x) = -x + 11$, domain: $x \geqslant 1$, range: $y \leqslant 10$

5. (a) $(f \circ g)(x) = \dfrac{1}{10 - x^2}$, domain: $x \neq \pm\sqrt{10}$, range: $y \neq 0$

   (b) $(g \circ f)(x) = 10 - \dfrac{1}{x^2}$, domain: $x \neq 0$, range: $y < 10$

6. (a) $h(x) = x + 3$, $g(x) = x^2$
   (b) $h(x) = x - 5$, $g(x) = \sqrt{x}$
   (c) $h(x) = \sqrt{x}$, $g(x) = 7 - x$
   (d) $h(x) = x + 3$, $g(x) = \dfrac{1}{x}$
   (e) $h(x) = x + 1$, $g(x) = 10^x$
   (f) $h(x) = x - 9$, $g(x) = \sqrt[3]{x}$
   (g) $h(x) = x^2 - 9$, $g(x) = |x|$
   (h) $h(x) = \sqrt{x - 5}$, $g(x) = \dfrac{1}{x}$

7. (a) (i) domain of f: $x \geqslant 0$
      (ii) domain of g: $x \in \mathbb{R}$
      (iii) $(f \circ g)(x) = \sqrt{x^2 + 1}$, domain: $x \in \mathbb{R}$
   (b) (i) domain of f: $x \neq 0$
      (ii) domain of g: $x \in \mathbb{R}$
      (iii) $(f \circ g)(x) = \dfrac{1}{x + 3}$, domain: $x \neq -3$
   (c) (i) domain of f: $x \neq \pm 1$
      (ii) domain of g: $x \in \mathbb{R}$
      (iii) $(f \circ g)(x) = \dfrac{3}{x^2 + 2x}$, domain: $x \neq 0, -2$
   (d) (i) domain of f: $x \in \mathbb{R}$
      (ii) domain of g: $x \in \mathbb{R}$
      (iii) $(f \circ g)(x) = x + 3$, domain $x \in \mathbb{R}$

Exercise 1.4

1. (a) 2 (b) 6
2. (a) -1 (b) b
3. 4
4. 6
5. (a) (i)
      (ii)

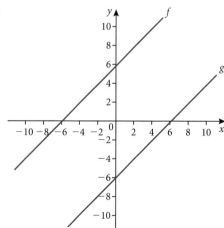

(b) (i)
(ii)

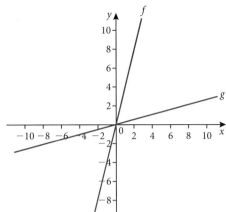

(c) (i)
(ii)

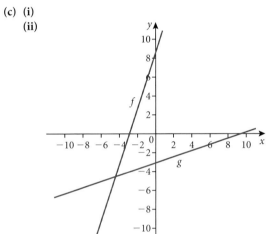

(d) (i)
(ii)

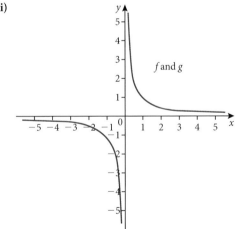

(e) (i)
(ii)

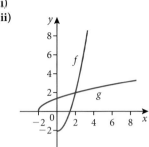

(f) (i)
(ii)

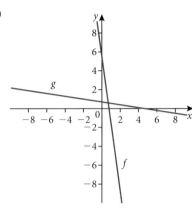

(g) (i)
(ii)

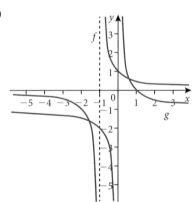

(h) (i)
(ii)

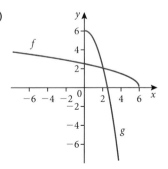

(i) (i)

(ii)

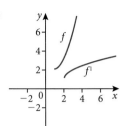

(j) (i)

(ii)

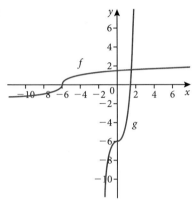

6. (a) $f^{-1}(x) = \frac{1}{2}x + \frac{3}{2}, x \in \mathbb{R}$

(b) $f^{-1}(x) = 4x - 7, x \in \mathbb{R}$

(c) $f^{-1}(x) = x^2, x \geqslant 0$

(d) $f^{-1}(x) = \frac{1}{x} - 2, x \in \mathbb{R}, x \neq 0$

(e) $f^{-1}(x) = \sqrt{4 - x}, x \leqslant 4$

(f) $f^{-1}(x) = x^2 + 5, x \geqslant 5$

(g) $f^{-1}(x) = \frac{1}{a}x - \frac{b}{a}, x \in \mathbb{R}$

(h) $f^{-1}(x) = -1 + \sqrt{x + 1}, x \geqslant -1$

(i) $f^{-1}(x) = \sqrt{\dfrac{1 + x}{1 - x}}, -1 \leqslant x < 1$

(j) $f^{-1}(x) = \sqrt[3]{x - 1}, x \in \mathbb{R}$

7. (a) $x > 1, f^{-1}(x) = \dfrac{x + 3}{x - 2}$

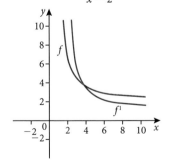

(b) $x > 2, f^{-1}(x) = \sqrt{x} + 2$

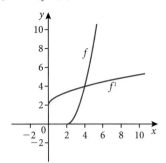

(c) $x > 0, f^{-1}(x) = \sqrt{\dfrac{1}{x}}$

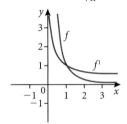

(d) $x > 0, f^{-1}(x) = \sqrt[4]{2 - x}$

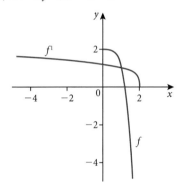

8. $x < -1, -1 \leqslant x \leqslant 1, x > 1$

9. (a) $\dfrac{3}{2}$ **(b)** 5

(c) -4 **(d)** $\dfrac{7}{2}$

(e) $g^{-1} \circ h^{-1} = \frac{1}{2}x - 1$ **(f)** $h^{-1} \circ g^{-1} = \frac{1}{2}x + \frac{1}{2}$

(g) $(g \circ h)^{-1} = \frac{1}{2}x + \frac{1}{2}$ **(h)** $(h \circ g)^{-1} = 2x + 2$

10. $f(f(x)) = f\left(\dfrac{a}{x + b} - b\right) = \dfrac{a}{\dfrac{a}{x + b} - b + b} - b = \dfrac{a}{\dfrac{a}{x + b}} - b$

$= \dfrac{a}{1} \cdot \dfrac{x + b}{a} - b = x + b - b = x$

Since $f(f(x)) = x$, then the function f is its own inverse.

Answers

Exercise 1.5

1. (a)

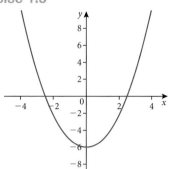

(b)

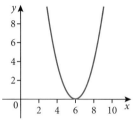

(c)

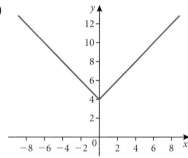

(d)

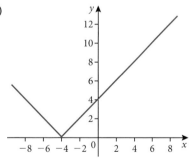

(e)

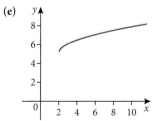

(f)

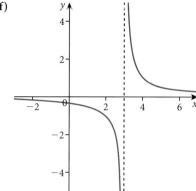

(g)

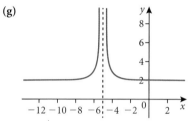

(h)

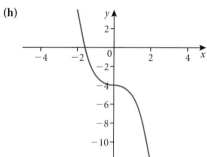

(i)

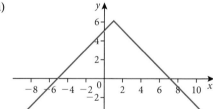

(j)

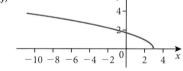

(k)

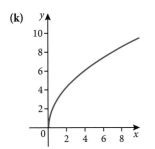

(l)

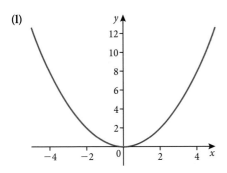

(m)

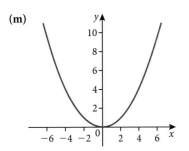

(n)

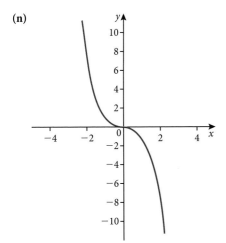

2. (a) $y = -x^2 + 5$ **(b)** $y = \sqrt{-x}$

  (c) $y = -|x + 1|$ **(d)** $y = \dfrac{1}{x - 2} - 3$

3. (a)

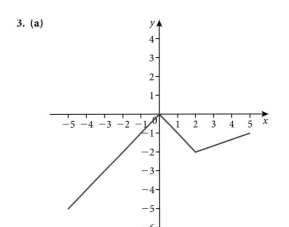

(b)

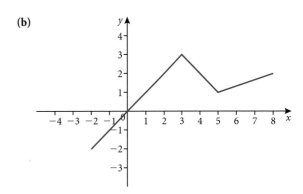

(c)

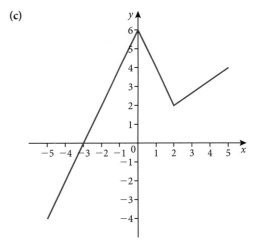

(d)

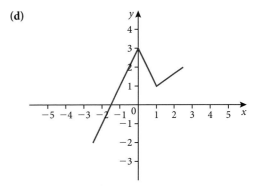

(e)

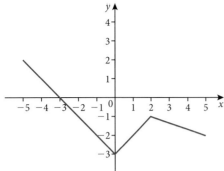

(f)

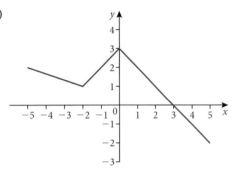

(g)

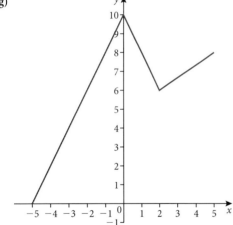

4. (a) horizontal translation 3 units right; vertical translation 5 units up (or reverse order)

(b) reflect over the x-axis; vertical translation 2 units up (or reverse order)

(c) horizontal translation 4 units left; vertical shrink by factor $\frac{1}{2}$ (or reverse order)

(d) horizontal shrink by factor $\frac{1}{3}$; horizontal translation 1 unit right; vertical translation 6 units down

5. (a) (i)

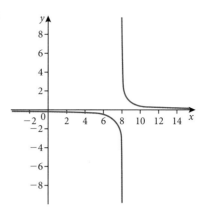

(ii)

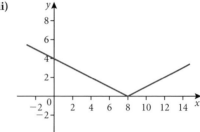

(iii)

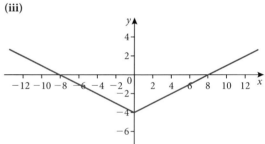

(b) (i)

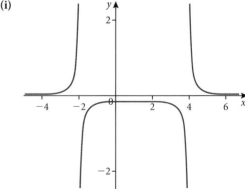

(ii)

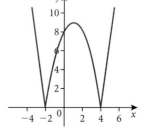

(iii)

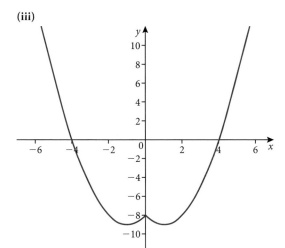

(c) (i)

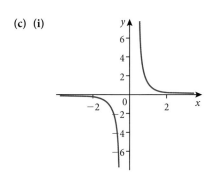

(ii)

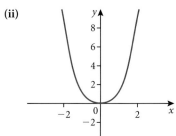

(iii)

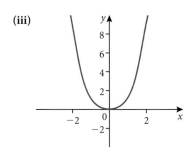

Chapter 1 practice questions

1. **(a)** $a = -3, b = 1$ **(b)** range: $y > 0$
2. **(a)** 5   **(b)** -9
3. **(a)** $g^{-1}(x) = -3x + 4$ **(b)** $x = \dfrac{2}{3}$
4. **(a)** $(g \circ h)(x) = 2x - 3$ **(b)** See Worked Solutions
5. See Worked Solutions
6. $(1, 3, 0)$

7. (a)

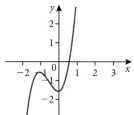

(b) maximum at $\left(-1, -\dfrac{1}{2}\right)$; minimum at $\left(0, -\dfrac{3}{2}\right)$

8. **(a)** $k = \dfrac{1}{2}$ **(b)** $p = -5$ **(c)** $q = 3$

9. (a)

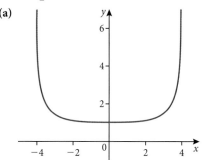

(b) $x = 4, x = -4$
(c) range: $y \geqslant 1$

10. (a)

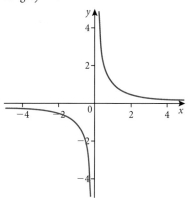

(b) $h(x) = \dfrac{1}{x + 4} - 2$

(c) (i) x-intercept: $\left(-\dfrac{7}{2}, 0\right)$; y-intercept: $\left(0, -\dfrac{7}{4}\right)$

    (ii) vertical asymptote: $x = -4$;
         horizontal asymptote: $y = -2$

    (iii)

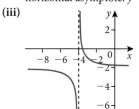

11. **(a) (i)** $\sqrt{11}$ **(ii)** 7   **(iii)** 0
     (b) $x < -3$
     (c) $(g \circ f)(x) = x - 2$
12. **(a)** 4   **(b)** $(g^{-1} \circ h)(x) = 2x^2 + 6$
     (c) $x = \pm 2\sqrt{2}$
13. **(a)** $f^{-1}(x) = \dfrac{1}{3}x + \dfrac{1}{3}$ **(b)** $(f \circ g)(x) = \dfrac{12}{x} - 1$

Answers

(c) $(f \circ g)^{-1}(x) = \dfrac{12}{x+1}$ **(d)** $(g \circ g)(x) = x$

14. (a) (i) $a = 8$ **(ii)** $b = -3$
   (b) reflection over x-axis

15. (a)

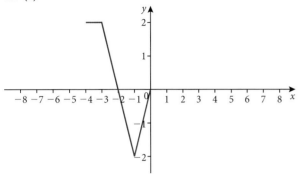

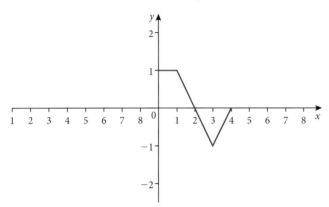

   (b) $A'(-3, -2)$

16.

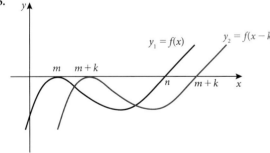

17. $(f \circ g)^{-1}(x) = \sqrt[3]{x-1}$

18. (a) $g(x) = \dfrac{x}{2x+1}$ **(b)** $\dfrac{2}{9}$

19. (a) $-\dfrac{\sqrt{2}}{2} \leqslant x \leqslant \dfrac{\sqrt{2}}{2}, x \neq 0$ **(b)** $f(x) \geqslant 0$

20. $f^{-1}(x) = \dfrac{x+1}{x-2}, x \neq 2$

21. (a) $-\dfrac{1}{2} < A < 2$ **(b)** $f^{-1}(x) = \dfrac{-2x-1}{x-2}$

22. (a) $g(x) = \sqrt[3]{x+1}$ **(b)** $g(x) = \sqrt[3]{x} + 1$

23. (a) $S = \{x: -\sqrt{3} < x < \sqrt{3}\}$ **(b)** $f(x) > \dfrac{\sqrt{3}}{3}$

24. $\dfrac{x}{4}$

25. (a) $A(1, 25), B(4, 0), C(7, -35), D(10, 0)$
   (b) $A(-1, -25), B(0, 0), C(1, 35), D(2, 0)$

Chapter 2

Exercise 2.2

1. (a) (i) $f(x) = (x-5)^2 + 7$; axis of symmetry is $x = 5$
      (ii) horizontal translation 5 units right; vertical translation 7 units up
      (iii) minimum: $(5, 7)$
   (b) (i) $f(x) = (x+3)^2 - 1$; axis of symmetry is $x = 3$
      (ii) horizontal translation 3 units left; vertical translation 1 unit down
      (iii) minimum: $(-3, -1)$
   (c) (i) $f(x) = -2(x+1)^2 + 12$; axis of symmetry is $x = -1$
      (ii) horizontal translation 1 unit left; reflection over x-axis; vertical stretch by factor 2; vertical translation 12 units up
      (iii) maximum: $(-1, 12)$
   (d) (i) $f(x) = 4\left(x - \dfrac{1}{2}\right)^2 + 8$; axis of symmetry is $x = \dfrac{1}{2}$
      (ii) horizontal translation $\dfrac{1}{2}$ unit right; vertical stretch by factor 4; vertical translation 8 units up
      (iii) minimum: $\left(\dfrac{1}{2}, 8\right)$
   (e) (i) $f(x) = \dfrac{1}{2}(x+7)^2 + \dfrac{3}{2}$; axis of symmetry is $x = -7$
      (ii) horizontal translation 7 units left; vertical shrink by factor $\dfrac{1}{2}$; vertical translation $\dfrac{3}{2}$ units up
      (iii) minimum: $\left(-7, \dfrac{3}{2}\right)$

2. (a) $x = 2, x = -4$ **(b)** $x = 5, x = -2$
   (c) $x = \dfrac{3}{2}, x = 0$ **(d)** $x = \dfrac{1}{3}, x = -4$
   (e) $x = 3, x = 2$ **(f)** $x = 2, x = \dfrac{1}{4}$

3. (a) $x = -2 \pm \sqrt{7}$ **(b)** $x = 5, x = -1$
   (c) no real solution   **(d)** $x = -4 \pm \sqrt{13}$
   (e) $x = 2, x = -4$ **(f)** $x = \dfrac{2 \pm \sqrt{22}}{2}$

4. (a) $x = 2 \pm \sqrt{5}$ **(b)** axis of symmetry: $x = 2$
   (c) minimum value of f is -5

5. (a) two real solutions   **(b)** no real solutions
   (c) two real solutions   **(d)** no real solutions

6. $p = \pm 2\sqrt{2}$

7. $k < 4$

8. $k < -1, k > 1$ **9.** $m < -3, m > 3$

10. $k > 12$

11. $x - 2 - x^2 \Rightarrow -(x^2 - x + 2) \Rightarrow -\left(x^2 - x + \dfrac{1}{4}\right) - \dfrac{7}{4}$
   $\Rightarrow -\left(x - \dfrac{1}{2}\right)^2 - \dfrac{7}{4} \leqslant -\dfrac{7}{4}$ for all x

12. (a) $y = -2x^2 + 6x + 8$
   (b) $y = \dfrac{2}{3}x^2 - \dfrac{7}{3}x + 1$

13. $-1 < k < 15$

14. $m < -2\sqrt{10}$ or $m > 2\sqrt{10}$

15. $f(x) = 3x^2 + 5x - 2$

16. $f(2) = 8$

17. $x < 1$ or $x > 3$

18. $\triangle - (2 - t)^2 - 4(2)(t^2 + 3) > 0 \rightarrow \ -7t^2 - 4t - 20 > 0,$
because $\triangle = -544$ for $-7t^2 - 4t - 20$ and leading coefficient is negative, then graph of $y = -7t^2 - 4t - 20$ is a parabola opening down and always below x-axis; hence $\triangle$ for original equation is always negative; thus, no real roots

19. product of roots $= \frac{c}{a}$, in this case $\frac{a}{a} = 1$, so they are reciprocals

20. (a) sum $= -3$, product $= -\frac{5}{2}$

(b) sum $= -3$, product $= -1$

(c) sum $= 0$, product $= -\frac{3}{2}$

(d) sum $= -a$, product $= -2a$

(e) sum $= 6$, product $= -4$

(f) sum $= \frac{1}{3}$, product $= -\frac{2}{3}$

21. $4x^2 + 5x + 4 = 0$

22. (a) $\frac{1}{9}$ **(b)** $\frac{1}{12}$ **(c)** $\frac{55}{27}$

23. (a) -2 and -6 **(b)** $k = 12$

24. (a) $-\frac{1}{4}$ **(b)** $4x^2 + x + 1 = 0$

25. (a) $x^2 - 19x + 25 = 0$ **(b)** $25x^2 + 72x - 5 = 0$

Exercise 2.3

1. (a) $3x^2 + 5x - 5 = (x + 3)(3x - 4) + 7$

(b) $3x^4 - 8x^3 + 9x + 5 = (x - 2)(3x^3 - 2x^2 - 4x + 1) + 7$

(c) $x^3 - 5x^2 + 3x - 7 = (x - 4)(x^2 - x - 1) - 11$

(d) $9x^3 + 12x^2 - 5x + 1 = (3x - 1)(3x^2 + 5x) + 1$

2. $(x - 7)(x - 1)(2x - 1)$

3. $(x - 2)(2x + 1)(3x + 2)$

4. $(x - 2)^2(x + 4)(3x + 2)$

5. (a) $Q(x) = x - 2, R = -2$

(b) $Q(x) = x^2 + 2, R = -3$

(c) $Q(x) = 3, R(x) = 20x + 5$

(d) $Q(x) = x^4 + x^3 + 4x^2 + 4x + 4, R = -2$

6. (a) $P(2) = 5$ **(b)** $P(-1) = 23$

(c) $P(-7) = -483$ **(d)** $P\left(\frac{1}{4}\right) = \frac{49}{64}$

7. $x = \frac{1 + \sqrt{5}}{2}$ or $x = \frac{1 - \sqrt{5}}{2}$

8. $k = 2\sqrt{3}$ or $k = -2\sqrt{3}$

9. $a = 5, b = 12$

10. (a) $x^3 - 3x^2 - 6x + 8$

(b) $x^4 - 3x^3 - 7x^2 + 15x + 18$

(c) $x^3 - 6x^2 + 12x - 8$

11. (a) $a = -1, b = -2$ **(b)** $3x + 2$

12. $a = \frac{4}{3}, b = \frac{1}{3}$ **13.** $a = -1, b = -4, c = 4$

14. $a = -5$ **15.** $b = 18$

16. (a) See Worked Solutions

(b) $R = 3$

17. (a) sum $= \frac{2}{3}$, product $= 5$

(b) sum $= 1$, product $= 7$

(c) sum $= \frac{1}{3}$, product $= -\frac{1}{2}$

18. See Worked Solutions

19. $-9, 3, 6$ **20.** $2, -4, 8$ **21.** $k = 3$ **22.** $k = -8$

Exercise 2.4

1. (a) vertical asymptote: $x = -2$
horizontal asymptote: $y = 0$

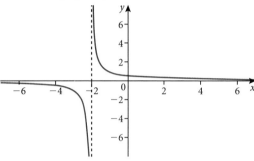

(b) vertical asymptote: $x = 2$
horizontal asymptote: $y = 0$

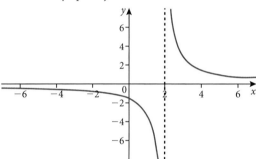

(c) x-intercept: $\left(\frac{1}{4}, 0\right)$, y-intercept: $(0, 1)$
vertical asymptote: $x = 1$
horizontal asymptote: $y = 4$

(d) x- and y-intercept: $(0, 0)$
vertical asymptote: $x = \pm 3$
horizontal asymptote: $y = 0$

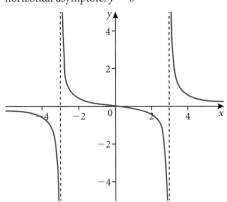

(e) x-intercept: none, y-intercept: $\left(0, -\dfrac{2}{3}\right)$

vertical asymptotes: $x = -3$, $x = 1$
horizontal asymptote: $y = 0$

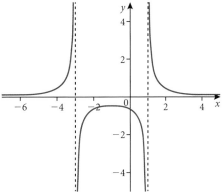

(f) oblique asymptote: $y = x$
vertical asymptote: $x = 0$

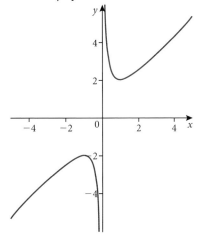

(g) x- and y intercept: $(0, 0)$
vertical asymptote: $x = -2$
horizontal asymptote: $y = 0$

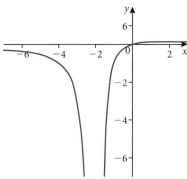

(h) x- and y-intercept: $(0, 0)$; x-intercept at $(-2, 0)$
vertical asymptote: $x = 1$
oblique asymptote: $y = x + 3$

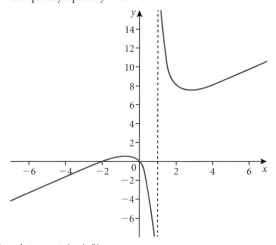

(i) x-intercept: $(-4, 0)$
y-intercept: $\left(0, -\dfrac{2}{3}\right)$
vertical asymptotes: $x = -3$ and $x = 4$
horizontal asymptote: $y = 0$

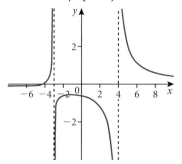

(J) *x*-intercept: $(2, 0)$
y-intercept: none
vertical asymptotes: $x = 0$ and $x = 4$
horizontal asymptote: $y = 0$

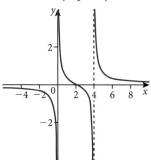

2. (a) domain $\{x : x \in \mathbb{R}, x \neq \pm 2\}$
range $\left\{ y : y \leqslant -\dfrac{5}{4} \text{ or } y > 2 \right\}$

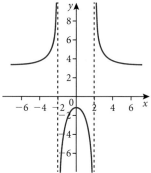

(b) domain $\{x : x \in \mathbb{R}, x \neq -4, 1\}$
range $\{y : y \in \mathbb{R}, y \neq 0\}$

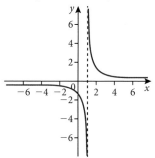

(c) domain $\{x : x \in \mathbb{R}\}$
range $\{y : 0 < y \leqslant 1\}$

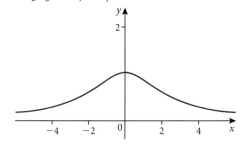

(d) domain $\{x : x \in \mathbb{R}, x \neq 1\}$
range $\{y : y \in \mathbb{R}, y \neq 0\}$

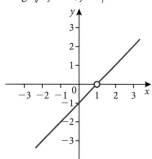

3. (a) *x*-intercept: $\left(\dfrac{5}{2}, 0\right)$

y-intercept: $\left(0, \dfrac{5}{18}\right)$

vertical asymptotes: $x = -6$ and $x = \dfrac{3}{2}$

horizontal asymptote: $y = 0$

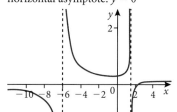

(b) *x*-intercept: none
y-intercept: $(0, -1)$
vertical asymptotes: $x = 1$
oblique asymptote: $y = x + 2$

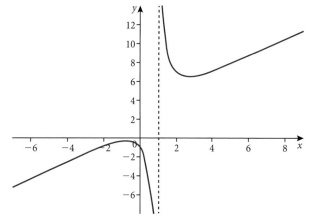

(c) *x*- and *y*-intercept: (0, 0)
horizontal asymptote: $y = 3$

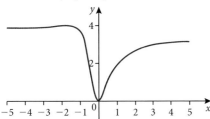

(d) *x*-intercept: none

y-intercept: $\left(0, \frac{1}{4}\right)$

vertical asymptotes: $x = -2$, $x = 1$ and $x = 2$
horizontal asymptote: $y = 0$

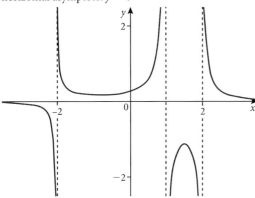

4. (a)

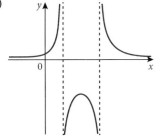

(b)

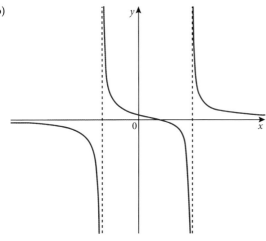

(c)

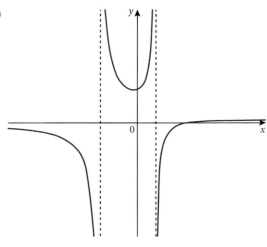

5. (a)

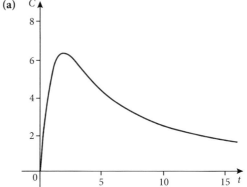

(b) at $t = 2$ minutes, concentration is 6.25 mg/l
(c) it continues to decrease and approaches zero as amount of time increases
(d) 50 minutes

Exercise 2.5

1. (a) $x = 3$ **(b)** $x = 9$
(c) $x = 5$ or $x = -2$ **(d)** $x = 11$ or $x = 3$
(e) $x = -5$ **(f)** $x = 1$ or $x = -2$
(g) $x = 2$ or $x = -2$ **(h)** $x = \frac{1}{2}$
(i) $x = \pm\sqrt{5}$ **(j)** $x = 27$ or $x = -\frac{125}{8}$
(k) $x = 3$ or $x = 2$ **(l)** $x = -\frac{1 \pm \sqrt{41}}{4}$
(m) $x = \frac{4}{3}$ or $x = -4$ **(n)** $x = 15$ or $x = \frac{9}{2}$
(o) no solution   **(p)** $x = 2$ or $x = -1$
(q) $x = 2$ or $x = \frac{1}{2}$ **(r)** $x = 9$
(s) $x = -5$ **(t)** $x = \frac{\pm 2\sqrt{5}}{5}$ or $x = \pm 1$
(u) $x = \frac{1 + \sqrt{41}}{2}$ **(v)** $x = \frac{49}{4}$ or $x = \frac{64}{9}$

2. (a) $-\frac{2}{3} < x < 2$ **(b)** $x < -2, x \geqslant 3$
(c) $-10 \leqslant x \leqslant 6$ **(d)** $x < \frac{3}{2}, x > 2$
(e) $x > \frac{17}{2}$ **(f)** $-4 \leqslant x \leqslant -1, 1 \leqslant x \leqslant 4$

(g) $x < -1, x > 2$ **(h)** $x < -1, -\dfrac{2}{3} < x < 3, x > 4$

3. $k < -\dfrac{3}{11}, k > 3$

4. (a) $p = \dfrac{9}{4}$ **(b)** $p < \dfrac{9}{4}$ **(c)** $p > \dfrac{9}{4}$

5. $k < -1, k > \dfrac{1}{3}$

6. (a) $m + \dfrac{1}{n} > 2 \Rightarrow mn + 1 > 2n \Rightarrow mn - 2n + 1 > 0$;
since $m > n \Rightarrow mn > n^2$ it follows that
$mn - 2n + 1 > n^2 - 2n + 1$ and since
$n^2 - 2n + 1 = (n - 1)^2 > 0$ then
$mn - 2n + 1 > 0 \Rightarrow m + \dfrac{1}{n} > 2$

(b) $(m + n)\left(\dfrac{1}{m} + \dfrac{1}{n}\right) > 4 \Rightarrow (m + n)\left(\dfrac{1}{m} + \dfrac{1}{n}\right)mn >$
$4mn \Rightarrow (m + n)(n + m) > 4mn \Rightarrow$
$m^2 + 2mn + n^2 > 4mn \Rightarrow m^2 - 2mn + n^2 >$
$0 \Rightarrow (m - n)^2 > 0$ which is true for all x and is
equivalent to original inequality – thus
$(m + n)\left(\dfrac{1}{m} + \dfrac{1}{n}\right) > 4$ is true for all x.

7. $x = \dfrac{-1 \pm \sqrt{13}}{2}$, $x = 1$ or $x = -2$

8. $(a + b + c)^2 < 3(a^2 + b^2 + c^2) \Rightarrow a^2 + b^2 + c^2 + 2ab$
$+ 2ac + 2bc < 3a^2 + 3b^2 + 3c^2$
$2a^2 + 2b^2 + 2c^2 - 2ab - 2ac - 2bc > 0 \Rightarrow a^2 - ab + b^2$
$- bc + c^2 - ac > 0$
Given $a \neq b \neq c$ are unequal then for any case,
e.g. $a > b > c > 0, (a^2 - ab) + (b^2 - bc) > (c^2 - ac)$
therefore $a^2 - ab + b^2 - bc + c^2 - ac > 0$ is true, and
hence $(a + b + c)^2 < 3(a^2 + b^2 + c^2)$ is true.

9. (a) $1 < x < 3$
(b) $x < -2, -1 < x < 1, x > 3$

10. If a and b have the same sign, then $|a + b| = |a| + |b|$; and
if a and b are of opposite sign, then $|a + b| < |a| + |b|$.

Exercise 2.6

1. (a) $\dfrac{3}{x + 2} + \dfrac{2}{x - 1}$ **(b)** $\dfrac{3}{x - 2} - \dfrac{2}{x}$

(c) $\dfrac{1}{2(x + 3)} + \dfrac{1}{2(x + 1)}$ **(d)** $\dfrac{9}{(x + 1)^2} - \dfrac{1}{x + 1} + \dfrac{6}{x}$

(e) $\dfrac{1}{x + 3} + \dfrac{3}{x + 2} - \dfrac{2}{x}$ **(f)** $\dfrac{1}{x + 1} - \dfrac{1}{x^2} + \dfrac{3}{x}$

(g) $\dfrac{1}{x - 1} - \dfrac{1}{x + 2}$ **(h)** $\dfrac{3}{2x - 1} - \dfrac{2}{x + 1}$

(i) $\dfrac{3}{x + 2} - \dfrac{2}{(x + 2)^2}$ **(j)** $\dfrac{1}{x - 2} - \dfrac{4}{x + 1} - \dfrac{6}{x^2} + \dfrac{3}{x}$

(k) $\dfrac{2}{x} - \dfrac{2x}{x^2 + 1}$ **(l)** $\dfrac{2}{3x} + \dfrac{3 - 2x}{3(x^2 + 3)}$

(m) $\dfrac{1}{3x} + \dfrac{9 - x}{3(x^2 + 6)}$ **(n)** $\dfrac{3}{8x} + \dfrac{16 - 3x}{8(x^2 + 8)}$

(o) $\dfrac{1}{3(x - 5)} + \dfrac{2}{3 \cdot (x + 1)} + \dfrac{1}{x}$

Chapter 2 practice questions

1. $x = a$ or $x = 3b$

2. $x \leq 4$

3. $c = 5$

4. $a = -\dfrac{1}{2}, b = 4, c = -2$

5. $\omega = -2, p = 2, q = -8$

6. (a) $m > -2$ **(b)** $-2 < m < 0$

7. $a = 2, b = -1, c = -2$

8. $x < 5, x > \dfrac{15}{2}$

9. $-1 < k < 15$

10. (a) $f(x) = 2 - \dfrac{3}{(x + 2)^2 + 1}$

(b) (i) $\lim\limits_{x \to +\infty} f(x) = 2$ **(ii)** $\lim\limits_{x \to -\infty} f(x) = 2$

(c) $(-2, -1)$

11. $k \in \mathbb{R}$ **12.** $a = -1$

13. $a = \dfrac{4}{3}, b = \dfrac{1}{3}$ **14.** $a = -6$

15. $a = 4$ **16.** $a = -2, b = 6$

17. $a = 1$ **18.** $k = 6$

19. $-2.80 < k < 0.803$ (3 s.f.)

20. $-3 \leq k \leq 4.5$ **21.** $-4 \leq m \leq 0$

22. $1 \leq x \leq 3$ **23.** $-2.30 < x < 0$ or $1 < x < 1.30$

24. $-3 \leq x \leq \dfrac{1}{3}$ **25.** $x < -1$ or $2 \leq x < 4$

26. $x < 3$ or $x > 27$ **27.** $-1 < x < \dfrac{1}{3}$

28. $\dfrac{-1}{x - 1} + \dfrac{3}{x + 2}$ **29.** $A = 9, B = -7$

30. $\dfrac{1}{(x - a)} - \dfrac{1}{(x - b)}$

Chapter 3

Exercise 3.1

1. (a) $-1, 1, 3, 5, 7$ **(b)** $-1, 1, 5, 13, 29$

(c) $\dfrac{3}{2}, \dfrac{3}{4}, \dfrac{3}{8}, \dfrac{3}{16}, \dfrac{3}{32}$ **(d)** $1, 7, -5, 19, -29$

(e) $5, 8, 11, 14, 17$ **(f)** $3, 7, 13, 21, 31$

2. (a) $-1, 1, 3, 5, 7, 97$

(b) $2, 6, 18, 54, 162, 4.786 \times 10^{23}$

(c) $\dfrac{2}{3}, -\dfrac{2}{3}, \dfrac{6}{11}, -\dfrac{4}{9}, \dfrac{10}{27}, -\dfrac{50}{1251}$

(d) $1, 2, 9, 64, 625, 1.776 \times 10^{83}$

(e) $3, 11, 27, 59, 123, 4.50 \times 10^{15}$

(f) $0, 3, \dfrac{3}{7}, \dfrac{21}{13}, \dfrac{39}{55}$, approx 1

(g) $2, 6, 18, 54, 162, 4.786 \times 10^{23}$

(h) $-1, 1, 3, 5, 7, 97$

3. (a) $u_n = \dfrac{1}{4}u_{n-1}, u_1 = \dfrac{1}{3}$

(b) $u_n = \dfrac{4a^2}{3}u_{n-1}, u_1 = \dfrac{1}{2}a$

(c) $u_n = u_{n-1} + a - k, u_1 = a - 5k$

4. (a) $u_n = n^2 + 3$ **(b)** $u_n = 3n - 1$

(c) $u_n = \dfrac{2n - 1}{n^2}$ **(d)** $u_n = \dfrac{2n - 1}{n + 3}$

5. (a) $2, \dfrac{3}{2}, \dfrac{5}{3}, \dfrac{8}{5}, \dfrac{13}{8}, \dfrac{21}{13}, \dfrac{34}{21}, \dfrac{55}{34}, \dfrac{89}{55}, \dfrac{144}{89}$

(b) Substitute $a_{n-1} = \dfrac{F_n}{F_{n-1}}$ and simplify.

Answers

6. (a) 1,1,2,3,5,8,13,21,34,55

(b) expand right side and simplify

(c) apply $F_n = \dfrac{1}{\sqrt{5}}\left(\dfrac{(1+\sqrt{5})^n - (1-\sqrt{5})^n}{2^n}\right)$ in

$F_n = F_{n-1} + F_{n-2}$, and then use result in **(b)** to simplify.

Exercise 3.2

1. $3, \dfrac{19}{5}, \dfrac{23}{5}, \dfrac{27}{5}, \dfrac{31}{5}, 7$

2. (a) arithmetic, $d = 2$, $a_{50} = 97$

(b) arithmetic, $d = 1$, $a_{50} = 52$

(c) arithmetic, $d = 2$, $a_{50} = 97$

(d) not arithmetic, *no common difference.*

(e) not arithmetic, *no common difference.*

(f) arithmetic, $d = -7$, $a_{50} = -341$

3. (a) (i) 26

(ii) $a_n = -2 + 4(n-1)$

(iii) $a_1 = -2$, $a_n = a_{n-1} + 4$ for $n > 1$.

(b) (i) 1

(ii) $a_n = 29 - 4(n-1)$

(iii) $a_1 = 29$, $a_n = a_{n-1} - 4$ for $n > 1$.

(c) (i) 57

(ii) $a_n = -6 + 9(n-1)$

(iii) $a_1 = -6$, $a_n = a_{n-1} + 9$ for $n > 1$.

(d) (i) 9.23

(ii) $a_n = 10.07 - 0.12(n-1)$

(iii) $a_1 = 10.07$, $a_n = a_{n-1} - 0.12$ for $n > 1$.

(e) (i) 79

(ii) $a_n = 100 - 3(n-1)$

(iii) $a_1 = 100$, $a_n = a_{n-1} - 3$ for $n > 1$.

(f) (i) $-\dfrac{27}{4}$

(ii) $a_n = 2 - \dfrac{5}{4}(n-1)$

(iii) $a_1 = 2$, $a_n = a_{n-1} - \dfrac{5}{4}$ for $n > 1$.

4. $13, 7, 1, -5, -11, -17, -23$

5. $299, 299\dfrac{1}{4}, 299\dfrac{1}{2}, 299\dfrac{3}{4}, 300$

6. $a_n = \dfrac{40}{9} + \dfrac{26}{9}(n-1) = \dfrac{14}{9} + \dfrac{26}{9}n$

7. $a_n = -\dfrac{142}{3} + \dfrac{11}{3}(n-1) = -51 + \dfrac{11}{3}n$

8. (a) 88 **(b)** 36 **(c)** 11 **(d)** 16 **(e)** 11

9. $9, 3, -3, -9, -15$ **10.** 99.25, 99.50, 99.75

11. $a_n = 4n - 1$ **12.** $a_n = \dfrac{19n - 277}{3}$

13. $a_n = 4n + 27$ **14.** Yes, 3271st term

15. Yes, 1385th term **16.** No

Exercise 3.3

1. (a) Geom., $r = 3^a$, $g_{10} = 3^{9a+1}$

(b) Arithmetic, $d = 3$, $a_{10} = 27$

(c) Geometric, $r = 2$, $b_{10} = 4096$

(d) Neither

(e) Geometric, $r = 3$, $u_{10} = 78732$

(f) Geometric, $r = 2.5$, $a_{10} = 7629.39453125$

(g) Geometric, $r = -2.5$, $a_{10} = -7629.39453125$

(h) Arithmetic, $d = 0.75$, $a_{10} = 8.75$

(i) Geometric, $r = -\dfrac{2}{3}$, $a_{10} = -\dfrac{1024}{2187}$

(j) Arithmetic, $d = 3$, $a_{10} = 79$

(k) Geometric, $r = -3$, $u_{10} = 19683$

(l) Geometric, $r = 2$, $u_{10} = 51.2$

(m) Neither

(n) Neither

(o) Arithmetic, $d = 1.3$, $a_{10} = 14.1$

2. (a) (i) 32

(ii) $-3 + 5(n-1)$

(iii) $a_1 = -3$, $a_n = a_{n-1} + 5$ for $n > 1$

(b) (i) -9

(ii) $19 - 4(n-1)$

(iii) $a_1 = 19$, $a_n = a_{n-1} - 4$ for $n > 1$

(c) (i) 69

(ii) $-8 + 11(n-1)$

(iii) $a_1 = -8$, $a_n = a_{n-1} + 11$ for $n > 1$

(d) (i) 9.35

(ii) $10.05 - 0.1(n-1)$

(iii) $a_1 = 10.05$, $a_n = a_{n-1} - 0.1$ for $n > 1$

(e) (i) 93

(ii) $100 - (n-1)$

(iii) $a_1 = 100$, $a_n = a_{n-1} - 1$ for $n > 1$

(f) (i) $-\dfrac{17}{2}$

(ii) $2 - 1.5(n-1)$

(iii) $a_1 = 2$, $a_n = a_{n-1} - 1.5$ for $n > 1$

(g) (i) 384

(ii) $3 \times 2^{n-1}$

(iii) $a_1 = 3$, $a_n = 2a_{n-1}$ for $n > 1$

(h) (i) 8748

(ii) $4 \times 3^{n-1}$

(iii) $a_1 = 4$, $a_n = 3a_{n-1}$ for $n > 1$

(i) (i) -5

(ii) $5 \times (-1)^{n-1}$

(iii) $a_1 = 5$, $a_n = -a_{n-1}$ for $n > 1$

(j) (i) -384

(ii) $3 \times (-2)^{n-1}$

(iii) $a_1 = 3$, $a_n = -2a_{n-1}$ for $n > 1$

(k) (i) $-\dfrac{4}{9}$

(ii) $972 \times \left(-\dfrac{1}{3}\right)^{n-1}$

(iii) $a_1 = 972$, $a_n = \left(-\dfrac{1}{3}\right)a_{n-1}$ for $n > 1$

(l) (i) $\dfrac{2187}{64}$

(ii) $a_n = -2\left(-\dfrac{3}{2}\right)^{n-1}$

(iii) $a_1 = -2$, $a_n = -\dfrac{3}{2}a_{n-1}$, $n > 1$

(m) (i) $\dfrac{390625}{117649}$

(ii) $a_n = 35\left(\dfrac{5}{7}\right)^{n-1}$

(iii) $a_1 = 35$, $a_n = \dfrac{5}{7}a_{n-1}$, $n > 1$

(n) (i) $-\dfrac{3}{64}$

(ii) $a_n = -6\left(\dfrac{1}{2}\right)^{n-1}$

(iii) $a_n = -6,\ a_n = \dfrac{1}{2}a_{n-1},\ n > 1$

(o) (i) 1216

(ii) $9.5 \times 2^{n-1}$

(iii) $a_1 = 9.5,\ a_n = 2a_{n-1},\ n > 1$

(p) (i) $69.833\,729\,609\,375 = \dfrac{893\,871\,739}{12\,800\,000}$

(ii) $a_n = 100\left(\dfrac{19}{20}\right)^{n-1}$

(iii) $a_1 - 100,\ a_n = \dfrac{19}{20}a_{n-1},\ n > 1$

(q) (i) $0.002\,085\,685\,73 = \dfrac{2187}{1\,048\,576}$

(ii) $a_n = 2\left(\dfrac{3}{8}\right)^{n-1}$

(iii) $a_1 = 2,\ a_n = \dfrac{3}{8}a_{n-1},\ n > 1$

3. 6,12,24,48

4. 35,175,875

5. -36

6. 21,63,189,567

7. $-24,\ 24$

8. $1.5, a_n = 24\left(\dfrac{1}{2}\right)^{n-1}$

9. $a_4 = \pm 3,\ r = \pm\dfrac{1}{2},\ a_n = 24\left(\pm\dfrac{1}{2}\right)^{n-1}$

10. $\dfrac{49}{3}$

11. 10th term

12. Yes, 10th term

13. Yes, 10th term

14. €2228.92

15. £945.23

16. €2968.79

17. 7745

18. $\dfrac{686}{27}$

19. 10th term

20. £2921.16

Exercise 3.4

1. 11 280

2. $-\dfrac{105\,469}{1024}$

3. 0.7

4. $\dfrac{10}{7}$

5. $\dfrac{16 + 4\sqrt{3}}{39}$

6. (a) $\dfrac{52}{99}$ (b) $\dfrac{449}{990}$ (c) $\dfrac{7459}{2475}$

7. £13 026.14

8. (a) 940 (b) 6578 (c) 42 625

9. $\dfrac{n(7 + 3n)}{2}$

10. 17 terms

11. 29 terms

12. $d = 4$

13. (a) 250, 125 250 (b) 83 501

14. $a = 1,\ d = 5$

15. (a) 2890 (b) 0.290 (c) -2.065

16. 11 400

17. 1.191

18. 49.2

19. $\dfrac{6}{5}$

20. $\dfrac{3 + \sqrt{6}}{2}$

21. (a) $3, \dfrac{18}{5}, \dfrac{93}{25}, \dfrac{468}{125}; \dfrac{15}{4}\left(1 - \dfrac{1}{5^n}\right)$

(b) $\dfrac{1}{6}, \dfrac{1}{4}, \dfrac{3}{10}, \dfrac{1}{3}; \dfrac{n}{2n + 4}$

(c) $\sqrt{2} - 1,\ \sqrt{3} - 1,\ 1,\ \sqrt{5} - 1 ; \sqrt{n + 1} - 1$

22. (a) 1.945 (b) 84.2

23. (a) 127 (b) 128

24. (a) $\dfrac{819}{128}$ (b) $\dfrac{32}{5}$

25. (a) 11 866 (b) 763 517

(c) 14 348 906 (d) ~ 150

Exercise 3.5

1. (a) $x^5 + 10x^4y + 40x^3y^2 + 80x^2y^3 + 80xy^4 + 32y^5$

(b) $a^4 - 4a^3b + 6a^2b^2 - 4ab^3 + b^4$

(c) $x^6 - 18x^5 + 135x^4 - 540x^3 + 1215x^2 - 1458x + 729$

(d) $16 - 32x^3 + 24x^6 - 8x^9 + x^{12}$

(e) $x^7 - 21bx^6 + 189b^2x^5 - 945b^3x^4 + 2835b^4x^3$
$- 5103b^5x^2 + 5103b^6x - 2187b^7$

(f) $64n^6 + 192n^3 + 240 + \dfrac{160}{n^3} + \dfrac{60}{n^6} + \dfrac{12}{n^9} + \dfrac{1}{n^{12}}$

(g) $\dfrac{81}{x^4} - \dfrac{216}{x^2\sqrt{x}} + \dfrac{216}{x} - 96\sqrt{x} + 16x^2$

2. (a) 56 (b) 0 (c) 1225

(d) 32 (e) 0

3. (a) $x^7 - 14x^6y + 84x^5y^2 - 280x^4y^3 + 560x^3y^4 - 672x^2y^5$
$+ 448xy^6 - 128y^7$

(b) $64a^6 - 192a^5b + 240a^4b^2 - 160a^3b^3 + 60a^2b^4$
$- 12ab^5 + b^6$

(c) $x^5 - 20x^4 + 160x^3 - 640x^2 + 1280x - 1024$

(d) $x^{18} + 12x^{15} + 60x^{12} + 160x^9 + 240x^6 + 192x^3 + 64$

(e) $2187x^7 - 5103bx^6 + 5103b^2x^5 - 2835b^3x^4 + 945b^4x^3$
$- 189b^5x^2 + 21b^6x - b^7$

(f) $64n^6 - 192n^3 + 240 - \dfrac{160}{n^3} + \dfrac{60}{n^6} - \dfrac{12}{n^9} + \dfrac{1}{n^{12}}$

(g) $\dfrac{16}{x^4} - \dfrac{96}{x^2\sqrt{x}} + \dfrac{216}{x} - 216\sqrt{x} + 81x^2$

(h) 112

(i) $1792\sqrt{3}$

4. (a) $x^{45} - 90x^{43} + 3960x^{41}$

(b) Does not exist as the powers of x decrease by 2 starting at 45. There is no chance for any expression to have zero exponent.

(c) $\dbinom{45}{43}x^2\left(\dfrac{-2}{x}\right)^{43} + \dbinom{45}{44}x\left(\dfrac{-2}{x}\right)^{44} + \left(\dfrac{-2}{x}\right)^{45}$

$= -\dbinom{45}{43}\dfrac{2^{43}}{x^{41}} + \dbinom{45}{44}\dfrac{2^{44}}{x^{43}} - \dfrac{2^{45}}{x^{45}}$

(d) $\dbinom{45}{21}x^{24}\left(\dfrac{-2}{x}\right)^{21} = -\dbinom{45}{21} \cdot 2^{21}x^3$

5. $\dbinom{n}{k} = \dfrac{n!}{k!(n - k)!} = \dfrac{n!}{(n - k)!k!} = \dfrac{n!}{(n - k)!(n - (n - k))!}$

$= \dbinom{n}{n - k}$

6. $(1 + 1)^n = \dbinom{n}{0} + \dbinom{n}{1} + \dbinom{n}{2}\cdots + \dbinom{n}{n}$

$2^n = 1 + \dbinom{n}{1} + \dbinom{n}{2}\cdots + \dbinom{n}{n} \Rightarrow 2^n - 1$

$= \dbinom{n}{1} + \dbinom{n}{2}\cdots + \dbinom{n}{n}$

7. (a) $k! = k(k - 1) \cdots \times 2 \times 1 = k((k - 1) \cdots \times 2 \times 1)$

(b) apply part **a**

(c) apply part **a**.

8. $\left(\dfrac{1}{3} + \dfrac{2}{3}\right)^6 = 1$

9. $\left(\dfrac{2}{5} + \dfrac{3}{5}\right)^8 = 1$

10. $\left(\dfrac{1}{7} + \dfrac{6}{7}\right)^n = 1$

11. 15

12. 90 720

13. 16 128

14. $1 + 10x + 45x^2$, 1.1045, 0.9045

Answers

15. Use definition of $\binom{n}{r}$, sum of an entry in the nth row plus twice the next entry plus the third entry is equal to the entry directly below the last entry but two rows below.

16. (a) $\dfrac{7}{9}$ **(b)** $\dfrac{19}{55}$ **(c)** $\dfrac{7952}{2475}$

17. $-145\,152$ **18.** $35a^3$ **19.** $96\,096$

20. $243n^5 - 810n^4m + 1080n^3m^2 - 720n^2m^3 + 240nm^4 - 32m^5$

21. $7\,838\,208$

22. $k = 3$

Exercise 3.6

1. (a) 120 **(b)** 120 **(c)** 20 **(d)** 336

2. (a) 1 **(b)** 1 **(c)** 120 **(d)** 120

3. (a) 70 **(b)** 70 **(c)** 330 **(d)** 330

4. (a) 0 **(b)** 39\,916\,800 **(c)** 0 **(d)** 10

5. (a) F **(b)** F **(c)** T

6. 24 **7.** 72 **8.** 312

9. 16\,777\,216 **10.** 262\,144 **11.** 1\,757\,600\,000

12. 81000 **13.** 40320,384 **14.** 40320,720

15. JANE, JAEN,JNAE,JNEA,JEAN,JENA,AJNE,AJEN, ANJE,ANEJ,AEJN,AENJ,NJAE,NJEA,NEJA,NEAJ, NAJE,NAEJ,EJAN,EJNA,EAJN,EANJ,ENJA,ENAJ

16. Mag,Mga,Mai,…(60 of them)

17. (a) 175\,760\,000 **(b)** 174\,790\,000

18. (a) 4080 **(b)** 1680 **(c)** 1050 **(d)** 1980

19. (a) 296 **(b)** 1460 **(c)** 504

20. (a) 125\,000 **(b)** 117\,600

(c) 2450 (or 2500) **(d)** 7350

21. 768

22. (a) 36 **(b)** 256

23. (a) 5985 **(b)** 2376 **(c)** 2475

(d) 5814; 2288; 1925

24. (a) 1\,192\,052\,400 **(b)** 4560, 0.00038%

(c) 265\,004\,096, 22.23%

25. (a) 74613 **(b)** 7560

26. $_{10}P_6 \times 9! = 54867456000.$

Chapter 3 practice questions

1. $D = 5, n = 20$

2. \$2098.63

3. (a) Nick: 20
Charlotte: 17.6

(b) Nick: 390
Charlotte: 381.3

(c) Charlotte will exceed the 40 hours during week 13.

(d) In week 11 Charlotte will catch up with Nick and exceed him

4. (a) loss for the second month = 1060 g
loss for the third month = 1123.6 g

(b) Plan A loss = 1880 g
Plan B loss = 1898.3 g

(c) (i) Loss due to plan A in all 12 months = 17 280 g
(ii) Loss due to Plan B in all 12 months = 16 869.9 g

5. (a) €895.42

(b) This is the future value of an annuity due = 6985.82

6. (a) 142.5 **(b)** 19003.5

7. (a) $\sqrt[3]{7}, 1, \sqrt[3]{7}, 1 \ldots$ **(b)** 0, 2, 0, 2, …

8. (a) On the 37th day **(b)** 407 km

9. (a) 1.5 **(b) (i)** 207 595 **(ii)** 2019

(c) 619 583 **(d)** Market saturation

10. $-4, 3006$

11. (a) $\sqrt{\dfrac{1}{4} + \dfrac{1}{4}} = \dfrac{\sqrt{2}}{2}$ **(b)** $\dfrac{1}{2}$

(c) (i) $\dfrac{1}{4}$ **(ii)** $\dfrac{1}{2}$ **(d) (i)** $\dfrac{1}{512}$ **(ii)** 2

12. (a) 1220 **(b)** 36920

13. (i) Area A = 1, Area B = $\dfrac{1}{9}$ **(ii)** $\dfrac{1}{81}$

(iii) $1 + \dfrac{8}{9}, 1 + \dfrac{8}{9} + \left(\dfrac{8}{9}\right)^2$ **(iv)** 0

14. (a) Neither, geometric converging, arithmetic, geometric diverging

(b) 6

15. (a) (i) Kell: 18 400, 18 800; YBO: 18 190, 19 463.3

(ii) Kell: 198 000; YBO: 234 879.62

(iii) Kell: 21 600; YBO: 31 253.81

(b) (i) After the second year

(ii) 4th year

16. (a) 62 **(b)** 936

17. (a) $7000(1 + 0.0525)^t$ **(b)** 7 years

(c) No, since 9912 < 10 015.0

18. (a) 11 **(b)** 2 **(c)** 15

19. 15, -8 **20.** $a = -2, b = -7$ **21.** 10300

22. (a) $a_n = 8n - 3$ **(b)** 50

23. 2 099 520 **24.** $6n - 5$ **25.** 72

26. 559 **27.** $-3, 3$ **28.** 9

29. 62 **30.** $-\dfrac{36}{5}$

31. (a) 4 **(b)** $16(4^n - 1)$

32. $|x| < \dfrac{5}{3}, 10$

33. 3168

34. (a) $\dfrac{n(3n + 1)}{2}$ **(b)** 30

35. -7

36. 1275 ln2

37. (a) 4, 8,16

(b) (i) $u_n = 2^n$ **(ii)** $2^{n+1} = 3 \times 2^n - 2 \times 2^{n-1}$

38. (a) $\dfrac{2}{3}$ **(b)** 9

39. $a = 2, b = -3$ **40.** 55 **41.** $-2, 4$

42. $\dfrac{\theta}{1 - \cos\theta}$

43. (a) 1, 5, 9 **(b)** $4n - 3$

44. (a) $32 + 80x + 80x^2 + 40x^3 + 10x^4 + x^5$;

(b) 32.8080401001

45. (a) \$5000(1.063)^n$ **(b)** \$6786.35

(c) 12

46. 7

47. $u_1 = 12, d = -1.5$

48. (a) $1, -nx, + \binom{n}{2}x^2, - \binom{n}{3}x^3$

(b) (i) $|u_3| - |u_2| = |u_4| - |u_3| \Rightarrow 3n^2 - 9n = n^3 - 6n^2 + 5n$

(ii) $n = 7$

49. (a) 330 **(b)** 150 **(c)** 325

50. (a) $81 + 108x + 54x^2 + 12x^3 + x^4$

(b) 92.3521

51. $\binom{7}{3}\binom{4}{2} = 210$

52. (a) 41   **(b)** $\sum_{n=1}^{41} 7 + 7n$ **(c)** 6314   **(d)** 287

53. (a) 34   **(b)** 81

54. (a) $6(4!) = 144$ **(b)** $4(3!)(3!) = 144$

55. (a) (i) $\dfrac{v_{n+1}}{v_n} = 2^d$ **(ii)** 2^a **(iii)** $v_n = 2^{a+(n-1)d}$

     (b) (i) $S_n = \dfrac{2^a(2^{dn} - 1)}{2^d - 1}$ **(ii)** $d < 0$.

         (iii) $S_\infty = \dfrac{2^a}{1 - 2^d}$ **(iv)** $d = -1$

Chapter 4

Exercise 4.1 & 4.2

1. (a) $y = b^x$

    (b) domain $\{x : x \in \mathbb{R}\}$, range $\{y : y > 0\}$

    (c) (i)

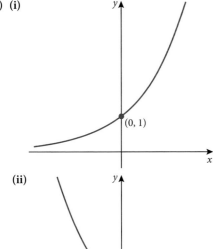

       (ii)

2. (a)

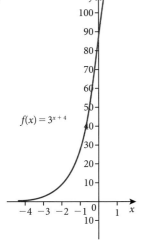

(i) y intercept: $(0, 81)$

(ii) horizontal asymptote: $y = 0$ (x-axis)

(iii) domain: $x \in \mathbb{R}$, range: $y > 0$

(b)

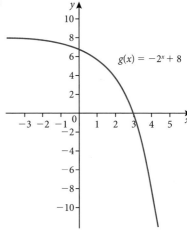

(i) x-intercept: $(3, 0)$, y-intercept: $(0, 7)$

(ii) horizontal asymptote: $y = 8$

(iii) domain: $x \in \mathbb{R}$, range: $y < 8$

(c)

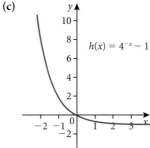

(i) x- and y-intercept: $(0, 0)$

(ii) horizontal asymptote: $y = -1$

(iii) domain: $x \in \mathbb{R}$, range: $y > -1$

(d)

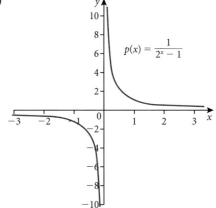

(i) x- and y-intercept: none

(ii) vertical asymptote: $x = 0$;
horizontal asymptote: $y = 0$ and $y = -1$

(iii) domain: $x \in \mathbb{R}$, $x \neq 0$, range: $y < -1$ or $y > 0$

Answers

(e)

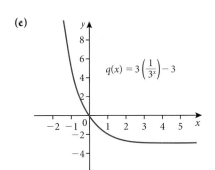

$$q(x) = 3\left(\frac{1}{3^x}\right) - 3$$

   (i) x- and y-intercept: $(0, 0)$
   (ii)   horizontal asymptote: $y = -3$
   (iii)   domain: $x \in \mathbb{R}$, range: $y > -3$

(f)

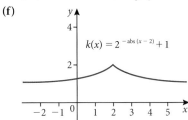

$$k(x) = 2^{-\text{abs}(x-2)} + 1$$

   (i) y-intercept: $\left(0, \dfrac{5}{4}\right)$
   (ii)   horizontal asymptote: $y = 1$
   (iii)   domain: $x \in \mathbb{R}$, range: $y > 1$

3. domain: $x \in \mathbb{R}$
   range: if $a > 0 \Rightarrow y > d$, if $a < 0 \Rightarrow y < d$
   y-intercept: $(0, a(b)^{-c} + d)$
   horizontal asymptote: $y = d$

4. $y = 2^{-x}$ $y = 4^{-x}$ $y = 8^{-x}$ $y = 8^x$ $y = 4^x$ $y = 2^x$

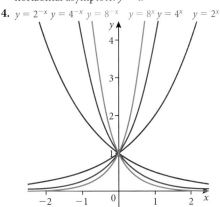

5. (a) $y = \left(\dfrac{1}{2}\right)^x$ **(b)** $y = \left(\dfrac{1}{4}\right)^x$ **(c)** $y = \left(\dfrac{1}{8}\right)^x$

6. $y = b^x$ is steeper

7. $P(t) = 100\,000(3)^{\frac{t}{25}}$ where t is number of years
   (a) $900\,000$ **(b)** $2\,167\,402$ **(c)** $8\,100\,000$

8. $N(t) = 10^4(2)^{\frac{t}{3}}$
   (a) $20\,000$ **(b)** $80\,000$ **(c)** $5\,120\,000$
   (d) $10\,485\,760\,000$

9. (a) $A(t) = A_0(2)^{\frac{t}{10}}$ **(b)** 7.18%

10. (a) $\$17\,204.28$ **(b)** $\$29\,598.74$ **(c)** $\$50\,922.51$

11. (a) $A(t) = 5000\left(1 + \dfrac{.09}{12}\right)^{12t}$

   (b) (c)   minimum number of years is 16

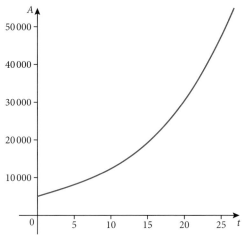

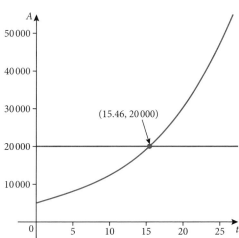

(15.46, 20 000)

12. (a) $\$16,850.58$ **(b)** $\$17,289.16$
    (c) $\$17,331.09$ **(d)** $\$17,332.47$

13. (a) $\$2$ **(b)** $\$2.61$ **(c)** $\$2.71$
    (d) $\$2.72$ **(e)** $\$2.72$

14. (a) $240\,310$ **(b)** $192\,759$

15. 8.90%

16. $0.0992A_0$ (or 9.92% of A_0 remains)

17. (a) $A(w) = 1000(0.7)^w$ **(b)** 20 weeks

18. $b > 0$ because if $b = 0$ then the result is always zero, and
   if $b < 0$ then b^x gives a positive result when x is an even
   integer and a negative result when x is an odd integer

19. Payment plan I: $\$465$; Payment plan II: $\$10,737,418.23$

20. (a) $a = 2, k = 3$

    (b) $a = \dfrac{1}{3}, k = 2$

    (c) $a = 3, k = -4$ or $a = -3, k = 4$

    (d) $a = 10, k = \dfrac{3}{2}$

Exercise 4.3

1.

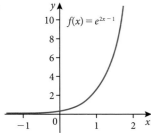

$f(x) = e^{2x-1}$

(a) (i) x-int.: none, y-int.: $\left(0, \dfrac{1}{e}\right)$

    (ii) horizontal asymptote: $y = 0$

    (iii) domain: $x \in \mathbb{R}$, range: $y > 0$

(b)

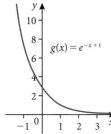

$g(x) = e^{-x+1}$

    (i) x-int.: none, y-int.: $(0, e)$ domain:

    (ii) horizontal asymptote: $y = 0$

    (iii) $x \in \mathbb{R}$, range: $y > 0$

(c)

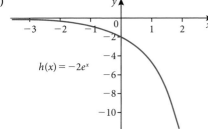

$h(x) = -2e^x$

    (i) x-int.: none, y-int.: $(0, -2)$

    (ii) horizontal asymptote: $y = 0$

    (iii) domain: $x \in \mathbb{R}$, range: $y < 0$

(d)

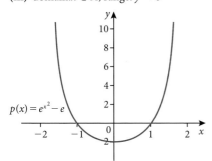

$p(x) = e^{x^2} - e$

    (i) x-int.: $(-1, 0)$ and $(1, 0)$, y-int.: $(0, 1 - e)$

    (ii) no asymptotes

    (iii) domain: $x \in \mathbb{R}$, range: $y > 1 - e$

(e)

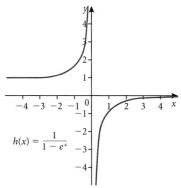

$h(x) = \dfrac{1}{1 - e^x}$

    (i) x-int.: none, y-int.: none

    (ii) horizontal asymptotes: $y = 1$ and $y = 0$

    (iii) domain: $x \in \mathbb{R}$, $x \neq 0$, range: $y < 0$, $y > 1$

(f)

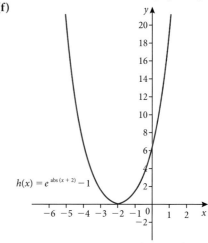

$h(x) = e^{\text{abs}(x+2)} - 1$

    (i) x-int.: $(-2, 0)$, y-int.: $(0, e^2 - 1)$

    (ii) no asymptotes

    (iii) domain: $x \in \mathbb{R}$, range: $y > 0$

2. (a) $e = \lim\limits_{n \to \infty} \left(1 + \dfrac{1}{n}\right)^n$

    (b) 0.3660323413, 0.3678610464, 0.3678792572

    (c) 0.36788; reciprocal of e, $\dfrac{1}{e} \approx 0.3678794412$

3. $y = \left(x + \dfrac{1}{x}\right)^x$ will not intersect $y = 2.72$

    because $\lim\limits_{x \to \infty} \left(x + \dfrac{1}{x}\right)^x = e \approx 2.718281828\ldots < 2.72$

4. (a) Bank A: earn 613.71 euros in interest

        Bank B: earn 614.06 euros in interest

    (b) Bank B account earns 0.36 euros more in interest

5. (a) Blue Star   **(b)** \$1358.42   **(c)** \$11.93

6. (a) 97.6%   **(b)** 78.7%   **(c)** 9.16%

    (d) 0.254%

7. (a) 5 kg   **(b)** 70.7%

    (c)   **(d)** 20 days

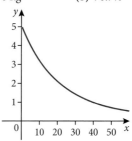

Answers

8. A

9. (a) $r \approx 1.07037$ (6 s.f.)   **(b)** 7.037% (4 s.f.)

10. (a) less than 1   **(b)** less than 1
    (c) greater than 1   **(d)** greater than 1

11. (a) £1568.31, £2459.60
    (b) 15.4 years
    (c) 15.4 years
    (d) same; doubling time is independent of initial amount

Exercise 4.4

1. (a) $2^4 = 16$ **(b)** $e^0 = 1$ **(c)** $10^2 = 100$

    (d) $10^{-2} = 0.01$ **(e)** $7^3 = 343$ **(f)** $e^{-1} = \dfrac{1}{e}$

    (g) $10^y = 50$ **(h)** $e^{12} = x$ **(i)** $e^3 = x + 2$

2. (a) $\log_2 1024 = 10$ **(b)** $\log_{10} 0.0001 = -4$

    (c) $\log_4\left(\dfrac{1}{2}\right) = -\dfrac{1}{2}$ **(d)** $\log_3 81 = 4$

    (e) $\log_{10} 1 = 0$ **(f)** $\ln 5 = x$

    (g) $\log_2 0.125 = -3$ **(h)** $\ln y = 4$

    (i) $\log_{10} y = x + 1$

3. (a) 6   **(b)** 3   **(c)** -3 **(d)** 5

    (e) $\dfrac{1}{2}$ **(f)** $\dfrac{1}{3}$ **(g)** -3 **(h)** 13

    (i) 0   **(j)** 6   **(k)** -3 **(l)** $\sqrt{2}$

    (m) 3   **(n)** $\dfrac{1}{2}$ **(o)** -2 **(p)** 88

    (q) $\dfrac{1}{2}$ **(r)** 18   **(s)** $\dfrac{1}{3}$ **(t)** π

4. (a) 0.239   **(b)** 0.549   **(c)** 1.40
    (d) 0.209   **(e)** 13.8

5. (a) $x > 2$ **(b)** $x > 0$ **(c)** $x > 0$ **(d)** $x < \dfrac{8}{5}$

    (e) $-2 \leqslant x < 3$ **(f)** $x < 1$

6. (a) domain $\{x : x > 0, x \neq 1\}$ range $\{y : y \in \mathbb{R}, y \neq 0\}$
    (b) domain $\{x : x\} > 1$ range $\{y : y > 0\}$
    (c) domain $\{x : x > 1\}$ range $\{y : y > 0\}$

7. (a) $f(x) = \log_4 x$ **(b)** $f(x) = \log_2 x$
    (c) $f(x) = \log_{10} x$ **(d)** $f(x) = \log_3 x$

8. (a) $\log_2 2 + \log_2 m = 1 + \log_2 m$
    (b) $\log 9 - \log x$
    (c) $\dfrac{1}{5}\ln x$
    (d) $\log_3 a + 3\log_3 b$
    (e) $\log 10x + \log(1 + r)^t = 1 + \log x + t\log(1 + r)$
    (f) $3\ln m - \ln n$

9. (a) $\log_b p + \log_b q + \log_b r$
    (b) $2\log_b p + 3\log_b q - \log_b r$
    (c) $\dfrac{\log_b p}{4} + \dfrac{\log_b q}{4}$
    (d) $\dfrac{\log_b q}{2} + \dfrac{\log_b r}{2} - \dfrac{\log_b p}{2}$
    (e) $\log_b p + \dfrac{1}{2}\log_b q - \log_b r$
    (f) $3\log_b p + 3\log_b q - \dfrac{1}{2}\log_b r$

10. (a) $\log x$ **(b)** $\log_3 72$ **(c)** $\ln\left(\dfrac{y^4}{4}\right)$

    (d) $\log_b 4$ **(e)** $\log\left(\dfrac{x}{yz}\right)$ **(f)** $\ln\left(\dfrac{36}{e}\right)$

11. (a) 9.97   **(b)** -5.32 **(c)** 2.06   **(d)** -0.179

12. (a) 4.32   **(b)** 1.86

13. $\log_b a = \dfrac{\log_a a}{\log_a b} = \dfrac{1}{\log_a b}$

14. $\log e = \dfrac{\ln e}{\ln 10} = \dfrac{1}{\ln 10}$

15. $dB = 10\log\left(\dfrac{I}{10^{-16}}\right) = 10(\log I - \log 10^{-16})$

      $= 10(\log I + 16) = 10\log I + 160$

    $10\log 10^{-4} + 160 = 10(-4) + 160 = 120$ decibels

16. See Worked Solutions

Exercise 4.5

1. (a) 0.699   **(b)** 2.5   **(c)** 7.99   **(d)** 3.64
    (e) -1.92 **(f)** 2.71   **(g)** 0.434   **(h)** 2.12
    (i) 4.42   **(j)** 0.225   **(k)** 0.642   **(l)** 22.0

2. (a) 3   **(b)** 0 or -1

    (c) $\dfrac{\ln(3/2)}{\ln 6}$ or $\dfrac{\ln(4/3)}{\ln 6}$ **(d)** 1 or -1

3. (a) $\$6248.58$ **(b)** $9\dfrac{1}{4}$ years

4. 12.9 years

5. 20 hours (≈ 19.93)

6. (a) 24 years (≈ 23.45)   **(b)** 12 years (≈ 11.9)
    (c) 9 years (≈ 8.04)

7. after 6 years

8. (a) 99.7%   **(b)** 127 000 years

9. (a) 37 dogs   **(b)** 9 years

10. (a) 459 litres
    (b) 8.89 minutes $\approx$ 8 min. 53 seconds
    (c) 39 minutes

11. (a) 5 kg   **(b)** 17.7 days

12. (a) $x = \dfrac{20}{3}$ **(b)** $x = 104$

    (c) $x = \dfrac{1}{e^3}$ **(d)** $x = 4$

    (e) $x = 98$ **(f)** $x = \pm\sqrt{e^{16}} \approx \pm 2980.96$

    (g) $x = 2$ or $x = 4$ **(h)** $x = 9$

    (i) $x = \dfrac{13}{5}$ **(j)** $x = 3$

    (k) $x = 1$ or $x = 100$

13. (a) $x > \dfrac{1}{\sqrt[5]{16}}$ **(b)** $x < 2$

    (c) $0 < x < \ln 6$
    (d) $0.161 < x < 1.14$ (approx. to 3 s.f.)

Chapter 4 practice questions

1. (a) $(8, 0)$ **(b)** $(0, 2)$ **(c)** $\left(-\dfrac{2}{3}, 3\right)$

2. (a) 183 g (3 s.f.)   **(b)** 154 years (3 s.f.)

3. $x = 2$

4. $y = 16$

5. $t = 0, \ln\left(\dfrac{1}{2}\right)$ or $\ln 2$

6. $x = e^{-1c}$ or e^{2c}

7. (a) $x = 3$ (b) $x = 6$

8. (a) $\log\left(\dfrac{a^2 b^3}{c}\right)$ (b) $\ln\left(\dfrac{e x^3}{\sqrt{y}}\right)$

9. 1900 years

10. $c = 42$

11. (a)

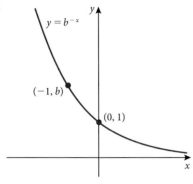

(b)

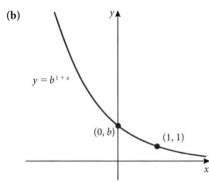

12. (a) $k \approx 0.0004332$ (b) 17.7% (3 s.f.)

13. $x \approx 1.28$

14. $1.52 < x < 1.79$

15. $-1 < x < -0.800 \cup x > 1$

16. (a) $x = -\dfrac{1}{2}$ or $x = 0$

    (b) (i) $x = \dfrac{1}{\ln a - 2}$ or $x = \dfrac{\log_a e}{1 - 2\log_a e}$

        (ii) $a = e^2$

17. $a = -2$, $b = 3$

18. $x = \sqrt{e}$, $x = e$

19. (a) $V = \$265.33$ (b) 235 months

20. $x = 5^{\frac{5}{3}}$ or $x = 5^{\frac{-5}{3}}$

21. $x = e - 3$ or $x = \dfrac{1}{e} - 3$

22. $x = -2.51, -1.51$ or 0.440 (3 s.f.)

23. $k = \dfrac{\ln 2}{20}$

24. (a) $f(x) = \ln\left(\dfrac{x}{x + 2}\right)$

    (b) $f^{-1}(x) = -\dfrac{2e^x}{e^x - 1}$ or $\dfrac{2e^x}{1 - e^x}$

25. (a) min. value of f is 0

    (b) from part (a) $f(x) > 0 \Rightarrow e^x - 1 - x \geqslant 0$
                $\Rightarrow e^x \geqslant 1 + x$

Chapter 5

Exercise 5.1

1. (a) Yes   (b) No   (c) No
   (d) Yes   (e) Yes

2. (a) $a = 7$ (b) No value   (c) $a = 7$
   (d) any value   (e) $a = 7$ (f) $a \neq 7$

3. antecedent: a; consequent: c
   (a) a, c (b) a, c (c) a, c
   (d) a, c (e) c, a (f) c, a
   (g) c, a

4. (a) **Converse**: If triangles have four sides, then quadrilaterals have three sides.
    Contrapositive: If triangles do not have four sides, then quadrilaterals do not have three sides

   (b) **Converse**: If $\sqrt{3}$ is a rational number, then the moon is made of butter.
    Contrapositive: If $\sqrt{3}$ is an irrational number, then the moon is not made of butter.

   (c) **Converse**: b divides 30 only if b divides 5.
    Contrapositive: b does not divide 30 only if b does not divide 5.

   (d) **Converse**: f to be continuous is sufficient for the differentiability of f.
    Contrapositive: f to be discontinuous is sufficient for the non-differentiability of f.

   (e) **Converse**: A sequence a is convergent whenever a is bounded.
    Contrapositive: A sequence a is not convergent whenever a is not bounded

   (f) **Converse**: A function f is integrable if f is bounded.
    Contrapositive: A function f is not integrable if f is not bounded.

   (g) **Converse**: $3 + 3 = 6$ is necessary for $3 + 2 = 5$.
    Contrapositive: $3 + 3 \neq 6$ is necessary for $3 + 2 \neq 5$.

5. (a) Q can be true or false.
   (b) Q must be true.
   (c) Q must be false.

6. (b), (c) & (d)

Exercise 5.2 & 5.3

1. (a) $20 = 4 + 6 + 10$
   (b) $2m = 2(p + q + r) = 2p + 2q + 2r$, m, p, q, and r are integers.
   (c) Not true: $(2n + 1) + (2k + 1) = 2(n + k + 1)$ which is even
   (d) Let $m = 2n + 1$ be any odd integer.
    $2n$ is even, thus it can be the sum of 3 even integers, so, $m = 2p + 2q + 2r + 1 = 2p + 1 + 2q - 1 + 2r + 1$
   (e) $n + n + 1 + n + 2 = 3n + 3 = 3(n + 1)$
   (f) $n + n + 1 + n + 2 + n + 3 = 4n + 6$. If $n = 1$, not true!
   (g) True. Every 3 consecutive numbers should have at least one even number and every 3 consecutive numbers should have 1 multiple of 3 (multiples of 3 are periodic with period 3!)
   (h) not true since $a = -b$ is another possibility.

Answers

2. **(a)** 1806

   (b) 126. Cannot be. The closest square to 2018 is 1936, they will be 44 years old, that means they are born 1892!

3. If $n \geq 2$, then $n!$ contains at least one even factor. Thus, $(n-1)\ldots 2.1 + 2 = 2((n-1).1 + 1)$ will be even.

4. **(a)** Needs to be proven in both directions.

   $5n + 3$ even $\Rightarrow$ if you add any odd number to it, the resulting number will be odd. So, $5n + 3 + 2n - 5 = 7n - 2$ is odd. Similarly, you can go in the opposite direction.

   (b) it is even. $5n + 3$ odd $\Rightarrow n$ is even $\Rightarrow 7n - 2$ is even.

5. $m^2 + n^2$ even $\Rightarrow m^2$ and n^2 are both even or both odd $\Rightarrow m$ and n have the same parity.

6. By contradiction:

   $\sqrt{x+y} = \sqrt{x} + \sqrt{y}$. Square both sides and simplify, $0 = xy$. Thus, one of x or y must be zero, which contradicts the fact that both are positive.

7. If $x = 0$, or $y = 0$, $xy = 0$ is obvious.

   In the opposite direction: If $xy = 0$. Assume that x and y are different from zero. But, there are no nonzero real numbers can have a product of zero. A contradiction.

8. **(a)** $\exists x \in O$ such that $x = k^2$ where, $k \in \mathbb{Z}$.

   (b) $\forall x \in O$ such that $x \neq k^2$ where, $k \in \mathbb{Z}$.

   (c) True. $x = 81$.

9. **(a)** $\forall x \in \mathbb{Z}^+, 13 | x. \exists x \in \mathbb{Z}^+, 13 \nmid x.$

   (b) False, a counter example: $x = 10, 13 \nmid 10$.

10. By contradiction: assume $a^2(b^2 - 2b)$ is odd, but at least a or b is even.

    Say a is even, then, $a = 2k$ and $a^2(b^2 - 2b) = 4k^2(b^2 - 2b)$ is even. Contradiction.

11. Contrapositive: $5 | m$ and $5 | n \Rightarrow m = 5r$ and $n = 5s \Rightarrow mn = 25rs \Rightarrow 25 | mn$.

12. Prove by cases: $m + n$ is even $\Rightarrow m$ and n must both be odd or both even.

    1. If both are odd: m^2 is odd and n^2 is odd $\Rightarrow m^2 + n^2$ is even.

    2. If both are even: m^2 is even and n^2 is even $\Rightarrow m^2 + n^2$ is even.

    Or, direct: $m + n$ is even $\Rightarrow m + n = 2k \Rightarrow m^2 + n^2 = (m+n)^2 - 2mn = 4k^2 - 2mn = 2N$.

13. If n is even, then $n = 2k \Rightarrow n^2 + 2n + 9 = 4(k^2 + k) + 9$, which is odd.

    If $n^2 + 2n + 9$ is odd, then $n^2 + 2n$ is even. But $2n$ is even, then, n^2 is even, and so, n must be even. (or by contrapositive method)

14. Let $a = 2k + 1$ and simplify.

    $a^2 + 3a + 5 = 2(2k^2 + 5k + 4) + 1 = 2N + 1$.

Exercise 5.4

Exercises in this section are proofs. Solutions provided are outlines or hints but not full solutions.

1. $S_n = n(n+1)$

   $2 + 4 + \cdots + 2k = k(k+1) \Rightarrow [2 + 4 + \cdots + 2k] + (2k+2) = k(k+1) + (2k+2) \cdots$

2. By MI

   $a_1 = 1 = 3^{1-1}$

   $a_k = 3^{k-1} \Rightarrow a_{k+1} = 3 \cdot 3^{k-1}$

3. By MI

   $a_2 = 1 + 4 = 4 \times 2 - 3$

   $a_k = 4k - 3 \Rightarrow a_{k+1} = 4k - 3 + 4 = 4(k+1) - 3$

4. By MI

   $a_2 = 2 + 1 = 2^2 - 1$

   $a_k = 2^k - 1 \Rightarrow a_k = 2(2^k - 1) + 1 = 2^{k+1} - 1$

5. By MI: $a_1 = \dfrac{1}{2} = \dfrac{1}{1+1}$,

   $a_k = \dfrac{k}{k+1} \Rightarrow a_{k+1} = \dfrac{k}{k+1} + \dfrac{1}{(k+1)(k+2)}$

   $= \cdots = \dfrac{k+1}{k+2}$

6. $1 - \left(\dfrac{1}{2}\right)^n$. By MI: $\dfrac{1}{2} = 1 - \left(\dfrac{1}{2}\right)^1$

   $\dfrac{1}{2} + \dfrac{1}{4} + \dfrac{1}{8} + \cdots + \dfrac{1}{2^k} = 1 - \left(\dfrac{1}{2}\right)^k$

   $\Rightarrow \dfrac{1}{2} + \dfrac{1}{4} + \dfrac{1}{8} + \cdots + \dfrac{1}{2^k} + \dfrac{1}{2^{k+1}} = 1 - \left(\dfrac{1}{2}\right)^k + \dfrac{1}{2^{k+1}}$

   $= 1 - \left(\dfrac{1}{2}\right)^k + \dfrac{1}{2}\left(\dfrac{1}{2}\right)^k = 1 - \left(\dfrac{1}{2}\right)^{k+1}$

7. MI: for $n = 0$, $1 = 2^{0+1} - 1$

   $1 + 2 + 2^2 + \cdots + 2^k = 2^{k+1} - 1 \Rightarrow 1 + 2 + 2^2 + \cdots + 2^k + 2^{k+1} = 2^{k+1} - 1 + 2^{k+1}$

8. $a_2 = a_1 r^{2-1}$

   $a_k = a_1 r^{k-1} \Rightarrow a_{k+1} = r \cdot a_k = \cdots = a_1 r^k$

9. $S_1 = a = \dfrac{a - ar}{1 - r}$

   $S_k = \dfrac{a - ar^k}{1 - r} \Rightarrow S_{k+1} = \cdots + ar^k$

   $= \cdots = \dfrac{a - ar^k + ar^k(1 - r)}{1 - r} = \dfrac{a - ar^{k+1}}{1 - r}$

10. $n = 4, 2^4 = 16 < 4! = 24$

    $2^k < k! \Rightarrow 2 \cdot 2^k < (k+1)k!$ since $2 < k + 1$

11. $n = 5, 2^5 = 32 > 5^2 = 25$

    $2^k > k^2 \Rightarrow 2 \cdot 2^k > 2k^2 = k^2 + k^2$, but $k > 4 \Rightarrow k^2 > 4k = 2k + 2k > 2k + 1$, thus,

    $2^{k+1} > 2k^2 = k^2 + k^2 > k^2 + 2k + 1 \cdots$

12. $n = 1, 1 \cdot 1! = (1+1)! - 1$

    $1 \cdot 1! + 2 \cdot 2! + 3 \cdot 3! + \cdots + k \cdot k! = (k+1)! - 1$

    $\Rightarrow 1 \cdot 1! \cdots + k \cdot k! + (k+1) \cdot (k+1)! = (k+1)! - 1 + (k+1) \cdot (k+1)!$

    $= (k+1)![1 + k + 1] - 1 = (k+2)! - 1$

13. $n = 1: \dfrac{1}{1 \cdot 2} = \dfrac{1}{1+1}$

    $\dfrac{1}{1 \cdot 2} + \dfrac{1}{2 \cdot 3} + \dfrac{1}{3 \cdot 4} + \cdots + \dfrac{1}{k \cdot (k+1)}$

    $= \dfrac{k}{k+1} \Rightarrow \dfrac{1}{1 \cdot 2} + \cdots + \dfrac{1}{k \cdot (k+1)} + \dfrac{1}{(k+1) \cdot (k+2)}$

    $= \dfrac{k}{k+1} + \dfrac{1}{(k+1) \cdot (k+2)} = \dfrac{k(k+2) + 1}{(k+1) \cdot (k+2)}$

    $= \dfrac{k+1}{k+2}$

14. $n = 1, 1 - 1 = 0$ is a multiple of 3.

    $k^3 - k = 3M \Rightarrow (k+1)^3 - (k+1)$

    $= k^3 + 3k^2 + 3k + 1 - k - 1 = k^3 - k + 3(k^2 + k)$

    $= 3M + 3N$

15. $n = 1, 1 - 1 = 0$ is divisible by 5.

    $k^5 - k = 5m \Rightarrow (k+1)^5 - (k+1) = k^5 + 5k^4 + 10k^3 + 10k^2 + 5k + 1 - k - 1$

    $= (k^5 - k) + 5(k^4 + 2k^3 + 2k^2 + k) = 5m + 5N$

16. We can prove this by several ways. We chose proof by cases.
$n^3 - n = (n - 1)n(n + 1)$
- If n is even, then it is divisible by 2, $n - 1$ or $n + 1$ must be divisible by 3.
- If n is odd, then $n - 1$ and $n + 1$ are divisible by 2. Either n is divisible by 3 or leaves a remainder of 1 or 2 when divided by 3. Thus either $n - 1$ or $n + 1$ would be divisible by 3.

17. We can prove this by several ways. We chose proof by cases:
$n^2 + n = n(n + 1)$.
- If n is even, then it is divisible by 2.
- If n is odd, then $n + 1$ is divisible by 2.

18. Using the binomial expansion
$$5^n = (4 + 1)^n = \sum_0^n \binom{n}{i} 4^i = 1 + \sum_1^n \binom{n}{i} 4^i \Rightarrow 5^n - 1$$
$$= \sum_1^n \binom{n}{i} 4 \cdot 4^{i-1} = 4 \sum_1^n \binom{n}{i} 4^{i-1}$$

19. MI
For $n = 1$: $\begin{pmatrix} a & 0 \\ 0 & b \end{pmatrix}^1 = \begin{pmatrix} a^1 & 0 \\ 0 & b^1 \end{pmatrix}$

$\begin{pmatrix} a & 0 \\ 0 & b \end{pmatrix}^k = \begin{pmatrix} a^k & 0 \\ 0 & b^k \end{pmatrix}$

$\Rightarrow \begin{pmatrix} a & 0 \\ 0 & b \end{pmatrix}^{k+1} = \begin{pmatrix} a & 0 \\ 0 & b \end{pmatrix}^k \begin{pmatrix} a & 0 \\ 0 & b \end{pmatrix} = \begin{pmatrix} a^k & 0 \\ 0 & b^k \end{pmatrix} \begin{pmatrix} a & 0 \\ 0 & b \end{pmatrix}$

$= \begin{pmatrix} a^{k+1} & 0 \\ 0 & b^{k+1} \end{pmatrix}$

20. (a) $\sum_{i=1}^1 (2i + 4) = 6 = 1 + 5$

$\sum_{i=1}^k (2i + 4) = k^2 + 5k$

$\Rightarrow \sum_{i=1}^{k+1} (2i + 4) = \sum_{i=1}^k (2i + 4) + 2(k + 1) + 4 = k^2 + 5k$
$+ 2(k + 1) + 4$
$= (k + 1)^2 + 5(k + 1)$

(b) $\sum_1^1 (2 \cdot 3^{1-1}) = 3^1 - 1$

$\sum_1^k (2 \cdot 3^{i-1}) = 3^k - 1$

$\Rightarrow \sum_1^{k+1} (2 \cdot 3^{i-1}) = \sum_1^k (2 \cdot 3^{i-1}) + 2 \cdot 3^k = 3^k - 1$
$+ 2 \cdot 3^k = 3 \cdot 3^k - 1$

(c) $\sum_1^1 \frac{1}{(2i - 1)(2i + 1)} = \frac{1}{(2 - 1)(2 + 1)} = \frac{1}{3} = \frac{1}{2 + 1}$

$\sum_1^k \frac{1}{(2i - 1)(2i + 1)} = \frac{k}{2k + 1}$

$\Rightarrow \sum_1^{k+1} \frac{1}{(2i - 1)(2i + 1)} = \sum_1^k \frac{1}{(2i - 1)(2i + 1)}$
$+ \frac{1}{(2(k + 1) - 1)(2(k + 1) + 1)}$
$= \frac{k}{2k + 1} + \frac{1}{(2k + 1)(2k + 3)} = \frac{2k^2 + 3k + 1}{(2k + 1)(2k + 3)}$
$= \frac{\cancel{(2k + 1)}(k + 1)}{\cancel{(2k + 1)}(2k + 3)}$

Chapter 5 practice questions

1. (a) $\{a_n\} = \{1, \sqrt[3]{7}, 1, \sqrt[3]{7}, \cdots\}$
(b) $\{a_n\} = \{2, 0, 2, 0, \cdots\}$

2. MI: for $n - 1$, $5^1 + 9^1 + 2 = 16$ is divisible by 4
Assume true for $n = k$: $5^k + 9^k + 2 = 4m$
Now, $5^{k+1} + 9^{k+1} + 2 = 5 \cdot 5^k + 9 \cdot 9^k + 2$, now add and subtract $4 \cdot 5^k$ and simplify
$5^{k+1} + 9^{k+1} + 2 = 5 \cdot 5^k + 9 \cdot 9^k + 2 = 5 \cdot 5^k + 4 \cdot 5^k +$
$9 \cdot 9^k - 4 \cdot 5^k + 2$
$= 9 \cdot 5^k + 9 \cdot 9^k - 4 \cdot 5^k + 2 = 9(5^k + 9^k) - 4 \cdot 5^k + 2$
$= 9(4m - 2) - 4 \cdot 5^k + 2$
With further simplification, the last result is a multiple of 4.

3. $S_n = 4n^2 - 2n \Rightarrow S_{n-1} = 4(n - 1)^2 - 2(n - 1)$
$u_n = S_n - S_{n-1} = 8n - 6$
$\Rightarrow u_2 = 10, u_m = 8m - 6, u_{32} = 250$
Geometric sequence:
$\frac{u_{32}}{u_m} = \frac{u_m}{u_2} \Rightarrow \frac{250}{8m - 6} = \frac{8m - 6}{10} \Rightarrow (8m - 6)^2 = 50^2$
$\Rightarrow 8m - 6 = \pm 50 \Rightarrow m = 7, \cancel{m = \frac{11}{2}}$

4. By M. Induction.
$n = 1 \Rightarrow 1^3 = \frac{1^2(1 + 1)^2}{4} = 1$, true
assume true for k: $1 + 2^3 + 3^3 + \cdots + k^3 = \frac{k^2(k + 1)^2}{4}$
Now, add $(k + 1)^3$ to the left side and simplify
$1 + 2^3 + 3^3 + \cdots + k^3 + (k + 1)^3$
$= \cdots = \frac{(k + 1)^2(k + 1 + 1)^2}{4}$

5. By MI: for $n = 0$, $24 \mid 0$. True
assume true for k: $24 \mid (5^{2k} - 1)$
Now, $24 \mid (5^{2k} - 1) \Rightarrow 5^{2k} = 24a + 1$
But $5^{2(k+1)} - 1 = 5^{2k} \cdot 5^2 - 1$ and by substituting
$5^{2k} = 24a + 1$ in and simplifying $5^{2(k+1)} - 1 = 24(25a + 1)$

6. By MI: for $n = 1$, $F_1^2 = F_1 F_{+1} \Rightarrow 1^1 = 1 \cdot 1$, true
Assume true for $n = k$: $F_1^2 + F_2^2 + F_3^2 + \cdots + F_k^2 = F_k F_{k+1}$
Add F_{k+1} and simplify
$F_1^2 + F_2^2 + F_3^2 + \cdots + F_k^2 + F_{k+1}^2 = F_k F_{k+1} + F_{k+1}^2$
$= \cdots = F_{k+1} F_{k+2}$

7. By MI: for $n = 1$: $\frac{1}{2!} = 1 - \frac{1}{(1 + 1)!} = \frac{1}{2}$, true
Assume true for $n = k$: $\frac{1}{2!} + \frac{2}{3!} + \frac{3}{4!} + \cdots + \frac{k}{(k + 1)!}$
$= 1 - \frac{1}{(k + 1)!}$,
Now, for $n = k + 1$
$\frac{1}{2!} + \frac{2}{3!} + \frac{3}{4!} + \cdots + \frac{k}{(k + 1)!} + \frac{k + 1}{(k + 2)!}$
$= 1 - \frac{1}{(k + 1)!} + \frac{k + 1}{(k + 2)!}$
$= 1 - \frac{k + 2 - k - 1}{(k + 2)!} = 1 - \frac{1}{(k + 2)!}$

Chapter 6

Exercise 6.1

1. (a) $\frac{\pi}{3}$ (b) $\frac{5\pi}{6}$ (c) $-\frac{3\pi}{2}$ (d) $\frac{\pi}{5}$

(e) $\frac{3\pi}{4}$ (f) $\frac{5\pi}{18}$ (g) $-\frac{\pi}{4}$ (h) $\frac{20\pi}{9}$

Answers

2. (a) $135°$ **(b)** $630°$ **(c)** $115°$ **(d)** $210°$

(e) $-143°$ **(f)** $300°$ **(g)** $15°$

(h) $89.95° \approx 90°$

3. (a) $390°, -330°$ **(b)** $\dfrac{7\pi}{2}, -\dfrac{\pi}{2}$ **(c)** $535°, -185°$

(d) $\dfrac{11\pi}{6}, -\dfrac{13\pi}{6}$ **(e)** $\dfrac{11\pi}{3}, -\dfrac{\pi}{3}$

(f) $3.25 + 2\pi \approx 9.5,\ 3.25 - 2\pi \approx -3.03$

4. (a) 12.6 cm **(b)** 14.7 cm

5. 1.5 radians, or approx. $85.9°$

6. $r \approx 7.16$

7. (a) area $\approx 13.96 \approx 14.0\text{ cm}^2$

(b) area $\approx 131\text{ cm}^2$

8. $\alpha = 3$ (radian measure), or $\alpha = 172°$

9. 32 cm

10. 6.77 cm

11. (a) 3π radians/second **(b)** 11.9 km/hr

12. 19.8 radians/second

13. $v = r\omega$

14. 23.1 cm

15. $20\,944$ sq metres

16. (a) $r \approx 30.6\text{ cm}$ **(b)** difference $\approx 0.0771\text{ cm}$

17. $150\sqrt{3}\text{ cm}^2$

18. Area of circle $= \left(\dfrac{4\pi}{\pi - 2}\right)A$

Exercise 6.2

1. (a) $t = \dfrac{\pi}{6}: \left(\dfrac{\sqrt{3}}{2}, \dfrac{1}{2}\right)$ $\quad t = \dfrac{\pi}{3}: \left(\dfrac{1}{2}, \dfrac{\sqrt{3}}{2}\right)$

2. (a) 0.6 **(b)** 1.0 **(c)** 0.5 **(d)** 0.5

(e) 2.7 **(f)** 0.1 **(g)** 0.3 **(h)** 1.6

3. (a) (i) I **(ii)** $\left(\dfrac{\sqrt{3}}{2}, \dfrac{1}{2}\right)$

(b) (i) IV **(ii)** $\left(\dfrac{1}{2}, -\dfrac{\sqrt{3}}{2}\right)$

(c) (i) IV **(ii)** $\left(\dfrac{\sqrt{2}}{2}, -\dfrac{\sqrt{2}}{2}\right)$

(d) (i) negative x-axis **(ii)** $(0, -1)$

(e) (i) II **(ii)** $(-0.416, 0.909)$

(f) (i) IV **(ii)** $\left(\dfrac{\sqrt{2}}{2}, -\dfrac{\sqrt{2}}{2}\right)$

(g) (i) IV **(ii)** $(0.540, -0.841)$

(h) (i) II **(ii)** $\left(-\dfrac{\sqrt{2}}{2}, \dfrac{\sqrt{2}}{2}\right)$

(i) (i) III **(ii)** $(-0.929, -0.369)$

4. (a) $\sin\dfrac{\pi}{3} = \dfrac{\sqrt{3}}{2}, \cos\dfrac{\pi}{3} = \dfrac{1}{2}, \tan\dfrac{\pi}{3} = \sqrt{3}$

(b) $\sin\dfrac{5\pi}{6} = \dfrac{1}{2}, \cos\dfrac{5\pi}{6} = -\dfrac{\sqrt{3}}{2}, \tan\dfrac{5\pi}{6} = -\dfrac{\sqrt{3}}{3}$

(c) $\sin\left(-\dfrac{3\pi}{4}\right) = -\dfrac{\sqrt{2}}{2}, \cos\left(-\dfrac{3\pi}{4}\right) = -\dfrac{\sqrt{2}}{2}, \tan\left(-\dfrac{3\pi}{4}\right)$
$= 1$

(d) $\sin\dfrac{\pi}{2} = \dfrac{1}{2}, \cos\dfrac{\pi}{2} = 0, \tan\dfrac{\pi}{2}$ is undefined

(e) $\sin\left(-\dfrac{4\pi}{3}\right) = \dfrac{\sqrt{3}}{2}, \cos\left(-\dfrac{4\pi}{3}\right) = -\dfrac{1}{2}, \tan\left(-\dfrac{4\pi}{3}\right) = -\sqrt{3}$

(f) $\sin 3\pi = 0, \cos 3\pi = -1, \tan 3\pi = 0$

(g) $\sin\dfrac{3\pi}{2} = -1, \cos\dfrac{3\pi}{2} = 0, \tan\dfrac{3\pi}{2}$ is undefined

(h) $\sin\left(-\dfrac{7\pi}{6}\right) = \dfrac{1}{2}, \cos\left(-\dfrac{7\pi}{6}\right) = -\dfrac{\sqrt{3}}{2}, \tan\left(-\dfrac{7\pi}{6}\right)$
$= -\dfrac{\sqrt{3}}{3}$

(i) $\sin(1.25\pi) = -\dfrac{\sqrt{2}}{2}, \cos(1.25\pi) = -\dfrac{\sqrt{2}}{2}, \tan(1.25\pi) = 1$

5. (a) $\sin\dfrac{13\pi}{6} = \sin\dfrac{\pi}{6} = \dfrac{1}{2}; \cos\dfrac{13\pi}{6} = \cos\dfrac{\pi}{6} = \dfrac{\sqrt{3}}{2}$

(b) $\sin\dfrac{10\pi}{3} = \sin\dfrac{4\pi}{3} = -\dfrac{\sqrt{3}}{2}; \cos\dfrac{10\pi}{3} = \cos\dfrac{4\pi}{3} = -\dfrac{1}{2}$

(c) $\sin\dfrac{15\pi}{4} = \sin\dfrac{7\pi}{4} = -\dfrac{\sqrt{2}}{2}; \cos\dfrac{15\pi}{4} = \cos\dfrac{7\pi}{4} = -\dfrac{\sqrt{2}}{2}$

(d) $\sin\dfrac{17\pi}{6} = \sin\dfrac{5\pi}{6} = \dfrac{1}{2}; \cos\dfrac{17\pi}{6} = \cos\dfrac{5\pi}{6} = -\dfrac{\sqrt{3}}{2}$

6. (a) $-\dfrac{\sqrt{3}}{2}$ **(b)** $-\dfrac{\sqrt{2}}{2}$ **(c)** undefined

(d) 2 **(e)** $-\dfrac{2\sqrt{3}}{3}$

7. (a) 0.598 **(b)** $-\dfrac{\sqrt{3}}{3}$ **(c)** $-\dfrac{\sqrt{2}}{2}$ **(d)** 1.04 **(e)** 0

8. (a) I, II **(b)** II **(c)** III **(d)** II

(e) I, IV **(f)** I **(g)** IV **(h)** II, IV

Exercise 6.3

1. (a)

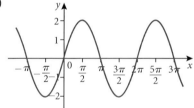

(b)

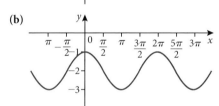

(c)

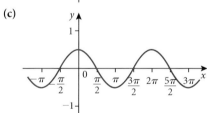

(d)

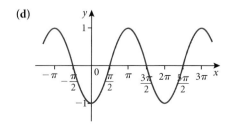

(e)

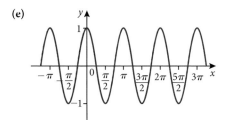

(f)

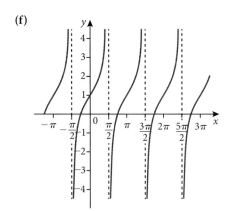

(g)

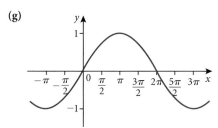

(h)

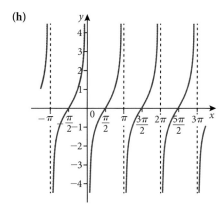

(i)

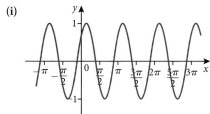

2. (a) (i)

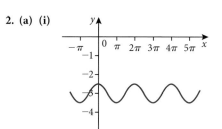

amplitude $= \dfrac{1}{2}$, period $= 2\pi$

(ii) domain: $x \in \mathbb{R}$, range: $-3.5 \leqslant y \leqslant -2.5$

(b) (i)

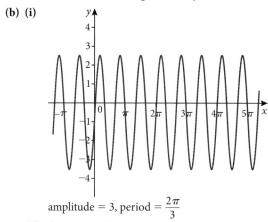

amplitude $= 3$, period $= \dfrac{2\pi}{3}$

(ii) domain: $x \in \mathbb{R}$, range: $-3.5 \leqslant y \leqslant 2.5$

(c) (i)

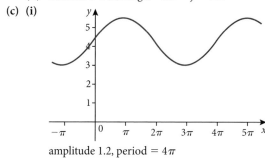

amplitude 1.2, period $= 4\pi$

(ii) domain: $x \in \mathbb{R}$, range: $3.1 \leqslant y \leqslant 5.5$

3. (a) $A = 3, B = 7$ **(b)** $A = 2.7, B = 5.9$

4. $A = 1.9, B = 4.3$

5. (a) $p = 8$ **(b)** $q = 6$

6. (a)

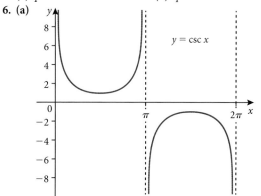

951

Answers

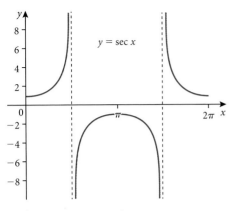

$y = \sec x$

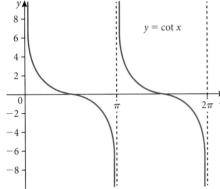

$y = \cot x$

(b) $y = \sec x$, domain: $x \neq k\pi$, k an integer range: $y \geq 1$, $y \leq -1$; $y = \csc x$, domain: $x \neq \dfrac{\pi}{2} + k\pi$, k an integer range: $y \geq 1$, $y \leq -1$; $y = \cot x$, domain: $x \neq 0, k\pi$, k an integer range: $y \in \mathbb{R}$

7. (a) $a = 2$, $b = 3$, $c = -1$ **(b)** $\dfrac{5\pi}{18}$

8. $a = 3$, $b = -\dfrac{\pi}{4}$, $c = -1$

Exercise 6.4

1. (a) $x = \dfrac{\pi}{3}, \dfrac{5\pi}{3}$ **(b)** $x = \dfrac{7\pi}{6}, \dfrac{11\pi}{6}$

(c) $x = \dfrac{\pi}{4}, \dfrac{5\pi}{4}$ **(d)** $x = \dfrac{\pi}{3}, \dfrac{2\pi}{3}$

(e) $x = \dfrac{\pi}{4}, \dfrac{3\pi}{4}, \dfrac{5\pi}{4}, \dfrac{7\pi}{4}$ **(f)** $x = \dfrac{\pi}{6}, \dfrac{5\pi}{6}, \dfrac{7\pi}{6}, \dfrac{11\pi}{6}$

(g) $x = \dfrac{\pi}{4}, \dfrac{3\pi}{4}, \dfrac{5\pi}{4}, \dfrac{7\pi}{4}$ **(h)** $x = \dfrac{\pi}{3}, \dfrac{2\pi}{3}, \dfrac{4\pi}{3}, \dfrac{5\pi}{3}$

(i) $x = 0, \dfrac{3\pi}{4}, \pi, \dfrac{7\pi}{4}, 2\pi$ **(j)** $x = 0, \dfrac{\pi}{2}, \pi, \dfrac{3\pi}{2}, 2\pi$

(k) $x = \dfrac{\pi}{3}, \dfrac{5\pi}{3}$ **(l)** $x = \dfrac{\pi}{4}, \dfrac{3\pi}{4}, \dfrac{5\pi}{4}, \dfrac{7\pi}{4}$

2. (a) $x \approx 0.412, 2.73$ **(b)** $x \approx 1.91, 4.37$

(c) $x \approx 1.11, 4.25$ **(d)** $x \approx 5.64, 3.78, 0.639, 2.50$

(e) $x \approx 2.961, 4.32$ **(f)** $x = \dfrac{\pi}{6}, \dfrac{5\pi}{6}, \dfrac{7\pi}{6}, \dfrac{11\pi}{6}$

(g) $x \approx 5.85, 5.01, 2.71, 1.87$ **(h)** $x \approx 3.43, 0.291$

3. (a) $\dfrac{5\pi}{2}, \dfrac{3\pi}{2}, \dfrac{\pi}{2}, -\dfrac{\pi}{2}, -\dfrac{3\pi}{2}, -\dfrac{5\pi}{2}$

(b) $\dfrac{\pi}{6}, -\dfrac{11\pi}{6}$ **(c)** $\dfrac{7\pi}{12}, \dfrac{19\pi}{12}$

(d) $0, \dfrac{\pi}{4}, \dfrac{\pi}{2}, \dfrac{3\pi}{4}, \pi, \dfrac{5\pi}{4}, \dfrac{3\pi}{2}, \dfrac{7\pi}{4}, 2\pi, \dfrac{9\pi}{4}, \dfrac{5\pi}{2}, \dfrac{11\pi}{4},$
$3\pi, \dfrac{13\pi}{4}, \dfrac{7\pi}{2}, \dfrac{15\pi}{4}$

4. (a) $x = \dfrac{5\pi}{6}, \dfrac{3\pi}{2}$ **(b)** $\theta = -\dfrac{3\pi}{4}, \dfrac{\pi}{4}$

(c) $x = 30°, 60°, 210°, 240°$ **(d)** $\alpha = -\dfrac{\pi}{6}, \dfrac{\pi}{6}$

(e) $\theta = \dfrac{2\pi}{3}, \dfrac{4\pi}{3}$ **(f)** $x = \dfrac{\pi}{6}, \dfrac{5\pi}{6}$

(g) $x = 7°, 97°$ **(h)** $\theta = \dfrac{\pi}{6}, \dfrac{5\pi}{6}$

5. $t \approx 1.5$ hours

6. (a) 80th day (March 21) and approximately 263rd day (September 20)

(b) 105th day (April 15) and approximately 238th day (August 26)

(c) 94 days – from 125th day to 218th day

7. (a) $x = \dfrac{\pi}{2}, \dfrac{2\pi}{3}, \dfrac{4\pi}{3}$ **(b)** $\theta = \dfrac{\pi}{2}, \dfrac{7\pi}{6}, \dfrac{11\pi}{6}$

(c) $x = -45°, 63.4°$ **(d)** $x \approx -1.87, 1.87$

(e) $x \approx 56.3°$ **(f)** $x = \dfrac{\pi}{4}, \dfrac{3\pi}{4}$

(g) no solution   **(h)** $x \approx 0°, 71.6°, 180°, 252°$

Exercise 6.5

1. (a) $\dfrac{\sqrt{2} - \sqrt{6}}{4}$ **(b)** $\dfrac{\sqrt{6} - \sqrt{2}}{4}$ **(c)** $2 - \sqrt{3}$

(d) $\dfrac{-\sqrt{6} - \sqrt{2}}{4}$ **(e)** $\dfrac{\sqrt{2} - \sqrt{6}}{4}$ **(f)** $2 - \sqrt{3}$

2. (a) $\dfrac{\sqrt{6} + \sqrt{2}}{4}$ **(b)** $\sqrt{\dfrac{\sqrt{6} + \sqrt{2} + 4}{8}}$

3. (a) $\tan\left(\dfrac{\pi}{2} - \theta\right) = \dfrac{\sin\left(\dfrac{\pi}{2} - \theta\right)}{\cos\left(\dfrac{\pi}{2} - \theta\right)} = \dfrac{\sin\dfrac{\pi}{2}\cos\theta - \cos\dfrac{\pi}{2}\sin\theta}{\cos\dfrac{\pi}{2}\cos\theta + \sin\dfrac{\pi}{2}\sin\theta}$

$= \dfrac{\cos\theta}{\sin\theta} = \cot\theta$

(b) $\sin\left(\dfrac{\pi}{2} - \theta\right) = \sin\dfrac{\pi}{2}\cos\theta - \cos\dfrac{\pi}{2}\sin\theta = \cos\theta$

(c) $\csc\left(\dfrac{\pi}{2} - \theta\right) = \dfrac{1}{\sin\left(\dfrac{\pi}{2} - \theta\right)}$

$= \dfrac{1}{\sin\dfrac{\pi}{2}\cos\theta - \cos\dfrac{\pi}{2}\sin\theta} = \dfrac{1}{\cos\theta} = \sec\theta$

4. (a) $\dfrac{4}{5}$ **(b)** $\dfrac{7}{25}$ **(c)** $\dfrac{24}{5}$

5. (a) $\dfrac{\sqrt{5}}{3}$ **(b)** $-\dfrac{4\sqrt{5}}{9}$ **(c)** $-\dfrac{1}{9}$

6. (a) $\sin 2\theta = -\dfrac{4\sqrt{5}}{9}$, $\cos 2\theta = \dfrac{1}{9}$, $\tan 2\theta = -4\sqrt{5}$

(b) $\sin 2\theta = \dfrac{24}{25}$, $\cos 2\theta = \dfrac{7}{25}$, $\tan 2\theta = \dfrac{24}{7}$

(c) $\sin 2\theta = \dfrac{4}{5}$, $\cos 2\theta = -\dfrac{3}{5}$, $\tan 2\theta = -\dfrac{4}{3}$

(d) $\sin 2\theta = -\dfrac{\sqrt{15}}{8}$, $\cos 2\theta = -\dfrac{7}{8}$, $\tan 2\theta = \dfrac{\sqrt{15}}{8}$

7. (a) $-\cos x$ **(b)** $-\cos x$ **(c)** $\tan x$ **(d)** $-\sin x$

8. (a) $\dfrac{1 + \sin\theta\cos\theta}{\cos\theta}$ **(b)** $\dfrac{1}{\sin^3\theta}$

(c) $\dfrac{\sin\theta + \cos\theta}{\sin 2\theta}$ **(d)** $\dfrac{1 + \sin^2\theta}{\cos^2\theta}$

9. (a) $\cos^3\theta$ **(b)** 1   **(c)** $\cos^2\theta$

(d) $2\tan^2\theta$ **(e)** $2\sin\alpha\cos\beta$ **(f)** $\cos^2 A$

(g) $2\cos\alpha\cos\beta$ **(h)** 1

10. (a)–(n) See Worked Solutions

11. $\tan\theta = \dfrac{5x}{x^2 + 14}$

12. (a) $x = \dfrac{\pi}{3}, \pi, \dfrac{5\pi}{3}$ **(b)** $x = -\dfrac{\pi}{3}, \dfrac{\pi}{3}$

(c) $x = 90°$ and $-90°$ **(d)** $x \approx 0.375, 2.77$

(e) $x \approx 0.615, 2.53, 3.76, 5.67$ **(f)** $x = \dfrac{3\pi}{4}, \dfrac{7\pi}{4}$

(g) $x = \dfrac{\pi}{3}, \dfrac{2\pi}{3}$ **(h)** $x = 0, \dfrac{\pi}{4}, \pi, \dfrac{5\pi}{4}$

(i) $x = 0, \dfrac{\pi}{3}, \dfrac{2\pi}{3}, \pi, \dfrac{4\pi}{3}, \dfrac{5\pi}{3}$

(j) $x = 30°, 90°, 105°, 150°, 165°$

13. $3\sin x - 4\sin^3 x$

14. (b) $x = \dfrac{\pi}{4}, \dfrac{3\pi}{4}, \dfrac{5\pi}{4}, \dfrac{7\pi}{4}$

Exercise 6.6

1. (a) $\dfrac{\pi}{2}$ **(b)** $\dfrac{\pi}{4}$ **(c)** $-\dfrac{\pi}{3}$

(d) $\dfrac{2\pi}{3}$ **(e)** 0   **(f)** $-\dfrac{\pi}{3}$

2. (a) $\dfrac{\pi}{3}$ **(b)** $\dfrac{3}{2}$ **(c)** 12

(d) not possible   **(e)** $\dfrac{\pi}{4}$ **(f)** not possible

(g) $\dfrac{3}{5}$ **(h)** $\dfrac{24}{25}$ **(i)** not possible

(j) $\dfrac{\pi}{3}$ **(k)** $\dfrac{2\sqrt{5}}{5}$ **(l)** $\dfrac{4}{5}$

(m) $\dfrac{63}{65}$ **(n)** $\dfrac{2\sqrt{20} - 3\sqrt{10}}{30}$

3. (a) $\sqrt{1 - x^2}$ **(b)** $\dfrac{\sqrt{1 - x^2}}{x}$ **(c)** $\dfrac{1}{\sqrt{x^2 + 1}}$

(d) $2x\sqrt{1 - x^2}$ **(e)** $\sqrt{\dfrac{1 - x}{1 + x}}$

(f) $\dfrac{-x^3 + x + 2x\sqrt{1 - x^2}}{x^2 + 1}$

4. $\cos\left(\arcsin\dfrac{4}{5} + \arcsin\dfrac{5}{13}\right) = \cos\left(\arccos\dfrac{16}{65}\right)$

$\cos\left(\arcsin\dfrac{4}{5}\right)\cos\left(\arcsin\dfrac{5}{13}\right)$

$- \sin\left(\arcsin\dfrac{4}{5}\right)\sin\left(\arcsin\dfrac{5}{13}\right) = \dfrac{16}{65}$

$\dfrac{3}{5}\cdot\dfrac{12}{13} - \dfrac{4}{5}\cdot\dfrac{5}{13} = \dfrac{36}{65} - \dfrac{20}{65} = \dfrac{16}{65}$ Q.E.D.

5. $\sin\left(\arctan\dfrac{1}{2} + \arctan\dfrac{1}{3}\right) = \sin\left(\dfrac{\pi}{4}\right)$

$\sin\left(\arctan\dfrac{1}{2}\right)\cos\left(\arctan\dfrac{1}{3}\right)$

$+ \cos\left(\arctan\dfrac{1}{2}\right)\sin\left(\arctan\dfrac{1}{3}\right) = \dfrac{\sqrt{2}}{2}$

$\dfrac{\sqrt{5}}{5}\cdot\dfrac{3\sqrt{10}}{10} + \dfrac{2\sqrt{5}}{5}\cdot\dfrac{\sqrt{10}}{10} = \dfrac{3\sqrt{50}}{50} + \dfrac{2\sqrt{50}}{50} = \dfrac{25\sqrt{2}}{50}$

$= \dfrac{\sqrt{2}}{2}$ Q.E.D.

6. $x = \dfrac{1}{2}$

7. (a) $x \approx 0.580, 2.56$ **(b)** $x \approx 2.21$

(c) $x \approx 1.11, 4.25$ **(d)** $x = \dfrac{\pi}{4}, \dfrac{5\pi}{4}; x \approx 2.82, 5.96$

(e) $x = \dfrac{\pi}{4}; x \approx 0\,464$ **(f)** $x \approx 1.37, 4.91$

(g) $x = \pi, 2\pi; x \approx 0.912, 2.23, 4.05, 5.37$

(h) $x = 0, \pi; x \approx 1.89$

8. $\theta = \arctan\left(\dfrac{2}{d}\right)$

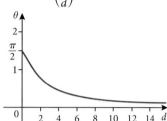

Chapter 6 practice questions

1. (a) 135 cm   **(b)** 85 cm

(c) $t = 0.5$ sec.   **(d)** 1 sec.

2. $x = \dfrac{\pi}{3}, \pi, \dfrac{5\pi}{3}$

3. $\theta \approx 2.28$ (radian measure)

4. (a) (i) -1 **(ii)** 4π **(b)**   four

5. (a) $p = 35$ **(b)** $q = 29$ **(c)** $m = \dfrac{1}{2}$

6. $x \approx 1.06, 2.05$

7. (a) $x = \dfrac{2\pi}{3}, \dfrac{4\pi}{3}$ **(b)** $x = \dfrac{\pi}{6}, \dfrac{\pi}{2}, \dfrac{5\pi}{6}, \dfrac{3\pi}{2}$

8. (a) $\sin x = \dfrac{1}{3}$ **(b)** $\cos 2x = \dfrac{7}{9}$

(c) $\sin 2x = -\dfrac{4\sqrt{2}}{9}$

9. (a) $1.6\sin\left(\dfrac{2\pi}{11}\left(t - \dfrac{9}{4}\right)\right) + 4.2$

(b) approximately 3.15 metres

(c) approximate 12:27 pm to 7:33 pm

10. $x \approx \dfrac{\pi}{4}, 1.89$

11. (a) arc length: 15 cm   **(b)** area ≈ 239cm^2

12. $k > 2.5, k < -2.5$

13. $k = 1, a = -2$

14. $\sec\alpha = -\dfrac{3}{2}$ or 4

15. (a) $\dfrac{84}{85}$ **(b)** $-\dfrac{13}{85}$ **(c)** $-\dfrac{84}{13}$

16. $\sin 2p = \dfrac{4}{5}, \sin 3p = \dfrac{11\sqrt{5}}{25}$

17. (a) $\dfrac{5}{13}$ **(b)** $-\dfrac{12}{13}$ **(c)** $-\dfrac{120}{169}$ **(d)** $\dfrac{119}{169}$

18. $\tan\theta = \dfrac{1}{3}$ or $\tan\theta = -3$

19. $\tan x = \dfrac{-(k + 1)}{k - 1}\tan\alpha$

20. $\theta = \pm\dfrac{3\pi}{8}, \pm\dfrac{\pi}{8}$

21. (a)

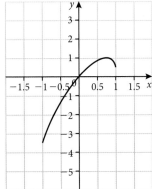

(b) $x \approx 0.412$ **(c)** $\cos(2) \leqslant g(x) \leqslant 1$

22. $24.1°$

23. $\dfrac{72}{\pi} \arccos \dfrac{8}{13}$ cm

Chapter 7

Exercise 7.1

1. (a) scalene   **(b)** equilateral
    (c) isosceles   **(d)** equilateral

2. $(0, 2, 0)$

3. (a) yes   **(b)** no

4. See Worked Solutions

5. (a) yes   **(b)** no   **(c)** no   **(d)** yes

6. (a) (i) 7,   **(ii)** 2,   **(iii)** 6,   **(iv)** 3
   (b) (i) 4,   **(ii)** 2,   **(iii)** $\sqrt{3}$,   **(iv)** 3

7. $S(2, 0, 16)$

8. (a) See Worked Solutions   **(b)** $\dfrac{\sqrt{899}}{2}$

9. surface area $= 116\pi$ units^2, volume $= \dfrac{116\pi\sqrt{29}}{3}$ units^3

10. 78 cm

11. surface area ≈ 262cm^2, volume ≈ 330 cm^3

12. $(-1, 6, -7), (3, 4, 5), (1, 2, 3)$

13. (a) $\dfrac{2}{3}$ **(b)** $\dfrac{2}{3}$

14. surface area ≈ 2170 m^2,   volume ≈ 6170 m^3

15. surface area $= 2052$ cm^2, volume $= 5832 + 324\sqrt{7}$ cm^3

Exercise 7.2

1. (a) (ii) $\cos\theta = \dfrac{4}{5}, \tan\theta = \dfrac{3}{4}, \cot\theta = \dfrac{4}{3}, \sec\theta = \dfrac{5}{4},$
    $\csc\theta = \dfrac{5}{3}$
    (iii) $\theta \approx 36.9°; 53.1°$

   (b) (ii) $\sin\theta = \dfrac{\sqrt{39}}{8}, \tan\theta = \dfrac{\sqrt{39}}{5}, \cot\theta = \dfrac{5\sqrt{39}}{39},$
    $\sec\theta = \dfrac{8}{5}, \csc\theta = \dfrac{8\sqrt{39}}{39}$
    (iii) $\theta \approx 51.3°; 38.7°$

   (c) (ii) $\sin\theta = \dfrac{2\sqrt{5}}{5}, \cos\theta = \dfrac{\sqrt{5}}{5}, \cot\theta = \dfrac{1}{2}, \sec\theta = \sqrt{5},$
    $\csc\theta = \dfrac{\sqrt{5}}{2}$
    (iii) $\theta \approx 63.4°; 26.6°$

   (d) (ii) $\sin\theta = \dfrac{3\sqrt{10}}{10}, \cos\theta = \dfrac{\sqrt{10}}{10}, \tan\theta = 3, \sec\theta = \sqrt{10},$
    $\csc\theta = \dfrac{\sqrt{10}}{3}$

(iii) $\theta \approx 71.6°, 18.4°$

(e) (ii) $\sin\theta = \dfrac{\sqrt{60}}{11}, \cos\theta = \dfrac{\sqrt{61}}{11}, \tan\theta = \dfrac{2\sqrt{915}}{61},$
    $\cot\theta = \dfrac{\sqrt{915}}{30}, \csc\theta = \dfrac{11\sqrt{60}}{60}$
    (iii) $\theta \approx 44.8°; 45.2°$

(f) (ii) $\sin\theta = \dfrac{7\sqrt{65}}{65}, \tan\theta = \dfrac{7}{4}, \cot\theta = \dfrac{4}{7}, \sec\theta = \dfrac{\sqrt{65}}{4},$
    $\csc\theta = \dfrac{\sqrt{65}}{7}$
    (iii) $\theta \approx 60.3°; 29.7°$

2. (a) $\theta = 60°, \dfrac{\pi}{3}$ **(b)** $\theta = 45°, \dfrac{\pi}{4}$ **(c)** $\theta = 60°, \dfrac{\pi}{3}$
   (d) $\theta = 60°, \dfrac{\pi}{3}$ **(e)** $\theta = 45°, \dfrac{\pi}{4}$ **(f)** $\theta = 30°, \dfrac{\pi}{6}$

3. (a) $x = 50\sqrt{3}, y = 100$ **(b)** $x \approx 8.60, y \approx 12.3$
   (c) $x \approx 20.6, y \approx 24.5$ **(d)** $x \approx 374, y \approx 299$
   (e) $x = 18, y = 18\sqrt{2}$ **(f)** $x = 200, y = 100\sqrt{3}$

4. (a) $\alpha = 60°, \beta = 30°$ **(b)** $\alpha \approx 67.4°, \beta \approx 22.6°$
   (c) $\alpha \approx 20.0°, \beta \approx 70.0°$ **(d)** $\alpha = 30°, \beta = 60°$

5. 114 metres   **6.** $67.4°$ **7.** 4.05 metres

8. 4105 m   **9.** $44°, 68°, 68°$ **10.** 5.76 km h^{-1}

11. 69.5 m   **12.** 28.7 m   **13.** 151 m

14. 59.2 m   **15.** $3\sqrt{5}$ **16.** -0.6

17. $\dfrac{|ap + bq + c|}{\sqrt{a^2 + b^2}}$ **18.** See Worked Solutions

19. $14°$

Exercise 7.3

1. (a) $\sin\theta = \dfrac{3}{5}, \cos\theta = \dfrac{4}{5}, \tan\theta = \dfrac{3}{4}$

   (b) $\sin\theta = \dfrac{12}{37}, \cos\theta = -\dfrac{35}{37}, \tan\theta = -\dfrac{12}{35}$

   (c) $\sin\theta = -\dfrac{\sqrt{2}}{2}, \cos\theta = \dfrac{\sqrt{2}}{2}, \tan\theta = -1$

   (d) $\sin\theta = -\dfrac{1}{2}, \cos\theta = -\dfrac{\sqrt{3}}{2}, \tan\theta = \dfrac{\sqrt{3}}{3}$

2. (a) $\sin 120° = \dfrac{\sqrt{3}}{2}, \cos 120° = -\dfrac{1}{2}, \tan 120° = -\sqrt{3},$
    $\cot 120° = -\dfrac{\sqrt{3}}{3}, \sec 120° = -2, \csc 120° = \dfrac{2\sqrt{3}}{3}$

   (b) $\sin 135° = \dfrac{\sqrt{2}}{2}, \cos 135° = -\dfrac{\sqrt{2}}{2}, \tan 135° = -1,$
    $\cot 135° = -1, \sec 135° = -\sqrt{2}, \csc 135° = \sqrt{2}$

   (c) $\sin 330° = -\dfrac{1}{2}, \cos 330° = \dfrac{\sqrt{3}}{2}, \tan 330° = -\dfrac{\sqrt{3}}{3},$
    $\cot 330° = -\sqrt{3}, \sec 330° = \dfrac{2\sqrt{3}}{3}, \csc 330° = -2$

   (d) $\sin 270° = -1, \cos 270° = 0, \tan 270° = $ undef.,
    $\cot 270° = 0, \sec 270° = $ undef., $\csc 270° = -1$

   (e) $\sin 240° = -\dfrac{\sqrt{3}}{2}, \cos 240° = -\dfrac{1}{2}, \tan 240° = \sqrt{3},$
    $\cot 240° = \dfrac{\sqrt{3}}{3}, \sec 240° = -2, \csc 240° = -\dfrac{2\sqrt{3}}{3}$

   (f) $\sin\dfrac{5\pi}{4} = -\dfrac{\sqrt{2}}{2}, \cos\dfrac{5\pi}{4} = -\dfrac{\sqrt{2}}{2}, \tan\dfrac{5\pi}{4} = 1,$
    $\cot\dfrac{5\pi}{4} = 1, \sec\dfrac{5\pi}{4} = -\sqrt{2}, \csc\dfrac{5\pi}{4} = -\sqrt{2}$

(g) $\sin\left(-\dfrac{\pi}{6}\right) = -\dfrac{1}{2}, \cos\left(-\dfrac{\pi}{6}\right) = \dfrac{\sqrt{3}}{2}, \tan\left(-\dfrac{\pi}{6}\right) = -\dfrac{\sqrt{3}}{3},$

$\cot\left(-\dfrac{\pi}{6}\right) = -\sqrt{3}, \sec\left(-\dfrac{\pi}{6}\right) = \dfrac{2\sqrt{3}}{3}, \csc\left(-\dfrac{\pi}{6}\right) = -2$

(h) $\sin\left(\dfrac{7\pi}{6}\right) = -\dfrac{1}{2}, \cos\left(\dfrac{7\pi}{6}\right) = -\dfrac{\sqrt{3}}{2}, \tan\left(\dfrac{7\pi}{6}\right) = \dfrac{\sqrt{3}}{3},$

$\cot\left(\dfrac{7\pi}{6}\right) = \sqrt{3}, \sec\left(\dfrac{7\pi}{6}\right) = -\dfrac{2}{\sqrt{3}}, \csc\left(\dfrac{7\pi}{6}\right) = -2$

(i) $\sin(-60°) = -\dfrac{\sqrt{3}}{2}, \cos(-60°) = \dfrac{1}{2}, \tan(-60°) = -\sqrt{3},$

$\cot(-60°) = -\dfrac{\sqrt{3}}{3}, \sec(-60°) = 2, \csc(-60°) = -\dfrac{2\sqrt{3}}{3}$

(j) $\sin\left(-\dfrac{3\pi}{2}\right) = 1, \cos\left(-\dfrac{3\pi}{2}\right) = 0, \tan\left(-\dfrac{3\pi}{2}\right)$

$= \text{undef.}, \cot\left(-\dfrac{3\pi}{2}\right) = 0, \sec\left(-\dfrac{3\pi}{2}\right) = \text{undef.},$

$\csc\left(-\dfrac{3\pi}{2}\right) = 1$

(k) $\sin\left(\dfrac{5\pi}{3}\right) = -\dfrac{\sqrt{3}}{2}, \cos\left(\dfrac{5\pi}{3}\right) = \dfrac{1}{2}, \tan\left(\dfrac{5\pi}{3}\right) = -\sqrt{3},$

$\cot\left(\dfrac{5\pi}{3}\right) = -\dfrac{\sqrt{3}}{3}, \sec\left(\dfrac{5\pi}{3}\right) = 2, \csc\left(\dfrac{5\pi}{3}\right) = -\dfrac{2\sqrt{3}}{3}$

(l) $\sin(-210°) = -\dfrac{1}{2}, \cos(-210°) = -\dfrac{\sqrt{3}}{2}, \tan(-210°)$

$= \dfrac{\sqrt{3}}{3}, \cot(-210°) = \sqrt{3}, \sec(-210°) = -\dfrac{2\sqrt{3}}{3},$

$\csc(-210°) = -2$

(m) $\sin\left(-\dfrac{\pi}{4}\right) = -\dfrac{\sqrt{2}}{2}, \cos\left(-\dfrac{\pi}{4}\right) = \dfrac{\sqrt{2}}{2}, \tan\left(-\dfrac{\pi}{4}\right) = -1,$

$\cot\left(-\dfrac{\pi}{4}\right) = -1, \sec\left(-\dfrac{\pi}{4}\right) = \sqrt{2}, \csc\left(-\dfrac{\pi}{4}\right) = -\sqrt{2}$

(n) $\sin\pi = 0, \cos\pi = -1, \tan\pi = 0, \cot\pi = \text{undef.},$
$\sec\pi = -1, \csc\pi = \text{undef.}$

(o) $\sin 4.25\pi = \dfrac{\sqrt{2}}{2}, \cos 4.25\pi = \dfrac{\sqrt{2}}{2}, \tan 4.25\pi = 1,$

$\cot 4.25\pi = 1, \sec 4.25\pi = \sqrt{2}, \csc 4.25\pi = \sqrt{2}$

3. $\sin\theta = \dfrac{15}{17}, \tan\theta = \dfrac{15}{8}, \cot\theta = \dfrac{8}{15}, \sec\theta = \dfrac{17}{8}, \csc\theta = \dfrac{17}{15}$

4. $\sin\theta = -\dfrac{6\sqrt{61}}{61}, \cos\theta = \dfrac{5\sqrt{61}}{61}$

5. $\cos\theta = -1, \tan\theta = 0, \cot\theta = \text{undef.}, \sec\theta = -1,$
$\csc\theta = \text{undef.}$

6. $\sin\theta = -\dfrac{\sqrt{3}}{2}, \cos\theta = \dfrac{1}{2}, \tan\theta = -\sqrt{3}, \cot\theta = -\dfrac{\sqrt{3}}{3},$

$\csc\theta = -\dfrac{2\sqrt{3}}{3}$

7. (a) (i) $30°$ **(ii)** $85°$
   (b) (i) $45°$ **(ii)** $7°$
   (c) (i) $60°$ **(ii)** $20°$

8. (a) $6\sqrt{3}$ **(b)** 88.9 **(c)** $675\sqrt{2}$

9. $28.5°$

10. (a) $236\,\text{cm}^2$ **(b)** $97.4\,\text{cm}^2$

11. (a) $9.06\,\text{cm}^2$ **(b)** $175\,\text{cm}^2$

12. $ab\sin\theta$

13. $x\sqrt{3}$

14. $\dfrac{2hf\cos\theta}{h+f}$

15. See Worked Solutions

16. (a) $A(x) = 24\sin x$
   (b) $0° < x < 180°$

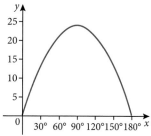

   (c) $(90°, 24)$, right-angled triangle, which will always give the maximum area because that is the maximum value of $\sin x$

17. (a) See Worked Solutions   **(c)** $7.02\,\text{m}$

18. $1740\,\text{km}$

19. (a) $\sec\theta = \dfrac{1}{\sqrt{1-x^2}}, 0 \leqslant x < \dfrac{\pi}{2}$

   (b) $\sin\beta = \dfrac{y\sqrt{1+y^2}}{1+y^2}$

20. $\cos\theta = OA, \tan\theta = PB, \cot\theta = CP, \sec\theta = OB, \csc\theta = OC$

21. (a) $45°$ **(b)** $33.7°$ **(c)** $60.3°$

22. (a) $71.6°$ **(b)** $45°$

Exercise 7.4

1. (a) infinite   **(b)** one triangle
   (c) one triangle   **(d)** one triangle
   (e) two triangles   **(f)** one triangle

2. (a) $BC \approx 17.9, AB \approx 27.0, A\widehat{C}B = 115°$
   (b) $AB \approx 18.1, BC \approx 22.5, B\widehat{A}C = 65°$
   (c) $AB \approx 74.1, B\widehat{A}C \approx 60.2°, A\widehat{B}C \approx 48.8°$
   (d) $B\widehat{A}C \approx 81.6°, A\widehat{B}C \approx 60.6°, A\widehat{C}B \approx 37.8°$
   (e) two possible triangles:
      (1) $B\widehat{A}C \approx 55.9°, A\widehat{C}B \approx 81.1°, AB \approx 40.6$
      (2) $B\widehat{A}C \approx 124.1°, A\widehat{C}B \approx 12.9°, AB \approx 9.17$
   (f) two possible triangles:
      (1) $A\widehat{B}C \approx 72.2°, A\widehat{C}B \approx 45.8°, AB \approx 0.414$
      (2) $A\widehat{B}C \approx 107.8°, A\widehat{C}B \approx 10.2°, AB \approx 0.102$

3. $10.8\,\text{cm}$ and $30.4\,\text{cm}$

4. $51.3°, 51.3°, 77.4°$

5. $71.6°$ or $22.4°$

6. $20.7°$

7. area $\approx 151\,\text{cm}^2$

8. (a) (i) $BC = 5\sin 36°$, or $BC > 5$
     (ii) $5\sin 36° < BC < 5$
     (iii) $BC < 5\sin 36°$
   (b) (i) $BC = 5\sqrt{3}$, or $BC > 10$
     (ii) $5\sqrt{3} < BC < 10$
     (iii) $BC < 5\sqrt{3}$

9. $x \approx 64.9\,\text{m}, y \approx 56.9\,\text{m}$

10. (a) $x = 5$
   (b) See Worked Solutions
   (c) $\dfrac{15\sqrt{3}}{14}$

11. $\dfrac{21\sqrt{15}}{4}$

12. (a) obtuse triangle
   (b) acute triangle
   (c) See Worked Solutions

Answers

13. 21.1

14. (a) 14   **(b)** $\cos\theta = \dfrac{3}{5}$, $WY = 2\sqrt{65}$

   (c) $2\sqrt{5}$ **(d)** 13.9°

15. 57.9°

16. (a) See Worked Solutions
   (b) See Worked Solutions
   (c) See Worked Solutions

17. (a) See Worked Solutions
   (b) See Worked Solutions
   (c) See Worked Solutions

18. (a) 1291.8 km   **(b)** 42.8°

Chapter 7 practice questions

1. $\sin A\widehat{O}B = \dfrac{24}{25}$

2. $\sin 2\theta = \dfrac{21}{29}$, $\cos 2\theta = \dfrac{20}{29}$

3. 101.5°

4. $\sin 2A = -\dfrac{120}{169}$

5. (a) 29.1 m   **(b)** 41.9 m

6. $C\widehat{A}B \approx 86.4°$

7. (a) 38.2°   **(b)** 17.3 cm²

8. (a) $A\widehat{C}B \approx 116°$ **(b)** 155 cm²

9. 78.5 km

10. $J\widehat{K}L \approx 31°$

11. (a) 3.26 cm   **(b)** 7.07 cm²

12. 70.5°

13. (a) 91 m   **(b)** $1690\sqrt{3}$
   (c) (i) See Worked Solutions
      (ii) $A_2 = 26x$
      (iii) $x = 40\sqrt{3}$
   (d) (i) supplementary angles have equal sines
      (ii) See Worked Solutions

14. (a) $2\sqrt{2} + 4$ **(b)** $5\sqrt{2} + 6\sqrt{3} + 5\sqrt{6} + 6$

15. 28.3 cm²

16. (a) $0 < \theta < 120°$
   (b) See Worked Solutions
   (c) 60°

17. (a) 120 cm²   **(b)** 2.16   **(c)** 161 cm²

18. (a) See Worked Solutions
   (b) $\sin\dfrac{\alpha}{2} = \pm\sqrt{\dfrac{1 - \cos\alpha}{2}}$
   (c) See Worked Solutions

19. $\cos\theta = \dfrac{b}{2a}$

20. 59.5 cm

21. $\triangle ABC = 72$ cm², $\triangle ABD = 24\sqrt{3} \approx 41.6$ cm²,
    $\triangle BCD \approx 34.6$ cm², $\triangle ACD \approx 69.3$ cm²

22. $D\widehat{E}F \approx 41.9°$

23. 43.0 metres

24. 52.26°

25. (a) $y = \dfrac{\sqrt{3}}{3}x$ **(b)** 56.6°

26. length ≈ 277 km, bearing $\approx 18.5°$

Chapter 8

Exercise 8.1

1. (a) $5+2i$ **(b)** $7 - \sqrt{7}\,i$ **(c)** $-6 + 0i$
   (d) $-7 + 0i$ **(e)** $0 + 9i$ **(f)** $0 - \dfrac{5}{4}i$

2. (a) $-1 - i$ **(b)** $-5 + 9i$ **(c)** $14 + 23i$
   (d) $-2 + 7i$ **(e)** $34 - 13i$ **(f)** $5 - i$
   (g) $\dfrac{16}{29} - \dfrac{11}{29}i$ **(h)** $\dfrac{4}{13} - \dfrac{7}{13}i$ **(i)** 1
   (j) $\dfrac{25}{36}$ **(k)** $-\dfrac{1}{13} - \dfrac{18}{13}i$ **(l)** $8 - i$
   (m) $-7 - 3i$ **(n)** $4 + 10i$ **(0)** $\dfrac{5}{13} + \dfrac{12}{13}i$
   (p) $\dfrac{48}{25} + \dfrac{36}{25}i$ **(q)** $2 + 9i$ **(r)** 68
   (s) $\dfrac{8}{13} - \dfrac{63}{26}i$ **(t)** $\dfrac{7}{65} + \dfrac{4}{65}i$ **(u)** $\dfrac{5}{169} + \dfrac{12}{169}i$
   (v) $\dfrac{12}{25} + \dfrac{8}{25}i$ **(w)** $\dfrac{498}{169} + \dfrac{553}{169}i$ **(x)** $-\dfrac{33}{25} - \dfrac{56}{25}i$

3. $a = \dfrac{17}{13}$, $b = -\dfrac{19}{13}i$

4. $x = -\dfrac{1}{2}$, $y = -2$ or $x = 1$, $y = 1$

5. (a) -8 **(b)** $\left((1 + i\sqrt{3})^3\right)^{2n} = 8_k^{2n}$ **(c)** $(-8)^{16}$

6. (a) $-4i$ **(b)** $\left(((-\sqrt{2} + i\sqrt{2})^2)^2\right)_k$ **(c)** $4(16)^{11}i$

7. $x^2 + y^2 = 4 \Rightarrow |z| = 2$

8. $\dfrac{9 - \sqrt{2}}{3} + \dfrac{2}{3}i$

9. $x = -\dfrac{2}{65}$, $y = \dfrac{29}{65}$

10. $\dfrac{1}{2}(1 + i)$

11. $5 + 12i$

12. $(x, y) = (2, -1)$ or $(x, y) = (-2, 1)$

13. (a) $(x, y) = (1, 3)$ or $(x, y) = (-1, -3)$
   (b) $2i$, $-1 - i$

14. $\left\{ -3i, \dfrac{3(\sqrt{3} + i)}{2}, \dfrac{3(-\sqrt{3} + i)}{2} \right\}$

15. $\dfrac{1}{2} - 2i$, 3

16. $f(x) = 2x^4 - 11x^3 + 15x^2 + 17x - 11$

17. $f(x) = x^4 + 2x^3 + 8x + 16$

18. $5 - 2i$, -3

19. $1 + i\sqrt{3}$, $-\dfrac{2}{3}$

20. $\dfrac{x + iy}{x - iy} \cdot \dfrac{x + iy}{x + iy} = \left(\dfrac{x^2 - y^2}{x^2 + y^2}\right) + \left(\dfrac{2xy}{x^2 + y^2}\right)i$

$\Rightarrow \left(\dfrac{x^2 - y^2}{x^2 + y^2}\right)^2 + \left(\dfrac{2xy}{x^2 + y^2}\right)^2 = 1$

21. (a) $k = \pm1$, $k = 0$ **(b)** $k = \pm\sqrt{3 \pm 2\sqrt{2}}$

22. $z_1 = 1 + i$, $z_2 = 2 - i$

23. $z_1 = \dfrac{7 - 4i}{3}$, $z_2 = \dfrac{1 + 6i}{3}$

Exercise 8.2

1. (a) $2\sqrt{2}\operatorname{cis}\dfrac{\pi}{4}$ **(b)** $2\operatorname{cis}\dfrac{\pi}{6}$ **(c)** $2\sqrt{2}\operatorname{cis}\dfrac{7\pi}{4}$
   (d) $2\sqrt{2}\operatorname{cis}\dfrac{11\pi}{6}$ **(e)** $4\operatorname{cis}\dfrac{5\pi}{3}$ **(f)** $3\sqrt{2}\operatorname{cis}\dfrac{3\pi}{4}$
   (g) $4\operatorname{cis}\dfrac{\pi}{2}$ **(h)** $6\operatorname{cis}\dfrac{7\pi}{6}$ **(i)** $\sqrt{2}\operatorname{cis}\dfrac{\pi}{4}$
   (j) $15\operatorname{cis}\pi$ **(k)** $\dfrac{1}{5}\operatorname{cis}(5.64)$ **(l)** $3\sqrt{2}\operatorname{cis}\dfrac{3\pi}{4}$
   (m) $\pi\operatorname{cis}0$ **(n)** $e\operatorname{cis}\dfrac{\pi}{2}$

2. (a) $\dfrac{-\sqrt{3}}{2} + \dfrac{i}{2}$, $\dfrac{\sqrt{3}}{2} + \dfrac{i}{2}$ **(b)** 1, $\dfrac{1}{2} - \dfrac{\sqrt{3}\,i}{2}$
   (c) $\dfrac{-\sqrt{3}}{2} + \dfrac{i}{2}$, $-i$ **(d)** $-i$, $\dfrac{-1}{2} + \dfrac{\sqrt{3}\,i}{2}$

(e) $\dfrac{\sqrt{6}+\sqrt{2}}{2}+i\dfrac{\sqrt{6}-\sqrt{2}}{2}, \dfrac{9(-\sqrt{6}+\sqrt{2})}{8}-i\dfrac{9(\sqrt{6}+\sqrt{2})}{8}$

(f) $-3\sqrt{3}-3+i(3\sqrt{3}-3), \dfrac{3\sqrt{3}-3}{4}-\dfrac{i(3\sqrt{3}+3)}{4}$

(g) $\dfrac{-\sqrt{2}}{2}(1+i), \dfrac{\sqrt{2}}{2}(1+i)$ **(h)** $6, \dfrac{-3}{4}-\dfrac{3\sqrt{3}\,i}{4}$

(i) $\dfrac{-5\sqrt{6}-15\sqrt{2}}{64}+i\dfrac{5\sqrt{6}-15\sqrt{2}}{64},$

$\dfrac{5\sqrt{6}-15\sqrt{2}}{48}-i\dfrac{5\sqrt{6}+15\sqrt{2}}{48}$

(j) $-3\sqrt{3}+3-i(3\sqrt{3}+3), \dfrac{3\sqrt{3}+3}{4}+\dfrac{i(3\sqrt{3}-3)}{4}$

3. (a) $z_1=2\text{cis}\dfrac{\pi}{6}, z_2=4\text{cis}\dfrac{-\pi}{3}, \dfrac{1}{z_1}=\dfrac{1}{2}\text{cis}\dfrac{-\pi}{6}, \dfrac{1}{z_2}=\dfrac{1}{4}\text{cis}\dfrac{\pi}{3},$

$z_1z_2=8\text{cis}\dfrac{-\pi}{6}, \dfrac{z_1}{z_2}=\dfrac{1}{2}\text{cis}\dfrac{\pi}{2}$

(b) $z_1=2\sqrt{2}\,\text{cis}\dfrac{\pi}{6}, z_2=4\sqrt{3}\,\text{cis}\dfrac{-\pi}{3}, \dfrac{1}{z_1}=\dfrac{\sqrt{2}}{4}\text{cis}\dfrac{-\pi}{6},$

$\dfrac{1}{z_2}=\dfrac{\sqrt{3}}{12}\text{cis}\dfrac{\pi}{3}, z_1z_2=8\sqrt{6}\,\text{cis}\dfrac{-\pi}{6}, \dfrac{z_1}{z_2}=\dfrac{\sqrt{6}}{6}\text{cis}\dfrac{\pi}{2}$

(c) $z_1=8\text{cis}\dfrac{\pi}{6}, z_2=3\sqrt{2}\,\text{cis}\dfrac{-3\pi}{4}, \dfrac{1}{z_1}=\dfrac{1}{8}\text{cis}\dfrac{-\pi}{6},$

$\dfrac{1}{z_2}=\dfrac{\sqrt{2}}{6}\text{cis}\dfrac{3\pi}{4}, z_1z_2=24\sqrt{2}\,\text{cis}\dfrac{-7\pi}{12}, \dfrac{z_1}{z_2}=\dfrac{4\sqrt{2}}{3}\text{cis}\dfrac{11\pi}{12}$

(d) $z_1=\sqrt{3}\,\text{cis}\dfrac{\pi}{2}, z_2=2\sqrt{2}\,\text{cis}\dfrac{-2\pi}{3}, \dfrac{1}{z_1}=\dfrac{\sqrt{3}}{3}\text{cis}\dfrac{-\pi}{2},$

$\dfrac{1}{z_2}=\dfrac{\sqrt{2}}{8}\text{cis}\dfrac{2\pi}{3}, z_1z_2=2\sqrt{6}\,\text{cis}\dfrac{-\pi}{6}, \dfrac{z_1}{z_2}=\dfrac{\sqrt{6}}{4}\text{cis}\dfrac{-5\pi}{6}$

(e) $z_1=\sqrt{10}\,\text{cis}\dfrac{\pi}{4}, z_2=2\sqrt{2}\,\text{cis}\dfrac{\pi}{2}, \dfrac{1}{z_1}=\dfrac{\sqrt{10}}{10}\text{cis}\dfrac{-\pi}{4},$

$\dfrac{1}{z_2}=\dfrac{\sqrt{2}}{4}\text{cis}\dfrac{-\pi}{2}, z_1z_2=4\sqrt{5}\,\text{cis}\dfrac{3\pi}{4}, \dfrac{z_1}{z_2}=\dfrac{\sqrt{5}}{2}\text{cis}\dfrac{-\pi}{4}$

(f) $z_1=2\text{cis}\dfrac{\pi}{3}, z_2=2\sqrt{3}\,\text{cis}0, \dfrac{1}{z_1}=\dfrac{1}{2}\text{cis}\dfrac{-\pi}{3}, \dfrac{1}{z_2}=\dfrac{\sqrt{3}}{6}\text{cis}0,$

$z_1z_2=4\sqrt{3}\,\text{cis}\dfrac{\pi}{3}, \dfrac{z_1}{z_2}=\dfrac{\sqrt{3}}{3}\text{cis}\dfrac{\pi}{3}$

4. (a) $x^2+(y-1)^2=x^2+(y+2)^2 \Rightarrow y=-\dfrac{1}{2}$

(b) $\arg(z_1)=\dfrac{-\pi}{6}, \arg(z_2)=\dfrac{-5\pi}{6}$

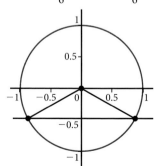

5.

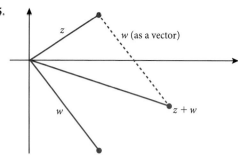

Note that the shortest distance between 0 and $z+w$ is the modulus of $z+w$. This is shorter in length than the path which goes from 0 to z to $z+w$. The total length of this second path is $|z|+|w|$.

6. (a) $\dfrac{\sqrt{3}}{2}-\dfrac{3i}{2},$ **(b)** $\dfrac{-2\sqrt{3}}{3},$ **(c)** $\sqrt{3}\,i$

7. $z_1=4, \arg(z_1)=\dfrac{-\pi}{6}, z_2=2\sqrt{2}, \arg(z_2)=\dfrac{\pi}{4}, z_3=8\sqrt{2},$

$\arg(z_3)=\dfrac{\pi}{12}$

8. $22-2\sqrt{3}\approx18.5$

9. (a) $\{(x,y): x^2+y^2=9\}$, the circle centre $(0,0)$ radius 3

(b) $\{(x,y): x=0\}$, the y-axis

(c) $\{(x,y): x=4\}$, the line $x=4$

(d) $\{(x,y): (x-3)^2+y^2=4\}$, the circle centre $(3,0)$ radius 2

(e) $\{(x,y): 1\leqslant x\leqslant3$ and $y=0\}$, the line segment between $(1,0)$ and $(3,0)$

10. (a) $\{(x,y): x^2+y^2<9\}$, the disk centre $(0,0)$ radius 3

(b) $\{(x,y): x^2+(y-3)^2>4\}$, all points excluding the interior of the disk centre $(0,3)$ radius 2

Exercise 8.3

1. (a) $-2-2i\sqrt{3}$ **(b)** 3 **(c)** 3i

(d) $\sqrt{2}-\sqrt{6}+i(\sqrt{2}+\sqrt{6})$ **(e)** $\dfrac{13}{2}+\dfrac{13i\sqrt{3}}{2}$

(f) $\dfrac{3e}{2}+\dfrac{3ei\sqrt{3}}{2}$

2. (a) $2\sqrt{2}\,e^{i\frac{\pi}{4}}$ **(b)** $2e^{i\frac{\pi}{6}}$ **(c)** $2\sqrt{2}\,e^{-i\frac{\pi}{6}}$

(d) $4e^{-i\frac{\pi}{3}}$ **(e)** $3\sqrt{2}\,e^{i\frac{3\pi}{4}}$ **(f)** $4e^{i\frac{\pi}{2}}$

(g) $6e^{i\frac{7\pi}{6}}$ **(h)** $3\sqrt{2}\,e^{i\frac{3\pi}{4}}$

(i) $\pi e^{2\pi i}$ or simply π **(j)** $e^{1+i\frac{\pi}{2}}$

3. (a) $32i$ **(b)** -64 **(c)** -10077696

(d) -262144 **(e)** 1296

(f) $17496(-1-i)$ **(g)** $\dfrac{1}{1296}$ **(h)** $\dfrac{1}{559872}(\sqrt{3}-i)$

(i) $-128\sqrt{3}-128i$

4. (a) $\sqrt{6}+i\sqrt{2}, -\sqrt{6}-i\sqrt{2}$ **(b)** $2e^{i\frac{\pi}{9}}; 2e^{i\frac{7\pi}{9}}; 2e^{i\frac{13\pi}{9}}$

(c) $\pm\dfrac{\sqrt{2}}{2}\pm i\dfrac{\sqrt{2}}{2}$

(d) $\left(-\dfrac{\sqrt{6}}{4}-\dfrac{\sqrt{2}}{4}\right)+i\left(\dfrac{\sqrt{2}}{4}-\dfrac{\sqrt{6}}{4}\right); \left(-\dfrac{\sqrt{6}}{4}+\dfrac{\sqrt{2}}{4}\right)$

$+i\left(-\dfrac{\sqrt{2}}{4}-\dfrac{\sqrt{6}}{4}\right);$

$\left(\dfrac{\sqrt{6}}{4}-\dfrac{\sqrt{2}}{4}\right)+i\left(\dfrac{\sqrt{2}}{4}+\dfrac{\sqrt{6}}{4}\right); \left(\dfrac{\sqrt{6}}{4}+\dfrac{\sqrt{2}}{4}\right)+i\left(\dfrac{\sqrt{6}}{4}-\dfrac{\sqrt{2}}{4}\right);$

$\left(-\dfrac{\sqrt{2}}{2}\right)+i\left(\dfrac{\sqrt{2}}{2}\right); \left(\dfrac{\sqrt{2}}{2}\right)-i\left(\dfrac{\sqrt{2}}{2}\right)$

(e) $\sqrt[5]{18}\,e^{i\frac{4\pi}{15}}; \sqrt[5]{18}\,e^{i\frac{10\pi}{15}}; \sqrt[5]{18}\,e^{i\frac{16\pi}{15}}; \sqrt[5]{18}\,e^{i\frac{22\pi}{15}}; \sqrt[5]{18}\,e^{i\frac{28\pi}{15}}$

Answers

5. **(a)** 2; $2e^{i\frac{2\pi}{5}}$; $2e^{i\frac{4\pi}{5}}$; $2e^{i\frac{6\pi}{5}}$; $2e^{i\frac{8\pi}{5}}$

  (b) $e^{i\left(-\frac{\pi}{16}\right)}$; $e^{i\frac{3\pi}{16}}$; $e^{i\frac{7\pi}{16}}$; $e^{i\frac{11\pi}{16}}$; $e^{i\frac{15\pi}{16}}$; $e^{i\frac{19\pi}{16}}$; $e^{i\frac{23\pi}{16}}$

  (c) $2e^{i\frac{5\pi}{18}}$; $2e^{i\frac{17\pi}{18}}$; $2e^{i\frac{29\pi}{18}}$

  (d) $\pm 2, \pm 2i$

  (e) $\sqrt{8}\,e^{i\left(\frac{3\pi}{20}\right)}$; $\sqrt{8}\,e^{i\left(\frac{11\pi}{20}\right)}$; $\dots$; $\sqrt{8}\,e^{i\left(\frac{35\pi}{20}\right)}$

  (f) $\left(-\dfrac{\sqrt{6}}{2}-\dfrac{\sqrt{2}}{2}\right)+i\left(\dfrac{\sqrt{2}}{2}-\dfrac{\sqrt{6}}{2}\right); \left(-\dfrac{\sqrt{6}}{2}+\dfrac{\sqrt{2}}{2}\right)$
  $+i\left(-\dfrac{\sqrt{2}}{2}-\dfrac{\sqrt{6}}{2}\right);$
  $\left(\dfrac{\sqrt{6}}{2}-\dfrac{\sqrt{2}}{2}\right)+i\left(\dfrac{\sqrt{2}}{2}+\dfrac{\sqrt{6}}{2}\right);\left(\dfrac{\sqrt{6}}{2}+\dfrac{\sqrt{2}}{2}\right)+i\left(\dfrac{\sqrt{6}}{2}-\dfrac{\sqrt{2}}{2}\right);$
  $-\sqrt{2}+i\sqrt{2};\sqrt{2}-i\sqrt{2}$

6. **(a)** $\cos(4\beta)+i\sin(4\beta)$ **(b)** $\cos(7\beta)+i\sin(7\beta)$
  (c) $\cos(3\beta)+i\sin(3\beta)$ **(d)** $\cos(2\beta)+i\sin(2\beta)$

7. $e^{i(\alpha+\beta)}=e^{i\alpha}e^{i\beta}\dots$ See Worked Solutions for full answer.

8. a, b, c Expand $(\cos\theta+i\sin\theta)^4$ using binomial theorem.

9. **(a)** simple addition and subtraction of z and $\dfrac{1}{z}$ in polar form

  (b) $2\cos 2n\alpha = z^n+\dfrac{1}{z^n}$; $2i\sin 2n\alpha = z^n-\dfrac{1}{z^n}$

10. 7

11. **(a)** See Worked Solutions **(b)** $1-i$
  (c) See Worked Solutions

12. **(a)** See Worked Solutions **(b)** $\dfrac{3+\sqrt{5}}{2}$
  (c) See Worked Solutions

13. 524 288

14. $\dfrac{3}{2}$

Chapter 8 practice questions

1. $x=2, y=-1$

2. **(a)** 0 **(b)** x^2+y^2-xy

3. **(a)** $2i$ **(b)** See Worked Solutions
  (c) 65536

4. **(a)** $z_1=\sqrt{2}\cos\left(-\dfrac{\pi}{6}\right)$; $z_2=\sqrt{2}\cos\left(-\dfrac{\pi}{4}\right)$
  (b) See Worked Solutions
  (c) $\dfrac{z_1}{z_2}=\dfrac{\sqrt{6}+\sqrt{2}}{4}+i\dfrac{\sqrt{6}-\sqrt{2}}{4}$; $\cos\dfrac{\pi}{12}=\dfrac{\sqrt{6}+\sqrt{2}}{4}$;
  $\sin\dfrac{\pi}{12}=\dfrac{\sqrt{6}-\sqrt{2}}{4}$

5. $\left(\dfrac{z_1}{z_2}\right)^4=\left(\dfrac{a^4}{2b^4}\right)+i\left(\dfrac{a^4\sqrt{3}}{2b^4}\right)$

6. $|z|=4$

7. $a=\dfrac{11}{5}, b=\dfrac{3}{5}$

8. $x=\sqrt{3}$

9. $a=0, b=-1$

10. **(a)** $z^5-1=(z-1)(z^4+z^3+z^2+z+1)$
  (b) $\text{cis}\left(\pm\dfrac{2\pi}{5}\right)$; $\text{cis}\left(\pm\dfrac{4\pi}{5}\right)$; $1\,\text{cis}0$
  (c) $\left(z^2-\left(2\cos\dfrac{2\pi}{5}\right)z+1\right)\left(z^2-\left(2\cos\dfrac{4\pi}{5}\right)z+1\right)$

11. **(a)** $8i=8\text{cis}\dfrac{\pi}{2}$
  (b) (i) $z=2\text{cis}\dfrac{\pi}{6}$ **(ii)** $z=\sqrt{3}+i$

12. **(a)** $|z|=1$; $\arg(z)=\dfrac{4\pi}{3}$ or $-\dfrac{2\pi}{3}$
  (b) $z^3=\text{cis}\,4\pi=1$ **(c)** $\dfrac{3}{2}+\dfrac{3\sqrt{3}}{2}i$

13. $-5-12i$

14. **(a)** True for $n=1$; assume true for $n=k$;
  Prove for $n=k+1$:
  $$z^{k+1}=(\cos k\theta+i\sin k\theta)(\cos\theta+i\sin\theta)\dots$$
  (b) substitute $n=-1$ and $-n$ into the equation and simplify
  (c) (i) $z^5+5z^3+10z+\dfrac{10}{z}+\dfrac{5}{z^3}+\dfrac{1}{z^5}$
  (ii) 1, 5, 10

15. $p=-\dfrac{2}{5}; q=\dfrac{6}{5}$

16. **(a)** See Worked Solutions
  (b) $z_1^2=4\text{cis}\dfrac{4\pi}{5}$; $z_1^3=8\text{cis}\dfrac{6\pi}{5}$; $z_1^4=16\text{cis}\dfrac{\pi}{5}$; $z_1^5=32$
  (c)

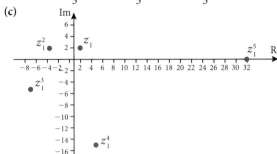

  (d) enlargement scale factor of 2 with $(0, 0)$ as centre,
  and a rotation of $\dfrac{2\pi}{5}$.

17. **(a)** See Worked Solutions
  (b) (i)

  (ii) Use $\tan^{-1}\left(\dfrac{1.5}{3}\right)=\dfrac{\pi}{6}$ **(iii)** $\dfrac{5\pi}{6}$
  (c) $k=4$

18. $a=3, b=1$

19. **(a)** $\cos n\theta+i\sin n\theta+\cos n\theta-i\sin n\theta$
  (b) See Worked Solutions

20. **(a)** $z=\dfrac{1}{2}e^{i\theta}$
  (b) $|z|=0.5$, hence less than 1
  (c) $S_\infty=\dfrac{e^{i\theta}}{1-\dfrac{1}{2}e^{i\theta}}$
  (d) (i) $S_\infty=\dfrac{\cos\theta+i\sin\theta}{1-\dfrac{1}{2}(\cos\theta+i\sin\theta)}$
  (ii) See Worked Solutions

21. $a=8; b=25; c=26$

22. $z=2+4i$

23. $z_1=1+4i; z_2=7-i$

24. **(a)** $z_1=2+2i; z_2=2-2i$ **(b)** $\dfrac{z_1^4}{z_2^2}=-8i$
  (c) See Worked Solutions **(d)** 0
  (e) $n=4k$, where $k\in\mathbb{Z}$

25. **(a)** $\left(\cos\dfrac{2\pi}{7}+i\sin\dfrac{2\pi}{7}\right)^7-1=\cos 2\pi+i\sin 2\pi-1=0$
  (b) By long division or synthetic division
  $$\dfrac{z^7-1}{z-1}=z^6+z^5+z^4+z^3+z^2+z+1$$

58

(c) Substitute $z = \cos\frac{2\pi}{7} + i\sin\frac{2\pi}{7}$ into $z^6 + z^5 + z^4 + z^3$

$+ z^2 + z + 1 = 0$ and simplify

$2\cos\frac{2\pi}{7} + 2\cos\frac{4\pi}{7} + 2\cos\frac{6\pi}{7} = -1$

26. (a) $z = \frac{2}{3}\text{cis}\frac{\pi + 2k\pi}{3} \Rightarrow z_1 = \frac{2}{3}\text{cis}\frac{\pi}{3}; z_2 = \frac{2}{3}\text{cis}\pi; z_3$

$= \frac{2}{3}\text{cis}\frac{5\pi}{3}$

(b) $3 \times \frac{1}{2} \times \frac{2}{3} \times \frac{2}{3} \times \frac{\sqrt{3}}{2} = \frac{\sqrt{3}}{3}$

27. (a) $(2 + 3i)^4 - 4(2 + 3i)^3 + 17(2 + 3i)^2 - 16(2 + 3i)$

$+ 52$

$= -119 - 120i + 184 - 36i - 85 + 204i - 32 - 48i$

$+ 52 = 0$

(b) $2 \pm 3i, \pm 2i$

28. (a) $z = \pm\frac{1 + i}{\sqrt{2}}$ **(b)** $2 + i$ or 1

29. $\cos 4\theta = $ real part of $(\cos\theta + i\sin\theta)^4$. Expand, using Pascal's triangle to get

$\cos 4\theta = \cos^4\theta - 6\cos^2\theta \sin^2\theta + \sin^4\theta$.

$\sin 4\theta = 4\cos^3\theta \sin\theta - 4\cos\theta \sin^3\theta$.

$\cos 5\theta = \cos^5\theta - 10\cos^3\theta \sin^2\theta + 5\cos\theta \sin^4\theta$

$\sin 6\theta = 6\cos^5\theta \sin\theta - 20\cos^3\theta \sin^3\theta + 6\cos\theta \sin^5\theta$.

30. $e^{\ln 2(1+i)} = e^{\ln 2}(\cos(\ln 2) + i\sin(\ln 2))$

$\Rightarrow$ real part $= 2\cos(\ln 2)$ and imaginary part $= 2\sin(\ln 2)$

Chapter 9

Exercise 9.1

1. (a) $\begin{pmatrix} \frac{5}{2} \\ -2 \\ 0 \end{pmatrix}$ **(b)** $\begin{pmatrix} 3 \\ 2\sqrt{3} \\ 0 \end{pmatrix}$ **(c)** $\begin{pmatrix} -1 \\ 2 \\ -2 \end{pmatrix}$ **(d)** $\begin{pmatrix} -2a \\ -a \\ -a \end{pmatrix}$

2. (a) $Q\left(-\frac{1}{2}, -3, 2\right)$ **(b)** $P\left(\frac{5}{2}, -2, 0\right)$

(c) $Q(0, -4a, 3a)$

3. (a) $(x, y, z) = (t, t, 5 - 5t)$,

or $(x, y, z) = (1 + t, 1 + t, -5t)$

(b) $(x, y, z) = (-1 + 4t, 5t, 1 - 3t)$

(c) $(x, y, z) = (2 - 4t, 3 - 6t, 4 + t)$

4. (a) $(7, -8, -1)$ **(b)** $(-1, \frac{11}{2}, \frac{29}{3})$

(c) $(2 - a, 4 - 2a, -b - 2)$

5. (a) $(-\frac{1}{3}, 1, \frac{1}{3})$ **(b)** $(1, -\frac{5}{3}, -16)$

(c) $\left(\frac{a + b + c}{3}, \frac{2a + 2b + 2c}{3}, a + b + c\right)$

6. (a) $D(1, 1, -6)$ **(b)** $D(-4\sqrt{2}, 3\sqrt{3}, -3\sqrt{5})$

(c) $D\left(\frac{5}{2}, 0, -4\right)$

7. $m = 5, n = 1$

8. (a) $v = \frac{2}{3}i + \frac{2}{3}j - \frac{1}{3}k$ **(b)** $v = \frac{3}{\sqrt{14}}i - \frac{2}{\sqrt{14}}j + \frac{1}{\sqrt{14}}k$

(c) $v = \frac{2}{3}i - \frac{1}{3}j - \frac{2}{3}k$

9. (a) $\frac{2}{3}(2i + 2j - k)$ **(b)** $\frac{2}{\sqrt{14}}(6i - 4j + 2k)$

(c) $\frac{5}{3}(2i - j - 2k)$

10. (a) $|u + v| = \sqrt{29}$ **(b)** $|u| + |v| = \sqrt{14} + \sqrt{5}$

(c) $|-3u| + |3v| = 3\sqrt{14} + 3\sqrt{5}$

(d) $\frac{1}{|u|}u = \frac{i}{\sqrt{14}} + \frac{3j}{\sqrt{14}} - \frac{2k}{\sqrt{14}}$ **(e)** $\left|\frac{1}{|u|}u\right| = 1$

11. (a) $(3, 4, -5)$ **(b)** $(0, -2, 5)$

12. (a) $(1, -\frac{4}{3})$ **(b)** $\sqrt{6}(4i + 2j - 2k)$

(c) $-\frac{2}{3}i + \frac{8}{3}j - 2k$

13. 0

14. $\pm\frac{\sqrt{14}}{14}$

15. None

16. $\pm\sqrt{\left(\frac{3}{13}\right)}$

17. (a) $a = (8,0,0), b = (8,8,0), c = (0,8,0)\ d = (0,0,8),$

$e = (8,0,8), f = (8,8,8)$

(b) $l = (8, 4, 8), m = 4, 8, 8), n = (8, 8, 4)$

(c) $\left(\overrightarrow{LM} + \overrightarrow{MN}\right) + \overrightarrow{NL} = \overrightarrow{LN} + \overrightarrow{NL} = \vec{0}$

18. (a) $c = (8,0,12), d = (0, 10, 12)$

(b) $f = (4, 5, 0), g = (4, 5, 12)$

(c) $\overrightarrow{AG} = (-4, 5, 12) = \overrightarrow{FD}$

19. $\pm\frac{\sqrt{6}}{3}$

20. $(\alpha, \beta, \mu) = \left(\frac{26}{7}, -\frac{11}{7}, \frac{3}{7}\right)$

21. $(\alpha, \beta, \mu) = (2, -1, 3)$

22. Not possible

23. (a) Rectangle

(b) $|u - v| = |u + v| \Rightarrow (u_1 - v_1)^2 + \cdots = (u_1 + v_1)^2 + \cdots$

24. $T_1 = 125(\sqrt{3} - 1); T_2 = 125\left(\frac{3\sqrt{2} - \sqrt{6}}{2}\right)$

25. $T_1 = 150; T_2 = 150\sqrt{3}$

Exercise 9.2

1. (a) $-16, 117.65°$ **(b)** $-20, 122.31°$

(c) $13, 40.24°$ **(d)** $-15, 151.74°$

2. (a) 6 **(b)** -6

3. (a) orthogonal   **(b)** acute   **(c)** orthogonal

4. (a) $v.u = 0 = w.u$;

(b) $\frac{3}{\sqrt{13}} + \frac{2}{\sqrt{13}}j, \frac{-3}{\sqrt{13}}i - \frac{2}{\sqrt{13}}j$

5. (a) $\cos\alpha = \frac{2}{\sqrt{14}}, \cos\beta = \frac{-3}{\sqrt{14}}, \cos\gamma = \frac{1}{\sqrt{14}};$

$\cos^2\alpha + \cos^2\beta + \cos^2\gamma = \frac{2^2}{14} + \frac{(-3)^2}{14} + \frac{1^2}{14} = 1;$

$\alpha \approx 58°, \beta \approx 143°, \gamma \approx 74°$

(b) $\cos\alpha = \frac{1}{\sqrt{6}}, \cos\beta = \frac{-2}{\sqrt{6}}, \cos\gamma = \frac{1}{\sqrt{6}};$

$\cos^2\alpha + \cos^2\beta + \cos^2\gamma = \frac{1^2}{6} + \frac{(-2)^2}{6} + \frac{1^2}{6} = 1;$

$\alpha \approx 66°, \beta \approx 145°, \gamma \approx 66°$

(c) $\cos\alpha = \frac{3}{\sqrt{14}}, \cos\beta = \frac{-2}{\sqrt{14}}, \cos\gamma = \frac{1}{\sqrt{14}};$

$\cos^2\alpha + \cos^2\beta + \cos^2\gamma = \frac{3^2}{14} + \frac{(-2)^2}{14} + \frac{1^2}{14} = 1;$

$\alpha \approx 37°, \beta \approx 122°, \gamma \approx 74°$

(d) $\cos\alpha = \frac{3}{5}, \cos\beta = 0\ \cos\gamma = \frac{-4}{5};$

$\cos^2\alpha + \cos^2\beta + \cos^2\gamma = \frac{3^2}{25} + \frac{0^2}{25} + \frac{(-4)^2}{25} = 1;$

$\alpha \approx 53°, \beta \approx 90°, \gamma \approx 143°$

6. $\begin{pmatrix} \frac{1}{2} \\ \frac{\sqrt{2}}{2} \\ -\frac{1}{2} \end{pmatrix}$

7. $\begin{pmatrix} \frac{3\sqrt{2}}{2} \\ \frac{3\sqrt{2}}{2} \\ 0 \end{pmatrix}$

8. (a) $m = -\dfrac{9}{8}$ (b) $m = 1$ or $-\dfrac{1}{4}$

9. $m = -14$

10. (a) $127°$ (b) $63°$ (c) $63°$

11. (a) $m = \dfrac{1}{3}$ (b) $m = -\dfrac{1}{4}$

12. m_A: $r = \begin{pmatrix} 4 \\ -2 \\ -1 \end{pmatrix} + m\begin{pmatrix} -1 \\ 0 \\ 3/2 \end{pmatrix}$; m_B: $r = \begin{pmatrix} 3 \\ -5 \\ -1 \end{pmatrix} + n\begin{pmatrix} 1/2 \\ 9/2 \\ 3/2 \end{pmatrix}$

m_C: $r = \begin{pmatrix} 3 \\ 1 \\ 2 \end{pmatrix} + k\begin{pmatrix} 1/2 \\ -9/2 \\ -3 \end{pmatrix}$; centroid $\left(\dfrac{10}{3}, -2, 0\right)$

13. (a) $90, 90, 82, 73, 60, 55, 53, 52, 47, 43, 38, 37$

   (b) 72.2

   (c) $103.3°, 133.5°, 46.5°$

   (d) 0

14. $k = 2$

15. $k = 0$ or $k = 4$

16. $x = -20, y = -14$

17. $x = 5$

18. (a) $117°$ (b) $\overrightarrow{AC} = \begin{pmatrix} 0 \\ 6 \\ 3 \end{pmatrix}$, $33°$

19. (a) $b = -\dfrac{1}{2}$ (b) $b = 0$ or $b = \dfrac{1}{2}$

   (c) $b = \dfrac{5}{2}$ or $b = 3$ (d) $b = \pm 4$

20. $(\boldsymbol{p} + \boldsymbol{q})(\boldsymbol{p} - \boldsymbol{q}) = \boldsymbol{p}^2 - \boldsymbol{q}^2 = |\boldsymbol{p}|^2 - |\boldsymbol{q}|^2 = 0$

21. $(-141.4, 141.4, 18)$

22. $t = 2$

23. $t = -\dfrac{1}{2}$

24. $t = 0$ or $t = \dfrac{1}{2}$

25. $90°$ or $\cos^{-1}\left(\dfrac{2}{\sqrt{6}}\right)$

26. $\boldsymbol{v} = |\boldsymbol{a}|\boldsymbol{b} + |\boldsymbol{b}|\boldsymbol{a}$; $\cos\alpha = \dfrac{\boldsymbol{a}\cdot\boldsymbol{b}}{|\boldsymbol{v}|} + \dfrac{|\boldsymbol{b}|\boldsymbol{a}}{|\boldsymbol{v}|}$; $\cos\beta = \dfrac{\boldsymbol{b}\cdot\boldsymbol{a}}{|\boldsymbol{v}|} + \dfrac{\boldsymbol{a}\boldsymbol{b}}{|\boldsymbol{v}|}$

27. $m = \dfrac{7}{4}, n = -\dfrac{1}{4}$

28. $\cos^2\dfrac{\pi}{4} + \cos^2\dfrac{\pi}{6} + \cos^2\dfrac{2\pi}{3} = \left(\dfrac{\sqrt{2}}{2}\right)^2 + \left(\dfrac{\sqrt{3}}{2}\right)^2 + \left(-\dfrac{1}{2}\right)^2$

$= \dfrac{1}{2} + \dfrac{3}{4} + \dfrac{1}{4} = \dfrac{3}{2} \neq 1$

29. $\dfrac{\pi}{3}, -\dfrac{2\pi}{3}$

30. $\cos^{-1}\left(\pm\dfrac{\sqrt{3}}{3}\right)$

31. $\pi - \alpha, \pi - \beta, \pi - \gamma$

32. $\pm\dfrac{1}{\sqrt{165}}(8\boldsymbol{i} + \boldsymbol{j} - 10\boldsymbol{k})$

Exercise 9.3

1. (a) $\boldsymbol{k} - \boldsymbol{j}$ (b) same

2. (a) $\boldsymbol{i} - \boldsymbol{k}$ (b) same

3. (a) $\boldsymbol{j} - \boldsymbol{i}$ (b)   same

4. $\boldsymbol{u} \times (\boldsymbol{v} + \boldsymbol{w}) = \begin{pmatrix} u_2(v_3 + w_3) - u_3(v_2 + w_2) \\ -u_1(v_3 + w_3) + u_3(v_1 + w_1) \\ u_1(v_2 + w_2) - u_2(v_1 + w_1) \end{pmatrix}$

$\boldsymbol{u} \times \boldsymbol{v} + \boldsymbol{u} \times \boldsymbol{w} = \begin{pmatrix} u_2(v_3 + w_3) - u_3(v_2 + w_2) \\ -u_1(v_3 + w_3) + u_3(v_1 + w_1) \\ u_1(v_2 + w_2) - u_2(v_1 + w_1) \end{pmatrix}$

Same for subtraction.

5. (a) $\begin{pmatrix} 13 \\ 0 \\ 13 \end{pmatrix}$ (b) $6\boldsymbol{i} - 8\boldsymbol{j} - 8\boldsymbol{k}$

   (c) $\begin{pmatrix} -5 \\ -1 \\ -7 \end{pmatrix}$ (d) $\boldsymbol{i} + \boldsymbol{j} - 3\boldsymbol{k}$

6. (a) $-2m^2 + 9m - 11$ (b) $-2m^2 + 9m - 11$

   (c) $-2m^2 + 9m - 11$

7. (a) $(-40, -115, 30)$ (b) $(-150, 60, 0)$

   (c) $(-80, -160, -640)$ (d) $(80, 160, 640)$

   (e) $(-40, -115, 30)$ (f) $(-150, 60, 0)$

8. $\dfrac{\sqrt{1774}}{1774}\begin{pmatrix} 19 \\ 33 \\ -18 \end{pmatrix}$

9. $\dfrac{\sqrt{6}}{6}\begin{pmatrix} 2 \\ 1 \\ 1 \end{pmatrix}$

10. (a) $\sqrt{209}$ (b) $\sqrt{139}$

11. $2\sqrt{43}$

12. Scalar triple product of $\left(\overrightarrow{PQ}, \overrightarrow{PR}, \overrightarrow{PS}\right) = \begin{vmatrix} 1 & 1 & -1 \\ 2 & 3 & -2 \\ 4 & 5 & -4 \end{vmatrix} = 0$

13. $m = 1$ or $m = \dfrac{11}{4}$

14. (a) $\dfrac{\sqrt{230}}{2}$ (b) $5\sqrt{29}$

15. (a) 128 (b) 21 (c) 1

16. (a) 78 (b) 63

17. (a) No   (b) Yes

18. (a) $-2, \dfrac{6}{5}$ (b) Not possible

19. (a) 49 (b) $7\sqrt{5}$

   (c) $\dfrac{7\sqrt{5}}{5}$ (d) $\sin^{-1}\left(\dfrac{7\sqrt{10}}{30}\right)$

20. (a) $\dfrac{49}{3}$, v(tetrahedron) $= \dfrac{1}{3}$v(parallelepiped)

   (b) $\dfrac{4}{3}$

21. $45°$

22. $\sqrt{|\boldsymbol{u}|^2 |\boldsymbol{v}|^2 - (\boldsymbol{uv})^2} = \cdots = |\boldsymbol{u}||\boldsymbol{v}|\sin\theta$

23. $\dfrac{|\overrightarrow{AP} \times \overrightarrow{AB}|}{|\overrightarrow{AB}|} = \dfrac{|\overrightarrow{AP}||\overrightarrow{AB}|\sin\theta}{|\overrightarrow{AB}|} = |\overrightarrow{AP}|\sin\theta = d$

24. (a) $\sqrt{\dfrac{564}{29}}$ (b) $\dfrac{6\sqrt{5}}{5}$ (c) $\sqrt{\dfrac{3}{2}}$

25. $2(\mathbf{u} \times \mathbf{v})$

26. $23(\mathbf{u} \times \mathbf{v})$

27. $(mp + nq)(\mathbf{u} \times \mathbf{v})$

28. (a) $o = \dfrac{1}{2}\left(\sqrt{(ab)^2 + (ac)^2 + (bc)^2}\right)$

   (b) $a = \dfrac{1}{2}ab$; $b = \dfrac{1}{2}bc$; $c = \dfrac{1}{2}ac$

   (c) result obvious

29. $\begin{pmatrix} 5t - \dfrac{1}{3} \\ -t - \dfrac{2}{3} \\ 3t \end{pmatrix}$

30. $\begin{vmatrix} \mathbf{i} & \mathbf{j} & \mathbf{k} \\ -1 & 2 & 3 \\ x & y & z \end{vmatrix} = \begin{pmatrix} 2z - 3y \\ z + 3x \\ -y - 2x \end{pmatrix} = \begin{pmatrix} 1 \\ 5 \\ 0 \end{pmatrix} \Rightarrow \begin{matrix} -3y + 2z - 1 \\ 3x + z = 5 \\ -2x - y = 0 \end{matrix} \Rightarrow$

No solutions.

Exercise 9.4

1. **(a)** $\mathbf{r} = \begin{pmatrix} -1 \\ 0 \\ 2 \end{pmatrix} + t\begin{pmatrix} 1 \\ 5 \\ -4 \end{pmatrix}$ $\begin{pmatrix} x \\ y \\ z \end{pmatrix} = \begin{pmatrix} -1 + t \\ 5t \\ 2 - 4t \end{pmatrix}$

   (b) $\mathbf{r} = \begin{pmatrix} 3 \\ -1 \\ 2 \end{pmatrix} + t\begin{pmatrix} 2 \\ 5 \\ -1 \end{pmatrix}$ $\begin{pmatrix} x \\ y \\ z \end{pmatrix} = \begin{pmatrix} 3 + 2t \\ -1 + 5t \\ 2 - t \end{pmatrix}$

   (c) $\mathbf{r} = \begin{pmatrix} 1 \\ -2 \\ 6 \end{pmatrix} + t\begin{pmatrix} 3 \\ 5 \\ -11 \end{pmatrix}$ $\begin{pmatrix} x \\ y \\ z \end{pmatrix} = \begin{pmatrix} 1 + 3t \\ -2 + 5t \\ 6 - 11t \end{pmatrix}$

2. **(a)** $\mathbf{r} = \begin{pmatrix} -1 \\ 4 \\ 2 \end{pmatrix} + t\begin{pmatrix} 8 \\ 1 \\ -2 \end{pmatrix}$ **(b)** $\mathbf{r} = \begin{pmatrix} 4 \\ 2 \\ -3 \end{pmatrix} + t\begin{pmatrix} -4 \\ -4 \\ 4 \end{pmatrix}$

   (c) $\mathbf{r} = \begin{pmatrix} 1 \\ 3 \\ -3 \end{pmatrix} + t\begin{pmatrix} 4 \\ -2 \\ 5 \end{pmatrix}$

3. **(a)** $\mathbf{r} = \begin{pmatrix} 3 \\ -2 \end{pmatrix} + t\begin{pmatrix} 2 \\ 3 \end{pmatrix}$ **(b)** $\mathbf{r} = \begin{pmatrix} 0 \\ -2 \end{pmatrix} + t\begin{pmatrix} 5 \\ 2 \end{pmatrix}$

4. $2x + 3y = 7$

5. $\mathbf{r} = 2\mathbf{i} - 3\mathbf{j} + t(4\mathbf{i} - 3\mathbf{j})$

6. $\mathbf{r} = \begin{pmatrix} -2 \\ 1 \\ 4 \end{pmatrix} + t\begin{pmatrix} 3 \\ -4 \\ 7 \end{pmatrix}$

7. **(a)** $(1, -1, 2)$ **(b)** $(-17, -1, 1)$
   (c) No **(d)** No

8. **(a)** $\mathbf{r} = \begin{pmatrix} 2 \\ -1 \end{pmatrix} + t\begin{pmatrix} 1 \\ 3 \end{pmatrix}$ $\begin{pmatrix} x \\ y \end{pmatrix} = \begin{pmatrix} 2 + t \\ -1 + 3t \end{pmatrix}$

   (b) $\mathbf{r} = \begin{pmatrix} 2 \\ -1 \end{pmatrix} + t\begin{pmatrix} -3 \\ 7 \end{pmatrix}$ $\begin{pmatrix} x \\ y \end{pmatrix} = \begin{pmatrix} 2 - 3t \\ -1 + 7t \end{pmatrix}$

   (c) $\mathbf{r} = \begin{pmatrix} 2 \\ -1 \end{pmatrix} + t\begin{pmatrix} 7 \\ 3 \end{pmatrix}$ $\begin{pmatrix} x \\ y \end{pmatrix} = \begin{pmatrix} 2 + 7t \\ -1 + 3t \end{pmatrix}$

   (d) $\mathbf{r} = \begin{pmatrix} 0 \\ 2 \end{pmatrix} + t\begin{pmatrix} 2 \\ -4 \end{pmatrix}$ $\begin{pmatrix} x \\ y \end{pmatrix} = \begin{pmatrix} 2t \\ 2 - 4t \end{pmatrix}$

9. **(a)** $t = \dfrac{3}{2}$ **(b)** No **(c)** $m = \dfrac{7}{2}$

10. **(a)** $\begin{pmatrix} 3 \\ -4 \end{pmatrix}, \begin{pmatrix} 7 \\ 24 \end{pmatrix}, 25$ **(b)** $\begin{pmatrix} -3 \\ 1 \end{pmatrix}, \begin{pmatrix} 5 \\ -12 \end{pmatrix}, 13$

   (c) $\begin{pmatrix} 5 \\ -2 \end{pmatrix}, \begin{pmatrix} 24 \\ -7 \end{pmatrix}, 25$

11. **(a)** $\begin{pmatrix} -96 \\ 128 \end{pmatrix}$ **(b)** $\begin{pmatrix} \dfrac{2040}{13} \\ \dfrac{-850}{13} \end{pmatrix}$

12. **(a)** $\begin{pmatrix} 24 \\ 18 \end{pmatrix}$ **(b)** $\mathbf{r} = \begin{pmatrix} 3 \\ 2 \end{pmatrix} + t\begin{pmatrix} 24 \\ 18 \end{pmatrix}$

   (c) in 10 minutes.

13. **(a)** $a = -3, b = -5$ **(b)** $-\dfrac{\sqrt{21}}{6}$

   (c) $\dfrac{\sqrt{15}}{6}, \dfrac{\sqrt{35}}{2}$

14. **(a)** $146.8°$ **(b)** 2.29

   (c) $L_1 : \mathbf{r} = \begin{pmatrix} 2 \\ -1 \\ 0 \end{pmatrix} + t\begin{pmatrix} 0 \\ 1 \\ 2 \end{pmatrix}; L_2 : \mathbf{r} = \begin{pmatrix} -1 \\ 1 \\ 1 \end{pmatrix} + s\begin{pmatrix} 1 \\ -3 \\ -2 \end{pmatrix}$

   No intersection.

15. **(a)** $\begin{pmatrix} x \\ y \\ z \end{pmatrix} = \begin{pmatrix} 1 + t \\ 3 - 2t \\ -17 + 5t \end{pmatrix}$ **(b)** $(4, -3, -2)$

16. **(a)** $\mathbf{r} = \begin{pmatrix} p \\ m \\ 0 \end{pmatrix} + t\begin{pmatrix} n \\ -m \end{pmatrix}$

   (b) **(i)** $bx - ay = bx_0 - ay_0$ **(ii)** slope $= \dfrac{b}{a}$

17. **(a)** $\mathbf{r} = \begin{pmatrix} t \\ t \\ 3t \end{pmatrix}, 0 \leq t \leq 1$

   (b) $\mathbf{r} = \begin{pmatrix} 2t-1 \\ t \\ 1-3t \end{pmatrix}, 0 \leq t \leq 1$

   (c) $\mathbf{r} = \begin{pmatrix} 1 - t \\ 3t \\ t-1 \end{pmatrix}, 0 \leq t \leq 1$

18. $\mathbf{r} = (2\mathbf{j} + 3\mathbf{k}) + 2t\mathbf{k};$ $\begin{cases} x = 0 \\ y = 2 \\ z = 3 + 2t \end{cases}$

19. $\mathbf{r} = (\mathbf{i} + 2\mathbf{j} - \mathbf{k}) + t(2\mathbf{i} - 3\mathbf{j} + \mathbf{k});$ $\begin{cases} x = 1 + 2t \\ y = 2 - 3t \\ z = -1 + t \end{cases}$

20. $\mathbf{r} = t(x_0\mathbf{i} + y_0\mathbf{j} + z_0\mathbf{k});$ $\begin{cases} x = tx_0 \\ y = ty_0 \\ z = tz_0 \end{cases}$

21. **(a)** $\mathbf{r} = (3\mathbf{i} + 2\mathbf{j} - 3\mathbf{k}) + t\mathbf{j};$ $\begin{cases} x = 3 \\ y = 2 + t \\ z = -3 \end{cases}$

   (b) $\mathbf{r} = (3\mathbf{i} + 2\mathbf{j} - 3\mathbf{k}) + t\mathbf{i};$ $\begin{cases} 3 + t \\ 2 \\ -3 \end{cases}$

22. $\dfrac{x - x_0}{x_0} = \dfrac{y - y_0}{y_0} = \dfrac{z - z_0}{z_0}$

23. **a** Intersect at $(1, 3, 1)$ **b** Parallel
    c Skew lines **d** Skew lines
    e Skew lines **f** Skew lines
    g $(4, -4, 8)$

24. $\left(\dfrac{16}{11}, \dfrac{35}{11}, \dfrac{13}{11} \right)$

25. $\left(\dfrac{1}{7}, \dfrac{25}{7}, \dfrac{37}{7} \right)$

26. $\left(\dfrac{43}{11}, \dfrac{58}{11}, -\dfrac{1}{11} \right)$

Exercise 9.5

1. B and C
2. None of them
3. **(a)** $\begin{pmatrix} 2 \\ -4 \\ 3 \end{pmatrix}\begin{pmatrix} x \\ y \\ z \end{pmatrix} = 26; 2x - 4y + 3z - 26 = 0$

   (b) $\begin{pmatrix} 2 \\ 0 \\ 3 \end{pmatrix}\begin{pmatrix} x \\ y \\ z \end{pmatrix} = -3; 2x + 3z + 3 = 0$

   (c) $\begin{pmatrix} 0 \\ 0 \\ 3 \end{pmatrix}\begin{pmatrix} x \\ y \\ z \end{pmatrix} = 3; 3z - 3 = 0;$

   $\mathbf{r} = \begin{pmatrix} 0 \\ 3 \\ 1 \end{pmatrix} + t\begin{pmatrix} 2 \\ -1 \\ 0 \end{pmatrix} + s\begin{pmatrix} 1 \\ 1 \\ 0 \end{pmatrix}$

   (d) $\begin{pmatrix} 5 \\ 1 \\ -2 \end{pmatrix}\begin{pmatrix} x \\ y \\ z \end{pmatrix} = 5; 5x + y - 2z - 5 = 0$

   (e) $\begin{pmatrix} 0 \\ 1 \\ -2 \end{pmatrix}\begin{pmatrix} x \\ y \\ z \end{pmatrix} = -2; y - 2z + 2 = 0$

   (f) $\begin{pmatrix} 1 \\ -6 \\ 2 \end{pmatrix}\begin{pmatrix} x \\ y \\ z \end{pmatrix} = 23; \mathbf{r} = \begin{pmatrix} 3 \\ -2 \\ 4 \end{pmatrix} + \lambda\begin{pmatrix} 2 \\ 1 \\ 2 \end{pmatrix} + \mu\begin{pmatrix} 2 \\ 0 \\ -1 \end{pmatrix}$

961

Answers

(g) $\begin{pmatrix} -2 \\ 2 \\ 1 \end{pmatrix}\begin{pmatrix} x \\ y \\ z \end{pmatrix} = -1; -2x + 2y + z = -1$

(b) $\begin{pmatrix} 18 \\ -3 \\ -11 \end{pmatrix}\begin{pmatrix} x \\ y \\ z \end{pmatrix} = 5 \ 18x - 3y - 11z = 5$

(i) $\begin{pmatrix} p \\ q \\ r \end{pmatrix}\begin{pmatrix} x \\ y \\ z \end{pmatrix} = p^2 + q^2 + r^2;$

$px + qy + rz = p^2 + q^2 + r^2$

(j) $4x - 2y + 7z = 14 ; \begin{pmatrix} 4 \\ -2 \\ 7 \end{pmatrix}\begin{pmatrix} x \\ y \\ z \end{pmatrix} = 14;$

$r = \begin{pmatrix} 1 \\ 2 \\ 2 \end{pmatrix} + m\begin{pmatrix} 2 \\ -3 \\ -2 \end{pmatrix} + n\begin{pmatrix} 4 \\ 1 \\ -2 \end{pmatrix}$

(k) $8x + 17y - 5z + 8 = 0 ; \begin{pmatrix} 8 \\ 17 \\ -5 \end{pmatrix}\begin{pmatrix} x \\ y \\ z \end{pmatrix} = -8;$

$r = \begin{pmatrix} 2 \\ -2 \\ -2 \end{pmatrix} + s\begin{pmatrix} 1 \\ 1 \\ 5 \end{pmatrix} + t\begin{pmatrix} -3 \\ 2 \\ 2 \end{pmatrix}$

(l) $\begin{pmatrix} 1 \\ -1 \\ 0 \end{pmatrix}\begin{pmatrix} x \\ y \\ z \end{pmatrix} = 3; x - y = 3$

(m) $\begin{pmatrix} 30 \\ 1 \\ -23 \end{pmatrix}\begin{pmatrix} x \\ y \\ z \end{pmatrix} = -86; 30x + y - 23z + 86 = 0$

(n) $\begin{pmatrix} 1 \\ 0 \\ -1 \end{pmatrix}\begin{pmatrix} x \\ y \\ z \end{pmatrix} = 1; x - z = 1;$

$r = \begin{pmatrix} 1 \\ 1 \\ 0 \end{pmatrix} + m\begin{pmatrix} 1 \\ 0 \\ 1 \end{pmatrix} + n\begin{pmatrix} 1 \\ -1 \\ 1 \end{pmatrix}$

All answers for 4(a), (d), (e) and (f) are to the nearest degree

4. (a) 64°   **(b)** 90°   **(c)** 45°
(d) 50°   **(e)** 15°   **(f)** 55°

5. (a) $(3, 6, -10)$ **(b)** $(2, -2, 6)$
(c) No intersection   **(d)** Plane contains line

6. (a) $r = \begin{pmatrix} 10 \\ -7 \\ 0 \end{pmatrix} + t\begin{pmatrix} 0 \\ -1 \\ 1 \end{pmatrix}$ **(b)** $r = \begin{pmatrix} 3 \\ 1 \\ 0 \end{pmatrix} + t\begin{pmatrix} 0 \\ 1 \\ 1 \end{pmatrix}$

(c) no intersection   **(d)** $r = \begin{pmatrix} \frac{7}{5} \\ 6 \\ \frac{5}{0} \end{pmatrix} + t\begin{pmatrix} 0 \\ -1 \\ 1 \end{pmatrix}$

7. $\begin{pmatrix} 1 \\ -1 \\ 1 \end{pmatrix}\begin{pmatrix} x \\ y \\ z \end{pmatrix} = 0$

8. $x + 6y + z = 16$

9. $\left(\frac{31}{21}, \frac{37}{21}, \frac{85}{21}\right)$

10. $\begin{pmatrix} 10 \\ 1 \\ -8 \end{pmatrix}\begin{pmatrix} x \\ y \\ z \end{pmatrix} = -32; 10x + y - 8z + 32 = 0$

11. $\begin{pmatrix} 4 \\ -3 \\ 2 \end{pmatrix}\begin{pmatrix} x \\ y \\ z \end{pmatrix} = 5; 4x - 3y + 2z - 5 = 0$

12. $(BC)x + (AC)y + (AB)z = ABC$

13. $r = \begin{pmatrix} 4 \\ -3 \\ -1 \end{pmatrix} + r\begin{pmatrix} 2 \\ -3 \\ 4 \end{pmatrix} + s\begin{pmatrix} 4 \\ 0 \\ -3 \end{pmatrix}$

14. $r = \begin{pmatrix} 2 \\ 3 \\ 0 \end{pmatrix} + m\begin{pmatrix} 2 \\ -3 \\ 4 \end{pmatrix} + n\begin{pmatrix} 1 \\ -2 \\ 1 \end{pmatrix}$

Chapter 9 practice questions

1. (a) $5i + 12j$ **(b)** $10i + 24j$

2. (a) $|\overrightarrow{OA}| = |\overrightarrow{OB}| = |\overrightarrow{OC}| = 6$
(b) $\overrightarrow{AC} = \begin{pmatrix} -1 \\ \sqrt{11} \end{pmatrix}$ **(c)** $\frac{\sqrt{3}}{6}$ **(d)** $3\sqrt{11}$

3. $a = 2, b = 8$

4. (a) 30 km/h, 39.4 km/h
(b) $(9, 12), (18, -8), \sqrt{481} = 21.9$ km   **(c)** 7 a.m.
(d) 24.4 km   **(e)** 54 minutes

5. (a) $\sqrt{16^2 + 12^2} = 20; \sqrt{12^2 + (-5)^2} = 13.$
(b) $y - 12 = \frac{-5}{12}(x - 16)$ **(c)** 90°
(d) (i) $12x - 5y = 301$ **(ii)** $(28, 7)$
(e) No. Air 1 passes the point at 13:00 but Air 2 at 14:00.

6. $2x + 3y = 5$

7. (a) $(6, 20)$ **(b)** $(6, -8); 10 \text{ km h}^{-1}$
(c) $4x + 3y = 84$ **(d)** collide at 15:00
(e) $\begin{pmatrix} x \\ y \end{pmatrix} = \begin{pmatrix} 18 \\ 4 \end{pmatrix} + (t - 1)\begin{pmatrix} 5 \\ 12 \end{pmatrix}$ **(f)** 26 km

8. (a) $2x^2 + 7x - 15 = 0$
(b) $x = \frac{3}{2}, x = -5.$

9. (a) (i) $\frac{1}{\sqrt{240^2 + 70^2}}\begin{pmatrix} 240 \\ 70 \end{pmatrix};$
(ii) $\begin{pmatrix} 288 \\ 84 \end{pmatrix};$ **(iii)** 50 minutes.
(b) 26.6°
(c) (i) $\begin{pmatrix} 99 \\ 168 \end{pmatrix};$ **(ii)** See Worked Solutions
(iii) $XY = 75$
(d) 180 km

10. (a) $\overrightarrow{ST} = \begin{pmatrix} 9 \\ 9 \end{pmatrix}, V(-4, 6)$ **(b)** $r = \begin{pmatrix} -4 \\ 6 \end{pmatrix} + \lambda\begin{pmatrix} 1 \\ 1 \end{pmatrix}$
(c) $\lambda = 5$
(d) (i) $a = 5$ **(ii)** 157°

11. $\begin{pmatrix} 2 \\ 3 \end{pmatrix}$

12. (a) $(3, -2)$ **(b)** $\overrightarrow{PQ} = \begin{pmatrix} 3 \\ -2 \end{pmatrix} \Rightarrow \cos\theta = -\frac{\begin{pmatrix} -7 \\ -3 \end{pmatrix} \cdot \begin{pmatrix} 3 \\ -2 \end{pmatrix}}{\sqrt{58}\sqrt{13}}$
(c) (i) supplementary angles
(ii) same sine as θ **(iii)** 23 square units.

13. (a) $\overrightarrow{OB} = \begin{pmatrix} -1 \\ 7 \end{pmatrix}; \overrightarrow{OC} = \begin{pmatrix} 8 \\ 9 \end{pmatrix}$ **(b)** $d = 11$
(c) $\overrightarrow{BD} = \begin{pmatrix} 12 \\ -3 \end{pmatrix}$
(d) (i) $\begin{pmatrix} x \\ y \end{pmatrix} = \begin{pmatrix} -1 \\ 7 \end{pmatrix} + t\begin{pmatrix} 12 \\ -3 \end{pmatrix}$ **(ii)** $t = 0$
(e) $t = \frac{2}{3}$
(f) scalar product = 0

14. (a) $\overrightarrow{BC} = -6i - 2j; \overrightarrow{OD} = -2i$ **(b)** 82.9°
(c) $r = i - 3j + t(2i + 7j)$ **(d)** $15i + 46j$

15. (a) $b = \begin{pmatrix} 0 \\ 0 \\ 5 \end{pmatrix} + 18 \times \frac{1}{\sqrt{3^2 + 4^2}}\begin{pmatrix} 3 \\ 4 \\ 0 \end{pmatrix}$
(b) (i) $(49, 32, 0)$ **(ii)** 53.7 km h^{-1}
(c) (i) $\frac{5}{6}$ hours   **(ii)** $(9, 12, 5)$

16. (a) (i) $\overrightarrow{AB} = \begin{pmatrix} 800 \\ 600 \end{pmatrix}$ **(ii)** $\frac{\overrightarrow{AB}}{\sqrt{800^2 + 600^2}}$
(b) (i) $250\begin{pmatrix} 0.8 \\ 0.6 \end{pmatrix}$ **(ii)** $\begin{pmatrix} -400 \\ -50 \end{pmatrix};$ **(iii)** at 16:00 hr.
(c) 27.8 km.

17. (a) $c = 1$ **(b)** $3i + 3k$
(c) $r = 3(1-t)i + (3 - t)j + (5 + 3t)k$
(d) $9x - 15y + 4z - 2 = 0$ **(e)** $\dfrac{15}{\sqrt{322}}$

18. (a) $\overrightarrow{AB} = -i - 3j + k; \overrightarrow{BC} = i + j$
(b) $-i + j + 2k$ **(c)** $\dfrac{\sqrt{6}}{2}$
(d) $-x + y + 2z = 3$ **(e)** $\begin{cases} 2 - t \\ -1 + t \\ -6 + 2t \end{cases}$
(f) $3\sqrt{6}$ **(g)** $\dfrac{1}{\sqrt{6}}(-i + j + 2k)$
(h) $E(-4,5,6)$

19. (a) $\begin{vmatrix} i & j & k \\ 1 & 2 & 3 \\ 2 & -1 & 2 \end{vmatrix} = 7i + 4j - 5k$
(b) Normal Vector $= 7i + 4j - 5k,$
$w = \lambda u + \mu v = (\lambda + 2\mu)i + (2\lambda - \mu)j + (3\lambda + 2\mu)k$
$(7i + 4j - 5k) \times ((\lambda + 2\mu)i + (2\lambda - \mu)j + (3\lambda + 2\mu)$
$k) = 0$

20. (a) $P(4, 0, -3), Q(3, 3, 0), R(3, 1, 1), S(5, 2, -1)$
(b) $3x + 2y + 4z = 0$ **(c)** 15

21. (a) $147°$ **(b)** $\dfrac{\sqrt{2}}{2} = 2.29$
(c) (i) $L_1 : \begin{cases} 2 \\ -1 + \lambda; L_2 : \begin{cases} -1 + \mu \\ 1 - 3\mu \\ 1 - 2\mu \end{cases} \\ 2\lambda \end{cases}$
(ii) no solution
(d) $\dfrac{9}{\sqrt{21}}$

22. (a) $(1, -1, 2)$ **(b)** $11i - 7j - 5k$
(c) $\mathbf{v.u} = 0$ **(d)** $r = \begin{pmatrix} 1 \\ -1 \\ 2 \end{pmatrix} + t\begin{pmatrix} 6 \\ 13 \\ -5 \end{pmatrix}$

23. (a) (i) $-5i + 3j + k$ **(ii)** $\dfrac{\sqrt{35}}{2}$
(b) (i) $-5x + 3y + z = 5$ **(ii)** $\dfrac{x - 5}{-5} = \dfrac{y + 2}{3} = z - 1$
(c) $(0,1,2)$ **(d)** $\sqrt{35}$

24. (a) $x - 2 = y - 5 = z + 1$ **(b)** $\left(\dfrac{1}{3}, \dfrac{10}{3}, -\dfrac{8}{3}\right)$
(c) $A'\left(-\dfrac{4}{3}, \dfrac{5}{3}, -\dfrac{13}{3}\right)$ **(d)** $\dfrac{\sqrt{654}}{3}$

25. (a) $3x - 4y + z = 6$
(b) (i) See Worked Solutions
(ii) $r = \begin{pmatrix} 1 \\ 2 \\ 11 \end{pmatrix} + t\begin{pmatrix} 1 \\ 4 \\ 13 \end{pmatrix}$
(c) $53.7°$

26. (a) $(3\mu - 2, \mu, 9 - 2\mu)$
(b) (i) $r = \begin{pmatrix} 4 \\ 0 \\ -3 \end{pmatrix} + \lambda\begin{pmatrix} 3 \\ 1 \\ -2 \end{pmatrix};$
(ii) $\overrightarrow{PM} = \begin{pmatrix} 3\mu - 6 \\ \mu \\ 12 - 2\mu \end{pmatrix}$
(c) (i) $\mu = 3$ **(ii)** $3\sqrt{6}$
(d) $2x - 4y + z = 5$ **(e)** Substitute in both

27. (a) $(1, -1, 2)$ **(b)** $2x - y + z = 5$
(c) $(3, 1, 3)$ and $(1, 2, 2)$

28. (a) (i) $\lambda = \mu$ **(ii)** $r = \begin{pmatrix} 2 \\ 1 \\ 1 \end{pmatrix} + t\begin{pmatrix} -1 \\ -2 \\ -1 \end{pmatrix}$
(b) $3x - 2y + z = 5$
(c) $r = \begin{pmatrix} 2 \\ 1 \\ 1 \end{pmatrix} + t\begin{pmatrix} -1 \\ -2 \\ -1 \end{pmatrix}$

29. (a) $\cos\theta = \sqrt{\dfrac{9}{13}}$
(b) (i) $\begin{pmatrix} 4 \\ 1 \\ 1 \end{pmatrix} \cdot \begin{pmatrix} -1 \\ 2 \\ 2 \end{pmatrix} = \begin{pmatrix} 4 \\ 3 \\ -1 \end{pmatrix} \cdot \begin{pmatrix} -1 \\ 2 \\ 2 \end{pmatrix} = 0$
(ii) $4 + 0 + 4 = 8, 4 + 0 - 4 = 0$
(iii) $r = \begin{pmatrix} 1 \\ 0 \\ 4 \end{pmatrix} + t\begin{pmatrix} -1 \\ 2 \\ 2 \end{pmatrix}$
(c) $a = 1, b = 3$
(d) $AB = \sqrt{0^2 + 3^2 + (-3)^2}$
(e) $P(1 - \sqrt{2}, 2\sqrt{2}, 4 + 2\sqrt{2})$ or $P(1 + \sqrt{2}, -2\sqrt{2}, 4 - 2\sqrt{2})$

30. (a) $(p, 2, 1)$ in vector form cannot be a multiple of $(1, 1, 3)$ in vector form
(b) $p = -5$ and $q = 4$.
(c) (i) If α is the acute angle between π and L, then
$\alpha = \sin\alpha = \dfrac{1}{\sqrt{11}} = \dfrac{\pi \cdot L}{|n||L|} = \dfrac{p + 5}{\sqrt{11}\sqrt{p^2 + 5}}$
(ii) $q = 10$

Chapter 10

Exercise 10.1

Note: Some answers may differ from one person to another because of variations in graph accuracy.

1. (a) Student, all students in a community, random sample of few students, qualitative
(b) Exam, 10th-grade students in a country, a sample from a few schools, quantitative.
(c) Newborns, heights of newborns in a city, sample from a few hospitals, quantitative
(d) Children, eye colour of children in a city, sample of children at schools, qualitative
(e) Working persons, commuters in a city, sample of a few districts, quantitative
(f) Country leaders, sample of few presidents, qualitative
(g) Students, origin countries of a group of international school students, qualitative

2. Answers are not unique.
(a) Skewed to the right as few players score very high
(b) Symmetric
(c) Skewed to the right
(d) Unimodal, or bi-modal, symmetric or skewed etc.

3. (a) (b) Quantitative
(c) (d) Qualitative

4. (a) Discrete
(b) Continuous
(c) Continuous
(d) Discrete
(e) Continuous
(f) Discrete (debatable!)

Answers

5.

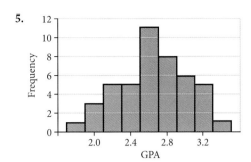

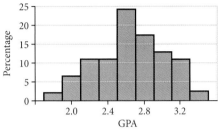

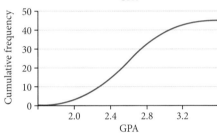

Relatively symmetric. No outliers.

6.

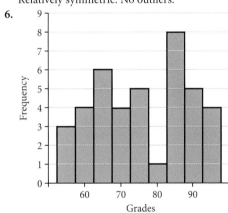

The grades appear to be divided into two groups, one with mode around 65 and the other around 85. No outliers are detected.

7. (a)

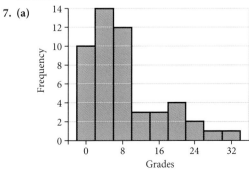

(b) The data is skewed to the right.

(c)

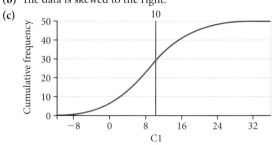

More than 35 out of 50 will lose the licence, about 70%.

8. (a)

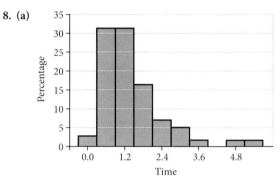

(b)

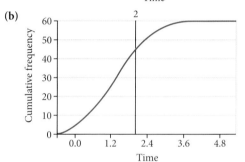

About 10 customers will have to wait more than 2 minutes.

9. (a) Skewed to the right, there is a mode at about 7 days stay, and a few extremes that stayed more than 20 days. A good proportion stayed for about 3 days.

(b)

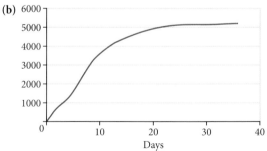

(c) Approximately 35% of the patients

10. (a) 40 minutes

(b) Approximately 30%

(c)
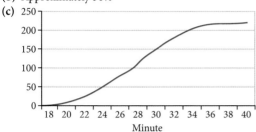

11. (a)

Speed	Frequency
$60 \leqslant$ speed < 75	20
$75 \leqslant$ speed < 90	70
$90 \leqslant$ speed < 105	110
$105 \leqslant$ speed < 120	150
$120 \leqslant$ speed < 135	40
$135 \leqslant$ speed < 150	10

(b), (c)
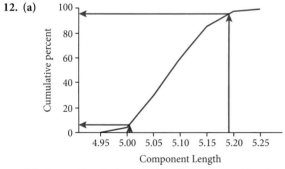

(d) As you see above, $\dfrac{25}{400} = 6.25\%$

12. (a)
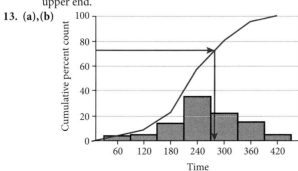

(b) about 5% at the lower end and also about 5% at the upper end.

13. (a),(b)
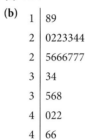

(c) As you see from diagram, about 275 seconds.

Exercise 10.2

1. (a) 6   **(b)** 6

(c) It appears to be symmetric as the mean and median are the same. A histogram supports this view.

2. (a) 7.8   **(b)** 7.5   **(c)** 7 or 8

3. Average = 1.16, median = 1. Median is more appropriate as the data is skewed to the right.

4. Mean = 307 036, median = 288 521. There are extreme values and it is skewed to the right. Median is more appropriate.

5. Mean = median = 430. It appears to be symmetric and hence either measure would be fine.

6. (a) €49.56   **(b)** €49.93

7. 2.05

8. (a) 9.96

(b)

1	89
2	0223344
2	5666777
3	34
3	568
4	022
4	66

Median is 27

(c)
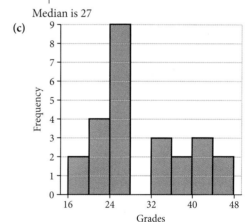

(d)
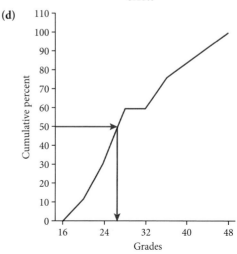

The median is approximately 27.

Answers

9. (a)

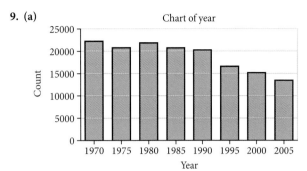

Chart of year

There appears to be a decline in the total number of injuries.

(b)

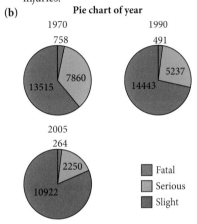

Pie chart of year

1970
758
7860
13515

1990
491
5237
14443

2005
264
2250
10922

Fatal
Serious
Slight

10. (a)

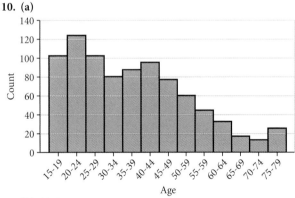

(b) 38.1

(c)

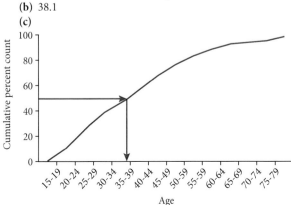

From the graph, the median is approximately 38.

11. Median = Approx. 7.5 days ; Mean = 9 days

12. Median = approx. 28 minutes; Mean = 27.7 minutes

13. Median = approx. 105; Mean = $103 \, \text{km h}^{-1}$

14. Median = approx. 5.075; Mean = 5.07

15. Median = approx. 225; Mean = 228.6

16. (a) 41.6   **(b)** 103.2

17. (a) 55.4   **(b)** 60.2

Exercise 10.3

1. (a) Mean = 71.47, S_{n-1} = 7.04

(b)

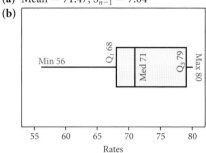

(c) No outliers

2. (a) Mean = 162.6, S_{n-1} = 23.35

(b) , Median = 162.5

11	79
12	567
13	089
14	123679
15	033445689
16	02334568
17	1344789
18	02255779
19	8
20	9
21	08

(c)

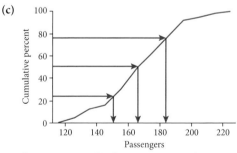

Q_1 approx. 150, Median approx. 165, Q_3 approx 182

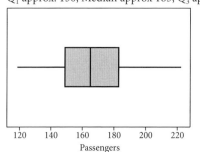

(d) Real Q_1 = 146.75, Q_3 = 179.25, IQR = 32.5. No outliers

(e) $\bar{x} \pm 3 s_{n-1}$ = (92.55, 232.65) No outliers

3. (a) and **(b)**

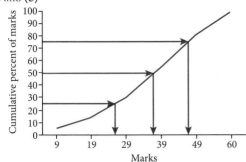

Q_1 = approx 26, Med = approx 37.5, Q_3 = approx 46.5

4. (a)

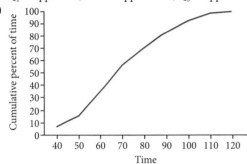

(b) Median = 63, IQR = 30   **(c)** about 72

5. 29.6

6. (a) Mean = 72.1, S_{n-1} = 6.1

(b) New mean = 85.1, S will not change.

7. (a)

Number of empty seats	$x \leqslant 10$	$x \leqslant 20$	$x \leqslant 30$	$x \leqslant 40$	$x \leqslant 50$	$x \leqslant 60$	$x \leqslant 70$	$x \leqslant 80$	$x \leqslant 90$	$x \leqslant 100$
Days	15	65	165	335	595	815	905	950	980	1000

(b)

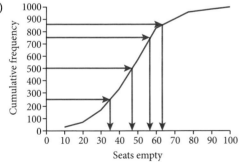

(c) (i) Median = 48   **(ii)** LQ = 36, UQ =56.5, IQR =20.5
(iii) 240 days   **(iv)** 64 seats empty

8. (a)

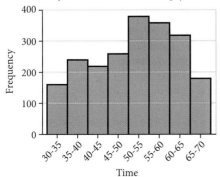

(b)

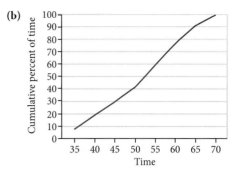

Median = 53, IQR = 19

(c) mean = 50.1 and S_{n-1} = 12.7

9. (a) Q_1 = 165.1, median = 167.64, Q_3 = 177.8, minimum = 152, maximum = 193

(b)

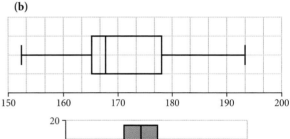

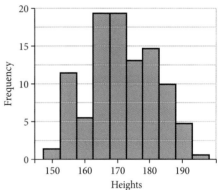

(c) Mean = 170.5, standard deviation = 9.61

(d) The heights are widely spread from very short to very tall players. Heights are slightly skewed to the right, bimodal at 165 and 170, no apparent outliers. The heights between the first quartile and the median are closer together than the rest of the data.

(e)

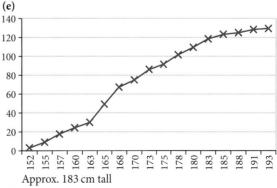

Approx. 183 cm tall

(f) 171.3

10. (a) 12   **(b)** 12   **(c)** 111

11. (a) 31   **(b)** Increase

12. 36.7

967

Answers

13. $x = 6$, $y = 11$

14. Mean = 11.12, Variance = 24.6 (calculating $\sigma^2 = 23.6$)

15. Std. dev. = 6.6, IQR ≈ 8

16. Std. dev ≈ 4.5, IQR ≈ 6

17. Std. dev ≈ 16.7, IQR ≈ 23

18. Std. dev ≈ 0.056, IQR ≈ 0.09

19. Std. dev ≈ 82.3, IQR ≈ 100

Exercise 10.4

1.

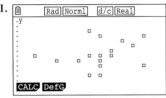

It appears that the data have a weak positive linear relationship. The correlation coefficient is 0.26 which confirms the weakness of the relationship.
The regression equation is: $y = 6.56 + 0.29\,x$. For every change of 1 unit in the x-values, the y-values will change, on average, by 0.29.

2. (a)

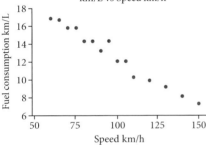

Scatterplot of Fuel Consumption km/L vs Speed km/h

(b) We chose the speed as the explanatory variable because the car must first run to cause a fuel consumption. Hence the speed helps explain the fuel consumption. The relationship appears to be negatively sloped because the consumption is measure by the distance travelled per liter of fuel.

(c) The relationship appears to be a relatively strong negative one without any apparent outliers. The correlation coefficient is -0.986 which is very close to -1. A very strong relationship.

(d) The regression equation is: Fuel cons.km/L = 24.1 − 0.116 Speed km h^{-1}. For every increase of 1 km h^{-1} in speed, the average number of km per liter will decrease by 0.116 km/L. i.e. consumption will increase.

3. (a)

Scatterplot of PPP vs GNI/Cap

(b) The relationship appears to be a positive one except for an outlier which can be traced to be Singapore. We chose the explanatory variable to be the Income because the income level dictates how willing are people to pay for goods.

(c) The relationship is relatively strong (weakened by Singapore's numbers). The correlation coefficient is 0.621. If we remove Singapore's data, then it becomes 0.886.

(d) The regression equation is: $PPP = 24383 + 0.351$ GNI/cap. For every increase of $1 in GNI/cap, the PPP will increase, on average by $0.351.

4. (a)

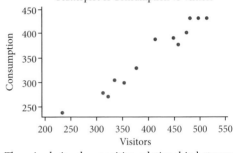

Scatterplot of Consumption vs Visitors

(b) There is obviously a positive relationship between the number of visitors and consumption. As the number of visitors increases the consumption will also increase.

(c) The relationship seems to be strong and there is an absence of outliers. The correlation coefficient is 0.978 which is very close to 1.

(d) The regression equation is:
Consumption = 40.0 + 0.777 Visitors. For every increase of 1 visitor, we expect, on average, that consumption will increase by 0.777.

5. See parts **(d)** in each of questions 1 to 4.

6.

The scatter plot shows a strong positive relationship. That is the higher the 'Before' score the higher the 'After' score is. The regression equation is: After = 20.2 + 1.03 Before. This means that, on average, for every change of 1 mark on the 'Before' test, the 'After' test is expected to change by 1.03. The correlation coefficient is 0.97 indicating a very strong linear relationship. For a student with 60 score on the 'Before' test, the model predicts, on average, a score of 81.90 on the 'After' test.

7. (a)

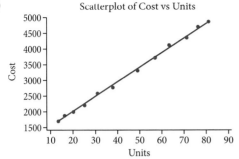

Scatterplot of Cost vs Units

(b) The regression equation is: Cost = 1066 + 47.1 units.

(c) For every increase of 1000 units in production, the cost, on average, will increase by 47 100 Euros. The correlation coefficient is 0.999, which is almost perfect association.

This is a strong linear relationship.

(d) Let number of 1000 units be x, then:

Cost = 1066 + 47.1x

$\dfrac{\text{Cost}}{x} = \dfrac{1066}{x} + 47.1 =$ cost per unit. If the cost is 105, then $105 = \dfrac{1066}{x} + 47.1 \Rightarrow x = 18.411$

Thus the number of units will be 18 400 units.

8. (a) $r = 0.493$. This is a relatively weak correlation between the two scores.

(b) The regression equation is:
Economics = 3.3 + 0.37 Physics

(c) 1.8. This shows the potential dangers of extrapolation.

9 (a)

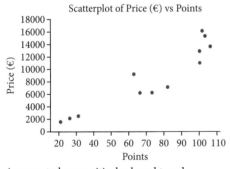

Scatterplot of Price (€) vs Points

Appears to be a positively sloped trend.

(b) The regression equation is:
Price (€) = −2689 + 154 points.

(c) The intercept is meaningless as zero is not in the domain of the explanatory variable. On average, for every increase of 1 point, we expect the price to increase by 154 Euros.

(d) $r = 0.93$ indicating a strong association between points and price.

(e) The average price of a 63-point diamond is predicted to be 7024 Euros.

(f) Residual = 2093.

Chapter 10 practice questions

1. (a) 12

(b) $\sqrt{30.83}$

2. 4

3. (a)

Time	1.6	2.1	2.6	3.1	3.6	4.1	4.6	5.1	5.6	6.1	6.6
Frequency	2	2	6	4	11	10	5	5	3	2	0

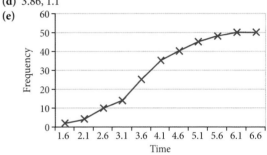

(b) 86%

(c) approx. 4

(d) 3.86, 1.1

(e)

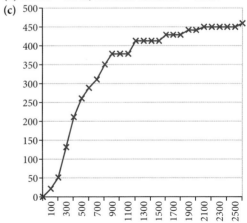

(f) Minimum = 1.6, Q_1 = 3, median = 4, Q_3 = 4.5, maximum = 6.2

4. (a) Median and IQR as the data is skewed with outliers.

(b) Mean = 682.6, standard deviation = 536.2

(c)

(d) Q_1 = 300, median = 500, Q_3 = 800, IQR = 500

(e) There are a few outliers on the right side. Outliers lie above Q_3 + 1.5IQR = 1550

(f) Data is skewed to the right, with several outliers from 1600 onwards. It is bimodal at 300–400.

5. (a) Spain, Spain **(b)** France

(c) On average, it appears that France produces the more expensive wines as 50% of its wines are more expensive than most of the wines from the other countries. Italy's prices seem to be symmetric while Frances' prices are skewed to the left. Spain has the widest range of prices.

Answers

6. (a) Mean = 52.65, standard deviation = 7.66

(b) Median = 51.34, IQR = 2.65

(c) Apparently, the data is skewed to the right with a clear outlier of 112.72! This outlier pulled the value of the mean to the right and increased the spread of the data. The median and IQR are not influenced by the extreme value.

7. (a) The distribution does not appear to be symmetric as the mean is less than the median, the lower whisker is longer than the upper one and the distance between Q_1 and the median is larger than the distance between the median and Q_3. Left skewed.

(b) There are no outliers as $Q_1 - 1.5\text{IQR} = 37 < 42$ and $Q_3 + 1.5\text{IQR} = 99 > 86$.

(c)

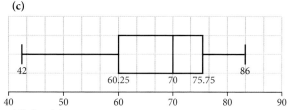

(d) See **(a)**

8. (a) 225

(b) $Q_1 = 205$, $Q_3 = 255$, 90th percentile = 300, 10th percentile = 190

(c) IQR = 50, since $Q_1 - 1.5\text{IQR} - 130 >$ minimum and $Q_3 + 1.5\text{IQR} = 330 < 400$ then there are outliers on both sides.

(d)

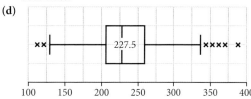

(e) The distribution has many outliers. Apparently skewed to the right with more outliers there. The middle 50% seem to be very close together while the whiskers appear to be quite spread.

9. (a)

Speed	Frequency
26–30	10
31–34	16
35–38	30
39–42	23
43–46	12
47–50	18
51–54	1

(b)

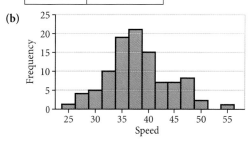

Data is relatively symmetric with possible outlier at 55. The mode is approximately 37

Histogram created from table:

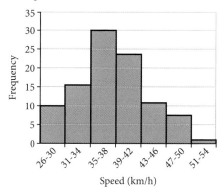

(c) Mean = 37.0, standard deviation = 7.9

(d)

Speed	Cumulative frequency
26–30	10
31–34	26
35–38	56
39–42	79
43–46	91
47–50	99
51–54	100

(e) Median = 37.6, $Q_1 = 34.5$, $Q_3 = 41.3$, IQR = 6.8

(f) There are outliers on the right since $Q_3 + 1.5\text{IQR} = 51.5 <$ maximum = 54.

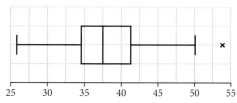

10. (a) Mean = 1846.9, median = 1898.6, standard deviation = 233.8, $Q_1 = 1711.8$, $Q_3 = 2031.3$, IQR = 319.5

(b) $Q_1 - 1.5\text{IQR} = 1232.55 >$ minimum, so there is an outlier on the left.

(c)

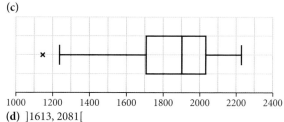

(d)]1613, 2081[

(e) The mean and standard deviation will get larger. The rest will not change much.

11. (a) 49.6 minutes   **(b)** 48.9 minutes

12. (a)

⩽10	⩽20	⩽30	⩽40	⩽50	⩽60	⩽70	⩽80	⩽90	⩽100
30	130	330	670	1190	1630	1810	1900	1960	2000

(b)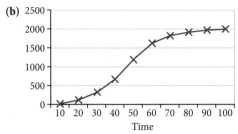

(c) (i) 47   **(ii)** About 500   **(iii)** Above 60

13. 1.74

14. (a) $\mu = 12$ **(b)** Standard deviation $= 5$

15. $k = 4$

16. (a) 97.2

(b)

30	60	90	120	150	180	210	240
5	20	53	74	85	92	97	100

(c)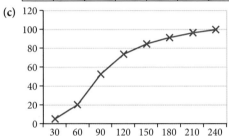

(d) Median $= 82$

     $Q_1 = 64$

     $Q_3 = 120$

17. (a) (i) 10   **(ii)** 24

(b) Mean $= 63$, standard deviation $= 20.5$

(c) Skew to the left   **(d)** 65

18. (a) 7.41

(b)

Weight	Number of packets
$w \leqslant 85$	5
$w \leqslant 90$	15
$w \leqslant 95$	30
$w \leqslant 100$	56
$w \leqslant 105$	69
$w \leqslant 110$	76
$w \leqslant 115$	80

(c) (i) Median $= 97$ **(ii)** $Q_3 = 101$

(d) 0   **(e)** 0.3

19. (a) 98.2

(b) (i) $m = 165, n = 275$

   (ii)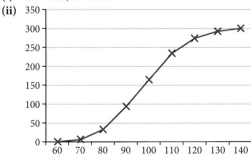

(c) (i) 67%   **(ii)** 115

20. (a) (i) 24   **(ii)** 158

(b) 40   **(c)** 7%

21. $a = 3$

22. (a)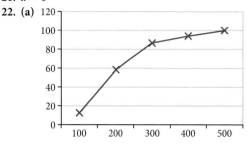

(b) IQR $= 110$ **(c)** $m = 7, n = 6$ **(d)** 199

(e) (i) 10   **(ii)** 0.36

23. (a) (i) 20   **(ii)** 24   **(b)** 10.

24. (a)

Mark	$[0, 20[$	$[20, 40[$	$[40, 60[$	$[60, 80[$	$[80, 100[$
Number of students	22	50	66	42	20

(b) Pass mark $= 43\%$

25. (a) 183   **(b)** 14

26. $a = 3, b = 7, c = 11, d = 11$.

27. (a) 100   **(b)** $a = 55, b = 75$

28. $x = 4, y = 10$.

29. (a) Apparently linear with two possible outliers: (7, 54) and (28, 78). It appears to be a linear relationship.

Scatterplot of Yield vs Rainfall

(b) 0.853. A relatively strong positive linear relationship.

(c) Yield $= 40.5 + 1.78$ Rainfall. On average, a change of 1 cm in rainfall corresponds to a change of 1.78 kg change in crop. The intercept is not useful in this case since 0 is not in the domain of the explanatory variable.

(d) 74.3   **(e)** 38°

Chapter 11

Exercise 11.1

Note: Some answers may differ from one student to another because of variations in graph accuracy.

1. (a) {left handed, right handed}

(b) all real numbers from (say) 50 cm to 210 cm.

(c) all real numbers from 0 to 720 (say).

2. {(1, h), (2, h), …, (1, t), …, (6, t)}

3. (a) {(1, Hearts), …, (King, Hearts), (1, Spades), …}

(b) {[(1, hearts), (King, Diamonds)], …,[(1, Spades), (10, Diamonds)],…}

(c) a: 52, b: 1326

Answers

4. (a) 0.47

(b) anywhere from 0 to 20.

(c) 10000.

5. (a) {(1, 1), (1,2), ..., (4, 4)}

(b) {3, 4, ..., 9}

6. (a) {(b, b), (b, g), (b, y), (g, b), (g, g), (g, y), (y, b), (y, g), (y, y)}

(b) {(y, y), (y, b), (y, g)}

(c) {(b, b), (g, g), (y, y)}

7. (a) {(b, g), (b, y), (g, b), (g, y), (y, b), (y, g)}

(b) {(y, b), (y, g)}

(c) ∅

8. (a) {(t, t, t), (t, t, h), (t, h, t), (h, t, t), (h, t, h), (h, h, t), (t, h, h), (h, h, h)}

(b) {(h, t, h), (h, h, t), (t, h, h), (h, h, h)}

9. {(I, fly), (I, dr), (I, tr), (H,dr), (H, b)}, {(I, fly)}

10. (a) {(1, g), (1, f), ..., (0, c}

(b) {(0, c), (0, s)}

(c) {(1, g), (1, f), (0, g), (0, f)}

(d) {(1, g), (1, f), (1, s), (1, c)}

11. (a) {(G_1, K_1, M_1), (G_1, K_2, M_1), (G_1, K_1, M_2), ...}

(b) A = all triplets containing G_2; B = all triplets Not containing K_1; C = all triplets containing M_2.

(c) (i) A∪B = All males or persons who drink

(ii) A∩C = All single males

(iii) C' = All non-single persons

(iv) A∩B∩C = All single males who drink

(v) A'∩B = All females who drink.

12. (a) {(R, L, L, S), (L, R, L, R), ...}, 81

(b) {(R, R, R, R), (L, L, L, L) (S, S, S, S)}

(c) (R, R, L, L), (R, L, R, S), ...}

(d) {(R, L, R, S), (S, S, R, L), ...}

13. (a) {(T, SY, O), (C, SN, O), ...}

(b) {(T, SY, O), (T, SY, F), (B, SY, O), ...}

(c) {(C, SY, O), (C, SN, O), (C, SY, F), ...}

(d) C∩SY = {(C, SY, O), (C, SY, F)}

C' = {(T, ..., ...), (B, ..., ...)}

C∪SY = all triplets containing C or SY.

14. (a) {(1, 1, 1), (1, 1, 0), (0, 1, 0), ...}

(b) X = {(1,1,0), (1,0,1), (0, 1, 1)}

(c) Y = {(1, 1, 1), (1, 1, 0), (1, 0, 1), (0, 1, 1)}

(d) Z = {(1, 1, 1), (1, 1, 0), (1, 0, 1)}

(e) (i) Z' = {(0, 1, 1), (0, 1, 0), (0, 0, 1), (0, 0, 0), (1, 0, 0)}

(ii) X∪Z = {(1, 1, 0), (1, 0, 1), (0, 1, 1), (1, 1, 1)}

(iii) X∩Z = { (1, 1, 0), (1, 0,1)}

(iv) Y∪Z ={(1, 1, 0), (1, 0, 1), (0, 1, 1), (1, 1, 1)}

(v) Y∩Z ={(1, 1, 1),(1, 1, 0), (1, 0, 1)}

15. (a) {1,2 , 31, 32, 41, 42, 51, 52, 341, 342, ...3452}

(b) {31, 32, 41, 42, 51, 52}

(c) all except {1, 2}

(d) {1, 31, 41, 51, 341, 351,4 31, 451, 531, 541, 3451, 4351, ...}

Exercise 11.2

1. (a) $\frac{3}{10}$ **(b)** $\frac{3}{4}$

2. (a) 0.63 **(b)** 1

3. (a) (i) $\frac{1}{52}$ **(ii)** $\frac{7}{26}$ **(iii)** $\frac{4}{13}$ **(iv)** $\frac{10}{13}$

(b) (i) $\frac{1}{51}$ **(ii)** $\frac{13}{17}$

(c) (i) $\frac{1}{52}$ **(ii)** $\frac{10}{13}$

4. (a) $\frac{4}{5}$ **(b)** $\frac{11}{30}$ **(c)** 1

5. (a) $\frac{1}{2}$ **(b)** $\frac{1}{12}$

6. (a) $\frac{1}{7}$ **(b)** $\frac{4}{7}$

7. (a) (i) {(1,1), (1,2)..., (6,6)}

(ii) $\frac{1}{6}$ **(iii)** $\frac{2}{9}$ **(iv)** $\frac{5}{6}$

(b) (i) 0 **(ii)** $\frac{1}{9}$ **(iii)** $\frac{5}{36}$ **(iv)** 0

8. (a) 0.04 **(b)** 0.55 **(c)** 0.1548

(d) 0.060372 **(e)** 0.104022

9. (a) Yes **(b)** No **(c)** No

10. (a) 0.06 **(b)** 0.42

(c) 0.3364 **(d)** 0.4084

11. (a) 0.183 **(b)** 0.69

12. $x > \frac{n-1}{2}$

13. (a) n = 20 **(b)** n = 12

14. (a) $\frac{5}{18}$ **(b)** $\frac{1}{3}$ **(c)** $\frac{5}{18}$

15. (a) $\frac{1}{3}$ **(b)** $\frac{5}{14}$

16. (a) $\frac{94}{54145} \simeq 0.00174$ **(b)** $\frac{143}{39984} \simeq 0.00358$

17. (a) $\frac{144}{3553} \simeq 0.0405$ **(b)** $\frac{943}{2261} \simeq 0.417$

18. (a) $\frac{1}{91}$ **(b)** $\frac{4}{91}$

19. (a) $\frac{78}{253} \simeq 0.308$ **(b)** $\frac{576}{1265} \simeq 0.455$

20. (a) 593775 **(b)** $\frac{608}{2639} \simeq 0.230$

(c) $\frac{2426}{7917} \simeq 0.306$ **(d)** $\frac{1045}{1131} \simeq 0.924$

21. (a) $\frac{10005}{29900492} \simeq 0.000335$ **(b)** $\frac{269265}{777412792} \simeq 0.000346$

(c) $\frac{777143527}{777412792} \simeq 0.9997$ **(d)** $\frac{85535486}{777412792} \simeq 0.110026$

22. (a) $\frac{3}{14} \simeq 0.214$ **(b)** $\frac{17}{95} \simeq 0.179$ **(c)** $\frac{41}{95} \simeq 0.432$

23. $\frac{7}{16} \simeq 0.4375$

24. $\frac{111}{400} \simeq 0.2775$

25. (a) $\frac{264}{1885} \simeq 0.140$ **(b)** $\frac{166}{84825} \simeq 0.00196$

(c) $\frac{2584}{39585} \simeq 0.0653$

26. (a) 0.096 **(b)** 0.008 **(c)** 0.512

Exercise 11.3

1. $\frac{7}{20}$

2. (a) $\frac{5}{10}$ **(b)** $\frac{4}{10}$ **(c)** $\frac{2}{10}$ **(d)** $\frac{1}{10}$ **(e)** $\frac{2}{3}$

3. P(A∩B) = $\frac{1}{9} \neq 0 \neq$ P(A)P(B)

4. $\frac{29}{35}$

5. 0.90

6. (a) 92%
  (b) (i) 0.64% **(ii)** 15.36% **(iii)** 14.72%
  (c) 48.68%
7. (a) 10 000 **(b)** $\dfrac{9}{10}$ **(c)** 0.3439 **(d)** $\dfrac{1000}{3439}$
8. (a) $\dfrac{15}{16}$ **(b)** $\dfrac{4}{5}$ **(c)** $\dfrac{1}{5}$
9. (a) {(1, 1) (1, 2), …, (6, 6)}
  (b)

x	2	3	4	5	6
P(x)	$\dfrac{1}{36}$	$\dfrac{1}{18}$	$\dfrac{1}{12}$	$\dfrac{1}{9}$	$\dfrac{5}{36}$
	7	8	9	10	11	12
	$\dfrac{1}{6}$	$\dfrac{5}{36}$	$\dfrac{1}{9}$	$\dfrac{1}{12}$	$\dfrac{1}{18}$	$\dfrac{1}{36}$

  (c) (i) $\dfrac{11}{36}$ **(ii)** $\dfrac{11}{12}$ **(iii)** $\dfrac{1}{3}$ **(iv)** $\dfrac{2}{3}$
10. (a) $\dfrac{7}{15}$ **(b)** $\dfrac{11}{75}$ **(c)** $\dfrac{9}{35}$ **(d)** $\dfrac{46}{75}$ **(e)** $\dfrac{11}{20}$
  (f) See Worked Solutions
11. (a) (i) 0.56 **(ii)** 0.15
  (b) $\dfrac{15}{56}$ **(c)** no

12.

P(A)	P(B)	Conditions for events A and B	P($A\cap B$)	P($A\cup B$)	P($A|B$)
0.3	0.4	Mutually exclusive	0.00	0.7	0.00
0.3	0.4	Independent	0.12	0.58	0.30
0.1	0.5	Mutually exclusive	0.00	0.60	0.00
0.2	0.5	Independent	0.10	0.60	0.20

13. (a) 0.30 **(b)** yes
14. (a) 65% **(b)** 35% **(c)** 52%
15. (a) 0.56 **(b)** 0.10
16. (a) $\dfrac{1}{216}$ **(b)** $\dfrac{91}{216}$ **(c)** $\dfrac{75}{216}$
17. (a) 0.21 **(b)** 0.441 **(c)** 0.657
18. (a) $\dfrac{23}{144}$ **(b)** $\dfrac{11}{144}$ **(c)** $\dfrac{15}{144}$ **(d)** $\dfrac{9}{23}$
19. (a) $A \cap B = \{(10, 5), (10, 4), …, (10, 1), (1, 10), …,$ (5, 10)\}, $p = 0.0694$
  (b) $A \cup B = \{(1, 12), … (1, 1), (2, 12), …, (3, 12), …, (4, 11),$ …, (5, 10), …\} $p = 0.778$
  (c) list, $p = 0.931$ **(d)** list, p = 0.222
  (e) same as **(c)** **(f)** same as **(d)**
  (g) This is $(A \cup B) - (A \cap B)$; $p = .709$
20. See Worked Solutions
21. (a) $\dfrac{1}{36}$ **(b)** $\dfrac{1}{28}$ **(c)** $\dfrac{91}{216}$ **(d)** 0.5
22. 0.5
23. (a) 0.103 **(b)** 0.0887 **(c)** 0.537
24. (a) 0.10 **(b)** 0.00001
25. (a) $\dfrac{21}{22} \approx 0.955$ **(b)** $\dfrac{23}{66} \approx 0.348$ **(c)** $\dfrac{13}{63} \approx 0.206$
26. (a) 0.36 **(b)** 0.64 **(c)** 0.75
  (d) 0.17 **(e)** 0.0455 **(f)** $\dfrac{5}{22} \approx 0.227$
27. (a) 0.8805 **(b)** 0.0471

Exercise 11.4

1. (a)

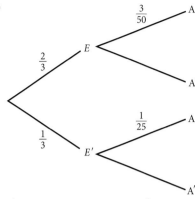

  (b) $\dfrac{4}{75}$ **(c)** $\dfrac{3}{4}$
2. (a) 0.061872 **(b)** 0.032777
3. (a) 0.52 **(b)** 0.692
4. (a) $\dfrac{1}{3}$ **(b)** $\dfrac{1}{2}$
5. (a) 0.55 **(b)** 0.555
6. 0.875
7. (a) 85.5% **(b)** 10.5%
8. (a) 0.64 **(b)** 0.703
9. (a) 0.7 **(b)** 0.50
10. 0.915
11. (a) 3.6% **(b)** 66.7% **(c)** 5.37% **(d)** 4.24%
12. 0.382
13. Antonio
14. (a) 0.1445 **(b)** 0.00058
15. (a) 0.375 **(b)** 0.333
16. (a) 0.93 **(b)** 0.108
17. (a) (i) P(F) = 0.4 **(ii)** P($F \cap T$) =0.224
  (iii) P($F \cup A'$) = 0.944 **(iv)** P($F'|A$) = 0.35
  (b) M is mutually exclusive with F. T is independent as P($F \cap T$) = P(T) P(F)
  (c) (i) 0.716 **(ii)** 0.704
18. (a) 0.012 **(b)** 0.030 **(c)** 0.40

Chapter 11 practice questions

1. (a) 0.30 **(b)** 0.72 **(c)** 0.7
2. (a) 0.0004 **(b)** 0.9996 **(c)** 0.0004
3. 0.99998
4. (a) (i) 0.85 **(ii)** 0.80 **(iii)** 0.15
  (b) 0.0833
5. (a) (i) 0.3405 **(ii)** 0.0108 **(iii)** 0.9622
  (iv) 0.30
  (b) Yes.
6. (a) 0.63 **(b)** 0.971
7. (a) 0.60 **(b)** Yes, P($B|A$) = P(B) = 0.60
  (c) 0.42
8. (a)

	Boys	Girls
Passed the ski test	32	16
Failed the ski test	14	12
Training, but did not take the test yet	20	16
Too young to take the test	4	6

  (b) (i) 0.6167 **(ii)** 0.56 **(iii)** 0.1463

9. (a) $\frac{3}{32}$ (b) $\frac{3}{4}$ (c) $\frac{5}{32}$

10. (a) 0.02   (b) 0.64

11. (a) 0.4   (b) 0.6

12. (a) 0.38   (b) 0.283

13. (a) Shaded area outside X and Y

     (b) (i) 2   (ii) $\frac{1}{18}$

     (c) No, $n(X \cap Y) \neq 0$

14. (a)

	Males	Females	Totals
Unemployed	20	40	60
Employed	90	50	140
Totals	110	90	200

     (b) (i) $\frac{1}{5}$ (ii) $\frac{9}{14}$

15. $\frac{44}{65}$

16. (a) 0.54   (b) 0.444

17. (a) $\frac{7}{12}$ (b) $\frac{11}{36}$ (c) $\frac{1}{3}$

18. (a) $\frac{1}{11}$ (b) $\frac{12}{121}$

19. (a) Ind.   (b) M   (c) N

20. (a) See Worked Solutions   (b) $\frac{47}{160}$

     (c) $\frac{35}{47}$

21. (a) See Worked Solutions

     (b) (i) 0.36   (ii) 0.84   (iii) 0.429

22. (a) (i) $\frac{8}{21}$ (ii) $\frac{1}{6}$

     (iii) No, $P(A \cap B) \neq P(A)P(B)$

     (b) $\frac{10}{17}$ (c) $\frac{200}{399}$

23. $\frac{1}{3}$

24. 0.80

25. $\frac{10}{19}$

26. $\frac{2}{5}$

27. (a) (i) $\frac{5}{36}$ (ii) $\frac{25}{216}$ (iii) $\frac{1}{6}\left(\frac{5}{6}\right)^{2n-2}$

     (b) See Worked Solutions

     (c) $\frac{5}{11}$ (d) 0.432

28. (a) 0.957   (b) 0.301

29. (a) 0.732   (b) $\frac{11}{61}$

30. (a) $\frac{2}{3}$ (b) $\frac{2}{9}$ (c) $\frac{3}{4}$

31. (a) $\frac{1}{10}$ (b) See Worked Solutions

     (c) $\frac{11}{90}$ (d) $\frac{3}{11}$

32. (a) $\frac{7}{3}$

     (b) (i) $\frac{7}{72}$ (ii) $\frac{2}{27}$

     (c) 0.559   (d) $n = 8$ (e) 3

Chapter 12

Exercise 12.1

1. (a) 4   (b) $3x^2$ (c) $2x$ (d) 6

2. (a) 0   (b) $\frac{5}{2}$

     (c) does not exist (increases without bound)

3. (a) $\frac{1}{8}$ (b) $\frac{3}{2}$ (c) $\frac{\sqrt{2}}{4}$ (d) $\frac{1}{4}$ (e) 1   (f) 3

4. e

5. (a) 3   (b) 3

6. As $x \to a$, $g(x) \to +\infty$

7. (a) horizontal: $y = 3$; vertical: $x = -1$

     (b) horizontal: $y = 0$; vertical: $x = 2$

     (c) horizontal: $y = b$; vertical: $x = a$

     (d) horizontal: $y = 2$; vertical: $x = \pm 3$

     (e) horizontal: $y = 0$; vertical: $x = 0, x = 5$

     (f) horizontal: none; vertical: $x = 4$

8. (a) $\frac{1}{3}$ (b) 4

9. See Worked Solutions

10. See Worked Solutions

Exercise 12.2

1. (a) $f'(x) = -2x$ (b) $g'(x) = 3x^2$

     (c) $h'(x) = \frac{1}{2\sqrt{x}}$ (d) $r'(x) = -\frac{2}{x^3}$

2. (a)

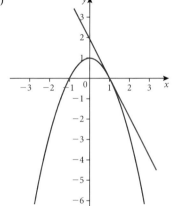

     (b)

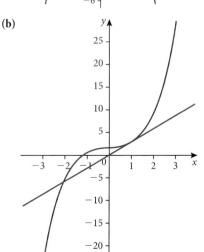

(c)

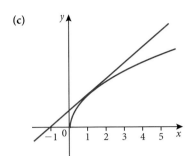

(d)

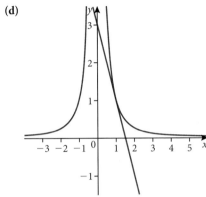

3. (a) (i) $y' = 6x - 4$ **(ii)** -4

  (b) (i) $y' = -2x - 6$ **(ii)** 0

  (c) (i) $y' = -\dfrac{6}{x^4}$ **(ii)** -6

  (d) (i) $y' = 5x^4 - 3x^2 - 1$ **(ii)** 1

  (e) (i) $y' = 2x - 4$ **(ii)** 0

  (f) (i) $y' = 2 - \dfrac{1}{x^2} + \dfrac{9}{x^4}$ **(ii)** 10

  (g) (i) $y' = 1 - \dfrac{2}{x^3}$ **(ii)** 3

4. $a = -5, b = 2$

5. (a) $(0, 0)$ **(b)** $(2, 8)$ and $(-2, -8)$

  (c) $\left(\dfrac{5}{2}, -\dfrac{21}{4}\right)$ **(d)** $(1, -2)$

6. (a) between A and B

  (b) (i) rate of change is positive at A, B and F

     (ii) rate of change is negative at D and E

     (iii) rate of change is zero at C

  (c) pair B & D, and pair E & F

7. $a = 1, b = 5$

8. $a = 1$

9. $(3, 6)$

10. (a) 12.61 **(b)** 12

11. (a) $f'(x) = 2ax + b$ **(b)** See Worked Solutions

12. (a) $4.\overline{6}$ degrees Celsius per hour

  (b) $C'(t) = 3\sqrt{t}$

  (c) $t = \dfrac{196}{81} \approx 2.42$ hours

13. (a) See Worked Solutions   **(b)** See Worked Solutions

14. See Worked Solutions

15. (a) $-\dfrac{1}{x^2}$ **(b)** $\dfrac{5}{(3-x)^2}$ $\left[\text{or } \dfrac{5}{(x-3)^2}\right]$

  (c) $-\dfrac{1}{2\sqrt{(x+2)^3}}$

16. See Worked Solutions

Exercise 12.3

1. (a) $(1, -7)$ **(b)** $\left(-\dfrac{3}{2}, 8\right)$ **(c)** $(3, 2)$

2. (a) (i) $y' = 2x - 5$

     (ii) increasing for $x > \dfrac{5}{2}$

     (iii) decreasing for $x < \dfrac{5}{2}$

  (b) (i) $y' = -6x - 4$

     (ii) increasing for $x < -\dfrac{2}{3}$

     (iii) decreasing for $x > -\dfrac{2}{3}$

  (c) (i) $y' = x^2 - 1$

     (ii) increasing for $x > 1, x < -1$

     (iii) decreasing for $-1 < x < 1$

  (d) (i) $y' = 4x^3 - 12x^2$

     (ii) increasing for $x > 3$

     (iii) decreasing for $x < 0, 0 < x < 3$

3. (a) (i) $(3, -130), (-4, 213)$

     (ii) $(3, -130)$ minimum because 2nd derivative is positive at $x = 3$

       $(-4, 213)$ maximum because 2nd derivative is negative at $x = -4$

     (iii)

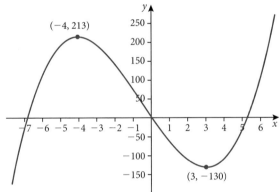

  (b) (i) $(0, -5)$

     (ii) stationary point is neither a maximum nor minimum because 1st derivative is always positive

     (iii)

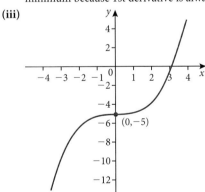

  (c) (i) $(1, 4), (3, 0)$

     (ii) $(1, 4)$ maximum because 2nd derivative is negative at $x = 1$

       $(3, 0)$ minimum because 2nd derivative is positive at $x = 3$

Answers

(iii)

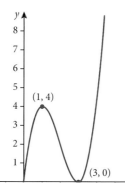

(d) (i) $(-1, 4)$, $(0, 6)$, $\left(\dfrac{5}{2}, -\dfrac{279}{16}\right)$

(ii) $(-1, 4)$ minimum because 2nd derivative is positive at $x = -1$

$(0, 6)$ maximum because 2nd derivative is negative at $x = 0$

$\left(\dfrac{5}{2}, -\dfrac{279}{16}\right)$ minimum because 2nd derivative is positive at $x = \dfrac{5}{2}$

(iii)

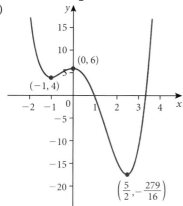

(e) (i) $\left(\dfrac{1}{4}, -\dfrac{1}{4}\right)$

(ii) $\left(\dfrac{1}{4}, -\dfrac{1}{4}\right)$ minimum because 2nd derivative is positive at $x = \dfrac{1}{4}$

(iii)

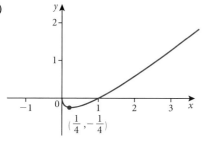

4. (a) $v(t) = 3t^2 - 8t + 1$; $a(t) = 6t - 8$

(b)

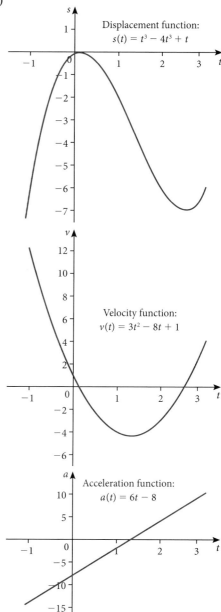

(c) $t \approx 0.131$, displacement ≈ 0.0646

(d) $t = 1.\overline{3}$, velocity $= -4.\overline{3}$

(e) object moves right at a decreasing velocity then turns left with increasing velocity then slowing down and turning right with increasing velocity

5. (a) (i) relative maximum at $(-2, 16)$; relative minimum at $(2, -16)$; **(ii)** inflection point at $(0, 0)$

(b) (i) absolute minima at $(-2, -4)$ and $(2, -4)$; relative maximum at $(0, 0)$; ii inflection points at $\left(-\dfrac{2\sqrt{3}}{3}, -\dfrac{20}{9}\right)$ and $\left(\dfrac{2\sqrt{3}}{3}, -\dfrac{20}{9}\right)$

(c) (i) relative maximum at $(-2, -4)$; relative minimum at $(2, 4)$

(ii) no inflection points

(d) **(i)** relative minimum at $(-1, -2)$; relative maximum at $(1, 2)$

  (ii) inflection points at $\left(-\dfrac{\sqrt{2}}{2}, -\dfrac{7\sqrt{2}}{8}\right)$, $(0, 0)$ and $\left(\dfrac{\sqrt{2}}{2}, \dfrac{7\sqrt{2}}{8}\right)$

(e) **(i)** relative minimum at $(-1, 0)$; absolute minimum at $(2, -27)$; relative maximum at $(0, 5)$;

  (ii) inflection points at $(1.22, -13.4)$ and $(-0.549, 2.32)$

6. (a) $v(0) = 27 \text{ m s}^{-1}$, $a(0) = -66 \text{ m s}^{-2}$

(b) $v(3) = 45 \text{ m s}^{-1}$, $a(3) = 78 \text{ m s}^{-2}$

(c) $t = \dfrac{1}{2}$ and $t = 2\dfrac{1}{4}$; where displacement has a relative maximum or minimum

(d) $t = \dfrac{11}{8} = 1.375$

(e) where acceleration is zero

7. $x \approx 5.77$ tonnes; $D \approx 34.6$ ($\$34{,}600$); this cost is a minimum because cost decreases to this value then increases (second derivative > 0)

8. $a - 3, b = 4, c = -2$

9. relative maximum at $\left(-2, -\dfrac{15}{4}\right)$, stationary inflection point at $(1, 3)$

$f(x) \rightarrow x$ as $x \rightarrow \pm\infty$

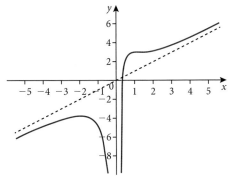

10. (a)

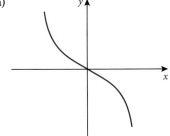

(b)

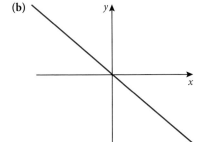

(c)

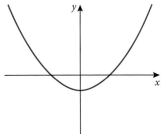

(d)

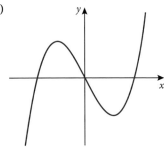

(f)

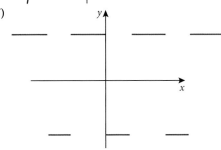

11. (a) **(i)** increasing on $1 < x < 5$; decreasing on $x < 1, x > 5$

  (ii) min. at $x = 1$; max. at $x = 5$

(b) **(i)** increasing on $0 \leq x < 1, 3 < x < 5$; decreasing on $1 < x < 3, x > 5$

  (ii) min. at $x = 3$; max. at $x = 1$ and $x = 5$

12. $x \approx 0.5$ and $x \approx 7.5$

13.

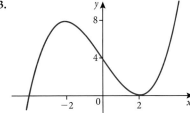

14. (a) right $1 < t < 4$; left $t < 1, t > 4$

(b) **(i)** $v_0 = -24$ **(ii)** $a_0 = 30$

(c) **(i)** $s_{max} = 16$ at $t = 4$

  (ii) $v_{max} = 13.5$ at $t = 2.5$

(d) velocity is max. at $t = 2.5$

15. (a) max. at $x \approx 6.50$, min. at $x \approx -0.215$

(b) max. is $\dfrac{7\pi}{4} + 1$, min. is $\dfrac{\pi}{4} - 1$

Exercise 12.4

1. (a) $y = -4x - 8$ **(b)** $y = \dfrac{4}{27}$

  (c) $y = -x + 1$ **(d)** $y = -2x + 4$

2. (a) $y = \dfrac{1}{4}x + \dfrac{19}{4}$ **(b)** $x = -\dfrac{2}{3}$

  (c) $y = x + 1$ **(d)** $y = \dfrac{1}{2}x + \dfrac{11}{4}$

3. at $(0, 0)$. $y = 2x$; at $(1, 0)$: $y = x + 1$; at $(2, 0)$: $y = 2x - 4$

4. $y = -2x + 1$

5. (a) $x = 1$

    (b) for $y = x^2 - 6x + 20$ eq. of tangent is $y = -4x + 19$
for $y = x^3 - 3x^2 - x$ eq. of tangent is $y = -4x + 1$

6. normal: $y = \dfrac{1}{2}x - \dfrac{7}{2}$; int. pt: $\left(-\dfrac{1}{2}, -\dfrac{15}{4}\right)$

7. eq. of tangent: $y = -3x + 3$; eq. or normal: $y = \dfrac{1}{3}x - \dfrac{1}{3}$

8. $a = -4, b = 1$

9. (a) $y = 2x + \dfrac{5}{2}$ **(b)** $\left(\dfrac{2}{3}, \dfrac{41}{27}\right)$

10. eq. of tangent: $y = -\dfrac{3}{4}x + 1$; eq. or normal: $y = \dfrac{4}{3}x - \dfrac{22}{3}$

11. (a) See Worked Solutions

    (b) See Worked Solutions

    (c)

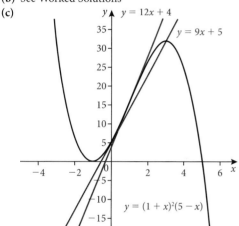

12. $y = 11x - 25$ and $y = -x - 1$

13. $y = (2\sqrt{2} - 2)x, y = -(2\sqrt{2} + 2)x$

14. (a) $y = \dfrac{1}{12}x + \dfrac{4}{3}$ **(b)** $\sqrt[3]{9} \approx 2.08$

15. $y = -\dfrac{1}{2\sqrt{a^3}}x + \dfrac{3}{2\sqrt{a}}$

16. $x_Q = -2x_P, y_Q = -8y_P$

17. $r\sqrt{3}$

18. See Worked Solutions

Chapter 12 practice questions

1. (a) gradient $= 3$

    (b) $y = 3x - \dfrac{9}{4}$

    (c)

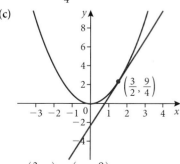

    (d) $Q\left(\dfrac{3}{4}, 0\right), R\left(0, -\dfrac{9}{4}\right)$

    (e) See Worked Solutions

    (f) $y = 2ax - a^2$

(g) $T\left(\dfrac{a}{2}, 0\right), U(0, -a^2)$

(h) x-coord.: $\dfrac{a + 0}{2} = \dfrac{a}{2}$; y-coord.: $\dfrac{a^2 - a^2}{2} = 0$

2. $A = 1, B = 2, C = 1$

3. (a) $x = 2$ or -2; $f'(1) = -6 < 0$ (decreasing) and
$f'(3) = \dfrac{10}{9} > 0$ (increasing) $\therefore f(2)$ is a turning point

    (b) vertical asymptote: $x = 0$ (y-axis);
oblique asymptote: $y = 2x$

4. $\left(\dfrac{1}{2}, 3\right)$

5. $a = 1$

6. (a) $y = 5x - 7$ **(b)** $y = -\dfrac{1}{5}x + \dfrac{17}{5}$

7. (a) $x = 1$

    (b) $-3 < x < -2, 1 < x < 3$

    (c) $x = -\dfrac{1}{2}$

    (d)

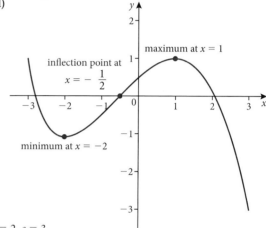

8. $b = 2, c = 3$

9.

function	diagram
f_1	c
f_2	d
f_3	b
f_4	a

10. (a) $\dfrac{2}{\pi}$ **(b)** $\dfrac{\sqrt{2}}{2}$ **(c)** $x \approx 0.881$

11. (a) (i) $x = 0$ **(ii)** $y = 3$

    (b) $\dfrac{dy}{dx} = \dfrac{2}{x^2}$

    (c) increasing for all x, except $x = 0$

    (d) no stationary points because $\dfrac{dy}{dx} = \dfrac{2}{x^2} \neq 0$

12. maximum at $(-1, 1)$, minimum at $(0, 0)$, maximum at $(1, 1)$

13. $a = \dfrac{8}{3}, b = \dfrac{16}{5}$

14. (a) 10 m s^{-1} **(b)** 10 sec.   **(c)** 50 metres

15. (a) $v = 14 - 9.8t, a = -9.8$

    (b) $t \approx 1.43$ sec.

    (c) velocity $= 0$, acceleration $= -9.8 \text{ m s}^{-2}$

16. $(-4, 120)$

17. (a) $y = (-\sqrt{3})x + \dfrac{\pi\sqrt{3}}{3} - 2$

    (b) $y = \left(\dfrac{\sqrt{3}}{3}\right)x - \dfrac{\pi\sqrt{3}}{9} - 2$

18. (a) $h = \dfrac{27 - r^2}{r}$; $V = \pi r(27 - r^2)$

(b) $r = 3$

19. $a = -2, b = 8, c = 10$

20. (a) $y = -7x + 1$ **(b)** $y = \dfrac{x}{7} + \dfrac{107}{7}$

21. (a) absolute minimum at $\left(\dfrac{3}{4}, -\dfrac{27}{256}\right)$

(b) domain: $x \in \mathbb{R}$, range: $y \geqslant -\dfrac{27}{256}$

(c) inflection points at $(0, 0)$ and $\left(\dfrac{1}{2}, -\dfrac{1}{16}\right)$

(d)

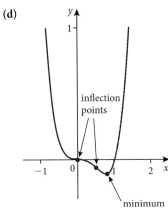

22. (a) $-\dfrac{5}{3}$ **(b)** $\dfrac{1}{4}$ **(c)** 3 **(d)** $\dfrac{1}{2\sqrt{x + 2}}$

23. (a) $f'(x) = \dfrac{3x - 4}{2\sqrt{x}}$ **(b)** $f'(x) = 3x^2 - \cos x$

(c) $f'(x) = -\dfrac{1}{x^2} + \dfrac{1}{2}$ **(d)** $f'(x) = -\dfrac{91}{3x^{14}}$

24. 3 solutions: $\left(\dfrac{11}{2}, \dfrac{1105}{8}\right)$ $(2, -15)$, and $(-2, 5)$

25. $\dfrac{17}{2}$ and $-\dfrac{47}{6}$

26. $\left(2, \dfrac{2}{3}\right)$, $\left(-2, -\dfrac{2}{3}\right)$

27. $(-1, -2)$

28. See Worked Solutions

29. (a) $(2, 20), (4, 16)$ **(b)** $(3, 18)$

30. (a)(b) changes direction at $t = 0$, then $v > 0$ for $0 < x \leqslant 2\pi$

(c) $t = 0, \pi, 2\pi$

(d) max. value of s is 2π

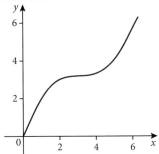

31. $a = \dfrac{1}{4}, b = \dfrac{3}{4}, c = -6, d = -\dfrac{5}{2}$; y-coord. is $-\dfrac{19}{2}$

32. absolute minimum points at $\left(-2, -\dfrac{1}{8}\right)$ and $\left(2, -\dfrac{1}{8}\right)$

33. (a) $y = -x + 2$ **(b)** $y = -x + \dfrac{\pi}{2}$

(c) See Worked Solutions

34. $y = 2x, (1, 2)$

35. (a) $v = 50 - 20t$

(b) $s = 1062.5\,\text{m}$

36.

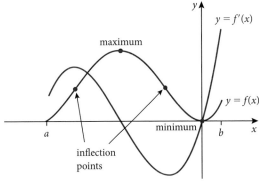

Chapter 13

Exercise 13.1

1. (a) $y' = 12(3x - 8)^3$ **(b)** $y' = -\dfrac{1}{2\sqrt{1 - x}}$

(c) $y' = \cos^2 x - \sin^2 x$ **(d)** $y' = \cos\left(\dfrac{x}{2}\right)$

(e) $y' = -\dfrac{4x}{(x^2 + 4)^3}$ **(f)** $y' = \dfrac{2}{(1 - x)^2}$

(g) $y' = \dfrac{-1}{2\sqrt{(x + 2)^3}}$ $\left[\text{or } \dfrac{-1}{(2x + 4)\sqrt{x + 2}}\right]$

(h) $y' = -2\sin x \cos x$

(i) $y' = \dfrac{-x + 2}{2\sqrt{(1 - x)^3}}$ $\left[\text{or } \dfrac{2 - 3x}{2\sqrt{1 - x}}\right]$

(j) $y' = \dfrac{-6x + 5}{(3x^2 - 5x + 7)^2}$

(k) $y' = \dfrac{2}{3\sqrt[3]{(2x + 5)^2}}$

(l) $y' = 2(2x - 1)^2(7x^4 - 2x^3 + 3)$

(m) $y' = \dfrac{x \cos x - \sin x}{x^2}$ **(n)** $y' = \dfrac{x^2 + 4x}{(x + 2)^2}$

(o) $y' = \dfrac{2 \cos x - 3x \sin x}{3\sqrt[3]{x}}$

2. (a) $y = -12x - 11$ **(b)** $y = \dfrac{9}{5}x - \dfrac{2}{5}$

(c) $y = 2x - 2\pi$ **(d)** $y = \dfrac{1}{2}x + \dfrac{1}{2}$

3. (a) $v(t) = -2t \sin(t^2 - 1)$

(b) velocity $= 0$

(c) $t = \sqrt{\pi + 1} \approx 2.04, t = 1$

(d) accelerating to the right then slowing down, turning around, accelerating to the left, slowing down, turning around again, then accelerating to the right

4. (a) (i) $y = -12x + 38$ **(ii)** $y = \dfrac{1}{12}x + \dfrac{7}{4}$

(b) (i) $y = \dfrac{2}{3}x - \dfrac{2}{3}$ **(ii)** $y = -\dfrac{3}{2}x + \dfrac{11}{3}$

(c) (i) $y = \dfrac{1}{4}x + \dfrac{1}{4}$ **(ii)** $y = -4x + \dfrac{9}{2}$

5. (a) $\dfrac{dy}{dx} = 2\sin(2x)$; $\dfrac{d^2y}{dx^2} = 4\cos(2x)$

(b) $\left(\dfrac{\pi}{4}, 0\right)$ and $\left(\dfrac{3\pi}{4}, 0\right)$

979

6. (a) (i) (0, 0) and (4, 0) **(ii)** $\left(\dfrac{4}{3}, \dfrac{256}{27}\right)$

(iii) $\left(\dfrac{8}{3}, \dfrac{128}{27}\right)$

(b)

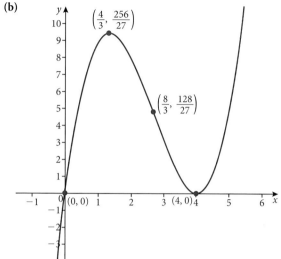

7. (a) See Worked Solutions
(b) See Worked Solutions
(c) $f''(3.8) = 0$ and $f''(3) = \dfrac{1}{3} > 0$, $f''(4) = -\dfrac{2}{625} < 0$, therefore graph of f changes concavity from up to down at $x = 3.8$ verifying that graph of f does have an inflection point at $x = 3.8$

8. $\dfrac{dy}{dx} = \dfrac{2a}{(x+a)^2}$; $\dfrac{d^2y}{dx^2} = \dfrac{-4a}{(x+a)^3}$

9. $\dfrac{d^ny}{dx^n} = \dfrac{(-1)^{n+1}n!}{(x-1)^{n+1}}$

10. (a) max. at (0, 2);
inflection pts. at $\left(-\dfrac{2\sqrt{3}}{3}, \dfrac{3}{2}\right)$ and $\left(\dfrac{2\sqrt{3}}{3}, \dfrac{3}{2}\right)$
(b) (i) none **(ii)** none **(iii)** all $x \in \mathbb{R}$
(c) (i) $\lim\limits_{x\to\infty} g(x) = 0$ **(ii)** $\lim\limits_{x\to-\infty} g(x) = 0$
(d)

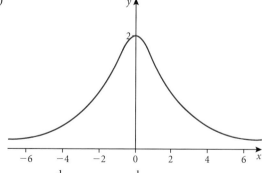

11. $\dfrac{d}{dx}(c \cdot f(x)) = \dfrac{d}{dx}(c) \cdot f(x) + c \cdot \dfrac{d}{dx}(f(x))$

$= 0 \cdot f(x) + c \cdot \dfrac{d}{dx}(f(x)) = c \cdot \dfrac{d}{dx}(f(x))$

12. $y = x^4(x^2 - 6) = 0$ when $x = 0$ and $x = \pm\sqrt{6}$; $y\left(\dfrac{1}{2}\right)$

$= -\dfrac{23}{16} < 0$, so $y < 0$ for $0 < x < 1$

$\dfrac{dy}{dx} = 4x(x^2 - 3) = 0$ when $x = 0$, $x = \pm\sqrt{3}$;

when $x = \dfrac{1}{2}$, $\dfrac{dy}{dx} = -\dfrac{11}{2} < 0$, so $\dfrac{dy}{dx} < 0$ for $0 < x < 1$

$\dfrac{d^2y}{dx^2} = 12(x^2 - 1) = 0$ when $x = 0$, $x = \pm 1$;

when $x = \dfrac{1}{2}$, $\dfrac{d^2y}{dx^2} = -9 < 0$, so $\dfrac{d^2y}{dx^2} < 0$ for $0 < x < 1$

$\dfrac{d^3y}{dx^3} = 24x > 0$ for $0 < x < 1$

Exercise 13.2

1. (a) $y' = x^2 e^x + 2xe^x$ **(b)** $y' = 8^x \ln 8$

(c) $y' = e^x \sec^2(e^x)$ **(d)** $y' = \dfrac{\cos x + x\sin x + 1}{(1 + \cos x)^2}$

(e) $y' = \dfrac{xe^x - e^x}{x^2}$ **(f)** $y' = 2\tan^3(2x)\sec(2x)$

(g) $y' = \left(\dfrac{1}{4}\right)^x \ln\left(\dfrac{1}{4}\right)$ **(h)** $y' = \cos x$

(i) $y' = \dfrac{-xe^x + e^x - 1}{(e^x - 1)^2}$

(j) $y' = -12\cos(3x)\sin(\sin(3x))$

(k) $y' = 2\ln 2(2^x)$ **(l)** $y' = \dfrac{\cos^3 x - \sin^3 x}{(\cos x - \sin x)^2}$

2. (a) $y = \dfrac{1}{2}x + \dfrac{3\sqrt{3} - \pi}{6}$ **(b)** $y = 2x + 1$

(c) $y = 16x + 4 - 2\pi$

3. (a) $x = \dfrac{\pi}{6}$, $x = \dfrac{5\pi}{6}$

(b) maximum at $\dfrac{\pi}{6}$, minimum at $\dfrac{5\pi}{6}$

4. $(0, -1)$ is an absolute maximum

5. (a) max. at $\left(\dfrac{\pi}{2}, 5\right)$; min. at $\left(\dfrac{3\pi}{2}, -3\right)$

(b) min. at $\left(\dfrac{3\pi}{4}, -1\right)$ and $\left(\dfrac{7\pi}{4}, -1\right)$

6. $y = 4$

7. (a) $f'(x) = e^x - 3x^2$; $f''(x) = e^x - 6x$
(b) $x \approx 3.73$ or $x \approx 0.910$ or $x \approx -0.459$
(c) decreasing on $(-\infty, -0.459)$ and $(0.910, 3.73)$ increasing on $(-0.459, 0.910)$ and $(3.73, \infty)$
(d) $x \approx -0.459$ (minimum); $x \approx 0.910$ (maximum); $x \approx 3.73$ (minimum)
(e) $x \approx 0.204$ or $x \approx 2.83$
(f) concave up on $(-\infty, 0.204)$ and $(2.83, \infty)$; concave down on $(0.204, 2.83)$

8. The two functions intersect for all x such that $\cos x = 1$, i.e. $x = k \cdot 2\pi, k \in \mathbb{Z}$. The derivatives for the two functions are $y' = -e^{-x}$ and $y' = -e^{-x}(\cos x + \sin x)$. The derivatives are equal whenever $x = k \cdot 2\pi, k \in \mathbb{Z}$. Therefore, the functions are tangent at all of the intersection points.

9. (a) $8\dfrac{m}{s^2}$ **(b)** $2.09\dfrac{m}{s}$

10. $y = ex$
11. (a) $f'(x) = 2^x \ln 2$ **(b)** $y = x\ln 2 + 1$
(c) $f'(x) = 2^x \ln 2 \neq 0$ for any x

12. (a) $(-1, -2e)$ and $\left(3, \dfrac{6}{e^3}\right)$

(b) $(-1, -2e)$ is a minimum; $\left(3, \dfrac{6}{e^3}\right)$ is a maximum

(c) (i) $\displaystyle\lim_{x \to \infty} h(x) = 0$
　　　(ii) as $x \to -\infty$, $h(x)$ increases without bound

(d) horizontal asymptote $y = 0$

(e)

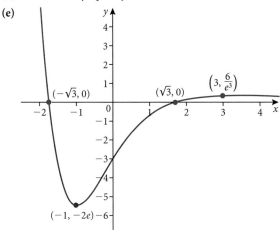

13. (a) $a = \dfrac{\pi}{2},\ b = \pi,\ c = \dfrac{3\pi}{2}$

(b) $\dfrac{d^{(n)}}{dx^{(n)}}(\sin x) = \sin\left(x + n \cdot \dfrac{\pi}{2}\right),\ n \in \mathbb{Z}^+$

14. (a) $\dfrac{d}{dx}(xe^x) = xe^x + e^x$;

$\dfrac{d^2}{dx^2}(xe^x) = xe^x + 2e^x$;　$\dfrac{d^3}{dx^3}(xe^x) = xe^x + 3e^x$

(b) $\dfrac{d^{(n)}}{dx^{(n)}}(xe^x) = xe^x + ne^x$

(c) See Worked Solutions

Exercise 13.3

1. (a) $\dfrac{dy}{dx} = -\dfrac{x}{y}$

(b) $\dfrac{dy}{dx} = \dfrac{-2xy - y^2}{x^2 + 2xy}$

(c) $\dfrac{dy}{dx} = \cos^2 y \left[\text{or } \dfrac{dy}{dx} = \dfrac{1}{1 + x^2}\right]$

(d) $\dfrac{dy}{dx} = \dfrac{-2x + 3y^2 - y^3}{-6xy + 3xy^2 - 2y}$

(e) $\dfrac{dy}{dx} = \dfrac{x^2 y + y^3}{x^3 + xy^2}$

(f) $\dfrac{dy}{dx} = \dfrac{-3xy - 2y^2}{2x^2 + 3xy}$

(g) $\dfrac{dy}{dx} = \dfrac{y - 1}{\cos y - x}$

(h) $\dfrac{dy}{dx} = \dfrac{4x^3 - 2xy^3}{3x^2 y^2 + 4y^3}$

(i) $\dfrac{dy}{dx} = \dfrac{-y}{x + e^y}$

(j) $\dfrac{dy}{dx} = \dfrac{x + 2}{y + 3}$

(k) $\dfrac{dy}{dx} = -\sin^2(x + y)$ $\left[\text{or } \dfrac{dy}{dx} = -\dfrac{x^2}{x^2 + 1}\right]$

(l) $\dfrac{dy}{dx} = \dfrac{18x^2 \sqrt{xy} - y}{x + 2\sqrt{xy}}$

2. (a) $y = -\dfrac{7}{5}x + \dfrac{4}{5}$;　$y = \dfrac{5}{7}x - \dfrac{24}{7}$

(b) $y = -2x + 4$;　$y = \dfrac{1}{2}x + \dfrac{3}{2}$

(c) $y = -\dfrac{\pi}{2}x + \pi$; $y = \dfrac{2}{\pi}x + \dfrac{\pi^2 - 4}{2\pi}$

(d) $y = -\dfrac{352}{23}x - \dfrac{32}{23}$;　$y = \dfrac{23}{352}x - \dfrac{5655}{176}$

3. $x^2 + y^2 = r^2 \Rightarrow \dfrac{dy}{dx} = -\dfrac{x}{y}$; at point (x_1, y_1), $m = -\dfrac{x_1}{y_1}$; centre of circle is $(0, 0)$; slope of line through (x_1, y_1) and $(0, 0)$ is $\dfrac{y_1}{x_1}$; because $-\dfrac{x_1}{y_1} \times \dfrac{y_1}{x_1} = -1$, the tangent to circle at (x_1, y_1) and line through (x_1, y_1) and $(0, 0)$ are perpendicular

4. (a) $(\sqrt{7}\ 0), (-\sqrt{7}, 0)$;

$\dfrac{dy}{dx} = \dfrac{-2x - y}{x + 2y}$, at both points $\dfrac{dy}{dx} = -2$

(b) $\left(\sqrt{\dfrac{7}{3}}, -2\sqrt{\dfrac{7}{3}}\right)$ and $\left(-\sqrt{\dfrac{7}{3}}, 2\sqrt{\dfrac{7}{3}}\right)$

(c) $\left(2\sqrt{\dfrac{7}{3}}, -\sqrt{\dfrac{7}{3}}\right)$ and $\left(-2\sqrt{\dfrac{7}{3}}, \sqrt{\dfrac{7}{3}}\right)$

5. $(3, -1)$

6. (a) $\dfrac{dy}{dx} = -\dfrac{4x}{9y}$, $\dfrac{d^2y}{dx^2} = \dfrac{-36y^2 - 16x^2}{81y^3}$

(b) $\dfrac{dy}{dx} = \dfrac{2 - y}{(x + 3)}$, $\dfrac{d^2y}{dx^2} = \dfrac{2y - 4}{(x + 3)^2}$

7. (a) $\dfrac{dy}{dx} = \dfrac{-1}{3x^{4/3}}$, $\dfrac{d^2y}{dx^2} = \dfrac{4}{9x^{7/3}}$

(b) $\dfrac{dy}{dx} = -\dfrac{y}{3x}$, $\dfrac{d^2y}{dx^2} = \dfrac{4y}{9x^2}$

8. $y = x + \dfrac{1}{2}$

9. (a) $\dfrac{dy}{dx} = \dfrac{3x^2}{x^3 + 1}$
(b) $\dfrac{dy}{dx} = \cot x$

(c) $\dfrac{dy}{dx} = \dfrac{2x}{(x^2 - 1)\ln 5}$
(d) $\dfrac{dy}{dx} = \dfrac{-2}{x^2 - 1}$

(e) $\dfrac{dy}{dx} = \dfrac{1}{2x \ln 10 \sqrt{\log x}}$
(f) $\dfrac{dy}{dx} = \dfrac{2a}{x^2 - a}$

(g) $\dfrac{dy}{dx} = -\sin x$
(h) $\dfrac{dy}{dx} = \dfrac{-1}{x \ln 3 (\log_3 x)^2}$

(i) $\dfrac{dy}{dx} = \ln x$
(j) $\dfrac{dy}{dx} = 0$

10. (a) $y = \left(\dfrac{1}{8 \ln 2}\right)x - \dfrac{1}{\ln 2} + 3$

11. See Worked Solutions

12. $x = \dfrac{1}{e^{\frac{3}{2}}}$

13. (a) $g'(x) = \dfrac{1 - \ln x}{x^2}$, $g''(x) = \dfrac{-3 + 2\ln x}{x^3}$

(b) $g'(x) = 0$ only at $x = e$; $g''(e) = -\dfrac{1}{e^3} < 0$, $\therefore$ abs. max at $x = e$, max value of g is $\dfrac{1}{e}$

14. (a) $\dfrac{dy}{dx} = \dfrac{1}{x^2 + 2x + 2}$

(b) $\dfrac{dy}{dx} = \dfrac{1}{x^2 + 1}$

(c) $\dfrac{dy}{dx} = \dfrac{6}{x\sqrt{x^4 - 9}}$

(d) $\dfrac{dy}{dx} = \left(\tan^{-1} x + \dfrac{x}{x^2 + 1} \right) e^{x \tan^{-1} x}$

15. $f'(x) = 0$; the graph of $f(x)$ is horizontal

16. (a)
(b)

17. $y = \left(\dfrac{\pi + 4}{2} \right) x + \dfrac{\pi - 4}{4}$

18. (a) For $0 \le x < \pi$, $f'(x) = -1$, therefore $f(x)$ is linear
(b) $y = -x + \dfrac{\pi}{2}$

19. $\sqrt{10} \approx 3.16$ m

20. (a) $\dfrac{1}{4}$ m s^{-1}, $\dfrac{1}{20}$ m s^{-1}
(b) $-\dfrac{1}{4}$ m s^{-2}, $-\dfrac{13}{800}$ m s^{-2}
(c) the particle initially is moving very fast to the right and then gradually slows down while continuing to move to the right
(d) $\lim\limits_{t \to \infty} s(t) = \dfrac{\pi}{2}$ m

Exercise 13.4

1. (a) $-18.1 \dfrac{\text{cm}}{\text{min}}$ **(b)** $-6.7 \dfrac{\text{cm}}{\text{min}}$

2. (a) $0.29 \dfrac{\text{cm}}{\text{sec}}$ **(b)** $0.439 \dfrac{\text{cm}}{\text{sec}}$

3. (a) $2\pi \dfrac{\text{cm}}{\text{hr}}$ **(b)** $8\pi \dfrac{\text{cm}}{\text{hr}}$

4. $\dfrac{d\theta}{dt} = \dfrac{3}{34} \approx 0.0882$ radians/min

5. 4.8 m/s

6. 8 ft/s

7. 69.6 km/hr

8. $\dfrac{dy}{dt} = \dfrac{12}{\sqrt{10}} \approx 3.79$

9. 0.01 m/s

10. $30 \dfrac{\text{mm}^3}{\text{sec}}$

11. 45 km/hr

12. $\dfrac{8\sqrt{3}}{3} \approx 4.62 \dfrac{\text{cm}}{\text{sec}}$

13. 1.5 units/s

14. $222.\overline{2} \dfrac{\text{m}}{\text{sec}} = 800$ km/hr

15. (a) 115 degrees/s   **(b)** 57 degrees/s
16. -485 km/hr

Exercise 13.5

1. $\sqrt{2}$ by $\dfrac{\sqrt{2}}{2}$

2. $13\dfrac{1}{3}$ cm by $6\dfrac{2}{3}$ cm

3. $\dfrac{\sqrt{5}}{2}$

4. (a) See Worked Solutions   **(b)** $S = 4x^2 + \dfrac{3000}{x}$

(c) 7.21 cm $\times$ 14.4 cm $\times$ 9.61 cm
5. $x = 5\sqrt{2\pi} \approx 12.5$ cm
6. $x \approx 5.51$ m
7. longest ladder ≈ 7.04 m
8. $d \approx 2.64$ km
9. $\dfrac{8}{5}$ units^2
10. 6 nautical miles
11. $h = R\sqrt{2}$, $r = \dfrac{R\sqrt{2}}{2}$
12. distance of point P from point X is $\dfrac{ac}{\sqrt{r^2 - c^2}}$
13. $x \approx 51.3$ cm, max volume ≈ 403 cm^3
14. See Worked Solutions

Exercise 13.6

1. (a) $\dfrac{1}{2}$ **(b)** 2   **(c)** $\dfrac{1}{2}$

2. (a) -1 **(b)** $-\dfrac{1}{3}$ **(c)** $\dfrac{1}{6}$ **(d)** $\dfrac{1}{3}$

     (e) $\ln 2$ **(f)** -1 **(g)** 1
3. 1

Chapter 13 practice questions

1.

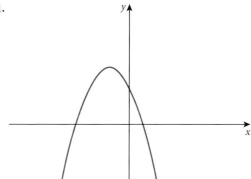

2. (a) (i) $a = -4$ **(ii)** $b = 2$
   (b) (i) $f'(x) = -3x^2 - 4x + 8$
       (ii) $\dfrac{-2 + 2\sqrt{7}}{3}, \dfrac{-2 - 2\sqrt{7}}{3}$
       (iii) $f(1) = 5$
   (c) (i) $y = 8x$ **(ii)** $x = -2$
3. (a) (i) $v(0) = 0$ **(ii)** $v(10) \approx 51.3$
   (b) (i) $a(t) = 0.99 e^{-0.15t}$
       (ii) $a(0) = 0.99$
   (c) (i) 66   **(ii)** 0
       (iii) as object falls it approaches terminal velocity
4. (a) $\left(-\dfrac{2}{3}, -\dfrac{149}{27} \right)$ is a minimum, $(-4, 13)$ is a maximum
   (b) $\left(-\dfrac{7}{3}, \dfrac{101}{27} \right)$ is an inflection point
5. (a) (i) $g'(x) = -\dfrac{3}{e^{3x}}$

       (ii) $e^{3x} > 0$ for all x, hence $-\dfrac{3}{e^{3x}} < 0$ for all x –

       therefore, $f(x)$ is decreasing for all x
   (b) (i) $e + 2$ **(ii)** $g'(x) = -3$

   (c) $y = -\dfrac{3}{e^{3e+6}} x + \left(-\dfrac{1}{e^{3e+6}} + e + 2 \right)$

6. (a)

(b) $f'(3) = 0$ and $f''(3) > 0 \Rightarrow$ stationary point at $x = 3$ and graph of f is concave up at $x = 3$, so $f(3)$ is a minimum

(c) $(4, 0)$

7. (a) $-\dfrac{4}{(2x+3)^3}$ **(b)** $5\cos(5x)\,e^{\sin(5x)}$

8. $A = 1, B = 2, C = 1$

9. $\dfrac{dy}{dx} = -1,\ \dfrac{d^2y}{dx^2} = -4$

10. (a) $\dfrac{dy}{dx} = \dfrac{-xe^x + e^x - 1}{(e^x - 1)^2}$

(b) $\dfrac{dy}{dx} = 2e^x\cos(2x) + e^x\sin(2x)$

(c) $\dfrac{dy}{dx} = 2x\ln x + 2x\ln 3 + x - \dfrac{1}{x}$

11. $y = -\dfrac{1}{2}x - \dfrac{3}{2},\ P(-3, 0),\ Q\left(0, -\dfrac{3}{2}\right)$

12. (a) $x = 3$; sign of $h''(x)$ changes from negative (concave down) to positive (concave up) at $x = 3$

(b) $x = 1$; $g)h'(x)$ changes from positive (h increasing to negative (h decreasing) at $x = 1$

13. $y = \dfrac{5}{7}x + \dfrac{11}{7}$

14. $h = 8\,\text{cm},\ r = 4\,\text{cm}$

15. maximum area is 16 square units; dimensions are 2 by 8

16. (a) E **(b)** A **(c)** C

17. $y = -\dfrac{1}{5}x + \dfrac{32}{5}$

18. (a) $y = 4x - 4$ **(b)** $y = -\dfrac{1}{4}x + \dfrac{1}{4}$

19. (a) absolute minimum at $\left(\dfrac{1}{\sqrt{e}}, -\dfrac{1}{2e}\right)$

(b) inflection point at $\left(\dfrac{1}{\sqrt{e^3}}, -\dfrac{3}{2e^3}\right)$

20. (a) $a = 16$ **(b)** $a = -54$

(c) $f'(x) = 2x - \dfrac{a}{x^2} = 0 \Rightarrow x = \sqrt[3]{\dfrac{a}{2}};\ f''(x) = 2 + \dfrac{2a}{x^3}$

$\Rightarrow f''\left(\sqrt[3]{\dfrac{a}{2}}\right) = 4 > 0$; hence, f is concave up at any critical point, so it cannot be a maximum

21. $y = -\dfrac{2}{3}x + 4$

22. $y = \left(\dfrac{\pi + 2}{2}\right)x - \dfrac{\pi^2}{8};\ y = \left(\dfrac{-2}{\pi + 2}\right)x + \dfrac{\pi}{2\pi + 4} + \dfrac{\pi}{4}$

23. (a) max at $\left(0, \dfrac{1}{\sqrt{2\pi}}\right)$, inflection points at $\left(-1, \dfrac{1}{\sqrt{2e\pi}}\right)$ and $\left(1, \dfrac{1}{\sqrt{2e\pi}}\right)$

(b) $\lim\limits_{x \to \pm\infty} f(x) = 0$; $y = 0$ (x-axis) is a horizontal asymptote

(c)

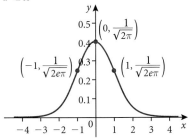

24. (a) min. at $x = 1$ because $f''(1) = \dfrac{1}{2} > 0$; max. at $x = 3$ because $f''(3) = -\dfrac{1}{6} < 0$

(b) inflection points at $x = -\sqrt{3}$ and $x = \sqrt{3}$ because $f''(x)$ changes sign at both values

25. $x = 20\sqrt{3} \approx 34.6\,\text{km/hr}$

26. incorrect; $\dfrac{\sin x}{1 - \cos x}$ is not of indeterminate form when $x = \pi$; $\lim\limits_{x \to \pi}\left(\dfrac{\sin x}{1 - \cos x}\right) = 0$

27. $\dfrac{dy}{dx} = \dfrac{2}{\sqrt{1 - x^2}}$

28. $-\dfrac{3}{4}$

29. $c = 2$

30. $a = -4,\ b = 18$

31. verify

32. $y = -x + 2$

33. (a) (i) $f'(x) = \dfrac{2(x^2 - 1)}{(x^2 + x + 1)^2}$

(ii) $A\left(1, \dfrac{1}{3}\right)$ $B(-1, 3)$ $\left(\text{or } A(-1, 3)\ B\left(1, \dfrac{1}{3}\right)\right)$

(b) (i)

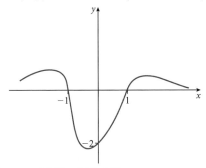

(ii) $x \approx -0.347, -1.53, 1.88$

(c) (i) range of f: $\left[\dfrac{1}{3}, 3\right]$

(ii) range of $f \circ f$: $\left[\dfrac{1}{3}, \dfrac{7}{13}\right]$

34. $\dfrac{1}{2\pi}\,\text{cm/s}$

35. (a) (i) See Worked Solutions

(ii) $f''(x) = \dfrac{x^2(\ln 2)^2 - 4x\ln 2 + 2}{2^x}$

(b) (i) $x = \dfrac{2}{\ln 2}$ **(ii)** $f''\left(\dfrac{2}{\ln 2}\right) < 0$ therefore, a maximum

(c) $x = \dfrac{2 + \sqrt{2}}{\ln 2} \approx 4.93,\ x = \dfrac{2 - \sqrt{2}}{\ln 2} \approx 0.845$

36. $240\,\text{km h}^{-1}$

37. (a) See Worked Solutions

(b) $\left(-\dfrac{1}{c}\ln b, \dfrac{a}{2b}\right)$

(c) See Worked Solutions

38. (a) $p = 2$ **(b)** $-\dfrac{4}{7}$

39. $\dfrac{1}{10}$ radians/s

40. $y = -\dfrac{5}{4}x + \dfrac{13}{2}$

41. $(-0.803, -2.08)$

42. (a) $k = \dfrac{\ln 2}{20}$ **(b)** 510 bacteria per minute

43. (a) $f'(x) = \dfrac{3}{3x+1}$ **(b)** $y = -\dfrac{7}{3}x + \dfrac{14}{3} + \ln 7$

44. See Worked Solutions

45. $\dfrac{dy}{dx} = \dfrac{e-1}{e}$

46. (a) See Worked Solutions

     (b) $b = -\dfrac{1}{\sqrt{6}}$

47. (a) $\dfrac{dy}{dx} = \dfrac{2-k}{2k-1}$ **(b)** $k = 2$

48. (a) $5\sqrt{5}\sqrt{x^2+4} + 5(2-x)$ minutes

     (b) See Worked Solutions

     (c) (i) $x = 1$

         (ii) 30 minutes

         (iii) $\dfrac{d^2 T}{dx^2} > 0$ for $x = 1$, therefore it's a minimum

49. $\dfrac{72}{\pi} \arccos \dfrac{8}{13}$ cm

Chapter 14

Exercise 14.1

1. (a) $\dfrac{x^2}{2} + 2x + c$ **(b)** $t^3 - t^2 + t + c$

     (c) $\dfrac{x}{3} - \dfrac{x^4}{14} + c$ **(d)** $\dfrac{2t^3}{3} + \dfrac{t^2}{2} - 3t + c$

     (e) $\dfrac{5u^{\frac{7}{5}}}{7} - u^4 + c$ **(f)** $\dfrac{4x\sqrt{x}}{3} - 3\sqrt{x} + c$

     (g) $-3\cos\theta + 4\sin\theta + c$ **(h)** $t^3 + 2\cos t + c$

     (i) $\dfrac{4x^2\sqrt{x}}{5} - \dfrac{10x\sqrt{x}}{3} + c$ **(j)** $3\sin\theta - 2\tan\theta + c$

     (k) $\dfrac{1}{3}e^{3t-1} + c$ **(l)** $2\ln|t| + c$

     (m) $\dfrac{1}{6}\ln(3t^2 + 5) + c$ **(n)** $e^{\sin\theta} + c$

     (o) $\dfrac{(2x+3)^3}{6} + c$

2. (a) $-\dfrac{5x^4}{4} + \dfrac{2x^3}{3} + cx + k$

     (b) $-\dfrac{x^5}{5} + \dfrac{x^4}{4} + \dfrac{x^2}{2} + 2x - \dfrac{11}{20}$

     (c) $\dfrac{4t^3}{3} + \sin t + ct + k$

     (d) $3x^4 - 4x^2 + 7x + 3$

     (e) $2\sin\theta + \dfrac{1}{2}\cos 2\theta + c$

3. (a) $\dfrac{(3x^2+7)^6}{36} + c$ **(b)** $-\dfrac{1}{18(3x^2+5)^3} + c$

     (c) $\dfrac{8\sqrt[4]{(5x^3+2)^5}}{75} + c$ **(d)** $\dfrac{(2\sqrt{x}+3)^6}{6} + c$

     (e) $\dfrac{\sqrt{(2t^3-7)^3}}{9} + c$ **(f)** $-\dfrac{(2x+3)^6}{18x^6} + c$

     (g) $-\dfrac{\cos(7x-3)}{7} + c$

     (h) $-\dfrac{1}{2}\ln(\cos(2\theta-1) + 3) + c$

     (i) $\dfrac{1}{5}\tan(5\theta - 2) + c$ **(j)** $\dfrac{1}{\pi}\sin(\pi x + 3) + c$

     (k) $\dfrac{1}{2}\sec 2t + c$ **(l)** $\dfrac{1}{2}e^{x^2+1} + c$

     (m) $\dfrac{1}{3}e^{2\sqrt{t}} + c$ **(n)** $\dfrac{2}{3}(\ln\theta)^3 + c$

Exercise 14.2

1. (a) $-\dfrac{1}{3}e^{-x^3} + c$

     (b) $-e^{-x}(x^2 + 2x + 2) + c$

     (c) $\dfrac{2}{9}x\cos 3x - \dfrac{2}{27}\sin 3x + \dfrac{1}{3}x^2 \sin 3x + c$

     (d) $\dfrac{1}{a^3}(2\cos ax - a^2 x^2 \cos ax + 2ax \sin ax) + c$

     (e) $\sin x(\ln(\sin x) - 1) + c$

     (f) $\dfrac{1}{2}x^2(\ln x^2 - 1) + c$

     (g) $\dfrac{1}{3}x^3 \ln x - \dfrac{1}{9}x^3 + c$

     (h) $2e^x + x^2 e^x - 2xe^x - \dfrac{1}{3}x^3 + c$

     (i) $\dfrac{1}{\pi^2}(\cos \pi x + \pi x \sin \pi x) + c$

     (j) $\dfrac{3}{13}\cos 2t\, e^{3t} + \dfrac{2}{13}e^{3t}\sin 2t + c$

     (k) $\sqrt{1-x^2} + x\arcsin x + c$

     (l) $e^x(x^3 - 3x^2 + 6x - 6) + c$

     (m) $-\dfrac{1}{4}e^{-2x}(\cos 2x + \sin 2x) + c$

     (n) $\dfrac{1}{2}x(\sin(\ln x) - \cos(\ln x)) + c$

     (o) $\dfrac{1}{2}x(\sin(\ln x) + \cos(\ln x)) + c$

     (p) $\ln(x+1) - 2x + x\ln(x^2 + x) + c$

     (q) $\dfrac{e^{kx}(k\sin x - \cos x)}{k^2 + 1} + c$

     (r) $\ln(\cos x) + x\tan x + c$

     (s) $\dfrac{1}{2}\sin x - \dfrac{1}{6}\sin 3x + c$

     (t) $\dfrac{1}{2}\arctan x(1 + x^2) - \dfrac{1}{2}x + c$

     (u) $2\sqrt{x}(\ln x - 2) + c$

2. Verification: First column represents u in "repeated by parts" and second column represents v.

1. (a) $-\dfrac{1}{15}\sqrt{(3-5t^2)^3} + c$ **(b)** $\dfrac{1}{3}\tan\theta^3 + c$

     (c) $-\cos\sqrt{t} + c$ **(d)** $\dfrac{1}{12}\tan^6 2t + c$

     (e) $2\ln(\sqrt{x} + 2) + c$ **(f)** $\dfrac{1}{10}\sec^5 2t + c$

     (g) $\dfrac{1}{2}\ln(x^2 + 6x + 7) + c$

     (h) $-\dfrac{k^3}{2a^4}\sqrt{a^2 - a^4 x^4} + c = -\dfrac{k^3}{2|a|^3}\sqrt{1 - a^2 x^4} + c$

     (i) $\dfrac{2}{5}(3x^2 - x - 2)\sqrt{x-1} + c$

     (j) $-\dfrac{1}{\pi}\cot\pi t + c$

     (k) $-\dfrac{2}{3}\sqrt{(1 + \cos\theta)^3} + c$

     (l) $\dfrac{2}{105}(15t^3 - 3t^2 - 4t - 8)\sqrt{1-t} + c$

     (m) $\dfrac{1}{15}(3r^2 + 2r - 13)\sqrt{2r-1} + c$

     (n) $\dfrac{1}{2}\ln(e^{x^2} + e^{-x^2}) + c$

3. (a) $-x^4\cos x + 4x^3\sin x + 12x^2\cos x - 24x\sin x$
$- 24\cos x + c$

(b) $x^5\sin x + 5x^4\cos x - 20x^3\sin x - 60x^2\cos x$
$+ 120x\sin x + 120\cos x + c$

(c) $e^x(x^4 - 4x^3 + 12x^2 - 24x + 24) + c$

4. No pattern in the second column

5. Use repeated "by parts" with $u = x^n$ and $dv = e^x\,dx$

6. "By parts" with $u = \ln x$

7. Repeated "by parts" to find the unknown integral

8. Repeated "by parts" to find the unknown integral

Exercise 14.3

1. (a) $\dfrac{\cos^5 t}{5} - \dfrac{\cos^3 t}{3} + c$

(b) $\dfrac{1}{192}\cos 6t - \dfrac{3}{64}\cos 2t + c$

(c) $\dfrac{\sin^4 3\theta}{12} + c$

(d) $\dfrac{1}{3}\cos^3\dfrac{1}{t} - \dfrac{2}{5}\cos^5\dfrac{1}{t} + \dfrac{1}{7}\cos^7\dfrac{1}{t} + c$

(e) $\sec x + \cos x + c$

(f) $\dfrac{1}{18}\tan^6 3x + c$

(g) $\dfrac{1}{24}(3\tan^4\theta^2 + 2\tan^6\theta^2) + c$

(h) $\dfrac{2\sec^5\sqrt{t}}{5} - \dfrac{2\sec^3\sqrt{t}}{3} + c$

(i) $\dfrac{1}{15}(\tan^3 5t - 3\tan 5t + 15t) + c$

(j) $\tan t - \sec t + c$

(k) $\csc t - \cot t + c$

(l) $-\ln(1 - \sin t) + c$

(m) $-2x - 3\ln(\sin x + \cos x) + c$

(n) $\arctan(\sec\theta) + c$

(o) $\dfrac{1}{2}(\arctan t)^2 + c$

(p) $\ln|\arctan| + c$

(q) $\arcsin(\ln x) + c$

(r) $\dfrac{-\cos x}{3}(\sin^2 x + 2) + c$ or $\dfrac{\cos^3 x}{3} - \cos x + c$

(s) $\dfrac{2}{5}(\cos^2 x\sqrt{\cos x} - 10\sqrt{\cos x}) + c$

(t) $\dfrac{-\cos\sqrt{x}}{3}(2\sin^2\sqrt{x} + 4) + c$ or $2\left(\dfrac{\cos^3\sqrt{x}}{3} - \cos\sqrt{x}\right) + c$

(u) $\dfrac{\sin(\sin t)}{3}(\cos^2(\sin t) + 2) + c$
or $\sin(\sin x) - \dfrac{\sin^3(\sin x)}{3} + c$

(v) $\ln(\sin\theta) + 2\sin\theta + c$

2. (a) $t\sec t - \ln|\sec t + \tan t| + c$

(b) $-\ln(2 - \sin x) + c$

(c) $\dfrac{1}{2}\ln(\cos(e^{-2x})) + c$

(d) $2\ln|\sec\sqrt{x} + \tan\sqrt{x}| + c$

(e) $\dfrac{1}{2}\tan x + c$

(f) $\dfrac{1}{6}(\arcsin 3x + 3x\sqrt{1 - 9x^2}) + c$

(g) $\dfrac{x}{4\sqrt{x^2 + 4}} + c$

(h) $2\ln(t + \sqrt{t^2 + 4}) + \dfrac{1}{2}t\sqrt{t^2 + 4} + c$

(i) $\dfrac{3}{2}\arctan\left(\dfrac{1}{2}e^t\right) + c$

(j) $\dfrac{1}{2}\arcsin\left(\dfrac{2}{3}x\right) + c$

(k) $\dfrac{1}{3}\ln\left|\dfrac{\sqrt{4 + 9x^2} + 3x}{2}\right| + c$

(l) $\ln(\sqrt{\sin^2 + 1} + \sin x) + c$

(m) $-\sqrt{4 - x^2} + c$

(n) $\dfrac{1}{2}\ln(x^2 + 16) + c$

(o) $-\arcsin\left(\dfrac{x}{2}\right) - \dfrac{\sqrt{4 - x^2}}{x} + c$

(p) $\dfrac{1}{9}\dfrac{x}{\sqrt{9 - x^2}} + c$

(q) $\dfrac{(x^2 + 1)^{\frac{3}{2}}}{3} + c$

(r) $\dfrac{(e^{2x} + 1)^{\frac{3}{2}}}{3} + c$

(s) $\dfrac{1}{2}(\sin^{-1}e^x + e^x\sqrt{1 + e^x}) + c$

(t) $\ln\left(\dfrac{1}{3}e^x + \dfrac{1}{3}\sqrt{e^{2x} + 9}\right) + c$

(u) $2\sqrt{x}(\ln x - 2) + c$

(v) $12\ln(x + 2) + \dfrac{8}{x + 2} + \dfrac{x^2}{2} - 4x + c$

3. $\dfrac{1}{2}\ln(x^2 + 9) + c_1; x = 3\tan\theta$ yields $\ln\left(\dfrac{\sqrt{x^2 + 9}}{3}\right)$
$+ c_2$; they differ by a constant

4. $x - 3\arctan\left(\dfrac{x}{3}\right) + c_1; x = 3\tan\theta$ yields $3(\tan\theta - \theta)$
$+ c_2 = 3\left(\dfrac{x}{3} - \arctan\dfrac{x}{3}\right) + c_2$

Exercise 14.4

1. (a) 24   **(b)** 40   **(c)** $\dfrac{24}{25}$ **(d)** 0

(e) $\dfrac{176\sqrt{7} - 44}{5}$ **(f)** 0   **(g)** 2

(h) -268 **(i)** $\dfrac{64}{3}$ **(j)** 2

(k) $\ln\left(\dfrac{11}{3}\right)$ **(l)** $\dfrac{44}{3} - 8\sqrt{3}$ **(m)** 3

(n) $\sqrt{\pi} + 1$

(o) (i)   6   **(ii)** 6   **(iii)**   12

(p) 1   **(q)** 4   **(r)** 0   **(s)** $\dfrac{\pi}{2}$

(t) $\dfrac{\pi}{6}$ **(u)** $\dfrac{\pi}{3}$ **(v)** $\dfrac{\pi}{8}$

2. (a) $\dfrac{14\sqrt{17} + 2}{3}$ **(b)** $\dfrac{1}{\pi}$ **(c)** $\ln(2)$

(d) $16\sqrt{2} - 5\sqrt{5}$ **(e)** $\sqrt{14} - \sqrt{10}$ **(f)** $\dfrac{3}{2}$

(g) $\pi^{3/2}\left(\dfrac{2\sqrt{3}}{27} - \dfrac{1}{12}\right)$ **(h)** $\dfrac{\pi}{6}$

(i) $-\dfrac{1}{2}\ln\left(\dfrac{37}{52}\right)$ **(j)** $-\arctan\left(\dfrac{\sqrt{15} - \sqrt{7}}{4}\right)$

(k) $\dfrac{2}{3}$ **(l)** 0   **(m)** -4 **(n)** $\dfrac{\pi}{6}$

(o) $\dfrac{1}{6}\arctan\left(\dfrac{4\sqrt{3}}{9}\right)$ **(p)** $\dfrac{\pi\sqrt{3} - 3\sqrt{3}\arctan\left(\dfrac{\sqrt{3}}{2}\right)}{18}$

(q) $\dfrac{1}{6}$ **(r)** $\dfrac{e - 1}{2}$

(s) $1 + \dfrac{e}{2}$ **(t)** $2\cos(1) + 2$

Answers

3. (a) $\dfrac{31}{5}$ (b) $\dfrac{2}{\pi}$ (c) $\dfrac{12 - 4\sqrt{3}}{\pi}$

   (d) $\dfrac{e^8 - 1}{8 e^8}$ (e) $\dfrac{\pi}{6 \ln 3}$

4. (a) $\dfrac{\sin x}{x}$ (b) $-\dfrac{\sin x}{x}$ (c) $-2x\dfrac{\sin x^2}{x^2}$

   (d) $2x\dfrac{\sin x^2}{x^2}$ (e) $\dfrac{\cos t}{1 + t^2}$ (f) $\dfrac{b - a}{5 + x^4}$

   (g) $-\csc\theta - \sec\theta$ (h) $\dfrac{1}{4x^{\frac{3}{4}}}\left(e^{x + 3x^{\frac{1}{2}}}\right)$

5. Yes

6. (a) $\dfrac{1}{3}\ln\left(\dfrac{3k + 2}{2}\right)$ (b) $k = \dfrac{2(e^3 - 1)}{3}$

7. Substitute $u = 1 - x$

8. (a) $-(1 - x)^{k+1}\left(\dfrac{1}{k + 1} + \dfrac{x - 1}{k + 2}\right)$

   (b) $\dfrac{1}{(k + 1)(k + 2)}$

9. (a) 0;   (b) $\sqrt{47}$;   (c) $\dfrac{15\sqrt{47}}{47}$

10. $f'(x) = 0$.

Exercise 14.5

1. (a) $\displaystyle\int\frac{5x + 1}{x^2 + x - 2}\,dx = \int\frac{3}{x + 2}\,dx + \int\frac{2}{x - 1}\,dx = 3\ln|x + 2| + 2\ln|x - 1| + c$

   (b) $\displaystyle\int\frac{x + 4}{x^2 - 2x}\,dx = \int\frac{3}{x - 2}\,dx - \int\frac{2}{x}\,dx = 3\ln|x - 2| - 2\ln|x| + c$

   (c) $\displaystyle\int\frac{x + 2}{x^2 + 4x + 3}\,dx = \int\frac{1}{2(x + 3)}\,dx + \int\frac{1}{2(x + 1)}\,dx = \frac{1}{2}\ln|x + 3| + \frac{1}{2}\ln|x + 1| + c$ $= \frac{1}{2}\ln|x^2 + 4x + 3| + c$

   (d) $\displaystyle\int\frac{5x^2 + 20x + 6}{x^3 + 2x^2 + x}\,dx = \int\frac{9}{(x + 1)^2}\,dx - \int\frac{1}{x + 1}\,dx + \int\frac{6}{x}\,dx = -\frac{9}{x + 1} - \ln|x + 1| + 6\ln|x| + c$

   (e) $\displaystyle\int\frac{2x^2 + x - 12}{x^3 + 5x^2 + 6x}\,dx = \int\frac{1}{x + 3}\,dx + \int\frac{3}{x + 2}\,dx - \int\frac{2}{x}\,dx = \ln|x + 3| + 3\ln|x + 2| - 2\ln|x| + c$

   (f) $\displaystyle\int\frac{4x^2 + 2x - 1}{x^3 + x^2}\,dx = \int\frac{1}{x + 1}\,dx - \int\frac{1}{x^2}\,dx + \int\frac{3}{x}\,dx = \ln|x + 1| + \frac{1}{x} + 3\ln|x| + c$

   (g) $\displaystyle\int\frac{3}{x^2 + x - 2}\,dx = \int\frac{1}{x - 1}\,dx - \int\frac{1}{x + 2}\,dx = -\ln|x + 2| + \ln|x - 1| + c = \ln\left|\frac{x - 1}{x + 2}\right| + c$

   (h) $\displaystyle\int\frac{5 - x}{2x^2 + x - 1}\,dx = \int\frac{3}{2x - 1}\,dx - \int\frac{2}{x + 1}\,dx = \frac{3\ln|2x - 1|}{2} - 2\ln|x + 1| + c$

   (i) $\displaystyle\int\frac{3x + 4}{(x + 2)^2}\,dx = \int\frac{3}{x + 2}\,dx - \int\frac{2}{(x + 2)^2}\,dx = 3\ln|x + 2| + \frac{2}{x + 2} + c$

   (j) $\displaystyle\int\frac{12}{x^4 - x^3 - 2x^2}\,dx = \int\frac{1}{x - 2}\,dx - \int\frac{4}{x + 1}\,dx - \int\frac{6}{x^2}\,dx + \int\frac{3}{x}\,dx = \ln|x - 2| - 4\ln|x + 1| + \frac{6}{x} + 3\ln|x| + c$

   (k) $\displaystyle\int\frac{2}{x^3 + x}\,dx = \int\frac{2}{x}\,dx - \int\frac{2x}{x^2 + 1}\,dx = 2\ln|x| - \ln(x^2 + 1) + c$

   (l) $\displaystyle\int\frac{x + 2}{x^3 + 3x}\,dx = \int\frac{2}{3x}\,dx + \int\frac{dx}{(x^2 + 3)} - \int\frac{2x}{3(x^2 + 3)}\,dx = \frac{2\ln|x|}{3} + \frac{\sqrt{3}}{3}\arctan\left(\frac{\sqrt{3}\,x}{3}\right) - \frac{\ln(x^2 + 3)}{3} + c$

   (m) $\displaystyle\int\frac{3x + 2}{x^3 + 6x}\,dx = \int\frac{1}{3x}\,dx + \int\frac{3}{x^2 + 6}\,dx - \int\frac{x}{3(x^2 + 6)}\,dx = \frac{\ln|x|}{3} + \frac{\sqrt{6}}{2}\arctan\left(\frac{\sqrt{6}\,x}{6}\right) - \frac{\ln(x^2 + 6)}{6} + c$

   (n) $\displaystyle\int\frac{2x + 3}{x^3 + 8x}\,dx = \int\frac{3}{8x}\,dx + \int\frac{2}{x^2 + 8}\,dx - \int\frac{3x}{8(x^2 + 8)}\,dx = \frac{3}{8}\ln|x| + \frac{\sqrt{2}}{2}\arctan\left(\frac{\sqrt{2}\,x}{4}\right) - \frac{3}{16}\ln(x^2 + 8) + c$

Exercise 14.6

1. (a) $\dfrac{125}{6}$ (b) $\dfrac{9\pi^2}{8} + 1$ (c) $4\sqrt{3}$

   (d) $\dfrac{10}{3}$ (e) $\dfrac{8}{21}$ (f) $\dfrac{125}{24}$

   (g) $\dfrac{13}{12}$ (h) 4π (i) $\dfrac{59}{12}$

   (j) 4.65 (k) $3\ln 2 - \dfrac{63}{128}$

   (l) (between $-\dfrac{\pi}{2}$ and $\dfrac{\pi}{6}$) $\sqrt{3}\ln\left(\dfrac{3}{4}\right) - 2\sqrt{3} + 4$

   (m) 18 (n) $\dfrac{32}{3}$ (o) $\dfrac{64}{3}$

   (p) 9 (q) $\dfrac{9}{2}$ (r) 19

   (s) $\dfrac{2\sqrt{3}}{3} + 2$ (t) $\dfrac{37}{12}$ (u) $\dfrac{1}{2}$

   (v) $\dfrac{2\sqrt{2}}{3}$

2. $\dfrac{269}{54}$

3. $\dfrac{e}{2} - 1$

4. $\dfrac{288\sqrt{3}}{35}$

5. $\dfrac{2\sqrt{2}}{3}$

6. $\dfrac{16}{3}$

7. 25.36

8. $m = 0.973$

9. $\dfrac{37}{12}$

Exercise 14.7

1. (a) $\dfrac{127\pi}{27}$ **(b)** $\dfrac{64\sqrt{2}\,\pi}{15}$ **(c)** $\dfrac{70\pi}{3}$

   (d) 6π **(e)** 9π **(f)** 2π

   (g) $\left(\dfrac{\sqrt{3}}{2} + 1\right)\pi$ **(h)** $\dfrac{512\pi}{15}$

   (i) approx. 5.937π **(j)** $\dfrac{32\pi}{3}$

   (k) $\pi(\sqrt{3} - 1)$ **(l)** $\dfrac{23\pi}{210}$ **(m)** $\dfrac{160\pi\sqrt{5}}{3}$

   (n) $\dfrac{64}{15}\pi$ **(o)** $\dfrac{\pi}{2}$ **(p)** $\dfrac{1778}{5}\pi$

   (q) $\dfrac{252}{5}\pi$ **(r)** $\dfrac{656}{3}\pi$ **(s)** $\dfrac{9}{8}\pi$

2. (a) $\dfrac{88}{15}\pi$ **(b)** $\dfrac{7}{6}\pi$

3. (a) $\dfrac{\pi}{2}$ **(b)** $2\pi(18 - 9\sqrt{2})$ **(c)** $\dfrac{32}{15}\pi$

   (d) $\dfrac{4}{5}\pi(121\sqrt{33} - 25\sqrt{15})$

   (e) $2\pi\left(\ln 2 - \dfrac{1}{4}\right)$ **(f)** $2\pi\left(\dfrac{11}{3}\sqrt{11} - \dfrac{2}{3}\sqrt{2}\right)$

   (g) $\dfrac{28}{3}\pi(\sqrt{34} - \sqrt{7})$ **(h)** $\pi\left(\dfrac{1}{2}\sqrt{2}\,\pi - \pi + 2\right)$

   (i) $\dfrac{284}{3}\pi$ **(j)** 2π **(k)** $\dfrac{256}{15}\pi$

Exercise 14.8

1. (a) $\dfrac{70}{3}$ m, 65 m   **(b)** 8.5 m to the left, 8.5 m.

   (c) 1 m, 1 m   **(d)** 2 m, $2\sqrt{2}$ m.

   (e) 18 m, 28.67 m   **(f)** $\dfrac{4}{\pi}$ m , $\dfrac{4}{\pi}$ m

2. (a) $3t$, 6 m, 6 m   **(b)** $t^2 - 4t + 3$, 0, 2.67 m

   (c) $1 - \cos t, \left(\dfrac{3\pi}{2} + 1\right)$ m, $\left(\dfrac{3\pi}{2} + 1\right)$ m

   (d) $4 - 2\sqrt{t + 1}$, 2.43 m, 2.91 m

   (e) $3t^2 + \dfrac{1}{2(1 + t)^2} + \dfrac{3}{2}$, 11.3 m, 11.3 m

3. (a) $4.9t^2 + 5t + 10$ **(b)** $16t^2 - 2t + 1$

   (c) $\dfrac{1}{\pi} - \dfrac{\cos \pi t}{\pi}$ **(d)** $\ln(t + 2) + \dfrac{1}{2}$

4. (a) $e^t + 19t + 4$ **(b)** $4.9t^2 - 3t$

   (c) $\sin(2t) - 3$ **(d)** $-\cos\left(\dfrac{3t}{\pi}\right)$

5. (a) 12; 20   **(b)** $\dfrac{13}{2}; \dfrac{13}{2}$

   (c) $\dfrac{9}{4}; \dfrac{11}{4}$ **(d)** $2\sqrt{3} - 6; 6 - 2\sqrt{3}$

6. (a) $-\dfrac{10}{3}; \dfrac{17}{3}$ **(b)** $\dfrac{204}{25}$ **(c)** $-6; \dfrac{13}{2}$

7. (a) $\dfrac{166}{5}$ **(b)** $\dfrac{166}{5}$ **(c)** $\dfrac{166}{5}$

8. (a) $50 - 20t$ **(b)** 1187.5

9. 1.0041 s

10. (a) 5 s   **(b)** 272.5 m   **(c)** 10 s

   (d) $-49\,\text{m s}^{-1}$ **(e)** 12.46 s   **(f)** $-73.08\,\text{m s}^{-1}$

Chapter 14 practice questions

1. (a) $p = 3$ **(b)** 3 square units

2. (a) $(0, 1)$ **(b)** $V = \pi\displaystyle\int_0^{\ln 2}\left(e^{\frac{x}{2}}\right)^2 dx$

   (c) See Worked Solutions

3. $a = e^2$

4. (a) $y = \dfrac{x}{e}$ **(b)** $\ln x + 1 - 1$

   (c) $\dfrac{1}{2} \cdot e \cdot 1 - \displaystyle\int_1^e \ln x\,dx$

5. (a) (i) 400 m

     (ii) $v = 100 - 8t$, 60 m/s

     (iii) 8 s

     (iv) 1344 m

   (b) Distance needed 625

6. (a) See Worked Solutions   **(b)** 2.31

   (c) $-\pi\cos x - \dfrac{x^2}{2} + c$; 0.944

7. $\ln 3$

8. (a) (i) See Worked Solutions

     (ii) $(1.57, 0)$; $(1.1, 0.55)$; $(0, 0)$, $(2, -1.66)$

   (b) $x = \dfrac{\pi}{2}$

   (c) (i) See Worked Solutions

     (ii) $\displaystyle\int_0^{\frac{\pi}{2}} x^2\cos x\,dx$

   (d) $\dfrac{\pi^2}{4} - 2 \approx 0.4674$

9. (a) 2π

   (b) range: $\{y\,|-0.4 < y < 0.4\}$

   (c) (i) $-3\sin^3 x + 2\sin x$ **(iii)** $\dfrac{2\sqrt{3}}{9}$

   (d) $\dfrac{\pi}{2}$

   (e) (i) $\dfrac{1}{3}\sin^3 x + c$ **(ii)** $\dfrac{1}{3}$

   (f) $\arccos\dfrac{\sqrt{7}}{3} \approx 0.491$

10. (a) (i) See Worked Solutions   **(ii)** See Worked Solutions

   (b) See Worked Solutions   **(c)** 3.69672

   (d) $\displaystyle\int_0^{\pi}(\pi + x\cos x)dx$ **(e)** $\pi^2 - 2 \approx 7.86960$

11. (a) (i) $10x - 1 - e^{2x}$ **(ii)** $\dfrac{\ln 5}{2} \approx 0.805$

   (b) (i) $f^{-1}(x) = \dfrac{\ln(x - 1)}{2}$ **(ii)** See Worked Solutions

   (c) $v = \pi\int_0^{\ln 2}(1 + e^{2x})^2\,dx$

12. $\pi\left(\dfrac{2}{15}a^5 + \dfrac{2}{3}a^3\right)$

13. $4\left(\dfrac{2}{5}\left(\dfrac{1}{2}x+1\right)^{\frac{5}{2}}-\dfrac{2}{3}\left(\dfrac{1}{2}x+1\right)^{\frac{3}{2}}\right)+c$

14. $a=-\dfrac{56}{27}$

15. $\dfrac{\pi}{2}(e^{2k}-1)$

16. $k=2$

17. $1800\,\text{m}$

18. $2a$ by $\dfrac{2}{3}a^2$

19. **(a)** $\ln x+1-k$ **(b)** $x>\dfrac{1}{e}$
  (c) (i) See Worked Solutions
   (ii) $(e^k,0)$
  (d) $\dfrac{e^{2k}}{4}$ **(e)** $y=x-e^k$
  (f) See Worked Solutions **(g)** Common ratio $=e$

20. $a=5$

21. $v=\sqrt{v_0^2+\dfrac{4k}{m}}$

22. **(a)**

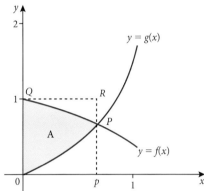

  (b) See Worked Solutions **(c)** 0.6937
  (d) $\displaystyle\int_0^p (e^{-x^2}-(e^{-x^2}-1))\,dx\approx\sim 0.467$

23. **(a)** See Worked Solutions
  (b) (i) $\dfrac{2\pi}{9}$ **(ii)** $\dfrac{4\pi}{9}$ **(iii)** $\dfrac{6\pi}{9}$
  (c) $\dfrac{n\pi}{9}(n+1)$

24. **(a)** $t=0,3,$ or 6
  (b) (i) $\displaystyle\left|\int_0^6 t\sin\left(\dfrac{\pi}{3}\right)t\,dt\right|$ **(ii)** $11.5\,\text{m}$

25. **(a)** 0.435 **(b)** $\dfrac{-2t}{(2+t^2)^2}$

26. **(a)** $\dfrac{dy}{dx}=\dfrac{2x^2}{\sqrt{1+x^2}}+2\sqrt{1+x^2}$
  (b) See Worked Solutions **(c)** $k=0.918.$

27. $6\,\text{m}$

28. 0.852

29. **(a)** See Worked Solutions
  (b) (i) $A=78;\ k=\dfrac{1}{15}\ln\dfrac{48}{78}$ **(ii)** 45.3

30. **(a)**

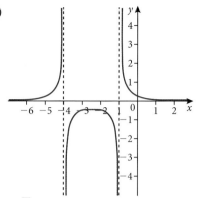

  (b) $\ln\sqrt[3]{\dfrac{8}{5}}$ **(c)** area $=2\ln\sqrt[3]{\dfrac{8}{5}}$

31. $\dfrac{(x+2)^2}{2}-6(x+2)+12\ln|x+2|+\dfrac{8}{x+2}+c$

32. **(a)**

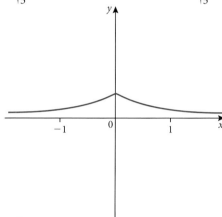

  (b) (i) $x=23$
   (ii) $x-\text{int}=e^2-3;\ y-\text{int}=\ln3-2$
  (c) $-1.34;\ 3.05$
  (d) (ii) $\displaystyle\int_0^{3.05}\left(4-(1-x)^2-(\ln(x+3)-2)\right)dx$
   (iii) 10.6
  (e) 4.63

33. **(a)** See Worked Solutions
  (b) $\ln x-\dfrac{1}{2}\ln(x^2+1)+c$ **(c)** $y=\dfrac{2e^\theta}{\sqrt{e^{2u}11}}$

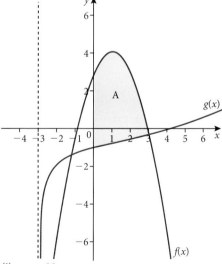

34. (a) $(1, 0)$, $\left(e^2, \dfrac{4}{e^2}\right)$ **(b)** $\dfrac{1}{3}$ **(c)** $\dfrac{\pi}{e}(24e - 65)$

35. (a) (i) See Worked Solutions

    (ii) $\dfrac{1}{2(e^x + e^{-x}) - (e^x - e^{-x})} = \cdots = \dfrac{e^x}{e^{3x} + 3}$; $\dfrac{\pi\sqrt{3}}{18}$

  (b) (i) $x = \ln\left(\dfrac{k \pm \sqrt{k^2 - n^2 + 1}}{2(n + 1)}\right)$

    (ii) 2 solutions $\Rightarrow k > \sqrt{k^2 - n^2 + 1}$
and $k^2 - n^2 + 1 > 0$

  (c) (i) See Worked Solutions

    (ii) Use quotient rule $f(x) > g(x)$ and result follows.

Chapter 15

Exercise 15.1

1. (a) discrete   **(b)** continuous   **(c)** continuous
  (d) discrete   **(e)** continuous   **(f)** continuous
  (g) discrete   **(h)** continuous   **(i)** continuous
  (j) discrete   **(k)** continuous   **(l)** continuous
  (m) discrete

2. (a) 0.4

  (b)

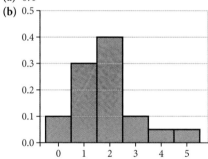

  (c) 1.85, 1.19   **(e)** 2.85, 1.19

  (f) $E(Z) = E(Y + b) = E(Y) + b$ and
    $V(Y) = V(Y + b) = V(Y)$

3. (a) 0.26   **(b)** 0.37   **(c)** 0.77
  (d) 16.29   **(e)** 8.1259
  (f) 4.145; 2.031475
  (g) $E(aX + b) = aE(X) + b$ and $V(aX + b) = a^2V(X)$

4. (a) 0.969   **(b)** 0.163   **(c)** 3.5
  (d) $\sum(x - 3.5)^2 \cdot P(x) = 1.048 \Rightarrow \sigma = \sqrt{1.048} \approx 1.02$
  (e) Empirical: 0.68, 0.95; approximately 0.68,
    approximately 0.90

5. $k = \dfrac{1}{30}$

x	12	14	16	18
$P(X = x)$	$6k$	$7k$	$8k$	$9k$

6. (a) $k = \dfrac{1}{10}$ **(b)** $\dfrac{37}{60}$ **(c)** $\dfrac{19}{30}$

  (d) $E(X) = 16$, SD $= 7$

  (e) $E(Y) = \dfrac{11}{5}$; $V(Y) = \dfrac{49}{25}$

7. (a) $\dfrac{1}{50}$

(b)

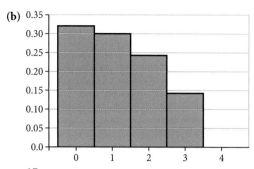

(c) $\dfrac{17}{25}$ **(d)** $\mu = 1.2$; Var $= 1.08$

8. (a) $P(X = 18) = 0.2$, $P(X = 19) = 0.1$, Symmetric
distribution.
  (b) $\mu = 17$, SD $= 1.095$

9. (a) $\mu = 1.9$, SD $= 1.338$ **(b)** between 0 and 5.

10. $k = 0.667$, $E(X) = 5.444$

11. (a) $k = 0.3$ or 0.7
  (b) for $k = 0.3$: $E(X) = 2.18$; for $k = 0.7$: $E(X) = 1.78$

12. (a)

x	0	1	2	3
$P(X = x)$	$\dfrac{1}{27}$	$\dfrac{2}{9}$	$\dfrac{4}{9}$	$\dfrac{8}{27}$

  (b) 2

13. (a) $k = \dfrac{1}{10}$ **(b)** $\dfrac{1}{2}$

14. (a) See table below.   **(b)** 0.85   **(c)** 0.15
  (d) 48.87   **(e)** 2.057   **(f)** 0.72

x	45	46	47	48	49	50	51	52	53	54	55
CDF	0.05	0.13	0.25	0.4	0.65	0.85	0.9	0.94	0.97	0.99	1

15. (a)

x	0	1	2	3	4	5	6
CDF	0.08	0.23	0.45	0.72	0.92	0.97	1

  (b) 0.72   **(c)** 0.97   **(d)** 2.63   **(e)** 1.440

16. (a) 0.9   **(b)** 0.09   **(c)** 0.009
  (d) (i) unacceptable   **(ii)** acceptable
  (e) $p(x) = (0.1^{x-1}) \times 0.9$

17. (a) 0   **(b)** 0.81   **(c)** 0.162
  (d) (i) either   **(ii)** acceptable
  (e) $(x - 1)(0.1^{x-2}) \times 0.9^2$, $x > 1$.

18. $n = 132$

19. (a) (i) $\dfrac{1}{9}$ **(ii)** $\dfrac{1}{81}$

  (b) (i) $\dfrac{73}{648}$ **(ii)** $\dfrac{575}{1296}$

  (c) (i) See Worked Solutions
    (ii)

X	1	2	3	4	5	6
CDF	$\dfrac{1}{1296}$	$\dfrac{15}{1296}$	$\dfrac{65}{1296}$	$\dfrac{175}{1296}$	$\dfrac{369}{1296}$	$\dfrac{671}{1296}$

    (iii) $\dfrac{6797}{1296}$

20. 9.3

Exercise 15.2

1. (a)

x	0	1	2	3	4	5
$P(X = x)$	0.01024	0.0768	0.2304	0.3456	0.2592	0.07776

Answers

(b)

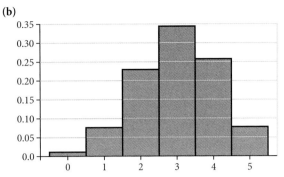

(c) (i) Mean = 3, SD = 1.095
(ii) Mean = 3, SD = 1.095
(d) between 2 and 4, and between 1 and 5.
(e) 0.8352, 0.898. Slightly more than the empirical rule.

2. (a) 0.001294494   **(b)** 0.000000011
(c) 0.99999999   **(d)** 0.99999966
(e) mean = 12, SD = 2.191

3. (a)

k	0	1	2	3	4	5	6
$p(x \leqslant k)$	0.11765	0.42018	0.74431	0.92953	0.98907	0.99927	1

(b)

Number of successes x	List the values of x	Write the probability statement	Explain it, if needed	Find the required probability
At most 3	0, 1, 2, 3	$p(x \leqslant 3)$	$p(x \leqslant 3)$	0.92953
At least 3	3, 4, 5, 6	$P(x \geqslant 3)$	$1 - p(x \leqslant 2)$	0.25569
More than 3	4, 5, 6	$p(x > 3)$	$1 - p(x \leqslant 3)$	0.07047
Fewer than 3	0, 1, 2	$p(x \leqslant 2)$	$p(x \leqslant 2)$	0.74431
Between 3 and 5 (inclusive)	3, 4, 5	$p(3 \leqslant x \leqslant 5)$	$p(x \leqslant 5) - p(x \leqslant 2)$	0.25496
Exactly 3	3	$P(x = 3)$	$P(x = 3)$	0.18522

4. (a)

k	0	1	2	3
$p(x \leqslant k)$	0.02799	0.15863	0.41990	0.71021
k	4	5	6	7
$p(x \leqslant k)$	0.90374	0.98116	0.99836	1

(b)

Number of successes x	List the values of x	Write the probability statement	Explain it, if needed	Find the required probability
At most 3	0, 1, 2, 3	$p(x \leqslant 3)$	$p(x \leqslant 3)$	0.71021
At least 3	3, 4, 5, 6, 7	$P(x \geqslant 3)$	$1 - p(x \leqslant 2)$	0.58010
More than 3	4, 5, 6, 7	$p(x > 3)$	$1 - p(x \leqslant 3)$	0.28979
Fewer than 3	0, 1, 2	$p(x \leqslant 2)$	$p(x \leqslant 2)$	0.41990
Between 3 and 5 (inclusive)	3, 4, 5	$p(3 \leqslant x \leqslant 5)$	$p(x \leqslant 5) - p(x \leqslant 2)$	0.56125
Exactly 3	3	$P(x = 3)$	$P(x = 3)$	0.290304

5. (a) p is not constant, trials are not independent.
(b) p becomes constant.
(c) $n = 3, p = \dfrac{5}{8}$

y	0	1	2	3
$P(Y = y)$	0.05273	0.26367	0.43945	0.24414

(d) 0.75586   **(e)** 1.875
(f) 0.703125   **(g)** 0.94727
6. (a) 0.10737   **(b)** 0.99363
(c) 0.89263   **(d)** 2
7. (a) 0.81707   **(b)** 1.00000   **(c)** 0.01618
8. (a) 0.033833   **(b)** 0.02449   **(c)** 0.78272
9. (a) 0.75   **(b)** 0.03251   **(c)** 0.17268
10. (a) 0.04317   **(b)** 0.00761   **(c)** 0.01125
(d) 0.13057   **(e)** 0.95683   **(f)** 10
(g) 3   **(h)** 4, 16
11. (a) 3   **(b)** 0.10131   **(c)** −0.00021
12. (a)

x	0	1	2	3	4	5
$P(x)$	0.03125	0.15625	0.3125	0.3125	0.15625	0.03125

(b) 0.03125   **(c)** 0.03125
(d) 0.96875   **(e)** 0.96875
(f) (a)

x	0	1	2	3	4	5
$P(x)$	0.32768	0.4096	0.2048	0.0512	0.0064	0.00032

(b) 0.32768   **(c)** 0.00032
(d) 0.67232   **(e)** 0.99968
14. 0.91296
15. (a) 0.10737   **(b)** 0.89263   **(c)** $1 - 0.8^n, n = 14$

Exercise 15.3

1. (a) $k = -\dfrac{3}{2}$ **(b)** 0.3125
(c) 0.6875   **(d)** 0.375; 0.3473; 0.2437
2. (a) $k = \dfrac{1}{6}$ **(b)** $\dfrac{1}{8}$
(c) $\dfrac{1}{2}$ **(d)** $\dfrac{7}{9}$, 0.697, 0.5329
3. (a) $k = \sqrt{2}$ **(b)** $\dfrac{49}{64}$ **(c)** $\dfrac{15}{64}$
(d) 0.7543; 0.7654; 0.3127
4. (a) $\dfrac{6}{37}$ **(b)** $\dfrac{133}{148}$
(c) $\dfrac{19}{74}$ **(d)** $\dfrac{50}{37}$, 1.5, 0.528
5. (a)

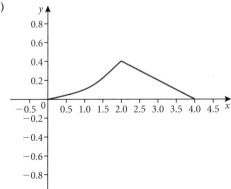

(b) $k = \dfrac{3}{53}$

(c) mean = 2.0283, median = 2.4176,
standard deviation = 1.58208

(d) 1.80819

6. (a) 24.7 hours   **(b)** 0.514   **(c)** 0.264

7. (a) 50 hours   **(b)** 50 hours

(c) 22.4 hours   **(d)** 0.104

(e) (i) 0.010816   **(ii)** 0.989184

8. (a)

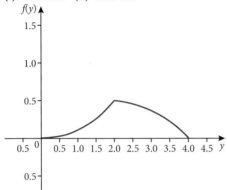

(b) $\dfrac{7}{3}$ **(c)** 0.694   **(d)** 155 barrels

9. (a) $\displaystyle\int_2^5 \dfrac{c}{(1-y)(y-6)}\,dy = 1$ **(b)** $\dfrac{7}{2}$; 0.916

10. (a) Area = 1 and $f(y) \geqslant 0$.

(b) $a = \dfrac{6}{125}$; $b = 5$ **(c)** 1.25

11. (a) $k = \dfrac{1}{b-a}$

(b) (i) & **(ii)** mean = median = $\dfrac{a+b}{2}$

(iii) variance = $\dfrac{(a-b)^2}{12}$

12. (a) 0.378   **(b)** 1.752   **(c)** 1.892   **(d)** 0.955

13. (a) $\dfrac{1}{8}$

(b) $f(x) = \begin{cases} 0 & 0 \leqslant x < 5 \\ \dfrac{3(x-7)^2}{8} & 5 \leqslant x \leqslant 7 \\ 0 & x > 7 \end{cases}$

(c) 5.4126

(d) 0.15

14. (a) $k = 3$ **(b)** $\dfrac{4}{5}$ **(c)** 0.8409

15. (a) $f(y) \geqslant 0, \displaystyle\int_0^\infty 2y\,e^{-y^2}\,dy = \left[-e^{-y^2}\right]_0^\infty = 0 + 1$

(b) 0.0183   **(c)** $\dfrac{\sqrt{\pi}}{2}$ **(d)** 0.8326

(e) 0.641   **(f)** 0.0769

16. (a) $\dfrac{5}{9}$ **(b)** 0.1944

(c) 0.1941   **(d)** 0.6207

17. (a) $\displaystyle\int_0^5 ky^2(5-y)\,dy = 1$ **(b)** 3, 3.1, 3.3

(c) 0.4752   **(d)** 1   **(e)** 0.64, no

18. (a) $\dfrac{10}{3}, \dfrac{15}{4}$ **(b)** See Worked Solutions

(c) 0.08032   **(d)** 0.8909

(e) (i) 0.9866   **(ii)** 0.99995   **(iii)** 0.01335

Exercise 15.4
(some answers are rounded)

1. (a) 0.5   **(b)** 0.499571   **(c)** 0.158655

(d) 0.682690   **(e)** 0.022750   **(f)** 0

2. (a) 0.76986   **(b)** 0.161514

(c) 0.656947   **(d)** 0.999944

3. (a) 0.008634   **(b)** 0.982732

4. 1.28

5. 1.96

6. (a) 0.066807   **(b)** 0.68269

(c) 678.16   **(d)** 134.90

7. (a) 1.76%   **(b)** 509.98   **(c)** 5.71

8. (a) 0.9696   **(b)** 0.546746

9. (a) 1 day   **(b)** 29 days   **(c)** 112 days

10. 1.56

11. 18.9192

12. 30.81

13. 100.28

14. 29.95

15. $\mu = 21.037, \sigma = 4.252$

16. $\mu = 18.988, \sigma = 0.615$

17. $\mu = 121.935, \sigma = 34.389$

18. (a) $\mu = 6.966, \sigma = 0.324$ **(b)** 0.252

19. (a) 0.655422   **(b)** 0.008198   **(c)** 82 bottles

20. (a) 22.73%

(b) 0.546%

(c) 29.678

(d) 229.183

21. (a) Not likely: chance is 0.135%

(b) 15.87%

(c) 68.27%   **(d)** 5396 km   **(e)** 43785

22. (a) 6.817   **(b)** 3.4315

(c) $\mu = 64.135, \sigma = 7.545$

23. 7.3%

24. (a) 216.06

(b) 15.31

25. (a) $\mu = 111.90, s = 17.09$ **(b)** 0.54

26. 9.1929

27. (a) (i) $\sigma = 1.355$

(ii) $\mu = 110.37$

(b) A = 108.64; B = 112.11

Exercise 15.5

1. (a) 0.5248; 0.8448   **(b)** 1.6; 0.96   **(c)** 5.8; 3.84

2. (a) 0.13; See Worked Solutions for histrogram.

(b) 0.48, 0.2   **(c)** 12.57, 1.4251

(d) (i) (ii) 25.14, 5.7004

(e) (i) (ii) 25.14, 2.28502

3. (a)

X	P(x)		Y	P(y)
1	0.166667		1	0.25
2	0.166667		2	0.25
3	0.166667		3	0.25
4	0.166667		4	0.25
5	0.166667			
6	0.166667			

Answers

(b) Mean (of X) = 3.5, variance = 2.917;
mean (of Y) = 2.5, variance = 1.25

(c)

x	$P(x)$
2	0.41667
3	0.083333
4	0.125
5	0.166667
6	0.166667
7	0.166667
8	0.125
9	0.083333

(d) (i) 6   **(ii)** 4.167

4. 3.5; 0.285

5. (a) (i) 0.1   **(ii)** 3.2   **(iii)** 1.68
   (b) (i) 16   **(ii)** 21.84

6. (a) 10; 3   **(b)** -4; 3   **(c)** 27; 17   **(d)** -15; 17

7. (a) $\sqrt{7} + \sqrt{13}$, 5   **(b)** $\sqrt{7} - \sqrt{13}$, 5
   (c) $2\sqrt{7} + 3\sqrt{13}$, 35   **(d)** $2\sqrt{7} - 3\sqrt{13}$, 35

8. (a) $2\sqrt{7} + 2$, 22   **(b)** $\sqrt{7} - 6$, 23
   (c) $2\sqrt{7} + 6$, 38   **(d)** $2\sqrt{7} - 6$, 38

9. (a) 0.98, 0.52405

(b)

x	$P(x)$
2.1	0.36
2	0.48
1.9	0.16

1.96, 1.0481

(c)

x	$P(x)$
2.85	0.064
2.95	0.288
3.05	0.432
3.15	0.216

2.94, $\dfrac{1}{57215}$

10. (a) 0.298   **(b)** 0.227   **(c)** 0.298

11. 0.007

Chapter 15 practice questions

1. (a) 34.5%   **(b)** 0.416   **(c)** 3325

2. (a) (i) 0.393   **(ii)** 0.656   **(b)** 50

3. (a) 0.1   **(b)** 10
   (c) See Worked Solutions   **(d)** 0.739

4. (a) (i) $a = -0.455$, $b = 0.682$
   (b) (i) 0.675   **(ii)** 0.428
   (c) See Worked Solutions   **(d)** $t = 62.6$

5. (a) 69.97%   **(b)** 0.00226

6. (a) 0.0808   **(b)** See Worked Solutions
   (c) $\mu = 25.5$, $s = 0.255$ **(d)** 12 500

7. (a) (i) 0.345   **(ii)** 0.115   **(iii)** 0.540
   (b) 0.119   **(c)** 737

8. (a) 15.9%   **(b)** 227 cm

9. (a) 0.0912   **(b)** $a = 251$, $b = 369$.

10. (a) $a = -1$, $b = 0.5$
   (b) (i) 0.841   **(ii)** 0.533
   (c) (i) See Worked Solutions   **(ii)** 0.647

11. $\mu = 66.6$, $\sigma = 22.6$

12. (a) 0.8
   (b) (i) See Worked Solutions
     (ii)

Y	0	1	2
$P(Y = y)$	$\dfrac{1}{15}$	$\dfrac{8}{15}$	$\dfrac{2}{5}$

   (c) $\dfrac{3}{10}$ **(d)** $\dfrac{1}{9}$

13. (a) 0.129886   **(b)** 0.676714   **(c)** 2

14. (a) (i) 0.002171   **(ii)** 0.00120
   (b) 0.8413

15. $\sigma = 0.00943 \text{ kg} \approx 9.4 \text{ g}$

16. (a) See Worked Solutions
   (b) $\dfrac{e}{4} - \sqrt{e} + \sqrt[4]{e}$
   (c) $\mu = \dfrac{e}{2} - 1$; $\sigma^2 = 1 + \dfrac{e}{3} - \dfrac{e^2}{4}$
   (d) $\sqrt{e} - \dfrac{e}{2}$
   (e) (i) $\left(\sqrt{e} - \dfrac{e}{2}\right)^3$
     (ii) $\dbinom{3}{2}\left(1 - \sqrt{e} + \dfrac{e}{2}\right)\left(\sqrt{e} - \dfrac{e}{2}\right)^2$

17. (a) 0.2212   **(b)** 0.125

18. (a) $x = 58.69$ **(b)** $\sigma = 3.41$
   (c) (i) Karl
     (ii) 0.00239

19. (a) $\mu = 1.63 \text{ nm}$
   (b) $0.44 \times 0.4013 + 0.56 \times 0.242 = 0.312$
   (c) 0.434   **(d)** \$6605.28

20. (a) 0.841
   (b) (i) 0.0681   **(ii)** 0.0312   **(iii)** 0.932

21. 0.164

22. (a) (i) 0.5, 0.13   **(ii)** 0.0828
   (b) 0.797

23. (a) 0.0548   **(b)** 0.908

24. (a) (i) $\dfrac{3}{2}$ **(ii)** $\dfrac{11}{9}$ **(iii)** $\dfrac{125}{36}$
   (b) 0.432

Chapter 16

Exercise 16.1

1. (a) $f(x) = \dfrac{1}{x}(\ln x + k) \Rightarrow \dfrac{dy}{dx} = \dfrac{1 - k - \ln x}{x^2}$

$x^2 \times \dfrac{1 - k - \ln x}{x^2} + x\left(\dfrac{1}{x}(\ln x + k)\right) - 1$

$= 1 - k - \ln x + \ln x + k - 1 = 0$

   (b) $\dfrac{dy}{dx} = \dfrac{1 - xy}{x^2}$

   (c) $f(x) = \dfrac{1}{x}(\ln x + 2)$, $f(x) = \dfrac{1}{x}\left(\ln x + 1 - \dfrac{\ln 2}{2}\right)$

2. (a) $y = \ln(e^x - C)$
   (b) $y^2 = Ce^{x^2} - 1$
   (c) $y = x + \ln\dfrac{1}{x + Ce^x + 1}$
   (d) $\dfrac{y - 1}{y + 1} = Ae^{(x-1)^2} + c$
   (e) $\ln(y + 1) + y\ln(y + 1) + 1 = (y + 1)(\ln(\ln x) + c)$

(f) $y^2 = C\tan^4\left(\dfrac{x}{2}\right) - \dfrac{1}{2}$

(g) $y + \ln|\sec y| = \dfrac{1}{3}x^3 + x + c$

(h) $\sqrt{(y^2 + 1)^3} = 3e^t(t - 1) + c$

(i) $e^{-y}(y + 1) = -\dfrac{1}{3}\sin^3\theta + c$

(j) $y = \tan(e^x + C)$

(k) $1 - \sqrt{(1 - x^2)}$

(l) $e^y = \ln(1 + e^x) + 1 - \ln(1 + e)$

(m) $y + \ln y = \dfrac{1}{3}x^3 - x - 5$

(n) $\ln y = \ln x + x^2 - 1$ or $y = xe^{(x^2 - 1)}$

(o) $\dfrac{1}{2}e^{(x^2)} = \ln y - \dfrac{1}{2}y^2 + 1$

3. (a) $y = 3e^{x^4}$ **(b)** $y = e^{\frac{1}{2}x^2}$

(c) $y = \dfrac{2}{2 - x^2}$ **(d)** $y = \dfrac{1}{3 - x}$

(e) $y = \ln\left(\dfrac{e}{1 - ex}\right)$

(f) $y^3 = \dfrac{3(x + 1)^2}{2} - \dfrac{1}{2}$

(g) $y = \dfrac{1}{\ln|x + 1| - 1}$

(h) $2y^3 + 6y = 3x^2 + 6x + 72$

(i) $y^2 = e^{x^2} - 1$

(j) $y + \ln|y| = \dfrac{x^2}{2} - x + 1$

(k) $\arcsin y = 1 - \sqrt{1 - x^2}$

(l) $y = \ln\left(\ln\dfrac{e(e^x + 1)}{1 + e}\right)$

(m) $y + \ln y = \dfrac{x^3}{3} - x - 5$

(n) $|y| = |x|e^{x^2 - 1}$

(o) $2\ln|y| - y^2 = e^{x^2} - 1$

4. (a) $|x| = C(x - y)^2$ **(b)** $|y^2 + 2xy - x^2| = C$

(c) $y = Ce^{-\frac{x^2}{2y^2}}$ **(d)** $e^{\frac{y}{x}} = 1 + \ln x^2$

(e) $x = e^{\sin\left(\frac{y}{x}\right)}$

(f) $\ln\left|\dfrac{(x + y)^2}{cx}\right| = \dfrac{y}{x}$ or $(x + y)^2 = cxe^{\frac{y}{x}}$

(g) $4x = y(\ln|x| - c)^2$

5. (a) See Worked Solutions

(b) (i) $C = 78, m = \dfrac{1}{15}\ln\dfrac{8}{13}$

    (ii) 45.3 (3 sf)

6. (a) $p = \dfrac{200}{1 + 7e^{-1/2\ln(39/23)}}$

(b) 70 tigers   **(c)** 7.37 yr

7. (a) 70   **(b)** 4.49 yr

(c) $\dfrac{dP}{dt} = 0.75P\left(1 - \dfrac{P}{2100}\right)$

8. (a) $\dfrac{dP}{dt} = kP \Rightarrow P(t) = 500e^{t\ln2} \Rightarrow P(12) = 2{,}048{,}000$

(b) $1000 = 500e^{t\ln2} \Rightarrow t = 1$ hour.

Exercise 16.2

1. (a) Not linear   **(b)** linear
    (c) linear   **(d)** Not linear

2. $y = x^2\ln|x| + Cx^2$.

3. $y = \dfrac{-x^2 + x + 2\sqrt{2}}{\sqrt{x^2 + 1}}$

4. $y = x - 1 + Ce^{-x}$

5. $y = \dfrac{1}{4}(1 + \sin x)(6x - \sin2x - 8\cos x + 12)$

6. (a) $y = e^{-x} + Ce^{-2x}$ **(b)** $y = \dfrac{-\cos x + C}{x^3}$

(c) $y = \dfrac{x^3 - 3x + C}{3(x - 1)^4}$ **(d)** $y = \csc x(\ln|\sec x| + C)$

(e) $y = x^2\ln|x| + Cx^2$ **(f)** $y = x - 1 + Ce^{-x}$

(g) $y = \dfrac{2}{3}\sqrt{x} + \dfrac{C}{x}$

(h) $y = \dfrac{1}{\sin x}\left(\int\sin(x^2)\,dx + C\right)$

(i) $u = \dfrac{1}{2(t + 1)}(t^2 + 2t + C)$

(j) $y = \dfrac{e^{x^2} + C}{2x^2}$

(k) $y = -\cos x + C\sec x$

(l) $y = e^{-x}\left(1 + \dfrac{C}{x}\right)$

7. (a) $y = \dfrac{x}{2} - \dfrac{1}{4} + \dfrac{5}{4}e^{-2x}$ **(b)** $y = x^2\left(\sec x + 2\left(\dfrac{9}{\pi^2} - 1\right)\right)$

(c) $y = \dfrac{1}{x}\ln x - \dfrac{1}{x} + \dfrac{3}{x^2}$ **(d)** $u = -t^2 + t^3$

(e) $y = -x\cos x - x$ **(f)** $y = (x^2 + 3)e^{x^2}$

8. $Q(t) = 3(1 - e^{-4t}), \quad I(t) = \dfrac{dQ}{dt} = 12e^{-4t}$

Exercise 16.3

1. 1.7616

2. (a) (i) 2.3928   **(ii)** 2.3701   **(iii)** 2.3681
  (b) $-3x^2e^{-x^2} = 6x^2 - 3x^2(e^{-x^3} + 2)$
  (c) (i) 0.0249   **(ii)** 0.0022
    (iii) 0.0002; error is divided approx. by 10

3. $y_1 = 2, y_2 = 2.75, y_3 = 3.5, y_4 = 4.25$

4. 0.4150

5. 1.8371

6. (a) $t = 0.5$:
    (i) 1.70308   **(ii)** 1.79548   **(iii)** 1.87734
    $t = 1.5$:
    (i) 2.44030   **(ii)** 2.432 92   **(iii)** 2.42672
    $t = 3$:
    (i) 1.11925   **(ii)** 1.12191   **(iii)** 1.12411
  (b) $t = 0.5$:
    (i) 2.30800   **(ii)** 2.30167   **(iii)** 2.29686
    $t = 1.5$:
    (i) 2.60023   **(ii)** 2.59352   **(iii)** 2.58830
    $t = 3$:
    (i) 2.73521   **(ii)** 2.73209   **(iii)** 2.72959
  (c) $t = 0.5$:
    (i) -1.48849 **(ii)** -1.46909 **(iii)** -1.45212
    $t = 1.5$:
    (i) 1.04687   **(ii)** 1.05351   **(iii)** 1.05941
    $t = 3$:
    (i) 1.51971   **(ii)** 1.50549   **(iii)** 1.49490

Answers

7. Exact equation: $y = \frac{1}{2}(\sin x - \cos x + e^x)$

x	0	0.2	0.4	0.6	0.8	1.0
$y(x)$ [exact]	0.0000	0.2200	0.4801	0.7807	1.1231	1.5097
$y(x)$ [$h = 0.2$]	0.0000	0.2000	0.4360	0.7074	1.0140	1.3561
$y(x)$ [$h = 0.1$]	0.0000	0.2095	0.4568	0.7418	1.0649	1.4273

Exercise 16.4

1. (a) $\sum_{k=0}^{\infty} \frac{x^{2k+1}}{(2k+1)!}$ **(b)** $\sum_{k=0}^{\infty} \frac{(-1)^k x^{4k}}{(2k)!}$

(c) $\sum_{k=0}^{\infty} \frac{(-1)^k 2^{2k+1} x^{2k+3}}{(2k+1)!}$ **(d)** $\sum_{k=1}^{\infty} \frac{(-1)^{k+1} 2^{2k-1} x^{2k}}{(2k)!}$

(e) $\sum_{k=0}^{\infty} \frac{(-1)^k x^{2k+2}}{(k+1)}$ **(f)** $\sum_{k=0}^{\infty} \frac{(-1)^k x^{2k+2}}{k!}$

(g) $1 + \sum_{k=1}^{\infty} \frac{(-1)^{k+1} x^k}{k+1}$

2. (a) 0.9802   **(b)** 0.0998   **(c)** -0.1054
    (d) 0.4969   **(e)** 0.7429

3. (a) $x^2 - \frac{x^4}{3!} + \frac{x^6}{5!}$ **(b)** $x + \frac{1}{3}x^3 + \frac{2}{15}x^5$

(c) $-x - \frac{3}{2}x^2 + \frac{4}{3}x^3$

4. $\sum_{k=0}^{\infty} k x^{k-1}$

5. $2x + \frac{2x^3}{3} + \frac{2x^5}{5} + \cdots = \sum_{k=1}^{\infty} \frac{2}{2k-1} x^{2k-1}$

6. (a) $f^{(n)}(x) = \frac{e^x + (-1)^n e^{-x}}{2}$

(b) $1 + \frac{x^2}{2} + \frac{x^4}{24} + \cdots$

(c) $f\left(\frac{1}{2}\right) \approx \frac{433}{384}$

7. $\sec^2 x = 1 + x^2 + \frac{2x^4}{3} + \frac{17x^6}{45} + \frac{62x^8}{315} + \cdots$

8. (a) $\lim_{x \to 0} \dfrac{\frac{x^2}{2!} - \frac{x^4}{4!} + \cdots}{\frac{x^2}{2!} + \frac{x^3}{3!} + \cdots} = 1$

(b) $\lim_{x \to 0} \dfrac{x - \arctan x}{x^3} = \frac{1}{3}$

9. (a) $y = a_0 \sum_{k=0}^{\infty} \frac{(-1)^k x^k}{k!} = a_0 e^{-x}$

(b) $y = a_0 \sum_{k=0}^{\infty} \frac{6^k x^k}{k!} = a_0 e^{6x}$

(c) $y = a_0 \sum_{k=0}^{\infty} \frac{x^{2k}}{k!} = a_0 e^{x^2}$

(d) $y = a_0 \sum_{k=0}^{\infty} \frac{(-1)^k 4^k x^{2k}}{k!} = e^{-4x}$

(e) $y = a_0 (x + 1)^3$

(f) $y = \sum_{k=0}^{\infty} (-1)^k x^{2k}$

10. (a) $1 + (1/4)x + \sum_{k=2}^{\infty} \frac{(-1)^{k+1} \cdot 3 \cdot 7 \cdot 11 \cdots (4k-5) x^k}{4^k k!}$

(b) $1 + \sum_{k=1}^{\infty} \frac{1 \cdot 4 \cdot 7 \cdots (3k-2) x^{2k}}{3^k k!}$

(c) $1 + \sum_{k=1}^{\infty} \frac{(-1)^k [1 \cdot 4 \cdot 7 \cdots (3k-2)] x^k}{3^k k!}$

(d) $3 + (x/27) + \sum_{k=2}^{\infty} \frac{(-1)^{k+1} \cdot 2 \cdot 5 \cdot 8 \cdot 11 \cdots (3k-4) x^k}{3^{4k-1} k!}$

11. (a) $\sqrt{1+x} = \sqrt{1 + 0.03} = 1 + \frac{0.03}{2} - \frac{0.03^2}{2^2 \cdot 2!} \approx 1.0149$

(b) $\sqrt{99} = \sqrt{100 - 1} = 10\sqrt{1 - 0.01}$

$$\approx 10\left(1 - \frac{0.01}{2} - \frac{0.01^2}{8}\right) = 9.9499$$

(c) $\int_0^{0.4} \sqrt[3]{1+x^4}\, dx = \int_0^{0.4} \left(1 + \frac{1}{3}x^4 - \frac{1}{9}x^8 + \cdots\right) dx$
$$\approx 0.4007$$

Chapter 16 practice questions

1. $y = \dfrac{e^{\sqrt{1+x^2}}}{e}$

2. $y = \arctan(1 - \cos x)$

3. $y = \dfrac{3x^3}{x^3 + 4}$

4. See Worked Solutions

5. 9.89 grams

6. (a) $2x^2 - y^2 = c$ **(b)** $y = \dfrac{x}{1 - cx}$

(c) $\ln(y - 1) - \ln y + c_1 = -\dfrac{1}{x}$ or $\dfrac{y}{y-1} = c_2 e^{\frac{1}{x}}$

(d) $x = c_1 \sin y$ or $y = \arcsin(c_2 x)$

(e) $y = c e^{\frac{x^2}{2}}$

(f) $y^2 = 2\sqrt{x^2 + 1} + c$

(g) $\ln\sqrt{\dfrac{y-1}{y+1}} = e^{-x} + c$

(h) $x = y \ln y - y + c$

7. $\int \frac{y+1}{y}\, dy = \int \frac{x+1}{x}\, dx \Rightarrow y + \ln|y| = x + \ln|x| + c$
$$\Rightarrow e^y e^{\ln y} = e^x e^{\ln x} e^c \Rightarrow y e^y = A x e^x$$

8. $y = \pm\sqrt{2\sin x + c}$
The constant c cannot be completely arbitrary because
$2\sin x + c \geqslant 0$.
If $c < -2$, then $2\sin x + c$ will be negative for all values of
x. If $-2 \leqslant c \leqslant 2$, then $2\sin x + c$ will be positive for some
values of x.

9. (a) $\dfrac{5}{2}$ **(b)** $\dfrac{5}{2}$

(c) Regardless of the initial value of the population, as time
increases, the population stabilizes at 2500.

10. $y = -\sqrt{x^2 + \tan x + 25}$

11. (a) See Worked Solutions   **(b)** $y = \dfrac{x+1}{x-1}$

12. $y = \dfrac{7x+1}{7-x}$

13. (a) $\dfrac{1}{3(x-2)} - \dfrac{1}{3(x+1)}$ **(b)** See Worked Solutions

14. (a) $y = c(x^2 - 1) + 1$

(b) $\dfrac{dy}{dx} + \left(\dfrac{2x}{1-x^2}\right) y = \dfrac{2x}{1-x^2}$; integrating factor
is $\left|\dfrac{1}{1-x^2}\right|$ leads to same solution as in part **(a)**.

15. (a) $y = x^4 + \dfrac{c}{x^2}$ **(b)** $y = c e^{\frac{x}{2}} - 1$

(c) $y = \dfrac{x^4}{3} + cx$ **(d)** $y = e^{\cos x}(x + c)$

(e) $y = e^{x^2}(x + c)$ **(f)** $y = x \ln|x| + cx$

16. $y = x \csc x + C \csc x$

17. (a) – (c) See Worked Solutions

    (d) $y = \dfrac{x}{2} + \dfrac{\arcsin x}{2\sqrt{1 - x^2}} + \dfrac{1}{\sqrt{1 - x^2}}$

18. (a) – (b) See Worked Solutions

    (c) $y = \tan x + C \sec x$

19. $y = \dfrac{1}{3}x^2 \ln x - \dfrac{1}{9}x^2 + \dfrac{10}{9x}$

20. $y = x - 1 + \dfrac{c}{e^x}$

21. $c = \dfrac{y - x}{(y + x)^2}$

22. (a) $y = c(x + 1)$ (b) $y = cx^2 - x$

    (c) $y = cx^3 - x$ (d) $2x^3 + 3xy^2 + 3y^3 = c$

    (e) $y^2 = \dfrac{x^2}{2} - \dfrac{c}{x^2}$ (f) $x^2 + 2xy - y^2 = c$

23. (a) See Worked Solutions   (b) $x^2 + 4xy - 3y^2 - 1 = 0$

24. See Worked Solutions

25. See Worked Solutions

26. (a) $\left| \dfrac{y}{y + 1} \right| = c|x|$ (b) $\left| \dfrac{y}{y + 1} \right| = \dfrac{1}{2}|x|$

    (c)
x_n	y_n
1.2	1.400
1.4	1.960
1.6	2.789
1.8	4.110

    (d)
x_n	approx. y_n	exact y_n	% error
1.2	1.400	1.5	6.7
1.4	1.960	2.3̇	16
1.6	2.789	4	30.3
1.8	4.110	9	54.3

27. $y \approx 1.5405$ at $x = 1$

28. $y \approx 5.9584$ at $x = 1$

29. $y^2 = Cx^3 - x^2$

30.
x_n	y_n
1.1	4.2
1.2	4.42543
1.3	4.67787
1.4	4.95904
1.5	5.27081

31. (a) See Worked Solutions

    (b) $y(1) \approx 0.32768$

    (c) $y(1) \approx 0.34868$

    (d) Actual value to 10 s.f. is $y(1) \approx 0.3678794412$; using more steps (and a smaller step size gives a better approximation.

32. $\alpha = 20 + 50e^{-\frac{t}{10}\ln\frac{5}{3}}$

33. (a) $y = (x + c)x^3$ (b) $y = (x + 1)x^3$

34. (a) $y = -2x + 12$ (b) $y = \dfrac{8x}{x + 1}$

35. (a) See Worked Solutions

    (b) $5x = \dfrac{y^2}{x^2} + 1$ or $y = \pm x\sqrt{5x - 1}$

36. (a) See Worked Solutions

    (b) $y = x - \dfrac{x^2}{2} + \dfrac{x^3}{6} - \dfrac{x^4}{12}$

    (c) $y = -x - \dfrac{x^2}{2} - \dfrac{x^3}{6} - \dfrac{x^4}{12}$

    (d) See Worked Solutions   (e) 0

Index

Index